Plant and Nematode Interactions

AGRONOMY

A Series of Monographs

The American Society of Agronomy and Academic Press published the first six books in this series. The General Editor of Monographs 1 to 6 was A. G. Norman. They are available through Academic Press, Inc., 111 Fifth Avenue, New York, NY 10003.

1. C. EDMUND MARSHALL: The Colloid Chemical of the Silicate Minerals. 1949
2. BYRON T. SHAW, *Editor*: Soil Physical Conditions and Plant Growth. 1952
3. K. D. JACOB, *Editor*: Fertilizer Technology and Resources in the United States. 1953
4. W. H. PIERRE and A. G. NORMAN, *Editors*: Soil and Fertilizer Phosphate in Crop Nutrition. 1953.
5. GEORGE F. SPRAGUE, *Editor*: Corn and Corn Improvement. 1955
6. J. LEVITT: The Hardiness of Plants. 1956

The Monographs published since 1957 are available from the American Society of Agronomy, 677 S. Segoe Road, Madison, WI 53711.

7. JAMES N. LUTHIN, *Editor*: Drainage of Agricultural Lands. 1957
8. FRANKLIN A. COFFMAN, Editor: Oats and Oat Improvement. 1961
9. A. KLUTE, *Editor*: Methods of Soil Analysis. 1986
   Part 1—Physical and Mineralogical Methods. Second Edition.
   A. L. PAGE, R. H. MILLER, and D. R. KEENEY, *Editors*: Methods of Soil Analysis. 1982
   Part 2—Chemical and Microbiological Properties. Second Edition.
10. W. V. BARTHOLOMEW and F. E. CLARK, *Editors*: Soil Nitrogen. 1965
    (Out of print; replaced by no. 22)
11. R. M. HAGAN, H. R. HAISE, and T. W. EDMINSTER, *Editors*: Irrigation of Agricultural Lands. 1967
12. FRED ADAMS, *Editor*: Soil Acidity and Liming. Second Edition. 1984
13. E. G. HEYNE, *Editor*: Wheat and Wheat Improvement. Second Edition. 1987
14. A. A. HANSON and F. V. JUSKA, *Editors*: Turfgrass Science. 1969
15. CLARENCE H. HANSON, *Editor*: Alfalfa Science and Technology. 1972
16. J. R. WILCOX, *Editor*: Soybeans: Improvement, Production, and Uses. Second Edition. 1987
17. JAN VAN SCHILFGAARDE, *Editor*: Drainage for Agriculture. 1974
18. G. F. SPRAGUE and J. W. DUDLEY, *Editors*: Corn and Corn Improvement. Third Edition. 1988
19. JACK F. CARTER, *Editor*: Sunflower Science and Technology. 1978
20. ROBERT C. BUCKNER and L. P. BUSH, *Editors*: Tall Fescue. 1979
21. M. T. BEATTY, G. W. PETERSEN, and L. D. SWINDALE, *Editors*: Planning the Uses and Management of Land. 1979
22. F. J. STEVENSON, *Editor*: Nitrogen in Agricultural Soils. 1982
23. H. E. DREGNE and W. O. WILLIS, *Editors*: Dryland Agriculture. 1983
24. R. J. KOHEL and C. F. LEWIS, *Editors*: Cotton. 1984
25. N. L. TAYLOR, *Editor*: Clover Science and Technology. 1985
26. D. C. RASMUSSON, *Editor*: Barley. 1985
27. M. A. TABATABAI, *Editor*: Sulfur in Agriculture. 1986
28. R. A. OLSON and K. J. FREY, *Editors*: Nutritional Quality of Cereal Grains: Genetic and Agronomic Improvement. 1987
29. A. A. HANSON, D. K. BARNES, and R. R. HILL, JR., *Editors*: Alfalfa and Alfalfa Improvement. 1988
30. B. A. STEWART and D. R. NIELSEN, *Editors*: Irrigation of Agricultural Crops. 1990
31. JOHN HANKS and J. T. RITCHIE, *Editors*: Modeling Plant and Soil Systems. 1991
32. D. V. WADDINGTON, R. N. CARROW, and R. C. SHEARMAN, *Editors*: Turfgrass. 1992
33. H. G. MARSHALL and M. E. SORRELLS, *Editors*: Oat Science and Technology. 1992
34. L. E. MOSER, D. R. BUXTON, and M. D. CASLER, *Editors*: Cool-Season Forage Grasses. 1996
35. ALBERT A. SCHNEITER, *Editor:* Sunflower Technology and Production. 1997
36. K. R. BARKER, G. A. PEDERSON, and G. L. WINDHAM, *Editors:* Plant and Nematode Interactions. 1998

Plant and Nematode Interactions

Coeditors
Kenneth R. Barker
Gary A. Pederson
Gary L. Windham

Managing Editor: J. M. Bartels

Editor-in-Chief ASA Publications: Jerry M. Hatfield

Editor-in-Chief CSSA Publications: P. S. Baenziger

Editor-in-Chief SSSA Publications: J. M. Bigham

Number 36 in the series
AGRONOMY

American Society of Agronomy, Inc.
Crop Science Society of America, Inc.
Soil Science Society of America, Inc.
Madison, Wisconsin, USA
Publishers
Madison, Wisconsin USA
1998

Related Society Publications

Introduction to Crop Protection

Microbial–Plant Interactions

For more information on these titles, please contact the ASA Headquarters Office: Attn: Marketing; 677 South Segoe Road; Madison, WI 53711-1086. Phone: (608) 273-8080, ext. 322. Fax: (608) 273-2021.

American Society of Agronomy, Inc.
Crop Science Society of America, Inc.
Soil Science Society of America, Inc.
677 South Segoe Road, Madison, WI 53711 USA

Library of Congress Catalog Card Number: 98-070082.

Printed in the United States of America

CONTENTS

FOREWORD

Nematodes are ubiquitous in soils, often a serious detriment, but sometimes beneficial to crop production. This monograph explores the general basis of nematode biology, systematics, and ecology, and provides specific information on the most important crop-nematode interactions. It is the most comprehensive treatise available on these complex nematode-plant relationships. Two score of the world's most knowledgeable experts on nematodes, plants, and their interrelationships have contributed to this volume, which should become an excellent reference book, textbook for college courses, and a definitive source book for biologists, plant and soil scientists and professionals in crop production. This monograph is a very important contribution to the agricultural sciences.

William W. McFee, *president*, ASA
Elizabeth L. Klepper, *president*, CSSA
D. Keith Cassel, *president*, SSSA

PREFACE

Nematodes are a complex, diverse group of roundworms that occur worldwide in essentially all environments. Many species are important parasites of plants and animals, whereas others are beneficial to agriculture and the environment. This monograph focuses primarily on plant-parasitic taxa, especially those attacking food and fiber crops. Annual crop losses to these obligate parasites have been estimated to be about $8 billion for U.S. growers and $78 billion worldwide. The purpose of this book is to provide an updated, in-depth treatment of these important plant parasites that would be of interest to established scientists and students in plant production and protection as well as fundamental research.

The soils in a hectare of all agroecosystem typically contains billions of plant-parasitic as well as beneficial nematodes. The damage to plants caused by nematodes often is overlooked because the associated symptoms, including slow growth, stunting, and yellowing also can be attributed to nutritional and water-related disorders. Nevertheless, the highly aggressive plant-parasitic nematodes which often severely restrict crop growth and yield have received much greater study than the nonparasitic, beneficial bacterivores, fungivores, omnivores and predacious forms. These beneficial nematodes often occur in greater numbers in soil, but their treatment in this monograph will be limited primarily to the ecology chapter. In addition to the important impact of nematodes on plant and soil health, certain species such as the bacteria-feeding *Caenorhabtitis elegans* are widely used as model biological research systems for molecular biology-genetics and related disciplines.

Of the 27 chapters in this monograph, the first 14 are discipline-oriented, and the remaining 13 focus on the key parasitic nematode species associated with the major food, feed, and fiber crops. The introductory chapter and appendices include extensive summary tables of the common and scientific names of plant species and given nematode species attacking the major crops. The discipline-oriented chapters range from basic nematode biology, anatomy and systematics, ecology, physiology and molecular biology, genetics, etiology of associated disease complexes, the use of genetic engineering for developing host resistance, to integrated management and information transfer. The commodity-oriented chapters focus primarily on the most important U.S. food-feed-fiber crops: alfalfa, corn, cotton, clover and other forage legumes, forage grasses, fruit and citrus trees, peanut, potato, rice and other cereals, soybean, small fruits, vegetables, and tobacco.

The editors are indebted to the authors of the respective chapters for their excellent contributions that resulted in this monograph. A special thank you is offered to Mrs. Judy Melton who assisted with the editing and the large number of scientists who reviewed selected sections and/or chapters of this book. Finally, we express our gratitude to the ASA, CSSA, and SSSA officials for supporting this endeavor and Jon Bartels and the ASA staff in Madison, Wisconsin, for extensive editing as well as facilitating the printing and distribution of this monograph.

Editors:

Kenneth R. Barker
Department of Plant Pathology
North Carolina State University
Raleigh, North Carolina

Gary A. Pederson
USDA-ARS
Mississippi State University, Mississippi

Gary L. Windham
USDA-ARS
Mississippi State University, Mississippi

CONTRIBUTORS

George S. Abawi	Professor, Department of Plant Pathology, New York State Agriculture Experiment Station, Cornell University, Geneva, NY 14456-0462
Kenneth R. Barker	Professor of Plant Pathology, Department of Plant Pathology, Box 7616, North Carolina State University, Raleigh, NC 27695-7616
J. Ole Becker	Assistant Cooperative Extension Specialist & Assistant Nematologist, Department of Nematology, University of California, Riverside, CA 92521-0415
Ernest C. Bernard	Professor, Department of Entomology and Plant Pathology, University of Tennessee, Knoxville, TN 37996-4500
Bill B. Brodie	Research Leader, USDA-ARS, and Professor, Department of Plant Pathology, Cornell University, Ithaca, NY 14853
Edward P. Caswell-Chen	Associate Professor, Department of Nematology, One Shields Ave., University of California, Davis, CA 95616-8668
Jianjun Chen	Postdoctoral Research Associate, Department of Nematology, University of California, Davis, CA 95616
Mark A. Conkling	Associate Professor, Department of Genetics, Box 7614, North Carolina State University, Raleigh, NC 27695-7614
Don W. Dickson	Professor, Department of Entomology and Nematology, Building 970, Hull Road, P.O. Box 110620, University of Florida, Gainesville, FL 32611-0620
Larry W. Duncan	Professor, Department of Entomology and Nematology, University of Florida, Citrus Research and Education Center, 700 Experiment Station Road., Lake Alfred, FL 33850
Jon D. Eisenback	Professor, Department of Plant Pathology, Physiology, and Weed Science, Virginia Polytechnic Institute and State University, Blacksburg, VA 24061
Joseph Esnard	Senior Research Associate, Department of Plant Pathology, Cornell University, Ithaca, NY 14853
John M. Ferris	Professor of Entomology, Department of Entomology, Purdue University, West Lafayette, IN 47907-1158
Virginia R. Ferris	Professor of Entomology, Department of Entomology, Purdue University, West Lafayette, IN 47907-1158
Diana Wall Freckman	Director, Natural Resource Ecology Laboratory, and Professor of Rangeland Ecosystem Science, Colorado State University, Fort Collins, CO 80523-1499
Gerald D. Griffin	Research Nematologist Emeritus, USDA-ARS, Forage and Range Research, Utah State University, Logan, UT 84322-6300
Kimberly D. Gwinn	Associate Professor, Department of Entomology and Plant Pathology, University of Tennessee, Knoxville, TN 37996
Richard S. Hussey	Professor, Department of Plant Pathology, University of Georgia, Athens, GA 30602-7274
Alva W. Johnson	Research Leader of Nematodes, Weeds and Crops Unit, USDA-ARS, P.O. Box 748, Tifton, GA 31793
Charles S. Johnson	Associate Professor of Plant Pathology, Southern Piedment Agricultural Research & Extension Center, Route 3, Box 60, Virginia Polytechnic Institute and State University, Blackstone, VA 23824

Robert A. Kinloch Associate Nematologist, University of Florida, West Florida Research and Education Center, 4253 Experiment Rd., Highway 182, Jay, FL 32565-9524

William C. Matthews Staff Research Associate, Department of Nematology, University of California, Riverside, CA 92521-0415

Edward C. McGawley Nematologist and Professor, Department of Plant Pathology and Crop Physiology, Louisiana State University, Baton Rouge, LA 70803

Robert McSorley Professor, Department of Entomology and Nematology, P.O. Box 110620, University of Florida, Gainesville, FL 32611-0620

Robert K. Niles Postdoctoral Research Associate, Natural Resource Ecology Laboratory, Colorado State University, Fort Collins, CO 80523

James P. Noe Associate Professor, Department of Plant Pathology, Miller Plant Sciences Bldg., University of Georgia, Athens, GA 30602

Joseph W. Noling Associate Professor, Department of Entomology and Nematology, University of Florida, Citrus Research and Education Center, University of Florida, 700 Experiment Station Road., Lake Alfred, FL 33850

Andrew P. Nyczepir Research Nematologist, USDA-ARS, Southeastern Fruit & Tree Nut Research Laboratory, 21 Dunbar Rd., Byron, GA 31008

Charles H. Opperman Associate Professor, Department of Plant Pathology, Box 7616, North Carolina State University, Raleigh, NC 27695-7616

Charles Overstreet Extension Nematologist, Louisiana Cooperative Extension Service, P.O. Box 25100, Baton Rouge, LA 70894-5100

Gary A. Pederson Research Geneticist, USDA-ARS, Forage Research Unit, P.O. Box 5367, Mississippi State, MS 39762-5367

Kenneth H. Quesenberry Professor of Agronomy, Agronomy Department, P.O. Box 110500, University of Florida, Gainesville, FL 32611-0500

Philip A. Roberts Professor, Nematologist, Department of Nematology, and Associate Dean, College of Natural and Agricultural Sciences, University of California, Riverside, CA 92521-0415

Donald P. Schmitt Professor of Plant Pathology, Department of Plant Pathology, University of Hawaii at Manoa, 3190 Maile Way, St. John 307, Honolulu, HI 96822

Brent S. Sipes Assistant Professor of Plant Pathology, Department of Plant Pathology, University of Hawaii at Manoa, 3190 Maile Way, St. John 307, Honolulu, HI 96822

James L. Starr Professor, Department of Plant Pathology and Microbiology, Texas A&M University, College Station, TX 77843-2132

John C. Veremis Post-Doctoral Researcher, Department of Nematology, University of California, Riverside, CA 92521-0415

Becky B. Westerdahl Associate Professor, Department of Nematology, One Shields Ave., University of California, Davis, CA 95616

Valerie M. Williamson Professor, Department of Nematology, University of California, Davis, CA 95616

Gary L. Windham Research Plant Pathologist, USDA-ARS, P.O. Box 5367, Mississippi State, MS 39762-5367

Lawrence D. Young Research Plant Pathologist, USDA-ARS, Soybean Research, 605 Airways Blvd., Jackson, TN 38301-3201

Bert M. Zuckerman Professor Emeritus, Department of Plant Pathology, University of Massachusetts, Amherst, MA 01003

Conversion Factors for SI and non-SI Units

Conversion Factors for SI and non-SI Units

To convert Column 1 into Column 2, multiply by	Column 1 SI Unit	Column 2 non-SI Units	To convert Column 2 into Column 1, multiply by
		Length	
0.621	kilometer, km (10^3 m)	mile, mi	1.609
1.094	meter, m	yard, yd	0.914
3.28	meter, m	foot, ft	0.304
1.0	micrometer, μm (10^{-6} m)	micron, μ	1.0
3.94×10^{-2}	millimeter, mm (10^{-3} m)	inch, in	25.4
10	nanometer, nm (10^{-9} m)	Angstrom, Å	0.1
		Area	
2.47	hectare, ha	acre	0.405
247	square kilometer, km^2 $(10^3\ m)^2$	acre	4.05×10^{-3}
0.386	square kilometer, km^2 $(10^3\ m)^2$	square mile, mi^2	2.590
2.47×10^{-4}	square meter, m^2	acre	4.05×10^3
10.76	square meter, m^2	square foot, ft^2	9.29×10^{-2}
1.55×10^{-3}	square millimeter, mm^2 $(10^{-3}\ m)^2$	square inch, in^2	645
		Volume	
9.73×10^{-3}	cubic meter, m^3	acre-inch	102.8
35.3	cubic meter, m^3	cubic foot, ft^3	2.83×10^{-2}
6.10×10^4	cubic meter, m^3	cubic inch, in^3	1.64×10^{-5}
2.84×10^{-2}	liter, L (10^{-3} m^3)	bushel, bu	35.24
1.057	liter, L (10^{-3} m^3)	quart (liquid), qt	0.946
3.53×10^{-2}	liter, L (10^{-3} m^3)	cubic foot, ft^3	28.3
0.265	liter, L (10^{-3} m^3)	gallon	3.78
33.78	liter, L (10^{-3} m^3)	ounce (fluid), oz	2.96×10^{-2}
2.11	liter, L (10^{-3} m^3)	pint (fluid), pt	0.473

	Mass		
2.20×10^{-3}	gram, g (10^{-3} kg)	pound, lb	454
3.52×10^{-2}	gram, g (10^{-3} kg)	ounce (avdp), oz	28.4
2.205	kilogram, kg	pound, lb	0.454
0.01	kilogram, kg	quintal (metric), q	100
1.10×10^{-3}	kilogram, kg	ton (2000 lb), ton	907
1.102	megagram, Mg (tonne)	ton (U.S.), ton	0.907
1.102	tonne, t	ton (U.S.), ton	0.907
	Yield and Rate		
0.893	kilogram per hectare, kg ha^{-1}	pound per acre, lb $acre^{-1}$	1.12
7.77×10^{-2}	kilogram per cubic meter, kg m^{-3}	pound per bushel, lb bu^{-1}	12.87
1.49×10^{-2}	kilogram per hectare, kg ha^{-1}	bushel per acre, 60 lb	67.19
1.59×10^{-2}	kilogram per hectare, kg ha^{-1}	bushel per acre, 56 lb	62.71
1.86×10^{-2}	kilogram per hectare, kg ha^{-1}	bushel per acre, 48 lb	53.75
0.107	liter per hectare, L ha^{-1}	gallon per acre	9.35
893	tonnes per hectare, t ha^{-1}	pound per acre, lb $acre^{-1}$	1.12×10^{-3}
893	megagram per hectare, Mg ha^{-1}	pound per acre, lb $acre^{-1}$	1.12×10^{-3}
0.446	megagram per hectare, Mg ha^{-1}	ton (2000 lb) per acre, ton $acre^{-1}$	2.24
2.24	meter per second, m s^{-1}	mile per hour	0.447
	Specific Surface		
10	square meter per kilogram, m^2 kg^{-1}	square centimeter per gram, cm^2 g^{-1}	0.1
1000	square meter per kilogram, m^2 kg^{-1}	square millimeter per gram, mm^2 g^{-1}	0.001
	Pressure		
9.90	megapascal, MPa (10^6 Pa)	atmosphere	0.101
10	megapascal, MPa (10^6 Pa)	bar	0.1
1.00	megagram, per cubic meter, Mg m^{-3}	gram per cubic centimeter, g cm^{-3}	1.00
2.09×10^{-2}	pascal, Pa	pound per square foot, lb ft^{-2}	47.9
1.45×10^{-4}	pascal, Pa	pound per square inch, lb in^{-2}	6.90×10^3

(continued on next page)

Conversion Factors for SI and non-SI Units (continued)

To convert Column 1 into Column 2, multiply by	Column 1 SI Unit	Column 2 non-SI Units	To convert Column 2 into Column 1, multiply by
	Temperature		
1.00 (K – 273)	Kelvin, K	Celsius, °C	1.00 (°C + 273)
(9/5 °C) + 32	Celsius, °C	Fahrenheit, °F	5/9 (°F – 32)
	Energy, Work, Quantity of Heat		
9.52×10^{-4}	joule, J	British thermal unit, Btu	1.05×10^{3}
0.239	joule, J	calorie, cal	4.19
10^{7}	joule, J	erg	10^{-7}
0.735	joule, J	foot-pound	1.36
2.387×10^{-5}	joule per square meter, $J\ m^{-2}$	calorie per square centimeter (langley)	4.19×10^{4}
10^{5}	newton, N	dyne	10^{-5}
1.43×10^{-3}	watt per square meter, $W\ m^{-2}$	calorie per square centimeter minute (irradiance), $cal\ cm^{-2}\ min^{-1}$	698
	Transpiration and Photosynthesis		
3.60×10^{-2}	milligram per square meter second, $mg\ m^{-2}\ s^{-1}$	gram per square decimeter hour, $g\ dm^{-2}\ h^{-1}$	27.8
5.56×10^{-3}	milligram (H_2O) per square meter second, $mg\ m^{-2}\ s^{-1}$	micromole (H_2O) per square centimeter second, $\mu mol\ cm^{-2}\ s^{-1}$	180
10^{-4}	milligram per square meter second, $mg\ m^{-2}\ s^{-1}$	milligram per square centimeter second, $mg\ cm^{-2}\ s^{-1}$	10^{4}
35.97	milligram per square meter second, $mg\ m^{-2}\ s^{-1}$	milligram per square decimeter hour, $mg\ dm^{-2}\ h^{-1}$	2.78×10^{-2}
	Plane Angle		
57.3	radian, rad	degrees (angle), °	1.75×10^{-2}

	Electrical Conductivity, Electricity, and Magnetism		
10	siemen per meter, S m^{-1}	millimho per centimeter, mmho cm^{-1}	0.1
10^4	tesla, T	gauss, G	10^{-4}
	Water Measurement		
9.73×10^{-3}	cubic meter, m^3	acre-inches, acre-in	102.8
9.81×10^{-3}	cubic meter per hour, m^3 h^{-1}	cubic feet per second, ft^3 s^{-1}	101.9
4.40	cubic meter per hour, m^3 h−1	U.S. gallons per minute, gal min^{-1}	0.227
8.11	hectare-meters, ha-m	acre-feet, acre-ft	0.123
97.28	hectare-meters, ha-m	acre-inches, acre-in	1.03×10^{-2}
8.1×10^{-2}	hectare-centimeters, ha-cm	acre-feet, acre-ft	12.33
	Concentrations		
1	centimole per kilogram, cmol kg^{-1}	milliequivalents per 100 grams, meq 100 g^{-1}	1
0.1	gram per kilogram, g kg^{-1}	percent, %	10
1	milligram per kilogram, mg kg^{-1}	parts per million, ppm	1
	Radioactivity		
2.7×10^{-11}	becquerel, Bq	curie, Ci	3.7×10^{10}
2.7×10^{-2}	becquerel per kilogram, Bq kg^{-1}	picocurie per gram, pCi g^{-1}	37
100	gray, Gy (absorbed dose)	rad, rd	0.01
100	sievert, Sv (equivalent dose)	rem (roentgen equivalent man)	0.01
	Plant Nutrient Conversion		
	Elemental	***Oxide***	
2.29	P	P_2O_5	0.437
1.20	K	K_2O	0.830
1.39	Ca	CaO	0.715
1.66	Mg	MgO	0.602

1 Introduction and Synopsis of Advancements in Nematology

KENNETH R. BARKER

North Carolina State University
Raleigh, North Carolina

Production of key food crops in several countries has more than doubled during the last 50 yr (Avery, 1995). World food production will likely need to double again during the next 30 to 50 yr to meet the projected population increase. New knowledge has been central to the unprecedented success of agriculture, and this must continue. Clearly, high-yielding, disease-resistant crop cultivars, as well as much improved agronomic practices and integrated pest/crop management, were key ingredients to past and current progress in food production (Roberts, 1993). Nematodes, including soilborne pathogenic and beneficial species, are an important component of this picture.

Nematodes, the most abundant multicellular organisms on earth, are found in all habitats and ecosystems of the biosphere. They occur in soil, decaying organic matter, all forms of plant life and most animals, including domesticated and wild species (Norton, 1978). Billions of nematodes may live in a hectare of farm land, or mud at ocean bottoms, as well as in forest floors. Marine, fresh water, and soil nematodes make up the majority of these organisms. To date, some 15 000 species of nematodes have been described, and the total number of species may exceed 500 000 (Poinar, 1983). Although many nematodes parasitize plants or animals, others are beneficial as they contribute to nutrient cycling or serve as biological control agents of plant pests (Stirling, 1991; Yeates, 1996).

Nematodes are defined as microscopic, roundworm invertebrates with a body cavity and complete digestive tract; they are nonsegmented, appendageless, and have bilateral symmetry (Hirschmann, 1971; Poinar, 1983). The digestive tract includes a stoma (mouth), alimentary canal (esophagus and intestine), and anus. They have a complex nervous system, a secretory-excretory system, a reproductive system and musculature system (longitudinal muscles), but have no specialized respiratory or circulatory systems. The cuticle-covered body of most nematodes species is transparent, which allows the delineation of sufficient anatomical detail to identify most genera under a stereoscopic microscope, typically at 20 to 60× magnification. The study of nematodes is addressed by numerous disciplines, including medicine, parasitology, ecology, entomology, biology, and plant pathology. During the last four decades, an independent, but limited discipline of *nematology* has developed in many countries throughout the world.

 Plant and Nematode Interactions, Agronomy Monograph no. 36.

Many types of nematodes are associated with all forms of plant life, including crop plants growing in the field, garden, orchard, turf, and the diverse plant species that occur in natural ecosystems (Fig. 1–1; Table 1–1). Plant-parasitic nematodes represent one of the most important groups of organisms that affect the growth and yield of crop plants (Table 1–2). Because of their microscopic size, however, most species of these pathogens were overlooked until effective nematicides and resistant crop cultivars were developed in the 1940s and 1950s. Unfortunately, they continue to be overlooked by many agriculturalists.

HISTORY

In light of the high population numbers of nematodes, N. A. Cobb (1915)—who is considered to be the father of American nematology—provided a dramatic description of the abundance of nematodes. He stated, "...if all the matter in the universe except the nematodes were swept away, our world still would be dimly recognizable—we would find its mountaintops, valleys, rivers, lakes and oceans represented by a film of nematodes." Two of the first records of parasitic nematodes are given by passages from the Chinese Yellow Emperor's Classic (2700 B.C.) and the Papyrus Ebers (dated around 1550 B.C.). Chinese "literature" as early as 235 B.C. includes descriptions that may refer to a white *Heterodera* female and an attached egg mass on soybean root (Noel, 1992). The Papyrus Ebers suggested a knowledge of the human intestinal parasite *Ascaris lumbricoides* and the tissue parasite *Dracunculus medinensis* (Poinar, 1983; Viglierchio, 1991). Both of these nematode species were quite large parasites, often exceeding 30 cm in length. In 1656, Borellus recorded his observations "Vinegar eels" the microbivorous or free-living nematode occurring in most vinegar at that time (Thorne, 1961). The statement "sowed cockle, reap'd no corn" in Shakespeare's "Love's labour's lost", Act 4, Scene 3, is suggested by Thorne (1961) as possibly our first record of plant-parasitic nematodes in 1594. The nematode that Thorne suspected to be in that reference actually was described by Needham in 1743. Thorne (1961) indicates that Needham, a clergyman, resolved the riddle of the "cockle" as he crushed shrunken, blackened wheat grains under his primitive microscope and discovered the first plant-parasitic nematode, now known as the wheat gall nematode, *Anguina tritici*. The following excerpt of Needham's description, as published by Thorne (1961), reflects the excitement involved in this discovery as well as the early stages of scientific thought in the mid-1700s:

Upon opening lately the small, black grains of smutty wheat, which they here distinguished from blighted corn, the latter affording nothing but a black dust, into which the whole substance of the ear is converted; perceived as soft, white fibrous substance, a small portion of which I placed upon my object-plate: it seemed to consist wholly of longitudinal fibers bundled together; and you will be surprised, perhaps that I should say, without any least sign of life or motion. I dropped a globule of water upon it, in order to try if the parts, when separated, might be viewed more conveniently; when to my great surprise these imaginary fibers, as it were, separated from each other, took life, moved irregular, not without a progressive, but twisting motion, and continued to do so, for the space of

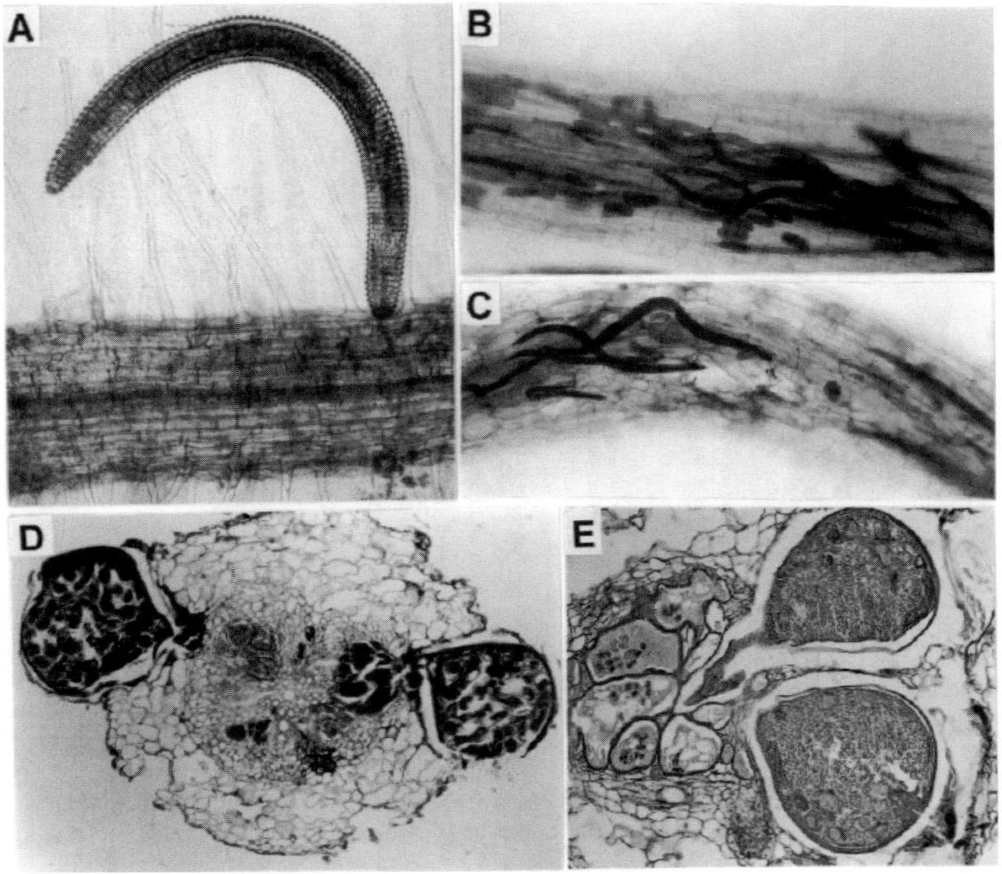

Fig. 1–1. Illustrations of types of plant-parasitic nematodes feeding on plant tissues: (*A*) Ectoparasite, *Criconemella xenoplax* feeding on a plant root (courtesy of S. W. Westcott, III and R. S. Hussey); (*B* and *C*) Migratory endoparasite, *Pratylenchus scribneri* in corn roots (courtesy of E. L. Davis and Luis Payan); (*D*) Mature adults (cysts) of *Heterodera glycines* on soybean rootlet; (*E*) *Meloidogyne incognita* adult (pear shape) females feeding on nurse or giant cells inside a galled tobacco root.

Table 1–1. Primary groups of nematodes, their common names and key hosts.†

Scientific name	Common name	Principal hosts
	Cyst-forming nematodes‡ (Nematodes that form cysts covering the eggs)	
	Lemon-shaped cysts	
Heterodera		
H. avenae	Oat nematode	Small grains, corn, grasses
H. carotae	Carrot-root nematode	Carrot
H. cruciferae	Cabbage-root nematode	Cabbage, other crucifers
H. glycines	Soybean-cyst nematode	Soybean
H. goettingiana	Pea-root nematode	Pea
H. humuli	Hop-nematode	Hop, hemp
H. schachtii	Sugar-beet nematode	Cole crops, celery, cress, kale, weeds, radish, spinach, turnip, sugar beet
H. trifolii	Clover-cyst nematode	Clover, legumes
H. zeae	Corn-cyst nematode	Corn
	Round cyst nematodes	
Globodera		
G. rostochiensis	Golden (potato cyst)	Irish potato, tomato, eggplant
G. pallida		Potato
G. tabacum	Tobacco cyst nematode	Tobacco, tomato, eggplant
Punctodera		
P. punctata	Grass cyst nematode	Wheat, other small grains
P. chalcoensis	Mexican corn cyst nematode	Corn
	Endoparasites [Nematodes that enter into the root tissue]	
Ditylenchus		
D. destructor	Potato rot nematode	Potato, dandelion
Meloidogyne	Root-knot nematodes	
M. arenaria	Peanut root-knot nematode	1700 hosts, including forages, small grains, grasses, fruit, vegetables, field crops, nursery, weeds
M. exigua	Coffee root-knot nematode	
M. hapla	Northern root-knot nematode	
M. incognita	Southern root-knot nematode	
M. javanica	Javanese root-knot nematode	
M. chitwoodi	Columbia root-knot nematode	Potato, alfalfa, carrot
Meloidodera	Cystoid nematode	Pine
Nacobbus	False root-knot nematode	
N. batatiformis		Sugar beet
N. aberrans		Potato, tomato
	Root-lesion nematodes	
Pratylenchus		
P. brachyurus	Smooth-headed nematode	Potato, corn, peanut, cotton, pineapple, avocado, tobacco
P. coffeae	Coffee nematode	Coffee, tea, abaca, sugarcane, banana, olive, apple
P. neglectus		Wheat, tobacco, corn
P. penetrans		Nursery, grasses, forages, oats, strawberry, apple, tobacco, tomato
P. pratensis		Grains, grasses, strawberry, lily, narcissus

(continued on next page)

Table 1–1. Continued.

Scientific name	Common name	Principal hosts
P. scribneri	Scribner's Meadow nematode	Potato, strawberry
P. thornei	Thorne's Meadow nematode	Wheat, grains, grasses
P. vulnus	Lesion nematode	Ornamentals, trees, forages, fruit, beans, walnut, grape, olive, fig, boxwood, stone fruits, citrus, raspberry, rose
P. zeae	Corn nematode	Corn, wheat
Radopholus		
R. oryzae	Rice nematode	Rice
R. similis	Burrowing nematode	Banana, citrus, ornamentals, avocado
Rotylenchulus		
R. reniformis	Kidney-shaped or Reniform nematode	Turf, cotton, peanut, tomamto weeds, ornamentals
Tylenchulus		
T. semipenetrans	Citrus nematode	Citrus, olive
	Ectoparasitic nematodes (Nematodes that feed from the root surface and normally do not bodily enter the root tissue)	
Belonolaimus	Sting nematodes	
B. gracilis		Turf, strawberry, corn, cotton,
B. longicaudatus		ornamentals, soybean, celery, vegetables, peanut, pine
Cacopaurus		
C. pestis	Walnut nematode	Walnut
Criconema	Pine nematodes	Pine trees
Criconemella (See *Mesocriconema*)	Ring nematodes	
Dolichodorus		
D. heterocephalus	Awl nematode	Celery, beans, corn, tomato, turf, water chestnut
Helicotylenchus‡	True spiral nematodes	
H. dihystera		Tomato, clover, soybean, turf
H. multicinctus		Banana
Hemicycliophora	Sheath nematode	Evergreens, trees, ornamentals
Hoplolaimus‡	Lance nematodes	Turf, nursery, corn,
H. galeatus		sugarcane, alfalfa, red
H. columbus	Columbia-lance nematodes	clover, soybean, cotton
Longidorus	Needle nematode	Grasses, peppermint, corn
Mesocriconema	Ring nematodes	
M. ornatum		Turf, cotton, peanut
M. xenoplax		Deciduous fruit, ornamentals
Paratylenchus	Pin nematodes	Celery, many vegetables, fig, trees
Rotylenchus	Spiral nematodes	Boxwood, ornamentals
Paratrichodorus		
(and *Trichodorus*)	Stubby-root nematodes	Turf, corn, celery, cole crops
P. minor		bean, pea, tomato, cotton, cowpea, beet, pepper, weeds
Tylenchorhynchus	Stylet nematodes	

(continued on next page)

Table 1–1. Continued.

Scientific name	Common name	Principal hosts
T. claytoni	Tobacco stunt nematode	Tobacco, corn, cotton, azalea, strawberry, alfalfa, peanut, bean, ornamentals
Xiphinema	Dagger nematodes	
X. americanum		Cotton, strawberry, oats, tobacco, peach, nursery
X. index		Grape
	Above-ground feeders (Nematodes that feed within plant tissue above ground)	
Anguina	Seed gall nematodes	
A. agrostis	Grass nematode	Grasses
A. tritici	Wheat nematode	Rye, wheat
Aphelenchoides	Foliar nematodes	
A. besseyi		Vanda orchid, causes white tip rice, strawberry, African violet
A. fragariae	Bud and leaf nematode	Flax, mint, onions, oats, sweet potato, nursery, begonia, fern, iris, lily
A. ritzemabosi	Bud and leaf nematode	Strawberry, white clover, ribes orchid, violet, chrysanthemum
Bursaphelenchus		
B. xylophilus	Pinewood nematode	Pine trees
Ditylenchus		
D. dipsaci	Stem and bulb, or Teasel nematode	Alfalfa, clover, onion, garlic, sweet potato, strawberry, nursery, corn, tobacco
Rhadinaphelenchus	Red ring nematode	
R. cocophilus		Coconut palm, oil palm

† In part, after "Nematode Chart" originally developed for early Nematode Workshops, sponsored by Shell Chemical Company.

‡ Often considered to be endoparasites or semi-endoparasites.

nine or ten hours when I threw them away. Now I am satisfied that they are a species of aquatic animals, and may be denominated worms, eels or serpents, which they very much resemble.

More than 100 yr elapsed before the second plant-parasitic nematode, the root-knot nematode, was discovered by another clergyman, Berkeley, in 1855. Many additional plant-parasitic nematodes were described in the mid-1800s through the early 1990s (Thorne, 1961; Nickle, 1991). During the 1930s to 1940s, Chitwood (see Chitwood & Chitwood, 1974) provided unmatched descriptions of the basic anatomy and general physiology of nematodes. In reference to nematode species causing damage on plants, the focus until 1951 was limited entirely to those that feed internally in plant tissues. At that time, Christie and Perry (1951) demonstrated that "ectoparasitic nematodes" could, in fact, also damage plants by inserting their stylets into plant roots. The majority of more than 1200 known species of plant-parasitic nematodes have been described during the last four decades of rapidly developing scientific information on all aspects of nematology. This includes initiation and establishment of many new journals such as

Table 1–2. Estimated annual losses due to nematodes for selected world crops.†

Crops‡	Number of estimates per crop	Food and Agriculture Organization production estimates	Estimated yield losses due to nematodes
		1000 metric t	%
Banana	78	2 097	19.7
Barley	49	171 635	6.3
Cassava	25	129 020	8.4
Citrus	102	56 100	14.2
Cocoa	13	1 660	10.5
Coffee	36	5 210	15.0
Corn	125	449 255	10.2
Cotton (lint only)	85	17 794	10.7
Field bean	70	19 508	10.9
Oat	37	43 355	4.2
Peanut	69	20 611	12.0
Potato	141	312 209	12.2
Rice	64	469 959	10.0
Sorghum	53	71 698	6.9
Soybean	91	89 893	10.6
Sugar beet	51	293 478	10.9
Sugar cane	65	935 769	15.3
Sweet potato	67	117 337	10.2
Tea	16	2 218	8.2
Tobacco	92	6 205	14.7
Wheat	89	521 682	7.0

† After Sasser and Freckman, 1987.
‡ See Appendix 2 for scientific names.

the *Journal of Nematology*, *Nematologica*, *Fundamental and Applied Nematology*, *Nematologia Mediterranea*, *Indian Journal of Nematology*, *Nematropica*, *The Pakistan Journal of Nematology*, and most recently, *The Russian Journal of Nematology*. An enumeration of some of the major developments in plant and soil nematology are presented in Table 1–3. As this compilation of advancements indicates, the study of nematodes could be considered to still be in an early phase.

An interesting treatment on the history and biology of nematodes is offered by Poinar (1983). He indicates that fossil records of nematodes are very sparse, and concludes that the available fossils are not old enough geographically to indicate anything about the origin of specific nematode groups. The authentic nematode fossils have been preserved in amber, a fossilized resin, dating back between 20 and 120 million years. Although amber arises from numerous tree species in many parts of the world, nematode fossils have been recovered primarily in amber deposits collected from the Baltic Sea area, Mexico, and the Dominican Republic (Poinar, 1983). For example, a piece of amber from Chiapas, Mexico, deposited about 20 million years ago, contained eggs, juveniles, and adults of the plant-parasitic Aphelenchoid, *Oligaphelenchoides atrebora*. Species of invertebrate parasitic nematodes in the Mermithidae have been found in Baltic and Dominican amber.

As indicated earlier, the development of nematology in the USA was catalyzed in large part by N. A. Cobb after joining the USDA in 1907 (Thorne, 1961). In addition to the many beautiful illustrations that he and his associate, W. E. Chambers, developed for various nematode taxa, he made major contributions

Table 1–3. Chronology of major advancements in the study of plant and soil nematodes with emphasis on developments in USA.

Advancement	Dates	Early contributors	Impact† (Immediate and/or delayed effects)
Discovery of first plant-parasitic nematode (*Anguina tritici*)	1743	Needham	Concept of disease causality; Andydrobiosis
Discovery of root-knot nematode	1855	Berkeley	Led to study of plant nematodes
Discovery of bulb-stem and sugar-beet nematodes	1857, 1859	Kühn; Schacht	Broadened concepts of plant parasites
Soil fumigation	1858–1880	Kühn et al.; Schmidt	Essential understanding of related crop damage
Disease complexes involving root-knot nematodes	1880s	Atkinson; Goeldi	Broadened concept of disease causality
Multiple tactics for nematode control	1880s–1911	Atkinson; Bessey	Set stage for integrated management
Early classical systematics/ taxonomy	1890s–1930s	Bastian; Cobb; deMan; Filipjev; T. Goodey; Micoletzky; Steiner; S. Stekhoven; Thorne	Established initial criteria for nematode types and taxa
Early Integrated nematode controls	1911–1930s	Bessy; Tyler	Early integrated nematode management
Concept of biological control	1930s	Duddington; Linford; Tyler	Natural management tactic
Nematode anatomy and physiology	1930s	Chitwood	Basis for much of current understanding/knowledge of nematodes
Discovery of the potato cyst nematode in the USA	1941	Betz; Buhrer; Chitwood; Feldmesser, Mai	Effective regulatory measures to limit spread
Discovery of nematicidal activity of "D-D"	1943	Carter	Definitive method to show and asses dramatic crop damage
Taxonomy of *Meloidogyne*	1930s–1940s	Chitwood; Christie	Provided sound criteria for species and host races
Discovery of the role of ectoparasites in plant disease	1940s–1950s	Christie & Perry	Broadened concepts of parasitism-pathogenesis
Agronomic importance of disease complexes	1950s	Hewitt; Mountain; Nusbaum; Pitcher; N. Powell; Weischer	Documented role of nematodes in predisposing plants to other parasites, including "breaking" of resistance
"Spreading" decline of citrus	1920s–1950s	Feldmesser; Rebóis & A. Taylor; Suit & DuCharme	Showed extreme damage potential of nematodes on major fruit trees
Regulatory programs	1940s–1990s	Esser; Hart; Mai; O'Bannon; Spears; A. Taylor	Contributed to containment of two or more major nematodes (Golden nematode of potato, burrowing nematode of citrus)
Discovery of soybean cyst nematode in USA	1954	Winstead & Sasser	Contributed much to study of nematodes as a science
Development of Federal-State research projects and workshops in USA	1950s	Allen; Cairns; Mai; Norton; Raski; Sasser; Throne	Established funding and means of collaborative research and graduate education, and training of first large group of nematologists in USA

"Axenic and monoxenic" culturing of nematodes	1914–1950s	Byars; Daugherty; Krusberg; Mountain; Nicholas; Tyler‡	Proved pathogenicity and enhanced study of physiology
Early host resistance to nematodes	1911–1950s	Barrons; Bingefors; Burkhart; Clayton; Ellenby; Mai & Peterson; Orton; Ross & Brim; Smith‡	Provided an ideal means of control
Methods of extraction and population assessment	1917–1960s	Baermann; Christie & Perry; Cobb‡	Provided basis for diverse studies on population biology and related crop-loss assessments
Development of educational centers for nematology	1950s	Allen; Christie; T. Goodey & J.B. Goodey; Jones; Mai; Oostenbrink; V. Perry; Raski; Sasser; Seinhorst; Sher; Steiner; Thomason; Thorne; Van Gundy‡	Advanced specialized education in nematology
Discovery of nematodes vectoring plant viruses	1956	Hewitt; Raski and Goheen; C.E. Taylor; Weischer	Further broadened role of nematodes in disease etiology
Concepts of population dynamics and ecology	1950s–1980s	Godfrey; Jones; Norton; Oostenbrink; Seinhorst; Wallace‡	Developed systems for ecology-based control/ management
Development of nonfumigant nematicides	1950s–1960s	Largely industry-oriented	Highly effective, short-term control
General and integrated management	1930s–1990s	Baines; Griffin; Jensen; McBeth; L.I. Miller; N.A. Minton; Overman; Reynolds;Tyler; Thomason‡	Interfacing management options on an ecologically sound basis
Biological control of plant-parasitic nematodes	1930s–1990s	Barron; Duddington; Mankau; Linford; Sayre; Zuckerman‡	Enhanced understanding of antagonists and related biology enhancing potential for practical biocontrol
Extensive systematics of many nematode groups	1940s–1990s	Allen; Andrássy; Baker; Coomans; DeConinck; Franklin; Geraert; Golden; B. Goodey; Hirschmann; Ichinohe; Jairajpuri; Khan; Kivjanova; Loof; Lordello; Luc; Maggenti; Meyl; Mulvey; Paramonov; Raski; Sher; Siddiqi; Steiner; Stone; Tarjan; Thorne‡	Provided concept and insight into diversity and enormity of nematode communities
Cytogenetics	1950s–1990s	Triantaphyllou‡	Enhanced understanding of evolution and taxa interrelationships
Modes of reproduction/ sexuality	1950s–1990s	Triantaphyllou‡	Provided information data base for genetics/molecular research
Survival mechanisms	1940s–1990s	Van Gundy‡	Fundamental knowledge and facilitated practical control

(continued on next page)

Table 1–3. Continued.

Advancement	Dates	Early contributors	Impact† (Immediate and/or delayed effects)
Genetic and physiological diversity in major species	1960s–1990s	Dropkin; Jones; Ludders; L.I. Miller; Ross; Sasser; Triantaphyllou‡	Facilitated development of durable resistant cultivars and recent molecular nematology
Damage functions/advisory programs	1960s–1990s	Oostenbrink; Seinhorst; Jones‡	Objective basis for nematode management
Population simulators	1960s–1990s	Jones‡	Facilitated effectiveness of research and management of nematodes
Tropical and international programs such as "International *Meloidogyne* Project"	1950s–1990s	Caveness; Jones; Oostenbrink; Sasser; Seinhorst‡	Expanded educational base of nematologists world wide and provided ecological-taxonomic data base
Physiology, behavior, and host-parasite relationships, including nature of feeding cells	1960s–1990s	A. Bird; Bergeson; Cohn; Croll; Dropkin; Krusberg; Shepherd; Van Gundy; Veech; Viglierchio‡	Database essential for molecular research and genetics engineering
Nematode ultrastructure	1960s–1990s	A. Bird; Endo; Hirschmann; Wright‡	Contributed to understanding of structure and function
Biological control of insects with nematodes	1960s–1990s	Dutky; Glaser; Nickle; Poinar; Steiner‡	Decades of research now resulting in effective control of some insects
Caenorhabditis elegans developmental biology and genetics-model system	1960s–1990s	Brenner; Dougherty; Nicholas‡	Providing fundamental information on cell lineage, behavior, gene function, aging, and overall genome for this model biological system
Molecular systematics/ taxonomy	1960s–1990s	‡	Provided new tools for identification and qualitative population assessments
Nutrient cycling (soil-plant system)	1980s–1990s	‡	Information being developed providing possible means of enhancing "soil health'" via practices that favor beneficial nematodes
Molecular markers for resistance genes	1990s	‡	Increasing efficiency of breeding for resistance
Molecular analysis of host-parasite interactions	1990s	‡	Providing fundamental knowledge on mechanisms of pathogenesis
Cloning of resistance genes	1990s	‡	Elucidation of the molecular mechanisms of resistance
Transgenic host resistance to plant-parasitic nematodes	1990s	‡	Great potential, but limited to model systems to date

† This compilation is somewhat subjective, but the kind reviews and inputs by many nematologists hopefully have minimized this problem.

‡ Active and recently retired scientists are not listed under their research activities (due to large numbers and risks of failing to include key individuals).

to areas of sampling and population assessments for plant-parasitic nematodes as well as developing a nematology section in the USDA (Sayre, 1994). G. F. Atkinson (1889) in the 1880s made numerous contributions to our early knowledge of root-knot nematodes, including the Fusarium root-knot disease complex on cotton (*Gossypium hirsutum* L.). Ernest R. Bessey (1911) also contributed greatly to the early information on plant-parasitic nematodes, especially through his 1911 review on these organisms. That review included options for nematode management such as destruction of infected roots after crop harvest.

Thorne (1961) provides a detailed description of the major USDA scientists and other nematologists contributing during the early part of this century. A number of developments during the 1940s and 1950s gave great impetus to the study of plant-parasitic nematodes. First, the discovery of an effective fumigant (DD; 1,3-dichloropropene-1,2-dichloropropane) by Carter in 1943 provided a means of effective nematode control as well as a method for demonstrating the losses caused by these plant pathogens. In 1941, the golden nematode was detected on Long Island, New York, and apparently had been introduced as military equipment was returned from Europe in World War I (Brodie, 1993). The demonstration of potential damage that ectoparasitic nematodes cause was published by Christie and Perry in 1950 and 1951.

The importance of predisposing plants to attack by fungi and bacteria in root-disease and wilt complexes, as well as their serving as vectors of certain plant viruses, did not receive emphasis until the 1950s and 1960s (Mountain, 1960; Lucas et al., 1955; Hewett et al., 1958). Atkinson (1889) and Goldi (1887) had published early findings that indicated root-knot nematodes predispose plants to fungi. The success of plant breeders in collaboration with nematologists in developing crop cultivars with resistance to nematodes such as root-knot, bulb and stem, and cyst nematodes also contributed greatly in documenting the detrimental effects of these parasites on the production of vegetable, fruit, and field crops (Bradley & Duffy, 1982; Evans, 1995; Roberts, 1993).

Additional and diverse treatments of the history and development of nematology include those by Evans (1995), Coomans (1992), Poinar (1988), Sayre (1994), and Viglierchio (1991). Evans (1995) provides a synopsis of the development of plant nematology and an interesting treatment of current developments in molecular nematology, including active scientists. The many important past and current contributions to nematology by scientists in the "low land countries" in Europe are discussed by Coomans (1992). Milestone developments related to various aspects of general science and nematology are outlined by Poinar (1988) and Viglierchio (1991). Finally, an interesting perspective of the importance of art in science, especially as utilized by N. A. Cobb and his associates, is provided by Sayre (1994).

In recent years, new technologies are bringing about a revolution in the nature of the study of all nematodes with the microbivorous species, *Caenorhabditis elegans,* serving as a model organism for many investigations on molecular genetics (Wood, 1988). In addition to advancing our understanding of all types of nematodes, molecular techniques offer great promise for genetically engineering nematode resistance in a wide range of crop plants, thereby making available new tactics for the management of these pathogens [see Chapters 6

(McSorley, 1998), 8–14(Noe, 1998; Sipes & Schmitt, 1998; Young, 1998; Roberts et al., 1998; Opperman & Conkling, 1998; Duncan & Noling, 1998; Westerdahl & Caswell-Chen, 1998)]. The potential of biological control also is receiving much study. In contrast, the options for traditional management of nematodes have been greatly reduced by the removal of a number of the most effective and cost-efficient nematicides, particularly the fumigants (Evans, 1995). Thus, as the scientific community strives to develop sustainable management systems for natural ecosystems (Lubchenco et al., 1991) and for all crop pests (Miller, 1993; Natl. Res. Council, 1991) including nematodes, integration of new technologies and traditional management tactics is being given a high priority (Evans, 1995; Persley, 1992: Roberts, 1993; Univ. California Div. of Natural Resour., 1992; Whitten, 1992).

GENERAL MORPHOLOGY

Nematodes vary greatly in shape and size. For example, a very small marine nematode, *Greeffiella minutum*, which is only 82 μm in length may be compared to another species, *Placentonema gigantissima*, which lives in the placenta of the sperm whale and may reach the length of 8 m (Poinar, 1983). In contrast, most species of plant-parasitic nematodes are 1 to 2 mm in length, but may vary in shape from filiform to fusiform, pear-shaped, lemon-shaped or kidney-shaped.

Most phytonematologists place nematodes in the phylum *Nematoda* (Nemata) and in two classes: the Adenophorea and the Secernentea (Maggenti, 1991). Nematodes also may be placed in the phylum Aschelminthes or Nemathelminthes. Some 18 orders make up the class Adenophorea, whereas the Secernentea has six orders (Maggenti, 1991). Plant-parasitic nematodes occur in only two orders: Dorylaimida (Adenophorea) and Tylenchida (Secernentea), with the majority being in the latter. Key characteristics of the Adenophorea include a one-cell, secretory-excretory system, a presence of hypodermal glands, males with two testes, females that produce limited numbers of eggs, an odontostyle (plant-parasitic species), setae on the lips, and a unique characteristic is the absence of a phasmid on the cuticle. In contrast, members of the Secernentea have a many-cell excretory system, no hypodermal glands, males usually with one testis, females that produce many eggs, a stomatostyle in plant-parasitic forms, lips with papillae, and phasmids present on the exterior cuticle. Many additional cuticular markings or modifications, such as striations, annulations, punctations, longitudinal incisures and openings are often characteristic of given nematode taxa. The morphology, taxonomy, and classification of nematodes are treated in Chapter 3 (Eisenback, 1998).

Traditionally, the taxonomy and systematics of nematodes have been based largely on morphological and physiological differences, and to a lesser extent, reproductive isolation and general ecological differences (Hirschmann, 1971; Luc et al., 1990). For example, given species will have two ovaries, whereas others have only one. The presence or absence of males also is helpful in nematode identification. More recently, biochemical and molecular techniques have come to the forefront in the taxonomic characterization of nematodes (Curran & Robinson, 1993; Ferris & Ferris, 1992). For disease diagnostics, in addition to the application of morphological characteristics and molecular biology, other criteria, including the use of differential host tests and associated host symptoms are often utilized (Barker & Davis, 1996).

DIVERSITY AMONG NEMATODES

Major Species of Plant-Parasitic Nematodes, Their Common Names and Hosts

Major trophic groups of nematodes include plant parasites, animal parasites, bacterivores, fungivores, predators, and omnivores (Norton, 1978). In addition to morphological differentiations, nematodes may be grouped on the basis of their feeding habits and general niches. In fact, nematodes with varying degrees of niche adaptation are found feeding and/or parasitizing all plants and animals. Some of the best known nematodes are animal parasites, including heartworms, pinworms, hookworms, and the guinea worm. The major plant-parasitic species of nematodes, including their scientific and common names, host niche and feeding habits, and their principal hosts are summarized in Table 1–1. Nematode common names may be based on associated symptoms and/or signs of infection, their general shape, or their host or combination of these characteristics. Most listings of nematodes are differentiated primarily on host-parasite relationships (i.e., endoparasites, ectoparasites) (Fig. 1–1). The separation into groups given in Table 1–1 is based on whether the adult females form cysts that protect eggs for long periods of time, true endoparasitic forms, ectoparasitic forms on roots, and aboveground feeders of plant tissues. These aspects of the host-parasite relationships are discussed for given crop hosts in several ensuing chapters.

Plant-parasitic nematodes generally have a similar and rather simple life cycle, including six life stages: the egg, four juvenile stages, and an adult stage. The developing embryo becomes the first-stage juvenile inside the egg. For most species, this juvenile molts before hatching from the egg and becomes the second-stage juvenile. The first-stage juvenile of a few species hatches from the egg. Upon hatching, the first- or second-stage juveniles commence feeding. If suitable plants are present, they generally continue feeding as they develop. In some instances, certain juvenile stages, including those of the soybean cyst nematode, *Heterodera glycines*, may enter a survival stage (diapause) when conditions are unfavorable for continued development.

Reproduction in plant-parasitic nematodes may be amphimictic or parthenogenetic. Most root-knot nematodes (*Meloidogyne* spp.) reproduce by the latter, whereas cyst nematodes (*Heterodera, Globodera*) typically increase via cross-fertilization. Some microbivorous nematodes, including *C. elegans,* may reproduce by hermaphroditism. The life cycle is completed with egg production, often involving 30 and up to 3000 eggs per female. Root-knot nematodes are very prolific with females often producing 300 to more than 1000 eggs each.

The number of life cycles per season or year varies greatly with nematode species. Females of *Meloidogyne* species may complete their life cycle within 3 to 4 wk, resulting in several life cycles (4–6 or more) per growing season. In contrast, the potato cyst nematode, *Globodera rostochiensis*, has only one life cycle per season. Temperature, soil conditions, and the type of plant host often impact the rate of nematode development and reproduction. The host range of nematodes is important to their reproduction and survival as well as in the selection of management tactics by the grower. For example, most *Meloidogyne* species have very wide host ranges and relatively poor survival mechanisms. In contrast, the cyst nematodes, *G. rostochiensis* and *H. glycines*, have more narrow host ranges and a more limited number of life cycles, but have a highly developed means of survival (diapause).

Many diverse species of nematodes often are found in given agroecosystems. They generally are considered to be the most abundant metazoans in soil, and are exceeded in species diversity only by arthropods (Bernard, 1992). Soil-inhabiting nematodes may be separated functionally into five trophic groups: microbivores, fungivores, plant parasites, predators, and omnivores. As indicated by Bernard (1992), these categories are not mutually exclusive. In fact, the feeding habit can vary with life stage or even generation with some nematode species. Numbers of species (species richness) may be as great as 100 in wheat (*Zea mays* L.) to soybean [*Glycine max* (L.) Merr.] rotations and up to 178 in some forest sites (Bernard, 1992). Currently, some scientists are using nematode maturity indices, relating to families and/or trophic groups and site stability to identify assemblages that colonize disturbed soils (see Chapter 4; Niles & Freckman, 1998).

Taxonomy/Systematics

The extreme biodiversity found in nematodes clearly poses challenges in advancing their taxonomy and systematics. This includes phylogenetic relationships, classification, continued descriptions of new species, and diagnosis of species and races/strains/genotypes of nematodes (Ferris & Ferris, 1992). Nematologists have been characterized as scientists who love to draw (Nicholas, 1975). This is reflected through the descriptions of hundreds of plant parasites and thousands of various trophic-group species being published over the last five decades. Nematode morphology will continue to serve as the primary basis of the taxonomy of these organisms. However, the application of DNA-based taxonomic methods are proving very useful in the identification of nematode species, subspecies, and other subspecific forms (races or pathotypes and genotypes) (Caswell-Chen et al., 1993; Curran & Robinson, 1991). This information also is proving to be invaluable in studying the phylogeny and general classification of nematodes.

Relationships of Plant-Parasitic Nematodes with Other Organisms

Crop phenology and economics, as related to plant-parasitic nematodes, often are considered independently of associated microflora, microfauna, and other plant pests. Nevertheless, the epidemiology of crop maladies associated with nematodes is often complex. For example, plant roots infected with the root-knot nematodes have an altered root geometry, morphology, and molecular biology, as well as physiology (Hussey, 1989; Sijmons, 1993), and thereby affected hosts are frequently predisposed to a wide range of associated fungi and bacteria (Hussey & McGuire, 1987; Lucas et al., 1955; Mountain, 1960) (see Chapter 7; Abawi & Chen, 1998). Also, infection by these and other nematodes may predispose plants resistant to specific fungi or bacteria to attack by these organisms. Certain nematode species in the Dorylaimida, primarily in the genera *Xiphinema, Paratrichodorus, Trichodorus, Longidorus*, and *Paralongidorus* often serve as vectors of a number of important plant viruses (Brown et al., 1995; Hewitt et al., 1958; Hussey & McGuire, 1987). Recent research with the soybean cyst nematode, *Heterodera glycines*, illustrates that nematodes also may enhance crop losses associated with insect as well as weed pests (Alston et al., 1991). An unusual relationship is found between a few plant-parasitic nematode species and insects. The pine wilt nematode, *Bursaphelenchus xylophilus*, is vectored by Cerambycid beetles, *Monochamus* spp., and the coconut red-ring nematode, *Rhadinaphelenchus cocophilus,* is vectored by the palm weevil, *Rhyncophorus palmarum*.

ECONOMIC IMPACT OF PLANT-PARASITIC NEMATODES

Severe infestations of given fields with nematodes such as *Meloidogyne* spp. or *H. glycines* often can result in annual yield losses of 10 to 50% (McSorley, 1987; Sasser & Freckman, 1987). The symptoms and signs of nematode infection and damage vary greatly with feeding habit and given species of nematode and host (Fig. 1–2). Expected economic losses for annual crops generally are inversely correlated with the level of infestation at the time the crop is planted. Yet, the development of reliable estimates for economic losses caused by nematodes has proven to be very difficult (Ferris, 1993; Roberts, 1993). This situation results from a number of factors, including the impact of environment on the activity of detrimental and beneficial nematodes (Yeates et al., 1993), as well as crop plants and the frequent involvement of nematode and disease complexes, and concomitant plant nematode species. Plant-parasitic nematodes, clearly, are a hidden enemy of crop plants, typically occurring in the soil and often overlooked. The differentiation of their direct effects on crop production and related economics vs. the enhanced detrimental effects caused by associated organisms offers great challenges in the field situation.

Experiments involving comparisons of effective nematode management treatments such as resistant cultivars, highly efficacious nematicides, or crop rotations have provided a means of developing damage functions and reliable estimates of yield loss resulting from given nematode infestations in given environments. For example, an economic analysis of the utilization of nematode-resistant cultivars shows that the returns over a period of a few years may be as great as 400 times the cost of development (Bradley & Duffey, 1982). Meaningful economic analysis of nematode damage (crop losses) for the regional, national, or world clearly is difficult. The most comprehensive calculation of crop loss estimates for given nematodes was presented by Sasser and Freckman in 1987. These data were based on a world survey with some 371 responses from nematologists and included a wide range of food and fiber crops. Annual yield loss estimates range from 20.6% for tomato to 3.3% for rye with an overall average of 12.3% (Table 1–2). The total estimated losses for the 21 crops given in Table 1–2 amounted to more than $77 billion. While the accuracy of these types of data will continue to be a source of debate, plant-parasitic nematodes clearly take their economic toll in production of the world's food, feed, and fiber supply.

Fig. 1– 2. Typical symptoms and signs of nematodes on crop plants: (*A*) Watermelon roots galled by the root-knot nematode *Meloidogyne incognita*; (*B*) Healthy (left) *Heterodera glycines*-resistant soybean roots versus diseased (infected) roots (right) [note tiny nematode "cysts" (mature females) on diseased roots]; (*C*) Enlarged portion of soybean root showing lemon-shaped cysts (sign of *H. glycines*); (*D*) Roots of three banana genotypes (first three root systems) depicting a range of root-necrosis symptoms induced by *Radopholus similis* (4th root system is nematode free) (photograph courtesy of Douglas Marin); and (*E*) Example of severe stunting and stand of soybean (foreground) caused by the Columbia Lance nematode, *Hoplolaimus columbus* (healthy plants in upper right were in a nematicide-treated plot; courtesy of D. P. Schmitt).

NEMATODE MANAGEMENT

Management practices designed to limit crop losses to plant-parasitic nematodes involve one or more tactics that focus on the strategies to reduce the initial inoculum, and/or to limit the rate of nematode population density increase (Roberts, 1993). The latter would involve host resistance as a control tactic. Other traditional management tactics include the prevention of spread of given nematodes, general land management, and cultural practices, such as crop rotation, physical treatments of infested planting materials or media, and the use of chemical soil treatments. Nematologists and consumers are indebted to geneticists and plant breeders for much of the progress in developing nematode-resistant crop cultivars (see Chapter 10; Young, 1998). This tactic is becoming more critical as nematicides continue to be removed from the market because of environmental and health concerns. Other promising treatments are biological controls (Stirling, 1991) and the development of host resistance through genetic engineering (Atkinson et al., 1994; Cui et al., 1997; Opperman et al., 1994). Undoubtedly, the practical applications of biological control and genetically engineered host resistance will increase as we gain basic information on soil biology, and molecular aspects of nematodes and their host plants. Modern computerized crop-management/harvest systems (Precision Agriculture) also offer new opportunities for more effective management of nematodes as well as general crop production (Copeland et al., 1996; Evans et al., 1996).

OUTLOOK—CURRENT TRENDS, RESEARCH INFORMATION AND TRANSFER

Nematology has a short, but very rich, history. Many of the advancements described for plant and soil nematology in this volume have come during the last four decades (see Table 1–3 for partial summary).

Although most of the nematological contributions to science and agriculture have been made in recent years, many fundamental discoveries were made in the 1900s. For example, Van Beneden's discovery and description of meiosis, as observed in studies of the animal parasite, *Parascaris equorum* (*Ascaras megalocephala*) were published in 1883 (see Chitwood & Chitwood, 1974). The new information age and molecular biology are clearly impacting agriculture and the management of crop pests, including plant-parasitic nematodes. In addition, world demands for a sustainable, yet competitive, agriculture also will impact the study and control of plant-parasitic nematodes on our food, fiber, and feed crops. Genetically engineered resistance to root-knot and cyst nematodes is on the horizon (Atkinson et al., 1994; Cai et al., 1997; Evans, 1995; Opperman et al., 1994). New diagnostics are being developed and should be available in the near future for a range of nematode species (Curran & Robinson, 1991). Genetic mapping and characterization of crop plants (Persley, 1992) should enhance the value of host resistance to nematodes and other plant pathogens. While much progress has been made in characterizing plant-nematode interactions, as described herein, many challenges remain. Opportunities for related advances include: "lessening the societal impact of plant-parasitic nematodes…"; "advancing our knowledge of fundamental nematode behavior…"; and "promoting the beneficial use of nematodes…" (Barker et al., 1994). The latter includes using beneficial nematodes for biological control of insects and other pests, developing a better understanding of the role of nematodes in soil nutrient cycling and environmental quality, and characterizing and manipulating nematode biodiversity to contribute to sustainable agriculture. Many of the other research activities pertaining to these goals are discussed in various chapters in this volume.

REFERENCES

Abawi, G.S., and J. Chen. 1998. Concomitant pathogen and pest interactions. p. 135–158. *In* K.R. Barker et al. (ed.) Plant and nematode interactions. Agron. Monogr. 36. ASA, CSSA, and SSSA, Madison, WI.

Alston, D.G., J.R. Bradley, Jr., H.D. Coble, and D.P. Schmitt. 1991. Impact of population density of *Heterodera glycines* on soybean canopy growth and weed competition. Plant Dis. 75:1016–1018.

Atkinson, G.F. 1889. A preliminary report upon the life history and metamorphoses of a root-gall nematode, *Heterodera marioni* (Greeff) Müll., and the injuries caused by it upon the roots of various plants. Sc. Contrib. Agric. Exp. Stn., Alabama Polytech. Inst. 1:177–226.

Atkinson, H.J., K.S. Blundy, M.C. Clarke, E. Hansen, G. Harper, V. Koritsas, M.J. McPherson, D. O'Reilly, C. Scollan, S.R. Turnbull, and P.E. Urwin. 1994. Novel plant defenses against nematodes. p. 197–210. *In* F. Lamberti et al. (ed.) Advances in molecular plant nematology. Plenus Press, New York and London.

Avery, D.T. 1995. Saving the planet with pesticides and plastic: The environmental triumph of high-yield farming. Hudson Inst., Indianapolis, IN.

Bessey, E.A. 1911. Root-knot and its control. USDA Bur. Plant Indust. Bull. 217.

Barker, K.R., and E.L. Davis. 1996. Assessing plant nematode infestations and infections. p. 103–136. *In* S.H. De Boer (ed.) Advances in botanical research. Vol. 23. Acad. Press, London.

Barker, K.R., R.S. Hussey, and L.R. Krusberg (ed.). 1994. Plant and soil nematodes: Societal impact and focus for the future. J. Nematol. 26:127–137.

Bernard, E.C. 1992. Soil nematode diversity. Biol. Fert. Soils 14:99–103.

Bradley, E.B., and M. Duffy. 1982. The value of plant resistance to soybean cyst nematode: A case study of Forrest soybeans. Rep. no. AGES820919. USDA, Natural Resour. Econ. Div., Econ. Res. Serv., Washington, DC.

Brodie, B.B. 1993. Probability of *Globodera rostochiensis* spread on equipment and potato tubers. J. Nematol. 25:291–296.

Brown, D.J.F., W.M. Robertson, and D.L. Trudgill. 1995. Transmission of viruses by plant nematodes. Annu. Rev. Phytopathol. 33:223–249.

Cai, D., M. Kleine, S. Kifle, H.-J. Harloff, N.N. Sandal, K.A. Marcker, R.M. Klein-Lankhorst, E.M.J. Salentijn, W. Lange, W.J. Stiekema, U. Wyss, F.M.W. Grundler, and C. Jung. 1997. Positional cloning of a gene for nematode resistance in sugar beet. Science (Washington, DC) 275:32–34.

Caswell-Chen, E.P., V.M. Williamson, and B.B. Westerdahl. 1993. Viewpoint: Applied biotechnology in nematology. J. Nematol. 25(4S):719–730.

Chitwood, B.G., and M.B. Chitwood. 1974. Introduction to nematology. Univ. Park Press, Baltimore, London, Tokyo.

Christie, J.R., and V.G. Perry. 1951. A root disease of plants caused by a nematode of the genus *Trichodorus*. Science (Washington, DC) 113:491–493.

Cobb, N. 1915. Nematodes and their relationships. p. 457–490. *In* USDA yearbook, 1914. Dep. Agric., Washington, DC.

Coomans, A. V. 1992. Development of nematology in the low land countries. p. 11–20. *In* F.J. Gommers and P.W.Th. Maas (ed.) Nematology from molecule to ecosystem. Europ. Soc. Nematol., Invergowrie, Dundee, Scotland.

Copeland, P.J., G.L. Malzer, J.G. Davis, J.A. Lamb, P.C. Robert, and T.W. Bruulsema. 1996. Using harvest index to locate environmental stress. p. 531–540. *In* P.C. Robert et al. (ed.) Precision agriculture. ASA, CSSA, SSSA, Madison, WI.

Curran, J., and M.P. Robinson. 1993. Molecular aids to nematode diagnosis. p. 545–564. *In* K. Evans et al. (ed.) Plant-parasitic nematodes in temperate agriculture. CAB Int., Wallingford, England.

Duncan, L.W., and J.W. Noling. 1998. Agricultural and sustainability and nematode integrated pest management. p. 251–287. *In* K.R. Barker et al. (ed.) Plant and nematode interactions. Agron. Monogr. 36. ASA, CSSA, and SSSA, Madison, WI.

Eisenback, J.D. 1998. Morphology and systematics. p. 37–63. *In* K.R. Barker et al. (ed.) Plant and nematode interactions. Agron. Monogr. 36. ASA, CSSA, and SSSA, Madison, WI.

Evans, K. 1995. Closing the gap between molecular biologists and traditional nematologists. Nematologica 41:385–394.

Evans, R.G., S. Han, M.W. Kraiger, and S.M. Schneider. 1996. Precision center pivot irrigation for efficient use of water and nitrogen. p. 75–84. *In* P.C. Robert et al. (ed.) Precision agriculture. ASA, CSSA, SSSA, Madison, WI.

Ferris, H. 1993. New frontiers in nematode ecology. J. Nematol. 25:374–382.

Ferris, V.R., and J.M. Ferris. 1992. Integration of classical and molecular approaches in nematode systematics. p. 92–100. *In* F. Gommers and P.W.Th. Maas (ed.) Nematology from molecule to ecosystem. Europ. Soc. Nematol., Wageningen, The Netherlands.

Goeldi, E.A. 1892. Relatorio sobre a molestia do cafeeiro na provincia do Rio de Janeiro. Arch. Mus. Nacl. Rio Janeiro 8:7–121.

Hewitt, W.B., D.J. Raski, and A.C. Goheen. 1958. Nematode vector of soil-borne fanleaf virus of grapevines. Phytopathology 48:586–595.

Hirschmann, H. 1971. Comparative morphology and anatomy. p. 11–63. *In* B.M. Zuckerman et al. (ed.) Plant-parasitic nematodes. Vol. 1. Acad. Press, New York.

Hussey, R.S. 1989. Disease-inducing secretions of plant-parasitic nematodes. Annu. Rev. Phytopathol. 27:123–141.

Hussey, R.S., and J.M. McGuire. 1987. Interaction with other organisms. p. 313–320. *In* R.H. Brown and B.R. Kerry (ed.) Principles and practice of nematode control in crops. Acad. Press, New York.

Lubchenco, J., A.M. Olson, L.B. Brubaker, S.R. Carpenter, M.M. Holland, S.P. Hubbell, S.A. Levin, J.A. MacMahon, P.A. Matson, J.M. Melillo, H.A. Mooney, C.H. Peterson, H.R. Pulliam, L.A. Real, P.J. Regal, and P.G. Risser. 1991. The sustainable biosphere initiative: An ecological research agenda. Ecology 72:371–412.

Luc, M.D., D.J. Hunt, and J.E. Machon. 1990. Morphology anatomy and biology of plant-parasitic nematodes—synopsis. p. 1–44. *In* M. Luc et al. (ed.) Plant-parasitic nematodes in subtropical and tropical agriculture. CAB, Int., Wallingford, England.

Lucas, G.B., J.N. Sasser, and A. Kelman. 1955. The relationship of root-knot nematodes to Granville wilt resistance in tobacco. Phytopathology 45:537–540.

Maggenti, A.R. 1991. Nemata: High classification. p. 147–187. *In* W.R. Nickle (ed.) Manual of agricultural nematology. Marcel Dekker, New York.

McSorley, R. (ed.). 1987. Bibliography of estimated crop losses in the United States due to plant-parasitic nematodes. Suppl. J. Nematol. 1:6–12.

McSorley, R. 1998. Population dynamics. p. 109–133. *In* K.R. Barker et al. (ed.) Plant and nematode interactions. Agron. Monogr. 36. ASA, CSSA, and SSSA, Madison, WI.

Miller, R.J. (ed.). 1993. The U.S. agriculture, food, and environmental system. Averting a crisis by building for future success: An imperative for the 1990s. Natl. Assoc. State Univ. Land-Grant Colleges, Washington, DC.

Mountain, W.B. 1960. (a) Theoretical considerations of plant-nematode relationships. (b) Acceptable standards of proof and approaches for evaluating plant-nematode relationships. (c) Mechanisms involved in plant-nematode relationships. p. 419–431. *In* J.N. Sasser and W.R. Jenkins (ed.) Nematology, fundamentals and recent advances with emphasis on plant-parasitic and soil forms. Univ. North Carolina Press, Chapel Hill.

National Research Council. 1991. Sustainable agriculture research and education in the field. Natl. Acad. Press, Washington, DC.

Nickle, W.R. (ed.). 1991. Manual of agricultural nematology. Marcel Dekker, Inc., New York.

Niles, R.K., and D.W. Freckman. 1998. Nematode ecology in bioassessment and ecosystem health. p. 65–85. *In* K.R. Barker et al. (ed.) Plant and nematode interactions. Agron. Monogr. 36. ASA, CSSA, and SSSA, Madison, WI.

Noe, J.P. 1998. Crop- and nematode-management systems. p. 159–171. *In* K.R. Barker et al. (ed.) Plant and nematode interactions. Agron. Monogr. 36. ASA, CSSA, and SSSA, Madison, WI.

Noel, G.R. 1992. History, distribution, and economics. p. 1–10. *In* R.D. Riggs and J.A. Wrather (ed.) Biology and management of the soybean cyst nematode. APS Press, St. Paul, MN.

Norton, D.C. 1978. Ecology of plant-parasitic nematodes. John Wiley & Sons, New York.

Nicholas, W.L. 1975. The biology of free-living nematodes. Clarendon Press, Oxford.

Opperman, C.H., G.N. Acedo, D.M. Saravitz, A.M. Skantar, W. Song, C.G. Taylor, and M.A. Conkling. 1994. Bioengineering resistance to sedentary endoparasitic nematodes. p. 221–230. *In* F. Lamberti et al. (ed.) Advances in molecular plant nematology. Plenum Press, New York and London.

Opperman, C.H., and M.A. Conkling. 1998. Bioengineering resistance to plant-parasitic nematodes. p. 239–250. *In* K.R. Barker et al. (ed.) Plant and nematode interactions. Agron. Monogr. 36. ASA, CSSA, and SSSA, Madison, WI.

Persley, G.L. 1992. Beyond Mendel's garden. p. 360–370. *In* A.A.S.A. Kadir and H.S. Barlow (ed.) CAB Int., Wallingford, England.

Poinar, G.O., Jr. 1983. The natural history of nematodes. Prentice Hall, Inc., Englewood, NJ.

Poinar, G.O., Jr. 1988. The impact of nematodes on mankind—a historical perspective. Nematologica 34:249–251.

Roberts, P.A. 1993. The future of nematology: Integration of new and improved management strategies. J. Nematol. 25:383–394.

Roberts, P.A., W.C. Matthews, and J.C. Veremis. 1998. Genetic mutations of host-plant resistance to nematodes. p. 209–238. *In* K.R. Barker et al. (ed.) Plant and nematode interactions. Agron. Monogr. 36. ASA, CSSA, and SSSA, Madison, WI.

Sasser, J.N., and D.W. Freckman. 1987. A world perspective of nematology: The role of the Society. p. 7–14. *In* J.A. Veech and D.W. Dickson (ed.) Vistas on nematology. Soc. Nematol., Hyattsville, MD.

Sayre, R.M. 1994. Art in phytopathology—portfolios of Nathan A. Cobb, nematologist. APS Press, St. Paul, MN.

Sijmons, P.C. 1993. Plant-nematode interactions. Plant Molec. Biol. 23:917–931.

Sipes, B.S., and D.P. Schmitt. 1998 Nematode-pesticide interactions. p. 173–185. *In* K.R. Barker et al. (ed.) Plant and nematode interactions. Agron. Monogr. 36. ASA, CSSA, and SSSA, Madison, WI.

Stirling, G.R. 1991. Biological control of plant-parasitic nematodes. CAB Int., Wallingford, England.

Thorne, G. 1961. Principles of nematology. McGraw-Hill Book Co., Inc., New York, Toronto, London.

University of California, Division of Agriculture and Natural Resources. 1992. Beyond pesticides: Biological approaches to pest management in California. ANR Publ., Univ. California, Oakland.

Viglierchio, D.R. 1991. The world of nematodes. David R. Viglierchio, Davis, CA.

Westerdahl, B.B., and E.P. Caswell-Chen. 1998. Information transfer. p. 289–315. *In* K.R. Barker et al. (ed.) Plant and nematode interactions. Agron. Monogr. 36. ASA, CSSA, and SSSA, Madison, WI.

Whitten, M.J. 1992. Pest management in 2000: What we might learn from the twentieth century. p. 9–44. *In* A.A.S.A. Kadir and H.S. Barlow (ed.) CAB Int., Wallingford, England.

Wood, W.B. (ed.). 1988. The nematode *Caenorhabditis elegans*. Cold Spring Harbor Lab., Cold Spring Harbor, NY.

Yeates, G.W. 1996. Nematode ecology. Russ. J. Nematol. 4:71–75.

Yeates, G.W., T. Bongers, R.G.M. de Goede, D.W. Freckman, and S.S. Georgiera. 1993. Feeding habits in soil nematode families and genera—an outline for soil ecologists. J. Nematol. 25:315–331.

Young, L.D. 1998. Breeding for nematode resistance and tolerance. p. 187–207. *In* K.R. Barker et al. (ed.) Plant and nematode interactions. Agron. Monogr. 36. ASA, CSSA, and SSSA, Madison, WI.

2 Biology of Plant-Parasitic Nematodes

JOHN M. FERRIS AND VIRGINIA R. FERRIS

Purdue University
West Lafayette, Indiana

Nematodes are worm-shaped invertebrates, and comprise one of the largest groups of animals. There are many different kinds of nematodes, and they are found practically everywhere. Nathan A. Cobb, often called the father of American nematology, once concluded a discussion about the abundance of nematodes in this way (Cobb, 1914):

> In short, if all the matter in the universe except the nematodes were swept away, our world would still be dimly recognizable, and if, as disembodied spirits, we could then investigate it, we should find its mountains, hills, vales, rivers, lakes, and oceans represented by a film of nematodes. The location of towns would be decipherable, since for every massing of human beings there would be a corresponding massing of certain nematodes. Trees would still stand in ghostly rows representing our streets and highways. The location of the various plants and animals would still be decipherable, and, had we sufficient knowledge, in many cases even their species could be determined by an examination of their erstwhile nematode parasites.

Although soil nematodes can rarely be seen with the naked eye, they are extremely important. Predacious nematodes prey on small invertebrates, including other nematodes, and microbial nematodes feed on soil fungi and bacteria. Plant-parasitic nematodes impact agricultural production in a dramatic fashion. Often referred to as the "unseen enemy," they attack all parts of growing plants. These different kinds of nematodes belong to well-defined taxonomic groups, and often most members of a group share similar feeding strategies and other life style characteristics (see Chapter 3; Eisenback, 1998). Plant-parasitic nematodes belong to one of two large groups of nematodes: the tylenchid group and the dorylaimid group. Their feeding, mostly by members of the tylenchid group, causes suppression of plant growth and, hence, losses in crop yields. They provide portals of entry and alter host physiology to enhance attack by other plant-parasitic organisms such as fungi and bacteria (Hussey & McGuire, 1987). In addition, members of the dorylaimid group can act as vectors of plant viruses (Hewitt et al., 1958). The total impact of nematodes on agriculture, even in the USA, has been difficult to assess (McSorley et al., 1987). On a worldwide scale, losses to parasitic nematodes of all kinds are estimated to be $78 billion per year for the major food and fiber crops, and about 12% or $8 billion per year in the USA (Barker et al., 1994). From studies carried out in a U.S. grassland biome

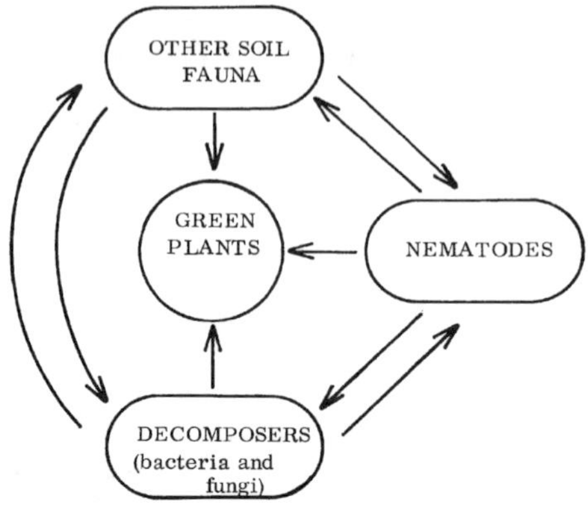

Fig. 2–1. Diagram showing relationships between nematodes and other soil microfauna and flora. Arrows point to prey in each instance (from Ferris & Ferris, 1974).

project, it was determined that the biomass of belowground feeding nematodes exceeded that of cattle grazing on aboveground plant parts (Smolik & Rogers, 1976).

A large number of nematode species are classified as "free-living" or microbivorous. Free-living nematode is the term applied to any soil or freshwa-

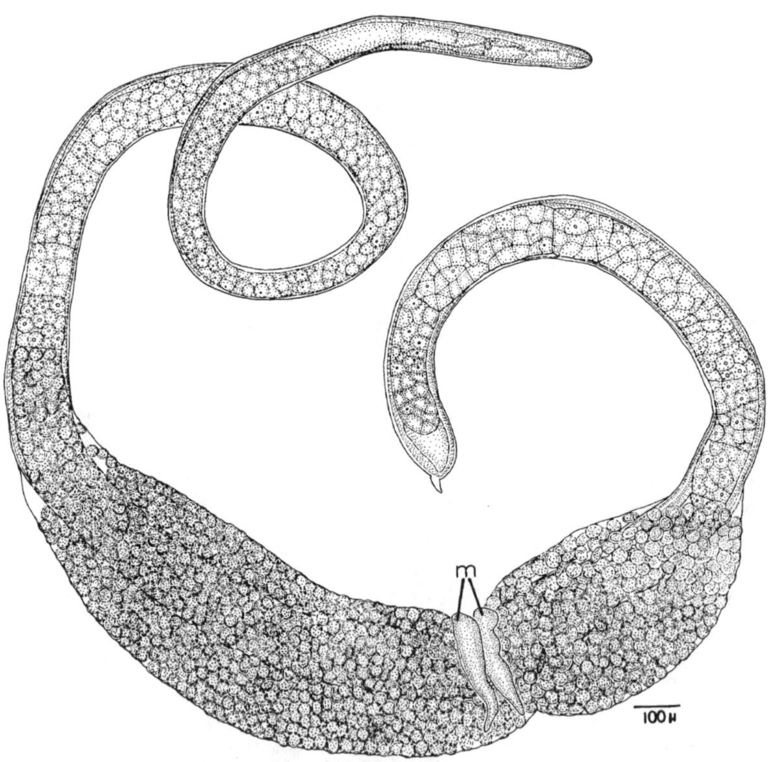

Fig. 2–2. Encapsulated female of *Tetradonema plicans*, m = males (from Ferris & Ferris, 1966).

ter nematode that does not feed on higher plants, animals (including man), or insects. These nematodes generally feed on other living microorganisms such as fungi and bacteria. They are an extremely important component of the ecosystem because they function in the recycling of nutrients, and they are often present in large numbers (Fig. 2–1). Other nematodes exist as parasites of insects, as well as man and other animals. Entomophilic nematodes occur in many different families of insects. These insect parasites belong to many systematic groups of nematodes and may vary greatly in size and shape, and in their life histories (Nickle & Welch, 1984). An example is *Tetradonema plicans*, which is a parasite of greenhouse fungus flies (*Bradysia coprophila*) (Ferris & Ferris, 1966). The fertilized female swells to enormous size and becomes a case for hundreds of eggs. After fertilization, tiny males remain attached to the female vulva until they die (Fig. 2–2). Some species of entomoparasitic nematodes are being used for control of larvae of the Japanese beetle (*Popillia japonica*), lepidopterous pests and other insects (Gaugler & Kaya, 1990).

Certain nematodes have been used as model systems in basic biological research. With the development of the light microscope, the large nematode *Ascaris*, an intestinal parasite of hogs (*Sus* spp.) and cattle (*Bos* spp.), was used in cytological investigations that provided basic contributions to genetics. These included the discovery that egg and sperm contribute equal numbers of chromosomes to the embryo (Van Beneden, 1883), and that the fate of cells in eggs is determined at cleavage in embryogenesis (Boveri, 1899). Today, the best known nematode used in basic research is *Caenorhabditis elegans* (Brenner, 1974; Marx, 1984; Wood, 1988). Originally, this nematode species was selected for detailed study of development because it was a small multicellular organism that could be examined with ease comparable to research with other microorganisms. Since the internal organs of *C. elegans* (and other microscopic nematodes) can be studied under the microscope while the animal is alive (Sulston et al., 1983), development has been traced from a single cell of a fertilized egg through all of the cells that make up the adult worm. Neurobiologists are using this nematode to determine how the various parts of a nervous system function, and geneticists are learning how genes control development and behavior in multicellular animals (Lewin, 1984; Wood, 1988).

COMMONALTIES AND DIFFERENCES AMONG NEMATODES

Nematodes are defined as triploblastic, noncoelomate, bilaterally symmetrical, internally nonsegmented round worms. They are vermiform (wormlike) in shape for at least part of their life cycle, tapering at the ends, and always circular in cross-section. Nematodes are often dioecious with separate male and female sexes, but some species contain hermaphrodites. Parthenogenesis also is common. Soil-dwelling nematodes are usually transparent when viewed through the microscope. Although they vary greatly in size, most plant-parasitic nematodes are microscopic, between 0.5 to 2 mm long and 0.05 mm in width. The body of nematodes, although flexible, shows little change in length and is adapted for undulatory movement. The body plan of a nematode is essentially a tube within a tube.

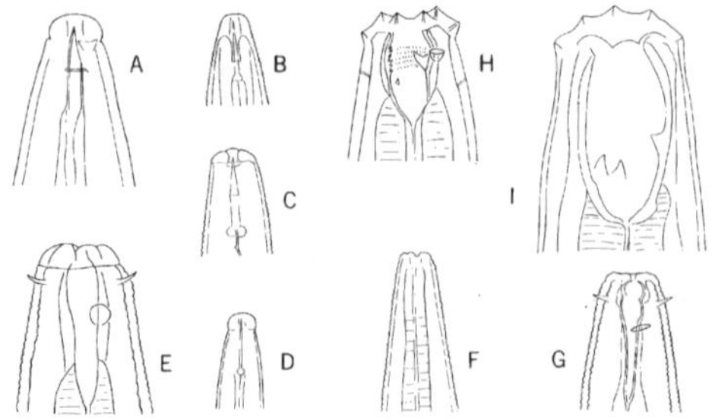

Fig. 2–3. Anterior ends of nematodes showing different kinds of modifications of the stoma. *A–D*, nematodes with stoma modified to form a stylet for piercing cells of plant hosts; *E–G*, nematodes with stoma modified for ingestion of microorganisms such as bacteria; *H* and *I*, nematodes with stoma modified to form teeth for ingesting soil microfauna, (see also Chapter 3, Eisenback, 1998).

Because nematodes are invertebrate animals, they have some features common to other invertebrates, such as insects. For example, both grow by a series of molts during their life cycles. Many species of plant-feeding nematodes are polyphagous, as are many insect species, feeding on weeds in the absence of a suitable agronomic crop. These and other similarities mean that control methods for economically important nematode pests are often similar to the approaches used to control insect pests. In a few instances, especially in the past, soil pesticides could control insect and nematode pests simultaneously.

Morphological features by which families, genera and species can be distinguished from each other include stoma, esophagus, reproductive system, size and cuticular ornamentation (see Chapter 3; Eisenback, 1998). The sexual characters of females (ovaries and reproductive openings) and males (secondary sexual characters, such as spicules) can be seen fairly easily with a low-power microscope. Adult shape may vary markedly from juvenile shape in some groups, with adult females swelling to become globoid or pear-shaped, rather than vermiform or worm-like. Modifications occur among nematodes in the shape of the stoma (or "mouth"), and these modifications usually relate to the food sources used by nematodes. For instance, the stoma of a plant-parasitic nematode is modified for piercing plant cells to ingest the contents, whereas the stoma of a free-living species will be modified for the ingestion of microorganisms (Fig. 2–3). Nematodes occasionally have distinctive tail shapes by which they may be recognized, and in some groups distinctive cuticular ornamentation may be diagnostic (see Chapter 3; Eisenback, 1998).

NEMATODE LIFE CYCLE AND BEHAVIOR

Life Cycle

Free-living nematodes can have a life cycle as short as 3 d. However, most plant-parasitic species fall into a 20- to 30-d life cycle (egg–egg) at temperatures between 18 to 27°C, although a few (dorylaimid nematodes) have life cycles as long as 1 to 2 yr. Egg development proceeds by division from a single cell to a multicellular worm encased in a cuticle. Nematodes generally undergo four molts, the first of which frequently occurs in the egg and leads to the emergence of the second stage infective juvenile (J2). The fourth stage, following the third molt, is pre-adult. Development is direct, without metamorphosis. Juveniles usually resemble adults in gross morphologic characters, with the exception of the reproductive system and secondary sexual characters.

Feeding

Most plant-parasitic nematodes have a cuticularized structure, called a stylet, found in the anterior end of the animal (Fig. 2–3A–D). It is hollow, with muscles attached to its posterior end. Contraction of these muscles protrudes the stylet to puncture cell walls. Once the stylet is inserted into a plant cell, muscles in the glandular/muscular esophagus produce a pumping force. Nematodes feed by sucking in plant fluids through the stylet. The material then passes through the lumen of the esophagus into the intestine.

Movement

Nematodes are too small to modify soil structure, so they are limited to the existing labyrinth of soil pore spaces. Because they are essentially aquatic organisms, nematodes require a film of moisture in which to "swim." They produce a sinusoidal wavelike motion by coordinated contraction of their longitudinal muscles. Nematode movement is governed by the relationships between the nematode's diameter, the pore size between soil particles and the depth of the water film. Nematode movement is greatest when the mean soil particle diameter is equal to about one-third to one-half the length of the nematode (Wallace, 1958a,b).

The vertical and horizontal distribution of nematodes in a cultivated field is usually irregular, but these obligate parasites are closely associated with plant roots and the area adjacent to the roots. Nematodes are concentrated mainly in the top meter of soil. Information is limited, but root-knot nematodes have been isolated from a depth of 5 m in a grape vineyard, and citrus nematodes (*Tylenchulus semipenetrans*) often have been found at depths of 2 to 3 m.

Reproduction and Survival

The reproductive system of nematodes in both sexes consists of one or two tubular gonads. The female reproductive system, located on the ventral side of the body, consists of ovary, oviduct, uterus and vagina, plus an opening called a vulva. The male gonad consists of testis, seminal receptacle, vas deferens and ejaculatory duct, and terminates in a common opening for the intestine and reproductive system located ventrally near the posterior end of the nematode. With some species of nematodes, males are not required for the production of viable eggs, although they may be present.

Female nematodes lay their eggs in whatever medium they occupy, including soil or host tissue. Females of root-knot nematode species are pear-shaped and have lost all motility. The female becomes an efficient egg-producing factory. Eggs are deposited outside the body in a gelatinous matrix, hatch, and the juveniles continue to invade nearby roots. With cyst nematodes, most eggs are retained in the swollen body of the female, although some may be deposited in a gelatinous matrix such as in the case of root-knot nematodes. When the swollen female dies, however, her body becomes a protective envelope ("cyst") for the remaining eggs for prolonged periods.

Many plant-parasitic nematodes can survive in soil for at least a year in the absence of a suitable host. During this period, they are assumed to be in a quiescent or inactive state that is associated with a lowered metabolic rate. The length of the quiescent period is limited by the amount of food reserves in the nematode and the environmental conditions. Nematodes tend to survive longer at low temperatures. Cyst nematodes are known to survive many years in the absence of a host plant and may require specific root exudates to induce hatching. Dormant states (termed anabiosis, anhydrobiosis, or diapause) of a few vermiform nematode species also enable them to withstand environmental conditions unfavorable for active development, thus permitting them to remain alive for as long as 10 yr. However, not all individuals in the population can survive these dormant periods. Generally, the population declines by a substantial percentage each year, thus making crop rotation a desirable control measure.

Some nematode species may also survive without a host, in or on pieces of root that remain alive in the soil. This means of survival is especially important in replanting fields having a history of perennial crops, as roots that are left in the soil may remain alive for as long as 4 to 5 yr. In some cases, nematodes may also feed on weeds when host crops are not available.

It is not likely that any major thermoregulation occurs in nematodes. As a group they tolerate rather wide temperature variations both in nature and under experimental conditions. However, all seem to have optimum temperature ranges that affect their geographic distribution. Heat tolerance depends on temperature and time. A free-living nematode, *Panagrellus* sp., and a root-knot nematode, *Meloidogyne incognita*, are both killed in about 1 h at a temperature of 46°C. At 43°C, the *Panagrellus* can survive for 6 h, whereas the root-knot species survives only 4 h (Santmyer, 1955). Heat resistance appears to be dependent on hydration of tissues, since desiccated worms withstand higher temperatures longer.

Cold resistance is affected by hydration and the medium in which nematodes are exposed. *Ditylenchus dipsaci* can survive –80°C for 28 d after lyophilization and storage in vacuo (Bosher & McKeen, 1954). Sayer (1963) found that external ice formation and super cooling were not lethal to root knot-nematodes, but internal ice formation was. A modern method for preserving nematodes in a living state for many years involves incubating vermiform nematodes in different concentrations of ethylene glycol for various time periods at room temperature and in a crushed ice bath before plunging them into liquid N for long-term cryopreservation. The procedure must be optimized for each species of nematode preserved in this manner (Triantaphyllou & McCabe, 1989).

RELATIONSHIP TO PLANT HOSTS

Aboveground Parasites

Some plant-parasitic nematodes invade aboveground portions of plants. Although sometimes found in soil, these nematodes migrate up the stems of plants in a film of water and enter leaves through open stomata in order to feed on leaf parenchyma cells. They may also invade stem tissue causing stunted, abnormal growth. When found in floral tissue or seeds, they cause galls and other distortions of healthy tissue.

Bulb, Stem and Tuber Nematodes

Species of the tylenchid genus *Ditylenchus*, which contains more than 80 recognized species (Sturhan & Brzeski, 1991), exhibit the feature called anabiosis. Certain juvenile stages can persist in a dry condition and remain viable for many years. One species, *D. dipsaci*, is an internal parasite of bulbs, stems and leaves, and rarely attacks roots. It causes heavy losses in onion (*Allium cepa* L. var. *cepa*), garlic (*Allium sativum* L.), alfalfa (*Medicago sativa* L. subsp. *sativa*), clover (*Trifolium* spp.) and oat (*Avena sativa* L.). Onions attacked by this nematode have a frosted appearance caused by dissolution of the middle lamella of cells as a result of nematode feeding. Because the outer layers of the onion bulb become soft, rots frequently develop during storage. Alfalfa attacked by *D. dipsaci* becomes shortened, thickened and clublike. Thus, plants do not grow tall enough to be mowed for hay. Later in the season, with the advent of high temperatures and dry conditions, nematode activity ceases and surviving alfalfa plants may resume fairly normal growth (Griffin, 1984). The nematode can survive many years in dried alfalfa stems. Although this species attacks many different plants, some populations will not transfer from one host plant to another. Since this nematode survives in plant parts and has the potential for being spread from one place to another in plant material, it has figured prominently in quarantine regulations.

Another species, *D. destructor*, the potato rot nematode, attacks only stolons and tubers of its host. It causes dry rot of potato tubers. This species differs from *D. dipsaci* in that it will feed on both higher plants and fungi.

Foliar Nematodes

A number of species of the genus *Aphelenchoides* are bud and foliage parasites, living primarily in leaves and buds, but they may also be found in soil. For many years, *A. ritzemabosi* has been a major pest of chrysanthemum (*Chrysanthemum* spp.) where it feeds on and destroys leaf parenchyma cells causing interveinal necrosis. It also affects many other plants including aster (*Aster*), dahlia (*Dahlia*), zinnia (*Zinnia*) and tobacco (*Nicotiana tabacum* L.) . Two species, *A. fragariae* and *A. besseyi*, are bud and leaf parasites of strawberry (*Fragaria* spp.), causing a major decline. The first species induces two different diseases. When the corynebacterium *Rhodococcus fasciens* is present, it is called cauliflower disease; it is called spring crimp or red plant when only the nematode is present. The other species causes summer crimp. Both affect leaf and bud development with symptoms of summer crimp, which includes reduction of leaf growth from mere rudiments to almost normal. The symptoms appear during hot weather and disappear with cool temperatures. Other members of this genus are heterogeneous with respect to feeding habits. Many species are associated with bark beetles and feed in frass of insect galleries (e.g., *Bursaphelenchus* species associated with pine wilt complex), whereas others are saprophagous in soil, living in part on fungi. Red-ring disease of the coconut palm (*Cocos nucifera* L.) is caused by the nematode *Rhadinaphelenchus cocophilus*, which can be vectored by the palm weevil *Rhynchophorus palmarum* (Fenwick, 1968).

Seed Gall Nematode

Anguina tritici is a parasite of wheat, causing leaf or stem galls, and seed galls. Dry seed galls contain quiescent juveniles of the nematode. When planted with normal seed, the nematodes revive with moisture and then attack seedlings developing from healthy seed. They infect shoots emerging from soil or move up the plant in a film of moisture to young leaves around the growing point. When primordia of flower buds form, the nematodes enter this tissue and stimulate development of galls instead of normal seeds. Cool, moist climates favor development of this nematode. It is effectively controlled by crop rotation and the use of clean seed. In the Pacific Northwest, another species, *A. agrostis*, is parasitic on bent grass. *Anguina agrostis* and *A. tritici* act as vectors for coryneform bacteria which induce seed galls that are highly toxic to cattle and sheep that feed on the plant galls (Bird, 1981).

Belowground Parasites

Ectoparasites/Semi-Endoparasites

Species of the genus *Hoplolaimus*, called lance nematodes, feed on many different kinds of plants. They often feed on roots of considerable size, injuring root cortex and endodermis. Sometimes they behave as ectoparasites, and at other times they may penetrate into the root and are often called semi-endoparasites. One common species, *H. galeatus*, is found with turf, corn (*Zea mays* L.) and soybean [*Glycine max* (L.) Merr.] in the midwestern USA (Alby et al. 1983), and can damage cotton (*Gossypium hirsutum* L.) in the coastal plains of the southeastern USA (Heald & Orr, 1984). The Columbia lance nematode, *H. columbus*, is associated with severe damage of cotton and soybeans in parts of southern USA.

Two genera of spiral nematodes, *Helicotylenchus* and *Rotylenchus*, are easily recognized under low microscopic magnification because, when killed by gentle heat, they look like a flat coil. Members of these genera are often numerous around the roots of grasses and generally do not cause very much damage. An exception is summer dormancy of Kentucky bluegrass (*Poa pratensis* L.) in Wisconsin where large numbers of *H. digonicus* were found to cause shallow, necrotic root systems (Perry et al., 1959). *Helicotylenchus* species were also found to be associated with grasses in the western Dakotas (Krupinsky et al.,

1981). The species *H. multicinctus* causes shallow necrotic lesions on banana roots. These lesions are then colonized by fungi belonging to the genera *Fusarium* and *Rhizoctonia* (Wehunt & Edwards, 1968).

Prior to 1970, approximately 100 species of stunt nematodes were placed in the genus *Tylenchorhynchus*, which became subdivided into smaller groupings that were described as separate genera. The agriculturally important species belong mainly to the three genera *Tylenchorhynchus*, *Merlinius* and *Amplimerlinius* (Tarjan, 1973). Species of these genera feed largely from outside the root, but some may become internal root parasites. Several species of stunt nematodes are associated with cotton, but the nematodes cause no measurable injury. *Tylenchorhynchus* species, including *T. dubius* and *T. nudus*, have been shown to damage grasses (Griffin, 1984).

Members of the genus *Belonolaimus*, sting nematodes, are widespread in the southeastern USA and occur in some western states. These large (at least 2 mm in length) nematodes cause severe stunting of plants, which frequently leads to early death of cotton, corn, sweetpotato [*Ipomoea batatas* (L.) Lam.] and other crops. The species *B. longicaudatus* not only causes extensive root injury to cotton, but also predisposes this crop to Fusarium wilt in varieties that are wilt-resistant in the absence of sting nematodes (Sasser, 1972; Smart & Nguyen, 1991). Another species of sting nematode, *B. nortoni*, has been associated with severely stunted corn in Nebraska (Kerr & Wysong, 1979).

Ring nematodes are short, fat nematodes with very heavy ringlike annulation (which gives them their common name) and are external parasites that feed at root tips and along sides of roots. They appear to be important parasites of trees and other woody ornamentals. Peach [*Prunus persica* (L.) Batsch var. *persica*] trees are damaged by at least two species of ring nematodes. *Criconemella curvata* causes extensive pits and lesions on roots of peach seedlings (Hung & Jenkins, 1969), which may be invaded by other microorganisms. *Criconemella xenoplax* limits the number of feeder roots on peach (Lownsbery et al., 1973), impairs root development and function (Wehunt, 1984), and has been implicated in peach tree short life (Nyczepir et al., 1983). Ring nematodes are often found in grasses of many kinds, including the turf of golf courses (Griffin, 1984; Safford & Riedel, 1976), and are now called *Mesocriconema* by some authorities (Andrássy, 1965; Mai & Mullin, 1996).

Pin nematodes, *Paratylenchus* spp., are very small with a very long stylet, in comparison to their length. Species of the genus feed from outside on root hairs or epidermal cells. They have a pre-adult stage that survives for long periods of time without feeding. In this stage, the nematode is very difficult to identify because it does not have a stylet. Although these nematodes can occur around plant roots in very large numbers, they are apparently of no economic importance except as members of the soil fauna. Pin nematodes are found in many parts of the world, but are easily underestimated because soil extraction efficiency of widely used techniques may be poor due to the small size of the nematodes.

Feeding by the needle nematode, *Longidorus breviannulatus*, causes dramatic stunting and yellowing of corn grown on light textured sandy soils of midwestern USA (Norton & Hoffman, 1975). This large dorylaimid nematode is especially active when weather after planting is cool and wet. Irrigation favors damage by needle nematodes. As the season becomes hotter and drier, feeding may be sufficiently reduced so that plants will resume normal growth. Yield, however, is reduced no matter how much the corn appears to have recovered. Other species of needle nematodes have been reported as economic pests of lettuce (*Lactuca* spp.), celery (*Apium* spp.) (Israel), sugar beet (*Beta vulgaris* L. subsp. *vulgaris*) (Europe), tobacco and pear (*Pyrus communis* L.) trees (Europe).

Dagger nematodes, large dorylaimid species, feed on succulent roots by inserting a long powerful stylet deep into the root. One dagger nematode species or species complex, *Xiphinema americanum*, is worldwide in distribution and thought to be a primary pathogen of fruit and nut trees. Another species, *X. diversicaudatum*, causes damage to roses (*Rosa* spp.), recognized by the small root galls it stimulates, and may cause decreased root growth of strawberries (Brown et al., 1993; Schindler, 1957; Schindler & Braun, 1957). Members of the genera *Xiphinema* and *Longidorus* are the only known nematode vectors of nepoviruses, which cause serious diseases of fruit trees, strawberries and other fruits, grapes (*Vitis* spp.) and many other plants (Brown et al., 1993; Nyczepir & Halbrendt, 1993).

Stubby-root nematodes belong to the genera *Trichodorus* and *Paratrichodorus*. Trichodorid nematodes transmit three tobraviruses (tobacco rattle, pea early browning and pepper ringspot) in various parts of the world (Decramer, 1991).

Migratory Endoparasites

All motile stages of lesion or meadow nematodes, *Pratylenchus* species, can enter and leave roots, so there is no single infective stage. Once inside the root, these nematodes migrate intercellularly, feeding and depositing eggs. *Pratylenchus penetrans* causes damage to many crops including vegetables, grasses, legumes, forest and fruit trees, and many ornamentals. This species is often found in corn fields in association with other species of lesion nematodes. Another species, *P. scribneri*, is associated with corn, soybean (Alby et al., 1983) and potato (*Solanum tuberosum* L.) (MacGuidwin, 1989); other species have been associated with a variety of crops. Injury by lesion nematodes can make plant roots particularly susceptible to invasion by other organisms. In potato, *Verticillium* wilt and *Rhizoctonia* disease increase when high densities of lesion nematodes are also present (Brodie, 1984). Some lesion nematode species are involved in disease complexes. Potato early death is an example of a synergistic interaction between *P. penetrans* and the fungus *Verticillium dahliae*. Population levels at which each of these two pathogens individually cause little or no damage, act synergistically to cause a reduction in potato yields (Rowe et al., 1985).

Burrowing nematodes, members of the genus *Radopholus*, generally occur in tropical or subtropical regions. Spreading decline of citrus in Florida, and root and rhizome necrosis of banana and pepper yellows disease, are caused by one or both of two putative sibling species, *R. citrophilus* and *R. similis* (Huettel et al., 1984; Loof, 1984; Loos, 1962; Suit & DuCharme, 1953). The closely related rice-root nematode, *Hirschmaniella oryzae*, is an important parasite in rice-growing regions.

Sedentary Endoparasites

Because of the conspicuous root galls stimulated by feeding of *Meloidogyne* species, these root-knot nematodes have been known for a very long time (Berkeley, 1855). Juveniles enter the roots of host plants and modify several host cells adjacent to vascular tissue into elaborate feeding sites called giant-cells. Associated local cell hyperplasia and hypertrophy result in gall formation. The disrupted vascular system of the roots leads to poor water and nutrient uptake and often secondary pathogen invasions. The collective species of root-knot nematode attack over 2000 plant species (Sasser, 1980). High population densities may kill seedlings. Although these nematodes are easily recognized by the presence of root galls, a number of hosts (cereals, corn, tubers, bulbs, and rhizomes) do not develop the characteristic galling in spite of attack by large numbers of parasites. At least four species are commonly found in the USA: the northern root-knot nematode, *M. hapla*; the southern root-knot nematode, *M. incognita*; the peanut root-knot nematode, *M. arenaria*; and the Javanese root-knot nematode, *M javanica*. A fifth species, *M. chitwoodi*, is found in potato-growing areas of the Pacific Northwest (Golden et al., 1980). Reproduction on differential hosts, morphology of perineal patterns (made by striae around the vulval and anal region on the posterior end of females), and isoenzyme analysis are used for diagnosing economically important species of root-knot nematodes (Barker et al., 1985; Esbenshade & Triantaphyllou, 1985; Taylor et al., 1955). Molecular methods are being developed (Powers & Harris, 1993).

Females of the cyst nematodes, *Heterodera* and *Globodera*, are also swollen, but have a very thick cuticle, which eventually forms a hard cyst with a variously patterned wall. Most cyst nematode species have a narrow host range, but the eggs within the cyst do not all hatch the following growing season. In fact, nematodes that belong to these genera are capable of surviving for many years in the absence of any food source. Cyst nematodes are serious pests because once they become established in a field, rotations with nonhost plants must be very long in order for the population to decline to a sufficiently low level to permit growing a susceptible host. Most cyst nematode species are capable of dramatic population increases with a single season of a favorable host. Important species in this group include: *H. glycines*, the soybean cyst nematode; *H. avenae*, the cereal cyst nematode; *H. schachtii*, the sugar beet cyst nematode; *H. trifolii*, the clover cyst nematode; and *G. rostochiensis* (also called the golden nematode) and *G. pallida*, the potato cyst nematodes.

Reniform nematodes, particularly *Rotylenchulus reniformis*, and false root-knot nematodes, *Naccobus* species, are important parasites of many crops. In both groups, the female nematode swells within the plant tissue and extrudes her eggs in a gelatinous matrix. The species *N. aberrans* induces galls on roots similar to those caused by root-knot species (Jatala, 1991).

The citrus nematode, *Tylenchulus semipenetrans*, is the cause of slow decline of citrus and occurs in most citrus-growing areas of the world causing considerable economic loss (Cohn, 1972). After a feeding site is established on a citrus root, the female nematode becomes immobile with its posterior portion exterior to the root becoming greatly swollen. Eggs are deposited in a gelatinous matrix to which soil particles become attached, making it difficult to see the nematode. Secondary microorganisms often invade infected tissue, causing dark, necrotic lesions in the root cortex and a gradual decline of the tree (Tarjan & O'Bannon, 1984).

OUTLOOK FOR THE FUTURE

Barker and Davis (1996) discussed the future outlook for research in the biology of plant-parasitic nematodes; repeated herein are several of their observations. As indicated, morphological examination will always be a component of nematode identification, although great strides are being made in the use of molecular data for nematode diagnostics. A breakthrough in the widespread use of molecular data for diagnostic purposes was isozyme analysis to distinguish among several of the root knot nematode species developed by Esbenshade and Triantaphyllou (1990). Serological methods to detect the potato cyst nematodes have been used in Europe (Robinson et al., 1993; Schomaker & Been, 1992; Schots et al., 1990). Molecular probes have been developed for a number of additional important plant parasitic species, including species of *Bursaphelenchus*, *Ditylenchus*, *Pratylenchus*, *Radopholus* and *Xiphinema*. At present, many of these probes are of limited use, owing to their cost and a need for special laboratory equipment, but progress will be made in developing molecular tools that are useful worldwide and which can be carried out by using simple kits that do not require specialized facilities. Improved methods for molecular diagnosis of pest nematodes will be integral to increased research on the molecular interactions between nematodes and their plant hosts. Such research is already underway, and will inevitably lead to improved methods for nematode control. Already, molecular markers have made it possible to track specific resistance genes to nematodes in plants and to deliver adapted host plant germplasm to plant breeders in record time (Vierling et al., 1996). An early goal for molecular researchers is the production of crop plants that are altered by genetic engineering in ways that prevent the growth and development of pest nematode species. More will be said on these topics in subsequent chapters.

REFERENCES

Alby, T., J. M. Ferris, and V. R. Ferris. 1983. Dispersion and distribution of *Pratylenchus scribneri* and *Hoplolaimus galeatus* in soybean fields. J. Nematol. 21:409–415.

Andrássy, I. 1965. Verzeichnis und Bestimmungsschlussel der Arten der Nematoden-Gattungen *Criconemoides* Taylor, 1936 und *Mesocriconema* n. gen. Opusc. Zool. Budapest 5:153–171.

Barker, K. R., C. C. Carter, and J. N. Sasser (ed.) 1985. An advanced treatise on *Meloidogyne*. Vol. 2. North Carolina State Univ. Graphics, Raleigh.

Barker, K. R., and E. L. Davis. 1996. Assessing plant-nematode infestations and infections. p. 103–104. *In* S. H. DeBoer (ed.) Potential of pathogen indexing technologies. Adv. Bot. Res. Vol. 23. Acad. Press.

Barker, K. R., R. S. Hussey, L. R. Krusberg, G. W. Bird, R. A. Dunn, H. Ferris, V. R. Ferris, D. W. Freckman, C. J. Gabriel, P. S. Grewal, A. E. MacGuidwin, D. L. Riddle, P. A. Roberts, and D. P. Schmitt. 1994. Plant and soil nematodes: Societal impact and focus for the future. J. Nematol. 26:127–137.

Beneden, E. van. 1883. Recherches sur la maturation de l'oeuf, la fecondation et la division cellulaire. Gand & Lepizig.

Berkeley, M. J. 1855. Vibrio forming excrescens on the roots of cucumber. Gard. Chron. 14:220.

Bird, A. F. 1981. The Anguina-Corynebacterium association. p. 303–323. *In* B. M. Zuckerman and R. A. Rohde (ed.) Plant parasitic nematodes. Vol. 3.

Bosher, J. E., and W. McKeen. 1954. Lyophilization and low temperature studies with the bulb and stem nematode *Ditylenchus dipsaci* (Kühn, 1858) Filipjev. Proc. Helminthol. Soc. Wash. 21:113–117.

Boveri, T. 1899. Die Entwickelung von *Ascaris megalocephala* mit besonderer Rucksicht auf die Kernverhaltnisse. Festschrift. v. Carl v. Kupffer, Jena.

Brenner, S. 1974. The genetics of the nematode *Caenorhabditis elegans*. Genetics 77:71–94.

Brodie, B. B. 1984. Nematode parasites of potato. p. 167–212. *In* W. R. Nickle (ed.) Plant and insect nematodes. Marcel Dekker, Inc., New York.

Brown, D. J. F., A. Dalmasso, and D. L. Trudgill. 1993. Nematode pests of soft fruits and vines. p. 427–462. *In* K. Evans et al. (ed.) Plant parasitic nematodes in temperate agriculture. CAB Int., Wallingford, Oxon, England.

Cobb, N. A. 1914. Nematodes and their relationships. p. 457–490. *In* Yearbook department agriculture 1914.

Cohn, E. 1972. Nematode diseases of citrus. p. 215–244. *In* J. M. Webster (ed.) Economic nematology. Acad. Press, New York.

Decramer, W. 1991. Stubby root and virus vector nematodes: *Trichodorus*, *Paratrichodorus*, and *Monotrichodorus*. p. 587–625. *In* W. R. Nickle (ed.) Manual of agricultural nematology. Marcel Dekker, Inc., New York.

Eisenback, J.D. 1998. Morphology and systematics. p. 37–63. *In* K.R. Barker et al. (ed.) Plant and nematode interactions. Agron. Monogr. 36. ASA, CSSA, and SSSA, Madison, WI.

Esbenshade, P. R., and A. C. Triantaphyllou. 1985. Use of enzymes for identification of *Meloidogyne* species. J. Nematol. 17:6–20.

Fenwick, D. W. 1968. Red-ring disease of the coconut palm. p. 38–48. *In* G. C. Smart, Jr., and V. G. Perry (ed.) Tropical nematology. Univ. Florida Press, Gainesville.

Ferris, J. M., and V. R. Ferris. 1966. Observations on *Tetradonema plicans*, an entomoparasitic nematode, with a key to the genera of the Family Tetradonematidae (Nematode: Trichosyringida). Ann. Entomol. Soc. Am. 59:964–971.

Gaugler, R., and H. R. Kaya (ed.). 1990. Entomopathogenic nematodes in biological control. CRC Press, Ann Arbor, MI.

Golden, A. M., J. H. O'Bannon, G. S. Santo, and A. M. Finley. 1980. Description and SEM observation of *Meloidogyne chitwoodi* n. sp. (Meloidogynidae), a root-knot nematode on potato in the Pacific Northwest. J. Nematol. 12:319–327.

Griffin, G. D. 1984. Nematode parasites of alfalfa, cereals and grasses. p. 243–321. *In* W. R. Nickle (ed.) Plant and insect nematodes. Marcel Dekker, Inc., New York.

Heald, C. M., and C. C. Orr. 1984. Nematode parasites of cotton. p. 147–166. *In* W. R. Nickle (ed.) Plant and insect nematodes. Marcel Dekker, Inc., New York.

Hewitt, W. B., D. J. Raski, and A. C. Goheen. 1958. Nematode vector of soil-borne fanleaf virus of grapes. Phytopathology 48:586–595.

Huettel, R. N., D. W. Dickson, and D. T. Kaplan. 1984. *Radopholus citrophilus* sp. n. (Nematoda), a sibling species of *Radopholus similis*. Proc. Helminthol. Soc. Wash. 74:412–418.

Hung, C.-L. P., and W. R. Jenkins. 1969. *Criconemoides curvatum* and the peach tree decline problem. J. Nematol. 1:12 (Abstr.).

Hussey, R. S., and J. M. McGuire. 1987. Interaction with other organisms. p. 293–328. *In* R. H. Brown and B. R. Kerry (ed.) Principles and practice of nematode control in crops. Acad. Press, Sydney, Australia.

Jatala, P. 1991. Reniform and false root-knot nematodes, *Rotylenchus* and *Naccobus* spp. p. 509–528. *In* W. R. Nickle (ed.) Manual of agricultural nematology. Marcel Dekker, Inc., New York.

Kerr, E. D., and D. S. Wysong. 1979. Sting nematode, *Belonolaimus* sp. in Nebraska. Plant Dis. Rep. 63:506–507.

Krupinsky, J. M., P. A. Donald, and R. E. Barber. 1981. *Helicotylenchus* spp. and *Tylenchorhynchus* spp. associated with grasses in the western Dakotas. J. Nematol. 13:446–447.

Lewin, R. 1984. Why is development so illogical? Science (Washington, DC) 224:1327–1329.

Loof, P. A. A. 1984. The family Pratylenchidae Thorne, 1949. p. 363–421. *In* W. R. Nickle (ed.) Plant and insect nematodes. Marcel Dekker, Inc., New York.

Loos, C. H. 1962. Studies on the life-history and habits of the burrowing nematode, *Radopholus similis*, the cause of blackhead of banana. Proc. Helminthol. Soc. Wash. 29:43–53.

Lownsbery, B. F., H. English, E. H. Moody, and F. J. Schick. 1973. *Criconemoides xenoplax* experimentally associated with a disease of peach tree decline problem. Phytopathology 63:994–997.

MacGuidwin, A. E. 1989. Distribution of *Pratylenchus scribneri* between root and soil habitats. J. Nematol. 21:418–426.

Mai, W.F., and P. G. Mullin. 1996. Plant-parasitic nematodes: a pictorial key to genera. Comstock Publ. Assoc., Ithaca, New York.

Marx, J. L. 1984. *Caenorhabditis elegans*: Getting to know you. Science (Washington, DC) 225:40–42.

McSorley, R., J. D. Arnett, S. C. Bost, W. W. Carter, S. Hafez, A. W. Johnson, T. Kirkpatrick, A. P. Nyczepir, J. D. Radewald, A. F. Robinson, and D. P. Schmitt. 1987. Bibliography of estimated crop losses in the United States due to plant-parasitic nematodes. Ann. Appl. Nematol. (Suppl. J. Nematol.) 1:6–12.

Nickle, W. R., and H. E. Welch. 1984. History, development and importance of insect nematology. p. 627–653. *In* W. R. Nickle (ed.) Plant and insect nematodes. Marcel Dekker, Inc., New York.

Norton, D. C., and J. K. Hoffman. 1975. *Longidorus breviannulatus* n. sp. (Nematoda: Longidoridae) associated with stunted corn in Iowa. J. Nematol. 7:168–171.

Nyczepir, A. P., and J. M. Halbrendt. 1993. Nematode pests of deciduous fruit and nut trees. p. 381–425. *In* K. Evans et al. (ed.) Plant parasitic nematodes in temperate agriculture. CAB Int., Wallingford, Oxon, England.

Nyczepir, A. P., E. I. Zehr, S. A. Lewis, and D. C. Harshman. 1983. Short life of peach trees induced by *Criconemella xenoplax*. Plant Dis. 67:507–508.

Perry, V. G., H. M. Darling, and G. Thorne. 1959. Anatomy, taxonomy and control of certain spiral nematodes attacking blue grass in Wisconsin. Univ. Wisconsin Res. Bull. 207.

Powers, T. O., and T. S. Harris. 1993. A polymerase chain reaction method for identification of *Meloidogyne* species. J. Nematol. 25:1–6.

Robinson, M. P., G. Butcher, R. H. Curtis, K. G. Davies, and K. Evans. 1993. Characterization of a 34D protein from potato cyst nematodes, using monoclonal antibodies with potential for species diagnosis. Ann. Appl. Biol. 123:337–347.

Rowe, R. C., R. M. Riedel, and M. J. Martin. 1985. Synergistic interactions between *Verticillium dahliae* and *Pratylenchus penetrans* in potato early dying disease. Phytopathology 75:412–418.

Safford, J., and R. M. Riedel. 1976. *Criconemoides* species associated with golf course greens in Ohio. Plant Dis. Rep. 60:405–408.

Santmyer, P. H. 1955. A comparison of the thermal death time of two dissimilar species of nematodes: *Panagrellus redivivis* (Linn, 1767) Goodey 1945, and *Meloidogyne incognita* var. *acrita*, Chitwood, 1949. Proc. Helminthol. Soc. Wash. 22:22–25.

Sasser, J. N. 1972. Nematode diseases of cotton. p. 187–214. *In* J. M. Webster (ed.) Economic nematology. Acad. Press, New York.

Sasser, J. N. 1980. Root-knot nematodes: A global menace to crop production. Plant Dis. 64:36–41.

Sayer, R. M. 1963. Effects of rapid cooling to subzero temperatures on *Meloidogyne hapla* and *M. incognita*. Phytopathology 53:888 (Abstr.).

Schindler, A. F. 1957. Parasitism and pathogenicity of *Xiphinema diversicaudatum*, an ectoparasitic nematode. Nematologica 2:25–31.

Schindler, A. F., and A. J. Braun. 1957. Pathogenicity of an ectoparasitic nematode, *Xiphinema diversicaudatum*, on strawberries. Nematologica 2:91–93.

Schomaker, C. H., and T. H. Been. 1992. Sampling strategies for the detection of potato cyst nematodes; developing and evaluating a model. p. 182–194. *In* F. Gommers and P. W. Th. Maas (ed.) Nematology from molecule to ecosystem. Europ. Soc. Nematol., Wageningen, The Netherlands.

Schots,. A. F., F. J. Gommers, J. Baker, and E. Egberts. 1990. Serological differentiation of plant-parasitic nematode species with polyclonal and monoclonal antibodies. J. Nematol. 22:16–23.

Smart, G. C., Jr., and K. B. Nguyen. 1991. Sting and awl nematodes: *Belonolaimus* spp. and *Dolichodorus* spp. p. 627–667. *In* W. R. Nickle (ed.) Manual of agricultural nematology. Marcel Dekker, Inc., New York.

Smolik, J. D., and L. E. Rogers. 1976. Effects of cattle grazing and wildfire on soil-dwelling nematodes of shrub-steppe ecosystem. J. Range Manag. 29:304–306.

Sturhan, D., and M. W. Brzeski. 1991. Stem and bulb nematodes, *Ditylenchus* spp. p. 423–464. *In* W. R. Nickle (ed.) Manual of agricultural nematology. Marcel Dekker, Inc., New York.

Suit, R. F., and E. P. DuCharme. 1953. The burrowing nematode and other parasitic nematodes in relation to spreading decline of citrus. Plant Dis. Rep. 37:379–383.

Sulston, J. E., E. Schierenberg, J. G. White, and J. N. Thomson. 1983. The embryonic cell lineage of the nematode *Caenorhabditis elegans*. Develop. Biol. 100:64–119.

Tarjan, A. C. 1973. A synopsis of the genera and species in the Tylenchorhynchinae (Tylenchoidea, Nematoda). Proc. Helminthol. Soc. Wash. 40:123–144.

Tarjan, A. C., and J. H. O'Bannon. 1984. Nematode parasites of citrus. p. 395–433. *In* W. R. Nickle (ed.) Plant and insect nematodes. Marcel Dekker, Inc., New York.

Taylor, A. L., V. H. Dropkin, and G. C. Martin. 1955. Perineal patterns of root-knot nematodes. Phytopathology 45:26–34.

Triantaphyllou, A. C., and E. McCabe. 1989. Efficient preservation of root-knot and cyst nematodes in liquid nitrogen. J. Nematol 21:423–426.

Vierling, R. A., J. Faghihi, V. R. Ferris, and J. M. Ferris. 1996. Association of RFLP markers with loci conferring broad-based resistance to the soybean cyst nematode (*Heterodera glycines*). Theor. Appl. Genet. 92:83–86.

Wallace, H. R. 1958a. Movement of eelworms. I. The influence of pore size and moisture content of the soil on the migration of larvae of the beet eelworm, *Heterodera schachtii* Schmidt. Ann. Appl. Biol. 46:74–85.

Wallace, H. R. 1958b. Movement of eelworms. III. The relationship between eelworm length, activity and mobility. Ann. Appl. Biol. 46:662–668.

Wehunt, E. J. 1984. Nematode parasites of peach and other tree crops. p. 435–455. *In* W. R. Nickle (ed.) Plant and insect nematodes. Marcel Dekker, Inc., New York.

Wehunt, E. J., and D. I. Edwards. 1968. *Radopholus similis* and other nematode species on banana. p. 1–19. *In* G. C. Smart, Jr., and V. G. Perry (ed.) Tropical nematology. Univ. Florida Press, Gainesville.

Wood, W. B. (ed.). 1988. The nematode *Caenorhabditis elegans*. Cold Spring Harbor Lab., New York.

3 Morphology and Systematics

JON D. EISENBACK

Virginia Polytechnic Institute and State University
Blacksburg, Virginia

Even though nematodes occupy nearly every habitat on earth, they are remarkably similar in morphology and life stages. These unsegmented roundworms form one clearly defined phylum that is conspicuously separated from other groups of organisms (Table 3–1). Members of the phylum Nemata are gonochoristic, triploblastic, bilateral, unsegmented, vermiform, aquatic pseudocoelomates (Maggenti, 1991b). Nematodes are circular in cross-section, and their overall body plan consists of a tube within a tube: the outer tube is formed by the body wall and the inner tube is composed of the alimentary canal (Figs. 3–1, 3–2). The area between the outer tube and the digestive system is not lined with an epithelium, but contains fluid and the tubelike gonads of the reproductive system (Fig. 3–2).

Nematode morphology is important for identification of species and for deriving information about their trophic category. To the untrained eye, nematodes appear to be very similar in morphology, but each species is morphologically distinct. Obviously, species within the same genus are very similar to each other, whereas species of different genera may be quite different, particularly if they also belong to different families or other higher taxonomic groups (Luc, 1988).

The morphology of nematodes often reveals their trophic group; function is derived from form (Lee & Atkinson, 1977). For example, bacterial feeding nematodes have a simple cylindrical or triangular buccal cavity; however, all nematodes that feed on plants possess a hollow, protrusile, hypodermic needlelike stylet (Fig. 3–3). Without a stylet, true plant parasitism is not possible because the stylet serves as the interface between the plant and the nematode.

MORPHOLOGY

The nematode body is made up of several distinct body systems: the body wall, nervous system, secretory-excretory system, digestive system, and reproductive system (Bird & Bird, 1991) (Fig. 3–1). Nematodes do not possess a specialized circulatory or respiratory system. The exchange of gases is thought to occur through the cuticle and circulation proceeds through the movement of fluids within the pseudocoelom and by simple diffusion across membranes.

Table 3–1. Characteristics of members of the phylum Nemata.

1. Inhabit marine, freshwater, and terrestrial environments as free-livers and parasites.
2. Gonochoristic, triploblastic, bilateral, unsegmented pseudocoelomates.
3. Vermiform, round in cross-section, covered with a three-layered cuticle.
4. Growth accompanied by molting of juvenile stages, usually four juvenile stages.
5. Oral opening surrounded by 6 lips and 16 sensory structures.
6. Possess unique cephalic sense organs called amphids.
7. Body wall contains only longitudinal muscles connected to longitudinal nerve chords by processes extending from each muscle.
8. Unique excretory system containing gland cells or a set of collecting tubes.
9. Longitudinal nerve cords housed within thickenings of the hypodermis.

Body Wall

Nematodes vary in size, but most plant-parasitic forms are 0.5 to 2 mm in length. Usually the body is vermiform or filiform (Fig. 3–1), but in some plant-parasitic nematodes the female becomes saccate, pyriform, reniform, or lemon-

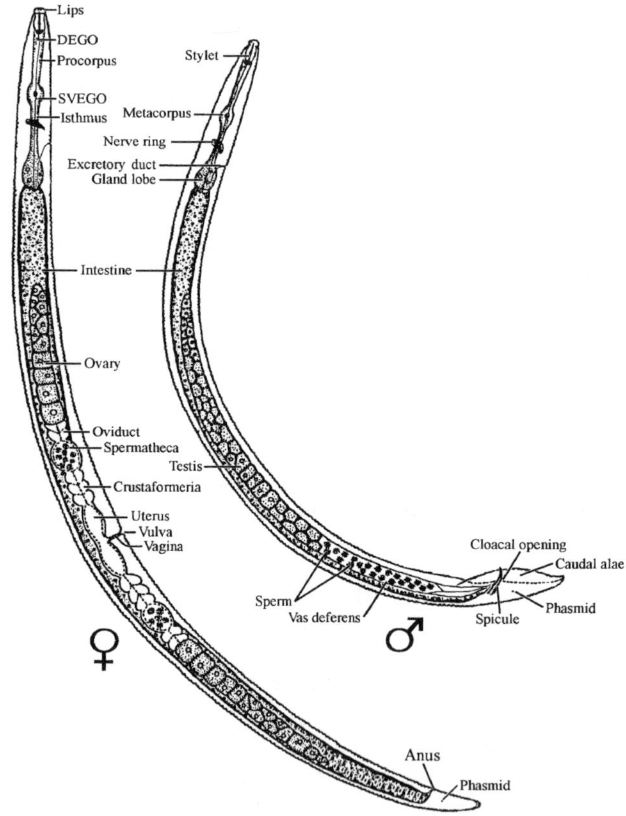

Fig. 3–1. Body morphology of a typical vermiform plant-parasitic nematode (lateral view); DEGO = dorsal esophageal gland orifice, SVEGO = subventral esophageal gland orifice (redrawn from Hirschmann, 1960).

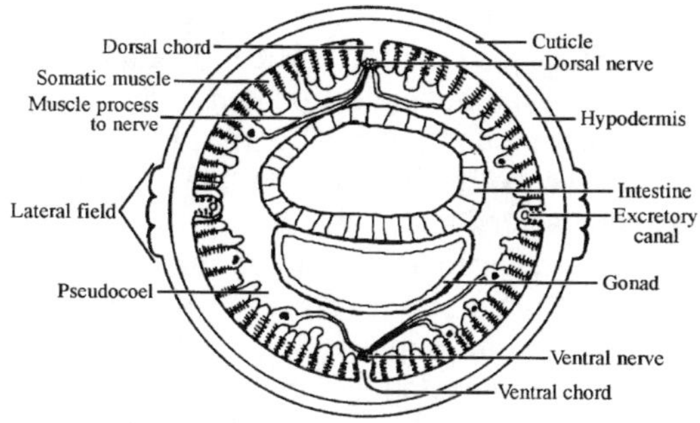

Fig. 3–2. Cross-section through the body of a typical plant-parasitic nematode.

shaped (Freckman & Baldwin, 1990) (Fig. 3–4). The swelling increases the reproductive potential of the organism, but results in the loss of mobility.

The body wall includes the cuticle, hypodermis, and somatic muscles (Figs. 3–1, 3–2). The body wall protects the nematode from the harsh external environment, serves as the exoskeleton, and provides the mechanism for movement of the organism through the soil or plant tissue. Also, the body wall contains much of the nervous and secretory-excretory systems, and it plays a role in the exchange of gases.

The cuticle is a nonliving, noncellular, triple-layered covering that is secreted by the underlying hypodermis. The cuticle is flexible. It covers the entire body and lines the esophagus, vulva, anus, cloaca, excretory pore, and sensory organs. The feeding stylet and copulatory spicules are formed from cuticle.

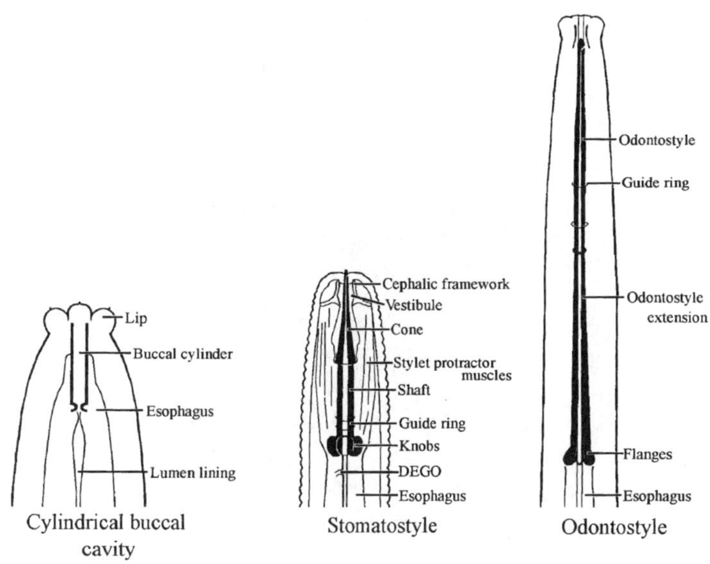

Fig. 3–3. Comparison of the anterior end of a microbivorous nematode having a cylindrical stoma, a typical plant-parasitic nematode possessing a stomatostylet, and a typical plant-parasitic nematode with an odontostylet (DEGO = dorsal esophageal gland orifice).

The composition and form of the cuticle is extremely variable, although it is composed of three primary zones: the cortical, medial, and basal layers. The cortical layer is amorphous and often electron dense; the medial layer often contains fluid, struts, and globular bodies; and the basal layer may hold laminae, striations, or fibers. These layers often form various markings and structures that are useful morphological characters.

The cuticle of many nematodes is marked with transverse indentions that occur at regular intervals along the body (Fig. 3–5). Striae are shallow, faint transverse markings; annulations are slightly deeper and easily seen with a light microscope; and apparent segmentations are very deep and pronounced. The annuli may also form scales, spines, and other outgrowths.

Likewise, the cuticle may also be marked with longitudinal markings (Fig. 3–5). Longitudinal ridges that run the entire length of the body are quite common on plant-parasitic nematodes. A combination of longitudinal and transverse markings form tessellation.

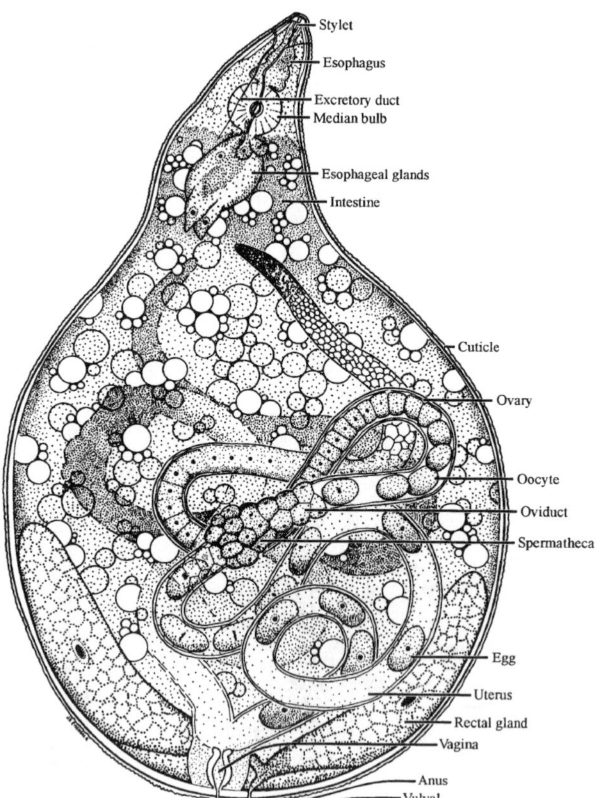

Fig. 3–4. Body morphology of a typical plant-parasitic nematode having a swollen female (lateral view).

Lateral fields are longitudinal ridges that occur on top of the hypodermal lateral chords (Fig. 3–5). They can be further divided by longitudinal incisures or transverse areolations. Longitudinal thickenings of the cuticle form various types of alae. Cervical alae occur near the anterior end, longitudinal alae are usually located in the lateral field, and caudal alae are situated near the tail. In the male, copulatory bursae are specialized caudal alae.

The hypodermis, which can be cellular, partially cellular, or syncytial, secretes the cuticle (Fig. 3–2). It lies beneath the cuticle and contains longitudinal thickenings between the somatic muscles that contain the nuclei, mitochondria, lipid droplets, endoplasmic reticulum, longitudinal nerves, and the canals of the secretory-excretory system. Most nematodes have four hypodermal chords: one dorsal, one ventral, and two lateral.

A single layer of obliquely arranged, spindle-shaped somatic muscle cells are grouped between the hypodermal chords (Fig. 3–2). The striated contractile portion of the muscle is tightly attached to the hypodermis, which in turn is tightly attached to the cuticle. A nonstriated, noncontractile portion of the muscle cell contains the nucleus and other cell organelles. It is connected to the nervous system by an elongated process of the noncontractile portion of the muscle cell. All of the muscle cells on the ventral side of the body are attached to the ventral nerve, and all of the cells on the dorsal side are connected to the dorsal nerve. Therefore, stimulation of muscles by the dorsal and ventral nerves causes contractions in the dorsal-ventral plane, and results in the characteristic sinusoidal movement of the nematode.

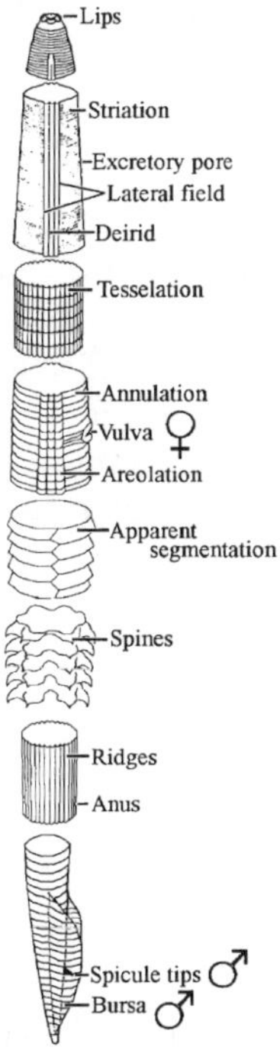

Fig. 3–5. Various external cuticular markings that are common to many plant-parasitic nematodes (Redrawn from Freckman & Baldwin, 1990).

Nervous System

The nervous system of nematodes is composed of a nerve ring surrounding the esophagus and is connected to longitudinal nerves that extend anteriorly and posteriorly. Additional nerve bundles may occur in the tail region of the nematode. Sensory organs are connected to the nervous system and function as tacto- and chemoreceptors (Fig. 3–6). These receptors appear as papillae, setae, or other sensilla. The typical arrangement and occurrence of sensory structures on the anterior end of the nematode is six inner labial sensilla, six outer labial sensilla, and four cephalic sensilla. These sensory organs may open to the external environment and function as chemoreceptors, or they may end just beneath the cuticle and function as tactoreceptors.

Nematodes also have large, paired, unique amphids near the anterior end and are neuro-secretory glands that function as chemoreceptors. The location of the amphidial openings on the anterior end is different for the two taxonomic classes of nematodes. In Secernentea, the amphids open on the anterior end, usually between the labial disc and lateral lips, but in Adenophorea they open posteriorly to the lips down on the body.

Small, paired, unique phasmids may be present on the posterior end of some nematodes, usually within the lateral field between the anus and tail tip (Fig. 3–1). Like the amphids, they appear to be neurosecretory organs that function as chemoreceptors. Phasmids are present in Secernentea, but are absent in Adenophorea.

Nematodes may also have other sensory organs, including deirids, ocelli, supplemental organs, and innervated setae, bristles, and warts. These additional sensory structures may occur on free-living nematodes; however, they are absent on plant-parasitic nematodes.

Secretory-Excretory System

The secretory-excretory system opens through the body wall ventrally, usually near the nerve ring. In Secernentea, the system is tubular and usually consists of a terminal duct that is distinctly lined with cuticle (Fig. 3–1). It is multicellular and composed of lateral excretory canals connected to a sinus. The excretory system is thought to function in osmoregulation and secretion. In some plant-parasitic nematodes, the secretory-excretory system produces a gelatinous matrix that is secreted around the eggs and is useful for protecting the eggs from the environment.

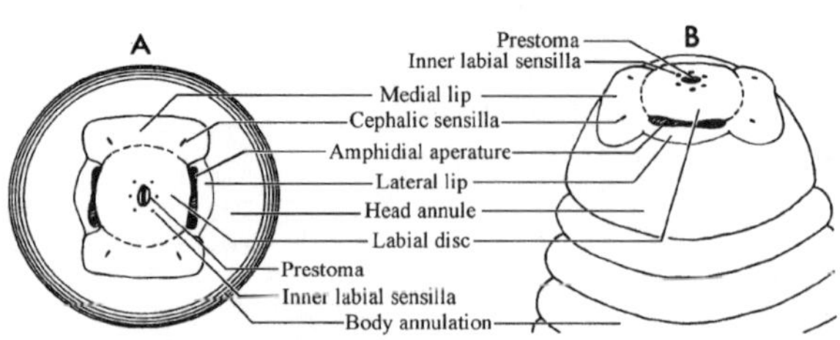

Fig. 3–6. External morphology of the head of a plant-parasitic nematode. *A*, face view; *B*, lateral view.

In Adenophorea, the secretory-excretory system is less complex and less variable. It is composed of a single gland cell connected to a duct that is not distinctly lined with cuticle. Unlike Secernentea, the secretory-excretory system functions only in secretion, and does not have an osmoregulatory role.

Digestive System

The digestive system of nematodes includes the stoma, esophagus, intestine, and posterior gut (Figs. 3–1 to 3–4). The stomatal opening is small and slitlike; it is usually surrounded by an oval-shaped prestoma that is encircled by the small pitlike openings of the six inner labial sensilla (Fig. 3–6). The stomatal opening occurs on the labial disc and is surrounded by six lips: two subdorsal, two subventral, and two lateral. Plant-parasitic nematodes are armed with a protrusile stylet which is usually hollow and functions like a hypodermic needle. In Secernentea, the stylet is thought to be derived from fusion of the stomatal lining and is, therefore, called a stomatostylet (Fig. 3–3). The stomatostylet consists of an anterior cone, a cylindrical shaft and three rounded knobs. The stylet protractor muscles attach to the knobs. However, in Adenophorea, the stylet is thought to be derived from a tooth and, therefore, is called an odontostylet (Fig. 3–3). The posterior end of the odontostylet may have an extension and three flanges that serve as points of attachment for the stylet protractor muscles (Fig. 3–3). In the plant-parasitic nematodes, *Trichodorus* and *Paratrichodorus*, the odontostylet is distinctly curved ventrally, lacks flanges, and is not hollow. It functions to pierce the cell wall of the root. The nematode secretes a hollow tube out of its stoma that connects it with the plant. This feeding tube serves as the interface between the nematode and the plant.

The stylet is contained within the cuticularized cephalic framework, which supports the anterior end of the nematode and provides sites for attachment of the stylet protractor muscles. The central cylinder of the framework, the vestibule, serves as a channel and guides the stylet through the stoma.

The esophagus is attached to the posterior portion of the stylet and is lined with cuticle. In the plant-parasitic nematodes in Adenophorea, the esophagus is divided into a narrow anterior procorpus and a broad posterior corpus (Fig. 3–7). Three to five esophageal gland cells empty into the lumen; one dorsal and two to four subventral glands. However, in Secernentea the esophagus is divided into distinct regions: a narrow procorpus, followed by a broad muscular median bulb or pump, a narrow isthmus, and a gland lobe (Fig. 3–7). The gland lobe may overlap the intestine in some genera and contains three gland cells; one dorsal gland, and two subventrals. In the order Tylenchida, the dorsal gland opens near the base of the stylet, but in the order Aphelenchida, it empties within the median bulb (Fig. 3–7).

The esophagus is connected to the intestine, which is an amorphous syncytium that fills the majority of the body cavity. Ingested nutrients that are absorbed by the intestine are circulated throughout the body by osmosis and movement within the fluid that fills the pseudocoelomic cavity. The reproductive system, which requires the greatest amount of energy, is conveniently located directly adjacent to the intestine. In juveniles and females, the intestine terminates

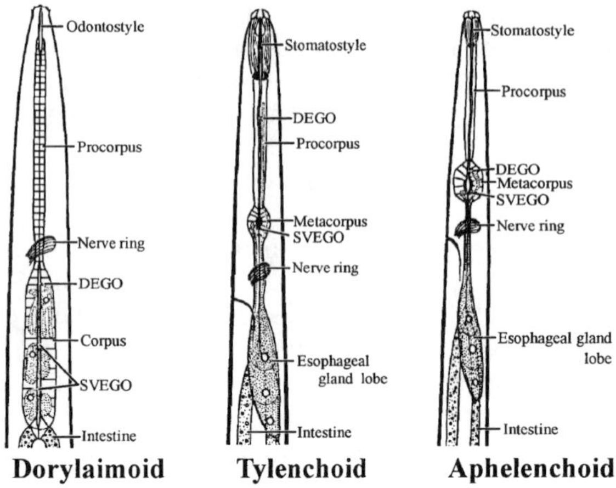

Fig. 3–7. Morphological comparison of the esophagi of the three orders of plant-parasitic nematodes: Tylenchida, Aphelenchida, and Dorylaimida (DEGO = dorsal esophageal gland orifice, SVEGO = subventral esophageal gland orifice; Hirschmann, 1960).

as a cuticle-lined rectum and passes through the body wall via the anus. In males, however, the intestine and testis form a common duct, the cloaca.

Reproductive System

The gonads of both the female and male are tubular invaginations within the pseudocoelom of the nematode, and both are similar in gross morphology (Figs. 3–1, 3–4). In the female, the gonad is made up of the ovary, oviduct, and uterus. The distal germinal zone of the ovary gives rise to small tightly packed oocytes that progressively enlarge in the proximal growth zone. The oocytes may be attached to each other centrally by a rachis. The oviduct may contain a spermatheca, an enlarged region for the storage of sperm. Following the spermatheca, the crustaformeria, a gland that produces the egg shell, connects the oviduct to the muscular uterus. There is a single uterus in monodelphic species which usually possesses a postvulvar uterine sac, or two uteri in didelphic species. The uteri are connected to the body wall by the vagina which opens to the outside as a slit-shaped vulva.

The male testis is divided into germinal and growth zones. It may be connected to the vas deferens by a vas efferens and an enlarged seminal vesicle that is used for the storage of mature sperm. The posterior end of the gonad may contain an ejaculatory duct and associated gland. Unlike the female, the male gonad shares a common duct with the intestine and opens to the outside through the cloacal opening. The male possesses a pair of cuticularized copulatory organs, called the spicules. They form a hollow tube when they are protracted through the cloacal opening and inserted into the vagina during copulation.

Nematode morphology has been described in detail in a large number of references (Andrássy, 1976; Bird, 1971; Bird & Bird, 1991; Chitwood & Chitwood, 1974; Crofton, 1966; Croll & Matthews, 1977; de Coninck, 1965;

Table 3–2. Genera of the most common plant-parasitic nematodes presented in alphabetical order according to their common name.

	Common name	Genus
1.	Awl nematode	(*Dolichodorus* spp.)
2.	Cyst nematodes	(*Globodera* and *Heterodera* spp.)
3.	Dagger nematode	(*Xiphinema* spp.)
4.	Foliar nematode	(*Aphelenchoides* spp.)
5.	Lance nematode	(*Hoplolaimus* spp.)
6.	Lesion nematode	(*Pratylenchus* spp.)
7.	Needle nematode	(*Longidorus* spp.)
8.	Pin nematode	(*Paratylenchus* spp.)
9.	Reniform nematode	(*Rotylenchulus* spp.)
10.	Ring nematode	(*Criconemella* spp.)
11.	Root-knot nematode	(*Meloidogyne* spp.)
12.	Sheath nematode	(*Hemicycliophora* spp.)
13.	Spiral nematodes	(*Helicotylenchus*, *Rotylenchus* and *Scutellonema* spp.)
14.	Sting nematode	(*Belonolaimus* spp.)
15.	Stubby-root nematodes	(*Paratrichodorus* and *Trichodorus* spp.)
16.	Stunt nematode	(*Tylenchorhynchus* spp.)

Freckman & Baldwin, 1990; Hirschmann, 1960, 1971; Hooper, 1978; Hyman, 1951; Lee & Atkinson, 1977; Luc et al., 1990; Maggenti, 1981, 1991a; Nicholas, 1984; Poinar, 1983; and Wharton, 1986).

SYSTEMATICS

Identification of nematode species is necessary to ascertain the cause of plant disease and to predict the impact of given population densities on crop production. Unfortunately, nematodes are difficult to identify by species, however the most common genera of plant-parasitic nematodes can be identified and counted under a dissecting microscope or an inverted microscope under low power (Table 3–2; Fig. 3–8).

In addition to morphology, molecular characters (including protein electrophoresis, immunological techniques, and DNA sequencing) may be useful for the identification of species, as well as subspecific forms such as races, pathotypes, and strains (Curran, 1991). Species-specific DNA probes have been developed for species of *Meloidogyne, Heterodera, Globodera*, and *Bursaphelenchus* (Curran, 1991; Baum et al., 1994).

The taxonomy of nematodes has been reviewed recently by Anderson and Mulvey (1979), Bongers (1988), Freckman and Baldwin (1990), Heyns (1971), Maggenti (1991b), Mai and Lyon (1975), Nickle (1991), and Siddiqi (1985). Baird and Bernard (1983) published a bibliography of keys to nematodes, and Ferris et al. (1976), Freckman and Baldwin (1990), Smart and Nguyen (1988), and Tarjan et al. (1977) have authored keys to various groups of terrestrial nematodes.

Nematodes belong to one phylum, Nemata, that is usually divided into two classes: Secernentea and Adenophorea. Two orders of Secernentea contain plant-parasitic nematodes, Tylenchida and Aphelenchida; whereas, Adenophorea has just one order with plant parasites, Dorylaimida. The groups that contain plant parasites will be described briefly, and the major morphological characters of the

Table 3–3. Morphological comparison of the gross morphological appearance of the most common genera of plant-parasitic nematodes.

Genus (common name)	Body length	Body width	Stylet length	Sclerotization of stylet
Awl nematodes	Long	Medium	Very long	Medium
Cyst nematodes				
Males	Long	Medium	Short	Medium
Juveniles	Short	Medium	Short	Medium
Dagger nematodes	Long	Thin	Very long	Medium
Foliar nematodes	Medium	Medium	Very short	Light
Lance nematodes	Long	Medium	Medium	Heavy
Lesion nematodes	Short	Medium	Short	Medium
Needle nematode	Very long	Thin	Very long	Light
Pin nematodes	Very short	Medium	Short	Medium
Reniform nematodes				
Males	Short	Medium	Short	Light
Juveniles	Short	Medium	Short	Light
Immature females	Short	Medium	Short	Light
Ring nematodes	Short	Stout	Long	Heavy
Root-knot nematodes				
Males	Long	Medium	Short	Light
Juveniles	Short	Medium	Very short	Light
Sheath nematodes	Medium	Medium	Long	Medium
Spiral nematodes	Medium	Medium	Short-Med.	Medium
Sting nematodes	Long	Thin	Very long	Medium
Stubby-root nematodes	Medium	Medium	Medium	Medium
Stunt nematodes	Medium	Medium	Short	Medium

† The following criteria describe the various attributes of the nematode genera listed in the table: Body length—short (<600 μm), medium (600–1500 μm), very long (>1500 μm); body width (based on body length/body width)—thin ($a > 50$), medium (20–50), and stout ($a < 20$); stylet length—very short (<15μm), short (15–30 μm), medium (30–80 μm), long (80–100 μm), and very long (>100 μm); and stylet sclerotization (subjective observation)—light, medium, and heavy.

common genera that occur in each group will be described below. A list of the most common genera, including their common and scientific names, are presented in Table 3–2, and the criteria that were used for describing the various subjective morphological attributes of each genus are listed in Table 3–3. The nematode genera are described in order of their relative economic importance to agricultural production.

Phylum Nemata: unsegmented roundworms, gonochoristic, triploblastic, bilateral, unsegmented, vermiform, aquatic pseudocoelomates; free-living, fungal feeders (fungivores), bacterial feeders (bacteriovores), onmivores, predators, parasites of invertebrates, plants, and animals; usually microscopic (<3 mm) (Fig. 3–8); rarely macroscopic (>7 m); transparent to pearly white, yellow, brown, to various tints dependent upon gut contents; alimentary canal tubular; terminal mouth; subterminal anus; body wall trilayered cuticle, hypodermis, and longitudinal somatic muscles; pseudocoelom filled with fluid under pressure forming a hydrostatic skeleton; locomotion typically in sinusoidal-shaped movements in the dorso-ventral plane; radially arranged lips; up to 16 sensilla in 3 whorls; two chemoreceptor amphids, anteriorly; sometimes two phasmids, posteriorly; secretory-excretory system glandular or tubular; nervous system a nerve ring, and

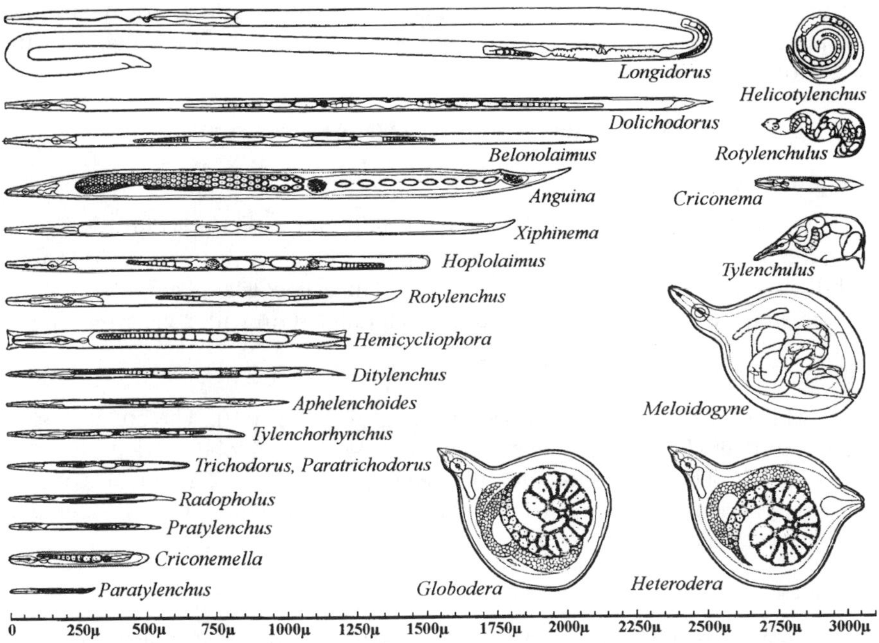

Fig. 3–8. Comparison of the size and morphology of 25 of the most common genera of plant-parasitic nematodes (modified from Agrios, 1969).

anterior and posterior longitudinal nerves; respiratory and circulatory systems absent; dioecious, reproduction by amphimixis, pseudogamy, hermaphroditism, and meiotic and mitotic parthenogensis; determinate cleavage of eggs; roughly 20 000 described species divided into two classes and 20 orders; many species remain undescribed (estimated to contain approximately 1 000 000 total species).

Class Secernentea: excretory system tubular, multicellular; hypodermal glands absent; phasmids present; amphids located between labial disc and lateral lips; setae on lips absent; many eggs produced per female; males usually with one testis; stylet, if present, a three-part stomatostylet (Table 3–4); this class has two orders that contain plant parasites, Tylenchida and Aphelenchida.

Order Tylenchida: Stomatostylet moderately to strongly sclerotized; three-part esophagus including a procorpus, metacorpus, and gland lobe; dorsal esophageal glands opening in procorpus near base of stylet; subventral esophageal gland openings in lumen of metacorpus; asymmetrical excretory system consist-

Table 3–4. Morphological comparison of the two classes of the phylum Nemata.

Character	Adenophorea	Secernentea
Excretory system	Glandular, one celled	Tubular, multicelled
Hypodermal glands	Present	Absent
Testes	Usually two	Usually one
Eggs produced	Few	Many
Stylet if present	Odontostylet	Stomatostylet
Setae on lips	Present	Absent
Phasmids	Absent	Present

ing of a single tube; sexual dimorphism common, soil dwelling, insect parasites, and plant parasites; the majority of plant-parasitic nematodes are contained within this order.

Root-Knot Nematodes (*Meloidogyne* spp.)

Second-Stage Juveniles

Second-stage, root-knot juveniles are short (400–600 μm) (Fig. 3–9A). The cephalic framework is weakly sclerotized and the stylet is very short (10–15 μm), lightly sclerotized, and has indistinct knobs. The esophageal gland lobe overlaps the intestine ventrally, and the tail tapers to a pointed tip with a clear terminus.

Males

Males of root-knot nematode are long (1.0–2.0 mm). The stylet is short (18–24 μm) and has distinct knobs. The esophageal gland lobe overlaps the intestine ventrally. The tail is short and rounded and lacks bursa. The spicules of root-knot males open a short distance from the tail tip, unlike those of cyst nematodes which open very near the terminus.

Females

Females of *Meloidogyne* are swollen, pear-shaped, pearly white, and sedentary (Figs. 3–4, 3–8, 3–9B). They deposit all of their eggs in a gelatinous mass, which usually protrudes from galled root tissue. Unlike cyst and reniform females, root-knot females usually remain completely endoparasitic.

Cyst Nematodes (*Globodera* and *Heterodera* spp.)

Second-Stage Juveniles

Second-stage juveniles of *Globodera* and *Heterodera* are short (400–600 μm). The head region of both is rounded, and the short (20–25 μm) stylet has distinct knobs. The esophageal gland lobe overlaps the intestine ventrally. The tail tapers to a pointed tip with a clear terminus. Juveniles of both genera are often found in soil together with cysts formed by females. The two genera cannot be distinguished from each other at low magnification.

Males

Males of *Globodera* and *Heterodera* are long (1.0–2.0 mm). The head region of both genera is rounded and the stylet is short (20–26 μm) with well-developed knobs. The esophageal gland lobe overlaps the intestine ventrally. The tail is short and rounded, and the spicules open very near the tail tip. Males of both genera inhabit soil and root tissue, and cannot be distinguished from each other at low magnification.

Females and Cysts

Live females of both *Globodera* and *Heterodera* are white to yellow. After death their bodies form cysts, which are light to dark brown. Most eggs are

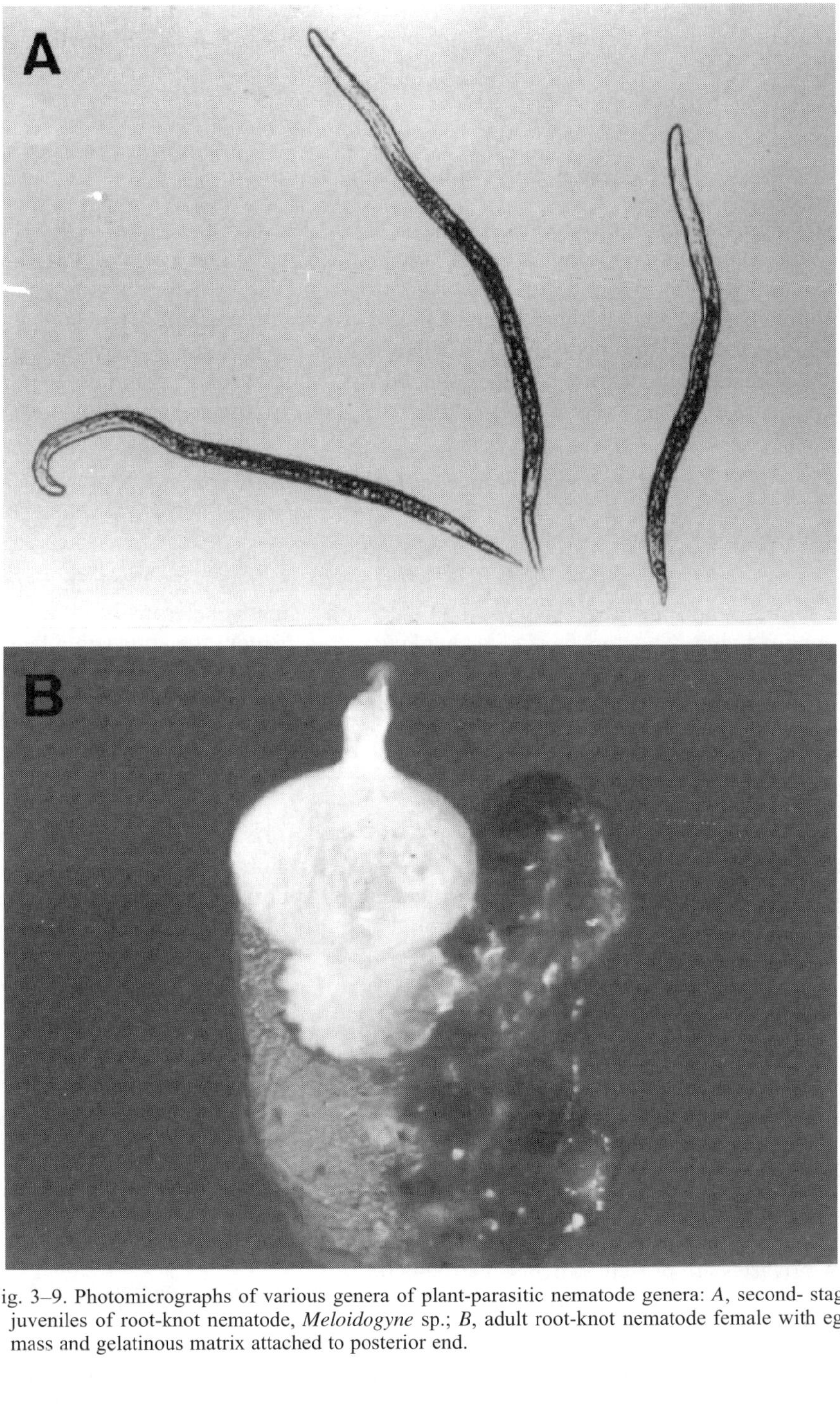

Fig. 3–9. Photomicrographs of various genera of plant-parasitic nematode genera: *A*, second- stage juveniles of root-knot nematode, *Meloidogyne* sp.; *B*, adult root-knot nematode female with egg mass and gelatinous matrix attached to posterior end.

retained within the body. Females of both genera have protruding necks, are immobile, and can be found attached to roots and in soil. Mature females and cysts of *Globodera* are round (Figs. 3–8, 3–9C). Females and cysts of *Heterodera* are lemon-shaped (Figs. 3–8, 3–9D).

Lesion Nematodes (*Pratylenchus* spp.)

Pratylenchus, the lesion nematodes, are small (0.45–0.70 mm) (Figs. 3–8, 3–9E). The anterior region is broad, and the head is blunt. The stylet is short (15–18 µm) with moderate to heavy sclerotization. The metacorpus is large and prominent, and the triradiate lumen lining is distinct but small. The esophagus overlaps the intestine ventrally. The vulva is near the tail, and one ovary is present. The tail tapers to a bluntly rounded tip. Males can be common or rare, depending upon the species. They have small, bluntly rounded bursae.

Reniform Nematodes (*Rotylenchulus* spp.)

Second-Stage Juveniles

Juveniles of reniform nematodes are short (0.35–0.45 µm). The head region is rounded and striated. The short (13–15 µm), lightly sclerotized stylet has rounded knobs. The dorsal esophageal gland orifice is more than one stylet length from the base of the stylet. The metacorpus is large, and the triradiate lumen lining is prominent. The esophagus overlaps the intestine ventrally. The tail is rounded.

Males

Males of *Rotylenchulus* are short (0.35–0.45 µm). The cephalic framework and esophagus are weakly developed. The stylet is short (12–15 µm) and lightly sclerotized. The metacorpus is conspicuous, but poorly developed and faintly lined with cuticle. The tail is bluntly pointed, and the bursae are small and do not reach the tip.

Pre-Parasitic Females

Unlike most other plant-parasitic nematodes, the preparasitic fourth-stage female of reniform nematode is the infective stage. They are short (0.35–0.45 mm), and the head region is rounded and striated. The stylet, like those of juveniles and males, is short (14–17 µm) with rounded knobs. The dorsal esophageal gland orifice is more than one stylet length from the base of the stylet. The triradiate lumen lining of the large metacorpus is prominent. The esophagus overlaps the intestine ventrally. The vulva occurs in the posterior region of the body and two ovaries are present. The tail is rounded.

Females

Females of *Rotylenchulus* are sedentary. Eggs are deposited into a gelatinous matrix which often covers the entire female. The anterior portion is usually

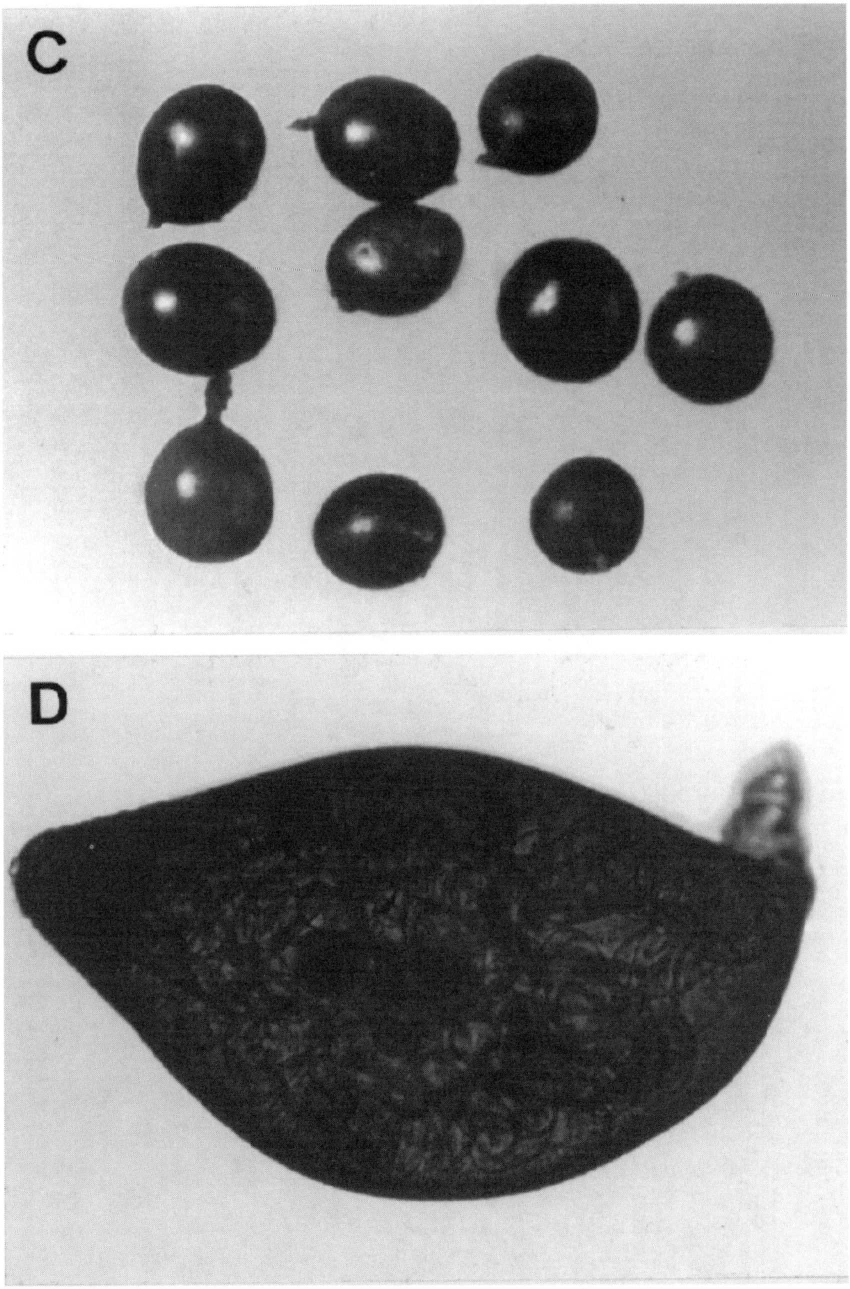

Fig. 3–9. Continued—Photomicrographs of various genera of plant-parasitic nematode genera: *C*, round cyst nematodes, *Globodera* sp.; *D*, lemon cyst of *Heterodera* sp.

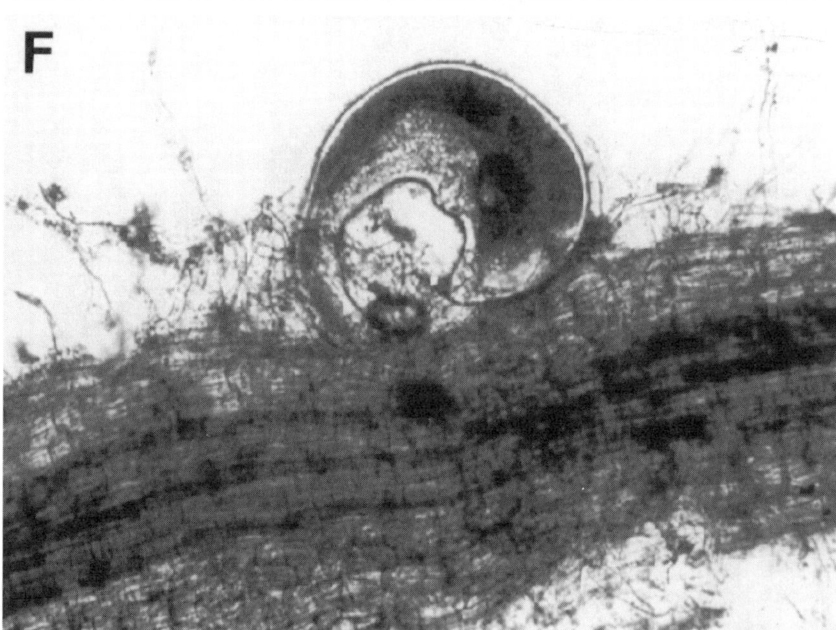

Fig. 3–9. Continued—Photomicrographs of various genera of plant-parasitic nematode genera: *E*, female lesion nematode, *Pratylenchus* sp.; *F*, mature female reniform nematode, *Rotylenchulus* sp. (*F* is courtesy of Edward McGrawley).

embedded in root tissue and the posterior, kidney-shaped portion remains external (Figs. 3–8, 3–9F). The vulval lips are prominent and protrude. The tail is small and pointed.

Sting Nematodes (*Belonolaimus* spp.)

Members of the genus *Belonolaimus* are long (2–3 mm), thin nematodes (Figs. 3–8, 3–10A). The head region is rounded and distinctly set off from the remainder of the body. Sting nematodes have a very long (100–140 μm) stylet with small, rounded knobs. The metacorpus is large and round, and the triradiate lumen lining is distinct. The esophagus overlaps the intestine dorsally. The vulva is located near the middle of the body, and two ovaries are present. The posterior end of the body is tapered, and the tail is rounded. Males of *Belonolaimus* have small curved spicules and a pointed tail that is surrounded by a small bursae. Males are slightly smaller than females.

Lance Nematodes (*Hoplolaimus* spp.)

Lance nematodes, which comprise the genus *Hoplolaimus*, are long (0.9–1.3 mm) and rounded at both ends (Figs. 3–8, 3–10B). Juveniles and adults frequently occur together in abundance. The head cap is prominent and set off from the remainder of the body. The stylet is heavily sclerotized and has large tulip-shaped knobs. The esophageal gland lobe overlaps the intestine dorsally. The vulva occurs near the middle of the body, and two ovaries are present. The tail is short and rounded in females and juveniles. Males are very common and appear similar to females, except that they have a pointed tail with small bursae and spicules.

Ring Nematodes (*Criconemella* or *Mesocriconema* spp.)

Criconemella or *Mesocriconema* spp. derive their common name from the coarse, ring-like appearance of the smooth-edged cuticular annulations (Figs. 3–8, 3–10C). Both mature and immature females are short (0.4–0.6 mm) and stout. Their stylets are long (65–90 μm) and heavily sclerotized and have anchor-shaped knobs. Like pin and sheath nematodes, the procorpus and metacorpus are fused, and the esophagus does not overlap the intestine. The triradiate lumen lining of the metacorpus is distinct and rounded. The vulva is located near the tail tip, and is difficult to detect. Only one ovary is present, and the posterior end of the body is blunt. Males of *Criconemella* or *Mesocriconema* are very rare.

Sheath Nematodes (*Hemicycliophora* spp.)

Species of *Hemicycliophora* appear to have a second cuticle that forms a loose sheath (Figs. 3–8, 3–10D). The cuticle is heavily annulated, and the head and/or tail may be withdrawn into the sheath. The stylet is long (80–95 μm) and slightly curved. Esophageal morphology is similar to that of pin and ring nematodes in that the procorpus and metacorpus are fused and the esophageal gland lobe does not overlap the intestine. The vulva is located near the rounded tail, the

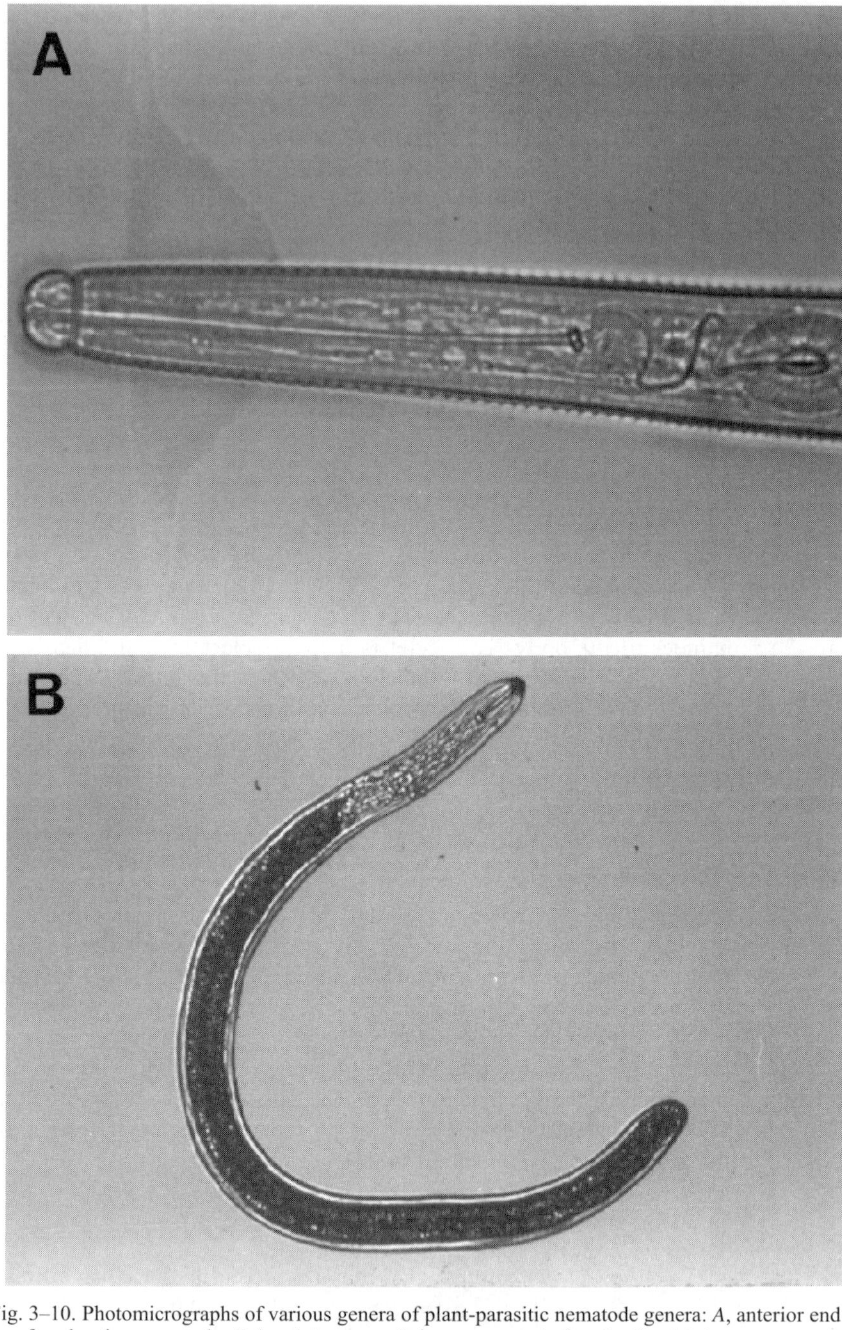

Fig. 3–10. Photomicrographs of various genera of plant-parasitic nematode genera: *A*, anterior end of a female sting nematode, *Belonolaimus* sp.; *B*, lance nematode female, *Hoplolaimus* sp.

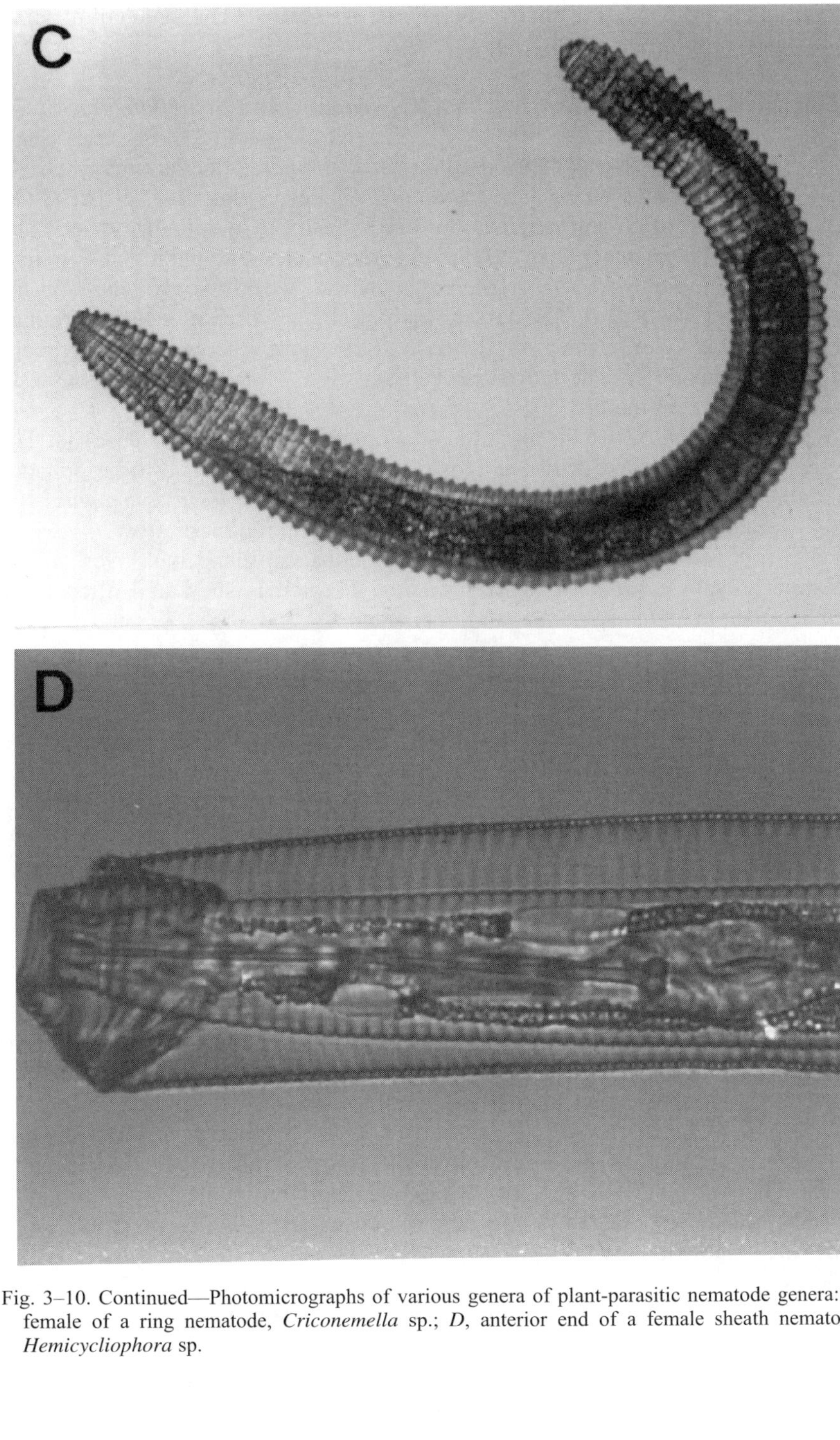

Fig. 3–10. Continued—Photomicrographs of various genera of plant-parasitic nematode genera: *C*, female of a ring nematode, *Criconemella* sp.; *D*, anterior end of a female sheath nematode, *Hemicycliophora* sp.

vulval lips protrude slightly, and a single ovary is present. The male tail is pointed and has a small bursae.

Spiral Nematodes (*Helicotylenchus*, *Rotylenchus*, and *Scutellonema* spp.)

Spiral nematodes, as their common name implies, often assume a loosely coiled shape in water when they are relaxed or dead (Figs. 3–8, 3–10E). The vulva of females of all spiral nematodes occurs near the middle of the body and two ovaries are present. Males are rare in some species, and unknown in others.

The lip region of *Helicotylenchus* is conical, weakly sclerotized, and the stylet is short (24–28 µm). The dorsal esophageal gland orifice occurs more than one-fourth of a stylet length from the base of the stylet. The esophagus overlaps the intestine ventrally. The tail is conical, usually offset, and often possesses a small projection on the tip.

In *Rotylenchus*, the lip region is also conical and weakly sclerotized. The dorsal esophageal gland orifice is close to the stylet knobs, and the esophagus overlaps the intestine dorsally. The tail is similar to that of *Helicotylenchus*.

Unlike *Helicotylenchus* and *Rotylenchus*, the lip region of *Scutellonema* is set off from the body and sclerotized. The esophageal gland lobe overlaps the intestine dorsally as it does in *Rotylenchus*. The tail is short and often more rounded than in *Helicotylenchus* and *Rotylenchus*.

Stunt Nematodes (*Tylenchorhynchus* spp.)

Tylenchorhynchus species have short (22–28 µm) stylets with backward sloping knobs (Figs. 3–8, 3–10F). The lip region is weakly developed, and the esophageal gland lobe does not overlap the intestine. The vulval occurs near the middle of the body, and two ovaries are present. Usually, the two uteri are clear and elongated. The tail is conical and tapers to a pointed tip. Males have small curved spicules and a busae that extends to the tail tip.

Pin Nematodes (*Paratylenchus* spp.)

Paratylenchus species, the pin nematodes, are very short (0.30–0.45 mm). The stylet is short (25–30 µm) with medially elongated knobs (in proportion to the body length, the stylet seems long). The procorpus is fused with the strong metacorpus, similar to that of ring and sheath nematodes. The vulva is located near the tail, and a single ovary is present. There is a distinct ventral curvature of the posterior portion of the body, and a marked postvulval reduction in body width. The tail tapers to a point. Males are common, and appear similar to females, but the esophagus is degenerate and the stylet is absent. Spicules of males are small and curved.

Awl Nematodes (*Dolichodorus* spp.)

Dolichodorus, commonly called the awl nematodes, are long (2–3 mm). The head is rounded and distinctly set off from the remainder of the body. The

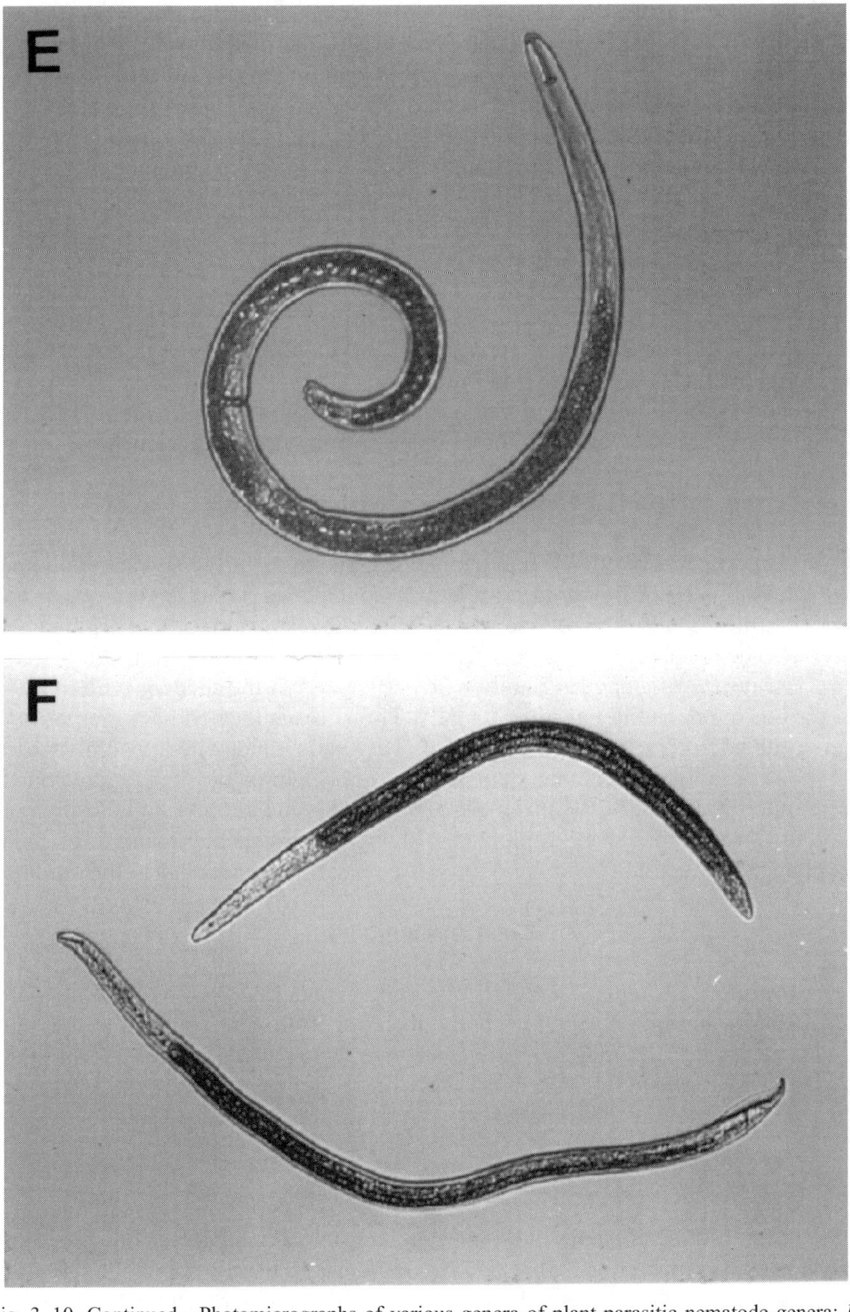

Fig. 3–10. Continued—Photomicrographs of various genera of plant-parasitic nematode genera: *E*, female spiral nematode, *Helicotylenchus* sp.; *F*, female and male stunt nematode, *Tylenchorhynchus* sp.

stylet is very long (90–120 µm) and has prominent knobs. The procorpus is swollen, and the metacorpus is large and rounded. The triradiate lumen lining of the metacorpus is distinct. The esophagus does not overlap the intestine. The vulva is located near the middle of the body, and two ovaries are present. The posterior end of the body is large and rounded, and the tail tip is long and pointed. Males have curved spicules and a pointed tail, which is surrounded by large bursae that project beyond the tail tip and form a deeply cleft lobe.

Order Aphelenchida

Stomatostylet present, but weakly sclerotized; three-part esophagus including a narrow procorpus, broad metacorpus, and gland lobe; both dorsal and subventral glands open into lumen in metacorpus (Fig. 3–7); cuticle nearly smooth or striated; soil dwelling fungal feeders, insect parasites, and plant parasites. *Aphelenchoides* is one of the common genera of the order Aphelenchida.

Foliar Nematodes (*Aphelenchoides* spp.)

Aphelenchoides, the foliar nematodes, move rapidly in water and may swim to the surface by rapid undulating movements. They have a very short (10–12 µm), lightly sclerotized stylet with inconspicuous knobs (Figs. 3–8, 3–11A). The metacorpus is large, often more than three-fourths of the body diameter. The dorsal esophageal gland orifice is located in the median bulb, and the esophagus overlaps the intestine dorsally. The vulva is located near the posterior end of the body. The female has a single ovary and an elongated postvulval uterine sac. The tail is conical and tapers to a blunt or pointed tip. It may have one or more mucro. The tail of the male is frequently curved ventrally and usually contains small rose-thorn shaped spicules. Although foliar nematodes are often found in foliage, flowers, and growing points of plants, they also occur in the soil.

Class Adenophorea

Excretory systems glandular, one cell; hypodermal glands present; phasmids absent; amphids located on body posterior to lips; setae on lips; few eggs produced per female; males usually with two testes; stylet if present, an odontostylet with extension (Table 3–4). Only one order of this class of nematodes, Dorylaimida, contains plant parasites.

Order Dorylaimida

Odontostylet present; two-part esophagus including a narrow procorpus and a broad corpus (Figs. 3–3, 3–7); body surface nearly smooth or striated; freshwater, soil-dwelling fungal feeders, nematode predators, and plant parasites; vectors of “tobra” and “nepo” plant viruses. The most common genera of plant parasites in this order include *Xiphinema* (Fig. 3–11B), *Longidorus* (Fig. 3–11C), *Trichodorus* (Fig. 3–11D), and *Paratrichodorus*.

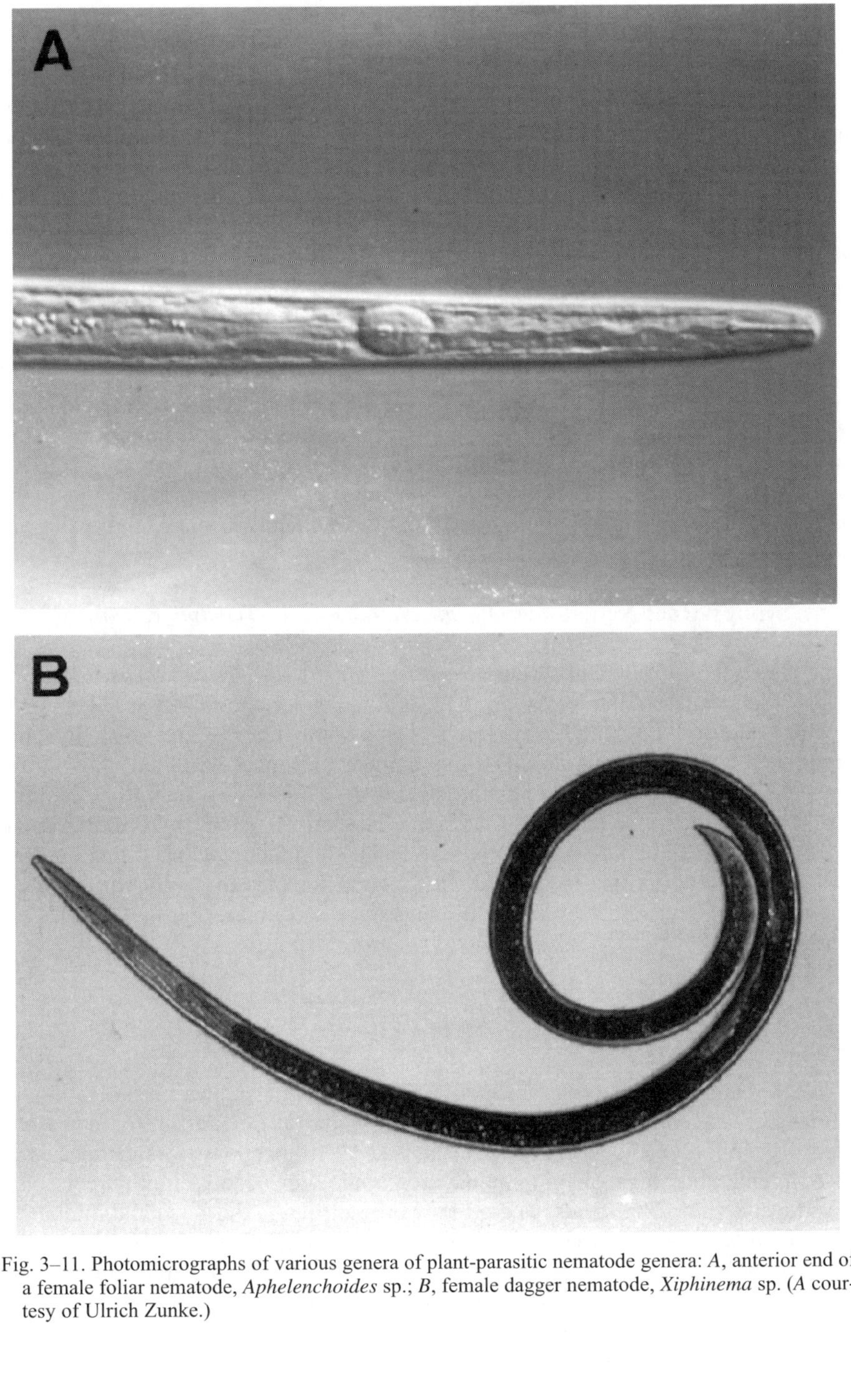

Fig. 3–11. Photomicrographs of various genera of plant-parasitic nematode genera: *A*, anterior end of a female foliar nematode, *Aphelenchoides* sp.; *B*, female dagger nematode, *Xiphinema* sp. (*A* courtesy of Ulrich Zunke.)

Dagger Nematodes (*Xiphinema* spp.)

Xiphinema, the dagger nematodes, are long (2–3 mm) and thin (Figs. 3–8, 3–11B). The stylet is very long (90–120 μm) and has no knobs. It has an extension that possesses three flanges. A guiding ring occurs in the posterior portion of the stylet. The procorpus is narrow and the corpus is broad, and the esophagus does not overlap the intestine. The vulva usually occurs near the middle of the body, and two ovaries are present. The tail is conical and tapers to a daggerlike point from which the genus gets its common name. Males are rare.

Needle Nematodes (*Longidorus* spp.)

Species of *Longidorus*, the needle nematodes, are very, very long (3.5–9 mm) and thin (Figs. 3–8, 3–11C). Their stylet is very long (80–120 μm) and lightly sclerotized, knobs are absent, and a guiding ring occurs in the anterior portion of the stylet. The anterior portion of the esophagus is narrow, and the posterior portion is broad and does not overlap the intestine. The vulva occurs near the middle of the body. Tail shape is variable, but frequently conical, and males are rare.

Stubby-Root Nematodes (*Paratrichodorus* and *Trichodorus* spp.)

The stubby-root nematodes, *Paratrichodorus* and *Trichodorus,* are most easily identified on the basis of their unique stylet morphology (Figs. 3–8, 3–11D). The stylet, which has no knobs, has a distinct dorsal curvature. In some specimens the outer layer of cuticle may appear loose and wrinkled. The esophagus has a narrow procorpus and a broad corpus. The vulva is near the middle of the body, and two ovaries are present. The tail is short and rounded. In *Paratrichodorus*, the vulva is lightly sclerotized, and the male has a small bursae and spicules that are nearly straight. In contrast, females of *Trichodorus* have a vulva that is heavily sclerotized, and males have no bursae, and the spicules are curved.

OUTLOOK

Nematode morphology is extremely prominent in the classification of nematodes and will likely remain the most important character for nematode identification. Identification of species and races will become more significant as recommendations for managing nematode populations become more precise and less dependent on chemicals. The effective use of plant resistance and biological control agents also requires an accurate identification of nematode populations. Unfortunately, identifications to species are often difficult and frequently end in a dilemma of uncertainty. These difficulties can be overcome with additional resources allocated to nematode taxonomy, by training additional taxonomists, and by the preparation of adequate descriptions and keys that illustrate all useful

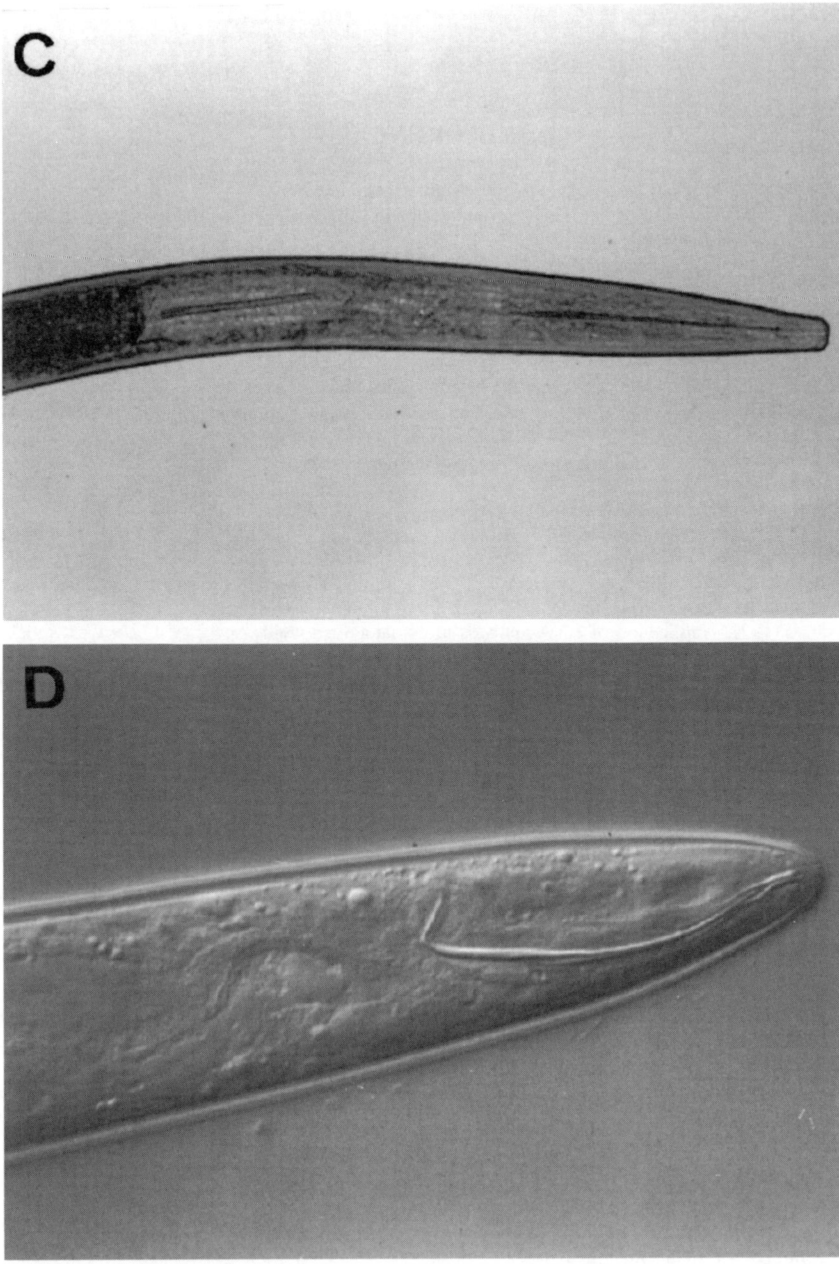

Fig. 3–11. Continued—Photomicrographs of various genera of plant-parasitic nematode genera: *C*, anterior end of a female needle nematode, *Longidorus* sp.; *D*, anterior end of a female stubby-root nematode, *Trichodorus* sp. (*D* courtesy of Ulrich Zunke.)

characters with proper recognition of the interspecific variation that occurs in nature.

REFERENCES

Agrios, G.N. 1978. Plant pathology. 2nd. ed. Acad. Press, New York.

Anderson, R.V., and R.H. Mulvey. 1979. Plant-parasitic nematodes in Canada. Part I. An illustrated key to the genera. Agric. Canada.

Baird, S.M., and E.C. Bernard. 1983. A bibliography of keys for the identification of plant-parasitic and free-living terrestrial nematodes. Nematol. Newsl. 29:9–18.

Baum, T.J., S.A. Lewis, and R.A. Dean. 1994. Isolation, characterization, and application of DNA probes specific to *Meloidogyne arenaria*. Phytopathology 84:489–494.

Bird, A.F. 1971. The structure of nematodes. Acad. Press, New York.

Bird, A.F., and J. Bird. 1991. The structure of nematodes. 2nd ed. Acad. Press, New York.

Bongers, T. 1988. De nematoden van nederland. Pirola Schoorl. Natuurhist. Biblioth. KNNV nr. 46. Wageningen Agric. Univ., the Netherlands.

Chitwood, B.G., and M. B. Chitwood. 1974. An introduction to nematology. Rev. Univ. Park Press, Baltimore, MD.

Crofton, H.D. 1966. Nematodes. Hutchinson Univ. Library, London.

Croll, N.A., and B.E. Matthews. 1977. Biology of nematodes. John Wiley & Sons Inc., New York. Stop

de Coninck, L. 1965. Classe de nematodes. p. 3–217. *In* P. P. Grasse (ed.) Traite de zoologie, Vol. 4. Masson et Cie, Paris.

Curran, J. 1991. Application of DNA analysis to nematode taxonomy. p. 125–143. *In* W. R. Nickle (ed.) Manual of agricultural nematology. Marcel Dekker, New York.

Ferris, V. R., L. M. Ferris, and J. P. Tjepkema. 1976. Genera of freshwater nematodes (Nematoda) of eastern North America. USEPA Aquat. Biol. Sect., Cincinnati, OH.

Freckman, D., and J. G. Baldwin. 1990. Nematoda. p. 155–200. *In* D. L. Dindal (ed.) Soil biology guide. John Wiley & Sons Inc., New York.

Heyns, J. 1971. A guide to the plant and soil nematodes of South Africa. A. A. Balkema, Cape Town.

Hirschmann, H. 1960. Gross morphology of nematodes. p. 125–135. *In* J. N. Sasser and H. R. Jenkins (ed.) Nematology. Univ. North Carolina Press, Chapel Hill.

Hirschmann, H. 1971. Comparative morphology and anatomy. p. 11–63. *In* B. M. Zuckermann et al. (ed.) Plant parasitic nemaotes. Vol. 1., Acad. Press, New York.

Hooper, D. J. 1978. Structure and classification of nematodes. p. 3–18. *In* S. F. Southey (ed.) Plant nematology. Tech. Bull. 7. Mins. Agric., Fisheries, and Food, London.

Hyman, L. H. 1951. The invertebrates: Acanthocephala, Aschelminthes and Entoprocta, the Pseudocoelomate Bilateria, Vol. III. McGraw-Hill Book Co., Inc., New York.

Lee, D.L. and H.J. Atkinson. 1977. Physiology of nematodes. McMillan Press, London.

Luc, M. 1988. Systematics and identification of plant-parasitic nematode genera. p. 1–7. *In* R. Fortuner (ed.) Nematode identification and expert system technology. Plenum Press, New York.

Luc, M., D.J. Hunt, and E. Machon. 1990. Morphology, anatomy and biology of plant parasitic nematodes—a synopsis. p. 1–44. *In* M. Luc et al. (ed.) Plant parasitic nematodes in subtropical and tropical agriculture. CAB Int. Inst. Parasitology, Wallingford, England.

Mai, W.F., P.G. Mullin, H.H. Lyons, and K. Loeffler. 1996. Plant-parasitic nematodes—a pictorial key to the genera. 5th ed. Cornell Univ. Press, Ithaca, NY.

Maggenti, A.R. 1976. Taxonomic position of nematoda among the pseudocoelomate bilateria. p. 3–10. *In* N. A. Croll (ed.) The organization of nematodes. Acad. Press, New York.

Maggenti, A.R. 1981. General nematology. Springer-Verlag, New York.

Maggenti, A.R. 1991a. General nematode morphology. p. 3–46. *In* W. R. Nickle (ed.) Manual of agricultural nematology. Marcel Dekker, Inc., New York.

Maggenti, A.R. 1991b. Nemata: Higher classification. p. 147–187. *In* W. R. Nickle (ed.) Manual of agricultural nematology. Marcel Dekker, New York.

Nicholas, W.L. 1984. The biology of free-living nematodes. Clarendon Press, Oxford, England.

Nickle, W.R. 1991. Manual of agricultural nematology. Marcel Dekker, New York.

Poinar, G.O. 1983. The natural history of nematodes. Prentice-Hall Inc., Englewood Cliffs, NJ.

Siddiqi, M.R. 1985. Tylenchida: parasites of plants and insects. Commonw. Agric. Bur., London.

Smart, G.C., and K.B. Nguyen. 1988. Illustrated key for the identification of common nematodes in Florida. Entomol. Nematol. Dep., Inst. Food Agric. Sci., Univ. Florida, Gainesville.

Wharton, D.A. 1986. A functional biology of nematodes. Johns Hopkins Univ. Press, Baltimore, MD.

4 From the Ground Up: Nematode Ecology in Bioassessment and Ecosystem Health

ROBERT K. NILES AND DIANA WALL FRECKMAN

Colorado State University
Fort Collins, Colorado

Biodiversity in nature and agriculture is represented by biotic units of organization that vary in number, kinds and function, within and between scales of biotic resolution (Solbrig, 1991). Put simply biodiversity is a unifying perspective in ecology that connotes biotic variation. Enumerating all the different nematode species inhabiting, for example, an agricultural field (Niles, 1991) represents one approach to assessing the nematode biodiversity in a habitat. However, biodiversity involves multiple resolutions of biotic organization. The variation inherent to biodiversity includes low orders of resolution, such as genes, populations, races and species, and high orders, such as feeding groups and communities. Thus, knowing the species (biodiversity) of plant-parasitic nematodes in an agricultural field conveys only limited information about the pathogenicity (biodiversity) of species populations or the structure (biodiversity) of the entire nematode community.

Concepts of scale and hierarchy are closely related, such that hierarchical relationships provide an analytic framework for understanding scalar phenomena (King, 1993). The scales of biodiversity conform to a structural hierarchy of nested subsystems, and biodiversity itself is simply one of the components comprising an ecosystem. Because a variety of hierarchical relationships link organisms and ecosystems (King, 1993), information about nematodes, at one scale of resolution, may confer information about the ecosystem at another scale.

Understanding the scales of biodiversity and their contributions to the structure and function of agroecosystems is basic to understanding the ecology of crop production and pest management. For example, biodiversity among plant-parasitic nematodes raises issues of their occurrence and consequence for crop growth. Low-order characteristics of nematode biodiversity can affect high-order attributes of agricultural systems. The presence in an agricultural field of a plant-parasitic species potentially influences, from the ground up, the management and profitability of the crop. Given the reality of these relationships in agriculture, nematode ecologists and other scientists are applying the principles learned from studying the biodiversity of plant-parasitic nematodes to understand, in general, the environmental consequences of human activities.

Nematode biodiversity may integrate environmental conditions to reflect the status and trends of ecosystems. Nematodes, as microscopic roundworms and distributed in high abundances throughout soil habitats, exhibit a sensitive relationship to their environment by responding to the spatial and temporal dynamics of resources and stressors in habitats. Resources are the sources of nutrition and characteristics of the habitat that enable nematodes to survive and reproduce. Stressors are the acute and chronic factors that challenge the physiological limits of nematode survival and reproduction. Resources and stressors tie nematode biodiversity to fundamental attributes of ecosystems: the plants and microorganisms in a habitat, upon which nematodes feed; natural and anthropogenic disturbances; physical and chemical properties of soil; and climate. These attributes individually and collectively influence the structure and function of nematode biodiversity (Fig. 4–1).

In many ecosystems, the structure of nematode biodiversity includes a large number of species, but the shifts and variation in nematode biodiversity encompass, as already described, scales of resolution from gene to community. A problem of nematode ecology is to distinguish when the properties of the habitat influence the structure of nematode biodiversity and when the activity of nematodes affects ecological processing (stocks and flows). Nematode activity, or function, involves the dynamics of soil foodwebs, in which nematodes feed only on living biota. Interactions between populations of nematodes and those of plants and microorganisms involve the flow of material and energy, contributing to soil processes of plant parasitism, herbivory, and decomposition. Agronomists recognize that plant-parasitic nematodes often constrain crop growth. Not as well understood, by agronomists and ecologists alike, is the contribution nematodes make to the primary function of soil biota, the decomposition of organic matter (Anderson, 1988; Freckman, 1988; Swift et al., 1979).

Contributing to the decomposition process, nematodes function as secondary decomposers that feed on bacteria and fungi, the primary decomposers. Under some conditions, nematode involvement in decomposition exerts a dispro-

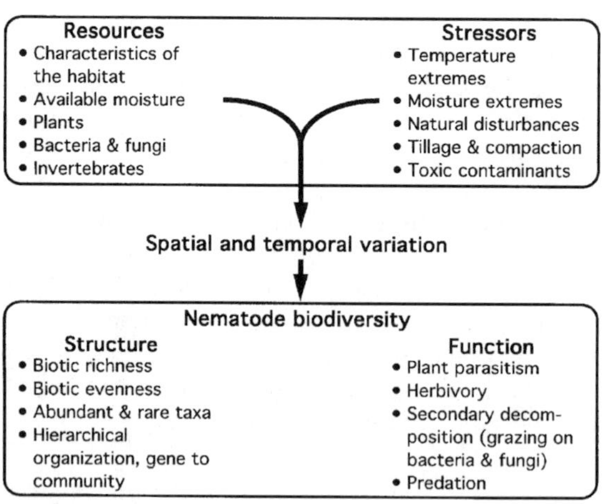

Fig. 4–1. The relationship of selected environmental resources and stressors to the structure and function of nematode biodiversity.

portionately strong and positive effect on nutrient mineralization (de Ruiter et al., 1993; Hunt et al. 1987), which could be one of the benefits of integrating decomposition into crop production. The sustained management of agroecosystems (optimizing crop productivity over the long term) may entail the conservation of detrital foodwebs to maintain and enhance the activity of soil biota, including that of nematodes, and thereby take advantage of organic-matter decomposition and subsequent nutrient release for crop growth (Beylich et al., 1995; Elliott, 1994). Current research suggests sustainability itself might be monitored by using the structural and functional attributes of nematode biodiversity to determine the ecological consequences of management practices (Elliott, 1994; Neher et al., 1995).

Using nematodes to assess sustainability relies on a body of knowledge about the changes in nematode biodiversity that occur when marine, aquatic, and soil habitats become polluted with toxic contaminants. Ecotoxicology focuses on toxic contaminants and their effects on organisms and ecosystems and thus informs our general understanding of environmental stressors and how they influence natural and managed ecosystems.

Evaluating the status and trends of ecosystems, whether to determine contaminant effect, sustainability, or some other attribute, involves issues of ecotoxicology (Amaral-Mendes et al., 1995; Schaeffer, 1991) and often entails the comparison of current conditions with those at a nominal reference site (one possessing pristine, uncontaminated, or another attribute useful for comparative purposes) (Cairns et al., 1993b; Clements & Kiffney, 1994; de Goede et al., 1993a; Schaeffer et al., 1988). The results of such comparisons can be interpreted by applying the concepts of ecosystem health, ecosystem integrity, or soil quality. In this review, we focus on ecosystem health and define it in the next section. However, differentiating the interrelated concepts of health, integrity, and quality warrants a more comprehensive discussion than can be detailed here (for descriptions of the ambiguity of these concepts see Doran & Parkin, 1994; Lemons & Westra, 1995; Shrader-Frechette, 1994). Still, the concepts of health, integrity, and quality originated, respectively, in the public health, water resource, and agricultural disciplines (Amaral-Mendes et al., 1995; Doran & Parkin, 1994; Karr, 1993; Rapport, 1995) and, therefore, differ in emphasis but share a common theme that environmental conditions relate to ecosystem structure and function (Amir & Hyman, 1993; Doran & Parkin, 1994; Elliott, 1994; Karr, 1993; Schaeffer, 1991).

Bioassessment strategies are one method of tracking the status and trends of ecosystems. By monitoring a group of organisms, bioassessment exploits linkages between organisms and ecosystems, enabling degradation by contaminates, changes in natural resources, and consequences of management to be detected and assessed (Cairns et al., 1993a,b; Clements & Kiffney, 1994; Doran & Parkin, 1994; Karr, 1993) (Fig. 4–2). In practice, organismal monitoring can be a powerful tool for evaluating ecosystems. The biotic properties of soil, as well as fine particulate organic matter (Elliott, 1994), reflect the condition of an ecosystem over space and time to a greater extent than the physical and chemical properties of soil or the concentration of a contaminant (McCarthy & Shugart, 1990). Thus, the responses of organisms to environmental stress are often a sensitive and reliable measure of ecosystem status and trends.

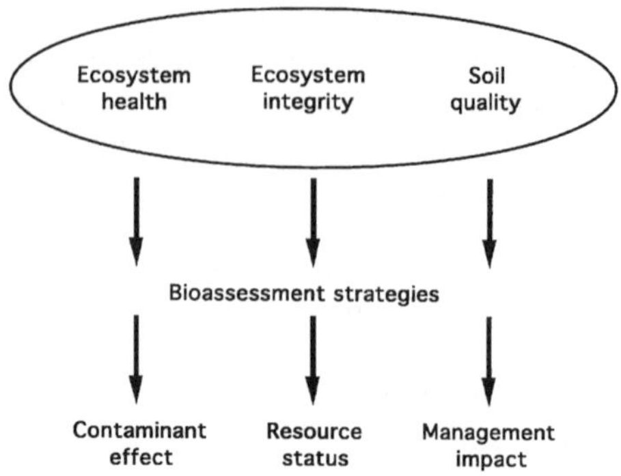

Fig. 4–2. The high-order concepts of ecosystem health, ecosystem integrity, and soil quality are interrelated in their respective emphases, yet each concept often relies on bioassessment strategies to evaluate the status and trends of ecosystems.

Nematodes are being evaluated as bioindicators of human activities that alter environmental conditions (Bongers, 1990; Freckman & Ettema, 1993; Neher et al., 1995). Over the last 10 yr, a number of reviews have outlined the fundamentals of nematode ecology (e.g., Bernard, 1992; Bouwman & Zwart, 1994; Freckman, 1988; Griffiths, 1989) and the relationship of nematodes to disturbance, soil quality, and sustainability (e.g., Blair et al., 1996; Elliott, 1994; Linden et al., 1994; Stork & Eggleton, 1992; Wasilewska, 1995). However, nematode ecotoxicology and its relationship to soil pollution, which represents an historic and current emphasis in nematode bioassessment, have received less attention (Parmelee, 1995; Wallwork, 1988). Thus, in this review of nematode ecology, we focus on the relationships between nematodes and ecosystem health. We begin by describing, from a nematologist's perspective, how bioassessment strategies integrate the health and contamination of ecosystems.

SOIL POLLUTION AND ECOSYSTEM HEALTH

Soil pollution constitutes a threat to ecosystem health that usually results from mixtures of toxic chemicals, rather than from single compounds (Poels et al., 1991). Mixture pollution of soil is a worldwide concern because, given current methods of remediation (Harris & Herbert, 1994), it may take centuries for polluted soils to recover (Kohsiek et al., 1994). Unsustainable practices in industry and agriculture pollute soils with heavy metals, organic compounds, pesticides and xenobiotics, endangering human health, degrading land use, and disrupting ecosystem function (Kohsiek et al., 1994). In the soil of agricultural fields, pesticide residues contaminating the plow layer often involve mixtures of residues, in which individual compounds decay at widely varying rates. Residue mixtures can adversely affect humans (if decay products leach to groundwater), wildlife, and microbial processes occurring in soil (Flury, 1996; Russell et al., 1995). Assessing the hazards posed by mixture pollution is a difficult problem because the toxicological properties of mixtures and the conditions of exposure in soil are usually unknown (Poels et al., 1991). As with chemical mixtures polluting water resources (Yang et al., 1989), new bioassessment strategies are needed to understand the environmental consequences of soil pollution.

Soil processes involve contributions from many nematode species, and inevitably, toxic contaminants in soil influence nematode biodiversity. Hence, nematode community structure reflects the pollutant-induced changes in ecosystems (Bongers, 1990). One perspective on biodiversity suggests species-rich groups of organisms, like nematodes, possess high levels of functional equivalency. Concomitant species may make equivalent, comparable, interchangeable, or redundant contributions to ecosystem function (Lawton & Brown, 1994; Lawton et al., 1996). One rationale for using nematodes in bioassessment contends that the structural components of an organismal community respond more rapidly to an environmental stressor than the functional consequences of organismal activity (King, 1993; Odum, 1985; Pratt & Bowers, 1992). Thus, impaired ecological processing, such as results from soil pollution, represents a heightened state of environmental degradation, but the structural components of biodiversity represent a more sensitive endpoint for assessing the status and trends of ecosystems (Schaeffer et al., 1988). Given this scenario, changes in biodiversity may represent an early warning sign of the ecosystem dysfunction that can occur, if a toxic mixture continues to pollute a habitat.

Ecosystem health emerges as a higher-order condition of the whole ecosystem, rather than of individual components comprising the ecosystem. Accordingly, losses of ecosystem health involve multiple dimensions that often include human judgments (Clements & Kiffney, 1994; King, 1993; Martinko, 1995; Rapport, 1995; Schaeffer et al., 1988). Different classes of environmental stressors (chemical, physical, and biological) hypothetically engender characteristic patterns of compromised health (Schaeffer, 1991). Conceptually, ecosystem health embraces the quantitative concerns of science and the socioeconomic dimensions of environmental policy, bringing together divergent perspectives (Shrader-Frechette, 1994), although not without misunderstanding about the hierarchical organization and properties of ecosystems (Rapport, 1995). One perspective equates degraded health with the dysfunction of an ecological process (Amir & Hyman, 1993; King, 1993; Rapport et al., 1985). This view, like many definitions of ecosystem health (Rapport, 1995), emphasizes biophysical rather than socioeconomic dimensions of ecosystems, possibly because the concept of ecosystem health is used ultimately for natural resource management (Martinko, 1995). By applying a biophysical and process-level concept of ecosystem health to soils contaminated with toxic chemicals, compromised health could correspond to inhibited decomposition (Linden et al., 1994; Moore & de Ruiter, 1993; Ramade, 1995). That is the definition of ecosystem health we adopt in this review. Because some microbial and secondary decomposers, including some nematodes, endure the toxicity of metal-contaminated soil (Bååth, 1989), decomposition proceeds slowly and dysfunctionally in contaminated soil, as evidenced by nutrient leaching, suppressed rates of substrate decay, reduced microbial activity, accumulation of litter, and changes in nematode biodiversity (Rapport et al., 1985; Yeates et al., 1994).

Given a top-down perspective on nematode biodiversity, which considers first the higher-order attributes of biodiversity and then the attributes at succeedingly finer scales of resolution, nematode biodiversity persists under the adverse conditions of polluted, disturbed, and extreme habitats. Persistence is a trait of nematodes that is independent of possible threats to ecosystem health. For example, in Antarctic deserts where climatic conditions are so extreme that the soils there were once thought to be sterile, one to two species of nematodes are the dominant animals in the habitat, with populations nearly as abundant as those in warm deserts (Freckman & Virginia, 1991). The survival of nematodes under conditions of environmental stress contrasts with the responses of other organisms, such as earthworms (Bisessar, 1982; Yeates et al., 1994), enchytraeids (Huhta et al., 1986; Salminen et al., 1995) and oribatid mites (Huhta et al., 1986), which fail to inhabit stressed environments or become extinct locally in polluted soil. Responses of nematodes to environmental stress, including to soil pollution, likely coincide with impaired decomposition, but the relationships among nematodes, contaminants, and decomposition remain speculative.

A challenge for nematode ecologists studying ecosystem health is to establish cause-and-effect relationships between soil pollution and nematode biodiversity by tying stress-induced changes in nematode biodiversity to the presence, bioavailability, and impact of soil contaminants (Clements & Kiffney, 1994; McCarthy & Shugart, 1990), thereby avoiding the ad hoc, descriptive nature of some top-down approaches to bioassessment (Cairns et al., 1993b). The relationship between nematodes and ecosystem health and the opportunities for using nematodes in bioassessment depend on: patterns of nematode succession (structure of biodiversity), contributions in detrital foodwebs (function of biodiversity), and knowledge of ecotoxicology (understanding the contaminant effect on structure and function).

NEMATODE BIODIVERSITY AND SUCCESSION

Nematodes are understood as a monophyletic group, one whose evolutionary history results from a single common ancestor. As a result, virtually all nematodes possess a similar set of characteristics that influences their responses to environmental conditions. This set of characteristics includes bilateral symmetry, an acellular cuticle supported by a hydrostatic skeleton, absence of body segmentation, a five-part life history (5 life stages), and the need for an aquatic habitat (Freckman & Baldwin, 1990). In addition, terrestrial species are often elongate and microscopic and display genomic to community levels of biotic organization. Genomic variation, as found among geographically separated populations of a single species (e.g., Tares et al., 1992), may confer heightened evolutionary advantage to populations possessing a particular trait (Caswell & Roberts, 1987). Community variation, which refers to the differences in structure that may be found among communities occupying similar habitats, probably occurs because individual species react independently to the balance between resources and stressors.

Succession in nematode communities is an expansive process, with biodiversity increasing across taxonomic and functional scales of richness. Mature nematode communities in many ecosystems include numerous species contributing in a variety of ways to ecosystem function. Benthic nematode communities in temperate streams can include over 150 species (Callahan et al., 1979). Communities in the soil of temperate forests can include 175 species (Johnson et al., 1972). Meadows can support more speciose communities than forests (Hánel, 1995). Even in agricultural fields, where nematodes are impacted by chemical, cropping, and physical stressors, communities possess a species richness that is only moderately reduced from that of mature communities in natural habitats (Niles, 1991). Structure in mature communities seldom includes a dominant species represented by most of the individuals in the community. Rather, the most abundant species may comprise only 25% of total individuals, and many species are represented by less than 1% of all individuals (Lawton et al., 1996; Niles, 1991).

Although most nematode species in a habitat might be considered rare, when functional richness is considered, taxonomic richness is subsumed into five functional groups that are defined by the species' feeding habits. Feeding groups (= trophic groups) are a high-order resolution of nematode biodiversity, one that includes plant-feeders (belowground herbivores and plant-parasites), bacterial-feeders, fungal-feeders, omnivores, and predators (Freckman & Baldwin, 1990; Yeates et al., 1993). Definitions of feeding groups can be controversial, based on difficulties of identification and arguments over the divergence of function among related species and the convergence of function among unrelated species (Bernard, 1992; Parmelee et al., 1995; Stork & Eggleton, 1992; Yeates, 1994). Still, analysis based on the structure and function of feeding groups reveals the magnitude of contributions that nematodes make to ecosystem processes (Beare et al., 1992).

Bacterial-feeding and fungal-feeding nematodes graze on microbial populations and influence organic matter decomposition. The nematodes involved in decomposition are often the most abundant of all nematode groups in the mature communities of ecosystems that differ from one another in structure and function as much as temperate forests, tropical forests, grasslands, deserts, and perennial crop fields (Coleman et al., 1991; Hánel, 1995; Hodda et al., 1995; Hunt et al., 1987; Juma & Mishra, 1988; Whitford, 1989).

Bacterial-feeding nematodes in the family Rhabditidae (rhabditid nematodes) predominate in the early stages of succession, when a surplus of labile substrates, such as those found in manure and fresh crop residue (possessing a narrow ratio of C to N), stimulates rhabditid species, in particular, to increase and become prominent (Bouwman et al., 1993; Griffiths et al., 1994). Chemical and physical stressors can induce a similar response (de Goede et al., 1993b; Weiss & Larink, 1991; Yeates et al., 1991). Over time, species in the Rhabditidae are replaced by species in the Cephalobidae (other bacterial-feeders in the order Rhabditida) (Bouwman & Zwart, 1994; de Goede et al., 1993b). The Cephalobidae are more tolerant of drought than the Rhabditidae (Bouwman & Zwart, 1994) and persist throughout the later stages of succession, often comprising the most abundant family of bacterial-feeding nematodes in the mature nematode community (Ettema & Bongers, 1993). Fungal-feeders, in the suborder Aphelenchina and the family Tylenchidae (aphelenchid and tylenchid nematodes), become more prominent as recalcitrant substrates (high in lignin and cellulose, with a wide ratio of C to N) accumulate in the habitat (Bouwman et al., 1993).

Secondary succession in nematode communities parallels plant succession and soil development (de Goede et al., 1993b). Long-lived nematodes, like those in the orders Dorylaimida and Mononchida (dorylaimid and mononchid nematodes), are associated with stable environments and are represented by a greater number of species in the mature community than in earlier stages of succession (de Goede et al., 1993b; Hánel, 1995; Wardle et al., 1995). Plant-feeding nematodes (herbivores and parasites) can vary as a proportion of the community by an order of magnitude between contrasting ecosystems, such as between meadows and forests (Hánel, 1995). In sand dune habitats in the Netherlands, plant-parasitic nematodes may determine the natural course of native plant succession (van der Putten et al., 1993). Many environmental stressors result in the increase in populations of plant-parasites (Wasilewska, 1995).

NEMATODE CONTRIBUTIONS IN DETRITAL FOODWEBS

The nematodes involved in decomposition include rhabditids, aphelenchids, and others (bacterial- and fungal-feeding nematodes) whose population dynamics may synchronize with those of the microorganisms upon which they feed (Freckman, 1988; Griffiths et al., 1994). Under varying environmental conditions, nematodes make direct and indirect contributions to the decomposition process and, at times, enhance the rates of C and nutrient mineralization (Anderson et al., 1981; de Ruiter et al., 1993; Hunt et al., 1987; Ingham et al., 1985; Trofymow et al., 1983). For example, upon death during the winter in a Saskatchewan grassland, nematodes represented an important source of labile P for uptake by plants (McKercher et al., 1979). A study of decomposition in a Polish rye (*Secale cereale* L.) field showed bacterial-feeding and fungal-feeding nematodes defecated 80% of the material they consumed (Wasilewska et al., 1981). Yet, despite these examples of direct nematode contributions to available nutrient pools, the indirect, or regulatory, effects that nematodes exert on microorganisms represent a greater influence on decomposition (Mishra & Mitchell, 1987; Sohlenius & Bostrom, 1984).

As indicated earlier, nematodes graze on the bacterial and fungal populations responsible for substrate decay. Relative abundances of bacterial-feeding and fungal-feeding nematodes can reflect the microbial pathway by which decomposition is proceeding in a habitat (Beare et al., 1992; Ingham et al., 1985; Sohlenius et al., 1988). A higher abundance of bacterial-feeders than fungal-feeders indicates that the bacterial pathway of decomposition is occurring to a greater extent than the fungal pathway. Nematode grazing affects bacterial and fungal activity by several mechanisms: (i) nematodes transport microorganisms to soil microhabitats that contain higher-quality resources, thus precipitating the increased growth of microbial populations; (ii) nematodes provide metabolic wastes and undigested food as a resource for use by microorganisms; (iii) fungal-feeding nematodes ingest senescent fungal hyphae, thereby removing secondary metabolites that inhibit fungal growth; and (iv) nematodes accelerate the growth of microbial populations by optimal grazing (Moore et al., 1991).

Optimal grazing occurs when nematode feeding maintains microorganisms in a logarithmic phase of population growth, which can enhance mineralization (Anderson et al., 1981; Brussaard et al., 1991; Moore et al., 1991). The link between optimal grazing and decomposition is an extension of the grazing optimization hypothesis that was originally formulated to explain the compensatory growth of plants in response to herbivore feeding (Hibert et al., 1981; Moore et al., 1991). The contributions made by nematodes to an optimized rate of decomposition have been observed in microcosm studies (e.g., Bouwman & Zwart, 1994; Brussaard et al., 1991; Griffiths & Caul, 1993; Ingham et al., 1985). Simulation modeling suggests bacterial-feeding nematodes in some ecosystems disproportionately influence nutrient dynamics, relative to their biomass (de Ruiter et al., 1993).

Other soil biota, such as protozoa and microarthropods, also participate in detrital foodwebs and influence decomposition by regulating microbial populations (Bouwman & Zwart, 1994; de Ruiter et al., 1993; Griffiths, 1989; Moore et al., 1991). However, in controlled experiments, nematodes were more successful than protozoa in exploiting low populations of bacteria in the rhizosphere (Griffiths, 1990). The nematodes' higher predation rate probably occurred because, unlike protozoa, nematodes migrated from surrounding soil to a concentrated bacterial resource (Griffiths & Caul, 1993).

When nematodes are especially abundant, grazing suppression may occur, and nematode predation may prevent the growth of microbial populations. This "overgrazing" by nematodes on microorganisms is thought to be an unlikely occurrence under agricultural field conditions (Bouwman & Zwart, 1994).

NEMATODE ECOTOXICOLOGY

The chemical stressors affecting nematode biodiversity and ecosystem health include the pollution of soil and sediment by heavy metals and organic compounds (e.g., Bisessar, 1982; Tietjen, 1980; Weiss & Larink, 1991) and the acidification of soil habitats (e.g., Huhta et al., 1986; Hyvonen & Persson, 1990; Ruess & Funke, 1992). In addition, air pollutants indirectly stress soil habitats through the increased incidence of plant disease that results from the interaction between pollutants and plant-parasitic nematodes (e.g., Khan & Khan, 1993).

Nematodes inhabit the interstitial water films surrounding soil particles. Chemical contaminants dissolved in soil water enter the nematode's body directly through the cuticle, which is the most important route for exposing nematodes to toxicants (Howell, 1983). However, when nematodes feed on bacteria and fungi that are themselves contaminated with heavy metals, the reproductive fitness of nematodes may be compromised (Doelman et al., 1984). Thus, a number of factors place nematodes in intimate contact with the toxicant activity of soil pollutants (Bongers, 1990).

Terrestrial nematodes in virtually all taxonomic groups are sensitive to toxic chemicals (Freckman et al., 1980). Under laboratory conditions, 12 nematode species, representing terrestrial variation in the phylum Nematoda, exhibited comparable sensitivities to Cd, but sensitivities to pentachlorophenol (PCP) appeared to correspond to phylogenic relationships. Representatives of the subphylum Penetrantia were more sensitive to PCP than were evolutionarily more advanced representatives of the subphylum Secernentia, which might have been related to differences in cuticle morphology (Kammenga et al., 1994). Despite a possible correlation between nematode phylogeny and sensitivity to toxicants, closely related species can differ markedly in their response to contaminants (Sturhan, 1986). For example, in comparing two species of marine nematodes under laboratory conditions, *Enoplus communis* accumulated nearly twice as much Cu and Zn as *E. brevis* (Howell, 1983).

Among communities of organisms, retrogression is a particularly common response to chemical contaminants and to other stressors (Rapport et al., 1985). Retrogression is an apparent reversal in the natural course of succession, as evidenced by communities that become simpler in structure, possessing reduced species richness, increased abundance of species with small body sizes and high reproductive capacities, and shifts in species dominance to early successional species (Bååth, 1989; Clements, 1991; Odum 1985). Gradient and other analyses (de Goede et al., 1993a; Pouyat et al., 1994; Wasilewska, 1995; Yeates et al., 1991) indicate that nematode communities in contaminated soil display increasingly prominent features of retrogression, as proximity to the source of contamination increases. In metal-contaminated soil surrounding a lead smelter (an example of soil pollution by a chemical mixture), nematodes and their microbial prey survived at metal concentrations of 28 000 ppm Pb, 972 ppm As, 599 ppm Cu, and 151 ppm Cd, with abundances decreasing as distance to the smelter lessened (Bisessar, 1982).

Heavy metals, organic compounds, and acidification cause many nematodes to decline in frequency, resulting in depressed species richness (Ruess & Funke, 1992; Sturhan, 1986; Tietjen, 1980; Yeates, 1994; Zullini & Peretti, 1986). Omnivorous dorylaimids seem particularly sensitive (Hánel, 1992; Parmelee et al., 1993; Zullini & Peretti, 1986). Predatory mononchids show differing sensitivities under differing conditions, exhibiting high sensitivity to some heavy metals (Parmelee et al., 1993; Sturhan, 1986), tolerance of other metals (Yeates et al., 1994), and early successional recolonization following fumigation with methyl bromide (Yeates et al., 1991). Possibly, as conjectured of plant-feeding nematodes (Yeates et al., 1994), mononchids are more responsive to the presence of a food resource than to the concentration of some contaminants. As the species sensitive to toxic chemicals are removed from the community, total nematode abundance may remain as great as before contamination (Parmelee et al., 1993; Weiss & Larink, 1991; Zullini & Peretti, 1986) or decline (Hyvonen & Persson, 1990; Kuperman, 1996; Ruess & Funke, 1992).

Fungal-feeding nematodes vary in their response to heavy metals and, under differing conditions, appear sensitive (Kuperman, 1996) or insensitive (Sturhan, 1986) to the metal contaminants in soil. Under experimental conditions, aphelenchid nematodes (fungal-feeders) were insensitive to acid deposition and maintained populations that were as abundant as in non-acidified soil (Hyvonen & Persson, 1990; Ruess & Funke, 1992). Still, in forests exposed to high ambient concentrations of SO_2, fungal-feeding nematodes declined in abundance (Hánel, 1992). In microcosms, fungal-feeding nematodes were more sensitive to PCP than bacterial-feeding nematodes (Salminen et al., 1995).

Many contaminants and other environmental stressors cause trophic structure to shift towards dominance by bacterial-feeding species, especially species in the families Rhabditidae and Cephalobidae (de Goede et al., 1993b; Hánel, 1992; Huhta et al., 1986; Weiss & Larink, 1991; Wasilewska, 1995; Yeates et al., 1991, 1994). Although bacterial-feeding species declined in abundance following artificial acidification (Hyvonen & Persson, 1990; Rucss & Funke, 1992), this could have been a short-term effect of the experiment (Hyvonen & Persson, 1990). Likewise, only a short-term decline in the abundance of bacterial-feeding nematodes was observed in acidified Scotch pine (*Pinus sylvestris* L.) forests (Huhta et al., 1986).

Plant-feeding nematodes inhabiting a soil contaminated with seven heavy metals appeared to be the only group of organisms in the fall that was unaffected by metal pollution (Kuperman, 1996). A similar response was observed in copper-contaminated microcosms, where plant-feeding nematodes increased in abundance, presumably upon release from the activity of predators (which in this experiment failed to survive in the contaminated soil) (Parmelee et al., 1993).

Altered structure and function in nematode communities, in response to soil pollution, depends on a range of species characteristics, such as the species' intrinsic rate of natural increase. A particularly useful trait, of unknown importance in nematode ecotoxicology, is the induction of metal-responsive genes, which produce metal-binding proteins, like metallothionein. Upon exposure to toxic metals, many organisms, including nematodes (Kugawa et al., 1994), induce metallothionein to sequester metal ions and possibly tolerate contaminated habitats (Garvey, 1990). The capacity or failure of nematodes in metal-contaminated soil to induce metallothionein or other stress-related proteins (Guven et al., 1995; van Leeuwen, 1995) is a mechanism for relating toxic contaminants to the altered structure and function in nematode communities and to the hazard of soil pollution.

The presence of metal responsive genes, conferring metal-tolerance for some nematode species, can presumably enhance the evolutionary fitness of species possessing such genes and cause such species to become more abundant in polluted soil. This mechanism of "pollution-induced community tolerance" caused differences in species composition between nematode communities inhabiting estuaries that were severely polluted and less polluted with Cu (Millward & Grant, 1995). Alternatively, changes in the composition of nematode species in polluted soil may reflect the differential growth rates of species under altered food regimes. Shifts in the structure and function of nematode biodiversity may parallel pollutant-induced changes in the composition of microbial communities (Platt et al., 1984). Thus, changes in nematode biodiversity could relate either to the differential fitness among species to toxic contaminants or to life-history traits enabling species to exploit a successional change in microbial populations. Whatever the mechanism(s) driving changes in nematode biodiversity under soil pollution, shifts in biodiversity at the species level can occur as a result of contaminant exposure (Arthington et al., 1986). For example, following acidification, *Acrobeloides buetschlii* replaced *Wilsonema schuurmansstekhoveni* as a dominant species among bacterial-feeding nematodes (Ruess & Funke, 1992). Similarly, at sites along a polluted stretch of river, *Tobrilus gracilis* increased in abundance under conditions of high chemical O_2 demand, whereas *T. husmanni* and *T. pellucidus* declined in abundance under the same conditions (Eder & Kirchengast, 1982).

Conceptually, responses of individual species to toxic contaminants can alter the structure and function of an entire ecosystem and thus influence ecosystem health (Ramade, 1995). This concept might be illustrated by following the hypothetical consequences of exposing an ecosystem to heavy metals (Fig. 4–3). Upon exposure some organisms, perhaps including some nematodes, induce the production of metal-binding proteins enabling those organisms to tolerate contaminated conditions. As a result, the composition and abundance of organismal communities would shift to tolerant species. Changes in species occurrence and dominance would affect the dynamics of belowground foodwebs (de Ruiter & Moore, 1993). In turn, altered biotic interactions in foodwebs would influence organic-matter decomposition. A shift in detrital processing to the bacterial pathway would probably occur. Changes in decomposition would be reflected in the relative abundances of nematode feeding groups; for example, as reflected in the increased relative abundance of bacterial-feeding nematodes (Wardle et al., 1995). Overall changes in nematode communities (retrogression) would indicate the loss of ecosystem health resulting from heavy metal contamination. We emphasize that only the individual steps in this cascade of events (Fig. 4–3) have been observed either in nature or under experimental conditions.

Relationships between nematode biodiversity and ecotoxicology show nematodes are a group of organisms that survives in polluted soil. We contend that the role of nematodes in decomposition and their persistence in stressed habitats link nematode biodiversity with ecosystem health. Given that nematodes can exert a mediating effect on decomposition, we suggest that nematodes influence decomposition under conditions of soil pollution, despite pollutant-induced retrogression in nematode communities. Decomposition in polluted soil might proceed at an intermediate rate when nematodes are present, a slower rate than in uncontaminated soil but a more rapid rate than if nematodes were absent. The key to this scenario is the influence exerted by nematodes on microbial activity, which indirectly affects rates of C and nutrient mineralization.

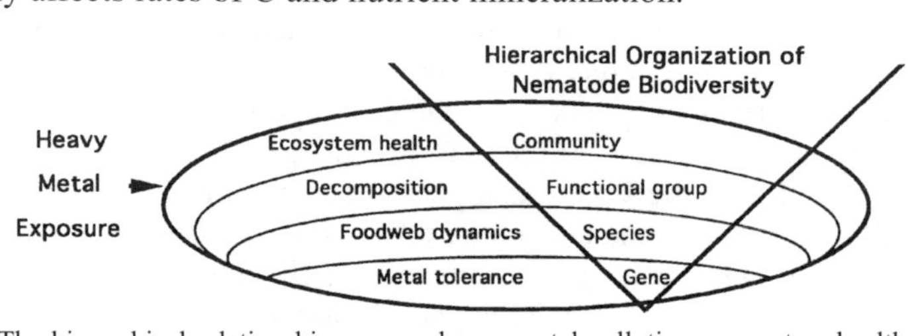

Fig. 4–3. The hierarchical relationships among heavy metal pollution, ecosystem health, and nematode biodiversity, where the loss of ecosystem health may result from the impaired decomposition of organic matter.

NEMATODES IN BIOASSESSMENT

Strategies of bioassay testing, which use single species, and of bioassessment, which use groups of species or whole communities (Hericks & Schaeffer, 1985), both adopt nematodes as indicators of contaminant effect and environmental degradation. Under controlled laboratory conditions, nematode bioassay tests evaluate the toxicity of chemical contaminants. One of the first developed of these tests is the *Panagrellus* assay, which has potential applications in monitoring marine and aquatic habitats and air quality (Samoiloff, 1987). Another group of bioassay tests adopts *Caenorhabditis elegans* to monitor soil and aquatic habitats (Donkin & Dusenbery, 1993; Guven et al., 1995). An *Acrobeloides* assay was developed to compare the toxicity of polychlorinated biphenyl (PCB) and carbofuran (Wasilewska et al., 1975).

Although bioassay tests are used routinely in ecotoxicology, they are criticized for their lack of ecological realism (Cairns et al., 1993a; McCarthy & Shugart, 1990). In nature, organisms take up toxicants at rates influenced by the varying physical and chemical properties of different soils. Thus, a particular shortcoming of bioassay tests is their failure to account for toxicant bioavailability (van Wensem et al., 1994). Another limitation of bioassay tests is the lack of information about the relationship between contaminant uptake by organisms and subsequent effect on ecosystem processes (Clements & Kiffney, 1994). Realistic, integrative methods for evaluating environmental contamination would recognize the dynamics of bioavailability and tie contaminant effect to ecosystem structure and function (Parmelee, 1995). To this end, current priorities in biomonitoring include the identification of sentinel species and biomarker assays and the establishment of cause-and-effect relationships between contaminants and ecosystem processes.

Sentinel species inhabit contaminated environments and display biomarker responses to toxic chemicals (genetic, physiologic, morphologic, or behavioral indicators of contaminant exposure) (McCarthy & Shugart, 1990). Invertebrate animals can serve as sentinel species (Lower & Kendall, 1990), but the biomarker approach has not been adopted using nematodes. Many species of nematodes are cosmopolitan in distribution, but identifying a sentinel nematode may prove difficult, given the biomonitoring experience in aquatic and marine habitats, where a single nematode species indicating water quality can be found (Arthington et al., 1986) but only with difficulty (Platt et al., 1984). Nonetheless, in terrestrial habitats, a multivariate analysis of 22 genera and other nematode groups inhabiting natural and agricultural ecosystems in the central region of Spain revealed fungal-feeding *Tylenchus* and plant-parasitic *Paratylenchus* were bioindicators of edaphic properties, especially of texture (Nombela et al., 1994). Similarly, high abundances of *Paratylenchus* indicated the degradation of environmental conditions in drained (= stressed) fen peat meadows in Poland (Wasilewska, 1991). As noted at the beginning of this review, species of plant-parasitic nematodes in agriculture provide numerous examples of how the abundance of a single nematode species can reflect and influence the status of entire ecosystems.

Unlike bioassay testing or sentinel species identification, bioassessment strategies monitor high-order shifts in biodiversity, to infer environmental conditions for purposes of surveillance, hazard assessment, regulatory assessment, and remediation (Cairns et al., 1993a; Herricks & Schaeffer, 1985). Nematode biodiversity at the high-order resolutions of feeding group and community has been recognized as sensitive to environmental stressors in terrestrial, aquatic, and marine habitats (Eder & Kirchengast, 1982; Elliott, 1994; Millward & Grant, 1995; Moore & de Ruiter, 1993; Parmelee et al., 1993; Platt et al., 1984; Tietjen, 1980; Wardle et al., 1995; Yeates et al., 1991). Thus, monitoring nematode biodiversity is advocated for bioassessment purposes, and this use of nematodes is further supported by a variety of nematode attributes and logistical considerations (Blair et al., 1996; Linden et al., 1994). Rationales for the use of nematodes in bioassessment include: (i) the organ level of biotic organization in nematodes integrates chemical stressors, rather than the cellular level in bacteria and protozoa; (ii) short generation times enable nematodes to respond quickly to many environmental stresses; (iii) nematodes survive in polluted soil; (iv) the limits to nematode migration (only a few centimeters in soil) enable the investigation of point sources of contamination; (v) nematodes influence soil processes of productivity and decomposition; (vi) nematode families and feeding groups (resolutions of biodiversity that are higher than the species level and more easily recognized by non-specialists) inhabit virtually all soils on earth and reflect changes in soil habitats; and (vii) methods of nematode sampling, extraction, and culturing are straightforward and easy to use (Bongers, 1990; Freckman, 1988).

A variety of ecological indices and summary statistics have been used in nematode bioassessment to describe the changes in nematode biodiversity that result from toxic contaminants (e.g., Platt et al., 1984). The variation in life-history strategies among plant-parasitic nematodes can be used in classifying genera into two groups: exploiter nematodes, characterized by unstable populations and the use of transitory resources; and persister nematodes, characterized by stable populations that inhabit predictable environments (Jones, 1980). Effectively expanding on this classification system, the recently introduced maturity index classifies the taxonomic families of nematodes into five groups, based on each family's life-history strategy and ecological requirements. This index thereby measures the successional maturity of nematode communities (Bongers, 1990; de Goede et al., 1993a; Yeates, 1994).

In computing the maturity index, variable weightings are assigned to the nematode abundances in each of five groups. Colonizer, *r*-selected families (producing many offspring and possessing high population growth rates; families found prominently in the early stages of succession) receive the least weighting. Persister, *K*-selected families (producing fewer offspring and possessing slower population growth rates; families found most abundantly when environmental conditions are constant during the later stages of succession) receive the highest weighting. As the intensity of a stressor increases, maturity index values usually decline (de Goede et al., 1993a,b; Ettema & Bongers, 1993; Freckman & Ettema, 1993; Yeates, 1994).

One shortcoming of strategies that assess high-orders of biodiversity, such as the maturity index, is the circumstantial relationship between changes in community structure and the environmental consequences of contaminant exposure (Clements, 1991; McCarthy & Shugart, 1990). For example, maturity index values do not prove the extent of a contaminant effect, a limitation recognized by researchers developing the index (de Goede & Bongers, 1994). Also, environmental variation across a region influences the higher-order structure of biodiversity and sometimes confounds the bioassessment of contaminant effect (Clements & Kiffney, 1994; Yeates, 1994; Yeates & Bird, 1994).

Researchers working with the Agroecosystem component of the Environmental Monitoring and Assessment Program (EMAP) are evaluating the ecological and statistical reliability of the maturity index to assess the status and trends of agroecological resources in the USA (Neher et al., 1995). Their analysis shows the maturity index is a more sensitive measure of environmental conditions than are ecological indices based on nematode-feeding groups. However, data from three agricultural sites in South Australia indicated that summary statistics, based on unweighted measures of species richness, were more discriminatory of management effect than was the maturity index (Yeates & Bird, 1994). Nonetheless, the maturity index possesses demonstrated merit for evaluating the status and trends of some ecosystems. Given the inherent limitations of the maturity index (Blair et al., 1996; Yeates, 1994), it may be most sensitive when assessing eutrophic habitats, which are characterized by high levels of microbial activity (stimulatory to bacterial-feeding, colonizer nematodes), or when assessing polluted habitats (where toxic contaminants kill omnivorous, persister nematodes) (Ruess & Funke, 1995).

Long-lived dorylaimid nematodes (many of which are the omnivorous, persister nematodes in a habitat) show particular sensitivity, in general, to environmental stressors (Freckman & Ettema, 1993; Niles, 1991; Wasilewska, 1991). When using the structural attributes of nematode biodiversity to assess environmental conditions, the presence and abundance of dorylaimid nematodes is often a sensitive indicator of deleterious change. This rule-of-thumb is refuted occasionally (Sohlenius & Sandor, 1989). Recently, study of the human impacts on tropical forests in Cameroon showed omnivorous nematodes (presumably dorylaimids) were more abundant in areas where the forest had been cleared, rather than in primary forest (Hodda et al., 1995). Nevertheless, ecological indices that incorporate information about biodiversity, such as the maturity index, can be robust tools for detecting the impact of environmental stressors. The discriminating power of these indices is primarily due to the high proportion of dorylaimid species that are sensitive to disturbance. When dorylaimid nematodes are absent from the habitat, values of these indices are often sharply lowered (Ettema & Bongers, 1993; Yeates & Bird, 1994), and the deleterious effect of a stressor can be inferred.

Relative absences of dorylaimid nematodes often reflect the impact of a toxic compound, but bioassessment strategies monitoring the nematodes involved in decomposition would rely on the relative abundance of a nematode group to determine environmental conditions. Such strategies might monitor the entire feeding group, but given the EMAP experience (which failed to discern environmental differences at the resolution of nematode feeding-groups) (Neher et al., 1995), a more focused approach monitoring a responsive subset of species would probably reveal more about the status of decomposition in an ecosystem (Wardel et al., 1995). Monitoring the nematodes involved in decomposition makes sense because these nematodes are tied to a key process occurring in soil (Moore & de Ruiter, 1993). We suggest that a bioassessment strategy linking nematodes to the status of decomposition has promise for establishing cause-and-effect relationships between soil pollution and its impact on ecosystems. The criteria for establishing cause-and-effect include consistency, responsiveness and mechanism, all of which can be determined by hypothesis testing, observation, and experimentation (Schreuder & Czaplewski, 1993).

OUTLOOK

The relationships among nematode biodiversity, decomposition, ecotoxicology, and ecosystem health apply broadly to contemporary problems of soil pollution and anthropogenic impacts on natural and agricultural ecosystems. Nematodes are perhaps uniquely suited for bioassessment because the structure and function of nematode biodiversity integrates ecological processes at contrasting scales of biodiversity resolution. Still, it is not sufficient to acknowledge some nematodes are particularly sensitive to disturbance. Rather, what is required is the explicit linkage of shifts in nematode biodiversity (changes in abundance, richness, physiologic or genetic response, or some other measure) to the presence, bioavailability, and impact of environmental stressors, such that altered ecological processing can be inferred with certainty from nematodes. In particular, this may involve the analysis and understanding of nematode contributions to detrital foodwebs and the ways those contributions are affected by soil contaminants (Crowder et al., 1995). Given predictable shifts in the successional patterns of nematodes, in addition to other exploitable aspects of nematode ecology, we believe future investigations of the nematode role in decomposition are likely to establish integrative, reliable measures of ecosystem health.

REFERENCES

Amaral-Mendes, J.J., E. Pluygers, and J. Farina. 1995. Environmental health and ecotoxicology: an indispensable link. p. 77–93. *In* D.J. Rapport et al. (ed.) Evaluating and monitoring the health of large-scale ecosystems. Springer-Verlag, Berlin, Germany.

Amir, S., and J. Hyman. 1993. Measures of ecosystem health and integrity. Water Sci. Technol. 27:481–488.

Anderson, J.M. 1988. Spatiotemporal effects of invertebrates on soil processes. Biol. Fert. Soils 6:216–227.

Anderson, R.V., D.C. Coleman, C.V. Cole, and E.T. Elliott. 1981. Effect of the nematodes *Acrobeloides* sp. and *Mesodiplogaster lheritieri* on substrate utilization and nitrogen and phosphorus mineralization in soil. Ecology 62:549–555.

Arthington, A.H., G.W. Yeates, and D.L. Conrick. 1986. Nematodes, including a new record of *Tobrilus diversipapillatus* in Australia, as potential indicators of sewage effluent pollution. Aust. J. Mar. Freshw. Res. 37:159–166.

Bååth, E. 1989. Effects of heavy metals in soil on microbial processes and populations (a review). Water Air Soil Pollut. 47:335–379.

Beare, M.H., R. W. Parmelee, P.F. Hendrix, W. Cheng, D.C. Coleman, and D.A. Crossley, Jr. 1992. Microbial and faunal interactions and effects on litter nitrogen and decomposition in agroecosystems. Ecol. Monogr. 62:569–591.

Bernard, E.C. 1992. Soil nematode biodiversity. Biol. Fert. Soils 14:99–103.

Beylich, A., H.-C. Frund, and U. Graefe. 1995. Environmental monitoring of ecosystems and bioindication by means of decomposer communities. Newsl. Enchytraeidae 4:25–34

Bisessar, S. 1982. Effect of heavy metals on microorganisms in soils near a secondary lead smelter. Water Air Soil Pollut. 17:305–308.

Blair, J.M., P.J. Bohlen, and D.W. Freckman. 1996. Soil invertebrates as indicators of soil quality. p. 273–292. *In* J.W. Doran and A.J. Jones (ed.) Methods for assessing soil quality. SSSA Spec. Publ. 49. SSSA, Madison, WI.

Bongers, T. 1990. The maturity index: An ecological measure of environmental disturbance based on nematode species composition. Oecologia 83:14–19.

Bouwman, L.A., G.H.J. Hoenderboom, A.C. Van Klinken, and P.C. de Ruiter. 1993. Effect of growing crops and crop residues in arable fields on nematode production. p. 127–131. *In* H.J.P. Eijsackers and T. Hamers (ed.) Integrated soil and sediment research: A basis for proper protection. Kluwer Acad. Publ., Dordrecht, the Netherlands.

Bouwman, L.A., and K.B. Zwart. 1994. The ecology of bacterivorous protozoans and nematodes in arable soil. Agric. Ecosyst. Environ. 51:145–160.

Brussaard, L., J.P. Kools, L.A. Bouwman, and P.C. de Ruiter. 1991. Population dynamics and nitrogen mineralization rates in soil as influenced by bacterial grazing nematodes and mites. p. 517–523. *In* G.K. Veeresh et al. (ed.) Advances in management and conservation of soil fauna. Oxford & IBH Publ. Co. Pvt. Ltd., New Delhi, India.

Cairns, J., Jr., P.V. McCormick, and S.E. Belanger. 1993a. Prospects for the continued development of environmentally-realistic toxicity tests using microorganisms. J. Environ. Sci. 5:253–268.

Cairns, J., Jr., P.V. McCormick, and B.R. Niederlehner. 1993b. A proposed framework for developing indicators of ecosystem health. Hydrobiologia 263:1–44.

Callahan, C.A., V.R. Ferris, and J.M. Ferris. 1979. The ordination of aquatic nematode communities as affected by stream water quality. p. 101–116. *In* J. Cairns Jr. et al. (ed.) Environmental biomonitoring, assessment, prediction, and management—certain case studies and related quantitative issues. Int. Co-op. Publ. House, Fairland, MD.

Caswell, E.P., and P.A. Roberts. 1987. Nematode population genetics. p. 390–397. *In* J.A. Veech and D.W. Dickson (ed.) Vistas on nematology. Soc. Nematol., Inc., Hyattsville, MD.

Clements, W.H. 1991. Community responses of stream organisms to heavy metals: A review of observational and experimental approaches. p. 363–391. *In* M.C. Newman and A.W. McIntosh (ed.) Metal ecotoxicology; concepts and applications. Lewis Publ., Chelsea, MI.

Clements, W.H., and P.M. Kiffney. 1994. Assessing contaminant effects at higher levels of biological organization. Environ. Toxicol. Chem. 13:357–359.

Coleman, D.C., A.L. Edwards, A.J. Belsky, and S. Mwonga. 1991. The distribution and abundance of soil nematodes in East African savannas. Biol. Fert. Soils 12:67–72.

Crowder, L.B., D.P. Reagan, and D.W. Freckman. 1995. Food web dynamics and applied problems. p. 327–336. *In* G.A. Polis and K.O. Winemiller (ed.) Food webs: Integration of patterns and dynamics. Chapman & Hall, New York.

de Goede, R.G.M., and T. Bongers. 1994. Nematode community structure in relation to soil and vegetation characteristics. Appl. Soil Ecol. 1:29–44.

de Goede, R.G.M., T. Bongers, and C.H. Ettema. 1993a. Graphical presentation and interpretation of nematode community structure: C-P triangles. Meded. Rijksfac. Landbouwwet. Gent 58/2b:743–750.

de Goede, R.G.M., S.S. Georgieva, B.C. Verschoor, and J.-W. Kamerman. 1993b. Changes in nematode community structure in a primary succession of blown-out areas in a drift sand landscape. Fund. Appl. Nematol. 16:501–513.

de Ruiter, P.C., and J.C. Moore. 1993. Simulation of the effects of contamination on the functioning of belowground food webs. p. 309–312. *In* H.J.P. Eijsackers and T. Hamers (ed.) Integrated soil and sediment research: A basis for proper protection. Kluwer Acad. Publ., Dordrecht, the Netherlands.

de Ruiter, P.C., J.C. Moore, K.B. Zwart, L.A. Bouwman, J. Hassink, J. Bloim, J.A. de Vos, J.C.Y. Marinissen, W.A.M. Didden, G. Lebbink, and L. Brussaard. 1993. Simulation of nitrogen mineralization in the below-ground food webs of two winter wheat fields. J. Appl. Ecol. 30:95–106.

Doelman, P., G. Nieboer, J. Schrooten, and M. Visser. 1984. Antagonistic and synergistic toxic effects of Pb and Cd in a simple food chain: Nematodes feeding on bacteria and fungi. Bull. Environ. Contam. Toxicol. 32:717–723.

Donkin, S.G., and D.B. Dusenbery. 1993. A soil toxicity test using the nematode *Caenorhabditis elegans* and an effective method of recovery. Arch. Environ. Contam. Toxicol. 25:141–151.

Doran, J.W., and T.B. Parkin. 1994. Defining and assessing soil quality. p. 3–21. *In* J.W. Doran et al. (ed.) Defining soil quality for a sustainable environment. ASA Spec. Publ. 35. ASA and SSSA, Madison, WI.

Eder, R., and M. Kirchengast. 1982. The nematode-fauna (Nemathelminthes, Nematoda) of a polluted part of the river Mur (Styria, Austria). Nematol. Medit. 10:127–134.

Elliott, E.T. 1994. The potential use of soil biotic activity as an indicator of productivity, sustainability and pollution. p. 250–256. *In* C.E. Pankhurst et al. (ed.) Soil biota management in sustainable farming systems. CSIRO Inform. Serv., East Melbourne, Australia.

Ettema, C.H., and T. Bongers. 1993. Characterization of nematode colonization and succession in disturbed soil using the maturity index. Biol. Fert. Soils 16:79–85.

Flury, M. 1996. Experimental evidence of transport of pesticides through field soils— a review. J. Environ. Qual. 25:25–45.

Freckman, D.W. 1988. Bacterivorous nematodes and organic-matter decomposition. Agric. Ecosyst. Environ. 24:195–217.

Freckman, D.W., and J.G. Baldwin. 1990. Nematoda. p. 155–200. *In* D.L. Dindal (ed.) Soil biology guide. John Wiley & Sons, Inc., New York.

Freckman, D.W., and C.H. Ettema. 1993. Assessing nematode communities in agroecosystems of varying human intervention. Agric. Ecosyst. Environ. 45:239–261.

Freckman, D.W., L.B. Slobodkin, and C.E. Taylor. 1980. Relation between species lists and tolerance to nematicides. p. 34–41. *In* D.L. Dindal Soil biology as related to land use practices. Proc. 7th Int. Soc. Soil Sci. Colloq. Soil Zool., Syracuse, NY. 29 July–3 Aug. 1979. EPA-560/13-80-038. USEPA, Office Pest. Toxic Subst., Washington, DC.

Freckman, D.W., and R.A. Virginia. 1991 Review. Nematodes in the dry valleys of southern Victoria Land. Antarct. J. 26:233–234.

Garvey, J.S. 1990. Metallothionein: A potential biomonitor of exposure to environmental toxins. p. 267–287. *In* J.F. McCarthy and L.R. Shugart (ed.) Biomarkers of environmental contamination. Lewis Publ., Boca Raton, FL.

Griffiths, B.S. 1989. The role of bacterial feeding nematodes and protozoa in rhizosphere nutrient cycling. Aspects Appl. Biol. 22:141–145.

Griffiths, B.S. 1990. A comparison of microbial-feeding nematodes and protozoa in the rhizosphere of different plants. Biol. Fert. Soils 9:83–88.

Griffiths, B.S., and S. Caul. 1993. Migration of bacterial-feeding nematodes, but not protozoa, to decomposing grass residues. Biol. Fert. Soils 15:201–207.

Griffiths, B.S., K. Ritz, and R.E. Wheatley. 1994. Nematodes as indicators of enhanced microbiological activity in a Scottish organic farming system. Soil Use Manage. 10:20–24.

Guven, K., J.A. Duce, and D.I. de Pomerai. 1995. Calcium moderation of cadmium stress explored using a stress-inducible transgenic strain of *Caenorhabditis elegans*. Comp. Biochem. Physiol. 110C:61–70.

Hánel, I. 1992. Soil nematodes in spruce forests with different degree of air pollution damage. p. 166–171. *In* J. Bohác (ed.) Bioindicatores deteriorisationis regionis. Inst. Landscape Ecol., Ceské Budejovice, Czechoslovakia.

Hánel, I. 1995. Secondary successional stages of soil nematodes in cambrisols of south Bohemia. Nematologica 41:197–218

Harris, M., and S. Herbert. 1994. Contaminated land; investigation, assessment and remediation. ICE design and practice guide. Thomas Telford Serv. Ltd., London.

Herricks, E.E., and D.J. Schaeffer. 1985. Can we optimize biomonitoring? Environ. Manage. 9:487–492.

Hibert, D.W., D.M. Swift, D.M. Detling, and M.I. Dyer. 1981. Relative growth rates and the grazing optimization hypothesis. Oecologia 51:14–18.

Hodda, M., G.F. Bloemers, F.R. Wanless, and P.J.D. Lambshead. 1995. Nematodes and carbon flux in a tropical forest in Cameroon: preliminary results. p. 131–137. *In* L.-O. Nilsson et al. (ed.) Nutrient uptake and cycling in forest ecosystems. Ecosyst. Res. Rep. 13. Comm. Europ. Commun., Brussels, Belgium.

Howell, R. 1983. Heavy metals in marine nematodes: Uptake, tissue distribution and loss of copper and zinc. Mar. Pollut. Bull. 14:263–268.

Huhta, V., R. Hyvonen, A. Koskenniemi, P. Vilkamaa, P. Kaasalainen, and M. Sulander. 1986. Response of soil fauna to fertilization and manipulation of pH in coniferous forests. Acta For. Fenn. 195:1–30.

Hunt, H.W., D.C. Coleman, E.R. Ingham, R.E. Ingham, E.T. Elliott, J.C. Moore, S.L. Rose, C.P.P. Reid, and C.R. Morley. 1987. The detrital food web in a shortgrass prairie. Biol. Fert. Soils 3:57–68.

Hyvonen, R., and T. Persson. 1990. Effects of acidification and liming on feeding groups of nematodes in coniferous forest soils. Biol. Fert. Soils 9:205–210.

Ingham, R.E., J.A. Trofymow, E.R. Ingham, and D.C. Coleman. 1985. Interactions of bacteria, fungi, and their nematode grazers: Effects on nutrient cycling and plant growth. Ecol. Monogr. 55:119–140.

Johnson, S.R., V.R. Ferris, and J.M. Ferris. 1972. Nematode community structure of forest woodlots. I. Relationships based on similarity coefficients of nematode species. J. Nematol. 4:175–182.

Jones, F.W.G. 1980. Some aspects of the epidemiology of plant parasitic nematodes. p. 71–92. *In* J. Palti and J. Kranz (ed.) Comparative epidemiology: A tool for better disease management. Centre Agric. Publ. Document., Wageningen, the Netherlands.

Juma, N.G., and C.C. Mishra. 1988. Effect of an annual and a perennial crop on trophic group dynamics of nematodes. Can. J. Soil Sci. 68:101–109.

Kammenga, J.E., C.A.M. Van Gestel, and J. Bakker. 1994. Patterns of sensitivity to cadmium and pentachlorophenol among nematode species from different taxonomic and ecological groups. Arch. Environ. Contam. Toxicol. 27:88–94.

Karr, J.R. 1993. Measuring biological integrity: Lessons from streams. p. 83–104. *In* S. Woodley et al. (ed.) Ecological integrity and the management of ecosystems. St. Lucie Press, Delray Beach, FL.

Khan, R. K., and M. W. Khan. 1993. The interaction of SO_2 and root-knot nematode on tomato. Environ. Poll. 81:91–102.

King, A.W. 1993. Considerations of scale and hierarchy. p. 19–45. *In* S. Woodley et al. (ed.) Ecological integrity and the management of ecosystems. St. Lucie Press, Delray Beach, FL.

Kohsiek, L.H.M., R. Franken, D. Fraters, J. Latour, A.M.A. van der Linden, R. Reiling, and W.J. Willems. 1994. The pollution of soils and groundwater in the European community. p. 35–93. *In* M.H. Donker et al. (ed.) Ecotoxicology of soil organisms. Lewis Publ., Boca Raton, FL.

Kugawa, F., H. Yamamoto, S. Osada, M. Aoki, M. Imagawa, and T. Nishihara. 1994. Metallothionein genes in the nematode *Caenorhabditis elegans* and metal inducibility in mammalian culture cells. Biomed. Environ. Sci. 7:222–231.

Kuperman, R.. 1996. A hierarchical approach to ecological assessment of contaminated soils at Aberdeen proving ground, USA. p. 197–212. *In* N.M. van Straalen and D.A. Krivolutskii (ed.) New approaches to the development of bioindicator systems for soil pollution. NATO Advan. Res. Workshop New Approach. Develop. Bioindicator Syst. Soil Pollut., Moscow, Russia. 1995. NATO ASI Ser. Environ. Vol. 16. Kluwer Acad. Publ., Dordrecht, the Netherlands.

Lawton, J.H., D.E. Bignell, G.F., P. Eggleton, and M.E. Hodda. 1996. Carbon flux and diversity of nematodes and termites in Cameroon forest soils. Biodiver. Conserv. 5:261–273.

Lawton, J.H., and V.K. Brown. 1994. Redundancy in ecosystems. p. 255–270. *In* E.-D. Schulze and H.A. Mooney (ed.) Biodiversity and ecosystem function. Springer-Verlag, New York.

Lemons, J., and L. Westra. 1995. Introduction to perspectives on ecological integrity. p. 1–11. *In* L. Westra and J. Lemons (ed.) Perspectives on ecological integrity. Kluwer Acad. Publ., Dordrecht, the Netherlands.

Linden, D.R., P.F. Hendrix, D.C. Coleman, and P.C.J. van Vliet. 1994. Faunal indicators of soil quality. p. 91–106. *In* J.W. Doran et al. (ed.) Defining soil quality for a sustainable environment. ASA Spec. Publ. 35. ASA and SSSA, Madison, WI.

Lower, W.R., and R.J. Kendall. 1990. Sentinel species and sentinel bioassay. p. 309–331. *In* J.F. McCarthy and L.R. Shugart (ed.) Biomarkers of environmental contamination. Lewis Publ., Boca Raton, FL.

Martinko, E. 1995. Rapporteur's report. p. 95–98. *In* D.J. Rapport et al. (ed.) Evaluating and monitoring the health of large-scale ecosystems. Springer-Verlag, Berlin, Germany.

McCarthy, J.F., and L.R. Shugart. 1990. Biological markers of environmental contamination. p. 3–14. *In* J.F. McCarthy and L.R. Shugart (ed.) Biomarkers of environmental contamination. Lewis Publ., Boca Raton, FL.

McKercher, R.B., T.S. Tollefson, and J.R Willard. 1979. Biomass and phosphorus contents of some soil invertebrates. Soil Biol. Biochem. 11:387–391.

Millward, R.N., and A. Grant. 1995. Assessing the impact of copper on nematode communities from chronically metal-enriched estuary using pollution-induced community tolerance. Mar. Pollut. Bull. 30:701–706.

Mishra, C.C., and M.J. Mitchell. 1987. Nematode populations in Adirondack forest soils and their potential role in sulfur cycling. Pedobiologia 30:277–283.

Moore, J.C., and P.C. de Ruiter. 1993. Assessment of disturbance on soil ecosystems. Vet. Parasitol. 48:75–85.

Moore, J.C., H.W. Hunt, and E.T. Elliott. 1991. Ecosystem perspectives, soil organisms, and herbivores. p. 105–140. *In* P. Barbosa et al. (ed.) Microbial mediation of plant-herbivore interactions. John Wiley & Sons, Inc., New York.

Neher, D.A., S.L. Peck, J.O. Rawlings, and C.L. Campbell. 1995. Measures of nematode community structure and sources of variability among and within agricultural fields. Plant Soil 170:167–181.

Niles, R.K. 1991. Relationship between agronomic management and nematode community structure. Ph.D. diss. Purdue Univ., West Lafayette, IN (Diss. Abstr. 91-32486).

Nombela, G., A. Navas, and A. Bello. 1994. Structure of the nematofauna in Spanish Mediterranean continental soils. Biol. Fert. Soils 18:183–192.

Odum, E.P. 1985. Trends expected in stressed ecosystems. BioScience 35:419–422

Parmelee, R.W. 1995. Soil fauna: linking different levels of the ecological hierarchy. p. 107–116. *In* C.G. Jones and J.H. Lawton (ed.) Linking species and ecosystems. Chapman & Hall, New York.

Parmelee, R.W., P.J. Bohlen, and C.A. Edwards. 1995. Analysis of nematode trophic structure in agroecosystems: functional groups versus high resolution taxonomy. p. 203–207. *In* H.P. Collins et al. (ed.) The significance and regulation of soil biodiversity. Kluwer Acad. Publ., Dordrecht, the Netherlands.

Parmelee, R.W., R.S. Wentsel, C.T. Phillips, M. Simini, and R.T. Checkai. 1993. Soil microcosm for testing the effects of chemical pollutants on soil fauna communities and trophic structure. Environ. Toxicol. Chem. 12:1477–1486.

Platt, H.M., K.M. Shaw, and P.J.D. Lambshead. 1984. Nematode species abundance patterns and their use in the detection of environmental perturbations. Hydrobiologia 118:59–66.

Poels, C.L.M., U. Gruntz, P. Isnard, D. Riley, M. Spiteller, W. ten Berge, W. Veerkamp, and W.J. Bontinck. 1991. Hazard assessment of chemical contaminants in soil. Chemosphere 21:3–24.

Pouyat, R.V., R.W. Parmelee, and M.M. Carreiro. 1994. Environmental effects of forest soil-invertebrate and fungal densities in oak stands along an urban-rural land use gradient. Pedobiologia 38:385–399.

Pratt, J.R., and N.J. Bowers. 1992. Variability of community metrics: Detecting changes in structure and function. Environ. Toxicol. Chem. 11:451–457.

Ramade, F. 1995. Qualitative and quantitative criteria defining a "healthy" ecosystem. p. 43–61. *In* D.J. Rapport et al. (ed.) Evaluating and monitoring the health of large-scale ecosystems. Springer-Verlag, Berlin, Germany.

Rapport, D.J. 1995. Ecosystem health: Exploring the territory. Ecosyst. Health 1:5–13.

Rapport, D.J., H.A. Regier, and T.C. Hutchinson. 1985. Ecosystem behavior under stress. Am. Nat. 125:617–640.

Ruess, L. 1995. Nematode fauna in spruce forest soils: A qualitative/quantitative comparison. Nematologica 41:106–124.

Ruess, L., and W. Funke. 1992. Effects of experimental acidification on nematode populations in soil cultures. Pedobiologia 36:231–239.

Ruess, L., and W. Funke. 1995. Nematode fauna of a spruce stand associated with forest decline. Acta Zool. Fenn. 196:348–351.

Russell, R.W., S.J. Hecnar, and G.D. Haffner. 1995. Organochlorine pesticide residues in southern Ontario spring peepers. Environ. Toxicol. Chem. 14:815–817.

Salminen, J., J. Haimi, A. Sironen, and J. Ahtiainen. 1995. Effects of pentachlorophenol and biotic interactions on soil fauna and decomposition in humus soil. Ecotoxicol. Environ. Saf. 31:250–257.

Samoiloff, M.R. 1987. Nematodes as indicators of toxic environmental contaminants. p. 433–439. *In* J.A. Veech and D. W. Dickson (ed.) Vistas on nematology. Soc. Nematol., Inc., Hyattsville, MD.

Schaeffer, D.J. 1991. A toxicological perspective on ecosystem characteristics to track sustainable development. VII. Ecosystem health. Ecotoxicol. Environ. Saf. 22:225–239.

Schaeffer, D.J., E.E. Herricks, and H.W. Kerster. 1988. Ecosystem health. I. Measuring ecosystem health. Environ. Manage. 12:445–455.

Schreuder, H.T., and R.L. Czaplewski. 1993. Long-term strategy for the statistical design of forest health monitoring system. Environ. Monit. Assess. 27:81–94.

Shrader-Frechette, K.S. 1994. Ecosystem health: A new paradigm for ecological assessment? Trends Ecol. Evol. 9:456–457.

Sohlenius, B., and S. Bostrom. 1984. Colonization, population development and metabolic activity of nematodes in buried barley straw. Pedobiologia 27:67–78.

Sohlenius, B., S. Bostrom., and A. Sander. 1988. Carbon and nitrogen budgets of nematodes in arable soil. Biol. Fert. Soils 6:1–8.

Sohlenius, B., and A. Sandor. 1989. Ploughing of a perennial grass ley—effect on the nematode fauna. Pedobiologia 33:199–210.

Solbrig, O.T. 1991. Biodiversity: Scientific issues and collaborative research proposals. MAB Digest 9. UNESCO, Paris, France.

Stork, N.E., and P. Eggleton. 1992. Invertebrates as determinants and indicators of soil quality. Am. J. Alter. Agric. 7:38–47.

Sturhan, D. 1986. Influence of heavy metals and other elements on soil nematodes. Rev. Nematol. 9:311.

Swift, M.J., O.W. Heal, and J.M. Anderson. 1979. Decomposition in terrestrial ecosystems. Blackwell Sci. Publ., Oxford, England.

Tares, S., P. Abad, N. Bruguier, and G. de Guiran. 1992. Identification and evidence for relationships among geographical isolates of *Bursaphelenchus* spp. (pinewood nematode) using homologous DNA probes. Heredity 68:157–164.

Tietjen, J.H. 1980. Population structure and species composition of the free-living nematodes inhabiting sands of the New York Bight Apex. Estuar. Coast. Mar. Sci. 10:61–73.

Trofymow, J.A., C.R. Morley, D.C. Coleman, and R.V. Anderson. 1983. Mineralization of cellulose in the presence of chitin and assemblages of microflora and fauna in soil. Oecologia 60:103–110.

van der Putten, W. H., C. van Dijk, and B.A.M. Peters. 1993. Plant-specific soil-borne diseases contribute to succession in foredune vegetation. Nature (London) 362:53–55.

van Leeuwen, M.A.W. 1995. Heat-shock and stress response of the parasitic nematode *Haemonchus contortus*. Parasitol. Res. 81:706–709.

van Wensem, J., J.J. Vegter, and N.M. van Straalen. 1994. Soil quality criteria derived from critical body concentrations of metals in soil invertebrates. Appl. Soil Ecol. 1:185–191.

Wallwork, J.A. 1988. The soil fauna as bioindicators. p. 203–215. *In* Biologia ambiental: Actas Congr. Biol. Ambiental Serv. (2nd Congr. Mundial Vasco), Univ. Pais Vasco, San Sebastian, Spain. 1987. Serv. Edit., Univ. Pais Vasco, San Sebastian, Spain.

Wardle, D.A., G.W. Yeates, R.N. Watson, and K.S. Nicholson. 1995. The detritus food-web and the diversity of soil fauna as indicators of disturbance regimes in agro-ecosystems. Plant Soil 170:35–43.

Wasilewska, L. 1991. Long-term changes in communities of soil nematodes on fen peat meadows due to the time since their drainage. Ekol. Pol. 39:59–104.

Wasilewska, L. 1995. Directions of changes in communities of soil nematodes in man-disturbed ecosystems. Acta Zool. Fennica 196:271–274.

Wasilewska, L., P.C. Oloffs, and J.M. Webster 1975. Effects of carbofuran and a PCB on development of a bacteriophagous nematode *Acrobeloides nanus*. Can. J. Zool. 53:1709–1715.

Wasilewska, L., E. Paplinska, and J. Zielinski. 1981. The role of nematodes in decomposition of plant material in a rye field. Pedobiologia 21:182–191.

Weiss, B., and O. Larink. 1991. Influence of sewage sludge and heavy metals on nematodes in an arable soil. Biol. Fert. Soils 12:5–9.

Whitford, W.G. 1989. Abiotic controls on the functional structure of soil food webs. Biol. Fert. Soils 8:1–6.

Yang, R.S.H., H.L. Hong, and G.A. Boorman. 1989. Toxicology of chemical mixtures: Experimental approaches, underlying concept, and some results. Toxicol. Lett. 49:183–197.

Yeates, G.W. 1994. Modification and quantification of the nematode maturity index. Pedobiologia 38:97–101.

Yeates, G.W., S.S. Bamforth, D.J. Ross, K.R. Tate, and G.P. Sparling. 1991. Recolonization of methyl bromide sterilized soils under four different field conditions. Biol. Fert. Soils 11:181–189.

Yeates, G.W., and A.F. Bird. 1994. Some observations on the influence of agricultural practices on the nematode faunae of some South Australian soils. Fund. Appl. Nematol. 17:133–145.

Yeates, G.W., T. Bongers, R.G.M. de Goede, D. W. Freckman, and S.S. Georgieva. 1993. Feeding habits in soil nematode families and genera—an outline for soil ecologists. J. Nematol. 25:315–331.

Yeates, G.W., V.A. Orchard, T.W. Speir, J.L. Hunt, and M.C.C. Hermans. 1994. Impact of pasture contamination by copper, chromium, arsenic timber preservative on biological activity. Biol. Fert. Soils 18:200–208.

Zullini, A., and E. Peretti. 1986. Lead pollution and moss-inhabiting nematodes of an industrial area. Water Air Soil Pollut. 27:403–410.

5 Physiological and Molecular Aspects of Nematode Parasitism

RICHARD S. HUSSEY

University of Georgia
Athens, Georgia

VALERIE M. WILLIAMSON

University of California
Davis, California

Interactions of plant-parasitic nematodes with their hosts are complex and dynamic and may involve, depending on the species, hatching stimuli, attraction to the host, penetration of and migration in host tissue, recognition of tissues suitable for feeding-site formation, and elaborate modification of host cells. Plant-parasitic nematodes are obligate parasites and, therefore, can only obtain nourishment for development and reproduction from living plant cells. This chapter reviews the feeding strategies and cellular alterations that root-parasitic nematodes induce on susceptible plants and the effects of this parasitism on the physiology of the host. The few nematodes that parasitize the aboveground parts of plants are not covered in this chapter. Since nematode parasitism of plants has been reviewed in the past (Dropkin, 1969; Fenoll et al., 1997; Grundler & Wyss, 1994; Hussey, 1985; Hussey, 1989; Hussey et al., 1994; Jones, 1981; Sijmons et al., 1994; Williamson & Hussey, 1996; Wyss, 1988; Zacheo & Bleve-Zacheo, 1994), emphasis in this chapter is on recent literature.

NEMATODE PARASITISM

Evolutionary adaptation of nematodes for plant parasitism involved development of a hollow, protrusible feeding apparatus, called a stylet, as well as marked morphological and physiological modifications of the esophagus (Fig. 5–1; Hussey, 1989). During plant parasitism, nematodes use the stylet to penetrate host cell walls and inject secretions into the cell, and to withdraw nutrients from the cytoplasm. Nematodes migrating in root tissue also use the stylet to cut openings in cell walls or to inject secretions between cells. Secretory glands in the esophagus are enlarged considerably in nematodes adapted to plant parasitism. Secretions produced by the esophageal glands also must have evolved to enable the nematodes to feed on or modify plant cells.

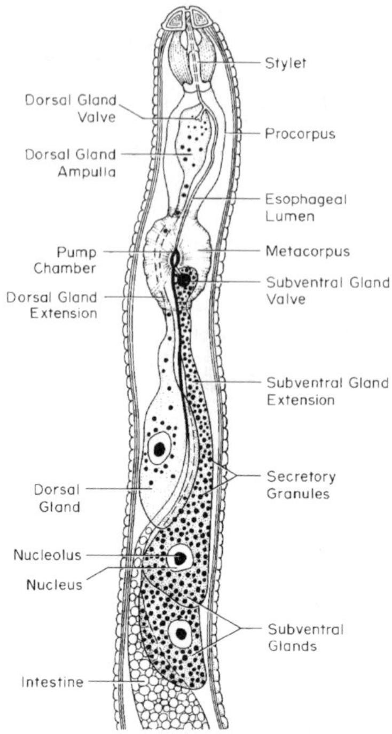

Fig. 5–1. The anterior end of a second-stage juvenile of a plant-parasitic nematode. (Reproduced with permission from the *Annual Review of Phytopathology*, Vol. 27 (1989), by Annual Reviews Inc.)

Chemical stimuli appear to be important for nematode orientation to host roots, finding appropriate feeding or penetration sites, migration within roots, and recognition of specific root tissue suitable for feeding site formation. Therefore, all nematodes possess an elaborate sensory system, with the primary chemosensory organs being structures called amphids. In plant-parasitic nematodes, the amphids are thought to have a role in recognition and infection of host plants. Disrupting the sensory functions involved in nematode interactions with plants to block their ability to find roots and establish feeding sites is an attractive target for an innovative management tactic (Dusenbery, 1987; Perry, 1994). However, little is known about the identity of chemical stimuli in the life cycle of plant-parasitic nematodes.

Feeding Strategies

The feeding relationships between parasitic nematodes and susceptible plants are diverse. Plant-parasitic nematodes can be separated into four general groups according to their feeding strategy: migratory ectoparasites, sedentary ectoparasites, migratory endoparasites, and sedentary endoparasites (Fig. 5–2). The amount of tissue destruction and the degree of plant response are often related to the type of feeding relationship. The *migratory ectoparasites* remain outside the root and use their stylet to feed either on epidermal cells or cells deeper

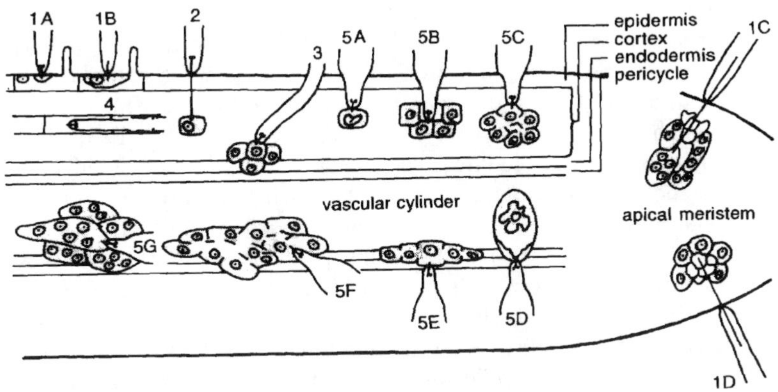

Fig. 5–2. Schematic representation of feeding sites and feeding structures of selected root parasitic nematodes. 1 = migratory ectoparasites: 1A, *Tylenchorhynchus dubius; 1B, Trichodorus* spp.; 1C, *Xiphinema index*; 1D, *Longidorus elongatus*. 2 = sedentary ectoparasites: *Criconemella xenoplax*. 3 = migratory ecto-endoparasites: *Helicotylenchus* spp. 4 = migratory endoparasites: *Pratylenchus* spp. 5 = sedentary endoparasites: 5A, *Trophotylenchulus obscurus*; 5B, *Tylenchulus semipenetrans*; 5C, *Verutus volvingentis*; 5D, *Cryphodera utahensis*; 5E, *Rotylenchulus reniformis*; 5F, *Heterodera* spp.; 5G, *Meloidogyne* spp. (Reproduced with permission from Sijmons et al. (1994, Vol. 32), *Annual Review of Phytopathology*, by Annual Reviews Inc.)

within the root. With the exception of species of a few genera that feed on root tips, e.g., *Paratrichodorus, Trichodorus* and *Belonolaimus,* nematodes with this type of feeding strategy generally cause little obvious tissue damage. *Sedentary ectoparasites* will feed from a single site or plant cell for a prolonged period of time while remaining outside the root. Feeding by *Criconemella xenoplax* causes little tissue damage, but other species like *Xiphinema index* and *Hemicycliophora arenaria* induce terminal galls when feeding at root tips. *Migratory endoparasites*, e.g., *Pratylenchus* and *Radopholus* species, enter the root and periodically feed as they migrate through the root tissue. The intracellular migration and extended feeding cause extensive destruction of root tissue along the path of the migrating parasite. *Sedentary endoparasites*, e.g., *Meloidogyne, Globodera,* and *Heterodera* species, have evolved very specialized and complex feeding relationships with their host plants. These nematodes enter roots as vermiform juveniles. Further development depends on modifying and obtaining nourishment from a few root cells in the vascular cylinder. These modified cells form specialized food cells that become the permanent feeding sites for the parasites. When feeding commences, their bodies grow and developing nematodes become immobile. In this feeding strategy, destruction of root tissue is limited to cells around the feeding site and the nematode.

Feeding Behavior

The feeding behavior of several parasitic nematodes has been studied in detail under laboratory conditions, including time-lapse video microscopy (Doncaster, 1971; Wyss, 1981). Due to differences in the nature of parasitism between ectoparasites and endoparasites, the feeding behavior of these two groups differs dramatically.

Feeding of ectoparasites can be divided into five distinct phases: exploration, insertion of the stylet into the host cell, injection of esophageal gland secretions through the stylet, ingestion of nutrients from the cytoplasm via the stylet, and retraction of the stylet from the host cell (Wyss, 1981). These phases are highly synchronized and involve coordination of different parts of the nematode's esophagus during feeding. The feeding of migratory and sedentary ectoparasites includes the same five phases, but each phase is considerably longer for the sedentary than for the migratory feeders. Feeding at diverse root sites by ectoparasites begins by the nematode exploring walls of epidermal cells with its lips and sometimes probing with its stylet (Doncaster & Seymour, 1973; Westcott & Hussey, 1992). When a suitable site is located by migratory ectoparasites, the cell wall is penetrated quickly by strong thrusting of the stylet. After wall penetration the stylet remains protruded and in contact with the cell cytoplasm during brief secretion and ingestion phases. However, for sedentary ectoparasites, which usually parasitize cells below the epidermis, stylet insertion can be a very deliberate process. For *C. xenoplax*, this process can take up to 80 min (Westcott & Hussey, 1992). Once the stylet of *C. xenoplax* is inserted to the final depth for feeding, it stays stationary for the remainder of the feeding episode, and a secretion phase that can last 1 to 3 h is initiated. Secretions from the esophageal glands pass through the stylet and accumulate around its tip in the cytoplasm of the parasitized cell. This secretory activity is followed by ingestion of nutrients from the cytoplasm, a continuous process that can last for several days (Westcott &Hussey, 1992). The ingestion phase is characterized by rhythmic maximum dilation of the pump chamber in the metacorpus. After the metacorpus ceases pumping and the ingestion phase is terminated, the stylet is retracted from the plant tissue to end the feeding cycle. During the feeding episode, juvenile nematodes usually increase in diameter and length, and in young adults the reproductive organs develop.

Until recently, observation of the feeding behavior of endoparasitic nematodes was limited because of problems associated with observing nematodes inside living roots. Now, high-resolution video-enhanced contrast light microscopy permits observations of nematode feeding behavior inside living root tissue. This technique was used to observe the behavior of the migratory endoparasite, *Pratylenchus penetrans*, in roots of several hosts (Zunke, 1990). This parasite usually enters a root by successively puncturing the walls of an epidermal cell and adjacent cortical cell with its stylet, although it also penetrates where lateral roots emerge or through entry sites of other nematodes. Once inside the root, the nematode migrates intracellularly through the cortical tissue by perforating cell walls with repeated stylet thrusts. While migrating nematodes may briefly feed from a cortical cell before passing through it, nutrients are primarily acquired by feeding from cortical cells for extended periods. In these feeding episodes and after the stylet is inserted into the cell, glandular secretions accumulate in the cytoplasm around the stylet tip. The ingestion of nutrients is accomplished by rapid pulsation of the metacorpal pump chamber which can last for several hours. Cell death always follows the extended feeding activity by *P. penetrans*.

The feeding behavior of sedentary endoparasites is well documented through video-microscopy studies with *Heterodera schachtii* (Wyss & Zunke, 1986; Wyss, 1992). Second-stage juveniles invade roots near the tip and migrate intracellularly in the cortical tissue toward the vascular cylinder by using their stylets to puncture and slit cell walls. Once a juvenile reaches the vascular cylinder, it stops migrating and starts to intensely probe several cells around its head with stylet thrusts. After the stylet carefully penetrates the wall of a parenchymous cell in the vascular cylinder, it remains protruded for several hours within the cell, which becomes the initial syncytial cell (Wyss, 1992). A permanent feeding site, called a syncytium, is formed from partial dissolution of walls of contiguous cells until several cells are incorporated into the feeding site (Jones, 1981). Syncytia are characterized by enlarged nuclei and nucleoli, dense cytoplasm with increased number of subcellular organelles, and ingrowths on walls adjacent to vascular tissue. After the syncytium is established, feeding behavior is divided into three continuous, cyclical phases. Esophageal gland secretions that pass through the stylet into the cytoplasm of the syncytium polymerize into a tubular structure called a feeding tube that remains associated with the stylet orifice during the food ingestion phase. Nutrient uptake is coordinated with continuous action of the metacorpal pump chamber and lasts 1 to 2 h. When the metacorpus stops pumping and the ingestion of nutrients ceases, the stylet is withdrawn from the initial syncytial cell. After a resting period of several minutes, the stylet tip is reinserted into the initial syncytial cell and a new feeding tube is formed to begin another feeding cycle. The site where the stylet penetrates the cell wall is covered with electron-dense material, called a feeding plug, that seals the syncytial wall around the stylet when it is withdrawn. The daily and overall food consumption for the different developmental stages varies, with an adult female *H. schachtii* consuming approximately 11×10^{-6} mL d (Müller et al., 1981).

Only the early stages of parasitism by the sedentary endoparasite *Meloidogyne* species have been studied using the video-enhanced contrast light microscopy, because gall formation precludes further observations due to root thickening. Infective juveniles of *M. incognita* are attracted to root tips where invasion occurs in the elongation region by repeated thrusting of the stylet until walls of epidermal cells are perforated (Wyss et al., 1992). Once inside the root, juveniles migrate intercellularly by separating the cells at the middle lamella with repeated stylet thrusts (Endo & Wergin, 1973; Wyss et al., 1992). Whether this process is entirely mechanical or involves enzymatic stylet secretions has not been resolved. In *Arabidopsis thaliana*, the infective juvenile usually enters the vascular cylinder by migrating intercellularly to the root tip, then turning around in the meristematic region and migrating back up in the center of the root. A permanent feeding site is then established in the vascular cylinder. The feeding site consists of large, modified cells, called giant-cells, that are characterized by many (50–100) enlarged, highly lobed nuclei with prominent nucleoli, increased numbers of organelles, dense cytoplasm with high rates of metabolism, and thickened walls with elaborate ingrowths. Although feeding from giant-cells has not been recorded in living roots, observations at the ultrastructural level suggests that

Meloidogyne adult females feed in cycles from the modified cells (Hussey & Mims, 1991).

Plant-parasitic nematodes withdraw nutrients from the cytoplasm of parasitized cells via their stylet directly and, in some instances, with the aid of a feeding tube. When parasitic nematodes establish prolonged feeding associations with a host cell, the plasma membrane in the parasitized cell becomes invaginated around the stylet tip (Endo, 1991; Hussey et al., 1992a; Rebois, 1980; Wyss et al., 1984). In this respect, parasitism of plant cells by nematodes is analogous to that by biotrophic fungi that breach the host cell wall and produce haustoria that invaginate the plasma membrane of the parasitized cell (Harder & Chong, 1984). Nematodes withdraw nutrients from the cytosol of the parasitized cell through a minute hole created in the plasma membrane at the stylet orifice (Fig. 5–3; Hussey et al., 1992b). In addition, invagination of the plasma membrane of the parasitized cell around a nematode's stylet tip appears to be associated with callose deposition between the plasma membrane and the stylet. Callose deposition may be a common wound response elicited when nematodes feed from plant cells. This response does not appear to inhibit feeding by nematodes and callose deposition is absent where the plasma membrane is tightly appressed against the stylet orifice.

Nematode-Gland Secretions

Many plant-parasitic nematodes possess large and complex esophageal secretory glands (Fig. 5–1). The complexity and secretory function of the three esophageal glands, one dorsal and two subventral, are revealed in ultrastructural studies of esophagi of nematodes in the order Tylenchida (Endo & Wergin, 1988; Hussey, 1989; Hussey & Mims, 1990). Each gland is a single large cell that connects to the esophageal lumen via an elaborate valve. The dorsal gland has a long cytoplasmic process extending anteriorly within the wall of the esophagus, to near the base of the stylet in most Tylenchida nematodes. The subventral glands, on the other hand, empty into the lumen of the esophagus just posterior to the metacorporal pump chamber. Secretory proteins synthesized in both types of glands are sequestered in spherical membrane-bound granules for export out of the gland. Nematode secretory granules vary in size, composition, and morphology among genera and species. They also differ between life stages of the same species and between the dorsal and subventral glands. Even within a gland in a specific life stage there can be different types of granules (Hussey, 1989). In sedentary endoparasites, the esophageal glands undergo marked morphological changes during parasitism. In *Meloidogyne* species, the subventral glands are the most active glands in infective second-stage juveniles. However, following the onset of parasitism, the dorsal gland is stimulated to increase production of secretory granules, and becomes the predominate gland in adult females (Bird, 1983; Hussey & Mims, 1990).

Identifying nematode gland secretions and determining their role in plant parasitism and pathogenesis remains one of the most intriguing and challenging areas of research in nematology. Secretions produced in the esophageal glands may function in egg hatching, penetration of and migration in plant tissue, induc-

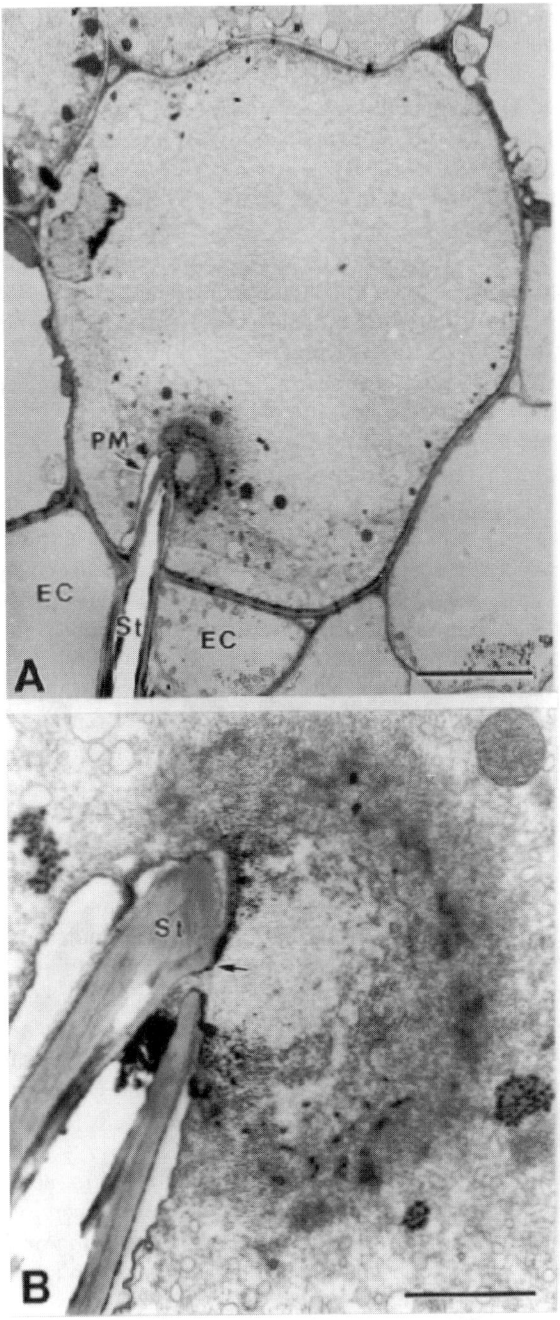

Fig. 5–3. Ultrathin sections of a stylet of *Criconemella xenoplax* feeding from a cortical cell in a tomato root. *A*. Longitudinal section of a stylet (St) that penetrated between two epidermal cells (EC) and was inserted in a cortical cell without penetrating the plasma membrane (PM) (bar = 5 μm). *B*. Detailed view of the minute hole created in the plasma membrane at the stylet (St) orifice (arrow) (bar = 1 μm). (Reprinted from Hussey et al., 1992b.)

tion and maintenance of discrete feeding sites, formation of feeding tubes, and intra- or extracorporeal digestion (Hussey, 1989). Although biochemical and cytochemical analyses have provided some general characterization of esophageal gland secretions (Hussey, 1989; Hussey et al., 1994), the identity and specific roles of these secretions in nematode-host interactions have not been elucidated.

Monoclonal antibody technology is used to investigate esophageal gland secretory proteins (Hussey et al., 1994). The specificities of these monoclonal antibodies reveal similarities and differences in proteins sequestered in secretory granules formed in the subventral and dorsal glands of *Meloidogyne* species (Davis et al., 1992). Furthermore, these antibodies reveal differences between the composition of stylet secretions of juveniles and adult females of *M. incognita* (Davis et al., 1994). Dorsal gland secretions have long been considered to have an important role in nematode-host interactions (Hussey, 1989). The location of the two subventral gland valves at the base of the metacorpal pump chamber has lead to speculation that secretions from the subventral glands can only pass posteriorly in the esophagus to the intestine and, thus, only function in digestion (Hussey, 1989; Wyss et al., 1992). However, monoclonal antibodies have been used to conclusively establish that secretions from subventral esophageal glands are secreted through the stylet of juveniles and adult females (Davis et al., 1994). Monoclonal antibody studies also have provided important information on the temporal synthesis of specific secretory components by the two types of esophageal glands in *M. incognita* and *Heterodera glycines* during parasitism of plants (Atkinson & Harris, 1989; Davis et al., 1994).

Even though various enzymes, e.g., cellulase, pectinases, and proteases, that could have a role in parasitism of plants are present in nematodes (Deubert & Rohde, 1971), their presence in esophageal glands or stylet secretions has not been firmly established. One secretory protein present in the esophageal glands and stylet secretions of *M. incognita* (Hussey et al., 1990; Davis et al., 1994) is known to be a large glycoprotein, but it has not been characterized further.

A monoclonal antibody specific for secretory granules in the subventral glands of adult females of *M. incognita* was used to isolate a cDNA clone that was used to clone a gene, *sec-1*, encoding a putative protein involved in esophageal gland secretion (Ray et al., 1994). The *sec-1* gene is specific to *Meloidogyne* species. Analogs were found in *M. arenaria*, *M. javanica*, *M. hapla,* but not in two other nematode species, *H. glycines*, and *Caenorhabditis elegans* (a nematode that feeds on bacteria). The protein encoded by the *sec-1* gene, SEC-1, was not detected with the monoclonal antibody in stylet secretions, but it has homologies with other proteins that suggest it is involved in movement of secretory granules. As additional genes encoding putative secretory proteins are cloned and characterized, we will increase our understanding of the basic biology of plant-parasitic nematodes and their interactions with plants.

Feeding Tubes

In addition to injecting stylet secretions that modify the recipient host cell, some plant-parasitic nematode species inject secretions that form unique tubelike structures called feeding tubes within the host cell. Feeding tubes are thought to facilitate withdrawal of nutrients from the food cell. The morphology of feeding tubes produced from secretions differs among nematode species (Rebois, 1980; Rumpenhorst, 1984; Hussey & Mims, 1991). The electron dense wall of the feeding tube may be thin, e.g., tubes formed by *Rotylenchulus reniformis* (Rebois, 1980); or thick and crystalline in nature, e.g., tubes formed by *Meloidogyne* species (Hussey & Mims, 1991). A detailed ultrastructural study has been conducted on feeding tubes formed in giant-cells by *M. incognita* (Fig. 5–4) (Hussey & Mims, 1991). The distal end of the feeding tube, which is 1 µm wide and up to 110 µm long with a lumen of 350-nm diam., is sealed with wall material. A compact membrane system, formed from rearrangement of the giant-cell endomembrane system, envelops newly formed and functional feeding tubes. The intimate association of the complex membrane system with the feeding tube indicates that it might function in synthesizing and/or transporting nutrients to the feeding tube for withdrawal by the parasite (Hussey & Mims, 1991). A new tube is formed each time the nematode reinserts its stylet into a food cell to initiate a new feeding cycle. Consequently, numerous feeding tubes are present in nematode-induced food cells. In giant-cells, the crystalline walls of feeding tubes no longer used by *M. incognita* persist in the cytoplasm, but the membrane systems that envelop newly formed tubes degenerate (Hussey & Mims, 1991).

The manner by which a nematode's stylet removes nutrients via the feeding tube is unclear. When the stylet of parasitic nematodes that produce feeding tubes is inserted in a root cell, the plasma membrane becomes invaginated around the stylet tip (Fig. 5–5) (Endo, 1991; Rebois, 1980; Wyss et al., 1984). Gland secretions injected through the stylet form the feeding tubes in the cytoplasm of the parasitized cell. A direct association of the stylet orifice on the outside of the plasma membrane with the lumen of the feeding tube located on the inside of this membrane would be necessary to facilitate withdrawal of nutrients from the cytoplasm of the parasitized cell via the feeding tube. Feeding tubes frequently project from short wall ingrowths through which the nematode's stylet is inserted, possibly facilitating alignment of the stylet orifice with the feeding tube lumen (Hussey & Mims, 1991). Wyss et al. (1984) considered feeding tubes to serve as a means for facilitating the flow of soluble nutrients in the cytosol of the food cells toward the stylet orifice. They also reported that the cytoplasm around feeding tubes appeared modified and suggested that enzymes from the tubes predigested the cytoplasm to promote entry of nutrients into the tubes. Since the distal ends of feeding tubes apparently are sealed and their walls do not have noticeable pores, the walls must be permeable to certain soluble assimilates. In fact, recent studies show that feeding tubes function as a molecular sieve during uptake of nutrients from the food cell by the nematode (Bökenhoff & Grundler, 1994). These microinjection studies with fluorescent probes of different molecular weights showed that dextrans of a molecular weight between 20 and 40 kDa were not taken up by feeding *H. schachtii*, but those less than 20 kDa were ingested. These authors concluded that the feeding tube formed by this nematode had a size exclusion limit between 20 to 40 kDa. Razak and Evans (1976) proposed that feeding tubes might function as cytoplasmic filters that prevent cell organelles from obstructing the stylet orifice (approximately 150-nm diam.). Regardless of

Fig. 5–4. Nomarski light micrograph of a cryosection of a giant-cell from a root of a tomato plant infected by *Meloidogyne incognita*. One long feeding tube (FT) and sections (arrows) of other feeding tubes are visible in the cytoplasm. A section of the head of the adult female nematode (N) that was feeding from the giant-cell is visible (bar = 20 μm). (Reprinted from Hussey & Mims, 1991.)

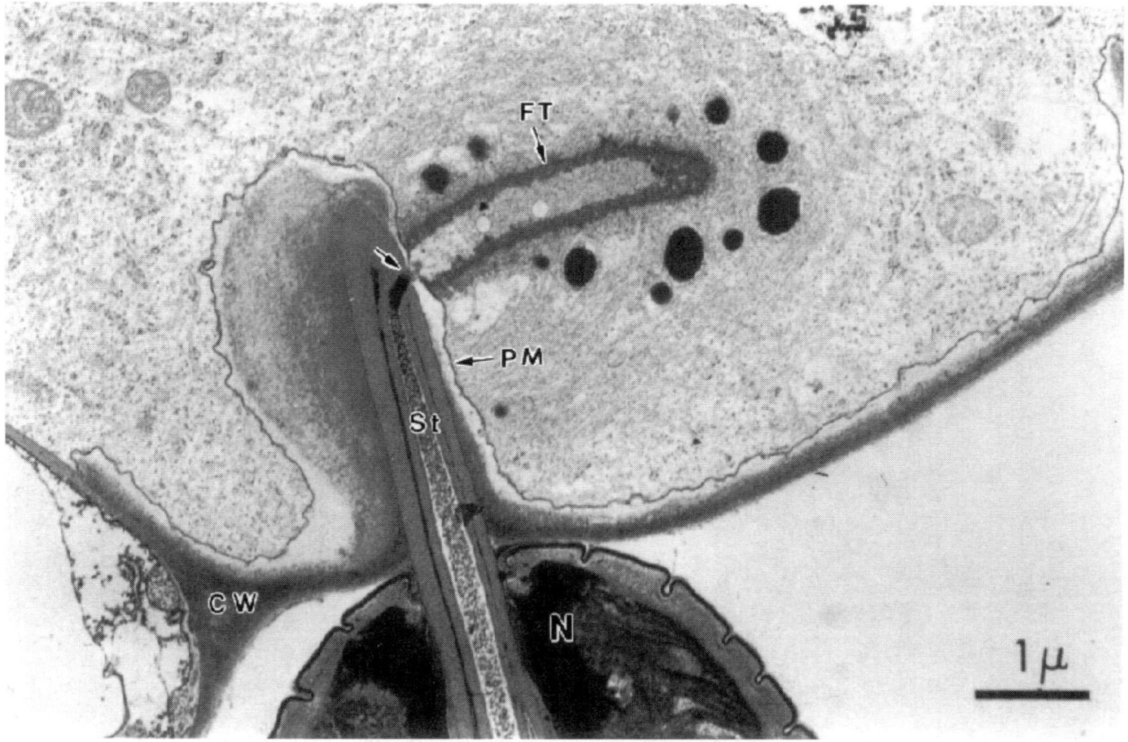

Fig. 5–5. Longitudinal section of stylet (St) of *Rotylenchulus reniformis* penetrating the wall (CW) of a cortical cell in a cotton root. The plasma membrane (PM) of the cortical cell is invaginated around inserted stylet tip. The stylet orifice (arrow) is aligned with an opening in the plasma membrane between the stylet and the lumen of a feeding tube (FT) formed from nematode secretions in the ctyoplasm of the parasitized cell. A section of the head of the nematode (N) is visible. (Reproduced with permission from Rebois, 1980.)

the manner in which feeding tubes function, more than likely they are essential for efficient withdrawal of nutrients by the parasites from the large food cells. In *M. incognita* parasitism, where the stylet is probably only inserted 2 to 3 μm through the cell wall and does not appear to penetrate the plasma membrane, the long feeding tube could function to provide the nematode access to a larger area of the giant-cell cytoplasm than accessible with the stylet alone. Conclusive evidence of an essential role for feeding tubes in nutrient acquisition by nematodes awaits experiments demonstrating that perturbation of feeding tube formation has a detrimental effect on nematode development.

Although the morphology of feeding tubes has been described, no data are available on the nature of the nematode secretions that contribute to feeding tube formation. Furthermore, it has not been definitively determined where these secretions are synthesized in the nematode, even though the dorsal esophageal gland has been implicated. An ultrastructural study showed that a crystalline secretory component, whose fine structure was similar to that of the feeding tube wall, accumulated in an open dorsal gland valve in an adult female of *M. incognita* (Hussey & Mims, 1990). In addition, other observations indicate that: (i) the fine structure of feeding tubes formed by *Meloidogyne* species is identical in giant-cells formed in different plant species (Hussey & Mims, 1991), (ii) subcellular structures similar to feeding tubes have not been reported in healthy cells of any plant species or plant cells parasitized by other organisms, (iii) feeding tubes form rapidly in parasitized cells following insertion of a nematode's stylet and initiation of the secretion phase of a feeding cycle (Wyss & Zunke, 1986), and (iv) the fine structure of feeding tubes formed by different nematode genera parasitizing the same plant species is dissimilar {e.g., *M. incognita* and *H. glycines* on soybean [*Glycine max* (L.) Merr.], R. S. Hussey, 1993, unpublished data). Together, these observations provide compelling evidence that feeding tubes form directly from nematode stylet secretions injected into the cytoplasm of parasitized cells. Stylet secretions produced in vitro by adult females of *Meloidogyne* species form a very viscous, insoluble precipitate that adheres to the lip region (Davis et al., 1994), but, when the nematode injects the secretions into the cytoplasm of giant-cells, these secretory components polymerize into the unique feeding tube. Apparently, interaction of the secretion with some component of the giant-cell cytoplasm, or perhaps the pH of the cytoplasm, induces the polymerization of the secretion into a tube.

HOST CELLULAR AND MOLECULAR RESPONSES TO NEMATODE PARASITISM

Cellular Modifications

Plant cells parasitized by different nematode species are modified in diverse ways. Cellular changes resulting from nematode parasitism can be categorized as either destructive or adaptive (Dropkin, 1969). Destructive cellular changes range from limited removal of cell contents by feeding nematodes to complete destruction of cells. By contrast, nematodes whose feeding activities are confined to one or a few plant cells induce elaborate modifications of these cells to form specialized feeding sites to support prolonged removal of nutrients by the parasite. The feeding sites result from nematode secretions modifying and regulating cell function, metabolism, and phenotype. These cells are modified by the nematode to function in increased uptake of solutes destined for the parasite. In these highly specialized cellular adaptations, intercellular transport of solutes is enhanced by development of elaborate wall ingrowths, e.g., *Heterodera* and *Meloidogyne* species feeding sites, or increases in number or modifications of plasmodesmata, e.g., *C. xenoplex* food cells. The feeding sites, which become the permanent nutrient source for the sedentary endoparasites, may take the form of a single uninucleate giant-cell (e.g., *Cryphodera utahensis*), several multinucleate giant-cells (e.g., *Meloidogyne* species), a multinucleate syncytium (e.g., *Heterodera* species and *R. reniformis*), or a group of discrete nurse cells (e.g., *Tylenchulus semipenetrans*). These specialized cellular adaptations are characterized by enlarged nuclei and nucleoli, increased cytoplasmic components, elevated metabolic rates, and, in some cells, thickened, remodeled cell walls. The development and morphology of the different types of cellular responses to nematode parasitism have been widely studied using light and electron microscopy. Since these classical studies have been extensively reviewed (Cohn & Spiegel, 1991; Dropkin, 1969; Endo, 1987; Jones, 1981; Wyss, 1988), we will focus this section on recent developments in understanding the biochemical and molecular bases of the cellular modifications due to nematode parasitism.

Biochemical and Molecular Responses

Biochemical and molecular changes associated with nematode infection have been mostly studied for sedentary endoparasites (Atkinson, 1994; Niebel et al., 1994; Sijmons, 1993). Responses in susceptible plants can be divided into two categories. The first category includes responses to the wounding or stress caused by nematode infection. The second involves changes directed toward initiation of feeding sites, such as giant-cells or syncytia, and the formation of root galls by *Meloidogyne* species. These changes are likely to be induced by nematode secretions and to be mediated by phytohormones. The focus of this section is on the responses in susceptible plants. Biochemical changes in resistant plants are discussed elsewhere (see Roberts et al., 1998, Chapter 11).

Biochemical responses have been observed both locally near the infecting nematode and throughout the plant. Root invasion by *Globodera rostochiensis* produces many changes in gene expression in potato (*Solanum tuberosum* L.) leaves by 6 h after infection (Hammond-Kosack et al., 1989). The induced proteins have properties of pathogenesis related (PR) proteins, a group of proteins that are induced systemically after infection of plants with various viral or fungal pathogens. Proteinase inhibitor proteins I and II, which are induced by wounding or by chewing insects (Ryan, 1990), were found to be induced in leaves, but not roots, upon nematode infection. Leaf protein changes in plants displaying susceptible and resistant responses to the nematode were not distinguishable. A gene encoding the enzyme catalase was shown to be induced in potato upon infection with *G. pallida* (Niebel et al., 1995). Induction occurred not only at the site of infection but also in uninfected roots, stems, and leaves of infected plants. Localized and systemic induction also occurred upon infection by *M. javanica* or root bacteria. Catalase is proposed to function to protect the plant from oxidative damage caused by the stress of nematode feeding (Niebel et al., 1995). The tomato [*Lycopersicon esculentum* (Mill) Karsten var. *lycopersicum*] gene *hmg2*, encoding 3-hydroxy-3-methylglutaryl CoA reductase, a key enzyme mediating expression in galling tissue after initiation of feeding by *M. incognita* or *M. hapla* (Cramer et al., 1993).

The gross and complex physiological changes that occur during the establishment of feeding sites of sedentary endoparasitic nematodes must require large and specific changes in gene expression. Changes in phytohormone levels have been observed in galled tissues. Auxin and ethylene levels increase greatly in *Meloidogyne javanica*-infected tomato roots (Glazer et al., 1986). Characterization of specific genes induced upon nematode infection is limited. Detailed analysis of expression of a gene encoding extensin after *M. javanica* and *Globodera tabacum* infection has been carried out in tobacco (*Nicotiana tabacum* L.) (Niebel et al., 1993). Extensins are hydroxyproline-rich glycoproteins that form a major protein component of plant cell walls. These proteins have been reported to be produced in plant-defense responses. Niebel et al. (1993) found that extensin mRNA levels were significantly increased in *M. javanica*-induced galls at 1 wk after infection and began to decrease by 3 wk post-infection. Increase in extensin mRNA levels is rapid: a large increase is seen in both resistant and susceptible tomato root tips by 12 h after infection with *M. javanica* (Lambert, 1995). Induction was seen in vascular parenchyma cells in contact with the nematodes, in pericycle, and cortical cells of galls (Niebel et al., 1993). Strong induction of extensin expression was not observed after cyst nematode infection. Since *Heterodera* and *Globodera* spp. induce more wounding than do *Meloidogyne* spp., these results suggest that the induction of extensin after *Meloidogyne* infection is not due to wounding. Another possible role for extensin is as a component of the nematode-induced alterations in plant cells. Localization studies showed that induction was primarily in the cortical cells of the gall (Niebel et al., 1993).

Changes in gene expression that are involved in feeding site establishment are likely to be localized to a few cells. It has been a challenge to investigate these changes because of the small amount of responding tissue. Recently, the ability to create transgenic plants containing "reporter genes," most commonly ß-glucuronidase (GUS), which can be easily visualized when expressed, has increased our ability to examine localized changes in gene expression. The expression of reporter genes under control of specific regulatory DNA regions, or promoters, has produced valuable information on localized nematode-induced changes in plant gene expression. Reporter gene constructs have been used to investigate the regulation of expression of *TobRB7* after *M. incognita* infection (Opperman et al., 1994). *TobRB7* is a tobacco gene encoding a water channel protein that, in uninfected plants, is expressed in root tips. The control regions of this gene also direct expression in giant-cells. Molecular dissection of the *TobRB7* control regions revealed that a small fragment of DNA immediately upstream from the gene could direct expression of other genes placed under its control to giant-cells. Interestingly, this control region does not increase gene expression in *G. tabacum*-induced syncytia. The use of such control regions for developing novel nematode control strategies is discussed elsewhere in this volume.

Other genes thought to be likely candidates for increased expression in feeding sites, especially for *Meloidogyne* species, are genes involved in cell division. A fusion of the promoter of CDC2, a protein kinase that is required for cell division, to the reporter gene GUS showed strong induction in transgenic tobacco near the head of the second-stage juvenile soon after it reached its feeding site (Niebel et al., 1994). After 3 d the expression was reduced. A similar induction pattern was seen near the site of syncytium initiation by *H. schachtii*, even though this nematode does not stimulate cell division.

Recently, experiments have been carried out to identify additional genes expressed in feeding cells induced by cyst and root-knot nematodes (Burrows, 1992; Bird, 1992). Polymerase chain reaction (PCR) amplification was used to construct a library highly enriched in syncytial material 26 d after infection of potato with *G. rostochiensis*, and a clone encoding a gene that appeared to be syncytium specific was identified (Gurr et al., 1991). Polymerase chain reaction and a novel subtraction technique were used to generate a cDNA library highly enriched for genes induced in giant-cells from tomato roots (Wilson et al., 1994). Fifty-eight genes expressed in tomato giant-cells induced by *M. incognita* infection have been characterized (Bird & Wilson, 1994). Some of these genes encode known sequences representing a diverse array of enzyme activities. Many of the genes are expressed in actively dividing and expanding tissues and in tissues other than the root. An apparent tomato analog of the *TobRB7* gene was present among these clones. The diversity of clones obtained indicates that there are complex changes in gene expression in nematode modified cells.

Transformation of plants with "promoter traps," reporter genes without promoters, has been used to identify other regulatory regions that may be affected by nematode infection. Several genes are reduced in expression locally upon nematode infection (Goddjin et al., 1993). An interesting hypothesis is that these genes are actively turned off by the nematode to repress defense responses. Alternatively, they may be turned off as a result of the stress produced by nematode infection.

WHOLE-PLANT RESPONSES TO NEMATODE DISEASES

Nematode parasitism of roots can have a dramatic impact on the physiology and growth of plants, leading to suppressed crop yields. The symptoms and yield losses caused by nematodes on susceptible host plants is variable and complex and involves several mechanisms that are directly related to the numbers of nematodes feeding on the roots. Nematodes remove plant nutrients during feeding, affect nutrient and water uptake and translocation by roots, modify or destroy root tissue, or retard root growth (Dropkin, 1979). The root damage from nematode parasitism results in disruption of physiological processes throughout the whole plant—causing stunted, usually chlorotic, crops producing low yields.

While visible symptoms of damage usually occur with high nematode numbers, some species can cause yield losses up to 30% on susceptible crops without obvious symptoms in high-yield environments (Niblack, 1993). In addition to direct effects on plant growth, nematode parasitism of roots causes indirect effects that also can suppress crop growth and yield. Nematodes feeding on legume roots can reduce the number or activity of nodules formed by the beneficial nitrogen-fixing bacteria that are necessary for plant growth (Barker et al., 1993). Nematode parasitism of roots also creates openings in the root surface that can serve as entry points for other soil-inhabiting microorganisms that can affect root function or destroy root tissue.

The physiological basis for suppressed growth and yield caused by nematodes has been reviewed by Melakeberhan and Webster (1993). Many of the major physiological processes in the host plant, including respiration, photosynthesis, nutrient absorption and translocation, water relations, and hormone balance, can be affected directly or indirectly by nematode parasitism. Most of the physiological studies of nematode-infected plants have involved sedentary endoparasitic nematodes, which dramatically modify host-root tissue.

Mineral Nutrition and Water Relations

The principal effect of nematode parasitism is damage to the host root system, which directly impacts the quantity and quality of nutrients and amount of water available to the shoot. The diminished function of nematode-infected roots is exhibited in the plant shoot through symptoms of nutrient deficiency, temporary wilting during periods of mild water stress or high temperatures, and suppressed biomass. Nutrients usually accumulate in the roots, and the amount translocated to the shoots of nematode-infected plants is reduced, though this effect varies with nematode feeding behavior and specific nutrients. Nematodes that parasitize cells in the vascular cylinder seem to have the greatest effect on plant nutrient and water relations since vascular parasitism results in disruption in the continuity of vessel elements and abnormal development of these elements. Nematode parasitism of roots affects the uptake of specific nutrients differently, depending on whether their radial movement through the root is via the symplastic pathway, through the cytoplasm, or the apoplastic pathway, through free space in the cell wall (Clarkson, 1991). For example, cyst nematodes feeding in the vascular tissue disrupt the continuity of endodermal tissue and enhance uptake of Ca which is transported apoplastically to the xylem (Price & Sanderson, 1984). Root branching and rate of root extension also can be suppressed by nematode infection. This abnormal root growth results in reduced surface area and limits the capacity of a root system to explore the soil and take up nutrients.

The effects of nematode infection on the concentration and distribution of nutrient elements in plants has been extensively studied and found to vary widely with nematode species and host, both temporally and spatially (Hussey, 1985; Melakeberhan & Webster, 1993). Melakeberhan et al. (1985a, 1986) have related the changes in nutrient concentrations over time with alterations of physiological processes in nematode-infected plants, and consider changes in concentration of nutrient elements one of the first effects that *Meloidogyne* species have on host physiology. Application of fertilizers can compensate for nematode effects on certain host nutrient levels (Melakeberhan et al., 1988), but the increased plant growth due to the fertilizer can result in higher nematode densities (Schmitt & Riggs, 1989).

Nematode parasitism of roots also directly affects the plant-water relations, which contribute significantly to the suppressed yields of nematode-infected crops (Wilcox-Lee & Loria, 1987). Water uptake and flow through roots is impeded by nematode infection (Dorhout et al., 1991) and decreases with increasing inoculum level and with decreasing soil moisture (Meon et al., 1978). The transpiration rate in infected plants is usually decreased and the diffusive resistance of stomata increased in nematode-infected plants (Haverkort et al., 1991; Schans, 1991), but in some interactions there is no change (Wilcox & Loria, 1986). Although nematode-induced water stress affects many plant physiological processes, supplemental irrigation does not compensate for the water deficit sufficiently to offset yield suppression (Koenning & Barker, 1995).

Photosynthesis and Respiration

The suppression of photosynthesis in nematode-infected plants has been documented for only a few nematode-host interactions. Nevertheless, the inability of roots damaged by any nematode species to supply sufficient nutrients, particularly N_2, and water to the shoot would affect the photosynthetic process. Most of the in-depth photosynthesis studies have been conducted on root-knot and cyst nematode-infected plants growing in growth chambers or greenhouses. A decrease of the photosynthetic rate in root-knot nematode infected plants has been reported for French bean (*Phaseolus vulgaris* L.) (Melakeberhan et al., 1985a,b, 1987), tomato (Meon et al., 1978), and grape (*Vitis* spp.) (Melakeberhan et al., 1990), and the impact is greater on younger than on older plants (Melakeberhan et al., 1986). A similar impact on photosynthesis has been reported for *Globodera pallida*-infected potato (Schans, 1991), and for *H. glycines*-infected soybean (Barker et al., 1993). The net photosynthetic rate of a highly resistant soybean cultivar was not suppressed by *H. glycines* (Barker et al., 1993), whereas *G. pallida* did restrict photosynthesis of partially resistant potato cultivars (Schans & Arntzen, 1991).

In the only study conducted under field conditions, the decrease in photosynthesis for *H. glycines*-infected soybean was greatest in soil types most favorable for nematode activity, but overall, the effect of nematode parasitism on photosynthesis was more variable than found in greenhouse studies (Koenning and Barker, 1995). This investigation also showed that only a portion of plant yield suppression could be attributed directly to restricted net photosynthetic rate caused by *H. glycines*. Suppression of individual leaf area and total plant biomass, which restricted the total plant photosynthetic area, is considered the primary impact of nematode parasitism on photosynthesis (Koenning & Barker, 1995; Poskuta et al., 1986; Melakeberhan et al., 1985a, 1986). In the *H. glycines*-soybean interaction, certain races of the nematode also suppress nodulation, causing N_2 deficiencies that would further contribute to the restricted net photosynthesis response of infected plants (Koenning & Barker, 1995).

In the few studies where respiration of nematode-infected plants was measured, respiration varied with time and nematode-host interaction. For example, respiration of grape was decreased by *M. incognita* infection (Melakeberhan et al., 1990), but in French bean infected by *M. incognita* respiration rate varied greatly (Melakeberhan et al., 1985a, 1986).

Nematode Consumption and Host-Assimilate Partitioning

In addition to altering the physiology of infected plants, nematodes feeding from plant cells consume nutrients normally utilized by the host and can, in some cases, affect photosynthate partitioning. Nematode species with high reproductive capacity require considerable amounts of nutrients (energy) for egg production and compete with the host for the pool of nutrients in roots. However, few studies have attempted to measure the energy demands of a parasitic nematode on a host plant (Atkinson, 1985; Powers & McSorley, 1993; Melakeberhan & Ferris, 1988). The energy demand of *M. incognita* for growth and reproduction on grape was equivalent to 15% of the total energy assimilated by the host and is considered to be a significant contributor to lower plant productivity (Melakeberhan & Ferris, 1988, 1989).

Feeding sites of sedentary nematodes are strong sinks for host photoassimilates and nutrients. As a consequence, energy is diverted from the host plant to the parasite. The increased metabolic activity of feeding-sites induced by *Meloidogyne* and *Heterodera* species stimulates mobilization of photosynthates from the shoots to the roots and symplastically or apoplastically from the phloem into the giant-cells where they are consumed by the nematode (Bird & Loveys, 1975; Bockenhoff et al., 1996; Dorhout et al., 1993; McClure, 1977). Free amino acids and glucose increase in feeding sites of *Heterodera* and *Meloidogyne* species (Gommers & Dropkin, 1977; Krauthausen & Wyss, 1982). Giant-cells take up solutes apoplastically from the xylem via a proton pump associated with the elaborate cell wall ingrowths and increased surface area of the plasma membrane (Jones, 1981; Dorhout et al., 1988, 1992).

OUTLOOK

We have entered an era where new tactics for managing plant-parasitic nematodes are needed. Clearly, development of new tactics will require an in-depth understanding of the interactions between nematodes and their hosts. These interactions appear to be mediated mainly by chemical signals. To date little is known about the chemicals involved in attraction of nematodes to their hosts or the signals that trigger the nematode to initiate a feeding response. The nature of the nematode compounds that initiate the complex changes that occur in their hosts are also largely unknown.

New tools in molecular biology have opened up the possibility of identifying the genes and molecular changes that are the basis for these interactions. Inroads are being made into identifying genes encoding stylet secretions, which are likely to be the main source of molecular signals that initiate the physiological and morphological changes in the host. The investigation of changes in host gene expression upon nematode infection is also a new and developing area. Many increases and decreases in gene expression have been observed, but it is not yet known how these changes are brought about and why. Do nematode secretions directly control changes in plant-gene expression? Are these changes secondary, perhaps due to hormonal changes or the formation of a sink? Do these genes have a role in feeding site development and maintenance? The next few years should bring answers to some of these questions and, perhaps, through better understanding of the interaction, lead us to new control tactics.

REFERENCES

Atkinson, H.J. 1985. The energetics of plant parasitic nematodes—a review. Nematologica 31:62–71.

Atkinson, H.J. 1994. Plant-nematode interactions: Molecular and genetic basis. p. 355–369. *In* K. Kohomoto et al. (ed.) Pathogenesis and host-parasite specificity in plant diseases: Histopathological, biochemical, genetic and molecular basis. Pergamon Press, Oxford, England.

Atkinson, H.J., and P.D. Harris. 1989. Changes in nematode antigens recognized by monoclonal antibodies during early infections of soya bean cyst nematode, *Heterodera glycines*. Parasitology 98:479–487.

Barker, K.R., S.R. Koenning, S.C. Huber, and J.-S. Huang. 1993. Biological and structural responses of plants to nematode parasitism with *Glycine max-Heterodera glycines* as a model system. p. 761–771. *In* D.R. Buxton (ed.) International crop science 1. CSSA, Madison, WI.

Bird, A.F. 1983. Changes in the dimensions of the esophageal glands in root-knot nematodes during the onset of parasitism. Int. J. Parasitol. 13:343–348.

Bird, A.F., and B.R. Loveys. 1975. The incorporation of photosynthates by *Meloidogyne incognita*. J. Nematol. 7:111–113.

Bird, D.Mck., and M.A. Wilson. 1994. DNA sequence and expression analysis of root-knot nematode-elicited giant cell transcripts. Mol. Plant-Microbe Interact. 7:419–424.

Bird, D.Mck. 1992. Mechanism of the *Meloidogyne*-host interaction. p. 51–59. *In* F.J. Gommers and P.W.Th. Maas (ed.) Nematology from molecule to ecosystem. Eur. Soc. Nematol., Inc., Dundee, Scotland.

Böckenhoff, A., and F.M.W. Grundler. 1994. Studies on the nutrient uptake by the beet cyst nematode *Heterodera schachtii* by *in situ* microinjection of fluorescent probes into the feeding structures in *Arabidopsis thaliana*. Parasitology 109:249–254.

Bockenhoff, A., D.A.M. Prior, F.M.W. Grundler, and K.J. Oparka. 1996. Induction of phloem unloading in *Arabidopsis thaliana* roots by the parasitic nematode *Heterodera schachtii*. Plant Physiol. 112:1421–1427.

Burrows, P.R. 1992. Initiation and development of the syncytial feeding sites of cyst nematodes: A molecular view. p. 32–40. *In* F.J. Gommers and P.W.Th. Maas (ed.) Nematology from molecule to ecosystem. Eur. Soc. Nematol., Inc., Dundee, Scotland.

Clarkson, D.T. 1991. Root structure and sites of ion uptake. p. 417–453. *In* Y. Waisel and A. Eshel (ed.) Plant roots—the hidden half. Marcel Dekker, Inc., New York.

Cohn, E., and Y. Spiegel. 1991. Root-nematode interactions. p. 789–805. *In* Y. Waisel and A. Eshel (ed.) Plant roots—the hidden half. Marcel Dekker, Inc., New York.

Cramer, C.L. D. Weissenborn, C.K. Cottingham, C.J. Denbow, J.D. Eisenback, D.N. Radin, and X. Xu. 1993. Regulation of defense-related gene expression during plant-pathogen interactions. J. Nematol. 25:507–518.

Davis, E.L., R. Allen, and R.S. Hussey. 1994. Developmental expression of esophageal gland antigens and their detection in stylet secretions of *Meloidogyne incognita*. Fund. Appl. Nematol. 17:255–262.

Davis, E.L., L.M. Aron, L.H. Pratt, and R.S. Hussey. 1992. Novel immunization procedures used to develop monoclonal antibodies that bind to specific structures in *Meloidogyne* spp. Phytopathology 82:1244–1250.

Deubert, K.H., and R.A. Rohde. 1971. Nematode enzymes. p. 73–90. *In* B.M. Zuckerman et al. (ed.) Plant parasitic nematodes. Vol. 2. Acad. Press, New York.

Doncaster, C.C. 1971. Feeding in plant parasitic nematodes: Mechanisms and behavior. p. 137–157. *In* B.M. Zuckerman et al. (ed.) Plant parasitic nematodes. Vol. 2. Acad. Press, New York.

Doncaster, C.C., and M.K. Seymour. 1973. Exploration and selection of penetration site by Tylenchida. Nematologica 19:137–145.

Dorhout, R., F. Gommers, and C. Kolloffel. 1991. Water transport through tomato roots infected with *Meloidogyne incognita*. Phytopathology 81:379–385.

Dorhout, R., F. Gommers, and C. Kolloffel. 1993. Phloem transport of carboxyfluorescein through tomato roots infected with *Meloidogyne incognita*. Physiol. Mol. Plant Pathol. 43:1–10.

Dorhout, R., C. Kolloffel, and F.J. Gommers. 1988. Transport of an apoplastic fluorescent dye to feeding sites induced in tomato roots by *Meloidogyne incognita*. Phytopathology 78:1421–1424.

Dorhout, R., C. Kolloffel, and F.J. Gommers. 1992. Alteration of distribution of regions with high net proton extrusion in tomato roots infected with *Meloidogyne incognita*. Physiol. Mol. Plant Pathol. 40:153–162.

Dropkin, V.H. 1969. Cellular responses of plants to nematode infection. Annu. Rev. Phytopathol. 7:101–121.

Dropkin, V.H. 1979. How nematodes induce disease. p. 219–238. *In* J.G. Horsefall and E.B. Cowling (ed.) Plant disease an advanced treatise. Vol. 4. Acad. Press, New York.

Dusenbery, D.B. 1987. Prospects for exploiting sensory stimuli in nematode control. p. 131–135. *In* J.A. Veech and D.W. Dickson (ed.) Vistas in nematology. Soc. Nematol., Hyattsvile, MD.

Endo, B.Y. 1987. Histopathology and ultrastructure of crops invaded by certain sedentary endoparasitic nematodes. p. 196–210. *In* J.A. Veech and D.W. Dickson (ed.) Vistas in nematology. Soc. Nematol. Hyattsvile, MD.

Endo, B.Y. 1991. Ultrastructure of initial responses of susceptible and resistant soybean roots to infection by *Heterodera glycines*. Rev. Nematol. 14:73–94.

Endo, B.Y., and W.P. Wergin. 1973. Ultrastructural investigation of clover roots during early stages of infection by the root-knot nematode, *Meloidogyne incognita*. Protoplasma 78:365–379.

Endo, B.Y., and W.P. Wergin. 1988. Ultrastructure of the second-stage juvenile of the root-knot nematode, *Meloidogyne incognita*. Proc. Helminthol. Soc. Wash. 55:286–316.

Fenoll, C., F.M.W. Grundler, and S.A. Ohl (ed.). 1997. Cellular and molecular aspects of plant-nematode interactions. Kluwer Acad. Publ., Dordrecht, the Netherlands.

Glazer, I., E. Epstein, D. Orion, and A. Apelbaum. 1986. Interactions between auxin and ethylene in root-knot nematode (*Meloidogyne javanica*) infected tomato roots. Physiol. Mol. Plant Pathol. 28:171–179.

Goddijn, O.J.M., K. Lindsey, F.M. Van der Lee, J.C. Klap, and P.C. Sijmons. 1993. Differential gene expression in nematode-induced feeding structures of transgenic plants harbouring promoter-*gusA* fusion constructs. Plant J. 4:863–873.

Gommers, F.J., and V.H. Dropkin. 1982. Quantitative histochemistry of nematode-induced transfer cells. Phytopathology 67:869–873.

Grundler, F.M.W., and U. Wyss. 1994. Strategies of root parasitism by sedentary plant parasitic nematodes. p. 309–319. *In* K. Kohomoto et al. (ed.) Pathogenesis and host-parasite specificity in plant diseases: histopathological, biochemical, genetic and molecular basis. Pergamon Press, Oxford.

Gurr, S.J., M.J. McPherson, C. Scollan, H.J. Atkinson, and D.J. Bowles. 1991. Gene expression in nematode-infected plant roots. Mol. Gen. Genet. 226:361–366.

Hammond-Kosack, K., H.J. Atkinson, and D.J. Bowles. 1989. Systemic accumulation of novel proteins in the apoplast of the leaves of potato plants following root invasion by the cyst-nematode *Globodera rostochiensis*. Physiol. Mol. Plant Pathol. 35:495–506.

Harder, D.E., and J. Chong. 1984. Structure and physiology of haustoria. p. 431–476. *In* W.R. Bushnell and A.P. Roelfs (ed.) The cereal rusts. Vol. 1. Acad. Press, New York.

Haverkort, A.J., T. Fasan, and M. Van De Waart. 1991. The influence of cyst nematodes and drought on potato growth. II. Effects on plant water relations under semi-controlled conditions. Netherlands J. Plant Pathol. 97:162–170.

Hussey, R.S. 1985. Host-parasite relationships and associated physiological changes. p. 144–153. *In* J.N. Sasser and C.C. Carter (ed.) An advanced treatise on *Meloidogyne*. Vol. 1. Biology and control. North Carolina State Univ. Graphics, Raleigh.

Hussey, R.S. 1989. Disease-inducing secretions of plant-parasitic nematodes. Annu. Rev. Phytopathol. 27:123–141.

Hussey, R.S., E.L. Davis, and C. Ray. 1994. *Meloidogyne* stylet secretions. p. 233–249. *In* F. Lamberti et al. (ed.) Advances in molecular plant nematology. Plenum Press, New York.

Hussey, R.S., and C.W. Mims. 1990. Ultrastructure of esophageal glands and their secretory granules in the root-knot nematode *Meloidogyne incognita*. Protoplasma 156:9–18.

Hussey, R.S., and C.W. Mims. 1991. Ultrastructure of feeding tubes formed in giant-cells induced in plants by the root-knot nematode *Meloidogyne incognita*. Protoplasma 162:99–107.

Hussey, R.S., C.W. Mims, and S.W. Westcott. 1992a. Ultrastructure of root cortical cells parasitized by the ring nematode *Criconemella xenoplax*. Protoplasma 167:55–65.

Hussey, R.S., C.W. Mims, and S.W. Westcott. 1992b. Immunocytochemical localization of callose in root cortical cells parasitized by the ring nematode *Criconemella xenoplax*. Protoplasma 171:1–6.

Hussey, R.S., O.R. Paguio, and F. Seabury. 1990. Localization and purification of a secretory protein from the esophageal glands of *Meloidogyne incognita* with a monoclonal antibody. Phytopathology 80:709–714.

Jones, M.G.K. 1981. Host cell responses to endoparasitic nematode attack: structure and function of giant cells and syncytial. Ann. Appl. Biol. 97:223–229.

Koenning, S.R., and K.R. Barker. 1995. Soybean photosynthesis and yield as influenced by *Heterodera glycines*, soil type and irrigation. J. Nematol. 27:51–62.

Krauthausen, H.-J. and U. Wyss. 1982. Influence of the cyst nematode *Heterodera schachtii* on relative changes in the pattern of free amino acids at feeding sites. Physiol. Plant Pathol. 21:425–436.

Lambert, K.N. 1995. Isolation of genes induced early in the resistance response to *Meloidogyne javanica* in *Lycopersicon esculentum*. Ph.D. diss. Univ. California, Davis.

McClure, M.A. 1977. *Meloidogyne incognita*: a metabolic sink. J. Nematol. 9:88–90.

Melakeberhan, H., R.C. Brooke, J.M. Webster, and J.M. D'Auria. 1985a. The influence of *Meloidogyne incognita* on the growth, physiology and nutrient content of *Phaseolus vulgaris*. Physiol. Plant Pathol. 26:259–268.

Melakeberhan, H., R.C. Brooke, and J.M. Webster. 1986. Relationship between physiological response of French beans of different age to *Meloidogyne incognita* and subsequent yield loss. Plant Pathol. 35:203–213.

Melakeberhan, H., and H. Ferris. 1988. Growth and energy demand of *Meloidogyne incognita* on susceptible and resistant *Vitis vinifera* cultivars. J. Nematol. 20:545–554.

Melakeberhan, H., and H. Ferris. 1989. Impact of *Meloidogyne incognita* on physiological efficiency of *Vitis vinifera*. J. Nematol. 21:74–80.

Melakeberhan, H., H. Ferris, and J.M. Dias. 1990. Physiological response of resistant and susceptible *Vitis vinifera* to *Meloidogyne incognita*. J. Nematol. 22:224–230.

Melakeberhan, H., and J.M. Webster. 1993. The phenology of plant-nematode interaction and yield loss. p. 26–41. *In* M. W. Khan (ed.) Nematode interactions. Chapman and Hall, New York.

Melakeberhan, H., J.M. Webster, and R.C. Brooke. 1985b. Response of *Phaseolus vulgaris* to a single generation of *Meloidogyne incognita*. Nematologica 31:190–202.

Melakeberhan, H., J.M. Webster, R.C. Brooke, and J.M. D'Auria. 1988. Effect of KNO_3 on CO_2 exchange rate, nutrient concentration and yield of *Meloidogyne incognita* infected beans. Rev. Nematol. 11:391–397.

Melakeberhan, H., J.M. Webster, R.C. Brooke, J.M. D'Auria, and M. Crackette. 1987. Effect of *Meloidogyne incognita* on plant nutrient concentration and its influence on the physiology of beans. J. Nematol 19:324–330.

Meon, S., H.R. Wallace, and J.M. Fisher. 1978. Water relations of tomato (*Lycopersicon esculentum* Mill. cv. Early Dwarf Red) infected with *Meloidogyne javanica* (Treub), Chitwood. Physiol. Plant Pathol. 13:275–281.

Müller, J., K. Rehbock, and U. Wyss. 1981. Growth of *Heterodera schachtii* with remarks on amounts of food consumed. Rev. Nematol. 4:227–234.

Niblack, T.L. (ed.). 1993. Protect your soybean profits: Manage soybean cyst nematode. Univ. Missouri Print. Services, Columbia.

Niebel, A., J A. Engler, C. Tire, G. Engler, M. Van Montagu and G. Gheysen. 1993. Induction patterns of an extensin gene in tobacco upon nematode infection. Plant Cell 5:1697–1710.

Niebel, A., G. Gheysen, and M. Van Montagu. 1994. Plant-cyst nematode and plant-root-knot nematode interactions. Parasitol. Today 10:424–430.

Niebel, A., K. Heungens, N. Barthels, D. Inze, M. Van Montague, and G. Gheysen. 1995. Characterization of a pathogen-induced potato catalase and its systemic expression upon nematode and bacterial infection. Mol. Plant-Microbe. Interact. 8:371–378.

Opperman, C.H., C.G. Taylor, and M.A. Conkling. 1994. Root-knot nematode-directed expression of a plant root-specific gene. Science (Washington, DC) 263:221–223.

Perry, R.N. 1994. Studies on nematode sensory perception as a basis for novel control strategies. Fund. Appl. Nematol. 17:199–202.

Poskuta, J.W., V.H. Dropkin, and C.J. Nelson. 1986. Photosynthesis, photorespiration, and respiration of soybean after infection with root nematodes. Photosynthetica 20:405410.

Powers, L.E., and R. McSorley. 1993. Energetics of *Meloidogyne incognita* on resistant and susceptible alyceclover genotypes. J. Nematol. 25:257–264.

Price, N.S., and J. Sanderson. 1984. The translocation of calcium from oat roots infected by the cereal cyst nematode *Heterodera avenae* (Woll.). Rev. Nematol. 7:239–243.

Ray, C., A.G. Abbott, and R.S. Hussey. 1994. *Trans*-splicing of a *Meloidogyne incognita* mRNA encoding a putative esophageal gland protein. Mol. Biochem. Parasitol. 68:93–101.

Razak, A.R., and A.A.F. Evans. 1976. An intracellular tube associated with feeding by *Rotylenchus reniformis* in cotton. Nematologica 26:396–405.

Rebois, R.V. 1980. Ultrastructure of a feeding peg and tube associated with *Rotylenchulus reniformis* in cotton. Nematologica 26:396–405.

Roberts, P.A., W.C. Matthews, and J.C. Veremis. 1998. Genetic mechanisms of host-plant resistance to nematodes. p. 209–238. *In* K.R. Barker et al. (ed.) Plant and nematode interactions. Agron. Monogr. 36. ASA, CSSA, and SSSA, Madison, WI.

Rumpenhorst, H.J. 1984. Intracellular feeding tubes associated with sedentary plant parasitic nematodes. Nematologica 30:77–85.

Ryan, C.A. 1990. Protease inhibitors in plants: Genes for improving defenses against insects and pathogens. Annu. Rev. Phytopathol. 28:425–449.

Schans, J. 1991. Reduction of leaf photosynthesis and transpiration rates of potato plants by second-stage juveniles of *Globodera pallida.* Plant Cell Environ. 14:707–712.

Schans, J., and F.K. Arntzen. 1991. Photosynthesis, transpiration and plant growth characters of different potato cultivars at various densities of *Globodera pallida.* Netherlands J. Plant Pathol. 97:297–310.

Schmitt, D.P., and R.D. Riggs. 1989. Population dynamics and management of *Heterodera glycines.* Agric. Zoo. Rev. 3:253–269.

Sijmons, P.C. 1993. Plant-nematode interactions. Plant Mol. Biol. 23:917–931.

Sijmons, P.C., H.J. Atkinson, and U. Wyss. 1994. Parasitic strategies of root nematodes and associated host cell responses. Annu. Rev. Phytopathol. 32:235–259.

Westcott III, S.W., and R.S. Hussey. 1992. Feeding behavior of *Criconemella xenoplax* in monoxenic culture. Phytopathology 82:936–940.

Wilcox, D.A., and R. Loria. 1986. Water relations, growth, and yield in two snap bean cultivars infected with root knot nematode, *Meloidogyne hapla* (Chitwood). J. Am. Soc. Hort. Sci. 111:34–38.

Wilcox-Lee, D., and R. Loria. 1987. Effects of nematode parasitism on plant-water relations. p. 261–266. *In* J.A. Veech and D.W. Dickson (ed.) Vistas in nematology. Soc. Nematol., Hyattsvile, MD.

Williamson, V.M., and R.S. Hussey. 1996. Nematode pathogenesis and resistance in plants. Plant Cell 8:1735–1745.

Wilson, M.A., D.McK. Bird, and E. van der Knaap. 1994. A comprehensive subtractive cDNA cloning approach to identify nematode-induced transcripts in tomato. Phytopathology 84:299–303.

Wyss, U. 1981. Ectoparasitic root nematodes: feeding behavior and plant cell responses. p. 325–351. *In* B.M. Zuckerman and R.A. Rohde (ed) Plant parasitic nematodes. Vol. 3. Acad. Press, New York.

Wyss, U. 1988. Pathogenesis and host-parasite specificity in nematodes. p. 417–432. *In* R.S. Singh et al. (ed.) Experimental and conceptual plant pathology. Gordon and Breach, New York.

Wyss, U. 1992. Observations on the feeding behavior of *Heterodera schachtii* throughout development, including events during molting. Fund. Appl. Nematol. 15:75–89.

Wyss, U., F.M.W. Grundler, and A. Munch. 1992. The parasitic behavior of second-stage juveniles of *Meloidogyne incognita* in roots of *Arabidopsis thaliana.* Nematologica 38:98.

Wyss, U., C. Stender, and H. Lehmann. 1984. Ultrastructure of feeding sites of the cyst nematode *Heterodera schachtii* Schmidt in roots of susceptible and resistant *Raphnus sativus* L. var. *oleiformis* cultivars. Physiol. Plant Pathol. 25:21–37.

Wyss, U., and U. Zunke. 1986. Observations on the behavior of second-stage juveniles of *Heterodera schachtii* inside host roots. Rev. Nematol. 9: 153–165.

Zacheo, G., and T. Bleve-Zacheo. 1994. Plant-nematode interactions: Histological, physiological, and biochemical interactions. p. 321–353. *In* K. Kohomoto et al. (ed.) Pathogenesis and host-parasite specificity in plant diseases: histopathological, biochemical, genetic and molecular basis. Pergamon Press, Oxford.

Zunke, U. 1990. Observations on the invasion and endoparasitic behavior of the root lesion nematode *Pratylenchus penetrans.* J. Nematol. 22:309–320.

6 Population Dynamics

ROBERT McSORLEY

University of Florida
Gainesville, Florida

Plant-parasitic nematodes occur in most cultivated soils, but associated damage usually results from high population densities, rather than from mere occurrence. In any cultivated field, estimates of current population density and future increase are critical in anticipating crop damage (see Duncan & Noling, 1998 , Chapter 13). Studies of nematode population dynamics seek to understand and predict nematode population growth, and to use this knowledge to improve nematode management (Duncan & McSorley, 1987; Ferris & Noling, 1987; McSorley & Phillips, 1993; Nusbaum & Barker, 1971; Seinhorst, 1970).

Population densities of plant-parasitic nematodes can increase quickly because reproduction rates are high and life cycles are short. Life cycles vary from a few days to over a year (Norton, 1978), but average about 3 to 6 wk for most species of plant parasites during the growing season. Temperature affects the length of the life cycle, which can often be expressed more consistently in degree days rather than in units of time (Duncan & McSorley, 1987).

Many nematodes—including the sting nematode (*Belonolaimus longicaudatus*), stubby-root nematode (*Trichodorus* spp., *Paratrichodorus* spp.), ring nematode (*Criconemella* spp.), and stunt nematode (*Tylenchorhynchus* spp.)—feed as ectoparasites throughout their life cycle and never enter the root system (Christie, 1959). Others, including lesion (*Pratylenchus* spp.) and burrowing nematodes (*Radopholus* spp.), migrate through root tissue while feeding. Root-knot (*Meloidogyne* spp.) and cyst (*Heterodera* spp., *Globodera* spp.) nematodes enter root systems as juveniles and establish a permanent feeding site where they live as sedentary endoparasites. A few others, such as the stem nematode (*Ditylenchus dipsaci*), enter aboveground plant parts. Despite the wide array of life cycles and parasitic habits among different kinds of plant nematodes, most species spend at least part of their cycle in the soil. Therefore, examination of a soil sample is the most convenient means for detecting and estimating the widest possible range of plant-parasitic nematodes.

ESTIMATING POPULATION DENSITY

Soil samples for nematode assay can be collected with many different types of equipment, including shovels, trowels, soil augers, or tubes (Ayoub, 1980;

Barker & Campbell, 1981; Southey, 1986). A typical cylindical sampling tube or cone-shaped modification is often used (Barker & Campbell, 1981).

Spatial Distribution

Vertical Distribution

In annual crops, most nematodes are often concentrated in the top 30 cm of soil, particularly the top 15 to 20 cm (Norton, 1978). Therefore, a sampling depth of 20 to 30 cm is probably adequate for most situations (McSorley, 1987a). Exceptions occur with shallow or deep-rooted crops, since nematode distribution may follow the root distribution of the host (Barker & Nusbaum, 1971; Ferris & McKenry, 1974). Deeper samples (0.5–1.0 m or more) may be more useful with some tree crops and other deep-rooted perennials (Barker & Campbell, 1981; Barker & Nusbaum, 1971; Ferris & McKenry, 1974). Seasonal migration in response to temperature, moisture, or changing root patterns can affect the vertical distribution of nematodes (Barker & Campbell, 1981; Barker & Nusbaum, 1971; McSorley, 1987a). Differences in soil texture with depth can also affect nematode distribution (Brodie, 1976). Even within the same crop and soil type, some nematode genera, such as *Longidorus* and *Xiphinema*, may show a preference for the deeper zones in the top 30 cm of soil (Flegg, 1968).

Horizontal Distribution

The irregular horizontal distribution of nematodes probably presents the greatest obstacle to obtaining reliable estimates of population density. Over a distance of a few meters, population densities can differ by an order of magnitude or more, resulting in clusters of peaks and valleys in population densities, which have been mapped and illustrated by several authors (Barker & Nusbaum, 1971; Barker et al., 1985; Goodell & Ferris, 1980). Stratification along plant rows is an important influence on horizontal distribution. Nematode distribution also may be correlated with a wide range of other biological factors and soil properties (Barker & Nusbaum, 1971; Francl, 1993), resulting in an irregular distribution which also is evident in fallow fields (McSorley & Parrado, 1982) and in crops without definite rows, such as alfalfa (*Medicago sativa*) (Goodell & Ferris, 1980). The clumped or contagious horizontal spatial distribution of nematodes can often be described statistically by the negative binomial distribution (Anscombe, 1950; Barker & Campbell, 1981; Barker et al., 1985; Goodell & Ferris, 1980; McSorley, 1987a).

Sampling and Sample Structure

Because of the clumped horizontal distribution of nematodes, soil for nematode analysis is collected from a number of locations within a field, plot, or row (Barker & Campbell, 1981; Barker & Nusbaum, 1971; McSorley, 1987a). At each location within the site (Fig. 6–1), a core is collected to the desired depth and position relative to any plants. The "core" could be collected with a trowel or shovel rather than a cylindrical sampling tube; the number of locations sampled is more critical than the size and shape of individual cores (McSorley, 1987a). All of the cores are then bulked and mixed to form a single sample representative of the site.

It is important that the cores collected are representative of the entire area sampled (Fig. 6–1). Example patterns for collecting cores are available for tree crops as well as row crops and fallow fields (Barker, 1985; Barker & Campbell, 1981). Most diagrams of this type illustrate a systematic, rather than a random, pattern of core collection (Barker & Campbell, 1981; McSorley, 1987a). In practice, systematic sampling seems to give more reliable results than random sampling, since core collection from high and low density areas is more likely (Anscombe 1950). Changing the pattern and direction of sampling had relatively little effect on sample precision (Goodell & Ferris, 1981).

Sample size is the number of cores that comprise the sample. Many studies have used sample sizes of 10 to 30 cores to estimate population densities in fields up to 1.6 to 2.0 ha in size (Barker, 1985; Barker & Campbell, 1981; Barker & Nusbaum, 1971; McSorley, 1987a). For several nematode species and field sizes, sample precision decreased rapidly as sample size decreased below 10 cores, but precision was not affected much as sample size increased above 30 cores (McSorley & Parrado, 1982). Larger numbers of cores may be needed in some situations, such as for quarantine samples or in other cases when density is believed to be extremely low (Barker & Campbell, 1981; Southey, 1986).

The reliability (precision) of the sample improves as the size of the area sampled decreases (McSorley, 1987a; McSorley & Parrado, 1982). Very large fields can be divided into randomly selected units of 2 ha or less for sampling (Barker & Campbell, 1981). Sites of any size which have known heterogeneity due to cropping history or soil type should be divided into smaller, more homogeneous units to be sampled individually (Barker & Campbell, 1981; Goodell & Ferris, 1980).

Since nematode population densities fluctuate seasonally, collection of samples can be timed to coincide with peaks in density, such as late in the growing season of an annual crop (Barker & Nusbaum, 1971; Barker et al., 1985; McSorley, 1987a). For annual crops, nematologists collect a sample at or shortly

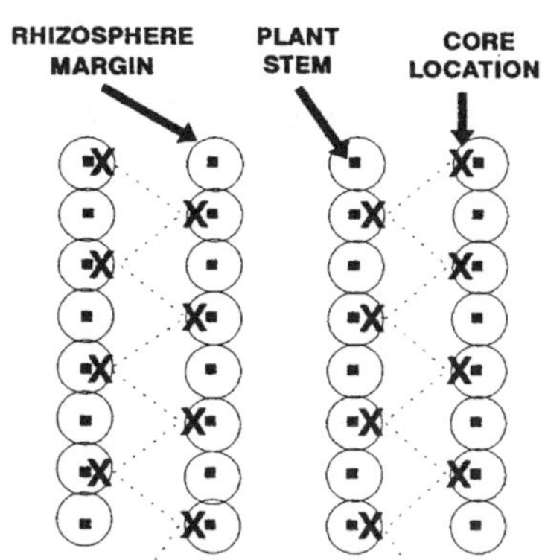

Fig. 6–1. Typical systematic pattern for sampling a row crop.

before planting to estimate initial population density (P_i), and a sample at or near harvest to estimate final population density (P_f) (Seinhorst, 1970). Some nematode stages may be more common at certain times of the year, and bioassays may be needed to estimate egg densities (Barker et al., 1985). To estimate population densities of nematode species which spend part of their life cycle in roots, it may be necessary to collect both root and soil samples. Although little attention has been given to the spatial distribution of nematodes in roots, a spatial pattern similar to that used for soil samples (e.g., Fig. 6–1) can be used (Barker, 1985). Sampling of nematode communities, cysts, perennials, tree crops, unusual substrates, and other specialized aspects of sampling are reviewed by various authors (Barker & Campbell,1981; Ferris & Noling, 1987; McSorley, 1987a; Southey, 1986).

Precision of Sampling

Often only one sample (of multiple cores) is collected from a plot or field. However, the collection of several samples (each of multiple cores) can greatly improve the precision of the estimated density measurement (Duncan et al., 1994; Goodell & Ferris, 1981; McSorley & Dickson, 1991). To evaluate sample precision, it is necessary to collect at least two samples from the site, so that estimates of the sample mean ($\bar{x}$) and standard deviation (s) can be obtained. If precision is defined in terms of the ratio between the standard error and the mean (C), then the level of precision associated with n samples can be estimated from the equation (Southwood, 1978)

$$n = [s/(\bar{x}\ C)]^2, \text{ or } C = s/(\bar{x}\ n^{1/2}) \qquad [1]$$

Alternatively, precision could be defined as the ratio between the half-width of the confidence interval and the mean (D) (Southwood, 1978), such that

$$n = [(t_{\propto[n-1]}s)/(\bar{x}\ D)]^2 \text{ or } D = (t_{\propto[n-1]}s)/(\bar{x}\ n^{1/2}) \qquad [2]$$

where $t_{\propto[n-1]}$ is the appropriate Student's t value for a confidence limit of $1 - \propto$ and $n - 1$ degrees of freedom.

In practice, coefficients of variation ($CV = 100s/\bar{x}$) of 50 or 100% are not uncommon in nematode sampling (McSorley & Dickson, 1991; Noe & Campbell, 1985). By substituting CV/100 for $s/\bar{x}$ in Eq. [1] or Eq. [2], estimates of precision associated with selected numbers of samples can be obtained (Table 6–1). The practical implications of using C or D as measures of precision are discussed elsewhere (McSorley et al., 1985; Southwood, 1978); for example, note that $D = 0.62$ corresponds to a 95% confidence interval of 38 to 162 around a mean of 100. The level of precision required depends on the purpose of the sampling program (Barker & Campbell, 1981), but large numbers of samples are needed to obtain precise estimates of population density (Table 6–1). Conversely, if relatively few samples are collected, precision is poor (Table 6–1); confidence intervals are especially broad, since the effect of $t_{\propto[n-1]}$ is particularly great with low sample numbers (Duncan et al., 1994). Possibly 80% confidence intervals are

Table 6–1. Precision of selected numbers of samples if the coefficient of variation (CV) is 50 or 100%.

Number of samples (n)	Ratio of standard error to mean (C)†		Ratio of half-width of 95% confidence interval to mean (D)‡	
	CV = 50%	CV = 100%	CV = 50%	CV =100%
2	0.35	0.71	4.49	8.99
3	0.29	0.58	1.24	2.48
4	0.25	0.50	0.80	1.59
5	0.22	0.45	0.62	1.24
9	0.17	0.33	0.39	0.77
16	0.12	0.25	0.27	0.53
25	0.10	0.20	0.21	0.41
100	0.05	0.10	0.10	0.20

† Calculated as $C = \mathrm{CV}/(100n^{1/2})$; note that Southwood (1978) uses the symbol E instead of C.
‡ Calculated as $D = (t_{0.05[n-1]}\ \mathrm{CV})/(100n^{1/2})$.

more practical than 95% confidence intervals in estimation of nematode densities (McSorley & Dickson, 1991). In any case, the levels of precision associated with typical nematode soil samples may often be quite low.

The above approach may be used to estimate precision of sampling in a single field or plot. However, if the variances (s^2) are plotted against the means for a series of similar fields or plots, they may be related by Taylor's power law (Taylor, 1961):

$$s^2 = a\ \bar{x}^b \text{ or } \log s^2 = \log a + b \log \bar{x} \qquad [3]$$

where a and b are constants. This relationship has been very useful in nematology in generalizing sampling plans for series of related sites (Duncan et al., 1994; Ferris, 1984b; McSorley et al., 1985; McSorley & Dickson, 1991). Ferris (1984b) discusses methodology for relating n and D if the parameters (a, b) of Taylor's power law are known.

EXTRACTION OF NEMATODES

Soil samples for nematode analysis are usually stored in plastic bags, and should be kept cool during transport (Hooper, 1986a). If samples must be stored prior to extraction, they should be kept cool (Hooper, 1986a). A storage temperature of 10 to 15°C is usually recommended (Barker & Campbell, 1981).

Extraction from Soil Samples

The many different methods for extracting nematodes from soil samples are summarized in recent laboratory manuals (Ayoub, 1980; Barker, 1985; Hooper, 1986a) and book chapters (Hooper, 1990; Hooper & Evans, 1993; McSorley, 1987a). Most soil extraction techniques can be divided into active methods, which depend on nematode behavior and motility, and passive methods, which do not (McSorley, 1987a; McSorley & Walter, 1991). The performance of active

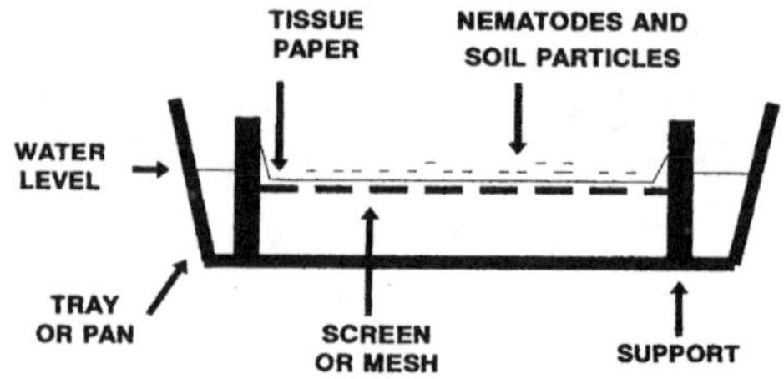

Fig. 6–2. Diagrammatic cross section of a Baermann tray for nematode extraction. Nematodes move through tissue paper and fall into water below.

methods depends on temperature (Robinson & Heald, 1989), which influences nematode activity.

Most active methods for nematode extraction from soil are based on the Baermann funnel (Baermann, 1917) or one of its many subsequent modifications (McSorley, 1987a; Robinson & Heald, 1989). The basic method involves incubation of a moist soil sample on a porous filter (paper or cloth), through which the nematodes migrate and fall into the clear water below (Fig. 6–2). Major improvements to the method have involved minimizing the thickness of the soil layer incubated or reducing the volume of soil incubated by sieving or elutriation (McSorley & Walter, 1991).

Passive methods for nematode extraction include mechanical separation of nematodes and soil by sieving and sedimentation (Cobb, 1918), elutriation (Cobb, 1924), centrifugation (Caveness & Jensen, 1955), or flotation (Byrd et al., 1966), or a combination of these techniques. Centrifugation is typically a two-step process in which the sample is centrifuged in water to precipitate nematodes and soil and remove organic debris in the supernatant. The pellet is then resuspended in a solution of sucrose or other solute of sufficient specific gravity to suspend nematodes and precipitate soil particles (Caveness & Jensen, 1955). A modification used by many laboratories is the preliminary sieving of the sample to reduce the volume of the soil that is actually centrifuged (Jenkins, 1964).

Numerous studies have compared the relative performance of two or more different extraction methods (Barker & Nusbaum, 1971; McSorley, 1987a). Reported estimates of extraction efficiency (nematodes recovered/nematodes present in sample) range from 0 to 100%, with most reported in the range 40 to 80% (McSorley, 1987a). Although extraction efficiencies can be useful for rating performance and comparing methods, they may vary with nematode life stages (Schneider & Ferris, 1987), and the methodology for their determination is subject to experimental error (Ferris, 1987; McSorley, 1987a; McSorley & Parrado, 1987). Therefore, attempts to "adjust" nematode count data through multiplication by estimated extraction efficiencies may introduce unknown and unwanted sources of error (McSorley & Parrado, 1987). Unfortunately, many different extraction methods and variations are in use today, and so it is difficult to compare quantitative data obtained from different laboratories. One approach is to divide a soil sample and submit a portion to two different laboratories for analysis. By submitting a series of divided samples, it is possible to develop a regression line relating counts obtained by the two laboratories (Fig. 6–3).

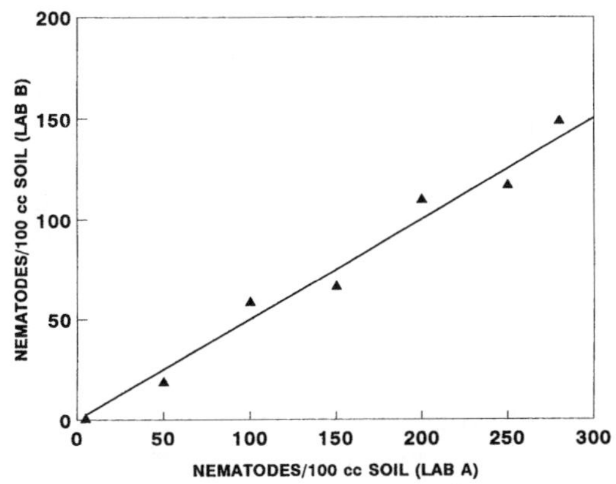

Fig. 6–3. Regression line relating nematode counts from soil samples divided between two hypothetical laboratories, A and B. Laboratory A consistently recovers more nematodes from a sample than Laboratory B.

This discussion pertains to extraction of vermiform stages of nematodes from soil. Bioassay or other methods may be required to estimate egg densities in soil or in cysts (Barker & Nusbaum, 1971; McSorley, 1987a).

Extraction from Plant Material

Many different methods and modifications are available for extracting nematodes from roots and other plant tissues (Ayoub, 1980; Barker, 1985; Hooper, 1986b; Hooper, 1990; Hooper & Evans, 1993; McSorley, 1987a). Migratory nematodes will emerge from roots incubated in a variety of chambers, from Baermann funnels to mist chambers (Ayoub, 1980). Immobile forms can sometimes be dislodged from the root surface by vigorous washing or freed by macerating the root tissue with a blender or with enzymatic digestion (Hooper, 1986b).

Cyst nematodes and galls or egg masses of root-knot nematodes can be examined and counted directly on roots. Often, the severity of galling or numbers of egg masses are rated on a numeric scale, such as the 0 to 5 scheme of Taylor and Sasser (1978) or the 0 to 10 scheme of Zeck (1971). Eggs of *Meloidogyne* spp. can be released from egg masses by briefly soaking infected root systems in a dilute solution of sodium hypochlorite (Hussey & Barker, 1973).

Transformations of Nematode Count Data

Since most nematode field data follows the negative binomial distribution, it may be necessary to log-transform nematode-count data to meet the assumptions of analysis of variance (Proctor & Marks, 1975). Square-root transformation is usually not appropriate, because nematode data rarely follow the Poison distribution (McSorley, 1987a). Root-knot rating scale data are usually not transformed.

DENSITY-INDEPENDENT POPULATION GROWTH

Population densities of many plant-parasitic nematodes are capable of increasing rapidly from low initial levels because life cycles are comparatively short and reproductive rates are high. Fecundity varies with species (Norton, 1978), with maximum numbers of eggs produced by root-knot and cyst nematodes. It is reported that over 1000 eggs may be produced by a single root-knot nematode (Taylor & Sasser, 1978).

Exponential Growth

The root system of most plants probably represents an almost limitless food supply to the first individual nematodes colonizing it. Under these conditions, classical models of exponential population growth in an unlimited environment (Andrewartha & Birch, 1954) may apply such that

$$P_{\mathrm{t}} = P_{\mathrm{i}}\exp(rt) \qquad [4]$$

where P_t = population density at time t, P_i = the initial population density, and r = the intrinsic rate of increase of the population (a constant). This pattern of exponential population growth (Fig. 6–4A) has been observed over a period of several months in the increase of *Ditylenchus destructor* in peanut (*Arachis hypogaea*) pegs and roots (Basson et al., 1991) and in the increase in densities of *Meloidogyne arenaria* juveniles in soil during a peanut crop (Rodriguez-Kabana et al., 1986).

If t represents the interval from planting to harvest of a crop, then P_{t} is usually represented by nematologists as P_{f}, the final population density on the crop (Seinhorst, 1970). The P_{f} is usually determined from a sample collected at harvest, and P_{i} from a sample collected at or shortly before planting. From Eq. [4], it is evident that the relationship between $\log(P_{\mathrm{f}})$ and $\log(P_{\mathrm{i}})$ is linear (Fig. 6–4B). Such linear relationships may occasionally be observed in the field at low population densities (McSorley & Dickson, 1989).

DENSITY-DEPENDENT POPULATION GROWTH

Although nematode populations may grow exponentially for short time intervals and at low densities, exponential growth cannot be sustained indefinitely. Eventually as population density increases, the food supply or some other aspect of the environment will limit the rate of population growth.

Logistic Growth

The logistic equation has been used to describe density-dependent population growth in many animal species (Andrewartha & Birch, 1954; Poole, 1974), including nematodes (Seinhorst, 1966). According to the logistic model, the rate of population growth (dP/dt) is related to population density (P)

$$dP/dt = rP[(E - P)/E] \quad [5]$$

where r is a constant (the intrinsic rate of increase for the population), and E is called equilibrium density by nematologists (Seinhorst, 1966, 1970), but is referred to as carrying capacity (with symbol K) by other authors (Andrewartha & Birch, 1954; Poole, 1974). The logistic model predicts rapid population growth at low densities, but a leveling-off at the equilibrium density over time (Fig. 6–5A). The solution of the differential equation (Eq. [5]) at time t is

$$P_t = E/[1 + \exp(c - rt)] \quad [6]$$

where c = a constant and $P < E$ (Seinhorst, 1966).

Although the logistic equation depends on a number of assumptions (Andrewartha & Birch, 1954; Poole, 1974), it appears to be descriptive of nematode population growth in a variety of situations (Seinhorst, 1970). For example,

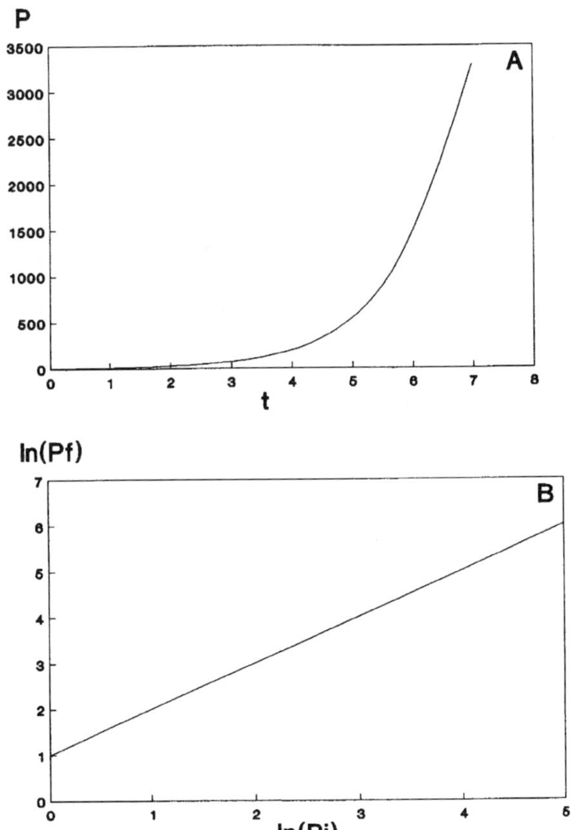

Fig. 6–4. Exponential population growth. A) Increase in population density (P_t) over time (t). B) Relationship between final population density (P_f) and initial population density (P_i) for a constant time interval.

the buildup of *M. arenaria* through several years of continuous peanut was described by this model (Rodriguez-Kabana & Ivey, 1986). In practice, population densities may fluctuate around the equilibrium density to some extent, and can drop below E if root growth declines due to nematode damage (Seinhorst, 1970).

Relationships Between the Initial and Final Population Densities

Seinhorst (1966) and Jones et al. (1967) have used Eq. [6] to derive an expression relating P_f to P_i after a specified time interval (usually the length of a crop)

$$P_f = aEP_i/[(a-1)P_i + E] \quad [7]$$

where $a = \exp(rt)$ is called the maximum rate of multiplication (Seinhorst, 1966). Under a specific set of conditions (nematode species, host cultivar, season, etc.),

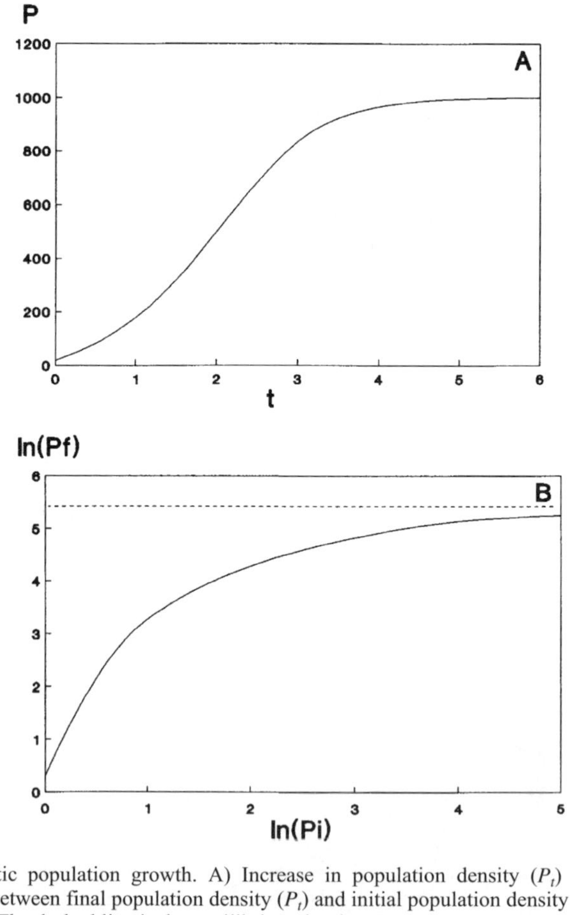

Fig. 6–5. Logistic population growth. A) Increase in population density (P_t) over time (t). B) Relationship between final population density (P_t) and initial population density (P_i) for a constant time interval. The dashed line is the equilibrium density, E.

a and E may be constant, and a curve relating P_f to P_i over a range of densities can be obtained (Fig. 6–5B). The ratio P_f/P_i represents the multiplication rate of the nematode at a given density (Seinhorst, 1966), but note that P_f/P_i may change with density (Fig. 6–5B). If a and E are known, then P_f can be estimated from Eq. [7] if P_i is measured. Relationships such as Eq. [7], in which measurement of P_i is critical and drives the model, are termed critical point models (Ferris, 1981).

Critical point models relating P_f and P_i have been applied to a number of nematode-host plant combinations (Duncan & McSorley, 1987; Ferris, 1985a; Ferris & Noling, 1987; McSorley & Gallaher, 1993; McSorley & Phillips, 1993; Seinhorst, 1966, 1970). The shape of the relationship between P_f and P_i (Fig. 6–5B) is influenced by a wide range of factors, including host condition and damage, condition of the nematode population, interspecific competition, temperature and other environmental factors, and agricultural practices (Ferris & Noling, 1987; McSorley & Phillips, 1993; Nusbaum & Barker, 1971; Seinhorst, 1970). Nevertheless, such models can be quite robust; for each of several nematode species on corn (*Zea mays*) or sorghum (*Sorghum bicolor*) in north Florida, relationships between P_f and P_i showed little variation with site, season, crop cultivar, or planting date (McSorley & Gallaher, 1993).

In addition to the logistic, several other types of models have been useful for expressing relationships between P_f and P_i, including quadratic (McSorley & Gallaher, 1993), increasing exponential (Noe et al., 1991), and Gompertz (Todd & Pearson, 1988) equations. Seinhorst (1967a) proposed an alternative model based on intraspecific competition, for use with sedentary nematodes having distinct generations. Seinhorst's (1967a) competition model served as the basis for models of population dynamics of *Globodera rostochiensis*, the potato cyst nematode (McSorley & Phillips, 1993).

Deriving Relationships Between the Final and Initial Population Densities

Relationships between P_f and P_i can be obtained from data collected in greenhouse, microplot, or field experiments (McSorley & Phillips, 1993). Density, age structure, and quality of nematode populations measured in greenhouse pots or microplots may differ from those measured in field plots (Barker & Noe, 1987; Ferris, 1984a). In the field, it is critical to have a range of P_i which is wide enough to define the entire shape of the curve up to and exceeding $P_i = E$ (Fig. 6–5B). If the range of P_i is limited to relatively low population densities, only a linear approximation of a portion of the relationship will be obtained (McSorley & Dickson, 1989). A greater range in P_i can sometimes be obtained by prior manipulation of population densities within the field to increase spatial heterogeneity (McSorley & Phillips, 1993), or by using a series of similar fields with the desired range of P_i (Ferris, 1985a).

If the range of P_i values is appropriate, a graph similar to Fig. 6–5B can be obtained by plotting ln (P_f) against ln (P_i) for each field or small plot (provided $P_i > 0$). However, determining the equation relating P_f and P_i (Eq. [7]) can be problematic if a and E are unknown (McSorley & Gallaher, 1993). Often, the regression of $\ln(P_f/P_i)$ on $\ln(P_i)$ is linear and of the form $\ln(P_f/P_i) = \ln(c) - b$

$\ln(P_i)$ (Ferris, 1985a; Ferris & Noling, 1987). Although the occurrence of P_i on both sides of the regression equation is a concern (McSorley & Gallaher, 1993), the method can be used to approximate E from the regression coefficients b and c, as $\ln(E) = [\ln(c)]/b$ (Ferris & Noling, 1987), or E could be estimated by visual inspection of the graphed data. The maximum rate of reproduction (a) can be estimated as the regression prediction for the lowest P_i in the data set (Ferris, 1985a), from the asymptote of the relationship of P_f and P_i at low density (Seinhorst, 1966), or by inspection of the data set (McSorley & Gallaher, 1993). These steps are presented in detail elsewhere (Ferris, 1985a; Ferris & Noling, 1987; McSorley & Gallaher, 1993; Seinhorst, 1966).

POPULATION DECLINE

Several authors have developed simulation models describing nematode population decline over the winter (Jeger & Starr, 1985) or during a nonhost crop (Jones et al., 1967). Often, population decline following harvest was described by simple critical point models (Ferris, 1985a; McSorley & Phillips, 1993). Exponential decline functions of the form $P = a \exp(-bt)$ or the equivalent form $\ln(P) = \ln(a) - bt$ have been used (Jeger & Starr, 1985; Windham & Barker, 1988) to relate population size (P) to number of days following harvest (t), where a and b are constant for a particular situation. The exponential decline model predicts a rapid population decline shortly after harvest, but a decreasing rate of decline as time goes on (Fig. 6–6). This model or others with similar form (MacGuidwin & Forge, 1991; Rodriguez-Kabana et al., 1986) have been descriptive of population declines over a wide range of locations and winter conditions, including plots with snow cover in Wisconsin (MacGuidwin & Forge, 1991), fallow soil above freezing in North Carolina (Windham & Barker, 1988), and a winter cover crop in Alabama (Rodriguez-Kabana et al., 1986).

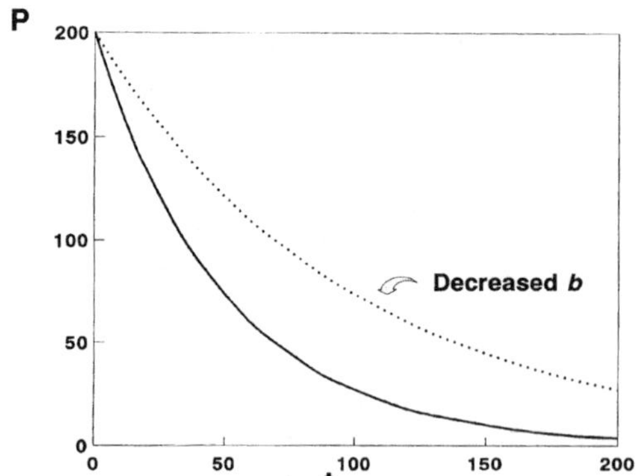

Fig. 6–6. Exponential decline of population density (P) over time (t). Solid line is a graph of the equation $P = a \exp(-bt)$, with $a = 200$ and $b = 0.02$; dotted line shows a similar equation, but with b decreased to $b = 0.01$.

Although the general shape of the exponential decline function (Fig. 6–6) applies to many situations, an equation with a particular set of *a* and *b* values will apply only to a very specific situation. Variation in environmental conditions, such as increased snow cover that could enhance nematode survival (MacGuidwin & Forge, 1991), could presumably decrease *b* and shift the decline curve upward (Fig. 6–6). If the population density following harvest ($t = 0$) is greater in one site than another, then *a* will of course be greater for the first site. The exponential model is thus sensitive to initial density, and recent data (MacGuidwin & Forge, 1991) support the hypothesis that winter survival is density-dependent. Survival of *M. incognita* over the winter in California showed strong density dependence, and methodology was presented for relating over-winter survival to population density in the fall (Ferris, 1985a). Nevertheless, there is some disagreement about whether winter survival is density dependent (Duncan & McSorley, 1987), and it is possible that alternative models that are not density dependent may apply in some situations. Finally, winter is not always synonymous with population decline, especially where a winter cover crop serves as a favorable host (McSorley & Duncan, 1995; McSorley & Phillips, 1993).

FACTORS AFFECTING POPULATION DYNAMICS

Numerous physical and biological factors directly affect nematode population dynamics. However, any factors that affect the health and cycle of the host crop will also have important indirect effects on the nematode parasites as well.

Host Crop

The host can have a drastic effect on nematode population dynamics (Seinhorst, 1970), and host-plant resistance is a very important method for managing plant-parasitic nematodes (see Young, 1998, Chapter 10; and Roberts et al., 1998, Chapter 11). Many nematologists use the term "resistance" to describe the effect of the host on nematode reproduction, especially for species that do not cause distinct host symptoms (Cook & Evans, 1987). Tolerance describes the amount of plant damage in response to a nematode population and is discussed in Chapter 13 (see Duncan & Noling, 1998). Nonhosts and immune cultivars are completely resistant and allow no nematode reproduction (Cook & Evans, 1987). However, resistance is usually not an "all or nothing" phenomenon. Several recent examples illustrate that hosts with some degree of resistance may allow low or even moderate levels of reproduction (Gallaher & McSorley, 1993; Pederson & Windham, 1992; Windham & Williams, 1987).

The multiplication rate P_f/P_i can give some indication of the relative efficiency of different hosts, provided one recognizes that P_f (Fig. 6–5B) and P_f/P_i vary with initial population density, P_i (Seinhorst, 1970). The pattern of the relationship between P_f and P_i for the same nematode species may often vary for different hosts or cultivars (Fig. 6–7), and a very wide range of patterns may be possible (Seinhorst, 1967b). In this context, terms such as "good host" or "weak host" may not be particularly meaningful; the shape of the relationship between

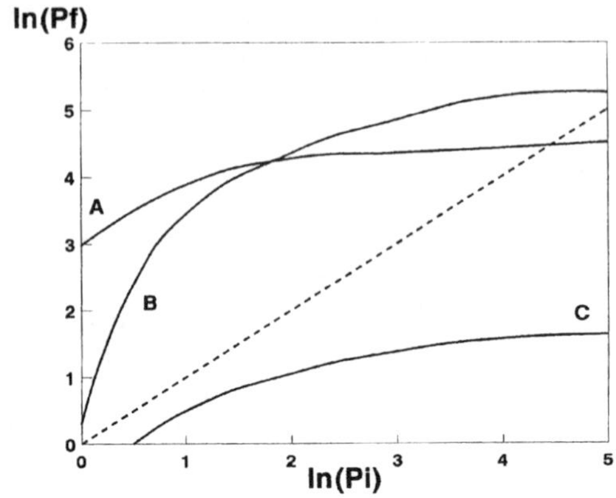

Fig. 6–7. Relationship between P_f and P_i for a nematode population on a series of host plant cultivars, A, B, and C. It is evident that cultivar C is more resistant than cultivars A or B, but it is difficult to tell whether A is more or less resistant than B (depends on P_i). The dashed line is the line $P_f = P_i$, where population size is just maintained.

P_f and P_i over the range of P_i commonly encountered is much more descriptive and quantitative (McSorley & Gallaher, 1993; Seinhorst, 1967b, 1970). Knowledge of the relationship between P_f and P_i on any crops or cultivars used is critical in anticipating the effect of crop rotation in nematode management (see Chapter 8, Noe, 1998).

Host Races and Population Genetics

Many nematologists, especially in the USA, use the term "race" to distinguish infraspecific forms (populations or even groups of individuals within the same population) that show physiological variation, such as differences in host preference, although use of the term is not technically correct in all instances (Dropkin, 1988; Triantaphyllou, 1987). "Races" of root-knot nematodes (*Meloidogyne* spp.) are distinguished by their ability to reproduce on a series of differential host plants (Taylor & Sasser, 1978), "races" of the soybean cyst nematode (*Heterodera glycines*) are distinguished by reproduction on four soybean (*Glycine max*) genotypes (Riggs & Schmitt, 1988), and pathotypes of potato cyst nematodes (*Globodera* spp.) are recognized by reproduction on a series of *Solanum* clones (Kort et al., 1977). "Races" exist in a number of other nematode species as well, especially *D. dipsaci* and *H. avenae* (Dropkin, 1988; Sidhu & Webster, 1981).

Although the race concept is often useful in distinguishing differences in behavior or population dynamics among relatively homogeneous populations, there are important limitations. The classification of a "+" or "–" response on a differential host may depend on slight differences in reproduction (Dropkin, 1988; Schmitt & Shannon, 1992). The dependence of P_f/P_i on initial inoculation density was discussed in the previous section. To minimize experimental error and variation in "race" tests, strict protocols must be followed (Riggs & Schmitt, 1991; Schmitt & Shannon, 1992). Expanding the number of differential hosts can result in detection of new "races" (Riggs et al., 1981). Several different "races" of soybean cyst nematode may occur in the same region and even in the same field, and the "race" structure of a population may shift within a few years, depending on the soybean cultivar grown (Anand et al., 1994). Although such shifts in "race" are probably of more concern with soybean cyst nematodes than with root-knot nematodes, differences in reproduction have also been observed within species and even "races" of root-knot nematodes, such as *M. arenaria* (Carpenter & Lewis, 1991). Finally, while the historical development of the race concept in nematology has often focused on nematode reproduction, cytological (Triantaphyllou, 1987) and biochemical methods (Dropkin, 1988) are becoming increasingly important for distinguishing "races" and other infraspecific forms.

From the above examples it is evident that the dynamics and behavior of a population consisting of mixed genotypes could differ from those of a more homogeneous "race." Future studies of nematode population genetics (Caswell & Roberts, 1987; Sidhu & Webster, 1981) will improve our understanding of how the genetic variability, structure, and shifts within a population will affect the dynamics of that population. At present, it is important that variability among, or even within, local populations be recognized (by "race" designations or perhaps eventually even by genotypes) before consistent predictions about their dynamics can be made.

Physical and Environmental Factors

Numerous physical factors directly affect nematode survival and rates of growth and reproduction, and therefore influence nematode population dynamics (Duncan & McSorley, 1987; Norton, 1978; Nusbaum & Barker, 1971; Wallace, 1973). Often it is difficult to separate the effects of individual physical factors on nematodes. Soil moisture and availability, for instance, usually depend on soil texture and structure (Heatherly & Young, 1991; Sultan & Ferris, 1991).

Temperature

Nematode growth and development are directly affected by temperature (Noe, 1991), and temperature or degree days are nearly always incorporated into simulation models of nematode population growth (Duncan & McSorley, 1987; Ferris, 1976; McSorley & Phillips, 1993). Temperature also may influence dormancy or diapause (Huang & Pereira, 1994), and subsequently affect the timing of events in the life cycle. High temperatures also may affect host plant resistance (Dropkin, 1969), resulting in increased reproduction on hosts resistant at lower temperatures.

Moisture and Humidity

In tropical climates with relatively little seasonal fluctuation in temperature, seasonal patterns in rainfall may have a major influence on fluctuations of nematode population densities (Caveness & Badra, 1980). Soil moisture is critical for nematode mobility and activity, but excessive levels can limit survival (Johnson et al., 1993; Sultan & Ferris, 1991; Wallace, 1973). The effect of moisture on population growth is so important that Jones (1975) suggested accumulating degree days in nematode population models only during times when soil moisture is adequate. Examples of effects of relative humidity on nematodes in soil are rare, but optimum humidity is important for foliar nematodes (Griffin, 1987).

Soil Texture and Edaphic Factors

Many nematode species show preference for specific soil types and textures (Norton, 1978; Wallace, 1973). For instance, it is fairly well established that root-knot nematode reproduction and damage are greater in sandy soils than in clay soils (Barker & Weeks, 1991). Soil texture also affects some nematode extraction methods, influencing population estimates (McSorley, 1987a).

Many other edaphic factors, which are influenced by soil texture, also may affect nematodes directly. In many cases, nematode population growth may be suppressed at extreme values of a given factor, but not at the normal ranges observed in the factor during crop growth. For instance, reproduction of *Pratylenchus penetrans* on alfalfa was greater at soil pH levels of 5.2 or 6.4 than at pH = 4.4 or 7.3 (Willis, 1972). Although many nematode species are affected by extremes in soil pH (Norton, 1978), there is relatively little evidence that nematodes are affected much within the normal range of pH observed for most agricultural crops (McSorley & Gallaher, 1994). Soil organic matter (Rodriguez-Kabana, 1986), salinity (Mashela et al., 1992), aeration (Van Gundy 1985), cation exchange capacity (Norton et al., 1971), bulk density (Noe & Barker, 1985), and several macro- and micronutrients (Francl, 1993; Noe & Barker, 1985) may affect distribution and dynamics of some plant-parasitic nematodes.

Numerous other environmental, physical, chemical, and edaphic factors can influence nematode dynamics and survival. The effects of these factors are discussed in more detail in a number of reviews (Ferris, 1976; Norton, 1978, 1979; Van Gundy, 1985; Wallace, 1971, 1973; Yeates, 1981).

Competition

Numerous examples in nematology have revealed that the presence of one species may interfere with the population growth of another species (Eisenback, 1985; Norton, 1978). In general, competition tends to be more severe between species with similar feeding habits (Eisenback, 1985). Nematode density, host suitability, and environmental factors (especially temperature) are probably the most important factors affecting the nature of nematode–nematode interactions.

The density dependence of nematode competition is recognized (Eisenback, 1985), but probably requires much more study. To fully evaluate effects of ranges in densities of both competing species on nematode population growth would require a large factorial experiment, similar to that used by Duncan and Ferris (1982) to determine effects of two root-knot nematode species on host yield. The size of such experiments (e.g., 8 levels of each of 2 species × 5 replications = 320 plots) may be the reason that there have been relatively few studies of this type. Research opportunities exist for the application of classical models of interspecific competition (Poole, 1974) to nematode population growth.

Predation and Parasitism

Many different predators and parasites have potential for some degree of biological control of plant-parasitic nematodes (Kerry, 1987; Stirling, 1991). Nematode predators and parasites are widespread in agricultural soils, and numerous examples of natural control are recognized (Stirling, 1991), particularly by fungi and bacteria (Kerry, 1990). One of the best-known examples is the decline of *H. avenae* under continuous cereals over a 14-yr period (Gair et al., 1969), attributed to the parasitism of nematode females and eggs by fungi (Kerry, 1987). An induction period of about 5 yr was required for fungal antagonists to build up and nematode population densities to decline (Kerry, 1987). Estimating and quantifying the levels of natural control present in soils can be a difficult task (Kerry, 1987), but several authors offer suggestions for enhancing and utilizing levels of naturally occurring biological control agents (Kerry, 1987; Kerry, 1990; Stirling, 1991).

In numerous cases, biological control agents have been mass-produced and released into greenhouse pots, microplots, or field sites (Kerry, 1990; Stirling, 1991). Efficacy was variable, and studies of the dynamics of the host-parasite relationship are few. The fungus *Hirsutella rhossiliensis* and the bacterium *Pasteuria penetrans* became established in microplots with nematode hosts within 2 yr (Oostendorp et al., 1991; Tedford et al., 1993). Nematode numbers were not affected by *H. rhossiliensis* (Tedford et al., 1993), however declines in nematode population densities and yield responses (in the third season) were reported in the experiment with *P. penetrans* (Oostendorp et al., 1991).

Agricultural Practices

Many agricultural practices are implemented with the intent to lower nematode population densities and improve plant growth. Crop rotation (see Noe, 1998, Chapter 8), nematicides (Sipes & Schmitt, 1998, Chapter 9), and resistant cultivars (Young, 1998, Chapter 10; Roberts et al., 1998, Chapter 11) are the most frequently used, but soil solarization (Heald & Robinson, 1987), manipulation of planting dates (Roberts et al., 1981), destruction of residual crop roots (Christie, 1959), and numerous other cultural practices may be applicable in some situations (McSorley & Duncan, 1995). Occasionally, practices such as use of fertilizers (Rodriguez-Kabana, 1986) may have some impact on nematode densities, although such effects are often inconsistent and unintentional. Changing from conventional tillage to a minimum or reduced tillage system may have variable effects on nematode population densities, depending on the nematode species and method used (McSorley & Duncan, 1995; Minton, 1986). Because the effects of tillage practices on plant-parasitic nematodes have often been inconsistent or nonsignificant, decisions to adopt such practices should be based on agronomic benefits rather than on expectations of nematode management. Effects of management practices against nematode pests of specific crops are discussed in subsequent chapters.

PREDICTION AND VARIABILITY

The many factors described in the previous section can individually or together affect population dynamics, and therefore make prediction difficult. Predictions can be made more realistic if the effects of external factors on nematode population growth can be described, or if statistical variability can be incorporated into predictions.

Modeling Nematode Population Dynamics

Simulation models can be used to describe the influence of external physical and biological factors on population growth over time within a plant-nematode system and simulation models of varying degrees of complexity have been developed for several nematode and host combinations (Duncan & McSorley, 1987; Ferris, 1981; McSorley & Phillips, 1993). Ferris (1976) outlines the many different data and assumptions needed to construct such a model. Unfortunately, the limited availability of specific data and equations describing the many interactions of external factors with nematode populations has probably limited the development and use of simulation models. Nevertheless, they remain an important means for conceptualizing the dynamics of plant-nematode systems (Ferris, 1981).

Critical point models predicting P_f from P_i are more widely used in practice in nematology (McSorley & Phillips, 1993), but they are more specific to a set of given conditions and therefore less adaptable than simulation models (Ferris, 1981). However, they are much easier to develop than simulation models. Even simulation models predicting potato-cyst nematode dynamics and damage over a period of years often incorporate critical point models for individual growing seasons (Jones et al., 1967; McSorley & Phillips, 1993). In some cases, critical point models can be generalized across similar sets of environmental and agronomic conditions (McSorley & Gallaher, 1993).

Stochastic Elements and Forecasting

Even if predictions can be improved by the use of simulation models or reliable critical point models, error due to statistical variation in measured and even derived parameters will remain. Some of the most important sources of error are imprecision of sampling in estimating P_i (Ferris, 1985b; McSorley, 1992), seasonal variation in weather data (Ferris & Noling, 1987; McSorley & Phillips, 1993), and variation around derived regression equations (McSorley, 1992). There have been relatively few attempts to incorporate variability from these sources into predictive models (Duncan & McSorley, 1987; McSorley & Phillips, 1993). In one instance (Perry, 1983), model predictions of *Globodera* spp. population densities were highly sensitive to spatial heterogeneity in P_i.

The effects of statistical error on population dynamics models and predictions can be estimated in several ways. Simulation models can be run many times to obtain an average result or frequency distribution of results (Goodell & Ferris, 1981; McSorley, 1992). Seasonal variation can be included by using average

Table 6–2. Distribution among logarithmic density classes of simulated P_i and P_f following a rye crop, if mean P_i = 15 nematodes/100 cm^3 soil with CV = 38%. Data are number of times out of 1000 simulations that values fell into each density class (McSorley, 1992).

Density class (nematodes/100 cm^3)	Number of simulated values in density class		
	P_i	P_f Single equation†	P_f Random equation‡
0–3	16	32	51
3–15	468	508	350
15–63	516	457	360
>63	0	3	239

† P_f predicted from a single regression equation.
‡ P_f predicted from one of two regression equations selected at random with equal probability.

weather data over several seasons (Ferris & Noling, 1987) or by randomly selecting weather data or equations from a series of different seasons (McSorley, 1992). The recognition that an individual nematode count is not absolute but the representative of a density class is a fundamental concept in visualizing patterns in variable data (Ferris, 1984a). Probability ranges in P_i values can be carried through models, resulting in ranges in predicted estimates, such as yield or P_f (Ferris, 1984a; McSorley & Phillips, 1993).

Estimating the effects of several interacting sources of error is particularly complex. Sampling error and confidence intervals around equations interact to produce probability ranges for predicted values (Ferris, 1984b). The combined effects of sampling error, seasonal variation, and variation around regression lines on predicted values of P_f were estimated through simulation from appropriate frequency distributions (McSorley, 1992). When P_f values were predicted from P_i using linear critical point models, relatively broad frequency distributions of P_f resulted, especially when CV values associated with P_i were high. For a typical scenario, estimates of P_i were distributed across several logarithmic density classes as a result of sampling error (Table 6–2). The distribution of P_f estimates predicted from this range of P_i spread more into the extreme density classes if P_f was predicted from a randomly selected equation (to simulate seasonal variation) than if P_f was always predicted from a single regression equation.

This example illustrates the feasibility of providing a statistically based forecast of expected population growth or decline. For instance, one might predict that P_f would be in the range 3 to 63/100 cm^3 96.5% of the time if seasonal variation was not considered, but only 71% of the time with seasonal variation (Table 6–2). Although a range of 3 to 63 might seem broad, this probably represents a more realistic situation than the prediction of some discrete P_f value, say $P_f = 32$.

OUTLOOK

Variability will probably always pose a serious problem for studies of nematode population dynamics. Despite this variability, differences in population densities among treatments can be measured in the field, as observed in hundreds of experiments. Realistic expectations are critical; it may be possible for example, to distinguish a population density level of 5 nematodes/100 cm^3 from 95/100 cm^3, but not 55/100 cm^3 from 60/100 cm^3. The resolution with which such distinctions can be made is unknown, but should be an important subject for future research. It is clear that much of the imprecision in nematode population density estimates is the result of sampling error (Barker & Campbell, 1981; Ferris, 1985b; McSorley, 1987a). Precision can be improved by increasing the intensity of sampling (more cores, smaller area sampled, etc.) (McSorley, 1987a) or by increasing number of replications for field tests (McSorley, 1987b). Repeated or replicated sampling of the same site (multiple samples of multiple cores) can greatly improve precision (Goodell & Ferris, 1981; McSorley, 1987a; McSorley & Dickson, 1991).

A list of research needed on nematode population dynamics would be very long. The dynamics of many different nematode species (and even "races" or genotypes) on hundreds of crops and thousands of cultivars must be determined. Each of these combinations could be affected by a range of physical and biological factors. The number of external factors affecting nematode dynamics is great, but fortunately many of these effects may be minor or could be generalized. Our poor resolution of nematode population density data will allow us to observe only the most drastic effects anyway. It is fortuitous that drastic effects have been particularly important in traditional nematode management (see Duncan & Noling, 1998, Chapter 13). Much additional research on nematode population dynamics is needed, but as progress in understanding the dynamics of more and more nematode and crop cultivar combinations increases, so will our ability to improve nematode management through effective use of novel crop rotation systems and other sustainable agriculture practices (see Noe, 1998, Chapter 8), nematicides (see Sipes & Schmitt, 1998, Chapter 9), host plant resistance (see Young, 1998, Chapter 10; Roberts et al., 1998, Chapter 11) including biotechnological approaches (see Opperman & Conkling, 1998, Chapter 12), biological control agents (Stirling, 1991), cultural practices (McSorley & Duncan, 1995), and other tactics (see Duncan & Noling, 1998, Chapter 13) for nematode management.

REFERENCES

Anand, S.C., S.B. Sharma, A.P. Rao-Arelli, and J.A. Wrather. 1994. Variation in parasitic potential of *Heterodera glycines* populations. Crop Sci. 34:1452–1454.

Andrewartha, H.G., and L.C. Birch. 1954. The distribution and abundance of animals. Univ. Chicago Press, Chicago, IL.

Anscombe, F.J. 1950. Soil sampling for potato root eelworm cysts. Ann. Appl. Biol. 37:286–295.

Ayoub, S.M. 1980. Plant nematology, an agricultural training aid. NemaAid Publ., Sacramento, CA.

Baermann, G. 1917. Eine einfache Methode zur Auffindung von *Ancylostomum* (Nematoden) Larven in Erdproben. Geneesk. Tijdschr. Ned. Ind. 57:131–137.

Barker, K.R. 1985. Soil sampling methods and procedures for field diagnosis. p. 11–20. *In* B. M. Zuckerman et al. (ed.) Plant nematology laboratory manual. Univ. Massachusetts Agric. Exp. Stn., Amherst, MA.

Barker, K.R., and C.L. Campbell. 1981. Sampling nematode populations. p. 451–474. *In* B.M. Zuckerman and R.A. Rohde (ed.) Plant parasitic nematodes. Vol. 3. Acad. Press, New York.

Barker, K.R., and J.P. Noe. 1987. Establishing and using threshold population levels. p. 75–81. *In* J. A. Veech and D. W. Dickson (ed.) Vistas on nematology. Soc. Nematol., Hyattsville, MD.

Barker, K.R., and C.J. Nusbaum. 1971. Diagnostic and advisory programs. p. 281–301. *In* B.M. Zuckerman et al. (ed.) Plant parasitic nematodes. Vol. 1. Acad. Press, New York.

Barker, K.R., D.P. Schmitt, and J.P. Noe. 1985. Role of sampling for crop-loss assessment and nematode management. Agric. Ecosyst. Environ. 12:355–369.

Barker, K.R., and W.W. Weeks. 1991. Relationships between soil and levels of *Meloidogyne incognita* and tobacco yield and quality. J. Nematol. 23:82–90.

Basson, S., D.G.M.A. DeWaele, and A.J. Meyer. 1991. Population dynamics of *Ditylenchus destructor* on peanut. J. Nematol. 23:485–490.

Brodie, B.B. 1976. Vertical distribution of three nematode species in relation to certain soil properties. J. Nematol. 8:243–247.

Byrd, D.W., Jr., C.J. Nusbaum, and K.R. Barker. 1966. A rapid flotation-sieving technique for extracting nematodes from soil. Plant Dis. Rep. 50:954–957.

Carpenter, A.S., and S.A. Lewis. 1991. Aggressiveness and reproduction of four *Meloidogyne arenaria* populations on soybean. J. Nematol. 23:232–238.

Caswell, E.P., and P.A. Roberts. 1987. Nematode population genetics. p. 390–397. *In* J.A.Veech and D.W. Dickson (ed.) Vistas on nematology. Soc. Nematol., Hyattsville, MD.

Caveness, F.E., and T. Badra. 1980. Control of *Helicotylenchus multicinctus* and *Meloidogyne javanica* in established plantain and nematode survival as influenced by rainfall. Nematropica 10:10–14.

Caveness, F.E., and H.J. Jensen. 1955. Modification of the centrifugal-flotation technique for the isolation and concentration of nematodes and their eggs from soil and plant tissue. Proc. Helminth. Soc. Wash. 22:87–89.

Christie, J.R. 1959. Plant nematodes. Their bionomics and control. Agric. Exp. Stn., Univ. Florida, Gainesville, FL.

Cobb, N.A. 1918. Estimating the nema population of the soil. USDA Agric. Tech. Circ. 1. USDA, Washington, DC.

Cobb, N.A. 1924. Removing nemas from soil by flotation. J. Parasit. 11:103–105.

Cook, R., and K. Evans. 1987. Resistance and tolerance. p. 179–231. *In* R.H. Brown and B.R. Kerry (ed.) Principles and practice of nematode control in crops. Acad. Press, Orlando, FL.

Dropkin, V.H. 1969. The necrotic reaction of tomatoes and other hosts resistant to *Meloidogyne*: Reversal by temperature. Phytopathology 59:1632–1637.

Dropkin, V.H. 1988. The concept of race in phytonematology. Annu. Rev. Phytopathol. 26:145–161.

Duncan, L.W., M.M. El-Morshedy, and R. McSorley. 1994. Sampling citrus fibrous roots and *Tylenchulus semipenetrans*. J. Nematol. 26:442–451.

Duncan, L., and H. Ferris. 1982. Interactions between phytophagous nematodes. p. 29–51. *In* D.W. Freckman (ed.) Nematodes in soil ecosystems. Univ. Texas Press, Austin, TX.

Duncan, L.W., and R. McSorley. 1987. Modeling nematode populations. p. 377–389. *In* J.A. Veech and D.W. Dickson (ed.) Vistas on nematology. Soc. Nematol., Hyattsville, MD.

Duncan, L.W., and J.W. Noling. 1998. Agricultural sustainability and nematode integrated pest management. p. 251–287. *In* K.R. Barker et al. (ed.) Plant and nematode interactions. Agron. Monogr. 36. ASA, CSSA, and SSSA, Madison, WI.

Eisenback, J.D. 1985. Interactions among concomitant populations of nematodes. p. 193–213. *In* J.N. Sasser and C.C. Carter (ed.) An advanced treatise on *Meloidogyne*. Vol. 1. North Carolina State Univ. Graphics, Raleigh.

Ferris, H. 1976. Development of a computer-simulation model for a plant-nematode system. J. Nematol. 8:255–263.

Ferris, H. 1981. Mathematical approaches to the assessment of crop damage. p. 405–420. *In* B.M. Zuckerman and R.A. Rohde (ed.) Plant parasitic nematodes. Vol. 3. Acad. Press, New York.

Ferris, H. 1984a. Nematode damage functions: The problems of experimental and sampling error. J. Nematol. 16:1–9.

Ferris, H. 1984b. Probability range in damage predictions as related to sampling decisions. J. Nematol. 16:246–251.

Ferris, H. 1985a. Density-dependent nematode seasonal multiplication rates and overwinter survivorship: A critical point model. J. Nematol. 17:93–100.

Ferris, H. 1985b. Population assessment and management strategies for plant-parasitic nematodes. Agric. Ecosyst. Environ. 12:285–299.

Ferris, H. 1987. Extraction efficiencies and population estimation. p. 59–63. *In* J.A. Veech and D.W. Dickson (ed.) Vistas on nematology. Soc. Nematol., Hyattsville, MD.

Ferris, H., and M.V. McKenry. 1974. Seasonal fluctuations in the spatial distribution of nematode populations in a California vineyard. J. Nematol. 6:203–210.

Ferris, H., and J.W. Noling. 1987. Analysis and prediction as a basis for management decisions. p. 49–85. *In* R.H. Brown and B.R. Kerry (ed.) Principles and practice of nematode control in crops. Acad. Press, Orlando, FL.

Flegg, J.J.M. 1968. The occurrence and depth distribution of *Xiphinema* and *Longidorus* species in southeastern England. Nematologica 14:189–196.

Francl, L.J. 1993. Multivariate analysis of selected edaphic factors and their relationship to *Heterodera glycines* population density. J. Nematol. 25:270–276.

Gair, R., P.L. Mathias, and P.N. Harvey. 1969. Studies of cereal nematode populations and cereal yields under continuous or intensive culture. Ann. Appl. Biol. 63:503–512.

Gallaher, R.N., and R. McSorley. 1993. Population densities of *Meloidogyne incognita* and other nematodes following seven cultivars of cowpea. Nematropica 23:21–26.

Goodell, P., and H. Ferris. 1980. Plant-parasitic nematode distributions in an alfalfa field. J. Nematol. 12:136–141.

Goodell, P.B., and H. Ferris. 1981. Sample optimization for five plant-parasitic nematodes in an alfalfa field. J. Nematol. 13:304–313.

Griffin, G.D. 1987. Effects of environmental factors and cultural practices on parasitism of alfalfa by *Ditylenchus dipsaci*. J. Nematol. 19:267–276.

Heald, C.M., and A.F. Robinson. 1987. Effects of soil solarization on *Rotylenchulus reniformis* in the lower Rio Grande Valley of Texas. J. Nematol. 19:93–103.

Heatherly, L.G., and L.D. Young. 1991. Soybean and soybean cyst nematode response to soil water content in loam and clay soils. Crop. Sci. 31:191–196.

Hooper, D.J. 1986a. Extraction of free-living stages from soil. p. 5–30. *In* J.F. Southey (ed.) Laboratory methods for work with plant and soil nematodes. Her Majesty's Stationery Office, London.

Hooper, D.J. 1986b. Extraction of nematodes from plant material. p. 51–58. *In* J.F. Southey (ed.) Laboratory methods for work with plant and soil nematodes. Her Majesty's Stationery Office, London.

Hooper, D.J. 1990. Extraction and processing of plant and soil nematodes. p. 45–68. *In* M. Luc et al. (ed.) Plant parasitic nematodes in subtropical and tropical agriculture. CAB Int., Wallingford, England.

Hooper, D.J., and K. Evans. 1993. Extraction, identification and control of plant parasitic nematodes. p. 1–59. *In* K. Evans et al. (ed.) Plant parasitic nematodes in temperate agriculture. CAB Int., Wallingford, England.

Huang, S.P., and A.C. Pereira. 1994. Influence of inoculum density, host, and low-temperature period on delayed hatch of *Meloidogyne javanica* eggs. J. Nematol. 26:72–75.

Hussey, R.S., and K.R. Barker. 1973. A comparison of methods of collecting inocula of *Meloidogyne* spp. including a new technique. Plant Dis. Rep. 57:1025–1028.

Jeger, M.J., and J.L. Starr. 1985. A theoretical model of the winter survival dynamics of *Meloidogyne* spp. eggs and juveniles. J. Nematol. 17:257–260.

Jenkins, W.R. 1964. A rapid centrifugal-flotation technique for separating nematodes from soil. Plant Dis. Rep. 48:692.

Johnson, A.B., H.D. Scott, and R.D. Riggs. 1993. Penetration of soybean roots by soybean cyst nematode at high water potentials. Agron. J. 85:416–419.

Jones, F.G.W. 1975. Accumulated temperature and rainfall as measures of nematode development and activity. Nematologica 21:62–70.

Jones, F.G.W., D.M. Parrott, and G.J.S. Ross. 1967. The population genetics of the potato cyst-nematode, *Heterodera rostochiensis*: Mathematical models to simulate the effects of growing eelworm-resistant potatoes bred from *Solanum tuberosum* spp. *andigena*. Ann. Appl. Biol. 60:151–171.

Kerry, B.R. 1987. Biological control. p. 233–263. *In* R.H. Brown and B.R. Kerry (ed.) Principles and practice of nematode control in crops. Acad. Press, Orlando, FL.

Kerry, B.R. 1990. An assessment of progress toward microbial control of plant-parasitic nematodes. Suppl. J. Nematol. 22:621–631.

Kort, J., H. Ross, H.J. Rumpenhorst, and A.R. Stone. 1977. An international scheme for identifying and classifying pathotypes of potato cyst-nematode *Globodera rostochiensis* and *G. pallida*. Nematologica 23:333–339.

MacGuidwin, A.E., and T.A. Forge. 1991. Winter survival of *Pratylenchus scribneri*. J. Nematol. 23:198–204.

Mashela, P., L.W. Duncan, J.W. Graham, and R. McSorley. 1992. Leaching soluble salts increases population densities of *Tylenchulus semipenetrans*. J. Nematol. 24:103–108.

McSorley, R. 1987a. Extraction of nematodes and sampling methods. p. 13–47. *In* R.H. Brown and B.R. Kerry (ed.) Principles and practice of nematode control in crops. Acad. Press, Orlando, FL.

McSorley, R. 1987b. Plot size and design for acquisition of field data in nematology. p. 52–58. *In* J.A. Veech and D.W. Dickson (ed.) Vistas on nematology. Soc. Nematol., Hyattsville, MD.

McSorley, R. 1992. Applied population modeling: Fact or fiction? p. 170–181. *In* F.J. Gommers and P.W.T. Maas (ed.) Nematology from molecule to ecosystem. Europ. Soc. Nematol., Invergowrie, Dundee, Scotland.

McSorley, R., W.H. Dankers, J.L. Parrado, and J.S. Reynolds. 1985. Spatial distribution of the nematode community on Perrine marl soils. Nematropica 15:77–92.

McSorley, R., and D.W. Dickson. 1989. Effects and dynamics of a nematode community on soybean. J. Nematol. 21:490–499.

McSorley, R., and D.W. Dickson. 1991. Determining consistency of spatial dispersion of nematodes in small plots. J. Nematol. 23:65–72.

McSorley, R., and L.W. Duncan. 1995. Economic thresholds and nematode management. Adv. Plant Pathol. 11:147–170.

McSorley, R., and R.N. Gallaher. 1993. Population dynamics of plant-parasitic nematodes on cover crops of corn and sorghum. J. Nematol. 25:446–453.

McSorley, R., and R.N. Gallaher. 1994. Effect of liming and tillage on soil nematode populations under soybean. Proc. Soil Crop Sci. Soc. Fla. 53:31–35.

McSorley, R., and J.L. Parrado. 1982. Estimating relative error in nematode numbers from single soil samples composed of multiple cores. J. Nematol. 14:522–529.

McSorley, R., and J.L. Parrado. 1987. Nematode losses during sieving from two soil types. Nematropica 17:125–146.

McSorley, R., and M.S. Phillips. 1993. Modelling population dynamics and yield loss and their use in nematode management. p. 6–85. *In* K. Evans et al. (ed.) Plant parasitic nematodes in temperate agriculture. CAB Int., Wallingford, England.

McSorley, R., and D.E. Walter. 1991. Comparison of soil extraction methods for nematodes and microarthropods. Agric. Ecosyst. Environ. 34:201–207.

Minton, N.A. 1986. Impact of conservation tillage on nematode populations. J. Nematol. 18:135–140.

Noe, J.P. 1991. Development of *Meloidogyne arenaria* on peanut and soybean under two temperature cycles. J. Nematol. 23:468–476.

Noe, J.P. 1998. Crop- and nematode-management systems. p. 159–171. *In* K.R. Barker et al. (ed.) Plant and nematode interactions. Agron. Monogr. 36. ASA, CSSA, and SSSA, Madison, WI.

Noe, J.P., and K.R. Barker. 1985. Relation of within-field spatial variation of plant-parasitic nematode population densities and edaphic factors. Phytopathology 75:247–252.

Noe, J.P., and C.L. Campbell. 1985. Spatial pattern analysis of plant-parasitic nematodes. J. Nematol. 17:86–93.

Noe, J.P., J.N. Sasser, and J.L. Imbriani. 1991. Maximizing the potential of cropping systems for nematode management. J. Nematol. 23:353–362.

Norton, D.C. 1978. Ecology of plant-parasitic nematodes. John Wiley & Sons, Inc., New York.

Norton, D.C. 1979. Relationship of physical and chemical factors to populations of plant- parasitic nematodes. Annu. Rev. Phytopathol. 17:279–299.

Norton, D.C., L.R. Frederick, P.E. Ponchillia, and J.W. Nyhan. 1971. Correlations of nematodes and soil properties in soybean fields. J. Nematol. 3:154–163.

Nusbaum, C.J., and K.R. Barker. 1971. Population dynamics. p. 303–323. *In* B.M. Zuckerman et al. (ed.) Plant parasitic nematodes. Vol. 1. Acad. Press, New York.

Opperman, M.A., and M.A. Conkling. 1998. Bioengineering resistance to plant-parasitic nematodes. p. 239–250. *In* K.R. Barker et al. (ed.) Plant and nematode interactions. Agron. Monogr. 36. ASA, CSSA, and SSSA, Madison, WI.

Oostendorp, M., D.W. Dickson, and D.J. Mitchell. 1991. Population development of *Pasteuria penetrans* on *Meloidogyne arenaria*. J. Nematol. 23:58–64.

Pederson, G.A., and G.L. Windham. 1992. Diallel analysis of resistance in white clover to southern root-knot nematode. Crop Sci. 32:1160–1162.

Perry, J.N. 1983. Effects of spatial heterogeneity on Jones' model for cyst-nematode population dynamics and crop root damage. J. Appl. Ecol. 20:849–856.

Poole, R.W. 1974. An introduction to quantitative ecology. McGraw-Hill, New York.

Proctor, J.F., and C.F. Marks. 1975. The determination of normalizing transformations for nematode count data from soil samples and of efficient sampling schemes. Nematologica 20:395–406.

Riggs, R.D., M.L. Hamblen, and L. Rakes. 1981. Infra-species variation in reactions to hosts in *Heterodera glycines* populations. J. Nematol. 13:171–179.

Riggs, R.D., and D.P. Schmitt. 1988. Complete characterization of the race scheme for *Heterodera glycines*. J. Nematol. 20:392–395.

Riggs, R.D., and D.P. Schmitt. 1991. Optimization of the *Heterodera glycines* race test procedure. J. Nematol. 23:149–154.

Roberts, P.A., S.D. Van Gundy, and H.E. McKinney. 1981. Effect of soil temperature and planting date of wheat on *Meloidogyne incognita* reproduction, soil populations, and grain yield. J. Nematol. 13:338–345.

Roberts, P.A., W.C. Matthews, and J.C. Veremis. 1998. Genetic mechanisms of host-plant resistance to nematodes. p. 209–238. In K.R. Barker et al. (ed.) Plant and nematode interactions. Agron. Monogr. 36. ASA, CSSA, and SSSA, Madison, WI.

Robinson, A.F., and C.M. Heald. 1989. Accelerated movement of nematodes from soil in Baermann funnels with temperature gradients. J. Nematol. 21:370–378.

Rodriguez-Kabana, R. 1986. Organic and inorganic nitrogen amendments to soil as nematode suppressants. J. Nematol. 18:129–135.

Rodriguez-Kabana, R., and H. Ivey. 1986. Crop rotation systems for the management of *Meloidogyne arenaria* in peanut. Nematropica 16:53–63.

Rodriguez-Kabana, R., C.F. Weaver, D.G. Robertson, and E.L. Snoddy. 1986. Population dynamics of *Meloidogyne arenaria* juveniles in a field with Florrunner peanut. Nematropica 16:185–196.

Schmitt, D.P., and G. Shannon. 1992. Differentiating soybean responses to *Heterodera glycines* races. Crop Sci. 32:275–277.

Schneider, S.M., and H. Ferris. 1987. Stage-specific population development and fecundity of *Paratrichodorus minor*. J. Nematol. 19:395–403.

Seinhorst, J.W. 1966. The relationships between population increase and population density in plant parasitic nematodes. I. Introduction and migratory nematodes. Nematologica 12:157–169.

Seinhorst, J.W. 1967a. The relationships between population increase and population density in plant parasitic nematodes. II. Sedentary nematodes. Nematologica 13:157–171.

Seinhorst, J.W. 1967b. The relationships between population increase and population density in plant parasitic nematodes. III. Definition of the terms host, host status and resistance. IV. The influence of external conditions on the regulation of population density. Nematologica 13:429–442.

Seinhorst, J.W. 1970. Dynamics of populations of plant parasitic nematodes. Annu. Rev. Phytopathol. 8:131–156.

Sidhu, G.S., and J.M. Webster. 1981. Genetics of plant-nematode interactions. p. 61–87. *In* B.M. Zuckerman and R. A. Rohde (ed.) Plant parasitic nematodes. Volume 3. Acad. Press, New York.

Sipes, B.S., and D.P. Schmitt. 1998. Nematode-pesticide interactions. p. 173–185. In K.R. Barker et al. (ed.) Plant and nematode interactions. Agron. Monogr. 36. ASA, CSSA, and SSSA, Madison, WI.

Southey, J.F. 1986. Principles of sampling for nematodes. p. 1–4. *In* J.F. Southey (ed.) Laboratory methods for work with plant and soil nematodes. Her Majesty's Stationery Office, London.

Southwood, T.R.E. 1978. Ecological methods. 2nd ed. Chapman and Hall, New York.Stirling, G.R. 1991. Biological control of plant parasitic nematodes. CAB Int., Wallingford, England.

Sultan, S.A., and H. Ferris. 1991. The effect of soil moisture and soil particle size on the survival and population increase of *Xiphinema index*. Revue Nematol. 13:345–351.

Taylor, A.L., and J.N. Sasser. 1978. Biology, identification and control of root-knot nematodes (*Meloidogyne* species). North Carolina State Univ. Graphics, Raleigh.

Taylor, L.R. 1961. Aggregation, variance and the mean. Nature (London) 189:732–735.

Tedford, E.C., B.A. Jaffee, A.E. Muldoon, C.E. Anderson, and B.B. Westerdahl. 1993. Parasitism of *Heterodera schachtii* and *Meloidogyne javanica* by *Hirsutella rhossiliensis* in microplots over two growing seasons. J. Nematol. 25:427–433.

Todd, T.C., and C.A.S. Pearson. 1988. Establishment of *Heterodera glycines* in three soil types. Ann. Appl. Nematol. (Suppl. J. Nematol.) 2:57–60.

Triantaphyllou, A.C. 1987. Genetics of nematode parasitism on plants. p. 354–363. *In* J.A. Veech and D.W. Dickson (ed.) Vistas on nematology. Soc. Nematol., Hyattsville, MD.

Van Gundy, S.D. 1985. Ecology of *Meloidogyne* spp.—emphasis on environmental factors affecting survival and pathogenicity. p. 177–182. *In* J.N. Sasser and C.C. Carter (ed.) An advanced treatise on *Meloidogyne*. Vol. I. North Carolina State Univ. Graphics, Raleigh.

Wallace, H.R. 1971. Abiotic influences in the soil environment. p. 257–280. *In* B.M. Zuckerman et al. (ed.) Plant parasitic nematodes. Vol. 1. Acad. Press, New York.

Wallace, H.R. 1973. Nematode ecology and plant disease. Edward Arnold, London.

Willis, C.B. 1972. Effects of soil pH on reproduction of *Pratylenchus penetrans* and forage yield of alfalfa. J. Nematol. 4:291–295.

Windham, G.L., and K.R. Barker. 1988. Winter survival of *Meloidogyne incognita* in six soil types. J. Nematol. 20:113–117.

Windham, G.L., and W.P. Williams. 1987. Host suitability of commercial corn hybrids to *Meloidogyne arenaria* and *M. incognita*. Ann. Appl. Nematol. (Suppl. J. Nematol.) 1:13–16.

Yeates, G.W. 1981. Nematode populations in relation to soil environmental factors: A review. Pedobiologia 22:312–338.

Young, L.D. 1998. Breeding for nematode resistance and tolerance. p. 187–207. In K.R. Barker et al. (ed.) Plant and nematode interactions. Agron. Monogr. 36. ASA, CSSA, and SSSA, Madison, WI.

Zeck, W.M. 1971. A rating scheme for field evaluation of root-knot nematode infestations. Pfl.-Nach. Bayer 24:141–144.

7 Concomitant Pathogen and Pest Interactions

GEORGE S. ABAWI AND JIANJUN CHEN

New York State Agricultural Experimental Station
Cornell University
Geneva, New York

Over a century ago, Atkinson (1892) observed that the severity of Fusarium wilt of cotton (*Gossypium hirsutum* L.) was increased in the presence of root-knot nematodes (*Meloidogyne* spp.). The involvement of plant-parasitic nematodes with bacterial diseases was first described in 1901 by Hunger with tomato wilt [*Pseudomonas solanacearum* (Smith) Smith] (Hunger, 1901). The first definitive documentation of plant-parasitic nematodes as vectors of plant viruses was published in 1958 with fanleaf disease of grape (*Vitis* spp.) (Hewitt et al., 1958). Since the first observation of Atkinson, numerous associations and specific interactions between plant-parasitic nematodes and pathogenic and nonpathogenic organisms have been documented on many crop plants. The voluminous information available in the literature has received numerous excellent reviews including those by Bergeson (1972), Eisenback and Griffin (1987), Lamberti and Roca (1987), Hirano (1975), Huang (1987), Mai and Abawi (1987), Norton (1978), Pitcher (1965, 1978), Powell (1971, 1979), Sikora and Carter (1987), Smith (1987), Taylor (1979), Wallace (1978), and Webster (1985). In addition, the recent book *Nematode Interactions*, edited by M. W. Khan (1993), presents a thorough and complete treatment of the reported interactions involving plant-parasitic nematodes and primarily other plant pathogens, but also includes interactions with rhizobial and mycorrhizal symbionts. Thus, this chapter will not attempt to review the vast, available information in-depth, but rather will focus on the underlaying principles and overall conclusions of the pertinent information on interactions involving plant-parasitic nematodes and plant pathogens as well as other pests, and will illustrate them with selected examples.

Plant-parasitic nematodes are widely distributed in natural and highly disturbed crop production areas of the world. Many species of nematodes [including the root-knot, cyst (*Heterodera glycines*), lesion (*Pratylenchus* spp.), citrus (*Tylenchulus semipenetrans*), and others] are highly damaging to host plants when present at high population densities (Brown & Kerry, 1987). In addition, the extensive studies on the transmission of plant viruses by nematodes and the interactions between root-knot nematodes and fungal soilborne pathogens (especially the Fusarium wilt fungi) have contributed greatly to the recognition and the

---

importance of plant-parasitic nematodes in microbial interactions in soil (Khan, 1993). Also, it is well accepted that plant-parasitic nematodes play a major role in the incidence and severity of root diseases of complex etiology, especially on crops exposed to a number of edaphic stress conditions such as water deficit, high temperature, nutrient deficiencies, and others (Wallace, 1978; Mai et al., 1994; Powelson & Rowe, 1993).

One square meter of high-fertility field soil may contain as many as 400 g of fungi, 300 g of bacterial cells, 39 g of protozoa, and 12 g of nematodes with the latter representing about 1×10^7 "worms" (Richards, 1976). Gottlieb (1976) reported that 1 g of surface soil contains 10^6 to 10^8 bacterial cells, 10^6 to 10^7 actinomycetes, 5×10^4 to 10^6 fungal colony forming units (cfu), 10^5 to 10^6 protozoa, and 10^4 to 5×10^5 algae. However, most soils do not have these high microbial densities, and plant pathogens represent only a small percentage of the total microorganisms present in any given soil. Nevertheless, the number of soil microorganisms present in the root zone of any crop plant is sufficient for innumerable and diverse interactions (Mai & Abawi, 1987). Thus, it is not surprising that many investigators have concluded that the majority of root diseases have a complex etiology and are greatly influenced by associated microorganisms (Wallace, 1978; Grogan, 1981; Fawcett, 1931). For example, Mayol and Bergeson (1970) described a synergistic interaction between root-knot nematodes and natural soil microflora on tomato. Tomato [*Lycopersicon esculentum* (Mill.) Karsten var. *lycopersicum*] growth was suppressed by 37 and 18% when the nematode and soil microflora, respectively, were present alone. However, tomato growth was suppressed by 75% when both the nematode and soil microflora were present.

Concomitant pathogen interactions have been observed and documented in both greenhouse and field experiments. For example, Fusarium wilt diseases were controlled by fumigation with nematicides in soils infested with the wilt pathogen and root-knot nematodes (Powell, 1971). Repeated documentation of concomitant pathogen interactions have been obtained largely from greenhouse or growth chamber experiments on numerous crop plants. Mai and Abawi (1987) summarized the interactions between Fusarium wilt fungi and *Meloidogyne* spp. on 13 crops. Approximately 70 described *Fusarium oxysporum* formae speciales are known to interact with nematodes (Armstrong & Armstrong, 1981; Beckman, 1989). More than 3000 parasitic and phoretic associations are known between nematodes and insects, but less than 10 involve plant-parasitic nematodes (Poinar, 1975). A recent literature survey showed that nematode-pathogen interactions occur on at least 45 crops (Evans & Haydock, 1993). Approximately 15 genera of plant-parasitic nematodes and 23 genera of fungi are involved in the observed interactions.

Clearly, multiple components are involved in the interaction between soilborne parasites and their hosts (Khan, 1993; Cook & Baker, 1983). The influence of a parasite on its host may well be to change its response to competitors and mutualists, alter its reaction to physical conditions of the environment and damage its ability to garner resources (Begon et al., 1990). The realism and predictive value of models of host-parasite interactions are increased by the incorporation of at least a third component such as another biological entity (Holt & Pickering,

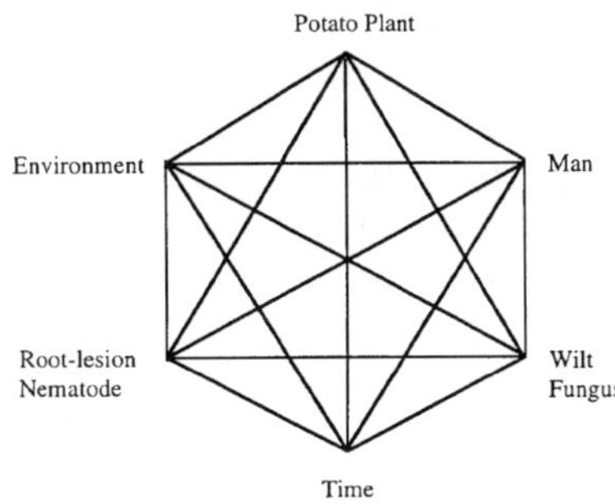

Fig. 7–1. *Pratylenchus penetrans* and *Verticillium dahliae* interactions associated with a *Solanum tuberosum* ecosystem (courtesy J. Chen, 1995).

1985; Anderson & May, 1986). However, the consideration of a triple component greatly increases the complexity of interactions. A conceptual illustration of a three-component interaction in an agroecosystem was recently presented (Fig. 7–1) by Chen (1995). The triangle of interaction among root-lesion nematode (*Pratylenchus* spp.), Verticillium wilt fungus, and potato (*Solanum tuberosum* L.) plant is coupled with the other 19 potential triangle interactions and numerous linear, quadrate, and polygonal interactions among the nematode, fungus, plant, environment, time, and man. Thus, extensive studies are needed to elucidate the biological interactions operating in such agroecosystems.

NEMATODE INVOLVEMENT/INTERACTION IN DISEASES OF COMPLEX ETIOLOGY

Plant-parasitic nematodes often cause plant diseases by themselves, resulting in significant losses in crop yield and quality. Specific symptoms and damage by nematode infections depend on the particular host and the nematode species involved. However, uneven growth, stunting, chlorosis, and poor quality and quantity of yield are general symptoms of severe infection and damage inflicted by plant-parasitic nematodes. Nevertheless, parasitic nematodes are only one component of the rhizosphere microbial community of crop plants, which also includes pathogenic and nonpathogenic fungi, bacteria, microarthropods, and other organisms. Thus, roots are exposed and ultimately attacked simultaneously or in a succession by various soilborne organisms. In addition, the potential exists for numerous interactions among the rhizosphere microbial community prior, during, and after invasion of roots. The latter often results in the development of root disease complexes with complex etiology and which are difficult to diagnose accurately (Agrios, 1988; Wallace, 1978). Plant-parasitic nematodes of various feeding habits have been recognized to play a major role in disease complexes (Mai et al., 1994; Powelson & Rowe, 1993; Powelson et al., 1993; Powell, 1971). Effective control of plant-parasitic nematodes involved in disease complexes has substantiated their significant role in such diseases and demonstrated the need for the integrated management of root diseases in general.

Replant diseases of fruit trees are good examples of diseases with a complex etiology, where plant-parasitic nematodes are major components of their etiology (Mai & Abawi, 1981; Mai et al., 1994; Jaffee et al., 1982; Wescott et al., 1986). These diseases have been encountered throughout the major temperate fruit-growing regions of the world for more than 200 yr. The diseases are manifested as uneven and poor growth as well as high mortality of fruit trees, especially those of pome and stone fruits, obtained on replanted old orchard sites (Fig. 7–2). Severely damaged root systems may lack feeder rootlets, be reduced in volume, and have numerous dead root segments. In addition, infected feeder roots are discolored, stunted, and may also exhibit witches' broom symptoms. However, the incidence and severity of replant diseases vary considerably within and between orchards and nurseries (Mai et al., 1994).

The etiology of replant diseases has been studied extensively, but still remains to be fully elucidated (Mai & Abawi, 1981). Most investigators have concluded that several soilborne organisms are involved in the complex etiology of these diseases and include plant-parasitic nematodes, pathogenic fungi and possibly bacteria as well as nonparasitic rhizosphere microorganisms. Other causal factors suggested in the literature include allelopathic compounds from old roots, unbalanced soil nutrition, impaired soil structure, cold injury and others. Although the latter factors may be involved in a number of orchards, they are considered as secondary contributing factors rather than primary causal factors. Preplant treatment of old orchard soil with fumigants has reduced tree mortality and increased growth, uniformity and yield of trees and their tolerance to stress factors. However, the responses to the broad-spectrum biocides, such as chloropicrin (trichloronitromethane), methyl bromide, and vorlex (methyl isothio-

Fig. 7–2. Symptoms of severe replant disease on pears in an orchard in western New York State (courtesy W. F. Mai, unpublished data).

cyanide), were greater than the responses to the strict nematicides such as 1,3-dichloropropene. These and many other soil treatments have strongly suggested the involvement of parasitic and nonparasitic soilborne organisms in the etiology of replant diseases.

Several nematodes, especially *Pratylenchus* species, play a major role in root destruction associated with replant diseases in several fruit production areas. *Pratylenchus penetrans* has been closely associated with replant failure and poor growth and yield of apple (*Malus domestica* Borkh.), cherry (*Prunus* spp.) and peach [*Prunus persica* (L.) Batsch var. *persica*], and to a lesser extent of other stone and pome fruits. Other species of plant-parasitic nematodes found associated with soils and roots of fruit trees include the dagger (*Xiphinema* spp.), the ring (*Criconemella* spp.), the pin (*Paratylenchus* spp.), and less frequently the root-knot (*Meloidogyne* spp.) nematodes (Mai & Abawi, 1981). Additionally, a number of fungi and other soilborne microorganisms known to be pathogenic to fruit trees are found in the soil of orchards with replant diseases in several production regions. Pathogenic fungi demonstrated to cause damage to feeder and other tree roots include *Rhizoctonia solani* on apple and cherry; *Pythium* spp. and *Phytophthora* spp. on peach, cherry, and apple; *Cylindocarpon lucidum* on apple; and *Thielaviopsis basicola* on cherry. Several *Fusarium* spp. and many other soilborne microorganisms also have been shown to be associated with roots and soils of poorly growing trees. Research in New York showed that nonparasitic rhizosphere microorganisms (possibly bacteria and actinomycetes), in addition to *P. penetrans*, are important causal agents of an apple replant disease (Jaffee, 1981). A rhizosphere wash obtained from apple seedlings growing in apple replant soil was shown to serve as an effective source of inoculum to induce replant disease on healthy seedlings. The causal agent(s) in the rhizosphere wash could be eliminated by filtration (0.2 μm) or heat treatment at 60°C for 10 min. Nevertheless, the exact etiology of the replant diseases is still not completely understood and the numerous interactions and their mechanisms among the various organisms need to be studied in detail.

TYPES AND CASE DOCUMENTATION OF NEMATODE-PATHOGEN PEST INTERACTIONS

Many reports document the interactions between plant-parasitic nematodes and other plant pathogen groups, including nonpathogenic organisms as well as symbiotic and mycorrhizal organisms. The following are brief summaries and conclusions regarding the reported interactions among plant-parasitic nematodes and fungal, bacterial, viral, other parasitic nematodes, and selected reported interactions with insects and weeds.

Interactions Between Nematodes and Fungal Pathogens

Interactions involving fungal plant pathogens and plant-parasitic nematodes have received much attention as suggested by the number of related publications. Various aspects of the nematode-fungal interactions have been reviewed

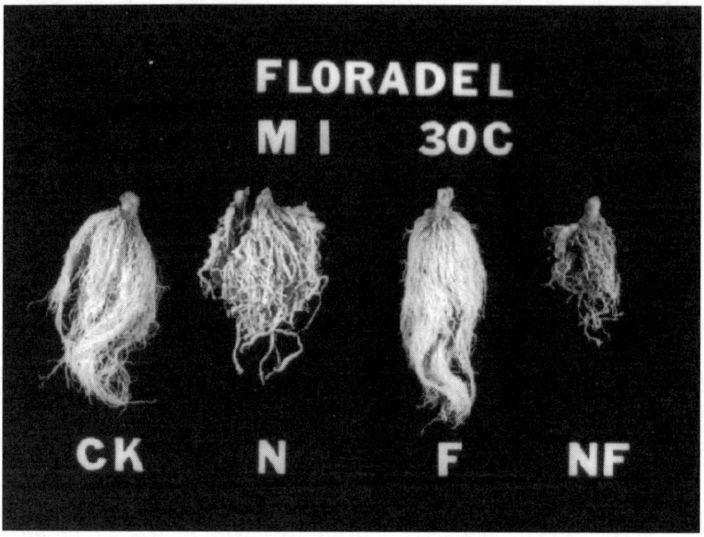

Fig. 7–3 Severity of root galling and root necrosis on root of 'Floradel' tomato that were (left to right) uninoculated (CK) or inoculated with *M. incognita* (N), *F. oxysporum* f. sp. *lycopersici* (F) or both (NF) at 30°C (Abawi & Barker, 1984).

previously (Powell, 1971; Bergeson, 1972; Webster, 1985; Mai & Abawi, 1987; Rowe et al., 1987; Evans & Haydock, 1993; Francl & Wheeler, 1993). Interactions between *Meloidogyne* spp. and Fusarium wilt fungi (several formae speciales of *F. oxysporum*) have received special attention and were documented on 20 crop species (Fig. 7–3). Interaction between these pathogens were especially obvious when the root-knot nematode infections preceded those of the Fusarium wilt pathogens by 3 to 4 wk. The majority of these studies have established that the presence of root-knot nematodes increases the incidence, rate of development, and severity of wilt or the mortality of *Fusarium*-susceptible and tolerant crop cultivars. In addition, growth parameters of such cultivars also are reduced and often synergistically in the presence of both pathogens. However, the role of the root-knot nematodes in the alteration or breakdown of the monogenic type of resistance to Fusarium wilt fungi (such as tomato cultivars with the dominant I-genes against *F. oxysporum* f. sp. *lycopersici*) remains controversial and requires further investigation (Mai & Abawi, 1987). Similarly, infections by *M. incognita* failed to alter the reaction of wilt resistant cowpea [*Vigna unguiculata* (L.) Walp. subsp. *unguiculata*] to *F. oxysporum* f. sp. *tracheiphilum* (Roberts et al., 1995). A recent report (France & Abawi, 1994) showed that the dry bean (*Phaseolus vulgaris* L.) cultivars Calima and Dark Red Kidney (resistant to the Brazilian race of *F. oxysporum* f. sp. *phaseoli*) developed symptoms of Fusarium wilt only in the presence of *M. incognita* (Fig. 7–4). Furthermore, the bean genotype A-107 (resistant to both *M. incognita* and *F. oxysporum* f. sp. *phaseoli*) exhibited Fusarium wilt symptoms and a moderately susceptible reaction to the Fusarium wilt pathogen only after the breakdown of its resistance to *M. incognita* was accomplished by elevated incubation temperature (27°C) (Fig. 7–5).

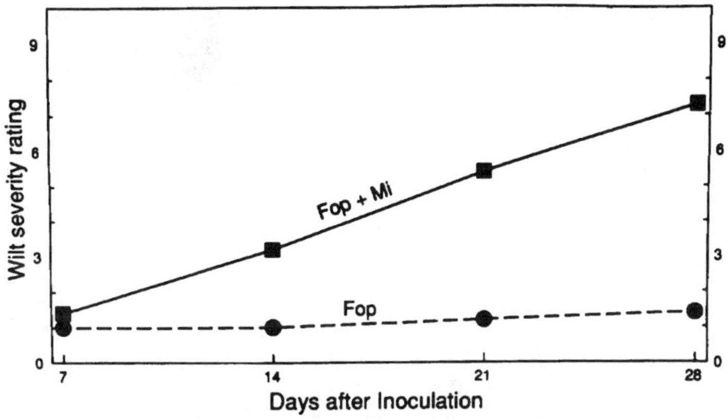

Fig. 7–4. Effects of *F. oxysporum* f. sp. *phaseoli* (FOP, broken line) alone and in combination with *M. incognita* (FOP + Mi, solid line) on Fusarium-wilt severity (rated on a scale of 1 = no visible symptoms to 9 = >95% of tissues affected) on 'Dark Red Kidney' bean (France & Abawi, unpublished data).

However, Fusarium wilt resistance to *F. oxysporum* f. sp. *ciceri* in 1, 4, and 48 genotypes of chickpea (*Cicer arietinum* L.) was reported to be ineffective, reduced, and not modified, respectively, by the presence of *M. javanica* (Maheshwari et al., 1995).

The interaction of lesion nematodes (*Pratylenchus* spp.) with Verticillium wilt fungi resulting in the potato early dying (PED) syndrome is another excellent and well-documented example of interactions between plant-parasitic nematodes and fungal pathogens (Powelson & Rowe, 1993). Potato early dying syndrome refers to the premature vine death and declining yields of potato, which has become a limiting factor in several potato production regions (Bird, 1981; Francl & Wheeler, 1993; Riedel et al., 1985; Rowe et al., 1987; Powelson et al., 1993; Wheeler & Riedel, 1994). The effective control of this disease under field conditions with fumigant and nonfumigant nematicides has long suggested the involvement of plant-parasitic nematodes. Results of detailed experiments conducted in field microplots clearly demonstrated a synergistic interaction between *P. penetrans* and *V. dahliae* resulting in PED syndrome. In these same tests, no

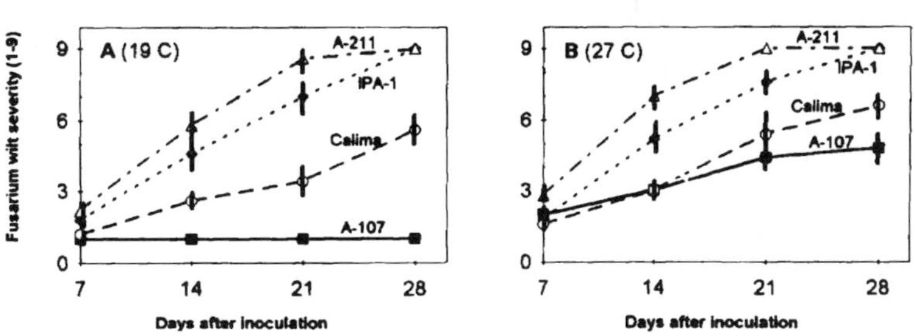

Fig. 7–5. A–B) Development of Fusarium-wilt in four bean genotypes inoculated with 10^6 spores/mL of *F. oxysporum* f. sp. *phaseoli* and 10^3 eggs of *M. incognita* as affected by incubation temperature of 19 (A) and 27°C (B) during the first 2 wk. Fusarium wilt severity was rated on a scale of 1 = no visible symptoms to 9 = >95% of tissues affected. Bars indicate SE (France & Abawi, 1994).

measurable effect on symptom development and potato yield was obtained when *P. penetrans* was present alone, whereas light to moderate effects were observed with *V. dahliae* alone. Information available also suggested that not all species of lesion nematodes interact equally with *Verticillium* on PED in potato. Actually, no interaction was observed with *P. crenatus* and *V. dahliae*, suggesting the specificity of this interaction. Similarly, Fusarium wilt of cotton was enhanced by *Belonolaimus longicaudatus*, but not by *Hoplolaimus galeatus* (Prot, 1993). It was suggested that the ectoparasitic *B. longicaudatus* has a very long stylet; thus it may provide access to the vascular tissues of the plant for the wilt pathogen. This is not the case with *H. galeatus*, which feeds in the cortical tissues. It also was reported that 'Moapa 90' alfalfa (*Medicago sativa* L. subsp. *sativa*) was predisposed to infection by *F. oxysporum* f. sp. *medicaginis* by *M. hapla*, but not by *Ditylenchus dipsaci* (Francl & Wheeler, 1993). Furthermore, early infection of broadleaf tobacco (*Nicotiana tabacum* L.) by *M.hapla* increased the incidence and severity of Fusarium wilt, but it was less than the increase caused by prior infections with *Globodera tabacum tabacum* (LaMondia, 1992).

Many examples also are available documenting the association and interaction of several plant-parasitic nematodes with various root-rot fungal pathogens (Evans & Haydock, 1993; Khan & Husain, 1989; Hasan & Khan, 1985; Khan, 1993; Powell, 1971). For example, *M. arenaria* and *Sclerotium rolfsii* cause a disease complex on peanut (*Arachis hypogaea* L.) due to their additive effects (Starr et al., 1996). Inoculation of cowpea with *M. incognita* and *Rhizoctonia solani* led to breakdown of resistance to both pathogens. Severe root rot of tomato was found to be associated with infection by *M. incognita* and *R. solani* as a result of the utilization of nutrients in gall tissues and root exudation by the latter pathogen (Wang & Bergeson, 1974). Infection of cotton by *R. solani* increased egg and juvenile densities of *Rotylenchulus reniformis* resulting in an antagonistic effect to plant growth (Sankaraligam & McGawley, 1994a,b). Another example is the predisposition of tobacco roots to severe decay incited by *Pythium ultimum* by a low level of *M. incognita*. The incidence and severity of sudden death syndrome of soybean [*Glycine max* (L.) Merr.] caused by a blue-pigmented form of *F. solani* was positively correlated with numbers of *Heterodera glycines*. However, population of the soybean cyst nematode (*Heterodera glycines*) was reduced by 47% in the presence of the fungal pathogen (McLean & Lawrence, 1995). Many other examples of nematode-fungal interactions results in an antagonistic effect on the nematode or the fungal pathogen have been documented (France & Abawi, 1994; Khan, 1993).

Interactions Between Nematode and Viral Pathogens

The efficient vectoring of several soilborne viruses by plant-parasitic nematodes has generated much interest and investigation concerning the nature of the nematode virus association and appropriate control strategies (Lamberti et al., 1975; Harrison, 1977, 1978; Harris, 1981; Taylor & Brown, 1981; Hussey & McGuire, 1987; Weischer, 1993). Only Dorylamoid nematode species are capable of vectoring certain plant viruses. The Longidorid nematodes species of *Xiphinema* (dagger), and *Longidorus* and *Paralongidorus* (needle) have been well documented to transmit several economically important nepoviruses (polyhedral-shaped viruses) including tobacco ring spot, tomato ring spot, grapevine fanleaf, cherry leaf roll, brome mosaic, and other viruses. The trichodorid nematodes species of *Trichodorus* and *Paratrichodorus* (stubby-root) have been shown to transmit the tobraviruses (rod-shaped viruses) including tobacco rattle and pea early browning viruses.

Several of the viruses that are transmitted by nematodes cause significant economic losses on major food crops such as apple (apple union necrosis and decline virus), peach (tomato and tobacco ringspot viruses), cherry (prunus stem pitting virus), grape (grapevine fanleaf, tomato ringspot, and peach rosette viruses), potato (tobacco rattle and tomato ringspot viruses), pea (pea early browning virus), raspberry (*Rubus* spp.) (raspberry ringspot, tomato ringspot, and arabic mosaic viruses), strawberry (*Fragaria chiloensis* Duchense) (tomato ringspot and arabic mosaic viruses), hop (*Humulus lupulus* L.) (arabic mosaic virus), and many others.

The vector nematodes have been found to be highly efficient in acquiring the virus from infected host plants or weed species and in their transmission efficiency to the roots of the same or related hosts. Transmission of several of these viruses is highly specific and may be accomplished by only one or a few species of nematodes. Particles of viruses transmitted by plant-parasitic nematodes are retained in association with the cuticular lining of the buccal cavity, lumen walls of the stylet or the cuticular lining of the esophageal lumen of the viruliferous nematodes. Depending on the nematode species and the particular virus, retention of the virus in the nematode may be for a period of a few weeks to a few years.

In the absence of a nematode-vector relationship, only a few reports suggest either stimulatory or antagonistic effects of virus infections on the infectivity, reproduction and damage caused by plant-parasitic nematodes. For example, a higher number of juveniles of *M. javanica* penetrated roots of tobacco ring spot-infected bean plants than in virus-free plants. In contrast, decreases in the populations of *D. dipsaci* and *Aphelenchoides ritzemabosi* were observed on tobacco infected with tobacco ringspot virus. However, it was suggested that the three pathogens interacted synergistically in reducing the growth of tobacco plants (Weischer, 1993).

Current molecular research approaches and the availability of biochemical techniques will undoubtedly contribute significantly to our understanding of nematode-virus-plant interactions. The double antibody sandwich and biotin-avidin ELISA (enzyme-linked immunosorbent assay) methods, as well as the molecular hybridization with homologous probes have been used to detect viruses in nematodes and to differentiate infection potential of several populations of the vector nematode (Esmenjaud et al., 1993). Enzyme-linked immunosorbent assay absorbances from greater than 10 *Xiphinema index* allowed the differentiation of the infection potential of grapevine fanleaf virus in several geographical populations of this nematode. In addition, investigations on the modifications of the coat protein of nematode-transmitted viruses and the expression of virus genes into transgenic host plants may well lead to understanding of the specificity and mechanisms of nematode-virus interactions (Grumet, 1994; Hackland et al., 1994; Fitchen & Beachy, 1993).

Interaction Between Nematodes and Bacterial Pathogens

Although only a few reports focus on interactions of plant-parasitic nematodes with bacterial plant pathogens, some are well documented (Powell, 1971, 1979; Norton, 1978; Taylor, 1979; El-Sherif, 1991; Sitaramaniah & Pathak, 1993; McKay & Ophel, 1993). The presence of root-knot nematodes was shown to play a major role in increasing the incidence and severity of bacterial wilts diseases incited by *Pseudomonas solanacearum* on various crops, including tomato, potato, eggplant (*Solanum melongena* L.) and tobacco. Ectoparasitic nematodes also increase bacterial wilt onset and severity, but not to the same level as the root-knot nematodes. Alfalfa wilt caused by *Clavibactor michiganensis* subsp. *insidiosum* (syn. *Corynebacterium insidiosum*) was similarly increased in the presence of *D. dipsaci*. In addition, several studies have documented the involvement of both *Rhodococcus faciens* (*C. fasciens*) and *A. ritzemabosi* in the expression of typical symptoms of cauliflower disease of strawberry. Crown gall of peach caused by *Agrobacterium tumefaciens* was reported to become more severe in the presence of *M. javanica* (Sitaramaniah & Pathak, 1993). In another study, *A. tumefaciens* stimulated the development and reproduction of *M. incognita* when added to the opposite split root of tomato (El-Sherif, 1991). In the majority of the reports on nematode-bacterial interactions, the nematodes act as vectors to the bacteria, provide wounds or entry to plant cells, and/or modify the physiology of host tissues to favor bacterial infections.

The association of *Anguina* spp. (especially *A. funesta*) with the bacterium *Clavibactor toxicus* in the toxicity of several grasses (especially Wimmera ryegrass, *Lolium rigidum*) to sheep and other animals further illustrates the importance and economic significance of these nematode-bacterial interactions (McKay & Ophel, 1993). The bacterium and its bacteriophage were shown to produce highly lethal glycolipid compounds, corynetoxins (3–5 mg kg^{-1} body weight in sheep), in seed galls. It has been well established that *C. toxicus* and the bacteriophage are carried into the developing seedheads of Wimmera ryegrass and others by *Anguina* spp. The bacterium attachment to the nematode cuticle was reported to be under the control of a single recessive gene. This important toxicity problem to sheep in Australia has stimulated extensive research on the development of new methods of control that include nematode-resistant grasses, biological control measures, animal protectants, and others.

Interactions Between Nematodal Pathogens

Polyspecific communities of plant-parasitic nematodes are usually found in agricultural soils as well as in undisturbed agroecosystems. Thus, it is not unusual for two or more nematodes to be found parasitizing the roots of the same host plant. Various interactions may result among the involved nematodes that may affect their reproductive capacities as well as modifying the etiology of particular disease exhibited by the invaded host plant (Eisenback, 1993; Eisenback & Griffin, 1987; Hirunsalee et al., 1995). Depending on the nematodes involved, their population density, and the suitability of the host to each, the resultant interaction may be detrimental, neutral, or beneficial to one or all nematode species involved. For example, Race 2 of *M. arenaria* depressed root infection, female development and reproduction of Race 1 of *M. arenaria* when they were concomitantly inoculated on 'Florigiant' peanut. However, *M. hapla* exhibited no significant effect against Race 1 of *M. arenaria* on the same peanut cultivar. In contrast, reproduction capacities and resultant root-galling severity of *M. hapla* and Race 2 of *M. arenaria* were restricted when either concomitantly inoculated with Race 1 of *M. arenaria* on 'McNair 373' tobacco (Hirunsalee et al., 1995). The growth of 'Ranger' alfalfa was synergistically reduced with concomitant inoculation with *D. dipsaci* and *M. hapla* (Eisenback & Griffin, 1987). However, the resistance of 'Speight G-28' tobacco to *M. incognita* was not altered with concomitant infections by *M. arenaria* and *M. incognita* (Ng'ambi et al., 1995). In a study involving *M. chitwoodi* and *Pratylenchus neglectus* on barley (*Hordeum vulgare* L.) and potato (Umesh & Ferris, 1994), it was concluded that the two species interact competitively and that the final outcome is influenced by soil temperature and the host plant. The authors also suggested that *P. neglectus* has the potential to suppress populations of *M. chitwoodi* and its damage to potato and barley. An inverse relationship between *P. neglectus* and *Heterodera avenae* on winter wheat (*Triticum aestivum* L.) was shown under field conditions and in vitro in split-root experiments (Lasserre et al., 1994). These examples illustrate the importance of nematode-nematode interactions and their potential impact on management strategies.

Interactions Between Nematodes and Insect Vectors

Several associations of plant-parasitic nematodes with insects have been shown to greatly increase the geographic distribution and survival of nematodes as well as the epidemic development of the diseases they cause (Giblin-Davis, 1993). The pine wood nematode (*Bursaphelenchus xylophilus*) is the causal agent of the pine wilt disease, which has become a major disease problem in native pine (*Pinus* spp.) forests in Japan, southern China and the USA. Fourth-stage dauer juveniles of *B. xylophilus* are transported to adjacent or distant new host trees on the body surface and within the trachae of several longhorn beetle species mostly in the genus *Monochamus* (Cerambycidae). The primary transmission of *B. xylophilus* to healthy trees occurs through the resin canals in wounds made by the vector beetles, usually in young branches. In highly susceptible tree host species and under favorable conditions, pine wood nematodes multiply rapidly reaching populations as high as 1000 nematodes/dry gram of wood. Depending on the age and physiological conditions, heavily infected trees may die within a few weeks to a year after initial infection. Such dead or dying pines become attractive breeding hosts for the next generation of vector beetles. Dauer juveniles of nematodes carried to the tree during oviposition of the beetles will hatch and feed on secondary fungi that become more abundant as the tree dies. Eggs of the beetle vector also hatch and undergo several larval instars and then pupate. Predauer juveniles of *B. xylophilus* are attracted and aggregate around the pupal chamber, then molt to the dispersal dauer stage and become again associated with the tracheae of the adult enclosed in the chamber. Internal host responses to nematode feeding and multiplication are the death of parenchymal cells and several chemical and physiological changes. An initial external symptom of pine wilt disease is the cessation of oleoresin exudation from wounds on the trunk, followed by wilting and yellowing of the foliage, and finally the reddish color of pine needles after tree death.

Another important and well-documented disease where a plant-parasitic nematode is closely associated with an insect vector is the red ring diseases of coconut (*Cocos nucifera* L.) and African oil palm (*Elaeis guineensis* Jacq.) (Giblin-Davis, 1993). These diseases are caused by *Rhadinaphelenchus cocophilus* which is vectored primarily by the African palm weevil, *Rhynchophorus palmarum* (Curculionidae: Coleoptera). The vectoring insects appear to acquire the dauer juveniles of *R. cocophilus* prior to metamorphosis and carry the nematodes internally to the adult stages. Thus, the red ring nematode appears to be parasitically associated with the insect vector and survives poorly outside the insect or its plant host. The vector insect deposits the nematodes into the leaf bases or wounds in susceptible palm trees during oviposition. The nematodes continue to invade the host tissues and feed on the parenchymal cells which can result in a lethal wilt in a few months. Other symptoms exhibited by infected coconut palms include premature nut fall, withering and necrosis of inflorescence, and yellowing, bronzing and death of progressively younger leaves. The diagnostic symptom is the presence of a brownish-red ring within the stem tissues, which contain high population densities of the dauer juveniles of *R. cocophilus*. Similar symptoms also occur on the African oil palm, but also include progressive premature yellowing and death of older leaves. A symptom on younger leaves is pale yellow-green chlorosis. However, the speed and the specific symptoms exhibited depend on the prevailing environmental conditions, the palm species and cultivar as well as the age of the trees. Losses in coconut and African oil palm have been estimated at 10 to 15%. The red ring and pine wilt diseases illustrate well the importance and damage potential of nematode-insect associations and the need to fully understand these interactions in order to formulate effective management options for these and similar complex diseases.

Nematode damage to host crops also may directly or indirectly influence the prevalence and damage caused by insect pests. Damage to soybean by moderate to high initial population densities of *H. glycines* resulted in delayed flowering and stunting, thus more open canopies and consequently higher infestation and damage by the corn earworm, *Helicoverpa zea* (Alston et al., 1991a). In another study, the reduction of the soybean canopy by *H. glycines* contributed to lower mortality of *H. zea* larvae due to parasitism and infection with entomopathogens (Alston et al., 1991b).

Interactions Between Nematodes and Weeds

It has been generally observed that fields heavily infested with plant-parasitic nematodes also have serious weed problems as compared to noninfested fields. High infestations with *H. glycines* were shown to restrict the canopy of soybean and result in increased weed populations and damage (Alston et al., 1991a,b). The competitiveness of soybean with common lambsquarters (*Chenopodium album* L.) was reduced in the presence of *H. glycines* (Chen et al.,

1995). Similarly, *Anguina amsinchia* intensified the competition between its weed host fiddleneck (*Amsinckia intermedia* Fish. & C.A. Mey) and wheat for the limited available resources for both plants (Pantone, 1995). In addition, weed-management strategies (herbicides, hand-hoeing, repeated spring-summer cultivation, and sawdust mulching) on asparagus (*Asparagus officinalis* L.) and corn (*Zea mays* L. subsp. *mays*) were shown to have different effects on the diversity of nematode taxa and trophic group composition (Yeates et al., 1993). Thus, it is important to integrate the management of weeds and nematodes and to study the various resultant interactions and their impact on soil biomass and crop yield. The nematicide fenamiphos [ethyl 3-methyl-4-(methylthio) phenyl (1-methylthyl) phosphophoramidate] and the herbicide EPTC (S-ethyl dipropylthiocarbamate) were found compatible and highly effective when applied together through a sprinkler irrigation system to control nematodes and weeds on snap bean (*Phaseolus vulgaris* L.) (Johnson et al., 1994).

Interactions Between Nematodes and Multiple Crop Pests

As suggested and discussed in the previous sections, interactions involving plant-parasitic nematodes and other plant pests as well as nonpathogenic organisms are common, especially in soil, on and in host plants. In addition, crop plants are frequently exposed to and attacked by several pests every growing season. Thus, the opportunity is high for the direct and/or indirect complex interactions involving plant-parasitic nematodes and several pests. Detailed information on these complex, multipest interactions are limited and difficult to obtain. The interaction of an insect-fungus-nematode pest complex was studied on soybean by Russin et al. (1989, 1990), which provides a good example of multiple pest interaction involving nematodes. They reported that the interrelationships between the soybean cyst nematode (*H. glycines*), stem canker fungus (*Diaporthe phaseolorum* var. *caulivora*), and soybean looper (*Pseudoplusia includens*) and their impact on soybean growth were additive. Juveniles in roots and cysts in soil of *H. glycines* were reduced on plants with stem canker. However, population of *H. glycines* were higher and proportionally related to the degree of defoliation caused by the soybean looper larvae. In addition, infections of soybean roots by *H. glycines* significantly reduced the length of stem cankers.

RESEARCH METHODOLOGIES EMPLOYED

The majority of the results reported in the literature documenting the numerous interactions and associations between plant-parasitic nematodes and other plant pathogens have been obtained from a general experiment with four treatments (nematode alone, pathogen alone, nematode plus pathogen, and non-infested control) conducted in greenhouses or growth chambers using pasteurized, sterilized, or fumigated soils (Khan, 1993). In addition, most of the early evidence provided on the interactions of plant-parasitic nematodes with other plant pathogens in disease development under field conditions was based on the use of fumigant-type nematicides. The advantages and disadvantages of these two approaches to study nematode-pathogen interactions have been reviewed recently, and several excellent suggestions for their modifications and improvements were made (Khan, 1993; Sikora & Carter, 1987). However, nonfumigant nematicides have been used in several studies to accurately document the role and interaction of plant-parasitic nematodes with other pathogens under field conditions such as those dealing with *H. glycines* and Fusarium wilt pathogen on soybean (Minton et al., 1985) and *Pratylenchus* spp. and *Verticillium* spp. on potato (Morsink, 1967; Rowe et al, 1987). Also, several studies have employed the use of field microplots of various dimensions utilizing natural field soils and various combinations of inoculum densities of both group of pathogens (Abawi & Barker, 1984; Martin et al., 1982; Rowe et al., 1985; Scholte & s'Jacob, 1989). Other techniques utilized to study nematode-pathogen interactions have included bridging, layering and grafting of roots or stem tissues (Sidhu & Webster, 1977) and double-root or split-root techniques (Faulkner et al., 1970; Conroy et al., 1972; Kotcon & Rouse, 1984). In tests employing such techniques, the nematode and the other plant pathogen were either in contact with the same root or were added to a separate portion of the divided root system.

Interactions between plant-parasitic nematodes and other plant pathogens have often been demonstrated by employing sequential inoculation of the pathogens involved, especially when root-knot nematodes were included (Powell, 1979; Webster, 1985; Mai & Abawi, 1987; Khan, 1993). In such tests, the particular host plants were usually inoculated with the nematodes 2 to 4 wk prior to inoculation with the other selected pathogen. The latter practice became widely used because of earlier findings that the incidence, rate of development, and severity of wilt on *Fusarium*-susceptible and *Fusarium*-tolerant crop cultivars was most pronounced when root-knot nematodes preceded the wilt pathogens by 2 to 4 wk. Although many reports on nematode-pathogen interactions have employed simultaneous inoculation procedures, this approach should be used more often as simultaneous infections of plants with various pathogens are probably more representative of natural conditions (Sikora & Carter, 1987). Selection of time of inoculations should be based on an analysis of disease development in the field. It has been proposed that such investigations should consider the use of factorial design experiments. These experiments should include ranges of inoculum densities of both pathogens, inoculation periods for both pathogens, ages of the host plant at inoculation time, and proper selections of the pre- and postinoculation conditions (Sikora & Carter, 1987; Evans & Haydock, 1993; Khan & Dasgupta, 1993).

The individual and combined effects of the interacting nematode and associated pathogen have been assessed by determining growth and yield parameters, nematode damage and reproduction, and plant-disease incidence and severity. However, limited quantitative data, if any, have been presented on the reproduction of the other interacting pathogen, the epidemiology of the disease and the physiological or biochemical characterization of the resultant interactions being studied.

Nematode-pathogen interactions can be synergistic, additive, antagonistic, or neutral (Chen & Bird, 1994; Evans & Haydock, 1993; Francl & Wheeler, 1993; MacGuidwin & Rouse, 1990; Sikora & Carter, 1987; Abawi & Jacobsen,

1984, Starr et al., 1996; Sankaralingam & McGawley, 1994a). Synergism and antagonism are terms describing quantitative plant disease interruptions in which the combined effect of a plant-parasitic nematode and another plant disease organism is either greater or less than the sum of the effects of the individual organisms. Thus, statistical evidence should be produced in all nematode-pathogen investigations to support the existence of quantitative interactions (Sikora & Carter, 1987). Chen and Bird (1994) reported that synergistic, additive, and antagonistic interactions between *P. penetrans* and *V. dahliae* associated with *S. tuberosum* in repeated greenhouse tests occurred once, nine times, and two times, respectively.

The recent availability and adaptations of molecular biology techniques to plant nematology will greatly advance our knowledge of nematode-host relationships and nematode-microbial interactions. Genetic and biochemical approaches to characterize nematode and host genes and their products using complementary DNA libraries, PCR (polymerase chain reaction), RFLP (restriction fragment length polymorphism), ELISA, monoclonal antibodies and others have already contributed greatly to our understanding of nematode parasitism of plants and host responses (Burrows, 1992; Goverse et al., 1994; Esmenjaud et al., 1993; Caswell-Chen et al., 1993). Such approaches and techniques are needed in studies dealing with interactions involving plant-parasitic nematodes and other pests.

MECHANISMS OF NEMATODE-PATHOGEN INTERACTIONS

In spite of the large number of studies conducted dealing with the interactions between numerous combinations of plant-parasitic nematodes and plant pathogens on a variety of crop plants, the exact mechanism(s) involved in the observed interactions and their impact on disease incidence and severity are not clearly understood (Pitcher, 1965; Bergeson, 1972; Webster, 1985; Powell, 1979; Mai & Abawi, 1987; Khan, 1993). For example, species of the root-knot nematode have been shown, and rather convincingly, to interact with *F. oxysporum* f. sp. *lycopersici* to enhance the onset of Fusarium wilt symptoms and the severity of wilt on *Fusarium*-tolerant tomato cultivars such as Rutgers. However, the exact mechanism involved in this interaction has not been fully elucidated, although the possible involvement of histopathological alterations of affected tissues, accumulation of certain metabolites in infected tissues and the rhizosphere have been advanced to explain this interaction. Another example that warrants study is the synergistic interaction between *V. dahliae* and *P. penetrans*, but not with *P. crenatus*, in the potato early dying syndrome. The following discussion is a summary of reported mechanisms that might be operative in various nematode-pathogen interactions. The suggested mechanisms have been basically arranged according to the role that plant-parasitic nematodes play in the development of the disease complexes.

Nematodes as Vectors of Plant Pathogens

The Longidorid (*Xiphinema*, *Longidorus* and *Paralongidorus*) and the Trichodorid (*Trichodorus* and *Paratrichodorus*) nematodes are efficient vectors of several important nepo- and tobraviruses, respectively (Hussey & McGuire, 1987, Weischer, 1993). The mechanism involved in virus transmission by nematodes is an ingestion-egestion type (Khan, 1993). Nematodes acquire the virus passively while feeding on cell sap of a virus-infected host plant. Ingested particles of the particular compatible virus are selectively absorbed on specific retention sites, depending on the nematode species. The structure of the virus coat protein appears to determine the retention site. Nematode transmitted viruses are noncirculative. Each virus is generally transmitted by one nematode specie, but there are a few exceptions. Transfer of the virus particles from retention sites into host plant cells occurs during feeding. It has been suggested that the dissociation of the virus particles from retention sites is accomplished through the changes in the pH of esophageal lumen by the esophageal gland secretions. This results in alteration of the surface charge of the virus particle, which is egested during feeding.

Plant-parasitic nematodes also are known to vector bacterial plant pathogens externally on their surface (Sitaramaiah & Pathak, 1993; Khan, 1993; McKay & Ophel, 1993). Examples include the vectoring of *Clavibacter tritici* (*Corynebacterium michiganense* pv. *tritici*) the causal agent of yellow ear rot of wheat by *Anguina tritici*; *Rodococcus faciens* (*C. faciens*) by *A. ritzemabosi* and *A. fragariae* resulting in the cauliflower disease syndrome of strawberry; *C. michiganensis* subsp. *isidiosum* (*C. michiganensis* pv. *isidiosum*) the causal agent of alfalfa wilt by *D. dipsaci*; and *C. toxicus* by *Anguina funestra* into developing seeds resulting in the annual ryegrass toxicity to livestock. Cells of bacterial pathogens have been shown attached to the respective nematode surfaces. Thus, the transport of the bacterial pathogens was considered necessary for the disease and the localized disseminations of these pathogens. An adhesin-receptor theory was suggested as the mechanism for the adhesion of bacterial cells and its specificity to the particular nematodes.

The possible vectoring of fungal pathogen propagules on nematode cuticles has not been substantiated, although a few reports suggest they occur (Khan & Pathak, 1993). Internal transport of fungal propagules by plant-parasitic nematodes has not been documented. Fungal and bacterial propagules probably cannot pass through the small stylet lumen, which is usually less than 1 μ in diameter. Accordingly, only viruses have been clearly documented to be carried internally by plant-parasitic nematodes.

Nematodes as Wound Agents

Plant-parasitic nematodes produce various type of wounds while feeding on or entering host plant roots. Wounds produced by ectoparasitic nematodes appear as small micropunctures, whereas the activities of migratory nematodes often result in the formation of necrotic lesions on the epidermal layer and ruptured cells in the form of tunnels within the cortical layer. Sedentary endoparasitic nematodes migrate intercellularly between cortical cells until they reach the vascular tissues and establish feeding sites known as giant cells or syncytia. Wounds caused by plant-parasitic nematodes are considered to be of major importance in interactions involving nematode-bacterial pathogens. Many plant pathogenic bacteria enter their host plants through natural or artificial wounds. Thus, the puncture and rupture of roots cells by nematodes and the availability of necrotic tissues aid in the rapid establishment of bacteria in host tissue and favor their subsequent invasion of healthy tissues. Wilt diseases incited by *P. solanacearum* in various hosts have been shown to be increased in incidence and severity as a result of nematode incited wounds and, to a lesser degree, by artificial wounds (Khan, 1993). The nature of injury caused by *P. penetrans* makes such nematode-infected roots particularly suitable as infection courts for weaker pathogens and secondary root invaders (Brodie, 1984; Mai et al., 1981). It also has been suggested that a number of Fusarium and Verticillium wilt pathogens may require wounds, such as those incited by *Meloidogyne* spp. or *Pratylenchus* spp., for enhanced infection of their host plants.

Nematodes as Substrate Modifiers

Interactions between plant-parasitic nematodes and other plant pathogens are usually explained by changes in the anatomical, physiological and biochemical parameters of host plants. Available data show that plant-parasitic nematodes cause extensive modifications of their host tissues, but the intensity and complexity of such modifications depend on the feeding habit of the nematode involved. Ectoparasitic nematodes generally caused the least modifications, whereas the greatest modifications were induced by the sedentary endoparasitic nematodes. Interactions between root-knot nematodes and Fusarium wilt pathogens have received the most attention in this regard (Mai & Abawi, 1987; Prot, 1993; Khan, 1993).

Root exudates may be altered by nematode infection and can lead to changes in the pH of the rhizosphere and a selective increase of competitors or antagonists (Wang & Bergeson, 1974; Mai & Abawi, 1987). The quantity and quality of root exudation significantly influence microbial activities in the rhizosphere and rhizoplane (Cook & Baker, 1983). Root exudates overcome soil fungistasis and stimulate chlamydospores and other resting propagules of *Fusarium* spp. and other pathogens to germinate and grow in the rhizosphere-rhizoplane region prior to their penetration of host tissues (Cook & Baker, 1983). Root exudates are relatively limited when a plant is young, but the nature and volume change as the plant grows older. Root-knot nematodes may change this pattern, resulting in considerably more damage to young plants (Mai & Abawi, 1987). Van Gundy et al. (1977) reported that exudates from *M. incognita*-infected tomato roots attracted hyphae of *Rhizoctonia solani*, enhanced initial sclerotial formation, and increased severity of root decay. Root exudates from root-knot infected roots were found to contain higher concentrations of Ca, Mg, Na, K, Fe, and Cu, amino acids and reducing compounds as compared to those from noninfected roots (Melakebrhan et al., 1985; Van Gundy et al., 1977; Khan, 1993). Bergeson et al. (1970) reported a consistent increase in the population density of *F. oxysporum* f. sp. *lycopersici* and a significant decrease in the population density of actinomycetes in the rhizosphere of *M. javanica*-infected tomato roots. It was concluded that infection by nematodes may have the dual effect of stimulating some plant pathogens while inhibiting antagonists.

Investigations of the interactions between root-knot nematodes and Fusarium wilt pathogens strongly suggest that host-tissue modifications are essential for increased wilt disease severity. This predisposition-phenomenon hypothesis became widely accepted among researchers with the evidence that the severity of wilt diseases was greatest when root-knot nematodes were inoculated 2 to 4 wk prior to the inoculation of host plants with the wilt pathogens, as compared to simultaneous inoculation with both pathogens (Porter & Powell, 1967; Powell, 1971, 1979). The hypothesis was further supported by results from experiments employing split-root, stem layering, and grafting techniques that showed predisposition to Fusarium wilt pathogens can be attained through systemically translocatable metabolites produced by the host at the site of nematode activity (Sidhu & Webster, 1977; Webster, 1985). Juveniles of root-knot nematodes establish their sedentary feeding sites in the xylem parenchyma tissues of the root, resulting in substantial changes in the morphology, anatomy, and biochemistry of the host plant. Since Fusarium wilt pathogens also are active in the xylem tissues, it is not surprising that these tissues were suspected to be the major site of interaction among these pathogens (Mai & Abawi, 1987; Khan, 1993; Webster, 1985).

Root-knot nematode infection sites (giant cells and surrounding xylem parenchyma cells) are maintained by the nematodes in a perpetual juvenile state with high metabolic activity (metabolic sink). This extended juvenile state of affected cells may delay the rate of maturation and suberization of other vascular tissues, a condition that may assist the Fusarium wilt pathogens in penetrating and becoming established in the xylem elements. Giant cells of root-knot nematodes have much higher concentrations of hemicellulose, organic acids, free amino acids, proteins, lipids, RNA, DNA, and photosynthates than cells of similar noninfected tissues. Fusarium wilt pathogens as well as other organisms obviously benefit from this enriched medium. The growth rate of Fusarium wilt pathogen hyphae and their colonization intensity was found to be greater in root-knot galls than those in nematode-free root tissues (Powell, 1979; Webster, 1985; Khan, 1993).

Toxins produced by Fusarium wilt fungi have been implicated in the development of wilt symptoms. The production of these toxins is known to be influenced by the level of nutrients in the growth medium. Thus, nutrient-rich host tissues that are modified by root-knot nematode infections may cause quantitative or qualitative changes in toxins produced in these tissues that contribute to increased wilt severity. It also was suggested that infections by root-knot nematodes may contribute to the inhibition of tylose formation (Webster, 1985). Such structures are formed in the xylem vessels by the expansion of cell walls of xylem parenchyma through the xylem pits and have been implicated as a host resistance response against wilt pathogens. Xylem parenchyma cells that are modified by root-knot nematodes are unlikely to produce tyloses. Many other examples exist on substrate modifications induced by plant-parasitic nematodes other than root-knot nematodes (Khan, 1993).

Nematodes as Modifiers of Host Resistance Reaction

Several authors have concluded that infection of host crops by plant-parasitic nematodes, especially the root-knot nematodes, reduces resistance to other plant pathogens, especially the Fusarium wilt pathogens (Powell, 1979; Mai & Abawi, 1987; Webster, 1985; Prot, 1993; Maheshwari et al., 1995). This conclusion was based on results showing that infections by root-knot nematodes increased the incidence, symptom development, and severity of the wilt diseases on *Fusarium*-susceptible and tolerant cultivars of several crop species. However, infection by root-knot nematodes did not alter the resistance of cultivars of cabbage (*Brassica oleracea* var. *capitata* L.), pea (*Pisum sativum* L.), chrysanthemum [*Dendranthema* × *grandiflorum* (Ramat.) Kitam] and other crops that possess single dominant genes for resistance against their respective wilt pathogen. The findings on the effects of root-knot nematodes on the development of Fusarium wilt diseases in tomato cultivars expressing the monogenic genes I-1 and I-2 against *F. oxysporum* f. sp. *lycopersici* have been variable. Failures of the monogenic *Fusarium*-resistant cultivars of tomatoes in the presence of both root-knot nematodes and the Fusarium wilt pathogens have not been reported under field conditions. It has been suggested that the contradictory results may be due to differences in inoculum densities of the pathogens used, aggressiveness of the pathogens, intervals and methods of inoculations, cultivars used, pre- and postinoculation environmental conditions, background soil microorganisms, assessment of wilting indices, and other factors. The loss of *Fusarium* resistance in 'Craigella GCR 161' tomato infected with *M. incognita* was associated with the lack of the antifungal substance rishitin, which was present in the healthy plants. Others have reported that the breakdown of resistance in the monogenic cultivars such as Chico III and Chesapeake is mediated by translocatable metabolites and is under genetic control.

The *Fusarium*-tolerant bean Calima exhibits wilt symptoms caused by *F. oxysporum* f. sp. *phaseoli* only in the presence of *M. incognita* (France & Abawi, 1994). Bean genotype A-107 (resistant to both *M. incognita* and the wilt pathogen) developed Fusarium wilt symptoms and a moderate susceptible reaction to the Fusarium wilt pathogen only after its resistance to *M. incognita* was broken by incubation at high temperature. Unfortunately, the inheritance of the resistance to the Fusarium wilt pathogen in A-107 is not known yet, but these results suggest that it might be controlled by polygenic factors.

Resistance to Fusarium wilt pathogens in crop plants can be either monogenic or polygenic (Bell & Mace, 1981; Beckman, 1989). Monogenic type of resistance is more stable and not as likely to be modified by temperature or infections by plant-parasitic nematodes as polygenic sources of resistance.

OUTLOOK

Significant progress has been made in characterizing and documenting important interactions and associations involving plant-parasitic nematodes and other plant pathogens. Knowledge of these interactions is essential to understanding the etiology and epidemiology of many root diseases and in the formulation of integrated management options for their control. The latter is of special importance in the development of biologically based integrated pest management programs as components of sustainable agricultural production systems.

Previous research has substantiated the interactions and/or associations between plant-parasitic nematodes and fungi, bacteria, viruses, other nematodes, injurious insects, and weed pests. Synergistic, additive, neutral and antagonistic interactions and/or associations have been reported and well documented. In the majority of these reports, the investigators have concluded that the presence of plant-parasitic nematodes, especially the root-knot nematodes, tends to increase the incidence and severity of disease or disease complex. However, interrelationships appear to be quite specific as to species and density of the nematode and other pathogens involved, cultivar of the host species, and test conditions. In addition, growth parameters of the host crops are also decreased, often synergistically, as a result of the interaction among the various pathogens and pests.

Detailed information of the mechanisms involved in the observed interactions among plant-parasitic nematodes, plant pathogens, and other pests remains very limited. Thus, extensive research efforts are needed to elucidate the exact mechanism(s) involved in the various interactions substantiated to date. Investigations on the specificity of interactions as to species of the pathogens and pests involved and cultivars of the host crop are warranted. Understanding how infections by plant-parasitic nematodes modify host tissues and alter or break down different resistance factors of a particular host cultivar to other pathogens is of great importance and is much needed. The various techniques and approaches being utilized in molecular biology and genetic engineering of host plants and characterization of pathogens and pests are needed and would greatly contribute to understanding the mechanisms of interactions involving nematodes. In addition, quantification of the population dynamics of the various pathogens and pests involved in the interaction also needs to be determined over several crop production cycles. Furthermore, such involved interaction studies should be conducted under field conditions and the particular host crops maintained to normal maturity stages in order to assess the impact of the interaction on the various parameters of marketable yield.

REFERENCES

Abawi, G.S., and K.R. Barker. 1984. Effects of cultivar, soil temperature, and population levels of *Meloidogyne incognita* on root necrosis and *Fusarium* wilt of tomatoes. Phytopathology 74:433–438.

Abawi, G.S., and B.J. Jacobsen. 1984. Effect of initial inoculum densities of *Heterodera glycines* on growth of soybean and kidney bean and their efficiency as hosts under greenhouse conditions. Phytopathology 74:1470–1474.

Agrios, G.N. 1988. Plant pathology. Acad. Press, New York.

Alston, D.G., J. R. Bradley, Jr., D.P. Schmitt, and H. D. Coble. 1991a. Response of *Helicoverpa zea* (Lepidoptera: Noctuidae) populations to canopy development in soybean as influenced by *Heterodera glycines* (Nematoda: Heteroderidae) and annual weed population densities. J. Econ. Entomol. 84:267–276.

Alston, D.G., J. R. Bradley, Jr., D.P. Schmitt, and H.D. Coble. 1991b. Relationship of *Heliothis zea* predators, parasitoids and entomopathogens to canopy development in soybean as affected by *Heterodera glycines* and weeds. Entomol. Exp. Appl. 58:279–288.

Anderson, R.M., and R.M. May. 1986. The invasion, persistence and spread of infectious diseases within animal and plant communities. Philos. Trans. R. Soc. London Ser. B 314:533–70.

Armstrong, G.M., and J.K. Armstrong. 1981. Formae speciales and races of *Fusarium oxysporum* causing wilt diseases. p. 391–399. *In* P.E. Nelson et al. (ed.) *Fusarium*: Disease, biology, and taxonomy. Pennsylvania State Univ. Press, Univ. Park.

Atkinson, G.F. 1892. Some diseases of cotton. Alabama Agric. Exp. Stn., Auburn Univ., Bull. 41.

Beckman, C.H. 1989. Recognition and response between host and parasite as determinants in resistance and disease development. p. 153–162. *In* E.C. Tjamos and C.H. Beckman (ed.) Vascular wilt diseases of plants. Springer-Verlag, Berlin.

Begon, M., J.L. Harper, and C.R. Townsend. 1990. Ecology — individuals, populations and communities. Blackwell Sci. Publ., Boston.

Bell, A.A., and M.E. Mace. 1981. Biochemistry and physiology of resistance. p. 431–486. *In* M.E. Mace et al. (ed.) Fungal wilt diseases of plants. Acad. Press, New York.

Bergeson, G.H. 1972. Concepts of nematode-fungus associations in plant diseases complexes: A review. Exp. Parasitol. 32:301–314.

Bergeson, G.H., S.D. Van Gundy, and I.J. Thomason. 1970. Effect of *Meloidogyne javanica* on rhizosphere microflora and *Fusarium* wilt of tomato. Phytopathology 60:1245–49.

Bird, G.W. 1981. Management of plant parasitic nematodes in potato production. p. 223–243. *In* J.H. Lashomb and R. Casagrande (ed.) Advances in potato pest management, Hutchinson Ross Publ. Co., Strundsburg, PA.

Brodie, B.B. 1984. Potato nematodes. p. 167–212. *In* W.R. Nickle (ed.) Plant and insect nematodes. Marcel Dekker. Inc., New York.

Brown, R.H., and B.R. Kerry. 1987. Principles and practice of nematode control in crops. Acad. Press, New York.

Burrows. P.R. 1992. Molecular analysis of the interactions between cyst nematodes and their hosts. J. Nematol. 24:338–342.

Caswell-Chen, E.P., V.M. Williamson, and B.B. Westerdahl. 1993. Viewpoint: Applied biotechnology in nematology. Suppl. J. Nematol. 25:719–730.

Chen, J. 1995. Feature, function, and nature of *Pratylenchus penetrans* and *Verticillium dahliae* interactions associated with *Solanum tuberosum*. Ph.D. diss. Michigan State Univ., East Lansing, MI (Diss. Abstr. AADAA-I9605848).

Chen, J., G.W. Bird, and K.A. Renner. 1995. Influence of *Heterodera glycines* on interspecific and intraspecific competition associated with *Glycine max* and *Chenopodium album*. J. Nematol. 27:63–69.

Chen, J., and G.W. Bird. 1994. Influence of *Pratylenchus penetrans* and *Verticillium dahliae* on *Solanum tuberosum* cultured with a split root-stolon system. J. Nematol. 26:541.

Conroy, J.J., R.J. Green, Jr., and J.M. ferris. 1972. Interaction of *Verticillium albo-atrum* and the root lesion nematode, *Pratylenchus penetrans*, in tomato roots at controlled inoculum densities. Phytopathology 62:362–366.

Cook, R.J., and K.F. Baker. 1983. The nature and practice of biological control of plant pathogens. Am. Phytopathol. Soc., St. Paul, MN.

Eisenback, J.D. 1993. Interactions between nematodes in cohabitance. p. 134–174. *In* M.W. Khan (ed.) Nematode interactions, Chapman and Hill, New York.

Eisenback, J.D., and G.D. Griffin. 1987. Interactions with other nematodes. p. 313–320. *In* J.A. Veech and D.W. Dickson (ed.) Vistas on nematology. Soc. Nematol., Inc., E.O. Painter Print. Co., deLeon Springs, FL.

El-Sherif, A.G. 1991. Interaction between *Meloidogyne incognita* and *Agrobacterium tumefaciens*, and *Fusarium oxysporum* f. sp. *lycopersici* on tomato. J. Nematol. 23:239–242.

Esmenjaud, D., B. Walter, J.C. Minot, R. Voisin, and P. Cornuet. 1993. Biotin-Avidin ELISA detection of grapevine fanleaf virus in the vector nematode *Xiphinema index*. J. Nematol. 25:401–405.

Evans, K., and P.P.J. Haydock. 1993. Interactions of nematodes with root-rot fungi. p. 104–133. *In* M.W. Khan (ed.) Nematode interactions. Chapman and Hall, New York.

Faulkner, L.R., W.J. Bolander, and C.B. Skotland. 1970. Interaction of *Verticillium dahliae* and *Pratylenchus minyus* in *Verticillium* wilt of peppermint: Influence of the nematode as determined by a double root technique. Phytopathology 60:100–103.

Fawcett, H.S. 1931. The importance of investigations on the effects of known mixtures of organisms. Phytopathology 21:545–550.

Fitchen, J.H., and R.N. Beachy. 1993. Genetically engineered protection against viruses in transgenic plants. Annu. Rev. Microbiol. 47:739–763.

France, R.A., and G.S. Abawi. 1994. Interaction between *Meloidogyne incognita* and *Fusarium oxysporum* f. sp. *phaseoli* on selected bean genotypes. J. Nematol. 26:467–474.

Francl, L.J., and T.A. Wheeler. 1993. Interaction of plant-parasitic nematodes with wilt-inducing fungi. p. 80–103. *In* M.W. Khan (ed.) Nematode interactions. Chapman and Hall, New York.

Giblin-Davis, R.M. 1993. Interactions of nematodes with insects. p. 302–344. *In* M.W. Khan (ed.) Nematode interactions. Chapman and Hall, New York.

Gottlieb, D. 1976. Production and role of antibiotics in soil. J. Antibiot. 29:987–1000.

Goverse, A., E.L. Davis, and R.S. Hussey. 1994. Monoclonal antibodies to the esophageal glands and stylet secretions of *Heterodera glycines*. J. Nematol. 26:251–259.

Grogan, R.G. 1981. The science and art of plant disease diagnosis. Annu. Rev. Phytopathol. 19:333–51.

Grumet, R. 1994. Development of virus resistant plants via genetic engineering. Plant Breed. Rev. 12:47–79.

Hackland, A.F., E.P. Rybicki, and J.A. Thomson. 1994. Coat protein-mediated resistance in transgenic plants. Arch. Virol. 139:1–22.

Hasan, A., and M.N. Khan. 1985. The effect of *Rhizoctonia solani, Sclerotium rolfsii*, and *Verticillium dahliae* on the resistance of tomato to *Meloidogyne incognita*. Nematol. Medit. 13:133–136.

Harris, K.F. 1981. Arthropod and nematode vectors of plant viruses. Annu. Rev. Phytopathol. 19:391–426.

Harrison. B.D. 1977. Ecology and control of viruses with soil-inhabiting vectors. Annu. Rev. Phytopathol. 15:331–360.

Harrison, B.D. 1978. The group of nematode-transmitted plant viruses, and molecular aspects of their variation and ecology. p. 255–264. *In* P.R. Scott and A. Bainbridge (ed.) Plant disease epidemiology, Blackwell, London.

Hewitt, W.B., D.J. Raski, and A.C. Goheen. 1958. Nematode vector of soil-borne fanleaf virus of grapevines. Phytopathology 48:586–595.

Hirano, K. 1975. Interrelationships between plant-parasitic nematodes and other plant pathogenic organisms. Rev. Plant Prot. Res. 8:55–68.

Hirunsalee, A., K.R. Barker, and M.K. Beute. 1995. Infection, reproduction potential, and root galling by root-knot nematode species and concomitant populations on peanut and tobacco. J. Nematol. 27:172–177.

Holt, R.D., and J. Pickering. 1985. Infectious disease and species coexistence: A model of Lotka-Voltera form. Am. Natur. 726:196–211.

Huang, J-S. 1987. Interactions of nematodes with Rhizobia. p. 301–306. *In* J.A. Veech and D. W. Dickson (ed.) Vistas on nematology. Soc. Nematol., Inc., E.O. Painter Print. Co., DeLeon Springs, FL.

Hunger, F.W.T. 1901. Een bacterie-ziekte den tomaat. Mededl. Plant., Batavia 48:4–57.

Hussey, R.S., and J.M. McGuire. 1987. Interaction with other organisms. p. 313–320. *In* R.H. Brown and B.R. Kerry (ed.) Principals and practice of nematode control in crops. Acad. Press, New York.

Jaffee, B.A. 1981. Etiology of an apple replant disease. Ph.D. diss. Cornell Univ., Ithaca, NY (Diss. Abstr. ADA81-19421).

Jaffee, B.A., G.S. Abawi, and W.F. Mai. 1982. Fungi associated with roots of apple seedlings grown in soil from an apple replant site. Plant Dis. 66:942–944.

Johnson, A.W., D.A. Smittle, D.R. Sumner, and N.C. Glaze. 1994. Efficacy and compatibility for fenamiphos and EPTC applied in irrigation water for nematode and weed control in snapbean production. J. Nematol. 26 (4S):690–696.

Khan, M.W. (ed). 1993. Nematode interactions. Chapman and Hall, New York.

Khan, M.W., and M.K. Dasgupta. 1993. The concept of interaction. p. 42–54. *In* M.W. Khan (ed.) Nematode interactions. Chapman and Hall, New York.

Khan, M.W., and K.N. Pathak. 1993. Nematodes as vectors of bacterial and fungal plant pathogens. p. 251–272. *In* M.W. Khan (ed.) Nematode interactions. Chapman and Hall, New York.

Khan, T.A., and S.I. Husain. 1989. Relative resistance of six cowpea cultivars as affected by the concomitance of two nematodes and a fungus. Nematol. Medit. 17:39–41.

Kotcon, J.B., and D.I. Rouse. 1984. Root deterioration in the potato early dying syndrome: causes and effects of root biomass reductions associated with colonization by *Verticillium dahliae*. Am. Potat. J. 61:557–568.

Lamberti, F., and F. Roca. 1987. Present status of nematodes as vectors of plant viruses. p. 321–328. *In* J.A. Veech and D.W. Dickson (ed.) Vistas on nematology. Soc. Nematol., Inc., E. O. Painter Print. Co., DeLeon Springs, FL.

Lamberti, F., C.E. Taylor, and J.W. Seihorst (ed.). 1975. Nematode vectors of plant viruses, Plenum Press, New York.

LaMondia, J.A. 1992. Predisposition of broadleaf tobacco to *Fusarium* wilt by early infection with *Globodera tabacum tabacum* or *Meloidogyne hapla*. J. Nematol. 24:425–431.

Lasserre, F., R. Rivoal, and R. Cook. 1994. Interactions between *Heterodera avenae* and *Pratylenchus neglectus* on wheat. J. Nematol. 26:336–344.

MacGuidwin, A.E., and D.I. Rouse. 1990. Role of *Pratylenchus penetrans* in the potato early dying disease of russet Burbank potato. Phytopathology 80:1077–1082.

Mai, W.F., I.A. Merwin, and G.S. Abawi. 1994. Diagnosis, etiology, and management of replant disorders in New York cherry and apple orchards. Acta Hortic. 363:33–41.

Mai, W.F., and G.S. Abawi. 1987. Interactions among root-knot nematodes and *Fusarium* wilt fungi on host plants. Annu. Rev. Phytopathol. 25:317–338.

Mai, W.F., and G.S. Abawi. 1981. Controlling replant diseases of pome and stone fruits in the Northeastern United States by replant fumigation. Plant Dis. 65:859–864.

Mai, W.F., B.B. Brodie, M.B. Harrison, and P. Jatala. 1981. Nematode diseases of potatoes. p. 93–101. *In* W.J. Hooker (ed.) Compendium of potato diseases. Am. Phytopathol. Soc., St. Paul, MN.

McKay, A.C., and K.M. Ophel. 1993. Toxigenic Clavibactor/Anguina associations infecting grass seedheads. Annu. Rev. Phytopathol. 31:151–167.

Maheshwari, T.U., S.B. Sharma, D.D.R. Reddy, and M.P. Haware. 1995. Co-infection of wilt-resistant chickpeas by *Fusarium oxysporum* f. sp. *ciceri* and *Meloidogyne javanica*. J. Nematol. 27:649–653.

Martin, M.J., R.M. Riedel, and R.C. Rowe. 1982. *Verticillium dahliae* and *Pratylenchus penetrans*: Interactions in the early dying complex of potato in Ohio. Phytopathology 72:640–647.

Mayol, P.S., and G.B. Bergeson. 1970. The role of secondary invaders in *Meloidogyne incognita* infection. J. Nematol. 2:80–83.

McLean, K.S., and G.W. Lawrence. 1995. Development of *Heterodera glycines* as affected by *Fusarium solani*, the causal agent of sudden death syndrome of soybean. J. Nematol. 27:70–77.

Melakebrhan, H., R.C. Brooke, J.M. Webster, and J.M. D'Auria. 1985. The influence of *Meloidogyne incognita* on the growth, physiology and nutrient content of *Phaseolus vulgaris*. Physiol. Plant Pathol. 26:259–68.

Minton, N.A., M.B. Parker, and D.R. Summer. 1985. Nematode control related to *Fusarium* wilt in soybean and root-rot and zinc deficiency in corn. J. Nematol. 17:314–21.

Morsink, F. 1967. *Pratylenchus penetrans*—its interaction with *Verticillium albo-atrum* in the Verticillium wilt of potatoes and its attraction by various chemicals. Ph.D. diss. Univ. New Hampshire, Durham (Diss. Abstr. AAD67-09958)

Ng'ambi, T.B.S., R.C. Rufty, and K.R. Barker. 1995. Response of root-knot resistant tobacco to concomitant populations of Meloidogyne species. Plant Dis. 79:1008–1013.

Norton, D.C. 1978. Ecology of plant-parasitic nematodes. John Wiley & Sons, Inc., New York.

Pantone, D.J. 1995. Replacement series of the competition between a weed and a crop as influenced by a plant parasitic nematode. Fund. Appl. Nematol. 18:93–97.

Pitcher, R.S. 1965. Interrelationships of nematodes and other pathogens of plants. Helminthol. Abstrs. 34:1–17.

Pitcher, R.S. 1978. Interactions of nematodes with other pathogens. p. 63–77. *In* J.F. Southy (ed.) Plant nematology. HM Stat. Office, London, England.

Porter, D.M., and N.T. Powell. 1967. Influence of certain *Meloidogyne* species on *Fusarium* wilt development in flue-cured tobacco. Phytopathology 57:282–85.

Poinar, G.O., Jr. 1975. Entomogenous nematodes. E.J. Brill, Leiden.

Powell, N.T. 1971. Interactions between nematodes and fungi in disease complexes. Annu. Rev. Phytopathol. 9:253–274.

Powell, N.T. 1979. Internal synergisms among organisms inducing disease. p. 113–133. *In* J.G. Horsfall and E.B. Cowling (ed.) Plant disease, an advanced treatise. Vol. 4. Acad. Press, New York.

Powelson, M.L., and R.C. Rowe. 1993. Biology and management of early dying of potato. Annu. Rev. Phytopathol. 31:111–126.

Powelson, M.L., K.B. Johnson, and R.C. Rowe. 1993. Management of diseases caused by soilborne pathogens. p. 149–158. *In* R.C. Rowe (ed.) Potato health management. APS Press, St. Paul, MN.

Prot, J. 1993. Biochemical and genetic basis of fungus-nematode interactions. p. 273–285. *In* M.W. Khan (ed.) Nematode interactions. Chapman and Hall, New York.

Richards, B.N. 1976. Introduction to the soil ecosystem. Longman, London.

Riedel, R.M., R.C. Rowe, and M.J. Martin. 1985. Differential interaction of *Pratylenchus crenatus, Pratylenchus penetrans*, and *Pratylenchus scribneri* with *Verticillium dahliae* in potato early-dying disease. Phytopathology 75:419–422.

Roberts, P.A., C.A. Frate, W.C. Mathews, and P.P. Osterli. 1995. Interactions of virulent *Meloidogyne incognita* and *Fusarium* wilt on resistant cowpea genotypes. Phytopathology 85:1288–1295.

Rowe, R.C., J.R. Davis, M.L. Powelson, and D.I. Rouse. 1987. Potato early dying: causal agents and management strategies. Plant Dis. 71:482–489.

Rowe, R.C., R.M. Riedel, and M.J. Martin. 1985. Synergistic interactions between *Verticillium dahliae* and *Pratylenchus penetrans* in potato early dying disease. Phytopathology 75:412–418.

Russin, J.S., M.B. Layton, D.J. Boethel, E.C. McGawley, J.P. Snow, and G.T. Berggren. 1989. Development of *Heterodera glycines* on soybean damaged by soybean looper and stem canker. J. Nematol. 21:108–114.

Russin, J.S., M.B. Layton, D.J. Boethel, E.C. McGawley, J.P. Snow, and G.T. Berggren. 1990. Growth, nodule development, and N_2-fixing ability in soybean damaged by an insect-fungus-nematode pest complex. Entomol. Soc. Am. 83:247–254.

Sankaralingam, A., and E.C. McGawley. 1994a. Interrelationships of *Rotylenchulus reniformis* with *Rhizoctonia solani* on cotton. J. Nematol. 26:475–485.

Sankaralingam, A., and E.C. McGawley. 1994b. Influence of *Rhizoctonia solani* on egg hatching and infectivity of *Rotylenchulus reniformis*. J. Nematol. 26:486–491.

Scholte, K., and J.J. s'Jacob. 1989. Synergistic interactions between *Rhizoctonia solani, Verticillium dahliae, Meloidogyne* spp. and *Pratylenchus neglectus* in potato. Potato Res. 32:387–95.

Sidhu, G., and J.M. Webster. 1977. Predisposition of tomato to the wilt fungus (*Fusarium oxysporum lycopesici*) by the root-knot nematode (*Meloidogyne incognita*). Nematologia 23:433–442.

Sikora, R.A., and W.W. Carter. 1987. Nematode interactions with fungal and bacterial plant pathogens-facts or fantasy. p. 307–320. In J.A. Veech and D.W. Dickson (ed.) Vistas on nematology. Soc. Nematol., Inc., E.O. Painter Print. Co., DeLeon Springs, FL.

Sitaramaiah, K., and K.N. Pathak. 1993. Nematode bacterial disease interactions. p. 222–250. *In* M.W. Khan (ed.) Nematode interactions. Chapman and Hall, New York.

Smith, G.S. 1987. Interactions of nematodes with mycorrhizal fungi. p. 292–300. *In* J.A. Veech and D.W. Dickson (ed.) Vistas on nematology. Soc. Nematol., Inc., E.O. Painter Print. Co., DeLeon Springs, FL.

Starr, J.L., M.-Y. Shim, T.A. Lee, Jr., and C.E. Simpson. 1996. Additive effects of *Meloidogyne arenaria* and *Sclerotium rolfsii* on peanut. J. Nematol. 28:99–106.

Taylor, C.E. 1979. *Meloidogyne* interrelationships with microorganisms. p. 375–398. *In* F. Lambeti and C.E. Taylor (ed.) Root-knot nematodes (*Meloidogyne* species), systematic, biology and control, Acad. Press, New York.

Taylor, C.E., and D.J.F. Brown. 1981. Nematode-virus interactions. p. 282–301. *In* B.T. Zuckerman and RA. Rohde (ed.) Plant parasitic nematodes. Vol. 3. Acad. Press, New York.

Umesh, K.C., and H. Ferris. 1994. Influence of temperature and host plant on the interaction between *Pratylenchus neglectus* and *Meloidogyne chitwoodi*. J. Nematol. 26:65–71.

Van Gundy, S.D., J.D. Kirkpatrick, and J. Golden. 1977. The nature and role of metabolic leakage from root-knot nematode galls and infection by *Rhizoctonia solani*. J. Nematol. 9:113–121.

Wallace, H.R. 1978. Diagnosis of plant diseases of complex etiology. Annu. Rev. Phytopathol. 16:379–402.

Wang, E.L.H.,and G.B. Bergeson. 1974. Biochemical changes in root exudate and xylem sap of tomato plants infected with *Meloidogyne incognita*. J. Nematol. 6:194–202.

Webster, J.M. 1985. Interaction of *Meloidogyne* with fungi on crop plants. p. 183–192. *In* J.N. Sasser and C.C. Carter (ed.) An advance treatise on *Meloidogyne*. Vol. 1. North Carolina State Univ. Graphics, Raleigh.

Weischer, B. 1993. Nematode-virus interactions. p. 251–231. *In* M.W. Khan (ed.) Nematode interactions. Chapman and Hall, New York.

Wescott, S.W. III, S.V. Beer, and W.C. Stiles. 1986. Infection of apple roots by actinomycetes associated with soils conducive to apple replant disease. Plant Dis. 70:1125–1128.

Wheeler, T.A., and R.M. Riedel. 1994. Interactions among *Pratylenchus penetrans*, *P. scribneri*, and *Verticillium dahliae* in the potato early dying disease complex. J. Nematol. 26:228–234.

Yeates, G.W., D.A. Wardle, and R.N. Watson. 1993. Relationships between nematodes, soil microbial biomass and weed-management strategies in maize and asparagus cropping systems. Soil Biol. Biochem. 25:869–876.

8 Crop- and Nematode-Management Systems

JAMES P. NOE

University of Georgia
Athens, Georgia

Plant-parasitic nematodes are soilborne microorganisms that lack the capacity for significant self-directed, long-distance movement. Usually, growers need only manage for the nematodes occurring in a particular field, especially at the beginning of the growing season. Further, all plant-parasitic nematodes are obligate parasites, i.e., they can feed only on a living host plant. For these reasons, crop rotation is an ideal potential tactic for nematode management in annual or short-term corps. A complete description of a cropping system includes more than simply the crop sequence over time, but rotations are of vital importance in the management of plant-parasitic nematodes, and have occupied a central place in the development of cropping systems.

A grower only has to alter the cropping sequence sufficiently to keep the nematode population densities in a given field below economic threshold levels. Nematodes do not move into a field from other infested fields during a single growing season, as do more mobile insect and fungal pests. If a grower does not plant a suitable host in a field, the endemic nematodes will starve or be limited to weed hosts. All of the previous arguments would lead to the conclusion that managing nematode parasites in an agricultural system is remarkably easy—simply rotate to nonhost crops. However, for a number of biological, tactical, and economic reasons, the implementation of a crop rotation system for nematode management is a difficult and complex undertaking. Crop rotations for nematode management also must be considered in the larger framework of cropping systems, which includes all aspects of the farming operation. A cropping system includes agronomic practices, management practices for other pests, and cultural and market constraints on the availability of management alternatives.

CROP ROTATION

Crop rotation has been, and continues to be, a management tactic of primary importance for limiting losses due to plant-parasitic nematodes. In 1911, Bessey (1911) suggested the use of certain crop rotations to minimize losses to root-knot nematodes in susceptible vegetable crops, providing one of the first reported use of this tactic specifically for nematode control. Long before this era,

---

 Plant and Nematode Interactions, Agronomy Monograph no. 36.

ancient agricultural crop sequencing patterns were developed empirically by growers, often to maintain the productivity of limited arable land holdings in the presence of nematodes and other soilborne pests and diseases. Rodriguez-Kabana and Canullo (1992) reported that evidence for the existence of specific crop sequences to enhance productivity in the presence of soilborne pathogens was found in ancient Chinese and Hindu writings, in the pre-Columbian Inca cultures, and in Medieval and Renaissance Europe. Since Bessey's more modern report in 1911, much attention has focused on crop rotations for nematode management. These efforts, however, declined in the 1960s to 1970s as chemical management and development of resistant cultivars became prominent. Nevertheless, with recent losses in chemical management tools and slower development of new sources for nematode resistance, renewed emphasis has been directed toward optimization of cropping systems for nematode management.

Crop rotation systems have been demonstrated to be useful against a limited number of nematode species, but variability in the outcomes and in the conditions under which specific systems were evaluated have so far precluded any broad conclusions as to the effectiveness of crop rotations in modern agriculture (Trivedi & Barker, 1986). In the following discussion of crop rotations, emphasis has been placed on those contributions that showed significant effects on nematode population densities. Enhancement of yield alone was not considered sufficient evidence for nematode suppression. Crop rotations may result in increased or decreased yields for a number of different reasons, including effects on other pests and on soil fertility. Assessment of nematode numbers is required to establish that nematode suppression was achieved.

A primary factor in the potential usefulness of crop rotation for nematode control is the host range of the target nematode species. Some species, such as the soybean cyst nematode (*Heterodera glycines*), have very limited host ranges, and it is relatively easy to suggest a nonhost crop. Other species, including root-knot nematodes (*Meloidogyne* spp.), have very wide host ranges, encompassing many common weeds, and rotation to a nonhost may be difficult. Another nematode-biology factor in designing crop rotations is the length of time a particular species may survive in the absence of a host. *Heterodera glycines* may survive for 7 to 10 yr in the absence of a host (Slack et al., 1972), whereas most *Meloidogyne* species usually decline 90 to 95% between growing seasons (Bergeson, 1959). Compensating mechanisms are present in both species to enable survival in the absence of the economically important host, although each species has developed a different tactic for survival. Soil texture, moisture, and temperature interact to affect nematode survival times, and also influence the rates of population increase under a susceptible crop. Plant-parasitic nematodes commonly occur in multiple-species communities, and a rotation that controls one species may enhance population densities of another. Interactions between nematode species in the same genus, and among pathogenic races of the same species, affect the success of a crop rotation system (Hirunsalee et al., 1995).

These factors must all be considered in determining the length of time required for an effective rotation system. Finally, a selection of suitable rotation crops needs to be identified to fill in the rotation scheme. This step has caused the most difficulty in modern agricultural production systems, where producers typically grow only one or two crops and are reluctant to rotate high value crops from prime arable land. If a required nonhost crop is not already being grown, it is often difficult to gain acceptance for a new crop, and alternative crops may have lower economic returns. Additional capitalization may be required for producing a new crop, both at the grower and wholesale buyer levels, and often these economic considerations far outweigh any perceived benefits from nematode control. For these reasons, only a few crop rotations have been widely implemented and accepted based solely on the need for management of plant-parasitic nematodes.

SPECIFIC ROTATION SCHEMES

A number of rotation schemes of potential use have been evaluated, leading to the development of crop rotations of practical significance for many of the most important nematode species (Table 8–1). A 1-yr rotation to cotton (*Gossypium hirsutum*) was shown to decrease subsequent population densities of the peanut root-knot nematode (*M. arenaria* race 1) on peanut (*Arachis hypogaea* L.) by 43% (Rodriguez-Kabana et al., 1987). This combination is a very attrac-

Table 8–1. Examples of potential rotation crops for important nematode species.

Nematode species	Primary hosts	Rotation crops
Belonolaimus longicaudatus	Peanut	Tobacco (*Nicotiana tabacum* L.), watermelon (*Citrullus lanatus*) (Robbins & Barker, 1973)
Criconemella ornata	Peanut	Cotton, soybean (Barker et al., 1982)
Globodera rostochiensis	Potato	Most crops except eggplant (*Solanum melongena* L.), tomato (Brodie et al., 1993; Mai & Lownsberry, 1952)
Ditylenchus dipsaci	Cereals, onion	Corn, barley, resistant oat (*Avena sativa* L.) (Rivoal and Cook, 1993)
Heterodera avenae	Cereals	Any nongraminaceous crop (Rivoal & Cook, 1993)
H. glycines	Soybean	Resistant cultivars, most nonleguminous crops (Riggs & Niblack, 1993)
H. schachtii	Sugar beet	Cereals, corn, potato, resistant oilseed crops (Cooke, 1993, Mai & Abawi, 1980)
Meloidogyne arenaria Race 1	Peanut	Cotton (Rodriguez-Kabana, et al. 1987)
Meloidogyne arenaria Race 2	Soybean, vegetables	Cotton, peanut
M. chitwoodi	Potato	Resistant alfalfa (O'Bannon & Santo, 1984)
M. hapla	Carrot [*Daucus carota* subsp. *sativus* (Hoffm.)], potato	Onion (*Allium cepa* L. var. *cepa*), radish (*Raphanus sativus* L.), lettuce (*Latuca sativa* L.) (Wilson, 1964)
M. incognita	Vegetables, cotton, others (very wide host range)	Resistant cultivars, peanut
M. javanica	Vegetables, legumes	Cotton, peanut
Rotylenchulus reniformis	Cotton, soybean, many tropical crops	Grain sorghum (Thames & Heald, 1974); corn (Braithwaite, 1974)

tive rotation at this time due to the high value and extensive planting of both crops in the southeastern USA. Peanut is not a host for the southern root-knot nematode (*M. incognita*), which is a major pathogen on cotton. Similar benefits for cotton from the peanut–cotton rotation were shown by Kirkpatrick & Sasser (1984). This rotation is widely practiced in regions where peanut is grown.

Southern root-knot nematode numbers can be reduced to nearly undetectable levels in cotton and corn (*Zea mays* L. subsp. *mays*) by various 4-yr rotations that included soybean [Glycine max (L.) Merr.] and peanut (Johnson et al., 1975). This extensive study also identified another potential problem in the identification of optimal crop rotations, in that population levels of other plant-parasitic nematode species were increased by the rotations that were most favorable for control of root-knot. Numbers of *M. hapla* juveniles increased on peanut in a 2-yr rotation with peanut, soybean, and vegetables (Johnson, 1985). The increase of *M. hapla* in this system was unexpected because only *M. incognita* had previously been detected in the plots. Yields of the second crop of peanut were reduced 51% in comparison to plots where nematicides were applied.

The nontarget species problem was again highlighted in a study, which showed that rotation to corn would decrease population densities of *H. glycines* on soybean, but the same rotation increased numbers of the peanut root-knot nematode, another damaging pathogen on soybean (Rodriguez-Kabana et al., 1988b). Kinloch (1983), however, found that rotation to corn for at least 2 yr decreased population densities of the southern root-knot nematode by as much as 50% on soybean. In consideration of the inverse system in southern Florida, root-knot population densities were lower in corn following sorghum (*Sorghum bicolor*) than in corn following soybean (McSorley & Gallaher, 1993), but numbers of the ring nematode (*Criconemella* spp.) were higher following sorghum. These results indicate the difficulty in designing crop rotations where plant-parasitic nematodes occur in multiple-species communities.

Forage crops may be useful rotation alternatives for root-knot susceptible vegetables. A 3-yr rotation to 'Coastal' bermudagrass [*Cynodon dactylon* (L.) Pers.] reduced population densities of *M. incognita* to extremely low levels in a vegetable production system (Johnson et al., 1995). This rotation had the additional advantage of reducing colony-forming units of the Fusarium-wilt pathogen (*Fusarium oxysporum*). Rotation to bahiagrass (*Paspalum notatum* Flugge) reduced population levels of *M. arenaria* to near zero, and increased yield of soybean by more than 100% (Rodriguez-Kabana et al., 1989b). In the midwestern USA's corn-producing region, rotation to alfalfa (*Medicago sativa* L. subsp. *sativa*) for 1 yr reduced populations of the sting nematode (*Belonolaimus* spp.) more than 90% as compared to continuous corn (Todd, 1991). Even in a multiple-crop per season system, such as upland rice (*Oryza sativa* L.), a simple rotation sequence may be helpful. In West Africa, alternating *Sesbania rostrata* Bremek & Oberm with rice decreased population levels of the rice root nematode (*Hirschmanniella oryzae*) by as much as 67% compared to continuous cultivation of rice (Germani et al., 1983).

Crop sequences also have been examined in combination with nematicide usage. A 6- to 8-yr rotation to barley (*Hordeum vulgare* L.) was required to provide acceptable yield of potato (*Solanum tuberosum* L.) in the presence of the

Golden potato-cyst nematode (*Globodera rostochiensis*), but this rotation was equally effective with only 2-yr duration when combined with nematicide treatment at the time of potato planting (Whitehead et al., 1991). In contrast, a 1-yr rotation to corn reduced numbers of the ring nematode (*C. ornata*) by as much as 50% on subsequent peanut crops, but treatment with nematicides in a peanut monoculture failed to reduce nematode numbers, and did not enhance yields (Ayers et al., 1989). Tillage has been shown to have no interaction with crop rotation in effects on population densities of *M. incognita* on corn, although tillage and rotation had interacting effects on the numbers of ring nematodes (McSorley & Gallaher, 1993). These types of combined management studies allow growers to evaluate management alternatives that clearly differ in cost and complexity.

TRAP CROPS AND GREEN MANURES

Other methods for the use of crop sequences to manage plant-parasitic nematodes include trap crops and green manures. A trap crop is most effective against sedentary endoparasitic nematodes that penetrate the host roots, establish a defined feeding site, and then lose the capacity for further movement or development. These nematodes must complete their life cycles on the host which they infect. A trap crop is planted, usually early in the growing season, and enough time is allowed for penetration of the roots by the nematodes. The trap crop is then destroyed, thus preventing reproduction by the nematodes. Trap crops have been evaluated for use against several plant-parasitic nematodes (Table 8–2). Wheat (*Triticum aestivum* L.) and rye (*Secale cereale* L.) may function as trap crops for *M. incognita* when grown as winter cover crops, since the nematodes typically would not have time to complete their life cycles during cooler winter temperatures (Johnson & Motsinger, 1990). The effectiveness of this practice, however, would depend on the temperatures experienced during the winter, and on planting dates in the spring. Steady decreases in nematode population levels have been observed where host crops were planted and then destroyed before nematode reproduction could occur (Godfrey & Hagan, 1934; Lamondia & Brodie, 1986), but this practice may be of limited usefulness due to associated costs and loss of crop-production time. Where a suitable range of potential trap crops can be identified, however, as for *Heterodera schachtii* on beet (*Beta vulgaris* L. subsp. *vulgaris*) (Gardner & Caswell-Chen, 1993), it is possible that

Table 8–2. Examples of trap crops with potential for control of plant-parasitic nematodes.

Nematode species	Primary host	Trap crop
Globodera rostochiensis	Potato	Potato (Lamondia & Brodie, 1986)
Heterodera avenae	Cereals	Oat (Stone, 1961)
H. schachtii	Beet	Buckwheat (*Fagopyrum esculentum* Moench), cabbage (*Brassica oleracea* var. *capitata* L.), oilseed radish, white mustard (*Sinapis alba* L.) (Gardner & Caswell-Chen, 1993)
Meloidogyne incognita	Vegetables	Wheat (Johnson & Motsinger, 1990); cowpea [*Vigna unguiculata* (L.) Walp. subsp. *unguiculata*], tomato (Godfrey & Hagan, 1934)

Table 8–3. Examples of green manure crops used for nematode control.

Nematode species	Primary host	Green manure crop
Hirschmaniella mucronata, H. oryzae	Rice	*Aeschynomeme afraspera* J. Leonard, *Sesbania rostrata* Bremek & Oberm (Prot et al., 1992)
Meloidogyne chitwoodi, *M. hapla*	Potato	Rapeseed (Mojtahedi et al., 1991), sudan grass (Mojtahedi et al., 1993)

growers would have the opportunity to adopt rotation crops that provide economic returns and still function as trap crops.

Green manures are crops grown specifically to provide biomass for incorporation into the soil. Some of these crops have been shown to have nematode-suppressive effects after incorporation (Table 8–3). Amendment of soil with sudangrass (*Sorghum* × *drummondii*) cultivars decreased numbers of *M. chitwoodi* 75 to 99%, with no phytotoxicity observed to tomato [*Lycopersicon esculentum* (Mill.) Karsten var. *lycopersicum*] (Mojtahedi et al., 1993). However, the amendments were applied at rates equivalent to 39 t ha^{-1}, and incorporated to a depth of 15 cm. *Meloidogyne chitwoodi* numbers also decreased dramatically in soil amended with rapeseed (*Brassica napus* L.), as compared to nonamended controls, after incubation for approximately 30 d in sealed plastic bags (Mojtahedi et al., 1991). Again, amendment rates were quite high, and phytotoxicity was observed on tomato. In field trials, the incorporation of crop residues from either *S. rostrata* or *Aeschynomene afraspera* significantly reduced population levels of *Hirschmanniella mucronata* and *H. oryzae*, and increased yields of subsequent rice crops (Prot et al., 1992). The mode of action for these green manure crops is not clearly known, but the release of toxic chemicals during the decomposition process has been suggested as a possible mechanism for nematode control. Green manures offer hope as another tool for managing nematodes, but as for any control method, the cost of growing the green manure crop and incorporating it into the soil must be considered, vs. the potential benefits of nematode control.

LIMITATIONS IN THE USE OF CROP ROTATION

Although many researchers have demonstrated the potential utility of crop sequence selection in the management of plant-parasitic nematodes, effective crop rotations have not yet been developed for many crop production systems. A primary problem in these systems has been the lack of a suitable nonhost crop. There is little value in recommending rotations to a turf cover crop, fallow, or some other nonprofitable alternative. The growers cannot afford the economic losses, and in most cases would be better off growing a cash crop even with the nematode problem. To address this problem, some studies have focused on the evaluation of exotic or uncommon crops as potential rotation alternatives (McSorley & Dickson, 1995; McSorley et al., 1994a,b; Rodriguez-Kabana et al., 1988a, 1989a). Rotation to such crops as castor (*Ricinus communis* L.) and velvetbean (*Mucuna pruriens* var. *utilis*) consistently lowered population densities of root-knot nematodes and increased yields of subsequent crops. These studies were useful in identifying numerous crops that were nonhosts for a number of important plant-parasitic nematode species, but the identified crops have not been introduced successfully into the desired production areas. Usually, adapting these exotic crops offers little or no economic incentive for a grower. Caution also is required in the introduction of new crops specifically to control a particular pest, since there may be undesirable side effects from the introduction. The lack of suitable, and profitable, rotation alternatives remains the largest obstacle in the use of crop rotation for nematode management. In a tropical cropping region, however, a selection of nine exotic and common crops were successfully arranged in a complex cropping sequence that kept population densities of *M. incognita* continuously below damaging levels (Hutton et al., 1983).

Crop-rotation research typically includes a rather limited range of all the possible rotations in a given system, and must be analyzed for limited periods of time, over only one, or perhaps a few, experimental locations. The reasons for these inherent limitations are all the same—limited funding and time available to the researcher to conduct the experiments. Crop-rotation experimental designs are extremely complex and labor intensive. In designing the field research, usually a promising series of rotations is based on previous information, and each rotation is implemented on a nematode-infested site. A properly randomized field design of this type will virtually exclude any use of large-scale equipment or extensive pesticide application procedures. Many replications are required, and the experiments should be continued for long periods of time. After experimental evaluation of the rotations has been completed, additional crops cannot be added to the analysis for further comparisons without repeating the experiments. However, the extreme effort required to gain useful information on cropping systems for nematode management can be minimized by the use of modeling and simulation procedures. The existence of robust models for describing the dependence of crop damage on nematode population densities at planting, and a well-developed base of information for modeling nematode population dynamics, provide a sound basis for the development of optimization models.

CROPPING SYSTEMS ANALYSIS

The management-oriented emphasis in crop rotation studies has been to reduce the population levels of plant-parasitic nematodes to below-threshold levels. This concept has been the unifying factor in nematode control tactics, and serves as the basis for integrating all available methods into a cropping system for nematode management. Damage to annual crops resulting from plant-parasitic nematodes usually is related to the population density of key species present in a field at planting (Noe, 1988; McSorley, 1998, see Chapter 6). These preplant nematode numbers are generally predictive of yield losses for annual crops, and can be used to determine damage thresholds for management (Barker et al., 1985). The emphasis on preplant nematode populations (P_i) in nematology is further strengthened by the fact that most nematode control tactics must be implemented before or at planting. Once the crop is in the ground, a grower typically has no further management alternatives for that growing season. New technologies, however, such as the application of nematicides through irrigation water, offer the possibility of management intervention during the growing season in limited situations. Other nematode-management alternatives, such as crop resistance, and biological control all operate by reducing the effective P_i of the target nematode species. Cultural practices such as fallowing, flooding, organic amendments, and even delayed planting times can serve to further reduce nematode P_i's at the time of planting a susceptible host. A detailed treatment of nematode population dynamics and pest management models are given in other chapters of this monograph.

End-of-rotation nematode counts, i.e., the number of nematodes on the next high-value susceptible host, has been a key variable of interest in most crop-rotation experiments. A population dynamic simulation model to predict the effect of a given crop sequence can be used to project and compare the outcomes of a large number of crop rotations. Single crop field studies to determine quantitative models for nematode population changes and crop yield relationships have been reported and are discussed in other chapters of this monograph. Typically, planting-time nematode assays and yields are measured in research plots and are used to determine the relationships of crop yields to initial nematode densities. These nematode-damage functions are usually represented by a negative logistic relationship, with a relatively level initial phase, then rapidly declining yield as nematode density increases, followed by another leveling off, or asymptotic approach to a minimum yield (Noe, 1988; Noe et al., 1991).

The next step in the development of cropping systems analysis for nematode management has been the merging of quantitative host-parasite relationships with crop-sequencing studies to simulate the effects of cropping sequences on nematode populations and crop yields (Noe, 1988; Noe et al., 1991). Simulation models developed from published information were used to predict population density oscillations and equilibrium levels of the cereal cyst nematode (*Heterodera avenae*) and the potato cyst nematode under multiple-year sequences of host and nonhost crops (Jones & Kempton, 1978). Logistic functions were fitted to field data that represented population changes during a single growing season, and the resulting models were linked together with an overwintering function to simulate multiple-year population dynamics. In both systems, the nematode population densities tended to converge on a stable oscillation, with higher and lower peaks representing host or nonhost crops, respectively. The mean level of the oscillations depended on the number of years of a nonhost crop. Kinloch (1986) used linear models to represent the relationships of initial to final population densities of *M. incognita* on susceptible and partially resistant soybean. Corn, which was a poor host for the field population of root-knot nematode, also was used to develop the models. That report also demonstrated an oscillating convergence to equilibrium densities over time, with the equilibrium level representative of the host status of the crop in a monoculture (Kinloch, 1986). Analysis of projected rotation sequences showed that a 2-yr rotation to corn or resistant soybean reduced nematode numbers by 90 to 95% as compared to the susceptible soybean. The models used by Kinloch (1986) also indicated that using corn in the 1st yr of rotation was more effective than a 1st yr of resistant soybean in providing for long-term suppression of root-knot nematode population densities.

In another study, nematode-damage functions, reproductive curves, overwintering functions, and crop enterprise budgets were assembled into a cropping sequence optimization model for the Columbia lance nematode (*Hoplolaimus columbus*) on cotton and soybean (Noe et al., 1991). The analytical method used in this study was an iterative process, proceeding in discrete steps from growing season to growing season, with the crop to be grown during each season represented by its characteristic mathematical models. A sequence of crops and P_i were input to the iterative portion of the program. The P_i was then used to project a final population density (P_f) and a crop yield, which was translated to a crop value based on market assumptions. The projected P_f for the first growing season, modified by appropriate between-crop survival functions, was then used as a P_i for the next crop in the sequence. The control structure of the computer program stepped through the entire cropping sequence, while calculating a measure of profitability. Nonlinear logistic models were used both for population dynamics and crop loss functions. Note that P_i provides the linkage between years and predicts the resulting effects on crop production values. When simulations were run over 3-yr rotations, a cotton–soybean–cotton sequence was predicted to earn \$63 ha^{-1} more than the least profitable sequence of soybean–cotton–soybean. Nematode population densities in the projected sequences oscillated in a manner similar to that reported by Jones and Kempton (1978), although both crops were hosts for the Columbia lance nematode. Relative differences among the levels of nematode reproduction provided adequate destabilization in the convergence to equilibrium densities. The discrete systems for optimization used in these studies can be improved by the use of quantitative decision-management tools to select only those sequences that are optimal for a given set of constraints. An improved method for the economic analysis of nematode management systems was derived from continuous decision models and was used to optimize returns and nematicide usage rates for control of root-lesion nematodes (*Pratylenchus* spp.) on corn (Osteen et al., 1988).

In perhaps the most detailed and complex attempt to simulate nematode population dynamics in a crop rotation system, Schmidt et al. (1993) used differential equations and transition probabilities to model a host-specific life cycle for the sugar beet cyst nematode (*H. schachtii*). Temperature-dependent and life-stage specific development of the nematodes was calculated for each host over the number of life cycles in a growing season. Transitional probabilities were used to predict maturation into more advanced life stages, and fecundity of the mature females was used to determine population density levels at the end of each crop. The final simulation model was remarkably accurate in predicting nematode population dynamics for actual field data from several cropping systems with sugar beet (*Beta vulgaris* L.) over a period of 15 to 18 yr. The detailed accuracy of such a model could be of tremendous benefit in the analysis and actual use of crop rotation systems, since nematode sampling data has a high level of imprecision, and the process of collecting nematode assay samples is laborious and expensive.

OUTLOOK

Much progress has been made in delineating the usefulness of cropping systems for management of plant-parasitic nematodes. Crop rotation will be of increased value in all agricultural systems as nematicides become less available and more expensive, but the most critical need for low-input management methods will still be found in low-income and subsistence agricultural systems. Much more information is needed on the complexity of tropical agricultural systems, where specialized techniques will be required to address the commonly used intercropping, relay cropping, and multicropping patterns (Noe, 1988). Tropical climates have profound effects on nematode distribution and on host-parasite relationships, which will subsequently affect the development of specialized crop rotations that are optimized for nematode control (Noe & Sikora, 1990). Much more information will be needed on the precise effects of climate and soil type variation in all agricultural regions. Clarification of the differences among crop cultivars and among pathogenic races of nematode species also will be critical to allow the formulation of general-use cropping systems for nematode management. To bring together a truly integrated management system, information on other nematode management techniques and on the effects of management practices used for other types of pests will need consideration, as will the basic agronomic needs and characteristics of each crop. Development of integrated management models will require greater cooperation among researchers and specialists across disciplines. Management-decision systems with a high enough level of complexity to be useful will have to be incorporated into expert-systems programming languages with sophisticated user interfaces before they will have any practical significance. However, the rapid rise in common, every day usage of more powerful microcomputers and higher-level software, combined with a virtual explosion in accessibility of information over the internet computer network, offers greater opportunities for information retrieval and dissemination. These developments also will enhance cooperation in a truly global community of agriculture.

Precision agriculture is another area of emerging technology that will be useful in the design of cropping systems for nematode management. Precision agriculture incorporates spatial variability into location-specific prescriptive management recommendations for individual fields (Vandenheuvel, 1996). Differences in soil texture, fertility, and other factors related to plant vigor within a field can be plotted with a high degree of spatial resolution. Ground-based management operations then can be linked to the geostationary-satellite Global Positioning System (GPS) to monitor and adapt location-specific application rates to precise coordinates. This technology will be beneficial by decreasing crop-management inputs to minimum effective levels, thus increasing profits, and decreasing adverse environmental impacts (Wallace, 1994). The most efficient means to obtain the required information on spatial variability within fields may arise from remote-sensing, particularly from the use of digital video-image analysis on aerial images (Beverly, 1996). Where plant vigor is being monitored, however, it will be essential to assess levels of soilborne plant pathogens and plant-parasitic nematodes before any conclusions are drawn regarding the nutritional status of the plant, or soil fertility. These compounding factors will make interpretation of remote sensing data difficult, but improvements in reliability of the data can be obtained from an understanding of regional distributions and spatial patterns of plant-parasitic nematodes. Recently adopted techniques using geostatistics to formulate predictive spatial models for nematodes (Wallace et al., 1993) offer hope that these factors can be included in remote-sensing data without requiring extensive ground-based sampling. It will be essential for researchers in cropping systems for management of plant-parasitic nematodes to interact and cooperate with farm-enterprise efforts in precision agriculture.

Although the effort required to develop economically sound, long-term cropping systems for limiting losses due to nematodes will be tremendous, the potential of this type of management is equally large, due to the unique biology and ecology of these parasites. Effective management protocols will require combinations of all the other available tactics along with crop rotation. These tactics will include chemical nematicides, biological control, crop resistance, cultural practices, and, hopefully, biologically engineered plant defenses. Previous experience would indicate that no one of these tactics will survive the continuing onslaught of these ubiquitous pests for long, but that successful future managers will carefully consider the best use of each, in a comprehensive cropping system. With the great promise of molecular methods for enhancing plant defenses against nematodes comes the likelihood that these defenses eventually will be overcome by development of new races of the nematodes. Crop rotation and other cultural methods for control of plant-parasitic nematodes will remain as important and active areas of future research and development.

REFERENCES

Ayers, A.R., H.E. Duncan, K.R. Barker, and M.K. Beute. 1989. Effects of crop rotation and nonfumigant nematicides on peanut and corn yields in fields infested with *Criconemella* species. J. Nematol. 21:268–275.

Barker, K.R., D.P. Schmitt, and V.P. Campos. 1982. Response of peanut, corn, tobacco, and soybean to *Criconemella ornata*. J. Nematol. 14:576–581.

Barker, K.R., D.P. Schmitt, and J.P. Noe. 1985. Role of sampling for crop-loss assessment and nematode management. Agric. Ecosyst. Environ. 12:355–369.

Bergeson, G.B. 1959. The influence of temperature on the survival of some species of the genus *Meloidogyne* in the absence of a host. Nematologica 5:344–354.

Bessey, E.A. 1911. Root-knot and its control. USDA Bull. 217. Bur. Plant Indust., Washington, DC.

Beverly, R.B. 1996. Video image-analysis as a nondestructive measure of plant vigor for precision agriculture. Commun. Soil Sci. Plant Anal. 27:607–614.

Braithwaite, C.W.D. 1974. Effect of crop sequence and fallow on populations of *Rotylenchulus reniformis* in fumigated and untreated soil. Plant Dis. Rep. 58:259–261.

Brodie, B.B., K. Evans, and J. Franco. 1993. Nematode parasites of potatoes. p. 87–132. *In* K. Evans et al. (ed.) Plant-parasitic nematodes in temperate agriculture. CAB Int., Wallingford, England.

Cooke, D. 1993. Nematode parasites of sugarbeet. p.133–169. *In* K. Evans et al. (ed.) Plant-parasitic nematodes in temperate agriculture. CAB Int., Wallingford, England.

Gardner, J., and E.P. Caswell-Chen. 1993. Penetration, development, and reproduction of *Heterodera schachtii* on *Fagopyrum esculentum*, *Phacelia tanacetifolia*, *Raphanus sativus Sinapis alpa*, and *Brassica oleracea*. J. Nematol. 25:695–702.

Germani, G., G. Reversat, and M. Luc. 1983. Effect of *Sesbania rostrata* on *Hirschmanniella oryzae* in flooded rice. J. Nematol. 15:269–271.

Godfrey, G.H., and H.R. Hagan. 1934. A study of the root-knot nematode trap crop under field conditions. Phytopathology 24:648–658.

Hirunsalee, A., K.R. Barker, and M.K. Beute. 1995. Effects of peanut-tobacco rotations on population dynamics of *Meloidogyne arenaria* in mixed race populations. J. Nematol. 27:178–188.

Hutton, D.G., P.L. Coates-Beckford, and S.A.E. Eason-Heath. 1983. Management of *Meloidogyne incognita* populations by crop rotation in a small-scale field trial and nematode pathogenic effects on selected cultivars (Jamaica). Nematropica 13:153–163.

Johnson, A.W. 1985. Specific crop rotation effects combined with cultural practices and nematicides. p.283–301. *In* J.N. Sasser and C.C. Carter (ed.) An advanced treatise on *Meloidogyne*: Biology and control. Vol. 1. North Carolina State Univ. Graphics, Raleigh.

Johnson, A.W., G.W. Burton, J.P. Wilson, and A.M. Golden. 1995. Rotations with coastal bermudagrass and fallow for management of *Meloidogyne incognita* and soilborne fungi on vegetable crops. J. Nematol. 27:457–464.

Johnson, A.W., C.C. Dowler, and E.W. Hauser. 1975. Crop rotation and herbicide effects on population densities of plant-parasitic nematodes. J. Nematol. 7:158–168.

Johnson, A.W., and R.E. Motsinger. 1990. Effects of planting date, small grain crop destruction, fallow, and soil temperature on the management of *Meloidogyne incognita*. J. Nematol. 22:348–355.

Jones, F.G.W., and R.A. Kempton. 1978. Population dynamics, population models, and integrated control. p.333–361. *In* J. F. Southey (ed.) Plant nematology. Her Majesty's Station. Office, London.

Kinloch, R.A. 1983. Influence of maize rotations on the yield of soybean grown in *Meloidogyne incognita* infested soil. J. Nematol. 15:398–405.

Kinloch, R.A. 1986. Soybean and maize cropping models for the management of *Meloidogyne incognita* in the coastal plain. J. Nematol. 18:451–458.

Kirkpatrick, T.L., and J.N. Sasser. 1984. Crop rotation and races of *Meloidogyne incognita* in cotton root-knot management. J. Nematol. 16:323–328.

Lamondia, J.A., and B.B. Brodie. 1986. The effect of potato trap crops and fallow on decline of *Globodera rostochiensis*. Ann. App. Biol. 108:347–352.

Mai, W.F., and G.S. Abawi. 1980. Influence of crop rotation on spread and density of *Heterodera schachtii* on a commercial vegetable farm in New York. Plant Dis. 64:302–305.

Mai, W.F., and B.F. Lownsbery. 1952. Crop rotation in relation to the Golden nematode population of the soil. Phytopathology 14:345–347.

McSorley, R.M. 1998. Population dynamics. p. 109–133. In K.R. Barker et al. (ed.) Plant and nematode interactions. Agron. Monogr. 36. ASA, CSSA, and SSSA, Madison,WI.

McSorley, R.M., D.W. Dickson, J.A. de Brito, T.E. Hewlett, and J.J. Frederick. 1994a. Effects of tropical rotation crops on *Meloidogyne arenaria* population densities and vegetable yields in microplots. J. Nematol. 26:175–181.

McSorley, R.M., D.W. Dickson, J.A. de Brito, and R.C. Hochmuth. 1994b. Tropical rotation crops influence nematode densities and vegetable yields. J. Nematol. 26:308–314.

McSorley, R.M., and D.W. Dickson. 1995. Effect of tropical rotation crops on *Meloidogyne incognita* and other plant-parasitic nematodes. J. Nematol. 27(4S):535–544.

McSorley, R.M., and R.N. Gallaher. 1993. Effect of crop rotation and tillage on nematode densities in tropical corn. J. Nematol. 25(4S):814–819.

Mojtahedi, H., G.S. Santo, A.N. Hang, and J.H. Wilson. 1991. Suppression of root-knot nematode populations with selected rapeseed cultivars as green manure. J. Nematol. 23:170–174.

Mojtahedi, H., G.S. Santo, and R.E. Ingham. 1993. Suppression of *Meloidogyne chitwoodi* with sudangrass cultivars as green manure. J. Nematol. 25:303–311.

Noe, J.P. 1988. Theory and practice of the cropping systems approach to reducing nematode problems in the tropics. J. Nematol. 20:204–213.

Noe, J.P., J.N. Sasser, and J.L. Imbriani. 1991. Maximizing the potential of cropping systems for nematode management. J. Nematol. 23:353–361.

Noe, J.P., and R.A. Sikora. 1990. Effects of tropical climates on the distribution and host-parasitic relationship of plant-parasitic nematodes. p. 583–597. *In* M. Luc et al. (ed.) Plant-parasitic nematodes in subtropical and tropical agriculture. CAB Int., Wallingford, England.

O'Bannon, J.H., and G.S. Santo. 1984. Effect of soil temperature on reproduction of *Meloidogyne chitwoodi* and *M. hapla* alone and in combination on potato and *M. chitwoodi* on rotation plants. J. Nematol. 16:309–312.

Osteen, C.D., L.J. Moffitt, and A.W. Johnson. 1988. Risk efficient action thresholds for nematode management. J. Prod. Agric. 1:332–338.

Prot, J.C., I.R.S. Soriano, D.M. Matias, and S. Savary. 1992. Use of green manure crops in control of *Hirschmanniella mucronata* and *H. oryzae* in irrigated rice. J. Nematol. 24:127–132.

Riggs, R.D., and T.L. Niblack. 1993. Nematode pests of oilseed crops and grain legumes. p. 209–258. *In* K. Evans et al. (ed.) Plant-parasitic nematodes in temperate agriculture. CAB Int., Wallingford, England.

Rivoal, R., and R. Cook. 1993. Nematode pests of cereals. p. 259–303. *In* K. Evans et al. (ed.) Plant-parasitic nematodes in temperate agriculture. CAB Int., Wallingford, England.

Robbins, R.T., and K.R. Barker. 1973. Comparisons of host range and reproduction among populations of *Belonolaimus longicaudatus* from North Carolina and Georgia. Plant Dis. Rep. 57:750–754.

Rodriguez-Kabana, R., P.S. King, D.G. Robertson, and C.F. Weaver. 1988a. Potential of crops uncommon to Alabama for management of root-knot and soybean cyst nematodes. Ann. Appl. Nematol. 2:116–120.

Rodriguez-Kabana, R., D.G. Robertson, P.A. Backman, and H. Ivey. 1988b. Soybean-peanut rotations for the management of *Meloidogyne arenaria*. Ann. App. Nematol. 2:81–85.

Rodriguez-Kabana, R., D.G. Robertson, L. Wells, P.S. King, and C.F. Weaver. 1989a. Crops uncommon to Alabama for the management of *Meloidogyne arenaria* in peanut. J. Nematol. 21(4S):712–716.

Rodriguez-Kabana, R., C.F. Weaver, R. Garcia, D.G. Robertson, and E.L. Carden. 1989b. Bahiagrass for the management of root-knot and cyst nematodes in soybean. Nematropica 19:185–193.

Rodriguez-Kabana, R., and G.H. Canullo. 1992. Cropping systems for the management of phytonematodes. Phytoparasitica 20:211–224.

Rodriguez-Kabana, R., H. Ivey, and P.A. Backman. 1987. Peanut-cotton rotations for the management of *Meloidogyne arenaria*. J. Nematol. 19:484–486.

Schmidt, K., R.A. Sikora, and O. Richter. 1993. Modeling the population dynamics of the sugar beet cyst nematode *Heterodera schachtii*. Crop Prot. 12:490–496.

Slack, D.A., R.D. Riggs, and M.L. Hamblen. 1972. The effect of temperature and moisture on the survival of *Heterodera glycines* in the absence of a host. J. Nematol. 4:263–266.

Stone, L.E.W. 1961. Oats as a trap-crop for cereal root eelworm. Plant Pathol. 10:164

Thames, W.H., and C.M. Heald. 1974. Chemical and cultural control of *Rotylenchulus reniformis* on cotton. Plant Dis. Rep. 58:337–341.

Todd, T.C. 1991. Effect of cropping regime on populations of *Belonolaimus* sp. and *Pratylenchus scribneri* in sandy soil. J. Nematol. 23(4S):646–651.

Trivedi, P.C., and K.R. Barker. 1986. Management of nematodes by cultural practices. Nematropica 16:213–236.

Vandenheuvel, R.M. 1996. The promise of precision agriculture. J. Soil Water Conserv. 51:38–40.

Wallace, A. 1994. High precision agriculture is an excellent tool for conservation of natural resources. Commun. Soil Sci. Plant Anal. 25:45–49.

Wallace, M. K., R.H. Rust, D.M. Hawkins, and D.H. MacDonald. 1993. Correlation of edaphic factors with plant-parasitic nematode population densities in a forage field. J. Nematol. 25:642–653.

Whitehead, A.G., R.M. Webb, and J. Beane. 1991. Effects of rotation length and oxamyl on potato yield and the potato-cyst nematode, *Globodera rostochiensis*, in a sandy loam soil. Ann. App. Biol. 118:371–380.

Wilson, J.D. 1964. Influence of preceding vegetable crop on nematode populations the following year. Plant Dis. Rep. 48:500–501.

9 Nematode-Pesticide Interactions

BRENT S. SIPES AND **DONALD P. SCHMITT**

University of Hawaii at Manoa
Honolulu, Hawaii

The use of synthetic chemicals, which are toxic to unwanted organisms and/or modify plant growth, has increased crop yields enormously, thus having a major effect on modern agricultural production. The introduction of nematicides provided growers an effective method of controlling plant-parasitic nematodes, thereby reducing associated crop losses (Johnson & Feldmesser, 1987). Since the 1950s, the popularity of pesticides has increased due to dramatic, rapid, and consistent control. Many pesticides are applied to soil and crops where nematodes are present. Consequently, interactions between nematodes and pesticides, both intentional and coincidental, are expected in some instances and probable in others.

Contemporary society is concerned about nontarget effects and the environmental impact of pesticides. Nontarget effects of a pesticide may reduce or enhance plant-parasitic nematode populations. The effect may be direct upon nematode processes or indirect through modification of the host plant. Additional confounding factors occur when two or more pesticides are applied to a crop sequentially or in combination. The activity of one pesticide may be significantly altered by the other and may produce an entirely unexpected effect (Putnam & Penner, 1974; Sine, 1993). Agriculturists are most concerned about those interactions that increase nematode damage to plants; however, antagonistic interactions warrant investigation as potential means of nematode control. In this chapter, discussion is limited to broad categories of pesticides, citing specific examples to demonstrate potential interactions between nematodes and pesticides.

BROAD-SPECTRUM BIOCIDES

Broad-spectrum biocides, commonly called fumigants, include halogenated hydrocarbons and methyl isothiocyanate (MIT) liberators. These compounds may affect a large portion of the soil biota or act more selectively against only certain organisms. Methyl bromide and MIT kill many seeds, bacteria, insects, fungi, and nematodes in the soil (Wright, 1981). Other halogenated hydrocarbons, such as 1,3-dichloropropene (1,3-D), affect fewer organisms than methyl bromide or MIT. The toxicity of 1,3-D is generally limited to nematodes, fungi (especially *Pythium* spp.), and wireworms (Elateridae). Many broad-spectrum fumigant pesticides, especially the nematicides ethylene dibromide, 1,2-dichloropropane-1,3-dibromopropene, and dibromochloropropane, have been banned in the USA because of their potential to contaminate the environment and their hazard to workers. Methyl bromide is scheduled for removal from the agricultural market in 2001 because it is implicated as an ozone depleter. These withdrawals of effective fumigant pesticides have left few effective broad-spectrum biocides registered for use in the USA.

The mode of action of halogenated hydrocarbons against nematodes can be represented by 1,3-D. Vermiform nematodes exposed to lethal doses of 1,3-D (LD_{50} cis-1,3-D = 21–33.7 $\mu g\ L^{-1}$ at 25°C) are initially hyperactive (McKenry & Thomason, 1974). Nematode activity gradually declines after a few hours of exposure, and the nematodes become paralyzed (Evans & Thomason, 1971). Death is believed to result from the oxidation of Fe in the cytochrome chain and the failure of respiration (Castro & Thomason, 1971).

Nematode-life stages and species vary in their susceptibility to halogenated hydrocarbon biocides. Higher rates of 1,3-D are required to kill eggs of *Rotylenchulus reniformis,* the reniform nematode, than juvenile or adult stages (Wright, 1981). Differential toxicity between eggs and juveniles appears to be related to differences in cuticle and egg shell permeability (Castro & Thomason, 1971; Evans & Thomason, 1971). The membrane surrounding the unhatched juvenile in the egg may be more restrictive to the chemical than the cuticle and membrane of the hatched second-stage juvenile. The level of nematode activity also may play a role in the differential toxicity of the compounds. First- and second-stage juveniles in eggs are comparatively less active than the infective juvenile stages. Consequently, the more active metabolism of hatched juveniles may facilitate uptake of the halogenated hydrocarbons.

No clear evidence exists on the mode of action of MIT on nematodes. These compounds can be very toxic in vitro, but field efficacy is variable.

Although MIT and halogenated hydrocarbons are nematicidal, nematodes frequently escape exposure and survive in treated soil (Lembright, 1990). Nematodes survive in areas where the fumigant has failed to penetrate or below the zone of fumigant placement in the soil (Lembright, 1990). Nematicide efficacy, therefore, depends upon inherent toxicity and application. The most toxic fumigant nematicide will be ineffective if not properly applied in the soil or retained in the soil for a sufficient period. The nematodes that survive after fumigation serve as an inoculum reservoir for subsequent infection of host plants.

Broad-spectrum biocides can create biological voids in the soil ecosystem where nematode competitors, predators, and parasites are eliminated (Van Gundy, 1985). Natural antagonists of nematodes in the soil, which may be eliminated by broad spectrum biocides, offer substantial control in perennial cropping systems or where nematode population densities have increased to large numbers. *Heterodera avenae*, the cereal cyst nematode, has been controlled by naturally occurring organisms in continuous monocrop wheat (*Triticum aestivum* L.) in Europe (Kerry et al., 1982). The absence of nematode competitors, predators, and parasites in the soil, as often occurs after the use of a broad-spectrum biocide, may allow plant-parasitic nematodes to increase to very large numbers. Problems may be especially severe where fumigation occurs frequently, preventing antagonists and predators from re-establishing in the soil.

INSECTICIDES AND NEMATICIDES

Nonfumigant insecticides and nematicides share similar chemistry and modes of action. Organophosphate and organocarbamate insecticides and nematicides are cholinesterase inhibitors that interfere with nematode sensory perception and neuromuscular behavior and, therefore, nematode development and reproduction (Nelmes et al., 1973; Opperman & Chang, 1991). Toxicity of these compounds, as with all pesticides, is a function of their concentration in the environment and the amount of time the nematode is exposed to the chemical (McKenry & Thomason, 1974). These compounds are rarely lethal to nematodes, and their effects are usually reversible, allowing nematode recovery after exposure. Consequently, these compounds would more accurately be termed nematistats rather than nematicides.

Nematode-life processes adversely affected by insecticide-nematicides include hatching, root penetration, and development (Hough & Thomason, 1975). Nematicides restrict the amount of hatch in water that occurs in *Heterodera glycines*, the soybean cyst nematode (Fig. 9–1) (Bostian et al., 1984). The inhibition of hatch is attributed to disruption of the necessary behavioral sequence of the hatching process (Wright, 1981). Penetration of plant roots by the nematode is impaired by these compounds, which may reflect disturbance of nematode sensory ability (Gourd et al., 1993). Males exposed to nematicides are unable to locate females, thus reducing copulation and fertilization, ultimately lowering nematode population densities (Wright, 1981). The organophosphate insecticide-nematicide fenamiphos [ethyl 3-methyl-4-(methylthio)phenyl(1-methylethyl)] retards the rate of *H. glycines* development in soybean [*Glycine max* (L.) Merr.] roots as compared to development in untreated plants (Fig. 9–2) (Sipes & Schmitt, 1989). Molecules of the nematicide may enter syncytia and be ingested by the nematode, thus, slowing or stopping nematode development.

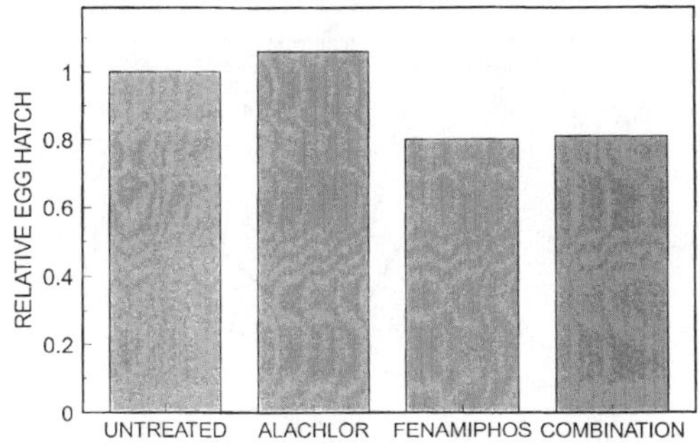

Fig. 9–1. Hatching of *Heterodera glycines* eggs exposed to the herbicide alachlor, the nematicide fenamiphos, or the combination of the two pesticides relative to an untreated control. Hatching in the treatments containing fenamiphos were lower than those without the nematicide ($P < 0.01$) (after Bostian et al., 1984).

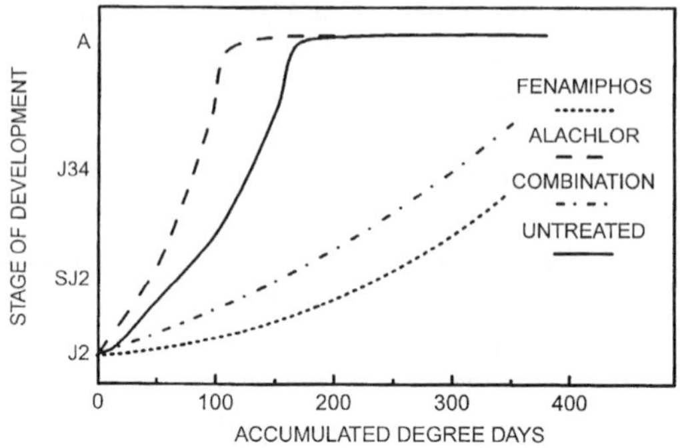

Fig. 9–2. Rate of *Heterodera glycines* development on soybean as affected by a herbicide (alachlor), nematicide (fenamiphos), or their combination. J2 = second-stage juvenile, SJ2 = swollen second-stage juvenile, J34 = third- or fourth-stage juvenile, A = adult male or female (after Sipes & Schmitt, 1989).

Nematode population densities often increase to larger numbers in fields following a nematicide treatment compared to untreated areas. Nematicides apparently protect the developing plant roots from nematode damage, resulting in a large root biomass accumulation. The nematicide may keep the nematode population density low all season. In other cases, as the nematicidal effect is reduced or eliminated, the surviving nematodes can produce large numbers of progeny because of the abundant available food supply (Fig. 9–3) (Schmitt et al., 1983).

Insecticides, in addition to having direct effects on nematodes, may have an indirect influence on nematode-plant damage and nematode population dynamics. The response of nematodes to insecticides is, in part, a chemical concentration effect. Most insecticides are applied at rates a quarter to a half required for nematicidal activity. Aldicarb [2-methy-2-(methylthio) propionaldehyde *O*-(methylcarbamoyl)oximine], an insecticide-nematicide, is labeled at 5.6 to 11.2 kg ha^{-1} and 11.2 to 22.4 kg ha^{-1} for insect and nematode control on soybean, respectively (Vroom, 1995). At higher rates, aldicarb is nematicidal, but at lower rates nematode activity is enhanced (Hough & Thomason, 1975). The insecticide

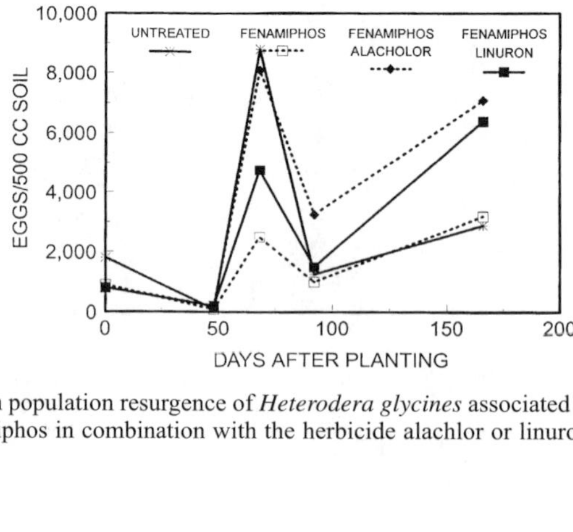

Fig. 9–3. Late season population resurgence of *Heterodera glycines* associated with application of the nematicide fenamiphos in combination with the herbicide alachlor or linuron (after Schmitt et al., 1983).

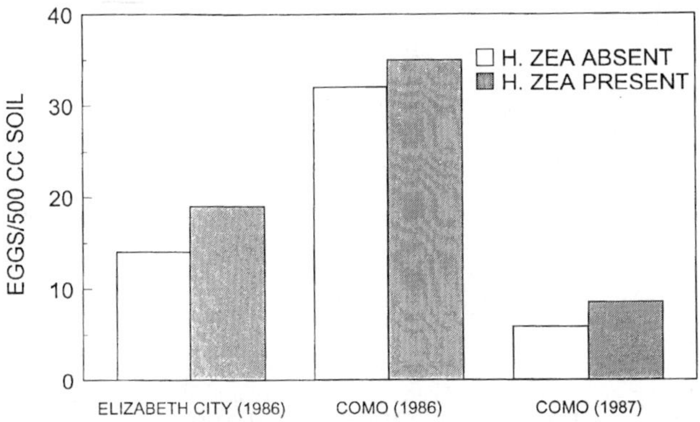

Fig. 9–4. The effects of *Helicoverpa zea* infestation in soybean on the number of eggs of *Heterodera glycines* in the soil at crop harvest (after Alston et al., 1993).

diflubenzuron (N-{[(4-chlorophenyl)amino]carbonyl}-2-6-difluorobenzamide), a chitin inhibitor, reduced the number of eggs that *Meloidogyne incognita*, the southern root-knot nematode, produced on cotton (*Gossypium hirsutum* L.) (Veech, 1978). Diflurobenzuron effects on nematodes may be limited because of differences in insect and nematode chitin synthesis or in the uptake and movement of the compound in the plant (Veech, 1978).

Insect feeding can enhance plant suitability for nematode reproduction (Alston et al., 1993); in contrast, controlling insect damage resulted in lower nematode population densities (Fig. 9–4). However, with severe insect damage, nematode population densities will decrease because a heavily damaged plant has insufficient roots to sustain large nematode populations.

Insecticides can adversely impact the movement of insect-vectored nematodes. The pine wilt nematode, *Bursaphelenchus xylophilus*, is vectored by a Cerambycid beetle (Cerambycidae) from diseased to healthy pine (*Pinus* spp.) trees (Linit, 1988). In Japan, an insecticide program was instituted to control the beetles, therefore limiting the movement of the nematode and spread of the disease. This method is costly, provides only limited control of the pine wilt nematode in Japan (Mamiya, 1988), and may not be practical for effective control of this nematode.

HERBICIDES

Herbicides are widely used on agricultural and uncultivated lands to control unwanted vegetation. Many species of weeds are good hosts of plant-parasitic nematodes and serve as inoculum sources in fields (Decker, 1972). Herbicides, which eliminate weed host of nematodes, are effective in reducing nematode population densities and controlling crop damage. Direct effects of herbicides on nematodes range from stimulatory to inhibitory. For example, Weischer and Muller (1985) exposed nematodes to 29 herbicides. Some herbicides enhanced nematode reproduction, whereas others adversely affected plant

Table 9–1. Selected herbicides with stimulatory or inhibitory action on nematode processes

Herbicide†	Process affected	Change in process	Nematode
		%	
Diallate	Reproduction	+17	*Heterodera schachtii*
EPTC	Galling	+44	*Meloidogyne incognita*
EPTC	Galling	+14	*Meloidogyne incognita*
Alachlor	Survival for 21 d	+6	*Heterodera glycines*
Vernolate	Reproduction	+474	*Helicotylenchus dihystera*
Vernolate	Hatch	−95	*Globodera rostochiensis*
Acifluorfen/bentazon	Eggs at 31 d	−17 to −35	*Heterodera glycines*
Acifluorfen	Hatch	−33 to −58	*Heterodera glycines*
Pebulate	Hatch	−96	*Heterodera schachtii*

† Literature cited for each example: diallate, Kraus and Sikora, 1983; EPTC, Johnson et al., 1994; alachlor, Bostian et al., 1984; vernolate, Schmitt and Corbin, 1981; acifluorfen/bentazon, Browde et al., 1994; acifluorfen, Wong et al., 1993; vernolate and pebulate, Perry and Beane, 1989.

resistance, and both cases resulted in increased nematode-incited plant damage. Two of the twenty-nine herbicides were toxic to the nematodes, consequently reducing plant damage. Conceptually, herbicides that affect nematode activity could be used to manage nematode populations. However, attempts to apply these concepts have not provided adequate control.

A variety of examples of herbicide influences on nematodes are summarized in Table 9–1. The mechanisms involved in these stimulatory and inhibitory processes are likely due to a multitude of phenomena. The impacts can be on nematode survival, hatching, root penetration, and reproduction. Alachlor [2-chloro-2′,6′-diethyl-N-(methoxymethyl)acetanilide] enhanced in vitro survival of second-stage juveniles by 6% over a 21-d period and stimulated hatching of *H. glycines* eggs (Bostian et al., 1984). Effects were similar between technical and formulated alachlor. Stimulatory herbicides, especially thiocarbamates, result in increased numbers of some nematodes and associated root galling. Diallate [*S*-(2,3-dichloroallyl)diisopropylthiocarbamate] enhanced the production of cyst nematodes 1.5- to 3-fold compared to untreated areas (Perry & Beane, 1989). This herbicide also has been associated with increased numbers of eggs and juveniles in the soil. Vernolate (*S*-propyl dipropylthiocarbamate), another thiocarbamate herbicide, increased the number of spiral nematodes, *Helicotylenchus dihystera*, 474% over those in untreated plots (Schmitt & Corbin, 1981). S-ethyl dipropylthiocarbamate (EPTC) substantially increased galling of snapbean (*Phaseolus vulgaris* L.) by *M. incognita* (Johnson et al., 1994).

The hatching stimulation induced by herbicides might be related to a change in the permeability of the egg shell due to a breakdown of the lipid membrane, the innermost layer of the three eggshell layers. In *Heterodera* spp., host-root exudates stimulate hatch by altering the permeability of the lipid membrane of the egg (Shepherd & Clarke, 1971). Kraus and Sikora (1983) suggest that thiocarbamate herbicides affect the lipid layer in the eggshell because these herbicides are known to interfere with fatty acid and lipid synthesis in plants, causing disintegration of membranes (Wilkenson, 1978). Preceeding hatching, membrane solubility may increase, therefore any compound that would increase membrane permeability also would likely facilitate hatch.

Table 9–2. Numbers of *Heterodera glycines* eggs and second stage juveniles (J2) per 100 cm^3 soil 31 d after application (69 d after planting) of acifluorfen plus bentazon in 1988 and 33 d after herbicide application (82 d after planting) in 1989 (data selected from Browde et al., 1994).

	Plant injury		Eggs		J2	
Herbicide rate	1988	1989	1988	1989	1988	1989
	—— % ——					
0	0	0	1438	5188	83	259
Low†	10	9	861	4472	52	296
High‡	14	19	1006	4841	59	222
Orthogonal contrasts (F; $P > F$)						
0 vs. low and high			4.75; 0.03	0.74; 0.39	4.78; 0.03	3.54; 0.06
Low vs. high			0.55; 0.46	0.01; 0.91	0.57; 0.45	1.44; 0.23

† Low = 1988; 0.14 kg a.i. ha^{-1} acifluorfen + 0.84 kg a.i. ha^{-1} bentazon; 1989; 0.28 kg a.i. ha^{-1} acifluorfen + 0.56 kg a.i. ha^{-1} bentazon.
‡ High = both years; 0.56 kg a.i. ha^{-1} acifluorfen + 0.84 kg a.i. ha^{-1} bentazon.

Penetration of plant roots by nematodes is enhanced by certain herbicides such as diallate (Kraus & Sikora, 1983). This increased amount of penetration could be related to greater hatching and, therefore, more juveniles available for infection. However, enhanced infectivity possibly may be related to a change in host status as a consequence of altered host physiology. Plant roots treated with herbicides might be more attractive to nematodes because of modified metabolism or may be more susceptible because of herbicide injury.

Several herbicides reduce nematode population densities. One of the most obvious means to account for this reduction in populations is injury to the host plant, as with the herbicide mixture aciflourfen {5-[2-chloro-4-(trifluorometyl) phenoxy]-2-nitrobenzoate} plus bentazon [3-isopropyl-1*H*-2,1,3-benzothiadiazin-4(3*H*)-one-2,2-dioxide] (Browde et al., 1994; Table 9–2). The numbers of eggs and juveniles of *H. glycines* were reduced by postemergence application of these two herbicides. Reductions in numbers of second-stage juveniles and eggs ranged from 16 to 40%. The reduction can be attributed to significant plant injury caused by these herbicides (Browde et al., 1994) (Table 9–2). A number of in vitro studies reveal other mechanisms that reduce nematode processes such as hatching. Field application of cycloate (*S*-ethyl N-ethylthicyclohexanecarbamate), pebulate (*S*-propyl butylethylthiocarbamate), vernolate (*S*-propyl dipropylcarbamothioate), and tri-allate [*S*-(2,3,3-trichloroallyl)diisopropylthiocarbamate] irreversibly inhibited hatching of eggs of *Globodera rostochiensis* (potato cyst nematode) and *Heterodera schachtii* (sugar beet cyst nematode) even when eggs were subsequently incubated in root diffusate (Perry & Beane, 1989) (Table 9–1).

PLANT-GROWTH REGULATORS

Nematodes have intimate relationships with their host plants, thus any alteration in plant metabolism likely will affect nematode behavior. The more intimate the host-parasite relationship, such as with sedentary parasitic nematodes, the more dramatic the plant-growth regulator effect likely will be on the nematode.

Nematode feeding and reproduction are most likely to be affected by plant-growth regulators because nematode life processes are closely linked to plant development. Nematode development may be hastened or slowed at different plant life stages. This fluctuation may be related to the quantity and quality of the food source. The rate of *H. glycines* reproduction on soybean increases as the host plant senesces (Hill & Schmitt, 1989). The mechanism of the nematode response is not understood, but plant metabolism has changed and may alter nematode behavior. Hatch of many nematodes, especially *G. rostochiensis*, is stimulated by host-root exudates (Shepherd & Clarke, 1971). Growth regulators that increase the amount of host-root exudates could inadvertently stimulate hatching. The greater number of juveniles would subsequently result in more plant damage.

Situations occur in which nematode reproduction is decreased by plant-growth regulators. Population densities of *R. reniformis* decreased immediately following application of ethylene to stimulate pineapple [*Ananas comosus* (L.) Merr.] flowering (Sipes & Schmitt, 1994). With flowering initiation, roots are no longer the primary sink in the plant, which could account for the decreases in nematode populations.

Aldicarb is an insecticide-nematicide that stimulates plant growth even in the absence of pests. At low rates, aldicarb has little direct effect on nematodes but increases the growth of soybean and tobacco (*Nicotiana tabacum* L.) (Barker et al., 1988; Barker & Powell, 1988). Additional root biomass is produced and provides more feeding sites for nematode population development. When aldicarb is applied at nematicidal (nematistatic) rates, it will reduce the number of early season nematode infections and stimulate plant growth, but a late season nematode population resurgence can be expected. The late-season population resurgence is attributed to the stimulated root-biomass accumulation. Nematode numbers could be larger with aldicarb than with fenamiphos, although the nematicidal effects are similar between the compounds.

FUNGICIDES

Fungicides appear to have fewer effects on plant-parasitic nematodes than other pesticides used in agriculture. However, several fungicides do affect nematode behavior and development. The thiocarbamates maneb (manganese ethylenebisdithiocarbamate) and nabam [disodium ethylenebis(dithiocarbamate)] stimulated hatch of *Globodera tabacum*, the tobacco cyst nematode (Miller, 1967). Pentachloronitrobenzene (PCNB) inhibited chitin formation in developing nematode eggs similar to the insecticide diflubenzuron (Wright, 1981). Reductions in nematode population densities are very slow with PCNB, suggesting an effect on nematode reproduction and not on penetration of juveniles into host roots (Wright, 1981).

Benomyl [methyl-(butylcarbamoyl)benzimidazol-2-ylcarbamate] and related thiabendazoles limited penetration of *G. tabacum* into roots of host plants due to a suppression of egg hatch and repulsion of nematodes from the roots (Miller, 1969). These compounds also slow juvenile development and increase the proportion of males in populations of *H. avenae* (Cook & York, 1972). Benzimidazole compounds interfere with spindle formation during cell division and DNA synthesis, thus affecting nematode embryonic development and reproduction (Wright, 1981).

Several fungicides, fosetyl-Al (ethyl hydrogen phosphonate), metalaxyl [methyl N-(2-methoxyacetyl)-N-(2,6oxylyl)-DL-alaninate], and triadimefon [1-(4-chlorophenoxy)-3,3-dimethyl-1-(1*H*-1,2,4-triazol-1-yl)butan-2-one], appear to induce effects similar to systemic acquired resistance (SAR) in plants (Kessmann et al., 1994). An SAR-type response may be induced against nematodes by benomyl. The reduced numbers of eggs and additional numbers of male *H. avenae* associated with exposure to benomyl-treated plants are similar to the lower population densities and death of females that occur on poor hosts or resistant plants (Triantaphyllou, 1973). Additional research into this area of SAR and nematodes is needed.

INTERACTIONS BETWEEN PESTICIDES

Many crops may receive two or more applications of pesticides during any particular crop cycle. These pesticides may be applied simultaneously, in tank mixes, or sequentially over that crop cycle. Whenever two or more pesticides are applied to the same site, whether simultaneously or sequentially, the potential for altering the behavior of one of the pesticides exists. Any alteration of nematicide activity could change its efficacy. The most obvious evidence of an interaction is the development of a precipitate when two or more chemicals are mixed together. Most pesticide interactions, however, are more subtle and produce synergistic, additive, or antagonistic responses in target or nontarget organisms (Putnam & Penner, 1974). Herbicides may enhance or reduce nematicide efficacy. Field plots treated with aldicarb and metribuzin (4-amino-6-*tert*-butyl-4,5-dihydro-3-methylthio-1,2,4-triazin-5-one), alachlor, or linuron [3-(3,4-dichlorophenyl)-1-methoxy-1-methylurea] had lower population densities of *H. glycines* than plots treated with the nematicide alone (Schmitt & Corbin, 1981). In contrast, population densities of *H. dihystera* were greater in plots treated with alachlor plus the insecticide-nematicides phorate {*O,O*-diethyl S-[(ethylthio)methyl] phosphorodithioate} or fensulfothion (*O,O*-diethyl *O*-4-methylsulfinylphenyl phosphorothioate) than in plots treated only with the insecticide-nematicides (Table 9–3).

The enhanced effectiveness of aldicarb in combination with herbicides may be partially related to the action of the herbicide on nematode life processes. If thiocarbamate herbicides cause nematode egg membrane disintegration and altered permeability, hatching may increase, resulting in more juveniles being exposed to aldicarb (Perry & Beane, 1989). An increase in hatch would therefore account for the enhanced efficacy of the nematicide. Since this response does not occur with all nematicides, aldicarb may be a critical component of the reaction. Juveniles may actually be affected while still within the eggshell. If this scenario is true, alteration of the egg membranes by the herbicide may facilitate the entry

Table 9–3. Effects of alachlor (herbicide), phorate (insecticide), fensulfothion (insecticide-nematicide), and their combinations on population densities of *Helicotylenchus dihystera* 60 and 180 d after treatment (after Schmitt & Corbin, 1981).

	No. nematodes 100 cm^{-3} soil	
Treatment	60 d	180 d
Untreated	65	65
Fensulfothion	55	138
Phorate	68	150
Alachlor	72	200
Fensulfothion + alachlor	330	588
Phorate + alachlor	242	188
LSD ($P = 0.05$)	178	203

of certain molecules, such as the active moiety of aldicarb. The nematode may be killed within the egg or incapacitated to the point at which it is unable to hatch.

The interference with or reduction of nematicide efficacy by other pesticides clearly is undesirable. Some herbicides reduce the efficacy of fenamiphos (Schmitt & Corbin, 1981). Greater nematode numbers occur following treatment with alachlor and fenamiphos than with the nematicide alone (Bostian et al., 1984; Sipes & Schmitt, 1989). The mechanism of this complex interaction may involve an alteration of host physiology. Soybean roots treated with the herbicide are morphologically different than those of untreated plants (Bostian et al., 1984) and, therefore, may be more suitable for nematode feeding and reproduction. An increase in root biomass could explain some of the nematode population-density differences associated with pesticides (Bostian et al., 1984). However, root masses were approximately equal in several treatment combinations involving insecticides, nematicides, and herbicides (Schmitt & Corbin, 1981). In experiments by Huang and Schmitt (1985–1986, unpublished data), alachlor-treated soybean produced less glyceollin, a phytoalexin associated with disease resistance, than untreated controls. Huang and Schmitt (1985–1986, unpublished data) hypothesized that the decrease in glyceollin may provide a more suitable plant in which the nematode can complete its life cycle. Furthermore, postinfection development of *H. glycines* was inhibited less by the combination of alachlor and fenamiphos than with the nematicide alone (Sipes & Schmitt, 1989) (Fig. 9–2). Other possible mechanisms may be related to reduced uptake of the nematicide or altered metabolism of the nematicide as a consequence of plant exposure to alachlor.

The effects of alachlor on fenamiphos vary from year-to-year, suggesting important environmental influences (Schmitt & Corbin, 1981; Schmitt et al., 1983). Soil moisture and temperature probably are critical factors. Moisture is required for activation and movement of most pesticides. With too much or too little moisture, the concentrations that trigger responses would not be present.

Two-, three-, and four-way pesticide-interactions probably occur simultaneously in agricultural fields. The consequence of an interaction is difficult to understand and, at this time, is unpredictable. A better understanding of interactions among pesticides will allow better pest and crop management systems to be designed.

PESTICIDE DYNAMICS IN THE SOIL

The soil environment, comprising both biological and physical aspects, represents the point of contact between the nematode and many pesticides. The dynamic nature of the soil system greatly influences all interactions between nematodes and pesticides, creating many possible scenarios. As illustrated herein, nematode activity may be enhanced, inhibited, or unaffected. Many interactions are probably so slight they are difficult to detect.

Conceptually, pesticides (other than nematicides) that are inhibitory or stimulatory to nematode activity could be used to manage nematode populations. Except for the use of herbicides to control weed hosts, these pesticides have not provided adequate nematode control to be used commercially. The failure to achieve sufficient activity against nematodes could be due to improper pesticide application, pesticide degradation in the soil environment, or a number of other factors.

An example illustrating the difficulty in exploiting herbicides for nematode control follows. Assumptions and calculations will be presented in an effort to demonstrate that herbicides have little opportunity to reduce nematode activity under field conditions, even if they demonstrate in vitro activity. Blazer (BASF Corp., Research Triangle Park, NC), a commercial formulation of acifluorfen, suppressed hatching of *H. glycines* at 50 μg mL^{-1} (10 μg a.i.) in vitro, a substantially higher rate than applied in the field (Wong et al., 1993). At any application rate, the quantity of chemical to which the nematode is exposed in soil will be much less than the amount applied. For example, the recommended foliar application rate for Blazer ranges from 0.13 to 0.41 kg a.i. ha^{-1} (equivalent to about 120–370 μg kg^{-1} soil if the herbicide is distributed in the top 10 cm of soil and soil bulk density is 1.1 g cm^3). We will assume the component responsible for the effect on nematodes is the active ingredient of Blazer, acifluorfen, even though the authors (Wong et al., 1993) did not state which component was responsible for the activity. This concentration is already considerably less than the effective concentration. Nevertheless, calculations estimating fate are important for interpretation of field activity. Given the additional assumptions that: (i) half of the chemical reaches the soil (photodegradation half-life is 4.5 d), (ii) the soil has an organic C content of 1% and gravimetric water content of 200 g kg^{-1} and (iii) the equilibrium adsorption coefficient K_{OC} = 113 L kg^{-1}, the acifluorfen concentration in the soil solution for the highest application rate given above is calculated to be about 0.28 μg mL^{-1}, or about 15% of what it would have been without any adsorption (R.E. Green, 1995, personal communication). In the period following application and movement of the herbicide into the soil, biodegradation and/or chemical degradation will reduce the concentration quickly (the half-life of acifluorfen in soil is about 14 d). Thus, the likelihood of acifluorfen having an effect on nematode processes is small, especially a few days after application. Most other herbicides used today would likely behave similarly. Likewise, calculations could be presented for other categories of pesticides that affect nematode activity in vitro. After dilution into the soil environment, the effects of some pesticides may be minute and have little effect on the nematode population.

OUTLOOK

In a natural environment, nematode population densities fluctuate greatly because of various environmental factors that affect nematode feeding and reproduction. Pesticides add to the complexity of the environment in which the nematode must survive. Although interactions between nematodes and pesticides are commonly observed in vitro, similar interactions are not always evident in the field. However, nematode damage to plants is often a matter of relative numbers and subtle changes in the behavior of nematodes.

For the near future, pesticides will remain an integral component of our agricultural production systems. Pesticides, whether they are synthetic or biologically based, provide food and fiber producers with dependable disease management tools. Therefore, the effects of pesticides on nematodes and their metabolism, on microorganism production of nematoxic by-products, or on plant physiology may still occur and inadvertently affect agricultural production. Pesticide-interaction studies should remain an area of focus and discovery. From a scientific viewpoint, we are interested in understanding how nematode behavior is affected by the wide array of pesticides that enter the environment. From a practical standpoint, it is important to recognize impacts of pesticides so that their use can be properly managed to avoid situations in which nematode damage will be aggravated in a crop.

REFERENCES

Alston, D.G., D.P. Schmitt, J.R. Bradley, and H.D. Coble. 1993. Multiple pest interactions in soybean: Effects on *Heterodera glycines* egg populations and crop yield. J. Nematol. 25:42–49.

Barker, K.R., S.R. Koenning, A.L. Bostian, and A.R. Ayers. 1988. Growth and yield responses of soybean to aldicarb. J. Nematol. 20:42–431.

Barker, K.R., and N.T. Powell. 1988. Influence of aldicarb on the growth and yield of tobacco. J. Nematol. 20:432–438.

Beane, J., and R. Perry. 1990. The influence of certain herbicides in pelletted form on the hatch and invasion of *Globodera rostochiensis*, *G. pallida*, and *Heterodera schachtii* Rev. Nematol. 13:275–281.

Bostian, A.L., D.P. Schmitt, and K.R. Barker. 1984. In vitro hatch and survival of *Heterodera glycines* as affected by alachlor and phenamiphos. J. Nematol. 16:22–26.

Browde, J.A., G.L. Tylka, L. Pedigo, and M.D.K. Owen. 1994. Response of *Heterodera glycines* population to a postemergence herbicide mixture and simulated insect defoliation. J. Nematol. 26:498–504.

Castro, C.E., and I.J. Thomason. 1971. Mode of action of nematicides. p. 289–296. *In* B.M. Zuckerman et al. (ed.) Plant parasitic nematode. Vol. 2. Acad. Press, New York.

Cook, R., and P.A. York. 1972. The effects of benomyl on *Heterodera avenae* on barley. Plant Dis. Rep. 56:261–264.

Decker, H. 1972. Plant nematodes and their control (Phytonematology). Kolos Publ., Moscow.

Evans, A.A.F., and I.J. Thomason. 1971. Ethylene dibromide toxicity to adults, larvae and moulting stages of *Aphelenchus avenae*. Nematologica 17:243–254.

Gourd, T.R., D.P. Schmitt, and K.R. Barker. 1993. Differential sensitivity of *Meloidogyne* spp. and *Heterodera glycines* to selected nematicides. J. Nematol. 25(4S):746–751.

Hill, N.S., and D.P. Schmitt. 1989. Influence of temperature and soybean phenology on dormancy induction of *Heterodera glycines*. J. Nematol. 21:361–369.

Hough, A., and I.J. Thomason. 1975. Effects of aldicarb on the behavior of *Heterodera schachtii* and *Meloidogyne javanica*. J. Nematol. 7:221–229.

Johnson, A.W., and J. Feldmesser. 1987. Nematicides—a historical review. p. 448–454 *In* J.A. Veech and D.W. Dickson (ed.) Vistas on nematology. Soc. Nematol., Hyattsville, MD.

Johnson, A.W., D.A. Smitte, D.R. Sumner, and N.C. Glaze. 1994. Efficacy and compatibility for fenamiphos and EPTC applied in irrigation water for nematode and weed control in snapbean production. J. Nematol. 26(4S):690–696.

Kessmann, H., T. Staub, J. Ligon, M. Oostendorp, and J. Ryals. 1994. Activation of systemic acquired disease resistance in plants. Eur. J. Plant Path. 100:359–369.

Kraus, R., and R.A. Sikora. 1983. Effects of the herbicide diallate, alone and in combination with aldicarb, on *Heterodera schachtii* population level on sugarbeet. Zschrf. Pflakran. Pflaschultz 90:132–139.

Kerry, B.R., D.H. Crump, and L.A. Mullen. 1982. Natural control of the cereal cyst nematode, *Heterodera avenae* Woll., by soil fungi at three sites. Crop Prot. 1:99–109.

Lembright, H.W. 1990. Soil fumigation: principles and application technology. J. Nematol. 22(4S):632–644.

Linit, M.J. 1988. Nematode-vector relationships in the pine wilt disease system. J. Nematol. 20:227–235.

Mamiya, Y. 1988. History of pine wilt disease in Japan. J. Nematol. 20:219–226.

McKenry, M.V., and I.J. Thomason. 1974. 1,3-dichloropropene and 1,2-dibromoethane compounds: II. Organism dosage-response studies in the laboratory with several nematode species. Hilgardia 42:422–438.

Miller, P.M. 1967. Stimulating emergence of larvae of *Heterodera tabacum* with fungicides. Plant Dis. Rep. 51:202–206.

Miller, P.M. 1969. Suppression by benomyl and thiabendazole of root invasion by *Heterodera tabacum*. Plant Dis. Rep. 53:963–966.

Nelms, A.J., D.L. Trudgill, and D.C.M. Carbide. 1973. Chemotherapeutic agents in the study of parasites. p. 95–112. *In* A.E.R. Taylor and R. Muller (ed.) Blackwell, Oxford.

Opperman, C.H., and S. Chang. 1991. Effects of aldicarb and fenamiphos on acetylcholinesterase and motility of *Caenorhabditis elegans*. J. Nematol. 23 :20–27.

Perry, R.N., and J. Beane. 1989. Effects of certain herbicides on the in vitro hatch of *Globodera rostochiensis* and *Heterodera schachtii*. Rev. Nematol. 12:191–196

Putnam, A.R., and D. Penner. 1974. Pesticide interaction in higher plants. Resid. Revs. 50:73–110.

Schmitt, D.P., and F.T. Corbin. 1981. Interaction of fensulfothion and phorate with preemergence herbicides on soybean parasitic nematodes. J. Nematol. 13:37–41.

Schmitt, D.P., F.T. Corbin, and L.A. Nelson. 1983. Population dynamics of *Heterodera glycines* and soybean response in soils treated with selected nematicides and herbicides. J. Nematol. 15:432–437.

Sine, C. 1993. Farm chemicals handbook. Meister Publ. Co., Willoughby, OH.

Sipes, B. S., and D. P. Schmitt. 1989. Effect of planting date, alachlor, and fenamiphos on *Heterodera glycines* development. J. Nematol. 21:33–41.

Sipes, B.S., and D.P. Schmitt. 1994. Population fluctuations of *Rotylenchulus reniformis* and its effects on pineapple yields. Plant Dis. 78:895–898.

Van Gundy, S.D. 1985. Biological control of nematodes: Status and prospects in agricultural IMP systems. p. 467–478. *In* M.A. Hoy and D.C. Herzog (ed.) Biological control in agricultural IMP systems. Acad. Press, Orlando, FL.

Veech, J.A. 1978. The effect of diflubenzuron on egg formation by the root-knot nematode. J. Nematol. 10:208–209.

Vroom, J.J. (ed.). 1995. Crop protection reference manual. 11th edition. p. 1680–1694. Chem. Pharmaceut. Press, Inc., New York.

Weischer, B., and J. Muller. 1985. Side effects of plant protection products on nematodes and their antagonists. Ber. Landu. 198:159–176.

Wilkenson, R.E. 1978. Physical response of lipid components to thiocarbamates, and antidotes. p. 85–108. *In* F.M. Pallos and J.E. Corsida (ed.) Chemistry and action of herbicidal antidote. Acad. Press, New York.

Wong, A.T.S., G.L. Tylka, and R.G. Hartzler. 1993. Effects of eight herbicides on in vitro hatching on *Heterodera glycines*. J. Nematol. 25:578–584.

Wright, D.J. 1981. Nematicides: Modes of action and new approaches. p. 421–449. *In* B.M. Zuckerman and R.A. Rohde (ed.) Plant parasitic nematodes. Vol. 3. Acad. Press, New York.

10 Breeding for Nematode Resistance and Tolerance

LAWRENCE D. YOUNG

USDA-ARS, Soybean Research
Jackson, Tennessee

Nematodes cause an estimated annual loss of about 12% ($8 billion) of the major food and fiber crops in the USA and more than $77 billion in the world (Sasser & Freckman, 1987). These economic losses would be much greater except for the many management practices employed to reduce losses. Use of nematode-resistant cultivars is a major management tactic used to suppress losses in crops. For example, soybean [*Glycine max* (L.) Merr.] cultivars resistant to *Heterodera glycines* (soybean cyst nematode) grown in infested soil yield 10 to 50% more than susceptible cultivars (Hartwig, 1981; Young & Hartwig, 1988). In fact, planting a single resistant cultivar increased profits for soybean producers over $400 million during a 6-yr period (Bradley & Duffy, 1982). Value of *H. glycines*-resistant cultivars can be enhanced because some of these cultivars also are resistant to *Meloidogyne* species and the reniform nematode, *Rotylenchulus reniformis*.

Resistant cultivars without nematicide treatment often yield as much as high-yielding susceptible cultivars treated with nematicides (Epps et al., 1981). In Florida, yields from soybean cultivars resistant to *Meloidogyne incognita* were five times greater than yields from highly susceptible cultivars (Kinloch et al., 1985). Over half of the hectarage planted to tobacco (*Nicotiana tabacum* L.) in the southeastern USA in 1986 was planted with cultivars resistant to Races 1 and 3 of *M. incognita* (Johnson, 1989).

Use of resistant cultivars has the following advantages: suppresses nematode reproduction, reduces need for toxic chemicals, shortens length of rotations, does not require use of specialized equipment, maintains cost of seed generally equal to that of susceptible cultivars (Boerma & Hussey, 1992), and may limit disease complexes associated with nematodes (Mai & Abawi, 1987). Planting resistant cultivars and using rotations are the primary practical methods of suppressing nematode damage to crops of low economic value.

Most of the nematode resistant cultivars are resistant to sedentary endoparasites such as *Globodera*, *Heterodera*, and *Meloidogyne* species. The stem nematode, *Ditylenchus dipsaci*, is an exception to this trend. Some success has been achieved with the migratory endoparasite, *Pratylenchus* species.

TERMINOLOGY

In nematology, resistance is used to describe the effects of the host on nematode reproduction and often relative host damage. It is a relative term describing a continuum of host-nematode interaction ranging from no nematode reproduction to much reproduction. A highly resistant genotype allows little reproduction, whereas a moderately resistant genotype supports an intermediate level of reproduction compared to a susceptible genotype that allows the nematode to reproduce freely. Although a plant that allows very little or no nematode reproduction is resistant using the above definition, few would utilize it if the plant was so intolerant to nematode infection that it produced little marketable product.

Tolerance has many definitions, but may describe the amount of host injury caused by the nematode (Wallace, 1987). Typically, a genotype is considered tolerant if it yields almost as well in the presence of nematode infection as it does in the absence of the nematode. Tolerance, like resistance, represents a continuum of reactions. Intolerant genotypes can be so severely injured that there is little plant growth. Since nematodes are obligate parasites, nematode reproduction can be greatly suppressed by intolerant plants due to lack of feeding sites for the nematode. Tolerance and resistance may be independent traits of a host (Boerma & Hussey, 1992).

Resistance also has been characterized by the number of genes controlling the trait. Monogenic [*M. incognita* resistance in tomato (*Lycopersicon esculentum* Miller)], oligogenic (*H. glycines* resistance in soybean), and polygenic describe resistance controlled by one, a few, and many genes, respectively. Size of the effects of genes on the phenotypic expression of resistance also has been used to categorize the reaction. Major genes have a large effect that usually is identified easily, whereas minor genes have only a small and possibly indiscernible effect on the expression of resistance.

SOURCES OF RESISTANCE

The first step in developing resistant cultivars by classical techniques to manage nematode populations is to locate a source of resistance. A modified listing of Fehr's (1987) priority for searching for resistance includes: adapted commercial cultivars, parents of hybrid cultivars, or parents of synthetic cultivars, adapted elite breeding lines or germplasms, unadapted cultivars or parents, unadapted elite breeding lines or germplasms, acceptable breeding lines, plant introductions (PI) of the cultivated species, interspecific hybrids, and transferred resistance genes from other species.

Generally, nematode resistance is available in commercial cultivars and elite breeding lines only after resistance to a particular nematode species has been the objective of a breeding program for several years. There have been a few exceptions. Resistance to the reniform nematode was found in 'Forrest' soybean, a byproduct of breeding for resistance to *H. glycines* (Rebois et al., 1970). Resistance to *M. incognita* was found in corn (*Zea mays* L.) inbreds (Windham

& Williams, 1988). The resistance in corn was probably the result of indirect selection when genotypes were grown in nematode-infested soil with direct selection for growth and vigor, not nematode resistance per se. Resistance also could have occurred because of independent segregation when introduced from germplasm selected for other traits. Progress in cultivar improvement is made much faster by using resistant commercial cultivars as parents as compared to using the original, less adapted source of resistance.

If resistance is not available in commercial cultivars or acceptable breeding lines, the next best source is the PI of the species. Development of *H. glycines*-resistant cultivars is illustrative of the use of PI from a germplasm collection (Fig. 10–1). Ross and Brim (1957) screened 2800 accessions in the U.S. soybean germplasm collection for resistance to this pest. They identified 'Ilsoy', 'Peking', PI 84751 and PI 90763 as resistant. Peking was used as the resistant parent to develop 'Pickett' (Brim & Ross, 1966), the first adapted cultivar with resistance to the nematode. Epps and Hartwig (1972) screened over 3000 accessions to find resistance to *H. glycines* populations reproducing on resistant cultivars like Pickett. They identified PI 88788 and seven other accessions as resistant. PI 88788 was used in the development of 'Bedford' (Hartwig & Epps, 1978). Anand and Gallo (1984) and Anand et al. (1985) tested all available accessions for resistance to *H. glycines* Races 3, 4 (the race was probably Race 14 instead of Race 4), and 5. They identified PI 437654 as resistant to all races and used it to develop 'Hartwig' (Anand, 1992). Young (1990a, 1995) evaluated approximately 2500 accessions added to the collection after Anand's work and identified PI 467312 as resistant to *H. glycines* Races 3, 5, and 14.

Fig. 10–1. Susceptible soybean (left) and resistant Hartwig soybean (right) grown in a field infested with *Heterodera glycines* Race 2 (courtesy of S. R. Koenning & K. R. Barker).

If resistance is not available in cultivars, breeding lines, or PI of the species, then closely related or distantly related species can be evaluated. Wild species have been used to develop nematode-resistant potato (*Solanum tuberosum* L.), tobacco, and tomato cultivars (Fassuliotis, 1987). Many undesirable traits are introduced along with resistance when wild species are used. The unacceptable traits must be eliminated through additional breeding effort. Problems of fertility and viability have been encountered with interspecific hybrids. Sometimes a bridge cross is required to transfer genes from one species to another. A bridge cross involves an interspecific cross between the donor species and a third species, which is more compatible with the cultivated species. The "bridge" interspecific hybrid is then crossed to the cultivated species. Embryo culture has been used to overcome viability problems (Fassuliotis & Nelson, 1988). Development of *M. incognita*-resistant tomato required embryo culture plus several years of breeding effort before the first cultivar was released (Boerma & Hussey, 1992).

Biotechnology techniques allow cloned genes to be transferred from one species to another. This may be a source of nematode resistance for cultivars in the future (Opperman & Conkling, 1998, see Chapter 12). Attempts are being made to clone the tomato *Mi* gene for *M. incognita* resistance (Williamson et al., 1994). Once cloned, the gene may be transferred to other species if a genotype of the recipient species is amenable to regeneration of transformed plants. Conventional plant breeding may be necessary to transfer the resistance gene from transformed plants to a commercial cultivar. Other biotechnology techniques, such as somaclonal variation and protoplast fusion between different species, may provide resistant germplasm to use in a conventional breeding program for the development of nematode-resistant cultivars. However, these approaches require the ability to regenerate whole viable plants from tissue culture.

INOCULUM PRODUCTION

It is essential for a successful breeding program to have the capability of evaluating a large number of progenies in a relatively short period of time. Two major components of a successful program that allow this are production of adequate inoculum and an effective evaluation technique. Inoculum production begins with selection of the nematode life stage to use, which is determined by the parasitic cycle of the nematode.

The second-stage juvenile is the infective stage for most sedentary endoparasites. These nematodes invade the root, establish a feeding site, and become sedentary. Since it is often difficult to obtain large numbers of infective juveniles, eggs are often used as inoculum, especially for *Meloidogyne*, *Heterodera*, and *Globodera* species. The vermiform female of *R.. reniformis* is the infective stage. However, inoculum for this nematode often includes a collection of all vermiform stages, juveniles, males, and females. Eggs of *R. reniformis* are not recommended for inoculum (Heald & Robinson, 1990) because of the frequent delay between egg hatch and development of the infective female. Most migratory endoparasites, such as *Pratylenchus* and *Ditylenchus*, deposit their eggs singly in either shoot or root tissues. In contrast, ectoparasites, including *Belonolaimus* and *Criconemella*, deposit eggs in soil. Therefore, inoculation with eggs is not a viable option, and juveniles or adults must be used as inoculum for these species.

Large quantities of inoculum are needed for most breeding programs. Infested field soil can be a source of inoculum for some species. However, the soil can be infested with other pathogens or pests that create problems during the evaluation procedures, especially in a greenhouse environment. Noel et al. (1990) described a procedure for producing *H. glycines* inoculum in outdoor ground beds. The infested soil was collected in the fall, placed in containers, and used in greenhouse evaluations. Use of soil infested in this manner was necessary to allow evaluation of a sufficient number of genotypes each year. Boerma and Hussey (1992) also supported use of infested field soil to evaluate large numbers of genotypes for *H. glycines* resistance.

Many programs use inoculum cultured on greenhouse-grown plants. Most endoparasite females deposit large quantities of eggs within an egg matrix or retain eggs within their bodies. Production of cultured inoculum is relatively easy for these species. Large quantities of eggs of *Meloidogyne* species are easily collected from galled roots with sodium hypochlorite (Hussey & Barker, 1973). Mature females of *H. glycines* can be washed from roots, and eggs can be released from the females and cysts by cracking them with a Ten-Broech homogenizer. Eggs are easily dispersed into soil. Other methods of inoculum production include tissue culture for *Pratylenchus* (Riedel et al., 1973) and *Ditylenchus* species and collecting infested seed or leaf tissues for *Aphelenchoides* species. *Aphelenchoides* can be stored in an anhydrobiotic state within seeds and dried leaf tissue (Griffin & Krusberg, 1990).

Much intraspecific genetic variation occurs in nematodes, especially in the endoparasites that have specialized interactions with their hosts. Sixteen races of *H. glycines* are possible, based on the four standard differentials. Twelve of these races have been identified (Riggs & Schmitt, 1988). Four host races have been described for *M. incognita* and two races for *M. arenaria* (Taylor & Sasser, 1978). Cytological races also are known for *M. arenaria* and *M. hapla* (Triantaphyllou, 1985). Thus, it is imperative that each nematode population used for evaluation of breeding material be properly characterized and constantly checked for purity.

Adequate sampling of within-race variation is critical to a successful evaluation program. Hussey and Boerma (1981) captured genetic diversity in *M. incognita* by using a mixture of three aggressive isolates collected from diverse geographic areas. The isolates were maintained separately in greenhouse culture, and the mixture was reconstituted at periodic intervals. Another technique to expose plant genotypes to diverse nematode isolates is to eliminate most susceptible genotypes with a preliminary evaluation using a representative isolate. The resistant genotypes identified in the preliminary evaluation are then exposed to multiple isolates of the nematode to confirm resistance to a broader array of nematode populations.

Aggressive nematode isolates can be used when a high level of resistance is available, but less aggressive isolates may be required if only low to moderate resistance is encountered. Too aggressive isolates can mask these lower levels of resistance (Windham & Pederson, 1989), which may be improved with recurrent selection breeding techniques. Moderate levels of resistance may provide adequate protection from nematode damage, particularly when cultural practices that decrease nematode population density are coupled with planting resistant cultivars.

EVALUATION METHODS

Resistance

Fassuliotis (1985) listed requirements for successful screening techniques. The technique must optimize the amount of inoculum needed to differentiate differences among host genotypes. The inoculum should be limited so that high dosages do not cause too much injury and mask potentially useful genetic material. The procedure should be simple, not laborious, and allow no plant to escape exposure to the infective stage of the nematode. Also, screening should be done under conditions conducive for good plant growth.

The screening procedure should be capable of evaluating thousands of genotypes in a reliable manner according to Boerma and Hussey (1992), who advocated a greenhouse environment for the procedure in order to obtain uniform nematode infestations between screenings and prevent seasonal restrictions. A greenhouse environment also allows evaluations in an area where the nematode is not indigenous. However, environmental conditions often vary sufficiently in greenhouses between seasons to appreciably affect results. Inclusion of standard resistant and susceptible genotypes in each screening allows normalization of variation in tests resulting from these changed environments.

In a screening program, either nematode reproduction or plant injury/symptomology is measured. One method of comparing nematode reproduction among genotypes is to calculate the ratio of nematodes present at the end of a test to the number present at the beginning of the test. Ratios less than one indicate the nematode population declined and that the plant genotype is resistant. If the level of resistance is not high, genotypes with ratios slightly higher than one may be retained for further breeding efforts. More often, final populations of nematodes occurring on breeding lines are compared with nematode populations on standard resistant cultivars. Direct enumeration of nematodes present on roots or in the soil also may be used to determine resistant genotypes when nematode densities are compared with those on the standard resistant checks included in each test.

Actual measurement of nematode numbers can be more laborious than is justified in the initial evaluation of breeding lines. A rating scheme based on a subjective assessment of nematode numbers or plant injury promotes rapid classification of plant response to the nematode. This is particularly true for *Heterodera* and *Meloidogyne* species. Plant response to *Meloidogyne* species is often measured by amount of root galling and/or number of egg masses (Table 10–1). These parameters are placed into one of 5 to 10 categories by visual inspection (Barker et al., 1986). Reproduction of *H. glycines* also is often placed in one of a few classes (Table 10–1). Low numbers of females on the roots require

Table 10–1. Rating scheme for *Heterodera glycines* and *Meloidogyne* nematode infection.

Scale	*H. glycines*	*Meloidogyne*
	no. cysts plant^{-1}	% root system galled
0	0	none†
1	1–5	trace
2	6–10	<25%
3	11–20	25–50%
4	21–40	50–75%
5	>40	>75%

† Kinloch, 1990.

visual enumeration of females, larger numbers of females are estimated. Only genotypes placed in the lowest classes have a practical level of resistance; therefore, accuracy is more important for these classes than it is for the more susceptible classes.

Although a greenhouse environment is often advocated for screening breeding material, the plant material also can be evaluated in the field for nematode reproduction as well as plant yield. Most nematode species have a patchy distribution in field sites. Kinloch (1990) suggested the following cultural procedures to enhance uniformity of nematode distribution in the field to an acceptable level: changing the direction of planting rows each time a crop is planted, manipulating soil with land-leveling equipment immediately before planting, avoiding cultivation and cropping in temperate regions during the off-season, planting a susceptible crop following evaluation in tropical regions to maintain the nematode population, and confining the screening test to the smallest area possible. He used hill plots to maximize the number of test genotypes within a small area.

Standard resistant and susceptible genotypes should be grown frequently among the genotypes in the field to help normalize the results. The test area should be examined for presence of other pathogens and pests that could affect results. For *Meloidogyne* species, galling is visually rated according to an index as illustrated in Table 10–1. Kinloch (1990) recommended that the most severely galled root portion within a hill be treated as representative of the hill. Also, if several replicates of a genotype are grown, the most severely galled replicate should be accepted as the rating for the genotype. This is more accurate for homozygous genotypes than it is for entries that are still segregating for nematode resistance. Entries with little or no nematode infection should be evaluated again to confirm that they are resistant and not escapes from infection.

Plants also have been evaluated for *H. glycines* resistance in the field. Ross and Brim (1957) grew glabrous susceptible soybean in double rows separated by 15 cm. Entries were dug approximately 30 d after planting while the females were still white or yellow in color but adhered to the roots. If an entry had few females on the root, the susceptible check was dug to determine if it had sufficient infection to accept the test entry as resistant. Evaluation in the field is the only way to obtain reliable data on yield of resistant genotypes. A genotype that is resistant could be intolerant or less tolerant of nematode infection than other genotypes. Field evaluation also allows selection of genotypes for other agronomic traits.

Evaluation techniques for screening different plant species for resistance to plant-parasitic nematodes recently have been published (Starr, 1990). These techniques may require adaptation to optimize differentiation between resistant and susceptible genotypes. Adjustments are needed to compensate for different environmental conditions and inoculum encountered by each evaluation program.

Tolerance

Tolerance in soybean to nematodes has been identified in different genotypes grown in infested fields by comparing yield of nematicide-treated plots to that of untreated plots (Anand & Koenning, 1986; Boerma & Hussey, 1984). Koenning et al. (1992) recommended that tolerance be measured by testing different genotypes against a range of nematode population densities and then comparing the slope of regression lines for yield against nematode density. They recommended using cultural practices in the previous year to create areas within a field with the range of nematode population densities required. However, this method of measuring tolerance might be limited to the later stages of evaluation in a breeding program because the method does not efficiently use labor and land.

Reese et al. (1988) found that genotypes in a breeding program could be selected for tolerance in untreated plots in infested fields. This procedure allows concurrent selection for high-yield potential. Elimination of treated plots for each genotype doubles the number of genotypes that can be evaluated with the same land area, thus improving chances for genetic gain. Dale et al. (1988) also found that potato could be directly selected for tolerance to *Globodera pallida* in infested fields. Genotypes identified as tolerant needed to be tested for nematode reproduction to confirm they were indeed tolerant and not resistant to nematode infection. Genotypes selected for tolerance can be tested in later stages of evaluation in nematicide-treated and untreated plots to more precisely measure tolerance.

BREEDING PROCEDURES

Selection of breeding procedures to utilize in developing nematode-resistant cultivars is dependent upon type of crop pollination, heritability of the trait, number of genes involved in resistance, and the productivity or agronomic adaptability of the donor parent. Recurrent selection, backcrossing, and pedigree breeding are commonly used in breeding for nematode resistance.

Recurrent selection is used primarily for population improvement, especially for cross-pollinating plants; however, the technique can be used for self-pollinating crops if limitations on crossing can be overcome. This breeding procedure is effective in increasing the level of resistance in a population when the initial frequency of resistance alleles is low. With this breeding technique, the most resistant plants are selected in sufficient number to retain ample genetic variation for other traits. The selected plants are intercrossed to form the improved population from progenies of the intercrosses. Members of the improved generation are then evaluated, selected and intercrossed to form the next improved population. Individuals can be extracted for evaluation for agronomic traits from the improved populations. Superior genotypes thus extracted can be used as cultivars, inbreds for crossing to produce hybrids, or synthetic populations. Recurrent selection in the population can continue until adequate resistance is obtained. Traits with heritability adequate to allow accurate phenotypic characterization by visual or simple tests greatly aid success via recurrent selection.

Recurrent selection has been used successfully to develop nematode-resistant germplasm. Barnes et al. (1990) developed two alfalfa (*Medicago sativa* L.) germplasms with resistance and tolerance to the root-lesion nematode, *Pratylenchus penetrans* using recurrent selection. Tolerance through development of many fibrous roots appeared to be the primary trait selected; however, the germplasms supported 20 to 30% fewer nematodes than the susceptible check. Pederson and Windham (1995) used recurrent selection to develop white clover (*Trifolium repens* L.) germplasm resistant to *M. incognita*. Quesenberry et al. (1989) used five cycles of restricted recurrent phenotypic selection in an infested field to increase *Meloidogyne* resistance in red clover (*T. pratense* L.). They then further increased resistance to *M. incognita*, *M. arenaria*, and *M. javanica* with two cycles of recurrent half-sib selection in the greenhouse.

The backcross technique is useful to transfer monogenic and oligogenic resistance from unadapted germplasm to adapted susceptible cultivars. Backcrossing is most often used for self-pollinating crop, although it also can be used to transfer traits to cross-pollinating crops. Some backcrossing procedures allow the resistance genes to be transferred without other adverse traits within 5 yr. One criticism of the backcross technique is that the backcross-derived cultivars are not superior to the recurrent parent. Boerma and Hussey (1992) indicated backcross-derived cultivars could be on the market within 3 yr of the recurrent parent when backcrossing was started 2 to 3 yr prior to release of the recurrent cultivar. The modified backcross technique is another way to improve traits other than nematode resistance during transfer of the nematode resistance genes from unadapted germplasm. In this procedure, crosses are made to more than one recurrent parent of similar maturity and agronomic characteristics during transfer of nematode resistance. Hartwig and Epps (1978) used modified backcrossing in the development of Bedford soybean. The "recurrent" parents were closely related. The modified backcross method requires more yield testing than the simple backcrossing method.

Pedigree breeding is the simple hybridization of parents and selection of resistant progenies. It is used when both parents are adapted, productive genotypes and is useful for both cross and self-pollinating crops when a homogeneous genotype is the desired product. The procedure is usually effective in producing competitive cultivars only after breeding for nematode resistance has been conducted intensively for several years and high-yielding resistant parents have been developed. The procedure is now being used extensively in development of nematode-resistant soybean cultivars.

SOME SUCCESSFUL BREEDING PROGRAMS

Nematode-resistant germplasm has been identified for a significant, but limited, number of crops (Roberts, 1982). Efforts to develop resistant cultivars for

Table 10–2.Major crops in which resistance to nematodes is available.†

Crop	Resistance to
Soybean	*Heterodera glycines, Meloidogyne incognita, M. arenaria, M. javanica, Rotylenchulus reniformis*
Alfalfa	*M. incognita, Ditylenchus dipsaci, Pratylenchus penetrans*
Clover (red, white)	*M. incognita*
Small grain (wheat, barley, oats)	*D. dipsaci, H. avenae*
Tobacco	*Globodera* spp., *M. incognita, M. arenaria, M. javanica*
Potato	*G. pallida, G. rostochiensis, M. chitwoodi*
Bean (snap, lima)	*M. incognita, M. javanica, P. scribneri*
Cowpea	*M. incognita*
Sweetpotato	*M. incognita, M. javanica, M. arenaria*
Tomato	*M. incognita, M. javanica, M. arenaria*
Apricot	*Meloidogyne* spp.
Citrus	*Tylenchulus semipenetrans, Radopholus citrophilus*
Grape	*Meloidogyne* spp.
Walnut	*Meloidogyne* spp.
Prunus rootstock (almond, nectarine, peach, plum)	*M. incognita, M. javanica, M. arenaria*

† Roberts, 1982, updated.

several crops have been made with varying degrees of success. Table 10–2 lists the major crops for which resistant cultivars have been released, and Table 10–3 gives the number of cultivars and germplasms released in the last 10 yr for those crops registered with the Crop Science Society of America (CSSA). Details follow for some of these crops, and more information is provided in chapters on individual crops.

Table 10–3. Total number of cultivars and germplasms and number of nematode-resistant ones for those crops registered by the Crop Science Society of America for which at least one cultivar or germplasm was referred to as resistant to a nematode.

Crop	Total	Nematode resistant
	no.	no.
Cultivars		
Alfalfa (*Medicago sativa*)	46	20
Alyceclover (*Alysicarpus vaginalis*)	1	1
Chickpea (*Cicer arietinum*)	10	1
Cotton (*Gossypium hirsutum*)	22	8
Cowpea (*Vigna unguiculata*)	2	2
Digitgrass (*Digataria eriantha*)	2	2
Hairy indigo (*Indigofera hirsuta*)	1	1
Red clover (*Trifolium pratense*)	3	1
Soybean (*Glycine max*)	147	50
Tobacco (*Nicotiana tabacum*)	18	3
Germplasms		
Alfalfa	80	3
Corn (*Zea mays*)	129	4
Cotton	365	27
Hairy indigo	2	2
Kura clover (*T. ambiguum*)	5	1
Peanut (*Arachis hypogaea*)	25	2
Soybean	70	12
Tobacco	18	3

Table 10–4. Soybean with cultivars resistant to *Heterodera* and/or *Meloidogyne* species and registered with the Crop Science Society of America during 1990 to 1994.

Maturity group	Cultivars registered	Cultivars resistant to *H. glycines*	*M. incognita*	*M. arenaria*	*M. javanica*
		no.			
00	1				
0	3				
I	12	2			
II	17	2			
III	20	2			
IV	11	5	1	1	
V	4	4	3	1	
VI	7	5	4	2	2
VII	4	1	3	2	2
VIII	3		2		1
IX	1				

Soybean

Nematode resistance has been incorporated into soybean cultivars probably with more success than in any other crop. Resistance to *H. glycines* has been incorporated into more cultivars than has resistance to *Meloidogyne* spp., but many cultivars grown in the southern USA have resistance to both nematodes. Table 10–4 lists the number of cultivars with nematode resistance registered with the CSSA from 1990 through 1994. Cultivars with resistance to *H. glycines* are available in all maturity groups (MG) from I to VII. One cultivar was resistant or tolerant to *Hoplolaimus columbus*, and one cultivar had resistance to *R.. reniformis*. Resistance to both *Meloidogyne* spp. and *Heterodera glycines* was more common in the later maturity groups. Many other cultivars with nematode resistance were released during this time by commercial seed companies that usually do not register their cultivars. A noteworthy germplasm release during this time was bred for tolerance to *Heterodera glycines* (Boerma et al., 1993).

The first resistant cultivars often did not yield as well as the most productive susceptible cultivars in noninfested soil. Today, the resistant cultivars are frequently the most productive cultivars in both infested and noninfested fields. In Tennessee in 1994, the two highest-yielding cultivars (3-yr avg.) in each of the MG IV and MG V yield trials were cultivars resistant to *Heterodera glycines*. Most of the yield trials were grown in noninfested soil (Graves et al., 1994).

Alfalfa

At least 25 alfalfa cultivars were registered from 1981 through 1994 that were either resistant or moderately resistant to the stem nematode *D. dipsaci*. One or more generations of phenotypic recurrent selection often was used to increase the percentage of plants with resistance to this nematode. Phenotypic recurrent selection also was used to develop alfalfa with resistance or tolerance to the lesion nematode, *Pratylenchus penetrans* (Barnes et al., 1990). A few cultivars were listed with moderate resistance to *M. hapla* (Kugler et al., 1991a,b).

Clovers and Other Forage Legumes

Only limited progress has been made in developing nematode-resistant cultivars in these crops. Pederson and Windham (1995) developed white clover germplasm with resistance to *M. incognita, M. arenaria, and M. graminicola*. 'Cherokee' red clover was described as having more resistance to *M. incognita, M. arenaria, M. hapla, and M. javanica* than other red clover cultivars (Quesenberry et al., 1993). Leath et al. (1987) developed red clover germplasm with resistance or tolerance to stunting caused by the clover cyst nematode, *Heterodera trifolii*. Quesenberry et al. (1989) selected red clover for tolerance to *Meloidogyne* spp. by selecting for early spring vigor during five cycles of recurrent restricted phenotypic selection. Kura clover (*Trifolium ambiguum* Bieb.) germplasm with resistance to *M. incognita* Race 4 has been released by Pederson et al. (1991). Alyceclover [*Alysicarpus vaginalis* (L.) DC] 'FL-3' was selected for resistance to *Meloidogyne* species (Baltensperger et al., 1990b). Hairy indigo (*Indigofera hirsuta* L.) cultivars with resistance to *M. incognita*, *M. arenaria*, and *M. javanica* have been developed (Baltensperger, 1989; Baltensperger et al., 1990a).

Cereals

Resistance to *D. dipsaci* in cultivars of both winter-sown and spring oat (*Avena sativa* L.) has been obtained from the land race 'Grey Winter' (Cook & Evans, 1987). Resistance to *D. dipsaci* also has been developed in some rye (*Secale cereale* L.) cultivars. Cultivars of barley (*Hordeum vulgare* L.), oat, and wheat (*Triticum aestivum* L. emend. Thell.) are available with resistance to *Heterodera avenae*, which is a serious pest in Europe and Australia. None of the cereal grain cultivars recently registered with the CSSA were listed as resistant to this nematode. However, most of these cultivars were developed in the USA.

Tobacco

In 1986 over half of the flue-cured tobacco hectarage from Virginia to Florida was planted with cultivars resistant to Races 1 and 3 of *M. incognita* (Johnson, 1989). Cultivars are now needed with resistance to Race 2 of *M. arenaria*, Races 2 and 4 of *M. incognita*, and *M. javanica* because these races and species are becoming an increasing problem (Barker, 1989). Cultivars with resistance to *Globodera tabacum solanacearum* have been developed. Although these cultivars can greatly suppress nematode increase, they are intolerant to nematode infection and do not yield adequately to compete with susceptible cultivars treated with a nematicide (Johnson, 1990). Some of the resistant flue-cured cultivars also have resistance to the related subspecies *G. tabacum* which parasitizes cultivars of shade grown tobacco (LaMondia, 1988).

Potato

Cultivars have been bred for resistance to the potato cyst nematodes, *Globodera rostochiensis* and *G. pallida*. Both species are found in Europe and

South America. Some cultivars have only partial resistance to *G. pallida* (Cook & Evans, 1987). Resistance to *G. rostochiensis* has been successful in the USA where the Ro1 pathotype has been found, and at least 16 cultivars with resistance have been released. In the USA, resistant cultivars are used with other management tactics to slow the spread of the nematode, which is subject to a quarantine program (Brodie & Brucato, 1993). Population densities of the nematode are maintained at a level much below the economic damage threshold to prevent spread of the nematode. Recently, much effort has been expended to locate and transfer resistance to *M. chitwoodi*. Resistance in *Solanum bulbocastanum* was transferred to cultivated potato using somatic hybridization (Austin et al., 1993).

Cotton

Many cotton (*Gossypium hirsutum* L.) cultivars have been selected for resistance to the Fusarium wilt-root-knot nematode complex. Some of this has been done in the field in the Regional Wilt Screening Test at Tallassee, Alabama (Baker, 1992). Shepherd (1986) indicated selection for field resistance to Fusarium wilt only resulted in cultivars with moderate resistance. However, selection for high resistance to the root-knot nematode, *M. incognita,* could give high field resistance to the Fusarium wilt-nematode complex, while selection for genetic resistance to Fusarium wilt alone would yield only moderate field resistance to the disease complex. Thus, selection for high resistance to *M. incognita* alone was expected to be effective to overcome any disease complex involving both organisms.

Vegetables

Several cultivars of southernpea [*Vigna unguiculata* (L.) Walp. subsp. *unguiculata*] or cowpea [*Vigna unguiculata* (L.) Walp.] and sweetpotato [*Ipomoea batatas* (L.) Lam.] have been released with resistance to *M. incognita*. An ongoing effort to maintain and increase resistance to this and other *Meloidogyne* species in these vegetables is apparent. Tomato is another crop with several cultivars with resistance to *Meloidogyne* species, especially *M. incognita*. Roberts (1982) indicated that of approximately 45 annual field and vegetable crops grown in California, *Meloidogyne* resistance in bean (large lima, *Phaseolus limensis* Macfad.), cowpea, sweetpotato, and tomato were the only cases where resistance was used as a nematode management tactic. Nematode resistance has been used in the development of snap bean, *Phaseolus vulgaris* L., (Wyatt et al., 1983) and hot pepper, *Capsicum frutescens* L. (Fery & Dukes, 1986). Some resistance to *Meloidogyne* species has been found in cucumber (*Cucumis sativus* L.) (Walters et al., 1993), okra [*Abelmoschus esculentus* (L.) Moench] (McLeod et al., 1983), and carrot [*Daucus carota* subsp. *sativus* (Hoffm.) Arcang.] (Yarger & Baker, 1981).

Fruit and Citrus Trees

Probably the most successful use of nematode resistance in fruit trees is the development of the 'Nemaguard' rootstock (Fig. 10–2). According to Roberts

(1982), approximately 75% of the almond [*Prunus dulcis* (Miller) Webb], nectarine [*Prunus persica* var. *nectarina* (L.) Batsch], peach [*Prunus persica* (L.) Batsch var. *persica*], and plum (*Prunus* spp.) plantings in California in 1979 were on Nemaguard rootstock. However, Nemaguard is susceptible to winter injury in the southeastern USA and is not used in that region (Ritchie & Clayton, 1981). Other *Meloidogyne*-resistant rootstocks have been released since Nemaguard: two for peach (Ramming & Tanner, 1983; Sherman et al., 1991) and two primarily for almond (Kester & Asay, 1986). The Nemaguard rootstock also is susceptible to the lesion nematode, *Pratylenchus vulnus*. Rootstocks have been identified that are resistant to this nematode (Culver et al., 1989). *Meloidogyne* resistance also is utilized in a large percentage of the walnut plantings in California (Roberts, 1982). Citrus rootstocks resistant to the citrus nematode, *Tylenchulus semipenetrans*, have been used successfully for 25 yr (Roberts, 1982). Limited success has been obtained from the use of citrus rootstocks resistant to *Radopholus citrophilus* (Roberts, 1992). Resistance to *Meloidogyne* spp. in grape (*Vitis vinifera* L.) rootstocks has been developed (Cook & Evans, 1987), and resistance to *Xiphinema index* has been identified.

Other Crops

Cultivars, germplasm lines, or parental lines of a few other crops with nematode resistance have been released within the last few years. These include strawberry (*Fragaria* × *ananass* Duch.) with resistance to *P. penetrans* (Potter, 1994), peanut (*Arachis hypogaea* L.) with resistance to *M. arenaria* (Simpson et

Fig. 10–2. Roots of resistant Nemaguard peach (right) compared to roots of a susceptible cultivar (left) grown in a *Meloidogyne* infested orchard (courtesy of K. R. Barker & C. N. Clayton).

al., 1993), and corn with resistance to *P. hexincisus* and *P. scribneri* (Wicks et al., 1990a,b).

LIMITATIONS ON USE OF RESISTANT AND TOLERANT CULTIVARS

Farmers select cultivars to improve profits through increased yield and/or quality. They must balance factors such as yield potential, market demands, cropping system, soil characteristics, and disease and pest problems. Nematode resistance is only one of these factors and usually is secondary to yield potential. Other factors a producer considers may eliminate selecting nematode-resistant cultivars for planting. Poor quality of the harvested product linked to nematode resistance may preclude growing some nematode-resistant cultivars of some crops. Flue-cured tobacco cultivars resistant to *G. tabacum solanacearum* effectively reduce nematode population densities. These cultivars have lower leaf quality and yield, however, and economic return is lower than when susceptible cultivars are treated with nematicides (Johnson, 1990). Infestations with multiple nematode species can pose difficulties for effective use of resistant cultivars. *Heterodera glycines*, *M. incognita*, and *M. arenaria* commonly occur together in soybean fields in southern Alabama (Weaver et al., 1989). Cultivars with effective resistance to *H. glycines* and *M. incognita* are available (Kinloch et al., 1985), but cultivars with resistance to some races of *H. glycines* and *M. arenaria* (Fig. 10–3) are not available (Kinloch et al., 1987). Therefore, *H. glycines* resistance cannot always be utilized effectively in fields infested with both pests.

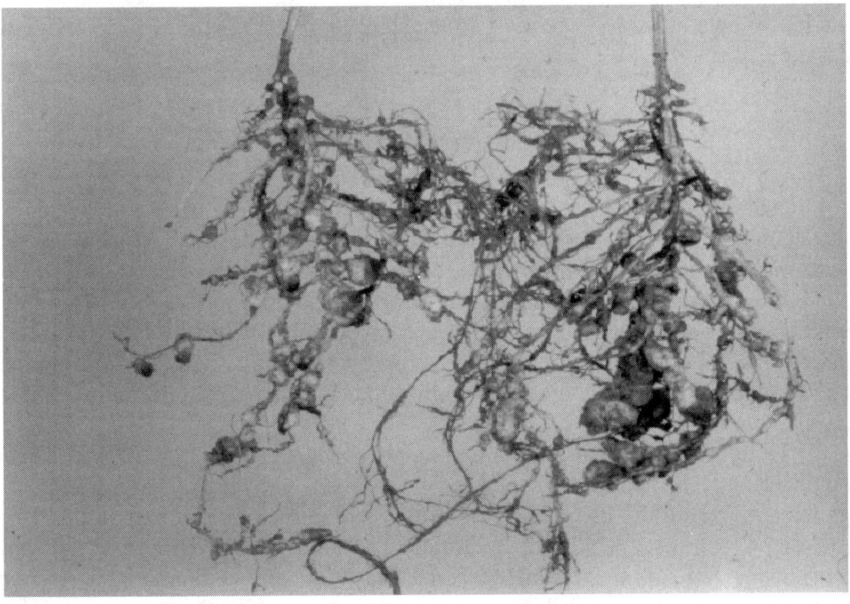

Fig. 10–3. Galled roots of Hartwig soybean (resistant to *Heterodera glycines*) after being grown in a field infested with *Meloidogyne arenaria* (courtesy of K. R. Barker).

Because plant-parasitic nematodes are obligate parasites, planting highly resistant cultivars places selection pressure on the nematode population for biotypes that can reproduce on the resistant cultivar. A prominent example of shifts in nematode biotypes resulting from planting resistant cultivars is the soybean and *H. glycines* interaction. In the southern USA, Race 3 was the prevalent race when the first *H. glycines*-resistant cultivar, Pickett, was released. Within a few years, Race 14 (then often referred to as Race 4) became the prevalent race in this region (Hartwig, 1981). Bedford soybean was developed for locations where Race 14 was a problem. Later, Race 5 became a prevalent race in Tennessee following several years of planting Bedford or related cultivars (Young, 1990b).

Shifts in pest species also can limit the effectiveness of nematode-resistant cultivars. For example, *M. incognita* Races 1 and 3 predominate in flue-cured tobacco fields in North and South Carolina, but *M. arenaria* and *M. javanica* are increasingly problematic (Barker, 1989; Johnson, 1989). Since cultivars resistant to the latter species are not available, the *M. incognita*-resistant cultivars may be damaged if grown. 'Bragg' soybean was grown in some areas of the southern USA because it was moderately resistant to *M arenaria*. Then *H. glycines* became a prevalent pest, and Bragg was no longer planted since it was susceptible to *H. glycines*.

Theoretically, planting tolerant cultivars may be an effective management tactic when combined with other control tactics. Hussey and Boerma (1992) described benefits of using a tolerant cultivar in place of susceptible cultivars in rotation schemes intended to reduce selection pressure exerted on *H. glycines* populations when resistant cultivars also are planted. They emphasized that tolerance is not a replacement for resistance in such rotations. Tolerant cultivars will allow buildup of high nematode population densities. Thus, other management tactics to lower nematode densities must be used in conjunction with planting tolerant cultivars.

OUTLOOK

Although some limitations are encountered on use of resistant cultivars in management of nematode populations, host-plant resistance is expected to be a vital management component in the future. In fact, with the increasing loss of available nematicides, resistance will become a more important tool in suppressing nematode damage. In order to maximize the effectiveness of host-plant resistance, it is important to integrate use of resistant cultivars with cultural practices that reduce nematode population densities and allow other crops and susceptible cultivars to be grown.

Hopefully, classical plant breeding procedures can be coupled with genetic engineering techniques to enhance the durability of native nematode resistance in plants and extend the use of host-plant resistance to crops and nematodes where resistance is now unavailable. Integration of restriction fragment length polymorphisms (RFLP) into plant breeding may expedite movement of desirable resistance genes among cultivars, allow easier transfer of resistance genes from related wild species, make possible analysis of complex polygenic resistance as ensembles of single genes, and allow cloning of genes whose products are unknown, such as genes governing tolerance to nematodes (Tanksley et al., 1989). Other molecular biology techniques, such as random amplified polymorphic DNA and DNA fingerprinting methods, along with new technologies, such as yeast artificial chromosomes and ribozymes and gene tagging, may aid in these efforts.

A likely imminent contribution of molecular biology to breeding for nematode resistance is marker-assisted selection through use of RFLPs (Young, 1992). Identification of genetic markers such as RFLPs closely linked to resistance genes can speed recognition of resistance alleles or phenotypes. Currently, genetic markers linked to nematode resistance loci have been identified in tomato for *M. incognita* (Williamson et al., 1994), in potato for *G. rostochiensis* (Kreike et al., 1993), and in soybean for *H. glycines* (Weisemann et al., 1992). Such markers will be used to identify resistant phenotypes when their use is more effective, cheaper, and/or easier to use than traditional techniques. Selection for traits not easily discerned with traditional methods could be greatly enhanced with use of closely linked molecular markers.

Opperman et al. (1994) outlined more futuristic ways that genetic engineering could be used to give plants resistance to nematodes in addition to transfer of nematode resistance genes from other species. Genes coding for preformed defenses could be introduced into agronomic crops. These defenses could include nematode-specific toxins under control of a constitutive promoter such as the BT genes, collagenase genes to disrupt nematode development, antibody genes to interfere with feeding or digestion, and proteinase inhibitors (Atkinson et al., 1994). The preformed defenses could impose selection pressure on the nematode community for nonsensitive nematode species, and nontarget species including humans could be exposed to the toxins through consumption of plants (Opperman et al., 1994). Selection pressure on non-sensitive species could result in some of the same drawbacks experienced with use of native resistance, i.e., species shifts and selection of variant biotypes of the target species. Nematode-induced expression of genes might be a better alternative to preformed defenses. Induced hypersensitivity through induction of a protein toxic to the feeding cells (giant cells of *Meloidogyne* spp. and syncytia of *Heterodera* and *Globodera* spp.) is a possibility (Opperman et al., 1994). The induced protein would need to be effective against the vast genetic variability experienced with these sedentary endoparasites and the native hypersensitive reaction of current resistant cultivars.

Host-plant resistance, whether developed by classical or molecular technology, will be an important nematode control tactic. Environmental and economic concerns demand resistance be put to its optimum use. Expansion of resistance as a control tactic will require concerted and long-term commitment of effort and resources.

REFERENCES

Anand, S.C. 1992. Registration of 'Hartwig' soybean. Crop Sci. 32:1069–1070.

Anand, S.C., and K.M. Gallo. 1984. Identification of additional soybean germplasm with resistance to race 3 of the soybean cyst nematode. Plant Dis. 68:593–595.

Anand, S.C., and S.R. Koenning. 1986. Tolerance of soybean to *Heterodera glycines*. J. Nematol. 18:195–199.

Anand, S.C., J.A. Wrather, and C.R. Shumway. 1985. Soybean genotypes with resistance to races of soybean cyst nematode. Crop Sci. 25:1073–1075.

Atkinson, H.J., K.S. Blundy, M.C. Clarke, E. Hansen, G. Harper, V. Koritas, M.J. McPherson, D. O'Reilly, C. Scollan, S.R. Turnbull, and P.E. Urwin. 1994. Novel plant defenses against nematodes. p. 197–210. *In* F. Lamberti et al. (ed.) Advances in molecular plant nematology. Plenum Press, New York.

Austin, A.J., J.D. Pohlman, C.R. Brown, H. Mojtahedi, G.S. Santo, D.S. Doucher, and J.P. Helgeson. 1993. Interspecific somatic hybridiztion between *Solanum tuberosum* L. and *S.bulbocastanum* Dun. as a means of transferring nematode resistance. Am. Potato J. 70:485–495.

Baker, S. 1992. Registration of 'Georgia King' cotton. Crop Sci. 32:493.

Baltensperger, D.D. 1989. Registration of 'FL-24' and 'FL-101' germplasm lines of hairy indigo. Crop Sci. 29:1331.

Baltensperger, D.D., H.E. Dominguez, W.H. Anderson, R.L. Stanley, L.S. Dunavin, and J.R. Rich. 1990a. Registration of 'Flamingo' hairy indigo. Crop Sci. 30:1366–1367.

Baltensperger, D.D., S.G. Taylor, W.H. Anderson, R.L. Stanley, L.S. Dunavin, M.J. Williams, R.A. Dunn, and R.J. Glennon. 1990b. Registration of 'FL-3' alyceclover. Crop Sci.30:1367–1368.

Barker, K.R. 1989. Yield relationships and population dynamics of *Meloidogyne* spp. on flue- cured tobacco. Suppl. J. Nematol.21:597–603.

Barker, K.J. L. Townshend, G.W. Bird, I.J. Thomason, and D.W. Dickson. 1986. Determining nematode population responses to control agents. p. 283–296. *In* K.D. Hickey (ed.) Methods for evaluating pesticides for control of plant pathogens. Am. Phytopathol. Soc., St. Paul, MN.

Barnes, D.K., J.A. Thies, D.L. Rabas, D.L. Nelson, and D.M.Smith. 1990. Registration of alfalfa germplasms with field resistance to the root-lesion nematodes. Crop Sci. 30:751–752.

Boerma, H., and R.S. Hussey. 1984. Tolerance to *Heterodera glycines* in soybean. J. Nematol. 16:289–296.

Boerma, H.R., and R.S. Hussey. 1992. Breeding plants for resistance to nematodes. J. Nematol. 24:242–252.

Boerma, H.R., R.S. Hussey, P.F. Reese, Jr., S.L. Finnerty, and E.D. Wood. 1993. Registration of a soybean cyst nematode tolerant soybean germplasm line G88-20092. Crop Sci. 33:216.

Brodie, B.B., and M.L. Brucato. 1993. Effect of inoculum density and egg age on establishment on *Globodera rostochiensis* populations. J. Nematol. 25:286–290.

Bradley, E.B., and M. Duffy. 1982. The value of plant resistance to soybean cyst nematode: A case study of Forrest soybeans. Rep. no. AGES820929. Natural Resour. Econ. Div., USDA, Washington, DC.

Brim, C.A., and J.P. Ross. 1966. Registration of Pickett soybeans. Crop Sci. 6:305.

Cook, R., and K. Evans. 1987. Resistance and tolerance. p. 179–231. *In* R.H. Brown and B.R. Kerry (ed.) Principles and practice of nematode control in crops. Acad. Press, New York.

Culver, D.J., D.W. Ramming, and M.V. McKenry. 1989. Procedures for field and greenhouse screening of *Prunus* genotypes for resistance and tolerance to root lesion nematode. J. Am. Hort. Soc. 114:30–35.

Dale, M.F.B., M.S. Phillips, R.M. Ayres, M. Hancock, M.Holliday, G.R. Mackay, and S.J. Tones. 1988. The assessment of the tolerance of partially resistant potato clones to damage by the potato cyst nematode *Globodera pallida* at different sites and in different years. Ann. Appl. Biol. 113:79–88.

Epps, J.M., and E.E. Hartwig. 1972. Reaction of soybean varieties and strains to race 4 of the soybean cyst nematode. J. Nematol. 4:222.

Epps, J.M., L.D. Young, and E.E. Hartwig. 1981. Evaluation of nematicides and resistant cultivar for control of soybean cyst nematode race 4. Plant Dis. 65:665–666.

Fassuliotis, G. 1985. The role of the nematologist in the development of resistant cultivars. p. 233–240. *In* J.N. Sasser and C.C. Carter (ed.) An advanced treatise on *Meloidogyne*. Vol. 1. North Carolina State Univ. Graphics, Raleigh, NC.

Fassuliotis, G. 1987. Genetic basis of plant resistance to nematodes. p. 364–371. *In* J.A. Veech and D.W. Dickson (ed.) Vistas on nematology. Soc. Nematol., Hyattsville, MD.

Fassuliotis, G., and B.V. Nelson. 1988. Interspecific hybrids of *Cucumis metuliferus* × *C. anguria* obtained through embryo culture and somatic embryogenesis. Euphytica 37:53–60.

Fehr, W.R. 1987. Principles of cultivar development. Vol. 1. Macmillan, New York.

Fery, R.L., and P.D. Dukes. 1986. 'Carolina Cayenne' pepper. Hortscience 21:330.

Graves, C.R., B.N. Duck, F. Allen, C.L. Click, R. Thompson, C. Brown, G. Percell, M. Smith, J. McClure, and B. Pitt. 1994. Performance of soybean varieties in 1994. Univ. Tenn. Res. Rep. 94–22.

Griffin, G.D., and L.R. Krusberg. 1990. Evaluating resistance to *Ditylenchus* and *Aphelenchoides* species. p. 58–66. *In* J.L. Starr (ed.) Methods for evaluating plant species for resistance to plant-parasitic nematodes. Soc. Nematol., Hyattsville, MD.

Hartwig, E.E. 1981. Breeding productive soybean cultivars resistant to the soybean cyst nematode for the southern United States. Plant Dis. 65:303–307.

Hartwig, E.E., and J.M. Epps. 1978. Registration of Bedford soybeans. Crop Sci. 18:915.

Heald, C.M., and A.F. Robinson. 1990. Screening for resistance to *Rotylenchulus* species. p. 42–50. *In* J.L. Starr (ed.) Methods for evaluating plant species for resistance to plant-parasitic nematodes. Soc. Nematol., Hyattsville, MD.

Hussey, R.S., and K.R. Barker. 1973. A comparison of methods of collecting inocula of *Meloidogyne* spp., including a new technique. Plant Dis. Rep. 57:1025–1028.

Hussey, R.S., and H.R. Boerma. 1981. A greenhouse screening procedure for root-knot nematode resistance in soybeans. Crop Sci. 21:794–796.

Hussey, R.S., and H.R. Boerma. 1992. Tolerance in soybean. p. 169–181. *In* R.D. Riggs and J.A. Wrather (ed.) Biology and management of the soybean cyst nematode. APS Press, St. Paul, MN.

Johnson, C.S. 1989. Managing root-knot on tobacco in the southern United States. Suppl. J. Nematol. 21:604–608.

Johnson, C.S. 1990. Control of *Globodera tabacum solanacearum* by rotating susceptible and resistant flue-cured tobacco cultivars. Suppl. J. Nematol. 22:700–706.

Kester, D.E., and R.N. Asay. 1986. 'Hansen 2168' and 'Hansen 536': Two new *Prunus* rootstock clones. Hortscience 21:331–332.

Kinloch, R.A. 1990. Screening for resistance to root-knot nematodes. p. 16–23. *In* J.L. Starr (ed.) Methods for evaluating plant species for resistance to plant-parasitic nematodes. Soc. Nematol., Hyattsville, MD.

Kinloch, R.A., C.H. Hiebsch, and H.A. Peacock. 1985. Comparative root-knot galling and yield responses of soybean cultivars to *Meloidogyne incognita*. Plant Dis. 69:334–336.

Kinloch, R.A., C.K. Hiebsch, and H.A. Peacock. 1987. Galling and yields of soybean cultivars grown in *Meloidogyne arenaria*-infested soils. J. Nematol. 19:233–239.

Koenning, S.R., S.C. Anand, and G.O. Myers. 1992. An alternative method for evaluating soybean tolerance to *Heterodera glycines* in field plots. J. Nematol. 24:177–182.

Kreike, C.M., J.R.A. de Koing, J.H. Vinke, J.W. van Ooijen, C. Gebhardt, and W.J. Stiekman. 1993. Mapping of loci involved in quantitatively inherited resistance to the potato cyst-nematode *Globodera rostochiensis* pathotype Ro1. Theor. Appl. Genet. 87:464–470.

Kugler, J.L., D.E. Huset, F.L. Bedard, and M.A. Peterson. 1991a. Registration of 'WL 317' alfalfa. Crop Sci. 31:1700.

Kugler, J.L., D.E. Huset, and M.A. Peterson. 1991b. Registration of 'WL 225' alfalfa. Crop Sci. 31:1700.

LaMondia, J.A. 1988. Tobacco resistance to *Globodera tabacum*. Suppl. J. Nematol. 2:77–80.

Leath, K.T., J.R. Bloom, R.R. Hill, Jr., and R.R. Smith. 1987. Registration of red clover germplasm line CCNR-1. Crop Sci. 27:1316.

Mai, W.F., and G.S. Abawi. 1987. Interactions among root-knot nematodes and Fusarium wilt fungi on host plants. Annu. Rev. Phytopathol. 25:317–338.

McLeod, J.M., W. Witcher, and W.M. Epps. 1983. Registration of okra plant introductions to root knot nematode and Fusarium wilt. Hortscience 18:249–250.

Noel, G.R., P. Jatala, and J. Franco. 1990. Screening for resistance to cyst nematodes, *Globodera* and *Heterodera* species. p. 24–32. *In* J.L. Starr (ed.) Methods for evaluating plant species for resistance to plant-parasitic nematodes. Soc. Nematol., Hyattsville, MD.

Opperman, C.H., G.N. Acedo, D.M. Saravitz, A.M. Skantar, W. Song, C.G. Taylor, and M.A. Conkling. 1994. Bioengineering resistance to sedentary endoparasitic nematodes. p. 221–230. *In* F. Lamberti et al. (ed.) Advances in molecular plant nematology. Plenum Press, New York.

Opperman, C.H., and M.A. Conkling. 1998. Bioengineering resistance to plant-parasitic nematodes. p. 239–250. *In* K.R. Barker et al. (ed.) Plant and nematode interactions. Agron. Monogr. 36. ASA, CSSA, and SSSA, Madison, WI.

Pederson, G.A., M.R. McLaughlin, and G.L. Windham. 1991. Registration of MS-2X, MS-4X, and MS-6X kura clover germplasms. Crop Sci. 31:1714.

Pederson, G.A., and G.L. Windham. 1995. Registration of MSNR4 root-knot nematode resistant white clover germplasm. Crop Sci. 35:1234–1235.

Potter, J.W. 1994. Wild and cultivated strawberries can tolerate or resist root-lesion nematodes. Hortscience 29:1074–1077.

Quesenberry, K.H., D.D. Baltensperger, R.A. Dunn, C.J. Wilcox, and S.R. Hardy. 1989. Selection for tolerance to root-knot nematodes in red clover. Crop Sci. 29:62–65.

Quesenberry, K.H., G.M. Prine, O.C. Ruelke, L.S. Dunavin, and P. Mislevy. 1993. Registration of 'Cherokee' red clover. Crop Sci. 33:208–209.

Ramming, D.W., and O. Tanner. 1983. 'Nemared' peach rootstock. Hortscience 18:376.

Rebois, R.V., J.M. Epps, and E.E. Hartwig. 1970. Correlation of resistance in soybeans to *Heterodera glycines* and *Rotylenchulus reniformis*. Phytopathology 60:695–700.

Reese, P.F., Jr., H.R. Boerma, and R.S. Hussey. 1988. Heritability of tolerance to soybean cyst nematode in soybean. Crop Sci. 28:594–598.

Riedel, R.M., J.G. Foster, and W.F. Mai. 1973. A simplied medium for monoxenic culture of *Pratylenchus penetrans* and *Ditylenchus dipsaci*. J. Nematol. 5:71–72.

Riggs, R.D., and D.P. Schmitt. 1988. Complete characterization of the race scheme for *Heterodera glycines*. J. Nematol. 20:392–395.

Ritchie, D.F., and C.N. Clayton. 1981. Peach tree short life: A complex of interacting factors. Plant Dis. 65:462–469.

Roberts, P.A. 1982. Plant resistance in nematode pest management. J. Nematol. 14:24–33.

Roberts, P.A. 1992. Current status of the availability, development, and use of host plant resistance to nematodes. J. Nematol. 24:213–227.

Ross, J.P., and C.A. Brim. 1957. Resistance of soybeans to the soybean cyst nematode as identified by a double-row method. Plant Dis. Rep. 41:923–924.

Sasser, J.N., and D.W. Freckman. 1987. A world perspective on nematology: The role of the society. p. 7–14. *In* J.A.Veech and D.W. Dickson (ed.) Vistas on nematology. Soc. Nematol., Hyattsville, MD.

Shepherd, R.L. 1986. Cotton resistance to the root knot-Fusarium wilt complex. II. Relation to root-knot resistance and its implications on breeding for resistance. Crop Sci. 26:233–237.

Sherman, W.B., P.M. Lyrene, and R.H. Sharp. 1991. Flordaguard peach rootstock. Hortscience 26:427–428.

Simpson, C.E., S.C. Nelson, J.L. Starr, K.E. Woodard, and O.D. Smith. 1993. Registration of TxAG-6 and TxAG-7 peanut germplasm lines. Crop Sci. 33:1418.

Starr, J.L. 1990. Methods for evaluating plant species for resistance to plant-parasitic nematodes. Soc. Nematol., Hyattsville, MD.

Taylor, A.L., and J.N. Sasser. 1978. Biology, identification and control of root-knot nematodes (*Meloidogyne* species). North Carolina State Univ., Raleigh.

Triantaphyllou, A.C. 1985. Cytogenetics, cytotaxonomy and phylogeny of root-knot nematodes. p. 113–126. *In* J.N. Sasser and C.C. Carter (ed.) An advanced treatise on *Meloidogyne*. Vol. 1. North Carolina State Univ. Graphics, Raleigh.

Wallace, H.R. 1987. A perception of tolerance. Nematologica 33:419–432.

Walters, S.A., T.C. Wehner, and K.R. Barker. 1993. Root-knot nematode resistance in cucumber and horned cucumber. Hortscience 28:151–154.

Weaver, D.B., R. Rodriguez-Kabana, and E.L. Carden. 1989. Long-term effect of crop rotation on soybean in a field infested with *Meloidogyne arenaria* and *Heterodera glycines*. Suppl. J. Nematol. 21:720–722.

Weisemann, J.M., B.F. Matthews, and T.E. Devine. 1992. Molecular markers located proximal to the soybean cyst nematode resistance gene, Rhg4. Theor. Appl. Genet. 85:136–138.

Wicks, Z.W., III, J.D. Smolik, M.L. Carson, and G.G. Scholten. 1990a. Registration of SD101 parental line of maize. Crop Sci. 30:242.

Wicks, Z.W., III, J.D. Smolik, M.L. Carson, and G.G. Scholten. 1990b. Registration of SD102 and SD103 parental lines of maize. Crop Sci. 30:242–243.

Williamson, V.M., K.N. Lambert, and I. Kaloshian. 1994. Molecular biology of nematode resistance in tomato. p. 211–219. *In* F. Lamberti et al. (ed.) Advances in molecular plant nematology. Plenum Press, New York.

Windham, G.L., and G.A. Pederson. 1989. Aggressiveness of *Meloidogyne incognita* host races on white clover. Nematropica19:177–183.

Windham, G.L., and W.P. Williams. 1988. Resistance of maize inbreds to *Meloidogyne incognita* and *M. arenaria*. Plant Dis. 72:67–69.

Wyatt, J.E., G. Fassuliotis, J.C. Hoffman, and J.R. Deakin. 1983. 'Nemasnap' snap bean. Hortscience 18:776.

Yarger, L.W., and L.R. Baker. 1981. Cultivar influence on response of carrot to northern root-knot nematode. Hortscience 16:69–71.

Young, L.D. 1990a. Soybean germplasm evaluated for resistance to races 3, 5, and 14 of soybean cyst nematode. Crop Sci. 30:735–736.

Young, L.D. 1990b. Survey of soybean cyst nematode races in Tennessee. Suppl. J. Nematol. 22:672–675.

Young, L.D. 1995. Soybean germplasm resistant to races 3, 5 or 14 of soybean cyst nematode. Crop Sci. 35:895–896.

Young, L.D., and E.E. Hartwig. 1988. Evaluations of soybeans resistant to *Heterodera glycines* race 5 for yield and nematode reproduction. J. Nematol. 20:38–40.

Young, N.D. 1992. Restriction fragment length polymorphisms (RFLPS) and crop improvement. Exp. Agric. 28:385–397.

11 Genetic Mechanisms of Host-Plant Resistance to Nematodes

PHILIP A. ROBERTS, WILLIAM C. MATTHEWS, AND JOHN C. VEREMIS

University of California
Riverside, California

Natural-host plant resistance is a major resource in agriculture for developing nematode-management tools and programs. This tactic has never been more clearly realized than at present, when the benefits of using chemical nematicides have been eroded or negated in many cases by their high risks to human and environmental health and safety. The potential for resistance has been achieved in some crops and situations but not in many others, reflecting that only a small portion of natural host-plant resistance has been utilized (Roberts, 1992). Several detailed reviews of nematode resistance are available that cover a full range of topics, from historical to current, from agriculturally applied to fundamental, and those that consider the host plant side of resistance, the plant-nematode interaction, and the matching nematode virulence characteristics (Bakker et al., 1993; Cook, 1991; Cook & Evans, 1987; Fassuliotis, 1982, 1987; Kaplan & Davis, 1987; Roberts, 1982, 1992, 1995; Sidhu & Webster, 1981; Starr, 1990; Triantaphyllou, 1987; Trudgill, 1986, 1991; Veech, 1982). The reader is referred to these for additional comprehensive information and extensive citation to other works.

Our approach in this chapter is to offer a synthesis of the genetic mechanisms involved in host-plant resistance to nematodes that incorporates the latest findings from current research programs. This treatment thus includes the exciting potential of the application of molecular biology techniques and knowledge that will undoubtedly help reveal the mechanistic processes of plant resistance and nematode interaction at the molecular and biochemical levels. Because the application of molecular techniques is just now being applied in a significant way, several major breakthroughs in our knowledge of nematode resistance and nematode-plant interaction are imminent (see Lamberti et al., 1994), and their realization could rapidly outdate some central areas of discussion offered here. Primary among these is the cloning, sequencing and gene-product analysis of nematode resistance genes, most likely root-knot nematode (*Meloidogyne*) or cyst nematode (*Globodera, Heterodera*) resistance genes from tomato (*Lycopersicon esculentum* Miller), potato (*Solanum tuberosum* L.), wheat (*Triticum aestivum* L.), sugarbeet (*Beta vulgaris* L. subsp. *vulgaris*) or wild relative, or *Arabidopsis*. In light of this timing of research activity, we have attempted to draw on analogies of nematode resistance in plants with resistance to other plant pathogen groups, in particular to bacteria, biotrophic fungi and viruses. Not only does this comparison demonstrate the striking mechanistic similarities between resistance-pathogen interactions in these different pathosystems, but also it provides insight into what can be expected for nematode resistance by review of the recent major advances. These developments include, for example, the cloning of the first R genes to bacteria in tomato and *Arabidopsis* (Bent et al., 1994; Martin et al., 1993; Mindrinos et al., 1994) to fungi in tomato and flax (*Linum usitatissimum* L.) (Jones et al., 1994; Lawrence et al., 1995), and to a virus in tobacco (*Nicotiana tabacum* L.) (Whitham et al., 1994), and the cloning and analysis of avirulence genes and their products in bacteria, fungi and viruses (coat proteins) (Joosten et al, 1994; Midland et al., 1993).

TERMINOLOGY

Definition of the important terms as used here are given to clarify meaning, particularly because certain terms applied to nematology are not quite the same as their use and meaning in classical plant pathology. These terms have been defined and described by several authorities in reviews, and the reader is referred to those references for additional descriptions (Cook & Evans, 1987; Roberts, 1982; Cook, 1991; Trudgill, 1986, 1991; Shaner et al., 1992).

Resistance is used to describe the ability of the plant to suppress development or reproduction of the nematode. It can range from *low* to *moderate* (*partial* or *intermediate*) resistance, to *high* resistance. The term *resistance* often also is used to refer to the capacity to suppress the disease, especially root-knot (Sasser et al., 1984). Complete resistance that prevents any nematode infection is referred to as *immunity*. *Susceptibility* is used as the opposite of resistance; thus a *susceptible* plant allows normal development and reproduction of the nematode to take place, and expression of any associated disease. *Tolerance* and its opposite, *intolerance*, are used to describe the ability of a plant to withstand the damage resulting from nematode infection; *tolerant* plants are able to grow and yield well despite the presence of infection, whereas *intolerant* plants are injured and grow less well when infected. Resistant plants are generally tolerant, but tolerance and resistance are not always coupled and have been shown to be under separate genetic control in some plant-nematode interactions (Evans & Haydock, 1990; Trudgill, 1991). The concept of tolerance to nematodes is used sometimes in a broader sense to describe general plant response to infections (Barker, 1993). Refer to other in-depth discussions on resistance and tolerance terminology and concepts in plant nematology and plant pathology by Barker (1993), Clarke (1986), and Wallace (1987). Resistance as it relates to the mode of inheritance can be *monogenic* (single gene), *oligogenic* (a few genes), or *polygenic* (many genes), depending on how many resistance genes (*R* genes) are involved. Resistance genes are further defined according to how much of a phenotypic effect they express, being either *major genes* (large effect) or *minor genes* (small effect).

Resistance may be further defined according to Vanderplank's (1978) classes of *vertical* resistance (*race-specific* or *qualitative*, differentiating interspecific variants—races, pathotypes or biotypes—of the pathogen) and *horizontal* resistance (*race-nonspecific* or *quantitative*, effective against all variants or populations of the pathogen). Vertical resistance is usually conferred by one or as many as three major genes and is identified with the gene-for-gene type of plant-pathogen interaction. Horizontal resistance tends to be polygenically inherited by several or more minor genes often with small additive effects that, depending on the number of genes present (i.e., during transfer in breeding programs), provide a quantitative level of resistance. In the nematode, genes for virulence are present that match with R genes in the host plant. *Virulence* is defined according to the ability of a nematode or other pathogen to reproduce on a host plant that possesses one or more resistance genes. *Virulent* nematodes are able to reproduce, whereas *avirulent* nematodes are unable to reproduce in the presence of the specific R gene(s). Populations of nematodes differ in the frequency of virulence to avirulence depending on the proportions of virulent or avirulent individuals present. In plant pathology the genes encoding this trait are typically called *avirulence* or *Avr* genes. Nematologists often refer to avirulence genes as *genes for parasitism* or *parasitism genes*.

MEASUREMENT OF THE RESISTANCE PHENOTYPE

Measurement of the host resistance and nematode virulence phenotypes currently often is based on an assessment of nematode reproduction on plants, following a standardized inoculation procedure that places nematodes (either eggs or infective vermiform stages) in proximity to host roots. Reproduction is assayed by extraction and counting of nematodes in a root system or in a unit weight of roots, and also sometimes by determining the population density of nematodes in soil around roots, either in greenhouse pots or field-planting rows (Starr, 1990). An index of symptoms induced by nematodes, such as the galls caused by root-knot nematodes, also is commonly used by many nematologists and plant breeders to measure resistance, particularly because of their ease of assessment (Starr, 1990; Sasser et al., 1984). Readers are referred to the in-depth compilation of techniques for measuring resistance in plants to the most important nematode genera (Starr, 1990), and to detailed treatment of measuring resistance and tolerance to nematodes by Trudgill (1986, 1991) that focuses primarily on cyst nematodes.

Nematodes produced in roots are enumerated by life stage, and for the sedentary endoparasitic forms by counting numbers of eggs produced, or numbers of egg-masses (a gelatinous matrix into which the female lays her eggs) of *Meloidogyne* spp., or numbers of adult females or cysts (dead females containing eggs) of cyst nematodes. Relative resistance is indexed against known standard susceptible and resistant controls, and designation of resistance is often based on nematode reproduction being below a given percentage (e.g., 5–10%) of that on susceptible plants. Quantitative resistance is more difficult to index and should be done by direct comparison with known susceptible and resistant internal standards in each resistance test (Trudgill, 1986, 1991). In segregating populations in breeding programs or in genetic analyses, the values of reproduction on both relevant susceptible and resistant parental genotypes are used as "book-end" controls for classifying individuals as susceptible, resistant or partially resistant, for example. An important and distinct exception to use in nematode reproduction is the indexing of root-galling symptom severity as a measure of resistance to *Meloidogyne* species (e.g., Luzzi et al., 1994). Root-galling is often closely correlated with nematode reproduction, but this is not always the case; it may be only a measure of resistance to root-galling reaction in interactions where galling and nematode reproduction are under separate genetic control.

RESISTANCE EXPRESSION IN PLANTS

Biochemical and cellular mechanisms and responses in nematode resistance reactions have been reviewed thoroughly elsewhere (Kaplan & Davis, 1987; Cook, 1991; Trudgill,1991; Veech, 1982). The overall lack of significant new information regarding these processes specifically for nematode-plant interactions directs us to present a very brief synopsis here, which is enough only to provide a working background as we explore the genetics of resistance and (a)virulence. Drawing on similarities and parallels with plant resistance to other pathogen groups is appropriate for mechanism analysis because understanding is further advanced, particularly for certain fungal and bacterial pathogens. While there are preinfection resistance processes operative against nematodes that prevent or suppress penetration and infection of host tissues, the great majority of resistance responses occur postinfectionally, following penetration into root or other host tissues by the specialized endoparasitic nematodes (e.g., *Ditylenchus, Heterodera, Globodera, Meloidogyne, Rotylenchulus, Tylenchulus*). Typically, resistant plants carrying major R genes are invaded, e.g., roots penetrated, to a similar extent as equivalent susceptible plants. Following invasion, the activation of incompatible reactions results in nematodes re-emerging from resistant roots and others dying within resistant roots (Forrest et al., 1986).

The expression of incompatibility or resistance after infection is typical of active resistance response found for other pathogens. It is further typified by being conferred by one or very few major R genes that operate in some form of gene-for-gene interaction process (see Keen, 1990). Kaplan and Davis (1987) reviewed plant incompatibility to nematodes according to this active resistance process and applied the concept of Keen (1990) for other plant pathogens that the resistance response can be divided into determinative and expressive stages. The determinative phase involves the detection of nematodes by plant cells through the recognition and binding of a nematode elicitor (ligand) by a plant receptor molecule. This recognition step generates directly or indirectly a signal to initiate the resistance or hypersensitive response, and involves gene activation or inactivation, m-RNA synthesis and finally DNA transcription. This recognition and signal transduction stage probably involves a complex number of steps in one or more biochemical pathways. The expressive phase includes the cascade of gene regulation events and biochemical changes that are initiated by the signaling process, with the net result of synthesis of compounds that directly or indirectly act to prevent the nematode or other pathogen from growth, development and regeneration. Nothing is known directly concerning the specific events and molecules involved in the signal process that determines resistance induction to nematodes. Understanding this system must await cloning and analysis of the plant R gene and its gene product, and the matching nematode avirulence gene and the elicitor role of its product. The nematode signal molecule(s) or elicitor is generally assumed to emanate from nematode esophageal gland secretions injected through the mouth stylet into host tissue. This gland-and-stylet route of nematode signaling would appear to fit more closely with the concept of specific elicitor-receptor signaling than the surface-to-surface type of interaction that has been considered as an alternative signal-recognition mechanism (see Kaplan & Davis, 1987). Herein, we discuss some of the findings of the first cloned gene products involved in other pathosystems for fungi, bacteria and viruses.

The secretions of esophageal glands of nematodes belonging to the highly specialized endoparasitic genera *Meloidogyne*, *Heterodera* and *Globodera* are intimately involved with the compatible or susceptible host response in which the invading nematode induces and then maintains a specialized feeding site of modified host cells (giant cells and syncytia, respectively, and commonly referred to as nurse or food cells) from which it withdraws nutrients (Hussey, 1989; Hussey et al., 1994; and see Hussey & Williamson, 1998, Chapter 5). In the expressive phase of the resistance response, the initiation and maintenance of the feeding site is prevented from following a normal path of progress. The timing of the incompatible response varies from as soon as 8 to 12 h after infection for *Meloidogyne* on tomato carrying the R gene *Mi* (Paulson & Webster, 1972), to about 48 h for *Globodera rostochiensis* infection of potato with gene *H1* (Rice et al., 1985), to as long as 2 wk for the citrus nematode, *Tylenchulus semipenetrans*, on resistant citrus rootstocks (Kaplan, 1981). Thus, the resistance reaction can actually prevent nurse cell initiation, or it can cause walling-off and degeneration of the feeding site at some point following its early development. Often in resistant roots a few nematodes may develop and reproduce, but they are often less fit with lower fecundity rates than those developing on equivalent susceptible plants, and the sex-ratio may shift in favor of males which, for cyst and root-knot nematodes at least, have smaller or no feeding requirements (Cook, 1991).

The nematode-resistance response exhibits many features of the hypersensitive reaction (HR) induced by bacteria and fungi (Kaplan & Davis, 1987; Keen, 1990), although differences are to be expected. The HR and associated biochemical responses in nematode resistant plants have been thoroughly reviewed elsewhere (Cook, 1991; Kaplan & Davis, 1987). A synopsis of primary features of the cellular reactions include cell necrosis, cell wall thickening, cell collapse and wound periderm formation, ribosome and endoplasmic reticulum proliferation, general loss of distinctness of cell membranes, and disappearance of mitochondria, Golgi bodies and nucleoplasm. Biochemical reactions include a wide array of enzyme changes; catalase, peroxidase and polyphenoloxidase levels are increased, as are phenylalanine ammonia lyase and tyrosine ammonia lyase levels, while superoxide dismutase (SOD) levels are lower than in susceptible responses. Chlorogenic acid and phenolic content can be elevated in resistance reactions, and enhanced activity of hydroxyproline in cell walls also has been recorded. The possible role of SOD-associated changes and the involvement of enzyme proteins in resistance reactions in relation to temperature effects on resistance expression is discussed later. Low molecular weight compounds with phytoalexin activity also can be produced in the hypersensitive response; glyceollin accumulated within 2 to 3 d after inoculation of incompatible soybean with *Meloidogyne incognita*, and gossypol and related terpenoid aldehydes accumulated in the stele of resistant cotton infected with *M. incognita* (Veech, 1982). Coumestans are produced in resistant Lima bean (*Phaseolus lunatus* L.) infected with the lesion nematode, *Pratylenchus scribneri* (see Veech, 1982). A group of papers presented at a recent colloquium entitled "Self Defense by Plants: Induction and Signaling Pathways" was published with a Foreword by Ryan and Jagendorf (1995). Although only limited mention is made of plant defenses to nematodes, the extensive discussion of detailed cellular and biochemical processes and modifications involved in plant defense include many of the elements and compounds summarized here that have been found associated with resistance responses to nematode infection. The reader is encouraged to refer to these studies for likely mechanisms in nematode resistance that await further analysis and description.

One additional resistance mechanism of important significance is the recent evidence for systemic acquired resistance induction by *Meloidogyne* spp. infection of tomato plants carrying gene *Mi* (Ogallo & McClure, 1995, 1996). This research has established clearly with split-root technique that infection with incompatible nematodes (*M. incognita*) induces resistance to compatible nematodes (*M. hapla*) subsequently inoculated onto a separate portion of the root system. Induced susceptibility also was found to occur by the reverse sequence of inoculation, i.e., initial inoculation with a compatible nematode (*M. hapla*) induced susceptibility to incompatible *M. incognita* that is normally controlled by *Mi*. Similar responses of induced resistance and induced susceptibility were found on pyrethrum by inoculations with *M. javanica* and *M. hapla*, to which pyrethrum is resistant and susceptible, respectively (Ogallo & McClure, 1995). These findings have important implications for the selection balance of virulence and avirulence within a nematode population, and for the community structure of mixed species populations of parasitic nematodes. Induced resistance and susceptibility are further examples of how nematode-plant interactions may be very similar to plant interactions with other groups of plant pathogens. Induced resistance involving bacterial, fungal and viral pathogens has been studied in some detail, including research on the signaling pathway that leads to systemic acquired resistance in aboveground pathosystems (Kessman et al., 1994; Ryals et al., 1995). However, little is known about the molecular basis of systemic acquired resistance and susceptibility to nematodes, and how nematode-induced responses relate to fungal and bacterial induced reactions.

GENETIC CONTROL OF RESISTANCE

Simple Versus Complex Inheritance

The focus herein is on major gene resistance of simple or fairly simple inheritance as one or a very few R genes primarily expressed in the dominant state. Both basic research and applied breeding efforts for nematode resistance usually involve such systems, and most identified resistance is conferred by simply inherited major genes. However, it should be noted that examples of quantitative nematode resistance of complex inheritance exist, and that attempts have been made to define the genetic basis of the resistance as completely as is possible.

The best understood case is the quantitative resistance to the potato cyst nematode *Globodera pallida* found in some wild potato relative species, particularly *Solanum vernei* Bitter & Wittm. and *S. spegazzinii* Bitter (Trudgill, 1986, 1991; Kreike et al., 1993). The quantitative nature of the resistance is expressed as a partially resistant phenotype in which nematode reproduction is suppressed to a moderate level. Those potato varieties bred for this resistance vary widely in their tolerance of nematode infection and clearly indicate the separate genetic nature of the tolerance and resistance traits (Trudgill, 1991). The number of genes involved in the resistance has not been clearly defined, but the phenotypic expression of wide variance among individuals of the same generation and breeder experience indicate that several genes of minor effect and with additive gene action play a role in the inheritance. Jones (1985) used the *S. vernei-G. pallida* resistance system as a basis for developing models for "multigenic" nematode resistance. The key points and simplifying assumptions made in his study provide useful working hypotheses for complex inheritance of nematode resistance in the absence of molecular analysis of such systems. Jones' model predictions are compatible with the preliminary findings from molecular mapping of quantitative resistance in the related interaction of *S. spegazzinii* and *G. rostochiensis* made by Kreike et al. (1993). They found that quantitative resistance involved at least two genes with additive effect that mapped to two different chromosomes. The fact that only 22% of the genetic variation could be accounted for with the DNA Restriction Fragment Length Polymorphism (RFLP) markers used strongly suggested that other minor R genes were probably operative in the resistance control, but they could not be detected within the limits of their mapping population (Kreike et al., 1993). Those results are compatible with the predictions made by Jones (1985), who determined that the pattern of genetic control could be explained by different numbers of minor genes each contributing partially to the resistance. He made the assumption that each of the minor genes involved was operating on a gene-for-gene basis of interaction with a series of matching avirulence genes in the nematode. This hypothesis was put forward for plant pathogens earlier by Parlevliet and Zadoks (1977). Jones (1985) developed predictions of the effects on female nematode fecundity of combinations of one major gene and up to five minor genes interacting, in addition to examining several minor genes only.

The scenario of a series of mostly independent minor genes with small additive effects is consistent with the difficulty encountered in breeding, whereby the transfer of the full level of parental resistance over successive generations of selection cannot be maintained. Thus, only some proportion of the full complement of genes is carried forward with the particular number or proportion reflecting the level of resistance phenotype transferred. Although this situation may well be an oversimplification of the complex nature of inheritance, including ignoring the likelihood of linked genes and pleiotropic and epistatic interactions in some cases, it provides a solid working hypothesis to explain the current experimental observations and a basis for molecular analysis of genetic mechanisms. This type of complex inheritance also has been encountered in some leguminous forage crops, such as white clover (*Trifolium repens* L.) and alfalfa (*Medicago sativa* L.) with resistance to cyst nematodes (*Heterodera trifolii*) and *M. hapla* or a related root-knot species (C. Mercer, 1995, personal communication). In these systems the same problems of being unable to maintain the parental level of resistance during transfer over several generations have been incurred.

Gene Families

Genetic and molecular evidence is growing and indicates or confirms that host-plant resistance genes often occur as clusters of genes in a multigene family or as multiple allelic series at a disease resistance locus, and in some cases as combinations of both "multi-gene" forms. Such complex gene arrangements thus can have intralocus or extralocus variation. They have been identified and, to some extent, characterized in a number of plant pathosystems as part of the defense against biotrophic fungi, bacteria, viruses, nematodes and certain insects (Hulbert et al., 1993; Martin et al., 1993; Staskawicz et al., 1995). Our knowledge of complex nematode resistance loci is limited, but increasing evidence for their existence in nematode-plant interactions allows some comparisons, analogies and inferences to be drawn from analysis of those involved in other disease-resistance systems.

Cyst Nematode Resistance Genes

The first evidence for complex nematode R gene arrangements was found in barley (*Hordeum vulgare* L.) for resistance genes controlling the cereal cyst nematode, *Heterodera avenae*. Cotton and Hayes (1969) determined through linkage analysis that at least three, and possibly more, *Ha* genes were grouped on barley Chromosome 2, and suggested that alleles representing different resistance genes also may be present. The resistance in soybean [*Glycine max* (L.) Merr.] to *Heterodera glycines* has not been fully characterized, and the number and nature of genes involved suggest a fairly complex pattern of inheritance of genes with major effect. Several genes reside in various combinations in different genotypes, and they show a range of differential specificities for reaction to at least 16 potential races of the nematode (Riggs & Schmitt, 1988; Caviness, 1992; Rao-Arelli et al., 1992). Of the several genes identified, the majority appear to be recessive, but dominant genes also are present. Certain genotypes carry a complement of recessive and dominant major genes. For example, the first commercially released resistant soybean, 'Picket', has resistance derived from the line Peking, which carries at least one dominant and two recessive genes (Rao-Arelli et al., 1992). Different soybean genotypes share some resistance genes, and genetic studies have suggested that multiple alleles and closely linked loci may be involved in the soybean R gene arrangements (Anand & Rao-Arelli, 1989; Rao-Arelli et al., 1992; Caviness, 1992). This pathosystem provides yet another likely example of the complex gene family type of R gene arrangement, but one that requires considerable further study.

Meloidogyne Resistance Genes in Cowpea

The *Rk* locus in cowpea [*Vigna unguiculata* (L.) Walp.] confers resistance to four root-knot nematode species, and it has some characteristics of a complex R gene locus. This gene was characterized by Fery and Dukes (1980) as being a single, completely dominant nuclear gene with major effect that expressed resistance to *M. incognita*, *M. javanica*, *M. hapla* and *M. arenaria*. The dominant resistance does not confer immunity but does inhibit nematode reproduction and galling to a very high degree when present in the heterozygous as well as homozygous state. In the recessive condition the plant is fully susceptible. Shortly after describing *Rk*, Fery and Dukes (1982) also described an intermediate or partial resistance in 'Pink Eye Purple Hull' that they determined to be governed by a different recessive allele at the *Rk* locus, and designated it Rk^i. The low level of resistance conferred by Rk^i was not of much practical use, but its presence suggested variability at the *Rk* locus.

Recently in two separate studies with different cowpea germplasm sources, additional resistance has been identified that is expressed at a higher level than original *Rk* (Fery et al., 1994; Roberts et al., 1996a,b). In each case this new, strong resistance has been found to be linked to the *Rk* locus. In our study, the new resistance was found to be effective against *M. incognita* and *M. javanica* isolates that are virulent to (i.e., not controlled by) the original gene *Rk*, in addition to providing a near-immune resistance response to *Rk*-avirulent nematode populations (Roberts et al., 1995, 1996a,b). This resistance (in accession #IT84S-2049) was found to be controlled by a single, completely dominant gene by analysis of F_1-F_3 and BC_1 and BC_1:F_2 progenies derived from crosses with susceptible cowpea. More importantly, in tests for allelism in crosses with genotypes that were homozygous for original gene *Rk*, the new resistance was found to be strongly associated with *Rk*. Analysis of large F_2 populations of more than 1000 plants revealed no susceptible recombinants. A testcross population of 1161 plants was produced by crossing the F_1 between the *Rk* and new-*Rk* plants, as the female parent, with a susceptible genotype, as male parent. This approach ensured that any possible selfing-derived progeny that might arise would be resistant. No susceptible recombinants were found in this TC_1 population, indicating that this new resistance is another allele at the *Rk* locus, or a separate locus tightly linked to *Rk* within approximately 0.17 map units (Roberts et al., 1996a). This new allele or locus conferring high resistance has been designated Rk^2.

Experiments with some small F_2 populations by Fery et al. (1994) of strongly expressed root-knot nematode resistance in three other cowpea accessions also suggest that in each case they are alleles of *Rk* or additional loci closely linked to *Rk*. However, they appear to differ from each other in terms of the expressed level of resistance according to suppression of nematode primary infection (e.g., numbers of egg masses on roots), rate of egg production (i.e., numbers of eggs per egg mass), and(or) galling. In a preliminary test (P.A. Roberts, 1995, unpublished data) at least two of these genotypes were different in resistance expression from that conferred by Rk^2. Thus, *Rk* is the first nematode resistance locus to be clearly revealed with complexity for resistance expression, whether as multiallelic variants at a single locus or possibly as a cluster of R gene loci. In addition to these allelic or locus variations at *Rk*, one or more independent modifier genes for the *Rk* locus have been found (J. D. Ehlers & P. A. Roberts, 1996, unpublished data), as discussed in the next section, that modifies resistance to a higher (in recessive form) or lower (in dominant form) level of expression.

Meloidogyne Resistance Genes in Tomato

In another example of nematode resistance involving complex multiple-gene arrangements, and also involving resistance to *Meloidogyne* spp., research on tomato and its wild relative species *Lycopersicon peruvianum* (L.) Mill. revealed the presence of several resistance factors (genes), some of which appear to be at least partially linked. The *Mi* gene, that confers resistance to *M. incognita*, *M. arenaria* and *M. javanica*, was first described as a single dominant gene in 1956 (Gilbert & McGuire, 1956), a few years after its successful introgression through backcrossing into *L. esculentum* from *L. peruvianum* (Smith, 1944). Additional research confirmed that it behaves as a single gene with major effect and that it is completely dominant, with plants heterozygous for *Mi* being as resistant as those homozygous for *Mi* (Bost & Triantaphyllou, 1982). For plant breeding purposes and in current molecular-marker research, *Mi* has behaved as a single dominant gene. This situation is further supported by the fact that all the current commercial tomato with root-knot resistance are derived from the same single plant source produced by Smith (1944). However, through breeding modification the length (size) of the introgressed *L. peruvianum* DNA region is variable in different tomato cultivars (Messeguer et al., 1991; Ho et al., 1992; Dixon et al., 1995). Thus, if additional resistance factors closely linked to *Mi* were present originally in the *L. peruvianum* donor, they now could be present in some, but not all, cultivars that possess *Mi*. This possibility is made more likely when considered that, despite the apparent recombination depression found in the *Mi* region (Messeguer et al., 1991; Ho et al., 1992; Dixon et al., 1995), early in the Hawaii resistance breeding program recombination occurred very close to *Mi* that unlinked the variant *L. peruvianum Aps-1* allele for acid phosphatase-1 from *Mi* (Medina-Filho & Stevens, 1980).

Some unconfirmed experiments have indicated that *Mi* or the *Mi* region on Chromosome 6 of tomato may have variant or additional linked *Meloidogyne* resistance factors. The first of these was reported by Sidhu and Webster (1973, 1975); in linkage and allelism tests they determined that *Mi* was, in fact, composed of two dominant genes that they designated *LMiR1* (same as *Mi*) and *LMiR2*, occurring in the varieties Nematex and Small Fry-1, respectively. In an F_2 population of 810 plants from the cross of the two resistant parents, they found one susceptible recombinant, and in a testcross of the F_1 of this cross with a susceptible cultivar, they reported 8 susceptible recombinants among 283 plants. From these data the two genes were determined to be closely linked, with *LMiR1* approximately 5.65 map units (centiMorgans) from *LMiR2* (Sidhu & Webster, 1975). Although no independent genetic analyses have confirmed these findings, additional indication for the existence of more than one gene or allele in the *Mi* locus region was provided by a demonstration that the same two cultivars,

Table 11–1. Summary of *Meloidogyne incognita* resistance genes in *Lycopersicon peruvianum* differentiated by temperature and nematode virulence (from Veremis & Roberts, 1996b).

	Nematode isolates		
	Project 77 (*Mi*-avirulent)		
Accession and clone	Heat-unstable	Heat-stable	557R (*Mi*-virulent)
PI 126440-9MH	Absent†	Absent	Absent
PI 128657-3R4	*Mi*‡	Absent	Absent
LA 1708-I	*Mi*	*Mi-4*	Absent
PI 270435-2R2	*Mi*	*Mi-2*	*Mi-8*
PI 270435-3MH	*Mi*	*Mi-6*	*Mi-7*
PI 126443-1MH	*Mi*	*Mi-5*	*Mi-3*

† Absent-susceptible reaction, no R gene expressed.
‡ All genes expressed in a dominant fashion.

Nematex and Small Fry-1, expressed a differential resistance reaction to a population of *M. incognita* from West Africa, that was virulent on Small Fry (*LMiR2*) but avirulent on Nematex (*LMiR1* or *Mi*) (Netscher, 1978). The differential reaction is indicative of different resistance specificities and, hence, different genetic factors (alleles or genes) in the two cultivars. A differential response between some other cultivars with *Mi*-based resistance to an Indian population of *M. javanica* also has been reported (Sikora et al., 1973), suggesting differential resistance factors may be present in tomato cultivars.

Recently, extensive genetic analyses of *Meloidogyne* spp. resistance genes in tomato other than *Mi* were undertaken. New sources of resistance have been identified in several accessions and clones of *L. peruvianum* (Ammati et al., 1985, 1986). Inheritance studies of these new resistance traits have revealed the presence of additional genes that segregate independently of *Mi* (Cap et al., 1993; Veremis & Roberts, 1996b). A single dominant gene designated *Mi-2* for resistance to *M. incognita* at high temperature, unlike the heat-sensitive *Mi*, was described from *L. peruvianum* PI 270435-clone 2R2 (Cap et al., 1993). In testcrossing experiments, additional single dominant resistance genes were found in several other accessions that segregated independently of *Mi*; these genes were found to be heat-stable for resistance expression or able to block certain *Mi*-virulent nematode isolates (Cap et al., 1993). An extensive testcrossing study revealed that several different *Meloidogyne* resistance genes occur in *L. peruvianum* (Veremis & Roberts, 1996a,b). These genes were distinguished according to linkage assessment and specificity of phenotype for expression at high temperature or expression to *Mi*-virulent nematode isolates. Based on these data, a summary of the relationships and nature of these genes has been developed (Table 11–1). Currently, we can recognize eight resistance genes in *L. peruvianum*—*Mi*, plus *Mi-2* to *Mi-8*.

Validation of the independence from *Mi* of the single dominant gene for resistance to *Mi*-virulent *M. incognita* in *L. peruvianum* PI 126443 clone 1MH was provided by mapping experiments that placed this new locus, designated *Mi-3*, in the telosomic region on the short arm of Chromosome 12 (Yaghoobi et al., 1995). In testcross experiments, the *Mi-2* and *Mi-3* loci were found to be distinct.

The *Mi-3* gene conferring the phenotype of resistance to *Mi*-virulent nematodes was found to be loosely linked to a heat-stable resistance locus (designated *Mi-5*) in the clone PI 126443 clone 1MH. Gene *Mi-5* was found to be independent of the heat-stable resistance locus *Mi-2* in the clone PI 270435-2R2. The PI 270435-2R2 clone also has a gene locus conferring resistance to *Mi*-virulent nematodes, designated *Mi-8*, that is loosely linked to *Mi-2*. Testcrossing revealed *Mi-3* to be distinct from *Mi-5*. In a third *L. peruvianum* clone, PI 270435-3MH, another linked pair of resistance genes was found, one conferring heat-stable resistance (*Mi-6*) and a second conferring resistance to *Mi*-virulent *M. incognit*a, designated *Mi-7* (Veremis & Roberts, 1996a,b). Testcrossing revealed that *Mi-6* is distinct from the other heat-stable resistance genes *Mi-2* and *Mi-5*, and *Mi-7* is distinct from the other genes (*Mi-3* and *Mi-8*) that confer resistance to *Mi*-virulent *M. incognita* isolates (Veremis & Roberts, 1996b). Yet another root-knot resistance gene was detected in *L. peruvianum* accession LA1708. This gene, designated *Mi-4*, also is inherited in a single dominant fashion and has the distinct phenotype of resisting certain *M. arenaria* populations that are not controlled by *Mi* to *Mi-3* and *Mi-5* to *Mi-8* (Veremis & Roberts, 1996a,b). The relationship of *Mi-4* to the other resistance genes requires genetic and molecular analyses.

To summarize the genetic background of resistance to *Meloidogyn*e spp. in tomato, eight R genes have been identified that are arranged in separate groups of linked loci. One of these regions is on Chromosome 6 (the *Mi* region, which also may house some other R genes as suggested by Sidhu and Webster (1975)), and a second group on Chromosome 12 (*Mi-3*, *Mi-5*). It is not known where the other two gene groups (*Mi-2* and *Mi-8*; *Mi-6* and *Mi-7*) reside. The pattern of R gene arrangements therefore appears as at least three multiple gene clusters representing complex regions of host-plant resistance, in addition to the *Mi* region on Chromosome 6. Molecular analysis will provide genetic definition of their physical and functional arrangements in these genomic regions.

Genomic Resistance Hot-Spots

Of particular interest in the tomato R gene system is that the *Mi* gene region on Chromosome 6 is now known to be a rich region or "hot-spot" for pest and disease resistance genes against a range of pest and pathogen groups. To date, in addition to the *Meloidogyne* R gene *Mi*, this region is known to house the resistance gene complex to the biotrophic leaf mold fungus *Cladosporium fulvum* (genes or alleles *Cf-2* and *Cf-5*) (Dixon et al., 1995), and an incompletely dominant gene (*Ol-1*) for resistance to another biotrophic fungus, *Oidium lycopersicum*, that causes powdery mildew (Van der Beek et al., 1994). A gene (*Ty-1*) that confers tolerance to tomato yellow leaf curl virus also maps to this region (Zamir et al., 1994). Resistance to the potato aphid, *Macrosiphum euphorbiae*, also has been found to be closely linked to *Mi* in this same region of tomato Chromosome 6 (Kaloshian et al., 1995). The factors behind this physical association of resistance genes in the same genomic region and their relationship to any functional commonalities remains to be revealed. However, this region of clustered genes points directly to an evolutionary association of development and divergence of resistance factors that share some significant homology, probably through some process of gene duplications. The emerging pattern of tomato *Mi* genes arranged in groups or clusters of genes in different parts (separate chromosomes) of the genome is directly comparable to the genomic arrangement of some other R gene-pathosystems. For example, the *C. fulvum* R genes also are arranged in separate clusters of linked genes; *Cf-1*, *Cf-4* and *Cf-9* on Chromosome 1 and *Cf-2* and *Cf-5* on Chromosome 6 (Jones et al., 1993). Furthermore, mutagenesis experiments revealed that another two genes (*Rcr-1* and *Rcr-2*), independent from each other and from the recently cloned *Cf-9*, are required for *Cf-9* resistance to be expressed fully (Hammond-Kosack et al., 1994), suggesting an even greater level of interactive gene complexity.

Complex resistance loci are known for other plant pathosystems that conform to gene-for-gene interactions. They should be expected to occur for highly specialized nematode parasites such as *Heterodera, Globodera* and *Meloidogyne* that also conform to gene-for gene type interactions with host plants and their resistance genes. The barley *Mla* locus for resistance to the powdery mildew fungus *Erysiphe graminis* f. sp. *hordei* has dual attributes of a complex R gene locus, with currently 28 characterized alleles, and another 16 race-specific powdery mildew R genes closely linked in coupling to named alleles in the *Mla* locus, giving a total of 44 mildew resistance genes at or near locus *Mla* (Jorgensen, 1994). In addition, another six race-specific resistance loci are located on the same Chromosome 5 as *Mla*. The maize (*Zea mays* L. subsp. *mays*) *Rp1* locus for resistance to the rust fungus *Puccinia sorghi* has been characterized as a complex R gene locus, composed of at least 14 alleles and genes within about 1 cM of *Rp1* and an additional three loci within one to three map units of *Rp1* (Hulbert et al., 1993; Sudupak et al., 1993). Thus, both features of multiple alleles and clustered loci are present. The tomato *Pto* gene that confers resistance to the bacterial speck pathogen *Pseudomonas syringae* pv. *tomato* has been shown by molecular cloning and DNA sequence analysis to be one of six members of a small multigene family based on modified tandem repeated sequences (Martin et al., 1993). One member of this family, *Fen*, is a gene that controls sensitivity to the herbicide fenthion {O,O-dimethyl O-[3-methyl-4-(methylthio)phenyl]-phosphorothioate} (Martin et al., 1994). The flax L^6 gene for resistance to the rust *Melampsora lini* also has a multiple allelic composition, and its recent molecular cloning (Lawrence et al., 1995) will enable an analysis of precisely what sequence changes and their resulting functional effects are present as a basis for allele specificity as first described by Flor (1942).

Multiple member gene families with alleles and linked genes appear to represent a common basis for race-specific major resistance genes, whereby multiple forms evolve and develop to combat new races or biotypes of the target pathogen. Unequal crossing-over via mispairing and, hence, unequal exchange between sequence duplications was indicated to be the primary cause for instability at the *Rp1* locus in maize (Sudupak et al., 1993). This recombination process occurring within the *Rp1* region would generate new sequence rearrangements, providing a mechanism for generating new R gene specificities that could keep pace with the shifting virulence specificities of the pathogen (Sudapak et al., 1993). Such rearrangements of pre-existing coding information to generate new specificities were considered by Staskawicz and others (1995) involved in molecular cloning of the first group of plant resistance genes to provide plausible working hypotheses for interpreting the primary results of the several R gene sequence arrangements of multigene families. At a late stage of preparing this chapter, the first nematode R gene, *Hs1*$^{pro-1}$ in sugarbeet conferring resistance to *Heterodera schachtii*, was cloned and sequenced (Cai et al., 1997). The gene encodes a 282-amino acid protein with some features similar to those of other pathogen R genes. Certainly, the current knowledge of the general biochemical and genetic function and arrangement of nematode R genes (e.g., hypersensitive reaction, conformity to gene-for-gene complementarity, complex loci, and genomic location within plant R gene hot-spots) supports further the contention that nematode R genes may be very similar to other pathogen R genes.

Modifier Genes

Evidence is limited for the existence of modifier genes that, through epistatic effects, alter the expression of nematode resistance genes. However, in at least two cases of resistance to *Meloidogyne* spp., the presence of modifiers has been studied. The resistance to *M. incognita* in common bean (*Phaseolus vulgaris*) line PI 165426 is conferred by two noncomplementary genes: *Me2* (which is dominantly expressed), and *me3* (which is recessive). Gene *me3* appears to be modified by a dominant suppresssor locus that is independent of the resistance gene itself (Omwega & Roberts, 1992). The presence of the modifier was indicated by a distortion of segregating family ratios in F_3 families derived from single F_2 plants that were tested at moderate and high temperatures known to influence resistance expression. In some F_3 families resistance-susceptibility segregation ratios at the different temperatures could be explained by a model that assumed the presence of a dominant suppressor locus for gene *me3* in some genotypes. Further genetic analysis is required to clarify the presence and action of this suppressive modifier of a resistance gene in common bean.

More conclusive evidence for resistance gene modification has been found for *Meloidogyne* spp. resistance in cowpea. Some genotypes of cowpea selected for heat tolerance at University of California, Riverside and referred to as H-lines, were found to possess elevated levels of resistance to *M. incognita* and *M. javanica*, including a broader based resistance effective against populations of these species that are virulent on gene *Rk* present in several current commercial cowpea cultivars (P.A. Roberts et al., 1996, unpublished data). The H-lines are known through selection and pedigree to possess gene *Rk*, and tests for allelism in crosses with genotypes with *Rk* confirmed that the H-lines carry *Rk*. However, from H-line crosses with genotypes carrying *Rk* and expressing the normal *Rk* level of resistance, evidence was obtained for the presence of one or two independent loci that modify *Rk* resistance expression (J.D. Ehlers & P.A. Roberts, 1996, unpublished data). The *Rk* modifier gene(s) promoted *Rk* resistance when present in the homozygous recessive condition. Conversely, in the heterozygous or homozygous dominant condition, *Rk* was expressed at the moderate level that has been used for many years in commercial cowpea. This modifier gene effect can be viewed in two ways. In our research work we have thought of it as a recessive enhancer or helper to resistance gene *Rk*. Alternatively, it is a dominant suppressor to *Rk*, similar to that indicated to be present in common bean and in many other resistance gene systems to different plant pathogens. An important characteristic of this modifier as a dominant suppressor is that it only partially suppresses *Rk* function against *Meloidogyne* populations; they are resisted quite effectively by *Rk* in the presence of the modifier. However, the modifier in the double recessive condition does facilitate *Rk* control of some normally *Rk*-virulent populations of both *M. incognita* and *M. javanica* (Roberts et al., 1996a). To summarize, the *Rk* gene modifier alters resistance expression to nematode populations that are controlled by *Rk*, and it broadens the spectrum of *Rk* resistance by modifying *Rk* to be effective against populations that are virulent to *Rk*.

Little is known concerning the biochemical mechanisms involved in the nematode resistance gene modification. However, modifier genes to resistance genes are not uncommon and have been reported in a number of crop species, including common bean, flax and wheat, to other pathogens such as stem and leaf rusts (Islam et al., 1989; Bai & Knott, 1992). In wheat, several suppressor genes are located on D-genome chromosomes that prevent expression of different leaf and stem rust resistance genes (Bai & Knott, 1992). In some cases suppression requires the presence of three modifiers on three different D-chromosomes, indicating that some complementary gene interaction is involved (Bai & Knott, 1992). Those authors speculated that the accumulation of resistance suppressors, which would appear to contradict the likely selective advantage of resistance genes and their primary role in plant defense, may reflect that resistance genes in the absence of disease cause a loss of fitness in plants due to small deleterious effects, an idea discussed earlier by Vanderplank (1978). Thus, some selective advantage would be present for retaining resistance-suppressor genes. Bai and Knott (1992) pointed to data of The et al. (1988) showing a lower yield potential of near-isogenic stem rust resistant wheat lines compared to the susceptibles in the absence of disease in support of the contention that resistance genes may have an associated cost to fitness. We have not accumulated enough reliable yield comparisons on cowpea lines to determine whether *Rk* has a measurable negative impact on yield in the absence of *Meloidogyne* infection.

Gene Linkage and Gene Markers

Genetic linkage of R genes to other factors due to proximity of location on the same linkage group or chromosome has both positive and negative aspects and connotations. On the negative side, plant breeders are faced with the task of incorporating R genes into genotypes of agronomically acceptable cultivars or rootstocks while pursuing yield and quality potential of the crop. This goal can be a significant challenge if undesirable traits are linked to the R gene, such that breaking linkage by recombination will only occur in very low frequency. Two examples of undesirable linkage involve resistance to *M. incognita* in cotton, and *Meloidogyne* spp. in tomato. The nematode resistance recently incorporated into upland Acala cotton has shown linkage or association with susceptibility to Verticillium wilt (*Verticillium dahliae*) (S. Oakley, 1995, personal communication). In tomato, the *Mi* resistance gene was linked to undesirable fruiting characters, including tough fruit attachment that affected mechanical harvesting suitability. Breeders worked for over a decade to break the linkage and provide good fruiting resistant varieties (Rick, 1967). Introgression of R genes into crops from wild species donors is especially challenging.

Advantageous linkages of R genes with useful traits do occur. As described earlier, the *Mi* gene region of tomato Chromosome 6 also contains several R genes to a range of fungal and viral pathogens and insects. Therefore, selection in breeding for one of these traits could result in selection for multiple disease and pest resistance. In fact, the aphid resistance was detected by field observation of its strong association with tomato cultivars carrying the *Mi* gene for nematode resistance (Kaloshian et al., 1995). The *Mi* gene in tomato is tightly linked to the acid phosphatase locus (*Aps-1*) that encodes acid phosphatase-1. A variant (slow-moving electrophoretic profile) allele at the *Aps-1* locus was introgressed with *Mi* from the wild donor species *L. peruvianum*. It has provided significant benefit since its detection in 1974 for the rapid selection of nematode resistance in breeding programs without the need for bioassaying plants for resistance (Medina-Filho & Stevens, 1980). Although electrophoretic detection of the *Aps-1*^{1} variant allele of this marker locus provides a convenient routine screen, in one group of Hawaii-derived breeding lines and cultivars, the *Mi-Aps-1*^{1} linkage was broken and therefore cannot be used (Medina-Filho & Stevens, 1980).

The recent surge in interest and application of molecular marker techniques for major gene traits in plants has led to the finding and development of several very useful DNA markers for nematode R gene loci. Efforts to pursue the map-based cloning of tomato gene *Mi* have led to a number of molecular markers linked to *Mi*, including RFLP and random amplified polymorphic DNA (RAPD)-based markers (Messeguer et al., 1991; Ho et al., 1992; Williamson et al., 1994a; Dixon et al., 1995). Williamson et al. (1994a) developed the RAPD-PCR (polymerase chain reaction)-based marker REX-1, which is a codominant marker that was converted into a stable and reliable form by cloning and sequencing its ends, which were then used to produce extended primers for marker detection. REX-1 appears especially useful for rapid resistance screening because it is more tightly linked to *Mi* than is *Aps-1*, so it can be used on all resistant tomato breeding materials including those which no longer have linkage to the *Aps-1*^{1} marker. Similar DNA marker developments have been made for some other nematode resistance genes. Isolation and characterization has been achieved of RAPD-based stable markers for the beet cyst nematode (*H. schachtii*) resistance gene loci *Hs1*$^{\text{pat-1}}$ and *Hs1*$^{\text{pro-1}}$ that have been incorporated into the sugarbeet as telosomic additions of Chromosomes 1 from *B. patellaris* and *B. procumbens*, respectively (Salentijn et al., 1995). RFLP markers also have been identified linked to the cereal cyst nematode (*H. avenae*) wheat resistance gene *Cre* (Williams et al., 1994), and to the potato cyst nematode (*G. rostochiensis*) potato resistance gene *H1* (Gebhardt et al., 1993). These and the many markers currently being developed will provide the dual advantage of reliable rapid screening protocols for resistance selection in breeding programs, and the starting points in some instances for chromosome walks as part of the strategy for map-based or positional cloning of resistance genes.

One resistance-gene marker study was made of quantitative, rather than major gene, resistance traits. Mapping of two quantitative trait loci (QTLs) involved in the quantitative resistance to *G. rostochiensis* Ro1 in the wild potato species *S. spegazzinii* was accomplished by means of RFLP analysis (Kreike et al., 1993). The two resistance loci (*Gro1.2* and *Gro1.3*) mapped to different chromosomes (10 and 11, respectively), were found to account for 22% of the genetic variation and appeared to have an additive gene action. The low percentage of genetic variation could be due to the distance of the RFLP markers from the R gene loci, and could be due to other undetected R genes being involved in the resistance expression (Kreike et al., 1993). The application of molecular techniques to study quantitative resistance is a high priority if breeders are going to effectively utilize quantitative resistance to nematodes for crop improvement.

TEMPERATURE SENSITIVITY OF RESISTANCE

A significant number of nematode resistance genes in plants are sensitive to high temperature, although not all nematode R genes are temperature sensitive. Typically, expression of resistance is lowered, weakened or broken at higher temperatures compared with expression at moderate temperatures, and the critical temperature regimes involved often show a distinct threshold at which transition from resistance to susceptibility occurs. Resistance to *Meloidogyne* spp. in a diverse range of host plants is temperature sensitive. For example, temperature-sensitive resistance has been found for *M. incognita* resistance in tomato (Dropkin, 1969; Ammati et al., 1986), sweetpotato [*Ipomoea batatas* (L.) Lam.] (Jatala & Russell, 1972), and common bean (Omwega & Roberts, 1992); for *M. hapla* resistance in alfalfa (Griffin, 1969); and for *M. naasi* resistance in oat (*Avena sativa* L.), white clover and potato (Gooris & D'Herde, 1972). Resistance to other nematodes also shows the same phenomenon, e.g., alfalfa resistance to *Ditylenchus dipsaci* (Chang et al., 1975). This phenomenon also has been reported for other disease or pest plant interactions, e.g., in rust diseases in wheat (Luig & Rajaram, 1972) and flax (Islam et al., 1989), and for Hessian fly (*Mayetiola destructor*) resistance in wheat (Tyler & Hatchett, 1983).

The underlying mechanism of temperature sensitivity is not clear. A number of researchers have described cellular and some biochemical changes and the timing of the changes induced by high temperature upon infection of resistant plants. The tomato R gene *Mi* for resistance to *Meloidogyne* was the first described nematode resistance gene interaction broken by high temperature (Dropkin, 1969). Resistance reversal occurs above a distinct threshold temperature of 28°C. Dropkin showed that exposure to high temperature (32°C) during the first 48 h after infection was adequate to break resistance of *Mi* permanently. When the plants were then maintained subsequently at 27°C for 4 wk, the nematodes completed their life-cycles and stimulated typical root-galling responses (Dropkin, 1969). The resistance response is initiated as early as 8 to 12 h after infection, when the invading juveniles have migrated through the cortical tissue of the root and begin to penetrate cells to initiate giant cell-feeding site formation (Paulson & Webster, 1972; Williamson et al., 1994b). These studies and the biochemical changes associated with them have led to the speculation that high temperature is interfering with or altering the normal protein-protein (enzymatic) interactions that are likely involved in the resistance signal transduction pathway process, such that at high temperature the resistance response is not triggered or "elicited." Vanderplank (1986) argued for a major role of protein association or polymerization as part of the specific susceptibility in gene-for-gene interaction requiring molecular association between host and pathogen. He offered that protein polymerization is an endothermic process resulting mainly from hydrophobic bonding, and this process will be favored by increased temperature, providing an underlying basis for transition from resistance to susceptibility at higher temperatures. Williamson et al. (1994b) speculated that after this early initial resistance induction period, the *Mi* gene product perhaps is no longer produced, or it may be in an inactive state, or that the elicitor signal molecules required to trigger the resistance response are not produced after the initial infection period of ±24 h.

The root ultrastructural studies by Paulson and Webster (1972) revealed a lack of significant hypersensitive reaction-associated changes in infected cells following heat-treating infected resistant roots at 33°C, and the formation of normal giant cells. A separate study was made of biochemical changes associated with the temperature-reversed resistance response of *Mi* in tomato at 5 d after infection at 34°C compared to 27°C (Zacheo et al., 1988). Polyphenoloxidase and peroxidase activity were not increased in infested roots at 34°C but were increased at 27°C, and activity of an oxygen-generating system and lipid peroxidation increased at 27°C but decreased at 34°C. Also, SOD and catalase activity were enhanced in roots at 34°C, but were unchanged in roots at 27°C. These researchers suggested that the heat-induced susceptibility resulted from a decline in the toxic superoxides and hydrogen peroxide due to their being scavenged effectively by SOD and catalase that showed increased activity. A continuing difficulty in interpretation of these events is that documenting the heat-induced susceptibility changes in infected resistant plants does not resolve the underlying mechanism, but tends to describe the changes associated with the compatible interaction. The precise identity of the molecules and reactions involved in the resistance-inducing signal transduction pathway events will be revealed through cloning of *Mi* or other nematode R genes, and the subsequent dissection of gene function. This opportunity already exists with the flax rust resistance L^6 gene due to its recent cloning (Lawrence et al., 1995).

Genetic analysis of R gene temperature sensitivity has been made in a limited number of cases. One striking component of this phenomenon is that not only do some genes remain expressed at high temperature, i.e., they are heat-stable or heat-insensitive, but the same host plant species and even the same individual plant genotype may contain both heat-sensitive and heat-stable genes for resistance to the same nematode. These gene differences indicate the high specificity of resistance involved in the nematode-plant interaction, in which different specific induction pathways must be involved. The series of *Mi* genes that have been identified in the tomato relative *L. peruvianum* exemplify this pattern of specificity (Ammati et al., 1986; Cap et al., 1993; Veremis & Roberts, 1996a,b). Several single dominant genes for resistance are present, some of which are heat sensitive (e.g., *Mi* and *Mi-8*), while others are heat-stable (Veremis & Roberts, 1996a,b). It remains to be determined whether the mechanisms governing heat-response are common within each of these two groups of genes, but that may be unlikely because the virulence specificities within the two groups differ, i.e., they are distinct genes within a group based on other characteristics.

Analysis of *Meloidogyne* resistance genes in common bean was made to determine the genetic basis of their sensitivity to temperature (Omwega & Roberts, 1992). It was revealed that the temperature at which the transition from resistance to susceptibility occurs was determined by whether the resistance gene is dominant or recessive. Recessive resistance conferred by gene *me3* and an additional one or two recessive genes was expressed to *M. incognita* at 26°C, but not at 28°C. However, resistance conferred by both dominant genes *Me1* to *M. javanica* and *Me2* to *M. incognita* has a transition to susceptibility when temperature is increased from 28°C (resistance expressed) to 30°C (susceptibility). An important additional finding was that these dominant genes are completely dominant at 26°C but show an allelic dosage response of incomplete dominance at 28°C (Fig. 11–1). The parent plants homozygous for *Me2* were completely resistant at 28°C, whereas the F_1 plants heterozygous for *Me2* expressed an intermediate level of resistance indicating incomplete dominance (Omwega & Roberts, 1992). The resistance ratios were changed in F_2 families from 3:1 to 1:2:1, further demonstrating the incomplete dominance in the heterozygotes at 28°C. These genetic relationships share features of heat-sensitive R genes to diseases and pests in other systems. For example, the wheat stem rust resistance gene *Sr6* was found to be dominant at 18°C but recessive at 25°C, and other wheat stem rust genes *Sr5*, *Sr8*, and *Sr9* became susceptible or moderately susceptible at or above 27°C (Luig & Rajaram, 1972). Resistance to Hessian fly in wheat showed allelic dosage response at 28°C, whereby heterozygous plants were less resistant than homozygous plants (Tyler & Hatchett, 1983). A common temperature threshold of about 28°C for resistance transition seems to be shared by many heat-sensitive dominant R genes, suggesting that at least in part a common process of temperature effect is at work in these host plant interactions.

Why would plants have resistance genes that fail to express resistance at their higher temperature ranges for growth? This would seem disadvantageous

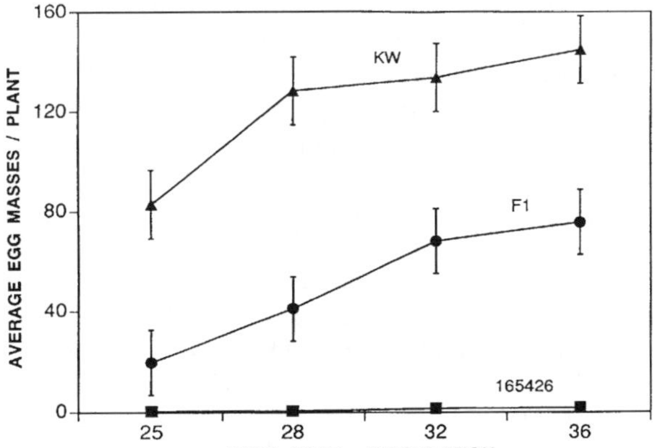

Fig. 11–1. Egg-mass production of *Meloidogyne incognita* on common bean genotypes Kentucky Wonder (KW), 165426, and their F_1, showing the incomplete dominance at 28°C of gene *Me2* that is in the homozygous recessive, homozygous dominant, and heterozygous condition in the three genotypes, respectively (from Omwega & Roberts, 1992).

for the plant because it provides an opportunity for the pathogen or pest. Molecular and biochemical dissection of R gene functional processes may reveal complexity of function for R genes that may influence additional phenotypic traits and provide explanation of the heat response, or that full R gene expression during infection has a high metabolic cost to the plant's energy budget and savings can be made by cessation of expression. However, a plausible speculation of direct selective advantage to the host plant from reversal of resistance at high temperature was offered by Cook and Evans (1987), who suggested that it could provide a mechanism of stabilizing selection by the plant for virulence to the R gene in the nematode population. They suggested that under most natural growing conditions, the early and main portions of the growing season would have temperatures moderate enough to be below the R gene transition temperature, thus allowing normal R gene expression. Later in the season when plants are established and less likely to be affected by damage of infection, high late season temperature would result in resistance reversal and plant infection. The advantage would come in the form of susceptibility removing the strong directional selection pressure that the R gene normally exerts for virulence increase in the nematode population. Avirulent nematodes would develop and dilute the selective force for virulence. Such a process would have long-term selective advantage for maintaining effectiveness of the R gene in the continuing presence of the resident parasite population.

GENETICS OF NEMATODE VIRULENCE

Gene-for-Gene Interactions

The gene-for-gene specificity model for plant-pathosystems is based on alleles at a resistance locus in the host plant interacting with alleles at an avirulence locus in the parasite or pathogen. This concept was originally developed by Flor (1942) based on the interaction between flax and its fungal pathogen flax rust. In most cases resistance is dominant over susceptibility at an R gene locus. At the avirulence locus, avirulence is dominant to recessive virulence (Keen, 1990). Based on the various combinations between the two sets of alleles, resistance (incompatibility) is expressed when the dominant R gene allele is matched by the dominant avirulence allele in the pathogen. In all other combinations, i.e., when either one or both dominant matching alleles is absent, a compatible (susceptible/virulent) interaction occurs. As discussed earlier, the determinative phase of the interaction is the signaling and recognition process between pathogen and host that has an elicitor (ligand)-receptor type of mechanism (Keen, 1990). It is in this phase that the specific avirulence gene action is involved.

In the two pathogen avirulence gene products deduced so far, very different compounds have been found. A *P. syringae* pv. *glycinea* avirulence gene *AvrD* acts indirectly in signaling by encoding enzymes involved in the synthesis of exported syringolides that elicit the HR resistance response in soybean carrying the matching R gene *Rpg4* (Smith et al., 1993). The *C. fulvum* fungus gene *Avr9* has been shown to encode a 28-amino acid secreted peptide that elicits a necrotic (HR) response when injected into plants carrying the matching R gene *Cf-9* (Jones et al., 1994; Joosten et al., 1994). The DNA sequence of *Cf-9* encodes a putative membrane-anchored extracytoplasmic glycoprotein with predicted homology to several receptorlike protein kinases, and to other members of the leucine-rich repeat family of proteins; this structure is consistent with that of a receptor that could bind the Avr9 peptide and activate resistance (Jones et al., 1994). Thus, generalizations over mechanism beyond involvement directly or indirectly in the signaling process are inappropriate. Nematode avirulence genes no doubt also fit into this general category of host-pathogen action, although gene product and detailed function must still be determined. One significant departure in nematodes from this pattern is that a few interactions involve recessive R genes and apparently recessive avirulence genes. Formal genetic proof of avirulence genes in nematodes participating in the classic gene-for-gene interaction process has been obtained as described later. The pattern of race, pathotype, and biotype specificities found within species of important nematode parasites suggests that the gene-for-gene process is widespread among the highly specialized nematode groups, even including interactions that involve quantitative, polygenic resistance (Jones, 1985).

Races, Pathotypes and Biotypes

Schemes for differentiating (a)virulence traits within and among closely related species have been primarily for some cyst nematodes (Dropkin, 1988). An international pathotype scheme for potato cyst nematodes recognizes three *G. pallida* and five *G. rostochiensis* pathotypes differentiated on potato clones with resistance genes (Bakker et al., 1993; Kort et al., 1977). Currently, 16 known and potential races of soybean cyst nematode (*H. glycines*) are differentiated, according to reaction on four distinct soybean resistance backgrounds (Riggs & Schmitt, 1988). Cereal cyst nematode (*H. avenae*) pathotypes are differentiated on small grain cereals including barley, oat, rye (*Secale cereale* L.) and wheat cultivars (see Cook & Evans, 1987). A scheme to differentiate beet cyst nematode (*H. schachtii*) (a)virulence characters for reaction to distinct resistance factors in sugar beet will be necessary as resistant sugar beets are released (Muller, 1992). Five citrus nematode (*T. semipenetrans*) biotypes are distinguished on differential hosts including citrus and olive germplasm (Inserra et al., 1980). These pathotype (biotype or race) schemes are relatively simple, based on one or very few host crop species, and can be expanded to incorporate new resistance differentials as they are identified. A limited race scheme is used to designate intraspecific variants of some *Meloidogyne* spp. (Hartman & Sasser, 1985), but it does not account for the large avirulence diversity within this complex genus. A biotype scheme for characterizing *Meloidogyne* spp. (a)virulence phenotypes has been proposed that would account for the range of avirulence factors that could be differentiated on R genes in any host plant (Roberts, 1995). This system was devised to form both a practical and conceptual guide to dealing with the vast complexity of avirulence factors present in *Meloidogyne* species.

Virulence in *Globodera* and *Heterodera*

The most comprehensive genetic analysis of nematode (a)virulence is that in the potato cyst nematodes, *G. rostochiensis* and *G. pallida*, which involves matching single dominant resistance genes *H1* and *H2*, respectively in potato (*S. tuberosum* or ex-*S. vernei*) lines (Bakker et al., 1993; Janssen et al., 1991; Jones et al., 1981). Mendelian proof has been obtained for (a)virulence being controlled by a single nuclear gene, with virulence recessive to avirulence, and fitting a gene-for-gene relationship for (in)compatibility between nematode parasite and plant host (Bakker et al., 1993; Janssen et al., 1991; Jones et al., 1981). The studies of *G. rostochiensis* virulence to potato resistance gene *H1* first required selection from several generations of nematode lines that were pure-breeding for virulence and avirulence, respectively, starting with single controlled matings from distinct populations that showed differences in ability to overcome resistance in potato (Janssen et al., 1990). Controlled crossings between the virulent and avirulent lines and selfing the F_1 progeny provided fifteen F_2 lines for analysis of virulence segregation. The observed numbers of juveniles conformed to the expected 3:1 segregation for avirulent (Genotypes AA and Aa) and virulent (Genotype aa) phenotypes for virulence inherited as a single recessive gene (Janssen et al., 1991). Reciprocal crossing comparisons gave no evidence of maternal (or sex-linked) inheritance (Janssen et al., 1991). These virulent and avirulent lines have provided the basis for developing a linkage map to pursue the clonal isolation of the avirulence gene. A controlled backcrossing program with the two lines has provided more than 100 progeny lines for use in detecting RAPD markers linked to the avirulence locus (Bakker et al., 1993; Rouppe van der Voort et al., 1994). Preservation of avirulence genes of *Globodera* spp. could be based on environmental sex determination and the lack of feeding requirement of males (Schouten, 1994).

A second program is underway to develop a linkage map of another cyst nematode, *H. glycines*, with a primary goal of identifying and isolating virulence genes that determine parasitic ability for several resistance genes in soybean genotypes (Opperman et al., 1994). This effort is based on three nematode lines inbred for more than 20 generations that each vary in ability to reproduce on designated resistance genes. Preliminary findings from virulence analysis of F_2 progeny derived from crossing two of the inbreds that each parasitize a single different resistant genotype suggest that virulence is controlled by a single unlinked dominant gene in each case of interaction with a specific resistance gene. Conformity to a Mendelian pattern of segregation (9:3:3:1) was found with no evidence of sex-linked inheritance (Opperman et al., 1994). An interesting departure of *H. glycines* from the *G. rostochiensis* virulence research is that virulence is dominant to avirulence for these two genes, and it is of compelling interest that most of the R genes in soybean are recessive rather than dominant. Thus, *H. glycines*-soybean interaction appears as a "reversed gene-for-gene interaction" of dominant virulence matching recessive resistance, counter to all reported gene-for gene interactions to date, including *G. rostochiensis*-potato interaction.

Virulence in *Meloidogyne*

Virulence in *Meloidogyne* has been reviewed in detail, including its description, processes of selection and fitness effects, and attempts at molecular characterization (Castagnone-Sereno, 1994; Roberts, 1995; Roberts &

Thomason, 1989). Lack of a functional characterization scheme has hindered understanding, interpretation and testing of the (a)virulence traits of *Meloidogyne* species matching specific host R genes (Roberts, 1995). Currently, the only scheme with intraspecific delineations is the widely used differential host test devised by Sasser (see Hartman & Sasser, 1985). This system was designed primarily to aid in the species differentiation of *M. arenaria, M. hapla, M. incognita* and *M. javanica*. The six differential hosts {designated cultivars of cotton, peanut (*Arachis hypogaea* L.), pepper (*Capsicum annuum* L.), tobacco, tomato, watermelon [*Citrullus lanatus* (Thunb.) Matsum. & Nakai var. *lanatus*]} also distinguish "races" within *M. incognita* and *M. arenaria*. Four races of *M. incognita* are separated by parasitic ability on tobacco ('NC-95' possessing a single dominant resistance gene to Races 1 and 3 of *M. incognita*) and cotton ('Deltapine 61' with unknown resistance factors); reactions of Races 1 to 4 to cotton/tobacco are –/–, –/+, +/–, and +/+, respectively. Two races of *M. arenaria* are separated by parasitic ability on peanut ('Florunner') (Hartman & Sasser, 1985).

These races are not aligned with any additional (a)virulence factors in other host plants, for example in cowpea and tomato. Both virulent and avirulent isolates for tomato gene *Mi* occur within *M. incognita* Race 1, and both virulent and avirulent isolates for cowpea gene *Rk* occur within each of *M. incognita* Races 1 and 3 (Roberts et al., 1995). Where additional resistance genes in the host have been identified, e.g., *Mi-3* in tomato, Rk^2 in cowpea, further differentiation of these *Meloidogyne* isolates can be made. Thus, an isolate can have differential (a)virulence interaction with two genes of the same host species. Similarly, a resistance gene can have a differential (in)compatibility with two isolates of the same *Meloidogyne* species. This situation is no doubt the common pattern to be expected for distribution of (a)virulence factors matching a broad range of resistance genes. It is unlikely that resistance gene homologies and, hence, avirulence gene homologies are that close at the host/parasite recognition level across the spectra of *Meloidogyne*-host interactions, even if underlying mechanisms of interaction may be similar (Roberts, 1995). Interactions of a nematode isolate with resistance genes from different plant taxa can be expected to vary widely. For example, *M. incognita* isolates virulent on cowpea gene *Rk* are avirulent on tomato genes *Mi* and *Mi2*, and vice versa. Resistance in Lima bean is effective against isolates virulent on cowpea gene *Rk*. Gene *Me1* in common bean is effective against *M. javanica* isolates that are virulent on cowpea gene *Rk*, and so on (P. A. Roberts, 1996, unpublished data).

Some speculation has emerged from these patterns, suggesting that each *Meloidogyne* species has an extensive series or bank of avirulence genes and alleles (Roberts, 1995). These genes match the array of resistance genes found within an extensive host range of diverse plant taxa. Presumably, each nematode isolate has a unique combination of the genes and alleles determining avirulence. The particular avirulence spectrum present in an individual will be the result of inheritance, and the mutation and selection forces acting upon the initial or ancestral complement of genes and alleles (Roberts, 1995). It is interesting to consider that a single base-pair substitution can change a fungal tomato pathogen (*C. fulvum*) gene from avirulent to virulent, resulting in loss of resistance in the tomato cultivar possessing the matching resistance gene and a new race designation (combination) for that pathogen isolate (Joosten et al., 1994). Changes from avirulence to virulence resulting in loss of resistance induction also have been shown to occur by complete loss of the fungal avirulence gene, and in phytopathogenic bacteria by deletion, mutation or insertional inactivation (Joosten et al., 1994; Keen et al., 1993). Similar events may occur in phytoparasitic nematodes, and in polyploid *Meloidogyne* species avirulence expression could be modified by gene action on homologous chromosomes.

Because traditional genetic analysis of (a)virulence is not possible in parthenogenetic *Meloidogyne* species, inference must be based on analysis of phenotype (Roberts, 1995). In applying the rule for a nongenetic demonstration of a gene-for-gene relationship between host and pathogen (parasite) (Robinson, 1987), current phenotype data for *Meloidogyne* does not provide the essential set of differential reactions. Differentials must belong to a single host species and a single parasite species (Robinson, 1987), and *M. incognita* isolates and tomato *Mi* resistance genes, and *M. incognita* isolates and cowpea *Rk* genes are close to indicating existence of gene-for-gene interaction. However, both cases fall short of the requirement for a transagonal quadratic check within a minimum 3 × 2 (or 2 × 3) set of differential interactions (Robinson, 1987). In order to have a transagonal fit, a minimum requirement is that at least one nematode isolate is virulent to resistance Gene 1 but avirulent to resistance Gene 2, and a second isolate is virulent to Gene 2 but avirulent to Gene 1, in addition to uniform virulence and(or) avirulence on a third host genotype. Within *M. incognita*, some isolates are avirulent to both resistance genes *Mi* and *Mi3* of tomato, some isolates are virulent to both genes, and some others are virulent to *Mi* but avirulent to *Mi3* (Roberts et al., 1990; Veremis, 1996a,b). However, isolates that are virulent to *Mi3* but avirulent to *Mi* have not yet been identified. Thus a true gene-for-gene relationship is indicated but not proven.

On cowpea, *M. incognita* isolates avirulent to both resistance genes *Rk* and Rk^2, and isolates virulent to *Rk* but avirulent to Rk^2 have been identified (Roberts et al., 1996a,b). However, isolates that are virulent to Rk^2 but avirulent to *Rk* have not yet been identified. If one steps outside the single species rule, the interaction between *M. incognita* and *M. javanica* isolates with resistance genes in common bean suggests a gene-for-gene relationship because a transagonal interaction is present. *Meloidogyne incognita* isolates are avirulent to gene combination *Me2me3* but virulent to gene *Me1*, whereas *M. javanica* isolates are virulent to *Me2me3* but avirulent to *Me1* (Omwega & Roberts, 1992). Thus, classical gene-for-gene interactions are indicated but not yet proven to exist for *Meloidogyne* species and their resistant hosts.

OUTLOOK

Crop production systems are being converted from a reliance on chemical control tactics with nematicides to a more active use of environmentally compatible strategies including host-plant resistance to control nematodes. The success of this transition will be founded upon increased understanding of the genetics of resistance in the host and the genetics of parasitism and virulence in the nematode. Both traditional and molecular genetics approaches are required to determine the mechanisms of resistance, virulence, and associated host-parasite interactions. Knowledge in these areas will provide a more useful framework to aid plant breeders and nematologists in breeding strategies that target nematode resistance and tolerance for crop improvement. The traditional approaches to gene transfer and introgression by plant breeders will be supplemented with novel plant transformation techniques for inserting cloned resistance genes. Once nematode resistance genes have been cloned, opportunities also will arise for modifying genes to extend their effectiveness to other nematodes and to a broader range of crop varieties and species. Most significantly, at a late stage of preparing this chapter, the first nematode R gene, *Hs1*$^{pro-1}$ in sugarbeet conferring resistance to the cyst nematode *Heterodera schachtii*, was cloned and sequenced (Cai et al., 1997). The gene has some features similar to those of other pathogen R genes. In addition, the first root-knot nematode R gene, *Mi* in tomato, has now been cloned and sequenced (V.M. Williamson, 1997, personal communication). These breakthroughs will greatly advance our understanding and use of host-plant resistance to nematodes.

Traditional and molecular genetics research on nematode traits controlling parasitism and virulence represents the "other half of the picture." Improved understanding of these processes will aid in the design of management programs for nematode-infested fields, in which nematode populations can be manipulated over both short-term and long-term horizons for crop protection. The effective deployment of resistance genes in cropping systems should aim to maximize their crop protection potential and at the same time promote the resistance gene durability. The expansion of knowledge of nematode parasitism and virulence mechanisms through genetics research should lead to ways of minimizing the selection for virulence to resistance in the nematode and offer up potential novel control methods. Although the mechanisms of nematode-plant interactions are not as well known as those of plant pathosystems involving fungi, bacteria, and viruses, the many apparent similarities between these pathosystems provide opportunities to share and direct research on nematode-plant interactions, including host-plant resistance.

REFERENCES

Ammati, M., I.J. Thomason, and H.E. McKinney. 1986. Retention of resistance to *Meloidogyne incognita* in *Lycopersicon* genotypes at high soil temperature. J. Nematol. 18:491–495.

Ammati, M., I.J. Thomason, and P.A. Roberts. 1985. Screening *Lycopersicon* spp. for new genes imparting resistance to root-knot nematodes (*Meloidogyne* spp.). Plant Dis. 69:112–115.

Anand, S.C., and A.P. Rao-Arelli. 1989. Genetic analysis of soybean genotypes resistant to soybean cyst nematode race 5. Crop Sci. 29:1181–1184.

Bai, D., and D.R. Knott. 1992. Suppression of rust resistance in bread wheat (*Triticum aestivum* L.) by D-genome chromosomes. Genome 35:276–282.

Bakker, J., R.T. Folkertsma, J.N.A.M. Rouppe van der Voort, J.M. De Boer, and F.J. Gommers. 1993. Changing concepts and molecular approaches in the management of virulence genes in potato cyst nematodes. Annu. Rev. Phytopathol. 31:169–190.

Barker, K.R. 1993. Resistance/tolerance and related concepts/terminology in plant nematology. Plant Dis. 77:111–113.

Bent, A.F., B.N. Kunkel, D. Dahlbeck, K.L. Brown, R. Schmidt, J. Giraudat, J. Leung, and B.J. Staskawicz. 1994. *RPS2* of *Arabidopsis thaliana*: A leucine-rich repeat class of plant disease resistance genes. Science (Washington, DC) 265:1856–1860.

Bost, S.C., and A.C. Triantaphyllou. 1982. Genetic basis of the epidemiologic effects of resistance to *Meloidogyne incognita* in the tomato cultivar Small Fry. J. Nematol. 14:540–544.

Cai, D., M. Kleine, S. Kifle, H.-J. Harloff, N.N. Sandal, K.A. Marcker, R.M. Klein-Lankhorst, E.M.J. Salentijn, W. Lange, W.J. Stiekema, U. Wyss, F.M.W. Grundler, and C. Jung. 1997. Positional cloning of a gene for nematode resistance in sugar beet. Science (Washington, DC) 275:832–834.

Cap, G.B., P.A. Roberts, and I.J. Thomason. 1993. Inheritance of heat-stable resistance to *Meloidogyne incognita* in *Lycopersicon peruvianum* and its relationship to the *Mi* gene. Theor. Appl. Genet. 85:777–783.

Castagnone-Sereno, P. 1994. Genetics of *Meloidogyne* virulence against resistance genes from solanaceous crops. p. 261–276. *In* F. Lamberti et al. (ed.) Advances in molecular plant nematology. Plenum Press, New York and London.

Caviness, C.E. 1992. Breeding for resistance to soybean cyst nematode. p. 143–156. *In* R.D. Riggs and J.A. Wrather (ed.) Biology and management of the soybean cyst nematode. APS Press, St. Paul, MN.

Chang, D.C.N., W.F. Campbell, and G.D. Griffin. 1975. Effects of temperature on the fine structural responses in the hypocotyl region of alfalfa lines to *Ditylenchus dipsaci*. J. Nematol. 7:239–246.

Clarke, D.D. 1986. Tolerance of parasites and disease in plants and its significance in host-parasite interactions. p. 161–197. *In* D.S. Ingram and P.H. Williams (ed.) Advances in plant pathology. Vol. 5. Acad. Press, London.

Cook, R. 1991. Resistance in plants to cyst and root-knot nematodes. Agric. Zool. Rev. 4:213–240.

Cook, R., and K. Evans. 1987. Resistance and tolerance. p. 179–231. *In* R.H. Brown and B.R. Kerry (ed.) Principles and practice of nematode control in crops. Acad. Press, Orlando, FL.

Cotton, J., and J.D. Hayes. 1969. Genetic resistance to the cereal cyst nematode (*Heterodera avenae*). Heredity 24:593–600.

Dixon, M.S., D.A. Jones, K. Hatzixanthis, M.W. Ganal, S.D. Tanksley, and J.D.G. Jones. 1995. High-resolution mapping of the physical location of the tomato *Cf-2* gene. Mol. Plant-Microbe Interact. 8:200–206.

Dropkin, V.H. 1969. The necrotic reaction of tomatoes and other hosts to *Meloidogyne*: Reversal by temperature. Phytopathology 59:1632–1637.

Dropkin, V.H. 1988. The concept of race in phytonematology. Annu. Rev. Phytopathol. 26:145–61.

Evans, K., and P.P.J. Haydock. 1990. A review of tolerance by potato plants of cyst nematode attack, with consideration of what factors may confer tolerance and methods of assaying and improving it in crops. Ann. Appl. Biol. 117:703–740.

Fassuliotis, G. 1982. Plant resistance to root-knot nematodes. p. 33–49. *In* R.D. Riggs (ed.) Nematology in the southern region of the United States. South. Coop. Ser. Bull. 276.

Fassuliotis, G. 1987. Genetic basis of plant resistance to nematodes. p. 364–371. *In* J.A. Veech and D.W. Dickson (ed.) Vistas on nematology. Soc. Nematol., Hyattsville, MD.

Fery, R.L., and P.D. Dukes. 1980. Inheritance of root-knot resistance in the cowpea (*Vigna unguiculata* (L.) Walp.). J. Am. Soc. Hort. Sci. 105:671–674.

Fery, R.L., and P.D. Dukes. 1982. Inheritance and assessment of a second root-knot resistance in southernpea [*Vigna unguiculata* (L.) Walp.]. HortScience 17:152.

Fery, R.L., P.D. Dukes, and J.A Thies. 1994. Characterization of new sources of resistance in cowpea to the southern root-knot nematode. HortScience 29:678–679.

Flor H.H. 1942. Inheritance of pathogenicity in *Melampsora lini*. Phytopathology 32:653–69

Forrest, J.M.S., D.L. Trudgill, and L.M. Cotes. 1986. The fate of juveniles of *Globodera rostochiensis* and *G. pallida* in roots of susceptible and resistant potato cultivars with gene H_1. Nematologica 32:106–114.

Gebhardt, C., D. Mugniery, E. Ritter, F. Salamini, and E. Bonnel. 1993. Identification of RFLP markers closely linked to the H1 gene conferring resistance to *Globodera rostochiensis* in potato. Theor. Appl. Genet. 85:541–544.

Gilbert, J.C., and D.C. McGuire. 1956. Inheritance of resistance to severe root-knot from *Meloidogyne incognita* in commercial type tomatoes. Proc. Am. Soc. Hort. Sci. 68:437–442.

Gooris, J., and C.J. D'Herde. 1972. Le cycle de développement de *Meloidogyne naasi* Franklin sur cereales de printemps et d'hiver et sur betteraves. Rev. Agric. Brux. 25:651–657.

Griffin, G.D. 1969. Effects of temperature on *Meloidogyne hapla* in alfalfa. Phytopathology 59:599–609.

Hammond-Kosack, K.E., D.A. Jones, and J.D.G. Jones. 1994. Identification of two genes required in tomato for full *Cf-9*-dependent resistance to *Cladosporium fulvum*. Plant Cell 6:361–374.

Hartman, K.M., and J.N. Sasser. 1985. Identification of *Meloidogyne* species on the basis of differential host test and perineal pattern morphology. p. 69–77. *In* K.R. Barker et al. (ed.) An advanced treatise on *Meloidogyne*. Vol. 2. North Carolina State Univ. Graphics, Raleigh.

Ho, J.-Y., R. Weide, H.M. Ma, M.F. van Wordragen, K.N. Lambert, M. Koorneef, P. Zabel, and V.M. Williamson. 1992. The root-knot nematode resistance gene (*Mi*) in tomato: Construction of a molecular linkage map and identification of dominant cDNA markers in resistant genotypes. Plant J. 2:971–982.

Hulbert S.H., M.A. Sudupak, and K.S. Hong. 1993. Genetic relationships between alleles of the *Rp1* rust resistance locus of maize. Mol. Plant-Microb. Interact. 6:387–392.

Hussey, R.S. 1989. Disease-inducing secretions of plant parasitic nematodes. Annu. Rev. Phytopathol. 27:123–141.

Hussey, R.S., E.L. Davis, and C. Ray. 1994. *Meloidogyne* stylet secretions. p. 233–249 *In* F. Lamberti et al. (ed.) Advances in molecular plant nematology. Plenum Press, New York and London.

Hussey, R.S., and V.M. Williamson. 1998. Physiological and molecular aspects of nematode parasitism. p. 87–108. *In* K.R. Barker et al. (ed.) Plant and nematode interactions. Agron. Monogr. 36. ASA, CSSA, and SSSA, Madison, WI.

Inserra, R.N., N. Vovlas, and J.H. O'Bannon. 1980. A classification of *Tylenchulus semipenetrans* biotypes. J. Nematol. 12:283–287.

Islam, M.R., K.W. Sheperd, and G.M.E. Mayo. 1989. Effect of genotype and temperature on the expression of *L* genes in flax conferring resistance to rust. Physiol. Mol. Plant Pathol. 35:141–151.

Janssen, R., J. Bakker, and F.J. Gommers. 1990. Selection of virulent and avirulent lines of *Globodera rostochiensis* for the H_1 resistance gene in *Solanum tuberosum* ssp. *andigena* CPC 1673. Rev. Nématol. 13:265–268.

Janssen, R., J. Bakker, and F.J. Gommers. 1991. Mendelian proof for a gene-for-gene relationship between virulence of *Globodera rostochiensis* and the H_1 resistance gene in *Solanum tuberosum* ssp. *andigena* CPC 1673. Rev. Nématol. 14:214–219.

Jatala, P., and C.C. Russell. 1972. Nature of sweet potato resistance to *Meloidogyne incognita* and the effects of temperature on parasitism. J. Nematol. 4:1–7.

Jones, D.A., M.J. Dickinson, P.J. Balint-Kurti, M. Dixon, and J.D.G. Jones. 1993. Two complex resistance loci revealed in tomato by classical and RFLP mapping of the *Cf-2, Cf-4, Cf-5* and *Cf-9* genes for resistance to *Cladosporium fulvum*. Mol. Plant-Microbe Interact. 6:348–357.

Jones, D.A., C.M. Thomas, K.E. Hammond-Kosack, P.J. Balint-Kurti, and J.D.G. Jones. 1994. Isolation of the tomato *Cf-9* gene for resistance to *Cladosporium fulvum* by transposon tagging. Science (Washington, DC) 266:789–793.

Jones, F.G.W. 1985. Modelling multigenic resistance to potato cyst nematodes. EPPO Bull. 15:155–166.

Jones, F.G.W., D.M. Parrott, and J.N. Perry. 1981. The gene-for-gene relationship and its significance for potato cyst nematodes and their solanaceous hosts. p. 23–36. *In* B.M. Zuckerman and R.A. Rohde (ed.) Plant parasitic nematodes. Vol. 3. Acad. Press, New York.

Joosten, M.H.A.J., T. J. Cozijnsen, and P.J.G.M. De Wit. 1994. Host resistance to a fungal tomato pathogen lost by a single base-pair change in an avirulence gene. Nature (London) 367:384–386.

Jørgensen, J.H. 1994. Genetics of powdery mildew resistance in barley. Crit. Rev. Plant Biol. 13:97–119.

Kaloshian, I., W.H. Lange, and V.M. Williamson. 1995. An aphid resistance locus is tightly linked to the nematode-resistance gene, *Mi*, in tomato. Proc. Natl. Acad. Sci. USA 92:622–625.

Kaplan, D.T. 1981. Characterization of citrus rootstock responses to *Tylenchulus semipenetrans* (Cobb). J. Nematol. 13:492–498.

Kaplan, D.T., and E.L. Davis. 1987. Mechanisms of plant incompatibility with nematodes. p. 267–276. *In* J.A. Veech and D.W. Dickson (ed.) Vistas on nematology. Soc. Nematol., Hyattsville, MD.

Keen, N.T. 1990. Gene-for-gene complementarity in plant-pathogen interactions. Annu. Rev. Genet. 24:447–463.

Keen, N.T., A.F. Bent, and B.J. Staskawicz. 1993. Plant disease resistance genes: Interactions with pathogens and their improved utilization to control plant diseases. p. 65–88. *In* I. Chet (ed.) Biotechnology in plant disease control. Wiley-Liss, New York.

Kessmann, H., T. Staub, C. Hofmann, T. Maetzke, and J. Herzog. 1994. Induction of systemic acquired disease resistance in plants by chemicals. Annu. Rev. Phytopathol. 32:439–459.

Kort, J., H. Ross, H.J. Rumpenhorst, and A.R. Stone. 1977. An international scheme for identifying and classifying pathotypes of potato cyst-nematodes *Globodera rostochiensis* and *G. pallida*. Nematologica 23:333–339.

Kreike, C.M., J.R.A. De Koning, J.H. Vinke, J.W. Van Ooijen, C. Gebhardt, and W.J. Stiekema. 1993. Mapping of loci involved in quantitatively inherited resistance to the potato cyst-nematode *Globodera rostochiensis* pathotype Ro1. Theor. Appl. Genet. 87:464–470.

Lamberti, F., C. De Georgi, and D.M. Bird (ed.). 1994. Advances in molecular plant nematology. Plenum Press, New York and London.

Lawrence, G.J., E.J. Finnegan, M.A. Ayliffe, and J.G. Ellis. 1995. The *L6* gene for flax rust resistance is related to the *Arabidopsis* bacterial resistance gene *RPS2* and the tobacco viral resistance gene *N*. Plant Cell 7:1195–1206.

Luig, N.H., and S. Rajaram. 1972. The effect of temperature and genetic background on host gene expression and interaction to *Puccinia graminis tritici*. Phytopathology 62:1171–1174.

Luzzi, B.M., H.R. Boerma, and R.S. Hussey. 1994. Inheritance of resistance to the southern root-knot nematode in soybean. Crop Sci. 34:1240–1243.

Martin, G.B., S.H. Brommonschenkel, J. Chunwongse, A. Frary, M.W. Ganal, R. Spivey, T. Wu, E.D. Earle, and S.D. Tanksley. 1993. Map-based cloning of a protein kinase gene conferring disease resistance in tomato. Science (Washington, DC) 262:1432–1436.

Martin, G.B., A. Frary, T. Wu, S.H. Brommonschenkel, J. Chunwongse, E.D. Earle, and S.D. Tanksley. 1994. A member of the tomato *Pto* gene family confers sensitivity to fenthion resulting in rapid cell death. Plant Cell 6:1543–1552.

Medina-Filho, H.P., and M.A. Stevens. 1980. Tomato breeding for nematode resistance: survey of resistant varieties for horticultural characteristics and genotype of acid phosphatase. Acta Hort. 100:383–391.

Messeguer, R., M. Ganal, M.C. de Vicente, N.D. Young, H. Bolkan, and S.D. Tanksley. 1991. High resolution RFLP map around the root knot nematode resistance gene (*Mi*) in tomato. Theor. Appl. Genet. 82:529–536.

Midland, S.L., N.T. Keen, J.J. Sims, M.M. Midland, M.M. Stayton, V. Burton, M.J. Smith, E.P. Mazzola, K.J. Graham, and J. Clardy. 1993. The structure of syringolides 1 and 2, novel C-glycosidic elicitors from *Pseudomonas syringae* pv. *tomato*. J. Org. Chem. 58:2940–2945.

Mindrinos, M., F. Katagiri, G-L. Yu, and F.M. Ausubel. 1994. The *A. thaliana* disease resistance gene *RPS2* encodes a protein containing a nucleotide-binding site and leucine-rich repeats. Cell 78:1089–1099.

Muller, J. 1992. Detection of pathotypes by assessing the virulence of *Heterodera schachtii*. Nematologica 38:50–64.

Netscher, C. 1978. Morphological and physiological variability of species of *Meloidogyne* in West Africa and implications for their control. Meded. LandbHoogesch. (Wageningen) 78–3:1–46.

Ogallo, J.L., and M.A. McClure. 1995. Induced resistance to *Meloidogyne hapla* by other *Meloidogyne* species on tomato and pyrethrum plants. J. Nematol. 27:441–447.

Ogallo, J.L., and M.A. McClure. 1996. Systemic acquired resistance and susceptibility to root-knot nematodes in tomato. Phytopathology 86:498–501.

Omwega, C.O., and P.A. Roberts. 1992. Inheritance of resistance to *Meloidogyne* spp. in common bean and the genetic basis of its sensitivity to temperature. Theor. Appl. Genet. 83:720–726.

Opperman, C.H., K. Dong, and S. Chang. 1994. Genetic analysis of the soybean-*Heterodera glycines* interaction. p. 65–75. *In* F. Lamberti et al. (ed.) Advances in molecular plant nematology. Plenum Press, New York and London.

Parlevliet, J.E., and J.C. Zadoks. 1977. The integrated concept of disease resistance: A new view including horizontal and vertical resistance in plants. Euphytica 26:5–21.

Paulson, R.E., and J.M. Webster. 1972. Ultrastructure of the hypersensitive reaction in roots of tomato, *Lycopersicon esculentum* L., to infection by the root-knot nematode, *Meloidogyne incognita*. Physiol. Plant Pathol. 2:227–234.

Rao-Arelli, A.P., S.C. Anand, and J.A. Wrather. 1992. Soybean resistance to soybean cyst nematode race 3 is conditioned by an additional dominant gene. Crop Sci. 32:862–865.

Rice, S.L., B.S.C. Leadbeater, and A.R. Stone. 1985. Changes in cell structure in roots of resistant potatoes parasitized by potato cyst nematodes. I. Potatoes with resistance gene H1 derived from *Solanum tuberosum* ssp. *andigena*. Physiol. Plant Pathol. 27:219–234.

Rick, C.M. 1967. Exploiting species hybrids for vegetable improvement. p. 217–229. *In* H.B. Tukey (ed.) Proc. 17th Int. Hort. Congr., Vol. 3, Univ. Maryland, College Park. 15–20 Aug. 1966. Univ. Maryland, College Park.

Riggs, R.D., and D.P. Schmitt. 1988. Complete characterization of the race scheme for *Heterodera glycines*. J. Nematol. 13:171–79.

Roberts, P.A. 1982. Plant resistance in nematode pest management. J. Nematol. 14:24–33.

Roberts, P.A. 1992. Current status of the availability, development and use of host plant resistance to nematodes. J. Nematol. 24:213–227.

Roberts, P.A. 1995. Conceptual and practical aspects of variability in root-knot nematodes related to host plant resistance. Annu. Rev. Phytopathol. 33:199–221.

Roberts, P.A., A. Dalmasso, G.B. Cap, and P. Castagnone-Sereno. 1990. Resistance in *Lycopersicon peruvianum* to isolates of *Mi* gene-compatible *Meloidogyne* populations. J. Nematol. 22:585–589.

Roberts, P.A., C.A. Frate, W.C. Matthews, and P.P. Osterli. 1995. Interactions of virulent *Meloidogyne incognita* and Fusarium wilt on resistant cowpea genotypes. Phytopathology 85:1288–1295.

Roberts, P.A., W.C. Matthews, and J.D. Ehlers. 1996a. New resistance to virulent root-knot nematodes linked to the *Rk* locus of cowpea. Crop Sci. 36:889–894.

Roberts, P.A., J.D. Ehlers, A.E. Hall, and W.C. Matthews. 1996b. Characterization of new resistance to root-knot nematodes in cowpea. p. 207–214. *In* B.B. Singh et al. (ed.) Advances in cowpea research. Jpn. Int. Res. Cent. Agric. Sci., and IITA, Ibadan, Nigeria.

Roberts, P.A., and I.J. Thomason. 1989. A review of variability in four *Meloidogyne* spp. measured by reproduction on several hosts including *Lycopersicon*. Agric. Zool. Rev. 3:225–252.

Robinson, R.A. 1987. Host management in crop pathosystems. Macmillan Publ., New York.

Rouppe van der Voort, J.N.A.M., J. Rosien, P.M. van Zandvoort, R.T. Folkertsma, E.L.J.G. van Enckevoort, R. Janssen, F.J. Gommers, and J. Bakker. 1994. Linkage mapping in potato cyst nematodes. p. 57–63. *In* F. Lamberti et al. (ed.) Advances in molecular plant nematology. Plenum Press, New York and London.

Ryals, J., K.A. Lawton, T.P. Delaney, L. Friedrich, H. Kessmann, U. Neuenschwander, S. Uknes, B. Vernooij, and K. Weymann. 1995. Signal transduction in systemic acquired resistance. Proc. Natl. Acad. Sci. USA 92:4202–4205.

Ryan, C.A., and A. Jagendorf. 1995. Self defense by plants. Proc. Natl. Acad. Sci. USA. 92:4075.

Salentijn, E.M.J., M.J.B. Arens-De Reuver, W. Lange, Th.S.M. De Bock, W.J. Steikema, and R.M. Klein-Lankhorst. 1995. Isolation and characterization of RAPD-based markers linked to the beet cyst nematode resistance locus ($Hs1^{pat-1}$) on chromosome 1 of *B. patellaris*. Theor. Appl. Genet. 90:885–891.

Sasser, J.N., C.C. Carter, and K.M. Hartman. 1984. Standardization of host suitability studies and reporting of resistance to root-knot nematodes. U.S. AID, and North Carolina State Univ., Raleigh.

Schouten, H.J. 1994. Preservation of avirulence genes of potato cyst nematodes through environmental sex determination: A model involving complete, monogenic resistance. Phytopathology 84:771–773.

Shaner, G., G.H. Lacy, E.L. Stromberg, K.R. Barker, and T.P. Pirone. 1992. Nomenclature and concepts of pathogenicity and virulence. Annu. Rev. Phytopathol. 30:47–66.

Sidhu, G.S., and J.M. Webster. 1973. Genetic control of resistance in tomato. I. Identification of genes for host resistance to *Meloidogyne incognita*. Nematologica 19:546–550.

Sidhu, G.S., and J.M. Webster. 1975. Linkage and allelic relationships among genes for resistance in tomato (*Lycopersicon esculentum*) against *Meloidogyne incognita*. Can. J. Genet. Cytol. 17:323–328.

Sidhu, G.S., and J.M. Webster. 1981. The genetics of plant-nematode parasitic systems. Bot. Rev. 47:387–419.

Sikora, R.A., K. Sitaramaiah, and R.S. Singh. 1973. Reaction of root-knot nematode-resistant tomato cultivars to *Meloidogyne javanica* in India. Plant Dis. Rep. 57:141–143.

Smith, M.J., E.P. Mazzola, J.J. Sims, S.L. Midland, N.T. Keen, V. Burton, and M.M. Stayton. 1993. The syringolides: Bacterial C-glycosyl lipids that trigger plant defense resistance. Tetrahed. Lett. 34:223–226.

Smith, P.G. 1944. Embryo culture of a tomato species hybrid. Proc. Am. Soc Hort. Sci. 44:413–416.

Starr, J.L. 1990. Methods for evaluating plant species for resistance to plant-parasitic nematodes. Soc. Nematol., Hyattsville, MD.

Staskawicz, B.J., F.M. Ausubel, B.J. Baker, J.G. Ellis, and J.D.G. Jones. 1995. Molecular genetics of plant disease resistance. Science (Washington, DC) 268:661–667.

Sudupak, M.A., J.L. Bennetzen, and S.H. Hulbert. 1993. Unequal exchange and meiotic instability of disease-resistance genes in the *Rp1* region of maize. Genetics 133:119–125.

The, T.T., B.D.H. Latter, and R.A. McIntosh. 1988. Grain yield of near-isogenic lines with added genes for stem rust resistance. p. 901–906. *In* Proc. 7th Int. Wheat Genet. Symp., 2.

Triantaphyllou, A.C. 1987. Genetics of nematode parasitism on plants. p. 354–363. *In* J.A. Veech and D.W. Dickson (ed.) Vistas on nematology. Soc. Nematol., Hyattsville, MD.

Trudgill, D.L. 1986. Concepts of resistance, tolerance and susceptibility in relation to cyst nematodes. p. 179–189. *In* F. Lamberti and C.E. Taylor (ed.) Cyst nematodes. Plenum Press, London and New York..

Trudgill, D.L. 1991. Resistance to and tolerance of plant parasitic nematodes in plants. Annu. Rev. Phytopathol. 29:167–192.

Tyler, J.M., and J.H. Hatchett. 1983. Temperature influence on expression of resistance to Hessian fly (Diptera: Cecidomyiidae) in wheat derived from *Triticum tauschii*. J. Econ. Entomol. 76:323–326.

Van der Beek, J.G., G. Pet, and P. Lindhout. 1994. Resistance to powdery mildew (*Oidium lycopersicum*) in *Lycopersicon hirsutum* is controlled by an incompletely-dominant gene *Ol-1* on chromosome 6. Theor. Appl. Genet. 89:467–473.

Vanderplank, J.E. 1978. Genetic and molecular basis of plant pathogenesis. Springer-Verlag, Berlin and Heidelberg.

Vanderplank, J.E. 1986. Specific susceptibility and specific feeding in gene-for-gene systems. Adv. Plant Pathol. 5:199–223.

Veech, J.A. 1982. Phytoalexins and their role in the resistance of plants to nematodes. J. Nematol. 14:2–9.

Veremis, J.C., and P.A. Roberts. 1996a. Relationships between *Meloidogyne incognita* resistance genes in *Lycopersicon peruvianum* differentiated by heat sensitivity and nematode virulence. Theor. Appl. Genet. 93:950–959.

Veremis, J.C., and P.A. Roberts. 1996b. Differentiation of *Meloidogyne incognita* and *M. arenaria* novel resistance phenotypes in *Lycopersicon peruvianum* and derived bridge-lines. Theor. Appl. Genet. 93:960–967.

Wallace, H.R. 1987. A perception of tolerance. Nematologica 33:419–432.

Whitham S., S.P. Dinesh-Kumar, D. Choi, R. Hehl, C. Corr, and B.J. Baker. 1994. The product of the tobacco mosaic virus resistance gene *N*: similarity to Toll and the interleukin-1 receptor. Cell 78:1101–1115.

Williams, K.J., J.M. Fisher, and P. Langridge. 1994. Identification of RFLP markers linked to the cereal cyst nematode resistance gene (*Cre*) in wheat. Theor. Appl. Genet. 89:927–930.

Williamson, V.M., J.-Y. Ho, F.F. Wu, N. Miller, and I. Kaloshian. 1994a. A PCR-based marker tightly linked to the nematode resistance gene, *Mi*, in tomato. Theor. Appl. Genet. 87:757–763.

Williamson, V.M., K.N. Lambert, and I. Kaloshian. 1994b. Molecular biology of nematode resistance in tomato. p. 211–219. *In* F. Lamberti et al. (ed.) Advances in molecular plant nematology. Plenum Press, New York and London.

Yaghoobi, J., I. Kaloshian, Y. Wen, and V.M. Williamson. 1995. Mapping a new nematode resistance locus in *Lycopersicon peruvianum*. Theor. Appl. Genet. 91:457–464.

Zacheo, G., G. Pricolo, and T. Bleve-Zacheo. 1988. Effect of temperature and biochemical changes in tomato inoculated with *Meloidogyne incognita*. Nematol. Medit. 16:107–112.

Zamir, D., I. Ekstein-Michelson, Y. Zakay, N. Navot, M. Zeiden, M. Sarfatti, Y. Eshed, E. Harel, T. Pleban, H. van-Oss, N. Kedar, H.D. Rabinowitch, and H. Czosnek. 1994. Mapping and introgression of the tomato yellow leaf curl virus tolerance gene, *Ty-1*. Theor. Appl. Genet. 88:141–146.

12 Bioengineering Resistance to Plant-Parasitic Nematodes

CHARLES H. OPPERMAN AND MARK A. CONKLING

North Carolina State University
Raleigh, North Carolina

Although plant-parasitic nematodes collectively cause estimated annual crop losses more than $77 billion on a worldwide scale, the vast majority of the damage is caused by sedentary endoparasitic forms, particularly *Meloidogyne* spp. (root-knot nematode) and *Globodera/Heterodera* spp. (cyst nematode) (Sasser & Freckman, 1987). All sedentary endoparasitic nematodes establish elaborate feeding sites within the host root. These feeding sites may range from slightly altered cortical cells to vascular cells that are developmentally altered in their fate and function (Jones, 1981). Feeding sites induced by the root-knot (*Meloidogyne* spp.) nematodes are the most elaborate. The specialized nature of the sedentary endoparasitic nematode-host relationship makes it vulnerable to identification and exploitation of weak points in the parasite's life cycle, and thus to designing genetically engineered resistance. This chapter will focus on the development of transgenic plants resistant to sedentary endoparasitic nematodes, although migratory forms will be discussed when appropriate.

Root-knot nematodes have a very broad host range, encompassing over 2000 plant species (Sasser, 1980). Most cultivated crops are attacked by at least one species of *Meloidogyne* (Sasser, 1980). The feeding site, or giant cells, of root-knot nematodes are initiated after the infective second-stage juvenile penetrates the host root, generally near the root tip, and migrates to the developing vascular cylinder. The induction of giant-cell formation is hypothesized to be due to glandular secretions injected via the nematode stylet into several root cells surrounding the head (Hussey, 1989), but no actual mechanisms for formation have been identified. The five to seven giant cells formed within or near the developing vascular cylinder become the permanent feeding site of root-knot nematode. The giant cells undergo repeated nuclear divisions without cytokinesis. Nuclei become enlarged and lobate, and may contain 14 to 16 times more DNA than do normal root tip nuclei (Jones, 1981; Wiggers et al., 1990). The giant cells become greatly enlarged, develop increased numbers of cellular organelles, and the cytoplasm appears dense and granular. The metabolically active giant cells act as nutrient sinks to provide food to the developing nematode (Huang, 1985). One characteristic feature of root-knot nematode giant cells is the highly invaginated

 Plant and Nematode Interactions, Agronomy Monograph no. 36.

and thickened cell wall, similar to cell walls observed in transfer cells (Jones, 1981).

Cyst nematodes also penetrate the root in the zone of elongation, but cause much more cellular damage than do root-knot nematodes during migration to the vascular cylinder. Cyst nematodes form a feeding site called a syncytium (Endo, 1992). The syncytium forms by a coalescence of cells surrounding the initial cell perturbed by the developing nematode. As the cell walls are partially dissolved, the syncytium incorporates the cytoplasm of the cell (Endo, 1992). Although syncytia also are multinucleate, it is exclusively because of the coalescence of cells. However, syncytia share many other characteristics with giant cells, including elaborate cell-wall ingrowths.

Root-knot and cyst nematode development is dependent upon feeding site formation. The infective second-stage juvenile is arrested in development while it is free in the soil. This motile nonfeeding stage also is highly resistant to detergent treatment (Opperman & Bird, 1996, unpublished results). Development of the second-stage juvenile is reinitiated after location of a food source, e.g., induction of a feeding site. Taken together, these factors strongly support the idea that second-stage juveniles are functionally homologous to diapaused dispersal stages observed in other parasitic and free-living species. Once development is reinitiated, the nematode passes through three additional molts to the adult. The major species of *Meloidogyne* (*M. arenaria, M. hapla, M. incognita*, and *M. javanica*) usually reproduce by parthenogenetic mechanisms; therefore most nematodes will develop into adult females under optimal conditions (Triantaphyllou, 1973). Males are generally formed under stress conditions on the nematode population, including overcrowding, high temperature, or poor nutrition. In contrast, cyst nematodes reproduce by amphimixis, so the sex ratio is generally 1:1. In resistant hosts, the ratio is often altered in favor of male nematodes due to differential survival under stress conditions. The life cycle of both nematodes varies with host and environment, but generally takes approximately 35 d to produce 300 to 1000 eggs.

THE NEED FOR NEW NEMATODE MANAGEMENT TACTICS

Traditional management of plant-parasitic nematodes has relied upon three primary tools; namely, crop rotation, nematicides, and host resistance. Crop rotation is hampered by the broad host range of many nematode species and may not be economically feasible for many growers. Use of nematicides is gradually being phased out due to health and environmental concerns. Host resistance is the most environmentally and economically sound method (Cook & Evans, 1987). Unfortunately, genetic resistance to plant-parasitic nematodes does not occur in all cultivated host species. Few sources of natural resistance have been identified for most crops. In those hosts where resistance is available, it has proved to be an extremely valuable commodity. For example, the introduction of the *Heterodera glycines*-resistant 'Forrest' has reportedly saved soybean [*Glycine max* (L.) Merr.] growers in the southern USA over $400 million during a 5-yr period (Bradley & Duffy, 1982). One of the most limiting aspects of currently available host resistance is the highly specific nature of the interaction. For example, resistance in tobacco (*Nicotiana tabacum* L.) is conferred by a single gene (*rk*), resulting in a hypersensitive response when infected by host Races 1 or 3 of *M. incognita* (Sasser, 1980; Slana & Stavely, 1981). The gene is not effective against the other races of *M. incognita*, nor against most populations of other *Meloidogyne* species. Tomato (*Lycopersicon esculentum* Miller) plants carrying the *Mi* gene may be resistant to *M. arenaria, M. incognita*, and *M. javanica*, but not *M. hapla* (Gilbert & McGuire, 1956). In both of these cases, resistance is manifested as a hypersensitive response (Kiraly, 1980). In other crop species, resistance to root-knot and cyst nematode is generally limited to specific genotypes (Triantaphyllou, 1987).

The lack of broad-spectrum resistance to the major plant-parasitic nematode species has complicated the development of integrated management strategies and tactics that do not rely on nematicides. A number of groups have been actively working on the isolation of nematode-resistance genes (Ho et al., 1992; Klein-Lankhorst et al., 1994; Ganal et al., 1995), and first gene has been cloned and sequenced (Cai et al., 1997). Its predicted protein is similar to that of disease-resistance genes previously cloned from higher plants (Cai et al., 1997, Dangl & Holub, 1997). It is conceivable that once nematode-resistance genes are molecularly characterized, they may be useful in transgenic approaches to gene transfer. For example, instead of having to rely on cumbersome crossing strategies between recalcitrant plant species, the resistance gene could be directly introduced via plant transformation. However, until the time that such genes are available it is impossible to speculate as to whether or not they will function in a heterologous host system. The remainder of this chapter will focus on approaches to design synthetic resistance via bioengineering.

GENE EXPRESSION DURING NEMATODE INFECTION

Plant-gene expression patterns are altered during nematode feeding site initiation (Sijmons, 1993; Bird & Wilson 1994). The most comprehensive study conducted to date was by Bird and Wilson (1994). They isolated over 200 cDNAs representing genes up-regulated in the giant cells of the root-knot nematode, *M. incognita*. Many of these genes normally are not expressed in root tissue, suggesting that the nematode is able to induce alterations in transcriptional control patterns. Preliminary reports indicate that structural or enzyme genes are up-regulated during either root-knot or cyst nematode feeding-site establishment. Many of these genes are classic wound-response genes that also are induced by invasion of other pathogens. For example, structural protein genes related to the extensin family have been isolated from root-knot nematode-infected tomato roots in a screen designed to identify up-regulated genes (Niebel et al., 1993). In another example, it was demonstrated that the well-characterized wound inducible *wun1* promoter was strongly up-regulated during cyst nematode invasion of potato (*Solanum tuberosum* L.), but was only slightly affected by root-knot nematode infection (Hansen et al., 1996). In one study, the promoter of the *hmg2* gene (hydroxymethylglutaryl CoA reductase) fused to GUS (bacterial reporter gene, β-glucoronidase) was observed to be strongly expressed in the developing giant cells and gall tissue shortly after infection by *M. incognita* or *M. hapla* (Cramer et al., 1993). The tomato *hmg*2 gene is a defense- related gene that is induced by both fungal and bacterial pathogens. Although a number of genes have been identified that are induced during nematode infection, very little data exist on the potential roles they may play in either defense responses or feeding site formation.

Information on control of gene expression in the nematode feeding site is of great significance to design genetically engineered resistance. However, few studies have focused on promoters involved in feeding-site gene expression. Results from one study suggested that many genes are down-regulated during feeding-site formation (Goddjin et al., 1993). In both transgenic *Arabidopsis* and tobacco plants, down-regulation of many constitutive promoters was observed during both cyst and root-knot nematode feeding site formation, including Cauliflower Mosaic Virus 35S, *rol* (root loci), A-D, nopaline synthase, γ-TIP, and T-*cyt* (Goddjin et al., 1993). Other groups, however, have not observed down-regulation of the CMV 35S promoter during root-knot nematode infection (Opperman & Conkling, 1994, unpublished results; Opperman et al., 1994a). These promoters represent an array of different regulatory sequences that confer strong expression of reporter gene constructs in uninfected root tissue.

A number of promoters of numerous plant organ-specific and pathogen-related genes have been characterized. Models emerging from such studies are that *cis*-acting sequences regulating gene expression in higher plants are often composites of several regions acting as transcriptional enhancers or silencers. Regulated transcription is a result of interaction among these sequences with various nuclear DNA-binding proteins and perhaps one another (Benfey et al., 1989). Computer comparisons of available regulatory region nucleotide sequences involved in nematode-induction or suppression do not reveal any obvious relationships (Goddjin et al., 1993).

The most comprehensive promoter analysis to date has been on the interaction of root-knot nematodes with the tobacco root-specific gene, *TobRB7*. Root growth and development are critically important to the success of a plant species, but relatively little characterization of root-specific gene expression has been reported. This limited information is due in part to a lack of easily characterized functions that are specific to root tissues and not other plant parts. *TobRB7* may be the most extensively characterized root-specific gene that has been isolated (Conkling et al., 1990; Yamamoto et al., 1990, 1991). The *TobRB7* gene encodes a protein of 250 amino acids and has conserved structural domains with several membrane spanning proteins (Yamamoto et al., 1990). Expression of *TobRB7* is limited to the root meristematic and immature vascular cylinder regions, as revealed by in situ hybridization studies (Yamamoto et al., 1991). Fusions were made between a deletion series of the 5′-flanking region and the bacterial reporter gene GUS (Jefferson, 1987) in order to identify sequences controlling the root-specific expression patterns. The deletion series experiments revealed that *cis*-acting elements necessary for root-specific expression of *TobRB7* are located between 636 and 299 nucleotides 5′ of the site of transcription initiation (Δ0.6) (Yamamoto et al., 1991) (Fig. 12–1). *TobRB7* expression was lost when the promoter was deleted to the –299 position (Δ0.3).

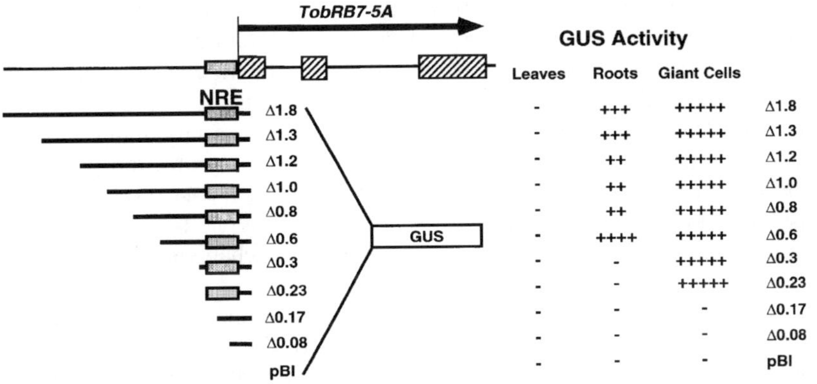

Fig. 12–1. A 5′ *TobRB7* promoter deletion series showing the different control regions fused to the reporter gene, β-glucuronidase (GUS). The nematode-responsive-element (NRE, grey box) is separable from the region controlling root expression, and may be used to specifically direct expression of a foreign gene in root-knot nematode giant cells. The numbers to the right depict the length of the truncated promoter constructs in kilobases. pBI is a vector control.

The GUS reporter-gene expression patterns in root-knot nematode-infected transgenic tobacco plants carrying the promoter deletion-reporter constructs demonstrated that *TobRB7* is one of the plant genes that is affected by nematode parasitism (Opperman et al., 1994a). Within 4 d after infection, significant levels of GUS activity in and around the developing feeding site were detected (Opperman et al., 1994a). This finding is further confirmed by results obtained with the deletion series of *cis*-acting sequences. Although no transgene expression is observed in uninfected plants with deletions containing only the Δ0.3 region, infection with either *M. arenaria*, *M. incognita*, *M. hapla*, or *M. javanica* significantly alters this pattern. In those plants, GUS accumulation was limited to the developing giant cells and appeared to be regulated by the nematode infection (Opperman et al., 1994a). The most significant aspect of these findings is that the nematode-responsive element (NRE) of the *TobRB7* promoter (Fig. 12–1) is not the same as the root-specific element, and can be uncoupled (Opperman et al., 1994a).

GENETICALLY ENGINEERED HOST DEFENSE

Rapid advances during the past 10 yr in our understanding of both plant molecular biology and host-pathogen interactions have made it possible to develop genetically engineered crop plants carrying disease-resistance traits. This strategy will prove particularly useful for crop-pest interactions in which no native resistance genes are known to occur. In addition, it is sometimes difficult or impossible to move resistance genes from a wild species into a cultivated one. In these cases, the best chance for developing resistant cultivars lies in genetic engineering.

The earliest—and probably most intensively—pursued approach has been plants genetically engineered to resist infection by viruses. Many of these strategies are based upon pathogen-derived genes as the foreign DNA sequence to be transformed into the plant. The first report of coat protein-mediated cross protection was tobacco plants transformed with the coat protein gene of tobacco mosaic virus (TMV) (Able et al., 1986). These experiments arose from observations that infection with a mild strain of virus could prevent or ameliorate later infection from a more pathogenic strain, the so-called cross-protection phenomenon (Fulton, 1986). Transgenic plants expressing the TMV coat protein gene were resistant to systemic infection by TMV. Many other plant-virus combinations have been subsequently studied. In no case is the mechanism of protection well understood at the molecular level.

A different approach to engineering virus resistance has been to construct a system of pathogen induced-hypersensitivity. In this strategy, a highly active cellular toxin gene (such as diphtheria toxin A fragment, or ricin) is transformed into plants under the control of a viral promoter (Wilson, 1993). Upon infection by the intact plant virus, the toxin gene is transcribed into mRNA during viral replication, resulting in production of the toxin and death of the infected cell. Resistance to potato virus X has been achieved through this approach, but the utility for field applications has been questioned (Wilson, 1993).

A number of genes involved in natural host-defense systems have been characterized and cloned, including so-called R genes (Dangl & Holub, 1997). The R genes may function in a signal transduction cascade, inducing the expression of other defense genes. Other host-defense genes cloned include chitinases, b-glucanases, proteinase inhibitors, and pathogenesis-related (PR) proteins (Graham et al., 1985; Metraux et al., 1989; Dixon & Lamb, 1990; Keen, 1992; Cramer et al., 1993). Several of these genes have been expressed in transgenic plants to provide either disease or insect resistance (see Lamb et al., 1992, for a review). Insect resistance may be produced by expression of proteinase-inhibitor genes. The cowpea [*Vigna unguiculata* (L.) Walp.] trypsin-inhibitor gene expressed in transgenic tobacco provided resistance to the tobacco budworm, *Heliothis virescens* (Hilder et al., 1987). The gene encoding the tomato- proteinase inhibitor II has been engineered into tobacco and provides protection from the tobacco hornworm, *Manduca sexta* (Johnson et al., 1989). Both of these proteinase inhibitors act with high specificity on trypsin, an enzyme that may be important in insect digestive processes (Plunckett et al., 1982). Approaches also have been developed that confer resistance to insect pests with transgenic plants that carry and express the Δ-endotoxin (BT) gene from the bacterium, *Bacillus thuringiensis* (Barton et al., 1987). Proper expression and function of the gene in plants, however, has required significant remodeling of the DNA sequence to compensate for plant-gene expression strategies (Perlak et al., 1991). Although insect-resistant crop cultivars have been developed using these approaches, several potential problems are associated with toxin-expressing plants. Constitutive expression of toxin genes in a host species may place strong selective pressures on the pest population to develop resistance. Another drawback is that nontarget pests or predators of target pests may be exposed to the protein toxin.

The previously mentioned studies point out how difficult and time consuming it is to design a strategy to engineer host resistance. Two of the major factors to be decided are what gene to insert (toxins, host defense genes, induced-hypersensitivity, etc.) and how to control expression of the foreign gene (i.e., what promoter to choose). Few plant-gene promoters characterized thus far respond to nematode infection, but a number of strong constitutive promoters are available. Some of these promoters have been used in the systems described earlier, but recent evidence (as follows) suggests that this approach may not be successful for nematodes. The relatively scant amount of information regarding plant-gene expression during nematode infection points out the difficulties in designing genetically engineered nematode resistance.

APPROACHES TO GENETIC ENGINEERING NEMATODE RESISTANCE

Several approaches are available for designing synthetic nematode-resistant transgenic crop cultivars. The simplest to implement is the expression of a nematode-specific toxin gene under the control of a constitutive promoter. As mentioned, this type of strategy has been used to design plants carrying the BT gene that are resistant to insect pests (Barton et al., 1987). However, very few protein toxins that act specifically upon plant parasitic nematodes have been identified. Even so, nematicidal plants could be produced if a suitable toxin gene and a constitutive promoter could be identified. Nematode natural enemies may be a source of toxin molecules. The fungal endoparasite *Nematoctonus* spp. secretes a nematoxic substance during spore germination that causes rapid paralysis and death of the nematode (Stirling, 1991). The predatory nematode *Seinura* injects a toxin into its nematode prey, causing immobilization within seconds (Stirling, 1991). Although the nature of these molecules is unknown, they may be potentially useful in the design of nematicidal transgenic plants if any of the toxins are peptides. One distinct advantage to this strategy is that it also would be effective on migratory forms of plant-parasitic nematodes. Because the migratory forms do not establish elaborate feeding sites, but rather browse on root cells, it is very hard to conceive of inducible strategies for engineering resistance.

Interference with either feeding or digestion also may be possible. For example, antibodies to nematode esophageal secretions may be utilized to disrupt either feeding site formation or feeding behavior (Hussey, 1989) via expression of functional antibodies in plants (Hiatt et al., 1989). Although it is too soon to tell whether this approach will yield resistance, transgenic plants expressing antibodies to stylet secretions have been produced and are being evaluated for response to nematode infection (Baum et al., 1996; Rosso et al., 1996). Proteinase inhibitors also may be used in an analogous way to the previously described insect resistance (Atkinson, 1994). In fact, this approach has resulted in the production of plants with altered response to cyst nematode infection (Urwin et al., 1995). In these experiments, tomato root cultures expressing the rice (*Oryza sativa* L.) oryzacystatin proteinase inhibitor gene did not support normal growth and development of the potato cyst nematode, *G. pallida* (Urwin et al., 1995).

The advantage of these strategies is that they are a preformed defense and do not require nematode-induction to be activated. Theoretically, any susceptible nematode that feeds for a suitable period upon cells containing the toxin moieties will not survive. There are several disadvantages to some of these approaches, however. Peptide toxins typically are narrow in their toxic spectra, as is the case with the BT toxins. The use of transgenic plants expressing toxins with activity against only certain nematode species may end up selecting for nonsensitive species or races, resulting in narrow resistance. The constitutive expression of any "toxin" gene may place upon the sensitive nematode population very strong selective pressure for resistance, placing the durability of this type of defense in doubt. In addition, the global, constitutive expression of toxin genes guarantees that nontarget species, including humans, will be exposed to the protein products. However, the preliminary results with the antifeedant proteinase inhibitors are encouraging, and the plantibody approach has yet to be fully explored. Clearly, there are numerous potential strategies in interactions as complex as those of sedentary endoparasitic nematodes with their hosts.

THE *TOBRB7-MELOIDOGYNE* INTERACTION

We have taken two independent approaches to designing transgenic nematode-resistant plants using the *TobRB7* NRE (Fig. 12–2). An approach using cell-specific cytotoxicity previously has been used to produce transgenic tobacco plants exhibiting male sterility (Mariani et al., 1990). In these experiments, expression of a gene encoding an RNase (*Barnase*) was directed by a tapetal cell-

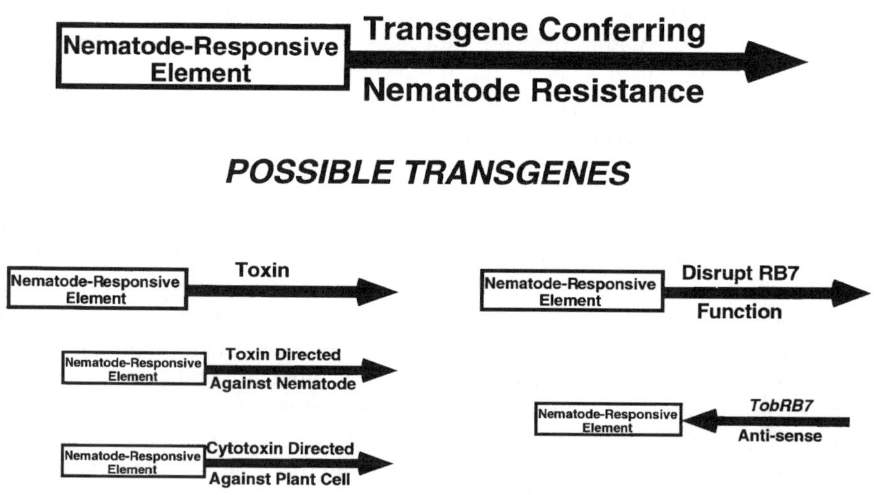

Fig. 12–2. Theoretical strategies to design transgenic resistant plants with a feeding site-specific promoter. The first approach is to drive expression of a peptide toxin gene directed against the developing nematode. Limitation of expression to the feeding site insures that only the target organism will be exposed to the toxin molecule. An alternate approach is to direct a toxin against the developing feeding site cells. It also may be possible to disrupt genes necessary for proper feeding site development and function. In this example, antisense constructs to genes up-regulated in feeding sites disrupt normal gene expression patterns. This approach has been used successfully with the *TobRB7* gene in tobacco against root-knot nematode species.

specific promoter. Expression of *Barnase* in the tapetal cells resulted in cell death, abortion of pollen development and, consequently, male sterility. Histochemical analysis of these transgenic plants showed that the cells surrounding the tapetum were unaffected. The apparent specificity of the *TobRB7* NRE suggested that we could use a similar approach to produce root-knot nematode-induced hypersensitivity. Constructs of *Barnase* driven by either the Δ0.6 or the Δ0.3 deletions were transformed into tobacco plants. As expected, the Δ0.6 plants did not form roots with high frequency, because this construct was probably expressed during vascular tissue development. In contrast, viable plants were obtained with the Δ0.3 NRE driving *Barnase*. These plants have no observable unique phenotype when uninfected. A screen of the transformants obtained revealed several lines with altered responses to root-knot nematode infection (Taylor, 1995). In one particular line, no root galls and two small egg masses were detected, in contrast to the heavy galling and large numbers (>300) of egg masses recovered on the control plants. Histochemical examination of the roots revealed that many second-stage juveniles had penetrated the Δ0.3 NRE-*Barnase* roots, but exhibited arrested development after 4 to 6 d. Control plants contained mostly adult female nematodes that had already deposited large numbers of eggs on the root surface. We have now repeated these transformations on two more occasions, and have recovered resistant plants from each independent experiment. We have, however, had difficulty recovering resistant lines from progeny of the barnase transformants. The highly cytotoxic barnase molecule has been reported in previous work from other laboratories to undergo a "gene silencing" phenomenon.

The second approach to develop transgenic plants resistant to root-knot nematodes has been to inhibit proper feeding site development or function using an antisense *TobRB7* construct. Full-length cDNA antisense constructs of *TobRB7* driven by either the Δ0.3 NRE or the 35S CaMV promoter were transformed into tobacco (Opperman et al., 1994b). Some plants containing constructs driven by the 35S promoter exhibited typical stresslike phenotypes, including pointed and narrow leaves, long internodal distances, and early flowering (Acedo & Conkling, 1994, unpublished data). When these plants were infected with root-knot nematode, however, a substantial reduction in galling and egg production was observed. Histochemical examination of roots revealed that numerous male nematodes had formed, and the feeding sites were small compared to the control plants. Transgenic plants containing the antisense construct driven by the Δ0.3 NRE showed no observable phenotype compared to the wild type. Greenhouse and field trials with the *TobRB*7 Δ0.3 NRE antisense plants provided strong evidence that nematode infection was substantially reduced. In both trials, root galling was reduced by approximately 70% compared to control treatments (Opperman & Conkling, 1996). The field trial indicated that protection lasted for the entire growing season (approximately 120 d). Galls that were observed on the roots of antisense plants were small and tended to appear as discrete entities compared to the large and clustered galls on the controls. Although these results provide only circumstantial evidence that *TobRB*7 is essential to giant-cell formation, they are the first indication that interference with feeding site establishment may be a viable approach to engineering transgenic nematode resistance.

OUTLOOK

The identification of the Nematode-Responsive-Element of *TobRB7* represents a significant step forward towards production of transgenic resistant crops to sedentary endoparasitic nematodes. As more genes are isolated that are involved in nematode-feeding-site formation, their control elements will be available for manipulation. The limitation of foreign gene expression to the developing nematode feeding site will allow the greatest flexibility and power in designing control strategies, and will overcome many of the problems associated with less specific constructs. The identification of genes necessary in a normal susceptible interaction is particularly significant. The control regions of these sequences coupled to proteins detrimental to the nematode feeding site should make it very difficult for the nematode to overcome the engineered resistance. Ultimately, the durability of transgenic sedentary endoparasitic nematode resistance may depend upon the specificity of the regions controlling foreign gene expression.

As more information is obtained regarding migratory genera of plant-parasitic nematodes, new methods of multiple nematode resistance will be possible. However, these strategies are necessarily more limited due to the less complex and interdependent interactions between these nematodes and their hosts. It may be possible to identify proteins that function as antifeedants or repellants to migratory forms, but as yet no such molecules have been isolated. Nevertheless, the rapid pace of progress on the sedentary endoparasitic forms provides optimism for the future. For example, it is not yet known whether proteinase inhibitors may function on migratory forms as well as the cyst nematodes. Renewed interest in these forms combined with the need for broad-spectrum protection in the absence of nematicides will almost certainly result in transgenic plants with broad and durable resistance to plant-parasitic nematodes in the future.

REFERENCES

Able, P.P., R.S. Nelson, N.B. De Hoffman, S.G. Rogers, R.T. Fraley, and R.N. Beachey, 1986. Delay of disease development in transgenic plants expressing the tobacco mosaic virus coat protein gene. Science (Washington, DC) 232:738–743.

Atkinson, H.J. 1994. Plant-nematode interactions: Molecular and genetic basis. p. 355–369. *In* K. Kohmoto et al. (ed). Pathogenesis and host-parasite specificity in plant diseases: Histopathological, biochemical, genetic and molecular basis. Pergamon Press, Oxford.

Baum, T.J., A. Hiatt, W.A. Parrott, L.H. Pratt, and R.S. Hussey. 1996. Expression in tobacco of a functional monoclonal antibody specific to stylet secretions of the root-knot nematode. Mol. Plant Microbe Interact. 9:382–387.

Barton, K.A., H.R. Whitely, and N.-S. Yang. 1987. *Bacillus thuringiensis* D-endotoxin expressed in transgenic *Nicotiana tabacum* provides resistance to lepidopterous insects. Plant Physiol. 85:1103–1109.

Benfey, P.N., L. Ren., and N.-H. Chua. 1989. The CaMV 35S enhancer contains at least two domains which can confer different developmental and tissue-specific expression patterns. EMBO J. 8:2195–2202.

Bird, D.M., and M.A. Wilson. 1994. DNA sequence and expression analysis of root-knot nematode-elicited giant cell transcripts. Mol. Plant-Microbe Interact. 7:419–424.

Bradley, E.B., and M. Duffy. 1982. The value of plant resistance to the soybean cyst nematode: A case study of Forrest soybean. USDA-ARS Staff Rep. AGES820929. U.S. Gov. Print., Washington, DC.

Cai, D., M. Kleine, S. Kifle, H.J. Harloff, N.N. Sandal, K.A. Marcker, R.M. Klein-Lankhorst, E.M.J. Salentijn, W. Lange, W.J. Stiekema, U. Wyss, F.M.J. Grundler, and C. Jung. 1997. Positional cloning of a gene for nematode resistance in sugar beet. Science (Washington, DC) 275: 832–834.

Conkling, M.A., C.-L. Cheng, Y.T. Yamamoto, and H.M. Goodman. 1990. Isolation of transcriptionally regulated root-specific genes from tobacco. Plant Physiol. 93:1203–1211.

Cook, R., and K. Evans. 1987. Resistance and tolerance. p. 179–232. *In* R.H. Brown and B.R. Kerry (ed.) Principles and practice of nematode control in crops. Acad. Press, Sydney.

Cramer, C.L., D.L. Weissenborn, C.K. Cottinghan, C.J. Denbow, J.D. Eisenback, D.N. Radin, and X. Yu. 1993. Regulation of defense-related gene expression during plant-pathogen interactions. J. Nematol. 25:507–518.

Dangl, J., and E. Holub. 1997. La dolce vita: A molecular feast in plant-pathogen ineractions. Cell 91:17–24.

Dixon, R.A., and C.J. Lamb. 1990. Molecular communication in interactions between plants and microbial pathogens. Annu. Rev. Plant Physiol. Plant Mol. Biol. 11:339–367.

Endo, B.Y. 1992. Cellular responses to infection. p. 37–49. *In* R.D. Riggs and J.A. Wrather (ed.) Biology and management of the soybean cyst nematode. APS Press, St. Paul, MN.

Fulton, R.W. 1986. Practice and precautions in the use of cross protection for plant virus disease control. Annu. Rev. Phytopath. 24:67–81.

Ganal, M.W., R. Simon, S. Brommonschenkel, M. Arnd, M.S. Phillips, S.D. Tanksley, and A. Kumar. 1995. Genetic mapping of a wide spectrum nematode resistance gene (Hero) against *Globodera rostochiensis* in tomato. Mol. Plant Microbe Interact. 8:886–891.

Gilbert, J.C., and D.C. McGuire. 1956. Inheritance of resistance to severe root-knot from *Meloidogyne incognita* in commercial type tomatoes. Proc. Am. Hort. Soc. 68:437–442.

Goddjin, O.J.M., K. Lindsey, F.M. van der Lee, J.C. Klap, and P.C. Sijmons. 1993. Differential gene expression in nematode-induced feeding structures of transgenic plants harbouring promoter-gusA fusion constructs. Plant J. 4:863–873.

Graham, J.S., G. Pearce, J. Merryweathers, K. Titani, L.H. Ericsson, and C.A. Ryan. 1985. Wound-induced proteinase inhibitors from tomato leaves. I. The cDNA-deduced primary structure of pre-inhibitor I and its post-translational processing. J. Biol. Chem. 260:6555–6561.

Hansen, E., G. Harper, M.J. McPherson, and H.J. Atkinson. 1996. Differential expression patterns of the wound-inducible transgene *wun1 - uidA* in potato roots following infection with either cyst or root knot nematodes. Physiol. Mol. Plant Pathol. 48:161–170.

Hiatt, A., R. Cafferkey, and K. Bowdish. 1989. Production of antibodies in transgenic plants. Nature (London) 342:76–78.

Hilder, V.A., A.M.R. Gatehouse, S.E. Sheerman, R.F. Barker, and D. Boulter. 1987. A novel mechanism of insect resistance engineered into tobacco. Nature (London) 330:160–162.

Ho, J.Y., R. Weide, H. Ma, M.F. van Wordragen, K.N. Lambert, M. Koornneef, P. Zabel, and V.M. Williamson. 1992. The root-knot nematode resistance gene (*Mi*) in tomato: Construction of a molecular linkage map and identification of dominant cDNA markers in resistant genotypes. Plant J. 2:971–982.

Huang, C.S. 1985. Formation, anatomy, and physiology of giant cells induced by root-knot nematodes. p. 155–164. *In* J.N. Sasser and C.C. Carter (ed.) An advanced treatise on Meloidogyne. Vol. 1. North Carolina State Univ. Graphics, Raleigh.

Hussey, R.S. 1989. Disease-inducing secretions of plant-parasitic nematodes. Annu. Rev. Phyopathol. 27:123–141.

Jefferson, R.A. 1987. Assaying chimeric genes in plants: The GUS gene fusion system. Plant Mol. Biol. Rep. 5:387–405.

Johnson, R., J. Narvaez, G. An, and C. Ryan. 1989. Expression of proteinase inhibitors I and II in transgenic tobacco plants: Effects on natural defense against *Manduca sexta* larvae. Proc. Natl. Acad. Sci. USA 86:9871.

Jones, M.G.K. 1981. Host cell responses to endoparasitic nematode attack: Structure and function of giant cells and syncytia. Ann. Appl. Biol. 97:353–372.

Keen, N.T. 1992. The molecular biology of disease resistance. Plant Mol. Biol. 19:109–122.

Kiraly, Z. 1980. Defenses triggered by the invader: hypersensitivity. p. 201–224. *In* J.G. Horsfall and E.B. Cowling (ed.) Plant disease—an advanced treatise. Acad. Press, New York.

Klein-Lankhorst, R.M., E.M.J. Salentijn, W.G. Dirkse, M. Arens-de Reuver, and W.J. Stikema. 1994. Construction of a YAC library from a *Beta vulgaris* fragment addition and isolation of a major satellite DNA cluster linked to the beet cyst nematode resistance locus *Hs1*$^{pat-1}$. Theor. Appl. Genet. 89: 426–434.

Lamb, C.J., J.A. Ryals, E.R. Ward, and R.A. Dixon. 1992. Emerging strategies for enhancing crop resistance to microbial pathogens. Bio/Technology 10:1436–1445.

Mariani, C., M. De Beuckeleer, J. Truettner, J. Leemans, and R.B. Goldberg. 1990. Induction of male sterility in plants by a chimeric ribonuclease gene. Nature (London) 347:737–741.

Metraux, J.P., W. Burkhart, M. Moyer, S. Dincher, W. Middlesteadt, S. Williams, G. Payne, M. Carnes, and J. Ryals. 1989. Isolation of a complementary DNA encoding a chitinase with structural homology to a bifunctional lysozyme/chitinase. Proc. Natl. Acad. Sci. USA 86:896–900.

Niebel, A., J. de Almeida Engler, C. Tiré, G. Engler, M. Van Montagu, and G. Gheysen. 1993 Induction patterns of an extensin gene in tobacco upon nematode infection. Plant Cell 5:1697–1710.

Opperman, C.H., C.G. Taylor, and M.A. Conkling. 1994a. Root-knot nematode directed expression of a plant root-specific gene. Science (Washington, DC) 263:221–223.

Opperman, C.H., and M.A. Conkling. 1996. Root-knot nematode induced *TobRB7* expression and antisense transgenic resistance strategies. Adv. Molec. Gen. Plant-Microbe Interact. 4:521–526.

Opperman, C.H., G.N. Acedo, D.M. Saravitz, A.M. Skantar, W. Song, C.G. Taylor, and M.A. Conkling. 1994b. Bioengineering resistance to sedentary endoparasitic nematodes. p. 221–232. *In* F. Lamberti et al. (ed.) Advances in molecular plant nematology. Plenum Press, New York.

Perlak, F.J., R.L. Fuchs, D.A. Dean, S.L. McPherson, and D.A. Fischhoff. 1991. Modification of the coding sequence enhances plant expression of insect control protein genes. Proc. Natl. Acad. Sci. USA 88:3324–3328.

Rosso, M.-N., A. Schouten, J. Roosien, T. Borst-Vrenssen, R.S. Hussey, F.J. Gommers, J. Bakker, A. Schots, and P. Abad. 1996. Expression and functional characterization of a single chain FV antibody directed against secretions involved in plant nematode infection process. Biochem. Biophys. Res. Commun. 220. 255–263.

Sasser, J.N. 1980. Root-knot nematodes: A global menace to crop production. Plant Dis. 64:36–41.

Sasser, J.N., and D.W. Freckman. 1987. A world perspective on nematology. p. 7–14. *In* D.W. Dickson and J.A. Veech (ed.) Vistas in nematology. Soc. Nematol., Hyattsville, MD.

Sijmons, P.C. 1993. Plant-nematode interactions. Plant Mol. Biol. 23:917–931.

Slana, L.J., and J.R. Stavely. 1981. Identification of the chromosome carrying the factor for resistance to *Meloidogyne incognita* in tobacco. J. Nematol. 13:61–66.

Stirling, G.R. 1991. Biological control of plant parasitic nematodes. CAB Int., Wallingford, England.

Taylor, C.G. 1995. Characterization of a tobacco promoter and its applications in creating nematode resistant transgenic plants. Ph.D. diss. North Carolina State University, Raleigh. Triantaphyllou, A.C. 1973. Environmental sex differentiation of nematodes in relation to pest management. Annu. Rev. Phytopathol. 11:441–462.

Triantaphyllou, A.C. 1987. Genetics of nematode parasitism on plants. p. 354–363. *In* D.W. Dickson and J.A. Veech (ed.) Vistas in nematology. Soc. Nematol., Hyattsville, MD.

Urwin, P.E., H.J. Atkinson, D.A. Waller, and M.J. McPherson, 1995. Engineered oryzacystatin-1 expressed in transgenic hairy roots confers resistance to *Globodera pallida*. Plant J. 8:121–131.

Wiggers, R.J., J.C. Starr, and H.J. Price. 1990. DNA content and variation in chromosome number in plant cells affected by the parasitic nematodes *Meloidogyne incognita* and *M. arenaria*. Phytopathology 80:1391–1395.

Wilson, T.M.A. 1993. Strategies to protect crop plants against viruses: Pathogen-derived resistance blossoms. Proc. Natl. Acad. Sci. USA 90:3134–3141.

Yamamoto, Y.T., C.-L. Cheng, and M.A. Conkling. 1990. Root-specific genes from tobacco and *Arabidopsis* homologous to an evolutionarily conserved gene family of membrane channel proteins. Nucl. Acids Res. 18:7449.

Yamamoto, Y.T., C. Taylor, G.N. Acedo, C.-L. Cheng, and M.A. Conkling. 1991. Characterization of cis-acting sequences regulating root-specific gene expression in tobacco. Plant Cell 3:371–382.

13 Agricultural Sustainability and Nematode Integrated Pest Management

LARRY W. DUNCAN AND JOSEPH W. NOLING

University of Florida
Lake Alfred, Florida

Many highly profitable modern agricultural practices cannot be sustained indefinitely, particularly in high-input agricultural systems (Allen et al., 1991). Major physical and energy inputs to agroecosystems are finite in quantity. The quality of soil is often compromised by modern farming practices (Elliott & Coleman, 1988), and agricultural chemicals pollute basic resources such as groundwater (Kaplan, 1988) and the atmosphere (UNEP, 1992). With increasing awareness of the cumulative adverse effects of many high-input agricultural practices, the paradigms called *integrated pest management* (IPM) and *sustainable agriculture* emerged to provide conceptual alternatives to past practices, as food and fiber-production systems evolve.

The use of nematicides historically has provided some of the strongest evidence for the need to reconsider how plant pests are managed. Most of the highly effective fumigant nematicides (primarily halogenated hydrocarbon-based) are among those agricultural chemicals that are now banned or scheduled for phase-out due to contamination of water, degradation of the atmosphere, and risk to farmworkers and consumers (McKenry et al., 1994; UNEP, 1992). The use of several organophosphate and carbamate-based nematicides has been discontinued in many parts of the world due primarily to groundwater contamination (Johnson & Feldmesser, 1987). The efficacies of several of the few remaining nematicides are compromised by accelerated microbial degradation (Davis et al., 1993) and/or by use restrictions designed to mitigate environmental hazards (Noling & Becker, 1994).

Growers have sustained increased nematode damage to crops as a result of these changes (Roberts, 1993). However, restricting the use of some soil-applied nematicides represented an important step toward developing more sustainable methods to manage nematodes. Moreover, a growing consensus among researchers and policy makers for the need to consider the sustainability of agricultural practices provides agricultural scientists and growers the opportunity to develop and implement pest management practices that emphasize long-term production rather than short-term profitability at the expense of future generations (Rajotte, 1993).

We shall consider nematode management in this chapter in the context of sustainability and IPM. Unfortunately, the integration of promising pest management methods is rudimentary or nonexistent in many agricultural systems today. Therefore, the various methods currently used to limit or prevent nematode damage to crops will be described in order to evaluate their potential as tactics to be used in IPM systems.

AGRICULTURAL SUSTAINABILITY

Few agricultural systems on earth exist in a state of equilibrium with respect to edaphic quality, nutrients and living communities. Agricultural systems have evolved to include numerous practices that affect nematode populations. Among these are reductions in the genetic base of crops (Gallaher & McSorley, 1993), in native habitats (Esser et al., 1993) and in use of crop rotation (Trivedi & Barker, 1986). The use of synthetic fertilizers (Noe & Imbriani, 1986) and pesticides (Thomason, 1987), irrigation (Duncan, 1995), monoculture (Kerry et al., 1982), mechanization (Fortnum & Karlen, 1985) and expansion into areas of low productivity (Avery, 1995) have all increased. All of these practices are being studied by nematologists in order to devise economic methods for nematode control that enhance agricultural sustainability by reducing the need for high-cost inputs. Increasingly, long-term environmental costs (e.g., resource depletion, air and water quality, wildlife diversity) are among those considered (Roberts, 1993; Thomason, 1987).

The myriad definitions of sustainability include broad concepts of balance (as described by Hawken, 1993), "...an economic state where the demands placed upon the environment by people and commerce can be met without reducing the capacity of the environment to provide for future generations...."; as well as more specific fundamental features pertaining to individuals, society and the environment, "A sustainable agriculture is one that equitably balances concerns of environmental soundness, economic viability and social justice among all sectors of society" (Allen et al., 1991). Achieving complete sustainability in agriculture is problematic, given finite resources and the limitations of current technology. Rather, "increased" sustainability is a valid concept to use when developing operational definitions for experimental purposes.

Consensus about what constitutes empirical evidence for the sustainability of a given activity will depend on the strength of a particular argument. Gliessman (1995) and others suggest that quantitative agroecology already provides many of the tools needed to compare agricultural activities with respect to sustainability. Some of these methods that have been used by nematologists to measure processes such as trophic structure (Yeates et al., 1993a), energy flow (Powers et al., 1993), nutrient cycling (Yeates et al., 1993b), food webs (Hunt et al., 1987) *r*-(rapid colonizers of ephemeral environments) and *k*-(adapted to stable environments) selection (Bongers, 1990), community structure (Showler et al., 1990), and succession (Ettema & Bongers, 1993) are routine and widely accepted. For example, among the soil-inhabiting nematodes, plant-parasitic species do not generally rate high on maturity indices (Bongers, 1990) of nematode community structure, based on *r*- and *k*-strategies. Since system stability increases with community succession (Altieri, 1994), the utility of indices of nematode community structure to predict system stability (and, by definition, natural regulation of some nematode species) is currently an active area of research (Niblack, 1989). Many recent concepts such as metapopulations, evolutionarily stable strategies and application of network theory to food webs await extensive exploration by nematologists (Ferris, 1993).

A tension that exists in determining the degree to which a particular activity is sustainable is that between the need for economic viability and other considerations such as those mentioned previously in the second definition of sustainability (Allen et al., 1991). Changes in agricultural practices that could benefit society in the long-term also may reduce profits to individual growers, decrease production per unit area and increase food prices. For example, there is a clear disparity in cereal return per unit input of fertilizers and pesticides among the eight countries shown in Fig. 13–1. For given amounts of these particular resources, some of the industries can be sustained for nearly twice as long as others, probably due partly to differences in climate and intrinsic soil productivity. Moreover, Fig. 13–1 does not reflect the fact that government subsidies permit the highest-producing industries to overproduce cereals, thereby disrupting international prices (e.g., the return of cereal per unit input is very low in those industries) and increasing environmental risk (Hamblin, 1995). Government subsidies currently foster artificially high levels of nematicide use in crops such as peanut (*Arachis hypogaea* L.) in southeastern USA (Rodriquez-Kabana et al., 1989).

This overly simplified analysis illustrates the extent to which agricultural industries could be disrupted if finite resources (i.e., phosphate fertilizers, water, petroleum-based products and activities) were allocated in the most efficient manner. Changes in practices that may cause significant environmental degradation can be similarly disruptive to growers. Therefore, social attitudes will play

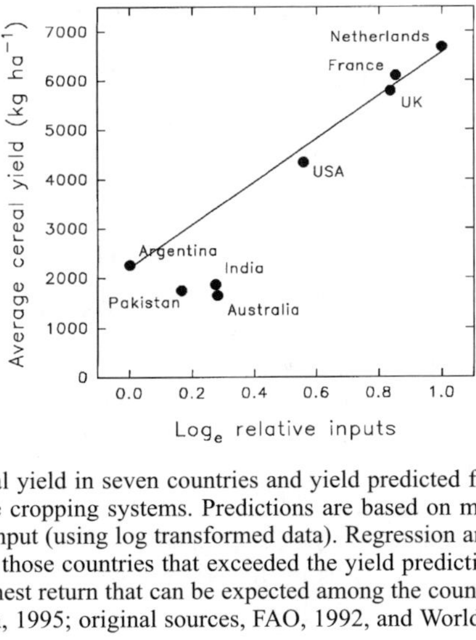

Fig. 13–1. Observed cereal yield in seven countries and yield predicted from levels of fertilizer and pesticides used in those cropping systems. Predictions are based on multiple linear regression of yield on both types of input (using log transformed data). Regression analysis was done twice, the second time using only those countries that exceeded the yield predictions from the first analysis, as a measure of the highest return that can be expected among the countries, using available methods (data from Hamblin, 1995; original sources, FAO, 1992, and World Resour. Inst., 1992).

an important role in efforts to achieve greater sustainability in agricultural systems (Rajotte, 1993). Fortunately, methods available in agroecology encompass social and cultural concerns, rather than treating agriculture as a separate entity (Maloney, 1995). Hamblin (1995) and others suggest the need for plant protection specialists to disengage from the mind-set that we must produce fail-safe methods to control each pest in high-input agricultural systems. Rather, we must consider the agroecosystem as a whole, which is likely to result in greater emphasis on increased product diversity, stability of production and profit per unit input.

NEMATODE INTEGRATED PEST MANAGEMENT

Ideally, nematode IPM would result from a holistic approach to managing an agroecosystem. Each intervention in that system by a grower would be based on knowledge of its interacting effects on the crop and the physical and biotic components of the environment, including other pests and diseases. The grower would know the impact on nematode population densities of inputs such as fertilizers (Castro et al., 1991), irrigation (Goodell & Ferris, 1989), tillage practices (Fortnum & Karlen, 1985), soil amendments (Kaplan & Noe, 1993), herbicides (Sipes & Schmitt, 1989), and fungicides (Meyer et al., 1991). Growers also would know many of the effects of nematode management practices on nontarget organisms (Rodriguez-Kabana & Curl, 1980), soil (Duncan, 1986) and water (Johnson & Feldmesser, 1987). Progress toward such management systems is rudimentary. Except where the involvement of other organisms in nematode-induced losses are obvious (Minton et al., 1985), IPM is frequently considered at the level of integrating various management methods exclusively to manage nematodes. As these systems are implemented, they will increasingly incorporate practices based on the interactions between nematodes and other diseases and pests (Alston et al., 1993) and other components of the cropping system.

In this context, IPM functions in terms of three processes (Duncan, 1991; Geier, 1966): (i) determining how the life system of the nematode must be modified to reduce it to tolerable levels, (ii) affecting the modification through biological knowledge combined with current technology, and (iii) developing new methods of management with current technology that are increasingly compatible with economic and environmental requirements.

Both concepts, IPM and sustainability, are interdependent (Duncan, 1991). Implementing an effective IPM program may require considerably more information, time and labor expense than does simple reliance on pesticides (Goodell & Ferris, 1981). Many management tactics are only partially effective, and incremental improvements from combinations of methods are often required to achieve economic control (Roberts, 1993). Grower reluctance to adopt many of these practices is well recognized (Riggs, 1995), so that the rate of progress toward IPM depends heavily on the level of social commitment to making agricultural systems more sustainable (Rajotte, 1993).

A frequently cited goal of IPM is the reduction of pesticide usage (McNamara, 1995). However, it is probably unwise to emphasize certain components of sustainability more than others. The judicious use of pesticides, based on biomonitoring and economic thresholds, is generally recognized as a valid IPM goal. Moreover, the elimination of pesticide usage is not necessarily intrinsic to sustainability. Citrus (*Citrus* spp.) trees grown on the deep sandy soils of central Florida suffer severe loss of roots caused by *Radopholus citrophilus* in the deeper soil profiles. Nematode management using the soil fumigant ethylene dibromide (EDB; 1,2-dibromoethane) was discontinued due to widespread contamination of groundwater (Kaplan, 1988). Growers now mitigate the effects of the nematode by frequent irrigation and fertilization of the shallow portion of the root system that is not damaged by the nematode; however, these growers are now in a situation comparable to that described previously (Fig. 13–1). Fruit yields per unit input of water and fertilizer are lower in these orchards than in orchards grown on shallower, heavier textured soils where *R. citrophilus* is less damaging to citrus. The nonpesticidal management practices currently used will become even less sustainable with increasing competition for water between citrus producers and growing urban communities. Pesticide-free agricultural systems also can reduce sustainability by increasing the area under cultivation (and no longer habitable to many species) in order to produce equivalent yields (Avery, 1995).

The working definition of IPM given in this section describes a systems approach to pest management in which specific types of information are processed in a manner designed to produce decisions that optimize how resources are employed and preserved (Bird, 1987). Given this approach to IPM, the concept of profitability may provide a useful operational definition of system sustainability at some levels of agricultural research. Profitability directly addresses the economic requirements to sustain a system. From the standpoint of conservation, the relative abundance of an input resource is reflected somewhat by its price. Waste of resources is minimized when nematode management practices are optimum to provide maximum profitability to the grower (Ferris, 1984a,b; Noe et al., 1991). On a broader level, agroecology research can help to provide the insights needed for society to recognize and quantify long-term costs associated with resource depletion and environmental degradation (Pimentel et al., 1980). If social mechanisms develop to reflect the true value of productive land and finite resources through land use regulations and cost-adjustment of input resources, then profitability will address the needs of both growers and society. Multidisciplinary studies could be instructive, if they involve nematologists, ecologists, economists and others to compare optimum nematode management practices at current prices against strategies based on estimates of true long-term profitability.

INTEGRATED PEST MANAGEMENT DECISIONS

Information and advice about nematode IPM are offered by private and government advisory services (Barker & Imbriani, 1984; McNamara, 1995), but individual growers usually make the final decisions. Regardless of the source of management decisions, growers currently must obtain information about the status of a field (type and extent of nematode infestation) by submitting soil or root samples to professional laboratories. Based on sample information, growers may respond to broadly defined action threshold densities of nematode populations (Barker & Imbriani, 1984; Ferris et al., 1986) with strategies and tactics designed to mitigate damage by nematodes to the crop(s) grown in the following season(s).

Large numbers of farmers make nematode IPM decisions at varying degrees of complexity. Certainly, the integration of pest management tactics to control one or more pests and diseases was routinely practiced before and after the advent of widespread nematicide usage, whenever such practices were perceived to be profitable (Tyler, 1933). Even as nematodes were first being recognized as plant pests, scientists recommended end-of-season root destruction to reduce population carry-over, nursery sanitation, clean fallow, crop rotation, use of trap crops, composts, and chemicals to manage nematodes (Atkinson, 1889). Moreover, deductive, non-empirical models (based on generation times and fecundity estimates), were employed to estimate the efficacy of practices such as early plow-down of crops to reduce end of season nematode densities (Atkinson, 1889). While all of these practices have continued in use, the development of complex, modern agricultural methods fostered increased specialization by growers. The discovery of effective fumigant nematicides in the 1940s often eliminated the need for management methods, such as use of resistant varieties and crop rotations, that interfered with the trend toward monoculture of the most profitable crops. The modern conception of IPM developed in reaction to that trend because of a recognized need for greater environmental stewardship. Management tactics, particularly those that pose the greatest risks to the environment, are to be employed only when needed. Consequently, the emphasis in nematode IPM has shifted toward promoting a reliance on biomonitoring coupled with predictive, empirically derived models as the basis for rational pest-management decisions.

Integrated pest management for all pest disciplines has progressed steadily in recent decades (Rajotte, 1993). Indeed, given some of the unique constraints to managing soilborne pests, grower acceptance of IPM for nematodes should be viewed as encouraging. Decisions affecting nematode control must be made well before planting annual crops (Ferris et al., 1986). Biomonitoring is more difficult for nematodes in soil than for aboveground pests and diseases (Duncan et al., 1994). Monitoring costs can be high because trained personnel working in laboratories are required to identify nematodes and to assess population densities (Goodell & Ferris, 1981). High costs can preclude efforts to identify patchiness of nematode-population densities within fields (Duncan et al., 1995; Schomaker & Been, 1992). Seasonal population fluctuations of nematode parasites of perennial crops that affect management timing decisions are more difficult to predict or monitor than are those for many aboveground diseases and arthropod pests (Duncan et al., 1993).

The special constraints to nematode IPM illustrate the importance of research to optimize the efforts required to make nematode IPM decisions. Improved mechanized soil-sampling devices are needed to reduce cost of sampling, increase the precision of density estimates, and permit more detailed mapping of population densities within fields (Duncan, 1991; Ferris et al., 1990). Serological and molecular methods to identify and quantify nematodes in soil and root samples are rapidly being developed (e.g., Hyman et al., 1990). No other area of research in nematology is more important to facilitate IPM decisions through improved diagnostic accuracy, expanded access to diagnostic methods for personnel not trained in nematology and, hopefully, reduced diagnostic costs. As these methods are combined with those of related disciplines, the value of soil sampling will increase by augmenting information gained from the cost of sampling.

The information gained from a soil or root sample is useful only in the context of a predictive decision-making paradigm. Based on the current nematode-infestation levels, growers require estimates of crop yields that will result from available management options and estimates of population densities that will remain to damage subsequent crops. Quantitative models, derived from conceptual models of host-parasite-environmental interactions, have been used to make inferences about empirical observations (Perry & Clark, 1983), test the validity of concepts (Jaffee et al., 1992) and to predict the outcome of management options (Ferris & Greco, 1990). From the standpoint of sustainable agriculture, models of sequential cropping systems (Noe, 1998, see Chapter 8) consistently demonstrate that optimum management strategies for the long-term do not result from a series of decisions about the crop at hand. The models most widely used to make IPM decisions about nematodes are derived from the research disciplines of population biology (McSorley, 1998, see Chapter 6) and crop loss assessment (crop-nematode damage functions).

Crop Loss Assessment

Crop loss assessment in nematology is used to identify economically important species (most plant-parasitic nematodes are not) and to derive density-dependent damage functions as the basis for rational management decisions. Controlled greenhouse studies are routinely used to establish nematode pathogenicity to a crop and to study the nature of the relationship between yield and nematode density (Seinhorst, 1965). However, parameter estimates for these relationships are often most useful when derived from field data, based on estimated nematode-population densities. The physical condition (Walters & Barker, 1993) and spatial patterns of nematodes inoculated in pots differ from those in the field (Noe & Campbell, 1985). Edaphic, biotic and climatic effects on plants and nematodes cannot be reproduced in pot studies, which mask the range of responses that can be expected. Containerized, field-microplot studies provide a compromise between the need for experimental control and natural conditions (McSorley et al., 1994).

Methods to conduct small-plot field trials involve establishing nematode-infested and nematode-free plots, or establishing plots with a randomized range of nematode densities, through the use of nematicides, or by previous cropping with host plants of variable susceptibility (Ferris, 1984a). Resistant and susceptible varieties of the same crop can be used to manipulate nematode densities in small plots with fewer residual differences between plots from nematicide effects or crop residues (Francl & Kenworthy, 1989). Because nematode populations typically exhibit patchiness, yield-nematode density relationships also are studied in small plots with naturally occurring ranges of nematode densities (Noe, 1993).

The latter approach may be especially useful to study cumulative effects of nematodes in orchards or other perennial crops where the cost of long-term studies can be prohibitive (Duncan et al., 1995).

Much care is required in the design and interpretation of data from the field. Whenever possible, the effects of practices to manipulate nematode densities should be measured in control treatments (Wheaton et al., 1985). Nematicides and cultural practices that affect nematodes may affect the subsequent crop directly (Barker & Powell, 1988) or indirectly by controlling other pests (Wheaton et al., 1985). Naturally occurring ranges of nematode densities are often due to variable edaphic or biotic factors that also influence crop growth and yield (Noe & Imbriani, 1986). Measurement of edaphic and soil biotic factors other than nematode densities can help to apportion putative nematode effects on yield in small plot studies, but the possibility of unmeasured factors affecting both variables cannot be eliminated.

Damage Functions

The preplant or initial population density (P_i) of nematodes is the usual predictor variable in various published damage functions (Barker, 1989; McSorley & Dickson, 1989; Noe, 1991, 1993; Seinhorst, 1965). The term *critical point model* has been applied to damage functions that predict yield based on population measurements from a single time point. *Multiple point models* have been proposed to integrate (similar to the heat unit concept) damaging effects of nematodes to perennial crops over time (Noling & Ferris, 1987). Because few options exist in annual cropping systems for managing nematodes after planting, simulation models to aid in nematode management are rare (Ferris, 1976).

A widely used damage function, $y = m + (1 - m)z^{P-T}$, for $P > T$ and $y = 1.0$ for $P \leq T$, where y = relative yield (yield/yield in the absence of nematodes), m = minimum yield obtained regardless of P_i, z = a rate-determining parameter, P = P_i and T = the tolerance threshold density, below which damage does not occur (Seinhorst, 1965) is illustrated in Fig. 13–2. The model was derived from classical competition theory of density dependent effects. This model and a variety of other damage functions (McSorley & Duncan, 1995) have been used to investigate management efficacy (Perry & Clark, 1983), overwinter survival (Jeger & Starr, 1985), race development (Jones et al., 1967), biological control of nematodes (Perry, 1978), and damage mechanisms (Trudgill, 1992).

The most direct application of damage functions is in estimation of *economic thresholds* or population densities at which the cost of management is equal to the value of the increased yield (Ferris, 1978; McSorley & Duncan, 1995). However, functions would best serve nematode IPM by providing growers with information about the likelihood of yield responses to various potential management practices. Unpredictable weather conditions (McSorley & Dickson, 1989), variable edaphic (Noe & Imbriani, 1986) and biotic (Duncan & Ferris, 1983b; Noe, 1991) conditions in fields and error associated with population estimates (Ferris, 1984a,b) limit the accuracy of individual predictions (McSorley, 1992). Less error is associated with yield estimates in irrigated dryland or Mediterranean climates (Ferris et al., 1986) than in temperate climates dependant

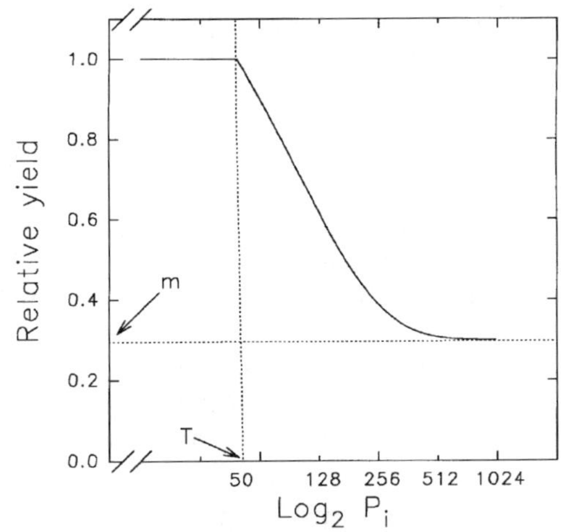

Fig. 13–2. Hypothetical nematode-crop damage function according to the model, $y = m + (1 - m) z^{P-T}$, for $P > T$ and $y = 1.0$ for $P \leq T$, where y = relative yield, m = minimum yield, z is a rate determining parameter, P = initial nematode population density and T = tolerance threshold population density, below which damage does not occur (Seinhorst, 1965).

on rainfall (McSorley & Dickson, 1989). Aggregation of nematodes in fields also typically results in overprediction of yield loss using damage functions and single P_i estimates (Perry, 1983). Methods to correct overpredictions have been proposed but not validated (Ferris et al., 1990; Noe & Barker, 1985a).

These limitations illustrate the need for multiple data sets to estimate the variability of the yield prediction and to identify key interacting factors. Factors that are easily identified by growers (Duncan & Ferris, 1983b) can be specified as part of the valid environmental bounds of the damage function or comprise terms of more complex formulae (Duncan & Ferris, 1983a). If yield responses to these factors are predictable based on initial measurements (i.e., soil texture, other pathogens), variation in yield estimates can be reduced. Probabilities associated with unpredictable factors such as rainfall and temperature are measurable. Theoretical foundations exist to optimize management practices based on variability of P_i estimation and yield responses (Ferris, 1984b) and can be extended to incorporate probabilities of important additional factors (McSorley & Phillips, 1993).

Cropping Systems

Pest management practices employed in a field in a given year influence the growth of subsequent crops due to their effects on residual populations of soil inhabitants and on the physical attributes of the soil matrix. The quantitative methods described previously are used increasingly to investigate the behavior of long-term cropping systems (Noe, 1998, see Chapter 8). The cropping systems approach to IPM is evolving in recognition of the complexity of management options and the importance of interactions over long periods of time (Ferris & Greco, 1992; Ferris et al. 1986; Jones et al., 1967; Kinloch, 1986; Noe et al.,

1991). For example, nearly 20 000 combinations of cropping/management sequences (3^9) are available during three seasons, to a grower considering production of any of six crops, if use of a resistant variety was an option for one crop, and if two crops were valuable enough to justify use of a nematicide. Optimizing the expected 3-yr profits, with respect to nematode management, is practical only by simulating the results of the various options.

In its most basic form, simulating long-term effects of nematodes, crop growth and profitability requires nematode population models, crop-nematode damage functions, management efficacy functions, and estimates of the costs of management tactics and the costs and value of the crops in question (Ferris & Greco, 1992). Computer algorithms are then derived to simulate, for a range of P_i, cumulative profits for all possible sequences of crops and management practices (Fig. 13–3). Essentially, the algorithms predict crop loss and profitability for

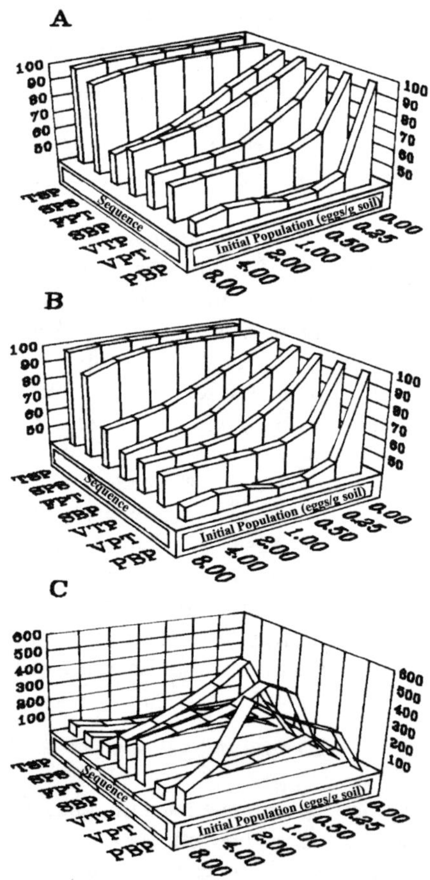

Fig. 13–3. Estimated profitability (relative) for (*A*) family farms or (*B*) commercial farms, of using various 3–yr cropping sequences with or without nematicides to control pea–cyst nematodes; and (*C*) estimated final (third year) nematode population densities (eggs/g soil), for a range of initial population densities (1st yr). [*B* = broad bean, *F* = pea (*Pisum sativum* L.) with D-D nematicide (1,3-dichloropropone), *P* = pea, *S* = sugarbeet [*Beta vulgaris* (L.) subsp. *vulgaris*], *T* = tomato, *V* = vetch (*Vicia sativa* L.); from Ferris & Greco, 1992].

each successive year in the sequence by predicting the nematode population densities at critical times.

The cropping systems approach represents the next logical step beyond the use of economic thresholds to advance the progress of IPM. It introduces complexity over time, by considering the nematode-host interaction for more than a single crop and season. Agricultural sustainability will be enhanced by the approach, because it optimizes the relationship between agricultural inputs and returns over a longer period of time. Moreover, as with simpler models, cropping systems models provide an eventual framework for incorporating intraseasonal complexity in the form of plant growth models and stochastic elements for factors that affect nematodes and plants.

Biomonitoring

Sampling methods to estimate the population density of nematodes in soil and plant material are described in Chapter 6 (McSorley, 1998). Special concerns about sampling pertaining to the practice of IPM include timing, estimating the population density of arthropod and disease organisms other than nematodes, and reliability of population density estimates.

Nematode samples differ from those of most other pest and disease organisms because they require intensive processing by trained personnel. To reduce high processing costs, numerous samples from throughout a field are typically pooled and subsampled. Disadvantages of this procedure are that they provide no variance estimate for the counts, and do not provide information about the spatial patterns of nematodes within fields. Growers must rely on researchers to estimate typical sample reliability, and until new methods are developed, they must make management decisions without very precise knowledge of where nematodes occur in their fields.

Lack of information about nematode spatial patterns can result in treatment of large areas of land when smaller applications would be more appropriate (Goodell & Ferris, 1981). This problem will not be resolved without improvements in sampling equipment and processing methods referred to previously. When developing sampling programs, researchers should acquire sufficient data to estimate sampling precision by assuming a known variability for the counts (Duncan et al., 1994). Published formulae that do not make this assumption probably overestimate the lengths of confidence intervals on population density estimates. It should be recognized that sampling plans for nematodes produce fairly imprecise population-density estimates that are still useful in making IPM decisions. This situation is because the majority of population estimates tend to be significantly below or above threshold densities that trigger management practices (Ferris, 1984b).

Although most nematode damage functions are used to make predictions based on preplant nematode population densities, samples collected at other points in the cropping cycle are sometimes more useful. After overwintering, population densities of root-knot nematodes often are below detection thresholds. In these cases, end-of-season population densities better predict damage to subsequent crops (Kinloch, 1986). Measurement of end-of-season plant symptoms, such as root-gall indices in carrot [*Daucus carota* subsp. *sativus* (Hoffm.) Arcang.], has proven useful to predict subsequent damage by *Meloidogyne hapla* and has the advantages of low sampling cost while providing information about nematode spatial patterns that permit directed treatments of fields (Belair & Boivin, 1988).

In most cases, sampling for nematodes is a continuing requirement to fulfill IPM objectives. Shifts in population densities of nematodes and other pests and pathogens occur continually in response to cropping sequences and climatic factors. Even in long-lived perennial crops, nematode densities fluctuate within seasons and between years. Population densities of *Tylenchulus semipenetrans* on citrus decline as soil moisture increases and respond to changes in nonstructural carbohydrate levels in roots (Duncan, 1995). Population levels are low during the summer rainy season of the subtropics, which also is when starch availability is lowest. Numbers of nematodes also increase in years when fruit production is low in response to increased allocation of carbohydrate to root production (Duncan et al., 1993). Thus, even in perennial crops, management decisions should reflect the time of year in which samples are collected, and estimates of nematode densities over several years may be necessary to understand the parasite load in an orchard.

The return on the investment in soil and root sampling can be increased by planning to acquire as much needed information as possible from each sample. Whenever possible, sampling should be scheduled and detection methods selected to optimize the types and quality of information obtained. Although population densities of *T. semipenetrans* are seasonal, propagule densities of *Phytophthora nicotianae* (a fungal root pathogen of citrus) isolated from soil are much less so. Samples acquired in spring or autumn, when nematode densities are highest, will generally provide the most precise density estimates for both organisms, since variance to mean ratios are inversely proportional to the mean (Taylor, 1961). However, serological detection of the fungus in root samples is most accurate during the warm summer months, when low nematode population densities reduce the precision with which they can be measured.

INTEGRATED PEST MANAGEMENT STRATEGIES

Strategies used to manage nematode populations derive from features of nematode biology, host parasite relationships, or available management tactics. Some of the common determinants of management strategies include the host range of nematode species (suitable rotation crop availability), survival mechanisms (e.g., host root diffusates are needed to stimulate hatch of some cyst nematodes), and the nature of biocide activity (fewer tactics are required with the use of some broad spectrum fumigants).

The concepts of *key pests* and *limiting factors* are often critical to successful strategy development (Thomason & Caswell, 1987). Management of *M. incognita* in cotton (*Gossypium hirsutum* L.) or *T. semipenetrans* in citrus does not enhance yield, if the wilt fungus *Verticillium alboatrum* is present in cotton or if citrus trees are stressed by poor water drainage. Indeed, for highly refined parasites such as *T. semipenetrans*, yield responses to nematode management occur most often in healthy-appearing, high-yielding orchards (Duncan, 1995; Thomason & Caswell, 1987). In many commonly encountered cases such as these, serious nematode pathogens are neither the key pests nor limiting factors to yield; correction of key deficiencies must occur before nematode management can be profitable.

Disease complexes and other interacting stress factors require specific management strategies. *Meloidogyne incognita* and other nematodes predispose resistant cultivars of cotton to infection by *Fusarium oxysporum*, resulting in synergistic damage to the plant (Powell, 1971). Therefore, the cotton-nematode damage function is essentially different depending on whether the fungus is present in nematode infested fields (Duncan & Ferris, 1983b). Salinity increases infection of susceptible and resistant citrus rootstocks by *T. semipenetrans* (Mashela et al., 1992). Thus, the use of resistant rootstocks in areas of high salinity may be questionable, considering the increased selection pressure against resistance that is likely to occur in the nematode gene pool.

Disease severity is often related to *plant phenology*. A common laboratory training exercise involves inoculating tomato (*Lycopersicon esculentum* Miller) seedlings of different ages with root-knot nematodes. Students observe that growth reduction of older seedlings is far less severe than that of younger plants. Numerous management tactics, such as manipulation of planting date or narrow-band soil fumigation, are used to delay nematode attack of young seedlings.

Some nematode management strategies are based on classic epidemiological principles from plant pathology. Exponential and logistic models of population growth (McSorley, 1998, see Chapter 6) also have been used to study the development of *monocyclic* and *polycyclic* epidemics in the field (Vanderplank, 1963). *Monocyclic* diseases develop slowly. Crops are infected mainly by organisms present in soil at planting. Disease progress curves generally appear linear during the growth phase of these diseases. Epidemics in which increasing numbers of plants are infected by successive generations of disease organisms during the cropping season are referred to as *polycyclic*. Sigmoidal curves, consisting of exponential, logistic and decay phases, describe the progress of polycyclic diseases. The basic parameters governing either type of disease process are initial inoculum density (P_i) and the disease growth rate (r).

Although nematodes often produce polycyclic population growth during a cropping cycle (Rodriguez-Kabana et al., 1986), the epidemiology of nematode diseases (the rate at which plants in a field become infected) is generally best described as monocyclic. Reproductive rates and motility are low for most nematode species compared to other pest and disease organisms. Therefore, the pattern of diseased plants in a field, throughout the cropping season, closely approximates the spatial pattern of the nematode P_i. However, when considered over several years (perennial hosts) or cropping cycles (annual hosts), nematode epidemics are often *polyetic*—requiring several years or cropping seasons for the rate of disease progress (sometimes exponential) to become apparent (Zadoks & Schein, 1979).

The simple growth models describing different types of epidemics can be used to demonstrate fundamental IPM principles such as the relative importance to disease progress of P_i vs. r, or general effects of management timing (e.g., dur-

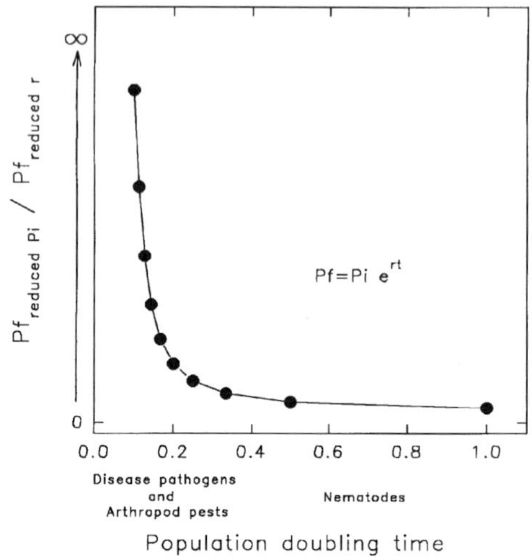

Fig. 13–4. Relative importance of r vs. P_i on population growth for organisms with different rates of increase. Disease doubling time is given as a proportion of the time available for uninterrupted population growth. The polycyclic disease model was solved twice for each doubling time, using the appropriate value for r and $t = 1.0$. The first solution reduced r to increase doubling time by two-fold. The second solution reduced P_i by 50%.

ing the exponential, logistic or decay portions of the disease cycle) on disease development. The first of the foregoing questions is an important element in nematode IPM, because with available methods (e.g., crop rotation, crop destruction, clean fallow, trap crops and some chemicals), it is often more practical to reduce P_i than r. Because most nematode-induced diseases are well described as monocyclic, the incidence of diseased plants is linear in time (a product of r and P_i) and proportional reduction of either P_i or r should have equivalent effects on the spread of disease in a single season.

It also is appropriate to consider the polycyclic nature of nematode population growth, as opposed to disease incidence. Population growth on individual plants determines the level of plant damage and the residual levels of nematodes available to damage subsequent crops. Polycyclic processes generally respond more to management that reduces r, rather than P_i, but this is highly dependent on the relationship of the population doubling time to the length of time available for uninterrupted population growth (Fig. 13–4). As the doubling time approaches the length of time available for population growth, the value of reducing P_i increases. For nematodes in annual crops, the time-frame available for growth often corresponds to the cropping season, after which populations may decline as a result of overwintering, or practices such as crop rotation, or fallow. Growth of nematode populations in perennial crops also is usually seasonal. Thus, reducing either P_i or r is generally successful in reducing nematode population growth, as well as the incidence of nematode-induced disease in a field.

Management timing can affect the relative advantage of reducing either P_i or r, by reducing the length of the population growth phase affected by a man-

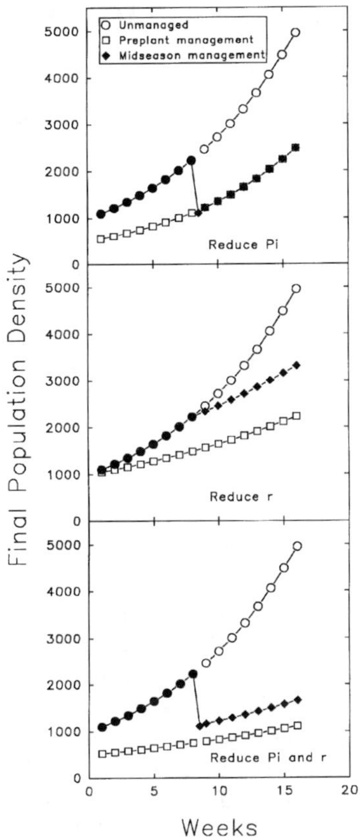

Fig. 13–5. Effects of management timing on pest population development for tactics that reduce P_i, r, or both in a proportional manner. Predictions from the exponential growth model, $P_f = P_i\, e^{rt}$.

agement tactic. For example, in perennial crops it is important to know when management practices should occur for best effect. Again, a simple exponential growth model can illustrate in a general way the importance of timing during the nematode population growth cycle for various types of management practices (Fig. 13–5). We might conclude from Fig. 13–5 that, all other things equal, management timing is more important when management affects r than P_i. However, from empirical data we know that management efficacy is governed by numerous additional fixed and stochastic variables. Nematode population growth rates, the fate of chemicals applied to soil, expression of resistance genes, as well as most other tactics, are affected by interactions between precipitation (Duncan, 1995), temperature (Omwega et al., 1990), microorganisms (Powell, 1971), root growth (Duncan et al., 1993) and numerous other factors that change during the time available for management. Again, more complex simulation models, incorporating these interacting fixed and stochastic variables, are needed to uncover the variety of expected responses to management and the probabilities of their occurrence (Ferris, 1976; McSorley, 1992).

INTEGRATED PEST MANAGEMENT TACTICS

A variety of tactics are available to manage plant-parasitic nematodes. Crop economics and knowledge of the biology and ecology of the crops, nematodes and other interacting organisms provides the basis to consider the use of a tactic. As noted previously, enhancing agricultural sustainability may require that growers select several management tactics to incrementally achieve their strategic aims (Roberts, 1993). It also should be apparent from the following examples of multiple-tactic management strategies that the complexity of nematode management is generally inversely related to crop value. Growers of low-value crops traditionally use a combination of methods to reduce nematodes below damaging levels, due to the high cost of chemical control. Conversely, in high-value commodity markets, growers are less likely to use management tactics that do not permit monoculture of the preferred crop. In this section we review some of the advantages and limitations of the most commonly used tactics to manage nematodes.

Sanitation

The major means of long-range movement of nematodes is through agricultural operations and commerce (McNamara, 1995). Thus, crop sanitation programs can be highly profitable endeavors (Maas, 1987). When used as part of an IPM system, sanitation may obviate the need for further management tactics in a substantial portion of an industry. Sanitation can be effectively utilized in everyday farming operations and in commerce involving plant and soil material.

Quarantine organizations target specific nematode pests and operate at regional, national and state levels. They generally function in three stages: (i) determination of key species for which introduction could be a major problem, (ii) assessment of the risk of introduction and the ability to establish in the new environment, and (iii) risk management (McNamara, 1995). Nematodes may be exogenous to the quarantine area or they may be present, but not widespread. A short list of species currently of most concern in North America include *Radopholus citrophilus, Globodera rostochiensis, G. pallida, Ditylenchus dipsaci, Heterodera glycines, H. trifolii, H. goettingiana, H. zeae,* and *Meloidogyne chitwoodi* (Lehman, 1995). In Europe, the majority of those species also are of quarantine concern as are species such as *Bursaphelenchus xylophilus, Naccobus abberans* and *Aphelenchoides besseyi* (McNamara, 1995). Spread of these and other nematodes is reduced or prevented through a combination of quarantine inspections and regulations governing the propagation and movement of commercial plant material (Maas, 1987).

Quarantine programs have proven to be immensely valuable to growers and to society. Commercial citrus nurseries in Florida have been legally required for the past 40 yr to sell only planting stock that is certified free of certain nematode species. It was estimated that the additional on-tree fruit value in a single year (1994) from restricting the spread of one of the nematodes, *Tylenchulus semipenetrans*, was $32.5 million, at a cost of $70 thousand (Lehman, 1995). This several hundredfold return on investment in the certification program does not take into account the additional economic and environmental costs that would have occurred for nematode management.

Common-sense sanitary practices also are useful at the farm level. Equipment sanitation (Esser, 1984), water filtration (Cohn, 1976), and paring or chemotherapy of planting material are some of the sanitation methods used in various industries (Maas, 1987). Roguing symptomatic plants can reduce the incidence of foliar nematodes such as *Aphelenchoides ritzemabosi* in ornamental nurseries (Brown, 1987). With rising interest in the use of municipal waste to amend soil, methods are being developed to sanitize composted materials of plant parasitic nematodes (O'Bannon et al., 1991).

Plant Resistance

The use of nematode-resistant varieties is the ideal management tactic to increase agricultural sustainability because it is safe, nonpolluting and requires no input by growers of additional resources. Indeed, plant resistance may be the only economic recourse to manage some nematodes in many low-value crops, and in less developed countries. Fortunately, the tactical use of host-plant resistance should rapidly increase, given the rate at which molecular methods are advancing our knowledge of the host parasite interaction and our ability to identify and manipulate genetic material (Opperman & Conkling, see Chapter 12).

Commercially available, nematode-resistant crop varieties are effectively used for nematode management in a large number of crops (Roberts, 1992). More than 130 varieties of soybean [*Glycine max* (L.) Merr.] with resistance to soybean cyst nematode have been developed and released in the USA (Anand et al., 1995b). Resistance to *Meloidogyne* spp. and *Tylenchulus semipenetrans* is available in the genera Prunus (e.g., 'Nemaguard') and Citrus (e.g. 'Swingle'), respectively. There are however, significant limitations to the availability of resistant varieties for many important nematode-crop combinations. For example, no commercially suitable cotton varieties are currently available with resistance to *Rotylenchulus reniformis* (Lawrence & McLean, 1994), or *M. incognita* (Jeffers & Roberts, 1993). No commercial peanut varieties resistant or tolerant to *M. arenaria* are available (Rodriguez-Kabana et al., 1989). No varieties resistant to *Meloidogyne* spp. are commercially available for cucurbits (squash, *Cucurbita* spp.; watermelon, *Citrullus* spp.; cucumber, *Cucumis* spp.), eggplant (*Solanum melongena* L.), lettuce (*Lactuca sativa* L.), or many other vegetable crops (Roberts, 1992).

Host plant resistance is commonly integrated with other nematode management tactics. Although the use of nematode-resistant varieties may outperform nematicide usage in nematode-susceptible monoculture such as tomato and soybean (Kinloch, 1983), the use of nematicides in combination with plant resistance can be very effective for heavily infested fields or when concomitant infestations of key nematodes occur. In the British potato industry, the use of plant resistance, crop rotation, and nematicides effectively control the potato cyst nematodes, *Globodera pallida* and *G. rostochiensis* (Roberts, 1993). Resistant varieties also are used in conjunction with tactics such as fallowing, tillage, addition of soil amendments, solarization, and/or use of nematode-tolerant crop varieties (Cook & Evans, 1987; Roberts, 1993; Schmitt & Imbriani, 1987).

Several problems are associated with effective implementation of host-plant resistance to nematodes. High soil temperatures can break resistance to *Meloidogyne* spp. and other nematodes in such crops as tomato, soybean, snapbean (*Phaseolus vulgaris* L.), alfalfa (*Medicago sativa* L.), and sweetpotato [*Ipomoea batatas* (L.) Lam.] (Omwega et al., 1990). Lack of multiple genes to manage complexes of key nematodes and pests may require use of additional tactics (Roberts, 1993). In annual crops, such as potato, it is often necessary to alternate the use of resistant and susceptible cultivars to avoid population shifts favoring resistance-breaking races (Anand et al., 1995a). Recently, blends of resistant and susceptible soybean cultivars were used to manage *H. glycines* while effectively reducing selection pressure on the population (Anand et al., 1995b). Use of plant resistance requires the ability to transfer resistance genes into horticulturally acceptable varieties (Cook & Evans, 1987). Some crop production systems, such as cotton, operate under strict market regulations to preserve uniform, high product quality standards for a region (Goodell, 1993). Resistant cultivars must conform to the quality standards to be useful in such systems. For this and other reasons, extensive nematode-resistant host germplasm has been identified, but not developed.

Crop Rotation

The use of crop rotation to minimize crop losses due to nematodes has been effectively used for centuries. The literature is replete with examples of nonhost crops used in rotation sequences to reduce nematode populations. For example, in the San Joaquin Valley of California, rotations using resistant cowpea [*Vigna unguiculata* (L.) Walp.], processing tomato, nonhost winter grains and alfalfa are used to manage *M. incognita* on cotton (Goodell, 1993). In the southeastern USA, *M. incognita* on cotton can be managed by rotating grain sorghum [*Sorghum bicolor* (L.) Moench] hybrids and resistant soybean (Kirkpatrick et al., 1995).

To be successful, there can be no ambiguity in the nematode host status of the rotation crop. This often requires careful assessment of the nematode community structure, because a major constraint to the success of crop rotation is the presence of more than one key nematode species in a field, particularly those with wide host ranges (Riggs, 1995). Use of crop rotation also is often constrained by the lack of crop or animal managerial skills, available land, specialized equipment to grow and harvest the crop, proximity of processing facilities or markets, or the scarcity of production-related information (Noling & Becker, 1994). Cotton is commonly used as a rotation crop to reduce *M. arenaria* in peanut (Rodriguez-Kabana et al., 1992). However, in some regions, cotton is not extensively used because many peanut farmers do not have the specialized equipment required for cotton production (Rodriguez-Kabana et al., 1989).

Several forages and pasture grasses have been identified as attractive rotation crops for producers with cattle operations (Rodriguez-Kabana et al., 1989). Generally, the inclusion of a forage crop in a rotation requires farm system diversity to include livestock. A current interest in forage legumes, like velvetbean

[*Mucuna pruriens* (L.) DC. var. *utilis* (Wallich ex Wright) Baker ex Burck], is related to their nitrogen-fixing potential, as well as their poor or nonhost status to root-knot and cyst nematodes, similar to that of many grasses. In some regions of the USA, livestock ranchers have initiated new short-term land leasing-resource management programs with vegetable growers. Ranchers lease pasture-land to growers for 2 to 3 yr and rotate back to bahiagrass (*Paspalum notatum* Flugge) sod for 6 to 8 yr (Coleman, 1996). Ranchers benefit from land leveling activities, increased productivity from residual nutrients in soil, and installation of irrigation wells. Growers benefit by enhanced crop production as a result of pest and disease suppression.

The influence of environment in dictating the success or failure of crop rotation cannot be overstated. In regions with hot, humid summers, the use of nematode-resistant rotation crops can be limited by stand establishment problems related to poor geographic adaptation of the plant to climatic conditions (McSorley & Gallaher, 1992). In Alabama, profitable production of corn (*Zea mays* L.) as a rotation crop with peanut is only possible in irrigated fields or in coastal areas where typical rainfall patterns are adequate for crop production (Rodriguez-Kabana et al., 1991). Rainfall patterns that support intense weed and disease pressures in subtropical areas mitigate the value of some rotation crops (Schroeder et al., 1993). During most winters, root-knot nematode populations decline under winter grains; however, the combination of a mild winter and a susceptible cultivar can increase populations high enough to damage the subsequent crop (Minton & Parker, 1987).

Crop rotations often do not reduce nematode population densities below economically damaging levels after a single nonhost crop (Johnson & Motsinger, 1990). Rotations of up to 7 yr are necessary to reduce densities of *H. schactii* below economic threshold levels in California. However, the necessity to extend the rotation beyond a single crop is often an economic constraint that causes growers to rely on chemical options (Goodell, 1993).

Cover Crops

Poor or nonhost cover crops, which are not grown for a harvestable product but turned into the soil, are used in crop rotation sequences for nematode control. To be effective, stands of cover crops should be established quickly and undesirable weeds, which can serve as alternative hosts, must be controlled. It also is desirable if cover crops serve to prevent erosion and otherwise improve the soil. Careful choice of cover crop is important in order to recover the cost of growing a nonmarketable crop through gains in pest management and soil quality.

The modes of action by which cover crops enhance yields of subsequent crops are complex. Poor or nonhost cover crops that do not support nematode reproduction during the growth of the crop do not always significantly reduce existing nematode population densities (Caswell-Chen & Goodell, 1992). In such cases, yield increases in successive crops are likely the result of improvements to soil fertility and water relations. However, substantial nematode population reduction can be achieved using crops that contain nematicidal chemicals or that act as trap crops which are infected by nematodes, but on which reproduction does not occur.

Sesbania rostrata Bremek & Oberm and *Aeschynomene afraspera* J. Leonard, used as green manure crops, reduced *Hirschmanniella mucronata* and *H. oryzae* and increased the yield of subsequent crops of irrigated rice (*Oryza sativa* L.) (Prot et al., 1992). Both plants appeared to act as trap crops since the nematodes entered, but did not exit, the roots. Rapeseed [*Brassica napus* (L.) var. *napus*] contains glucosinolate compounds, which degrade to form fungicidal, bactericidal, and nematicidal isothiocyanates (Gamliel & Stapleton, 1993). However, the results of using rapeseed as a cover crop to manage *Meloidogyne spp.* have been variable (Johnson et al., 1992; Mojtahedi et al., 1993). In Florida, the summer cover crops, sorghum, sudangrass [*Sorghum* × *drummondii* (Steudel) Millsp. & Chase], American jointvetch (*Aeschynomene americana* L.), or hairy indigo (*Indigofera hirsuta* L.) may limit damage due to *Meloidogyne* spp. (McSorley et al., 1994). Hairy indigo and American jointvetch are two nitrogen-fixing cover crops adaptable for managing sting (*Belonolaimus longicaudatus*) or root-knot nematodes (Rhoades, 1984). Sorghum has been used extensively in Florida for restoring large amounts of soil organic matter to soil, but while it is a poor host for *Meloidogyne* spp., it is a good host for the sting nematode, *B. longicaudatus* (Overman & Martin, 1978). Most small grains commonly used as winter cover crops in the southeastern USA can support limited and sometimes significant reproduction of root-knot nematode under suitable environmental conditions (Minton & Parker, 1987). Likewise, cover crop rotations with some pasture land grasses have significantly reduced, but not eliminated, root-knot nematodes (Caswell-Chen & Goodell, 1992; Kinloch, 1973). Prompt incorporation of some green-manure cover crops before flowering and seed production is necessary to avoid introduced weed problems or crop-seed contamination (Mojtahedi et al., 1993).

Planting Date

The use of planting date is a tactic to limit early season nematode reproduction and subsequent crop damage for a number of crops. Nematode activity thresholds have been established to use planting date for nematode management in crops such as wheat (*Triticum aestivum* L.) and other small grains (Johnson & Motsinger, 1990). Jeffers and Roberts (1993) showed that delayed planting of cotton could be an important tactic in a strategy to manage the disease complex involving *Fusarium oxysporum* and *M. incognita.* Even though temperatures were not low enough to restrict root penetration and subsequent development of *M. incognita,* delayed planting reduced wilt incidence and severity.

The effect of planting date on final population densities of *H. glycines* varies considerably among locations as well as from year to year (Koening et al., 1993). Risks are likely to be minimized in northern latitudes where the occurrence of threshold activity temperatures are more predictable, based on more consistent seasonal temperatures. Unfortunately, growers in northern climates may have little flexibility in planting dates due to shorter growing seasons.

Seasonal demand for many high-value agricultural products often dictates planting and harvesting schedules by increasing the value of early harvested crops (Van Sickle et al., 1995). Therefore, adjustment of planting date appears to be most effectively used in agronomic crops where commodity price market fluctuations are generally less severe.

The maturity groupings of soybean cultivars affect nematode population densities by defining the cropping length and the time available for nematode reproduction. Early maturing cultivars can restrict final nematode densities, thus increasing yields in subsequent crops (Koenning et al., 1993). The modest, but incremental, reduction in nematode populations that can be achieved with this tactic provides an opportunity to combine the use of early maturity group with other tactics (e.g., rotations with nonhost and resistant cultivars) to achieve nematode control.

Fallowing and Conservation Tillage

In some locations, clean fallow can be used as a simple, cost-effective nematode management tool, particularly when the interim period between successive crops is reasonably short and precludes the use of other tactics. Without food, rates of nematode survival depend on prevailing environmental conditions (primarily temperature and moisture) and physical processes such as tillage (Johnson & Motsinger, 1990). Clean fallow is not successful in cases where nematodes hatch in response to host root exudates, in dry soil where nematodes are well adapted for anhydrobiosis, or for nematodes that survive on lipid reserves over long periods (Roberts et al., 1981). For polyphagous nematodes, fallowing often is a viable tactic only if weed-free conditions are maintained during the fallow period. The use of herbicides can result in slower rates of population decline and may not be as effective as clean fallow practices utilizing tillage (Overman et al., 1970).

In irrigated areas of the arid western USA, a wet fallow system is used to manage nematodes (Goodell & Ferris, 1989). Irrigation water is applied to fallow soils to stimulate larval activity, contributing to the starvation death of nematodes in the absence of a host. In the northwestern USA, summer fallow is used to reduce damage by the cereal cyst nematode (*Heterodera avenae*), and by fungal pathogens of nonirrigated wheat (Smiley et al., 1994). The efficacy of fallow is enhanced when coupled with other cultural practices, such as early crop destruction.

The unfavorable effects of fallowing on soil organic matter, soil structure and beneficial soil biota also must be considered (Duncan, 1986). Often, fallow cannot be implemented due to potentially serious problems of soil erosion. However, where appropriate, clean fallowing can be as cost-effective for nematode management as other tactics, including crop rotation, cover cropping or use of nematicide (Duncan, 1986; Johnson et al., 1992; Kinloch & Dunavin, 1993).

The effects of conservation tillage practices on nematode survival are difficult to predict (Minton, 1986), and studies report conflicting results. The severity of root damage to wheat caused by *H. avenae* was greater in minimum till than plowed soil, an effect opposite to that reported from Australia (Smily et al.,

1994). Generally, conservation-tillage systems negate the benefits of discing or plowing up root debris to the surface where harsh environmental conditions reduce nematode survival. However, crop residue left intact in the soil may provide substrates needed by nematode antagonists. Stubble also increases water infiltration and reduces soil temperature, thereby affecting egg hatch or nematode mobility.

Soil Amendments

Various organic and inorganic soil amendments have been added to soil and investigated for their suppressive effects on plant-parasitic nematodes (Muller & Gooch, 1982). Crop-yield responses in these amended soils, particularly for amendments high in N, are frequently attributed to improved soil fertility and water-holding capacity as well as suppression of plant-parasitic nematodes. Although the mode of action is not demonstrated in many studies, most explanations suggest that amendments or their degradation products are either directly toxic to nematodes, or serve to enhance the proliferation of soilborne antagonists that kill or weaken nematodes.

Chitin and inorganic fertilizers that release ammoniacal nitrogen into soil have been shown to suppress nematode populations due to plasmolysing effects and to selective proliferation of microbial antagonists (Rodriguez-Kabana, 1986). Yield enhancement and nematode control have been achieved in some, but not all cases with these materials (Spiegel et al., 1989; Rodriguez-Kabana, 1986).

The addition to soil of composted chicken litter and other animal manures stimulated plant growth and was suppressive to *Meloidogyne* spp. populations in soil (Kaplan & Noe, 1993; Natl. Res. Counc., 1989). However, tomato root-gall severity caused by *M. incognita* and incidence of Fusarium crown and root rot caused by *Fusarium oxysporum* f. sp. *radicoli* were significantly increased following soil applications of solid municipal composted waste (Noling, 1995, unpublished data). Marine kelp derivatives are mildly nematicidal, while other algal extracts are totally ineffective against *Meloidogyne* spp. (Paracer et al., 1987).

At present, the use of soil amendments for nematode management has not proceeded beyond the exploratory research phase. These strategies often are described as potential crop and pest management alternatives to chemical methods, particularly suitable for organic, low-input or sustainable agricultural systems (Natl. Res. Counc., 1989). High rate recommendations, high cost, and inconsistent effects on nematodes are factors that may limit the utility of many of these materials to homeowner or small farm use (Rodriguez-Kabana, 1986). New federally mandated guidelines (U.S.) currently under development regulating the contamination of agricultural food products with biological mycotoxins also may preclude or limit the utility of some soil amendments.

Physical Methods of Nematode Control

The application of various forms of energy, primarily thermal and electromagnetic, have been evaluated to control nematodes in soil and plant materials.

Several excellent reviews of physical methods of nematode control have been published (e.g., Maas, 1987). Because of the relatively low thermal sensitivities of nematodes and other soilborne pests, physical approaches that utilize heat have generally proved to be the most effective and widely used in some sectors of agriculture.

Steam is used in commercial greenhouse, shadecloth, and nursery operations as an effective broad spectrum pasteurization treatment for soilborne pest and disease control, but has not been widely adopted in production agriculture. Hot water dips are frequently used in sanitation programs to disinfest plant material of nematodes and have been highly effective in reducing regional and international spread of nematode-infested plant materials when used in quarantine and regulatory programs (Mass, 1987). Recent attempts to use hot water to treat soil in open fields have shown some promise, if the method can be used as a synergist in combination with other tactics such as solarization, or with low dosages of nematicides (Saleh et al., 1988; Noling, 1994).

Soil solarization has been successfully used in some parts of the world to achieve noncropped soil temperatures lethal to many soilborne pests and can sometimes be used to reduce soilborne pathogens in established perennial fruit and nut crops (Chellemi et al., 1993). Efficacy of soil solarization against nematodes has ranged from excellent to incomplete (Stapleton et al., 1987). Despite high soil temperatures that can be achieved, lack of efficacy against some pests may be related to depth limitations or to the heat resistance of the pest organism. In these cases, the combined use of soil solarization with solar heated water (Saleh et al., 1988), pesticides (Stapleton et al., 1987), organic amendments or biological antagonists (Gamliel & Stapleton, 1993b), may improve efficacy and eventually foster an increased use of solarization.

Biological Control

The primary approaches to biological control of nematodes have been augmentation of natural control and inundation with biological organisms. To date, only the former approach has met with consistent, albeit very limited, success. A growing body of evidence suggests that monoculture or long rotations of crops susceptible to a given nematode species can eventually establish a microflora that regulates some plant-parasitic nematodes at levels below economic thresholds for chemical control (McSorley & Duncan, 1995). The best-known example of this is widespread control of *Heterodera avaenae* in cereal monocultures in Europe by the fungi *Nematophthora gynophila* and *Verticillium chlamydosporium* (Kerry, 1995). Similar observations in the southeastern USA suggest the potential to manage *Meloidogyne spp.* with the bacterium *Pasturia penetrans* by selective cropping to maintain a continuous prey population density (McSorley & Duncan, 1995).

Although some commercial products are available, there are currently no systems in which soil-borne nematodes are managed by inundation with biological control agents. Because nematode-suppressive soils often result from complex interactions between the rhizosphere environment and combinations of antagonistic agents, it is unclear how niche requirements can be created for inoculated organisms, given the extreme edaphic and biological variability of agricultural soils (Stirling, 1991). Commercialization of effective biological control products also has been constrained by the high cost of producing large quantities of the most promising fastidious parasites of nematodes.

In the near term, exploitation of biological control probably will develop in an IPM context (Kerry, 1995). The use of nematode-tolerant crop varieties in cropping systems can help in exploiting natural control of nematodes by avoiding excessive crop loss during the early epidemic phase in which nematode antagonists require continuous prey availability. This will allow development of adequate population densities to provide control (McSorley & Duncan, 1995). The use of soil amendments and solarization has usefully augmented biological control of nematodes and may be useful in small farms and in the more stable environments created by horticultural crops. The use of seed treatments with organisms has been suggested as an IPM strategy to protect seedlings during the most susceptible period of growth (Oostendorp & Sikora, 1989). Low rates of nematicides may be necessary to complement the effects of biological control agents in reducing nematode populations sufficiently (Stirling, 1991). Selection of hosts that permit only limited development of *Meloidogyne* spp. can increase biological control, because heavy infection by the nematode results in larger root galls containing many nematodes deep within the tissue where they are inaccessible to parasites and predators (Kerry, 1995). Epidemiological models of epizootics are being used to elucidate environmental requirements for effective biological control (Jaffee et al., 1992). Finally, restrictions on chemical management have the effect of promoting the development of biological control by increasing its commercial attractiveness (Becker & Schwinn, 1993).

Chemical Control

Prior to 1940, soilborne pest and disease management was achieved almost exclusively through nonchemical approaches, primarily crop rotation (Bewick, 1989). The subsequent development of effective soil fumigants and narrower-spectrum nematicides fostered awareness of plant-parasitic nematodes by providing evidence of their involvement in crop loss either directly, as virus vectors, or as predisposing agents in disease complexes (Powell, 1971). Growers responded, replacing former management tactics with chemical usage in order to use the same fields to produce a preferred crop each year, taking advantage of financial investments in property, land improvements, and site location relative to climate and marketing opportunities (Bewick, 1989; Heald, 1987; Johnson & Feldmesser, 1987; Overman & Martin, 1978). Moreover, multiple weed, disease, arthropod and nematode pest problems often were managed with a single operation. Family owned, diversified farms were increasingly replaced by highly specialized, corporate-owned businesses whose large capital investments in land, labor and equipment were supported by the profitable use of chemical pest control. Against this background, multiple-tactic IPM strategies will gain grower acceptance not by merely minimizing the undesirable side effects of chemical usage, but by addressing, as nearly as possible, the broad range of economic benefits obtained through chemical management.

Chemical nematicides can be classed by volatility. Fumigant compounds (primarily halogenated hydrocarbons or isothiocyanate precursors) generally provide superior broad-spectrum control of soil pests and disease pathogens. Nonfumigant compounds (primarily organophosphate or carbamate inhibitors of acetylcholine) have a narrower spectrum of target organisms and either kill or disorient nematode behavior by disrupting the nervous system. The spectra and modes of activity, and the high phytotoxicity of most fumigants compared to nonfumigant nematicides, affect many of the strategic possibilities for using the two types of compounds (Johnson & Feldmesser, 1987; Perry & Clark, 1983).

Benefits of nematicides include not only higher yield, but more uniformity of crop quality and maturity, so that harvesting is completed in fewer pickings at lower total cost. The availability of the reliable, economical, broad-spectrum soil fumigant methyl bromide was a key factor in the development of many high-value, high-input, multiple cropping systems (Ferguson & Padula, 1994). The value of fumigant and nonfumigant compounds for nematode control and crop yield enhancement also has been demonstrated repeatedly in major agronomic and horticultural crops such as banana (*Musa* × *paradisiaca* L. var. *paradisiaca*), citrus, corn, cotton, grape (*Vitis vinifera* L.), peanut, pineapple [*Ananas comosus* (L.) Merr.], potato, soybean and tobacco.

Key factors affecting the efficacy of nematicides include timing (see IPM strategies), placement (Duncan, 1989), distribution in soil (Noling, 1991), retention in soil of volatile compounds (Noling, 1991), leaching (Thomason, 1987), and accelerated microbial degradation with repeated application (Davis et al., 1993). Research and grower efforts to optimize practices affecting those factors are important not only to achieve high efficacy, but to reduce the risks associated with nematicide usage (Thomason, 1987). Research on nematicide efficacy should focus on high-value crops, because management intensity is directly correlated with risk of investment (Noling & Becker, 1994). Growers are reluctant to use other, less lucrative, management tactics when producing these high-input commodities where resources are available for innovation. For example, the use-efficiency of nematicides might be increased in sandy soils by using narrower, covered planting beds and closer drip emitter spacings, and by placing plants closer to drip emitters. Strategic considerations of such a system are the poor lateral dispersion of water soluble nematicides in these soils, the requirement of many plants for protection from nematodes only during the early seedling stage, and broader spectrum pest control (weeds) from the bed-coverings.

Adoption of IPM methods can reduce the use of nematicides not only by increasing the use of other management tactics, but by providing rational guidelines for management decisions. In the absence of decision guidelines based on damage functions, economic thresholds and biomonitoring, growers of many crops are unwilling to face the risk of not using chemical treatments (UNEP, 1992). They have access to no remedial control methods if nematode problems develop in their crop. There is no possibility of treating higher-than-expected nematode densities between crops in some double-cropping systems, where features such as bedding and tarping but no drip-irrigation exist (Noling, 1991). Banks often require soil treatment as a form of insurance for farm loans. However, when sound sampling and decision-making protocols became available, use of nonfumigant nematicides in peanut in Virginia declined by 35% (Phipps, 1993). A similar program resulted in an 80% reduction in methyl bromide use in the Netherlands for control of potato-cyst nematodes (Duncan, 1991).

CURRENT AND FUTURE CONSIDERATIONS FOR INTEGRATED PEST MANAGEMENT

Current Integrated Pest Management Systems

Most of the principles of nematode IPM are well-illustrated by current practices in the soybean producing areas of the southeastern USA (Table 13–1). In this region, soybean seed production can be significantly limited by *H. glycines*, *M. incognita*, *M. arenaria*, and *H. columbus*. Monitoring of nematode community structure is commonly practiced and damage functions are available, relating soybean growth and seed yield to preplant population density of many damaging species (e.g., Kinloch, 1986; Noe, 1993; Schmitt et al., 1987). Damage functions, reproduction functions and economic thresholds have been described under a variety of edaphic and climatic conditions for soybean and other crops used for rotation. Cropping-system models of the influence of tactics such as rotation, resistance and nematicides on *H. glycines* and *M. incognita* are developed to some extent (Kinloch, 1986; Noe et al., 1991). Viable, regionalized cropping systems have been developed to manage combinations of these nematodes with strategies designed to reduce P_i in subsequent, susceptible crops. A number of state and private organizations are available to aid growers in using this information within the context of IPM (Barker & Imbriani, 1984).

Management strategies depend largely on whether one or more key nematode pests are present in a field. Management is based primarily on rotation of soybean varieties and nonhost crops, when only *H. glycines* is present (Anand et al., 1995a). In North Carolina, 3-yr rotations of a nonhost crop, a resistant and a susceptible variety of soybean have been used for nearly two decades to manage *H. glycines* (Koenning et al., 1993). Four-year rotations in which rice is alternated with susceptible and resistant soybean has effectively controlled *H. glycines* in areas where rice can be grown (Riggs, 1995). Rotations of resistant, tolerant, and susceptible soybean varieties are practiced to stabilize yields and minimize race specific shifts of *H. glycines* field populations. This combination of tactics also is sometimes necessary, because there are insufficient levels of varietal resistance to *H. glycines* to prevent yield losses on heavily infested soils (Kinloch & Dunavin, 1993). The narrow host range of *H. glycines* makes it possible to utilize a wide number of nonhost crops between crops of soybean, including corn, wheat, rye (*Secale cereale* L.), barley (*Hordeum vulgare* L.), grain sorghum, peanut and tobacco. In general, the yield of a susceptible soybean crop increases in proportion to the number of intervening nonhost crops.

The wider host range of *Meloidogyne* spp. reduces the number of useful rotation crops for soybean production. Tobacco is unsuitable when root-knot nematodes are present, as is peanut for managing populations of *M. arenaria*. Some soybean varieties are resistant to both *H. glycines* and *M. incognita*.

Table 13–1. Availability of IPM tactics and management decision models in selected agronomic crops (sources cited in text).

	IPM decisions		IPM tactics								
Crop nematode	Damage/ function	Cropping system models	Crop rotation	Resistant varieties	Tolerant varieties	Planting date	Maturity group	Crop destruction	Certified plant material	Chemical	Stress avoidance
					Soybean						
Heterodera glycines	X	X	X	X	X	X	X		X	X	X
Hoplolaimus columbus	X	X	X		X	X	X		X	X	X
Meloidogyne spp.	X	X	X	X	X	X	X		X	X	X
					Cotton						
Rotylenchulus reniformis	X		X					X		X	X
H. columbus	X	X	X		X			X		X	X
Belonolaimus longicaudatus	X		X					X		X	X
M. incognita	X		X	X				X		X	X
					Peanut						
M. arenaria	X		X							X	X
					Tomato						
Meloidogyne spp.	X		X	X	X		X	X	X	X	X
					Potato (*Solanum tuberosum* L.)						
Globodera spp.	X	X	X	X	X		X		X	X	X
M. hapla	X		X			X	X	X	X	X	X
Pratylenchus penetrans	X		X		X		X			X	X
M. chitwoodi	X		X				X		X	X	X
					Tobacco (*Nicotiana tabacum* L.)						
M. javanica	X		X	X				X	X	X	X
M. incognita	X		X	X				X	X	X	X
M. arenaria	X		X	X				X	X	X	X

Hoplolaimus columbus has such a wide host range that economic thresholds are available for nematicide use against this nematode in cotton when it is rotated with tolerant soybean varieties in Georgia (Noe, 1993). Planting date adjustments and/or use of early maturing soybean varieties also are sometimes used to minimize nematode reproduction, survivorship, and yield losses due to *H. glycines* (Pacumbaba & Tadesse, 1991) and *H. columbus* (Mueller & Sanders, 1987). Although not as well characterized, use of nonhost cover crops, no-till planting practices, and avoidance of moisture, nutritional, and other pest stresses are used by growers for nematode management. Except for management of *H. columbus*, nematicide use is generally uneconomical or of little utility, particularly as a sole tactic (Schmitt & Imbriani, 1987; Sasser & Uzzell, 1991).

Combinations of the previous tactics have been developed, suitable in terms of region and nematode community structure. For example, a commonly recommended system for growers in Arkansas (R. D. Riggs, 1995, personal communication) with fields infested by *H. glycines* and *M. incognita* consists of grain sorghum and winter wheat in the 1st yr. Winter wheat requires that soybean (resistant to both species) be planted late during the 2nd yr. The late planting may help to further reduce *H. glycines* (Koenning & Anand, 1991), although *M. incognita* may increase to higher levels during this time. Following resistant soybean in Year 2, either winter wheat, winter fallow, or a nonhost cover crop is recommended. A high-yielding tolerant or susceptible soybean cultivar is then planted in the final year of the rotation.

Growers who employ this recommended rotation sequence clearly pay a high cost for nematode management. During periods when soybean prices are favorable, the preferred soybean crop is grown only 2 yr out of 3. Moreover, resistant and tolerant varieties often yield less than the best susceptible varieties. It may be necessary to sample soil with higher intensity, to insure that nematode populations behave as expected (prior to planting the susceptible variety). Thus, it is not uncommon to find growers who grow resistant soybean in monoculture for as long as possible, trusting in plant breeders to keep ahead of the inevitable resistance-breaking nematode races created by the practice (Riggs, 1995). Similar economic reasons are often responsible for heavy use of chemical management in high-value cropping systems.

Table 13–1 lists several additional agronomic crops and a comparison of the IPM practices that can be used for nematode management. The degree to which IPM is practiced in these crops varies, generally in relation to crop value and public policy. Growers of fresh-market tomatoes have no reason to forgo the profits from growing tomato in monoculture, because the crop value supports the use of methyl bromide for broad spectrum pest control (UNEP, 1994). Growers of lower-value processing tomatoes often use tactics, such as rotation and resistance in addition to nematicides, due to high nematicide costs and local nematicide-use restrictions. Similarly, cotton prices usually support nematicide economic thresholds at nematode densities that are commonly encountered, whereas prices of soybean and peanut do not. The economics of nematicide usage in potato is similar to that of processing tomato. In Europe, however, local-use restrictions due to public concern about environmental safety have resulted in several highly developed IPM systems (McSorley & Phillips, 1993). Other factors influencing the implementation of the practices in Table 13–1 include regional market niches (e.g., suitable rotation crop availability) and climatic and edaphic factors affecting pests, management practices (e.g., suitable resistant varieties) and environmental risk (e.g., nematicide leachability in sandy soils).

Future Considerations for Integrated Pest Management

We have tried to emphasize in this chapter the need for broader public support to achieve greater sustainability of agricultural endeavors. Indeed, given the will and resources that non-agricultural segments of society can bring to bear, legislative and educational initiatives may be as important as agricultural research in promoting the development of IPM. Therefore, it is necessary to reflect beyond the traditional modes of research, development and extension, to understand how IPM is likely to develop in the future.

Integrated pest management will be increasingly affected by legislation concerned with safety and the environment. As concern about the sustainability of agriculture increases, more consideration may be given to distributing the attendant costs and economic risks of IPM more equally between growers and the public (Rajotte, 1993). Methods that have been suggested include: (i) provision of compensation programs as insurance against crop-failure due to IPM, (ii) subsidizing insurance programs for private sector IPM specialists, (iii) requiring grower compliance with IPM practices to qualify for other government aid or subsidy programs, (iv) greater subsidization of the cost of biomonitoring through state-operated laboratories, and (v) implementing new methods to classify chemicals to permit faster registration of low-risk, highly targeted compounds. Possible roles for education in furthering IPM include: (i) reversing the current trend of decreased support for agricultural extension, (ii) requiring certification and the attendant education for an increasing number of pest management functions, (iii) instituting university programs to supply personnel needed by private IPM companies, (iv) reinstituting the use of grower demonstration fields to highlight IPM in the way chemical management was once promoted, and (v) relying more heavily on experts in the social sciences to develop linkages between IPM experts, growers and the public.

New ways must be found to cultivate public awareness of the need to support not only the benefits, but the costs of IPM. Basic research to develop novel management tactics utilizes an increasing portion of reduced research budgets. Therefore, the informational research needs for IPM (e.g., development and validation of cropping systems and the need for multiple-year experiments to estimate variability of models) will be sustained only through a sound public-science partnership. For many citizens today, it does not seem inconsistent to favor low food prices, reduced use of pesticides and reduced public funding of seemingly esoteric programs in agricultural research and extension. Without a greater awareness of the complexity involved in managing agroecosystems and the vital role of agriculture in national economies, such attitudes are not surprising. Non-agriculturists would have a more realistic context in which to judge the considerable progress of IPM, if they were more aware of constraints that exist in the form of grower risk (lack of information needed for decisions) and the added costs associated with many IPM practices. For example, the cropping-systems model component of IPM, can be used to provide some of the information (e.g., Fig. 13–3) needed for public policy decisions on sustainable agriculture, by helping economists to project the relative use and cost of resource inputs, resultant crop yields and food prices for traditional and IPM systems.

Future research in IPM should focus on practicality as well as innovation. More attention must be paid to the factors that constrain the adoption of multiple-tactic nematode control. Quantifying the risk associated with management practices will better enable growers to make IPM decisions, but requires collecting numerous data in long-term experiments under a wide variety of conditions (Ferris, 1984b, McSorley & Duncan, 1995). The need for methods to better determine the levels and distribution of nematodes in fields has been discussed. Application of geostatistical methods and global positioning technology to develop precision management methods awaits the development of more efficient methods to collect and process soil samples. New molecular-based methods will provide answers to the problem of sampling (Powers & Harris, 1993), and help overcome taxonomic problems associated with species and race determination (Thomason & Caswell, 1987). The promise of molecular biology to expand the tactical use of novel forms of resistance is profound. (Gommers et al., 1995). Compared to entomology, nematology has much to discover about interfering with biochemical and physiological processes of nematodes such as pheremones, growth regulators, and host-parasite communication (Dussenberry, 1987). Finally, the development of efficient methods for growers to understand and use new information, in whatever media are most appropriate, will do much to determine the rate at which new IPM practices are adopted.

We did not contrast, in this chapter, differences in the requirements and prospects of nematode IPM in high- vs. low-input agricultural systems (Noe, 1988). Many of the IPM goals for high-input systems (e.g., increased genetic diversity, complex cropping systems) already exist in various forms in small-plot, subsistence agricultural systems. It has been suggested that high-input growers may have more to learn from the study of traditional agriculture than do small-plot farmers from the study of modern agriculture (Altieri, 1994). As efforts are made to address the serious need for increased food production in developing countries, many problems encountered in high-input systems can be avoided, if the necessary social and political conditions exist. In many countries, the pressures placed by growing populations on the amount and quality of arable land is already enormous. Clearly, on a worldwide basis, advances in research must be matched by education and commitment by many sectors of society in order to develop sustainable systems for nematode management.

REFERENCES

Allen, P.A., D. Van Dusen, J.L. Lundy, and S.R. Gliessman. 1991. Integrating social, environmental, and economic issues in sustainable agriculture. Am. J. Altern. Agric. 6:34–39.

Alston, D.G., D.P. Schmitt, J.R. Bradley, Jr., and H.D. Coble. 1993. Multiple pest interactions in soybean: Effects on *Heterodera glycines* egg populations and crop yield. J. Nematol. 25:42–49.

Altieri, M.A. 1994. Biodiversity and pest management in agroecosystems. Food Prod. Press. Binghamton, NY.

Anand, S.C., S.R. Koenning, and S.B. Sharma. 1995a. Effect of temporal deployment of different sources of resistance to soybean cyst nematode. J. Prod. Agric. 8:119–123.

Anand, S.C., S.R. Koenning, and S.B. Sharma. 1995b. Performance of blends of soybean cyst nematode resistant and susceptible cultivars. Crop Sci. 35:524–528.

Atkinson, G.F. 1889. Nematode root galls. Agric. Exp. Stn., Alabama Polytech. Inst., Bull. no. 9.

Avery, D.T. 1995. Saving the plant with pesticides and plastic: The environmental triumph of high-yield farming. Hudson Inst., Indianapolis, IN.

Barker, K.R. 1989. Yield relationships and population dynamics of *Meloidogyne* spp. on flue-cured tobacco. Suppl. J. Nematol. 21: 597–603.

Barker, K.R., and J.L. Imbriani. 1984. Nematode advisory programs—status and prospects. Plant Dis. 68:735–741.

Barker, K.R., and N.T. Powell. 1988. Influence of aldicarb on the growth and yield of tobacco. J. Nematol. 20:432–438.

Becker, J.O., and F.J. Schwinn. 1993. Control of soil-borne pathogens with living bacteria and fungi: Status and outlook. Pest. Sci. 37:355–363.

Belair, G., and G. Boivin. 1988. Spatial pattern and sequential sampling plan for *Meloidogyne hapla* in muck-grown carrots. Phytopathology 78:604–607.

Bewick, T.A. 1989. Use of soil sterilants in Florida vegetable production. Acta Hort. 25:61–72.

Bird, G.W. 1987. Role of nematology in integrated pest management programs. p. 114–121. *In* J.A. Veech and D.W. Dickson (ed.) Vistas on nematology. Soc. Nematol., Hyattsville, MD.

Bongers, T. 1990. The maturity index: An ecological measure of environmental disturbance based on nematode species composition. Oecologia 83:14–19.

Brown, R.H. 1987. Control strategies in low-value crops. p. 351–388. *In* R.H. Brown and B.R. Kerry (ed.) Principles and practice of nematode control in crops. Acad. Press, London.

Castro, C.E., H.E. McKinney, and S. Lux. 1991. Plant protection with inorganic ions. J. Nematol. 23:409–413.

Caswell-Chen, E.P., and P.B. Goodell. 1992. Potential cover crops for reduction of *Meloidogyne incognita* in California. J. Nematol. 24:585.

Chellemi, D.O., S.M. Olson, J.W. Scott, D.J. Mitchell, and R. McSorley. 1993. Reduction of phytoparasitic nematodes on tomato by soil solarization and genotype. Suppl. J. Nematol. 25:800–805.

Cohn, E. 1976. Report of investigations on nematodes of citrus and subtropical fruit crops in South Africa. Citrus Subtrop. Fruit Res. Inst., Nelspruit.

Coleman, R.G. 1996. Tomato producers decrease due to NAFTA & marketplace. Fla. Grow. Ranch. 89:36, 40.

Cook, R., and K. Evans. 1987. Resistance and tolerance. p. 179–231. *In* R.H. Brown and B.R. Kerry (ed.) Principles and practice of nematode control in crops. Acad. Press, Orlando, FL.

Davis, R.F., A.W. Johnson, and R.D. Wauchope. 1993. Accelerated degradation of fenamiphos and its metabolites in soil previously treated with fenamiphos. J. Nematol. 25:679–685.

Duncan, L.W. 1986. Effects of bare fallow on plant-parasitic nematodes in the Sahelian zone of Senegal. Revue. Nématol. 9:75–81

Duncan, L.W. 1989. Effect of fenamiphos placement on *Tylenchulus semipenetrans* and yield in a Florida citrus orchard. Suppl. J. Nematol. 21:703–706.

Duncan, L.W. 1991. Current options for nematode management. Annu. Rev. Phytopathol. 29:469–490.

Duncan, L. 1995. Environmental determinants of crop damage by *Tylenchulus semipenetrans* and other nematode parasites of citrus. p. 204–212. *In* Progr. Proc. Int. Congr. Trop. Nematol., Rio Quente, Brasil. 4–9 June 1995. Rio Quente, Brasil.

Duncan, L.W., and H. Ferris. 1983a. Validation of a model for prediction of host damage by two nematode species. J. Nematol. 15:227–234.

Duncan, L.W., and H. Ferris. 1983b. Effects of *Meloidogyne incognita* on cotton and cowpeas in rotation. p. 22–26. *In* Proc. Beltwide Cotton Prod. Res. Conf. Natl. Cotton Council, Memphis, TN.

Duncan, L.W., J.H. Graham, and L.W. Timmer. 1993. Seasonal patterns associated with *Tylenchulus semipenetrans* and *Phytophthora parasitica* in the citrus rhizosphere. Phytopathology 83:573–581.

Duncan, L.W., P. Mashela, J. Ferguson, J. Graham, M.M. Abou-Setta, and M.M. El-Morshedy. 1995. Estimating crop loss in orchards with patches of mature citrus trees infected by *Tylenchulus semipenetrans*. Nematropica 25:43–51.

Duncan, L.W., J.D. Toole, R.N. Inserra, W.S. Castle, and J.H. O'Bannon. 1994. Comparing two sampling methods to detect *Radopholus citrophilus* in Florida Citrus. Proc. Soil Crop Sci. Soc. Fla. 53:22–25.

Dusenberry, D.B. 1987. Prospects for exploring sensory stimuli in nematode control. p. 131–136. *In* J.A. Veech and D.W. Dickson (ed.) Vistas on nematology. Soc. Nematol., Hyattsville, MD.

Elliott, E.T., and D.C. Coleman. 1988. Let the soil work for us. Ecol. Bull. 39:23–32.

Esser, R.P. 1984. How nematodes enter and disperse in Florida nurseries via vehicles. Fla. Dep. Agric. Consum. Serv., Div. Plant Ind., Nematol. Circ. 109:1–2.

Esser, R.P., G.T. Smith, and J.H. O'Bannon. 1993. An eleven year phytoparasitic nematode survey of Florida citrus groves and their environs. Fla. Dep. Agric. Consum. Serv., Div. Plant Ind., Nematol. Bull. 15.

Ettema, C.H., and T. Bongers. 1993. Characterization of nematode colonization and succession in disturbed soil using the Maturity Index. Biol. Fert. Soils 16:79–85.

Ferguson, W., and A. Padula. 1994. Economic effects of banning methyl bromide for soil fumigation. USDA Econ. Res. Serv., Agric. Econ. Rep. 677.

Ferris, H. 1976. Development of a computer-simulation model for a plant-nematode system. J. Nematol. 8:255–263.

Ferris, H. 1978. Nematode economic thresholds: Derivation, requirements, and theoretical considerations. J. Nematol. 10:341–350.

Ferris, H. 1984a. Nematode damage functions: The problems of experimental and sampling error. J. Nematol. 16:1–9.

Ferris, H. 1984b. Probability range in damage predictions as related to sampling decisions. J. Nematol. 16:246–51.

Ferris, H. 1993. New frontiers in nematode ecology. J. Nematol. 25:374.

Ferris, H., D.A. Ball, L.W. Beem, and L.A. Gudmundson. 1986. Using nematode count data in crop management decisions. Calif. Agric. 40:12–14.

Ferris, H., and N. Greco. 1990. Management strategies for *Heterodera goettingiana* in a vegetable cropping system in Italy. Rev. Nématol. 15:25–33.

Ferris, H., T.A. Mullens, and K.E. Ford. 1990. Stability and characteristics of spatial description parameters for nematode populations. J. Nematol. 22:427–39.

Fortnum, B.A., and D.L. Karlen. 1985. Effect of tillage system and irrigation on population densities of plant nematodes in field corn. J. Nematol. 17:25–28.

Francl, L.J., and W.J. Kenworthy. 1989. Heterogeneity in population density of soybean cyst nematodes after evaluation of soybean germplasm. J. Nematol. 21:561.

Gallaher, R.N., and R. McSorley. 1993. Population densities of *Meloidogyne incognita* and other nematodes following seven cultivars of cowpea. Nematropica 23:21–26.

Gamliel, A., and J.J. Stapleton. 1993. Characterization of antifungal volatile compounds evolved from solarized soils amended with cabbage residuals. Phytopathology 83:899–905.

Geier, P.W. 1966. Management of insects pests. Annu. Rev. Entomol. 11:471–490.

Gliessman, S.R. 1995. Sustainable agriculture: An agroecological perspective. p. 45–57. *In* J. H. Andrews and I. C. Tommerup (ed.) Advances in plant pathology. Vol. 11. Acad. Press, San Diego, CA.

Gommers, F.J., F.A. van Engelen, A. Schouten, J. Roosien, J.M. de Boer, J.P.W.G. Stokkermans, A. Schots, W.J. Stiekema, and J. Bakker. 1995. Plantbodies: A new route towards resistance against plant parasitic nematodes. p. 237–243. *In* Progr. Proc. Int. Congr. Trop. Nematol., Rio Quente, Brazil. 4–9 June 1995. Rio Quente, Brazil.

Goodell, P.B. 1993. Integrated mange for cotton nematode in the irrigated west. p. 182–183. *In* D.J. Herber and D. A. Richter (ed.) Proc. Beltwide Cotton Prod. Res. Conf., New Orleans, LA. 10–14 Jan. 1994. Natl. Cotton Counc. Am., Memphis, TN.

Goodell, P.B., and H. Ferris. 1981. Sample optimization for five plant-parasitic nematodes in an alfalfa field. J. Nematol. 13:304–313.

Goodell, P.B., and H. Ferris. 1989. Influence of environmental factors on the hatch and survival of *Meloidogyne incognita*. J. Nematol. 21:328–334.

Hamblin, A. 1995. The concept of agricultural sustainability. p. 1–19. *In* J. H. Andrews and I. C. Tommerup (ed.) Advances in plant pathology. Vol. 11. Acad. Press, San Diego, CA.

Hawken, P. 1993. The ecology of commerce. Harper Colling, New York.

Heald, C.M. 1987. Classical nematode management practices. p. 100–104. *In* J.A. Veech and D.W. Dickson (ed.) Vistas on nematology. Soc. Nematol., Hyattsville, MD.

Hunt, H.W., D.C. Coleman, E.R. Ingham, R.E. Ingham, E.T. Elliott, J.C. Moore, S.L. Rose, C.P.P. Reid, and C.R. Morley. 1987. The detrital food web in a shortgrass prairie. Biol. Fert. Soils 3:57–68.

Hyman, B.C., J.J. Peloquin, and E.G. Platzer. 1990. Optimization of mitochondrial DNA based hybridisation assays to diagnosis in soil. J. Nematol. 22:273–278.

Jaffee, B., R. Phillips, A. Muldoon, and M. Mangel. 1992. Density-dependent host-pathogen dynamics in soil microcosms. Ecology 73:495–506.

Jeffers, D.P., and P.A. Roberts. 1993. Effect of planting date and host genotype on the root-knot nematode-Fusarium will disease complex of cotton. Phytopathology 83:645–654.

Jeger, M.J., and J.L. Starr. 1985. A theoretical model of the winter survival dynamics of *Meloidogyne* spp. eggs and juveniles. J. Nematol. 17:257–260.

Johnson, A.W., and J. Feldmesser. 1987. Nematicides: A historical review. p. 448–454. *In* J.A. Veech and D.W. Dickson (ed.) Vistas on nematology. Soc. Nematol., Hyattsville, MD.

Johnson, A.W., A.M. Golden, D.L. Auld, and D.R. Sumner. 1992. Effects of rapeseed and vetch as green maure crops and fallow on nematodes and soil-borne pathogens. J. Nematol. 24:117–126.

Johnson, A.W., and R.E. Motsinger. 1989. Suitability of small grains as hosts of *Meloidogyne* species. Suppl. J. Nematol. 21:650–653.

Johnson, A.W., and R.E. Motsinger. 1990. Effects of planting date, small grain crop destruction, fallow, and soil temperature on the management of *Meloidogyne incognita*. J. Nematol. 22:348–355.

Jones, F.G.W., D.M. Parrott, and G.J.S. Ross. 1967. The population genetics of the potato cyst-nematode, *Heterodera rostochiensis*: Mathematical models to simulate the effects of growing eelworm-resistant potatoes bred from *Solanum tuberosum* spp. *Andigena*. Ann. Appl. Biol. 60:151–171.

Kaplan, D.T. 1988. Future considerations for nematode management in citrus. p. 969–975. *In* R. Goren and K. Mendel (ed.) Proc. Int. Citrus Congr., 6th, Tel-Aviv, Israel. 6–11 March. Balaban, Rehovot, Israel.

Kaplan, M., and J.P. Noe. 1993. Effects of chicken-excrement amendments on *Meloidogyne arenaria*. J. Nematol. 25:71–77.

Kerry, B.R. 1990. An assessment of progress toward microbial control of plant-parasitic nematodes. Suppl. J. Nematol. 22:621–631.

Kerry, B.R. 1995. Microbial agents for the biological control of plant parasitic nematodes. p. 244–250. *In* Progr. Proc. Int. Congr. Trop. Nematol., Rio Quente, Brazil. 4–9 June. Rio Quente, Brasil.

Kerry, B.R., D.H. Crump, and L.A. Mullen. 1982. Natural control of the cereal cyst nematode, *Heterodera avenae* Woll., by soil fungi at three sites. Crop Prot. 1:99–109.

Kinloch, R.A. 1973. Nematode and crop responses to short-term rotations of corn and soybean. Proc. Fla. Crop Soil Sci. Soc. 33:86–88.

Kinloch, R.A. 1983. Influence of maize rotations on the yield of soybean grown in *Melodogyne incognita* infested soil. J. Nematol. 15:398–404.

Kinloch, R.A. 1986. Soybean and maize cropping models for the management of *Meloidogyne incognita* in the Coastal Plain. J. Nematol. 18:451–58.

Kinloch, R.A., and L.S. Dunavin. 1993. Summer cropping effects on the abundance of *Meloidogyne arenaria* race 2 and subsequent soybean yields. Suppl. J. Nematol. 25:806–809.

Kirkpatrick, T.L., P.D. Colyer, and T.A. Wheeler. 1995. Interactions with Fusarium diseases. p. 199. *In* D.A. Richter and J. Armour (ed.) Proc. Beltwide Cotton Prod. Res. Conf. Natl. Cotton Counc., San Antonio, TX. 4–7 January. Natl. Cotton Counc. Am, Memphis, TN.

Koenning, S.R., and S.C. Anand. 1991. Effects of wheat and soybean planting date on *Heterodera glycines* population dynamics and soybean yield with conventional tillage. Plant Dis. 75:301–304.

Koenning, S.R., D.P. Schmitt, and K.R. Barker. 1993. Effects of cropping systems on population density of *Heterodera glycines* and soybean yield. Plant Dis. 77:780–786.

Lehman, P.S. 1995. Role of plant protection organizations in nematode management. p. 137–148. *In* Progr. Proc. Int. Congr. Trop. Nematol., Rio Quente, Brazil. 4–9 June. Rio Quente, Brasil.

Maas, P.W.Th. 1987. Physical methods and quarantine. p. 265–293. *In* R.H. Brown and B.R. Kerry (ed.) Principles and practice of nematode control in crops. Acad. Press, London.

Maloney, A.P. 1995. Three sources for non-chemical management of plant disease: Towards an ecological framework. p. 103–130. *In* J.H. Andrews and I.C. Tommerup (ed.) Advances in plant pathology. Vol. 11. Acad. Press, San Diego, CA.

Mashela, P., L.W. Duncan, and R. McSorley. 1992. Salinity reduces resistance to *Tylenchulus semipenetrans* in citrus rootstocks. Nematropica 22:7–12.

McKenry, M., T. Buzo, J. Kretsch, S. Kaku, E. Otomo, R. Ashcroft, A. Lange, and K. Kelley. 1994. Soil fumigants provide multiple benefits: Alternatives give mixed results. Calif. Agric. 48:22–28.

McNamara, D.G. 1995. The role of the European and Mediterranean plant protection organization in nematode management. p. 122–127. *In* Progr. Proc. Int. Congr. Trop. Nematol., Rio Quente, Brazil. 4–9 June 1995.

McSorley, R. 1992. Applied population modeling: fact or fiction. p. 170–181. *In* F.J. Gommers and P.W.Th. Maas (ed.) Nematology from molecule to ecosystem. Dekker and Huisman, Wildervank, The Netherlands.

McSorley, R. 1998. Population dynamics. p. 109–133. *In* K.R. Barker et al. (ed.) Plant and nematode interactions. Agron. Monogr. 36. ASA, CSSA, and SSSA, Madison, WI.

McSorley, R., and D.W. Dickson. 1989. Effects and dynamics of nematode community on soybean. J. Nematol. 21:490–499.

McSorley, R., D.W. Dickson, J.A. De Brito, and R.C. Hochumuth. 1994. Tropical rotation crops influence nematode densities and vegetable yields. J. Nematol. 26: 308–314.

McSorley, R., and L.W. Duncan. 1995. Economic thresholds and nematode management. p. 147–171. *In* J.H. Andrews and I.C. Tommerup (ed.) Advances in plant pathology. Vol. 11. Acad. Press, San Diego, CA.

McSorley, R., and R.N. Gallaher. 1992. Comparison of nematode population densities on six summer crops at seven sites in north Florida. Suppl. J. Nematol. 24:699–706.

McSorley, R., and M.S. Phillips. 1993. Modeling population dynamics and yield losses and their use in nematode management. p. 61–85. *In* K. Evans et al. (ed.) Plant parasitic nematodes in temperate agriculture. CAB Int., Wallingford, England.

Meyer, S.L.F., R.M. Sayre, and R.N. Huettel. 1991. Benomyl tolerance of ten fungi antagonistic to plant-parasitic nematodes. J. Nematol. 23:402–408.

Minton, N.A. 1986. Impact of conservation tillage on nematode populations. J. Nematol. 18:135–140.

Minton, N.A., and M.B. Parker. 1987. Root-knot nematode management and yield of soybean as affected by winter cover crops, tillage systems, and nematicides. J. Nematol. 19:38–43.

Minton, N.A., M.B. Parker, and D.R. Sumner. 1985. Nematode control related to fusarium wilt in soybean and root rot and zinc deficiency in corn. J. Nematol. 17:314–321.

Mojtahedi, H., G.S. Santo, and R.E. Ingham. 1993. Suppression of *Meloidogyne chitwoodi* with sudangrass cultivars as green manure. J. Nematol. 25:303–311.

Mosjidis, J.A., R. Rodríguez-Kábana, C.F. Weaver, and P.S. King. 1994. Reaction of nine vicia species to *Meloidogyne arenaria* race 2 and *Heterodera glycines* race 4. Nematropica 24:1–5.

Mueller, J.D. 1994. Environmental effects on nematode management. p. 232–234. *In* D.J. Herber and D.A. Richter (ed.) Proc. Beltwide Cotton Prod. Res. Conf., San Diego, CA. 5–8 January. Natl. Cotton Council, Memphis, TN.

Mueller, J.D., and G.B. Sanders. 1987. Control of *Hoplolaimus columbus* on late-planted soybean with aldicarb. Ann. Appl. Nematol.(Suppl. J. Nematol.) 1:123–126.

Muller, R., and P.S. Gooch. 1982. Organic amendments in nematode control. An examination of the literature. Nematropica 12:319–326.

National Research Council. 1989. Alternative agriculture. Natl. Acad. Press, Washington, DC.

Niblack, T.L. 1989. Applications of nematode community structure research to agricultural production and habitat disturbance. J. Nematol. 21:437–443.

Noe, J.P. 1988. Theory and practice of the cropping systems approach to reducing nematode problems in the Tropics. J. Nematol. 20:204–213.

Noe, J.P. 1991. Variability among populations of *Meloidogyne avenaria* race 1 with respect to reproduction and pathogenicity on various crops. J. Nematol. 23:544.

Noe, J.P. 1993. Damage functions and population changes of *Hoplolaimus columbus* on cotton and soybean. J. Nematol. 25:440–445.

Noe, J.P. 1998. Crop- and nematode-management systems. p. 159–171. In K.R. Barker et al. (ed.) Plant and nematode interactions. Agron. Monogr. 36. ASA, CSSA, and SSSA, Madison, WI.

Noe, J.P., and K.R. Barker. 1985. Overestimation of yield loss of tobacco caused by the aggregated spatial pattern of *Meloidogyne incognita*. J. Nematol. 17:245–251.

Noe, J.E., and C.L. Campbell. 1985. Spatial pattern analysis of plant-parasitic nematodes. J. Nematol. 17:86–93.

Noe, J.P., and J.L. Imbriani. 1986. Yield-loss relationships and population dynamics of *Hoplolaimus columbus* on cotton, as modified by edaphic parameters. J. Nematol. 18:624.

Noe, J.P., J.N. Sasser, and J.L. Imbriani. 1991. Maximizing the potential of cropping systems for nematode management. J. Nematol. 23:353–361.

Noling, J.W. 1991. Metham sodium application via drip irrigation systems. Fla. Dep. Agric. Consum. Serv., Div. Plant Ind., Nematol. Circ. no. 196.

Noling, J.W. 1994. Use of hot water, broadspectrum fumigants, and soil solarization for nematode control. J. Nematol. 25:559.

Noling, J.W., and J.O. Becker. 1994. The challenge of research and extension to define and implement alternatives to methyl bromide. Suppl. J. Nematol. 26:573–586.

Noling, J.W., and H. Ferris. 1987. Nematode-degree days, a density-time model for relating epidemiology and crop losses in perennials. J. Nematol. 19:108–118.

O'Bannon, J.H., R.N. Inserra, R.P. Esser, and P.S. Lehman. 1991. Guidelines for compost sanitation. Fla. Dep. Agric. & Consum. Serv., Div. Plant Ind., Nematol. Circ. no. 193.

Omwega, C.O., I.J. Thomason, and P.A. Roberts. 1990. Effect of temperature on expression of resistance to *Meloidogyne* spp. in common bean *(Phaseolus vulgaris*). J. Nematol. 22:446–451.

Oostendorp, M., and R.A. Sikora. 1989. Seed treatment with antagonistic rhizobacteria for the suppression of Heterodera schactii early root infection of sugar beet. Rev. Nematol. 12:77–83.

Opperman, C.H., and M.A. Conkling. 1998. Bioengineering resistance to plant-parasitic nematodes. p. 239–250. In K.R. Barker et al. (ed.) Plant and nematode interactions. Agron. Monogr. 36. ASA, CSSA, and SSSA, Madison, WI.

Overman, A.J., H.H. Bryan, and R.W. Harkness. 1970. Effect of off-season culture on weeds, nematodes, and potato yields on marl soils. Proc. Fla. State Hort. Sci. Soc. 84:135–139.

Overman, A.J., and F.G. Martin. 1978. A survey of soil and crop management practices in the Florida tomato industry. Proc. Fla. State Hort. Sci. Soc. 91:294–297.

Pacumbaba, R.P., and W. Tadesse. 1991. The effect of different planting dates on the infestation of soybean by *Heterodera glycines* race 5. J. Agron. Crop Sci. 166:278–284.

Paracer, S., A.C. Tarjan, and L.M. Hodges. 1987. Effective use of marine algal products in the management of plant-parasitic nematodes. J. Nematol. 19:194–200.

Perry, J.N. 1978. A population model for the effect of parasitic fungi on numbers of the cereal cyst-nematode, *Heterodera avenae*. J. Appl. Ecol. 15:781–787.

Perry, J.N., and S.J. Clark. 1983. Use of population models to estimate the proportion of cyst-nematode eggs killed by fumigation. J. Appl. Ecol. 20:857–864.

Phipps, P.M. 1993. IPM in peanuts: Developing and delivering working IPM systems. Plant Dis. 77:307–309.

Pimentel, D., D. Andow, R. Dyson-Hudson, D. Gallahan, S. Jacobson, M. Irish, S. Kroop, A. Moss, I. Schreiner, M. Shepard, T. Thompson, and B. Vinzant. 1980. Environmental and social costs of pesticides: A preliminary assessment. Oikos 34:126–140.

Powell, N.T. 1971. Interactions between nematodes and fungi in disease complexes. Annu. Rev. Phytopathol. 9:253–274.

Powell, N.T. 1979. Internal synergisms among organisms inducing disease. p. 113–133. *In* J.G. Horsfall and E.B. Cowling (ed.) Plant disease. Vol. 4. Acad. Press, New York.

Powers, L.E., R. McSorley, and R.A. Dunn. 1993. Effects of mixed cropping on a soil nematode community in Honduras. J. Nematol. 25:666–673.

Powers, T.O., and T.S. Harris. 1993. A polymerase chain reaction method for identification of five major *Meloidogyne* species. J. Nematol. 25:1–6.

Prot, J.C., I.R.S. Soriano, D.M. Matias, and S. Savary. 1992. Use of green manure crops in control of *Hirschmanniella mucronat* and *H. oryzae* in irrigated rice. J. Nematol. 24:127–132.

Rajotte, E.G. 1993. From profitability to food safety and the environment: Shifting the objectives of IPM. Plant Dis. 77:296–299.

Rhoades, H.L. 1984. Effects of fallowing, summer cover crops, and fenamiphos on nematode populations and yields in cabbage-field corn rotation in Florida. Nematropica 14:131–138.

Riggs, R.D. 1995. Management of races of SCN. p. 107–110. *In* Progr. Proc. Int. Congr. Trop. Nematol., Rio Quente, Brazil. 4–9 June 1995.

Roberts, P.A. 1992. Current status of the availablility, development, and use of host plant resistance to nematodes. J. Nematol. 24:213–227.

Roberts, P.A. 1993. The future of nematology: Integration of new and improved management strategies. J. Nematol. 25:383–294.

Roberts, P.A., S.D. Van Gundy, and H.E. McKinney. 1981. Effects of soil temperature and planting date of wheat on *Meloidogyne incognita* reproduction, soil populations, and grain yield. J. Nematol. 13:338–345.

Rodriguez-Kabana, R. 1986. Organic and inorganic nitrogen amendments to soil as nematode suppressants. J. Nematol. 18:129–135.

Rodriguez-Kabana, R., and E.A. Curl. 1980. Nontarget effects of pesticides on soilborne pathogens and disease. Annu. Rev. Phytopathol. 18:311–332.

Rodríguez-Kábana, R., P.S. King, D.G. Robertson, and C.F. Weaver. 1989. Crops uncommon to Alabama for the management of *Meloidogyne arenaria* in peanut. Suppl. J. Nematol. 21:712–716.

Rodriguez-Kabana, R., D.G. Robertson, C.F. Weaver, and L. Wells. 1991. Rotations of bahiagrass and castorbean with peanut for the management of *Meloidogyne arenaria*. Suppl. J. Nematol. 23:658–661.

Rodriguez-Kabana, R., C.F. Weaver, D.G. Robertson, and E.L. Snoddy. 1986. Population dynamics of *Meloidogyne arenaria* juveniles in a field with florunner peanut. Nematropica 16:185–196.

Saleh, H., W.I. Abu-Gharbieh, and L. Al-Banna. 1988. Effect of solarization combined with solar-heated water on *Meloidogyne javanica*. Nematologica 34:290–291.

Sasser, J.N., and G. Uzzell. 1991. Control of the soybean cyst nematode by crop rotation in combination with a nematicide. J. Nematol. 23:344–347.

Schmitt, D.P. and J.L. Imbriani. 1987. Management of *Hoplolaimus columbus* with tolerant soybean and nematicides. Ann. Appl. Nematol. 1:59–63.

Schmitt, D.P., H. Ferris, and K.R. Barker. 1987. Response of soybean to *Heterodera glycines* race 1 and 2 in different soil types. J. Nematol. 19:240–250.

Schomaker, C.H., and T.H. Been. 1992. Sampling strategies for the detection of potato cyst nematodes; developing and evaluating a model. p. 182–194. *In* F.J. Gommers and P.W.Th. Maas (ed.) Nematology from molecule to ecosystem. Dekker and Huisman, Wildervank, The Netherlands.

Schroeder, J., S.H. Thomas, and L. Murray. 1993. Yellow and purple nutsedge and chile peppers host southern root-knot nematode. Weed Sci. 41:150–156.

Seinhorst, J.W. 1965. The relationship between nematode density and damage to plants. Nematologica 11:137–154.

Showler, A.T., T.E. Reagan, and R.M. Knaus. 1990. Sugar-cane weed community interactions with arthropods and pathogens. Insect Sci. Appl. 11:1–11.

Sipes, B.S., and D.P. Schmitt. 1989. Effect of planting date, alachlor, and fenamiphos on *Heterodera glycines* development. J. Nematol. 21:33–41.

Smiley, R.W., R.E. Ingham, W. Uddin, and G.H. Cook. 1994. Crop sequences for managing cereal cyst nematode and fungal pathogens of winter wheat. Plant Dis. 78:1142–1149.

Spiegel, Y., E. Cohn, and I. Chet. 1989. Use of chitin for controlling *Heterodera avenae* and *Tylenchulus semipenetrans*. J. Nematol. 21:419–422.

Stapleton, J.J., B. Lear, and J.E. DeVay. 1987. Effect of combining soil solarization with certain nematicides on target and nontarget organisms and plant growth. Ann. Appl. Nematol. 1:107–112.

Stirling, G.R. 1991. Biological control of plant parasitic nematodes: progress, problems and prospects, CAB Int., Wallingford, England.

Taylor, L.R. 1961. Aggregation, variance and the mean. Nature (London) 189:732–735.

Thomason, I.J. 1987. Challenges facing nematology: Environmental risks with nematicides and the need for new approaches. p. 469–476. *In* J.A. Veech and D.W. Dickson (ed.) Vistas on nematology. Soc. Nematol., Hyattsville, MD.

Thomason, I.J., and E.P. Caswell. 1987. Principals of nematode control. p. 87–130. *In* R.H. Brown and B.R. Kerry (ed.) Principles and practice of nematode control in crops. Acad. Press, London.

Trivedi, P.C., and K.R. Barker. 1986. Management of nematodes by cultural practices. Nematropica 16:213–236.

Trudgill, D.L. 1992. Mechanisms of damage and of tolerance in nematode infested plants. p. 133–145. *In* F.J. Gommers and P.W.Th. Mass (ed.) Nematology from molecule to ecosystem. Europ. Soc. Nematol., Invergourie, Dundee.

Tyler, J. 1933. The root-knot nematode. Univ. Calif. College Agric. Circ. 330. Agric. Exp. Stn., Berkeley.

United Nations Environmental Programme (UNEP). 1992. Synthesis report of the methyl bromide interim scientific assessment and methyl bromide interim technology and economic assessment. p. 33. *In* Montreal Protocol Assess. Suppl. June 1992. UNEP, Nairobi, Kenya.

United Nations Environmental Programme (UNEP). 1994. Montreal Protocol on substances that deplete the ozone layer. UNEP, Nairobi, Kenya.

Vanderplank, J.E. 1963. Plant diseases: Epidemics and control. Acad. Press, New York.

Van Sickle, J.J., E. Belibasis, D. Cantliff, G. Thompson, and N. Oebker. 1994. Competition in the winter fresh vegetable industry. USDA Econ. Res. Serv. Agric. Econ. Rep. 691. U.S. Gov. Print. Office, Washington, DC.

Walters, S.A., and K.R. Barker. 1993. Comparison of two inoculum preparation methods of *Rotylenchulus reniformis*. Suppl. J. Nematol. 25:778–784.

Wheaton, T.A., C.C. Childers, L.W. Timmer, L.W. Duncan, and S. Nikdel. 1985. Effects of Aldicarb on yield, fruit quality, and tree condition of Florida citrus. Proc. Fla. State Hort. Sci. Soc. 98:6–10.

Yeates, G.W., T. Bongers, R.G.M. de Goede, D.W. Freckman, and S.S. Georgieva. 1993a. Feeding habits in soil nematode families and genera-an outline for soil ecologists. J. Nematol. 25:315–331.

Yeates, G.W., D.A. Wardle, and R.N. Watson. 1993b. Relationships between nematodes, soil microbial biomass and weed-management strategies in maize and asparagus cropping systems. Soil Biol. Biochem. 25:869–876.

Zadoks, J.C., and R.D. Schein. 1979. Epidemiology and plant disease management. Oxford Univ. Press, New York.

14 Information Transfer

BECKY B. WESTERDAHL AND EDWARD P. CASWELL-CHEN

University of Califonia
Davis, California

The number of research scientists developing new information is at a record high. An often-heard generalization is that "the majority of the research scientists that have *ever* lived are alive today" (Asimov, 1982). Science is experiencing an information explosion, and the quantity of new information offered is enormous. The ramifications of such information, and the demand and need for information also have increased. The information era is quickly becoming the era of electronic information. Still, the need for information quality must not be sacrificed to quantity.

Information transfer has changed dramatically during the past 20 yr. The quantity and availability of information is increasing, and the time lag between information gathering and dissemination is decreasing. The advent of networked computers and satellite relays has made information available on a worldwide basis. The availability of personal computers allows the storage, retrieval, and analysis of information on a scale unprecedented in the history of man.

The sources and means for generating and analyzing information now include such diversified techniques as aerial and satellite imaging (Denison et al., 1996; Skole & Tucker, 1993), global positioning systems (Herring, 1996), global information systems (Burrough, 1986), electron microscopy (Eisenback, 1985; Hall, 1995), confocal microscopy (Czymmek et al., 1994), scanned-probe microscopes (Wickramasinghe, 1989), automated remote data collection, computer databases (Eeckman & Durbin, 1995), DNA cloning, sequence analysis, genome projects (Caswell-Chen et al., 1993; Hodgkin et al., 1995), and bioinformatics (Marshall, 1996).

In the past, information transfer was relatively simple and accomplished via two primary means—verbal and written. The primary means for information transfer remain the same, but the technological means for effecting transfer are new.

The subject addressed herein is the development and transfer of information in nematology. An overview of information transfer at the basic and applied levels is provided. The objectives are to improve the reader's access to information on nematology, and to introduce new ideas.

The primary focus herein is on information related to nematode management. The types of information available for transfer and the techniques for infor-

 Plant and Nematode Interactions, Agronomy Monograph no. 36.

mation transfer are included. Also provided are caveats for successful transfer and use of information, and the needs of the audience or clientele to whom the information transfer is directed. Future directions are mentioned. Traditional methods, recently developed methods, and those on the cusp of widespread availability are given. Some of the types of information transfer considered herein apply to pertinent topics, such as the geographic distribution of nematode species, the host status of cultivars for specific nematodes, nematode-control strategies and tactics, crop rotations, DNA sequences, phylogenetic relationships among nematodes, and nematode biology.

No one best method for transferring information is given, because there are many types of information, many uses for the information, and many audiences for whom the information is intended. Therefore, the challenges of information transfer are endless and fluid. The capacity to effectively transfer information improves with experience and practice. With experience, one learns to better judge the level of presentation appropriate for different audiences and the most effective means of information presentation.

NEMATODES AND NEMATODE MANAGEMENT

Nematology was accepted as a separate discipline in the early 1900s following the demonstration that nematodes were significant pathogens (Dowler & Van Gundy, 1984; Caswell & Apt, 1989). Entomopathogenic nematodes were embraced within nematology with the realization that some nematodes had potential as biological control agents for insects and other invertebrates of agricultural importance (Nickle & Welch, 1984).

Future management strategies for plant-parasitic nematodes in sustainable agriculture will require combinations of management tactics. Sustainable agriculture will require information on nematode ecology, especially fundamental knowledge of how different tactics influence nematode population dynamics and genetics. As the research community defines the important details of nematode population and community dynamics and the influence of genetic variability on nematode fitness, sustainable agricultural systems based on integrated nematode management strategies using cover crops, resistant cultivars, or biological control agents will be implemented. Because nematicides continue to become less available, individuals directly involved in nematode management have an increasing need for access to detailed information on available strategies and tactics. Thus, information transfer becomes more important.

From very applied beginnings, in which traditional management principles and classical taxonomy were the primary emphasis, the study of nematology has grown until today it is a diverse and complex science. Nematodes are utilized as model systems for many disciplines such as molecular and biochemical systematics, genetics, developmental biology, physiology, behavior, and ecology. Today, development of new management techniques for plant-parasitic nematodes depends heavily on discoveries in the many research specialties within nematology (Caswell-Chen et al., 1993).

AUDIENCE AND CLIENTELE

When preparing to transfer information, one must consider the intended audience. This consideration will affect the content and depth of the information that is presented, as well as the most appropriate means for presentation. The major audiences for information on nematology include researchers, farmers or growers, pest control advisers, agriculturally related industries, government agencies, universities or colleges, the media, and the general public.

The Teachable Moment

The optimal time to make a point to a student or client is at the time when the information is needed (Dunn, 1987). It is important that appropriate information be available when that moment arises. In nematology this has been traditionally accomplished via written materials, such as textbooks, journals, and extension publications, and in verbal forms such as formal classes, short courses, extension presentations to agricultural groups, and one-on-one consultations. With the development of the personal computer, the "teachable moment" may be expanded in ways that were not previously possible.

Clientele—Farmers and Pest Control Advisors

Frequently, growers not farming nematode-infested land have little knowledge of nematodes. Such growers require introductory information, including pictures of nematode damage symptoms, guidelines on how to sample for nematodes and information on how to prevent nematode movement and introduction. Growers and corporations that farm large areas frequently hire pest control advisors (PCAs). Pest control advisors are professionals who earn their living advising growers on pest management strategies. Often, they are less knowledgeable about nematodes than about other pathogens and pests, although PCAs are sufficiently knowledgeable to suspect nematode infestations based on crop symptoms. Typically, they are required to obtain a prescribed number of hours of continuing education each year by attending courses, and so are frequently a focus of information transfer. Pest control advisors know how to take soil samples for analysis, can provide advice on nematode management, and know where to obtain further information on nematodes. Growers who have dealt with nematode problems for an extended period are often interested in more advanced topics, such as new chemical control techniques, crop rotation, and the availability of resistant cultivars.

Audiences at extension meetings typically include a mixture of growers and PCAs with varying levels of knowledge about nematodes. The challenge is to present sufficient introductory information as not to confuse the novice, and enough advanced information to keep the more informed interested. Growers and PCAs also are a primary audience for written guidelines and bulletins on pest management. One of the axioms of extension work remains the importance of one-on-one field consultation with growers or PCAs. Although such contact is expensive because of the time and travel involved, such interaction is invaluable for information transfer.

Clientele—Industry

Food processors may take an active role in commodity production and employ field personnel to work with growers on crop production practices. Processors may be more involved in nematode management than one might anticipate. They establish planting and harvest dates for particular fields, and often dictate acceptable cultivars, pesticides, and lengths of rotations to nonhost crops.

Farm-equipment manufacturers need to know the factors affecting movement of various nematicides through soil in order to develop appropriate implements. Equipment should be constructed to minimize the spread of nematodes.

Pesticide manufacturers often employ fewer nematologists than professionals representing the other pest management disciplines. As a result, nematologists frequently must communicate with entomologists, plant pathologists, or weed scientists as related to the need for new nematicides and appropriate methods for evaluating potential products. Information on markets, pesticide-use data, and crop loss figures are of great interest to pesticide manufacturers. In some cases, markets have not been developed because of the lack of registered products, and because the only yield- and quality-loss data available for many crops are rough estimates.

Seed producers, plant producers, and breeders also have diverse educational backgrounds and may need encouragement to develop nematode-resistant or tolerant varieties. They must have information on the culture and maintenance of pure lines of various nematode species, and methods for screening large numbers of plants for host status to nematodes. Because of the divergence in definitions for resistance and tolerance, clarification on the way cultivars of various plants are labeled may be needed. For example, a nematode-resistant tomato (*Lycopersicon esculentum* Miller) is resistant to only some of the many root-knot nematode species, but the labels do not specify which species. Importantly, a resistant tomato limits reproduction of those species to which it is resistant, and so is valuable for use in crop rotation programs (Roberts & Thomason, 1989). In contrast, nematode resistant alfalfa (*Medicago sativa* L.) varieties may actually be tolerant, and are not damaged by the feeding of one or more nematode species even though they do allow significant nematode reproduction (Roberts et al., 1998, see Chapter 11).

Clientele—Government

Pesticide Registration

Nematologists may be asked by state or federal agencies to assist with the preparation of information needed to register, or otherwise evaluate, nematicides. The individuals within government agencies responsible for collecting such information may not be trained in nematology. The expertise of individuals in the agency should be determined, as it will affect the types of information provided. Often when a government agency needs information, it is an urgent need for all possible information on a topic. For example, the need for information to evaluate nematicides relative to environmental impact statements has become so great in recent years, that agencies have resorted to employing private consultants to prepare the reports. The contracts are typically given to companies or individuals with no nematology background, with the rationale that nematologists cannot prepare unbiased reports on nematicides. The contracted consultants then phone nematologists soliciting the needed information. The result is that in addition to providing the requested information, the nematologist also must try to provide a basic education in nematology over the telephone. This interface is part of the challenge of conveying information.

The registration of new nematicidal active ingredients is relatively rare, but when it occurs, someone from a government agency is likely to call university and/or USDA nematologists for information on the product. Questions asked typically concern the wording of various parts of the label with respect to application timing or methodology. Nematologists also are asked by growers or county agencies to assist with the preparation of applications for Section 18 Emergency Registrations and 24C or Special Local Need (SLN) Registrations (USEPA) (see your state department of agriculture for further information on these programs). Special registrations require, for example, information on the nature of the pest problem and historical information on nematode management. In some states, city and county agencies have the responsibility for enforcing pesticide regulations and, at times, need advice on various subjects involving nematodes and nematicides.

Often, governmental agencies are referred to as if they were a single, monolithic entity. A typical government agency employs hundreds or even thousands of individuals, often acting and thinking individually within a broad set of guidelines provided by the agency. This is worth considering, because in some instances one only needs to communicate with a single individual within an agency, and that individual will act on the information or will present it to others in the agency.

Judicial Proceedings

Nematologists also may be called to testify at public hearings or judicial proceedings. Testimony can be presented verbally, in writing, or both. Hearing officers and justices are typically chosen to be unbiased arbitrators and have little knowledge of nematodes. If 35-mm slides are to be used as part of the testimony, copies should be made beforehand, because they should be submitted to become a part of the permanent record. Printed or hard copies of slides (e.g., color laser copies) or handouts may be a better choice than slides in such situations. Also, remember that the audience is the person or persons making the decision based on the testimony presented, rather than the other individuals testifying.

Clientele—University Setting

Education

Dickson and Meredith (1987) and Abrantes and Hague (1992) have surveyed and summarized the universities and colleges involved in teaching nematology. Some 22 universities in the USA and 16 in other countries offer at least one graduate-level nematology course. Only a limited number of institutions offer more than a single course devoted to nematodes or principles of nematology. Most courses in nematology are designed to supplement curricula in plant pathology, entomology, or closely allied disciplines. The majority of graduate courses in nematology are offered in alternate years and have average enrollments of 5 to 10 students. Some students entering nematology today do so because of an interest in pest management, but have primary training in plant pathology or entomology. Still others have their primary training in disciplines such as crop science, horticulture, biochemistry, physiology, microbiology, genetics, or molecular biology.

In addition to traditional university courses, several universities periodically offer nematology short-courses that last from a few days to a few weeks. The 10-d Nematode Identification Course for Professional Consultants offered annually at Clemson University in South Carolina is perhaps the oldest and best known. An International Nematology Course is offered at the University of California at Davis for 4 wk approximately every other summer, and a 5-d course in nematode taxonomy is offered in June each year by the University of California at Riverside.

Professors who teach nematologists recognize that their students will probably enter careers where they will be the only nematologist in a group of plant pathologists, entomologists, or related biologists. Approximately 50% of nematology graduates find careers with universities, 25% are hired by industry and the government, and less than 10% go into private business or administration.

Institutions that have a single faculty position devoted to nematology are the rule rather than the exception. This nematologist is typically housed in a department devoted to plant pathology or entomology, and frequently has teaching, research, and extension responsibilities. In many land-grant institutions, the nematologist also will be affiliated with the agricultural experiment station.

A range of course materials for teaching nematology are available. The need for self-study materials and for students to be able to interact with faculty in various specialties can be greatly facilitated through computers. A cooperatively published laboratory manual (Zuckerman et al., 1990) addresses a range of issues within nematology. In addition to formal course work, a great deal of training in nematology is gained through practical experience, with nematode identification serving as a particularly germane example. Although the basic principles can be learned in a course, the ability to routinely identify nematodes extracted from soil samples requires a year or more of practical experience gained in close collaboration with a trained nematologist.

Most of the early nematology teachers had their primary training in plant pathology or entomology and were raised in agricultural backgrounds. Many contemporary teachers still have their primary training in plant pathology or entomology, although fewer come from an agricultural background. Nematology is increasingly encompassing individuals with primary training in other disciplines.

Teaching nematology at the university or college level requires a balance between teaching historical developments in nematology (to minimize the tendency for students to "reinvent the wheel") and, more recently, developments and specialized concepts. To maintain strong programs within universities and colleges, nematologists need to describe nematology research and extension activities in terms consistent with the mission of the university.

Cooperative Extension

Associated with U.S. land-grant universities is an organization known as Agricultural, or Cooperative Extension (CE) (Radewald, 1969). Programs and responsibilities of CE vary from state to state. Within some states, specialists in disciplines such as nematology, plant pathology, irrigation, or crops, are housed on university campuses and work in collaboration with researchers of the agricultural experiment station. The university may have offices in all of the counties in the state desiring to participate in the program. Funding for the county offices is provided cooperatively by the county, the university, and the federal government (via the USDA-ARS). Farm advisors or county agents serve as sources of information for farmers, PCAs, the general public, and 4-H youth programs. Specialists are the link between the county personnel and the university. They have primary responsibility for transmitting research information developed by the university to county agents, keeping university researchers apprised of problems that require further research, keeping advisors abreast of new developments in nematology, and assisting the advisors with the diagnosis of nematode problems. Specialists and advisors also conduct research in growers' fields. Advisors typically have expertise in the particular crop of interest, while the nematology specialist provides nematode expertise across a variety of crops. Cooperative extension administers the 4-H youth program to educate youth in agricultural principles and the Master Gardener program to educate a volunteer force of adults to assist home gardeners. Training in nematology is included where applicable in these programs. Moens and Maas (1992) discuss the similarities and differences in extension nematology in Europe and the USA.

Clientele—The Media

When media representatives need information, they want answers quickly and may print what they are told (or their interpretation of it) the same day. Writers from the general press typically lack experience with nematodes, while farm press writers may be quite familiar with nematodes. When contacted by a writer, the challenge is to quickly focus thoughts about the topic, think carefully, and then answer. Asking the press representative some questions may buy time in which to formulate appropriate answers. When providing information, short, nonspeculative, objective answers are the best and the safest. Always request that a copy of the article be faxed or mailed to you so it can be reviewed for accuracy, but don't expect to get one.

The employing organization also may have a media office that will call looking for stories to put on the news wires. Their deadlines are usually not critical, and there may be an opportunity to edit the resulting article for accuracy. News releases have a way of reappearing periodically for several months as they are picked up by various trade publications. This is one means of conveying the importance of particular nematode problems across broad areas, for some period of time.

Clientele—General Public

The home gardener frequently becomes aware of nematodes when they experience a crop failure due to root-knot nematodes in their own backyard. A recent publication by the Society of Nematologists Extension Committee (1988) lists publications written specifically for home gardeners. Some older publications that are still found in libraries may be outdated because they contain recommendations for nematicides that are no longer available for homeowner use. Current publications emphasize a diversity of management strategies relative to nematode biology, including fallow, crop rotation and the use of resistant cultivars (Flint, 1990).

The adverse publicity surrounding nematicides also can bring nematologists into contact with the general public. The challenges here are similar to those discussed in dealing with the media.

Because nematodes are small, and microscopes are rare in elementary and middle schools, nematodes are more difficult than many other organisms to incorporate into a science curriculum. Still, many nematologists participate in programs that bring scientists in contact with young students. Some visit classrooms to discuss nematodes and to demonstrate the disease problems they can cause. Others participate in outreach programs, and have students working in their laboratories during the summer. Some nematologists participate in programs that bring school teachers to campuses in the summer to develop new science curricula.

The use of computers in schools has increased tremendously in recent years. The use of computer networks, which we will discuss in later sections, can be used to provide students access to educational material on nematodes, including visual images and video clips of nematodes.

EQUIPMENT AND TECHNIQUES FOR INFORMATION TRANSFER

Solutions to nematode problems are frequently of a regional nature. Therefore, the best source of information may reside with an extension nematologist at the nearest land-grant university (Table 14–1), USDA research station, or state department of agriculture. Within the USA, an expert can usually be located with a telephone call to the closest land-grant university. If the university does not have a department of nematology, resource personnel may be housed with departments of entomology or plant pathology.

Publications

Many of the classic nematology textbooks are out of print, difficult to find, or are available only in libraries (Caveness, 1964; Chitwood & Chitwood, 1950;

Table 14–1. Extension nematologists employed at various land-grant universities.

State	Nematologist	University
Alabama	W. S. Gazaway	Auburn Univ.
California	B. B. Westerdahl	UCD
California	J. O. Becker	UCR
California	M. V. McKenry	UCR
California	A. T. Ploeg	UCR
Florida	D. W. Dickson	Univ. of Florida
Florida	R. A. Dunn	Univ. of Florida
Florida	J. W. Noling	Univ. of Florida
Florida	G. C. Smart	Univ. of Florida
Georgia	R. F. Davis	Univ. of Georgia
Idaho	S. A. Hafez	Univ. of Idaho
Iowa	Greg L. Tylka	Iowa State Univ.
Louisiana	C. Overstreet	Louisiana State Univ.
Michigan	G. W. Bird	Michigan State Univ.
Mississippi	Joe A. Fox	Mississippi State Univ.
North Carolina	S. R. Koenning	North Carolina State Univ.
North Carolina	T. A. Melton	North Carolina State Univ.
Ohio	R. M. Riedel	Ohio State Univ.
Oregon	G. B. Newcomb	Oregon State Univ.
Pennsylvania	J. M. Halbrendt	Pennsylvania State Univ.
South Carolina	J. D. Mueller	Clemson Univ.
Virginia	C. S. Johnson	Virginia Polytechnic Inst. and State Univ.
Washington	G. S. Santo	Washington State Univ.

Christie, 1959; Dropkin, 1980; Goodey, 1963; Jenkins & Taylor, 1968; Maggenti, 1981; Peachey & Chapman, 1966; Sasser & Jenkins, 1960; Southey, 1965, 1970, 1982; Tarjan, 1960, 1967; Taylor, 1967; Thorne, 1961; Wallace, 1963). Many of the more recent texts are edited efforts, composed of chapters written by acknowledged experts in the field with chapters organized by nematodes or by crops (Evans et al., 1993; Luc et al., 1990; Nickle, 1984, 1991). Some of the classic texts are periodically revised, for example, an updated edition of Mai et al. (1996) is now available.

A number of scientific journals are devoted to presenting research on nematodes, including: *Fundamental and Applied Nematology* (formerly *Revue de Nematologie*), *Indian Journal of Nematology*, *Japanese Journal of Nematology*, *Journal of Nematology* (published by the Society of Nematologists), *Nematologica*, *Nematologica Mediterranea*, *Nematropica* (published by the Organization of Tropical American Nematologists), *Pakistan Journal of Nematology*, and the *Russian Journal of Nematology*. Research articles on nematodes may be found in other journals, such as *Phytopathology*, *Plant Disease* (formerly *Plant Disease Reporter*), and the *Annals of Applied Biology*. The contents of each of these journals can be retrieved through *Current Contents*. *Current Contents* is available as a paper publication, on diskette, or CD-ROM (see http://www.isinet.com). *Nematological Abstracts* are published by the Centre for Agriculture and Biosciences (CAB) (http://www.cabi.org/) in the United Kingdom, and contain abstracts from the major nematology journals.

Libraries, Museums, and Nematode Collections

Most major universities have libraries with collections of nematology books and journals. Most of the major educational centers for nematology (see Table 14–1) have libraries that contain many of the major nematology texts. Excellent documented collections of older reprints are available at the USDA in Beltsville, Maryland, and in the Department of Nematology at the University of California at Davis. The Society of Nematologists Archives are located on the campus of Iowa State University under the curatorship of G.L. Tylka.

Extensive collections of preserved nematodes exist in several locations. The Society of Nematologists Systematics Resources Committee published a list of voucher depositories in 1992 (Robbins et al., 1992). A major nematode repository for the USA is the USDA Nematode Collection in Beltsville, Maryland, established in June 1960 by A. Morgan Golden. The USDA collection includes over 33 700 permanent slides and vials of preserved mass collections, and approximately 19 000 species entries. The collection includes designated types for more than 1500 species. The USDA Gerald Thorne Collection includes 6600 slides, while the Steiner Mermithid Collection includes over 3400 slides with both collections including many original types. The curator of the USDA collection is Z. A. Handoo (e-mail address: NEMATAX@asrr.arsusda.gov).

The University of California Davis Nematode Collection, which was started in 1944 (UCDNC) (http://ucdnema.ucdavis.edu), has a large collection of plant-parasitic, free-living soil, freshwater, and marine nematodes. The UCDNC consists of a type specimen collection, a general slide collection, mass collections, and collections and voucher specimens. The type specimen collection includes nematodes from 16 orders and a total of approximately 640 species. Approximately 300 primary type specimens, and 10 000 secondary type specimens, are included in the collection. The general slide collection contains approximately 53 000 specimens, while the mass collections include almost 6000 samples from 96 localities. A special California Academy of Sciences Nematode Collection is included within the UCDNC. The current curator of the UCDNC is S. A. Nadler (e-mail address: sanadler@ucdavis.edu).

The University of California Riverside Nematode Collection (UCRNC) is among the world's largest museums of plant-parasitic and soil nematodes. The UCRNC includes approximately 15 000 vials of preserved mass collections, and 175 000 mounted specimens on 35 000 microscope slides. The curator is J. G. Baldwin (e-mail address: james.baldwin@ucr.edu).

Scientific Meetings and Societies

Some of the largest annual gatherings of nematologists are found at the annual meetings of the Society of Nematologists (SON) in the USA, the Organization of Tropical American Nematologists (ONTA) held in the Western Hemisphere, and the biennial meeting of the European Society of Nematologists (ESN). Each of these organizations produces a newsletter that contains announcements and current information on the activities of the respective scientific societies. Nematology societies also sponsor periodic international nematology congresses.

Meetings of other scientific societies, such as the American Phytopathological Society (APS), the American Society of Parasitologists (ASP), Society for Invertebrate Pathology (SIP), the Entomological Society of America (ESA), and the Ecological Society of America (ECSA) are attended by nematologists. Research papers presented at these meetings may be given in separate sessions devoted to nematodes, or may be scattered throughout the program.

Nematology Films and Video Tapes

High-quality films and videotapes of nematodes in action are available. A videocassette recording for teaching identification of the most common genera of plant-parasitic nematodes was prepared by J. D. Eisenback and E. C. McGawley in 1991, and is a valuable teaching tool for the identification of nematodes. Using time-lapse photography and video-enhanced contrast microscopy, U. Wyss has been involved in developing more than 13 excellent films of the feeding and development of *Heterodera schachtii* (sugarbeet-cyst nematode), *Trichodorus similis* (stubby-root nematode), *Tylenchorhynchus dubius* (stunt nematode), *Xiphinema index* (dagger nematode). These films have been produced and published in cooperation with the Institut für den Wissenschaftlichen Film, Göttingen, Germany, and are used worldwide in research and teaching. A trio of recent video tapes on the soybean cyst nematode are available: "Soybean Cyst Nematode Management" from the University of Kentucky; "How to Take a Good Nematode Sample" from the University of Missouri; and "Testing Soils for Soybean Cyst Nematode" from the University of Illinois.

The Telephone

The telephone has gone from being a simple real-time, point-to-point communication device, to an important and essential tool for information transfer via messaging, conference communication, and data transmission. Portable, or cellular, phones have made travel with a telephone simple. Recently available features, such as call waiting, call forwarding, voice mail, and paging, have all enhanced the value of the telephone. Telephone lines now serve as transmission lines for data, allowing distance learning, video conferencing, and computer-to-computer communication.

Written documents can be transferred via phone lines using facsimile (FAX) machines or FAX boards in computers, or by electronic mail (e-mail, see below). FAX machines and e-mail are valuable for exchanging information because they transmit a hard copy of the actual document over long distances, and they do not require that the recipient be available to acknowledge receipt of the message. Many computers with FAX boards also have optical character recognition (OCR) software that is capable of turning FAX documents into ASCII text files that can be manipulated and processed by computer word processing and database applications.

Teleconferencing, a multiway telephone conversation between several people in different locations, can be a time and cost saving means to exchange information in a group setting. This may be necessary to achieve consensus when all the participants are in distant locations and can not possibly get together in one location for a meeting. Video conference via satellite or by the Internet (worldwide network of computers) is becoming commonplace, and in the future will probably replace the telephone conference call.

Photography and Video

Photographs are used to document research and convey visual information, such as disease symptoms. Photographs are used as illustrations in books and journals, and color slides are used as illustrations in oral presentations. The standard photographic approach is being augmented by computer digitization and manipulation of images, and emulsion-based cameras are being replaced by digital cameras (Grotta & Grotta, 1996).

At the present time, a presentation illustrated with 35-mm slides is a popular method for information transfer at both professional and extension meetings, although projection of computer-based images is starting to replace photographic slides. The slide projector will probably remain the tool of choice for some time, as almost every room at formal scientific meetings is equipped with a slide projector. Documenting research as it progresses with a 35-mm camera and slide film increases the clarity of information presented, and simplifies the preparation of presentations.

Many factors should be considered when purchasing a camera. High-quality inexpensive cameras are readily available. If one will be involved in field work in a dusty environment with the possibility of inclement weather, one of the smaller, lighter, less expensive, auto-focus, auto-exposure cameras with a built-in zoom lens should be considered.

For permanent storage, slides and negatives can be transferred to a CD-ROM disk. This service is offered commercially, and approximately 80 slides or negatives can be placed on a disk, but this depends on the resolution and format of the saved images. To preserve several thousand slides, it becomes economical to consider the purchase of computer equipment to do this. With the pace at which computer technology is developing, their direct utilization at meetings may become standard within a few years.

Digital cameras have been improved substantially during the past 8 yr, and although the affordable models (less than $10 000) are not yet ready to replace emulsion-based cameras, the currently available cameras represent a range of options and quality (Grotta & Grotta, 1996; Hogan, 1996). The advantage of the digital cameras is that the images they save are ready for storage and manipulation on the computer, and hence can be transferred as standard digital information with no loss of quality. Digital images can be used in computer graphics, allowing presentations to be made using computer presentation software.

Video cameras represent another means for recording information. As noted previously, many different commercial nematology videos and films are available for research and teaching. Individual researchers can use video cameras to record field experiments or techniques in the laboratory. Video cameras can be connected to microscopes to record images of nematodes in action. Relatively inexpensive "frame grabbers" are now available that will digitally capture individual video frames. The captured frames can be treated as individual digital photographs and manipulated and stored using the computer.

Remote Sensing

The benefits to be derived from airplane and satellite imaging have barely been sampled by nematologists. Some dramatic photographs of crops known to be infested with nematodes have been taken from airplanes (e.g., Radewald, 1978, p. 41), demonstrating that nematode damage to some crops can be discerned and delimited from several hundred feet above the ground. Such photographs are excellent teaching examples to demonstrate the uneven distribution of nematodes within a field and nematode dissemination, and from which to illustrate the merits of various sampling strategies. Aerial and satellite images can be used to monitor vegetation status, and have potential use in assessing geographic prevalence of nematode disease (e.g., Denison et al., 1996; Skole & Tucker, 1993).

The federal government has established a $10 billion fleet of satellites and ground control stations (Herring, 1996; Kinal, 1996). We benefit from these everyday with respect to communications and weather forecasting. Some of these satellites have been placed in geosynchronous orbits (consistently orbiting in the same location in relationship to the earth) and are used by the military for navigation. The government also allows civilians to utilize this Global Positioning System (GPS), although the positioning accuracy is degraded slightly for civilian uses (Herring, 1996). For example, small hand-held GPS receiver positioning devices are commercially available that receive signals from several satellites simultaneously, and report the user's location (latitude, longitude, and altitude) on a screen. Airplane pilots and boat operators utilize these devices for navigation. Such positioning devices also will assist the user to return to a previous location whose coordinates have been stored in the device, or, to move to a new location entered by the user. A limitation of the civilian access to the GPS is that the accuracy is generally limited to ±50 m, although the inaccuracy can be circumvented using "differential GPS" so that ±1 m accuracy is attainable (Herring, 1996).

Recently, GPS receivers on a PC (personal computer) card have become available. By plugging the card into a portable computer, field position can be defined and computations and mapping in the field can be pursued (Kinal, 1996). In the near future such GPS receivers will be placed in automobiles, and in conjunction with onboard computer maps, will provide drivers real-time navigational directions. It will be possible to navigate by keying in coordinates and letting the computer guide.

In the future, detailed information on the spatial distribution of nematodes in relation to various environmental parameters will be digitally collated using geographical information systems (GIS) (Burrough, 1986). Such compilations of information will eventually allow historical assessment of nematode disease prevalence, variability, and dissemination.

For example, highly accurate GPS information will allow researchers to return to exactly the same spot in a given field, enabling long-term monitoring of nematode densities in research plots. As fields are surveyed, growers will be able to create detailed maps of soil type, drainage, nutrient status (Denison et al., 1996), and nematode infestation levels. Farm equipment linked to the survey database and real-time GPS apparatus will allow precision placement of fertilizer and nematicides as the machinery traverses the field.

Genome Analysis

The exciting analysis of nematode and plant genomes continues unabated, and the data generated are stored in a number of easily accessible computer databases. The *Caenorhabditis elegans* genome project is making great strides (Hodgkin et al., 1995). The data available on *C. elegans* are being managed through databases, such as "A *Caenorhabditis elegans* Database" (ACeDB) (Eeckman & Durbin, 1995). This database covers a wide range of information, and can be accessed via the World Wide Web (WWW) (http://probe.nalusda.gov:8000/acedocs), and the ACeDB source and binary code can be downloaded via anonymous ftp (file transfer protocol) at "cele.mrc-lmb.cam.ac.uk" from "pub/acedb." Computer-based transfer systems (WWW and ftp) are discussed later.

Nucleotide and protein sequences from plants, nematodes, and other organisms can be accessed through several databases, including the NIH database "GenBank." GenBank is the NIH database of all known nucleotide and protein sequences, and can be accessed via the WWW (http://www.ncbi.nlm.nih.gov/). The database entries include the name and taxonomy of the source organism, a concise description of the sequence, and a table of characteristics specifying coding regions and other sites of biological significance.

Computers

The computer is the most important tool for information transfer to emerge in recent years. The following discussion is limited to the basic equipment and jargon needed to relate computers to information transfer in nematology.

The physical computer equipment is called hardware. The electronic instructions used by computers to accomplish various tasks are called programs, applications, or software. Currently available computer technology is revolutionizing information transfer. With this technology, persons needing information on nematodes can obtain information from formal classes, textbooks, slides, and videos via computers, and can ask questions and receive answers via electronic mail, Internet relay chat, or on-line video conference, at any time of the day or night.

Sources of Information

Hundreds of books are available on learning to use a computer and on different applications. Many are devoted to specific programs and are frequently more helpful than the manuals actually provided with the software.

Hardware

Computer technology is changing so fast that today's state-of-the-art computer will be antiquated in approximately 2 yr. Since the beginnings of the PC revolution in the late 1970s, the "best" type of personal computer to purchase has been controversial. Selection should be based on the tasks required, the software preferred, computers used by colleagues and support available. Over time, the differences among the various brands are decreasing, and users now consider economics and reliability as primary determinants in selection. The ability to share information among all computers has improved recently.

Portable computers have been getting progressively smaller and more powerful, and have longer battery life. Some people use only a portable computer, and don't have a desktop model. The main disadvantages at the time of this writing of a portable compared to a larger desktop model would be in price and smaller screen-size and quality.

It is standard today that computers have modems that allow communication with other computers via cables, infrared transmission, or over telephone lines, and the capacity to send and receive faxes. Several models allow one to make drawings on the screen that can be saved or printed, and some will even convert handwriting into printed text. One company manufactures a watch that will hold the address book and calendar from a desktop computer. Many of the hand-held computers utilize a thin card slightly thicker and bigger than a credit card, called a PCMCIA card, that is inserted into a slot in the computer. The PCMCIA cards that function as modems, network cards, or data storage devices are available. Hand-held computers can be connected to desktop computers for data exchange. Limitations of hand-held computers at the present time include the minimal size of the keyboard and screen, limited memory, and relatively slow speed. However, it is important to maintain some perspective, as almost every one of these hand-held devices exceeds the power of state-of-the-art computers of 40 yr ago—computers that would have occupied an entire room!

Computer Printers

Printers used with computers are of three basic types known as dot matrix, ink jet, or laser printers. Ink jet and laser printers are available that produce high resolution black and white or color copy, and they can be used to prepare high-quality documents with print clarity equal to that of a book.

Important Extras for Information Transfer

A number of additional devices are available, through a computer store, magazine, or catalog, that can be plugged into a computer to increase the effectiveness of information transfer. A few of the most common and most useful follow.

A modem allows a computer to connect to and utilize a telephone line for the transfer of information. The user can send information to, or retrieve information from, another computer located anywhere in the world. Computers of different types can exchange data in a format useable by both. The Internet is accessible and is described in more detail in a later section.

Video projection units allow computer output to be projected onto a screen for an audience. Several different types of projection units are available (Chinnock, 1996). A projection panel plugs into the monitor port of a computer, and is placed on the platform of an overhead projector to allow the projection of the image on the monitor for viewing. Video projectors that do not require overhead projectors also are available. Projection panels, or projectors, are useful for demonstrating computer software, such as databases or population dynamics models, and can be used in place of 35-mm slides. Computer graphical presentation programs are being used more frequently by speakers to present their information. Computers combined with the new projection technology allow the development of standard presentations, but also allow speakers to develop presentations involving still images, video, and sound.

Many new, excellent forms of projection are available. The cable connections can be somewhat confusing if not used often, so it is advisable to give the equipment a trial run prior to the actual presentation. Although this equipment can be invaluable for demonstrations to small groups, the resolution of many of the available systems is limited by the projector light intensity and may not be adequate for presentations to large groups.

The CD ROM stands for compact disc read only memory. These drives are in wide use. Approximately 35 million CD ROM drives will be sold in 1996 (Bell, 1996). The CDS are the same thin, circular pieces of plastic that have become popular in recent years for playing recorded music. The CD ROMs covering a wide range of subjects and applications are commercially produced, and soon CDS with information on nematodes will be available. Recordable CD drives now cost less than $1000. These are useful for archival purposes, and can be used for storage of images such as photomicrographs of nematodes. The CD ROM disks can hold approximately 600 MB (megabytes) of information, facilitating storage of large files.

The next generation of CD ROM, termed DVD-ROM (digital versatile disc), will appear in 1997 (Bell, 1996). The DVD ROM will hold approximately 14 times more data (approximately 4.7 gigabytes, GB) than current CDS, and anticipated innovations will allow development of 17-GB DVD ROMs (Bell, 1996). Recordable DVD drives will soon follow.

Software

Software programs, or applications, refer to the instructions that a user loads onto the computer. Software is purchased, or is sometimes distributed for a minimal fee ("shareware"), and is available from a variety of sources on a variety of media. Programs also can be downloaded over a telephone line or network, or provided via CD ROM.

Operating Systems

The most common operating systems for personal computers are Macintosh DOS (disk operating system), MS DOS, and Windows 95. A system called UNIX is common on mainframe computers, and a microcomputer version called Linux is available. Operating systems, as well as software, are periodically upgraded and improved.

Applications for Information Transfer

For ease of explanation, the major uses of computers are divided here into several broad categories of tasks performed by the software. Some programs may address several tasks.

Word processing programs accomplish the equivalent of typing text with a typewriter with the added advantages of automated editing, spell checking, and grammar checking.

Spreadsheets provide a screen that resembles a sheet of paper composed of rows and columns and are used for entering and analyzing data. They provide a large number of automatic functions that act on the rows and columns to sort, summarize, and allow statistical analysis of data.

Statistics programs typically operate on data entered in a spreadsheet, with the treatments from an experiment listed in one column and individual variables that have been measured (such as nematode survival, yield, or gall rating) placed in succeeding columns. The amount of time saved by using a computer becomes immediately evident the first time one runs an analysis of variance program. Results that might take hours of hand calculation can literally be obtained in a matter of seconds.

Database programs are used to organize and manipulate large sets of data. Previous compilations of nematode parasites and their hosts have been available (Goodey et al., 1965), but they were only in book form. Database formats allow the user to enter, coordinate, sort, extract, easily rearrange, and print information. Data are separated into categories or "fields." For example, an address database would consist of separate fields for last name, first name, street, city, state, zip code, and telephone number. Each address in the database would be stored in a separate "record" similar to a card in a rolodex file. An example in nematology would be a database such as "NEMABASE," developed at the UCD (Caswell-Chen et al., 1995). NEMABASE contains summaries of nematode-host associations and related data such as soil type, study location, and reference.

Graphics or drawing programs provide a variety of tools and special effects for making publication quality illustrations, drawings, or 35-mm slides. Such programs often include a slide-show feature allowing entire presentations to be given directly from the computer.

Developing Your Own Software

Although most users will find an application ready for whatever they need to do, a number of programming languages are available that more experienced operators can use to write their own programs. Learning the basics of one of these computer languages can take time, so before making a decision to learn a language, the objectives of the programming effort should be carefully defined.

"Macro" and script writing commands are available within many spreadsheet and database programs. These are easier to learn than formal languages and

Table 14–2. Selected World Wide Web (WWW) sites that have information on nematodes.

Institution	WWW-URL
USDA	www.ars-grin.gov:80/ars/Beltsville/barc/psi/nem/home-pg.html
Univ. of Arizona	ag.arizona.edu/PLP/plphome.html
UCD	ucdnema.ucdavis.edu
UCR	nemweb.ucr.edu/nemhp.html
Univ. of Florida	www.ifas.ufl.edu/~ENTWEB/entomolo.htm
Univ. of Nebraska	ianrwww.unl.edu/ianr/plntpath/nematode/wormhome.htm
North Carolina State Univ.	www2.ncsu.edu/unity/lockers/project/plantpath/
Texas A&M Univ.	cygnus.tamu.edu/PLPA/dept_toc.html

will accomplish many of the tasks that previously could only be accomplished through formal program writing.

A number of population dynamics models have been developed for various nematodes. Descriptions of many of these are available in the literature.

Integrated Pest Management/Nematology Programs

Computer resources in nematology are literally increasing as we write. NEMABASE is a database on the host status of plants to plant-parasitic nematodes and contains over 35 000 records (Caswell-Chen et al., 1995). NEMAPLEX is a phytonematology teaching aid for the PC (Ferris, 1996, personal communication). A database for teaching general nematology using the WWW has been developed, and can be accessed on the UCD Nematology WWW server. Nematology home pages on the WWW are available for many nematology institutions (Table 14–2), and will provide an up-to-date guide of what is available.

A number of pest management programs have sections devoted to plant-parasitic nematodes, for example, the UC IPM Calex Cotton Model (Goodell et al., 1990). The INTERACT computer system run by the UC Integrated Pest Management Project (UCIPM) is accessible over the Internet and the WWW (http://www.ipm.ucdavis.edu). It contains nematode pest management guidelines, programs to calculate degree days, a nematode-grape model, a nematode population-dynamics model, weather data from throughout California, and a database of pesticide use in California.

Computer Networks

In 1966, the U.S. Department of Defense Advanced Research Projects Agency (ARPA) began using the phone system to link computers around the country, forming the ARPANet. Soon, nonmilitary computers were using the ARPANet approach and began to be Internet worked with the ARPANet, forming the Internet. The Internet now crosses international boundaries and, although the numbers change every day, there are approximately 3 million computers on the Internet connecting nearly 20 million users around the globe (Levine & Baroudi, 1994).

Access to the Internet provides the user with the following basic capacities: (i) sending and receiving e-mail, (ii) connecting to remote computers (Telnet),

(iii) transferring files from one computer to another (FTP), (iv) communicating with Internet users around the world in real time using Internet Relay Chat, (v) accessing news groups and bulletin boards via USENET, and (vi) searching the Internet for information on various subjects using WAIS, GOPHER, VERONICA, or ARCHIE (Levine & Baroudi, 1994). Each user on the Internet has an electronic address that allows him or her to be found on the Internet. This address is associated with a mailbox, which receives and holds e-mail. Many nematologists worldwide are active users of the Internet and communicate regularly via e-mail.

If Internet access is not provided, a number of private subscription services are available. The WWW allows Internet information to be viewed in a graphical format via a language (hyper text markup language, or HTML), that allows specification of links among documents on the Internet. Programs called "web browsers" (e.g., Mosaic, Netscape, Internet Explorer) allow users access to the WWW. Each site on the WWW has an address, or a "Uniform Resource Locator" (URL). Using a browser, you can go to specific sites by opening a location using the appropriate URL. Nematologists and other biologists are in the process of placing a variety of information on nematodes onto the WWW. An advantage of information presented on the WWW is that it can be continually upgraded with the most current data. This system allows users to easily monitor and obtain the most recent findings.

Easy graphical access to the Internet is a relatively recent development whose benefits are only just beginning to be realized. The explosion of interest in the WWW is nothing short of amazing. The worldwide availability of information that can be continually updated represents a knowledge transfer revolution. It is currently possible to track the stock market, go shopping, check satellite weather photographs, and consult physicians via the WWW. It is now commonplace that companies, magazines, and publishing houses have a presence on the WWW, and they include their WWW address in their advertising.

A number of institutions with nematology projects have information available on the WWW. The addresses of some nematology WWW sites are included in Table 14–2, and most of these servers have links to other sites.

DIAGNOSTIC LABORATORIES AND CLINICS

Importance to Information Transfer

Because of the microscopic nature of plant-parasitic nematodes, diagnostic laboratories are probably a more essential element in information transfer in nematology than in any other pest management discipline (Barker & Imbriani, 1984). Although symptoms observable in the field may indicate the presence of a nematode problem, the presence of plant-parasitic nematodes must be confirmed by microscopic examination. A possible exception would be if adult female cyst nematodes are visible on roots. Although the basics of nematode taxonomy and identification can be learned in regular courses or "short courses" offered by several universities, the ability to do this efficiently and accurately is a skill learned on the job while working for several years with a skilled nematologist. Consequently, some 150 people, or less, worldwide consider themselves skillful enough to diagnose nematode problems for others.

Diagnostic Providers

Some states have diagnostic laboratories associated with the state department of agriculture or a land-grant university (Barker & Imbriani, 1984; Imbriani, 1985). The services provided in each state and the laboratory policies regarding who may submit samples, costs for processing samples, and the time required vary widely. Several states (e.g., California, Oregon, Washington) have one or more privately run diagnostic labs. Charges are typically higher than for the state run laboratories. A number of laboratories run other types of agricultural samples (e.g., salinity or soil fertility) that will accept nematode samples and subcontract them to a nematology laboratory for processing. The local university, state nematologist, farm advisor or county agricultural commissioner will likely know the closest laboratories, approximate charges, and reputation.

There is no single best method for processing nematode samples, and it is likely that no two nematology diagnostic laboratories use identical procedures. The procedures used within a single lab vary depending on the crop and nematodes involved, and facilities available. Differences in nematode size and motility are only two reasons for multiple extraction methods. Because of this, inquiry should be made regarding which methods were used if they are not listed on the laboratory report. Some laboratories provide recommendations based on the nematodes found, and others will not. If it is necessary to go to another nematologist to discuss the results, information regarding which laboratory actually ran the sample and the extraction methods utilized is important. Preferred methods for taking nematode samples vary by crop and location. It is best to consult with the laboratory to determine the preferred method for the locality. Some important concepts regarding sampling are: (i) nematodes are not usually evenly distributed in a field—a sample should be made up of a number of subsamples; (ii) most plant-parasitic nematodes will be found within the root zone of the current or previous crop; and (iii) samples should be kept cool (from room to refrigerator temperatures) from the time they are taken until they arrive at the laboratory (Barker, 1985; Goodell & Ferris, 1981).

Diagnostic Services

Imbriani (1985) contains an excellent summary of the operation of a nematode diagnostic laboratory. Soil samples comprise the largest workload of such laboratories. If roots are included in the sample, a separate method will typically be utilized to recover endoparasitic species. In both of these types of samples, extracted nematodes are identified using dissecting or compound microscopes. Some laboratories also will conduct differential host-range tests as an aid in determining what root-knot nematode species are present, or to analyze for genetic diversity in a population, which might affect the use of resistant varieties. Isozyme electrophoresis is used by some laboratories to identify root-knot nematode species. Molecular techniques also are being developed for species identification of root-knot, cyst and lesion nematode. In some instances, molecular techniques are sensitive enough that they can be used to identify a single juvenile root-knot nematode to species (Powers & Harris, 1993).

VERBAL PRESENTATIONS

General Considerations

With the technology available today for remote or distance learning, it is interesting to consider the intangibles that still bring people together to listen to presentations by others. Why will we travel for a distance and pay to see a live theater performance when we could view the "same" performance on television or videotape in our own homes? It is interesting to contemplate some of the issues arising from distance learning; however, it is clear that personal appearances allow interaction between speaker and audience. The audience is provided an opportunity to ask questions in an interactive setting. This benefits the speaker and the audience, because the speaker is provided feedback on the efficacy of his or her presentation. In addition, growers and PCAs obtain credit for continuing education or for pesticide training. It is questionable whether impersonal video presentations can function in the same regard as live presentations. A few practical suggestions for successful presentations are worth consideration (Table 14–3).

Specific Considerations for Nematology

Presenting information on nematodes to audiences is difficult because nematodes are microscopic. Providing the audience with a hands-on demonstration of nematodes requires much more effort than showing the same thing with

Table 14–3. The following suggestions deal with aspects of a presentation that can be addressed to help assure a successful presentation.

Important points to consider	Specific suggestions
Visuals should not contain too much information	A maximum 10 lines of information per graphic—present the data on multiple visuals if necessary
Assess colors and fonts to assure visibility of graphics	Red-green color blindness is common, so avoid using these on same graphic
Avoid visuals that are too big for screen	Prepare visuals in horizontal or "landscape" format
Limit your presentation to the time allotted	Prepare the presentation to allow at least 5 minutes for questions
When speaking direct your attention to the audience	Avoid talking to the screen or blackboard
Speak clearly, at a moderate pace, and with sufficient volume	Repeat questions so all the audience can hear them and summarize key points
Check the room or venue prior to speaking	Determine if microphones, pointers, slide projector remote controls, lectern lighting, or other aids you may need are present
Preview your visuals in the venue	Make sure slides are properly oriented and that the carousel functions properly

35-mm slides or a drawing on the blackboard. The increased impact, however, is worth the extra effort. Virtually any dissecting microscope with a magnification of 60 to 100 × and substage illumination can be used. For more detailed viewing, Swift (Swift Instrum. Int. S.A., Japan), for example, makes a small (approximately 10 × 15 × 6 cm), relatively inexpensive field microscope with a battery-operated light. It has excellent resolution at magnifications of 40, 100 and 400. Spare batteries and bulbs are recommended accessories as are paper towels, dissecting needles, a small bottle of water, small petri dishes, and an extension cord. Nematodes extracted from any soil sample can be stored in a vial and later poured into a petri dish for display under a dissecting microscope. A slide made with water and sealed with nail polish the day before works well for high magnification observation. A demonstration with one or two microscopes can be set up in the back of a meeting room in a few minutes. Larger displays utilizing a number of microscopes and plants with symptoms take considerably more effort to put together, but always receive high ratings from participants in meetings. Plants with root-knot galls or cyst nematodes are effective for demonstrations.

Entire meetings or portions of them can be devoted to hands-on demonstrations rather than a series of verbal presentations. Other possibilities include a display of sampling equipment and methodology, extraction techniques, nematicide application equipment, insect-parasitic nematodes, computer databases, video tapes or films, and nematode biological control agents. At larger meetings, demonstrations can be set up in several different rooms or different parts of the same room and the group divided into subgroups, which rotate on a time schedule through the various sections.

Outdoor Presentations

Field days and agricultural tours require a considerable amount of coordination and preparation. However, taking the participants to the local experts and problems can be a more effective learning tool than bringing the experts and their 35-mm slides to the group. Standing in a field heavily damaged by nematodes while listening to a grower discuss the economic loss and the lack of adequate controls leaves a lasting impression. In addition to nematode-infested field sites, participants enjoy visits to local food processing industries. This is a situation where the local farm advisors can be an invaluable resource. If responsible for a field tour of several days' duration, planning 1 to 2 yr in advance is recommended.

An easel with a large tablet and colored markers are the usual substitute for a blackboard in the field. Basic information is frequently written on the tablet ahead of time and then added to during the presentation. Handouts also can be utilized to help present the information and taken home for later review. A shovel is useful for digging roots and looking for symptoms.

If the weather permits, setting up a microscope demonstration in the field is only slightly more difficult than in a meeting room. Usually enough natural light is available for a dissecting microscope, and higher power microscopes are available that utilize a battery-operated light. A table for the demonstration is useful, although the hood or trunk of a car or the tailgate of a truck can be used if needed. Handouts or pictures should be anchored in case of wind.

Research Plots

Research plots can be excellent ways to transfer technology and information to growers. Some argue that a plot established for demonstration does not need replicated treatments. However, nothing is as reassuring or convincing as being able to look around a plot and see the same difference in various replicates. Consider giving each grower a plot map without the treatments labeled and asking them to rate the visual differences on a numerical scale for possible use in a future publication.

Still or video pictures taken during the course of a field experiment can be invaluable in explaining the significance of the results. Many growers will be more convinced by a photograph of the results than by seeing tables of data. Experienced farm advisors often sacrifice randomization by organizing treatments in a progression that will clearly demonstrate treatment affects to growers (e.g., low rate–high rate in a chemical efficacy trial or similar plant varieties in a cover crop trial). Field days should be coordinated with cooperating growers in advance to avoid any misunderstandings (e.g., irrigation or spray schedules that prevent entry).

WRITTEN PRESENTATIONS

The Society of Nematologists Extension Committee has recently compiled a list of publications available in various states (1988).

For more than 50 yr, nematologists have prepared nematode-management recommendations for various crops and locations. In some states these are now available on computer, as well as in a written format. For example, the UC Pest Management Guidelines for different crops are available on the IMPACT computer system. These publications are typically oriented towards a particular crop rather than an individual nematode species. Such a publication may contain a listing of the nematodes likely to cause problems, symptoms and signs of nematode damage, directions on how to take soil and plant tissue samples, and general management recommendations. This electronic format has the advantage of being easily upgraded when new practices are developed or when chemical recommendations change, and the disadvantage of not being able to provide the depth of information frequently needed for nonchemical means of control. Growers are typically more aware of the presence of insects and some other plant pests than they are of nematodes. Guidelines for managing nematodes may be included with those for other pests on the same crop.

The Commonwealth Institute of Helminthology (CIH, St. Albans, Herts, United Kingdom) has published an excellent series, called Descriptions of Plant Parasitic Nematodes, that is available by mail order. These are black- and white-publications describing individual nematode species. Each description includes drawings and photographs of the nematodes, morphological measurements, type host and locality, systematic position, distribution and hosts, biology and life-history, host-parasite relationships, and control. Some contain photographs of symptoms as well.

The APS (Minneapolis, Minnesota) publishes an excellent series of compendia on plant pathogens on particular crops. They include color photographs of damage symptoms and literature citations, and each contains a chapter on nematodes.

The UC Integrated Pest Management Program publishes manuals on IPM for various commodities. Over the years the Florida Department of Agriculture and Consumer Services, Division of Plant Industry, has published an excellent series of circulars on nematodes and nematode diseases.

Grant Proposals

The opportunities for technology transfer that can take place through the grant process are easily overlooked. Some agencies providing research funding (e.g., commodity organizations) include reviewers who are not researchers. Most commodity groups have a grower committee that reviews proposals and provides advice on what should be funded. Presentations should be prepared with this in mind, as this situation could be a "teachable moment." Some organizations prepare and distribute research summaries to the media, various government, and nonprofit agencies and may use proposals in preparing them.

OUTLOOK

The dramatic developments in information acquisition and dissemination are sure to change and improve nematology teaching, research, and extension. It is clear that several new technologies are becoming more widely used in nematology. Faster computers connected to the WWW will allow the rapid exchange of information. For research and extension, this will allow access to databases or electronic manuscripts that might otherwise be obscure. For teaching, this will allow the delivery of text, images and video to multiple, distant locations. This will enhance learning, and expand the audience of teachers and extension personnel. The new technology provides new tools to transfer information about nematode diseases and their management to our clientele. The linkage of large volumes of related data will be facilitated by the use of high-density digital storage. Integrated data sets on nematode taxonomy, host range, and genetics may aid the identification, diagnosis, and management of nematode diseases.

Molecular markers will assist the characterization and accurate identification of nematode isolates, allowing management to be matched to characteristics of the isolate. Such information will advance the development and application of unique nematode management tactics.

The GPS and GIS technology will foster development of more detailed information on the distribution and abundance of parasitic nematodes in the field. This information will be used in conjunction with new farm machinery that will allow precise application of fertilizers, nematicides, or even resistant cultivars within certain areas of a field where they are needed.

All of these new information technologies will advance the abilities of agricultural professionals to identify and monitor nematode problems in the field, research a wide range of possible solutions, and apply appropriate management tactics in a site-specific manner.

REFERENCES

Abrantes, I.M.De O., and N.G.M. Hague. 1992. Seminar: Nematology teaching. Nematologica 38:455–458.

Asimov, I. 1982. Asimov's biographical encyclopedia of science and technology. Doubleday, New York.

Barker, K.R. 1985. Sampling nematode communities. p. 3–17. *In* K.R. Barker et al. (ed.) An advanced treatise on *Meloidogyne*. Vol. 2. North Carolina State Univ. Graphics, Raleigh.

Barker, K.R., and J. L. Imbriani. 1984. Nematode advisory programs—status and prospects. Plant Dis. 68:735–741.

Bell, A.E. 1996. Next-generation compact discs. Sci. Am. 275:42–46.

Burrough, P.A. 1986. Principles of geographical information systems for land resources assessment. Oxford Univ. Press, Oxford.

Caswell, E.P., and W.J. Apt. 1989. History of pineapple nematode research in Hawaii: Past, present, and future. J. Nematol. 21:147–157.

Caswell-Chen, E.P., V.M. Williamson, and B.B. Westerdahl. 1993. Applied biotechnology in nematology. Suppl. J. Nematol. 25:719–730.

Caswell-Chen, E.P., H. Ferris, B.B. Westerdahl, and R.L. Sloan. 1995. A PC/MAC-platform database on the host status of crop and weed species to plant-parasitic nematodes. Nematol. Newsl. 41:7–8.

Caveness, F.E. 1964. A glossary of nematological terms. F. E. Caveness, Ibadan.

Chinnock, C.B. 1996. Aurevoir, overheads. Byte 21:34.

Chitwood, B.G., and M.B. Chitwood. 1950. An introduction to nematology. Section 1. rev. ed. B.G. Chitwood, Washington, D.C.

Christie, J.R. 1959. Plant nematodes: their bionomics and control. Univ. Florida Press, Gainesville.

Czymmek, K.J., J.H. Whallon, and K.L. Klomparens. 1994. Confocal microscopy in mycological research. Exp. Mycol. 18:275–293.

Denison, R.F., R.O. Miller, D. Bryant, A. Abshahi, and W.E. Wildman. 1996. Image processing extracts more information from color infrared aerial photos. Calif. Agric. 50:9–13.

Dickson, D.W., and J.A. Meredith. 1987. Perspectives on nematology teaching. p. 22–27. *In* J.A. Veech and D.W. Dickson (ed.) Vistas on nematology: A commemoration of the twenty-fifth anniversary of the Society of Nematologists. E. O. Painter Print. Co., DeLeon Springs, FL.

Dowler, W.M., and S.D. Van Gundy. 1984. Importance of agricultural plant nematology. p. 1–12. *In* W.R. Nickle (ed.) Plant and insect nematodes. Dekker, New York.

Dunn, R.A. 1987. Perspectives on extension nematology. p. 32–37. *In* J.A. Veech and D.W. Dickson (ed.) Vistas on nematology: A commemoration of the twenty-fifth anniversary of the Society of Nematologists. E. O. Painter Print. Co., DeLeon Springs, FL.

Dropkin, V.H. 1980. Introduction to plant nematology. John Wiley & Sons, Inc., New York.

Eeckman, F.H., and R. Durbin. 1995. ACeDB and Macace. p. 583–605. *In* H.F. Epstein and D.C. Shakes (ed.) *Caenorhabditis elegans*: Modern biological analysis of an organism. Acad. Press, New York.

Eisenback, J.D. 1985. Techniques for preparing nematodes for scanning electron microscopy. p. 79–105. *In* K.R. Barker et al. (ed.) An advanced treatise on *Meloidogyne*. Vol. 2. Methodology. North Carolina State Univ. Graphics, Raleigh.

Evans, K., D.L. Trudgill, and J.M. Webster (ed.). 1993. Plant parasitic nematodes in temperate agriculture. CAB Int., London.

Flint, M.L. 1990. Pests of the garden and small farm: A growers guide to using less pesticide. Statewide Integrat. Pest Manage. Program Publ. 3332.

Goodell, P., and H. Ferris. 1981. Sample optimization for five plant-parasitic nematodes in an alfalfa field. J. Nematol. 13:304–313.

Goodell, P., R.E. Plant, T.A. Kerby, J.F. Strand, L.T. Wilson, L. Zelinski, J.A. Young, A. Corbett, R.D. Horrocks, and R.N. Vargas. 1990. Calex-Cotton: An integrated expert system for cotton production and management. Calif. Agric. 44:18–21.

Goodey, J.B., M.T. Franklin, and D.J. Hooper. 1965. The nematode parasites of plants catalogued under their hosts. 3rd ed. Tech. Commun. Commonw. Bur. Helminth., no. 30, London.

Goodey, T. 1963. Soil and freshwater nematodes. 2nd ed. Methuen and Co. Ltd., London.

Grotta, D., and S.W. Grotta. 1996. Digital cameras: The death of film? PC Magaz. 15:145–171.

Hall, D.H. 1995. Electron microscopy and three-dimensional image reconstruction. p 395–436. *In* H.F. Epstein and D.C. Shakes (ed.) *Caenorhabditis elegans*: Modern biological analysis of an organism. Acad. Press, New York.

Herring, T.A. 1996. The global positioning system. Sci. Am. 274:44–50.

Hodgkin, J., R.H.A. Plasterk, and R.H. Waterston. 1995. The nematode *Caenorhabditis elegans* and its genome. Science (Washington, DC) 270:410–414.

Hogan, K. 1996. Digital cameras 101: Advice from the pro. Imag. Mag. 5:52–60.

Imbriani, J.L. 1985. The practical application of a nematode advisory service. p. 309–323. *In* J.N. Sasser and C.C. Carter (ed.) An advanced treatise on *Meloidogyne*. Vol. 1. North Carolina State Univ. Graphics, Raleigh.

Jenkins, W.R., and D.P. Taylor. 1968. Plant nematology. Reinhold Publ. Co., New York.

Kinal, G.V. 1996. Where on earth am I? Byte 21:175–178.

Levine, J.R., and C. Baroudi. 1994. The internet for dummies. 2nd ed. Int. Data Group Books, Foster City, CA.

Luc, M., R.A. Sikora, and J. Bridge (ed.). 1990. Plant parasitic nematodes in subtropical and tropical agriculture. CAB Int., London.

Mai, W.F., P.G. Mullin, H.H. Lyon, and K. Loeffler. 1996. Plant-parasitic nematodes: A pictorial key to genera. 5th ed. Cornell Univ. Press, Ithaca, NY.

Maggenti, A. 1981. General nematology. Springer-Verlag, New York.

Marshall, E. 1996. Hot property: Biologists who compute. Science (Washington, DC) 272:1730–1732.

Moens, M., and P.W.Th. Maas. 1992. Seminar: A view on extension nematology and its relation to research in Europe—results of an inquiry. Nematologica 38:474–481.

Nickle, W.R. (ed.). 1984. Plant and insect nematodes. Marcel Dekker, Inc., New York.

Nickle, W.R. (ed.). 1991. Manual of agricultural nematology. Marcel Dekker, Inc., New York.

Nickle, W.R., and H.E. Welch. 1984. History, development, and importance of insect nematology. p. 627–653. *In* W.R. Nickle (ed.) Plant and insect nematodes. Dekker, New York.

Peachey, J.R., and M.R. Chapman. 1966. Chemical control of plant nematodes. Tech. Commun. Commonw. Bur. Helminth., no. 36. Commonw. Bur. Helminthol., St. Albans, England.

Powers, T.O., and T.S. Harris. 1993. A polymerase chain reaction method for identification of five major *Meloidogyne* species. J. Nematol. 25:1–6

Radewald, J.D. 1969. The role of agricultural extension in nematology—past, present and future. p. 333–340. *In* J. E. Peachey (ed.) Nematodes of tropical crops. CAB Int., London.

Radewald, J.D. 1978. Nematode diseases of food and fiber crops of the southwestern United States. p. 41. Univ. Calif. Div. Agric. Sci. Publ. 4083.

Robbins, R.T., R.V. Anderson, W.I. Decraemer, Z.A. Handoo, M. Mundo-Ocampo, and E.M. Noffsinger. 1992. Why "voucher" specimens are important to you, the researcher. Nematol. News. 38:14–19.

Roberts, P.A., W.C. Mathews, and J.C. Veremis. 1998. Genetic mechanisms of host-plant resistance to nematodes. p. 209–238. In K.R. Barker et al. (ed.) Plant and nematode interactions. Agron. Monogr. 36. ASA, CSSA, and SSSA, Madison, WI.

Roberts, P.A., and I.J. Thomason. 1989. A review of variability in four *Meloidogyne* spp. measured by reproduction on several hosts including *Lycopersicon*. Agric. Zool. Rev. 3:225–252.

Sasser, J.N. and W.R. Jenkins (ed.). 1960. Nematology. Fundamentals and recent advances with emphasis on plant parasitic and soil forms. Univ. North Carolina Press, Chapel Hill.

Skole, D., and C. Tucker. 1993. Tropical deforestation and habitat fragmentation in the Amazon: Satellite data from 1978 to 1988. Science (Washington, DC) 260:1905–1910.

Society of Nematologists Extension Committee. 1988. Bibliography of U.S. extension and extension-related publications on nematodes. Ann. Appl. Nematol. 2:121–136.

Southey, J.F. (ed.). 1965. Plant nematology. no. 7. 2nd ed. Tech. Bull. Min. Agric. Fish. Food, London.

Southey, J.F. (ed.). 1970. Laboratory methods for work with plant and soil nematodes. no. 2. 5th ed. Tech. Bull. Min. Agric. Fish. Food, London.

Southey, J.F. (ed.). 1982. Plant nematology. Ref. Book 407. Min. Agric., Fish. Food. HMSO, London.

Tarjan, A.C. 1960. Check list of plant and soil nematodes. Univ. Florida Press, Gainesville.

Tarjan, A.C. 1967. Supplement to check list of plant and soil nematodes. Univ. Florida Press, Gainesville.

Taylor, A.L. 1967. Introduction to research on plant nematology. An FAO guide to the study and control of plant-parasitic nematodes. Food Agric. Organiz., U.N., Rome.

Thorne, G. 1961. Principles of nematology. McGraw-Hill, New York.

Wallace, H.R. 1963. Biology of plant parasitic nematodes. Edward Arnold, London.

Wickramasinghe, H.K. 1989. Scanned-probe microscopes. Sci. Am. 261:98–105.

Zuckerman, B.M., W.F. Mai, and L.R. Krusberg (ed.). 1990. Plant nematology laboratory manual. rev. ed. Univ. Massachusetts, Amherst.

15 Soybean

ROBERT A. KINLOCH
University of Florida
Jay, Florida

Soybean, *Glycine max* (L.) Merr., is an annual herbaceous legume that provides for half of the world's oilseed and protein meal. Though originating and grown in eastern Asia for many centuries, the crop has been amenable to breeding for adaptive traits and is now grown throughout many latitudes from temperate to subtropical. The bulk of soybean production is now in the Americas, with the USA accounting for slightly less than half of the world's supply. Brazil, China, and Argentina are the next major producers. Soybean growth and maturity respond to day length, and cultivars are assigned to 12 maturity groups according to their responsiveness. Group 00, 0, and I are adapted to the longer days in higher latitudes of southern Canada and the northern states of the USA. Succeeding groups are adapted further south. Group VIII cultivars are the latest grown in the USA, primarily in Florida, southern Alabama, and southern Georgia. Groups IX and X are grown in the subtropics (Hartwig, 1973).

Throughout most of its range, soybean is susceptible to an assemblage of soilborne nematode pathogens that feed both endo- and ectoparasitically on the plant. Plant-parasitic nematodes often become limiting factors of production in many areas when crop or disease management practices are employed that favor particular nematode species.

SOYBEAN CYST NEMATODE

Soybean cyst nematode, *Heterodera glycines*, was described in 1952 and was determined to be the causal agent of Yellow Dwarf, a soybean disease that had been recognized for many decades in northern China, Japan, and Korea. Infestations have since been identified in Canada, Colombia, Indonesia, and eastern Siberian Russia (Young, 1992). In the USA, the nematode was first detected in North Carolina in 1954. It is now known to occur in 26 eastern states from Minnesota to Florida. Recent surveys have found this nematode to be widely distributed in Maryland (Sindermann et al., 1993), Missouri (Niblack et al., 1993) and Michigan (Warner et al., 1994).

Biology

The soybean cyst nematode is a genetically heterogeneous species with field populations exhibiting variability in their ability to parasitize certain soybean cultivars. Early efforts resulted in the identification of soybean germplasms with resistance to specific populations of this pathogen (Ross, 1962; Smart, 1964). Four physiological races of the nematode were eventually accepted based on their ability to reproduce on four differential soybean germplasms (Golden et al., 1970). To accommodate subsequent populations that did not fit the initial four-race scheme, the classification was expanded to describe 16 possible races based on the same soybean germplasm differentials (Riggs & Schmitt, 1988: Niblack, 1992). To date, eight of these additional races have been reported (MacDonald et al., 1980; Riggs et al., 1981; Lehman & Dunn, 1987; Schmitt & Barker, 1988).

Two to six hundred eggs (100 by 47 μm) of this nematode typically are retained within the encysted body wall of the dead female. However, up to one-third of the eggs may be found within a gelatinous matrix exuded from the posterior end of the cyst. Following a molt within the egg, the second-stage juvenile nematode (450 by 18 μm) hatches and moves through the soil to penetrate the soybean 1 cm or more behind the root tip. Following migration to the vascular tissue, the juvenile starts feeding on an induced syncytium (food cell) and becomes sedentary. Three additional molts are passed as the nematode enlarges to an adult. The adult males (1300 by 30 μm) revert to a vermiform shape and move from their feeding site through the rhizosphere in search of the females. Reproduction is bisexual. The females remain sedentary and, while erupting through the root surface, enlarge to lemon-shaped proportions of 470 to 790 by 210 to 580 μm. Males are pheromonally attracted to the sedentary females and are found around the posterior of the latter during copulation (Rende et al., 1982). The cyst color changes from white to yellow as it matures and on death becomes dark brown and filled with eggs. The life cycle from root infection to egg is completed within temperatures of 10 to 35°C and optimally takes approximately 22 d at 21 to 24°C. A few generations, predominantly from eggs or juveniles that have been released from the cyst, may occur within a crop season. The majority of the eggs remain within the cyst. Their hatching is slow, and only half may be released by the next crop season. A genetically induced egg dormancy is implicated (Hill & Scmitt, 1989).

Dormancy could be induced by declining temperatures, and host phenology may be involved (Yen et al., 1995). Dormancy may be partially broken by increasing soil temperatures and exposure to host root exudates. Survival of eggs within the cyst is favored by cool moist conditions, and eggs may remain viable for three or more yr in the absence of a host (Moore, 1984). Fungal contaminants in the cysts are implicated for a shorter endurance in the more southerly ranges of this nematode (Morgan-Jones et al., 1981; Kim & Riggs, 1991). Cyst protection and egg endurance has considerable bearing on dispersion of the species and management of disease. Cysts are readily spread in contaminated soil that clings to farm machinery. Movement of soil by wind, water, seed stocks, and animals are implicated in the spread of this disease (Epps, 1969a,b).

Fig 15–1. Early season symptoms of soybeans grown in soil heavily infested with soybean cyst nematodes, *Heterodera glycines*.

Symptoms

Disease symptoms are nonspecific, and their severity varies with nematode population density and ameliorating conditions such as nutrient supply and ample water. When soybean is grown under ample moisture conditions, the syncytia induced by the soybean cyst nematode are more likely to be found in the root cortex rather than the stele, hence, susceptible cultivars may exhibit considerable tolerance to infection in well-irrigated fields (Johnson et al., 1993). Damage is usually greater in sandier soils. Stunted plants may exhibit foliar chlorosis and roots that exhibit varying degrees of brown discoloration to necrosis. Bacterial root nodulation can be restricted, and severely infected plants die before normal maturation. Patches of stunted plants may be observed as early as 6 wk following seeding into dense infestations (Fig. 15–1). Though a hand lens is recommended, eruptive white females can be readily seen on the roots with unaided vision (Fig 15–2). Several may be observed on a 1-cm length of root in dense infections. Cysts become difficult to see as they mature to the brown cyst stage. Confirmation of an infestation is best provided by extraction of cysts and infective second-stage juveniles from soil samples. For diagnostic purposes, samples may be taken from specific sites once symptoms are suspected. However, for management decisions, it is preferable to remove several samples from a wide area in the field after the crop has been harvested.

ROOT-KNOT NEMATODES

Root-knot disease of soybean in the USA is largely confined to the southeastern states, south of 35°N lat. Thirty-five percent of soybean fields in North

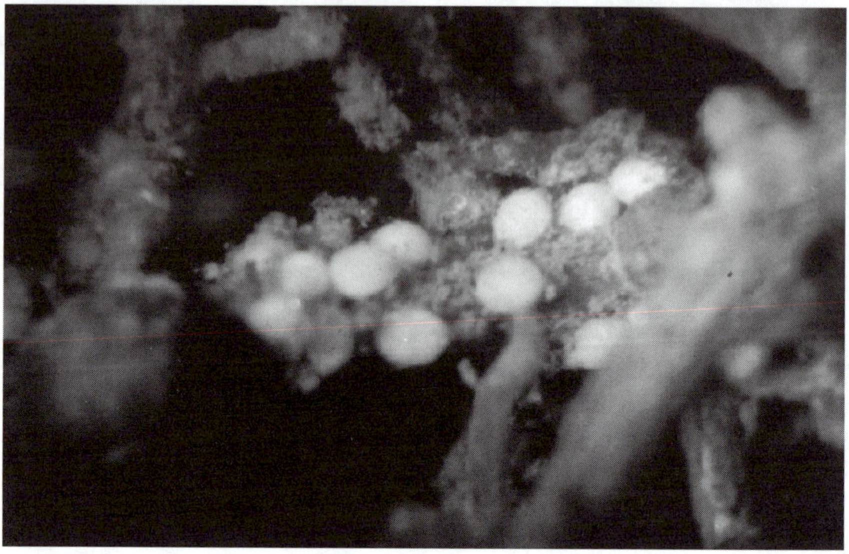

Fig 15–2. Portion of soybean roots infected with a cluster of white females of the soybean cyst nematode, *Heterodera glycines*.

Carolina (Schmitt & Barker, 1988) and 28% in South Carolina (Lewis et al., 1993) were found to be infested with root-knot nematodes. Similar to higher frequencies of infestation are found in Georgia, Alabama, Florida, Mississippi, and Louisiana. Several species of root-knot nematodes are known to infect soybean. *Meloidogyne incognita* is the most prevalent, but other species may be more common in certain regions. *M. javanica* is the predominant species found on soybean produced in fields that had been traditionally planted to tobacco (*Nicotiana tabacum* L.) in north central Florida (Garcia & Rich, 1985). *Meloidogyne hapla* is likely to be the causal species in the more northerly reaches of root-knot disease in North Carolina and Virginia. Concomitant infestations can be frequent and the recovered species may be a result of recent cropping. Of the two races of *M. arenaria*, Race 2 (nonpathogenic to peanut, *Arachis hypogaea* L.) appears to be the more pathogenic to soybean (Carpenter & Lewis 1991; Pedrosa et al. 1994). During the 1970s, possibly as a result of the use of *M. incognita*-resistant soybean cultivars and the decline in the use of fumigant nematicides, *M. arenaria* became the more prevalent species in certain localities (Kinloch, 1980). A reverse in this trend may be imminent with the current increase in the planting of cotton (*Gossypium hirsutum* L.), which is a host to only *M. incognita*. Although both *M. incognita* and *M. javanica* are prevalent on soybean in Argentina (March et al., 1985a,b), the latter is the predominant species in Brazil.

Biology

Though survival in the egg stage may be prevalent in the northern reaches of its range, *M. incognita* survives the winter months predominantly as a second-stage infective juvenile. These juveniles are inactive, and their soil densities

Fig 15–3. Severely galled roots of susceptible soybean 'Hartz 8112' infected with *Meloidogyne incognita* (left) and roots of relatively gall free resistant soybean 'Centennial' (right).

remain static in the cool winter months with numbers ranging from a few to as high as 3000 per 100 cm^3 soil. Juveniles become active as soil temperatures climb in the spring. A decline in nematode populations occurs if host plants are absent. Populations can become exceedingly low such that they may escape detection in May to June. Juveniles can be expected to occur below normal sampling depths in deep sandy soils. Penetration of the developing soybean root is at or immediately behind the root tip. Juveniles migrate only short distances before settling and feeding. Host response to feeding is the development of multinucleate giant cells from which the nematode gains sustenance. Surrounding parenchymatous cells become greatly enlarged. The plant root swells to produce galls, which are the characteristic symptoms of infection (Fig 15–3). Nematode development to an adult proceeds by molting through third- and fourth-juvenile stages. Though some juveniles become redundant males and revert to a vermiform shape, the majority of them become females that, when fully grown, are pear- to lemon-shaped (600–800 μm long and 300–500 μm wide) with their posteriors protruding through the root surface. One to several females may reside within a gall. Reproduction is parthenogenetic, and eggs are deposited into a posteriorly protruded egg sac within 4 wk of root invasion. Eggs hatch in the soil, and root reinfection is rapid enough that several overlapping generations occur within the crop season.

Symptoms

Shoot growth symptoms to root-knot nematode invasion are not determinative of soybean infection. Associated stunting and chlorosis are typical of general nutrient deficiencies. Root galling is, however, characteristic. Small galls

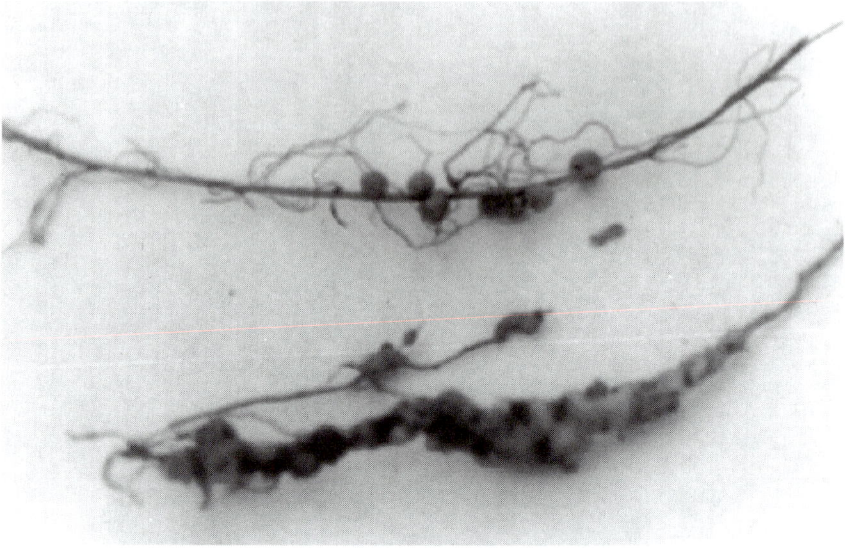

Fig 15–4. Healthy soybean rootlet showing rhizobial nodules and lateral roots (top) and root-knot nematode infected rootlet showing root-knot nematode galling and a reduced number of lateral roots (bottom).

may appear within 2 to 3 wk of planting (Fig. 15–4) but become large and very evident by August. Rhizobial nodulation is considerably reduced when roots become densely infected with root-knot nematodes. The southern blight fungus (*Sclerotium rolfsii*) is often associated with root-knot nematode infection. Its presence may cause root gall deterioration and early plant death. Plants heavily infected with root-knot nematodes can be expected to mature 1 to 2 wk earlier than normal.

RENIFORM NEMATODE

The reniform nematode, *Rotylenchulus reniformis*, is widely distributed throughout the tropics and subtropics. It is an obligate parasite of many agricultural crops, particularly in the Leguminosae, Malvaceae, Cucurbitaceae, and Solanaceae. Host range studies of specific crops have produced varying results due to the existence of populations with distinctive host preferences (Dasgupta & Seshadri, 1971). A serious pest of soybean and cotton in the southern USA (Heald & Thames, 1982), reniform nematode is found in North Carolina, South Carolina, Georgia, Florida, Alabama, Mississippi, Louisiana, Texas, Arkansas, and Missouri.

Biology

Vermiform second-stage juveniles of this nematode are found free in the soil. They do not feed, but rapidly molt through third and fourth stages to become adults. Females partially or totally penetrate the plant root-cortex where they become sedentary. The female swells to a characteristic kidney shape and deposits usually less than a hundred eggs into a gelatinous sac, which protrudes into the soil. The vermiform males do not enter plant tissue and are found congregating in and around the egg sacs. Both amphimictic and parthenogenetic reproducing populations are known (Nakasono, 1978). The life cycle lasts slightly less than 4 wk, and several overlapping generations can occur within one growing season.

Symptoms

Though females induce the development of a syncytium (food cell) at the feeding site, the root does not swell in response to feeding as with root-knot nematode but becomes necrotic and stunted when heavily infected. Parasitized plants exhibit varying degrees of stunting and chlorosis. Initial field infestations may exhibit patchy areas of poor growth. This nematode, however, can resist desiccation (Birchfield & Martin, 1967) and is rapidly spread. Hence, damaged plants are generally distributed throughout a field. Uprooted infected plants frequently display soil particles adhering to the gelatinous egg sacs and may give the root a grainy or rough appearance. The nematode is best detected by extraction from soil samples taken just prior to and for a few months after soybean harvest. Numbers exceeding 10 000 per 100 cm^3 soil are frequently encountered from established infestations.

COLUMBIA LANCE NEMATODE

The Columbia lance nematode, *Hoplolaimus columbus*, is a particularly troublesome pest of soybean in South Carolina (Lewis et al., 1993). Infestations also occur in North Carolina, Georgia, and Alabama. In addition to soybean, the nematode reproduces on bermudagrass [*Cynodon dactylon* (L.) Pers. var. *dactylon],* common bean (*Vicia faba* L.), cotton, lima bean (*Phaseolus vulgaris* L.), maize [*Zea mays* (L.) subsp. *mays*], millet (*Panicum miliaceum* L.), sugarcane (*Saccharum officinarum* L.), and wheat [*Triticum aestivum* (L.) emend. Thell.] (Fassuliotis, 1974b; Lewis & Smith, 1976).

Biology

Hoplolaimus columbus is a relatively large vermiform nematode that can reach 1750 μm in length. It is a migratory ecto- and endoparasite, and all of its life stages can be found in both the root and rhizosphere. All stages are infective. It reproduces parthenogenetically, and males are not found in the field though they have been recovered from laboratory cultures (Fassuliotis, 1974a). Females repeatedly lay single eggs. Greater numbers of this nematode are found in root tissue than in the soil during the growing season. However, as soybean approaches maturity, most of the nematode population returns to the soil (Perez et al., 1994) sending population densities to as high as 300 per 100 cm^3 soil.

Symptoms

Damage to the plant is primarily due to the destruction of cortical and epidermal structures as the nematode migrates through the root (Lewis, 1989). Root shedding and a restriction of lateral roots are pronounced in heavy infestations. Plants become stunted and chlorotic. Since there are no characteristic aboveground symptoms by which the presence of the nematode can be determined, it must be identified from root samples taken prior to plant maturity or from soil samples taken after harvest. This nematode is considerably resistant to desiccation and can survive for several months in the absence of a host.

STING NEMATODE

The sting nematode, *Belonolaimus longicaudatus*, is a severe pathogen of soybean. It has many other hosts including agronomic, vegetable, forage, and turf crops (Robbins & Barker, 1973). It has been found in Alabama, Arkansas, Kansas, Mississippi, Missouri, Louisiana, Oklahoma, New Jersey, Virginia, and Texas. However, it is particularly troublesome in localized areas characterized by deep sandy soils in the Atlantic Coastal Plain from North Carolina to Florida.

Biology

The sting nematode feeds ectoparasitically. It is one of the largest plant-parasitic nematodes and can reach up to 2600 μm in length. The nematode reproduces bisexually and approximately 40% of the population are males. Reproduction of sting nematode is greatest from 25 to 30°C when the life cycle takes approximately 28 d. Several generations can occur through the summer months. Its geographic distribution is governed largely by its temperature requirements and its sensitivity to soil type, preferring soils with a minimum of 80% sand and a maximum of 10% clay (Robbins & Barker, 1974).

Symptoms

Shoot symptoms of damaged plants display nondescriptive stunting that varies in intensity with the soil population densities of the nematode. Often clearly defined spots of damaged plants can be found in a field. Infected roots show varying combinations of stubby root, necrosis, and lesions. Although this can be a very damaging pest of soybean due to its confined distribution, the sting nematode is probably the least significant of the major nematode pathogens attacking soybean in the southeastern USA.

OTHER NEMATODE ASSOCIATES

Lesion nematodes, *Pratylenchus* spp., are rather stout vermiform nematodes less than 1000 μm long. All species are mobile endoparasites and their life stages may be found in both root tissue and soil. Depending on their level of infestation, host roots exhibit varying degrees of dark lesions caused by cell necrosis and invasion by fungi and bacteria. These nematodes are frequently found as associates of soybean but are generally considered to be minor pests of the crop. In North Carolina, *P. brachyurus* has been found to cause up to 30% soybean yield losses (Schmitt & Barker, 1980). However, damage to soybean appears to be very erratic and dependent on variations in environmental conditions between locations, seasons, and cultivars (Koenning et. al., 1985). Similar host reactions are attributed to *P. penetrans* (Schmitt & Barker, 1980). Management of *P. hexincisus* with a nematicide had no influence on soybean cultivar yields in Missouri (Niblack, 1992). However, *P. sefaensis* is implicated in reducing rhizobial nodulation and soybean yield in Senegal (Germani et al., 1984).

Soybean is recognized as an excellent host for the spiral nematodes, particularly *Helicotylenchus dihystera* and *H. pseudorobustus*. However, there is a lack of evidence to show concern over this relationship (Kinloch et al., 1971; Niblack, 1992). Similarly, populations of stubby-root nematodes, *Paratrichodorus* spp., stunt nematodes, *Tylenchorhynchus* spp., and ring nematodes, *Criconemella* spp., that are frequently found in the soybean rhizosphere, have not been shown to influence soybean yields (Riggs, 1992).

ECONOMIC IMPACT

The soybean cyst nematode is probably the most damaging nematode to soybean worldwide. A recent compilation of information concerning this nematode has been presented (Riggs & Wrather, 1992). In the north central region of the USA, where the bulk of the nation's soybean is grown, soybean cyst nematode causes more yield reduction than any other disease (Doupnik, 1993). An estimated yield loss of 4% is credited to soybean cyst nematode in Illinois, Indiana, and Missouri (Noel, 1992). This nematode is only slightly less important in the southern states. The more southern the latitude in the USA the more important become the root-knot, reniform, lance, and sting nematodes. For example, the estimated yield loss to soybean cyst nematode in North Carolina for 1991 was 7.0% as opposed to 1.5% for all other nematodes. Soybean cyst nematode accounted for 3% and other nematodes caused 7.0% of soybean yield losses in South Carolina. In Missouri a 7% soybean yield loss is credited to the soybean cyst nematode and negligible losses are assigned to other nematodes. Further south, soybean yield losses of 4% are credited to the soybean cyst nematode and 4% to other nematode pathogens in Louisiana (Sciumbato, 1993). Plant shoot symptoms produced by parasitic nematodes are not definitive. Cursory field observations may ascribe factors other than nematodes, such as low soil fertility levels and inadequate soil moisture, as the cause of poor soybean growth. It is likely that considerably more yield losses occur due to nematodes than is generally accepted.

MANAGEMENT

Crop Rotation

Crop rotation has long been recognized as an effective nematode management practice. Its utility is dependent on the availability of economically attractive alternative crops that are nonhosts to the target nematode. A grower's proactive adoption of rotation practices, suitable for the soybean parasitic nematodes indigenous to the area, is generally considered to be the best nematode management defense. Monoculturing, or the too frequent planting of soybean, can lead to a buildup of nematodes to such damaging levels that resort to planting possibly less profitable nonhost crops for a number of years may be the only recourse.

Since the soybean cyst nematode has a narrow host range among agricultural crops, crop rotation can be used effectively for its management. Recommended rotations vary depending on the availability of alternative crops and the density of the cyst population in the soil. Years out of soybean production is largely dependent on latitude. In the more southerly states of the USA, considerable hatching of the soybean cyst nematode occurs through the winter months. This, and the frequent presence of cyst parasites, reduces the overwintering populations. In addition, a dormancy factor may be involved that delays hatching in more northerly latitudes (Hill & Schmitt, 1989). Hatching ceases in the central and northern states during the fall, as soil temperature drops, but continues throughout the winter in the southernmost areas of the USA. In the northern USA, up to 3 yr of nonhost cropping may be required. Generally, two consecutive years of nonhost crops, such as maize or sorghum [*Sorghum bicolor* (L.) Moench], can reduce the cyst population in the soil to below damaging levels (Riggs & Schmitt, 1989). Cyst populations have been found to decline by 75% in 1-yr rotations and up to 92% in 2-yr rotations (Slack et al., 1981). In the southern states, peanut or cotton may be used as the rotational management crop and 1 yr of rotation may be sufficient in more southerly areas. The use of soybean cyst resistant soybean cultivars in a crop rotation is an effective tactic for managing soybean cyst nematode (Schmitt & Noel, 1984). However, to reduce the pressure on the nematode population and thus delay the selection of resistance-breaking populations (Young, 1982), inclusion of a susceptible cultivar with a good degree of tolerance in the rotation has been recommended (Anand & Koenning, 1986). Tolerant cultivars include Coker 488 (maturity Group VIII), Deltapine 417 (VII), and Northrup-King S72-60 (VII). Even where soil populations of cyst nematode are low, planting a tolerant rather than a resistant soybean cultivar may be more advantageous since there would be less selection pressure on the nematode population (Hussey & Boerma, 1989).

The root-knot nematodes are pests of soybean in the more southerly ranges of the crop's production area. Though there is a larger selection of alternative field crops that can be grown in this region, root-knot nematodes have broad host ranges, and alternative cropping choices can be complicated by possible multiple nematode pest infestations. These can include lance nematode, reniform nematode, soybean cyst nematode, and other root-knot nematode species (Weaver et al, 1995). This requires that a considerable emphasis be placed on frequent assays for soil nematode constituents. Usually populations of *M. incognita* can be reduced to, or maintained at, below damaging levels with 1 yr out of soybean production and cropping with nonhost peanut, or with lesser hosts as sorghum and maize (Kinloch, 1986). Cotton, a host for both *M. incognita* and *R. reniformis*, cannot be used in rotation with soybean for managing either nematode. Fifty commercial maize hybrids maintained *R. reniformis* below initial population levels indicating they were relatively poor hosts for the nematode (Windham & Lawrence, 1992) and are thus good alternative crops for rotational management of both reniform and root-knot nematodes on soybean. Among the major agronomic crops, peanut and cotton are nonhosts of *M. arenaria* Race 2 and can be grown in rotation with soybean for management of this nematode (Kinloch & Dunavin, 1993).

Hoplolaimus columbus has a wide host range including many crops that are normally rotated with soybean. Thus, maize and cotton cannot be recommended. Of the major nonhost crops that may be used as alternatives for management are ryegrass (*Lolium multiflorum* Lam.), bahiagrass (*Paspalum notatum* Flugge), and peanut. Caution is emphasized with the latter because of fungal pathogens common to both crops. Small hectarage infestations may be rotated to sweet potato [*Ipomoea batatas* (L.) Lam.], tomato (*Lycopersicon esculentum* Miller), watermelon [*Citrullus lanatus* (Thunb.) Matsum & Nakai var. *lanatus*], and pepper (*Capsicum annuum* L.); all of which are nonhosts to this lance nematode.

The use of winter wheat preceding a soybean crop has been shown to reduce populations of soybean cyst nematode (Koenning & Anand, 1991) and lesion nematode (Koenning et al., 1985). In both instances reductions in the populations of the nematodes attacking soybean was due to the delayed planting of soybean.

Resistance

Over the last 30 yr, considerable public and private research has been placed in developing cultivars resistant to the nematodes that attack soybean. In soybean, genetic resistance to the soybean cyst nematode is complex and controlled by multiple genes, both dominant and recessive (Anand, 1994). A recently compiled list of soybean cyst resistant cultivars includes 18 from maturity Group I, 42 from Group II, 129 from Group III, 79 from Group IV, 48 from Group V, 15 from Group VI, 15 from Group VII, and 4 from Group VIII (Anonymous, 1994). The preponderance of these cultivars have resistance to soybean cyst nematode Races 3 and 4. However, resistance to Races 1, 9, and 14 are represented. One public cultivar, Hartwig in maturity Group V, is resistant to all known races of this nematode. Caution is recommended against the monoculture of resistant cultivars due to the pressure such a practice exerts for the selection of resistance breaking races within heterogeneous soybean cyst nematode populations (Young, 1994).

Resistance in soybean against the parthenogenetic root-knot nematode, *M. incognita*, is more quantitative in nature than it is against the soybean cyst nematode. Consequently there is less restriction against monoculturing root-knot resistant cultivars. Nonetheless, the widespread prevalence of soybean cyst nematode

Fig 15–5. Susceptible soybean Hartz 8112 (left) and resistant soybean 'Kirby' (right) growing in soil infested with the root-knot nematode, *Meloidogyne incognita*.

and other soilborne pathogens of soybean argues against the monoculture of soybean regardless of cultivar resistance. *Meloidogyne incognita*-resistant cultivars are mostly in maturity Group VI and above, being adapted to the warmer latitudes where root-knot nematodes are most prevalent (Fig 15–5). Some slight galling and partial yield loss of *M. incognita*-resistant cultivars can be expected when planted in heavily infested soil (Kinloch, 1992). Resistance selection against *M. arenaria* has not been so thoroughly developed, and cultivars considered as resistant to *M. arenaria* do not have the comparable resistance as those for *M. incognita* (Kirkpatrick & May, 1989). Recent findings of soybean germplasm in PI 12300538, PI 230977, and 'Jackson', with considerable resistance to this nematode, may provide agronomically acceptable resistant cultivars in the future (Luzzi et al., 1995).

Several resistant cultivars are available for the management of *R. reniformis* (Robbins et al.,1994a). Since many of these resistant cultivars also are resistant to *H. glycines*, some unknown relationship is suspected between these resistances in certain cases. However, some *H. glycines*-resistant cultivars, such as Mack, Lloyd, and Walters, are very susceptible to *R. reniformis* (Robbins et al., 1994b).

Several soybean cultivars have been reported as more tolerant to *H. columbus* than others. These include Hartwig (maturity Group V), Centennial (VI), Davis (VI), Deltapine 506 (VI), Northrup-King RA 680 (VI), Hagood (VII), Maxcy (VIII), Asgrow 6785 (VI), Deltapine 726 (VI), Young (VI), Duocrop (VII), Northrup-King Coker 6847 (VII), Stonewall (VII), Kirby (VIII), Northrup-King Coker 368 (VIII), Northrup-King Coker 488 (VIII) and Northrup-King S83-3' (VIII) (Anonymous, 1992). Reproduction of Columbia lance nematode is as

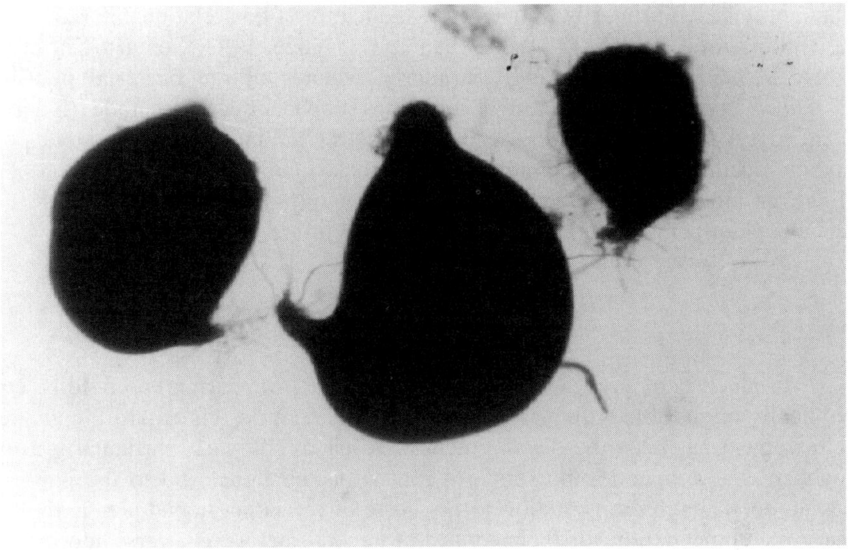

Fig 15–6. Mature cysts of soybean cyst nematode, *Heterodera glycines*, within emanating fungal hyphae.

prolific on these as on the more susceptible cultivars; hence, their use does not provide any durable influence on the management of the nematode.

Pesticides

The use of pesticides for the management of nematodes on soybean is not as prevalent as in years prior to the removal of the fumigant nematicides DBCP (1,2-dibromo-3-chloropropane) and EDB (ethylene dibromide). Available nematicides are of questionable economic value in soybean production under current market price conditions. Though nematicides such as 1,3-dichloropropene, carbofuran (2,3-dihydro-2,2-dimethyl-7-bensofuranylmethyl-carbamate), and aldicarb [2-methyl-2-(methylthio)proprianaldehydeO-(methyl-carbomoyl)oxime], are available and cleared for use on soybean, alternative nematode management strategies such as the use of resistant cultivars and crop rotation are more widely employed. However, management of Columbia lance nematode or sting nematode may be exceptions. The wide host ranges of these nematodes do not make them amenable for management by crop rotation and there are no resistant soybean cultivars. Although both nematodes are relatively easier to control chemically than cyst, root-knot, and reniform nematodes, the use of a tolerant soybean cultivar was a necessary adjunct to nematicides to provide profitable yields in the management of *H. columbus* (Schmitt & Imbriani, 1987).

Biological

No biological management practices are currently being employed against the nematodes parasitic on soybean. Soybean cyst nematode appears to be readily parasitized by several soil-borne fungi (Fig. 15–6), and research has been initiated in this area primarily to determine the nature of this relationship (Godoy et al, 1982; Kim & Riggs, 1991; McLean & Lawrence, 1993; Senyu Chen et al., 1994; Meyer & Meyer, 1995). Similarly, isolates of the bacterial parasites *Pasteuria* spp. have been found infecting soybean cyst nematode (Noel & Stanger, 1994) and specific root-knot nematodes (Narabu, 1995). Nematode population declines have been attributed to these parasites (Nishazawa, 1987). However, attempts to utilize these and other suppressive parasitic organisms to manage plant-parasitic nematodes awaits in the future.

OUTLOOK

Production of soybean in many parts of the world currently would be economically impossible without nematode management. In this regard the greatest achievement has been the development of resistant cultivars, particularly to the soybean cyst nematode, the southern root-knot nematode, and to the reniform nematode. Though crop rotation to nonhosts of the target nematodes is the primary recommendation for their control, a general lack of growers' adoption of rotation practices has been due to the availability of many resistant soybean cultivars and is testament to the achievements made in soybean breeding. This is a continuing process, and future releases of nematode resistant soybean cultivars can be expected. These releases, with other improved soybean traits including herbicide tolerance and yield enhancement, ensure that the use of resistant cultivars will be the mainstay of nematode management in soybean for many years. Of immediate concern is the development of adaptable cultivars with resistance, or tolerance, to *M. javanica* in Brazil, and to *M. arenaria*, and *H. columbus* in the USA.

Developing avenues of research into the parasites of plant-parasitic nematodes and the chemistry involved in soybean cyst nematode sex attractants (Jaffe et al., 1989) and egg hatching stimulants (Masamune et al., 1982) will hopefully provide new approaches to nematode management and complement the use of resistant cultivars.

REFERENCES

Anand, S.C. 1994. Genetic diversity for resistance to *Heterodera glycines* race 5 in Soybean. J. Nematol. 26:76–79.

Anand, S.C., and S.R. Koenning. 1986. Tolerance of soybean to *Heterodera glycines*. J. Nematol. 18:195–199.

Anonymous. 1992. Characteristics of soybean varieties for South Carolina. Clemson Univ. Coop. Ext. Serv. Circ. 545.

Anonymous. 1994. Soybean varieties with soybean cyst nematode resistance. Univ. Illinois Coop. Ext. Serv., Livingston County Ext. Unit.

Birchfield, W., and W.J. Martin. 1967. Reniform nematode survival in air-dried soil. Phytopathology 57:804.

Carpenter, A.S., and S.A. Lewis. 1991. Aggressiveness and reproduction of four *Meloidogyne arenaria* populations on soybean. J. Nematol. 23:232–238.

Dasgupta, D.R., and A.R. Seshadri. 1971. Races of the reniform nematode *Rotylenchulus reniformis* Linford and Oliveira, 1940. Ind. J. Nematol.1:21–24.

Doupnik, B., Jr. 1993. Soybean production and disease loss estimates for north central United States from 1989 to 1991. Plant Dis. 77:1170–1171.

Epps, J.M. 1969a. Recovery of soybean cyst nematode (*Heterodera glycines*) from the digestive tract of blackbirds. J. Nematol. 3:417–419.

Epps, J.M. 1969b. Survival of the soybean cyst nematode in seed stocks. Plant Dis. Rep. 53:403–405.

Fassuliotis, G. 1974a. A description of males of *Hoplolaimus columbus*. J. Nemtol. 6:116–118.

Fassuliotis, G. 1974b. Host range of the Columbia lance nematode, *Hoplolaimus columbus*. Plant Dis. Rep. 58:1000–1002.

Garcia M.R., and J.R. Rich. 1985. Root-knot nematodes in North-Central Florida soybean fields. Nematropica 15:43–48.

Germani, G., J. Mugnier, and Y. Dommergues. 1984. Influence of pathogenic nematodes on nodulation and seed yield of soybeans in Senegal. Rev. Nematol. 7:335–340.

Godoy, G.R., R. Rodriguez, and G. Morgan-Jones. 1982. Parasitism of eggs of *Heterodera glycines* and *Meloidogyne arenaria* by fungi isolated from cysts of *H. glycines*. Nematropica 12:111–119.

Golden, A.M., J.M. Epps, R.D. Riggs, L.A. Duclos, J.A. Fox, and R.L. Bernard. 1970. Terminolgy and identity of infraspecific forms of the soybean cyst nematode (*Heterodera glycines*). Plant Dis. Rep. 54:544–546.

Hartwig, E.E. 1973. Varietal development. p. 187–210. *In* B.E. Caldwell (ed.) Soybeans: Improvement, production, and uses. Agron. Monogr. 16. ASA, Madison, WI.

Heald, C.M., and W.H. Thames. 1982. The reniform nematode, *Rotylenchulus reniformis*. p.139–143. *In* R.D. Riggs (ed.) Nematology in the southern region of the United States. Southern Coop. Ser. Bull. 276. Arkansas Agric. Exp. Son.

Hill, A.S., and D.P. Schmitt. 1989. Influence of temperature and soybean phenology on dormancy induction of *Heterodera glycines*. J. Nematol. 21:361–369.

Hussey, R.S., and H.R. Boerma, 1989. Tolerance in maturity groups V–VII soybean cultivars to *Heterodera glycines*. J. Nematol. 21(4S):686–692.

Jaffe, H., R.N. Huettel, A.B. Demilo, D.K. Hayes, and R.V. Rebois. 1989. Isolation and identification of a compound from soybean cyst nematode, *Heterodera glycines*, with sex pheromone activity. J. Chem. Ecol. 15:2031–2043.

Johnson, A.B., K.S. Kim, R.D. Riggs, and H.D. Scott. 1993. Location of *Heterodera glycines*-induced syncytia in soybean as affected by soil water regimes. J. Nematol. 25:422–426.

Kim, D.G., and R.D. Riggs. 1991. Characteristics and efficacy of a sterile hyphomycete (ARF18), a new biocontrol agent for *Heterodera glycines* and other nematodes. J. Nematol. 23:275–282.

Kinloch, R.A. 1980. The control of nematodes injurious to soybean. Nematropica 10:141–153.

Kinloch, R.A. 1986. Soybean and maize cropping models for the management of *Meloidogyne incognita* in the Coastal Plain. J. Nematol. 18:451–458.

Kinloch, R.A. 1992. Management of root-knot nematodes in soybean. p.147–154. *In* L.G. Copping et al. (ed.) Pest management in soybean. Elvisier Appl. Sci., London.

Kinloch, R.A., M.C. Lutrick, and W.K. Robertson. 1971. The reproduction of *H. dihystera* on soybeans. Proc. Soil Crop Sci. Soc. Fla. 31:262–263.

Kinloch, R.A., and L.S. Dunavin. 1993. Summer cropping effects on the abundance of *Meloidogyne arenaria* race 2 and subsequent soybean yield. J. Nematol. 25(4S):806–808.

Kirkpatrick, T.L., and M.L. May. 1989. Host suitability of soybean cultivars for *Meloidogyne incognita* and *M. arenaria*. J. Nematol. 21(4S):666–670.

Koenning, S.R., D.P. Schmitt, and K.R. Barker. 1985. Influence of planting date on population dynamics and damage potential of *Pratylenchus brachyurus* on soybean. J. Nematol. 17:428–434.

Koenning, S.R., D.P. Schmitt, and K.R. Barker. 1985. Influence of selected cultural practices on winter survival of *Pratylenchus brachyurus* and subsequent effects on soybean yield. J. Nematol. 17:464–469.

Koenning, S.R., and S.C. Anand. 1991. Effects of wheat and soybean planting date on *Heterodera glycines* population dynamics and soybean yield with conventional tillage. Plant Dis. 75:301–304.

Lehman, P.S., and R.A. Dunn. 1987. Distribution of Florida populations of the soybean cyst nematode with previously undescribed genetic variation. Plant Dis. 71:68–70.

Lewis, S.A. 1989. Lance nematodes. p. 67–68. *In* J.B. Sinclair and P.A. Blackmon (ed.) Compendium of soybean diseases. 3rd ed. APS Press, St. Paul, MN.

Lewis S.A., and F.H. Smith. 1976. Host plants, distribution, and ecological associations of *Hoplolaimus columbus*. J. Nematol. 8:264–270.

Lewis, S.A., C.E. Dyre, J.A. Saunders, E.R. Shipe, and J.M. Halbrendt. 1993. Plant-parasitic nematodes on soybean in South Carolina. J. Nematol. 25(4S):890–894.

Luzzi, B.M., H.R. Boerma, and R.S. Hussey. 1995. Inheritance of resistance to the peanut root–knot nematode in soybean. Crop Sci. 35:50–53.

Mac Donald, D.H., G.R. Noel, and W.E. Lueschen. 1980. Soybean cyst nematode, *Heterodera glycines* populations in Minnesota. Plant Dis. 64:319–321.

March, G.J., J.A. Ornaghi, J.E. Beviacqua, E.M. Astorga, A.B. Lopez, G.T. Boito, and J. Marcellino. 1985a. Efectos de la aplicacion de carbofuran en soja para el control del "nematodo de la galla" *Meloidogyne incognita* (Kofoid & White) Chitwood. Gaceta Agron. 5:424–435.

March, G.J., J.A. Ornaghi, J.E. Beviacqua, E.M. Astorga, and J. Marcellino. 1985b. Comportamiento de cultivares de soja frente al nematodo causanta de agallas *Meloidogyne javanica* (Treub) Chitwood. Informativo de Invesigacion Agropecuaria 444:70–77.

Masamune, T., M. Anetai, M. Takasugi, and N. Katsui. 1982. Isolation of a natural hatching stimulus, glycinoeclepin A, for the soybean cyst nematode. Nature (London) 297:495–496

McLean, K.S., and G.W. Lawrence. 1993. Interrelationship of *Heterodera glycines* and *Fusarium solani* in sudden death syndrome of soybean. J. Nematol. 25:434–439.

Moore, W.F. (ed.). 1984. Soybean cyst nematode. Mississippi Coop. Ext. Serv.

Morgan–Jones, G., B.O. Gintis, and R. Rodriguez–Kabana. 1981. Fungal colonization of *Heterodera glycines* cysts in Arkansas, Florida, Mississippi, and Missouri soils. Nematropica 11:155–163.

Nakasono, K. 1978. Sexual attraction of *Rotylenchulus reniformis* females and some differences in the attractiveness between amphimictic and parthenogenetic populations (Nematoda: Nacobbidae). Helminthol. Abstr. 47:31.

Meyer, S.L.F., and R.J. Meyer. 1995. Effects of a mutant strain and a wild type strain of *Verticillium lecanii* on *Heterodera glycines* populations in the greenhouse. J. Nematol. 27:409–417.

Nakasono, K. 1978. Sexual attraction of *Rotylenchulus reniformis* females and some differences in attractiveness between amphimictic and parthenogenetic populations (Nematoda: Nacobbidae). Helminthol. Abstr. 47:31.

Narabu, T. 1995. Attachment specificity of *Pasteuria penetrans* endospores to *Meloidogyne* spp. J. Nematol. 27:511.

Niblack, T.L. 1992. *Pratylenchus, Paratylenchus, Helicotylenchus*, and other nematodes on soybean in Missouri. J. Nematol. 24(4S):738–744.

Niblack, T.L. 1992. The race concept. p. 73–86. *In* R.D. Riggs and J.A. Wrather (ed.) Biology and management of the soybean cyst nematode. APS Press, St. Paul, MN.

Niblack, T.L., R.D. Heinz, G.S. Smith, and P.A. Donald. 1993. Distribution, density, and diversity of *Heterodera glycines* in Missouri. J. Nematol. 25(4S):880–886.

Noel, G.R. 1992. History, distribution, and economics. p. 1–13. *In* R.D. Riggs and J.A. Wrather (ed.) Biology and management of the soybean cyst nematode. APS Press, St. Paul, MN.

Noel, G.R., and B.A. Stanger. 1994. First report of *Pasteuria* sp. attacking *Heterodera glycines* in North America. J. Nematol. 26(4S):612–615.

Nishizawa, T. 1987. A decline phenomenon in a population of the upland cyst nematode, *Heterodera elachista*, caused by a bacterial parasite, *Pasteuria penetrans*. J. Nematol. 19:546.

Pedrosa, E.M.R., R.S. Hussey, and H.R. Boerma. 1994. Response of resistant soybean plant introductions to *Meloidogyne arenaria* races 1 and 2. J. Nematol. 26:182–187.

Perez, E.E., J.D. Mueller, and S.A. Lewis. 1994. Population dynamics of *Hoplolaimus columbus* on soybean. J. Nematol. 26:115.

Rende, J.F., P.M. Tefft, and L.W. Bone. 1982. Pheromone attraction in the soybean cyst nematode *Heterodera glycines* race 3. J. Chem. Ecol. 8:981–991.

Riggs, R.D. 1992. Management of nematode problems on soybean in the United States of America. p. 128–136. *In* L.G. Copping et al. (ed.) Pest management in soybean. Elvisier App. Sci., London.

Riggs, R.D., M.L. Hamblen, and L. Rakes. 1981. Infra-species variation in reaction to hosts in *Heterodera glycines* populations. J. Nematol. 13:171–179.

Riggs, R.D., and D.P. Schmitt. 1988. Complete characterization of the race scheme for *Heterodera glycines*. J. Nematol. 20:392–395.

Riggs, R.D., and D.P. Schmitt. 1989. Soybean cyst nematode. p. 65–67. *In* J.B. Sinclair and P.A. Blackmon (ed.) Compendium of soybean diseases. 3rd ed. APS Press, St Paul, MN.

Riggs, R.D., and J.A. Wrather (ed.). 1992. Biology and management of the soybean cyst nematode. APS Press, St. Paul, MN.

Robbins, R.T., and K.R. Barker. 1973. Comparisons of host range and reproduction among populations of Belonolaimus longicaudatus from North Carolina and Georgia. Plant Dis. Rep. 57:750–754.

Robbins, R.T., and K.R. Barker. 1974. The effects of soil type, particle size, temperature and moisture on reproduction of *Belonolaimus longicaudatus*. J. Nematol. 6:1–6.

Robbins, R.T., L. Rakes, and C.R. Elkins. 1994a. Reniform nematode reproduction and soybean yield of four soybean cultivars in Arkansas. J. Nematol. 26(4S):656–658.

Robbins, R.T., L. Rakes, and C.R. Elkins. 1994b. Reproduction of the reniform nematode on thirty soybean cultivars. J. Nematol. 26(4S):659–664.

Ross, J.P. 1962. Physiological strains of *Heterodera glycines*. Plant Dis. Rep. 46:766–769.

Schmitt, D.P., and K.R. Barker. 1980. Damage and reproductive potentials of *Pratylenchus brachyurus* and *Pratylenchus penetrans* on soybean. J. Nematol. 13:327–332.

Schmitt, D.P., and J.L. Imbriani. 1987. Management of *Hoplolaimus columbus* with tolerant soybean and nematicides. Ann. Appl. Nematol. (Suppl. J. Nematol.) 1:59–63

Schmitt, D.P., and K.R. Barker. 1988. Incidence of plant-parasitic nematodes in the Coastal Plain of North Carolina. Plant Dis. 72:107–110.

Schmitt, D.P., and G.R. Noel. 1984. Nematode parasites of soybean. p. 13–59. *In* W.R. Nickle (ed.) Plant and insect nematodes. Marcel Dekker, New York.

Sciumbato, G.L. 1993. Soybean disease loss estimates for the southern United States during 1988–1991. Plant Dis. 77:954–956.

Senyu Chen, D.W. Dickson, J.W. Kimbrough, R. McSorley, and D.J. Mitchell. 1994. Fungi associated with females and cysts of *Heterodera glycines* in a Florida soybean field. J. Nematol. 26:296–303.

Sindermann, A., G. Williams, S. Sardanelli, and L.R. Krusberg. 1993. Survey for *Heterodera glycines* in Maryland. J. Nematol. 25(4S):887–889.

Slack, D.A., R.D. Riggs, and M.L. Hamblen. 1981. Nematode control in soybeans: Rotation and population dynamics of soybean and other nematodes. Arkansas Agric. Exp. Stn. Rep. Ser. 263.

Smart, G.C. 1964. Physiological strains and one additional host of the soybean cyst nematode, *Heterodera glycines*. Plant Dis. Rep. 48:542–543.

Warner, F., R. Mather, G. Bird, and J. Davenport. 1994. Nematodes in Michigan 1. Disrtribution of *Heterodera glycines* and other plant-parasitic nematodes in soybean. J. Nematol. 26(4S):720–726.

Weaver, D.B., R. Rodriguez-Kabana, and E.L. Carden. 1995. Comparison of crop rotation and fallow for management of *Heterodera glycines* and *Meloidogyne* spp. in soybean. J. Nematol. 27:585–591.

Windham, G.L., and G.W. Lawrence. 1992. Host status of commercial maize hybrids to *Rotylenchulus reniformis*. J. Nematol. 24(4S):745–748.

Yen, J.H., T.L. Niblack, and W.J. Wiebold. 1995. Dormancy of *Heterodera glycines* in Missouri. J. Nematol. 27:153–163.

Young, L.D. 1982. Reproduction of Tennessee soybean cyst nematode population on cultivars resistant to race 4. Plant Dis. 66:251–252.

Young, L.D. 1992. Management of the soybean cyst nematode, *Heterodera glycines*, in soybeans. p. 137–146. *In* Coping et al. (ed.) Pest management in soybean. Elvisier Appl. Sci., London.

Young, L.D. 1994. Changes in reproduction of a *Heterodera glycines* race 5 isolate culture on ‘Cordell’ and ‘Bedford’ soybean. J. Nematol. 26(4S):653–655.

16 Corn

GARY L. WINDHAM

USDA-ARS, Crop Science Research Laboratory
Mississippi State, Mississippi

Corn (*Zea mays* L.) is the major feed grain grown in the USA leading all other crops in both dollar value and number of hectares harvested. In 1994, more than 31 million hectares of corn with a value of over $22 billion was harvested (Anonymous, 1995). A wide range of plant-parasitic nematodes is commonly found in association with corn. Over 60 species of plant-parasitic nematodes have been documented as parasites of corn in North America (Norton, 1983). Because corn has such an extensive fibrous root system, yield losses may not be noticeable until high nematode populations develop or until the crop is stressed by environmental conditions.

Parasitism of corn by plant-parasitic nematodes ranges from weak to strong. Several nematode species have a localized distribution and are severe pathogens on corn. Their effects on yield are easy to establish. Other nematode species are found in most corn fields usually in mixtures with other plant-parasitic nematodes, and with some difficulty have been implicated in yield losses. And then other nematodes rarely cause yield losses in corn, but increase in number rapidly on corn and damage subsequent susceptible crops planted in rotation with corn.

Because plant-parasitic nematodes are often found in polyspecific communities and the additional presence of other soil-infesting pests, yield losses caused by nematodes have been difficult to quantify. Nevertheless, scientists in several states have documented yield losses in corn due to plant-parasitic nematodes (Soc. Nematol. Crop Loss Assess. Comm., 1987). Yearly yield losses in corn were 4.2% in Iowa, and 7% in Georgia and South Carolina. Without the use of nematicides, estimates of yield losses would increase to 5.5, 7.8, and 22.8% in Iowa, South Carolina, and Georgia, respectively.

This chapter covers nematodes commonly associated with corn that cause yield losses in the USA. Other reviews should be consulted for information on nematodes that are minor pests on corn in the USA, but more important in other parts of the world (Norton, 1984; Rivoal & Cook, 1993; Swarup & Sosa-Moss, 1990).

IMPORTANT NEMATODE PARASITES

Root-Knot Nematodes

Species and Races

Meloidogyne incognita (southern root-knot nematode), *M. arenaria* (peanut root-knot nematode, and *M. javanica* (javanese root-knot nematode) are important parasites of corn in the southeastern USA (Baldwin & Barker, 1970b; Windham & Williams, 1987, 1988a,b). Corn is a host to the four host races of *M. incognita* and both races of *M. arenaria* (Ibrahim et al., 1993; Windham & Williams, 1987). *Meloidogyne chitwoodi* (Columbia root-knot nematode) is a pest of corn in the Pacific Northwest (Santo & O'Bannon, 1981). Corn is not considered a host for *M. hapla* (northern root-knot nematode).

Symptoms

Root-knot nematode damage to corn plants is easily confused with other problems such as nutrient deficiencies and compacted or acid soils. Aboveground symptoms may include stunting, unthrifty growth, and signs of moisture and nutrient deficiencies. Belowground symptoms may include heavily galled, stunted roots; however, root-knot galls are commonly very small. Plants supporting a large root-knot population may have numerous small galls and show very little sign of a stunted root system or necrosis due to secondary parasites. Damaged plants are generally localized in clusters in fields.

Biology

The life cycle of root-knot nematodes on corn is typical of that found in other susceptible crops. Infective second-stage juveniles hatch from eggs and penetrate available root tips. When the juveniles begin to feed, giant cells develop only from cells in the stele. As the expanding giant cells expand into the cortex, the endodermis persists and at least partially encloses the giant cells (Baldwin & Barker, 1970a). Giant-cell walls in corn may not thicken as early in development as those observed in most other hosts. Globose preadults and adult females are found in susceptible plants as early as 15 to 18 d postinoculation at 26°C (Windham & Williams, 1994a). Egg-producing females have been detected at 25 d after inoculation.

Root-knot development may be affected greatly in resistant corn. Penetration of some resistant corn genotypes is much less than susceptible genotypes in some instances (Baldwin & Barker, 1970a), whereas others have reported that second-stage juveniles entered resistant plants at rates similar or greater than those for susceptible corn genotypes (Windham & Williams, 1994a). Development of root-knot nematodes in resistant plants is much slower than in susceptible plants, and total number of nematodes in the roots of resistant plants may decline rapidly at 27 d postinoculation after peaking at 15 d postinoculation. The status of corn as a good or poor host of root-knot nematode is not reflected in the ability of their root leachates to elicit nematode egg hatch (Idowu & Fawole, 1990).

In the southeastern USA, *Meloidogyne* spp. population densities may increase rapidly from May to August (Johnson et al., 1990). Root-knot populations are at highest levels in July and August when the crop is close to harvest. Root-knot nematodes are commonly found in corn fields at a soil depth of 0 to 15 cm, but during the summer months greater numbers may be found at a depth of 15 to 45 cm (Rich & Schenck, 1981).

Disease Complexes

Several reports have concerned the interactions of *M. incognita* with fungi on corn. In a greenhouse test, corn root and shoot growth was more severely stunted in pots infested with both *Fusarium moniliforme* and *M. incognita* than in pots infested with either pathogen alone (Palmer & MacDonald, 1974). In field microplots, the effect of *M. incognita* on aflatoxin production by *Aspergillus flavus* was determined (Windham et al., 1994). Aflatoxin production in *A. flavus*-susceptible hybrids grown in nematode infested plots did not differ from plants grown in the absence of the nematode. However, *A. flavus*-resistant hybrids grown in nematode infested plots had aflatoxin production equal to the susceptible hybrids. *Meloidogyne incognita* may have altered the physiology of the resistant plants or, simply, the added stress caused by the nematode allowed aflatoxin to accumulate in the resistant plants.

Interactions of *M. incognita* with maize mosaic virus on corn also have been documented (Khurana et al., 1970). Symptom severity was the greatest when the nematode and virus were both present. *Meloidogyne incognita* reproduction was greater in plants infected with maize mosaic than in plants free of the virus. The nematode failed to transmit the virus.

Economic Importance

Commercial corn hybrids (Windham & Williams, 1987, 1988b), tropical hybrids (Windham & Williams, 1994b), and open-pollinated cultivars (Aung et al., 1990) have all shown to be excellent hosts for *M. incognita, M. arenaria,* and *M. javanica*. Because of its extensive root system, corn may exhibit tolerance to moderate levels of *Meloidogyne* spp., but heavily infested fields may have yield losses of 30% (Gazaway et al., 1991). The economic importance of *Meloidogyne* species may be greater on root-knot susceptible crops planted in rotation with corn. Root-knot nematodes increase rapidly on corn and cause substantial yield losses in subsequent crops in the rotation.

Meloidogyne spp. are commonly found associated with corn in the southeastern USA (Barker, 1974; Gazaway et al., 1991; Nelson, 1955; Schmitt & Barker, 1988). In a survey in North Carolina, root-knot nematodes were present in 51% of the corn fields sampled (Schmitt & Barker, 1988). In an earlier survey in North Carolina, frequency of occurrence of root-knot nematode in corn fields in the Mountains, the Piedmont, and the Coastal Plain was 54, 59, and 77%, respectively (Barker, 1974).

The relationship between initial numbers of *Meloidogyne* spp. and corn growth has been studied. In greenhouse experiments, initial populations of 0.5 to 1024 eggs/g of soil were used to determine the effect of *M. incognita* on corn growth (Di Vito et al., 1980). The relation between fresh weight of shoots and initial density suggested a tolerance limit of 10 eggs/g of soil. In field microplot tests, the threshold density for *M. incognita* was found to be greater than 400 nematodes/100 cm^3 of soil (Barker & Olthof, 1976). In North Carolina, the probability of nematode damage to corn by the *M. incognita* and *M. arenaria* is rated, for advisory purposes, as “slight” when 0 to 4900 nematodes per 0.47 L of soil are found in fall samples and “moderate” when greater than 5000 nematodes per 0.47 L of soil are found (Koenning et al., 1990).

Lesion Nematodes

Species

Pratylenchus spp. are some of the most important nematode parasites of corn in the USA, mainly because they are found more often associated with corn than any other plant-parasitic nematode, and they also can cause significant yield losses. *Pratylenchus hexincisus, P. penetrans and P. scribneri* are important parasites of corn in the midwestern USA (Ferris & Bernard, 1967; Norton, 1983; Norton et al., 1985; Taylor & Schleder, 1959). *Pratylenchus zeae* and *P. brachyurus* are often associated with declining corn yields in the southeastern USA (Endo, 1959; Johnson et al., 1975). Other lesion nematodes commonly found in corn fields include *Pratylenchus agilis, P. alleni, P. crenatus, P. flakkensis, P. neglectus, P. pratensis, and P. sefaensis* (Afolami & Fawole, 1991; Biggs et al., 1994; Dickerson et al., 1964; Rebois & Huettel, 1986; Thomas & Norton, 1986).

Symptoms

The main characteristic of corn roots infected with *Pratylenchus* spp. is necrotic lesions (Fig. 16–1). Dark brown discrete lesions may develop in fibrous and coarse corn roots. Lesions may cover the entire root system. Pruning of fibrous roots and sloughing of the cortical tissue in infected roots may occur in heavily infested fields. Roots invaded by *Pratylenchus* spp. may develop more lateral roots than found on nematode-free plants (Ogiga & Estey, 1975). Plants may be severely stunted with yield losses near 100%. Stunted plants may be localized in fields, or a general suppression of growth may be evident throughout the entire field. Chlorotic plants may be present in severely infested fields (Norton, 1984).

Biology

The life cycle of *Pratylenchus* spp. is similar to other plant-parasitic nematodes. Adult females lay eggs singly in root tissue or in the soil. The first-stage juvenile molts to the second stage within the egg. The second-stage juvenile hatches from the egg, and then molts three times to become an adult. The presence of corn roots have been shown to stimulate egg hatch (DeWaele et al., 1988). The optimum temperature for reproduction of *P. penetrans* on corn is 24°C (Dickerson et al., 1964), whereas the optimum temperature for *P. brachyurus, P. hexincisus, and P. zeae* is 30°C (Olowe & Corbett, 1976; Zirakparvar et al.,

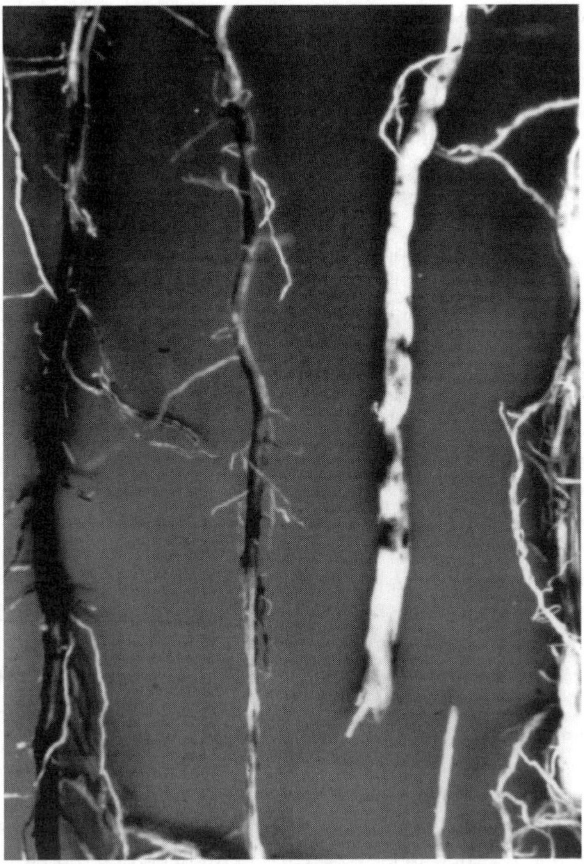

Fig. 16–1. Corn roots with necrotic lesions due to penetration and feeding by *Pratylenchus* sp. (photo courtesy of T. A. Melton).

1980). The life cycle of *Pratylenchus* spp. on corn may be completed after 3 to 4 wk.

All *Pratylenchus* juvenile stages outside the egg and adults can infect corn roots. Although classified as migratory endoparasites, initially lesion nematodes feed ectoparasitically on all zones of the root tip except the root cap before entering the root (Ogiga & Estey, 1975). Invasion of roots by *Pratylenchus* spp. may occur at the root tip, the root hair region, and occasionally in young lateral root junctions (Olowe, 1977). Although penetration of the root appears to be mechanical, there is often discoloration of cells in advance of the nematodes, suggestive of an enzymatic action (Ogiga & Estey, 1975). After puncturing the epidermal cell with its stylet, the nematode will force itself directly through the opening and into cells beneath the epidermis. Groups of lesion nematodes may enter the root simultaneously and migrate around the stelar region in the cortex to produce large cavities. *Pratylenchus* spp. vary in the amount of root damage they produce in corn. *Pratylenchus zeae* causes greater mechanical breakage of cells but less necrosis and less stunting of root growth than *P. brachyurus* (Olowe, 1977). Root penetration by *P. brachyurus* is primarily at the root tip, which may severely limit root growth.

Pratylenchus spp. are commonly found at greater numbers in corn fields at a depth of 0 to 15 cm, followed by 15 to 30 cm, with the fewest at 30 to 45 cm (Rich & Schenck, 1981). Vertical distribution of these nematodes is closely related to root growth, and lesion populations migrate vertically in the soil depending on where the most feeding sites are found. *Pratylenchus* populations in the midwestern USA peak on corn in late August or early September and remain at high levels through November (Thomas, 1978). In the southeastern USA, *Pratylenchus* populations on corn peak in July and August, and remain high through October (Johnson et al., 1975; McSorley & Dickson, 1989).

Disease Complexes

A number of reports have been made of interactions of *Pratylenchus* spp. with other pathogens on corn. *Pratylenchus scribneri* interacted synergistically with *Helminthosporium pedicellatum* to enhance the invasion of corn roots by the fungus, with a subsequent increase in severity of root rot disease and a decline in productivity (Egunjobi et al., 1986). In combination with *Fusarium moniliforme*, *P. brachyurus* and *P. zeae* suppressed corn growth more than when the nematodes were applied separately (Jordaan et al., 1987). Corn seedlings were more vulnerable to damage by *F. moniliforme* and *Pratylenchus* spp. than older growth stages. The presence of *F. moniliforme* made corn roots more attractive and increased penetration by lesion nematodes. *Pratylenchus penetrans* in combination with *F. moniliforme, Rhizoctonia solani,* and *Helminthosporium sativum, and Gibberella zeae* was implicated in root rot of corn in New York (Miller et al., 1963). Numbers of *P. penetrans* were directly correlated with the severity of the root rot.

Interactions of *Pratylenchus zeae* with a virus of corn also have been reported (Nath et al., 1978). Symptoms were more severe with concomitant infections of corn by *P. zeae* and maize mosaic virus than when either pathogen was inoculated singly. *Pratylenchus zeae* numbers were 10-fold higher when the virus was inoculated 1 wk prior to the nematode.

Economic Importance

Pratylenchus spp. are considered the most important plant-parasitic nematodes affecting corn in the USA Corn Belt. These nematodes also are important pests of corn in the northeastern and southeastern USA. *Pratylenchus* spp. are more often associated with corn than any other plant-parasitic nematode. In a recent survey in Mississippi (Windham & Williams, 1990) and in a survey of corn in the Coastal Plain of North Carolina (Nelson, 1955), *Pratylenchus* spp. were found in over 90% of the fields sampled. *Pratylenchus hexincisus* and *P. scribneri* are the most common and, therefore, the most important lesion nematodes in the midwestern USA (Norton, 1983). *Pratylenchus scribneri* as well as *P. brachyurus* and *P. zeae* are important parasites of corn in the southeastern USA (Koenning et al., 1990; Osteen et al., 1982).

Yield losses of corn caused by *Pratylenchus* spp. have been well documented. Using data collected in Georgia, yield losses were estimated at 1243 kg ha^{-1} when 100 *Pratylenchus* spp. per 150 cm^3 of soil were present at harvest (Osteen et al., 1982). Yield losses increased to 2284 kg ha^{-1} when *Pratylenchus* spp. increased to 200 nematodes per 150 cm^3 of soil at harvest. In the absence of other pests, nematicides applied to fields infested with *P. hexincisus* increased corn yields by 26% (Norton & Hinz, 1976). When numbers of *P. hexincisus* reach 1000 to 4000 per gram of dry fibrous root at midseason, there is a marginal chance of this nematode causing yield losses (Norton, 1983). As the numbers increase to 5000 per gram of dry root or higher, the probability of having significant yield losses is much greater. *Pratylenchus penetrans* reduced corn root growth, stalk height, and stalk diameters in Wisconsin (Dickerson et al., 1964). When numbers of *P. scribneri* exceeded 8000 nematodes per gram of dry root at midseason, yield losses exceed 240 kg ha^{-1} (Smolik & Evenson, 1987). In a Kansas field containing *P. scribneri* and *P. neglectus*, a 1% loss in seed yield for each 1000 nematodes/g root was reported (Todd & Oakley, 1996). *Pratylenchus zeae* inhibited corn shoot growth by 38% in the greenhouse and restricted yields in the field by 20% (Tarte, 1971). In a greenhouse study, 6000 *P. zeae* were necessary to suppress corn shoot growth (Chevres-Roman et al., 1971). For advisory purposes in North Carolina, fall counts of 0 to 490 *P. scribneri* or *P. brachyurus* per 0.47 L of soil indicate a slight probability of damage, whereas there is a moderate probability of damage when fall counts exceed 500 lesion nematodes per 0.47 L of soil (Koenning et al., 1990).

Sting Nematodes

Species

The sting nematode (*Belonolaimus* spp.) can have a devastating effect on the growth and yields of field and sweet corn. Several *Belonolaimus* spp. have been documented as pathogens of corn. *Belonolaimus longicaudatus* is the sting nematode commonly found associated with corn in the southeastern USA (Rau, 1958). Early reports (Christie et al., 1952; Christie, 1953) of this pathogen attacking corn in the southeastern USA were attributed to *B. gracilis*, but *B. longicaudatus* was most likely the pathogen involved. A sting nematode closely resembling *B. nortoni* has been isolated from fields of stunted corn in the midwestern USA (Kerr & Wysong, 1979; Todd, 1989).

Symptoms

Corn plants parasitized by *Belonolaimus* spp. are stunted and may have an appearance of nutrient deficiency. Seedlings in heavily infested areas may be severely stunted and devoid of a root system, which can lead to plant death (Fig 16–2). Severely stunted plants may tassel when less than a foot high. Plants may appear drought stressed and wilt during the heat of the day. When sweet corn is severely stunted, the foliage frequently turns yellow. Occasionally the leaf will be yellow while the veins are green. Below-ground symptoms include roots with coarse, stubby branches, and necrotic lesions may develop on the sides of the roots. Lateral roots may develop when the root tips have been severely damaged and stop growing. Root damage by a *Belonolaimus* sp. in Nebraska was most noticeable at about 25 cm below the soil surface (Kerr & Wysong, 1979).

Fig. 16–2. Corn damaged by the sting nematode (*Belonolaimus longicaudatus*) (photo courtesy of K. R. Barker).

Biology

The sting nematode is a large, slender worm that reaches a length of nearly 2.5 mm and has a very long stylet, which may be 60 to 150 μm. As with many ectoparasites, the life cycle of the sting nematode has not been thoroughly characterized. The life cycle is similar to that of other plant-parasitic nematodes. Reproduction is bisexual with males comprising 40% of the population. Eggs are deposited by adult females in the soil, and the first-stage juvenile molts once before leaving the egg. The nematode has three juvenile stages outside the egg and undergoes three molts before becoming an adult. *Belonolaimus* spp. feed on root tips and along the sides of succulent roots by inserting its long slender stylet. The nematode has only occasionally been found inside the root.

Belonolaimus spp. are only found in very sandy soils that have a sand content of 80% or higher. Symptomatic corn damaged by these parasites may be clustered or uniformly distributed in the field. These nematodes are most commonly found in the top 30 cm of soil, but migrate to depths of 45 to 60 cm as the growing season progresses presumably to feed on root tips (McSorley & Dickson, 1990; Rich & Schenck, 1981).

Economic Importance

Corn fields heavily infested with a *Belonolaimus* sp. can have greatly reduced yields or even a complete yield loss. Even in a polyspecific community including other plant-parasitic nematodes, the damage caused by *Belonolaimus* spp. is so severe that it is the only nematode that should be monitored and managed (McSorley & Dickson, 1989). Treating infested fields with nematicides commonly increases yields by 75% or more (Dickson & Hewlett, 1987; Johnson & Dickson, 1973).

Any detectable number of sting nematodes may pose a serious problem. Corn fields with a fall count of 10 or more sting nematodes per 0.47 L of soil have a high probability of sustaining substantial damage (Koenning et al., 1990). Corn yields decrease rapidly as the nematode population increases, and virtually no yield is predicted in fields with densities of 30 sting nematodes per 100 cm^3 of soil (McSorley & Dickson, 1989).

Needle Nematode

Species

Needle nematodes have only relatively recently been shown to be pathogens of corn. *Longidorus breviannulatus* was first found associated with stunted corn in three counties of southeastern Iowa in 1971 and 1972 (Norton & Hoffmann, 1975). This nematode also has been collected from corn fields in Illinois, Indiana, and Delaware (Malek et al., 1980; Norton & Hoffmann, 1975).

Symptoms

Damage by *L. breviannulatus* is often evident 2 wk after seedling emergence. Symptomatic plants are often localized in fields in oval to oblong patches up to 2 ha in size. Plants are stunted, chlorotic, and may have a purple discoloration characteristic of phosphorus deficiency. Stunted plants often appear to be under drought stress. Occasionally, plants may die during the first 6 to 8 wk after planting. Damaged plants may approach the height of unaffected plants; however, stalk diameter and ears are much reduced in size. Root symptoms include yellow discoloration, swollen root tips, stubby roots and root pruning. Feeder roots may be scarce causing the root system to have a coarse appearance. The root injury caused by *L. breviannulatus* closely resembles damage caused by *Belonolaimus* and *Paratrichodorus* spp. When soil moisture levels are high, brushlike crown roots may proliferate near the soil surface.

Biology

Longidorus breviannulatus is one of the largest plant-parasitic nematodes with a length of 4 to 5 mm. The nematode has a long odonto style ranging from 81 to 88 µm. Very little information is available on the life cycle of *L. breviannulatus*. Males are found only occasionally. This parasite has only been found in soils with 49% sand content or higher (Malek et al., 1980). Highest populations of *L. breviannulatus* and the most severe damage to corn occur in soils with a sand content >90%. Greatest numbers of *L. breviannulatus* are found at a depth of 0 to 15 cm from early to midseason. After midseason, these nematodes are found at a depth of 15 to 30 cm. Populations of *L. breviannulatus* drop significantly through winter (Norton & Edwards, 1988). Corn is an excellent host for this nematode. In the greenhouse, an initial population of 100 nematodes increased to over 4000 after 322 d (Norton & Hoffmann, 1975).

Economic Importance

Although *L. breviannulatus* is only distributed in sandy areas in the midwestern USA and Delaware, this nematode can have a devastating effect on corn yields. Heavily infested fields can have yield losses as high as 62% (Malek et al., 1980). Stunting of plants and root symptoms have been observed in areas where populations were as few as 10 needle nematodes per 100 cm^3 of soil. When populations exceed 100 nematodes per 100 cm^3 of soil, seedling death is common.

Stubby-Root Nematodes

Species

Paratrichodorus minor (syn. *Trichodorus minor, T. christiei, P. christiei, T. obesus*) is considered an important parasite of corn in the coastal plain region of the southeastern USA (Barker & Olthof, 1976; Christie, 1953; Christie & Perry, 1951; Johnson, 1975). *Paratrichodorus porosus* also has been reported as a serious parasite of corn (Chevres-Roman et al., 1971).

Symptoms

The most characteristic symptom of damage on corn by *Paratrichodorus minor* is a stunted, stubby root system. Stubby-root nematodes may be particularly damaging to seedlings. As the first roots of the seedling develop, these nematodes feed on root tips, which become devitalized and stop growing. Root

Fig. 16–3. Corn planted in a field heavily infested with *P. minor* resulting in a poor stand and stunted plants (photo courtesy of K. R. Barker).

systems usually show little discoloration or necrosis as a direct reaction to the nematode. In later stages, root systems may become necrotic after secondary invaders enter the plant. The aboveground symptoms are similar to plants deprived of a root system. Plants are stunted and chlorotic, wilt easily, and have very little ability to withstand drought (Fig. 16–3).

Biology

Paratrichodorus minor is an ectoparasite and feeds almost exclusively at the root tips. When feeding, its head very rarely becomes embedded in the root. Feeding by stubby-root nematodes is very different from other plant-parasitic nematodes in that food is not ingested through the stylet. These nematodes have a dorsally curved stylet referred to as an "onchiostyle." *Paratrichodorus minor* is most widely distributed in sandy soils but has been found in muck soils (Christie, 1952). Soil populations can increase rapidly and then decline with equal abruptness. High numbers of these parasites are usually found associated with corn in July (Johnson et al., 1975; McSorley & Dickson, 1990). Population densities of *Paratrichodorus minor* have been found to be concentrated at a soil depth of 30 to 45 cm early in the growing season, but become more equally distributed through the soil horizons as the growing season progresses (McSorley & Dickson, 1990). These nematodes have commonly been found at deep soil levels, which may create sampling problems and limit detection of these parasites. *Paratrichodorus minor* is considered a poor competitor in polyspecific communities of plant-parasitic nematodes (Johnson et al., 1974).

Economic Importance

Since it was first found to be a parasite of corn in the early 1950s, *Paratrichodorus minor* has been considered a major pest of corn in the coastal plain region of the southeastern USA. Although it is capable of causing severe injury to corn, stubby-root nematode may be equally important in areas where corn has moderate damage, which may not attract the attention of growers. Initial populations of *P. minor* exceeding 20 nematodes per 100 cm^3 of soil can suppress corn yields (Barker & Olthof, 1976). Populations of 1000 to 2000 nematodes per 0.47 L of soil were needed for *P. porosus* to damage corn (Chevres-Roman et al., 1971). Growers in North Carolina are advised that fields with fall counts of 10 to 490 *P. minor* per 0.47 L of soil stand a slight chance of being damaged, and fields with 500 or more nematodes stand a moderate chance of being damaged (Koenning et al., 1990).

In some areas, *Paratrichodorus minor* is no longer considered a major problem on corn. In the early 1970s, stubby-root nematodes were found in 74, 57, and 21% of the corn fields sampled in the Coastal Plain, the Piedmont, and the Mountain regions of North Carolina, respectively (Barker, 1974). When corn fields in the North Carolina were surveyed in the mid-1980s, *P. minor* was found in only 9% of the fields (Schmitt & Barker, 1988). Schmitt and Barker (1988) suggested that the widespread use of nonfumigant insecticides/nematicides resulted in the drastic decline of this nematode.

OTHER NEMATODE PARASITES

Awl Nematode

The awl nematode, *Dolichodorus heterocephalus*, is similar to the sting nematode in size, shape, and general appearance. A female awl nematode has a bluntly pointed tail, and the male has a short, wide bursa. The sting female has a blunt rounded tail, and the male has a long narrow bursa. Awl and sting nematodes have similar feeding habits and inflict the same type damage on corn roots. Awl nematodes feed mainly at root tips, but may feed along the sides of roots and also at the base of the hypocotyl. Seed may not germinate if awl nematodes feed on the seed embryo. Plants may be severely stunted and have greatly depleted root systems. Roots may have numerous necrotic lesions. No feeder roots may be present, and secondary roots also may be absent. The awl nematode can cause damage as severe as sting nematodes, but is more localized and has a less extensive distribution than the sting nematode. If awl and sting nematodes are found in the same field, sting nematodes are more aggressive and usually predominate (Rhoades, 1985). The awl nematode has been primarily a problem on sweet and field corn in Florida in areas that have high soil moisture (Christie, 1953; Perry, 1953; Rhoades, 1985).

Cyst Nematode

The corn cyst nematode, *Heterodera zeae*, is an important parasite of corn in Egypt, India, and Pakistan (Bajaj & Gupta, 1994; Koshy et al., 1970; Shahina et al., 1989). This nematode was first found in the USA in 1981 in Kent County, Maryland (Sardanelli et al., 1981). *Heterodera zeae* was most recently found in Cumberland County, Virginia, in a field with severely stunted corn plants (Eisenback et al., 1993). The optimum temperature for reproduction for the corn cyst nematode is from 33 to 36°C at which the life cycle can be completed in 15 to 18 d (Hutzell & Krusberg, 1990; Hashmi et al., 1993b). Minimal reproduction occurred at 24°C (Hashmi et al., 1993). The *H. zeae*-infested field in Virginia produced 7525 kg ha^{-1} of corn in 1991 when a nematicide was applied, and produced only 1254 kg ha^{-1} in 1992 when no nematicides were applied. Yield losses cannot be directly attributed to the corn cyst nematode since other plant-parasitic nematodes were present. Corn-yield losses due to *H. zeae* have not been demonstrated in Maryland. The high temperatures required by *H. zeae* to complete its life cycle indicate it could be a potential problem for corn production in the southern USA.

Dagger Nematode

Although dagger nematodes are important parasites of many agriculturally important plants, very limited information has been collected on the relationship of this nematode with corn. The dagger nematode is slender, up to 2 mm long, has a dorylaimoid esophagus, and has a flanged odontostyle that may be 130 µm long. *Xiphinema americanum* is the most commonly found dagger species associated with corn in the USA (Barker, 1974; Griffin 1964; Taylor & Schleder, 1959). Other dagger species found in corn fields include *X. chambersi* (Barker, 1974) and *X. rivesi* (Biggs et al., 1994). Maximum numbers of dagger nematodes generally occur on corn at a depth of 0 to 15 cm in August and September (Norton & Edwards, 1988; Thomas, 1978). Numbers of *X. americanum* have been negatively correlated with corn yields (Norton et al., 1978).

Lance Nematode

Hoplolaimus galeatus and *H. columbus* (Columbia-lance nematode) have been associated with yield losses of corn in the USA (Ahmad & Chen, 1980; Fassuliotis, 1974). *Hoplolaimus indicus* is an important parasite of corn in India (Haider et al., 1978). Stunting is the most common symptom caused by lance nematodes. Plants attacked by *H. galeatus* and *H. indicus* have discolored root systems and a reduction in the number of feeder roots. In fields heavily infested with *H. indicus*, sloughing off of epidermal and cortical tissues has been observed. Also, these plants may be chlorotic and have leaf tips that are starting to dry.

Hoplolaimus spp. can be classified as semiendoparasites or as migratory endoparasites. Semiendoparasites are partially embedded in the root, whereas migratory endoparasites migrate in and out of the root and feed in the cortex and the stelar area. *Hoplolaimus indicus* is a migratory endoparasite, induces root lesions, and produces extensive tunneling in the cortex. Lance nematode life stages (eggs, juveniles, and adults) may be found in the soil or inside the root tissues.

Field and sweet corn are excellent hosts of *Hoplolaimus galeatus* and *H. columbus* (Ahmad & Chen, 1980; Fassuliotis, 1974; Lewis & Smith, 1976). *Hoplolaimus galeatus* is not considered a serious parasite of corn in the southeastern USA, but has been associated with yield losses in the midwestern USA (Norton et al., 1978). Corn yields in fields infested with *H. galeatus* were increased by as much as 26% when nonfumigant and fumigant nematicides were applied (Norton & Hinz, 1976). *Hoplolaimus columbus*, a serious pest of cotton (*Gossypium hirsutum* L.) and soybean [*Glycine max* (L.) Merr.] in the coastal plain regions of Georgia, South Carolina, and North Carolina, poses a moderate problem for corn growers when populations exceed 500 nematodes per 0.47 L of soil (Koenning et al., 1990). As with other nematodes that reproduce freely on corn, the economic importance of *H. columbus* may not be realized until cotton or soybean follow corn in a rotation system.

Ring Nematode

Criconemella ornata and *C. sphaerocephala* are commonly associated with corn in the southeastern USA (Barker, 1974; McSorley & Dickson, 1990). Ring nematodes are named for the prominent annulations found on their cuticle. Although ring nematodes are not very large (0.2–1.0 mm in length), they have a massive stylet. Ring nematode numbers usually peak on corn in July (Johnson et al., 1974, 1975). Studies of the vertical distribution of *C. sphaerocephala* in corn fields have shown that the majority of these nematodes are found in the soil at a depth of 15 to 45 cm (McSorley & Dickson, 1990). Sweet and field corn are excellent hosts for *Criconemella* spp. (Johnson, 1975; McSorley & Gallaher, 1991, 1992). Although huge increases in nematode numbers on corn have been observed over a growing season (Barker et al., 1982), it has been difficult to establish the pathogenicity of ring nematode on corn. In field microplots, *C. ornata* had a negative effect on corn growth at initial populations of about 200 to 300 nematodes per 500 cm^3 of soil; however, corn yields were not affected. In some field experiments, nematicide treatments were inconsistent in demonstrating yield losses while other showed yield increases of more than 30% (Ayers et al., 1989; Johnson & Chalfant, 1973).

Spiral Nematode

Spiral nematodes (*Helicotylenchus dihystera, H. digonicus, H. pseudorobustus,* and *Scutellonema brachyurus*) are commonly found in association with corn in the USA (Barker, 1974; Fortnum & Karlen, 1985; Griffin, 1964; Norton et al., 1985; Taylor & Schleder, 1959). Generally, spiral nematodes are considered semiendoparasites, e.g., the nematodes are only partially embedded in the roots in which they are feeding. Spiral nematodes also may be found completely embedded in corn roots. Plants attacked by spiral nematodes may have numerous small, light to dark brown lesions. Infected areas in the roots are characterized by damage to cortical cell walls and often by the presence of a ruptured epidermis (Taylor, 1961). In most cases, these nematodes are considered weak to moderate pathogens; however, high numbers of spiral nematodes have been associated with declining corn yields. In greenhouse tests, 10 000 *H. digonicus* were necessary to suppress corn root and shoot growth (Griffin, 1964). The threshold density for *H. dihystera* on corn in field microplots was found to be greater than 200 nematodes per 100 cm^3 of soil (Barker & Olthof, 1976). Sweet corn, which is an excellent host for *H. dihystera,* has stunted growth and restricted ear size and number when grown in a field heavily infested with *H. dihyste*ra (Johnson, 1975). *Helicotylenchus pseudorobustus* can increase rapidly on corn. Initial populations of 20 000 nematodes per pot can increase to over several hundred thousand over a 12-mo period (Norton, 1977). Fields infested with *H. pseudorobustus* and treated with granular nematicides have shown yield increases of 15 to 20% compared to untreated fields (Norton et al., 1978).

Stunt Nematodes

Stunt nematodes (*Tylenchorhynchus* spp.) are ectoparasites that are common in corn fields in the USA. *Tylenchorhynchus claytoni* and *T. maximus* are important parasites in corn in the southeastern and the midwestern USA, respectively (Barker, 1974; Griffin, 1964; Nelson, 1955; Taylor & Schleder, 1959). Other *Tylenchorhynchus* spp. associated with corn are *Tylenchorhynchus clarus, T. cylindricus, T. fischeri, T. latus, T. nudus*, and *T. striatus* (Taylor & Schleder, 1959). Stunt nematodes are easily identified by their tylenchoid esophagus, which has a well-developed medium bulb and a prominent nonoverlapping glandular basal bulb. More information about the pathogenicity of *T. claytoni* on corn is available than for any other *Tylenchorhynchus* sp. These nematodes feed almost exclusively at the root surface although they may be found partially embedded or completed inside the root. Under gnotobiotic conditions, feeding by *T. claytoni* changed the size and shape of nuclei in corn roots (Deubert et al., 1967). In greenhouse tests, 50 *T. claytoni* per pot had no effect on corn growth, 200 caused some damage to corn, and 500 severely stunted corn (Nelson, 1956). In another greenhouse test, 6000 to 8000 *T. claytoni* were necessary to cause damage to corn (Chevres-Roman et al., 1971). In field microplots, the threshold density for *T. claytoni* on corn was found to be greater than 200 nematodes per 100 cm^3 of soil (Barker & Olthof, 1976). *Tylenchorhynchus maximus* also has been shown to be pathogenic to corn. Corn grown in pots in the greenhouse infested with 1000 *T. maximus* had stunted shoot and root growth (Griffin, 1964)

CONTROL

Host Resistance

For most field crops, resistance is the only practical means for controlling plant-parasitic nematodes. Resistant crops are economical in that little or no additional costs are assessed to the grower, and no special equipment is needed to utilize resistant genotypes. Also, from the standpoint of impact on the environment, resistant cultivars are considered the best choice of control practices. Progress has been made in recent years in developing corn germplasm with resistance to root-knot and lesion nematodes.

In the mid 1980s, mass screenings of corn were conducted to evaluate material for root-knot nematode resistance (Windham & Williams, 1987, 1988a,b). Resistance to *M. incognita* was not found in commercial hybrids (Windham & Williams, 1987). However, several commercial hybrids were identified with resistance to *M. arenaria* and *M. javanica* (Windham & Williams, 1987, 1988b). Commercial hybrids, 'Northrup King Brand 508' and 'Pioneer Brand 3147', were poor hosts for *M. javanica* (Windham & Williams, 1988b) and for *M. arenaria* Races 1 (Ibrahim et al., 1993) and 2 (Windham & Williams, 1987). These hybrids were excellent hosts for *M. incognita*. Other materials were then evaluated in an effort to identify sources of resistance to *M. incognita*, and consequently, high levels of resistance were identified in corn inbreds and in open-pollinated cultivars (Aung et al., 1990; Windham & Williams, 1988a). *Meloidogyne arenaria* and *M. javanica* reproduction on corn inbreds and hybrids were found to be positively correlated (Windham & Williams, 1988b). However, reproduction of *M. incognita* on inbreds and hybrids was not positively correlated with the reproduction of *M. arenaria* or *M. javanica*. These sources of resistance were used to make diallel crosses, which effectively limited reproduction of *Meloidogyne* spp. in greenhouse and field tests (Poerba et al., 1990; Williams & Windham, 1988, 1990, 1992). Hybrids developed from these root-knot resistant materials were successfully used in Florida to suppress populations of *M. incognita* (McSorley & Dickson, 1995). Researchers in Mississippi released inbred lines Mp709, Mp710, Mp711, and Mp712 which have resistance to *M. incognita* and *M. arenaria* (W. P. Williams & G. L. Windham, 1996, personal communication).

Numerous efforts have been made to identify sources of resistance in corn to lesion nematodes (Edwards & Norton, 1988; Georgi et al., 1983; Norton et al., 1985; Smolik & Wicks, 1987; Waudo & Norton, 1983). Researchers have been successful in identifying resistance to *Pratylenchus hexincisus* and *P. scribneri* (Smolik & Wicks, 1987). Wicks et al. (1990a,b) released inbred lines SD101, SD102, and SD103 which have resistance to *P. hexincisus*. SD101 also has resistance to *P. scribneri*.

Efforts have been made to find nematode resistance in corn and close relatives of corn to additional plant-parasitic nematodes. Corn cultivars and lines have been evaluated for resistance to the corn cyst nematode (Hashmi et al., 1993a; Shahina et al., 1989). In greenhouse evaluations, one line limited *Heterodera zeae* reproduction after 8 wk to 30 cysts while the susceptible check had over 8000 cysts (Hashmi et al., 1993a). Perennial teosinte (*Zea diploperennis*), which is a relative to maize, was found to be an excellent source of resistance to the spiral nematode *Helicotylenchus pseudorobustus* (Norton et al., 1985). No progress has been made in identifying resistant corn genotypes to ectoparasitic nematodes.

In the past, host resistance has not been utilized to manage plant-parasitic nematodes attacking corn for several reasons. The first reason is simply that until relatively recently no material was available for commercial corn seed companies to use to incorporate nematode resistance into their material. With the release of corn germplasms resistant to lesion nematodes (Wicks et al., 1990a,b) and the release of germplasms resistant to root-knot nematodes (W. P. Williams & G. L. Windham, 1996, personal communication), corn hybrids can be developed with nematode resistance. The second reason that resistance has not been utilized is that plant-parasitic nematodes are not considered by commercial corn seed companies to be a major problem on corn in the USA; therefore, limited time and resources have been delegated to this problem. As with many soilborne pests, nematode damage to corn is often subtle, yet significant, and may go unrecognized. In addition to harming the corn crop, nematodes that increase on corn may have more of an economic importance on susceptible crops subsequently planted in the rotation with corn. Development of corn hybrids with nematode resistance probably has the greatest unrecognized potential of any crop for controlling and managing plant-parasitic nematodes. This is because corn is widely grown throughout the USA, and corn is often grown in rotation with crops that are susceptible to plant-parasitic nematodes. As the number of nematicides registered for use on corn continue to decline, host resistance in corn may become an invaluable tool in controlling and managing nematodes damaging a variety of crops.

Cultural

The effects of various tillage methods on nematode population dynamics in corn fields have been studied extensively. Lesion nematode populations were significantly higher in conventionally tilled corn plots compared to minimum till or no-till plots (Bergeson & Ferris, 1986; Fortnum & Karlen, 1985; McSorley & Gallaher, 1993). It was suggested that no-till plots had more compact soil, which would limit root growth and restrict nematode movement (Bergeson & Ferris, 1986). In some cases, lesion nematode (Yeates & Hughes, 1990) and spiral nematode (*S. brachyurus*) numbers (Fortnum & Karlen, 1985) have been less in conventional tilled plots than in no-till plots (Yeates & Hughes, 1990). Minimum tillage has not been shown to affect root-knot nematodes (Fortnum & Karlen, 1985). A no-till ridge treatment had greater numbers of *Helicotylenchus pseudorobustus* and *X. americanum* than other tillage treatments (Thomas, 1978). Fall tillage has been effective in suppressing nematode populations in corn fields (Thomas, 1978). In Tennessee, numbers of root-knot nematodes were reduced by 73% by fall tillage (Southards, 1971). Spring cultivation also has been used to reduce numbers of lesion nematodes that overwinter in corn fields (Norton & Edwards, 1988). Subsoiling increased corn yields in a field infested with *B. longicaudatus*, *Hoplolaimus galeatus*, *Paratrichodorus minor*, and *Pratylenchus* spp. (Rich et al., 1985). Tillage effects on plant-parasitic nematodes may vary with the nematode species present, soil type, host plant, and location.

Crop rotations have been effective in limiting the buildup of selected plant-parasitic nematodes on corn. A rotation including cotton and peanut (*Arachis hypogaea* L.) suppressed reproduction of *Pratylenchus* spp. (Johnson et al., 1975). Cotton and soybean rotated with corn provided control of the ring nematode *C. ornata* (Johnson et al., 1974). A 3- to 4-yr rotation of continuous alfalfa (*Medicago sativa* L.) was comparable to fallow in reducing numbers of sting and lesion nematodes (Todd, 1991). Sorghum [*Sorghum bicolor* (L.) Moench] has proven effective in suppressing numbers of *M. incognita* (McSorley & Gallaher, 1991), but sorghum rotations have not been adequate in keeping *M. arenaria* (Rodriguez-Kabana & Touchton, 1984), *Paratrichodorus minor* (McSorley & Gallaher, 1991) and *Criconemella* sp. (McSorley & Gallaher, 1992) below damaging levels. Problems associated with using crop rotations are that nonhost crops may be of a lower value than corn, and nontarget plant-parasitic nematodes may increase.

Fallowing land has limited value for controlling plant-parasitic nematodes attacking corn. Very few growers can afford to leave land vacant. Also, fallowed land is susceptible to wind and water erosion. Fallowed land must be maintained weed free to prevent nematode reproduction on weed hosts. Attempts to control sting and lesion nematodes in corn fields by fallowing have not been successful (Todd, 1991). When corn followed 1 yr of fallow, sting and lesion nematodes increased to densities equal to areas planted in continuous corn. Although nematode populations decreased rapidly in fallowed plots, nematodes were still at detectable levels after 2 yr without a cultivated crop.

Chemical

The use of nematicides is probably not economically feasible unless there is a high probability of damage by a severe pathogen such as *Belonolaimus* sp. or moderately pathogenic nematodes are present in high numbers in the field at planting. Numerous examples are available of nematicides effectively controlling plant-parasitic nematodes attacking corn (Johnson & Chalfant, 1973; Zirakparvar, 1979). *Belonolaimus longicaudatus* numbers were reduced, and yields of sweet and field corn were increased after nematicide treatments (Rhoades, 1978, 1979). Nonfumigant nematicides reduced the number of lesion nematodes in corn roots by as much as 98% and increased yields (Bergeson, 1978). Corn yields in fields containing a polyspecific community of nematodes including *Pratylenchus* spp., a *Helicotylenchus* sp., a *Hoplolaimus* sp., and a *Xiphinema* sp. increased following treatment with nonfumigant nematicides (Norton et al., 1978). In corn fields where nematicides are applied at an insecticidal rate, nematodes have not been controlled and yields have not increased (Smolik, 1978).

Treating corn seed with a nematicide was effective in reducing numbers of the lesion nematode *Pratylenchus penetrans* (Chiba et al., 1993). Coating seeds with a nematicide would have several advantages over conventional application methods. It would be less expensive, easier to treat the crop, and less harmful to the environment.

Unfortunately, a number of problems are associated with using nematicides for controlling nematodes in corn fields. Few available nematicides can be applied to corn fields. Ethoprop (O-ethyl S,S-dipropylphosphorodithioate) and terbufos {S-[(1,1-dimethylethyl) thiomethyl] O,O-diethylphosphorodithioate} are the only nematicides recommended for use in many states. In many cases, there is a resurgence of nematode numbers following nematicide treatment (Biggs et al., 1994). Presumably, nematicide-treated corn plants have an increase in root mass, which provides more feeding sites for surviving nematodes. This leads to a significant late-season increase in the nematode population. Nematicides also are considered an environmental liability.

Biological

Very little information is available on the use of biological control agents in controlling nematodes on corn. In a greenhouse study, *Trichoderma harzianum* and *T. koningii* were used to control *M. arenaria* on two commercial corn hybrids (Windham et al., 1989). Shoot and root growth was significantly enhanced, regardless of the presence of the nematode. When a root-knot susceptible hybrid was grown in soil infested with either *Trichoderma* spp., nematode reproduction was suppressed.

OUTLOOK

Controlling plant-parasitic nematodes on corn in the USA will be challenging in the future. With increased pressure from environmentalists, chemical treatments of corn to manage nematode populations will probably be greatly reduced. The number of scientists and resources allocated to corn nematology also have diminished in past years. Recent dramatic increases in corn production in the southeastern USA may lead to an increase in nematode populations and associated yield losses. All of these factors will have an major impact on corn nematology in coming years.

A number of important developments in host-plant resistance should have an impact on the management of nematodes on corn. Germplasm has been released with resistance to root-knot and lesion nematodes. Molecular techniques may be used to move genes that are associated with nematode resistance in corn or other crops into commercially acceptable material. Once commercial hybrids with nematode resistance are available to growers, they may become part of a crop rotation sequence which will manage nematodes below damaging levels. Host-plant resistance may become the most important tool for controlling nematodes in corn and other crops in the future.

REFERENCES

Afolami, S.O., and B. Fawole. 1991. Effect of *Pratylenchus sefaensis* Fortuner 1973 on growth and yield of Zea mays L. cv. FARZ-7 under continuous cropping. Plant Soil 138:133–138.

Ahmad, M., and T.A. Chen. 1980. Effect of certain environmental factors and host plants on reproduction of *Hoplolaimus galeatus*. Plant Dis. 64:479–480.

Anonymous. 1995. The world of corn, 1995. Natl. Corn Growers Assoc., Natl. Corn Develop. Found., St. Louis, MO.

Aung, T., G.L. Windham, and W.P. Williams. 1990. Reproduction of *Meloidogyne incognita* on open-pollinated maize varieties. Suppl. J. Nematol. 22 (4S):651–653.

Ayers, A.R., H.E. Duncan, K.R. Barker, and M.K. Beute. 1989. Effects of crop rotation and nonfumigant nematicides on peanut and corn yields in fields infested with *Criconemella* species. J. Nematol. 21:268–275.

Bajaj, H.K., and D.C. Gupta. 1994. Existence of host races in *Heterodera zeae* Koshy et al. Fund. Appl. Nematol. 17:389–390.

Baldwin, J.G., and K.R. Barker. 1970a. Histopathology of corn hybrids infected with root knot nematode, *Meloidogyne incognita*. Phytopathology 60:1195–1198.

Baldwin, J.G., and K.R. Barker. 1970b. Host suitability of selected hybrids, varieties and inbreds of corn to populations of *Meloidogyne* spp. J. Nematol. 2:345–350.

Barker, K.R. 1974. Influence of geographic area and previous crop on occurrence and densities of plant-parasitic nematodes in North Carolina. Plant Dis. Rep. 58:991–995.

Barker, K.R., and T.H. Olthof. 1976. Relationships between nematode population densities and crop responses. Annu. Rev. Phytopathol. 14:327–353.

Barker, K.R., D.P. Schmitt, and V.P. Campos. 1982. Response of peanut, corn, tobacco, and soybean to *Criconemella ornata*. J. Nematol. 14:576–581.

Bergeson, G.B. 1978. Control of the lesion nematode (*Pratylenchus* spp.) in corn with carbofuran. Plant Dis. Rep. 62:295–297.

Bergeson, G.B., and J.M. Ferris. 1986. Influence of tillage methods on *Pratylenchus* spp. two soil types. Plant Dis. 70:326–328.

Biggs, A.R., J.B. Kotcon, T.A. Baugher, A.R. Collins, D.M. Glenn, H.W. Hogmire, R.E. Byers, A.J. Sextone, and G.W. Lightner. 1994. Comparison of corn and fescue rotations on pathogenic nematodes, nematode biocontrol agents, and soil structure and fertility on an apple replant site. J. Sust. Agric. 4:39–56.

Chevres-Roman, R., H.D. Gross, and J. N. Sasser. 1971. The influence of selected nematode species and number of consecutive plantings of corn and sorghum on forage production, chemical, composition of plant and soil, and water use efficiency. Nematropica 1:40–41.

Chiba, M., G.J. Fulop, B.D. McGarvey, and J.W. Potter. 1993. Distribution and persistence of oxamyl in relation to root-lesion nematode control following seed treatment of corn. J. Agric. Food Chem. 41:2160–2163.

Christie, J.R. 1952. Some new nematode species of critical importance to Florida growers. Soil Sci. Soc. Fla. Proc. 12:30–39.

Christie, J.R. 1953. Ectoparasitic nematodes of plants. Phytopathology 43:295–297.

Christie, J.R., A.N. Brooks, and V.G. Perry. 1952. The sting nematode, *Belonolaimus gracilis*, a parasite of major importance on strawberries, celery, and sweet corn in Florida. Phytopathology 42:173–176.

Christie, J.R., and V.G. Perry. 1951. A root disease of plants caused by a nematode of the genus *Trichodorus*. Science (Washington, DC) 113:491–493.

Deubert, K.H., R.L. Norgens, S.M. Paracer, and B.M. Zuckerman. 1967. The influence of *Tylenchus agricola* and *Tylenchorhynchus claytoni* on corn roots under gnotobiotic conditions. Nematologica 13:56–62.

DeWaele, E., G.C. Loots, and J. Heyns. 1988. Observations on the effect of maize roots on the hatching of *Pratylenchus zeae* and *P. brachyurus*. Phytophylactica 20:135–137.

Dickerson, O.J., H.M. Darling, and G.D. Griffin. 1964. Pathogenicity and population trends of *Pratylenchus penetrans* on potato and corn. Phytopathology 54:317–322.

Dickson, D.W., and T.E. Hewlett. 1987. Effect of two nonfumigant nematicides on corn grown in two adjacent fields infested with different nematodes. Ann. Appl. Nematol. 1:89–93.

Di Vito, M., N. Vovlas, and R.N. Inserra. 1980. Influence of *Meloidogyne incognita* on growth of corn in pots. Plant Dis. 64:1025–1026.

Edwards, J., and D.C. Norton. 1988. Relative host efficiencies of races of maize from Colombia to the nematodes *Pratylenchus scribneri* and *Hoplolaimus galeatus*. Fitopatol. Colomb. 12:21–23.

Egunjobi, O.A., D.C. Norton, and C. Martinson. 1986. Interaction of *Pratylenchus scribneri* and *Helminthosporium pedicellatum* in the etiology of corn root rot. Phytoparasitica 14:287–295.

Eisenback, J.D., D.M. Reaver, and E.L. Stromberg. 1993. First report of corn cyst nematode (*Heterodera zeae*) in Virginia. Plant Dis. 77:647.

Endo, B.Y. 1959. Responses of root-lesion nematodes, *Pratylenchus brachyurus* and *P. zeae*, to various plants and soil types. Phytopathology 49:417–421.

Fassuliotis, G. 1974. Host range of the Columbia lance nematode, *Hoplolaimus columbus*. Plant Dis. Rep. 58:1000–1002.

Ferris, V.R., and R.L. Bernard. 1967. Population dynamics of nematodes in fields planted to soybeans and crops grown in rotation with soybeans. I. The genus *Pratylenchus* (Nemata: Tylenchida). J. Econ. Entomol. 60:405–410.

Fortnum, B.A., and D.L. Karlen. 1985. Effect of tillage system and irrigation on population densities of plant nematodes in field corn. J. Nematol. 17:25–28.

Gazaway, W.S., A. Hagan, P.L. Mask, and R.T. Gudauskas. 1991. Corn diseases in Alabama. Alabama Coop. Ext. Ser. Circ. ANR-601.

Georgi, L., J.M. Ferris, and V.R. Ferris. 1983. Population development of *Pratylenchus hexincisus* in eight corn inbreds. J. Nematol. 15:243–252.

Griffin, G.D. 1964. Association of nematodes with corn in Wisconsin. Plant Dis. Rep. 48:458–459.

Haider, M.G., R.P. Nath, and S.S. Prasad. 1978. Studies on the lance nematode *Hoplolaimus indicus*. I. Pathogenicity and histopathogenesis on maize. Indian J. Nematol. 8:9–12.

Hashmi,G., S. Hashmi, L.R. Krusberg, and R.N. Huettel. 1993a. Resistance in *Zea mays* to *Heterodera zeae*. Suppl. J. Nematol. 25:820–823.

Hashmi, S., L.R. Krusberg, and S. Sardanelli. 1993b. Reproduction of *Heterodera zeae* and its suppression of corn plant growth as affected by temperature. J. Nematol. 25:55–58.

Hutzell, P.A., and L.R. Krusberg. 1990. Temperature and the life cycle of *Heterodera zeae*. J. Nematol. 22:414–417.

Ibrahim, I.K., S.A. Lewis, and D.C. Harshman. 1993. Host suitability of Graminaceous crop cultivars for isolates of *Meloidogyne arenaria* and *M. incognita*. Suppl. J. Nematol. 25:858–862.

Idowu, A.A., and B. Fawole. 1990. Effects of cowpea and maize root leachates on *Meloidogyne incognita* egg hatch. J. Nematol. 22:136–138.

Johnson, A.W. 1975. Resistance of sweet corn cultivars to plant-parasitic nematodes. Plant Dis. Rep. 59:373–376.

Johnson, A.W., and R.B. Chalfant. 1973. Influence of organic pesticides on nematode and corn earworm damage and on yield of sweet corn. J. Nematol. 5:177–180.

Johnson, A.W., C.C. Dowler, N.C. Glaze, D.R. Sumner, R.B. Chalfant, A.M. Golden, and J.E. Epperson. 1990. Effects of nematicides on nematode population densities and crop yields in a turnip-corn-pea cropping system. Phytoprotection 71:55–63.

Johnson, A.W., C.C. Dowler, and E.W. Hauser. 1974. Seasonal population dynamics of selected plant-parasitic nematodes on four monocultured crops. J. Nematol. 6:187–190.

Johnson, A.W., C.C. Dowler, and E.W. Hauser. 1975. Crop rotation and herbicide effects on population densities of plant-parasitic nematodes. J. Nematol. 7:158–168.

Johnson, J.T., and D.W. Dickson. 1973. Evaluation of methods and rates of application of three nematicide-insecticides for control of the sting nematode on corn. Soil Crop Sci. Soc. Proc. Fla. 32:171–173.

Jordaan, E.M., G.C. Loots, W.J. Jooste, and D. DeWaele. 1987. Effects of root-lesion nematodes (*Pratylenchus brachyurus* Godfrey and *P. zeae* Graham) and *Fusarium moniliforme* Sheldon alone or in combination, on maize. Nematologica 33:213–219.

Kerr, E.D., and D.S. Wysong. 1979. Sting nematode, *Belonolaimus* sp., in Nebraska. Plant Dis. Rep. 63:506–507.

Khurana, S.M., B.K. Goswami, and S.P. Raychaudhuri. 1970. Interaction of maize mosaic with root-knot nematode *Meloidogyne incognita* (Kofoid & White) Chitwood in maize (*Zea mays* L.). Phytopathol. Zeitschr. 69:267–272.

Koenning, S.R., H.E. Duncan, J.E. Bailey, K.R. Barker, and J.L. Imbriani. 1990. Nematode thresholds for soybeans, corn, cotton, and peanuts. North Carolina Agric. Ext. Ser. Publ. AG-394.

Koshy, P.K., G. Swarup, and C.L. Sethi. 1970. *Heterodera zeae* n. sp. (Nematoda: Heteroderidae), a cyst-forming nematode on *Zea mays*. Nematologica 16:511–516.

Lewis, S.A., and F.H. Smith. 1976. Host plants, distribution, and ecological associations of *Hoplolaimus columbus*. J. Nematol. 8:264–270.

Malek, R.B., D.C. Norton, B.J. Jacobsen, and N. Acosta. 1980. A new corn disease caused by *Longidorus breviannulatus* in the Midwest. Plant Dis. 64:1110–1113.

McSorley, R., and D.W. Dickson. 1989. Effects and dynamics of a nematode community on maize. J. Nematol. 21:462–471.

McSorley, R., and D.W. Dickson. 1990. Vertical distribution of plant-parasitic nematodes in sandy soil under maize. Plant Soil 123:95–100.

McSorley, R., and D.W. Dickson. 1995. Effect of tropical rotation crops on *Meloidogyne incognita* and other plant-parasitic nematodes. Suppl. J. Nematol. 27:535–544.

McSorley, R., and R.N. Gallaher. 1991. Nematode population changes and forage yields of six corn and sorghum cultivars. Suppl. J. Nematol. 23:673–677.

McSorley, R., and R.N. Gallaher. 1992. Comparison of nematode population densities on six summer crops at seven sites in north Florida. Suppl. J. Nematol. 24(4S):699–706.

McSorley, R., and R.N. Gallaher. 1993. Effect of crop rotation and tillage on nematode densities in tropical corn. Suppl. J. Nematol. 25:814–819.

Miller, R.E., C.W. Boothroyd, and W.F. Mai. 1963. Relationship of *Pratylenchus penetrans* to roots of corn in New York. Phytopathology 53:311–315.

Nath, R.P., G. Swarup, S.S. Prasad, and A.B. Prasad. 1978. Interaction of *Pratylenchus zeae* and maize mosaic virus on maize plants. Indian Phytopathol. 31:307–309.

Nelson, R.R. 1955. Nematode parasites of corn in the coastal plain of North Carolina. Plant Dis. Rep. 39:818–819.

Nelson, R.R. 1956. Resistance to the stunt nematode in corn. Plant Dis. Rep. 40:635–639.

Norton, D.C. 1977. *Helicotylenchus pseudorobustus* as a pathogen on corn, and its densities on corn and soybean. Iowa State J. Res. 51:279–285.

Norton, D.C. 1983. Maize nematode problems. Plant Dis. 67:253–256.

Norton, D.C. 1984. Nematode parasites of corn. p. 61–94. *In* W.R. Nickle (ed.) Plant and insect nematodes. Marcel Dekker, New York.

Norton, D.C., and J. Edwards. 1988. Age structure and community diversity of nematodes associated with maize in Iowa sandy soils. J. Nematol. 20:340–350.

Norton, D.C., J. Edwards, and P.N. Hinz. 1985. Nematode populations in maize and related species. Maydica 30:67–74.

Norton, D.C., and P. Hinz. 1976. Relationship of *Hoplolaimus galeatus* and *Pratylenchus hexincisus* to reduction of corn yields in sandy soils in Iowa. Plant Dis. Rep. 60:197–200.

Norton, D.C., and J.K. Hoffman. 1975. *Longidorus breviannulatus* n. sp. (Nematoda: Longidoridae) associated with stunted corn in Iowa. J. Nematol. 7:168–171.

Norton, D.C., J. Tollefson, P. Hinz, and S.H. Thomas. 1978. Corn yield increases relative to nonfumigant chemical control of nematodes. J. Nematol. 10:160–166.

Ogiga, I.R., and R.H. Estey. 1975. Penetration and colonization of *Brassica rapa* and *Zea mays* root tissues by *Pratylenchus penetrans*. Phytoprotection 56:23–30.

Olowe, T. 1977. Histological changes in maize root induced by *Pratylenchus brachyurus* and *P. zeae* in the absence of other micro-organisms. Niger. J. Plant Prot. 3:41–51.

Olowe, T., and D.C. Corbett. 1976. Aspects of the biology of *Pratylenchus brachyurus* and *P. zeae*. Nematologica 22:202–211.

Osteen, C., A.W. Johnson, and C.C. Dowler. 1982. Applying the economic threshold concept to control lesion nematodes on corn. USDA ERS Tech. Bull. no. 1670. Natl. Tech. Inform. Serv., Springfield, VA.

Palmer, L.T., and D.H. MacDonald. 1974. Interaction of *Fusarium* spp. and certain plant parasitic nematodes on maize. Phytopathology 64:14–17.

Perry, V.G. 1953. The awl nematode, *Dolichodorus heterocephalus*, a devastating plant parasite. Proc. Helminthol. Soc. Wash. 20:21–26.

Poerba, Y.S., G.L. Windham, and W.P. Williams. 1990. Resistance of maize hybrids to *Meloidogyne javanica*. Nematropica 20:169–172.

Rau, G.J. 1958. A new species of sting nematode. Proc. Helminthol. Soc. Wash. 25:95–98.

Rebois, R.V., and R.N. Huettel. 1986. Population dynamics, root penetration, and feeding behavior of *Pratylenchus agilis* in monoxenic root cultures of corn, tomato, and soybean. J. Nematol. 18:392–397.

Rhoades, H.L. 1978. Effect of nonfumigant nematicides on the sting nematode, *Belonolaimus longicaudatus*, and yield of sweet corn. Soil Crop Sci. Soc. Fla. Proc. 37:218–220.

Rhoades, H.L. 1979. Evaluation of nematicides and methods of their application for control of nematodes on field corn. Nematropica 9:43–47.

Rhoades, H.L. 1985. Effects of separate and concomitant populations of *Belonolaimus longicaudatus* and *Dolichodorus heterocephalus* on *Zea mays*. Nematropica 15:171–174.

Rich, J.R., J.T. Johnson, and C.H. Hodge. 1985. Corn response to subsoiling and nematicide application. J. Nematol. 17:404–407.

Rich, J.R., and N.C. Schenck. 1981. Seasonal variations in populations of plant-parasitic nematodes and vesicular-arbuscular mycorrhizae in Florida field corn. Plant Dis. 65:804–807.

Rivoal, R., and R. Cook. 1993. Nematode pests of cereals. p. 259–303. *In* K. Evans et al. (ed.) Plant parasitic nematodes in temperate agriculture. CAB Int., Wallingford, England.

Rodriguez-Kabana, R., and J.T. Touchton. 1984. Corn and sorghum as rotational crops for management of *Meloidogyne arenaria* in peanut. Nematropica 14:26–36.

Santo, G.S., and J.H. O'Bannon. 1981. Pathogenicity of the Columbia root-knot nematode (*Meloidogyne chitwoodi*) on wheat, corn, oat, and barley. J. Nematol. 13:548–550.

Sardanelli, S., L.R. Krusberg, and A.M. Golden. 1981. Corn cyst nematode, *Heterodera zeae*, in the United States. Plant Dis. 65:622.

Schmitt, D.P., and K.R. Barker. 1988. Incidence of plant-parasitic nematodes in the coastal plain of North Carolina. Plant Dis. 72:107–110.

Shahina, F., M. Abid, and M.A. Maqbool. 1989. Screening for resistance in corn cultivars against *Heterodera zeae*. Pak. J. Nematol. 7:75–79.

Smolik, J.D. 1978. Influence of previous insecticidal use on ability of carbofuran to control nematode populations in corn and effect on corn yield. Plant Dis. Rep. 62:95–99.

Smolik, J.D., and P.D. Evenson. 1987. Relationship of yields and *Pratylenchus* spp. population densities in dryland and irrigated corn. Ann. Appl. Nematol. 1:71–73.

Smolik, J.D., and Z.W. Wicks, III. 1987. Reproduction of *Pratylenchus hexincisus* and *P. scribneri* in corn inbreds. Ann. Appl. Nematol. 1:29–31.

Society of Nematologists Crop Loss Assessment Committee. 1987. Bibliography of estimated crop losses in the United States due to plant-parasitic nematodes. Ann. Appl. Nematol. 1:6–12.

Southards, C.J. 1971. Effect of fall tillage and selected hosts on the population density of *Meloidogyne incognita* and *Pratylenchus zeae*. Plant Dis. Rep. 55:41–44.

Swarup, G., and C. Sosa-Moss. 1990. Nematode parasites of cereals. p. 109–135. *In* M. Luc et al. (ed.) Plant parasitic nematodes in subtropical and tropical agriculture. CAB Int., Wallingford, England.

Tarte, R. 1971. The relationship between preplant populations of *Pratylenchus zeae* and growth and yield of corn. J. Nematol. 3:330–331.

Taylor, D.P. 1961. Biology and host-parasite relationships of the spiral nematode, *Helicotylenchus microlobus*. Proc. Helminthol. Soc. Wash. 28:60–66.

Taylor, D.P., and E.G. Schleder. 1959. Nematode associated with Minnesota crops. II. Nematodes associated with corn, barley, oats, rye, and wheat. Plant Dis. Rep. 43:329–333.

Thomas, S.H. 1978. Population densities of nematodes under seven tillage regimes. J. Nematol. 10:24–27.

Thomas, S.H., and D.C. Norton. 1986. Influence of selected maize hybrids on nematode populations under differing edaphic conditions. Plant Dis. 70:234–237.

Todd, T.C. 1989. Population dynamics and damage potential of *Belonolaimus* sp. on corn. Suppl. J. Nematol. 21:697–702.

Todd, T.C. 1991. Effect of cropping regime on populations of *Belonolaimus* sp. and *Pratylenchus scribneri* in sandy soil. Suppl. J. Nematol. 23:646–651.

Todd, T.C., and T.R. Oakley. 1996. Seasonal dynamics and yield relationships of *Pratylenchus* spp. in corn roots. Suppl. J. Nematol. 28:676–681.

Waudo, S.W., and D.C. Norton. 1983. Population changes of *Pratylenchus hexincisus* and *P. scribneri* in maize inbred lines. Plant Dis. 67:1369–1370.

Wicks, Z.W., III, J.D. Smolik, M.L. Carson, and G.G. Scholten. 1990a. Registration of SD101 parental line of maize. Crop Sci. 30:242.

Wicks, Z.W., III, J.D. Smolik, M.L. Carson, and G.G. Scholten. 1990b. Registration of SD102 and SD103 parental lines of maize. Crop Sci. 30:242–243.

Williams, W.P., and G.L. Windham. 1988. Resistance of corn to southern root-knot nematode. Crop Sci. 28:495–496.

Williams, W.P., and G.L. Windham. 1990. Resistance of maize to *Meloidogyne arenaria* and *Meloidogyne javanica*. Theor. Appl. Genet. 80:810–812.

Williams, W.P., and G.L. Windham. 1992. Reaction of a diallel cross of maize to *Meloidogyne incognita* under field conditions. Field Crops Res. 30:167–171.

Windham, G.L., and W.P. Williams. 1987. Host suitability of commercial corn hybrids to *Meloidogyne arenaria* and *M. incognita*. Ann. Appl. Nematol. 1:13–16.

Windham, G.L., and W.P. Williams. 1988a. Resistance of maize inbreds to *Meloidogyne incognita* and *M. arenaria*. Plant Dis. 72:67–69.

Windham, G.L., and W.P. Williams. 1988b. Reproduction of *Meloidogyne javanica* on corn hybrids and inbreds. Ann. Appl. Nematol. 2:25–28.

Windham, G.L., and W.P. Williams. 1990. Plant-parasitic nematodes associated with maize in Mississippi. Phytopathology 80:439.

Windham, G.L., and W.P. Williams. 1994a. Penetration and development of *Meloidogyne incognita* in roots of resistant and susceptible corn genotypes. J. Nematol. 26:80–85.

Windham, G.L., and W.P. Williams. 1994b. Reproduction of *Meloidogyne incognita* and *M. arenaria* on tropical corn hybrids. Suppl. J. Nematol. 26:753–755.

Windham, G.L., M.T. Windham, and W.P. Williams. 1989. Effects of *Trichoderma* spp. on maize growth and *Meloidogyne arenaria* reproduction. Plant Dis. 73:493–495.

Windham, G.L., N. Zummo, and W.P. Williams. 1994. Aflatoxin accumulation in maize infected with *Aspergillus flavus* and *Meloidogyne incognita*. Phytopathology 84:778.

Yeates, G.W., and K.A. Hughes. 1990. Effect of three tillage regimes on plant and soil nematodes in an oats/maize rotation. Pedobiologia 34:379–387.

Zirakparvar, M.E. 1979. Population changes of *Pratylenchus hexincisus* as influenced by chemicals in fibrous and coarse roots of corn. Plant Dis. Rep. 63:55–58.

Zirakparvar, M.E., D.C. Norton, and C.P. Cox. 1980. Population increase of *Pratylenchus hexincisus* on corn as related to soil temperature and type. J. Nematol. 12:313–318.

17 Cotton

JAMES L. STARR

Texas A&M University
College Station, Texas

Cotton is the major fiber crop grown in the USA with an annual production of 5.0 to 6.3 million hectares. The production is concentrated in 16 southern and western states (North Carolina–California) (Fig. 17–1A), with California, Georgia, and Texas accounting for more than 50% of the total. Total production (bales per state) has more than doubled in some states since 1983, especially in the Southeast. North Carolina, South Carolina, and Georgia increased total hectares of cotton (*Gossypium hirsutum* L.) by 60, 50, and 70%, respectively, from 1994 to 1995. Upland cotton accounts for nearly 98% of total U.S. production. The remainder is in long staple or pima cotton (*G. barbadense*). This chapter will be restricted to upland cotton.

Cotton is produced across a variety of climates and soil types, and is a relatively long season crop. Most cultivars are specifically adapted to a given region; e.g., the Southeast, Midsouth, Southwest, or Far West. Cultivars common to the high plains of Texas mature in about 140 d, whereas cultivars adapted to California require about 195 d to produce a crop. Cotton fruit formation begins with the development of the floral bud (square) at 5 to 8 wk after planting (El-Zik et al., 1989). Although flowering may continue over a period of several weeks, 60 to 85% of the total bolls are set in the first 3 wk of flowering. Approximately 55 d are required for a flower to develop to a mature boll.

Nematodes have long been recognized as pathogens of cotton, but in recent years there has been a renewed appreciation of yield losses due to these plant pathogens. Total losses due to nematodes were estimated at 3.5×10^7 kg (1.6×10^5 bales) in 1982. By 1992 this figure had increased more than threefold to 11.4×10^7 kg (5.3×10^5 bales) (Goodell, 1993a). It is generally accepted that this increase is due to more complete data on the distribution frequencies of important nematode species and on the damage potentials of these pathogens rather than any actual increase in losses (Blasingame, 1993). In addition to nematodes, cotton is attacked by more than 40 arthropod pests, chief of which are *Heliothis* spp., pink bollworm (*Pectinophora gossypiella*), and the boll weevil (*Anthonomus grandis*) (Ridgeway, 1984). The most important plant pathogens are the seedling disease pathogens, Verticillium (*Verticillium dahlia*) and Fusarium (*Fusarium oxysporum* f. sp. *vasinfectum*) wilts, and boll rot pathogens (*Alternaria* spp., *Ascochyta gossypium*, *Collectrotichum* spp., *Diplodia gossyp-*

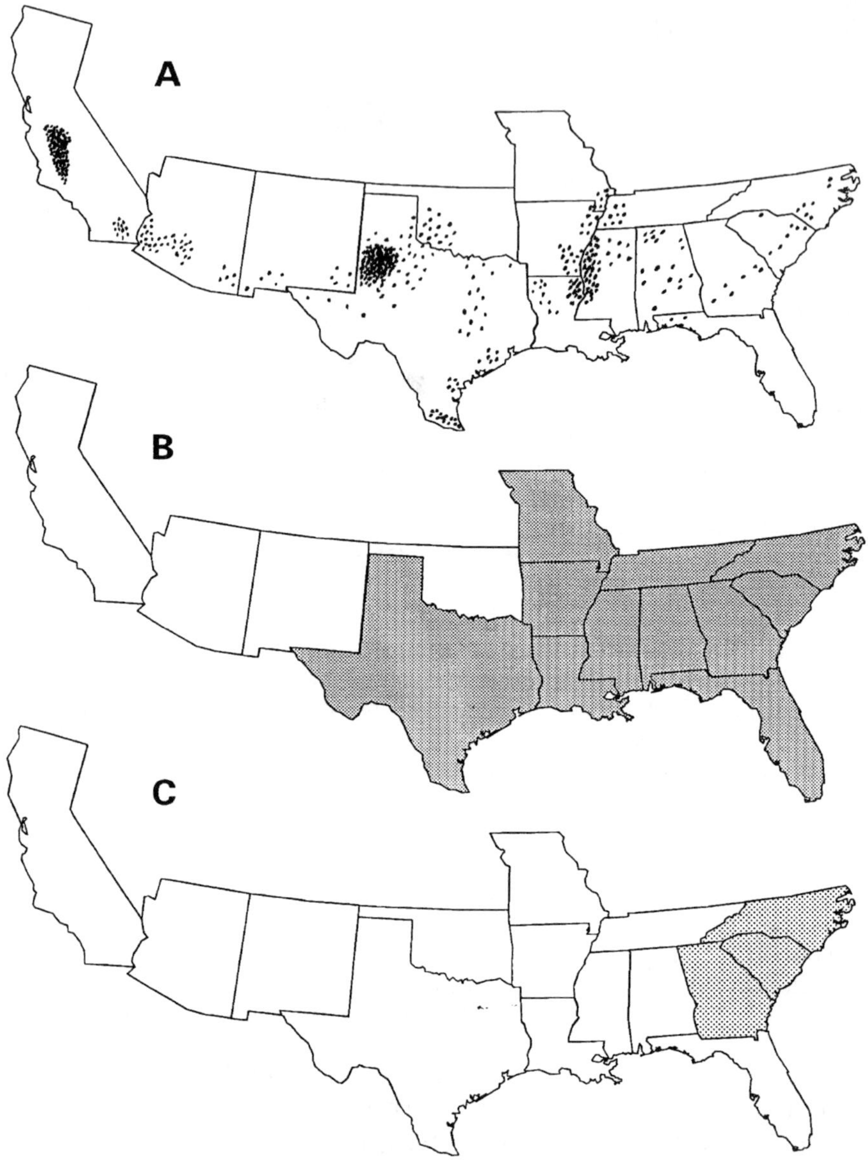

Fig. 17–1. Distribution of cotton and associated plant parasitic nematodes in the USA: (*A*) Principle areas of cotton production. The root-knot nematode, *Meloidogyne incognita*, is distributed across the entire Cotton Belt but is most prevalent in coastal plains of the Southeast, the Panhandle of Texas, and the San Joaquin Valley of California. (*B*) States with confirmed infestations of the reniform nematode, *Rotylenchulus reniformis*. This nematode is most prevalent in the midsouth along the Mississippi River and in the Lower Rio Grand Valley of Texas, but is being found with increasing frequency in other areas of the cotton belt. (*C*) The sting nematode, *Belonolaimus longicaudatus*, and the Columbia lance nematode, *Hoplolaimus columbus*, are found associated with cotton primarily in the coastal plain regions of the Carolinas and Georgia. The sting nematode is found in other states (California, Florida, and Texas) associated with other crops, especially turfgrasses.

ina) (Bell, 1984). Disease loss estimates by the Cotton Disease Council places plant-parasitic nematodes, as a group, second in importance behind seedling diseases in terms of total losses caused (Blasingame, 1992).

Four major nematode pathogens of cotton occur in the USA (Fig. 17–1). These are, in order of importance, the root-knot nematode (*Meloidogyne incognita*), reniform nematode (*Rotylenchulus reniformis*), Columbia lance nematode (*Hoplolaimus columbus*) and sting nematode (*Belonolaimus longicaudatus*). Each of these species will be discussed in detail. Species of *Longidorus* and *Xiphinema* have been associated with severely stunted cotton having poorly developed root systems in Africa (Starr & Page, 1990) but confirmation of their status as pathogens of cotton is lacking. Numerous other nematode species are frequently associated with cotton and probably parasitize cotton without causing detectible yield suppression. In this discussion, parasitism refers to the ability of the nematode species to complete its life cycle when utilizing a given plant species as a nutrient source, whereas pathogenesis is the development of plant disease as a result of the nematode's feeding activities. An aggressive pathogen is a nematode species or population that causes relatively greater amounts of plant disease with greater yield suppression than that of nonaggressive nematodes.

ROOT-KNOT NEMATODES

General Biology

Among the 60 recognized species of *Meloidogyne* (Eisenback & Triantaphyllou, 1991), two are parasitic and pathogenic on cotton. They are *M. acronea* and some populations of *M. incognita*. *Meloidogyne acronea* has been reported only from the Republic of South Africa and Malawi (Bridge et al., 1976), whereas *M. incognita* is the most widely distributed of the root-knot nematodes and is distributed worldwide (Taylor et al., 1982).

Four host races are recognized within *M. incognita* (Hartman & Sasser, 1985) based on their differential ability to parasitize cotton ('Deltapine 16') and root-knot nematode-resistant tobacco (*Nicotiana tobacum* L. 'NC 95'). Only Host Races 3 and 4 are able to parasitize cotton, apparently with similar levels of aggressiveness (Starr & Veech, 1986). These two host races are worldwide in distribution, with Host Race 3 being more common (Hartman & Sasser, 1985) and is most prevalent in regions with a history of cotton production. Host Race 4 is becoming more prominent in areas where tobacco also is produced (K. R. Barker, 1994, personal communication). Host Races 3 and 4 can be found occasionally in areas were cotton is not routinely grown.

All species of root-knot nematodes are classified as sedentary endoparasites because they lose their motility after successful infection of the host and because the invading second-stage juveniles (J2) are completely within the host tissues during initial stages of nematode development. All root-knot species are capable of inducing galling of host roots through hypertrophy and hyperplasia of root cortical tissues; however, galling may be minimal or indistinct on some host species. *Meloidogyne acronea* causes limited galling of cotton, and mature females are frequently exposed on the root surface (Page, 1983).

The preferred infection court for root-knot nematodes is near the root tip. Second-stage juveniles are the only infective stage, typically migrating intercellularly through the cortical cells and establishing a permanent feeding site in the region of the protophloem and protoxylem, just behind the root apex (Endo & Wergin, 1973). Development of the nematode is dependent on its being able to stimulate the host cells to differentiate into specialized nurse cells called giant cells. The giant cells are hypermetabolic, multinucleate transfer cells (Jones & Northcote, 1972) that facilitate the flow of photosynthates from the host to the parasitic nematode (McClure, 1977). The giant cells are the parasite's permanent feeding site (Bird, 1972). Each giant cell develops from a single cell by mitosis without cytokinesis (Jones & Northcote, 1972; Starr, 1993b). Development of the giant cells disrupts the continuity of the vascular system at the infection site, increasing resistance to water transport (Meon et al., 1978). The development of these additional metabolic sinks (i.e., giant cells and developing nematodes) and the disruption of normal root function with regard to water and mineral transport are responsible for the pathogenic effects of parasitism by root-knot nematodes (Hussey, 1985). Root wounding is not a significant factor in pathogenesis by *Meloidogyne* spp.

Development of the invading, preparasitic J2 nematodes can be detected within 2 to 3 d of root penetration as the nematode begins to swell from the vermiform shape (tubular with tapering ends) to the sausage-shaped, advanced, parasitic J2. The nematode goes through four juvenile stages (Triantaphyllou & Hirschmann, 1960). The first juvenile stage develops within the egg and molts to the J2 prior to hatch. Under favorable conditions, the J2 will molt to a J3 within 10 to 14 d after successful infection of cotton. Development of the J3, through the J4, to the adult stage occurs rapidly (4–6 d), and the nematode does not feed during this period of development. After the last molt, the adult females commence feeding again and undergo a period of rapid growth, achieving their characteristic enlarged pear-shaped bodies within a few days (Bird, 1972). The giant cells also undergo rapid enlargement to their final size during this time. Egg production begins 5 to 7 d after the final molt (Starr & Veech, 1986). Adult males, when present, lose the swollen sausage shape with the final molt and are again vermiform and motile. The males do not feed after the final molt and egress from the roots. Total time from root invasion to egg production in a favorable host with optimal temperatures (approximately 28°C) is 21 to 28 d (Starr & Veech, 1986).

Meloidogyne incognita reproduces by an obligatory mitotic parthenogenic mechanism (Triantaphyllou, 1985), with each female producing typically 750 eggs (Starr, 1993a). The eggs are deposited in a gelatinous matrix, which is produced by rectal glands, to form a distinct egg mass. The egg mass typically ruptures through the root cortex and is visible on the root surface, but may be embedded within the gall. Initially the egg masses are clear to creamy in color and sticky. The egg masses harden and become progressively darker in color over a period of several days until they become dark brown. The egg masses help protect the eggs from desiccation (Godfrey & Morita, 1929) and may have antimicrobial activity (Orion & Kritzman, 1991). Development of *M. acronea* is similar to that of *M. incognita* except that it reproduces by typical amphimixis, requiring sperm from males to fertilize the oocytes (Page, 1983).

Root-knot nematodes have no specialized survival stages but are able to survive adverse environmental conditions by a variety of mechanisms. The egg masses help protect the eggs from desiccation, and J2 in the soil may enter an anhydrobiotic state as the soil dries (Towson, 1977). Intact egg masses containing viable eggs can be extracted from soil after approximately 90 d of simulated winter survival (Starr, 1993a). *Meloidogyne incognita* is sensitive to freezing temperatures (Vrain, 1978), hence the species distribution is limited by winter soil temperatures. Both eggs and J2 contribute to winter survival (Starr & Jeger, 1985). Soil populations may be 90% eggs in the fall at the time of cotton crop maturity. Viable eggs can be extracted as late as March, but at the time of planting the populations are almost entirely composed of J2. Percentage winter survival is a density-dependent phenomenon (Ferris, 1985; Starr & Jeger, 1985), ranging from 37 to 1%, with a mean of 9%.

Symptoms of Root-Knot Nematode on Cotton

Foliar symptoms of root-knot nematode damage to cotton are similar to those of many other nematode-incited diseases in that they are not diagnostic. Infected plants may be slightly to severely stunted and may be chlorotic. Symptoms of nematode damage may be mistaken for those of nutrient deficiencies. Affected plants are usually distributed in an irregular, clustered pattern over the field. Affected areas may be elliptical, with the long axis parallel to the cotton rows.

Root symptoms can be diagnostic for root-knot nematodes, with large galls present on the tap and lateral roots late in the growing season, especially where the initial nematode population density was high (Fig. 17–2A). When initial

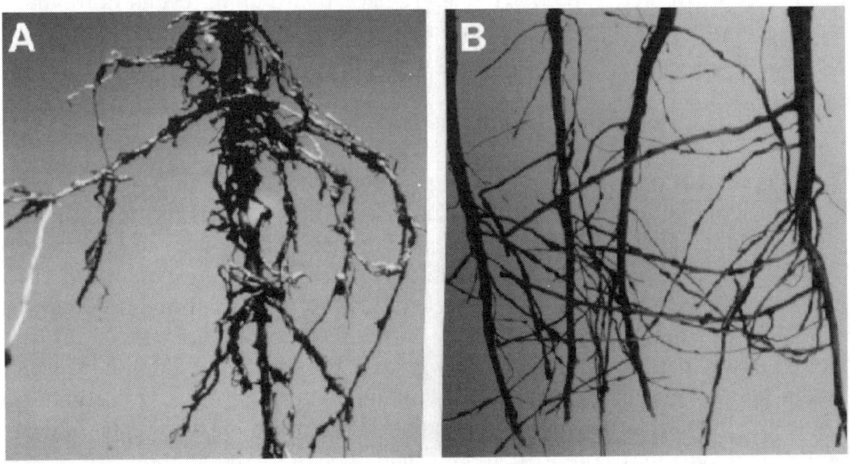

Fig. 17–2. Root galling of cotton by *Meloidogyne incognita* host Races 3 and 4: (*A*) Severe galling caused by high initial nematode population densities (>100 J2/500 cm^3 soil at planting). (*B*) Moderate root galling that is sufficient to suppress cotton yields by 10 to 20% and caused by initial population densities of <20 J2/500 cm^3 soil at planting.

nematode populations are lower, root galls may be small and indistinct (Fig 17–2B). Even low infection levels causing limited galling are sufficient to suppress cotton yield, especially if the crop is subjected to other stresses.

Ecology and Distribution

Meloidogyne incognita is favored by coarsely textured soils and mild winter temperatures. Several studies have documented the effects of soil texture on distribution and activity of root-knot nematodes (Taylor et al., 1982; Robinson et al., 1987; Starr et al., 1993). In surveys of Texas cotton production areas, *M. incognita* was rarely detected in soils with greater than 40% clay content, but was frequently found in samples of more coarsely textured soil (Starr et al., 1993). Distribution of this species is further limited by winter temperatures. It does not occur in the field in the USA above the −1°C isotherm for average January temperatures (Taylor et al., 1982). *Meloidogyne incognita* has a host range of hundreds of plant species; it parasitizes nearly all major crop species of the South, with the exception of peanut (*Arachis hypogaea* L.).

On a worldwide basis, *M. incognita* Host Race 1, which does not parasitize cotton, has the greatest frequency distribution, accounting for 71% of all populations from agricultural soils. Host Races 3 and 4, which do parasitize cotton, account for only 17% of the populations (Hartman & Sasser, 1985). In the Cotton Belt, the two host races affecting cotton are more prevalent than in areas without a history of cotton production (Fig. 17–1A). In some regions of the Cotton Belt, more than 50% of the cotton fields are infested with Host Races 3 or 4, especially in regions characterized by coarsely textured soils and a history of cotton production (e.g., southern Alabama, the southern high plains of Texas, and the southern San Joaquin Valley of California) (Fig. 17–1A) (Blasingame, 1993; Starr et al., 1993). Within individual fields, nematode distribution is usually highly aggregated (Noe & Barker, 1985a). This aggregated distribution results in the patchy occurrence of symptomatic plants and necessitates use of composite soil samples from a small area (≤5 ha) when estimating the nematode's population densities.

Population Dynamics and Damage Functions

Population densities of *Meloidogyne* spp. fluctuate greatly over the course of a year. Populations densities are usually at a maximum as the cotton matures in the autumn and at a minimum in the spring immediately prior to planting (Fig. 17–3). Differences between the minimum and maximum population densities may exceed four orders of magnitude. Males are not usually considered when estimating population densities as they do not feed and are not needed for reproduction by the parthenogenetic *M. incognita*. Eggs and J2 are the life stages present in the soil and can be extracted with some ease, but usually not with high efficiency. In many instances, only population densities of J2 are reported despite the fact that eggs may comprise greater than 90% of the population from early in the growing season until immediately after harvest (Fig. 17–3) (Barker et al., 1987). The low numbers of J2 present in the soil during the growing season is probably because the J2 invade the host roots rapidly upon hatching. Indeed,

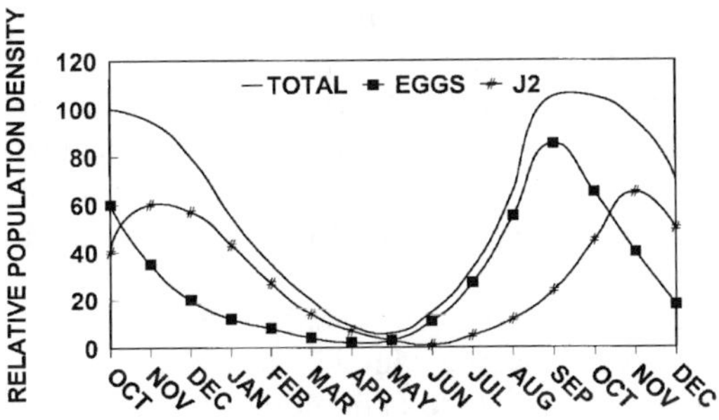

Fig. 17–3. A schematic of the shifts in the population densities of *Meloidogyne incognita* on cotton. Note that most of the extractable soil population is present as eggs during the growing season. As cotton reaches maturity and the plant senesces, the proportion of the population present as eggs declines because of a lack of reproductive females. J2 accumulate in the soil as the crop matures because of the absence of suitable host roots to invade. J2 are the predominant stage in the soil in the late winter and early spring.

there is little incentive for the J2 to remain in the soil as opposed to in the host roots, where it presumably has a competitive advantage over other soil microfauna. Other life stages (J3, J4, and adult females) are present only within the roots and are not usually considered when estimating nematode population densities. This omission is justified when the purpose of the population density estimate is to determine inoculum present and/or winter survival.

The relationship between initial population densities of eggs and J2 present at planting and cotton yield have been determined from field studies in California (Roberts & Matthews, 1984) and from artificially infested microplots in Texas (Starr et al., 1989). In these studies, the damage threshold in sandy soils ranged from 5 to 10 J2/100 cm^3 soil. This population density is near the lower limit for detection by most sampling schemes and soil extraction methods. Thus, it is generally recommended that samples be collected in the autumn when population densities are greatest and easier to detect. Consequently, assuming a 90% winter mortality rate, a fall population density of 50 to 100 eggs and J2/100 cm^3 represents a damage threshold for most sandy soils. In more finely textured, loamy soils, the damage threshold is expected to be at a greater nematode population density. These nematodes rarely cause yield loss in clayey soils. It is important to consider the effects of the clustered distribution pattern of root-knot nematodes on the relationship between population densities and crop growth. In most cases, because the nematode is not uniformly distributed, these damage thresholds tend to overestimate associated losses (Noe & Barker, 1985b).

Disease Complexes

The soil is a complex biological system, and the activities of most soil inhabitants are affected by other organisms present. Thus, disease complexes involving root-knot nematodes, cotton, and numerous soilborne fungal pathogens are common. The most widely recognized complex is that between *M. incognita* and the vascular wilt pathogen *Fusarium oxysporum* f. sp. *vasinfectum* (Roberts et al., 1985; Starr et al., 1989). Other interactions between the nematode and several seedling disease pathogens [*Pythium* spp., *Rhizoctonia solani* (Brodie & Cooper, 1964), and more recently *Thielaviopsis basicola* (Kirkpatrick & Rothrock, 1995)] have been described. In each case, the incidence and severity of the disease caused by the fungal pathogen is greater in the presence of the nematode. This has been especially evident in the case of Fusarium wilt, where coinfection of the plant by the nematode enables the wilt pathogen to overcome the plant's genetic resistance to wilt.

Separate studies (Roberts et al., 1985; Starr et al., 1989) indicate that the Fusarium wilt/root-knot nematode complex does not alter the damage threshold relationship between the nematode and cotton yields, but the disease complex does result in greater yield losses at each nematode population density above the threshold level. Increased early season plant mortality in fields infested with both *M. incognita* and *F. oxysporum* f. sp. *vasinfectum* suppress the population densities of the nematode, an obligate biotroph requiring the living host for reproduction. Similar research of how the nematodes affect the epidemiology of seedling diseases have not been reported.

A common misconception is that these disease complexes arise due to root-wounds caused by the nematodes. Yet, as discussed previously, root-knot nematodes do not cause significant physical damage to cotton roots during root penetration and establishment of permanent feeding sites (Endo & Wergin, 1973). It is more likely that physiological changes within the root gall interfere with normal host resistance mechanisms (Huang, 1985). No data are available that suggests that wounds caused by the nematodes provide a means of ingress for the fungal pathogens. In the cotton seedling disease complex involving *M. incognita* and *R. solani*, the nematode parasitizes the roots, whereas the fungus often colonizes the hypocotyl (Brodie & Cooper, 1964). Thus, even if the nematode wounded the root, these wounds would be unlikely to provide a pathway for fungal invasion of the hypocotyl.

RENIFORM NEMATODES

General Biology

Of the 10 species of the reniform nematode, only *Rotylenchulus reniformis* and *R. parvus* parasitize cotton (Jatala, 1991). *Rotylenchulus reniformis* is widely distributed in tropical and subtropical regions, whereas *R. parvus* has been found primarily in Africa. Both species have wide host ranges, but little is known of the biology or host-parasite relations of *R. parvus.*

Reniform nematodes are sedentary semi-endoparasites in that the anterior third of the body is embedded into the host roots while the posterior two-thirds remains exterior to the root (Fig. 17–4). The nematode usually parasitizes the pericycle, endodermis, and phloem tissues of younger roots (Rebois et al., 1975). *Rotylenchulus reniformis* does not have as restricted an infection court as do root-knot nematodes. The infection court for reniform nematodes is not restricted to the root tip, it invades feeder roots at any point along their length. The reniform nematode also induces host cells to differentiate into specialized nurse cells (the

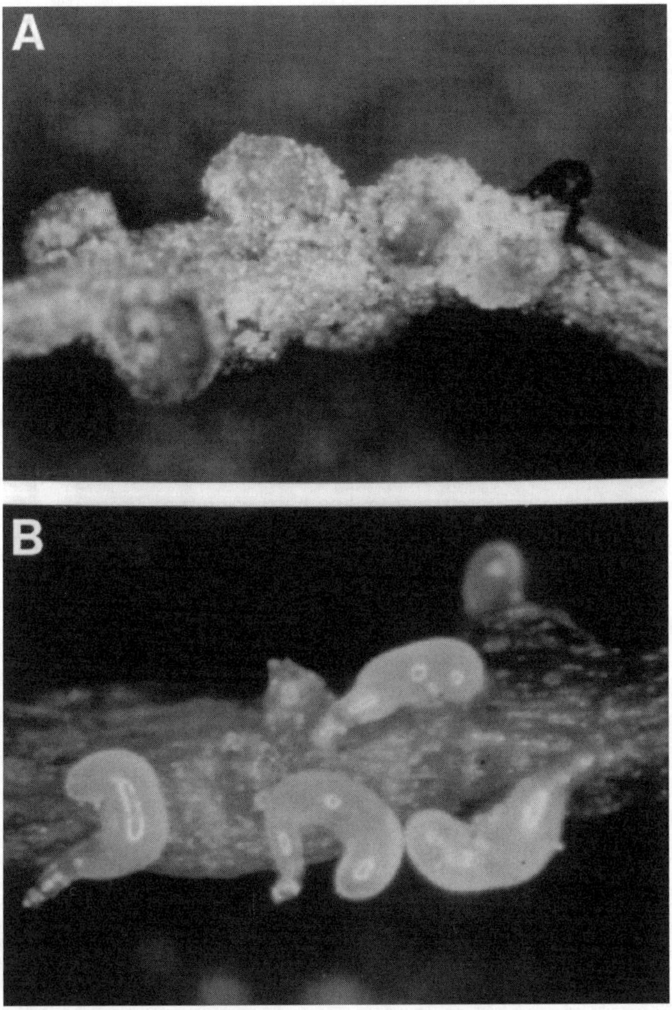

Fig. 17–4. Signs of *Rotylenchulus reniformis* on cotton roots: (*A*) Egg masses of the reniform nematode with adhering soil particles that give roots a "dirty" appearance. (*B*) Mature females of reniform nematode exposed on the root surface after removal of the egg masses. Note distinct kidney shape (photographs provided by J. D. Mueller).

syncytium) at the nematode's permanent feeding site. The syncytium is a hypermetabolic, multinucleate cell that develops through the dissolution of portions of the cell walls of adjacent cells and a coalescence of cytoplasm (Rebois et al., 1975; Jones & Dropkin, 1975). Nurse cells that develop via dissolution of cell walls and coalescence of cytoplasm from several cells are referred to as syncytia, whereas a giant cell induced by *Meloidogyne* spp. develops from a single cell (Jones & Northcote, 1972; Starr, 1993b).

Reniform nematodes also have four juvenile stages, with the J1 occurring only within the egg. Development of the reniform nematode is unusual in that the

J2, which hatches from the egg, progresses through the J3 and J4 stages to the immature adult (also referred to as preadults) in the soil prior to infection of the host. These preparasitic stages of the nematode do not feed (Birchfield, 1962). It is the vermiform, immature adult female that is the infective stage and not the J2 as is the case for root-knot nematodes. Males develop fully in the soil, have a degenerative esophagus, and apparently do not feed. Development in the soil from freshly hatched J2 to infective female takes about 10 d. Following penetration of the root cortex with the anterior portion of the body and transformation of root cells to syncytia at the permanent feeding site, the posterior portion of the female body exterior to the root swells resulting in the characteristic kidney shape of the mature female (Fig. 17–4b). The adult females become sedentary soon after they begin feeding, whereas the males remain motile. *Rotylenchulus reniformis* is an amphimixitic species, and the males are attracted to the females soon after the females develop to their full size, with egg production beginning 5 to 7 d after infection of the cotton root (Birchfield, 1962). As with root-knot nematodes, the eggs of the reniform nematode are deposited into a gelatinous matrix. The egg mass usually covers the female's body and may contain up to 100 eggs.

Symptoms

Cotton infected with reniform nematodes generally lacks diagnostic symptoms. A microscopic sign of the pathogen is the female protruding from the root (Fig 17–4B). Soil particles generally adhere to the egg mass and cause roots with a high level of infection to appear "dirty" (Fig. 17–4A). Infected cotton plants

Fig. 17–5. Field symptoms of damage to cotton by *Rotylenchulus reniformis*. Plants in background growing in fumigated soil. Note uniform stunting of plants in nonfumiganted soil in the foreground (photograph by C. M. Heald).

also can be stunted and, in severe cases, may exhibit symptoms of potassium deficiency (Gazaway, 1993). Fields infested with reniform nematodes may not exhibit the patchy appearance of affected plants as is characteristic of root-knot nematodes. Rather, fields are more uniformly infested (J. L. Starr & T. A. Wheeler, 1994, unpublished data) and the plants more uniformly stunted (Gazaway, 1993) (Fig. 17–5). Yield losses due to reniform nematodes in such fields are frequently overlooked because there are no obvious or striking symptoms, other than suppressed yield.

Ecology and Distribution

Rotylenchulus reniformis is not as well characterized as *M. incognita* with respect to its biology and ecology, but it also is restricted to warm temperate and tropical climates. This nematode is not known to occur in the USA above the southern boot-heel of Missouri or the southern panhandle of Texas (Heald & Robinson, 1990; Blasingame, 1993) (Fig. 17–1B). Reniform nematodes have not been found in the far western cotton-producing states of New Mexico, Arizona, or California. The known distribution of reniform nematodes associated with cotton in the South has increased substantially since 1990. Prior to 1990, the known distribution of reniform nematodes in Texas was limited to four counties of the Lower Rio Grande Valley and one county in the panhandle region. By 1995, the known distribution was 17 counties distributed across the state (Starr et al., 1993; J. L. Starr & T. A. Wheeler, 1995, unpublished data). Similar increases in the known distribution of reniform nematodes are occurring in several southern states. It is not known whether the nematode is continuing to spread to new fields or whether increased awareness and survey efforts are providing better documentation of existing distribution.

Rotylenchulus reniformis can be found in numerous soils types (Heald & Robinson, 1990). In Texas, however, reniform nematodes are rarely found in soil with greater than 40% sand and are more prevalent in finely textured silty soils (Robinson et al., 1987; Starr et al., 1993). Although root-knot and reniform nematodes have widely overlapping host ranges, with common hosts including cotton, soybean [*Glycine max* (L.) Merr.], sweetpotato [*Ipomoea batatas* (L.) Lam.], tomato (*Lycopersicon esculentum* Miller), and melon (*Cucumis melo* L.), they are seldom found inhabiting the same soil (Starr et al., 1993). This observation suggests significant interspecific competition occurs between *M. incognita* and *R. reniformis*. The apparent effects of soil texture on the distribution of these two species may affect the outcome of this interspecific competition.

Several reports indicate that reniform nematodes survive desiccation better than most plant-parasitic nematode species. They are reported to survive for up to 2 yr in dry soil in the absence of a host (Apt, 1976). Reniform nematodes survive in dry soils by entering an anhydrobiotic state. A slow loss of water from the environment enhances survival by anhyrobiosis (Womersley & Chung, 1987).

Population Dynamics and Damage Function

The population dynamics of *R. reniformis* are probably similar to the generalized scheme presented for root-knot species with dramatic fluctuations over the course of 1 yr (Fig. 17–3). Maximum population densities of *R. reniformis* on cotton may exceed 30 000 individuals/500 cm^3 soil. Because reniform nematodes develop through all juvenile stages prior to invasion of cotton roots, population densities of the vermiform stages (J2, J3, J4, and immature females) in the soil during the growing season tend to be greater than those observed for root-knot species. Eggs still constitute a substantial portion of the total soil population during the latter third of the growing season, with eggs comprising as much as 60% of the total population. The numbers of recoverable eggs decline to less than 1% of the total population soon after harvest (J. L. Starr, 1994, unpublished data). Because of the rapid decline in egg numbers, it is likely (but not yet tested) that eggs do not play a significant role in the winter survival of reniform nematodes.

The damage threshold for reniform nematodes is probably at a greater population density than it is for root-knot nematodes. Some states do not recommend implementation of specific management options until preplant populations of reniform nematodes exceed 1000 individuals/500 cm^3 soil (Goodell, 1993b). Cotton, in field tests, has failed to exhibit a positive yield response to nematicide treatment unless preplant nematode populations densities were greater than 500 individuals/500 cm^3 soil (Starr & Page, 1990).

Disease Complexes

Rotylenchulus reniformis increases the incidence and severity of Fusarium and Verticillium wilts of cotton (Neal, 1954; Prasad & Padeganur, 1980). In contrast to the situation with root-knot nematodes and Fusarium wilt, reniform nematodes have higher population densities when associated with cotton that also is infected with *Verticillium dahlia* (Verticillium wilt) than when cotton is not infected with this wilt pathogen (Prasad & Padeganur, 1980). Root-knot nematodes are not known to affect Verticillium wilt of cotton or other crops. This differential response may be because both reniform nematodes and *V. dahlia* are favored by finely textured soils whereas both root-knot nematodes and *F. oxysporum* f. sp. *vasinfectum* are favored by coarsely textured soils. Reniform nematodes also have been associated with increased incidence and severity of cotton seedling diseases (Brodie & Cooper, 1964).

LANCE NEMATODES

General Biology

Four species of lance nematodes (*Hoplolaimus* spp.) parasitize cotton with varying degrees of pathogenesis (Mueller, 1993, Starr & Page, 1990). The most important and most aggressive of these is the Columbia lance nematode, *H. columbus*. *Hoplolaimus galeatus* is probably the mostly widely distributed of the lance nematodes parasitizing cotton but is not aggressive and is rarely associated with economic loss. *Hoplolaimus aegypti* and *H. indicus* parasitize cotton in Egypt (Koura, 1986) and in India (Gura & Mishra, 1981), respectively. A fifth species, *H. magnistylus*, has been found associated with cotton in the mid-South (Mueller, 1993) but data on its parasitism or pathogenesis of cotton is lacking.

All lance nematodes are migratory semi-endoparasites in that they remain motile during all stages of development and because they feed both at the root surface and internally. Three juvenile stages (J2–J4) and adults are capable of invading cotton roots and feeding. The lance nematodes do not induce specialized, permanent feeding sites as do reniform or root-knot nematodes but feed on different root cells as they migrate along the length of the root. They are most frequently associated with younger feeder roots and probably do not migrate more than a few centimeters along the root-length. The preferred feeding site is the root cortex (Lewis et al., 1976).

The lance nematodes *H. aegypti*, *H. galeatus*, and *H. indicus* are amphimictic species, reproducing by cross-fertilization, whereas *H. columbus* reproduces by parthenogenesis (Krall, 1978). All species deposit their eggs singly within the root-cortex or along the root surface; they do not produce a gelatinous egg mass.

Symptoms

Feeding activities of lance nematodes in the root cortex cause substantial tissue destruction and cellular necrosis. Necrosis often includes the pericycle, endodermis and phloem even though the nematodes seldom invade these tissues (Lewis et al., 1976; Koura, 1986). Although *H. columbus* is recognized as being more aggressive than *H. galeatus*, these differences are not readily apparent based on root invasion and cortical cell damage.

Cotton plants infected with lance nematodes tend to be stunted with poorly developed root systems. Parasitized roots may have extensive necrosis. Foliar chlorosis is common in more severely infested fields (Mueller, 1993). Yield suppression can be severe and losses in the range of 5 to 10% may occur in the absence of recognizable symptoms.

Distribution, Population Dynamics, and Damage Functions

Although lance nematodes in general are widely distributed in cotton-production areas, the individual species may be quite restricted in their distribution. *Hoplolaimus galeatus* is widely distributed across the Cotton Belt but *H. columbus* is restricted to the sandy soils of the coastal plain of the Carolinas and Georgia (Fig. 17–1C) (Blasingame, 1993). Lance nematodes have wide host ranges and numerous crop species support sufficient nematode reproduction to maintain the population. Although the Columbia lance nematode is generally associated with coarsely textured sandy soils, *H. galeatus* can be found in finely textured soils (J. L. Starr, 1994, unpublished data).

As with most plant-parasitic nematodes, population densities of lance nematodes can fluctuate greatly during the year and are generally lowest in the spring just prior to planting and at the maximum at about the time of crop harvest. Nematodes can be readily extracted from the soil and the roots, but increases in the population densities in the soil usually lag behind increases in numbers in the roots (Kraus-Schmidt & Lewis, 1979). In contrast with reniform nematodes, *H. columbus* is frequently found in the same fields with *M. incognit*a. The lance nematode usually suppresses populations of the root-knot nematode (Bird et al., 1974; Kraus-Schmidt & Lewis, 1981). Although these two species do not compete for the same feeding site, the cortical necrosis associated with feeding activities of the lance nematode may result in the accumulation of secondary metabolites that are toxic to the giant cells necessary for development of the root-knot nematode (Kraus-Schmidt & Lewis, 1981).

Hoplolaimus columbus is considered an aggressive pathogen with relatively low population densities frequently associated with detectible yield losses. Population densities of 200 to 500 *H. columbus* per gram root fresh weight at 42 d after planting were associated with yield losses of 100 kg lint per hectare, whereas population densities of 2000 nematodes per gram roots were associated with losses of greater than 200 kg lint per hectare (Mueller & Sullivan, 1988). In other research, a maximum lint yield suppression of 20% was observed with initial populations densities of approximately 250/500 cm^3 soil (Noe et al., 1991). Damage thresholds for *H. columbus* in the southeastern portion of the Cotton Belt have been set at about 250 nematodes/500 cm^3 soil for spring samples (Goodell, 1993b). For samples collected in the fall, the damage threshold is approximately 500 nematodes/500 cm^3 soil, reflecting high rates of winter survival.

STING NEMATODE

Of the nine recognized species of sting nematodes, only *Belonolaimus longicaudatus* is recognized as a pathogen of cotton (Smart & Nguyen, 1991). The nematode is a very aggressive pathogen with a recognized host range of more than 20 crop and weed species. Variation in host range among different geographic isolates of this species has been reported, but distinct races or biotypes have not been described (Abu-Gharbeih & Perry, 1970; Robbins & Barker, 1973). There are no known reports of populations of *B. longicaudatus* not being pathogenic to cotton, but cotton has been reported as a poor host in some tests (Robbins & Barker, 1973). There is some reproductive isolation among isolates with differing host ranges, suggesting that they may be different species. The sting nematode is an amphimictic species, reproducing by cross-fertilization (Robbins & Hirschmann, 1974).

The horizontal and vertical distribution of the sting nematode is closely related to soil texture. The nematode will not survive in soils with less than 85% sand content (Robbins & Barker, 1974). Population densities of the nematode declined with depth as the sand content decreased with soil depth (Brodie, 1976). This nematode species also is favored by moderate to warm soil temperatures (22–25°C). *Belonolaimus longicaudatus* is sparsely distributed in the coarsely textured, sandy soils of the coastal plains of Southeast and is present in Texas and California; however, it has been associated with cotton only in North Carolina, South Carolina, and Georgia (Fig. 17–1C). Less than 2% of the fields in this region have detectible populations of the nematode (J. L. Starr, 1980, unpublished data; Schmitt & Barker, 1988). If cotton production continues to expand in the sandy soils of the coastal plains of the southeastern USA, it is expected that the incidence of cotton fields infested with the sting nematode will increase.

The sting nematode is a migratory ectoparasite that feeds exclusively from the root surface, utilizing its long stylet to feed upon cortical cells several cell layers deep in the root. The feeding activities of the sting nematode result in extensive necrosis of the root cortex, leading to a suppression of root development. Affected plants are stunted to various degrees, depending on nematode population densities. Population densities as low as 10 nematodes/500 cm^3 soil at planting is the usual damage threshold (Wrather, 1993). In severely infested fields (nematode population densities >100/500 cm^3 soil) cotton may not survive beyond the seedling stage, resulting in uneven plant stands.

The sting nematode has been observed to increase the incidence of Fusarium wilt of cotton (Holdeman & Graham, 1954; Yang et al., 1976) and control of the nematode with nematicides reduces the incidence of wilt (Brodie & Hauser, 1970). *Belonolaimus longicaudatus* has a greater affect on Fusarium wilt, in terms of disease incidence, than do root-knot nematodes (Yang et al., 1976).

MANAGEMENT OF NEMATODES ON COTTON

Several methods are available to producers for managing nematodes on cotton to reduce yield losses, and current research suggests that other methods will become available in the future. Only two primary management methods, however, are widely used today. These are the use of the nematicides aldicarb (Temik) [2-methyl-2-(methylthio)propioaldehyde 0-(methylcarbamyl)oxime] and 1,3-dichloropropene (Telone II) and, to a lesser degree, crop rotation. Host resistance has real potential for improved management of root-knot nematodes and may eventually be an option for other species. Biological control, based on current technology, is a more remote possibility for large acreage, annual crops like cotton.

Unfortunately, few cotton producers have management programs aimed specifically at plant-parasitic nematodes. In a survey of producers and crop consultants at the 1995 Beltwide Cotton Production Research Conferences (San Antonio, Texas), most growers (89%) indicated that they were aware of nematodes as potential pathogens of cotton, but less than 45% regularly sample fields to obtain estimates of specific nematode problems (Mueller & Starr, 1995, unpublished data). Less than 50% of the respondents indicated that they were willing to spend more than $50/ha to control nematodes. Admittedly, this was not an extensive survey (137 respondents) but producers and crop consultants from 12 states were included. Nonetheless, these data suggests that most producers are probably unaware of the extent of losses incurred by the cotton industry from plant-parasitic nematodes. Additionally, most producers seem unwilling to invest significant time or other resources in nematode management. This attitude is often seen with respect to nematodes, reflecting the unseen nature of these soilborne pathogens and the often subtle damage they inflict on a crop.

Most of the aldicarb used on cotton is applied at rates most effective for control of early season insects, especially thrips (*Frankliniella* spp.). Only 37 of 137 producers and crop consultants indicated that they used aldicarb for nematode control on a regular or occasional basis. Available data indicate that aldicarb at the lowest recommended rate for nematode control (1.1 kg a.i. ha^{-1}, applied in-furrow) will provide some yield enhancement in fields infested with reniform or root-knot nematodes (Mueller & Sullivan, 1988). More consistent yield responses in nematode-infested fields are observed with application of 1.8 to 2.3 kg a.i. ha^{-1}. The higher rates of aldicarb usually result in a suppression of nematode population densities at crop maturity, whereas nematode population densities may be unaffected by lower application rates. This lack of population suppression is generally believed to be a reflection of the mode of action of aldicarb, in that it interferes with the nematode's sensory system to inhibit motility, host penetration, feeding activity, and mating of amphimictic species rather than having immediately lethal effects on the nematode (Hague & Gowen, 1987). Although aldicarb provides a real economic benefit when nematode population densities are moderate to high, most producers are reluctant to use the chemical at necessary rates because of low profit margins.

Few cotton producers use the fumigant nematicide 1,3-dichloropropene because it must be applied preplant and it requires additional time to dissipate from the soil prior to planting. Only 17 of 137 respondents to the survey of cotton producers indicated that they used a fumigant for nematode management on a regular or occasional basis (J. D. Mueller & J. L. Starr, 1995, unpublished data). Most producers are unwilling to commit this effort to nematode management. However, the fumigant is effective in suppressing nematode population densities with a corresponding yield enhancement in fields with a moderate to heavy infestation of aggressive nematodes (Mueller & Sullivan, 1988; Thomas & Smith, 1993).

Crop rotations can effectively control many plant-parasitic nematode species. Despite the generally broad host ranges of the nematodes attacking cotton, rotations that effectively manage different nematode problems have been developed. In California, a rotation of cotton with root-knot resistant cowpea [*Vigna unguiculata* (L.) Walp.] suppresses population densities of root-knot nematodes (Duncan & Ferris, 1984), and in Texas and the Southeast a cotton-peanut rotation effectively controls root-knot nematodes on both crops (Kirkpatrick & Sasser, 1983; Rodriguez-Kabana, 1991). Sorghum [*Sorghum bicolor* (L.) Moench] is an effective rotation crop for management of reniform nematodes on cotton (Thames & Heald, 1974). A 1-yr rotation to a crop unfavorable to root-knot nematode generally improves yields of cotton, but greater yield responses are usually obtained with longer rotation schemes. Nematode population densities, which decline in the absence of a favorable host, can increase rapidly to near maximum levels on susceptible crops like cotton. For crop rotation to be an acceptable management system for cotton producers, the alternative crop(s) must be at least as profitable as cotton. Second, the alternative crop must be compatible with cotton production in terms of equipment, herbicide usage, water requirements, and climatic conditions. No one system will be acceptable across the entire Cotton Belt, but alternatives for each region are possible. Crop rotation must have benefits beyond nematode control to be acceptable to most cotton producers.

Much recent effort has been expanded to identify, characterize, and develop systems for the biological control of important plant-parasitic nematodes. Two principal areas of research have been on soilborne fungi that parasitize nematode eggs (e.g., *Paecilomyces lilacinus*) and on the bacterial parasite of vermiform nematodes *Pasteuria penetrans* (Kerry, 1990). Organic amendments to the soil that suppress nematode population densities also have potential (Kaplan & Noe, 1993). These approaches may be used as either primary or secondary components of a nematode management system. To be truly effective, improved systems for propagation and delivery of the biocontrol agents need to be developed. An effective biological control system should provide either suppression of early season nematode activity or become established in a field to effect a long-term suppression of the target nematode population.

Host resistance is yet another approach to managing nematode populations. Resistance to plant-parasitic nematodes is often defined as an inhibition of nematode reproduction and is not based on crop yield response (Cook & Evans, 1987). Cotton cultivars with improved yield in nematode-infested fields may not be resistant to the nematodes but may have improved tolerance to parasitism. Although improved yield is the ultimate objective of a nematode management system and tolerance may achieve this objective as well as resistance, these two mechanisms should not be confused. Nematode populations are not suppressed with tolerance and represent a threat to subsequent crops. In cases where both resistance and tolerance are available, resistance is the best approach. Where resistance is not available, tolerance is an acceptable alternative.

High levels of resistance to root-knot nematodes are available among advanced generation cotton breeding lines (Shepherd et al., 1988) and are being incorporated into cotton genotypes with high yield potential and fiber quality (Starr & Smith, 1993; Ogallo et al., 1997). Currently available cotton cultivars with resistance to the Fusarium wilt/nematode complex have improved resistant to the wilt pathogen in the presence of the nematode, but may still be susceptible to the nematode (Starr & Smith, 1993).

Resistance to reniform nematodes has not yet been identified in genotypes cross-compatible with *G. hirsutum*, but efforts are in progress to introgress resistance to reniform nematodes from *G. longicalyx* into *G. hirsutum* via hexaploid bridges (A. A. Bell, 1995, personal communication) and by monosomic additions (Frerich, 1995). Cotton lines with improved tolerance to reniform nematodes have been reported (Cook & Namkin, 1994).

No source of resistance to the Columbia lance or sting nematodes has yet been identified. As these two species are migratory parasites with less complex host-parasite relationships than exists between root-knot or reniform nematodes and their hosts, it is possible that naturally occurring resistance may not exist or may be sufficiently rare as to be difficult to identify with current levels of effort. New approaches to host resistance, especially genetic engineering of novel resistance mechanisms (Opperman et al., 1994), may be the best hope for developing resistance to lance and sting nematodes. Efforts to identify tolerance to the Columbia lance nematode may have a greater potential for success than efforts to identify resistance (Bowman & Schmitt, 1994).

Cultural practices also can assist in the management of nematode problems. In the sandy soils of the coastal plains of the Southeast, a soil hard pan or plow pan frequently develops that limits cotton root development to the upper soil profile. Disruption of this barrier to root penetration by subsoiling suppresses nematode population densities in the upper soil profile and improves cotton yields (Hussey, 1977). This may be related to lower root density in the upper profile as the cotton roots penetrate deeper into the soil.

With the possible exception of resistance to root-knot nematodes, no single management alternative is likely to be applicable across all cotton-production environments in the near future. A sufficient number of effective alternatives, however, exist so that producers should be able to develop an integrated system that is compatible with their production system and achieve higher yields through improved management of nematode problems. The first step is to recognize that nematodes are pathogens that suppress cotton yields and then to determine the extent of the problem in individual fields. Nematode pathogens of cotton are widespread, and significant problems will likely increase with increased monoculture of cotton.

REFERENCES

Abu-Gharbieh, W.I., and V.G. Perry. 1970. Host differences among Florida populations of *Belonolaimus longicaudatus* Rau. J. Nematol. 2:209–216.

Apt, W.J. 1976. Survival of reniform nematodes in desiccated soils. J. Nematol. 11:316.

Barker, K.R., J.L. Starr, and D.P. Schmitt. 1987. Usefulness of egg assays in nematode population-density determinations. J. Nematol. 19:130–134.

Bell, A.A. 1984. Cotton protection practices in the USA and world. B: Diseases. p. 288–309. *In* R.J. Kohel and C.F. Lewis (ed.) Cotton. Agron. Monogr. 24. ASA, CSSA, and SSSA, Madison, WI.

Birchfield, W. 1962. Host-parasite relations of *Rotylenchulus reniformis* on *Gossypium hirsutum*. Phytopathology 52:862–865.

Bird, A.F. 1972. Quantitative studies on the growth of syncytia induced in plants by root-knot nematodes. Int. J. Parasitol. 2:157–170.

Bird, G.W., O.L. Brooks, and C.E. Perry. 1974. Dynamics of concomitant field populations of *Hoplolaimus columbus* and *Meloidogyne incognita*. J. Nematol. 6:190–194.

Blasingame, D. 1992. Cotton disease loss estimate committee report. p. 172. *In* 52nd Cotton Dis. Counc., Proc. Beltwide Cotton Prod. Res. Conf., New Orleans. 10–14 January. Natl. Cotton Counc., Memphis, TN.

Blasingame, D. 1993. Nematode distribution and density. p. 4–6. *In* Cotton nematodes: Your hidden enemies. Cotton Found., Memphis, TN.

Bowman, D.T., and D.P. Schmitt. 1994. Screening cotton for tolerance to *Hoplolaimus columbus*. Plant Dis. 78:695–697.

Bridge, J., E. Jones, and L.J. Page. 1976. *Meloidogyne acronea* associated with reduced growth of cotton in Malawi. Plant Dis. Rep. 60:5–7.

Brodie, B.B. 1976. Vertical distribution of three nematode species in relation to certain soil properties. J. Nematol. 8:243–247.

Brodie, B.B., and W.E. Cooper. 1964. Relation of plant parasitic nematodes to post-emergence damping-off of cotton. Phytopathology 54:1023–1027.

Brodie, B.B., and E.W. Hauser. 1970. Multiple pest control in cotton with mixtures of selective pesticides. Phytopathology 60:1609–1612.

Cook, R., and K. Evans. 1987. Resistance and tolerance. p. 179–232. *In* R.H. Brown and B.R. Evans (ed.) Principles and practice of nematode control in crops. Acad. Press, London.

Cook, C.G., and L.N. Namkin. 1994. Influence of reniform nematodes on upland cotton cultivars and breeding lines. p. 256–257. *In* 54th Cotton Disease Council. Proc. Beltwide Cotton Conf., San Diego, Ca. 5–8 January. National Cotton Council, Memphis, TN.

Duncan, L.W., and H. Ferris. 1984. Effects of *Meloidogyne incognita* on cotton and cowpeas in rotation. p. 22–26. *In* 44th Cotton Disease Council, Proc. Beltwide Cotton Prod. Res. Conf., Atlanta, GA. 8–12 January. National Cotton Council, Memphis, TN.

Eisenback, J.D., and H.H. Triantaphyllou. 1991. Root-knot nematodes: *Meloidogyne* species and races. p. 191–274. *In* W.R. Nickle (ed.) Manual of agricultural nematology. Marcel Dekker, Inc., New York.

El-Zik, K.M., D.W. Grimes, and P.M. Thaxton. 1989. Cultural management and pest suppression. p. 11–36. *In* R.E. Frisbie et al. (ed.) Integrated pest management systems and cotton production. John Wiley & Sons, Inc., New York.

Endo, B.Y., and W.P. Wergin. 1973. Ultrastructural investigations of clover roots during early stages of infection by the root-knot nematode *Meloidogyne incognita*. Protoplasma 78:365–379.

Ferris, H. 1985. Density-dependent nematode seasonal multiplication rates and overwinter survivorship: A critical point model. J. Nematol. 17:93–100.

Frerich, S.J. 1995. Evaluation of *Gossypium longicalyx* monosomic addition lines of *Gossypium hirsutum* for resistance to reniform. M.S. thesis. Texas A&M Univ., College Station.

Gazaway, W.S. 1993. Reniform nematodes. p. 13–16. *In* Cotton nematodes: your hidden enemies. Cotton Found., Memphis, TN.

Godfrey, G.H., and H.T. Morita. 1929. Effects of some environmental factors on the root-knot nematode. Phytopathology 19:83–84.

Goodell, P.B. 1993a. Cotton Nematodes. p. 1–3. *In* Cotton nematodes: Your hidden enemies. Cotton Found., Memphis, TN.

Goodell, P.B. 1993b. Sampling for nematodes. p. 23–25. *In* Cotton nematodes: your hidden enemies. Cotton Found., Memphis, TN.

Gura, H.S., and S.D. Misha. 1981. Pathogenicity of the lance nematode, *Hoplolaimus indicus* to cotton *Gossypium hirsutum*. Ind. J. Nematol. 11:87–88.

Hague, N.G.M., and S.R. Gowen. 1987. Chemical control of nematodes. p 131–178. *In* R.H. Brown and B.R. Evans (ed.) Principles and practice of nematode control in crops. Acad. Press, London.

Hartman, K.M., and J.N. Sasser. 1985. Identification of *Meloidogyne* species on the basis of differential host test and perineal pattern morphology. p. 69–77. *In* K.R. Barker et al. (ed.) An advanced treatise on *Meloidogyne*. Vol. 2. methodology. North Carolina State Univ. Graphics, Raleigh.

Heald, C.M., and A.F. Robinson. 1990. Survey of current distribution of *Rotylenchulus reniformis* in the United States. Suppl. J. Nematol. 22:695–699.

Holdeman, O.L., and T.W. Graham. 1954. Effect of the sting nematode on expression of Fusarium wilt of cotton. Phytopathology 44:683–685.

Huang, C.S. 1985. Formation, anatomy, and physiology of giant cells induced by root-knot nematodes. p. 155–164. *In* J.N. Sasser and C.C. Carter (ed.) An advanced treatise on *Meloidogyne*. Vol. 2. North Carolina State Univ. Graphics, Raleigh.

Hussey, R.S. 1977. Influence of subsoiling on cotton yield and vertical distribution of soil-borne microorganisms. J. Nematol. 9:83–86.

Hussey, R.S. 1985. Host-parasite relations and associated physiological changes. p. 144–153. *In* J.N. Sasser and C.C. Carter (ed.) An advanced treatise on *Meloidogyne*. Vol. 2. North Carolina State University Graphics, Raleigh.

Jatala, P. 1991. Reniform and false root-knot nematodes, *Rotylenchulus* and *Nacobbus* spp. p. 509–528. *In* W.R. Nickle (ed.) Manual of agricultural nematology. Marcel Dekker, Inc., New York.

Jones, M.K.G., and V.H. Dropkin. 1975. Cellular alterations induced in soybean roots by three endoparasitic nematodes. Physiol. Plant Pathol. 5:119–124.

Jones, M.K.G., and D.H. Northcote. 1972. Multinucleate transfer cells induced in *Coleus* roots by the root-knot nematode *Meloidogyne arenaria*. Protoplasma 75:381–395.

Kaplan, M., and J.P. Noe. 1993. Effects of chicken-excrement amendments on *Meloidogyne arenaria*. J. Nematol. 25:71–77.

Kerry, B. 1990. An assessment of progress toward microbial control of plant-parasitic nematodes. Suppl. J. Nematol. 22:621–631.

Kirkpatrick, T.L., and J.N. Sasser. 1983. Crop rotation and races of *Meloidogyne incognita* in cotton root-knot management. J. Nematol. 16:323–328.

Kirkpatrick, T.L., and C.S. Rothrock. 1995. Interaction of *Thielaviopsis basicola* and *Meloidogyne incognita* on cotton. p. 211. *In* 53rd Cotton Disease Council. Proc. Beltwide Cotton Conf., San Antonio, TX. 4–7 January. Natl. Cotton Counc., Memphis, TN.

Koura, F.H. 1986. Comparative histopathogenesis of some Egyptian cotton cultivars infected with *Hoplolaimus aegypti*. Egypt. J. Phytopathol. 18:133–141.

Krall, E.L. 1978. Root parasitic nematodes—Family Hoplolaimidae. (Translated from Russian). Nauka Publ., Leningrad, USSR.

Kraus-Schmidt, H., and S.A. Lewis. 1979. Seasonal fluctuations of various nematodes in cotton fields in South Carolina. Plant Dis. Rep. 63:859–853.

Kraus-Schmidt, H., and S.A. Lewis. 1981. Dynamics of concomitant populations of *Hoplolaimus columbus*, *Scutellonema brachyurus*, and *Meloidogyne incognita* on cotton. J. Nematol. 13:41–48.

Lewis, S.A., F.H. Smith, and W.M. Powell. 1976. Host-parasite relations of *Hoplolaimus columbus* on cotton and soybean. J. Nematol. 8:141–145.

McClure, M.A. 1977. *Meloidogyne incognita*, a metabolic sink. J. Nematol. 9:88–90.

Meon, S., H.R. Wallace, and J.M. Fisher. 1978. Water relations of tomato (*Lycopersicon esculentum* Mill. Early Red Drawf) infected with *Meloidogyne javanica* (Treub), Chitwood. Physiol. Plant Pathol. 13:275–281.

Mueller, J.D. 1993. Lance nematodes. p. 17–20. *In* Cotton nematodes: Your hidden enemies. Cotton Found., Memphis, TN.

Mueller, J.D., and M.J. Sullivan. 1988. Response of cotton to *Hoplolaimus columbus*. Ann. App. Nematol. 2:86–89.

Neal, C.D. 1954. The reniform nematode and its relationship to the incidence of Fusarium wilt of cotton at Baton Rouge. Phytopathology 44:447–450.

Noe, J.P., and K.R. Barker. 1985a. Relation of within field variation of plant-parasitic nematode population densities to edaphic factors. Phytopathology 75:247–292.

Noe, J.P., and K.R. Barker. 1985b. Overestimation of yield loss of tobacco caused by the aggregated spatial pattern of *Meloidogyne incognita*. J. Nematol. 17:245–251.

Noe, J.P., J.N. Sasser, and J.L. Imbriani. 1991. Maximizing the potential of cropping systems for nematode management. J. Nematol. 23:353–361.

Ogallo, J.L., P.B. Goodell, J. Eckert, and P.A. Roberts. 1997. Evaluation of NemX, the first cultivar of upland cotton with high resistance to *Meloidogyne incognita*. J. Nematol. 29:531–537.

Opperman, C.H., C.G. Taylor, and M.A. Conkling. 1994. Root-knot nematode-directed expression of a plant root-specific gene. Science (Washington, DC) 263:221–223.

Orion, D., and G. Kritzman. 1991. Antimicrobial activity of *Meloidogyne javanica* gelatinous matrix. Rev. Nematol. 14:481–483.

Page, S.L.J. 1983. Biological studies of the African cotton root nematode *Meloidogyne acronea*. Ph.D. diss., London Univ.

Prasad, K.S., and G.M. Padeganur. 1980. Observations on the association of *Rotylenchulus reniforms* with Verticillium wilt of cotton. Ind. J. Nematol. 10:91–92.

Rebois, R.V., P.A. Madden, and B.J. Eldridge. 1975. Some ultrastructural changes induced in resistant and susceptible soybean roots following infection by *Rotylenchulus reniformis*. J. Nematol. 7:122–139.

Ridgeway, R.L. 1984. Cotton protection practices in the USA and world. A: Insects. p. 266–287 *In* R.J. Kohel and C.F. Lewis (ed.) Cotton. Agron. Monogr. 24. ASA, CSSA, and SSSA, Madison, WI.

Robbins, R.T., and K.R. Barker. 1973. Comparison of host-range and reproduction among populations of *Belonolaimus longicaudatus*. Plant Dis. 750–754.

Robbins, R.T., and K.R. Barker. 1974. The effects of soil type, particle size, temperature, and moisture on reproduction of *Belonolaimus longicaudatus*. J. Nematol. 6:1–6.

Robbins, R.T., and H. Hirschmann. 1974. Variation among populations of *Belonolaimus longicaudatus*. J. Nematol. 6:87–94.

Roberts, P.A., and W.C. Matthews. 1984. Cotton growth responses to root-knot nematode infection. p. 17–19. *In* 44th Cotton Disease Counc., Proc. Beltwide Cotton Prod. Res. Conf., Atlanta, GA. 8–12 January. Natl. Cotton Counc., Memphis, TN.

Roberts, P.A., S.N. Smith, and W.C. Matthews. 1985. Quantitative aspects of the interaction of *Meloidogyne incognita* with Fusarium wilt on Acala cotton in field plots. p. 21. *In* 45th Cotton Disease Counc., Proc. Beltwide Cotton Prod. Res. Conf., Natl. Cotton Council, Memphis, TN.

Robinson, A.F., C.M. Heald, S.L. Flannagan, W.H. Thames, and J. Amador. 1987. Geographic distributions of *Rotylenchulus reniformis, Meloidogyne incognita*, and *Tylenchulus semipenetrans* in the Lower Rio Grande Valley as related to soil texture and land use. Suppl. J. Nematol. 1:20–25.

Schmitt, D.P., and K.R. Barker. 1988. Incidence of plant-parasitic nematodes in the coastal plain of North Carolina. Plant Dis. 72:107–110.

Shepherd, R.L., J.C. McCarty, Jr., W.L. Parrott, and J.N. Jenkins. 1988. Resistance of cotton cultivars and elite breeding lines to root-knot nematodes. Mississippi Agric. For. Exp. Stn. Tech. Bull. 158.

Smart, G.C., and K.B. Nguyen. 1991. Sting and awl nematodes: *Belonolaimus* and *Dolichodorus* spp. p. 627–668. *In* W.R. Nickle (ed.) Manual of agricultural nematology. Marcel Dekker, Inc., New York.

Starr, J.L. 1993a. Recovery and longevity of egg masses of *Meloidogyne incognita* during simulated winter survival. J. Nematol. 25:244–248.

Starr, J.L. 1993 b. Dynamics of the nuclear complement of giant cells induced by *Meloidogyne incognita*. J. Nematol. 25:416–420.

Starr, J.L., C.M. Heald, A.F. Robinson, R.G. Smith, and J.P. Krause. 1993. *Meloidogyne incognita* and *Rotylenchulus reniformis* and associated soil textures from some cotton production areas of Texas. Suppl. J. Nematol. 25:895–899.

Starr, J.L., and M.J. Jeger. 1985. Dynamics of winter survival of eggs and juveniles of *Meloidogyne incognita* and *M. arenaria*. J. Nematol. 17:252–256.

Starr, J.L., M.J. Jeger, R.D. Martyn, and K. Schilling. 1989. Effects of *Meloidogyne incognita* and *Fusarium oxysporum* f. sp. *vasinfectum* on plant mortality and yield of cotton. Phytopathology 79:640–646.

Starr, J. and S.L.J. Page. 1990. Nematode parasites of cotton and other fiber tropical fiber crops. p. 539–556. *In* M. Luc et al. (ed.) Plant parasitic nematodes in subtropical and tropical agriculture. CAB Int., London.

Starr, J.L., and C.W. Smith. 1993. Root-knot nematode and Fusarium wilt—resistance to both pathogens. p. 178–180. *In* 53rd Cotton Disease Counc., Proc. Beltwide Cotton Res. Conf., New Orleans, LA. 10–14 January. Natl. Cotton Counc., Memphis, TN.

Starr, J.L., and J.A. Veech. 1986. Comparison of development, reproduction and aggressiveness of *Meloidogyne incognita* races 3 and 4 on cotton. J. Nematol. 18:413–415.

Taylor, A.L., J.N. Sasser, and L.A. Nelson. 1982. Relationship of climate and soil characteristics to geographical distribution of *Meloidogyne* species in agricultural soils. North Carolina State Univ. Graphics, Raleigh.

Thames, W.H., and C.M. Heald. 1974. Chemical and cultural control of *Rotylenchulus reniformis* on cotton. Plant Dis. Rep. 58:337–341.

Thomas, S.H., and D.W. Smith. 1993. Effects of 1,3-dichloropropene for *Meloidogyne incognita* management on cotton under furrow irrigation. Suppl. J. Nematol. 25:752–757.

Towson, A.J. 1977. Survival of *Meloidogyne javanica* in a desiccated soil. M.S. thesis. Univ. Hawaii.

Triantaphyllou, A.C. 1985. Cytogenetics, cytotaxomony, and phylogeny of root-knot nematodes. p. 113–126. *In* J.N. Sasser and C.C. Carter (ed.) An advanced treatise on *Meloidogyne*. Vol 1. North Carolina State Univ Graphics, Raleigh.

Triantaphyllou, A.C., and H. Hirschmann. 1960. Post infection development of *Meloidogyne incognita* Chitwood, 1949. Ann. Inst. Phytopathol. Benaki 3:1–11.

Vrain, T.C. 1978. Influence of chilling and freezing temperatures on infectivity of *Meloidogyne incognita* and *M. hapla*. J. Nematol. 10:177–180.

Womersley, C., and C. Chung. 1987. Natural dehydration regimes as a prerequisite for the successful induction of anhydrobiosis in the nematode *Rotylenchulus reniformis*. J. Exp. Biol. 143:359–372.

Yang, H., N.T. Powell, and K.R. Barker. 1976. Interactions of concomitant species of nematodes and *Fusarium oxysporum* f. sp. *vasinfectum* on cotton. J. Nematol. 8:74–80.

18 Alfalfa

GERALD D. GRIFFIN

USDA-ARS, Forage and Range Research Laboratory
Logan, Utah

Several species of plant-parasitic nematodes are associated with cultivated and native vegetation throughout the world. The number and kind of species found are differentiated by host species, and climatic and environmental conditions. Nematodes have a relatively low reproductive rate, and movement is constrained by such soil parameters as texture and water. Nematode-disease symptoms are generally slow in developing in plants when compared to those of other pathogens including insects, bacteria, and fungi. This explains why symptoms of nematode disease may not be immediately recognized (Webster, 1987). Some nematode symptoms may be similar to or identified as being due to other causes. Thus, it is often difficult to accurately determine the importance of nematodes on plant growth and survival.

Plant-parasitic nematodes invade and parasitize all forages including alfalfa (*Medicago sativa* L.). Nematodes affect the growth and persistence of forages in both irrigated and nonirrigated cultivated and rangeland soils (Griffin, 1984). Many plant-parasitic nematodes reduce the growth of alfalfa. The most important are the alfalfa stem nematode, *Ditylenchus dipsaci*; root-knot nematodes, *Meloidogyne* spp.; and the lesion nematode, *Pratylenchus penetrans* (Griffin, 1984; Leath et al., 1988; Santo et al., 1980; Barnes et al., 1990; Thies et al., 1992). Other nematodes, including the foliar nematode (*Aphelenchoides ritzemabosi*), other *Pratylenchus* spp., and ectoparasitic nematodes including *Helicotylenchus* spp., *Tylenchorhynchus* spp., *Merlinius* spp., and *Xiphinema americanum* (Griffin, 1984, 1996), have been found associated with suppressed growth and yields of alfalfa. Nematodes also play an important role in disease complexes with other plant pathogens, and affect crops grown in rotation with alfalfa.

IMPORTANT NEMATODE PARASITES

Alfalfa Stem Nematode

History and Distribution

The alfalfa stem nematode, *Ditylenchus dipsaci*, is the most important nematode pathogen of alfalfa and is found in most areas where alfalfa is grown.

This pathogen was first reported on alfalfa in Germany in 1881 and was discovered in the USA in 1923 (Thorne, 1961). The stem nematode is a serious pest in areas of high rainfall and in irrigated fields. It is usually found in heavy soils, but it can damage alfalfa in all soil textures if sufficient soil water is present for nematode movement.

Several biological races of *D. dipsaci* have been separated, and several reports have shown race variability. Races can only be separated by hosts and not morphological differences. Green (1981) found several weeds and wild plants to be hosts of *D. dipsaci*, which may be sources of the different biological races. The alfalfa race found in the USA is limited mainly to alfalfa, but Edwards and Taylor (1963) found a race in the midwestern USA that was able to reproduce on several plant species. Sturhan (1971) stated that more than 20 biological races have been described that attack more than 300 plant species. Although *D. dipsaci* may be host specific, it is able to infect and parasitize a wide range of nonhost plants. The alfalfa stem nematode race found in the Intermountain Region of the USA, host specific to alfalfa and sainfoin (*Onobrychis viciifolia* Scop.) (Griffin et al., 1975), caused characteristic symptoms on nonhost seedlings of Indian sweet clover [*Melilotus indicus* (L.) Mill.], onion (*Allium cepa* L.), tomato (*Lycopersicon esculentum* Miller), sugarbeet (*Beta vulgaris* L. subsp. *vulgaris*), and wheat (*Triticum aestivum* L. emend. Thell) (Griffin, 1975); the nematode caused seedling mortality ranging from 20% on sugarbeet and tomato to 100% on onion.

The major avenue of nematode dissemination and distribution was initially through soil particles and plant debris in unclean seed lots (Thorne, 1961). Such spread has been lessened by the use of certified and clean seed. However, anything that moves soil or plant tissue will move the nematode. Nematodes are readily spread by animals, machinery during harvest, and rain and irrigation water. The reuse of waste water is a major method of nematode dissemination under an irrigation system (Faulkner & Bolander, 1970).

Biology and Symptomology

Both males and females of *D. dipsaci* are 1.0 to 1.3 mm long. All vermiform nematode stages are able to invade and parasitize alfalfa. The nematode enters primordial crown bud tissue and migrates into the developing crown bud. Nodes and internodes of parasitized stems are short and swollen and usually discolored. Pathological effects of *D. dipsaci* on alfalfa are due to enzymatic action of the nematode and a physiological imbalance of auxins produced by the plant (Griffin, 1984). Growing stems may succumb to nematode parasitism or outgrow the swelling and grow normally. Stem necrosis may occur after an extended period of parasitism associated with high humidity at moderate soil temperatures. Necrosis may extend up the stem for more than 30 cm under certain conditions. Great numbers of nematodes (several thousand/g of fresh tissue) can be found in blackened stems. The numbers of stems per crown become fewer from increased nematode parasitism and invasion of the alfalfa crown by opportunistic fungi and bacteria. This leads to rapid deterioration of the crown tissue and subsequent death of the plant. The stand of alfalfa is thinned and becomes invaded by weeds (Griffin, 1984, 1987; Thorne, 1961).

Temperature influences the damage potential of *D. dipsaci* on alfalfa (Griffin, 1987, 1968). Under warm, humid, weather conditions the nematode may migrate into leaf tissue causing distortion and curling of the leaves. Leaves of individual stems may show "white flagging," resulting from destruction of the chloroplast precursor (Campbell & Griffin, 1973). Optimum temperature for invasion and reproduction of *D. dipsaci* in alfalfa is 15 to 20°C. A life cycle can be completed in 19 to 23 d in a susceptible alfalfa cultivar at the optimum temperature range. The stem nematode is unique in its ability to invade and parasitize plants over a temperature range of 5 to 30°C. *Ditylenchus dipsaci* is able to acclimatize to temperature, and there is a direct correlation between the acclimation temperature and the temperature at which maximum penetration of plant tissue occurs (Croll, 1967; Griffin, 1974).

Under conditions not completely understood, *D. dipsaci* can invade the alfalfa seed if the inflorescence is infected. Roots are occasionally infected, resulting in internal cavities (Krusberg, 1961). Gall-like outgrowths that girdle the root crown may be present. Large numbers of nematodes can survive periods of environmental stress in alfalfa crown tissue, while only small numbers have been recovered from alfalfa field soil at any time of the year (Tseng et al., 1968). This is similar to data obtained from *D. dipsaci* on onion and oat (*Avena sativa* L.) (Lewis & Mai, 1960; Wallace, 1962).

Reports differ on the ability of *D. dipsaci* to survive outside of host tissue (Chitwood et al., 1940; Edwards, 1951). This nematode can survive less than 3 yr under a nonhost cropping of grain under heavy rainfall or irrigation, although fourth-stage nematodes can survive for many years in an anhydrobiotic state in fallow soil where the soil water remains very low (Griffin, 1987). *Ditylenchus dipsaci* has been known to survive for 20 yr in a dormant stage in soil particles in unclean seed (Fielding, 1951). Since the nematode is able to remain in a dormant state until activated by water, longevity depends on the availability of adequate host tissue in the soil when activation occurs.

Disease Complexes

Ditylenchus dipsaci interacts with other plant pathogens in disease complexes, and is involved in the development of other pathogens on alfalfa including bacterial wilt caused by *Clavibacter michiganensis* ssp. *insidiosus* (Hawn & Hanna, 1966), Verticillium wilt caused by *Verticillium albo-atrum* (Vrain, 1987), and Fusarium wilt caused by *Fusarium oxysporum* f. sp. *medicaginis* (Griffin, 1990). Another possible disease complex involves *D. dipsaci* and the chrysanthemum foliar nematode, *A. ritzemabosi* (Grundbacher & Stanford, 1962). Recent research has shown both nematodes to be cohabitants of alfalfa stem and crown tissue of plants collected from several western states of the USA (Gray et al., 1994)

Ditylenchus dipsaci and *F. oxysporum* f. sp. *medicaginis* synergistically affected the mortality and plant growth of Ranger alfalfa, a cultivar susceptible to stem nematode and Fusarium wilt. The nematode-fungus relationship had an additive effect on mortality and plant growth of 'Lahontan' (nematode resistant and Fusarium wilt susceptible) and of 'Moapa 69' (nematode susceptible and

Fusarium wilt resistant). Combination of the two pathogens reduced the mean weights of Ranger shoots, but not the mean weights of Lahontan and Moapa 69 below those of plants inoculated with either organism alone. Suppressions of shoot growth from combined inoculations of Lahontan and Moapa 69 were similar to those for *D. dipsaci* alone. No differences occurred between simultaneous and sequential pathogen inoculations of any cultivar. Inoculation with *D. dipsaci* restricted root growth of Ranger and Moapa cultivars, but not Lahontan. *Fusarium oxysporum* f. sp. *medicaginis* suppressed the root growth of Ranger and Lahontan, but not Moapa 69 (Griffin, 1990).

Soil water at time of cutting affected persistence and growth of alfalfa in soil infested with *D. dipsaci* alone, or with *D. dipsaci* + *F. oxysporum* f. sp. *medicaginis*. Persistence of Ranger and Moapa 69 was reduced by *D. dipsaci* when soil water was maintained at the high soil water level, whereas *D. dipsaci* did not affect stand persistence where soil water declined to the low soil water level before cutting. Shoot and root growth of all plant cultivars was suppressed by the nematode + fungus combination at the high soil water level. Also shoot and root growth of Ranger was restricted below uninoculated control plants by single inoculations of *D. dipsaci* or *F. oxysporum* f. sp. *medicaginis*, the growth of Moapa 69 was inhibited by *D. dipsaci*, and the growth of Lahontan was limited by *F. oxysporum* f. sp. *medicaginis*. At the low soil water level, root growth of Ranger and Lahontan was suppressed only by the combined inoculation, whereas Moapa 69 root growth was not affected by any treatment. *Rhizobium* nodulation was suppressed by single and combined inoculations of *D. dipsaci* and *F. oxysporum* f. sp. *medicaginis* at the high soil water level but not at the low soil water level; the one exception was *D. dipsaci* on Lahontan at high soil water level. Fusarium wilt of the three cultivars was greater in the high soil water level than in the low soil water level with the exception of the fungus alone on Moapa 69. Combined inoculation increased wilt symptoms above the single inoculation of *F. oxysporum* f. sp. *medicaginis* in Ranger in the high soil water level treatment but not in the low soil water level treatments. However, combined inoculation of Lahontan and Moapa 69 alfalfa with the nematode + fungus did not increase Fusarium wilt over that of the fungus alone at either water level (Griffin, 1992).

Root-Knot Nematodes

History and Distribution

The most important root-knot nematode species pathogenic to alfalfa are the northern root-knot nematode (*Meloidogyne hapla*), the southern root-knot nematode (*M. incognita*), the Javanese root-knot nematode (*M. javanica*), and the peanut root-knot nematode (*M. arenaria*). A race of another root-knot nematode species, *M. chitwoodi*, recently discovered in the western USA, also parasitizes and reproduces on alfalfa (Griffin & Thomson, 1988; Santo & Pinkerton, 1985).

The northern root-knot nematode, *M. hapla*, is the most widely distributed root-knot nematode species in alfalfa in the USA (Griffin, 1984). It is found in the northern hemisphere where soil temperatures drop to 0°C or below and summer soil temperatures seldom exceed 25 to 30°C. It occurs worldwide and is found in the northern and southern USA, Africa, Asia, Canada, Europe, and South America (Taylor & Sasser, 1978).

Meloidogyne incognita is probably the most important root-knot species on cultivated plant species worldwide. Unlike *M. hapla*, it cannot survive harsh temperate climates, and is only found at optimum soil temperatures ranging from 25 to 30°C. *M. incognita* eggs usually cannot survive soil temperatures below freezing, and juveniles cannot survive soil temperatures below 5°C. Since less than 10% of the alfalfa hectarage is found in areas where *M. incognita* is climatically adapted, this nematode is a less important pathogen on alfalfa than *M. hapla*.

The Javanese root-knot nematode, *M. javanica*, like *M. incognita* is limited in its distribution in the alfalfa-producing regions of the USA. It is common in sandy loam soil in the arid southwestern regions of the USA where average January temperatures approach 7°C (Taylor & Sasser, 1978).

Two other species of *Meloidogyne* of lesser importance on alfalfa are *M. chitwoodi* and *M. arenaria*. The predominance of *M. hapla* in the Pacific Northwest and Intermountain Region of the USA has been challenged by the identification of *M. chitwoodi* (Golden et al., 1980). Alfalfa is considered a nonhost of *M. chitwoodi*, but a race from Utah and Washington parasitizes and reproduces on *M. hapla*-resistant and susceptible cultivars (Griffin & Thomson, 1988; Santo & Pinkerton, 1985). Alfalfa is not as suitable a host for *M. chitwoodi* as it is for *M. hapla* (Griffin et al., 1986). *Meloidogyne arenaria* is similar to *M. incognita* in geographical limitations, and its economic importance on alfalfa is minor.

Plants parasitized by *Meloidogyne* spp. may be stunted, may exhibit excessive adventitious root growth, and may have a reduced stand. The size and degree of galling are dependent on the nematode species.

Biology and Symptomology

Meloidogyne hapla is less pathogenic on alfalfa than *D. dipsaci* (Griffin, 1984). Invasion of alfalfa seedlings with large numbers of *M. hapla* juveniles can result in a high mortality of seedlings that severely reduces alfalfa yields. A tolerance level of less than 1 nematode/cm^3 of soil has been reported for *M. hapla* on 'Washoe' alfalfa under greenhouse conditions (Inserra et al., 1980).

Meloidogyne hapla may increase to a very high population on alfalfa but is less virulent than *D. dipsaci*. High populations of *M. hapla* can, however, result in severe invasion and destruction of *Rhizobium* nodules, resulting in a reduction in nitrogen fixation (Gray et al., 1990). Although *M. hapla* causes yield losses in alfalfa, greater losses may occur on susceptible vegetable and field crops grown in rotation with alfalfa.

Meloidogyne hapla can predispose resistant alfalfa plants to other plant pathogens. This nematode increased the incidence of bacterial wilt, Fusarium wilt, and seedling damping off and root infection (caused by *Phytophthora megasperma* f. sp. *medicaginis* Kuan & Erwin) in resistant and susceptible alfalfa (Hunt et al., 1971; Griffin & Hunt, 1972a; Griffin & Thyr, 1988; Gray et al., 1990; Griffin & Gray, 1994). In addition, *M. hapla* also interacts with *D. dipsaci* to increase the mortality rate and plant-growth suppression of resistant and susceptible alfalfa cultivars (Griffin, 1987).

The life cycles are similar for all root-knot nematode species including *M. hapla* (Taylor & Sasser, 1978). The first molt occurs within the egg, and infective stage of *Meloidogyne* spp. is the second-stage juvenile (approximately 0.40–0.50 mm long). Juveniles are attracted to and invade metabolically active alfalfa tissue. The primary invasion sites are undeveloped tissues near the root tip, where nematode feeding may cause cellular hypertrophy. Early invasion also occurs in *Rhizobium* nodules (Gray et al., 1990). Nematode juveniles become oriented toward the root plerome, feed on this tissue and prohibit its differentiation into vascular tissue. The cells enlarge, and a cell wall may form during cell division but soon disappears. The adjoining cells coalesce, forming giant cells that become a continual food source for the nematode. Juveniles become sedentary and increase in size and girth.

Nematodes undergo a total of four molts before becoming adults. The female becomes oval in shape and remains sedentary throughout life, feeding on the giant cells. The male, after molting the fourth cuticle, becomes a cylindrical filiform nematode approximately 1.2 mm long and migrates from the root. The pear-shaped female grows to approximately the size of the head of a pin. One female deposits 100 to 350 or more eggs in a protective gelatinous matrix that adheres to the outer surface of the root. When temperature and soil water are favorable, the eggs hatch and the juveniles escape into the soil. The role of the male is not well understood since reproduction generally occurs without the egg being fertilized. Males are often found, however, in large numbers in the soil, in egg masses, and occasionally in the root tissue. Food availability impacts the sex ratio within root-knot nematode species; the ratio of males to females increases as inoculum density increases or as the food supply decreases.

The life cycle of root-knot nematodes on alfalfa differs among species. The optimum temperature for hatching, mobility, invasion, and growth is 25, 20, 15 to 20, and 20 to 25°C, respectively, for *M. hapla*, and 30, 25, 15 to 30, and 25 to 30°C, respectively, for *M. javanica*. The life cycle of *M. hapla* is approximately 30 d at 25°C. The life cycle for *M. incognita* is similar to that of *M. javanica*.

Gall size differs depending on the root-knot nematode species. Galls of *M. hapla* are small and usually oval, whereas those of *M. incognita* are large and more cylindrical. A distinct characteristic of *M. hapla* galls is multiple lateral root growth from the galls.

Meloidogyne incognita is not as damaging on alfalfa as *M. hapla* and *M. javanica*. Shoot growth of 'Atlantic' alfalfa was suppressed more by *M. hapla* than *M. incognita* (Chapman, 1960, 1963). However, reproduction of *M. incognita* was greater than that of *M. hapla*. A similar relationship occurred between *M. incognita* and the Javanese root-knot nematode, as Reynolds and O'Bannon (1960) found *M. javanica* to be more damaging than *M. incognita* on nondormant alfalfa varieties. The major impact of *M. incognita* is on economically important plant cultivars such as those of tomato and cotton (*Gossypium hirsutum* L.) that are grown in rotation with alfalfa.

Meloidogyne hapla and *M. incognita* have been associated with other pathogens in disease complexes on alfalfa, and these nematodes increased the severity of Fusarium wilt on alfalfa.

Root-Lesion Nematodes

History and Distribution

Root-lesion nematodes, *Pratylenchus* spp., occur worldwide. They attack and parasitize a wide range of plant species, are important pathogens on cultivated and noncultivated plants, and are one of the most important groups of plant-parasitic nematodes. Symptoms of root-lesion nematode are hard to diagnose since they are similar to those caused by many other root pathogens. Roots invaded by *Pratylenchus* spp. are often invaded by secondary microorganisms; nematode invasion creates avenues for easy entry by other pathogens. Hence, *Pratylenchus* spp. are involved in important disease complexes with other plant pathogens.

Biology and Symptomology

Lesion nematodes are migratory endoparasites. They are found in root tissue and in the soil. Eggs are deposited in root tissue and soil since juveniles and adults migrate in and out of plant roots. Both juveniles and adults invade host root tissue where they break down tissues by feeding and forcing their way through and between the epidermal and cortical cells. No increase in host-cell development occurs as with *Meloidogyne* spp. Kable and Mai (1968) reported that alfalfa roots were invaded by only the fourth-stage juveniles, whereas Townshend (1978) stated that roots of 'DuPuit' alfalfa also were invaded by third-stage juveniles. The nematode may penetrate the entire length of the roots (Townshend & Stobbs, 1981). *Pratylenchus* spp. usually feeds only in the root cortex, but also penetrates and feeds in the root stele tissue. The nematode produces enzymes, resulting in water-soaked areas of the root that later become necrotic. These areas may coalesce, the discoloration intensifies, and the entire root may become dark brown to blackened and totally necrotic.

Cortical cells penetrated and fed upon by *P. penetrans* contain only cytoplasmic debris (Townshend et al., 1989). Proximal cells had increased in tannin deposits, degenerated mitochondria, numbers of ribosomes, and had no internal membrane structure. The endodermis was usually collapsed and contained massive tannin deposits on the inner cell wall and cell lumen. Effects on the stele were similar to those on the epidermis, except tannin deposits were not as prominent. Multivesiculate structures were observed both in the endodermis and in the stele.

Several species of lesion nematodes attack and parasitize alfalfa in North America (Griffin,1984; Griffin, 1993a; Kimpinsky & Willis, 1981; Sontirat & Chapman, 1970; Thorne, 1961; Townshend, 1984). *Pratylenchus* spp. severely damaged alfalfa sown in winter wheat fields, reportedly because of competition with wheat under moist field conditions, heavy nematode invasion, and destruction of alfalfa seedling roots; minimal damage occurred following summer fallow (Chapman, 1954).

Pratylenchus penetrans is the most important lesion nematode of alfalfa. It has a wide host range and is known to reduce establishment and production of alfalfa in the USA and Canada (Chapman, 1954; Griffin, 1993b; Nelson et al.,

1985). Nematode populations approaching 100/cm^3 soil have been observed in Minnesota alfalfa fields (Thies et al., 1992).

Nematode parasitism affects alfalfa cold hardiness (Suzuki & Willis, 1974). Plant age affects nematode invasion (Olthof, 1982), and the efficiency of root penetration by lesion nematodes is dependent on the nematode stage (Sonitrat & Chapman, 1970). Townshend (1978) found that *P. penetrans* females have a wider penetration temperature range than do the juvenile stages. Temperature also affects nematode invasion, root penetration increases with an increase in soil temperature (Kimpinsky & Willis, 1981).

Pratylenchus pratensis was associated with poor stands and low yields of alfalfa in Rhode Island (Stessel, 1960). *Pratylenchus neglectus* is endemic in both cultivated and virgin soil in the USA (Thorne, 1961), and is associated with poor growth of alfalfa (Griffin & Gray, 1990). Populations of *P. neglectus* as high as 14000/root system have been collected from alfalfa and wheat (Griffin & Gray, 1990). Although considered mildly pathogenic, Griffin and Gray (1990) found a pathotype of *P. neglectus* that reduced the persistence and growth of alfalfa. It differs from other *P. neglectus* populations in the degree of virulence and reproductive capability (Griffin, 1991b).

Soil temperature affects host-parasite relationships including those involving *Pratylenchus* spp. (Griffin, 1984). Temperature limitations depended more on the host plant than on the nematode by affecting host susceptibility (Dickerson, 1979). Lesion nematodes are active over a wide temperature range. Invasion of roots by *P. penetrans* females may occur over a temperature range of 5 to 30°C (Townshend, 1978); this aids the nematode in surviving severe climatic changes that may occur in the soil. The optimum temperature for lesion nematode reproduction appears to be approximately 30°C (Dunn, 1973; Griffin & Gray, 1990). Nelson et al.(1985), however, found the optimum temperature for *P. penetrans* reproduction was 25°C. Soil texture affects the host-parasite relationship between lesion nematodes and plants. Endo (1959) and Griffin (1996) found that the pathological relationship of lesion nematodes to host plants also was affected by soil type. Nematodes were more virulent in sandy-loam than in clay-loam soils.

Comparable to stem and root-knot nematodes, lesion nematodes interact with other plant pathogens on alfalfa. Mauza and Webster (1982) and Chapman and Turner (1975) found positive interactions with *P. penetrans* and two species of *Fusarium* on alfalfa, and between *M. incognita* and *P. penetrans* on alfalfa.

Other Nematodes

Foliar Nematode

The foliar nematode, *Aphelenchoides ritzemabosi*, is a major parasite of a wide range of ornamental plants. It was first discovered on chrysanthemum [*Dendranthema* × *grandiflorum* (Ramat.) Kitam.] in the USA by Halsted (1891, p. 366–370). Grundbacher and Stanford (1962) found *A. ritzemabosi* associated with the alfalfa stem nematode, *D. dipsaci*, on alfalfa. Recent work has focused on its importance and distribution on alfalfa (Gray et al., 1994). Initial work showed that *A. ritzemabosi* is usually associated with *D. dipsaci* in alfalfa, but environmental conditions have a greater affect on *A. ritzemabosi* than on *D. dipsaci* (Williams et al., 1991). *Aphelenchoides ritzemabosi* invasion causes damping off, general stunting, and "blind shooting" (seedling development is arrested in the cotyledon stage). The effects from *D. dipsaci*—tissue swelling and shortened internodes. Additional study is needed to further elucidate the host-parasite relationship between *A. ritzemabosi* and alfalfa, and the relationship of *D. dipsaci* to *A. ritzemabosi*.

Ectoparasitic Nematodes

Several genera of ectoparasitic nematodes are associated with alfalfa. The dagger nematode, *Xiphinema americanum*, caused serious losses to alfalfa and red clover (*Trifolium pratense* L.) in the North Central Region of the USA (Norton, 1963). This agrees with the findings of Taylor et al. (1958) who found *X. americanum* widely distributed in alfalfa fields in Minnesota along with *Criconemella, Helicotylenchus, Hoplolaimus, Paratylenchus, Paratrichodorus* and *Tylenchorhynchus* genera. The spiral nematode, *Helicotylenchus digonicus*, associated with the poor growth of alfalfa in Canada, increased significantly in numbers during the growing season (Townshend & Potter, 1973). The stunt nematode, *Tylenchorhynchus clarus*, suppressed growth of 'Moapa' alfalfa, resulting in reduced growth of feeder roots, although neither lesions nor abnormal root growth were observed (Noel & Lownsbery, 1978). Control strategies and tactics have not been developed for these nematodes.

Several viruses that infect alfalfa are vectored by nematodes. For example, the pea early browning virus of alfalfa is transmitted by *Paratrichodorus* spp.

NEMATODE CONTROL

Alfalfa Stem Nematode

Plant resistance is the major method of controlling *D. dipsaci*. Resistance to the alfalfa race of *D. dipsaci* was first discovered by Gerald Thorne in a Turkestan alfalfa selection growing in a nematode infested field in northern Utah. The alfalfa cultivar Nemastan was developed from this selection (Thorne, 1961). Lahontan was selected from Nemastan and also is resistant to bacterial wilt. Germplasm resistant to several plant pathogens, including *D. dipsaci*, is available to plant breeders (Hartman et al., 1979; Peaden et al., 1976). Many proprietary alfalfa cultivars have multiple-pest resistance to plant pathogens, including *D. dipsaci* (Anonymous, 1996).

Only the rate of invasion, the degree of damage, and nematode reproduction differs between resistant and susceptible alfalfa seedlings. Although greater numbers of *D. dipsaci* invade susceptible than resistant alfalfa seedlings, similar ultrastructural changes occur within tissues of both resistant and susceptible alfalfa. Susceptible plants, however, demonstrate a greater degree of swelling and disorganization of cellular structures. The greatest injury occurs at 20 to 30°C in both resistant and susceptible alfalfa. As resistant plants age, they overcome the damage and grow normally (Griffin, 1968, 1987).

Rotation of alfalfa with nonhost crops such as durum wheat (*Triticum durum* Desf.) or sugarbeet, for 2 or 3 yr usually reduces the nematode population to an undetectable level (Griffin, 1975). However, the field can become recontaminated from machinery, animals, or waste irrigation water. In irrigated areas of the arid and semiarid western USA, the alfalfa stem nematode is usually a problem only in the first cutting of alfalfa. Reinvasion occurs in subsequent cuttings mainly from the nematodes in the soil. Invasion can be restricted if alfalfa is cut when the top 5 to 10 cm of the soil is dry and irrigation is delayed until new shoot growth appears. When the soil water is 50% of field capacity, immediately before or after cutting, nematodes cause severe stunting (Griffin, 1987).

The timing of burning of alfalfa for weed control appears to affect the incidence of nematode-infected plants. Nematode invasion of primordial stem tissue is significantly reduced in spring plant growth when fall burning is used for weed control. However, spring burning is not recommended because it increases invasion of alfalfa stem tissue by *D. dipsaci*. Two- to three-year-old alfalfa fields have been completely devastated by this nematode because of spring burning (Griffin, 1987). It appears that fall burning destroys alfalfa stubble or weed hosts and nonhost plants that may be reservoirs for *D. dipsaci*. An increase in soil temperature resulting from spring burning may increase nematode metabolism, resulting in an increase in nematode invasion of primordial stem tissue.

Invasion by *D. dipsaci* has been associated with winterkill of alfalfa (Hawn & Hanna, 1966). *Ditylenchus dipsaci* reduced nonstructural carbohydrate (TNC) in alfalfa and contributed to winterkill (Boelter et al., 1985). Persistence of dormant Ranger and nondormant Moapa alfalfa, both susceptible to *D. dipsaci*, varied with stand age and harvest frequency (Griffin, 1991a). Increasing the harvest of alfalfa from three to four cuttings significantly decreases the storage of nonstructural carbohydrate in the crown and roots and reduces alfalfa persistence and yields. Similar results occur if the last alfalfa cutting is postponed until immediately before frost (Griffin, 1991a). Winterkill of mature alfalfa plants has been shown to be caused by *D. dipsaci*, and the mortality rate of infected plants was attributed to interference in the plants ability to either produce or store nonstructural carbohydrate (Boelter et al., 1985).

A fair to good alfalfa stand that may persist for 4 to 5 yr is often found when soil infestation by *D. dipsaci* results from infected seed or contaminated machinery. However, if the nematode source is from irrigation water or if the number of nematodes in the soil at the time of planting is high, the alfalfa stand can be decimated in less than 2 to 3 yr. There is a more rapid decline in nondormant alfalfa than in dormant alfalfa; nondormant types decline more rapidly since there is a longer growing period and longer association of the nematode with the plant, resulting in greater nematode invasion and reproduction (Griffin, 1996, unpublished data).

Several systemic nematicides are effective in controlling *D. dipsaci* (Gray & Soh, 1989; Griffin, 1967; Nigh et al., 1969). Both resistant and susceptible alfalfa seedlings are invaded by *D. dipsaci*. Seed treatment should provide protection during the succulent seedling stage (Gray & Soh, 1989). Carbofuran (2,3-

dihydro-2,2-dimethyl-7-benzofuranyl methylcarbamate) application around the seed at planting, or spraying aldicarb [2-methy-2-(methylthio)propionaldehyde O-(methylcarbamoyl)-oxime] on alfalfa stubble reduced the degree of parasitism and increased alfalfa yields (Whitehead & Tite, 1988). Chemicals are only a temporary control, cost plus the strong potential of reinfestation of the alfalfa should be considered before selecting chemicals as a control methodology. The use of nematicides in conjunction with a resistant cultivar may be justified under certain conditions when nematode populations are high. Recently all nematicides were removed from the registered list of chemicals that can be used on alfalfa.

Root-Knot Nematodes

Because of the wide host range of *M. hapla*, crop rotation is not an effective control measure. Resistance to this nematode is the most practical control methodology and is found in several dormant cultivars. Resistant germplasm (Nevada Synthetic XX) from 'Vernal' selections (Peaden et al., 1976), has been used to develop resistant cultivars (Anonymous, 1996). Since pathotypes may be found that parasitize and reproduce on resistant germplasm and cultivars (Griffin & McKenry, 1989), it is important to include geographically isolated populations in breeding and screening programs.

Biological control may be useful in controlling root-knot nematodes on alfalfa. Townshend et al. (1989) found that 3-wk-old alfalfa seedlings grown from seeds coated with a spore suspension of *Meria coniospora* Dreschler and planted in *M. hapla*-infested soil had 58 and 60% fewer galls in two experiments than did seedlings from untreated seeds. Treatment of *M. hapla*-infested soil with rape (*Brassica napus* L. var. *napus*) seed meal, a soil enhancer, increased shoot growth of resistant, but not susceptible, alfalfa (Hafez et al, 1988).

Nematicides have proven efficacy in controlling root-knot nematodes. Townshend (1989), using a polymer sticker to coat alfalfa seed with oxamyl {S-methyl N',N'-dimethyl-N-[(methylcarbamoyl)oxy]-1-thio-oxamimidate}, reduced the number of *M. hapla* in the soil and roots. Since seedlings of both susceptible and resistant alfalfa cultivars are parasitized by *M. hapla* (Griffin & Hunt, 1972b), nematicides could be effective in protecting the alfalfa plant during the vulnerable seedling stage until the plant's resistance system becomes active. Nematicides, however, are not registered for use on alfalfa.

Resistance is the major method of control of *M. incognita* and *M. javanica* in alfalfa. Resistance is found in several nondormant cultivars including African, Moapa, and Sonara alfalfa selections, and Nevada Synthetic YY alfalfa germplasm (Anonymous, 1996; Reynolds et al., 1970; Hartman et al., 1979). Because of the wide host range of root-knot nematodes, crop rotation is not a satisfactory method of control.

Root-Lesion Nematodes

Because of the extensive host range of lesion nematodes, and since more than one species may be involved in a nematode-host relationship, it has been difficult to breed for resistance or tolerance. However, significant strides have been made recently in breeding alfalfa for resistance, and alfalfa germplasm resistant to *P. penetrans* has been released (Barnes et al., 1990; Thies et al.,1994). Thies et al. (1994) reported that when alfalfa populations selected for resistance to *P. penetrans* were compared with susceptible cultivars, they showed the greatest stand density and fewest nematodes/gram fresh weight. Nelson et al. (1985) found variability among alfalfa clones and progenies for resistance and susceptibility to *P. penetrans*. This agrees with the report of Hafez and Miller (1990) who found nematode tolerance in an experimental alfalfa selected for resistance to *P. penetrans*. Thies et al. (1994) found that additive effects resulting from combining ability effects should be effective in developing alfalfa cultivars resistant to *P. penetrans*.

Willis and Thompson (1979) illustrated reduction of *P. penetrans* invasion of alfalfa by soil fallowing prior to seeding. However, the nematode population increased on alfalfa to its pre-fallow population after only 1 yr. Chemical control of root lesion nematodes has been reported with carbofuran (Thies et al., 1992; Willis & Thompson, 1979). Although expensive, carbofuran (though no longer registered for use on alfalfa) usage may have been justified when alfalfa was sown in fields with a high nematode population. Often, the expense involved in such short range chemical control is prohibitive.

Griffin (1994) found that interplanting of alfalfa with resistant crested wheatgrass [*Agropyron cristatum* (L.) Gaertn.] restricted the invasion of alfalfa roots by *P. penetrans*. Alfalfa growth and persistence were not affected by interplanting alfalfa and grass in *P. neglectus*-infested soil since both cultivars were hosts of the nematode.

Foliar Nematode

There has been no conscious effort to develop a control scenario for *A. ritzemabosi* in alfalfa. Recent studies, however, that show an association of *A. ritzemabois* with *D. dipsaci* in alfalfa over a large geographic region indicate that control practices for *D. dipsaci* may control *A. ritzemabosi* (Gray et al., 1994). Preliminary studies indicate alfalfa genotypes and varieties do exist that have resistance to *A. ritzemabosi*. Varietal resistance could have originated from field selections or under control conditions in which *A. ritzemabosi* was unknowingly present in the field or in inoculum used in selection studies.

Other Nematodes

Little or no research has been focused on the control of less important nematodes associated with the growth of alfalfa, including most ectoparasitic nematodes. Since ectoparasitic nematodes are known to parasitize most plants, it appears that the only method of control would be through plant resistance and/or tolerance. Until now, however, no attempt has been made to screen alfalfa for resistance to plant-parasitic nematodes except for *D. dipsaci, Meloidogyne* spp., and *Pratylenchus* spp.

OUTLOOK

Plant-parasitic nematodes occur wherever alfalfa is grown and negatively impact the growth and persistence of this important forage crop. The host-parasite relationship between nematodes and alfalfa may be significantly different between geographical regions, depending on climatic and environmental differences (Griffin, 1984, 1987, 1991a). Studies also have shown the importance of nematodes in disease complexes on alfalfa (Gray et. al., 1990; Griffin & Thyr, 1988; Hawn & Hanna, 1966).

Development and use of resistant germplasm is the most desirable type of disease and nematode control. Alfalfa cultivars and germplasms that are resistant to lesion, stem and root-knot nematodes have been released, but new cultivars still must be developed. (Anonymous, 1996; Barnes et al., 1990; Hartman et al., 1979; Peaden et al., 1976; Thorne, 1961).

A control strategy that may positively impact nematode control is interplanting of a resistant and susceptible cultivar in nematode-infested soil. A preliminary study has shown that an interplanting of resistant and susceptible alfalfa and grass cultivars extended the growth and longevity of the susceptible cultivar in either *M. hapla*- or *M. chitwoodi*-infested soil (Griffin et al., 1992). Other nematode control strategies and tactics including biological control have potential for various field crops. They may or may not be effective in controlling nematodes and disease complexes on perennial forage crops.

ACKNOWLEDGMENTS

Cooperative investigation by USDA-ARS and the Utah Agricultural Experiment Station.

REFERENCES

Anonymous. 1996. Alfalfa variety characterization. Cert. Alfalfa Seed Counc., Davis, CA.

Barnes, D.K., J.A. Thies, D.L. Rabas, D.L. Nelson, and D.M. Smith. 1990. Registration of two alfalfa germplasms with field resistance to the root-lesion nematode. Crop Sci. 30:751–752

Boelter, R.H., F.A. Gray, and R.H. Delaney. 1985. Effect of *Ditylenchus dipsaci* on alfalfa mortality, winterkill, and yield. J. Nematol. 17:140–144.

Campbell, W.F., and G.D. Griffin. 1973. Fine structure analyses of stem-nematode induced white flagging in *Medicago sativa*. J. Nematol. 5:123–126.

Chapman, R.A. 1963. Development of *Meloidogyne hapla* and *M. incognita* in alfalfa. Phytopathology 53:1003–1005.

Chapman, R.A. 1960. The effect of *Meloidogyne incognita* and *M. hapla* on the growth of Kenland red clover and Atlantic alfalfa. Phytopathology 50:181–182.

Chapman, R.A. 1954. Meadow nematodes associated with failure of spring-sown alfalfa. Phytopathology 44:542–545.

Chapman, R.A., and D.R. Turner. 1975. Effect of *Meloidogyne incognita* on *Pratylenchus penetrans* in red clover and alfalfa. J. Nematol. 7:6–10.

Chitwood, B.G., A.C. Newhall, and R.L. Clement. 1940. Onion bloat or eelworm rot, caused by the bulb or stem nematode, *Ditylenchus dipsaci* (Kühn) Filipjev. Proc. Helminthol. Soc. Washington 7:44–51.

Croll, N.A. 1967. Acclimatization in the eccritic thermal response of *Ditylenchus dipsaci*. Nematologica 13:385–389.

Dickerson, O.J. 1979. The effect of temperature on *Pratylenchus scribneri* and *P. alleni* populations on soybean and tomatoes. J. Nematol. 11:23–26.

Dunn, R.A. 1973. Extraction of eggs of *Pratylenchus penetrans* from alfalfa callus and relationship between age of culture and yield of eggs. J. Nematol. 5:73–74.

Edwards, G.R. 1951. Insect pests of onions. J. Dep. Agric. S. Austral. 54:558–561.

Edwards, D.I., and D.P. Taylor. 1963. Host range of an Illinois population of the stem nematode *Ditylenchus dipsaci* isolated from onion. Nematologica 9:305–312.

Endo, B.Y. 1959. Responses of root lesion nematodes, *Pratylenchus brachyurus* and *P. zeae* to various plant and soil types. Phytopathology 49:417–421.

Eriksson, K.B. 1972. Nematode diseases of pasture legumes and turf grasses. p. 66–96. *In* J.M. Webster (ed.) Economic nematology. Acad. Press, New York.

Faulkner, L.R., and O.J. Bolander. 1970. Acquisition and distribution of nematodes in waterways of the Columbia Basin in Eastern Washington. J. Nematol. 2:362–367.

Fielding, M.J. 1951. Observations on the length of dormancy in certain plant infecting nematodes. Proc. Helminthol. Soc. Washington 18:110–112.

Golden, A.M., J.H. O'Bannon, G.S. Santo, and A.M. Finley. 1980. Description and SEM observation of *Meloidogyne chitwoodi* n. sp., (Meloidogynidae) a root-knot nematode on potato in the Pacific Northwest. J. Nematol. 12:109–112.

Gray, F.A., and D.H. Soh. 1989. A nematicide seed treatment to control *Ditylenchus dipsaci* on seedling alfalfa. J. Nematol. 21:184–88.

Gray, F.A., G.D. Griffin, D.A. Johnson, J.W. Eckert, and J.E. Kazimir. 1990. Interrelationship between *Meloidogyne hapla* and *Phytophthora megasperma* f. sp. *medicaginis* in seedling damping-off and root infection in alfalfa. Phytopathology 80:228–232.

Gray, F.A., J.L. Williams, G.D. Griffin, and T.E. Wilson. 1994. Distribution in the western U. S. and cultivar reaction to mixed populations of *Ditylenchus dipsaci* and *Aphelenchoides ritzemabosi* in alfalfa. J. Nematol. 26:705–719.

Green, C.D. 1981. The effects of weeds and wild plants on the reinfestation of land by *Ditylenchus dipsaci*. p. 217–224. *In* J. M. Thresh (ed.) Pests, pathogens, and vegetation. Pitman, London.

Griffin, G.D. 1996. Importance of soil texture on the relationship of plant parasitic nematodes to range grasses. Nematropica 26:27–37.

Griffin, G.D. 1994. Effect of single and interplantings on virulence of *Pratylenchus penetrans* and *P. neglectus* to alfalfa and crested wheatgrass. J. Nematol. 26:460–466.

Griffin, G.D. 1993a. Comparative response of alfalfa to *Pratylenchus penetrans* populations. J. Nematol. 25:461–465.

Griffin, G.D. 1993b. Differential host response of alfalfa to *Pratylenchus penetrans* and *P. neglectus* populations. Nematropica 23:27–33.

Griffin, G.D. 1992. Stem nematode-Fusarium wilt complex in alfalfa as related to irrigation management at harvest time. J. Nematol. 24:315–320.

Griffin, G.D. 1991a. The relationship of *Ditylenchus dipsaci* and harvest practices to the persistence of alfalfa. J. Nematol. 23:306–315.

Griffin, G.D. 1991b. Differential pathogenicity of four *Pratylenchus neglectus* populations on alfalfa. J. Nematol. 23:380–385.

Griffin, G.D. 1990. Pathological relationship of *Ditylenchus dipsaci* and *Fusarium oxysporum* f. sp. *medicaginis* on alfalfa. J. Nematol. 22:333–336.

Griffin, G.D. 1987. The importance of environmental factors and cultural practices on parasitism of alfalfa by *Ditylenchus dipsaci*. J. Nematol. 19:267–276.

Griffin, G.D. 1984. Nematode parasites of alfalfa, cereals, and grasses. p. 243–321. *In* W.R. Nickle (ed.) Plant and insect nematodes. Marcel Dekker, New York.

Griffin, G.D. 1975. Parasitism of nonhost cultivars by *Ditylenchus dipsaci*. J. Nematol. 7:236–238.

Griffin, G.D. 1974. Effect of acclimation temperature on infection of alfalfa by *Ditylenchus dipsaci*. J. Nematol. 6:57–59.

Griffin, G.D. 1968. The pathogenicity of *Ditylenchus dipsaci* to alfalfa and the relationship of temperature to plant infection and susceptibility. Phytopathology 58:929–932.

Griffin, G.D. 1967. Chemical control of the stem nematode *Ditylenchus dipsaci*. Plant Dis. Rep. 51:973–974.

Griffin, G.D., and K.H. Asay. 1996. Biology and pathogenicity of three ectoparasitic nematode species on crested wheatgrasses. Nematropica 26:15–25.

Griffin, G.D., K.H. Asay, and W.H. Horton. 1996. Factors affecting population trends of plant parasitic nematodes on rangeland grasses. J. Nematol. 28:107–114.

Griffin, G.D., and F.A. Gray. 1990. Biology and pathogenicity of *Pratylenchus neglectus* on alfalfa. J. Nematol. 22:546–551.

Griffin, G.D., and O.J. Hunt. 1972a. Effect of temperature and inoculation timing on the *Meloidogyne hapla/Corynebacterium insidiosum* complex in alfalfa. J. Nematol. 4:70–71.

Griffin, G.D., and O. Hunt. 1972b. Effect of plant age on resistance of alfalfa to *Meloidogyne hapla*. J. Nematol. 4:87–90.

Griffin, G.D., and M.V. McKenry. 1989. Susceptibility of Nevada Synthetic XX germplasm to a California race of *Meloidogyne hapla*. J. Nematol. 21:292–293.

Griffin, G.D., and S.V. Thomson. 1988. The Columbia root-knot nematode, *Meloidogyne chitwoodi*, discovered in the state of Utah. Plant Dis. 72:363.

Griffin, G.D., and B.D. Thyr. 1988. Interaction of *Meloidogyne hapla* and *Fusarium oxysporum* f. sp. *medicaginis* on alfalfa. Phytopathology 78:421–425.

Griffin, G.D., and W.W. Waite. 1971. Attraction of *Ditylenchus dipsaci* and *Meloidogyne hapla* by resistant and susceptible alfalfa. J. Nematol. 3:215–219.

Griffin, G.D., O.J. Hunt, and J.J. Murray. 1975. Pathogenicity of *Ditylenchus dipsaci* to sainfoin (*Onobrychis viciaefolia* Scop.). J. Nematol. 7:93–93.

Griffin, G.D., R.N. Inserra, N. Volvas, and D.V. Sisson 1986. Differential reaction of alfalfa cultivars to *Meloidogyne chitwoodi* and *M. hapla* populations. J. Nematol. 18:347–352.

Griffin, G.D., M.D. Rumbaugh, and K.H. Asay. 1992. Effect of *Meloidogyne chitwoodi* and *M. hapla* on competitive growth of alfalfa and grasses. J. Nematol. 24:593.

Grundbacher, F.J., and E.H. Stanford. 1962. Genetic factors conditioning resistance in alfalfa to the stem nematode. Crop Sci. 2:211–217.

Hafez, S.L., and D.R. Miller. 1990. The effect of *Pratylenchus penetrans* on different alfalfa cultivars. p. 18. *In* G.R. Bauchen (ed.) Rep. 32nd Alfalfa Improv. Conf., Pasco, WA. 19–24 August. N. Am. Alfalfa Improv. Conf., Beltsville, MD.

Hafez, S.L., D.R. Miller, and K. Hara. 1988. The effect of different soil enhancer materials on different alfalfa varieties growing in root-knot nematode infested soil. p. 6. *In* J.B. Moutray (ed.) Rep. 31st Alfalfa Improv. Conf., Beltsville, MD. 19–23 June. N. Am. Alfalfa Improv. Conf., Beltsville, MD.

Halsted, B.D. 1891. Nematodes as enemies of plants. Ann. Rep. New Jersey Agric. Exp. Stn. Bull.

Hartman, B.J., O.J. Hunt, R.N. Peaden, H.J. Jensen, B.D. Thyr, L.R. Faulkner, and G.D. Griffin. 1979. Registration of Nevada Synthetic YY Alfalfa. (Reg. no. GP 99). Crop Sci. 19:416–417.

Hawn, E.J., and M.R. Hanna. 1966. Influence of stem nematode infestation on bacterial wilt reaction and forage yield of alfalfa varieties. Can. J. Plant Sci. 47:203–208.

Hunt, O.J., G.D. Griffin, J.J. Murray, M.W. Pederson, and R.N. Peaden. 1971. The effect of root-knot nematodes on bacterial wilt in alfalfa. Phytopathology 61:256–259.

Inserra, R.N., J.H. O'Bannon, and G.S. Santo. 1980. The effect of *Meloidogyne hapla* on growth of alfalfa seedlings. Nematologica 26:379–381.

Kable, P.F., and W.F. Mai. 1968. Influence of soil moisture on *Pratylenchus penetrans*. Nematologica 14:101–122.

Kimpinsky, J.M., and C.B. Willis. 1981. Influence of soil temperature and pH on *Pratylenchus penetrans* and *P. crenatus* in alfalfa and timothy. J. Nematol. 13:333–338.

Krusberg, L.R. 1961. Studies on the culturing and parasitism of plant parasitic nematodes in particular *Ditylenchus dipsaci* and *Aphelenchoides ritzembosi* on alfalfa tissues. Nematologica 6:181–200.

Leath, K.T., D.C. Erwin, and G.D. Griffin. 1988. Diseases and nematodes. p. 621–670. *In* A.A. Hanson et al. (ed.) Alfalfa and alfalfa improvement. Agron. Monogr. 29. ASA, CSSA, and SSSA, Madison, WI.

Lewis, G.D., and W.F. Mai. 1960. Overwintering and migration of *Ditylenchus dipsaci* in organic soils of southern New York. Phytopathology 50:341–343.

Mauza, B.E., and J.M. Webster. 1982. Suppression of alfalfa growth by concomitant populations of *Pratylenchus penetrans* and two *Fusarium* species. J. Nematol. 14:364–367.

Nelson, D.L., D.K. Barnes, and D.H. McDonald. 1985. Field and growth chamber evaluation for root-lesion nematode resistance in alfalfa. Crop Sci. 25:35–39.

Nigh, E.L., Jr., K.C. Ellis, and R.B. Hine. 1969. Systemic control of the alfalfa stem nematode. Progr. Agric. Ariz. 21:10–11.

Noel, G.R., and B.F. Lownsbery. 1978. Effect of temperature on the pathogenicity of *Tylenchorhynchus clarus* to alfalfa and observations on feeding. J. Nematol. 10:195–198.

Norton, D.C. 1963. Population fluctuations of *Xiphinema americanum* in Iowa. Phytopathology 53:66–68.

Olthof, Th.H.A. 1982. Effect of age of plant on penetration by *Pratylenchus penetrans*. J. Nematol. 14:100–105.

Peaden, R.N., O.J. Hunt, L.R. Faulkner, G.D. Griffin, H.J. Jensen, and E.H. Stanford. 1976. Registration of a multiple-pest resistant alfalfa germplasm. (Reg. no. GP 51). Crop Sci. 16:125–126.

Reynolds, H.W., W.W. Carter, and J.H. O'Bannon. 1970. Symptomless resistance of alfalfa to *Meloidogyne incognita acrita*. J. Nematol. 2:131–134.

Reynolds, H.W., and J.H. O'Bannon. 1960. Reaction of sixteen varieties of alfalfa to two species of root-knot nematodes. Plant Dis. Rep. 44:441–443.

Santo, G.S., D.W. Evans, and D.B. Bowers. 1980. Reaction of three alfalfa cultivars to several species of plant-parasitic nematodes. Plant Dis. 64:404–405.

Santo, G.S., and J.N. Pinkerton. 1985. A second race of *Meloidogyne chitwoodi* discovered in Washington State. Plant Dis. 69:361.

Sontirat, S., and R.A. Chapman. 1970. Penetration of alfalfa roots by different stages of *Pratylenchus penetrans*. J. Nematol. 2:270–271.

Stessel, G.J. 1960. Effect of nematocides on *Pratylenchus pratensis* and alfalfa yields. Phytopathology 50:656.

Sturhan, D. 1971. Biological Races. p. 51–71. *In* B.M. Zuckerman et al. (ed.) Plant parasitic nematodes. Acad. Press, New York.

Suzuki, M., and C.B. Willis. 1974. Root lesion nematodes affect cold tolerance of alfalfa. Can. J. Plant Sci. 50:577–581.

Taylor, A.L., and J.N. Sasser. 1978. Biology, identification, and control of root-knot nematodes (*Meloidogyne* species). Int. *Meloidogyne* Project. North Carolina State Univ. Graphics, Raleigh.

Taylor, D.P., R.V. Anderson, W.A. Haglund. 1958. Nematodes associated with Minnesota crops. I. Preliminary survey of nematodes associated with alfalfa, flax, peas, and soybeans. Plant Dis. Rep. 42:195–198.

Thies, J.A., D. Basigalup, and D.K. Barnes. 1994. Inheritance of resistance to *Pratylenchus penetrans* in alfalfa. J. Nematol. 26:452–459.

Thies, J.A., D.K. Barnes, D.L. Rabas, C.C. Sheaffer, and R.D. Wilcoxson. 1992. Seeding date, carbofuran, and resistance to root-lesion nematode affect alfalfa stand establishment. Crop Sci. 32:786–792.

Thorne, G. 1961. Principles of nematology. McGraw-Hill, New York.

Townshend, J.L. 1989. Efficacy of oxamyl coated on alfalfa seed with a polymer sticker in *Pratylenchus* and *Meloidogyne* infested soils. J. Nematol. 21:242–246

Townshend, J.L. 1984. Inoculum densities of five plant parasitic nematodes in relation to alfalfa seedling growth. Can. J. Plant Pathol. 6:309.

Townshend, J.L. 1978. Infectivity of *Pratylenchus penetrans* on alfalfa. J. Nematol. 10:318–322.

Townshend, J.L., and J.W. Potter. 1973. Nematode numbers under cultivars of forage legumes and grasses. Can. Plant Dis. Surv. 53:194–195.

Townshend, J.L., and L. Stobbs. 1981. Histopathology and histochemistry of lesions caused by *Pratylenchus penetrans* in roots of forage legumes. Can. J. Plant Pathol. 3:123–128.

Townshend, J.L., M. Meskine, and G.L. Barron. 1989. Biological control of *Meloidogyne hapla* on alfalfa with the fungus *Meria coniospora*. J. Nematol. 21:178–183.

Townshend, J.L., L. Stobbs, and R. Carter. 1989. Ultrastructural pathology of cells affected by *Pratylenchus penetrans* in alfalfa roots. J. Nematol. 21:530–539

Tseng, S.T., K.R. Allred, and G.D. Griffin. 1968. A soil population study of *Ditylenchus dipsaci* (Kühn) Filipjev in an alfalfa field. Proc. Helminthol. Soc. Wash. 35:57–62.

Vrain, T.C. 1987. Effect of *Ditylenchus dipsaci* and *Pratylenchus penetrans* on Verticillium wilt of alfalfa. J. Nematol. 19:379–383.

Wallace, H.R. 1962. Observations on the behavior of *Ditylenchus dipsaci* in soil. Nematologica 7:91–101.

Webster J.W. 1987. Introduction. p. 1–11. *In* R.H. Brown and B.R. Kerry (ed.) Principles and practice of nematode control in crops. Acad. Press, New York.

Whitehead, A.G., and D.J. Tite. 1988. Effects of seed row and foliar applied pesticides on the yield of lucerne in clay soil, in the presence or absence of lucerne stem nematode, *Ditylenchus dipsaci*. J. Agric. Sci. 110:279–283

Williams, J.L., F.A. Gray, and G.D. Griffin. 1991. Biology of the chrysanthemum foliar nematode and its association with the alfalfa stem nematode in irrigated alfalfa in western U.S. J. Nematol. 23:555.

Willis, C.B., and L.S. Thompson. 1979. Effects of phenamiphos, methyl bromide, and fallowing on *Pratylenchus penetrans*, yield of *Medicago sativa*, and *Fusarium* infections. J. Nematol. 11:265–269.

19 Clovers and Other Forage Legumes

GARY A. PEDERSON

USDA-ARS, Forage Research Unit
Mississippi State, Mississippi

KENNETH H. QUESENBERRY

University of Florida
Gainesville, Florida

Clovers and other forage legumes are important components of forage and pasture systems throughout the world. Legumes added to a grass pasture generally increase the quality of forage feed, resulting in improved animal performance over grass alone. Legumes also fix atmospheric N, thereby reducing the need for fertilizer N in forage systems. White clover (*Trifolium repens* L.) is an excellent grazing crop with its stems (stolons) running along the ground, producing nutritious leaves for animals to ingest. Red clover (*T. pratense* L.) is commonly grown as a hay crop and tolerates acid or poorly drained soils better than alfalfa (*Medicago sativa* L.). Birdsfoot trefoil (*Lotus corniculatus* L. var. *corniculatus*) is a nonbloating legume that can be managed to reseed. Numerous other forage legumes are important components of pasture and hay systems, often in niche environments where other species are not well adapted. This chapter concentrates on nematode pathogens of clover and forage legumes except for alfalfa, which will be covered in another chapter. For additional information on nematodes of clovers and other forage legumes, reviews by Eriksson (1972), Cook and York (1980), Leath (1985), Skipp and Gaynor (1987), Cook and Yeates (1993), Griffin et al. (1996), Mercer and Watson (1996), and Taylor and Quesenberry (1996) should be consulted.

Nematode infection has a significant impact on yield and persistence of clover and other forage legumes. Since many forage legumes are used in hay and pasture systems, often as one component of a legume/grass system and are fed directly to animals rather than sold as a cash crop, it is difficult to determine the economic value of forage legumes as well as the impact that nematodes have on forage legume production. Hague (1980) estimated that nematodes caused a 6% loss in U.S. clover pastures of 5.4 million hectares, resulting in an estimated loss of $33 million. This estimate only includes clover and does not provide any estimate of losses in birdsfoot trefoil, lespedeza (*Lespedeza* and *Kummerowia* sp.), lupine (*Lupinus* sp.), and other temperate and tropical forage legumes. Nematode infestation also limits N_2 fixation, which has been estimated to have a value of $11 to 48 million (New Zealand dollars) for each 1% increase in N from white clover in New Zealand (Mercer, 1994).

IMPORTANT NEMATODE PARASITES

Root-Knot Nematodes

Species and Races

Root-knot nematode species that are important pathogens of clovers and other forage legumes include the northern root-knot nematode (*Meloidogyne hapla*), the southern root-knot nematode (*M. incognita*; Fig. 19–1), the javanese root-knot nematode (*M. javanica*), the peanut root-knot nematode (*M. arenaria* Race 1), and *M. arenaria* Race 2 (Table 19–1). In New Zealand, an undescribed root-knot nematode species previously thought to be *M. hapla* has been reported throughout the country and was the species used in much of the clover root-knot

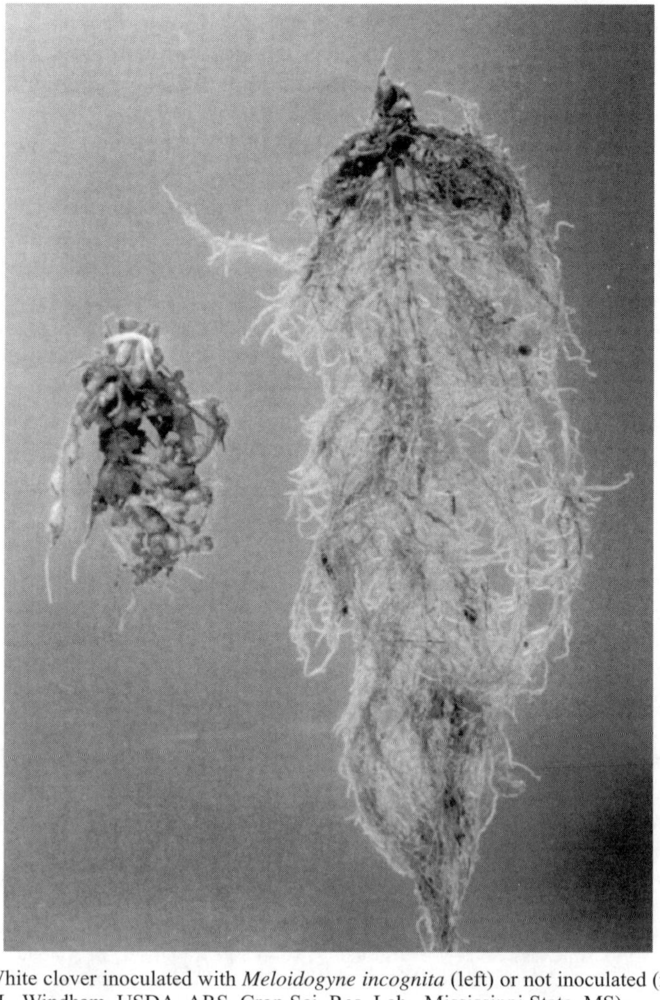

Fig. 19–1. White clover inoculated with *Meloidogyne incognita* (left) or not inoculated (right) (courtesy of G.L. Windham, USDA, ARS, Crop Sci. Res. Lab., Mississippi State, MS).

Table 19–1. Some recent forage legume/nematode interactions.

Forage legume	Nematode	Parameter	Reference
Alysicarpus vaginalis (L.) DC	*Meloidogyne arenaria, M. incognita*	Reproduction	Powers et al., 1992
Desmodium spp.	*M. arenaria, M. hapla, M. incognita, M. javanica*	Reproduction	Sasser et al., 1987
Desmodium ovalifolium Guillemin & Perrottet	*Pterotylenchus cecidogenus*	Pathogenicity	Lehman, 1991
Kummerowia stipulacea	*Heterodera lespedezae*	Reproduction	Fagbenle et al., 1986
K. striata	*H. lespedezae*	Reproduction	Fagbenle et al., 1986
Lespedeza cuneata (Dum-Cours.) G. Don	*H. lespedezae*	Reproduction	Fagbenle et al., 1986
Lotus corniculatus L. var. *corniculatus*	*Pratylenchus crenatus, P. penetrans*	Pathogenicity	Kimpinski et al., 1984
Lupinus termis Forsskal	*M. arenaria, M. incognita, M. javanica*	Pathogenicity	Ibrahim et al., 1983
Melilotus officinalis Lam.	*H. lespedezae*	Reproduction	Fagbenle et al., 1986
Trifolium alexandrinum L.	*M. arenaria, M. incognita, M. javanica*	Pathogenicity	Baltensperger et al., 1985
	M. graminicola, M. incognita	Reproduction	Windham & Pederson, 1992
	H. daverti, M. incognita, Rotylenchulus reniformis	Pathogenicity	Massoud & Ghorab, 1988
	D. dipsaci	Reproduction	Whitehead, 1992
	Tylenchorhynchus vulgaris, Helicotylenchus dihystera	Pathogenicity	Hasan & Jain, 1987
T. ambiguum M. Bieb.	*M. graminicola, M. incognita*	Reproduction	Windham & Pederson, 1992
	H. trifolii, M. trifoliophila	Reproduction	Mercer, 1988
	D. dipsaci	Pathogenicity	Whitehead, 1992
T. fragiferum L.	*M. arenaria, M. incognita, M. javanica*	Pathogenicity	Quesenberry et al., 1986
	M. graminicola, M. incognita	Reproduction	Windham & Pederson, 1992
T. hybridum L.	*H. lespedezae*	Reproduction	Fagbenle et al., 1986
	M. incognita	Pathogenicity	Pederson & Windham, 1989
	H. trifolii, M. trifoliophila	Reproduction	Mercer, 1988

(continued on next page)

Table 19–1. Continued.

Forage legume	Nematode	Parameter	Reference
T. incarnatum L.	*M. arenaria, M. incognita, M. javanica*	Pathogenicity	Quesenberry et al., 1986
	M. graminicola, M. incognita	Reproduction	Windham & Pederson, 1992
T. nigrescens Viv.	*H. trifolii, M. trifoliophila*	Reproduction	Mercer, 1988
	M. incognita	Pathogenicity	Pederson & Windham, 1989
T. pratense L.	*M. arenaria, M. incognita, M. javanica*	Pathogenicity	Quesenberry et al., 1986
	M. arenaria, M. javanica	Reproduction	Call et al., 1996b; 1997a
	M. graminicola, M. incognita	Reproduction	Windham & Pederson, 1992
	H. lespedezae	Reproduction	Fagbenle et al., 1986
	D. dipsaci	Reproduction	Whitehead et al., 1987
	P. crenatus, P. flakkensis, P. neglectus, P. penetrans	Pathogenicity	Kimpinski et al., 1984; Kornobis, 1987
T. repens L.	*M. arenaria, M. incognita, M. javanica*	Pathogenicity	Quesenberry et al., 1986
	M. graminicola, M. incognita	Reproduction	Windham & Pederson, 1992
	H. lespedezae	Reproduction	Fagbenle et al., 1986
	H. trifolii, M. trifoliophila	Reproduction	Mercer, 1988
	D. dipsaci	Reproduction	Whitehead et al., 1987
T. resupinatum L.	*M. graminicola, M. incognita*	Reproduction	Windham & Pederson, 1992
T. subterraneum L.	*M. arenaria, M. incognita, M. javanica*	Pathogenicity	Kouame et al., 1989
	M. graminicola, M. incognita	Reproduction	Windham & Pederson, 1992
	H. daverti	Reproduction	Nordmeyer & Sikora, 1983
	H. trifolii, M. trifoliophila	Reproduction	Mercer, 1988
T. vesiculosum Savi	*M. arenaria, M. incognita, M. javanica*	Pathogenicity	Baltensperger et al., 1985
	M. graminicola, M. incognita	Reproduction	Windham & Pederson, 1992
Vicia villosa Roth subsp. *villosa*	*H. lespedezae*	Pathogenicity	Fagbenle et al., 1986

nematode research in that country since 1985 (Van Den Bosch & Mercer, 1996). This species was recently identified as *M. trifoliophila* and also has been found in Tennessee (Bernard & Eisenback, 1997). Though described as *M. hapla* in most references cited (Mercer & Watson, 1996), this species will be described as *M. trifoliophila* throughout this chapter for clarity. The rice root-knot nematode (*M. graminicola*) was isolated from white clover and has shown potential to be an important pathogen of *Trifolium* spp. (Windham & Pederson, 1992). Root-knot nematodes cause damage to forage legumes and other crops by infecting and stopping growth of root tips, inducing plants to form galls that are important nutrient sinks, and disrupting the vascular system of the plant affecting translocation of water and nutrients (Fig. 19–2). *Meloidogyne* spp. have been detected on white clover in 58% of New Zealand pastures (Skipp & Christensen, 1983), 77% of Australian pastures (McLeish et al., 1997), and 4% (*M. hapla*) of England

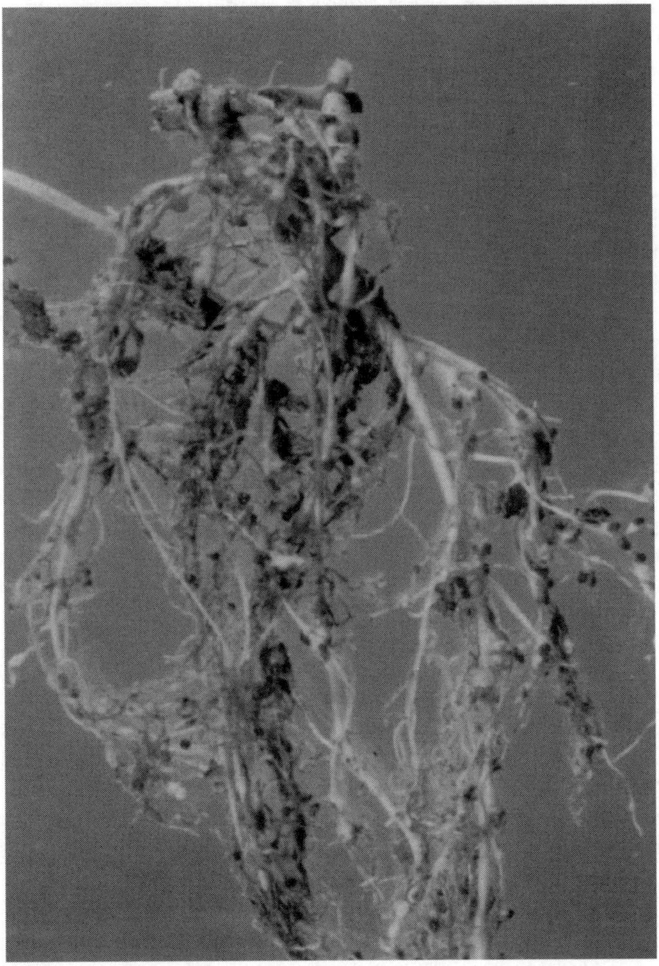

Fig. 19–2. White clover infected with *Meloidogyne incognita* (courtesy of G.L. Windham, USDA, ARS, Crop Sci. Res. Lab., Mississippi State, MS).

and Wales pastures (Cook et al., 1992b). *Meloidogyne arenaria* was found on subterranean clover (*T. subterraneum* L.) in all pastures sampled in Western Australia (Pung et al., 1988).

Nematode host race and population differences are not important in root-knot nematode parasitism of white clover, as little difference in damage potential among races has been reported in *M. incognita* (Windham & Pederson, 1989) and *M. trifoliophila* (Mercer & Grant, 1993). Use of less aggressive nematode populations, however, may enable identification of white clover with moderate levels of resistance (Windham & Pederson, 1989).

Symptoms

Root-knot nematodes can have their greatest effects on forage legume growth and development when infection occurs during the seedling or early stage of growth. Greater numbers of galls were formed by *M. arenaria* on subterranean clover when plants were inoculated as seedlings rather than as more mature plants (Pung et al., 1991). Seasonal rains stimulate both subterranean clover germination and hatching of *M. arenaria* eggs, which favors significant infection of seedling roots (Pung et al., 1992a). This infection results in a shallow root system with limited water and nutrient uptake by the plants. Infected plants wilt more rapidly during periods of moisture stress, and may be chlorotic due to reduced N_2 fixation and N uptake.

Nematode population density is a major factor in determining the relative extent of damage to the plant. An increase in initial population density (inoculation rate) causes greater suppression in plant growth (dry weight) in white clover by *M. incognita* (Shepherd, 1965; Windham & Pederson, 1992), *M. graminicola* (Windham & Pederson, 1992), and *M. trifoliophila* (Mercer, 1990). A similar response of an increase in initial population density to greater plant growth suppression was observed in five species of sweet clover (*Melilotus* spp.) by *M. incognita* (Azmi, 1988) and in subterranean clover by *M. arenaria* (Pung et al., 1988). Nematode density effects can interact with temperature. The greatest suppression in subterranean clover root growth by *M. arenaria* was at the intermediate (3000 eggs pot^{-1}) egg density at 20°C, whereas at 25°C it was at the highest level (6000 eggs pot^{-1}; Pung et al., 1988).

Biology

The life cycle of root-knot nematodes in forage legumes is similar to that noted in other susceptible crop species (Taylor & Sasser, 1978). Second-stage juveniles, which are the only infective stage of root-knot nematodes, enter roots at or near the root tip. They penetrate through epidermal cells, establishing feeding sites in the area of differentiating xylem tissue. Cells adjacent to the head of the nematode do not differentiate, but begin to enlarge. Through repeated division without the usual cytokinesis and possibly some cell-wall breakdown, nurse or giant cells are formed (Hussey & Williamson, 1998, see Chapter 5). The female nematode feeds on surrounding cells and swells into a large oval shape. Galls are formed from enlargement of the nematode and from hypertrophy and hyperplasia of the cells surrounding the giant cells. Eggs are produced in a gelatinous egg mass, which adheres to the posterior of the female, usually at the root surface. When environmental conditions are favorable, the eggs hatch and second-stage juveniles infect roots of the same or adjacent plants. For *M. trifoliophila* in white clover, 281 degree days above 10°C (29 d) were required for development from eggs to females producing eggs, and 407 degree days (42 d) were required to complete one generation (Mercer, 1990). Nematode development is temperature dependent, as development is slowest during the cool period of the year (Pung et al., 1992a). Optimum temperature for root-knot nematode invasion is 20 to 25°C and for nematode reproduction is 25°C.

Disease Complexes

Infection of plants by root-knot nematodes may alter the responses to other disease organisms. These effects may be due to wounding, caused by entry of nematodes into the roots, which exposed root tissue to infection by fungi and bacteria or by nematode-induced physiological changes in the roots (Taylor, 1990). *Meloidogyne incognita* combined with *Fusarium roseum, F. oxysporum, Rhizoctonia solani,* or *Nigrospora* spp. inoculations resulted in greater pathogenic effects on white clover than either the nematode or fungus alone (Halpin, 1963). Other researchers have found no increase in infection by *F. oxysporum* or bacterial pathogens in white clover roots infected with *M. hapla* (Sarathchandra et al., 1995) or in predisposition to more severe infection by *F. oxysporum* in subterranean clover infected previously with *M. arenaria* (Pung et al., 1991). The damage potential of pathogens in studies of fungal and nematode interactions may be dependent upon the methods used in the study (Pung et al., 1992b). Concomitant infection by *M. incognita* and a lesion nematode (*Pratylenchus penetrans*) in red clover resulted in suppressed reproduction by *P. penetrans* (Chapman & Turner, 1975). Root-knot nematode and virus infections may have a synergistically negative effect on forage legume growth. In the greenhouse, white clover root growth and rooting stolon nodes were restricted more by mixed infection of *M. incognita* and peanut stunt virus than with either pathogen alone (McLaughlin et al., 1993). However, in a field study, *M. incognita* and peanut stunt virus reduced white clover productivity and persistence in an additive manner (McLaughlin & Windham, 1996).

Economic Importance

Infection by root-knot nematodes has a number of detrimental effects on forage legume productivity and persistence. Dry-matter yield losses range from: 40 to 55% in white clover, red clover, and birdsfoot trefoil infected with *M. hapla* (Townshend & Potter, 1978); 30% in red clover infected with a mixture of *M. arenaria, M. javanica,* and *M. hapla* (Call et al., 1996a); 50% in red clover infected with *M. javanica* (Call et al., 1996a); 17% in white clover infected with *M. incognita* (Pederson et al., 1991b); and 11% in berseem clover (*T. alexandrinum* L.) infected with *M. incognita* (Windham & Pederson, 1988). Stolon growth (dry weight) may be suppressed by 36 to 83% in white clover infected with *M. incognita* (Brink & Windham, 1990). Persistence has been reduced by 74% in birdsfoot trefoil infected with *M. hapla* (Townshend & Potter, 1978) and 12 to 20% in white clover infected with *M. incognita* (Pederson et al., 1991b; Fig. 19–3A,B).

Fig. 19–3. Two-year-old stands of white clover not infested (*A*) or infested with *Meloidogyne incognita* (*B*) (courtesy of G.L. Windham, USDA, ARS, Crop Sci. Res. Lab., Mississippi State, MS).

Flowering has been delayed in red clover infected with *M. javanica* and with a mixture of *M. arenaria, M. javanica,* and *M. hapla* (Call et al., 1996a).

Root-knot nematodes also affect plant nutrition. Average total N has been reduced 27 to 60% in *M. javanica* infected-, and 43 to 48% in *M. hapla* infected-white clover plants, though the mechanism by which the nematode affects N_2 fixation has not been determined (Taha & Raski, 1969; Yeates et al., 1977). Phosphorus and potassium concentrations in red clover stems have been reduced up to 15% by infection of roots with *M. incognita* or *M. hapla* (Pokharel & Kruchina, 1991).

Clover Cyst Nematode

Species and Races

The main cyst nematode affecting forage legume production in the world is the clover cyst nematode (*Heterodera trifolii*). This parasite occurs in at least 30 states in the USA (Miller, 1986). *Heterodera trifolii* has been found on white clover in 62% of England and Wales (Cook et al., 1992b), 75 to 81% of New Zealand (Skipp & Christensen, 1983; Mercer & Woodfield, 1986), 77% of Australian (McLeish et al., 1997), and 74% of Prince Edward Island (Kimpinski et al., 1993) pastures. On red clover, it has been found in 22% of pastures in Poland (Kornobis, 1987). Other *Heterodera* spp. that have been shown to be important pathogens of forage legumes include *H. lespedezae* and *H. daverti* (Table 19–1).

A number of host races and isolate differences have been noted in cyst nematodes, especially in *H. trifolii*. Singh and Norton (1970) reported variability in host reactions to six North American isolates of *H. trifolii* as well as to eight isolates from Iowa. They concluded that different biological forms occur within *H. trifolii* and that any given isolate may not be representative of a particular region. Similarly, variation was reported in host reaction to five European populations of *H. trifolii* (Sikora & Maas, 1986), though few differences were noted in white clover cultivar reaction to seven Netherlands populations of *H. trifolii* (Dijkstra, 1971). Mercer and Grant (1993) reported variation in reaction of resistant and susceptible white clover lines to different New Zealand *H. trifolii* populations. Resistance to *H. trifolii* would need to be developed against several populations of the nematode. Significant differences in host reaction of striate lespedeza [*Kummerowia striata* (Thunb.) Schindler], red clover, and alsike clover (*T. hybridum* L.) to two different isolates of *H. lespedezae* have been reported (Fagbenle et al., 1986).

Symptoms

Infection by *H. trifolii* may cause damage to root tips in young seedlings; however, more severe root damage occurs after several generations of the nematodes have reduced the root growth rate (Mankau & Linford, 1960). Development of syncytia disrupt translocation of nutrients throughout the root system reducing root growth of infected plants.

An increase in *H. trifolii* population density causes restricted plant growth (dry weight) in white clover (Stelter & Meinl, 1972; Mercer, 1990) and red clover (Stelter & Meinl, 1972), reductions in white clover persistence (Hidding et al., 1963), and lower crude protein levels in white clover (Hidding et al., 1963). Yeates (1978), however, found that low infection by *H. trifolii* initially stimulated white clover growth, but this effect decreased over time and with increased P application.

Biology

The biology and life cycle for *H. trifolii* on white clover was described by Mankau and Linford (1960), and differences among *Heterodera* spp. were investigated by Sikora and Maas (1986). Second-stage juveniles generally enter the root in the root hair region though they can enter the root almost anywhere. The nematodes either reside permanently near the entry site or invade throughout the root, destroying cells along their path. Juveniles may damage or kill the root tip by moving through and destroying differentiating tissue or the apical meristem.

When the juvenile begins to feed at its permanent site within the root, it causes the formation of enlarged nurse cells called the syncytia. A syncytium is formed by dissolution of cell walls and enlargement of the nuclei (Mankau & Linford, 1960). Continual enlargement and elongation of syncytia occur as additional cell walls dissolve. As it feeds and grows, the female nematode enlarges, destroying surrounding cortex tissue with the posterior of the nematode reaching the root surface. The enlarging body of the female nematode often splits the root or destroys cortex tissue and lies exposed on the root surface.

Adult females are formed 17 to 19 d following penetration of the root, begin producing eggs within their body cavity by 22 d, and become filled with eggs within 26 to 28 d (Mankau & Linford, 1960; Mercer, 1990). The body cavity becomes entirely filled with eggs forming a spherical, lemon shape. Some eggs also are deposited in a gelatinous matrix near the posterior of the female and hatch within a short time. The first-stage juvenile molts within the egg to become a second-stage juvenile (J2). These J2 either hatch or remain dormant within the egg shell. The female eventually dies, and the body cavity forms a dark brown cyst. This cyst may persist in the soil for a number of years, but survival of the juveniles and eggs therein is dependent on moisture and temperature conditions (Yeates & Visser, 1979). Plant resistance may affect cyst nematode development as more adults, cysts per plant, and eggs per cyst were produced on susceptible than on resistant white clover genotypes (Grant et al., 1996).

Heterodera trifolii in white clover requires 244 degree days above 10°C (26 d) for development from eggs to females producing eggs, and 310 degree days (33 d) to complete one generation (Mercer, 1990). Temperature may have a greater effect on generation time of cyst nematodes, as low temperatures of 10 to 20°C resulted in increased generation times to 56 d or 840 degree days (Sikora & Maas, 1986). Optimum temperature for development of *H. trifolii* in white clover is 25°C (Norton, 1967a) and for development of *H. lespedezae* in red clover and striate lespedeza is 22 to 26°C (Fagbenle et al., 1989).

Females of *H. trifolii* were most abundant in white clover roots during the spring and fall periods of plant growth in New Zealand (Yeates, 1977). Two generations per year were produced in the field, with most cysts and juveniles found between 2.5 and 5 cm from the soil surface (Yeates, 1977). Cysts of *H. trifolii* may move into uninfected pastures by a number of manners including wind, water, farm equipment, bird migration, or in soil attached to potato (*Solanum tuberosum* L.) tubers (Sikora & Maas, 1986; Kimpinski et al., 1993).

Disease Complexes

Cyst nematodes may interact with fungi or other disease organisms causing increased damage to forage legumes. White clover roots damaged by *H. trifolii* had more infections by *Fusarium oxysporum* and other fungal species than undamaged roots (Skipp & Christensen, 1981). Simultaneous inoculation of subterranean clover with *H. daverti* and *F. avenaceum* or *F. oxysporum* resulted in additive inhibition in plant growth (Nordmeyer & Sikora, 1983). Inoculation of *H. daverti* 1 to 2 wk following fungal inoculation resulted in a synergistic effect on clover growth. Interactions between nematodes have not been observed for simultaneous infection of red clover with *H. trifolii* and *Pratylenchus penetrans*

(Freckman & Chapman, 1972), of berseem clover with *H. daverti* and *M. incognita* or *Rotylenculus reniformis* (Massoud & Ghorab, 1988), and of white clover with *H. trifolii* and *M. hapla* (Mankau & Linford, 1960).

Economic Importance

Infection by cyst nematodes has been shown to limit productivity of forage legumes. Shoot growth in the greenhouse has been suppressed 14 to 37% in white clover and 13 to 50% in red clover infected with *H. trifolii* and *H. galeopsidis* (syn. *H. trifolii*; Stelter & Meinl, 1972), 24 to 52% in striate lespedeza infected with *H. lespedezae* (Fagbenle et al., 1989), and 17% in subterranean clover infected with *H. daverti* (Nordmeyer & Sikora, 1983). Root growth may be suppressed 19 to 50% in striate lespedeza infected with *H. lespedezae* (Fagbenle et al., 1989). In the field, total dry-matter production was reduced 14 to 46% in white clover and 27% in red clover infected with *H. trifolii* (Norton, 1967a). Nitrogen content has been reduced 27 to 63% in white clover inoculated with *H. trifolii* (Taha & Raski, 1969; Yeates, 1977; Yeates et al., 1977). Nodule number per gram of root was not reduced due to *H. trifolii*, but the entire root system was stunted (Taha & Raski, 1969). White clover crude protein levels were decreased 6% due to *H. trifolii* infection (Hidding et al., 1963).

Stem Nematode

Species and Races

The stem nematode, *Ditylenchus dipsaci*, is a major pathogen of red and white clover in temperate regions of the world (Table 19–1). Damage to red clover was observed as early as the 19th century in northern Europe. In the 1920s, selections for resistance to *D. dipsaci* in red clover were made in Scandinavia, which resulted in the resistant cultivars, Merkur in 1937 and Resistenta in 1941 (Bingefors, 1957). *Ditylenchus dipsaci* has been found on white clover in 41% of England and Wales pastures (Cook et al., 1992b) and 77% of Australian pastures (McLeish et al., 1997).

At least 21 host races of *D. dipsaci* have been reported, and isolate differences have been reported within these host races (Viglierchio et al., 1973; Whitehead et al., 1987). In an evaluation of 121 *D. dipsaci* populations from 12 different host species, Whitehead et al. (1987) observed that alfalfa was the only host that differentiated a host race as only alfalfa stem nematode populations multiplied on alfalfa. Red clover was host to *D. dipsaci* populations from white clover (2 of 4) and white clover was host to *D. dipsaci* populations from onion (*Allium cepa* L. var. *cepa*) (3 of 14), field bean (*Phaseolus vulgaris* L.) (1 of 13), and alfalfa (1 of 13). Attempts to develop a host range test to identify the host races of *D. dipsaci* populations have not been successful (Whitehead et al., 1987).

Symptoms

When red clover seedlings are infected, the hypocotyl at the base of the petiole begins to swell within 1 or 2 d (Bingefors, 1957). Young leaves are deformed, and the seedling may die. Severe infection of red clover plants results in galls at the base of the plant, swollen leaf petioles, deformed leaves, reduced plant vigor, and stunting. Shoots will have large basal galls at the crown with leaves bunched near the top of the shoot and flowers stunted and distorted. Reproduction by nematodes continues until the plant dies. Heavily infested red clover fields will have patches of severely infected plants with the pattern of infection moving down the slope or in the direction of machinery movement (Bingefors, 1957).

Biology

The life cycle of *D. dipsaci* in forage legumes is similar to that in other crops. *Ditylenchus dipsaci* is an endoparasitic nematode entering legume leaf buds, young leaves, or shoot apices often as a fourth-stage juvenile. Infection occurs usually during periods of cool temperatures and high humidity when plants are covered with a film of moisture. Upon molting, the nematode matures to an adult in the intercellular area feeding on parenchyma cells within the cortex. The female lays 200 to 500 eggs within the host with the complete life cycle taking 19 to 25 d. Host cells about the nematode divide and enlarge causing a characteristic swelling of the stem.

Active movement by *D. dipsaci* itself is quite limited. The nematode generally spreads via passive movement with soil, plant materials, water, machinery, animals, man, or by seed and debris with seed (Bingefors, 1957; Courtney, 1962). Viable stem nematodes have been reported on red clover seeds covered with honeydew from clover aphids (Courtney, 1962). Slugs feeding on infected white clover plants spread adults, juveniles, and eggs of stem nematode (Cook et al., 1989). *Ditylenchus dipsaci* may survive for three or more years by remaining in diapause within dried red clover hay or soil until being activated by moisture (Courtney, 1962; Evans, 1987). Infectivity of *D. dipsaci* following diapause is dependent on the season of the year (Evans, 1987).

Economic Importance

Forage legume productivity has been shown to be limited by *D. dipsaci*. Dry-matter yields of three white clover cultivars were reduced 77% by *D. dipsaci* infestation in adjacent areas (Cook et al., 1992a). Treatment of a *D. dipsaci*-infested area with nematicide resulted in a 38% increase in dry-matter yield and a 29% increase in petiole length of five white clover cultivars (Williams & Barclay, 1972). Red clover cultivars resistant to *D. dipsaci* averaged 136 to 151% greater hay yields in the 2nd yr and 205 to 360% greater hay yields in the 3rd yr than a susceptible standard cultivar in 36 naturally infested field trials in Sweden (Bingefors, 1985).

Lesion Nematodes

Species and Races

The lesion nematode most commonly associated with damage to forage legumes is *Pratylenchus penetrans*. Other species also found on forage legumes include *P. brachyurus, P. coffeae, P. crenatus, P. flakkensis, P. neglectus, P. pratensis,* and *P. scribneri* (Chapman, 1958; Kornobis, 1987; Skipp & Gaynor, 1987). *Pratylenchus* spp. are common pathogens in forage legume stands throughout the world, and are found on red clover in 57% of fields in USA (Lau & Reed, 1960), 52% of fields in Poland (Kornobis, 1987), and 100% of fields in Prince Edward Island (Kimpinski et al., 1993). On white clover, *Pratylenchus* spp. have been found in 92% of pastures in England and Wales (Cook et al., 1992b), 94% of Australian (McLeish et al., 1997), and 48% of New Zealand pastures (Skipp & Christensen, 1983).

Symptoms

The main symptom of lesion nematodes is necrosis of root cortical cells, which follows the tunnels of the nematodes forming dark, elongate lesions of decaying tissue. These lesions first appear as water-soaked areas that become elliptical in shape and rusty-red to black in color in birdsfoot trefoil and red clover (Townshend & Stobbs, 1981). The lesions increase in size and coalesce causing splitting of the primary root cortex.

Biology

Pratylenchus spp. are migratory endoparasitic nematodes, and their life cycle has been well documented on other crop hosts. Most studies of the pathology of *Pratylenchus* spp. in clovers and other forage legumes has been with *P. penetrans* in red clover (Freckman & Chapman, 1972; Turner & Chapman, 1972; Townshend & Stobbs, 1981). Juveniles and adults of *P. penetrans* may enter anywhere along the length of the root, though they tend to penetrate main roots where lateral roots have broken the cortex, often 3 to 10 mm from the root tip (Freckman & Chapman, 1972; Townshend & Stobbs, 1981). Once within the root, lesion nematodes feed and reproduce within epidermal and parenchyma cells, but do not enter the endodermis or stele. *Pratylenchus penetrans* completes a generation in 54 to 65 d and produces 16 to 35 eggs female^{-1} at a rate of 1 to 2 eggs d^{-1}. In red clover, 8 to 12 eggs were produced per *P. penetrans* female in 17- to 19-d experiments (Chapman & Turner, 1975). Eggs are usually produced in the root tissue, and the first molt occurs within the egg. Second-stage juveniles move throughout the root and mature. Juveniles or adults may leave the root and initiate new lesions on other roots.

Relative increase of *P. penetrans* is inoculum density dependent in red clover roots (Freckman & Chapman, 1972; Turner & Chapman, 1972). No differences were found for red clover shoot and root weight between three inoculation rates of a population of *P. penetrans, P. coffeae,* and *P. neglectus* (Chapman, 1958). An increase in population density of *P. penetrans* resulted in a suppression in shoot growth for red clover and birdsfoot trefoil, but not for white clover (Willis & Thompson, 1969).

Disease Complexes

Due to the necrotic lesions formed by lesion-nematode infection, secondary pathogens often cause additional rotting of the surrounding root tissue. This relationship does not appear to be synergistic in clovers as simultaneous infection of red clover with *P. penetrans* and *Fusarium avenaceum* resulted in an additive suppression in shoot growth (Jin et al., 1991). When *P. penetrans* infection was reduced by nematicide treatment, fewer infections with *Fusarium* spp. and *Rhizoctonia solani* were observed in red clover and birdsfoot trefoil tap roots (Thompson & Willis, 1975). No increase in fungal disease was found in white clover roots damaged by *Pratylenchus* spp. (Skipp & Christensen, 1981).

Economic Importance

Shoot growth (dry weight) in the greenhouse was suppressed by 4 to 50% in birdsfoot trefoil infected with *P. penetrans* (Willis & Thompson, 1969), 28% in kura clover (*T. ambiguum* Bieb.) infected with *P. penetrans* (Thies et al., 1995), 7 to 12% in white clover infected with *P. penetrans* (Willis & Thompson, 1969; Thies et al., 1995), 14 to 27% in red clover infected with *P. penetrans* (Willis & Thompson, 1969; Thies et al. 1995), and 31 to 41% in red clover infected with *P. penetrans, P. coffeae,* and *P. neglectus* (Chapman, 1958). Other forage legume species having shoot growth suppression due to *P. penetrans* infection include sainfoin (*Onobrychis viciifolia* Scop., 25%), crownvetch (*Coronilla varia* L., 16%), and white sweetclover (*Melilotus alba* Medikus; 15%; Thies et al., 1995). Yield losses in birdsfoot trefoil caused by *P. penetrans* infection were related to relative soil moisture. Birdsfoot trefoil maintained at 70 to 100% field capacity had a 55% suppression in shoot growth due to nematode infection, whereas plants maintained at 50% field capacity had only a 28% suppression (Thompson & Willis, 1970a).

Other Nematode Pests

Though semi-endoparasitic and ectoparasitic nematodes such as *Helicotylenchus* spp., *Paratylenchus* spp., *Tylenchorhynchus* spp., and others are commonly found in forage legume/grass pastures (Hasan & Jain, 1987; Kornobis, 1987; Cook et al., 1992b; McLeish et al., 1997; Table 19–1), little is known of their effects on growth of clovers and other forage legumes. An increase in population density of *Tylenchorhynchus vulgaris* from 10 to 10 000 nematodes per 10-cm pot caused 15 to 67% suppression in shoot growth of berseem clover compared to uninoculated control plants (Hasan & Jain, 1987). Red clover dry-matter yields were reduced 20% in the field and 15% in the greenhouse by *Xiphinema americanum* (Norton, 1967b).

CONTROL

Host Resistance

Host resistance is the most economical method for controlling nematodes in forage legumes as the perceived value of forages is often not great enough to justify more expensive methods of control. The greatest success in breeding for nematode resistance in forage legumes has been in red clover for stem nematode resistance in Scandinavia. The first resistant cultivar, Merkur, was released over 50 yr ago and most red clover cultivars sold in Scandinavia are resistant (Andersson, 1985; Bingefors, 1985).

Selection for resistance has been conducted for root-knot nematode species in alyceclover [*Alysicarpus vaginalms* (L.) DC] (Powers et al., 1992), lespedeza (Minton & Donnelly, 1971), white clover (Mercer et al., 1992; Pederson & Windham, 1995), and red clover (Quesenberry et al., 1989). Selection programs for resistance to *D. dipsaci* and *H. trifolii* have been conducted in white clover (Dijkstra, 1971; Williams, 1972; Cook & Mizen, 1989; Van Den Bosch et al., 1993; Mercer, 1994) and red clover (Bingefors, 1957; Armstrong et al., 1977; Andersson, 1985; Leath et al., 1987). Selection for resistance to *P. penetrans* has been conducted in red clover (Kimpinski et al., 1992). These selection programs have resulted in a number of resistant or tolerant cultivars and germplasms (Table 19–2).

In the USA, recurrent selection has been successfully used in the development of root-knot nematode resistance in red and white clover (Quesenberry et al., 1993; Pederson & Windham, 1995) and clover cyst nematode resistance in red clover (Leath et al., 1987). In cross-pollinated species such as red and white clover, the recombining of resistant phenotypes in recurrent selection can develop highly resistant populations even though the initial populations have very little nematode resistance (Windham & Pederson, 1991). Five cycles of restricted recurrent phenotypic selection were used in a naturally infested (*Meloidogyne* spp.) field to develop 'Cherokee' red clover which has less root galling and reproduction by *M. arenaria, M. hapla, M. incognita,* and *M. javanica* than other red clovers (Quesenberry et al., 1989). Subsequently, six cycles of recurrent phenotypic selection for reduced gall and egg-mass scores in Cherokee red clover infested with *M. arenaria, M. incognita*, or *M. javanica* were conducted in the greenhouse to produce FLMR6 (Fig. 19–4). This population shows no significant yield loss due to infestation with *M. javanica* or a mixture of *M. arenaria, M. javanica*, and *M. hapla* (Call et al., 1996a) and reduced penetration, maturation, and fecundity by *M. arenaria* and *M. janvanica* (Call et al., 1996b, 1997a). Four cycles of phenotypic recurrent selection for resistance to root galling and reproduction by *M. incognita* were conducted in the greenhouse to develop MSNR4 white clover germplasm that has resistance to *M. arenaria, M. graminicola*, and *M. incognita* (Pederson & Windham, 1995; Fig. 19–5). Two cycles of phenotypic recurrent selection for resistance to stunting by *H. trifolii* were conducted in the greenhouse to develop CCNR-1 red clover germplasm that shows less yield suppression due to clover cyst nematode infestation than other red clover cultivars (Leath et al., 1987).

Techniques such as interspecific hybridization, chromosome doubling, and mutation breeding have been used to produce nematode-resistant cultivars and breeding lines in forage legumes. Four vetch cultivars derived from interspecific hybrids of *Vicia sativa* L. subsp. *sativa* × *V. cordata* Wulf and *V. sativa* × *V. narbonensis* L. were released with resistance to *M. incognita* and *M. javanica* (Table 19–2; Donnelly, 1979). Studies of white clover interspecific hybrids and related species showed that *Trifolium nigrescens* Viv. had resistance to *M. incognita* and *H. trifolii*, but not *M. trifoliophila* (Pederson & Windham, 1989; Mercer, 1994).

Table 19–2. Forage legumes released with nematode resistance or tolerance.

Forage legume	Name	Nematode species	Reference
Alyceclover	FL-3	*Meloidogyne arenaria, M. incognita, M. javanica*	Baltensperger et al., 1990b
	FL-5	*M. arenaria, M. incognita, M. javanica*	Baltensperger et al., 1989
Hairy indigo	Flamingo	*M. arenaria, M. incognita, M. javanica*	Baltensperger et al., 1990a
	FL-24 and FL-101	*M. arenaria, M. incognita, M. javanica*	Baltensperger, 1989
Kura clover	MS-2X, MS-4X, MS-6X	*M. incognita*	Pederson et al., 1991a
White clover	SC-1	*M. incognita*	Gibson, 1973
	MSNR4	*M. arenaria, M. graminicola, M. incognita*	Pederson & Windham, 1995
	Grasslands Prestige	*Ditylenchus dipsaci*	Cooper & Chapman, 1993
Red clover	CCNR-1	*Heterodera trifolii*	Leath et al., 1987
	Cherokee	*M. arenaria, M. hapla, M. incognita, M. javanica*	Quesenberry et al., 1993
	Bjorn	*Ditylenchus dipsaci*	Andersson, 1985
	Sara	*Ditylenchus dipsaci*	Jonsson, 1985
Sericea lespedeza	AU Lotan	*M. incognita*	Donnelly, 1981
Vetch	Cahaba White, Vantage, Nova II, Vanguard	*M. incognita, M. javanica*	Donnelly, 1979

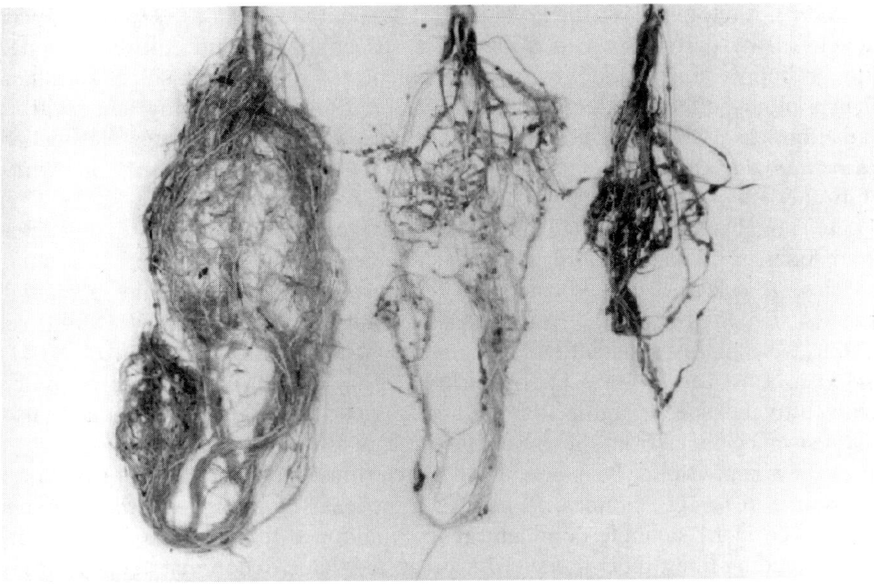

Fig. 19–4. Improvement in root growth of red clover following recurrent selection for reduced galling and egg-mass production by *Meloidogyne* species.

About 75% of progeny from backcrosses of white clover × *T. nigrescens* hybrids with white clover had less than 10% of the root system galled by *M. incognita*, though root gall ratings were equivalent to susceptible white clover (Pederson et

Fig. 19–5. MSNR4 root-knot nematode resistant white clover (right) and a susceptible check cultivar (left) inoculated with *Meloidogyne incognita* (courtesy of G.L. Windham, USDA, ARS, Crop Sci. Res. Lab., Mississippi State, MS).

al., 1990). A tetraploid red clover cultivar, Ulva, with resistance to *D. dipsaci* was developed by chromosome doubling of a susceptible diploid cultivar, Ultuna, though improved nematode resistance was reported to be due to selection rather than a ploidy effect (Bingefors & Ellerstrom, 1964). Gamma ray and neutron radiation was used to produce advanced generations of annual Korean lespedeza [*Kummerowia stipulacea* (Maxim.) Makino] with resistance to *M. incognita* (Offutt & Riggs, 1970).

Though most cultivated *Trifolium* spp. have low resistance to root-knot nematodes, some native North American species are quite resistant. The annual species, *T. carolinianum* Michx. and *T. bejariense* Moric., and the perennial species, *T. calcaricum* J.L. Collins & Wieboldt and *T. stoloniferum* Muhl. ex Eaton, were highly resistant to *M. arenaria, M. incognita, M. javanica,* and *M. hapla* in most cases (Quesenberry et al., 1997). Due to sexual incompatibility with cultivated species, molecular methods of gene transfer will be needed to use the resistance genes from these North American clovers.

Few studies have been conducted to determine the inheritance of nematode resistance in forage legumes. In sericea lespedeza [*Lespedeza cuneata* (Dum.-Cours.) G. Don], a single dominant gene conditioned resistance to *M. hapla* in three breeding lines, though one cross suggested that two dominant genes conditioned resistance in one of the lines (Adamson et al., 1974). In white clover, additive genetic effects were found to be most important in determining resistance to *M. incognita,* and direction of crossing was not important as no maternal effects were observed (Pederson & Windham, 1992). High narrow-sense heritabilities for resistance to *M. arenaria, M. incognita,* and *M. javanica* have been reported in red clover (Quesenberry et al., 1989). A combining ability analysis study of red clover parents varying in resistance to *M. incognita* Race 3 found that general combining ability effects were important for reduced galling and egg mass production (Call et al., 1997b). Resistance to *H. trifolii* in white clover is determined by more than one dominant gene (Dijkstra, 1971). Resistance to *D. dipsaci* in red clover is determined by two dominant genes, one of which is closely linked to the self-sterility locus (Nordenskiold, 1971).

Plant type in white clover appears to be related to stem nematode resistance as numerous studies have reported that resistance was found in Ladino-type white clovers (Williams, 1972; Caradus et al., 1990). Rapid spread by stolons also may allow white clover to escape stem nematode infestations as opposed to species such as red clover where death of the crown kills the entire plant (Cook et al., 1992a).

Cultural

Crop rotation has rarely been proposed as a measure to control nematodes in forage legumes, since legumes are often components of permanent grass pastures or pastures on marginal land that is not easily rotated with other crops. Control of *D. dipsaci* in red clover fields can be obtained by rotation with nonhost species, though recommendations suggest that intervals between red clover plantings be at least 5 yr (Courtney, 1962). Several researchers of forage legume crop rotation and nematode control have attempted to find forage legumes that could be used as N_2-fixing green manure crops that reduce nematode density. This green manure crop would be of most benefit to a high-value cash crop following in the rotation. These studies have not been very successful in identifying species that adversely affect nematode population density. Increases in nematode population density have been reported for *M. incognita* on crimson clover (*T. incarnatum*; McSorley & Gallaher, 1992), *M. hapla* on white clover (Guyton et al., 1989), *P. penetrans* on red clover (Kimpinski et al., 1984), and others. Some success has been noted using the summer forage legumes, American jointvetch (*Aeschynomene americana* L.) and hairy indigo (*Indigofera hirsuta* L.), to reduce *M. incognita, M. arenaria,* and *Heterodera glycines* populations in a soybean [*Glycine max* (L.) Merr.] production system (Rodriguez-Kabana et al., 1988). Recent findings using the FLMR6 red clover breeding population suggest that the high level of root-knot nematode resistance in this material could be used in crop rotation to suppress soil nematode populations (Call et al., 1996a,b).

Sanitation is an excellent method of preventing the introduction of stem nematodes into new red clover fields. Contaminated red clover seed, hay, and machinery should be kept out of any new red clover seedings (Courtney, 1962).

Other cultural methods have rarely been effective in controlling nematodes or overcoming the effects of nematode parasitism in forage legumes. Sod-seeded red clover had fewer juvenile *H. trifolii* and *Pratylenchus* spp. in the soil than conventional seeded red clover, though no difference in root populations of *Pratylenchus* spp. was observed (Kunelius et al., 1988). Increasing white clover plant density had no effect on *M. incognita* reproduction (Brink & Windham, 1990). Interplanting white clover with perennial ryegrass (*Lolium perenne* L.) or tall fescue (*Festuca arundinacea* Schreber) had no beneficial effect on clover growth in *M. hapla* and *H. trifolii*-infested soil (Watson, 1990). Increasing the P nutrition of white clover could not overcome the effect of *M. hapla* and *H. trifolii* infestation (Widdowson et al., 1973) and increased *Pratylenchus* spp. populations (Yeates, 1976). Increases in sheep stocking rate and lime on white clover resulted in increased numbers of *H. trifolii* and *Pratylenchus* spp. juveniles after 6 yr (Yeates, 1976), while lime application and sheep stocking (as opposed to cutting) had no effect on *D. dipsaci* infestation in white or red clover (Courtney, 1962; Cook et al., 1992a). More severe defoliation treatments of birdsfoot trefoil resulted in greater yield losses due to *P. penetrans* (Thompson & Willis, 1970a).

Chemical

Chemical control of nematodes in forage legumes is rarely economically feasible, though it has been shown to be effective in controlling nematode infestation. Registration of nematicides for use on forage legumes has been rare due to concerns with potential residues in forages that may be eaten by lactating animals. Nematicide treatment effectively eliminated *M. arenaria* infections on subterranean (Pung et al., 1991) and berseem clovers, and reduced *Tylenchorhynchus* spp. populations on berseem clover (Kassab & Taha, 1990). Three different nematicides reduced *M. hapla* and *H. trifolii* populations on white clover and resulted in increased total yields of 6 to 15% (Yeates et al., 1975). In six trials using three nematicides, nematicide treatment reduced *H. trifolii* cyst populations

32 to 77% and increased white clover yields 20 to 84% (Yeates & Risk, 1991). Nematicide seed coatings have given variable results in protecting red and white clover seedlings from infestation with *M. hapla, H. trifolii,* and *D. dipsaci* (Mercer & Campbell, 1986; Barker et al., 1993). Nematicide treatment restricted initial infection by *D. dipsaci* in white clover (Williams & Barclay, 1972; Cook et al., 1992a). Treatment with nematicides reduced *P. penetrans* populations 78% in red clover (Thompson & Willis, 1970b), 20% in white clover (Thies et al., 1995), and 18 to 43% in birdsfoot trefoil (Thompson & Willis, 1970b; Thies et al., 1995) and increased dry-matter yields 22% in red clover and 40% in birdsfoot trefoil in the 1st yr (Thompson & Willis, 1970b). Three-year average dry-matter yields were increased 23 to 67% in red clover and 54% in white clover when the soil was treated with a nematicide prior to seeding (Mislevy et al., 1994). Nematicide effects may diminish over time as nematicide treatment of *P. penetrans* increased birdsfoot trefoil dry-matter yields 14 to 40% in the first two growing seasons but had no effect on yield by 27 mo (Willis & Thompson, 1975).

Biological

Few fungal species and bacterial strains have been evaluated for control of nematodes in forage legumes. Most research on biological agents that may impact nematode control in forage legumes is currently being conducted with the same or related nematode species on other hosts. For additional information on these studies, reviews by Stirling (1991) and Dickson et al. (1994) should be consulted.

Trichoderma harzianum reduced *M. incognita* reproduction 38 to 58% on white clover, but did not affect *M. arenaria* reproduction (Windham et al., 1993). A number of fungal species isolated from *H. trifolii* cysts will parasitize eggs of *H. trifolii* and *M. hapla* (Morgan-Jones & Rodriguez-Kabana, 1988; Hay & Skipp, 1993; Mercer, 1994). The parasitic fungus, *Verticillium balanoides*, was carried into white clover plants by its host, *D. dipsaci*, and infected and sporulated on the nematode inside the plant tissue (Hay & Regnault, 1995). Preinoculation of plants with vesicular-arbuscular mycorrhizal (VAM) fungi (mainly *Glomus* spp.) lessened plant-growth suppression by *M. incognita* in berseem clover and by *M. hapla* in white clover, but nematode reproduction was not affected (Cooper & Grandison, 1986; Taha & Abdel-Kader, 1990). However, when plants were inoculated with the nematode prior to or simultaneously with VAM, the nematode suppressed fungal development (Taha & Abdel-Kader, 1990). For *Tylenchorhynchus* spp. and *T. vulgaris* infection on berseem clover, plant-growth suppression by the nematode was eliminated by either preinoculation with VAM fungi or simultaneous inoculation; however, nematode populations were only suppressed by preinoculation with VAM fungi (Hasan & Jain, 1987; Kassab & Taha, 1990).

Few bacteria have been evaluated for biological control of nematodes in forage legumes. Strains of two *Pseudomonas fluorescens* and one *Bacillus* spp. inhibited galling by *M. incognita* in white clover by 21 to 36% and resulted in increased root and shoot weights (Becker et al., 1988). *Pasteuria penetrans* strains showed little adherence to *M. trifoliophila* (Mercer & Watson, 1996).

OUTLOOK

Nematode control in clover and other forage legumes in the USA has taken a positive step forward recently with releases of root-knot nematode resistance in red and white clover (Quesenberry et al., 1993; Pederson & Windham, 1995). As these materials are incorporated into other cultivars, better control may be obtained on root-knot nematodes in hay fields and pastures. Resistance breeding efforts to these and other nematodes in forage legumes will continue to provide producers with nematode control at a cost they can afford. Though chemical control will probably never be a component of nematode control in forage legumes due to cost and registration restrictions, biological agents selected to control nematode species in other crops may be useful for nematode control in pastures. As transformation protocols are perfected in clover and other forage legumes, molecular biological methods may be used to move resistance genes from unrelated species into forage legumes. Other possible gene transfers that have been proposed for nematode control in forage legumes include genes that code for production of collagenase, *Bacillus thuringiensis* endotoxins, bacterial chitinase, or proteinase inhibitors (Mercer & Watson, 1996).

ACKNOWLEDGMENTS

Contribution of the USDA-ARS in cooperation with the Mississippi Agricultural and Forestry Experiment Station. Journal Article no. BC-8967 of the Mississippi Agricultural and Forestry Experiment Station.

REFERENCES

Adamson, W.C., E.D. Donnelly, N.A. Minton, and J.D. Miller. 1974. Inheritance of resistance to *Meloidogyne hapla* in *Lespedeza cuneata*. J. Hered. 65:365–368.

Andersson, B. 1985. Some results of selection for disease resistance in herbage legumes at Svalof, Sweden. p. 230–231. *In* T. Okubo (ed.) Proc. 15th Int. Grassl. Congr., Kyoto, Japan. 24–31 August. Sci. Counc. Jpn. and Jpn. Soc. Grassl. Sci., Nishinasuno, Japan.

Armstrong, J.M., J.N. Pinkerton, and H.J. Jensen. 1977. Responses of red clover germplasm to stem nematodes in greenhouse trials. Plant Dis. Rep. 61:1060–1063.

Azmi, M.I. 1988. Effect of root-knot nematode, *Meloidogyne incognita*, on sweet clovers. Indian J. Nematol. 18:347.

Baltensperger, D.D. 1989. Registration of 'FL-24' and 'FL-101' germplasm lines of hairy indigo. Crop Sci. 29:1331.

Baltensperger, D.D., H.E. Dominguez, W.H. Anderson, R.L. Stanley, L.S. Dunavin, and J.R. Rich. 1990a. Registration of 'Flamingo' hairy indigo. Crop Sci. 30:1366–1367.

Baltensperger, D.D., K.H. Quesenberry, R.A. Dunn, and M.M. Abd-Elgawad. 1985. Root-knot nematode interaction with berseem clover and other temperate forage legumes. Crop Sci. 25:848–851.

Baltensperger, D.D., S.G. Taylor, and R. Glennon. 1989. Registration of 'FL-5' germplasm line of alyceclover. Crop Sci. 29:1330–1331.

Baltensperger, D.D., S.G. Taylor, W.H. Anderson, R.L. Stanley, L.S. Dunavin, M.J. Williams, R.A. Dunn, and R.J. Glennon. 1990b. Registration of 'FL-3' alyceclover. Crop Sci. 30:1367–1368.

Barker, G.M., P.J. Addison, and R.N. Watson. 1993. Seed treatment for protection of seedling white clover from nematodes. *In* A.J. Popay (ed.) Proc. 46th N. Z. Plant Protection Conf., Christchurch and Wellington, New Zealand. 10–13 August 1992. N.Z. Plant Prot. Soc., Palmerston North, NZ.

Becker, J.O., E. Zavaleta-Mejia, S.F. Colbert, M.N. Schroth, A.R. Weinhold, J.G. Hancock, and S.D. Van Gundy. 1988. Effects of rhizobacteria on root-knot nematodes and gall formation. Phytopathology 78:1466–1469.

Bernard, E.C., and J.D. Eisenback. 1997. *Meloidogyne trifoliophila* n. spp. (Nemata: Meloidogynidae) a parasite of clover from Tennessee. J. Nematol. 29:43–54.

Bingefors, S. 1957. Studies on breeding red clover for resistance to stem nematodes. Inst. Plant Husband. Publ. 8. R. Agric. Coll. Sweden, Uppsala.

Bingefors, S. 1985. Improved resistance to stem nematodes in Swedish cultivars of lucerne and red clover and its value. Zeit. Pflanzenzucht. 95:164–172.

Bingefors, S., and S. Ellerstrom. 1964. Polyploidy breeding in red clover. The tetraploid variety Svalof's Ulva compared with some diploid and tetraploid varieties. Zeit. Pflanzenzucht. 51:315–334.

Brink, G.E., and G.L. Windham. 1990. White clover response to nematode infestation and plant density. Crop Sci. 30:1295–1298.

Call, N.M., R.A. Dunn, and K.H. Quesenberry. 1996a. Effects of root-knot nematodes on red clover grown in microplots. Proc. Soil Crop Sci. Soc. Fla. 55:77–81.

Call, N.M., K.H. Quesenberry, and R.A. Dunn. 1996b. *Meloidogyne arenaria* development in three clover populations. Crop Sci. 36:1393–1397.

Call, N.M., K.H. Quesenberry, and R.A. Dunn. 1997a. *Meloidogyne javanica* development in three clover populations. Crop Sci. 37:265–269.

Call, N.M., K.H. Quesenberry, D.S. Wofford, and R.A. Dunn. 1997b. Combining ability analysis of resistance to southern root-knot nematode in red clover. Crop Sci. 37:121–124.

Caradus, J.R., M.B. Forde, S. Wewala, and A.C. Mackay. 1990. Description and classification of a white clover (*Trifolium repens* L.) germplasm collection from southwest Europe. N.Z. J. Agric. Res. 33:367–375.

Chapman, R.A. 1958. The effect of root-lesion nematodes on the growth of red clover and alfalfa under greenhouse conditions. Phytopathology 48:525–530.

Chapman, R.A., and D.R. Turner. 1975. Effect of *Meloidogyne incognita* on reproduction of *Pratylenchus penetrans* in red clover and alfalfa. J. Nematol. 7:6–10.

Cook, R., D.R. Evans, T.A. Williams, and K.A. Mizen. 1992a. The effect of stem nematode on establishment and early yields of white clover. Ann. Appl. Biol. 120:83–94.

Cook, R., and K.A. Mizen. 1989. Resistance to plant parasitic nematodes in ryegrasses and white clover. p. 737–738. *In* R. Desroches (ed.) Proc. 16th Int. Grassl. Congr., Nice, France. 4–11 October. French Grassl. Soc., Versailles Cedex.

Cook, R., K.A. Mizen, R.A. Plowright, and P.A. York. 1992b. Observations on the incidence of plant parasitic nematodes in grassland in England and Wales. Grass For. Sci. 47:274–279.

Cook, R., B.J. Thomas, and K.A. Mizen. 1989. Dissemination of white clover mosaic virus and stem nematode, *Ditylenchus dipsaci*, by the slug, *Deroceras reticulatum*. p. 107–112. *In* I. Henderson (ed.) Slugs and snails in world agriculture. Monogr. 41. Br. Crop Protect. Counc., Croydon, England.

Cook, R., and G.W. Yeates. 1993. Nematode pests of grassland and forage crops. p. 305–350. *In* K. Evans et al. (ed.) Plant parasitic nematodes in temperate agriculture. CAB Int., Wallingford, England.

Cook, R., and P.A. York. 1980. Nematodes and herbage improvement. p. 177–207. *In* Rep. Welsh Plant Breeding Station for 1979. Welsh Plant Breed. Stn., Aberystwyth, England.

Cooper, B.M., and D.F. Chapman. 1993. Grasslands Prestige (G39), a white clover cultivar originating from northern New Zealand. p. 458–459. *In* M.J. Baker et al. (ed.) Proc. 17th Int. Grassl. Congr., Palmerston, NZ, and Rockhampton, Australia. 8–21 February. N.Z. Grassl. Soc., Palmerston, NZ.

Cooper, K.M., and G.S. Grandison. 1986. Interaction of vesicular-arbuscular mycorrhizal fungi and root-knot nematode on cultivars of tomato and white clover susceptible to *Meloidogyne hapla*. Ann. Appl. Biol. 108:555–565.

Courtney, W.D. 1962. Stem nematode of red clover in the Pacific Northwest. Washington Agric. Exp. Stn. Bull. 640.

Dickson, D.W., M. Oostendorp, R.M. Giblin-Davis, and D.J. Mitchell. 1994. Control of plant-parasitic nematodes by biological antagonists. p. 575–601. *In* D. Rosen et al. (ed.) Pest management in the subtropics, biological control—a Florida perspective. Intercept Ltd., Andover, England.

Dijkstra, J. 1971. Breeding for resistance to *Heterodera* in white clover. Euphytica 20:36–46.

Donnelly, E.D. 1979. Registration of Cahaba White, Vantage, Nova II, and Vanguard vetch. Crop Sci. 19:414.

Donnelly, E.D. 1981. Registration of AU Lotan serecia lespedeza. Crop Sci. 21:474.

Eriksson, K.B. 1972. Nematode diseases of pasture legumes and turf grasses. p. 66–96. *In* J.M. Webster (ed.) Economic nematology. Acad. Press, London, England.

Evans, A.A.F. 1987. Diapause in nematodes as a survival strategy. p. 180–187. *In* J.A. Veech and D.W. Dickson (ed.) Vistas on nematology. Soc. Nematol., Hyattsville, MD.

Fagbenle, H.H., D.I. Edwards, and R.B. Malek. 1986. Cyst production by two geographical isolates of *Heterodera lespedezae* on selected legumes. Plant Dis. 70:643–644.

Fagbenle, H.H., D.I. Edwards, and R.B. Malek. 1989. Effects of temperature on pathogenicity and population development of two isolates of *Heterodera lespedezae*. J. Nematol. 21:173–178.

Freckman, D.W., and R.A. Chapman. 1972. Infection of red clover seedlings by *Heterodera trifolii* Goffart and *Pratylenchus penetrans* (Cobb). J. Nematol. 4:23–28.

Gibson, P.B. 1973. Registration of SC-1 white clover germplasm. Crop Sci. 13:131.

Grant, J.L., C.F. Mercer, and T.M. Stewart. 1996. Effects of resistance in white clover (*Trifolium repens*) on *Heterodera trifolii*. J. Nematol. 28(4S):492–500.

Griffin, G.D., E.C. Bernard, G.A. Pederson, G.L. Windham, K.H. Quesenberry, and R.A. Dunn. 1996. Nematode pathogens of American pasture/forage crops. p. 257–284. *In* S. Chakraborty et al. (ed.) Pasture and forage crop pathology. ASA, CSSA, and SSSA, Madison, WI.

Guyton, R.F., J.R. Thompson, R. Kimoto, B.A. Kratky, O.V. Holtzmann, B.D. Thyr, and W.W. Miller. 1989. Root-knot nematode populations and carrot yield following five forage legumes and continuous carrots. HortScience 24:71–73.

Hague, H.G.M. 1980. Nematodes of legume crops. *In* R.J. Summerfield et al. (ed.) Advances in legume science. p. 199–205. *In* (ed.) Proc. Int. Legume Conf., Kew, England. 31 July–4 August 1978. R. Bot. Gardens, Kew, England.

Halpin, J.E. 1963. The effect of soil fungi and root-knot nematodes on the growth of white clover in field bins. Phytopathology 53:877.

Hasan, N., and R.K. Jain. 1987. Parasitic nematodes and vesicular-arbuscular mycorrhizal (VAM) fungi associated with berseem (*Trifolium alexandrinum* L.) in Bundelkhand region. Ind. J. Nematol. 17:184–188.

Hay, F.S., and L. Regnault. 1995. Introduction of the endoparasitic nematophagous fungus *Verticillium balanoides* into white clover (*Trifolium repens*) by its host, stem nematode (*Ditylenchus dipsaci*). Austral. Plant Pathol. 24:249–251.

Hay, F.S., and R.A. Skipp. 1993. Fungi and actinomycetes associated with cysts of *Heterodera trifolii* Goffart (Nematoda: Tylenchida) in pasture soils in New Zealand. Nematologica 39:376–384.

Hidding, J., M.J. Hijink, and M. Oostenbrink. 1963. Opbrengst-en kwaliteitsverlies van witte klaver in een grasklavermengsel door het klaversystenaaltje, *Heterodera trifolii*. Meded. Landb. Opzoekstn. Gent. 28:679–684.

Hussey, R.S., and V.M. Williamson. 1998. Physiological and molecular aspects of nematode parasitism. p. 87–108. *In* K.R. Barker et al. (ed.) Plant and nematode interactions. Agron. Monogr. 36. ASA, CSSA, and SSSA, Madison, WI.

Ibrahim, I.K.A., M.A. Rezk, and H.A.A. Khalil. 1983. Resistance of some plant cultivars to root-knot nematodes, *Meloidogyne* spp. Nematol. Medit. 11:189–192.

Jin, X., J.B. Kotcon, and J.B. Morton. 1991. Interactions between *Pratylenchus penetrans* and *Fusarium avenaceum* in red clover. Nematropica 21:105–109.

Jonsson, H.A. 1985. Red clover (*Trifolium pratense*) Sara. Agric. Hort. Genet. 43:43–51.

Kassab, A.S., and A.H.Y. Taha. 1990. Interaction between plant-parasitic nematodes, vesicular-arbuscular mycorrhiza, rhizobia, and nematicide on Egyptian clover. Ann. Agric. Sci. (Cairo) 35:509–520.

Kimpinski, J., B.R. Christie, and T.M. Choo. 1992. Evaluation of red clover for nematode resistance. Forage Notes 36:33–35.

Kimpinski, J., H.T. Kunelius, and C.B. Willis. 1984. Plant parasitic nematodes in temperate forage grass and legume species in Prince Edward Island. Can. J. Plant Pathol. 6:160–164.

Kimpinski, J., G. Plumas, and M.C. MacDonald. 1993. Occurrence of the clover cyst nematode, *Heterodera trifolii*, in Prince Edward Island soils. J. Nematol. 25(4S):876–879.

Kornobis, S. 1987. Plant parasitic nematodes associated with poor growth of red clover in the region Wielkopolska. Prace Nauk. Inst. Ochrony Roslin 29:109–117.

Kouame, C.N., D.D. Baltensperger, G.M. Prine, and R.A. Dunn. 1989. Screening subterranean clover (*Trifolium* spp.) germplasm for resistance to *Meloidogyne* species. J. Nematol. 21:379–383.

Kunelius, H.T., M.R. Carter, J. Kimpinski, and J.B. Sanderson. 1988. Effect of seeding method on alfalfa and red clover establishment and growth, soil physical condition and nematode populations. Soil Till. Res. 12:163–175.

Lau, N.E., and J.P. Reed. 1960. Nematodes associated with red clover in its second growth year. Plant Dis. Rep. 44:402–404.

Leath, K.T. 1985. General diseases. *In* N.L. Taylor (ed.) Clover science and technology. Agron. Monogr. 25. ASA, CSSA, and SSSA, Madison, WI.

Leath, K.T., J.R. Bloom, R.R. Hill, Jr., and R.R. Smith. 1987. Registration of red clover germplasm line CCNR-1. Crop Sci. 27:1316.

Lehman, P.S. 1991. *Pterotylenchus cecidogenus*, a nematode causing stem galls on *Desmodium ovalifolium*. Univ. Florida Nematol. Cir. 194.

Mankau, R., and M.B. Linford. 1960. Host-parasite relationships of the clover cyst nematode, *Heterodera trifolii* Goffart. Illinois Agric. Exp. Stn. Bull. 667.

Massoud, S.I., and A.I. Ghorab. 1988. Histopathological responses of clover roots infected with *Meloidogyne incognita, Heterodera daverti*, and *Rotylenchulus reniformis*. Egypt. J. Phytopathol. 20:79–85.

McLaughlin, M.R., and G.L. Windham. 1996. Effects of peanut stunt virus, *Meloidogyne incognita*, and drought on growth and persistence of white clover. Phytopathology 86:1105–1111.

McLaughlin, M.R., G.L. Windham, and A.S. Heagle. 1993. Response of *Trifolium repens* clones to infection by *Meloidogyne incognita* and peanut stunt virus. J. Nematol. 25(4S):869–875.

McLeish, L.J., G.N. Berg, J.M. Hinch, L.V. Nambiar, and M.R. Norton. 1997. Plant parasitic nematodes in white clover and soil from white clover pastures in Australia. Aust. J. Exp. Agric. 37:75–82.

McSorley, R., and R.N. Gallaher. 1992. Managing plant-parasitic nematodes in crop sequences. Proc. Soil Crop Sci. Soc. Fla. 51:42–45.

Mercer, C.F. 1988. Reaction of some species of *Trifolium* to *Meloidogyne hapla* and *Heterodera trifolii*. p. 275–280. *In* P.P. Stahle (ed.) Proc. 5th Australasian Conf. Grassl. Invert. Ecol., Melbourne, Australia. 15–19 August. Univ. Melbourne, Australia.

Mercer, C.F. 1990. Development of the nematodes *Meloidogyne hapla* Chitwood and *Heterodera trifolii* Goffart in white clover. Nematologica 36:227–236.

Mercer, C.F. 1994. Plant-parasitic nematodes in New Zealand. N.Z. J. Zool. 21:57–65.

Mercer, C.F., and B.D. Campbell. 1986. Early invasion of overdrilled red clover seedlings by nematodes. N.Z. J. Agric. Res. 29:495–499.

Mercer, C.F., B.M. Cooper, and J.L. Grant. 1992. Selection for resistance and tolerance of white clover to root-knot (*Meloidogyne hapla*) and clover cyst (*Heterodera trifolii*) nematodes. N.Z. J. Agric. Res. 35:219–224.

Mercer, C.F., and J.L. Grant. 1993. Reproduction of *Meloidogyne hapla* and *Heterodera trifolii* from several sites in New Zealand on resistant and susceptible lines of white clover. Nematologica 39:312–321.

Mercer, C.F., and R.N. Watson. 1996. Nematode pathogens of New Zealand pastures. p. 241–256. *In* S. Chakraborty et al. (ed.) Pasture and forage crop pathology. ASA, CSSA, and SSSA, Madison, WI.

Mercer, C.F., and D.R. Woodfield. 1986. A survey of root knot and clover cyst nematodes in dry hill country. N.Z. J. Agric. Res. 29:129–135.

Miller, L.I. 1986. Economic importance of cyst nematodes in North America. p. 373–385. *In* F. Lamberti and C.E. Taylor (ed.) Cyst nematodes. Plenum Press, New York.

Minton, N.A., and E.D. Donnelly. 1971. Reaction of field-grown sericea lespedeza to selected *Meloidogyne* spp. J. Nematol. 3:369–373.

Mislevy, P., J. Noling, F.G. Martin, and K.H. Quesenberry. 1994. Influence of methyl bromide on clover production in subtropical Florida. p. 160. *In* Agronomy abstracts. ASA, Madison, WI.

Morgan-Jones, G., and R. Rodriguez-Kabana. 1988. Fungi colonizing cysts and eggs. p. 39–58. *In* G.O. Pointer, Jr., and H.B. Jansson (ed.) Diseases of nematodes. Vol. 2. CRC Press, Boca Raton, FL.

Nordenskiold, H. 1971. The genetic background of the resistance to nematodes (*Ditylenchus dipsaci*) in red clover (*Trifolium pratense*). Hereditas 69:301–302.

Nordmeyer, D., and R.A. Sikora. 1983. Studies on the interaction between *Heterodera daverti, Fusarium avenaceum* and *F. oxysporum* on *Trifolium subterraneum*. Rev. Nematol. 6:193–198.

Norton, D.C. 1967a. Relationships of *Heterodera trifolii* to some forage legumes. Phytopathology 57:1305–1308.

Norton, D.C. 1967b. *Xiphinema americanum* as a factor in unthriftiness of red clover. Phytopathology 57:1390–1391.

Offutt, M.S., and R.D. Riggs. 1970. Radiation-induced resistance to root-knot nematodes in Korean lespedeza. Crop Sci. 10:49–50.

Pederson, G.A., M.R. McLaughlin, and G.L. Windham. 1991a. Registration of MS-2X, MS-4X, and MS-6X kura clover germplasms. Crop Sci. 31:1714.

Pederson, G.A., and G.L. Windham. 1989. Resistance to *Meloidogyne incognita* in *Trifolium* interspecific hybrids and species related to white clover. Plant Dis. 73:567–569.

Pederson, G.A., and G.L. Windham. 1992. Diallel analysis of resistance in white clover to southern root-knot nematode. Crop Sci. 32:1160–1162.

Pederson, G.A., and G.L. Windham. 1995. Registration of MSNR4 root-knot nematode resistant white clover germplasm. Crop Sci. 35:1234–1235.

Pederson, G.A., G.L. Windham, M.M. Ellsbury, M.R. McLaughlin, R.G. Pratt, and G.E. Brink. 1991b. White clover yield and persistence as influenced by cypermethrin, benomyl, and root-knot nematode. Crop Sci. 31:1297–1302.

Pederson, G.A., G.L. Windham, and M.R. McLaughlin. 1990. Incorporation of nematode and virus resistance from *Trifolium* interspecific hybrids into white clover. *In* Proc. 1990 Forage Grassl. Conf., Blacksburg, VA. 6–9 June 1990. Am. Forage Grassl. Council, Georgetown, TX.

Pokharel, N.P., and S.N. Kruchina. 1991. Effects of *Meloidogyne incognita* (Kofoid & White, 1919) Chitwood, 1949 and *Meloidogyne hapla* Chitwood, 1949 on the macro-nutrient content of *Trifolium pratense* L. Arch. Phytopathol. Pflanzenschutz 27:41–44.

Powers, L.E., R.A. Dunn, R. McSorley, D.D. Baltensperger, and D.S. Wofford. 1992. Effects of resistance in alyceclover (*Alysicarpus* spp.) on root-knot nematode (*Meloidogyne* spp.) populations. Trop. Grassl. 26:30–39.

Pung, S.H., M.J. Barbetti, and K. Sivasithamparam. 1988. Association of *Meloidogyne arenaria* with root rot of subterranean clover in Western Australia. N.Z. J. Exp. Agric. 16:91–96.

Pung, S.H., M.J. Barbetti, and K. Sivasithamparam. 1991. The effect of aldicarb and benomyl on root-knot nematode and root rot in subterranean clover. Austral. Plant Pathol. 20:130–133.

Pung, S.H., M.J. Barbetti, and K. Sivasithamparam. 1992a. Effect of soil environment on infection of subterranean clover by *Meloidogyne arenaria*. Aust. J. Agric. Res. 43:87–104.

Pung, S.H., K. Sivasithamparam, and M.J. Barbetti. 1991. Effect of placement and sequence of inoculation with *Fusarium oxysporum* on its interaction with *Meloidogyne arenaria* on subterranean clover. J. Phytopathol. 132:273–286.

Pung, S.H., K. Sivasithamparam, and M.J. Barbetti. 1992b. Interactions between *Fusarium oxysporum* and *Meloidogyne arenaria* as root pathogens of subterranean clover affected by soil sterilisation and inoculum substrate. N.Z. J. Agric. Res. 35:83–91.

Quesenberry, K.H., D.D. Baltensperger, and R.A. Dunn. 1986. Screening *Trifolium* spp. for response to *Meloidogyne* spp. Crop Sci. 26:61–64.

Quesenberry, K.H., D.D. Baltensperger, R.A. Dunn, C.J. Wilcox, and S.R. Hardy. 1989. Selection for tolerance to root-knot nematodes in red clover. Crop Sci. 29:62–65.

Quesenberry, K.H., N.M. Call, D.E. Moon, and R.A. Dunn. 1997. Response of native eastern North American *Trifolium* spp. to root-knot nematodes. Crop Sci. 37:270–274.

Quesenberry, K.H., G.M. Prine, O.C. Ruelke, L.S. Dunavin, and P. Mislevy. 1993. Registration of 'Cherokee' red clover. Crop Sci. 33:208–209.

Rodriguez-Kabana, R., P.S. King, D.G. Robertson, C.F. Weaver, and E.L. Carden. 1988. New crops with potential for management of soybean nematodes. Nematropica 18:45–52.

Sarathchandra, S.U., M.E. di Menna, G. Burch, J.A. Brown, R.N. Watson, N.L. Bell, and N.R. Cox. 1995. Effects of plant-parasitic nematodes and rhizosphere microorganisms on the growth of white clover (*Trifolium repens* L.) and perennial ryegrass (*Lolium perenne* L.). Soil Biol. Biochem. 27:9–16.

Sasser, J.N., K.M. Hartman, and C.C. Carter. 1987. Summary of preliminary crop germplasm evaluations for resistance to root-knot nematodes. North Carolina State Univ. and U.S. AID, Raleigh, NC.

Shepherd, R.L. 1965. Reaction of *Trifolium repens* L. to root-knot nematodes, *Meloidogyne incognita acrita*. Ph.D. diss. Auburn Univ., Auburn, AL. (Diss. Abstr. 65-9638).

Sikora, R.A., and P.W.T. Maas. 1986. An analysis of the *Heterodera trifolii* complex and other species in the *schachtii* group attacking legumes. p. 293–313. *In* F. Lamberti and C.E. Taylor (ed.) Cyst nematodes. Plenum Press, New York.

Singh, N.D., and D.C. Norton. 1970. Variability in host-parasite relationships of *Heterodera trifolii*. Phytopathology 60:1834–1837.

Skipp, R.A., and M.J. Christensen. 1981. Invasion of white clover roots by fungi and other soil microorganisms. I. Surface colonisation and invasion of roots growing in sieved pasture soil in the glasshouse. N.Z. J. Agric. Res. 24:235–241.

Skipp, R.A., and M.J. Christensen. 1983. Invasion of white clover roots by fungi and other soil microorganisms. IV. Survey of root-invading fungi and nematodes in some New Zealand pastures. N.Z. J. Agric. Res. 26:151–155.

Skipp, R.A., and D.L. Gaynor. 1987. Pests—nematodes. p. 493–512. *In* M.J. Baker and W.M. Williams (ed.) White clover. CAB Int., Wallingford, England.

Stelter, H., and G. Meinl. 1972. The effects of the infestation of red and white clover by *Heterodera trifolii* and *Heterodera galeopsidis*. Arch. Pflanzenschutz 8:463–470.

Stirling, G.R. 1991. Biological control of plant parasitic nematodes. CAB Int., Wallingford, England.

Taha, A.H.Y., and K.M. Abdel-Kader. 1990. The reciprocal effects of prior invasion by root-knot nematode or by endomycorrhiza on certain morphological and chemical characteristics of Egyptian clover plants. Ann. Agric. Sci. (Cairo) 35:521–532.

Taha, A.H.Y., and D.J. Raski. 1969. Interrelationships between root-nodule bacteria, plant-parasitic nematodes and their leguminous host. J. Nematol. 1:201–211.

Taylor, A.E., and J.N. Sasser. 1978. Biology, identification, and control of root-knot nematode (*Meloidogyne* species). North Carolina State Univ. Graphics, Raleigh.

Taylor, C.E. 1990. Nematode interactions with other pathogens. Ann. Appl. Biol. 116:405–416.

Taylor, N.L., and K.H. Quesenberry. 1996. Advances in red clover science. Kluwer Acad. Publ., Dordrecht, The Netherlands.

Thies, J.A., A.D. Petersen, and D.K. Barnes. 1995. Host suitability of forage grasses and legumes for root-lesion nematode *Pratylenchus penetrans*. Crop Sci. 35:1647–1651.

Thompson, L.S., and C.B. Willis. 1970a. Reproduction of *Pratylenchus penetrans* and growth of birdsfoot trefoil as influenced by soil moisture and cutting management. Can. J. Plant Sci. 50:499–504.

Thompson, L.S., and C.B. Willis. 1970b. Effect of nematicides on root lesion nematodes and forage legume yields. Can. J. Plant Sci. 50:577–581.

Thompson, L.S., and C.B. Willis. 1975. Influence of fensulfothion and fenamiphos on root lesion nematode numbers and yield of forage legumes. Can. J. Plant Sci. 55:727–735.

Townshend, J.L., and J.W. Potter. 1978. Yield losses among forage legumes infected with *Meloidogyne hapla*. Can. J. Plant Sci. 58:939–943.

Townshend, J.L., and L. Stobbs. 1981. Histopathology and histochemistry of lesions caused by *Pratylenchus penetrans* in roots of forage legumes. Can. J. Plant Pathol. 3:123–128.

Turner, D.R., and R.A. Chapman. 1972. Infection of seedlings of alfalfa and red clover by concomitant populations of *Meloidogyne incognita* and *Pratylenchus penetrans*. J. Nematol. 4:280–286.

Van Den Bosch, J., and C.F. Mercer. 1996. Variation in white clover for root-knot nematode resistance. N.Z. J. Agric. Res. 39:137–148.

Van Den Bosch, J., C.F. Mercer, J.L. Grant, and I.K. Black. 1993. Clover cyst nematode resistance in white clover. p. 928–929. *In* M.J. Baker et al. (ed.) Proc. 17th Int. Grassl. Congr., Palmerston North, NZ, and Rockhampton, Australia. 8–21 February. N.Z. Grassland Soc., Palmerston North, NZ.

Viglierchio, D.R., I.A. Siddiqui, and N.A. Croll. 1973. Culturing and population studies of *Ditylenchus dipsaci* under monoxenic conditions. Higardia 42:177–213.

Watson, R.N. 1990. Effects of plant nematodes and *Acremonium* endophyte on white clover establishment with ryegrass or tall fescue. p. 347–351. *In* Proc. 43rd N.Z. Weed Pest Control Conf. N.Z. Weed and Pest Control Soc., Palmerston North, NZ.

Whitehead, A.G. 1992. Sources of resistance to stem nematode, *Ditylenchus dipsaci* (Kuhn) Filipjev, in species of *Medicago* and *Trifolium*. Ann. Appl. Biol. 120:73–81.

Whitehead, A.G., J.E. Fraser, and A.J.F. Nichols. 1987. Variation in the development of stem nematodes, *Ditylenchus dipsaci*, in susceptible and resistant crop plants. Ann. Appl. Biol. 111:373–383.

Widdowson, J.P., G.W. Yeates, and W.B. Healy. 1973. The effect of root nematodes on the utilisation of phosphorus by white clover on a yellow-brown loam. N.Z. J. Agric. Res. 16:77–80.

Williams, W.M. 1972. Laboratory screening of white clover for resistance to stem nematode. N.Z. J. Agric. Res. 15:363–370.

Williams, W.M., and P.C. Barclay. 1972. The effect of clover stem eelworm on the establishment of pure swards of white clover. N.Z. J. Agric. Res. 15:356–362.

Willis, C.B., and L.S. Thompson. 1969. Effect of the root-lesion nematode on yield of four forage legumes under greenhouse conditions. Can. J. Plant Sci. 49:505–509.

Willis, C.B., and L.S. Thompson. 1975. Influence of carbofuran and benomyl on yield and persistence of birdsfoot trefoil. Can. J. Plant Sci. 55:95–99.

Windham, G.L., and G.A. Pederson. 1988. Effects of *Meloidogyne incognita* on forage yields of four annual clovers. Plant Dis. 72:152–154.

Windham, G.L., and G.A. Pederson. 1989. Aggressiveness of *Meloidogyne incognita* host races on white clover. Nematropica 19:177–183.

Windham, G.L., and G.A. Pederson. 1991. Reaction of *Trifolium repens* cultivars and germplasms to *Meloidogyne incognita*. J. Nematol. 23(4S):593–597.

Windham, G.L., and G.A. Pederson. 1992. Comparison of reproduction by *Meloidogyne graminicola* and *M. incognita* on *Trifolium* species. J. Nematol. 24:257–261.

Windham, G.L., M.T. Windham, and G.A. Pederson. 1993. Interaction of *Trichoderma harzianum, Meloidogyne incognita,* and *Meloidogyne arenaria* on *Trifolium repens*. Nematropica 23:99–103.

Yeates, G.W. 1976. Effect of fertiliser treatment and stocking rate on pasture nematode populations in a yellow-grey earth. N.Z. J. Agric. Res. 19:405–408.

Yeates, G.W. 1977. Soil nematodes in New Zealand pastures. Soil Sci. 123:415–422.

Yeates, G.W. 1978. Influence of *Heterodera trifolii* cyst inoculum on white clover growth and final cyst count in pots of a yellow-brown pumice soil. N.Z. J. Agric. Res. 21:149–151.

Yeates, G.W., W.B. Healy, J.P. Widdowson, N.A. Thomson, and B.N. MacDiarmid. 1975. Influence of nematicides on growth of plots of white clover on a yellow-brown loam. N.Z. J. Agric. Res. 18:411–416.

Yeates, G.W., and W.H. Risk. 1991. Effect of clover cyst nematode (*Heterodera trifolii*) on white clover production in Southland. DSIR Land Resour. Tech. Rec. 37. Lower Hutt, New Zealand.

Yeates, G.W., D.J. Ross, B.A. Bridger, and T.A. Visser. 1977. Influence of the nematodes *Heterodera trifolii* and *Meloidogyne hapla* on nitrogen fixation by white clover under glasshouse conditions. N.Z. J. Agric. Res. 20:401–413.

Yeates, G.W., and T.A. Visser. 1979. Persistence of *Heterodera trifolii* (Nematoda) cysts in the absence of host plants. N.Z. J. Agric. Res. 22:649–651.

20 Forage Grasses

ERNEST C. BERNARD AND KIMBERLY D. GWINN

University of Tennessee
Knoxville, Tennessee

GERALD D. GRIFFIN

Utah State University
Logan, Utah

The amount of land in the USA available as forage for grazing animals is nearly 2.4×10^8 ha (Barnes & Baylor, 1995); worldwide, the figure is more than 10 times as large (Bula et al., 1977). Most of this immense area consists of indigenous grasses, planted forage grasses, or a mixture of grasses and legumes, such as clover (*Trifolium* spp.). "Forage" is a general term used for the edible parts of plants, other than separated grain, that serve as food of grazing animals. Pasture is grazing land planted to specific grasses or grass-legume mixtures by landowners. Pasture is usually fenced and may receive periodic improvement, such as N fertilization, weed control, and overseeding. Rangeland usually is composed of self-seeding indigenous grasses. Range usually receives little or no supplemental care, although the deteriorating condition and productivity of rangelands (Schmautz et al., 1980) is spurring research into renovation approaches (Asay et al., 1985). [Definitions are adapted from Barnes et al. (1995).] This portion of this chapter deals mainly with research on forage grass nematodes in the USA and Canada, with some relevant data from other world regions, and excludes grass species grown primarily for silage. Several reviews of forage nematodes in other regions of the world, as well as North America, have appeared over the past 15 yr, the most recent being that of Cook and Yeates (1993). It should be noted that the division of Poaceae into recreational turf, forage grasses, silage crops, and cereals is arbitrary from a nematode's point of view. Valuable insights into nematode biology, behavior, and management relevant to forage grass production can be obtained by study of these organisms attacking nonforage grasses.

Pasture and rangeland in the USA provide forage for nearly 10^8 cattle and more than 10^7 sheep. The health of forage grasslands is vital to the continued efficient production of animal protein, but because forage land has a low unit value (compared, for example, to residential and recreational turf), relatively little research effort has been expended on forage grass nematodes. Effects of parasitic

 Plant and Nematode Interactions, Agronomy Monograph no. 36.

nematodes on forage grasses may be very subtle and difficult to separate from other problems, such as drought, grazing pressure, and root diseases. Nevertheless, plant-parasitic nematodes clearly have the potential to hamper the establishment of rangeland and pasture grasses. They not only can decrease productivity and yields, but also may affect plant forage quality. A major problem in studying effects of nematodes on grasses is variability observed in pathogenicity. Inherent biological differences in nematode reproductive potential, effects of environmental factors on their biological activity, and effects on growth of plant tissues necessary for population maintenance greatly influence the importance of individual species of these parasites. The usual presence of multiple plant-parasitic species also complicates the characterization of nematode-grass relationships (McGlohon et al., 1961; Santerre & Levesque, 1982; Smolik, 1974; Townshend & Potter, 1973; Willis et al., 1976). Controlled field and greenhouse experiments have shown that significant reductions in the persistence and growth of many grass species can be attributed to these parasites (Griffin, 1992; Griffin & Asay, 1992). Damage potential may be partially related to reproductive potential. Certain nematodes, such as root-knot nematodes (*Meloidogyne* spp.) and cyst nematodes (*Heterodera* spp.), may produce several generations per year, whereas the dagger nematode (*Xiphinema americanum*) only has one generation per year. Host-parasite relationships between nematodes and grasses also are affected by differences in temperature and soil texture. Hence, differences in the reproductive potential of nematodes, and their differential reactions to climatic and edaphic conditions, can greatly affect seasonal and yearly population trends and nematode pathogenicity. Population densities of *Tylenchorhynchus acutoides* were approximately six times those of *X. americanum*, apparently because of differences in fecundity. Although little is known of the biology and fecundity of *T. acutoides*, it appears to have at least two reproductive life cycles per year, compared to only one for *X. americanum* (Griffin & Darling, 1964; Jaffee et al., 1987; Norton, 1963).

SOIL FACTORS AFFECTING NEMATODES

Soil water differences affect the population dynamics of plant-parasitic nematodes on grasses (Griffin et al., 1988). Although little data have been generated on grass root phenology, crested wheatgrass [*Agropyron desertorum* (Fischer ex Link) Schultes] may initiate adventitious roots in early spring and fall when soil water is more optimal or abundant (Thorgeirsson, 1985). Also, root growth is closely related to nematode population increase. The greatest numbers of these parasites occurred on all grass selections during the spring of 1984, when low soil temperatures and adequate soil water favored plant growth (Griffin et al. 1988). Ectoparasitic nematodes were affected more by changes in soil water and temperature than were endoparasitic species (Griffin et al. 1988).

The effects of other soil factors on pathogenicity of nematodes to forage grasses have been investigated extensively (Griffin, 1992, 1995). Growth of crested wheatgrasses was significantly reduced by the ectoparasitic nematodes

Merlinius brevidens, *Tylenchorhynchus acutus*, *X. americanum*, and the root lesion nematode, *Pratylenchus neglectus*, in controlled experimental field and greenhouse studies. The latter was the most damaging and had the greatest rate of reproduction. *Merlinius brevidens* and *T. acutus* were similar in their reproduction on the grasses, while *X. americanum* was the least aggressive and failed to reproduce on any grass. Temperature affected the host-parasite relationships, with the greatest suppression in shoot growth and nematode reproduction occurring at 20 to 25°C. Soil texture also affected nematode-grass relationships. Nematode pathogenicity and reproduction was greatest in Kidman fine sandy loam soil [coarse-loamy mixed mesic Calcic Haploxeroll (84% sand, 8% silt, 8% clay; pH 7.4, 1.0% organic matter)]; *P. neglectus* was less affected by soil texture than were the ectoparasitic nematodes. Reproduction of *P. neglectus* was greatest among all nematode species, whereas *X. americanum* did not increase in any soil texture on any cultivar, and declined by 70% in fine silty soil.

Wallace et al. (1993) studied the population densities of plant-parasitic nematodes in a Minnesota forage grass field dominated by reed canarygrass (*Phalaris arundinacea* L.). Numbers of individual species varied with many soil factors, including soil textures, moisture content, and acidity.

ECOLOGICAL ROLES OF NEMATODES IN FORAGE GRASS ECOSYSTEMS

Total nematode biomass in temperate grasslands typically is low, in one study reaching only 3.8 g fresh weight per surface square meter (Sohlenius, 1980). Despite this small contribution, nematodes may account for a considerable part of total herbivore grazing in range ecosystems. Smolik and Lewis (1982) estimated that the biomass of plant-feeding nematodes on South Dakota rangeland was 20% that of a hypothetical cattle population, and five times the biomass of herbivorous arthropods. Plant-parasitic nematodes consumed 6 to 12% of annual net root production in a little bluestem [*Schizachyrium scoparium* (Michaux) Nash]-western wheatgrass [*Pascopyrum smithii* (Rydb.) Löve] prairie in South Dakota (Ingham & Detling, 1984). Consumption was more pronounced in areas heavily grazed by prairie dogs. In another study at the same site, Ingham and Detling (1990) concluded that root-feeding nematodes reduced aboveground net primary production by 16 times more than could be accounted for by consumption. In a Colorado shortgrass prairie, primarily blue grama [*Bouteloua gracilis* (Willd. ex Kunth) Lagasca ex Griffiths], both plant-parasitic and microbivorous nematodes were most abundant in sandy loam soils (Ingham et al., 1982). Bacterial-feeding nematodes greatly increased the availability of N to *B. gracilis*, compared to the presence of bacteria alone, due to excretion of N compounds by the nematodes (Ingham et al., 1985b). Plant shoots also contained more P in treatments that included nematodes. In other work, nematode parasitism contributed to blue grama die-off by restriction of net primary production, but was not believed to be the sole cause (Stanton et al., 1984).

CURRENT STATUS OF FORAGE GRASS-NEMATODE RELATIONSHIPS

Among the potentially most significant parasites of forage grasses are root-knot nematodes, *Meloidogyne* spp.; lesion nematodes, *Pratylenchus* spp.; and seedhead parasites, *Anguina* spp. It is important to remember, however, that most basic nematode research on forage grasses has been conducted in the greenhouse to determine reproductive capacity. Pot experiments are important for determining the compatibility of nematode and host, but the results may be difficult to apply to the field, where innumerable factors influence nematode population dynamics. Similarly, nematode effects on forage growth and yield usually have been estimated with nematicide trials, which cause many other changes in the soil environment. A list of many of the known nematode-forage grass interactions is presented in Table 20–1. A worldwide bibliography of nematode-forage grass interaction studies from 1961 to 1985 has been published (Stanton & Rizo P., 1989).

Plant-parasitic nematodes likely invade and parasitize all species of forage grasses. However, the study of nematode host ranges and pathogenicity on forage grasses has received only a small fraction of the attention devoted to higher value field and truck crops. The available evidence strongly suggests that these parasites may impose significant constraints on forage grass production. Much basic research still needs to be done on all aspects of the relationships between nematodes and forage grasses. Some extrapolation can be made from the greater knowledge base on nematode pathogenicity to turfgrasses and cereals. These findings strongly suggest that pasture and range grasses are likely to be just as susceptible to nematode parasitism and production losses.

IMPORTANT NEMATODE PARASITES

Root-Knot Nematodes

Species and Races

Root-knot nematodes that cause significant losses in legumes and vegetable crops (*Meloidogyne arenaria*, *M. chitwoodi*, *M. hapla*, *M. incognita*, *M. javanica*) are infrequently reported as parasites of grasses, although all except *M. hapla* may reproduce well on some grasses experimentally (McGlohon et al., 1961). A large number of *Meloidogyne* spp. parasitic on grasses are placed in the "*graminis*" group, including *M. graminicola*, *M. graminis*, *M. marylandi*, and *M. naasi*. Host races of *M. naasi* have been designated (Michell et al., 1973; Ediz & Dickerson, 1976), but their relevance to forage grasses is unknown. Root-knot nematode species are identified by examination of many morphological characters. The most important character is the pattern of striae surrounding the anus and vulva in mature females (perineal pattern).

Symptoms

Root galls are the characteristic belowground symptom of *Meloidogyne* spp. infection. Root-knot nematodes induce the formation of greatly enlarged

Table 20–1. Some reported forage grass/nematode interactions.†

Forage Grass	Nematode	Parameter	Reference
Agropyron cristatum (L.) Gaertner (crested wheatgrass)	*Meloidogyne chitwoodi*	Pathogenicity	Griffin & Asay, 1989
	Pratylenchus neglectus	Pathogenicity	Griffin, 1994
A. desertorum (Fisher ex Link) Schultes (standard crested wheatgrass)	*Meloidogyne chitwoodi*	Reproduction	Griffin et al., 1984
		Pathogenicity	Griffin & Asay, 1989
Agrostis avenacea J.F. Gmelin (blown grass)	*Anguina* sp.	Pathogenicity	McKay & Ophel, 1993
Andropogon gerardii Vitman var. *gerardii* (big bluestem)	*Belonolaimus* sp.	Survey	Kerr & Wysong, 1979
A. virginicus L. (broomsedge)	*Tylenchulus graminis*	Reproduction	Inserra et al., 1988
Arrhenatherum elatius (L.) P. Beauv. ex J.S. Presl. & C. Presl. (tall oatgrass)	*Meloidogyne arenaria, M. incognita, M. javanica*	Reproduction	McGlohon et al., 1961
	Paratylenchus projectus	Reproduction	Townshend & Davidson, 1989
Bouteloua curtipendula (Michaux) Torrey (sideoats grama)	*Tylenchorhynchus claytoni*	Pathogenicity	Ingham & Detling, 1986
B. gracilis (Willd. ex Kunth) Lagasca ex Griffiths (blue grama)	*Belonolaimus* sp.	Survey	Kerr & Wysong, 1979
	Mixed species	Ecology	Ingham et al., 1982
	Mixed species	Food webs	Ingham et al., 1985b
	Mixed species	Vertical distribution	Ingham et al., 1985a
	Mixed species	Pathogenicity	Stanton et al., 1984
Bromus inermis Leysser (smooth bromegrass)	*Meloidogyne arenaria, M. incognita, M. javanica*	Reproduction	McGlohon et al., 1961
	M. chitwoodi	Reproduction	Griffin et al., 1984
	M. microtyla	Reproduction	Townshend et al., 1984b
	Paratylenchus projectus	Reproduction	Townshend & Davidson, 1989
	Pratylenchus penetrans	Root populations	Petersen et al., 1991
Buchloe dactyloides (Nutt.) Engelm. (buffalograss)	*Tylenchorhynchus robustoides*	Pathogenicity	Smolik, 1982
Chloris gayana Kunth (Rhodes grass)	*Belonolaimus longicaudatus*	Resistance	Boyd & Perry, 1969
	Meloidogyne javanica	Resistance	York, 1989, 1990
C. pectinata Benth	*Belonolaimus longicaudatus*	Resistance	Boyd & Perry, 1969
Cynodon dactylon (L.) Pers. var. *dactylon* (bermuda grass)	*Meloidogyne arenaria, M. incognita*	Reproduction	Windham & Brink, 1991

(continued on next page)

Table 20–1. Continued.

Forage Grass	Nematode	Parameter	Reference
Dactylis glomerata L. (orchardgrass)	*Helicotylenchus digonicus*	Population dynamics	Townshend & Potter, 1973
	Hoplolaimus galeatus	Reproduction	McGlohon et al., 1961
	Meloidogyne arenaria, M. incognita, M. javanica	Reproduction	McGlohon et al., 1961
	M. chitwoodi	Reproduction	Griffin et al., 1984
	M. graminis	Reproduction	Williams & Laughlin, 1968
	M. microtyla	Reproduction	Townshend et al., 1984b
	M. naasi	Reproduction	Michell et al., 1973
	Paratylenchus projectus	Reproduction	Townshend & Davidson, 1989
	P. penetrans	Root populations	Petersen et al., 1991
	Tylenchorhynchus agri	Reproduction	Coates-Beckford & Malek, 1978
Elytrigia intermedia (Host) Nevski subsp. *intermedia* (intermediate wheatgrass)	*Meloidogyne chitwoodi*	Reproduction	Griffin et al., 1984
Eremochloa ophiuroides (Munro) Hackel (centipede grass)	*Tylenchulus graminis*	Reproduction	Inserra et al., 1988
Festuca arundinacea Schreber (tall fescue)	*Helicotylenchus dihystera*	Survey	Pedersen et al., 1988
	H. pseudorobustus	Reproduction	Kimmons et al., 1990
	Hoplolaimus galeatus	Reproduction	McGlohon et al., 1961‡
	Meloidogyne chitwoodi	Reproduction	Griffin et al., 1984‡
	M. marylandi	Population dynamics	Gwinn & Bernard, 1993
	M. naasi	Growth, reproduction	Person-Dedryver Fischer, 1987‡
	Mixed species	Pathogenicity	Hoveland et al., 1975‡
	Paratrichodorus minor	Survey	West et al., 1988
	Paratylenchus projectus	Survey	Bernard et al., 1993 (unpubl.)
		Survey	West et al., 1988
	Pratylenchus penetrans	Reproduction	Townshend et al., 1984a‡
	P. scribneri	Population dynamics	Gwinn & Bernard, 1993
		Reproduction	Kimmons et al., 1990
		Reproduction	Minton, 1965‡
	Tylenchorhynchus acutus	Survey	West et al., 1988
	T. hopperi	Survey	West et al., 1988
	T. maximus	Survey	Bernard et al., 1993 (unpubl.)

F. pratensis Hudson (meadow fescue)	*Meloidogyne naasi*	Reproduction	Michell et al., 1973
	Pratylenchus penetrans	Root populations	Petersen et al., 1991
F. rubra L. subsp. *rubra* (red fescue)	*Meloidogyne marylandi*	Population dynamics	Gwinn & Bernard, 1993
Leymus cinereus (Scribner & Merr.) A. Löve (Great Basin wild ryegrass)	*M. chitwoodi*	Reproduction	Griffin et al., 1984
Lolium multiflorum Lam. (annual or Italian ryegrass)	*Meloidogyne arenaria, M. incognita, M. javanica*	Reproduction	McGlohon et al., 1961
	M. microtyla	Reproduction	Townshend et al., 1984b
	M. naasi	Reproduction	Michell et al., 1973
		Growth, reproduction	Person-Dedryver & Fischer, 1987
	Paratylenchus projectus	Reproduction	Townshend & Davidson, 1989
L. perenne L. (perennial ryegrass)	*Meloidogyne marylandi*	Population dynamics	Gwinn & Bernard, 1993
	M. naasi	Reproduction	Cook et al., 1991
		Reproduction	Michell et al., 1973
		Growth, reproduction	Person-Dedryver & Fischer, 1987
	Amplimerlinius icarus, A. macrurus, Merlinius brevidens, M. nothus, Tylenchorhynchus dubius, T. lamelliferus, T. maximus	Feeding behavior	Bridge & Hague, 1974
	A. icarus, T. maximus	Pathogenicity	Bridge & Hague, 1974
L. rigidum Gaudin (Wimmera ryegrass)	*Anguina funesta*	Pathogenicity	McKay & Ophel, 1993
Pascopyrum smithii (Rydb.) Löve (western wheatgrass)	*Anguina agropyronifloris*	Host range	Norton, 1965; Collins, 1966
	Meloidogyne chitwoodi	Reproduction	Griffin et al., 1984
	Mixed species	Ecology	Ingham & Detling, 1984
	Tylenchorhynchus robustoides	Pathogenicity	Smolik, 1982
Paspalum notatum Flugge (bahiagrass)	*Meloidogyne arenaria*	Rotation for control	Rodríguez-Kábana et al., 1991
Phalaris arundinacea L. (reed canarygrass)	*Meloidogyne microtyla*	Reproduction	Townshend et al., 1984b
	Mixed species	Ecology	Wallace et al., 1993
	Mixed species	Pathogenicity	Hoveland et al., 1975
	Paratylenchus projectus	Reproduction	Townshend & Davidson, 1989
	Pratylenchus penetrans	Root populations	Petersen et al., 1991
	Tylenchorhynchus agri	Reproduction	Coates-Beckford & Malek, 1978

(continued on next page)

Table 20–1. Continued.

Forage Grass	Nematode	Parameter	Reference
Phleum pratense L. (timothy)	*Helicotylenchus digonicus*	Population dynamics	Townshend & Potter, 1973
	Meloidogyne microtyla	Reproduction	Townshend & Potter, 1973
	Pratylenchus projectus	Reproduction	Townshend & Potter, 1973
		Reproduction	Townshend & Davidson, 1989
	Pratylenchus crenatus	Reproduction, survival	Kimpinski & Willis, 1981
	P. neglectus	Reproduction	Townshend & Potter, 1973
	P. penetrans	Reproduction, survival	Kimpinski & Willis, 1981
	Tylenchorhynchus agri	Reproduction	Coates-Beckford & Malek, 1978
Polypogon monspeliensis (L.) Desf. (annual beardgrass)	*Anguina* sp.	Pathogenicity	McKay & Ophel, 1993
Pseudoroegneria spicata (Pursh) A. Löve (bluebunch wheatgrass)	*Meloidogyne chitwoodi*	Reproduction	Griffin et al., 1984
Schizachyrium rhizomatum (Swallen) Gould	*Tylenchulus graminis*	Reproduction	Stokes, 1969
S. scoparium (Michaux) Nash (little bluestem)	Mixed species	Ecology	Ingham & Detling, 1984
	Mixed species	Pathogenicity, control	Ingham & Detling, 1990

† Grass list is composed of many of the species grown for forage in the USA and/or Canada. Nematode reports and references are included for non-North American species that appear to pose a potential threat to forage grass production in North America.

‡ Endophyte status not reported for tall fescues used in these studies.

vascular cells (giant cells) that serve as "nurse cells" for the nematode. In addition, cortical cells surrounding the feeding site usually multiply to form the swollen gall. In many grass-parasitic root-knot nematodes, little or no cortical expansion occurs, and thus galling is negligible (Griffin et al., 1984). Detection of nematode infection by root examination is more difficult without the presence of galls (Fig. 20–1). In the *M. graminis* group, females and their egg masses may protrude enough from the roots to give a gall-like appearance (Figs. 20–2, 20–3). Root-staining techniques are useful for detecting nematodes inside roots (Byrd et al., 1983). Aboveground symptoms, such as chlorosis and stunting, are not nematode-specific. Analysis of soil and root samples is a necessity for accurate determination of nematode problems.

Biology

Root-knot nematodes are the most studied of all plant-parasitic nematodes, and the basic life cycle is well understood. The life cycle of these pathogens on grasses is identical to that of the better-studied species (Taylor & Sasser, 1978). Infective second-stage juveniles migrate through soil to roots and enter them near their tips (Fig. 20–1), where they promote the growth of the giant cells in the vascular tissue. Typically, a female induces development of two to eight giant cells (Fig. 20–4), which are maintained for the life of the nematode. If the nematode dies, the giant cells it has maintained also die. Selection of nematode-resistant cultivars often is based on the inability of root-knot nematodes to establish giant cells.

After a feeding site is formed, the juvenile nematode enlarges as it feeds and molts. Mature females produce a gelatinous secretion from rectal glands, into which they deposit up to 2000 eggs (Figs. 20–2, 20–5). Most economically important root-knot nematodes reproduce parthenogenetically, and males are rare. Species parasitic on herbaceous crops usually are capable of several generations per year.

The barley root-knot nematode, *M. naasi*, a major cereals pest in Europe, has been reported from scattered localities across the USA, and has been implicated in poor establishment of *Lolium* spp. in Oregon (Cook & Yeates, 1993). This nematode has a wide host range in the Poaceae as well as in several dicotyledonous families (Michell et al., 1973). In French studies, *M. naasi* reproduced better on Italian ryegrass (*Lolium multiflorum* Lam.) than on perennial ryegrass (*L. perenne* L.), and reproduced poorly on tall fescue (*Festuca arundinacea* Schreber) and orchardgrass (*Dactylis glomerata* L.) (Person-Dedryver & Fischer, 1987). Big bluestem (*Andropogon gerardii* Vit.) and buffalograss [*Buchloe dactyloides* (Nutt.) Engelm.] also were poor hosts (Ediz & Dickerson, 1976). Nematicide tests have provided evidence that *M. naasi* may restrict plant productivity up to 22% (Cook & Yeates, 1993).

Meloidogyne microtyla, a recently described Canadian species (Mulvey et al., 1975), parasitizes a wide range of grasses (Townshend et al., 1984b), among them some forage species. At present, this nematode appears to be of concern only on turfgrasses and golf courses.

The Columbia root-knot nematode (*M. chitwoodi*) is a potentially important pathogen of forage grasses, although its distribution in range and pasture is not well known. Crested wheatgrass, western wheatgrass, smooth bromegrass (*Bromus inermis* Leysser), and orchardgrass were all found to be good hosts of *M. chitwoodi*, but bluebunch wheatgrass [*Pseudoroegneria spicata* (Pursh) Löve], Great Basin wild ryegrass [*Leymus cinereus* (Scribner & Merr.) A. Love], and intermediate wheatgrass [*Elytrigia intermedia* (Host) Nevski subsp. *intermedia*] were poor hosts or nonhosts (Griffin et al., 1984). Three crested wheatgrass

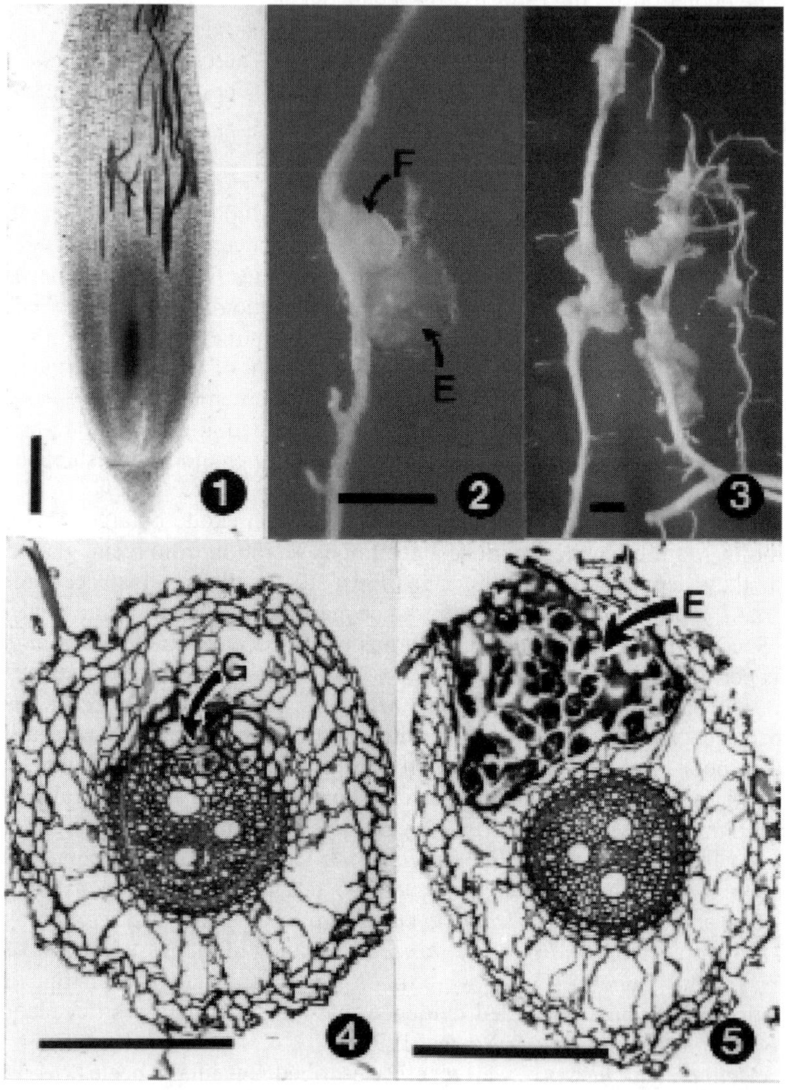

Figs. 20–1 to 20–5. A root-knot nematode, *Meloidogyne marylandi*, on roots of tall fescue (*Festuca arundinacea* Schreber): 1) Infective juveniles near root tip (scale bar = 0.5 mm). 2) Female (F) and egg mass (E) on root (note lack of root swelling) (scale bar = 0.5 mm). 3) Multiple infection of root segments (scale bar = 0.5 mm). 4) Cross-section of infected root with giant cells (G) and disruption of endodermis (scale bar = 0.5 mm). 5) Cross-section of root with partially embedded egg mass (E)(scale bar = 0.5 mm).

cultivars were all susceptible to *M. chitwoodi*, but the three geographically distinct nematode isolates used exhibited some variability in virulence and reproduction (Griffin & Asay, 1989).

Meloidogyne graminis and *M. marylandi* are similar species with overlapping host ranges (Jepson, 1987; Williams & Laughlin, 1968) and wide distributions in the eastern USA, and *M. graminicola* occurs in several southeastern states. They may experimentally suppress the growth of turfgrasses, but have not been implicated in field losses.

Disease Complexes

Root-knot nematodes are components of disease complexes, especially wilts, that attack dicotyledonous crops (Mai & Abawi, 1987), but are not known to participate similarly in forage-grass diseases.

Economic Importance

The economic importance of root-knot nematodes on forage grasses has not been adequately documented. *Meloidogyne chitwoodi* inhibited growth of crested wheatgrass in a growth chamber experiment, suggesting that the pathogen damage potential could be more severe in the more stressful natural environment (Griffin & Asay, 1989). Since some root-knot nematodes are significant cereal pathogens [e.g., *M. naasi* on barley (*Hordeum vulgare* L.) and *M. chitwoodi* on wheat (*Triticum aestivum* L. emend. Thell.)], there is no reason to doubt that forage grasses could be more generally affected than is currently recognized.

Lesion Nematodes

Species and Races

Numerous species are recognized in the genus *Pratylenchus*. Many are important species on cereals, but fewer species have been recorded from forage grasses. Those most commonly reported are *P. crenatus*, *P. neglectus*, *P. penetrans*, and *P. scribneri*. Lesion nematodes are similar in appearance and require careful microscopic observation for separation of species.

Symptoms

Lesion nematodes are migratory endoparasites of plant roots. They live in the cortex of their hosts, feeding and moving from cell to cell. This activity often leads to secondary invasion of roots by other soilborne pathogens, sometimes resulting in root necrosis. However, root necrosis is not a definitive indication of the presence of lesion nematodes. Soil and root analysis is necessary to determine the presence of these parasites.

Biology

Reports of lesion nematodes on forage grasses in North America have largely come from surveys, but *P. neglectus*, a widespread species, may suppress shoot and root growth of crested wheatgrass (Griffin, 1994). Another common species, *P. penetrans*, may have little or no effect on grass growth. Petersen et al. (1991) found that *P. penetrans* was unable to maintain itself on any of 12 forage grass species, but Kimpinski and Willis (1981) reported that *P. penetrans* and *P. crenatus* reproduced well on timothy (*Phleum pratense* L.), especially as soil temperature increased.

Disease Complexes

Like root-knot nematodes, lesion nematodes may be involved in wilt disease complexes on dicotyledonous crops, and also may facilitate the entry of root-rot fungi (*Fusarium* spp.) (Powelson & Rowe, 1993; Ronacadori et al., 1974). On wheat, *P. neglectus* and various root-rot fungi in combination caused an increase in root lesions (Taheri et al., 1994). On maize [*Zea mays* (L.) subsp. *mays*], the presence of *P. hexincisus* in roots increased the severity of anthracnose leaf blight (*Colletotrichum graminicola*) (Nicholson et al., 1985), and two *Pratylenchus* spp. on maize increased the severity of *Fusarium* root rot (Jordaan et al., 1987). Even though disease complexes involving lesion nematodes have not been documented on forage grasses, additional research likely will reveal examples on pasture and range grasses.

Economic Importance

Experimental evidence suggests that lesion nematodes can suppress the growth, and therefore the forage quantity, of some grasses (Griffin, 1994). Further research is needed to firmly establish lesion nematodes as serious pests of forage grasses.

Stunt Nematodes

Species and Races

Stunt nematodes, a large complex of several genera, are common in grass rhizospheres. *Merlinius* spp. and *Tylenchorhynchus* spp. are the most common stunt nematode genera parasitizing grasses. There are numerous species in each genus. Races have not been recognized in any of the genera.

Symptoms

Stunt nematodes generally feed ectoparasitically on roots and readily move among roots. Some species on perennial ryegrass may be sedentary ectoparasites or semi-endoparasites (Bridge & Hague, 1974). The common name is largely a misnomer, since most species cause little noticeable damage to plants. However, some species on forage grasses may be more significant. For example, *T. maximus* damaged root tips and limited main root growth, but stimulated lateral root growth (Bridge & Hague, 1974).

Biology

In a growth chamber experiment, *T. claytoni* significantly restricted root biomass, shoot biomass, and tiller production of sideoats grama [*Bouteloua cur-*

tipendula (Michaux) Torrey], but net photosynthesis and transpiration were not significantly affected (Ingham & Detling, 1986). Mechanical defoliation limited biomass and tiller production as well, but also suppressed photosynthesis and transpiration. *Tylenchorhynchus robustoides*, a very common species on prairie (Smolik & Lewis, 1982), limited growth of western wheatgrass and buffalo grass, especially at higher temperatures (Smolik, 1982). This parasite also inhibited foliar production of these grasses up to 64%. Many other species recorded from forage grass ecosystems, including *T. maximus* and *T. nudus*, need to be studied more intensively.

Disease Complexes

Interactions of stunt nematodes and other pathogens on forage grasses have not been reported. However, *M. brevidens* and *Olpidium brassicae*, a root-parasitic fungus, may together severely restrict wheat yields (Anderson & Potter, 1991).

Economic Importance

The impact of stunt nematodes is largely unexplored. Some rangeland grasses appear to be damaged by *Tylenchorhynchus* spp. (Ingham & Detling, 1984; Smolik, 1982), but actual impact on net production cannot yet be estimated.

Sting Nematodes

Species and Races

Sting nematodes comprise a small, closely related group of three or four species in the USA, with another small group of species in South America and Australia (Smart & Nguyen, 1991). The most significant species is *Belonolaimus longicaudatus*. This nematode is a major constraint on the production of many crops grown in the southeastern USA Coastal Plain, and is a serious parasite of grasses in Florida and California (Mundo-Ocampo et al., 1994; Smart & Nguyen, 1991). Another species of uncertain identity occurs in sandy loam soils of Kansas and Nebraska (Dickerson et al., 1972; Kerr & Wysong, 1979), but is probably more widespread than current records indicate.

Symptoms

Symptoms of sting nematodes on grasses include an iron-deficiency chlorosis of the foliage, stunting, and failure of stolons to root successfully at nodes (Boyd & Perry, 1969). Plant death often follows, and the affected area has the appearance of circular or irregular dead patches, with a narrow border of chlorotic plants (Fig 20–6). Sting nematodes feed primarily on root tips, destroying the meristem and preventing further root growth. As adventitious roots grow from the blind root, their root tips are attacked in turn. This type of feeding results in the symptom termed "stubby-root" and accounts for the high degree of damage usually associated with *B. longicaudatus*.

Fig. 20–6. Sting nematode (*Belonolaimus longicaudatus*) damage to bermudagrass [*Cynodon dactylon* (L.) Pers. var. *dactylon*] turf (courtesy of M. H. Newman).

Biology

Sting nematodes are highly destructive to many turfgrasses. The damage threshold for *B. longicaudatus* on Florida turfgrasses has been estimated at only 10 nematodes/100 cm^3 soil (Dunn, 1987). Some 120 forage grass introductions to Florida were susceptible to *B. longicaudatus* (Boyd & Perry, 1969). In Nebraska, *Belonolaimus* sp. was found in the root zones of big bluestem and blue grama growing in a native prairie (Kerr & Wysong, 1979). This same species severely stunts corn in Kansas (Todd, 1989). In expanded testing of pangola grass [*Digitaria decumbens* (Stent)] and Rhodes grass (*Chloris gayana* Kunth) introductions, many highly resistant plants were identified (Boyd & Perry, 1969). The use of resistant selections is recommended as the most practical approach to sting nematode management (Boyd & Perry, 1969).

Disease Complexes

Sting nematodes interact with wilt fungi (*Fusarium* spp.) to make wilt symptoms more severe on dicotyledons, and may break resistance in *Fusarium* wilt-resistant cotton (*Gossypium hirsutum* L.) (Cooper & Brodie, 1962; Holdeman & Graham, 1954). Sting nematodes also increase the severity of *Pythium* root rot on dicotyledons (Johnson & Littrell, 1970). They are not known to interact with root-rot pathogens of grasses.

Economic Importance

Sting nematodes should be considered a significant threat to plant health wherever they and a host of value coincide. Grasses of many species are highly susceptible to root damage. Pastures or other forage land with symptoms of dieback or expanding areas of dead plants should be assayed for sting nematodes.

Floral and Leaf Gall Nematodes

Species and Races

Species of *Anguina* are obligate parasites in leaves or florets, causing galls on affected parts. Krall (1991) has reviewed some of the more important species. In North America, the most important species is *A. agrostis*, the bentgrass nematode. This species limits seed production on bentgrass (*Agrostis tenuis* Sibth.) and several other *Agrostis* spp. (Courtney & Howell, 1952; Pinkerton & Alderman, 1994). Norton (1965) described *A. agropyronifloris* from western wheatgrass, apparently the same nematode species noted by Collins (1966) on western wheatgrass in Montana. The status of *A. agropyronifloris* is unsettled. The species has been synonymized with *A. agrostis* (see Krall, 1991), but morphological differences exist (Norton, 1965), and Norton was unable to demonstrate parasitism of western wheatgrass by *A. agrostis* from bentgrass. Host preference variability in *A. agrostis* suggests the presence of host races or a species complex (Southey, 1973). Several other *Anguina* spp. are known from forage grasses, but there is no information regarding their impact on natural reseeding or seed harvest [see Griffin (1984) for review]. *Anguina graminis* is a parasite of several fescues (Griffin, 1984). Continued taxonomic confusion regarding species delimitations hampers study of *Anguina*/grass relationships, but application of molecular diagnostic techniques could help resolve this problem (Cenis, 1993).

Symptoms

Anguina spp. on grasses usually reproduce within the developing florets of grasses, although leaves and stems also may be parasitized. Their activity converts the developing seed into a hollow, brown or black cockle in which nematode eggs and juveniles are contained (Bird et al., 1980; Courtney & Howell, 1952; McKay et al., 1993).

A potentially far more serious problem is the association of some *Anguina* spp. with bacteria, especially *Clavibacter toxicus* (= *Corynebacterium rathayi*), extensively reviewed by McKay and Ophel (1993). This bacterium produces corynetoxin, which induces convulsions in grazing mammals. Mortality rates can be very high, especially for sheep; many thousands of sheep and cattle have died from corynetoxin poisoning since 1970 (McKay & Ophel, 1993; McKay et al., 1993). The problem is most frequently seen in annual Wimmera ryegrass (*Lolium rigidum* Gaudin) fields in Australia and is termed annual ryegrass toxicity. The same disease complex also may be serious in valley paddocks of blown grass, *Agrostis avenacea* C. Gmelin, and annual beardgrass, *Polypogon monspeliensis* (L.) Desf.; on these two grasses the syndrome is termed "flood plain staggers."

Biology

Anguina spp. are disseminated by sowing cockles along with healthy seed. Juvenile nematodes emerging from the cockles locate new plants to continue the life cycle (Courtney & Howell, 1952). Typically, *Anguina* spp. can be controlled by fallowing land for one growing season, since the juveniles cannot survive that long without invading a host. Effective methods exist for separation of cockles from viable seed by flotation in a brine solution, or by other mechanical methods (Christie, 1959).

Disease Complexes

The association of *Anguina* spp. and bacteria, especially *Clavibacter* spp., is widespread in Australia, but also occurs in the USA. In Australia, *Anguina funesta* is the principal vector of *C. toxicus* on *L. rigidum*, while several other species appear to be vectors on the other grasses. The nematodes carry *C. toxicus* on their cuticles as they migrate to new plants. This relationship is not mutualistic, since the bacterium can degrade the nematode cuticle (Bird & Riddle, 1984). The toxin is produced by the bacterium inside the developing seed, where the nematode reproduces. Numerous approaches are currently being studied for successful management of this disease complex. The most important approach is desiccation of the seedheads before anthesis with herbicides (McKay et al., 1982). However, annual ryegrass is becoming increasingly resistant to herbicides used to promote desiccation; therefore, annual ryegrass toxicity incidence is likely to increase as well (Riley & Gill, 1994).

Several additional *Anguina* spp. are associated with *Clavibacter* spp. [= *Corynebacterium* spp. on grasses (Davis et al., 1984)] (McKay & Ophel, 1993). Corynetoxin-like symptoms in grazing animals have been reported from Oregon (Galloway, 1961). Other associations of bacteria with *Anguina* spp. have been summarized (Riley & McKay, 1990). In a survey of bacteria within *A. agrostis*-induced galls on bentgrass, Wen and Viglierchio (1992) found six genera of bacteria, none of which were identical to *C. toxicus*.

Economic Importance

Cockle formation, with a corresponding loss of 3 to 9% of seed (Alderman, 1988, 1991), obviously is of most importance in seed fields of susceptible grasses. This problem currently exists only in seed production for lawn and recreational turf. Increased interest in native range grasses probably will lead to more seed-production fields, and with this increase may come greater infestations of seed-limiting species of *Anguina*.

Annual ryegrass toxicity and floodplain staggers have been devastating agents of livestock mortality in certain areas of Australia (McKay & Ophel, 1993; McKay et al., 1994). The distribution of *Anguina* spp. in North America is imperfectly known, but the frequent association of *Anguina* spp. with *Clavibacter* spp. must be considered a major potential problem for livestock production in the USA, even though such a problem has not yet appeared in epidemic proportions. Research on *Anguina-Clavibacter* relationships in the USA is warranted by the extreme toxicity of corynetoxin and the widespread occurrence of *Anguina* spp. We must know definitively how widespread this relationship is in North America, and whether it can develop on our important forage and range species.

Other Nematodes of Potential Significance

A pin nematode (*Paratylenchus projectus*) parasitizes many turf and forage grasses (Townshend & Davidson, 1989). In a field survey in Ontario, *P. projectus* reproduced well on timothy cultivars, but not on orchard grass (Townshend & Potter, 1973). Pin nematodes may increase to enormous numbers in the rhizospheres of host plants, but seldom are pathogenic.

Tylenchulus graminis is a sedentary endoparasite of broomsedge (*Andropogon virginicus* L.), centipede grass [*Eremochloa ophiuroides* (Munro) Hackel], and *Schizachyrium rhizomatus* (Swallen) Gould (Inserra et al., 1988; Stokes, 1969). A closely related nematode (*T. semipenetrans*) is a major pathogen of citrus (*Citrus* spp.). Although *T. graminis* is not considered a pest of any economically significant grass, its close taxonomic relationship to the citrus nematode (*Tylenchulus semipenetrans*) suggests that further host range studies are needed.

Numerous additional nematode species have been reported as parasites of turfgrasses and cereals (Griffin, 1984), and some perhaps should be considered as potential problems on forage grasses. Among this group are: ring nematodes (*Criconemella* spp.), spiral nematodes (*Helicotylenchus* spp.), and cyst nematodes (*Heterodera* spp.). However, little or no data are available on the direct effects of most nematodes on forage grasses. Surveys and research on turf and recreational grasses have demonstrated that some nematode species may increase to very high numbers on the roots of their grass hosts, but often without much effect on foliar growth. This situation may be due to heavy watering and fertilization regimes used on recreational and residential turf. The normal habit of forage grasses may actually involve suboptimal performance because of the pressure of unfavorable growing conditions, pests, and pathogens, including nematodes. Certainly, additional pathogenicity studies are indicated for this very important component of agricultural production.

MANAGEMENT OF PLANT-PARASITIC NEMATODES ON FORAGE GRASSES

The immense area and low unit value of pastoral lands make intensive nematode control strategies unsuitable and economically nonviable. Therefore, except for seed production fields, approaches that use chemical nematicides or require intensive mechanical input usually are not realistic options (Brown, 1987). When economically feasible, nematicides are as effective on new grasslands as they are for other crops. However, increasingly severe restrictions on the use of both fumigant and nonfumigant nematicides, as well as their high cost of purchase and application, make their use very unlikely. Crop rotations may be useful in certain high-value situations. For example, bahiagrass (*Paspalum notatum* Flugge)—when grown in rotation with peanut (*Arachis hypogaea* L.)—can greatly reduce numbers of *M. arenaria* (Rodríguez-Kábana et al., 1991). Such use of a forage grass in tandem with a high-value field crop obviously limits its use as a long-term pasture.

The most feasible approach to most nematode problems on forage grasses is the use of resistant or tolerant selections. Nematode tolerance has rarely been reported in forage grasses. Still, 'Hycrest' wheatgrass (*Agropyron cristatum* 'Fairway' × *A. desertorum* 'Nordan') is more tolerant of nematode feeding than are its parents (Griffin, 1995). Many native forage grasses on rangeland have not been examined for resistance, but there is some evidence that good sources for resistance exist. The search for nematode resistance in forage grasses probably is limited by the perceived lack of symptoms. When forage grass selections or introductions are screened for resistance to individual nematode species, an abundance of highly resistant or immune plants may be identified. For instance, in one screening 30% of western wheatgrass plants and 86% of Great Basin wild ryegrass [*Leymus cinereus* (Scribn. & Merr.) A. Löve] plants appeared to be resistant to *M. chitwoodi* (Griffin et al., 1984). Many introductions of Rhodes grass and pangola grass are resistant to *B. longicaudatus* (Boyd & Perry, 1969), and several Rhodes grass accessions from southern Africa are resistant to *M. javanica* (York, 1989, 1990). Reproduction of a lance nematode (*Hoplolaimus galeatus*) was greatly limited on a tall fescue genotype with roots of small diameter, compared to a genotype with roots of a larger diameter (Rodríguez-Kábana et al., 1978; Williams et al., 1983). Yu et al. (1990) succeeded in transferring genes for resistance to *M. naasi* from *Aegilops variabilis* to wheat, suggesting that intergeneric gene manipulations may be possible among other Poaceae. As the maintenance and improvement of rangelands become an issue of higher priority, the need for enlargement and increased screening of germplasm collections will become more acute.

A seemingly very different approach to nematode resistance is exhibited by some species of Poaceae, especially in the genera *Festuca* (fescues) and *Lolium* (ryegrasses), which harbor symbiotic, endophytic fungi that confer multiple resistance to insects, pathogens, and drought. The potential importance of this symbiosis is significant, since there is experimental evidence that endophytic fungi can be induced to colonize nonendophytic species, such as wheat. Because nematode-endophyte-grass interrelationships are a recently recognized, but potentially very important, phenomenon they are detailed extensively in the next section.

FORAGE GRASSES WITH ENDOPHYTIC FUNGI

Many grasses house symptomless fungal endophytes of the genus *Neotyphodium* (White et al., 1993). *Neotyphodium* endophytes and several genera of fungi in the ascomycetous family Clavicipitaceae [e.g., *Epichloe typhina* (causal agents of "choke" disease), *Claviceps* spp. (the ergot fungi), and *Balansia* spp.] are closely related (Clay, 1989). Two endophytes are particularly well studied because of their association with toxicoses in grazing animals: *Neotyphodium coenophialum* on tall fescue and *N. lolii* on perennial ryegrass. Tall fescue (primarily 'Kentucky-31') is planted on 15 million hectares in the USA. Perennial ryegrass is a primary forage species in New Zealand.

Endophyte-infected (E+) grasses are nutritionally comparable to their endophyte-free (E–) counterparts, but animals raised in E+ pastures typically have poor performance and exhibit fescue toxicosis, a disease syndrome that includes reductions in weight gain, reproductive capacity, and lactation (Bacon et al., 1986; Stuedemann, 1988). In the USA, fescue toxicosis causes $600 million in cattle production losses annually (Fribourg et al., 1991). Horse and sheep production also is affected (Stuedemann, 1988). In New Zealand, endophyte infection of perennial ryegrass induces a serious neuromuscular disease of sheep, ryegrass staggers. Animals with staggers exhibit muscle immobilization, which is amplified by stress (Fletcher, 1993). Interestingly, reed canarygrass produces alkaloids and other toxins without the aid of endophytes; ingestion of high-alkaloid canarygrass causes symptoms in cattle similar to those from eating endophyte-infected grasses (Hovin & Buckner, 1983).

Endophytes were successfully cultured from perennial ryegrass and tall fescue in the early 1940s (White et al., 1993), but otherwise went virtually unstudied until Bacon et al. (1977) isolated an endophytic fungus, *Epichloe typhina*, from tall fescue plants found in pastures where grazing animals were exhibiting symptoms of fescue toxicosis. Morgan-Jones and Gams (1982) reclassified the endophyte as *A. coenophialum*. A ryegrass endophyte was first observed by Sampson (1933), but was not named until 1984 when Latch et al. (1984) named the endophyte *Acremonium loliae*, corrected to *A. lolii* (Latch et al., 1987). These fungi are true symbionts, endophytes spend their entire life cycle within the host and do not rely on sporulation for dissemination. Fungal hyphae have been found in all plant parts except the roots and leaf blades (Hinton & Bacon, 1985), and the only means of dissemination is via seed (Siegel et al., 1987; Welty et al., 1987). These endophytes are now placed in *Neotyphodium* (Glenn et al., 1996).

Since ingestion of endophyte-infected grasses results in animal toxicoses, removal of endophytes was assumed to be a viable solution for elimination of toxicoses. However, endophyte-free cultivars of tall fescue often produce sparse, unthrifty stands; in years of severe drought, endophyte-free stands typically are lost to drought stress. Ryegrass pastures with a low percentage of *N. lolii* infestation reverted within 2 to 3 yr to a high endophytic status due to selective grazing on *N. lolii*-free plants (Prestidge & Ball, 1993). In New Zealand, the impact of *N. lolii* infection on perennial ryegrass is so great that yield differences imparted by the presence or absence of the endophyte have often been mistaken as improved genetic performance in new perennial ryegrass cultivars (Easton, 1984). Consequently, many researchers have begun to examine the interactions between grasses and their endophytes.

Tall fescue and its endophyte coexist in a mutualistic relationship (Bacon et al., 1986; Gwinn et al., 1992; Stuedemann, 1988). The fungus is protected, disseminated, and nourished by the plant. Endophyte-infected (E+) plants are more resistant to drought, pathogens, and herbivores than are endophyte-free (E–) plants (Arachevaleta et al., 1989; Bernard et al., 1990; Blank et al., 1993; Clay & Cheplick, 1989; Gwinn et al., 1991; Johnson et al., 1985; West et al., 1993). Despite the fundamental effects of endophytes on growth and survival of fescues and ryegrasses, some refereed publications still appear in which the endophyte status of the host is not identified.

What is the significance of endophyte infection to nematode management? The use of endophytes to confer resistance to pathogens, including nematodes, is an exciting possibility. If endophytic fungi can be selected or engineered to eliminate deleterious effects on mammals, but still retain their beneficial characteristics, it should be possible to insert them into nonendophytic grasses. Already, endophytic fungi have been successfully transferred to wheat (Marshall et al., 1993). Many fundamental questions remain, however, regarding the ability of nematode species to parasitize endophyte-infected grasses, and the ability of the endophyte-grass symbiosis to affect nematode populations. Extensive study of nematodes on *F. arundinacea* and *L. perenne* provides an indication of the questions that must be answered in this complicated system.

Information concerning the relationship of plant-parasitic nematodes to tall fescue is confused and contradictory before 1987, because the endophyte status of fescue was not recognized as a potential factor in the nematode-plant reaction. Several ectoparasitic nematodes can be successfully reproducing parasites or pathogens of *F. arundinacea* (Hoveland et al., 1975; McGlohon et al., 1961; Pedersen & Rodríguez-Kábana, 1984, 1985), but endophyte status rarely has been known. Some recent experiments and surveys have begun to clarify the relationship of nematodes and the endophyte-fescue symbiosis. Population densities of *Helicotylenchus dihystera* and *Paratrichodorus minor* were reduced in pastures with high endophyte infestation levels (Pedersen et al., 1988; West et al., 1988). In greenhouse experiments (Kimmons et al., 1990), *Helicotylenchus pseudorobustus* population densities declined slightly on E+ fescue. Numbers of *T. acutus* under E+ fescue were only 25% of those under E− fescue (West et al., 1988). On the other hand, some have questioned whether *Neotyphodium* endophytes had any real effects on nematodes in turf soils, considering that growing conditions and sampling methods were more important determinants of nematode densities (Halisky & Myers, 1989). Considerable variation may exist in the endophyte-grass symbiosis. Nevertheless, it is now generally recognized that E+ fescue can suppress several nematode species. *Paratylenchus* sp. (pin nematodes) numbers were reduced in endophyte-infested plots in Missouri (O'Day et al., 1990). In a replicated field experiment in Tennessee, numbers of *P. projectus* were reduced to nondetectable levels in E+ fescue but increased in E− plots, whereas *T. maximus* was unaffected by endophyte status of the grass (E. C. Bernard et al., 1993, unpublished data). In an Alabama experiment, population densities of *P. projectus* were greatly decreased in E+ fescue plots, but *Tylenchorhynchus ewingi* was not affected (West et al., 1988) . Nematode species, especially ectoparasites, appear to have differential sensitivities to endophyte-infested fescue.

Migratory endoparasitic nematodes often have been reported as parasites of tall fescue. In greenhouse pot experiments, *H. galeatus* was highly fecund on Kentucky 31 tall fescue (McGlohon et al., 1961, Rodríguez-Kábana et al., 1978), and was common in Alabama pastures (Hoveland et al., 1975). Species of *Pratylenchus* may be suppressed by tall fescue, although endophyte status has seldom been reported. Populations of *P. penetrans* were reduced 20 to 40% after 8 wk on tall fescue of unknown endophyte status (Townshend et al., 1984a). *Pratylenchus scribneri* increased fivefold on tall fescue of unknown endophyte status (Minton, 1965), but *P. scribneri* populations in fescue pastures with high endophyte infestation levels were only 1% of their levels in E– pastures (West et al., 1988). This reduction may be a partial explanation for drought tolerance in E+ fescue. In greenhouse experiments, *P. scribneri* declined to nondetectable levels in the roots and rhizosphere of E+ plants, but maintained itself near initial inoculum levels on E– plants (Kimmons et al., 1990).

Little research has been performed on parasitism of tall fescue by root-knot nematodes. *Meloidogyne hapla* and *M. incognita* were unable to parasitize seven tall fescue cultivars (Chapman, 1973), but in another experiment, one *M. incognita* isolate reproduced on tall fescue (McGlohon et al., 1961). The endophyte status in these studies was not reported. *Meloidogyne marylandi* increased up to fivefold on E– fescue in greenhouse experiments, but did not increase on E+ fescue (Kimmons et al., 1990). Egg production on E+ fescue was less than 10% of that on E– fescue. Hatching and invasion of E+ and E– roots by *M. marylandi* were investigated over a 17-d period (Gwinn & Bernard, 1993). Up to Day 8, hatching (juveniles in soil + roots) was similar in both E+ and E– treatments. Root invasion was consistently higher in E– root systems after Day 6, and rose dramatically by Day 17. Total hatch (soil + root) was much higher in E– treatments. The ability of *M. marylandi* to reproduce on four clone pairs of differing genotypes also was investigated. Sixty days after infestation, egg masses were absent from roots of all E+ plants, but were abundant on all E– plants (E.C. Bernard & K. D. Gwinn, 1993, unpublished data). Kirkpatrick et al. (1990) observed a similar phenomenon with *M. graminis* on E+/E– fescue clones.

Although other Poaceae harbor endophytic fungi, little research has been focused on their relationships with nematodes. Parasitism of perennial ryegrass by *M. naasi* was not affected by presence or absence of *N. lolii* (Cook et al., 1991). *Meloidogyne marylandi* invasion and reproduction on E+ and E– red fescue and perennial ryegrass were unaffected by endophyte status, and *P. scribneri* invasion and reproduction even appeared to be enhanced in E+ roots of perennial ryegrass and red fescue (K. D. Gwinn & E. C. Bernard, 1993, unpublished data).

It should be clear that use of endophytic fungi for nematode management in grasses has great potential. This approach would be advantageous in providing a nonpesticidal alternative similar in most respects to classical resistance, and could provide additional benefits such as drought tolerance and pest resistance. Mechanisms of endophyte-mediated resistance must be identified first so that plant-fungus symbioses antagonistic to pests, but harmless to vertebrate grazers, can be identified and evaluated.

OUTLOOK

Future progress in forage grass-nematode research will be dependent on the resources devoted to it. These resources are far less than for many other important crops. In the USDA Current Research Information System (CRIS) for 1991 to the present, only 32 projects have a forage grass-nematode component, and some of these are only peripheral to the problems of nematode effects on productivity. By contrast, 174 projects have a soybean [*Glycine max* (L.) Merr.]-nematode component, while more than 200 deal with nematodes on vegetables.

Symptoms of nematode parasitism on grasses are nonspecific and subtle, which makes the perception of plant damage more difficult. Because of reduced perception, grasses are sometimes thought to be less affected by nematodes than are dicotyledonous crop plants. This chapter should provide an idea of the possible damage that nematodes can inflict on grasses, but it does not tell the whole story. Research on forage grass nematodes has tended to be spotty and isolated because of the low unit value of forage land. In reality, nematodes could be exacting a significant toll on pasture and range, but we perceive these affected grasses as being in their normal state. Recent research (Griffin, 1992, 1995; Ingham & Detling, 1984, 1990) clearly demonstrates the effects nematodes can have on native grass stands. In some cases, a significant increase in forage productivity might result simply from introduction of nematode-resistant seed during renovation. The following areas should be pursued for the protection and enhanced productivity of pasture and forage grass crops. These suggestions partially reflect those recently listed by Skipp (1996).

1. Germplasm resources should be maintained and further developed to include collections of genotypes from all potential forage grass regions. Chemical nematode control is not practical on forage land. Enhanced germplasm resources will increase the chance of locating nematode-resistant selections.

2. Determination of the extent of grass parasitism by *Anguina* spp., *Clavibacter* spp., and other bacteria should be a major priority. These parasites occur in North America and may already have caused toxicity to some animals in Oregon (Galloway, 1961). Repetition of the serious problem now occurring in Australia would be a serious blow to the North American beef industry.

3. Comprehensive surveys of *Neotyphodium* endophyte status in pasture and range grasses are urgently needed. *Neotyphodium* spp. are to date known only from the subfamily Pooideae, which also contains many of our widespread forage species (Clay, 1989). Some of the research cited in this chapter could be erroneously interpreted if the plants contained an endophyte, and if the particular endophyte provided resistance to nematodes. Better understanding of the presence of endophytes in forage grasses is essential for interpretation of previous research.

4. More effort also is needed in determining the mechanisms and biochemistry by which E+ tall fescue is resistant to root-knot and cyst nematodes. Further progress in this area could result in discovery of useful approaches to resistance not previously recognized.

5. The relationship of grass adaptability to nematodes in a new geographic environment has scarcely been considered. Extensive data are available for prediction of adaptability to new environments (Vallentine, 1989), but these data do not extend to pests and pathogens. The relationship of indigenous nematodes to introduced grasses should be examined whenever possible.

6. Basic research on the population dynamics and damage potential of nematodes associated with pasture and range must continue. Increasing interest in pasture renovation or range improvement (Vallentine, 1989) will require a

more complete approach to pest management, similar to that for monocultured food crops. An adequate understanding of nematode population behavior will be important for grass establishment and maintenance.

REFERENCES

Alderman, S.C. 1988. Distribution of *Gloeotinia temulenta*, *Claviceps purpurea*, and *Anguina agrostis* among grasses in the Williamette Valley of Oregon in 1988. J. Appl. Seed Prod. 6:6–10.

Alderman, S.C. 1991. Distribution of *Gloeotinia temulenta*, *Claviceps purpurea*, and *Anguina agrostis* among grasses in the Williamette Valley of Oregon in 1989. J. Appl. Seed Prod. 9:44–48.

Anderson, R.V., and J.W. Potter. 1991. Stunt nematodes: *Tylenchorhynchus*, *Merlinius*, and related genera. p. 529–586. *In* W.R. Nickle (ed.) Manual of agricultural nematology. Marcel Dekker, New York.

Arachevaleta, M., C.W. Bacon, and C.S. Hoveland. 1989. Effect of the tall fescue endophyte on plant response to environmental stress. Agron. J. 81:83–90.

Asay, K.H., W.H. Horton, and W.T. Hansen, II. 1985. New grasses for intermountain rangelands. Utah Sci. 46:119–123.

Bacon, C.W. 1995. Toxic endophyte-infected tall fescue and range grasses: Historical perspectives. J. Anim. Sci. 73:861–870.

Bacon, C.W., P.C. Lyons, J.K. Porter, and J.D. Robbins. 1986. Ergot toxicity from endophyte infected grasses: A review. Agron. J. 78:106–116.

Bacon, C.W., J.K. Porter, J.D. Robbins, and E.S. Luttrell. 1977. *Epichloe typhina* from toxic tall fescue grasses. Appl. Environ. Microbiol. 34:576–581.

Barnes, R.F, and J. E. Baylor. 1995. Forages in a changing world. p. 3–13. *In* R.F Barnes et al. (ed.) Forages. Vol. 1. Iowa State Univ. Press, Ames.

Barnes, R.F, D.A. Miller, and C.J. Nelson (ed.). 1995. Forages. Vol. 1. Iowa State Univ. Press, Ames.

Bernard, E.C., A.M. Cole, J.B. Oliver, and K.D. Gwinn. 1990. Survival and fecundity of *Folsomia candida* (Collembola) fed tall fescue tissues or ergot peptide-amended yeast. p. 125–127. *In* R. Joost and S. Quisenberry (ed.) Proc. Int. Symp. *Acremonium*/Grass Interactions, New Orleans, LA. 3–5 November. Louisiana Agric. Exp. Stn., Baton Rouge.

Bird, A.F., and D.L. Riddle. 1984. Effect of attachment of *Corynebacterium rathayi* on movement of *Anguina agrostis* larvae. Int. J. Parasitol. 14:503–511.

Bird, A.F., B.A. Stynes, and W.W. Thomson. 1980. A comparison of nematode and bacteria-colonized galls induced by *Anguina agrostis* in *Lolium rigidum*. Phytopathology 70:1104–1109.

Blank, C.A., K.D. Gwinn, and A.M. Gavin. 1993. Tolerance of tall fescue to soilborne pathogens is influenced by *Acremonium coenophialum*. p. 145–147. *In* D. E. Hume et al. (ed.) Proc. 2nd Int. Conf. *Acremonium*/Grass Interactions, Palmerston North, New Zealand. 24–27 February. AgRes., Palmerston North, New Zealand.

Box, T.W., R.J. Bula, R.J., L.V. Crowder, D.A. Holt, L.R. Humphries, and V.L. Lechtenberg. 1977. Potential of the world's forages for ruminant animal production. Winrock Int., Morrilton, AR.

Boyd, F., and V.G. Perry. 1969. The effect of sting nematodes on establishment, yield, and growth of forage grasses on Florida sandy soils. Soil Crop Sci. Soc. Fla. Proc. 29:288–300.

Bridge, J., and N.G.M. Hague. 1974. The feeding behaviour of *Tylenchorhynchus* and *Merlinius* species and their effect on growth of perennial ryegrass. Nematologica 20:119–130.

Brown, R.H. 1987. Control strategies in low value-crops. p. 351–387. *In* R.H. Brown and B.R. Kerry (ed.) Principles and practice of nematode control in crops. Acad. Press, Sydney, Australia.

Byrd, D.W., Jr., T. Kirkpatrick, and K.R. Barker. 1983. An improved technique for clearing and staining plant tissues for detection of nematodes. J. Nematol. 15:142–143.

Cenis, J.L. 1993. Identification of four major *Meloidogyne* spp. by random amplified polymorphic DNA (RAPD-PCR). Phytopathology 83:76–78.

Chapman, R.A. 1973. Southern and northern root-knot nematodes did not reproduce on tall fescue. p. 94. *In* 85th Ann. Rep. Univ. Kentucky Agric. Exp. Stn., Kentucky Agric. Exp. Stn., Lexington, KY.

Cheeke, P.R. 1995. Endogenous toxins and mycotoxins in forage grasses and their effects on livestock. J. Anim. Sci. 73:909–918.

Christie, J.R. 1959. Plant nematodes. Their bionomics and control. Univ. Florida, Gainesville.

Clay, K. 1989. Clavicipitaceous endophytes of grasses: Their potential as biocontrol agents. Mycol. Res. 92:1–12.

Clay, K., and G.P. Cheplick. 1989. Effect of ergot alkaloids from fungal endophyte-infected grass on fall armyworm (*Spodoptera frugiperda*). J. Chem. Ecol. 15:169–182.

Coates-Beckford, P.L., and R.B. Malek. 1978. Host preferences of the stunt nematode *Tylenchorhynchus agri*. Plant Dis. Rep. 62:793–796.

Collins, D.D. 1966. The occurrence of seed-gall nematodes in a native population of western wheatgrass (*Agropyron smithii*) in Montana. Plant Dis. Rep. 50:45.

Cook, R., G.C. Lewis, and K.A. Mizen. 1991. Effects on plant-parasitic nematodes of infection of perennial ryegrass, *Lolium perenne*, by the endophytic fungus, *Acremonium lolii*. Crop Prot. 10:403–407.

Cook, R., and G.W. Yeates. 1993. Nematode pests of grassland and forage crops. p. 305–350. *In* K. Evans et al. (ed.) Plant parasitic nematodes in temperate agriculture. CAB Int., Wallingford, England.

Cooper, W.E., and B.B. Brodie. 1962. Correlation between *Fusarium* wilt indices of cotton varieties with root-knot and sting nematodes as predisposing agents. Phytopathology 52:6.

Courtney, W.D., and H.B. Howell. 1952. Investigation of the bent grass nematode, *Anguina agrostis* (Steinbuch 1799) Filipjev 1936. Plant Dis. Rep. 36:75–83.

Davis, M.J., A.G. Gillaspie, A.K. Vidaver, and R.W. Harris. 1984. *Clavibacter*: a new genus containing some phytopathogenic coryneform bacteria, including *Clavibacter xyli* subsp. *xyli* sp. nov., subsp. nov. and *Clavibacter xyli* subsp. *cynodontis* subsp. nov., pathogens that cause ratoon stunting disease of sugarcane and bermudagrass stunting disease. Int. J. Syst. Bacteriol. 34:107–117.

Dickerson, O.J., W.G. Willis, F.J. Dainello, and J.C. Pair. 1972. The sting nematode, *Belonolaimus longicaudatus*, in Kansas. Plant Dis. Rep. 56:957.

Dunn, R.A. 1987. Turf nematodes: Diagnosis and control for turf maintenance professionals. Nematol. Univ. Florida Plant Prot. Point. 18. Univ. Florida, Gainesville.

Easton, H.S. 1984. Ryegrasses. *In* G. S. Wratt and H.C. Smith (ed.) Plant breeding in New Zealand. Butterworths/DSIR, Auckland, NZ.

Ediz, S.A., and O.J. Dickerson. 1976. Life cycle, pathogenicity, histopathology, and host range of race 5 of the barley root-knot nematode. J. Nematol. 8:228–232.

Fletcher, L.R. 1993. Grazing ryegrass/endophyte association and their effect on animal health and performance. p. 115–120. *In* D.E. Hume et al. (ed.) Proc. 2nd Int. Conf. *Acremonium*/Grass Interactions, Palmerston North, New Zealand. 3–5 February. AgRes., Palmerston North, New Zealand.

Fribourg, H.A., C.S. Hoveland, and K.D. Gwinn. 1991. Tall fescue and the fungal endophyte—a review of current knowledge. Tennessee Farm Home Sci. 160:30–37.

Galloway, J.H. 1961. Grass seed nematode poisoning in livestock. J. Am. Vet. Med. Assoc. 139:1212–1214.

Glenn, A.E., C.W. Bacon, R. Price, and R.T. Hanlin. 1996. Molecular phylogeny of *Acremonium* and its taxonomic implications. Mycologia 88:369–383.

Griffin, G.D. 1984. Nematode parasites of alfalfa, cereals, and grasses. p. 243–321. *In* W.R. Nickle (ed.) Plant and insect nematodes. Marcel Dekker, New York.

Griffin, G.D. 1992. Pathological effects of *Pratylenchus neglectus* to wheatgrasses. J. Nematol. 24:442–449.

Griffin, G.D. 1994. Effect of single and interplantings on virulence of *Pratylenchus penetrans* and *P. neglectus* to alfalfa and crested wheatgrass. J. Nematol. 26:460–466.

Griffin, G.D. 1996. Importance of soil texture to the pathogenicity of plant parasitic nematodes on rangeland grasses. Nematropica 26:27–37.

Griffin, G.D., and K.H. Asay. 1989. Pathological reaction of crested wheatgrass cultivars to four *Meloidogyne chitwoodi* populations. J. Nematol. 21:446–452.

Griffin, G.D., and K.H. Asay. 1992. Importance of soil texture on the pathogenicity of plant parasitic nematodes to range grasses. Phytopathology 82:1096.

Griffin, G.D., K.H. Asay, D.L. Crebs, and W.H. Horton. 1988. Population trends of plant parasitic nematodes on crested wheatgrass. J. Nematol. 20:638.

Griffin, G.D., and H.M. Darling. 1964. An ecological study of *Xiphinema americanum* Cobb in an ornamental spruce nursery. Nematologica 10:471–479.

Griffin, G.D., R.N. Inserra, and N. Vovlas. 1984. Rangeland grasses as hosts of *Meloidogyne chitwoodi*. J. Nematol. 16:399–402.

Gwinn, K.D., and E.C. Bernard. 1993. Interactions of endophyte-infected grasses with the nematodes *Meloidogyne marylandi* and *Pratylenchus scribneri*. p. 156–160. *In* D.E. Hume et al. (ed.)

Proc. 2nd Int. Conf. *Acremonium*/Grass Interactions. 3–5 February. AgRes, Palmerston North, NZ.

Gwinn, K.D., E.C. Bernard, and C.D. Pless. 1992. Pathogen and pest resistance in endophyte-infected tall fescue. p. 139–145. *In* P.M. Gresshoff (ed.) Plant biotechnology and development. CRC Press, Boca Raton, FL.

Gwinn, K.D., C.A. Blank, A.M. Cole, and C.D. Pless. 1991. Resistance of endophyte-infected tall fescue seedlings to pathogens and pests. Tennessee Farm Home Sci. 160:72–77.

Halisky, P.M., and R.F. Myers. 1989. Effect of *Acremonium* endophytes on parasitic nematodes in *Festuca* and *Lolium* grown under turf maintenance. p. 124–131. *In* Rutgers Turfgrass Proc.

Hinton, D.M., and C.W. Bacon. 1985. The distribution and ultrastructure of the endophyte of toxic tall fescue. Can. J. Bot. 63:35–42.

Holdeman, Q.L., and T.W. Graham. 1954. Effect of the sting nematode on expression of *Fusarium* wilt in cotton. Phytopathology 44:683–685.

Hoveland, C.S., R. Rodríguez-Kábana, and C.D. Berry. 1975. Phalaris and tall fescue forage production as affected by nematodes in the field. Agron. J. 67:714–717.

Hovin, A.W., and R.C. Buckner. 1983. Alkaloids in tall fescue and reed canarygrass: A review. p. 241–248. *In* M. Rechsigl (ed.) CRC handbook of naturally occurring food toxicants. CRC Press, Boca Raton, FL.

Ingham, R.E., R.V. Anderson, W.D. Gould, and D.C. Coleman. 1985. Vertical distribution of nematodes in a shortgrass prairie. Pedobiologia 28:155–160.

Ingham, R.E., and J.K. Detling. 1984. Plant-herbivore interactions in a North American mixed-grass prairie III. Soil nematode populations and root biomass on *Cynomys ludovicianus* colonies and adjacent uncolonized areas. Oecologia (Berlin) 63:307–313.

Ingham, R.E., and J.K. Detling. 1986. Effects of defoliation and nematode consumption on growth and leaf exchange in *Bouteloua curtipendula*. Oikos 46:23–28.

Ingham, R.E., and J.K. Detling. 1990. Effects of root-feeding nematodes on aboveground net primary production in a North American grassland. Plant Soil 121:279–281.

Ingham, R.E., J.A. Trofymow, R.V. Anderson, and D.C. Coleman. 1982. Relationships between soil type and soil nematodes in a shortgrass prairie. Pedobiologia 24:139–144.

Ingham, R.E., J.A. Trofymow, E.R. Ingham, and D.C. Coleman. 1985b. Interactions of bacteria, fungi, and their nematode grazers: Effects on nutrient cycling and plant growth. Ecol. Monogr. 55:119–140.

Inserra, R., N. Vovlas, J.H. O'Bannon, and R.P. Esser. 1988. *Tylenchulus graminis* n. sp. and *T. palustris* n. sp. (Tylenchulidae) from native flora of Florida, with notes on *T. semipenetrans* and *T. furcus*. J. Nematol. 20:266–287.

Jaffee, B.A., M.B. Harrison, R.L. Shaffer, and M.B. Strang. 1987. Seasonal population fluctuations of *Xiphinema americanum* in New York and Pennsylvania orchards. J. Nematol. 19:369–378.

Jepson, S. 1987. Identification of root-knot nematodes (*Meloidogyne* spp.). CAB Int., Wallingford, England.

Johnson, A.W., and R.H. Littrell. 1970. Pathogenicity of *Pythium aphanidermatum* to chrysanthemum in combined inoculations with *Belonolaimus longicaudatus* or *Meloidogyne incognita*. J. Nematol. 2:36–41.

Johnson, M.C., D.L. Dahlman, M.R. Siegel, L.P. Bush, G.C.M. Latch, D.A. Potter, and D.R. Varney. 1985. Insect feeding deterrents in endophyte-infected tall fescue. Appl. Environ. Microbiol. 49:568–571.

Jordaan, E.M., G.C. Loots, W.J. Jooste, and D. De Waele. 1987. Effects of root-lesion nematodes (*Pratylenchus brachyurus* Godfrey and *P. zeae* Graham) and *Fusarium moniliforme* Sheldon alone or in combination, on maize. Nematologica 33:213–219.

Kerr, E.D., and D.S. Wysong. 1979. Sting nematode, *Belonolaimus* sp., in Nebraska. Plant Dis. Rep. 63:506–507.

Kimmons, C.A., K.D. Gwinn, and E.C. Bernard. 1990. Nematode reproduction on endophyte-infected and endophyte-free tall fescue. Plant Dis. 74:757–761.

Kimpinski, J.M., and C.B. Willis. 1981. Influence of soil temperature and pH on *Pratylenchus penetrans* and *P. crenatus* in alfalfa and timothy. J. Nematol. 13:333–338.

Kirkpatrick, T.L., J.D. Barham, and J.R. Bateman. 1990. Host status for *Meloidogyne graminis* of tall fescue selections and clones with and without the endophyte *Acremonium coenophialum*. p. 154–156. *In* R. Joost and S. Quisenberry (ed.) Proc. Int. Symp. *Acremonium*/Grass Interactions, Baton Rouge, LA. 3–5 November. Louisiana Agric. Exp. Stn., Baton Rouge.

Krall, E.I. 1991. Wheat and grass nematodes: *Anguina*, *Subanguina*, and related genera. p. 721–760. *In* W.R. Nickle (ed.) Manual of agricultural nematology. Marcel Dekker, New York.

Latch, G.C.M., M.J. Christensen, and G.J. Samuels. 1984. Five endophytes of *Lolium* and *Festuca* in New Zealand. Mycotaxon 20:535–550.

Latch, G.C.M., L.R. Potter, and B.F. Tyler. 1987. Incidence of endophytes in seeds from collections of *Lolium* and *Festuca* species. Ann. Appl. Biol. 11:59–64.

Mai, W.F., and G.S. Abawi. 1987. Interactions among root-knot nematodes and *Fusarium* wilt fungi on host plants. Annu. Rev. Phytopath. 25:317–338.

Marshall, D., L.R. Nelson, and B. Tunali. 1993. The occurrence of *Acremonium* and other endophytic fungi in the indigenous wild cereals of Turkey. p. 8–10. *In* D.E. Hume et al. (ed.) Proc. 2nd Int. Conf. *Acremonium*/Grass Interactions, Palmerston North, New Zealand. 24–27 February. AgRes., Palmerston North, New Zealand.

McGlohon, N.E., J.N. Sasser, and R.T. Sherwood. 1961. Investigations of plant-parasitic nematodes associated with forage crops in North Carolina. North Carolina State Univ. Exp. Stn.. Tech. Bull. 148:1–39.

McKay, A.C., J.M. Fisher, and A.J. Dube. 1982. Control of the nematode associated with annual ryegrass toxicity. Aust. J. Agric. Res. 34:403–413.

McKay, A.C., and K.M. Ophel. 1993. Toxigenic *Clavibacter*/*Anguina* associations infecting grass seedheads. Annu. Rev. Phytopathol. 31:151–167.

McKay, A.C., K.M. Ophel, T.B. Reardon, and J.M. Gooden. 1993. Livestock deaths associated with *Clavibacter toxicus*/*Anguina* sp. infection in seedheads of *Agrostis avenacea* and *Polypogon monspeliensis*. Plant. Dis. 77:635–641.

Michell, R.E., R.B. Malek, D.P. Taylor, and D.I. Edwards. 1973. Races of the barley root-knot nematode, *Meloidogyne naasi*, I. Characterization by host preference. J. Nematol. 5:41–44.

Minton, N.A. 1965. Reaction of white clover and five other crops to *Pratylenchus scribneri*. Plant Dis. Rep. 49:856–859.

Morgan-Jones, G., and W. Gams. 1982. Notes on hyphomycetes. XLI. An endophyte of *Festuca arundinacea* and the anamorph of *Epichloe typhina*, a taxon in one of two new sections of *Acremonium*. Mycotaxon 15:311–318.

Mulvey, R.H., J.L. Townshend, and J.W. Potter. 1975. *Meloidogyne microtyla* sp. nov. from southwestern Ontario, Canada. Can. J. Zool. 53:1528–1536.

Mundo-Ocampo, M., J.O. Becker, and J.G. Baldwin. 1994. Occurrence of *Belonolaimus longicaudatus* on bermudagrass in the Coachella Valley. Plant Dis. 78:529.

Nicholson, R.L., G.B. Bergeson, F.P. DeGennaro, and D.M. Viveiros. 1985. Single and combined effects of the lesion nematode and *Colletotrichum graminicola* on growth and anthracnose leaf blight of corn. Phytopathology 75:654–661.

Norton, D.C. 1963. Population fluctuations of *Xiphinema americanum* in Iowa. Phytopathology 53:66–68.

Norton, D.C. 1965. *Anguina agropyronifloris* n. sp., infecting florets of *Agropyron smithii*. Proc. Helminthol. Soc. Wash. 32:118–122.

O'Day, M.H., W.C. Bailey, and T.L. Niblack. 1990. Phytonematode communities in tall fescue varieties with varying levels of endophyte (*Acremonium coenophialum*) infection. p. 163–169. *In* R. Joost and S. Quisenberry (ed.) Proc. Int. Symp. *Acremonium*/Grass Interactions, Baton Rouge, LA. 3–5 November. Louisiana Agric. Exp. Stn., Baton Rouge.

Pedersen, J.F., and R. Rodríguez-Kábana. 1984. Differences among nematode populations in tall fescue pastures in north, central, and south Alabama. Crop Sci. 24:819–821.

Pedersen, J.F., and R. Rodríguez-Kábana. 1985. Regional differences in nematode populations of tall fescue pastures in Alabama. Alabama Agric. Exp. Stn.. Circ. 276:1–12.

Pedersen, J.F., R. Rodríguez-Kábana, and R.A. Shelby. 1988. Ryegrass cultivars and endophyte in tall fescue affect nematodes in grass and succeeding soybean. Agron. J. 80:811–814.

Person-Dedryver, F., and J. Fischer. 1987. Grasses as hosts of *Meloidogyne naasi* Franklin: I. Variation in host status of species and varieties grown in France. Nematologica 33:61–71.

Petersen, A.D., D.K. Barnes, and J.A. Thies. 1991. Preference of root-lesion nematode for alfalfa and forage grasses growing in binary mixtures. Crop Sci. 31:567–570.

Pinkerton, J.N., and S.C. Alderman. 1994. Epidemiology of *Anguina agrostis* on Highland Colonial bentgrass. J. Nematol. 26:315–323.

Powelson, M.L., and R.C. Rowe. 1993. Biology and management of early dying of potatoes. Annu. Rev. Phytopathol. 31:111–126.

Prestidge, R.A., and O.J.-P. Ball. 1993. The role of endophyte in alleviating plant biotic stress in New Zealand. p. 141–151. *In* D.E. Hume et al. (ed.) Proc. 2nd Int. Conf. *Acremonium*/Grass Interactions, Palmerston North, New Zealand. 24–27 February. AgRes., Palmerston North, New Zealand.

Riley, I.T., and G.S. Gill. 1994. Relationship between herbicide resistance in *Lolium rigidum* and populations of *Anguina funesta*, the nematode vector in annual ryegrass toxicity. Aust. J. Exp. Agric. 34:633–635.

Riley, I.T. and A.C. McKay. 1990. Specificity of the adhesion of some plant pathogenic micro-organisms to the cuticle of nematodes in the genus *Anguina* (Nematoda: Anguinidae). Nematologica 36:90–103.

Rodríguez-Kábana, R., R.L. Haaland, C.B. Elkins, and C.S. Hoveland. 1978. Relationship of root diameter and number of roots to damage of tall fescue by lance nematodes. J. Nematol. 10:279.

Rodríguez-Kábana, R., D.G. Robertson, C.F. Weaver, and L. Wells. 1991. Rotations of bahiagrass and castorbean with peanut for the management of *Meloidogyne arenaria*. J. Nematol. (Suppl.) 23:658–661.

Roncadori, R.W., P.S. Lehman, ans S.M. McCarter. 1974. Effect of *Pythium irregulare* on cotton growth and yield, and joint action with other soil-borne pathogens. Phytopathology 64:1303–1306.

Sampson, K. 1933. The systemic infection of grasses by *Epichloe typhina* (Pers.) Tul. Trans. Br. Mycol. Soc 21:84–97.

Santerre, J., and R. Levesque. 1982. Inventaire de nematodes phytoparasites dans des cultures de plantes fourrageres au Quebec: 1973 a 1978. Can. Plant Dis. Surv. 62:13–18.

Schmautz, J.E., M.D. Belinger, and R.W. Harris. 1980. Range. p. 246–315. *In* An assessment of the forest and range land situation in the United States. USDA-FS-345. U.S. Gov. Print. Office, Washington, DC.

Siegel, M.R., G.C.M. Latch, and M.C. Johnson. 1987. Fungal endophytes of grasses. Annu. Rev. Phytopathol. 25:239–315.

Skipp, R.A. 1996. The future of pasture and forage crop pathology. p. 647–649. *In* S. Chakraborty et al. (ed.) Pasture and forage crop pathology. ASA, CSSA, and SSSA, Madison, WI.

Smart, G.C., Jr., and K.B. Nguyen. 1991. Sting and awl nematodes: *Belonolaimus* spp. and *Dolichodorus* spp. p. 627–667. In W. R. Nickle (ed.) Manual of agricultural nematology. Marcel Dekker, New York.

Smolik, J.D. 1974. Nematode studies at the Cottonwood site. Grassl. Biome Ecosyst. Stud. Tech. Rep. 251. U.S. Int. Biome Proj., Colorado State Univ., Fort Collins.

Smolik, J.D. 1982. Effect of *Tylenchorhynchus robustoides* on growth of buffalo grass and western wheatgrass. J. Nematol. 14:585–588.

Smolik, J.D., and J.K. Lewis. 1982. Effect of range condition on density and biomass of nematodes in a mixed ecosystem. J. Range Manage. 35:657–663.

Sohlenius, B. 1980. Abundance, biomass and contribution to energy flow by soil nematodes in terrestrial ecosystems. Oikos 34:186–194.

Southey, J.F. 1973. *Anguina agrostis*. CIH Descript. Plant-Parasitic Nematodes. Set 2, no. 20. Commonweal. Agric. Bur., Wallingford, England.

Stanton, J.M., and N. Rizo P. 1989. Nematodes of forage legumes and grasses: Catalogue and bibliography 1961–1985. CAB Int., Wallingford, England.

Stanton, N.L., D. Morrison, and W.A. Laycock. 1984. The effect of phytophagous nematode grazing on blue grama die-off. J. Range Manage. 37:447–450.

Stokes, D.E. 1969. *Andropogon rhizomatus* parasitized by a strain of *Tylenchulus semipenetrans* not parasitic to four citrus rootstocks. Plant Dis. Rep. 53:882–885.

Stuedemann, J.A. 1988. Fescue endophyte: history and impact on animal agriculture. J. Prod. Agric. 1:39–44.

Taheri, A., G.J. Hollamby, and V.A. Vanstone. 1994. Interaction between root lesion nematode, *Pratylenchus neglectus* (Rensch 1924) Chitwood and Oteifa 1952, and root rotting fungi of wheat. N.Z. J. Crop. Hort. Sci. 22:181–185.

Taylor, A.L., and J.N. Sasser. 1978. Biology, identification, and control of root-knot nematode (*Meloidogyne* species). North Carolina State Univ. Graphics, Raleigh.

Thorgeirsson, H. 1985. Temporal and spatial partitioning of the soil water resource between two *Agropyron* bunchgrasses and *Artemisia tridentata*. M.S. thesis, Utah State Univ., Logan.

Todd, T.C. 1989. Population dynamics and damage potential of *Belonolaimus* sp. on corn. J. Nematol. 21 (Suppl.):697–702.

Townshend, J.L., R.A. Cline, V.A. Dirks, and C.F. Marks. 1984a. Assessment of turfgrasses for the management of *Pratylenchus penetrans* and *Paratylenchus projectus* in orchards. Can. J. Plant Sci. 64:355–360.

Townshend, J.L., and T.R. Davidson. 1989. Grass and legume hosts of *Paratylenchus projectus*. Nematologica 35:128–131.

Townshend, J.L., and J.W. Potter. 1973. Nematode numbers under cultivars of forage legumes and grasses. Can. Plant Dis. Surv. 194–195.

Townshend, J.L., J.W. Potter, and T.R. Davidson. 1984b. Some moncotyledonous and dicotyledonous hosts of *Meloidogyne microtyla*. Plant Dis. 68:7–10.

Vallentine, J.F. 1989. Range development and improvements. 3rd ed. Acad. Press, San Diego, CA.

Wallace, M.K., R. H. Rust, D.M. Hawkins, and D.H. MacDonald. 1993. Correlation of edaphic factors with plant-parasitic nematode population densities in a forage field. J. Nematol. 25:642–653.

Welty, R.E., M.D. Azevedo, and T.M. Cooper. 1987. Influence of moisture content, temperature, and length of storage on seed germination and survival of endophytic fungi in seeds of tall fescue and perennial ryegrass. Phytopathology 77:893–900.

Wen, Y.C., and D.R. Viglierchio. 1992. Bacterial associates of *Anguina* species isolated from galls. Fund. Appl. Nematol. 15:231–234.

West, C.P, and K.D. Gwinn. 1993. Role of *Acremonium* in drought, pest and disease tolerances of grasses. p. 131–140. *In* D.E. Hume et al. (ed.) Proc. 2nd Int. Conf. *Acremonium*/Grass Interactions. 3–5 February. AgRes., Palmerston North, NZ

West, C., E. Izekor, D.M. Oosterhuis, and R.T. Robbins. 1988. The effect of *Acremonium coenophialum* on growth and nematode infestation of tall fescue. Plant Soil 112:3–6.

White, J.F., G. Morgan-Jones, and A.C. Morrow. 1993. The ecology and evolution of endophytes. Agric. Ecosyst. Environ. 44:13–37.

Williams, A.S., and C.W. Laughlin. 1968. Occurrence of *Hypsoperine graminis* in Virginia and additions to the host range. Plant Dis. Rep. 52:162–163.

Williams, C.B., C.B. Elkins, R.L. Haaland, C.S. Hoveland, and R. Rodríguez-Kábana. 1983. Effects of root diameter, nematodes, and soil compaction on forage yield of two tall fescue genotypes, *Festuca arundinacea*. p. 121–124. *In* J.A. Smith and V.W. Hays (ed.) Proc. 14th Int. Grass. Congr., Lexington, KY. 15–24 1981. Westview Press, Boulder, CO.

Willis, C.B., J.L. Townshend, R.V. Anderson, J. Kimpinski, R.H. Mulvey, J.W. Potter, J. Santerre, and L.Y. Wu. 1976. Species of plant-parasitic nematodes associated with forage crops in eastern Canada. Plant Dis. Rep. 60:207–210.

Windham, G.L., and G.E. Brink. 1991. Host efficiency of bermuda grass to *Meloidogyne incognita* and *Meloidogyne arenaria*. Nematropica 21:88–96.

York, P.A. 1989. Resistance to *Meloidogyne javanica* (root-knot nematode) in *Chloris gayana* (Rhodes grass). Nematologica 35:225–233.

York, P.A. 1990. Range of susceptibility within and between diploid and tetraploid strains of *Chloris gayana* (Rhodes grass) to *Meloidogyne javanica*. Rev. Nematol. 13:45–50.

Yu, M.Q., F. Pewrson-Dedryver, and J. Jahier. 1990. Resistance to root knot nematode, *Meloidogyne naasi* (Franklin) transferred from *Aegilops variabilis* Eig to bread wheat. Agronomie 6:451–456.

21 Rice and Other Cereals

EDWARD C. MCGAWLEY AND CHARLES OVERSTREET

Louisiana State University
Baton Rouge, Louisiana

On a worldwide basis, rice (*Oryza sativa* L.) is consumed daily by more people than any other single food crop. The economic value of losses, even minor ones such as those caused by the stem rot (*Sclerotium oryzae* Cattaneo) and leaf scald fungi [*Microdochium oryzae* (Hashioka & Yokogi) Samuels and I.C. Hallett], translate into millions of dollars on a global basis. A conservative estimate of the annual yield loss caused to this important and life-sustaining crop by plant-parasitic nematodes, a major category of pest, is 10% (Sasser & Freckman, 1987). An approximation of the magnitude of the loss in rice yields caused by nematodes in 1995 can be obtained by multiplying the 10% loss estimate by USA and world production figures of 7.9 and 371.0 Tg (Fig. 21–1), respectively (USDA Natl. Agric. Stat. Serv., Jan., 1995; USDA Econ. Res. Serv., 1995).

IMPORTANT NEMATODES

Aphelenchoides Species

Species and Races

Aphelenchoides besseyi, which was first described by Christie in 1942, is by far the most common species of this genus associated with cultivated rice. This foliar-feeding parasite causes the disease commonly called "white-tip" because of the tattered and whitish appearance of the tips of infected leaf blades.

Differences in rice cultivar reactions to *A. besseyi* among years and locations, as well as differences in the host range of the nematode, have been noted by various investigators (Atkins & Todd, 1959; Noegel & Perry, 1963). To date, however, the question of whether these differences in cultivar reaction result from environmental conditions or from changes in the nematode genome remains unanswered. The white-tip nematode has a known host range that includes 35 genera of higher plants and many fungi, some of which are plant pathogens (Fortuner & Williams, 1975; McGawley et al., 1984). Recently, two new species, *A. nechaleos* and *A. paranechaleos*, have been described from rice in Vietnam and Sierra Leone (Hooper & Ibrahim, 1994). The analysis of enzymes and non-specific esterase patterns by native and sodium dodecylsulfate polyacrylanide gel

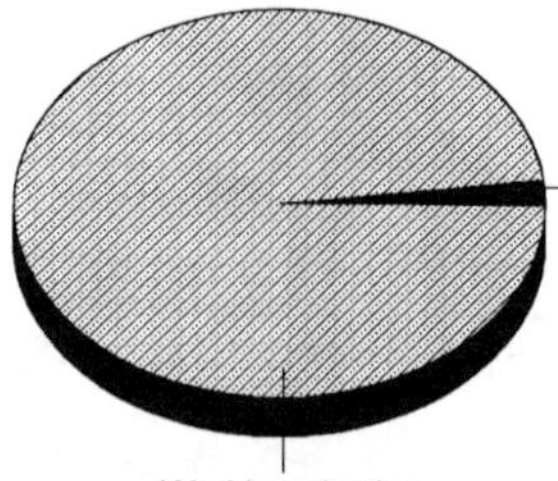

Fig. 21–1. 1995 world and U.S. rice production.

electrophoresis (SDS-PAGE) (Ibrahim et al., 1994) have been shown to be a highly useful and reproducible tool for the differentiation of these three rice-parasitic species of *Aphelenchoides* as well as *A. bicaudatus* and *A. arachidis* from the Pit-Pit plant (*Setaria palmaefolia*) and ground nut (*Arachis hypogaea* L.), respectively.

Symptoms

Rice white tip disease is known to occur in Africa, Northern, Southern, and Central America, most of Asia, Eastern Europe, and the Pacific. The primary symptom of rice white tip disease is the appearance of tattered, stringy-white tips on infested leaf blades (Ou, 1985; Prot, 1992). Plants infected with the nematode may be stunted, sometimes lodge, and frequently produce small, often sterile, panicles with reduced numbers of spikelets. Ou (1985) indicates that, under field conditions, symptoms of the disease are usually apparent during the elongation period of growth. The upper leaves, especially the flag leaf, of severely diseased plants are markedly twisted, and emergence of the panicles from the boot may be incomplete. It should be noted, however, that as early as 1959 Atkins and Todd observed that resistant rice cultivars rarely display the classic "white-tip" foliar symptoms although the nematodes can be isolated from infected plants and yields are markedly reduced.

Biology

Aphelenchoides besseyi is normally amphimictic, although parthenogenetic reproduction of the nematode has been reported to occur in nature (Sudakova & Stoyakov, 1967). The nematode is seed-transmitted and can survive in a quiescent state on infested seed for long periods of time (Cralley, 1949; Yoshii & Yamamoto, 1950), which on unhulled grain can exceed 5 yr (McGawley, 1995, unpublished data). *Aphelenchoides besseyi* feeds ectoparasitically in the young folded leaves during the early growth stages, enters the spikelets after ear formation, and occurs in the glumes when the grain matures (Yoshi & Yamamoto, 1950). The nematode can be extracted (by Baerman funnel) from infested grain, in which its population density can reach levels as high as 100 per grain (Barat et al., 1969; McGawley et al., 1984). The white tip nematode, like many other related "aphelenchoid" nematodes, is able to feed on and reproduce in association with a wide spectrum of fungi (Iyatomi & Nishizawa, 1954; Todd & Atkins, 1959). In association with the fungus *Gliocladium virens*, the white tip nematode, very much like the pinewood nematode *Bursaphelenchus xylophilus* (McGawley et al., 1985), can reach population levels in excess of several hundred thousand per 90-mm diam. Petri dish and actually inhibit sporulation of the fungus.

Disease Complexes

Disease complexes involving the interrelationship of *A. besseyi* with the rice pathogenic fungi *Sclerotium oryzae* (Nishizawa et al., 1953; McGawley et al., 1984) and *Pyricularia oryzae* (Tikhanova & Ivanchenko, 1968) have been described. Nishizawa et al. (1953) found that the presence of the nematode limited the severity of rice stem rot disease. In contrast, McGawley et al. (1984) reported an increase in the severity of stem rot disease in the presence of the nematode. They also documented enhanced nematode reproduction in the presence of the fungus, and a synergistic interaction between the two pathogens with respect to suppression of plant growth. In all three of these studies, yield losses reported for both organisms together have been greater than the sum of those observed for each pathogen alone (a synergistic relationship).

Economic Importance

Overall, rice white tip disease is currently of only minor importance in the USA. The utilization of hot water seed treatment, coupled with the application of insecticidal chemicals during storage of seed that also are toxic to the nematode (McGawley et al., 1984), minimizes dissemination. Additionally, the availability of resistant cultivars has markedly lowered yield losses as compared with estimates of 40 yr ago (Atkins & Todd, 1959), which averaged about 15 to 20% in susceptible cultivars vs. only less than 1% currently (Hollis & Keoboonrueng, 1984). Internationally, however, white tip disease loss estimates as high as 71% were reported from the USSR for 1966 (Tikhonova, 1966) and 60% from India in 1974 (Muthukrishnan et al., 1974). The disease also is of serious economic concern in several of the major rice-producing regions of Africa (Fomba, 1984).

Ditylenchus angustus

Species and Races

Ditylenchus angustus is the only species of this genus reported to parasitize rice. No reports to date suggest the existence of divergent biotypes of this nematode.

Symptoms

In 1913 Butler first described *D. angustus* as the causal agent of "Ufra" disease of rice in Bangladesh (Butler, 1913). The disease also is known as "Tiem Dot San" in Vietnam, "Akhet-pet" in Burma, and "yad-ngo" in southern Thailand. This nematode, like *A. besseyi*, is an ectoparasitic, foliar-feeding nematode species. As such, it frequently causes malformed plant parts, especially leaf sheath tissue and lower nodes, wilting, and stunting of plants, which assume a typical "patchy-brown" pattern under field conditions. White patches or white speckles occur in a splash pattern at the bases of young leaves of rice plants infected with *D. angustus* during the vegetative growth stages. Symptoms on older plants include partially emerged panicles with empty grain and twisted and distorted flag leaves.

Biology

Individuals of *D. angustus*, like the white tip nematode, have evolved the ability to survive for several months in a quiescent or dehydrated state, although they show no intrinsic ability to control water loss and survive severe desiccation (Ibrahim & Perry 1993a). Recent data indicate that survival of this nematode is dependent upon high humidity and/or protection by rice plant tissue (Ibrahim & Perry, 1994). Moreover, it is probable that the fourth-stage juvenile of the nematode is the principal stage that survives between crop seasons and invades new hosts (Ibrahim & Perry, 1994). Infection can take place either from soil or in water. Nematodes in water probably locate rice plants more by chance than by attraction, but can invade roots in as little as 1 h. In either case, nematodes migrate up the rice plant stem to feed in developing leaf tissue if relative humidity requirements of at least 75% exist. The nematodes accumulate and feed on the developing panicles and, at harvest, are coiled within the dried glumes of the lower spikelets on each panicle, but not within the grains (Bridge et al., 1990). This nematode can be grown in monoxenic culture (Plowright & Akehurst, 1992).

Disease Complexes

In 1969, Vuong indicated lesions associated with the feeding activities of the nematode could serve as entry portals for fungi such as *Cladosporium* and *Fusarium*, and coworkers (1986) suggested that parasitism of rice plants by *D. angustus* increased their susceptibility to blast disease.

Economic Importance

Ufra disease is economically important in both deep-water and lowland rice production and is particularly devastating in areas with high annual rainfall or in particularly wet years. Loss estimates as high as 60 to 100% have been reported (Cuc & Kinh, 1981; Mondal & Miah, 1987), although damage has a tendency to be scattered and irregular in occurrence because of the high relative humidity requirements for mobility of the nematode.

Meloidogyne Species

Species and Races

Races 2 and 4 of *M. incognita* and Race 1 of *M. arenaria* reproduce on rice (Ibrahim et al., 1983). *Meloidogyne oryzae* was described by Maas et al. in 1972, and *M. salasi* was described by Lopez in 1984. Bridge et al. (1990) listed two root-knot species (*M. salasi* and *M. oryzae*) that usually occur on irrigated rice only; three species (*M. arenaria*, *M. javanica*, and *M. incognita*) that occur on both upland and irrigated rice; and only one species (*M. graminicola*) that occurs on upland, irrigated, lowland, and deep water rice. Recently, another root-knot species, *M. triticoryzae*, which apparently parasitizes both wheat (*Triticum aestivum* L.) and rice, has been described from India (Gaur et al., 1993).

Symptoms

Root-knot nematode disease of rice was first described in Arkansas in 1934 (Tullis, 1934). Root-knot nematodes cause conspicuous, malformed, galled roots on most plant hosts. Serious infestations by the nematode can cause dwarfing and death of plants with an associated range of nondiagnostic aboveground symptoms, which simply reflect a damaged and dysfunctional root system. Infection of rice roots by *M. graminicola* and *M. oryzae* causes the tips of affected roots to take on a characteristic swollen, hooked appearance. In all cases, however, diagnosis as root-knot disease should be based on observation of females and egg masses protruding from or embedded within galled root tissue. As many as 50 females (Segeren et al., 1984) can be embedded within a single gall. Additionally, some rice plants can support as many as 1785 galls without any visible symptoms under normal conditions (Patnik, 1969).

Biology

The biology of root-knot nematode is very similar on most hosts with only a few exceptions as regards *M. oryzae* and *M. graminicola*. These two species of *Meloidogyne* have been more widely associated with and more extensively studied on rice than any of the other species. One of the main factors that makes *M. oryzae* and *M. graminicola* more of a concern in rice production is their ability to survive in poorly drained, flooded soil. *Meloidogyne graminicola*, in particular, is able to survive for extended periods of time in flooded soil in a completely endoparasitic mode within submerged rice root tissue. Eggs can hatch within galled root tissue, and the juveniles can migrate within the same root system to establish new feeding sites in which they themselves mature and produce offspring (Bridge & Page, 1982). Similarly, eggs in soil produced on a previous rice crop or on intercrop weed species can hatch prior to establishment of the permanent flood and infect newly transplanted seedlings. In glasshouse studies, Plowright and Bridge (1990) simulated conditions necessary to evaluate the effect of *M. graminicola* on the rice cultivar IR36 grown under upland conditions. They found that an initial population of 80 juveniles per milliliter of soil caused seedling death within 10 d, and by 32 d after inoculation there was 80% seedling mortality. They further noted the importance of this root-knot species in deepwater rice production since seedlings, established under upland conditions, failed to elongate above rising flood water. Plant-to-plant infection by root-knot juveniles or from eggs that hatch after establishment of the permanent flood has not been reported. The life cycle duration for *M. graminicola* is apparently highly dependent upon environmental conditions since some investigators indicate a generation time of as little 19 d (Bridge & Page, 1982), others estimate it at 23 to

27 d (Yik & Birchfield, 1979) and still others indicate the life cycle can require as long as 51 d to complete (Rao & Israel, 1973).

Most known species of root-knot nematodes have very extensive host ranges. The host range for *M. arenaria*, for example, includes more than 300 plant species, whereas those for *M. incognita* and *M. javanica* number more than 700 (Prot, 1992).

Disease Complexes

Historically, root-knot nematodes have been widely cited as interactants in disease complexes, especially those involving other rhizosphere-inhabiting pathogens. The lack of such reports for this crop, undoubtedly, is related to the fact that the low O_2 status of flooded soil constitutes an inhospitable environment for many microorganisms. A few reports, however, detail interactions involving *M. javanica* and the foliar pathogen, *Pyricularia oryzae*, and between *M. javanica* and *Fusarium* spp. (Prot, 1992).

Economic Importance

Root-knot nematodes are certainly a limiting factor in rice production. Estimates of losses, especially to upland rice, have been as high as 50 to 60% (Diomande, 1981; Babatola, 1984). *Meloidogyne graminicola* is of particular economic importance since it is pathologically associated with upland, irrigated, lowland, and deep-water rice production.

Hirschmaniella Species

Species and Races

In their discussion of this nematode genus, Bridge et al. (1990) indicate that seven species of *Hirschmaniella* are known to damage rice, six species are endoparasites of rice roots, and four are associated with weed species commonly found in rice fields. The 1968 key by Sher remains the best source for separation of species within the genus *Hirschmaniella*, and is current with respect to separation of the seven species known to damage rice. Survey work by Cuc and Prot (1992) in the Mekong Delta of Vietnam was designed to identify nematode species associated with deep water rice production that also damage irrigated rice. They found that two species of rice root nematode, *H. mucronata* and *H. oryzae*, were present in virtually all deep water rice fields surveyed. These two nematodes species are currently highly suspect as causal agents of damage to irrigated rice.

Symptoms

As described for root-knot nematode herein, rice root nematode causes no diagnostic symptoms on aerial plant parts. Foliar chlorosis, plant stunting, restricted numbers of tillers, and delays in flowering have all been attributed to parasitism by nematodes in this genus (Hollis & Keoboonrueng, 1984; Ichinohe, 1988). Discrete root symptoms also are lacking for this group of nematodes.

Adults, juveniles, and eggs can usually be observed by microscopic examination of infested root tissue.

Biology

The migratory "tunneling" activities of individuals in this genus as they move through, feed on, and reproduce in rice root tissues can result in a severely deteriorated root system. Most species that parasitize rice have moderately wide host ranges. Bridge et al. (1990) list 32 weed hosts, 13 of which support high reproduction, and eight crop hosts for *Hirschmaniella* spp. In addition to alternate crop and weed hosts and rice stubble, individuals of *Hirschmaniella* spp. are well equipped to survive unfavorable temperature and moisture conditions. Additionally, in the absence of rice plants they may act as facultative anaerobes (Babatola, 1981) or enter a quiescent state (Fortuner & Merny, 1979).

Disease Complexes

Only one reference details the involvement of *Hirschmaniella* spp. in a disease complex on rice. This interrelationship is commonly referred to as the rice root browning disease complex (Lee & Park, 1975).

Economic Importance

Across *Hirschmaniella* species and environments, loss estimates for rice root nematode, which occurs in fields in slightly more than half of the rice-producing areas of the world (Hollis & Keoboonrueng, 1984; Ichinohe, 1988), average 30%.

Heterodera Species

Species and Races

Four species of cyst nematodes, *Heterodera elachista*, *H. oryzicola*, *H. oryzae*, and *H. sacchari*, are known to attack rice. The existence of aberrant biotypes for any of these four species has not been reported. These four species can easily be distinguished on the basis of non-specific esterase banding patterns (Nobbs et al., 1992). Separation on the basis of morphological characters is possible although some characters, especially the lip region, are very similar among the four species.

Symptoms

Root damage, and associated browning of heavily infected roots, leads to the non-specific, aboveground symptoms discussed above for other nematode genera. Diagnosis as cyst nematode damage is based upon observing the white, yellow, or brown female life stages attached to rice roots.

Biology

All four of these cyst nematode species occur in the soil and roots of both upland and irrigated rice. *Heterodera sacchari* reproduces parthenogenetically, and *H. elachista*, *H. oryzicola*, and *H. oryzae* are amphimictic. Additionally, females of these latter three species deposit eggs into large egg masses from which second-stage juveniles hatch freely in water. Eggs of *H. elachista*, *H. oryzicola*, and *H. oryzae*, which are retained within the cyst, however, must be stimulated to hatch by rice root diffusate (Jayaprakash & Rao, 1982). Conversely, *H. sacchari* does not usually produce an egg mass; and, eggs within the cyst do not require root diffusate to induce hatching (Bridge et al., 1990) since juveniles can survive long periods of time in a desiccated state (Ibrahim & Perry, 1992). Recent data of Ibrahim et al. (1993a) prompts a slight modification in thought with regard to the effect of root exudate on hatching of eggs of *H. sacchari*. They conducted a study of the hatching behavior of *H. sacchari* and *H. oryzicola* as it relates to host age. Their data confirmed previous reports that *H. oryzicola* is dependent on host diffusates to stimulate hatch, regardless of host age. With *H. sacchari*, however, there was some evidence of a root exudate effect on egg hatch. The authors contend that, if diffusates are involved, those from older plants are more stimulatory than those from young plants. The apparent necessity of rice root exudates for hatch of eggs of *H. elachista*, *H. oryzicola*, and *H. oryzae* probably relates to the fact that they, unlike *H. sacchari*, have a rather narrow host range. Hence, the emergence of infective juveniles from eggs is restricted to a time when there is maximum probability for host contact.

Economic Importance

Rice cyst nematodes, probably because of their restricted distribution, are of only minor pathological concern. Nevertheless, Bridge et al. (1990) list several localized reports of rice cyst nematode damage estimates in the range of 7 to 42%.

Pratylenchus Species

Species and Races

Reports from all major rice production areas of the world attest to frequent association of approximately 10 different species of lesion nematodes, especially *P. zeae*, with upland rice (Plowright et al., 1990; Prot, 1990; Prot et al., 1992a). An excellent compendium of the genus *Pratylenchus*, by Frederick and Tarjan (1989), is available and covers all species currently reported to be parasitic on rice.

Symptoms

Root damage by lesion nematodes can cause brown to black lesions on heavily infected roots. Such damage can lead to the nonspecific, aboveground symptoms reported by Plowright et al. (1990) for *P. zeae*. Diagnosis of lesion-nematode damage must be based upon finding life stages of the nematode associated within roots and soil around visibly stunted rice plants.

Biology

Lesion nematodes are primarily migratory endoparasites but they are known to migrate into soil from heavily damaged or decaying roots. They can exist for extended periods under bare fallow or they can survive cropless periods by infecting several alternate weed hosts (Fortuner, 1976). No relationship between population densities of *P. zeae* and yield has been established since a constant level of yield reduction was observed across a range of nematode population densities (Plowright et al., 1990). In subsequent work designed to address this problem, Prot and Savary (1993) evaluated the potential, under both field and greenhouse conditions, of using correspondence analysis for assessing this nematode density:plant yield relationship. They determined *P. zeae* densities and yields of the rice cultivar UPL Ri5 at 490 spots in a field and found that this analysis revealed meaningful associations related primarily to the nematode-carrying capacity of the root system.

Economic Importance

Lesion nematode species such as *P. zeae* and *P. indicus* have been reported to cause damage estimated at 13 to 29% (Plowright et al., 1990) and 33% (Prasad & Rao, 1978), respectively. Evaluation of the damage potential of other species of *Pratylenchus* which are parasitic on rice awaits further research.

Other Nematodes

Various species in the genera *Criconemella, Criconema, Hoplolaimus, Paralongidorous, Xiphinema, Tylenchorhynchus, Helicotylenchus*, and *Caloosia* are listed by Bridge et al. as known rice parasites in their 1990 discussion, and in a recent report by Prot et al. (1992). Very little new information about these nematodes has been generated since then.

CONTROL

Aphelenchoides besseyi

Prasad and Varaprasad (1992) conducted studies designed to eliminate *A. besseyi* from infested seed. They evaluated 16 seed treatments on 10 rice entries heavily infested with the nematode. They found that soaking seed in a 0.2% solution of mancozeb (Zn ion and manganese ethylene bisdithiocarbamate) and monocrotophos (dimethyl phosphate of 3-hydroxy-n-methyl-cis-crotonamide) followed by vacuum fumigation with methyl bromide for 2 h at 30°C completely eliminated the nematode with no adverse effect on seed viability. Gergon and Prot (1993) evaluated the efficiency of benomyl [methyl 1-(butylcarbomoyl)-2-benzimidazolecarbamate] applied as a seed treatment and spray, and carbofuran (2-3-dihydro-2, 2-dimethyl-7-benzofuranyl methylcarbamate) applied in floodwater to prevent the nematode from infecting rice seedlings. They found that seedlings from benomyl-treated seed that were transplanted and sprayed with benomyl either 1 or 15 d later were protected from the disease. Carbofuran broadcast into floodwater either 1 or 30 d after transplanting nematode-infested seedlings to the field failed to control the disease. Other, more conventional, methods for controlling white tip disease include the use of resistant/tolerant cultivars, burning of rice stubble, seed treatment with hot water, as well as cultural methods such as seeding and transplanting into water and reducing the plant population density per field.

Ditylenchus angustus

Elimination of Ufra nematode-infested plant material by burning or burying can be an effective management technique. Other research has noted the utility of measures such as controlling Ufra by manipulating water flow (Sein & Zan, 1977), employing resistant or tolerant rice cultivars, or by manipulating rice cropping patterns and methods of cultivation (McGeachie & Rahman, 1983).

Meloidogyne Species

As indicated earlier, most species of root-knot nematodes, except *M. graminicola* and *M. oryzae,* are intolerant of flooded conditions. Hence, water management is a key element in root-knot control. The use of root-knot resistant cultivars, as always, is the most desirable management tactic. Developing resistance to *M. graminicola*, however, seems to be the greatest challenge. Bridge et al. (1990) list 32 rice cultivars from India, Thailand, and the USA that are reported to contain resistance to *M. graminicola*. Tandingan et al. (1994) evaluated the tolerance of 15 rice cultivars to *M. graminicola* under upland and irrigated conditions. They found that for several of the cultivars, IR29 and IR74 in particular, although infestation by the juveniles of *M. graminicola* was greater under irrigated than upland conditions, yields of irrigated cultivars were 20% greater.

Hirschmaniella Species

Prot et al. (1992b) evaluated the effects of green manure crops of *Sesbania rostrata* Bremek. & Oberm. and *Aeschynomene afraspera* J. Leonard in a rice production system. Populations of both *Hirschmaniella mucronata* and *H. oryzae* were reduced significantly. However, since these two crops do not generate direct income, their use was considered not to be economical. Other methods for controlling this nematode that show promise include crop rotation, bare fallow and the use of resistant cultivars.

Heterodera Species

Work reported by Nishizawa et al. some 25 yr ago (1972) suggested that rotation of rice with nonhost crops showed potential for cyst nematode control. This recommendation seems to be applicable even today since cyst-resistant cultivars tend to be susceptible to damage by root-knot nematodes.

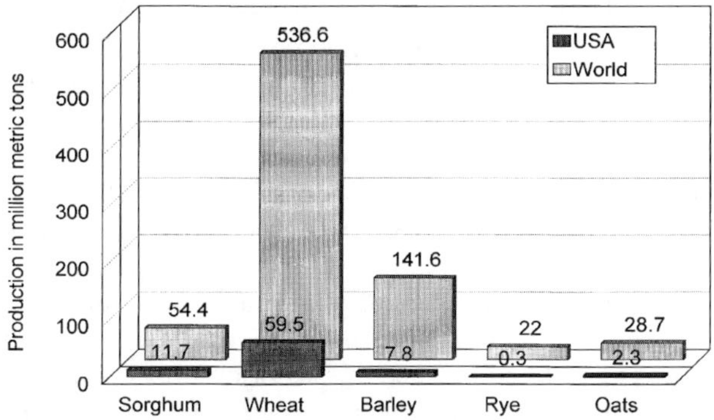

Fig. 21–2. 1995 world and U.S. production of sorghum, wheat, barley, rye, and oat.

Pratylenchus Species

Currently, no satisfactory control measures are available for lesion nematodes on rice. The use of nematicides currently is too expensive. Crop rotation is unreliable because of the wide host range and number of lesion nematode species that are found associated with the crop. Rice cultivars resistant to lesion nematodes have not yet been identified.

OTHER CEREAL CROPS

Aside from rice, the cereal crops include sorghum [*Sorghum bicolor* (L.) Moench], wheat [*Triticum aestivum* (soft wheat) or *T. durum* Desf. (durum wheat)], barley (*Hordeum vulgare* L.), rye (*Secale cereale* L.), and oat (*Avena sativa* L.). These crops constitute an important source of protein, starch, and amino acids essential, both directly and indirectly, in human nutrition. Production in teragrams for the USA and world for 1995 was 11.7 and 54.4 for sorghum; 59.5 and 536.6 for wheat; 7.8 and 141.6 for barley; 0.3 and 22.0 for rye; and 2.3 and 28.7 for oat (Fig. 21–2.) (USDA Natl. Agric. Stat. Serv., Jan., 1995; USDA Econ. Res. Serv., 1995). A wide spectrum of both foliar and root parasitic nematode species are known from these crops. Yield losses they cause across these five crops average about 6%/yr (Sasser & Freckman, 1987) with the greatest losses occurring on wheat.

SORGHUM

Important Nematodes

Meloidogyne Species

Species and Races. The major species of root-knot nematodes that attack sorghum include *M. incognita, M. naasi*, and *M. acronea*. *Meloidogyne incogni-*

ta was first reported on sorghum by Orr in 1967 in west Texas. Of the four described races of *M. incognita*, only Race 3 has been evaluated extensively (Fortnum & Currin, 1988; Ibrahim et al., 1993b; Thomas & Murray, 1987) with regard to its effects on sorghum growth. Several researchers working with *M. incognita* and sorghum did not specify which race of *M. incognita* was under investigation (McSorley & Gallaher, 1991; Birchfield, 1983; Gallaher et al., 1988). However, McSorley et al. (1987), who conducted a rotation study with *M. incognita* Race 1, found low population levels of this race associated with sorghum and concluded that it was not a favorable host for reproduction. The work of Thomas and Murray (1987) with Race 3 of the *M. incognita* reveals the importance of race identification since this work described sorghum yield losses up to 45% and nematode population increases over the course of the experiment that ranged from 30- to 42-fold. In contrast, Orr and Morey (1978) considered sorghum to be a poor host of the cotton root-knot nematode (*M. incognita acrita*), which may be either Race 3 or Race 4. They also noted, however, that there was abundant reproduction by the nematode.

Michell et al. (1973a) described five races within *M. naasi*. However, only Race 5 of this species develops and reproduces on sorghum (Michell et al., 1973b).

Symptoms. Visual damage by *M. incognita* was first reported on sorghum as stunted plants which bloomed late and had suppressed yield (Orr, 1967). Root systems were small, and numerous galls were present on lateral roots. When compared at similar nematode population levels, damage to cotton was far greater than on sorghum (Orr, 1967). The cotton root-knot nematode caused elongated swellings, discrete knots, and abundant secondary root proliferation when infecting sorghum (Orr & Morey, 1978). Lateral roots were found to be more heavily infected than brace roots, and no galls were observed on the brace roots.

Meloidogyne naasi caused aboveground symptoms that were visible as irregular areas containing severely stunted sorghum plants with severely galled root systems (Aytan & Dickerson, 1969; Ediz & Dickerson, 1976). Additionally, the latter authors observed significantly suppressed shoot and root growth and restricted plant height, which resulted in smaller stem diameters.

Biology. *Meloidogyne naasi* reproduces by facultative meiotic parthenogenesis. The nematode completes its life cycle in 31 d at 30°C under greenhouse conditions (Aytan & Dickerson, 1969). The optimum soil temperature for *M. naasi* to complete its life cycle on sorghum was 26°C and required 34 d (Ediz & Dickerson, 1976). When compared at the same temperature, nematode maturation was more rapid on sorghum than on barley. Michell et al. (1973b) suggest that a better host-parasite relationship exists for *M. naasi* on sorghum than on barley.

Disease Complexes. The 1958 report by Norton was the first to suggest an interaction between a sorghum-associated nematode, *Pratylenchus hexincisus*, and another plant pathogen, *Macrophomina phaseolina*. This work showed an additive interaction between the two organisms, especially under drought conditions. Orr and Morey (1978) later reported an association involving *M. incognita* and *Pythium* spp. on sorghum. In greenhouse tests, they reported that more extensive root necrosis and injury to sorghum plants occurred with both organisms together.

Economic Importance. On a worldwide basis, *M. incognita* apparently is the most damaging root-knot species on sorghum. Loss estimates in greenhouse, microplot, and field tests of 15, 45, and 33%, respectively, have been reported for this species (Orr, 1967; Thomas & Murray, 1987).

Pratylenchus Species

Species and Races. Eight species in this genus including *P. zeae, P. hexincisus, P. crenatus, P. penetrans, P. brachyurus, P. coffeae, P. sudanensis,* and *P. thornei* have been found associated with sorghum. See the 1989 compendium of the genus *Pratylenchus*, by Frederick and Tarjan, for identification of many lesion nematode species.

Symptoms. Diagnosis as damage by lesion nematode must be based upon finding life stages of the nematode associated within roots and soil around visibly stunted sorghum plants. Unfortunately, lesion nematodes, like many other plant pathogens, cause nondiagnostic lesions on the roots of susceptible plants. These lesions may vary from small, lightly colored areas to dark, necrotic areas depending upon population density of the nematode (Cuarezma-Teran & Trevathan, 1985; Acosta & Malek, 1981).

Biology. Lesion nematodes are migratory endoparasites of which approximately 50% of the described species are parthenogenetic. Norton (1958) found that *P. hexincisus* was abundant in cortical parenchyma tissue of sorghum rootlets and that feeding individuals oriented themselves parallel to the longitudinal axis of the root. Norton further noted that nematodes were rarely found in older roots but were common in herbaceous rootlets. He also made the observation that drought stress augmented the damage to sorghum caused by *P. hexincisus*. In greenhouse experiments, McDonald and Van Den Berg (1993) more critically examined the influence of water regime on damage to sorghum caused by the lesion nematodes *P. zeae* and *P. brachyurus*. In their system, in which soil was infested with either 9000 *P. zeae* or 6300 *P. brachyurus* per 4-L pot, neither species was pathogenic to sorghum, and nematode reproduction was unaffected by the degree of water stress imposed on sorghum plants. Moreover, they found that inoculation with *P. brachyurus* actually stimulated root development and plant height. These results conflict with data of Cuarezma-Teran and Trevathan in Mississippi (1985). They associated damaged sorghum plants with the incidence of *P. zeae* and the stunt nematode, *Quinisulcius acutus*. The 1985 data, based upon greenhouse experimentation using nematode infestation levels and durations comparable to those of McDonald and Van Den Berg (1993), showed significant damage to sorghum by *P. zeae*. These investigators also had reported previously that there was a negative correlation between *P. zeae* population density and soil pH and percentage of base saturation (1984). In his discussions of lesion nematodes associated with sorghum, Starr (1986, 1992) indicates that only

P. zeae, P. hexincisus, and *P. thornei* have been shown to be pathogenic to sorghum.

Disease Complexes. The combination of *P. hexincisus* and the charcoal rot fungus (*Macrophomina phaseolina*) increased sorghum disease ratings and decreased top weights under a watering regime that allowed plants to temporarily wilt between waterings (Norton, 1958). Effects of the nematode and fungus together were additive with respect to disease rating and plant growth. Population development of *P. hexincisus* was not influenced by the occurrence of charcoal rot disease. The various interactions of *P. zeae* and the fungi *Curvularia* spp., *Fusarium moniliforme*, *Rhizoctonia solani*, and *Macrophomina* spp were mostly antagonistic with respect to effects on foliage and growth of sorghum (Bee-Rodriguez & Ayala, 1977). The most damaging combination involved the nematode and *Curvularia* spp.

Economic Importance. Lesion nematodes are very widespread and frequently associated with sorghum. Estimates of percentage of damage to the crop by *Pratylenchus* species, however, are lacking. The percentage of loss to the sorghum caused by all nematodes was estimated at 6.9% for the USA in 1987 (Sasser & Freckman). Currently, insufficient data are available to judge what proportion of this loss is attributable to lesion nematodes.

Tylenchorhynchus and *Quinisulcius* Species

Species and Races. Several species within these two related genera are known pathogens of sorghum including *T. annulatus, T. nudus,* and *Q. acutus* (Starr, 1986, 1992). A report by Chevres-Roman et al. (1971) also indicated that *T. claytoni* was pathogenic to sorghum and corn (*Zea mays* L.), but this observation has not been confirmed by subsequent research. References for identification of species in the genera *Tylenchorhynchus* and *Quinisulcius* are available (Anderson & Potter, 1991; Tarjan, 1973).

Symptoms. Root damage and possible nondiagnostic browning of heavily infected roots lead to the nonspecific aboveground symptoms discussed herein for other nematode genera.

Biology. Cuarezma-Teran et al. (1984) found a positive correlation between soil P and K and population density of *Q. acutus*. Cuarezma-Teran and Trevathan (1985) indicated root parasitism by *Q. acutus* suppressed plant growth in greenhouse tests while causing discrete root lesions and discoloration. With *T. nudas*, temperature apparently plays a key role in both reproduction and damage capability by the nematode (Smolik, 1977). Maximum injury to sorghum was apparent at 15 to 20°C, and maximum reproduction occurred at 30°C. The only visible damage symptoms were restricted root size. In this system, soil fertility also played an important role since nutrient stress resulted in enhanced pathogenicity by the nematode.

Economic Importance. Comments made earlier for lesion nematodes and sorghum also apply here.

Other Nematodes

Paratrichodorus allius is associated with sorghum (Smolik, 1977) in South Dakota. Microplot studies by Wenefrida et al. (1994) in Louisiana examined single and combined effects of *Macrophomina phaseolina* and ring (*Criconemella xenoplax*) and stunt (*Tylenchorhynchus annulatus*) nematodes on several cultivars of sorghum. Effects of the three organisms on growth limitation were antagonistic with regard to sorghum growth, nematode reproduction, and colonization of roots.

CONTROL

Host Resistance

Ten sorghum cultivars were evaluated for their reaction to *M. incognita* Race 3, and Races 1 and 2 of *M. arenaria* and *M. javanica* (Fortnum & Currin,1988). All were considered to be either poor hosts or, in the case of Pioneer B815, a nonhost of all three root-knot species. In related field tests over 2 yr, soil populations of root-knot nematodes following sorghum were suppressed to 1.4 and 4.6% of those which occurred on tobacco (*Nicotiana tabacum* L.). These data strongly suggest that sorghum can be an effective rotation crop for management of some root-knot nematodes. Findings by Birchfield (1983) with sorghum and root-knot nematodes are in contrast to those of Fortnum and Currin (1988). In Birchfield's work eight out of nine cultivars of sorghum were very susceptible to *M. incognita*. The cultivar Funk G 499BR was the only one that Birchfield considered to be moderately resistant. Work by Ibrahim et al. (1993b) supports the findings of Fortnum and Currin (1988) with respect to *M. arenaria* and those of Birchfield relative to *M. incognita* and sorghum.

Cultural

Few generalizations can be made relative to the utility of sorghum as a rotation crop for nematode control. Reports by some (as discussed earlier under Biology") indicate sorghum supports abundant root-knot nematode reproduction, whereas those by others show the crop to be either a poor host (Bridge, 1987) or a virtual nonhost. A corn-vetch (*Vicia villosa* Roth)-sorghum rotation significantly reduced (about 3-fold) the population density of *M. incognita* at crop harvest when compared with a corn-vetch-corn scheme (Gallaher et al, 1988). Additionally, when nematode numbers from the sorghum-vetch-sorghum scheme were compared with those from corn-vetch-corn, the difference was approximately sevenfold. Other work by McSorley and Gallaher (1991) supported previous observations of the comparative effectiveness of corn and sorghum against *M. incognita* in cropping systems.

Rodriguez-Kabana and Touchton (1984) found that sorghum reduced populations of *M. arenaria*, but populations recovered rapidly on susceptible peanut (*Arachis hypogaea* L.). The use of sorghum alone was not considered economical as a rotational crop with peanut.

Corn and soybean have been endorsed as effective rotation crops for management of *M. naasi* on sorghum although *M. naasi* infestation is currently very limited in the USA (Ediz & Dickerson, 1976).

Chemical

Currently, the nematicides aldicarb [2-methyl-2-(methylthio)propionaldehyde 0-(methylcarbamoyl)oxime], terbufos (S-{[(1,1-dimethylethyl) thio] methyl} 0,0-diethyl phosphorodithioate), 1,3-dichloropropene, and a combination of 1,3 dichloropropene and chloropicrin (trichloronitromethane) are labeled for use on sorghum in the USA (Anonymous, 1995). However, environmental concerns and the economics of sorghum production usually preclude their usage.

WHEAT

Important Nematodes

Heterodera Species

Species and Races. The cereal cyst nematode, *Heterodera avenae,* is the most important nematode on wheat. Publications by Baldwin and Mundo-Ocampo (1991) and Golden (1986) are excellent references to genera of cyst nematodes and species of the genus *Heterodera*, respectively. *Heterodera avenae* was first reported on oat in Germany in 1874 and is widely distributed throughout many of the wheat-growing areas of the world. Infestations are known in 33 countries (Meagher, 1977; Sarena et al., 1988). In North America, *H. avenae* is a serious problem in Canada and has been reported in the USA from California, Michigan, Oregon, Idaho, and Washington (Jensen et al., 1975; Hafez & Golden, 1984, 1985; Hafez et al., 1992; Granez, 1985). Most references list at least 10 host races (pathotypes) of *H. avenae*, and some list as many as 20 (Baldwin & Mundo-Ocampo, 1991). A number of other *Heterodera* species cause damage to wheat. These include *H. bifenestra, H. hordecalis, H. latipons, H. mani, H. pakistanensis*, and *H. zeae*. *Punctodera punctata* is another cyst nematode specie that also parasitizes wheat.

Symptoms. Root damage and associated browning of heavily infected roots lead to the nonspecific aboveground symptoms characteristic of infestation by many other plant-parasitic nematodes. Diagnosis as cyst nematode is based upon observing the white, yellow, or brown female life stages attached to wheat roots. With *H. avenae*, leaf discoloration is common, and tillering may be greatly reduced. In severely infested fields, yield may be drastically reduced.

Biology. *Heterodera avenae* is a sexually dimorphic species. The nematode has a life cycle similar to other cyst species although there is usually only one generation of new individuals produced per year (Decker, 1989; Valdeolivas & Romero, 1985). Juveniles of *H. avenae* survive best in a low moisture regime at soil temperatures of 5 to 10°C (Davies & Fisher, 1976). Invasion and development of the nematode are temperature dependent and usually slow during winter months (Kerry & Hague, 1974). The optimum temperature for egg hatch of *H.*

avenae is 10°C in Australia (Brown, 1987), but it has been reported to occur at temperatures as high as 20 to 22°C (Winslow, 1955). Egg-bearing females develop within 6 to 9 wk after invasion. Hosts of this nematode include cereals such as wheat, oat, rye, and barley (*Hordeum vulgare* L.) as well as numerous grasses (Williams & Siddiqi, 1972; Decker, 1989). Even though corn is considered to be a poor host of *H. avenae* (Johnson & Fushtry, 1966), the nematode is capable of causing significant damage to this crop. Injury to corn roots is caused by penetration of juveniles, even though they usually do not fully mature (Miller, 1986). Similarly, the corn cyst nematode (*H. zeae*), is capable of penetrating and potentially damaging wheat roots (Bajaj et al., 1986) when the temperature approaches 30 to 32°C.

Both low and moderate population densities of *H. avenae* were found to delay maturity of wheat (Zancada & Althofer, 1994). Rivoal et al. (1990) found that the damage to wheat resulted from additive detrimental effects throughout the growing season. They also contended that *H. avenae* damage to wheat was similar to that of water stress.

Disease Complexes. The work of Meagher and Chambers (1971), Meagher et al. (1978), and Brown (1984b) details the relatively few studies that have been conducted with *H. avenae* and other wheat pests. These investigators reported a synergistic interaction between *H. avenae* and *Rhizoctonia solani*, which affected both growth and yield of wheat. Recent work by Lasserre et al. (1994) details the interaction between *H. avenae* and the lesion nematode (*Pratylenchus neglectus*). These investigators report that, in long-term field experiments, population development of lesion nematode on wheat that followed *H. avenae*-susceptible oat was inhibited in the presence of *H. avenae*. No inhibition of lesion reproduction on wheat occurred when the *H. avenae* population had been reduced by previous cultivation of oat cultivars resistant to *H. avenae*. Another study by Smiley et al. (1994) involving the incorporation of wheat into a crop rotation scheme found that coinfestation of wheat with *H. avenae* and *Gaeumannomuyces graminis* var. *tritici* had the greatest negative impact on yield. This work also documented a reduction in the numbers of roots produced in the presence of both the nematode and *Pythium* spp.

Economic Importance. The limited number of reports describing damage to wheat by *H. avenae* indicates that it is a minor pest in the USA. The nematode, however, is widespread throughout Europe, India, and Australia and is considered to be a major pest. At least 2 million hectares of land in Australia are infested with *H. avenae*, and losses are estimated at $72 million (Australian) annually (Brown, 1984b). Other annual loss estimates for *H. avenae* include £3 million sterling for Europe and $9 million in India (Wallace, 1965; Van Berkum & Seshadri, 1970). Damage estimates as high as 50% under field conditions have been attributed to *H. avenae* in Pakistan with an overall yield loss estimate for the country of 15 to 20% (Maqbool, 1987).

Meloidogyne Species

Species and Races. Worldwide, 10 to 15 root-knot nematode species are known to attack wheat (Sikora, 1988). *Meloidogyne naasi* and *M. artiellia* are the two species of major economic importance on a worldwide basis, although neither occurs with great frequency in the USA. *Meloidogyne naasi* is distributed widely throughout Europe in the wheat-producing areas of England, France, the Netherlands, Belgium, Germany, and Italy (Jepson, 1987). In the USA, the nematode is reported from California, Illinois, Kansas, Kentucky, and Oregon (Anonymous, 1984). *Meloidogyne naasi* and *M. artiellia* are described in a current work by Eisenback and Triantaphyllou (1991). Reports of the incidence of *M. artiellia* on wheat have come from Europe, Italy, Spain, and Syria (Greco & DiVito, 1988).

In the USA, *M. incognita, M. javanica, M. arenaria*, and *M. chitwoodi* are the primary root-knot nematode species associated with wheat. Kaloshian et al. (1989a), in a study of the comparative reproduction of *M. javanica* and *M. incognita* on wheat, found that reproduction by *M. javanica* was greater than that by Races 1, 2, or 3 of *M. incognita*. On six wheat cultivars, females of Race 2 of *M. incognita* produced more eggs than did females of Race 3 or females of *M. javanica* (Kaloshian et al., 1989b). Currently, only *M. chitwoodi* is thought to pose a serious threat to wheat production, especially in the Pacific Northwest (Santo et al., 1980). At least three races of *M. chitwoodi* are known (see "Biology" section below). The distribution of *M. chitwoodi* appears to be somewhat limited having been detected only in the USA, Argentina, Holland, South Africa, Australia, and Mexico (Eisenback & Triantaphyllou, 1991). *Meloidogyne triticoryzae*, which apparently parasitizes both wheat and rice, was recently described from India (Gaur et al., 1993).

Symptoms. Galls produced by root-knot nematode species on wheat are generally small. Kilpatrick et al. (1976) associated yellow and stunted wheat seedlings, missing plants in the rows, and small and poorly developed roots with infestation by *M. naasi*.

Griffin (1992) found that *M. chitwoodi* reduced the tillering of all wheat cultivars tested. He also noted significant reductions in plant height, shoot weight, root weight, head set, head length, and head dry weights of all the spring wheat cultivars. A direct correlation was made between P_i (the initial population density) and root-galling indices when plants were inoculated with 2, 10, and 20 eggs/cm^3 of soil.

Biology. *Meloidogyne chitwoodi* can reproduce and cause injury to wheat, corn, barley, and oat (Santo & O'Bannon, 1981). Under greenhouse conditions, the suppression in growth of 'Nugaines' winter wheat by *M. chitwoodi* was related directly to nematode population density (Nyczepir et al., 1984). Subsequent microplot experiments showed that nematode population densities of 0.75 and 9.0 eggs/cm^3 of soil affected only the head weights of 'Felder' spring wheat, and that Nugaines was unaffected by the nematode at any level and may be tolerant. Development of *M. chitwoodi* on wheat required 105, 51, 36, and 21 d at 10, 15, 20, and 25°C, respectively (Inserra et al., 1985). Second-stage juveniles of *M. chitwoodi* invaded root tips or axial roots, and giant cells were observed after 10 d at 25°C. Egg-laying females were produced within 25 d at this temperature. Reproduction index was highest when an infestation level of 2 eggs/cm^3 of soil was used, and reproduction of *M. chitwoodi* varied considerably among the winter wheat cultivars tested.

Santo and Pinkerton (1985) reported that potato (*Solanum tuberosum* L.) following alfalfa (*Medicago sativa* L.) was damaged by *M. chitwoodi*. Subsequent greenhouse research conducted with this isolate of *M. chitwoodi* showed good reproduction on several cultivars of alfalfa compared with that observed for the wild race" of *M. chitwoodi*. Griffin (1988) also found a population of *M. chitwoodi* that was virulent on potato and could reproduce on several cultivars of alfalfa. Because of these findings, at least two races of *M. chitwoodi* have been described based on their ability to reproduce on alfalfa (Griffin & Thomson, 1988; Santo & Pinkerton, 1985). Mojtahedi et al. (1988) showed that Race 2 of *M. chitwoodi* also reproduced well on oat, barley, and both spring and summer wheat. They proposed adding 'Red-Cored Chantenay' carrot [*Daucus carota* subsp. *sativus* (Hoffm.) Arcang.] and 'Thor' alfalfa to the differential host list for *M. chitwoodi*. Mojtahedi et al. (1988) also noted that Race 1 of *M. chitwoodi* reproduced well on carrot but not on the alfalfa, and that the inverse was true for Race 2. Differences in aggressiveness also have been observed between with the two alfalfa" isolates of *M. chitwoodi*. The Utah population of *M. chitwoodi* had greater reproduction indices on Nugaines and Wander' winter wheat than did the Idaho population (Griffin, 1992). A third race of *M. chitwoodi* has recently been described from California that can, unlike Races 1 and 2, reproduce on a wild potato, *Solanum bulbocastanum* Dunal (Mojtahedi & Santo, 1994).

Meloidogyne incognita, *M. javanica*, and *M. arenaria* are known to infect, develop, and reproduce on wheat, barley, oat, and rye (Roberts & Van Gundy, 1981; Johnson & Motsinger, 1990). The ability of these three root-knot nematode species, especially *M. incognita*, to damage wheat is minimal. Some investigators (Patel et al., 1986; Sharma, 1981) suggest that, under conditions favorable for reproduction, *M. javanica* can suppress wheat growth and yield. Others (Roberts & Van Gundy, 1981) indicate that neither *M. javanica* nor *M. incognita* adversely affect root or shoot growth or tillering of wheat across a range of inoculum densities and soil temperatures. Temperature requirements for reproduction may account, in part, for the lack of reports of damage to wheat by *M. incognita* since it does not reproduce at temperatures below 15 to 18°C (Thomason, 1962; Roberts et al., 1981).

In full-season microplot studies, Opperman et al. (1988) found that final population densities of *M. incognita*, *M. arenaria*, and *M. javanica* declined over time to levels below the original infestation level, and none affected wheat yield. They attributed the low reproduction and lack of effect on wheat growth to temperature effects rather than genetic resistance.

Disease Complexes. Research in France with *M. naasi* and several species of lesion nematode indicates a negative correlation between the incidence of the two genera of nematodes in wheat roots (Esmenjaud et al., 1990). They further stated, however, that infection of wheat roots with lesion nematodes did not inhibit penetration by infective juveniles of *M. naasi*.

Economic Importance. Root-knot nematode is much more important in other wheat- growing areas of the world than it is in the USA. *Meloidogyne naasi* has been shown to cause losses to wheat in southern Italy in 17% of fields sampled (Lamberti, 1979).

Pratylenchus Species

Species. Species of lesion nematodes, including *Pratylenchus neglectus* (syn. *P. minyus*), *P. thornei, P. penetrans,* and *P. fallax* have been found widely associated with wheat in both the USA and overseas (Anonymous, 1987).

Symptoms. Reports by several researchers describe the chlorotic, stunting symptoms associated with infection of wheat roots by lesion nematodes. Van Gundy et al. (1974) found *P. thornei* associated with reduced wheat stands and stunted plants. Mojtahedi and Santo (1988) identified *P. thornei* as a pathogen of wheat causing stunting and chlorotic plants. Both internal and external symptoms of wheat root invasion by lesion nematodes are similar to those described for other cereal crops. There is little doubt the infection portals and cortical tunnels caused by lesion nematode feeding and migration can serve as entry points for other rhizosphere microorganisms.

Biology. Benedict and Mountain (1956) were among the first to implicate lesion nematodes as a limiting factor in wheat production in Canada. The pathogenicity and histopathology of wheat infected with *P. thornei* were described by Baxter and Blake in 1967 and 1968 and are similar to that described for lesion nematodes on other crops. Mojtahedi and Santo (1992) found that *P. neglectus* was a weak pathogen on Stephens' wheat in Washington.

Disease Complexes. Very little work has been published in the area of lesion nematode disease complexes of wheat since the first association between the lesion nematode (*P. neglectus*) and *Rhizoctonia solani* was noted (Benedict & Mountain, 1956).

Economic Importance. A current estimate of the economic importance of lesion nematodes on wheat is lacking. Data from Mexico in 1974 (Van Gundy et al.) showed a 19 to 30% yield increase across six wheat cultivars when populations of *P. thornei* were controlled by application of the fumigant nematicide 1,3-dichloropropene. Similarly, a 1989 report from Canada showed a 10 to 19% yield response as the result of controlling *P. penetrans* with aldicarb (Kimpinski et al., 1989).

Other Nematodes

An excellent treatment of the biology, ecology, and pathology of the stem nematode, *Ditylenchus dipsaci*, and the seed gall nematode, *Anguina tritici,* is given in the review by Rivoal and Cook (1993). The recent review by Khan and Pathak (1993), which describes the vector potential of nematodes for dissemination of plant pathogenic bacteria and fungi, also contains interesting information on several wheat disease associations involving *A. tritici* and *Clavibacter tritici, Ustilago tritici, Tilletia foetida*, and *Neovossia indica*.

The needle nematode, *Longidorus leptocephalus,* has been shown to damage both wheat and barley (Sykes, 1979). Wheat also is reported to be damaged by several stunt nematodes including *Merlinius brevidens* and *Tylenchorhynchus* (Langdon et al., 1961; Jones, 1979).

CONTROL

Host Resistance

Many investigators have attempted, with little success, to identify resistance to root-knot nematodes in wheat over the years (Pedersen & Rodriguez-Kabana, 1987; Opperman et al., 1988) The difficulty is exemplified by the work of Roberts et al. (1982). They screened 29 wild and commercial cultivars of durum and bread wheat and found that all were all hosts of *M. incognita, M. javanica*, and *M. naasi*. Their work did identify one accession of *Aegilops squarrosa* L., which contained resistance to the four host races of *M. incognita* and four populations of *M. javanica*. Subsequent research showed that the resistance from *A. squarrosa* was more effective than that contained in most wheat cultivars developed for use in the southeastern USA (Kalsohian et al., 1989b).

In other work, Kaloshian et al. (1989a) also found that another *A. squarrosa* accession, G 3489, and the durum wheat/G 3489 cross, synthetic hexaploid G 4299 could give suppression of field populations of *M. chitwoodi* over time. Wheat cultivars that have the $6M^v$ chromosome derived from another *Aegilops* spp., *A. ventricosa* Tausch, also have been shown to have partial resistance against the cereal cyst nematode, *H. avenae* (Rivoal et al., 1993). This chromosome apparently allows only limited reproduction by *H. avenae* and could theoretically reduce the selection pressure placed on cultivars with monogenic resistance (Young, 1992). Several other sources of resistance to *H. avenae* in wheat have been noted recently (Rivoal & Cook, 1993). Unfortunately, some cultivars such as Katyl for example, which have been bred specifically for resistance to *H. avenae*, do not yield satisfactorily and have not been readily accepted by producers (Brown, 1984a,b). Additionally, cultivars like Katyl may be resistant in one environment (Australia) and susceptible in another (India) as the result of differences in pathotypes of the nematode.

Other research has attempted to identify and incorporate tolerance rather than resistance to nematodes into wheat cultivars (Rivoal et al., 1990). Resistant cultivars by definition permit minimal, if any, nematode reproduction. With tolerant cultivars, the focus is on satisfactory plant performance in the presence of the nematode, irrespective of the amount of reproduction that occurs. From the standpoint of nematode reproduction, then, tolerant cultivars act like susceptible cultivars in producing a pool of inoculum that may adversely impact subsequent plantings of the same or other crops.

Cultural

Wheat plants infected by *M. incognita* have rarely been observed to be seriously injured. However, the fact that wheat harbors the nematode is of great concern when it is employed as a rotation crop that may provide nematode inoculum to a subsequent crop. Findings of some studies, such as those by Johnson and Motsinger (1990) and Minton and Bondari (1994), demonstrate that the use of wheat (and selected cultivars of oat and rye) as a rotation crop impacts reproduction of *M. incognita* in a manner similar to that observed under fallow conditions. The situation becomes more complicated when evaluating rotation effects in fields containing more than one nematode species. Lasserre et al. (1994) showed an inverse relationship between *P. neglectus* and *H. avenae* on winter wheat. On wheat crops grown after either resistant ('Panama') or susceptible ('Peniarth') oat, populations of *P. neglectus* were higher where *H. avenae* had been controlled than in plots where *H. avenae* was still present at moderate levels. Similarly, reductions in *P. neglectus* populations coincided with increased reproduction by *H. avenae*. Esmenjaud et al. (1990) found a similar relationship for *M. naasi* and *Pratylenchus* spp. They further noted that sugarbeet (*Beta vulgaris* L. subsp. *vulgaris*) is an effective rotation crop for reduction of populations of lesion nematode. The findings of Corbett (1972) and Van Gundy et al. (1974) are in line with those of Esmenjaud et al. (1990) regarding the inhibitory effects of sugarbeet on lesion nematode reproduction. They found that sugarbeet negatively affected population development of *P. neglectus, P. fallax*, and *P. thornei*.

Wheat grown under field conditions where soil temperatures measured approximately 1°C during the winter appeared to escape damage by *M. chitwoodi* (Inserra et al., 1985). Plants were well developed by the time nematode infection occurred, and corresponding effects on wheat growth and yield were insignificant. Wheat, barley, corn, oat, potato, sugarbeet, and tomato (*Lycopersicon esculentum* L.) are hosts of *M. chitwoodi* but alfalfa, peanut, black pepper (*Piper nigrum* L.), and strawberry (*Fragaria* × *ananass*) are not (Golden et al., 1980; Santo et al., 1980). Both groups recommend the use of alfalfa to reduce populations of *M. chitwoodi* although it is now known that Race 2 of this nematode parasitizes alfalfa. Although resistance to *M. chitwoodi* in wheat may be of limited value to improve the yield performance, its true value may be in managing the residual population, which will confront a more susceptible rotation crop such as potato (Roberts, 1992).

Oat and corn are not considered good hosts of *M. artiellia* (Greco & Di Vito, 1988).

Chemical

Kerr (1967) observed greater root weights and increased tillering and height of wheat plants grown in soil containing *Merlinius brevidens* (syn. *Tylenchorhynchus brevidens*) and treated with steam, chloropicrin, or 1,3 dichloropropene than in nontreated plots. Soil fumigation with 1,3 dichloropropene produced a 19 to 30% yield increase across six wheat cultivars in tests against the lesion nematode *P. thornei* (Van Gundy et al., 1974). Soil fumigants were found to be more effective than nonfumigants in reducing populations of *M. chitwoodi* (Santo & Ponti, 1981). A significant correlation was made between yield response to aldicarb and egg counts and disease ratings for *H. avenae* (Simon, 1980). Orion and Shlevin (1989) reported control of both *H. avenae* and *Pratylenchus mediterranus* and higher wheat grain yield when seed dressings of oxamyl [*N,N*-dimethyl-2-methylcarbamoyloxyimino-2-(methylthio)acetamide], furathiocarb (*N,N*'-dimethyl-*N,N*'-thiodicarbamate), or carbofuran were used. Brown (1984a,b) reported similar results with the placement of low rates of nematicides into the seed drill.

Biological

A number of soilborne fungi such as *Catenaria auxiliaris, Nematophthora gynophila*, and *Verticillium chlamydosporium* have been found to suppress reproduction of *H. avenae* by acting themselves as parasites of eggs, females, and cysts (Kerry, 1981; Kerry & Crump, 1977). Initially, observations such as these seem promising but are probably still a long way from implementation for nematode suppression under field conditions.

Other

Low population levels of *H. avenae* have been shown to stimulate early wheat growth (Rivoal & Sarr, 1987). This is probably the result of increased uptake of fertilizer due to the formation of adventitious roots in response to nematode parasitism.

BARLEY, OAT, AND RYE

Important Nematodes

All three of these crops are subject to damage by most of the same species of *Heterodera* as those that attack wheat. Barley is slightly less susceptible to *H. avenae* than wheat, and oat is apparently slightly more susceptible (Decker, 1989). Volkmar (1989) reported that the main effect of *H. avenae* infestation of oat was the suppression of root elongation. Volkmar (1989) found that seminal root length was reduced more in NZC, an oat cultivar tolerant to *H. avenae*, than in two intolerant cultivars.

Generally, the same root-knot nematode species that attack wheat parasitize barley, oat, and rye, although oat is apparently a poor host for *M. artiellia* (Greco & Di Vito, 1988). All five races of *M. naasi* have been shown to attack barley, oat, and rye (Michell et al., 1973b), and the nematode has been reported from most parts of the world where they are cultivated. Michell et al. (1973a) compared the life cycle duration of the five known races of *M. naasi* on barley and sorghum. On barley, Race 5 individuals matured much quicker than individuals representing the other four races. Griffin (1992) found that *M. chitwoodi* affected tillering, height, shoot and head weights, number of heads per plant, and head length of either spring or winter barley. With *M. incognita, M. arenaria*, and *M. javanica*, Opperman et al. (1988) found considerable variation in egg production among six cultivars of oat tested.

The lesion nematode (*Pratylenchus fallax*) was found associated with stunted barley plants in 1972 (Corbett). *Pratylenchus neglectus* was reported from barley in 1973 (Siddiqui et al.). Survey work (Kimpinski et al., 1989) indicated that *P. penetrans* also was a common nematode parasite of barley. Work by Umesh and Ferris (1992) focused on the effect of temperature on the pathogenicity of *P. neglectus*. In subsequent studies of competition between *P. neglectus* and the Columbia root-knot nematode (*M. chitwoodi*), on barley across a range of temperatures data indicated that there was a competitive and reciprocally inhibitory reproductive interaction between the two nematode species (Umesh et al., 1994; Umesh & Ferris, 1994). The species that parasitized the roots first inhibited subsequent penetration by the other species. Weight of barley shoots was reduced by *M. chitwoodi* alone at 25°C but not when combined with *P. neglectus* at this temperature. Prior invasion of roots by *P. neglectus* reduced egg production, final population level, and reproductive index for *M. chitwoodi* in a manner linearly related to the initial levels of *P. neglectus*.

Barley also is reported to be damaged by *Longidorus elongatus* (Brown & Sykes, 1975), the stunt nematode *Tylenchorhynchus dubius* (Jones, 1979), and *Paratrichodorus anemones* (Spaull, 1980). Boag et al. (1990) found *Longidorus elongatus* the most prevalent species present in a survey of cereal fields in Scotland.

CONTROL

Management of nematodes of barley, oat, and rye relies heavily on the use of resistant cultivars. Cultivars of spring barley resistant to *H. avenae* consistently out yielded susceptible ones in infested fields (Cotten, 1970). In the absence of the nematode, no differences in yield were observed between *H. avenae* resistant and susceptible cultivars. Work with six oat cultivars and *M. incognita, M. javanica* and *M. arenaria* indicated considerable variation in reproduction among the three root-knot species (Opperman et al., 1988). Cultivar screening data with multiple cultivars of barley, oat, and rye (Johnson & Motsinger, 1989) and the same three root-knot species showed high rates of nematode reproduction on most plant cultivars.

The employment of cultural practices such as the use of winter cover crops, crop rotation, and delayed planting also can aid in reducing nematode damage to barley, oat, and rye. Juveniles of *M. incognita* were more abundant in the spring in fallow soil than in soil following a crop of rye grown during the winter (Minton & Parker, 1987). Townshend (1989) evaluated oat as a rotation crop for tobacco (*Nicotiana tabacum* L.) fields infested with several lesion nematodes. He found that ‘Saia’ oat was more resistant to *P. penetrans* and *P. neglectus* than ‘Woodstock’, and that *P. crenatus* did not develop on either cultivar. Juveniles and adults of *M. incognita* were more abundant on rye planted 1 October than on rye planted either 1 November or 1 December (Johnson & Motsinger, 1990). Population density of *M. incognita* was not measurably influenced by rye preceding a minimum-till soybean crop although root-knot gall indices were lower and yield of soybean was enhanced following rye (Minton, 1992).

Comparative studies of the effectiveness of nematicides labeled for control of nematode parasites of barley, oat, and rye are limited. Nematicides currently labeled for use on barley, oat, and rye in the USA include 1,3-dichloropropene and a combination of this compound and chloropicrin (Anonymous, 1995). Aldicarb reduced root-lesion nematode populations in wheat and barley and increased yields during 2 of 4 yr (Kimpinski et al., 1989).

OUTLOOK

Current world population is estimated to be about 5.8 billion. This figure is forecast to climb to 6.2 billion by the year 2000 and to 9.8 billion by 2050 (Anonymous, 1995). An increased demand for food and fiber will undoubtedly accompany this substantial increase in human population density. Minimizing losses sustained from major agricultural pest species such as nematodes will remain a major responsibility of plant protection scientists. The challenge will be to employ traditional pest management strategies and tactics including the use of pesticides, in an environmentally responsible manner. At the same time, agricultural production systems must be developed that employ the most current technologies available, exploit biological competition among microorganisms for the suppression of pathogens, and focus on long-term crop sustainability instead of short-term profitability.

REFERENCES

Acosta, N., and R.B. Malek. 1981. Symptomatology and histopathology of soybean roots infected by *Pratylenchus scribneri* and *P. alleni*. J. Nematol. 13:6–12.

Anderson, R.V., and J.W. Potter. 1991. Stunt nematodes: *Tylenchorhynchus*, *Merlinius*, and related genera. p. 529–586. *In* W.R. Nickle (ed.) Manual of agricultural nematology. Marcel Dekker, Inc., New York.

Anonymous. 1984. The distribution of plant parasitic nematode species in North America. Soc. Nematol.

Anonymous. 1987. Diseases caused by nematodes. *In* M.V. Wiese (ed.) Compendium of wheat diseases. APS Press, St. Paul, MN.

Anonymous. 1995. Crop protection chemicals reference. 11th ed. Chem. Pharmaceut. Publ. Corp., New York.

Anonymous. 1995. Concise report on the world population situation in 1995. U.N. Publ. no. E.95.XIII.14. Chem. Pharmaceut. Publ. Corp., New York.

Atkins, J.G., and E.H. Todd. 1959. White tip diseases of rice. Phytopathology 48:189–191.

Aytan, S., and O.J. Dickerson. 1969. *Meloidogyne naasi* on sorghum in Kansas. Plant Dis. Rep. 53:737.

Babatola, J.O. 1981. Effect of pH, oxygen, and temperature on the activity and survival of *Hirschmaniella* spp. Nematologica 27:285–291.

Babatola, J.O. 1984. Rice nematode problems in Nigeria, their occurrence, distribution, and pathogenesis. Trop. Pest Manage. 30:256–265.

Bajaj, H.K., D.C. Gupta, and R.S. Dashiya. 1986. Development of *Heterodera zeae* Koshy et al. on wheat and maize. Nematologica 32:209–215.

Baldwin, J.G., and M. Mundo-Ocampo. 1991. Heteroderinae, cyst- and non-cyst-forming nematodes. p. 275–362. *In* W.R. Nickle (ed.) Manual of agricultural nematology. Marcel Dekker, Inc., New York.

Barat, H., M. Delassus, and H.H. Vuong. 1969. The geographical distribution of white tip disease of rice in tropical Africa and Madagascar. p. 269–273. *In* J.E. Peachey (ed.) Nematodes of tropical crops. Commonw. Bureau Helminthol. Tech. Comm. 40. Commonw. Agric. Bur., Farnham Royal, Bucks, England.

Baxter, R.I., and C.D. Blake. 1967. Invasion of wheat roots by *Pratylenchus thornei*. Nature (London) 215:1168–1169.

Baxter, R.I., and C.D. Blake. 1968. *Pratylenchus thornei*: A cause of root necrosis in wheat. Nematologica 14:351–361.

Bee-Rodriguez, D., and A. Ayala. 1977. Interaction of *Pratylenchus zeae* with four soil fungi on sorghum. J. Agric. Puerto Rico 61:501–506.

Benedict, W.G., and W.B. Mountain. 1956. Studies on the etiology of a root rot of winter wheat in southwestern Ontario. Can. J. Bot. 34:159–174.

Birchfield, W. 1983. Wheat and grain sorghum varietal reaction to *Meloidogyne incognita* and *Rotylenchulus reniformis*. Plant Dis. 67:41–42.

Boag, B., S. Bowen, A.M. Spaull, G. Wright, and B.F.L. Smith. 1990. Migratory plant-parasitic nematodes associated with cereals in Scotland. Ann. Appl. Biol. 117:399–406.

Bridge, J. 1987. Control strategies in subsistance agriculture. p. 389–420. *In* R.H.Brown and B.R. Kerry (ed.) Principles and practice of nematode control in crops. Acad. Press, Orlando, FL.

Bridge, J., M. Luc., and R.A. Plowright. 1990. Nematode parasites of rice. p. 69–108. *In* M. Luc et al. (ed.) Plant parasitic nematodes in subtropical and tropical agriculture. CAB Int. Inst. Parasitol. Wallingford, England.

Bridge, J., and S.J.L. Page. 1982. The rice root nematode, *Meloidogyne graminicola*, on deepwater rice (*Oryza sativa* subsp. *Indica*). Rev. Nematol. 5:225–232.

Brown, R.H. 1984a. Ecology and control of cereal cyst nematode. (*Heterodera avenae*) in southern Australia. J. Nematol.16:216–222.

Brown, R.H. 1984b. Cereal cyst nematode I: Chemical control in Australia. Plant Dis. 68:922–928.

Brown, R.H. 1987. Control strategies of low value crops. p. 351–387. *In* R.H. Brown and B.R. Kerry (ed.) Principles and practice of nematode control in crops. Acad. Press, Orlando, FL.

Brown, E.B., and G.B. Sykes. 1975. Studies on the relation between density of *Longidorus elongatus* and yield of barley and potatoes. Plant Pathol. 24:221–223.

Butler, E.J. 1913. Diseases of rice. Pusa Agric. Res. Bull. 34.

Chevres-Roman, R., H.D.Gross, and J.N. Sasser. 1971. The influence of selected nematode species and number of consecutive plantings of corn and sorghum on forage production, chemical composition of plant and soil, and water use efficiency. Nematropica 1:40–41.

Corbett, D.C.M. 1972. The effect of *Pratylenchus fallax* on wheat, barley and sugar beet roots. Nematologica 18:303–308.

Cotten, J. 1970. Field experiments with spring barley resistant to cereal cyst nematode; 1965–1968. Ann. Appl. Biol. 65:163–168.

Cralley, E.M. 1949. White tip of rice. Phytopathology 39:5.

Cuarezma-Teran, J.A., L.E. Trevathan, and S.C. Bost. 1984. Nematodes associated with sorghum in Mississippi. Plant Dis. 68:1083–1085.

Cuarezma-Teran, J.A., and L.E. Trevathan. 1985. Effects of *Pratylenchus zeae* and *Quinisulcius acutus* alone and in combination on sorghum. J. Nematol. 17:169–174.

Cuc, N.T.T., and D.N. Kinh. 1981. Rice stem nematodes in Vietnam. Int. Rice Res. Newsl. 6(6):14–15.

Cuc, N.T.T., and J.C. Prot. 1992. Root-parasitic nematodes of deepwater rice in the Mekong Delta of Vietnam. Fund. Appl. Nematol. 15:575–577.

Davies, K.A., and J.M. Fisher. 1976. Duration of infectivity of second stage larvae of *Heterodera avenae*. Nematologica 22:163–168.

Decker, H. 1989. Cyst-forming nematodes of oats. p. 223–231. *In* N.M. Sveshnikova (ed.) Plant nematodes and their control (phytonematology). E.J. Brill Publ. Co., Leiden.

Diomande, M. 1981. Effets direct et combine des engrais et de *Meloidogyne incognita* sur le riz pluvial. Rev. Nematol. 4:71–74.

Ediz, S.A., and O.J. Dickerson. 1976. Life cycle, pathogenicity, histopathology, and host range of race 5 of the barley root-knot nematode. J. Nematol. 8:228–232.

Eisenback, J.D., and H.H. Triantaphyllou. 1991. Root-knot nematodes: *Meloidogyne* species and races. p. 191–274. *In* W.R. Nickle (ed.) Manual of agricultural nematology, Marcel Dekker, Inc., New York.

Esmenjaud, D., R. Rivoal, and H. Marzin. 1990. Numbers of *Pratylenchus* spp. (Nematoda) in the field on winter wheat in different cereal rotations. Nematologica 36:217–226.

Fomba, S.N. 1984. Rice disease situation in mangrove and associated swamps in Sierra Leone. Trop. Pest Manage. 30:73–81.

Fortnum, B.A., and R.E. Currin, III. 1988. Host suitability of grain sorghum cultivars to *Meloidogyne* spp. Ann. Appl. Nematol. 2:61–64.

Fortuner, R. 1976. *Pratylenchus zeae*. C.I.H. Descript. Plant-Parasitic Nematodes. Set 6, no. 77. Commonw. Agric. Bur., St. Albans, Herts, England.

Fortuner, R., and G. Merny. 1979. Root parasitic nematodes of rice. Rev. Nematol. 2:79–102.

Fortuner, R., and K.J.O. Williams. 1975. Review of the literature on *Aphelenchoides besseyi* Christie, 1942, the nematode causing 'White-tip' disease in rice. Helminthol. Abstr. B44:1–40.

Frederick, J.J., and A.C. Tarjan. 1989. A compilation of the genus *Pratylenchus*, Filipjev, 1936 (Nemata: Pratylenchidae). Rev. Nematol. 12:243–256.

Gallaher, R.N., D.W. Dickson, J.F. Corella, and T.E. Hewlett. 1988. Tillage and multiple cropping systems and population dynamics of phytoparasitic nematodes. Ann. Appl. Nematol. 2:90–94.

Gaur, H.S., M. Saha and E. Khan. 1993. *Meloidogyne triticoryzae* (Nematoda: Meloidogynidae), a root-knot nematode damaging wheat and rice in India. Ann. Plant Protect. Sci. 1:18–26.

Gergon, E.B., and J.-C. Prot.1993. Effect of benomyl and carbofuron on *Aphelenchoides besseyi* on rice. Fund. Appl. Nematol.16(6):563–566.

Golden, A.M. 1986. Morphology and identification of cyst nematodes. p. 75–99. *In* F. Lamberti and C.E. Taylor (ed.) Cyst nematodes. Plenum Press, New York.

Golden, A.M., J.H. O'Bannon, G.S. Santo, and A.M. Finley. 1980. Description and SEM observations of *Meloidogyne chitwoodi* n. sp. (Meloidgynidae), a root-knot nematode on potato in the Pacific Northwest. J. Nematol. 12:319–327.

Granez, L.S.O. 1985. Observations on the morphology of *Heterodera carotae* and *Heterodera avenae* in Michigan USA. J. Nematol. 17:519.

Greco, N., and M. DiVito. 1988. The importance of plant parasitic nematodes in food legume production in the Mediterranean region. p. 28–45. *In* M.C. Sarena et al. (ed.) Nematodes parasitic to cereals and legumes in temperate semi-arid regions. Proc. Workshop, Larnaca, Cyprus. 1–5 Mar. 1987. Int. Cent. Agric. Res., Allepo, Syria.

Griffin, G.D. 1992. Comparative effects of two populations of *Meloidogyne chitwoodi* on *Triticum aestivum* and *Hordeum vulgare*. Nematropica 22:65–74.

Griffin, G.D., and S.V. Thomson. 1988. The Columbia root-knot nematode, *Meloidogyne chitwoodi*, discovered in the state of Utah. Plant Dis. 72:363.

Hafez, S.L., and A.M. Golden. 1984. First report of oat cyst nematode in eastern Washington. Plant Dis. 68:351.

Hafez, S.L., and A.M. Golden. 1985. First report of oat cyst nematode. (*Heterodera avenae*) on barley in Idaho. Plant Dis. 69:360.

Hafez, S.L., A.M. Golden, F. Rashid and Z. Handoo. 1992. Plant-parasitic nematodes associated with crops in Idaho and eastern Oregon. Nematropica 22:193–204.

Hollis, J.P., and S. Keoboonrueng. 1984. Nematode parasites of rice. p. 95–146. *In* W.R. Nickle (ed.) Plant and insect nematodes. New York & Basel, Marcel Dekker.

Hooper, D.J., and S.K. Ibrahim. 1994. *Aphelenchoides nechaleos* n. sp. and *A. paranechaleos* n. sp. (Nematoda: Aphehenchoididae) from rice plants. Fund. Appl. Nematol. 17:106–114.

Ibrahim, I.K.A., M.A. Rezk, and H.A.A. Khalil. 1983. Resistance of some plant cultivars to root-knot nematodes, *Meloidogyne* spp. Nematol. Medit. 11:189–192.

Ibrahim, S.K., S.A. Lewis, and D.C. Harshman. 1993b. Host suitability of graminaceous crop cultivars for isolates of *Meloidogyne arenaria* and *M. incognita*. J. Nematol. 25:858–862.

Ibrahim, S.K., and R.N. Perry. 1992. Observations on the desiccation survival of 2nd stage juveniles of the rice cyst nematodes, *Heterodera sacchari*. Nematologica 38:328–334.

Ibrahim, S.K., and R.N. Perry. 1993a. Use of esterase patterns of females and galled roots for identification of species of *Meloidogyne*. Fund. Appl. Nematol. 16:187–191.

Ibrahim, S.K., and R.N. Perry. 1993b. Desiccation survival of the rice stem nematode *Ditylenchus angustus*. Fund. Appl. Nematol.16:31–38.

Ibrahim, S.K., and R.N. Perry. 1994. Infectivity and population dynamics of the rice stem nematode, *Ditylenchus angustus*. Nematologica 40:412–422.

Ibrahim, S.K., R.N. Perry, R.A. Plowright, and J. Rowe. 1993a. Hatching behavior of the rice cyst nematodes *Heterodera sacchari* and *H. oryzicola* in relation to age of host plant. Fund. Appl. Nematol.16:23–29.

Ibrahim, S.K., R.N. Perry, and D.J. Hooper. 1994. Use of esterase and protein patterns to differentiate two new species of *Aphelenchoides* on rice from other species of *Aphelenchoides* and from *Ditylenchus angustus* and *Ditylenchus myceliophagus*. Nematologica 40:267–275.

Ichinohe, M. 1988. Current research on the major nematode problems in Japan. J. Nematol. 20:184–190.

Inserra, R.N., N. Voulas, J.H. O'Bannon, and G.D. Griffin. 1985. Development of *Meloidogyne chitwoodi* on wheat. J. Nematol. 17:322–326.

Iyatomi, K., and T. Nishizawa. 1954. Artificial culture of the strawberry nematode *Aphelenchoides fragariae*, and the rice white tip nematode, *Aphelenchoides besseyi*. Jpn. J. Appl. Zool. 19:8–15.

Jayaprakash, A., and Y.S. Rao. 1982. Hatching behavior of the cyst nematode *Heterodera oryzicola*. Ind. J. Nematol. 12:400–402.

Jensen, H.J., H. Eshtighi, P.A. Koepsell, and N. Goetze. 1975. The oat cyst nematode, *Heterodera avenae,* occurs in oats in Oregon. Plant Dis. Rep. 59:1–3.

Jepson, S.B. 1987. Identification of the root-knot nematodes (*Meloidogyne* species). Commonw. Agric. Bureau, Farnham Royal, England.

Johnson, A.W., and S.G. Fushtey. 1966. The biology of the oat cyst nematode *Heterodera avenae* in Canada II. Nematode development and related anatomical changes in roots and corn. Nematologica 12:630–636.

Johnson, A.W., and R.E. Motsinger. 1989. Suitability of small grains as hosts of *Meloidogyne* species. J. Nematol. 21:650–653.

Johnson, A.W., and R.E. Motsinger. 1990. Effects of planting date, small grain crop destruction, fallow, and soil temperature on the management of *Meloidogyne incognita*. J. Nematol. 22:348–355.

Jones, R.K. 1979. Migratory plant parasitic nematodes as pests of cereals. Ann. Appl. Biol. 92:257–262.

Kaloshian, I., P.A. Roberts, and I.J. Thomason. 1989a. Resistance to *Meloidogyne* spp. in allohexaploid wheat derived from *Triticum turgidum* and *Aegilops squarrosa*. J. Nematol. 21:42–47.

Kaloshian, I., P.A. Roberts, and I.J. Thomason. 1989b. Resistance in *Triticum* and *Aegilops* spp. to *Meloidogyne chitwoodi*. J. Nematol. 21:632–634.

Kerr, E.D. 1967. Population dynamics and fumigation studies of certain nematodes associated with roots of wheat in Nebraska. Plant Dis. Rep. 51:637–641.

Kerry, B.R. 1981. Fungal parasite: A weapon against cyst nematodes. Plant Dis. 65:390–393.

Kerry, B.R., and D.H. Crump. 1977. Observations on fungal parasites of females and eggs of the cereal cyst-nematode, *Heterodera avenae*, and other cyst nematodes. Nematologica 23:193–201.

Kerry, B.R., and N.G.M. Hague. 1974. The invasion and development of the cereal cyst-nematodes, *Heterodera avenae,* in the roots of autumn- and spring-grown cereals. Ann. Appl. Biol. 78:319–330.

Khan, M.W., and K.N. Pathak. 1993. Nematodes as vectors of bacterial and fungal plant pathogens. p. 251–272. *In* M.W. Khan (ed.) Nematode interactions. Chapman and Hall, London.

Kilpatrick, R.A., L. Gilchrist and A.M. Golden. 1976. Root-knot on wheat in Chile. Plant Dis. Rep. 60:135.

Kimpinski, J., R.V. Anderson, H.W. Johnston, and R.A. Martin. 1989. Nematodes and fungal diseases in barley and wheat on Prince Edward Island. Crop Prot. 8:412–416.

Lamberti, F. 1979. Economic importance of *Meloidogyne* spp. in subtropical and Mediterranean climates. p. 342–357 *In* F. Lamberti and C.E. Taylor (ed.) Root-knot nematodes (*Meloidogyne* species) systematics, biology, and control. Acad. Press, London.

Langdon, K.R., F.B. Struble and H.C. Young, Jr. 1961. Stunt of small grains, a new disease caused by the nematode *Tylenchorhynchus brevidens*. Plant Dis. Rep. 45:248–252.

Lasserre, F., R. Rivoal, and R. Cook. 1994. Interactions between *Heterodera avenae* and *Pratylenchus neglectus* on wheat. J. Nematol. 26:336–344.

Lee, Y.B., and J.S. Park. 1975. Influence of the rice root nematode, *Hirschmaniella oryzae*, on the root browning of rice. Kor. J. Plant Prot. 14:35–36.

Lopez, R. 1984. *Meloidogyne salasi* sp. nov. (Nematoda: Meloidogynidae), a new parasite of rice (*Oryza sativa* L.) from Costa Rica and Panama. Turrialba 34:275–286.

Maas, P.W.T., H. Sanders, and J. Dede. 1972. *Meloidogyne oryzae* n. sp. (Nematoda: Meloidogynidae) infesting irrigated rice in Surinam (South America). Nematologica 24:305–311.

Maqbool, M.A. 1987. Present status of research on plant parasitic nematodes in cereals and food and forage legumes in Pakistan. p. 173–180. *In* M.C. Sarena et al. (ed.) Nematodes parasitic to cereals and legumes in temperate semi-arid regions. ICARDA, Aleppo, Syria.

McDonald, A.H., and E.H. Van Den Berg. 1993. Effect of watering regimen on injury to corn and grain sorghum by *Pratylenchus* species. J. Nematol. 25:654–658.

McGawley, E.C., M.C. Rush, and J.P. Hollis. 1984. Occurrence of *Aphelenchoides besseyi* in Louisiana rice seed and its interaction with *Sclerotium oryzae* in selected cultivars. J. Nematol. 16:65–68.

McGawley, E.C., K.L. Winchell, J.P. Jones, W. Birchfield, and G.T. Berggren. 1985. Population development and influence of *Bursaphelenchus xylophilus* on *Gliocladium virens*. J. Nematol. 16:69–76.

McGeachie, I., and L. Rahman. 1983. Ufra disease: a review and a new approach to control. Trop. Pest Manage. 29:325–332.

McSorley, R., and R.N. Gallaher. 1991. Nematode population changes and forage yields of six corn and sorghum cultivars. J. Nematol. 23:673–677.

McSorley, R., J.L. Parrado, R.V.Tyson, V.H. Waddill, M.L. Lamberts, and J.S. Reynolds. 1987. Effect of sorghum cropping practices on winter potato production. Nematropica 17:45–60.

Meagher, J.W. 1977. World dissemination of the cereal cyst nematode (*Heterodera avenae*) and its potential as a pathogen of wheat. J. Nematol. 9:9–15.

Meagher, J.W., R.H. Brown, and A.D. Rovira. 1978. The effects of cereal cyst nematode (*Heterodera avenae*) and *Rhizoctonia solani* on the growth and yield of wheat. Aust. J. Agric. Res. 29:1127–1137.

Meagher, J.W., and S.C. Chambers. 1971. Pathogenic effects of *Heterodera avenae* and *Rhizoctonia solani* and their interaction in wheat. Aust. J. Agric. Res. 22:189–194.

Michell, R.E., R.B. Malek, D.P. Taylor, and D.I. Edwards. 1973a. Races of the barley root-knot nematode, *Meloidogyne naasi*. I. Characterization by host preference. J. Nematol. 5:41–43.

Michell, R.E., R.B. Malek, D.P. Taylor, and D.I. Edwards.1973b. Races of the barley root-knot nematode, *Meloidogyne naasi*. II: Developmental rates. J. Nematol. 5:44–46.

Miller, L.I. 1986. Economic importance of cyst nematodes in North America. p. 373–385. *In* F. Lamberti and C.E. Taylor (ed.) Cyst nematodes. Plenum Press, New York.

Minton, N.A. 1992. Nematode management in minimum-till soybean with resistant cultivars, rye rotation, and aldicarb. Nematropica 22:21–28.

Minton, N.A., and K. Bondari. 1994. Effects of small grain crops, aldicarb, and *Meloidogyne incognita* resistant soybean on nematode populations and soybean production. Nematropica 24:7–15.

Minton, N.A., and M.B. Parker. 1987. Root-knot nematode management and yield of soybean as affected by winter cover crops, tillage systems, and nematicides. J. Nematol. 19:38–43.

Mojtahedi, H., and G.S. Santo. 1988. First report of *Pratylenchus thornei* on dryland wheat in Washington state. Plant Dis. 72:175.

Mojtahedi, H., and G.S. Santo. 1992. *Pratylenchus neglectus* on dryland wheat in Washington. Plant Dis. 76:323.

Mojtahedi, H., and G.S. Santo. 1994. A new host race of *Meloidogyne chitwoodi*. Plant Dis.. 78:1010.

Mojtahedi, H., G.S. Santo, and J.H. Wilson. 1988. Host tests to differentiate *Meloidogyne chitwoodi* races 1 and 2 and *M. hapla*. J. Nematol. 20:468–473.

Mondal, A.H., and S.A. Miah. 1987. Ufra problem in low lying areas of Bangladesh. Int. Rice Res. Newsl. 12:29–30.

Mondal, A.H., L. Rahman, H.J. Ahmed, and S.A. Miah. 1986. The causes of increasing blast susceptibility of Ufra-infected rice plants. Banglad. J. Agric.11:77–79.

Muthukrishnan, T.S., G. Rajendran, and J. Chandrasekara. 1974. Studies on the white tip nematode of rice *Aphelenchoides besseyi* in Tamil Nadu. Ind. J. Nematol. 7:8–16.

Nishizawa, T., S. Yamomoto, and H. Mizuta. 1953. Studies on the varietal resistance of rice plant to the rice nematode disease "senchu shingare byo" (VII). Bull. Kyushu Agric. Exp. Stn. 2:71–80.

Nishizawa, T., K. Shimizu, and T. Nagashima. 1972. Chemical and cultural control of the rice nematode, *Heterodera oryzae* Luc et Berdon Brizuela, and hatching responses of the larvae to some extracts. Jpn. J. Nematol. 2:27–32.

Nobbs, J.M., S.K. Ibrahim, and J. Rowe. 1992. A morphological and biochemical comparison of the four cyst nematode species, *Heterodera elachista, H. oryzicola, H. oryzae* and *H. sacchari* (Nematoda: Heteroderidae) known to attack rice (*Oryza sativa*). Fund. Appl. Nematol.15:551–562.

Noegel, K.A., and V.G. Perry. 1963. A foliar disease of chrysanthemum incited by the strawberry summer crimp nematode. Proc. Soil Crop Soc. Fla. 22:162–166.

Norton, D.C. 1958. The association of *Pratylenchus hexincisus* with charcoal rot of sorghum. Phytopathology 48:335–358.

Nyczepir, A.P., R.N. Inserra, J.H. O'Bannon, and G.S. Santo. 1984. Influence of *Meloidogyne chitwoodi* and *M. hapla* on wheat growth. J. Nematol. 16:162–165.

Opperman, C.H., J.R. Rich, and R.H. Dunn. 1988. Reproduction of three root-knot nematodes on winter small grains. Plant Dis. 72:869–871.

Orion, D., and E. Shlevin. 1989. Nematicide seed dressing for cyst and lesion nematode control in wheat. J. Nematol. 21:629–631.

Orr, C.C. 1967. Observations on cotton root-knot nematode in grain sorghum in west Texas. Plant Dis. Rep. 51:29.

Orr, C.C., and E.D. Morey. 1978. Anatomical response of grain sorghum roots to *Meloidogyne incognita acrita*. J. Nematol. 10:48–53.

Ou, S.H. 1985. Rice diseases. 2nd ed. Commonw. Mycol. Inst. Kew, England.

Patel, G.J., D.J. Patel, D.K. Jogani, and S.J. Patel. 1986. Wheat–a new host of *Meloidogyne javanica* in India. Ind. J. Nematol.16:134.

Patnik, N.C. 1969. Pathogenicity of *Meloidogyne graminicola* (Golden and Birchfield, 1965) in rice. All India Nematol. Symp., New Delhi.

Pedersen, J.F., and R. Rodriguez-Kabana. 1987. Nematode response to cool season annual graminaceous species and cultivars. Ann. Appl. Nematol. 1:116–118.

Plowright, R., and J. Bridge. 1990. Effect of *Meloidogyne graminicola* (Nematoda) on the establishment, growth and yield of Rice CV IR36. Nematologica 36:81–89.

Plowright, R.A., and T.E. Akehurst. 1992. Monoxenic culture of the Ufra nematode, *Ditylenchus angustus*. Fund. Appl. Nematol. 15:327–330.

Plowright, R.A., T.A. Matias, and T.W. Mew. 1990. The effect of *Pratylenchus zeae* on upland rice. Rev. Nematol. 13:283–291.

Prasad, J.S., and S.Y. Rao. 1978. Potentiality of *Pratylenchus indicus*, the root lesion nematode, as a new pest of upland rice. Ann. Zool. Ecol. Anim. 10:627–634.

Prasad, J.S., and K.S. Varaprasad. 1992. Elimination of white-tip nematode, *Aphelenchoides besseyi*, from rice seed. Fund. Appl. Nematol. 15:305–308.

Prot, J.-C. 1990. Population density of the nematode *Pratylenchus zeae* at harvest and yield of upland rice UPL Ri 5. Int. Rice Res. Newsl. 15:34.

Prot, J.-C. 1992. Diseases caused by nematodes. p. 46–50. *In* R.K. Webster and P.S. Gunnell (ed.) Compendium of rice diseases. APS Press, St. Paul, MN.

Prot, J.-C., M. Herman, and A. Ahmadin. 1992a. Plant parasitic nematodes associated with upland rice in Sitiung, West Sumatra, Indonesia. Int. Rice Res. Newsl. 17:27–28.

Prot, J.-C., and S. Savary. 1993. Interpreting upland rice yield and *Pratylenchus zeae* relationships: Correspondence analysis. J. Nematol. 25:277–285.

Prot, J.-C., I.R.S. Soriano, D.M. Matias, and S. Savary. 1992b. Use of green manure crops in control of *Hirschmaniella mucronata* and *H. oryzae* in irrigated rice. J. Nematol. 24:127–132.

Rao, Y.S., and P. Israel. 1973. Life history and bionomics of *Meloidogyne graminicola*, the rice root-knot nematode. Ind. Phytopathol. 26:333–340.

Rivoal, R., and R. Cook. 1993. Nematode pests of cereals. p. 259–303. *In* K. Evans et al. (ed.) Plant parasitic nematodes in temperate agriculture. CAB Int., Wallingford, England.

Rivoal, R., G. Doussinault, and M. Hulle. 1990. Influence of *Heterodera avenae* on winter wheat in France: experiments with resistant and susceptible varieties. Ann. Appl. Biol. 116:537–548.

Rivoal, R., J. Jakier, and M. Hulle. 1993. Partial resistance to *Heterodera avenae* in wheat lines with the 6 M^V chromosome from *Aegilops ventricosa*. J. Nematol. 25:265–269.

Rivoal, R., and E. Sarr. 1987. Field experiments on *Heterodera avenae* in France and implications for winter wheat performance. Nematologica 33: 460–479.

Roberts, P.A. 1992. Current status of the availability, development, and use of host plant resistance to nematodes. J. Nematol. 24:213–227.

Roberts, P.A., and S.D. Van Gundy. 1981. The development and influence of *Meloidogyne incognita* and *M. javanica* on wheat. J. Nematol. 13:345–352.

Roberts, P.A., S.D. Van Gundy, and H.E. McKinney. 1981. The effects of soil temperature and planting date of wheat on *Meloidogyne incognita* reproduction, soil population, and grain yield. J. Nematol. 13:338–345.

Roberts, P.A., S.D. Van Gundy, and J.G. Waines. 1982. Reaction of wild and domesticated *Triticum* and *Aegilops* species to root-knot nematodes (*Meloidogyne*). Nematologica 28:182–191.

Rodriguez-Kabana, R., and J.T. Touchton. 1984. Corn and sorghum as rotational crops for management of *Meloidogyne arenaria* in peanut. Nematropica 14:26–36.

Santo, G.S., and J.H. O'Bannon. 1981. Pathogenicity of the Columbia root-knot nematode (*Meloidogyne chitwoodi*) on wheat, corn, oat, and barley. J. Nematol. 13:548–550.

Santo, G.S., J.H. O'Bannon, A.M. Finley, and A.M. Golden. 1980. Occurrence and host range of new root-knot nematode (*Meloidogyne chitwoodi*) in the Pacific Northwest. Plant Dis. 64:951–952.

Santo, G.S., and J.N. Pinkerton. 1985. A second race of *Meloidogyne chitwoodi* discovered in Washington state. Plant Dis. 69:361.

Santo, G.S., and R.P. Ponti. 1981. Control of root-knot nematodes on Russet Burbank potato in Washington with fumigant and nonfumigant nematicides. J. Nematol. 13:460.

Sarena, M.C., R.A. Sikora, and J.P. Srivastava. 1988. p. 69–84. *In* M.C. Sarena et al. (ed.) Nematodes parasitic to cereals and legumes in temperate semi-arid regions. ICARDA, Aleppo, Syria.

Sasser, J.N., and D.W. Freckman. 1987. A world perspective on nematology: The role of the society. p. 7–14. *In* J.A.Veech and D.W. Dickson (ed.) Vistas on nematology. Soc. Nematol., Hyattsville, MD.

Segeren, V.D., H. Oever, and M.L. Sanchit-Bekker. 1984. Observations on *Meloidogyne oryzae* Maas, Sanders and Dede 1978 in irrigated rice in Suriname. Surin. Landbuow. 32:51–59.

Sein, T., and K. Zan. 1977. Ufra disease spread by water flow. Intl. Rice Res. Newsl. 2:5.

Sharma, R.D. 1981. Patogenicidade do nematoide *Meloidogyne javanica* ao trigo *(Triticum aestivum* L.). Publ. Soc. Brasil. Nematol. 5:104–118.

Sher, S.A. 1968. Revision of the genus *Hirschmaniella* Luc and Goodey, 1963 (Nematoda: Tylenchoidea). Nematologica 14:243–275.

Siddiqui, I.A., S.A. Sher, and A.M. French. 1973. Distribution of plant parasitic nematodes in California. California Dep. Food Agric., Sacramento.

Sikora, R.A. 1988. Plant parasitic nematodes of wheat and barley in temperate and semiarid regions—a comparative analysis. p. 46–68. *In* M.C. Sarena et al. (ed.) Nematodes parasitic to cereals and legumes in temperate semi-arid regions. ICARDA, Aleppo, Syria.

Simon, A. 1980. A plant assay of soil to assess potential damage to wheat by *Heterodera avenae*. Plant Dis. 64:917–919.

Smiley, R.W., R.E. Ingham, W. Uddin, and G.H. Cook. 1994. Crop sequences for managing cereal cyst nematodes and fungal pathogens of winter wheat. Plant Dis. 78:1142–1149.

Smolik, J.D. 1977. Effects of *Trichodorus allius* and *Tylenchorhynchus nudus* on growth of sorghum. Plant Dis. Rep. 61:855–858.

Spaull, A.M. 1980. Effect of *Paratrichodorus anemones* on growth of spring wheat and barley. Nematologica 26:163–169.

Starr, J.L. 1986. *Pratylenchus* spp. p. 53. *In* R.A. Frederickson (ed.) Compendium of sorghum diseases. APS Press, St. Paul, MN.

Starr, J.L. 1992. Nematode pathogens of sorghum. p. 179–185 *In* W.A.C. de Milliano et al. (ed.) Sorghum and millet diseases: A second world review. ICRISAT, Patancheru, India.

Sudakova, M.I., and A.V. Stoyakov. 1967. Reproduction and duration of life of *Aphelenchoides besseyi* Christie, 1942. Zool. Z. 46:1079–1099.

Sykes, G.B. 1979. Yield losses in barley, wheat, and potatoes associated with field populations of 'large form' *Longidorus leptocephalus*. Ann. Appl. Biol. 91:237–241.

Tandingan, I.C., J.-C. Prot, and R.G. Davide. 1994. Tolerance to *Meloidogyne graminicola* in rice cultivars under two water regimes. J. Nematol. 26:567.

Tarjan, A.C. 1973. A synopsis of the genera and species in Tylenchorhynchinae (Tylenchoidea, Nematoda). Proc. Helminthol. Soc. Wash. 40:123–144.

Thomas, S.H., and L. Murray. 1987. Yield reduction in grain sorghum associated with injury by *Meloidogyne incognita* race 3. J. Nematol. 19:559.

Thomason, I.J. 1962. Reaction of cereals and sudan grass to *Meloidogyne* spp. and relation of soil temperatures to *M. javanica* populations. Phytopathology 52:787–791.

Tikhanova, L.V., and Y.N. Ivanchenko. 1968. The combined infestation of rice with a fungus, *Pyricularia,* and a nematode, *Aphelenchus*. Trudy Vsesoyuznogo Inst. Gel' mintol.. K.I. Skryabina 14:287–290.

Todd, E.H., and J.G. Atkins. 1959. White tip disease of rice. I. Symptoms laboratory culture of nematodes, and pathogenicity test. Phytopathology 48:632–637.

Townshend, J.L. 1989. Population densities of four species of root-lesion nematodes (*Pratylenchus*) in the oat cultivars Saia and OAC Woodstock. Can. J. Plant Sci. 69:903–905.

Tullis, E.C. 1934. The root-knot nematode of rice. Phytopathology 24:938–942.

Umesh, K.C., and H. Ferris. 1992. Effects of temperature on *Pratylenchus neglectus* and on its pathogenicity to barley. J. Nematol. 24:504–511.

Umesh, K.C., and H. Ferris. 1994. Influence of temperature and host plant on the interaction between *Pratylenchus neglectus* and *Meloidogyne chitwoodi*. J. Nematol. 26:65–71.

Umesh, K.C., H. Ferris, and D.E. Bayer. 1994. Competition between the plant-parasitic nematodes *Pratylenchus neglectus* and *Meloidogyne chitwoodi*. J. Nematol. 26:286–295.

U.S. Department of Agriculture, National Agricultural Statistics Service. January 1995. http://www.usda.gov/census/census92/agrimenu.html.

U.S. Department of Agriculture, Economic Research Service. 1995. http://www.fred.ifas.ufl.edu/EconFAQ/node57.html.

Valdeolivas, A., and M.D. Romero. 1985. The biology of *Heterodera avenae* in Spain. p. 187–290. *In* F. Lamberti and C.E. Taylor (ed.) Cyst nematodes. Plenum Press, New York.

Van Gundy, S.D., J. G. Perez, L.H. Stolzy, and I.J. Thomason. 1974. A pest management approach to the control of *Pratylenchus thornei* on wheat in Mexico. J. Nematol. 6:107–116.

Van Berkum, J.A., and A.R. Seshadri. 1970. Some important nematode problems in India. p. 136–137. *In* 10th Int. Nematol. Symp., Pescara, Italy.

Volkmar, K.M. 1989. Cereal cyst nematode (*Heterodera avenae*) on oats. II. Early root development and nematode tolerance. J. Nematol. 21:384–391.

Vuong, H.H. 1969. The occurrence in Madagascar of the rice nematodes *Aphelenchoides besseyi* and *Ditylenchus angustus*. p. 274–288. *In* J.E. Peachey (ed.) Nematodes of tropical crops. Commonw. Bur. Helminthol. Tech. Comm. 40. St. Albans, England.

Wallace, H.R. 1965. The ecology and control of the cereal root nematode. J. Aust. Inst. Agric. Sci. 31:178–186.

Wenefrida, I., J.S. Russin, and E.C. McGawley. 1994. Interrelationships between *Macrophomina phaseolina* and nematodes associated with sorghum in Louisiana. Phytopathology 84:1090.

Winslow, R.D. 1955. The hatching responses of some root eelworms of the genus *Heterodera*. Ann. Appl. Biol. 45:19–36.

Yik, C.P., and W. Birchfield. 1979. Host studies and reactions of cultivars to *Meloidogyne graminicola*. Phytopathology 69:497–499.

Yoshi, H., and S. Yamamoto. 1950. A rice nematode disease 'Sencha Shingane Byo'. II. Hibernation of *Aphelenchoides oryzae*. J. Fac. Agric., Kyusha Univ. 9:223–233.

Young, L.D. 1992. Problems and strategies associated with long-term use of nematode-resistant cultivars. J. Nematol. 24:228–233.

Zancada, M., and M.V. Althofer. 1994. Effect of *Heterodera avenae* on the yield of winter wheat. Nematologica 40:224–248.

22 Tobacco

CHARLES S. JOHNSON

Virginia Polytechnic Institute and State University
Blackstone, Virginia

Tobacco (*Nicotiana tabacum* L.) is the most widely grown commercial nonfood crop in the world (Akehurst, 1981). World production was estimated at 8.3 Tg in 1993. The most important tobacco-producing countries are China, USA, India, Brazil, Turkey, the former USSR, Zimbabwe, Greece, and Italy (Johnson & Reed, 1994). The specific leaf characteristics required by manufacturers tend to make tobacco a high-value crop. Most tobacco types thrive on coarse-textured, inherently infertile soils (Akehurst, 1981). The potential to secure relatively large economic returns from tobacco in geographic regions poorly suited to alternative crops has historically meant that production areas of the world tend to be very dependent upon this one crop. In addition to supporting large numbers of farmers throughout the world, cured leaf and tobacco products also generate tax income important to many governments.

Plant-parasitic nematodes, especially *Meloidogyne* species, may be the primary pathogens of tobacco worldwide. Nematodes have been estimated to cause yield losses of approximately 15% in the world's tobacco crop, but losses in the USA are generally much less (Sasser & Freckman, 1987; Shew & Lucas, 1991). Initial symptoms of injury to tobacco roots by plant-parasitic nematodes are often seen in host foliage as apparent nutrient deficiencies, particularly of N, K, and P (Barker & Lucas, 1984; Ramakrishnayya & Hussaini, 1987). Boron deficiencies also have been associated with root-knot nematode infection of burley tobacco (Barker & Lucas, 1984). Losses in tobacco yield and poor quality result from stunting and delayed crop maturity, as well as from increased incidence of other tobacco diseases often enhanced by nematode parasitism (Powell, 1979). Plant-parasitic nematodes occur virtually everywhere tobacco is grown, although the extent of the damage that they cause depends on the climate and soil environment of each tobacco production area (Shepherd & Barker, 1990). In addition, the tobacco plant's ability to partially compensate for pest injury and it's relatively short growing season often allow the crop to escape serious economic injury when environmental conditions favor the crop (Graham et al., 1964). Generally, nematodes cause economically significant yield losses in tobacco yield and quality in regions with warm climates and light sandy or sandy loam soils (Van Gundy, 1981). Consequently, flue-cured tobacco is the tobacco type most severely affected by plant-parasitic nematodes.

Root-knot nematodes are the most common nematodes damaging tobacco around the world (Shepherd & Barker, 1990). In addition to *Meloidogyne* spp., lesion (*Pratylenchus* spp.), cyst (*Globodera* spp.), reniform (*Rotylenchulus* spp.), stem and bulb (*Ditylenchus* spp.), and bud and leaf (*Aphelenchoides* spp.) nematodes cause significant damage to tobacco in various tobacco-production areas. Certain species of *Longidorus*, *Paratrichodorus*, *Trichodorus*, and *Xiphinema* also can transmit viruses to tobacco. Ring (*Criconemella*), spiral (*Helicotylenchus*, *Rotylenchus*, and *Scutellonema*), and stunt (*Tylenchorhynchus*) nematodes have been reported from tobacco fields, but are not considered major pests of the crop. Some nematode parasites of tobacco (*G. tabacum tabacum*) tend to occur by themselves in tobacco fields, and to inhibit other tobacco nematodes (*P. penetrans*) (Lucas, 1975). More commonly, however, various nematode genera and species occur concomitantly. These nematode communities may interact in various ways. Individual genera may not be able to increase to as great an extent when they must compete with other nematodes for infection sites (Barker & Lucas, 1984). However, plant-parasitic nematodes may induce root-disease complexes with each other (*Meloidogyne* and *Pratylenchus* spp.) (Johnson & Nusbaum, 1970; Patel & Patel, 1991) or with soil fungi or bacteria (Shepherd & Barker, 1990; LaMondia, 1992).

IMPORTANT NEMATODE PARASITES

Root-Knot Nematodes

Species and Races

Meloidogyne arenaria, *M. incognita*, and *M. javanica* are the most common of the important root-knot nematode species occurring on tobacco (Shepherd & Barker, 1990). *Meloidogyne incognita* is better adapted to tobacco-growing regions with temperate climates, whereas *M. javanica* predominates in tropical and subtropical areas, and *M. hapla* occurs in colder areas of temperate regions. The average temperature during the coldest part of the year appears to be the primary climactic factor impacting distribution of the four main root-knot nematode species on tobacco (Taylor et al., 1982). In the USA, *M. incognita* has historically predominated in the Carolinas, whereas *M. javanica* has been found more often on tobacco in Florida (Shepherd & Barker, 1990). All of the four known host races of *M. incognita* parasitize tobacco, but the predominant race varies in differing tobacco production areas (Barker, 1989; Shepherd & Barker, 1990). *Meloidogyne arenaria* and *M. hapla* also often parasitize tobacco in fields infested by *M. incognita* (Schmitt & Barker, 1988). Incidence of Race 2 of *M. arenaria* has increased in the Carolinas with increased use of *M. incognita*-resistant cultivars (Barker, 1989). Other species of *Meloidogyne* also may parasitize tobacco, but are not known to be significant pests in tobacco-producing regions of the world (Rammah & Hirschmann, 1990; Shepherd & Barker, 1990).

Symptoms

The first noticeable symptoms of root-knot nematode attack are usually an uneven pattern of poor color, premature wilting, and stunting (Fig. 22–1) of

Fig. 22–1. Irregular stunting of flue-cured tobacco caused by *Meloidogyne arenaria* in South Carolina (photograph courtesy of B. A. Fortnum, Clemson Univ.).

Fig. 22–2. Galling of flue-cured tobacco roots by *Meloidogyne* spp.

tobacco plants. Plants in affected areas will tend to wilt, especially in the afternoon on hot days, despite the presence of adequate soil moisture. Affected plants also usually exhibit a more yellowish color than those in more vigorous parts of the field. Leaves may show chlorosis or necrosis along their margins or tips. Foliar symptoms reflect deficiencies in N and K, respectively, caused by nematode infection. Marginal foliar necrosis usually increases with time. Stunting of growth is often distributed very unevenly, but tends to worsen within noticeable "hot spots" or foci. These areas tend to be oriented along tobacco rows more so than across rows of the crop (Noe & Campbell, 1985). Weed cover may be extensive in significantly stunted areas due to increased exposure to sunlight and reduced competition for soil nutrients.

The diagnostic symptom of root-knot nematode infection is a localized abnormal swelling of tobacco roots to form a gall or "knot" (Fig. 22–2). Galls are formed as parasitic nematodes stimulate swelling of root tissue around their feeding sites. Galls can be shaped irregularly, but tend to be spherical or spindle shaped. Galls also vary in size, often due to the species involved, from that of a pinhead (2 mm) for *M. hapla* to diameters of over 2.5 cm for *M. javanica*. Small galls may contain a single nematode, while large galls may contain numerous nematodes in varying stages of development. Careful dissection of a gall will reveal an extremely small, pear shaped, pearly white female nematode. The entire root system can be distorted in severely infected plants, but infections are often found most easily on small fibrous roots, where galls may appear similar to "beads on a string." *Meloidogyne incognita* and *M. javanica* can cause numerous, large galls distorting 90% of the root system (Fig. 22–3) (Shepherd & Barker, 1990). Infection by *M. arenaria* typically results in somewhat smaller galls that also may be distributed in a beadlike manner throughout a large proportion of the

Fig. 22–3. Effect of *Meloidogyne* spp. on roots of burley tobacco.

root system (Barker & Lucas, 1984). Infection by *M. hapla*, however, is associated with production of very small galls. Root cortical tissue often ruptures open as female nematodes mature and galls grow in size. These ruptures allow egg masses of the nematode to be deposited on the surface of infected roots, where the egg masses are barely visible as extremely small, light brown areas or masses on the surface of roots. Root decay also is often associated with root knot of tobacco, although such decay is very limited when *M. hapla* is involved (Shepherd & Barker, 1990). Root necrosis is caused by secondarily invading fungi and bacteria that frequently cause important root disease complexes with root-knot nematodes (Powell, 1979).

Biology

Prior to transplanting of the tobacco crop, root-knot nematode populations exist as eggs and juveniles in the soil, mostly in the plow layer (Fig. 22–4) (Shew & Lucas, 1991). *Meloidogyne* spp. undergo their first molt to form the infective, second-stage juveniles (J2) while still within the egg, and hatch in response to increasing soil temperature. Root-knot infection of tobacco is negligible at soil temperatures below 16°C or above 36°C, but severe at 22 to 32°C (Lucas, 1975). Optimum soil temperatures for *M. hapla* are 15 to 25°C, whereas *M. javanica* reproduces best at 25 to 30°C (Taylor & Sasser, 1978).

At temperatures typical for flue-cured tobacco production, *M. arenaria* and *M. javanica* penetrate tobacco roots and progress through their life cycle more quickly than *M. incognita* (Arens et al., 1981; Barker et al., 1981). Populations of *M. incognita* build up more quickly on tobacco than do those of *M. hapla* (Barker et al., 1981). *Meloidogyne javanica* penetrates tobacco roots more quickly than

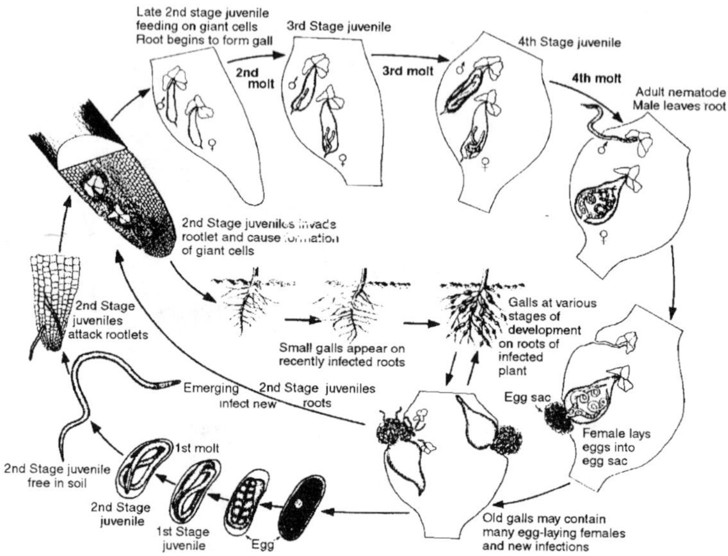

Fig. 22–4. Life cycle of root-knot nematodes (*Meloidogyne* spp.) (reprinted, by permission, from *Plant Pathology*, by G. N. Agrios, 1978).

either *M. arenaria* or *M. incognita* (Arens et al., 1981). Compared to other species, fewer J2 of *M. hapla* penetrate roots of flue-cured tobacco cultivars, regardless of the cultivar reaction to Races 1 and 3 of *M. incognita* (Sosa-Moss et al., 1983). After contacting a susceptible host root, J2 penetrate and then migrate through the region of cell differentiation and elongation until they reach the vascular cylinder. There they establish feeding sites, apparently by injecting "secretory substances" into three to five host cells via their stylets. These nematode secretions are thought to stimulate formation of hypertrophied, multinucleate "giant cells" on which the nematode feeds (Hussey, 1989). Giant cells initiated by *M. incognita* and *M. hapla* in tobacco roots are more rounded and compact than the extensive, highly lobed syncytia associated with *M. arenaria* and *M. javanica* (Sosa-Moss et al., 1983). Hyperplasia of other host cells surrounding the feeding site also occurs.

As feeding progresses, J2 enlarge to become sausage shaped. Feeding ceases when the nematodes molt through their very short (3–4 d) third- and fourth-developmental stages, but is resumed by adult female nematodes. Adult females enlarge to a pyriform or pear shape as they continue to feed. Upon maturity, the posterior of adult females may protrude from the exterior of roots. Between 300 and 2000 eggs may be deposited by a single female in a gelatinous matrix on the root surface. Fecundity of individual female root-knot nematodes is density dependent, generally decreasing as population densities increase. Females may continue to produce eggs for weeks, and may live for some time after egg production stops (Taylor & Sasser, 1978).

Root-knot nematodes can complete their life cycle in about 20 d when soil temperatures are near 25°C (Lucas, 1975). At these temperatures, root-knot nematodes can complete from 5 to 10 generations within a single tobacco growing season (Lucas, 1975). However, *M. javanica* required 56 d to complete a generation at 14°C, 67 d at 18°C, 21 d at 26°C, and 24 d at 30°C (Barker & Lucas, 1984; Madulu & Trudgill, 1994). Soil moisture levels have little influence on development of root-knot nematodes on tobacco, as long as soil moisture remains between 40 and 80% of the moisture-holding capacity of the soil (Lucas, 1975).

Meloidogyne arenaria, *M. incognita*, and *M. javanica* reproduce parthenogenetically, whereas *M. hapla* is facultatively parthenogenetic (Barker & Lucas, 1984). Emerging males migrate to the root surface and tend to cluster themselves in and around egg masses. The proportion of male nematodes in root-knot populations is usually very low, but may increase when penetration sites are limiting or due to unfavorable physiological changes in the host (Barker & Lucas, 1984).

The four most common root-knot nematode species differ in their reproduction and damage potential on tobacco (Arens & Rich, 1981; Barker et al., 1981). *Meloidogyne incognita* populations can increase by more than 500 times during the first 6 to 8 wk of the growing season, when tobacco growth is most pronounced and leaves are formed (Barker et al., 1981). In contrast, populations of *M. arenaria* and *M. javanica* can increase about 600- and 1100-fold during the first 6 to 8 wk after transplanting, while *M. hapla* may increase 75-fold early in the growing season (Arens & Rich, 1981; Barker et al., 1981). *Meloidogyne incognita* galls tobacco roots 2 to 2.5 times as extensively as *M. hapla* (Barker, 1989; Barker et al., 1981).

Increase of root-knot nematode populations on tobacco is density dependent (Barker & Lucas, 1984). As a result, root-knot population densities peak at different periods of time, depending upon initial population levels. High initial nematode population densities suppress root growth, decreasing the number and suitability of available penetration sites for subsequent generations. Under such conditions, penetration sites become limiting, causing nematode populations to peak before the end of the tobacco production season. The more normal growth of tobacco roots when initial nematode population densities are lower provides sufficient infection sites to allow nematode reproduction throughout the growing season. Peak nematode population densities are then observed at or soon after final harvest of tobacco (Rich et al., 1984).

Survival of root-knot nematodes between tobacco growing seasons is primarily influenced by soil temperature and moisture. *Meloidogyne hapla* can survive freezing temperatures and remain infective in significant numbers for up to 90 d, while eggs of *M. incognita* and *M. javanica* were killed within 11 d (Taylor & Sasser, 1978). Populations of *M. javanica* from tropical areas are even more sensitive to cold than isolates from temperate climates (Barker & Lucas, 1984). Sunlight and drying together can kill nematode eggs within 30 min, but juveniles can survive up to 40 wk below the surface of fallow soil (Lucas, 1975).

Meloidogyne species occur in a wide range of soil types, but primarily damage tobacco grown in sandy and sandy loam soils. (Barker & Weeks, 1991; Noe & Barker, 1985a). *Meloidogyne incognita* reproduction on tobacco is higher in loamy sands than in clay or organic soils (Barker & Weeks, 1991). In loamy sand soil, however, higher population densities of *Meloidogyne* were associated with higher percentage clay content, lower concentrations of Cu, slower cycling of soil moisture between wet and dry, higher organic matter content, and greater Na concentrations of the soil (Noe & Barker, 1985a). Increased root galling also has been associated with lower soil pH and higher soil K levels (Kincaid et al., 1970; Marks & Sayre, 1964). Increased survival between growing seasons also was correlated with lower population densities in the fall.

Disease Complexes

Disease complexes are an integral part of the ecology of root-knot nematodes on tobacco. Extensive decay is common in tobacco roots severely infected by root-knot nematodes. Root-knot nematodes enhance the incidence and severity of tobacco black shank [caused by *Phytophthora parasitica* Dast. var. *nicotiana* (B. de Haan) Tucker], Fusarium wilt (caused by *Fusarium oxysporum* Schlecht. var. *nicotianae*), and bacterial wilt [also called Granville wilt, caused by *Pseudomonas solanacearum* (Smith) Smith] (Shepherd & Barker, 1990; LaMondia, 1992). Root-knot nematodes also often negate the effectiveness of host resistance and chemical treatments for these diseases (Shepherd & Barker, 1990). Resistance to root-knot nematodes limited crop losses to black shank, but host resistance to black shank does not similarly limit damage caused by root-knot nematodes (Csinos et al., 1986). Although *Rhizoctonia solani* Kühn does not normally attack older tobacco roots, such infection was observed after penetration by root-knot nematodes (Shepherd & Barker, 1990). Increased stem damage by *R. solani* was not associated with root-knot infection. Parasitism by *Meloidogyne* species can even facilitate extensive root necrosis by soil-inhabiting fungi, such as *Aspergillus*, *Botrytis*, *Curvularia*, *Penicillium*, *Pythium*, and *Trichoderma,* that do not usually infect tobacco (Powell, 1979). Much of the crop loss associated with root-knot nematodes may be due to root decay caused by such secondary invaders (Barker et al., 1981).

Root-knot nematode infection also increases foliar disease problems caused by *Alternaria alternata* (Fr. ex Fr.) Kiessl., and air pollution (O_3) (Bisessar & Palmer, 1984; Shepherd & Barker, 1990). Host infection by tobacco mosaic virus and tobacco ring spot virus also have been reported to influence development of root-knot nematodes (Bird, 1969; Goswami & Raychaudhuri, 1973). Resistance to Races 1 and 3 of *M. incognita* in flue-cured tobacco also has been associated with varying reactions to host strains of potato virus Y (Rufty et al., 1983). The necrotic reaction of root-knot resistant cultivars to certain strains of potato virus Y (PVY) may be pleiotropic (Rufty et al., 1983; Yung et al., 1991). In contrast, root-knot nematodes were found not to interact with the parasitic plant *Orobanche cernua* (Bridge et al., 1984). Interactions between root-knot nematodes and other plant pathogens involves more than simply changing the availability of infection courts. Root-knot nematode infection changes the structure and function of the tobacco root system, as well as the physiology of the entire plant (Powell et al., 1971). However, changes in tobacco physiology caused by these nematodes do not always cause systemic changes in host reaction to other pathogens (Moorman et al., 1980).

Root-knot nematodes occur as populations of a single species or as communities of several species, and must often compete with other plant-parasitic nematodes (Fortnum et al., 1984; Schmitt & Barker, 1988). Reproduction and development of *M. arenaria* and *M. incognita* may be suppressed when these species parasitize roots simultaneously (Ibrahim, 1987; Ng'ambi et al., 1995). Infection by other root-knot species has been reported to predispose resistant flue-cured tobacco cultivars to increased parasitism by Races 1 and 3 of *M. incognita* (Ibrahim, 1987; Shepherd & Barker, 1990). Recent work, however, indicates that mixed populations of *Meloidogyne* spp. do not increase nematode parasitism of *M. incognita*-resistant cultivars (Ng'ambi et al., 1995). Root-knot and lesion nematodes also may compete with each other for penetration sites on tobacco roots (Lucas, 1975). Reproduction of *M. hapla* can be inhibited by concomitant populations of *M. incognita* or *Pratylenchus brachyurus*. Population increases of both *M. incognita* and *P. brachyurus* also may be suppressed when these two species occur together. Similar interactions have been observed among *M. incognita, M. javanica*, and *Rotylenchulus reniformis* (reniform nematode) (Patel & Patel, 1991).

Economic Importance

Root-knot nematodes are the most important nematode parasites of tobacco (Shepherd & Barker, 1990). However, losses due to root knot may be routinely underestimated because reduced vigor is not as eye-catching as the plant mortality caused by other tobacco diseases. Estimated losses in U.S. tobacco caused by root-knot nematodes averaged approximately $11 million from 1991 to 1992 (Anonymous, 1993). Their economic importance results from a number of factors other than the stunting and delayed maturity associated with root-knot nematode infection alone. Disease complexes associated with root-knot nematodes may be of equal or greater importance compared to infection by root-knot nematodes alone (Barker & Lucas, 1984). Because tobacco is a highly taxed commodity, losses at the farm gate are often magnified by equivalent reductions in tax revenue (Shepherd & Barker, 1990). The high economic value of tobacco, the consistent effectiveness of pesticides, and the poor reliability of methods for predicting damage caused by root-knot nematodes also increase the economic importance of these nematodes by persuading tobacco growers to routinely apply nematicides for control of root knot (Johnson, 1989a). Annual root-knot nematode control costs for U.S. flue-cured tobacco were over $20 million in 1991 and 1992 (Anonymous, 1993).

Tobacco plants infected by root-knot nematodes grow more slowly, an effect often also manifested as delayed flowering of the crop and reduced sucker growth (Rich & Barker, 1984). Root-knot nematodes decrease nutrient uptake, increase transpiration, and reduce the water use efficiency of infected tobacco plants (Patel et al., 1988; Rahi et al., 1988). Peroxidase activities and levels of phenols and polyphenols also increase in infected tobacco plants, particularly in root-knot resistant varieties (Ibrahim, 1991; Shukla & Chakraborty, 1988). Resistant cultivars may be stunted during the first 60 d of the growing season, but recover before the harvest season (Barker & Melton, 1990). Susceptible cultivars, however, may exhibit progressive stunting throughout the growing season. Delayed host growth results in harvest of fewer, lighter tobacco leaves with inferior quality characteristics. The initial stunting of resistant cultivars has been associated with the hypersensitive nature of resistance and to root decay caused by soil microorganisms (Barker & Melton, 1990; Sosa-Moss et al., 1983). Recovery of resistant cultivars may be due to less root decay resulting from suppressed nematode reproduction.

Considerable research has been conducted to quantify direct losses in tobacco yield and value caused by root-knot nematodes (Barker, 1989). As few as two eggs of *M. incognita* per cubic centimeter of soil at transplanting can significantly suppress tobacco yield (diVito et al., 1983). Yields of susceptible cultivars of flue-cured tobacco may drop 5 to 10% for each 10-fold increase in the initial population of *M. incognita* (Barker et al., 1981). Resistant cultivars lost 3% of their yield for each 10-fold increase in initial *M. incognita* populations. Yield losses may approach 13 to 19% for each 10-fold increase in the initial populations of *M. arenaria* and *M. javanica*, while yield losses of 3 to 5% have been estimated for each 10-fold increase in the initial population of *M. hapla*. Under Florida conditions, *M. javanica* suppressed tobacco yield twice as much as *M. incognita* (Arens & Rich, 1981). Although nematode population increase may be dramatic late in the growing season (when tobacco is harvested), high nematode population densities and root galling are usually more correlated with lowered yield if nematode data is obtained near the time of the first harvest, approximately 8 to 10 wk after transplanting (Fig. 22–5) (Ferris, 1974; Rich et al., 1984). However, the optimum sample time for obtaining such correlations varies from year to year, depending upon initial nematode population densities and environmental parameters such as soil temperature and moisture (Barker, 1989; Rich et al., 1984). Root necrosis indicates the combined effects of root-knot nematodes and associated soil microorganisms (Barker, 1989).

The patchy distribution of nematode populations also needs to be remembered when considering losses in yield and quality caused by root-knot nematodes (Noe & Barker, 1985b). Nematode effects on tobacco yield can be overestimated by as much as 288% when it is assumed that low (50 juveniles/500 cm^3 of soil) populations of root-knot nematodes are distributed uniformly. Such levels are in the range of economic thresholds used in flue-cured tobacco production (Barker & Imbriani, 1984; Melton et al., 1994). Higher initial nematode population densities tend to be less aggregated, and under these conditions, crop loss estimates based upon average conditions in a field improve dramatically.

Leaf chemistry parameters are important aspects of tobacco quality. Root-knot infection increases the nicotine content of tobacco roots, but has been associated with lower sugar and nicotine content in tobacco leaves (Davis & Rich, 1987; Barker & Weeks, 1991). Lower nicotine levels in cured leaf associated with these pathogens have been attributed to suppression of root growth, resulting in fewer root tips (the site of nicotine synthesis in tobacco), or to disrupted translocation of nicotine from roots to leaves by root-knot nematodes (Barker & Weeks, 1991). Lower levels of total amino acids have been reported for leaves of *M. incognita*-infected tobacco, but higher amino acid levels were found in infected roots (Zhang & Zhou, 1989).

Soil fertility, moisture, texture, water-holding capacity, temperature, and biota significantly influence the relationship between root-knot nematodes and

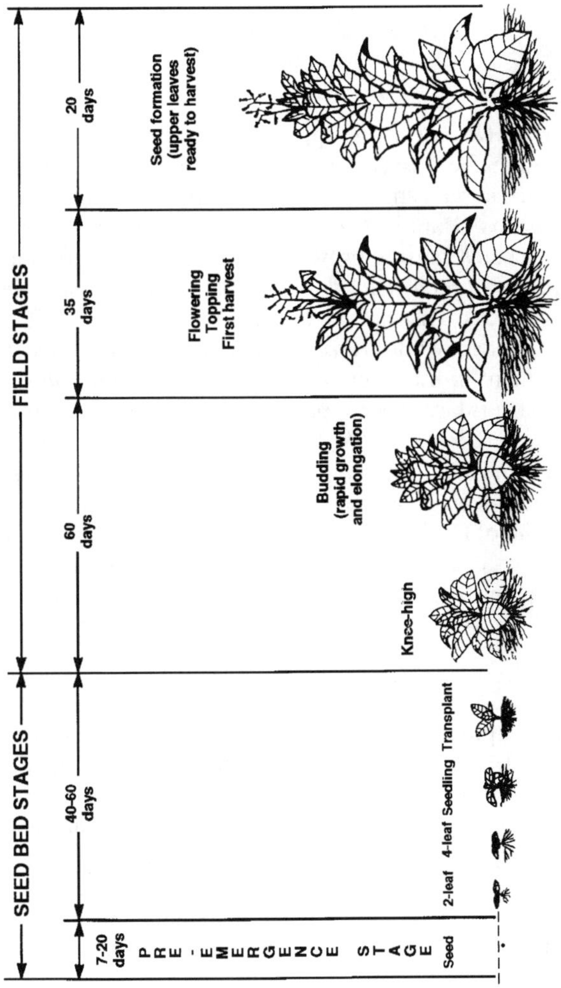

Fig. 22–5. Growth stages in the development of flue-cured tobacco (courtesy of H. D. Shew, Dept. of Plant Pathology, North Carolina State Univ.).

tobacco yield and quality (Barker, 1989). Although soil type may have slight to moderate influences on reproduction of *M. incognita*, nematode effects on tobacco yield and quality are smaller in clay soils compared to sandy, loamy, or muck soils (Barker & Weeks, 1991). Low populations of *M. incognita* may even increase yield slightly in clay soils (Barker & Weeks, 1991). Yield effects of root knot were greater under conditions of moisture stress or with high irrigation than under an intermediate level of soil moisture (Wheeler et al., 1991).

Lesion Nematodes

Species

Lesion nematodes (*Pratylenchus* spp.) are the primary cause of brown root rot of tobacco, although the disease may often be a complex involving lesion nematodes and soil microorganisms such as *Fusarium*, *Pythium*, *Macrophomina*, *Mucor*, and *Thielaviopsis* (Barker & Lucas, 1984). *Pratylenchus pratensis* has been reported as the species causing brown root rot in North Carolina, Kentucky, and Virginia; whereas *P. brachyurus* damaged tobacco in South Carolina and elsewhere (Barker & Lucas, 1984). *Pratylenchus brachyurus* and *P. zeae* tend to occur in warmer climates; while *P. penetrans*, *P. thornei*, and *P. neglectus* (*minyus*) are predominant in more temperate regions. *Pratylenchus crenatus*, *P. neglectus*, and *P. penetrans* have been reported from tobacco fields in Canada and New Zealand (Shepherd & Barker, 1990). Although *P. zeae* has been reported on tobacco in South Carolina and Trinidad, it did not reproduce on or damage tobacco in greenhouse experiments (Shepherd & Barker, 1990). Evidence is equivocal on the damage potential of *P. alleni*, *P. brachyurus*, and *P. scribneri* on flue-cured tobacco in North Carolina (Shepherd & Barker, 1990). *Pratylenchus hexincus*, *P. thornei*, and *P. vulnus* also have been reported on tobacco (Barker & Lucas, 1984).

Symptoms

Aboveground symptoms of tobacco caused by lesion nematodes are very similar to those caused by root-knot and other nematodes (Fig. 22–6) (Lucas, 1975). Symptoms involve stunted plants that tend to wilt when soil moisture conditions should be adequate, indicating deficiencies in moisture and nutrients. Symptomatic plants may occur singly or in groups distributed throughout the field in no obvious pattern. Infected seedlings may never exceed a height of 30 cm (Shew & Lucas, 1991). Leaves on infected plants may become chlorotic and then necrotic ("rimfire"). Macroscopic root symptoms are very similar to those of black root rot, caused by the fungus *Thieloviopsis basicola* (Fig. 22–7). Lesions first appear as discrete, elliptical, colorless, water-soaked areas that may occur anywhere from just behind the root tip to mature tissue (Lucas, 1975). Most lesions are found in the region of root elongation. As lesions mature, they attain a pale yellow color that darkens to dark brown, to almost black. Lesions coalesce to girdle infected roots, causing cortical tissue to slough-off, with only the vascular cylinder remaining. These root symptoms are common diagnostic indicators of brown root rot. Severely infected feeder roots shrivel and die. Large nematode

Fig. 22–6. Stunting of burley tobacco by lesion nematodes (*Pratylenchus* spp.) in Ontario, Canada.

populations cause extensive root pruning that gives the root system a "stubby" look. Secondary feeder roots often proliferate above infected areas, forming a mass of small roots near the soil surface and close to the root crown (Lucas, 1975). The extensive destruction of feeder roots in severely infected plants enables them to be pulled quite easily from the soil, even in midsummer. Eggs,

Fig. 22–7. Black root rot lesions on tobacco roots, similar to symptoms caused by lesion nematodes (*Pratylenchus* spp.). Note discrete necrotic regions on small roots.

juveniles, and adults can be observed in root tissues using appropriate extraction or staining techniques (Daykin & Hussey, 1985).

Biology

Lesion nematodes begin their life cycle as eggs in root tissue or soil (Fig. 22–8). Eggs of lesion nematodes can hatch within 6 to 17 d of being laid, depending upon the species and environmental conditions (Shew & Lucas, 1991). Root exudates attract lesion nematodes to host roots, where they congregate and feed (Zunke, 1990a). Upon contact with a host root, lesion nematodes begin feeding almost immediately (Zunke, 1990b).

Lesion nematodes feed ectoparasitically and endoparasitically (Zunke, 1990b). Nematodes select a feeding site by rubbing their lips against the surface of host cells. After inserting their stylet, lesion nematodes inject saliva into a host cell for several minutes before ingesting cytoplasm (Zunke, 1990a,b). Lesion nematodes may feed for several seconds or many hours. Individual nematodes may feed at several sites on a single root hair. Lesion nematodes penetrate intracellularly by pushing epidermal cells apart, and intercellularly after feeding on epidermal cells (Zunke, 1990a). Penetration wounds attract other lesion nematodes and provide an excellent infection court for secondary microorganisms (Lucas, 1975; Zunke, 1990a). In addition to penetrating roots in the zone of elongation like root-knot nematodes, lesion nematodes also penetrate small, relatively mature roots in areas where root hairs are dense (Zunke, 1990a,b). All active stages of the nematode are capable of penetration, but juveniles tend to infect smaller, lateral roots and feed on cells for shorter periods of time (Zunke, 1990a).

Lesion nematodes begin migrating through host roots by puncturing, penetrating, and partially digesting the walls of neighboring cells (Zunke, 1990a).

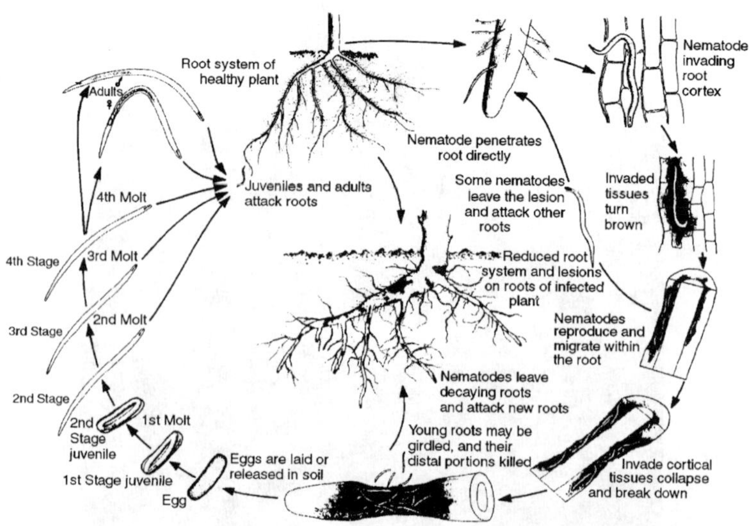

Fig. 22–8. Life cycle of lesion nematodes (*Pratylenchus* spp.) (reprinted, by permission, from *Plant Pathology*, by G. N. Agrios, 1978).

These parasites feed by depleting ruptured cells of their cytoplasm. Feeding and migration is frequently suspended by rest phases where nematodes remain coiled within host cells for several hours. Lesion nematodes restrict their activity to the root cortex (Zunke, 1990a). Brief feeding causes little noticeable change in host cells, but extended feeding causes rapid cytoplasmic streaming and pronounced hypertrophy of root hair nuclei. Root hairs may survive for 10 or more hours after brief feeding, but die within 1 h of extended feeding. Host cells that eventually die due to extended feeding and nematode migration may live for hours after nematodes have moved elsewhere (Zunke, 1990a). Disruption of the root cortex by nematode migration and feeding results in the water-soaked appearance of early lesions. Oxidation of polyphenols in infected cells results in the gradual discoloration of infected host tissue. Oxidized polyphenols inhibit nematode respiration and can exert a repellent effect, perhaps explaining the migratory behavior of lesion nematodes. (Lucas, 1975).

Nematodes molt through a third and fourth stage as they feed, either inside cortical cells or in the soil (Zunke, 1990a). Populations of *P. penetrans* and *P. vulnus* generally contain abundant males and reproduce by cross-fertilization or amphimixis (Barker & Lucas, 1984). Populations of *P. brachyurus*, *P. neglectus* (*minyus*), and *P. zeae* are monosexual and reproduce by parthenogenesis. Females may begin laying eggs after several hours of continuous feeding (Zunke, 1990a). Female lesion nematodes usually deposit eggs singly in infected roots at an average rate of 1 egg/d, for up to 30 d (Lucas, 1975). Eggs also may be laid in soil, near the surface of host roots (Zunke, 1990a). Eggs in root tissue hatch within 6 to 17 d, depending upon the species. The length of the life cycle varies across species of *Pratylenchus* and with environmental conditions, particularly soil temperature and moisture (Lucas, 1975). *Pratylenchus penetrans* can complete its life cycle in 28 d, whereas *P. brachyurus* may require 35 to 40 d, and other species in the genus may take up to 65 d to complete a generation, even under optimum environmental conditions (Lucas, 1975).

Freshly hatched juveniles may feed in the same root in which they hatch, or migrate to another (Lucas, 1975). Lesion nematodes can move through the soil as far as 2 cm in a 7-d period (Townshend & Webber, 1971). Reproduction and migration of lesion nematodes is greater in coarse-textured soils than in fine-textured soils (Townshend, 1971). Movement of *P. penetrans* through soil tends to be greatest when soil moisture is close to field capacity (Townshend & Webber, 1971).

Lesion nematodes can survive for some time in soil without a host (Lucas, 1975). The mechanism for this survival is unknown. Winter mortality of *P. penetrans* ranges from 40 to 65% in tobacco-producing areas of Canada (Barker & Lucas, 1984). Survival has been related to soil aeration and to increasing moisture tension in the soil, but appears unrelated to soil type or texture (Townshend & Webber, 1971). Anhydrobiosis (the ability to survive after losing almost all body water) may increase the ability of lesion nematodes to survive freezing temperatures (Townshend, 1984). Lesion nematodes that become anhydrobiotic can survive from one to several years depending upon the rate at which soil dries, soil texture, and soil moisture (Townshend, 1984, 1987).

Disease Complexes

In addition to damaging tobacco roots themselves, lesion nematodes provide entry sites for plant pathogenic fungi such as *Fusarium*, *Pythium*, *Macrophomina phaseolina*, *Thielaviopsis basicola*, and *Verticillium* (Lucas, 1975). Simultaneous inoculation with *Pratylenchus brachyurus* and *P. parasitica* var. *nicotianae* enhanced development of tobacco black shank on a black shank susceptible flue-cured tobacco cultivar (Lucas, 1975). However, nematode penetration 3 wk before fungal infection delayed appearance of black shank symptoms. Resistance to black shank was not altered by lesion nematode infection. Penetration of a black shank-susceptible Connecticut shade tobacco cultivar by *P. penetrans* 4 or 7 d before fungal inoculation also delayed black shank development (McIntyre & Miller, 1978). Black shank was severe when nematodes and fungus were applied simultaneously. Apparent protection from black shank by lesion nematode infection also increased with initial nematode numbers. Under wet conditions, *P. hexincus* has been associated with increased development of black root rot of tobacco, caused by *T. basicola* (syn. *Chalara elegans*) (Shepherd & Barker, 1990). Increased reproduction of some populations of *P. penetrans* also has been observed in the roots of some tobacco cultivars infected with the black root rot fungus in Canada (Lucas, 1975).

Lesion nematodes also interact with other nematode genera. Concomitant populations of *M. incognita*, *M. hapla*, and *Pratylenchus brachyurus* can sometimes mutually suppress reproduction (Barker & Lucas, 1984; Lucas, 1975). Interactions between lesion and root-knot nematodes on tobacco can differ significantly among cultivars, often depending upon the resistance characteristics of each cultivar to the individual nematode species comprising the nematode complex (Lucas, 1975). Attack by *Tylenchorhynchus claytoni* can restrict numbers of *P. penetrans* in tobacco roots (McIntyre & Miller, 1976). *Globodera tabacum tabacum* suppresses infection and survival of *P. penetrans* (Lucas, 1975). Reproduction of tobacco cyst nematodes also was suppressed by antagonism with lesion nematodes, but population densities of *G. tabacum tabacum* increased or remained stable over time, while those of *P. penetrans* often decreased to low levels.

Economic Importance

Crop losses caused by lesion nematodes in tobacco have been estimated at about 1%, but have been so low in recent years that numerical estimates are difficult to obtain (Shepherd & Barker, 1990). They are considered to be of minor to moderate (or local) importance in most tobacco-producing countries around the world (Shepherd & Barker, 1990). Widespread use of cultural and chemical control practices effective against most plant-parasitic nematodes may account for the apparent decline in the importance of lesion nematodes in tobacco (Barker & Lucas, 1984). The relatively poor host suitability of tobacco for *Pratylenchus* spp. and the apparent increase in resistance to these nematodes in cultivars bred for resistance to certain races of *M. incognita* also may explain these trends (Shepherd & Barker, 1990). Some also have suggested that a limited ability to migrate through soil, compared to root-knot nematodes, may be involved (Barker & Lucas, 1984).

Tobacco plants infected by lesion or root-knot nematodes transpire up to twice as much water as healthy plants during the first 8 wk of the growing season (Barker & Lucas, 1984). However, healthy plants transpire more than nematode-infected plants later in the growing season (12 or more wk after transplanting). The effects of *P. brachyurus* on transpiration of tobacco are less persistent than those of *M. hapla* or *M. incognita*. Nematode infection also restricts the water-use efficiency of tobacco, but may actually stimulate growth of tobacco when soil moisture is not a limiting factor. Under Canadian conditions, yield and quality of flue-cured tobacco decrease with preplant population densities of 2000 or more *P. penetrans*/kg of soil (Lucas, 1975). Although the numbers of *P. penetrans* are lower during midseason compared to the beginning or end of the growing season, soil population densities between 500 and 2000/ kg of soil during midseason (approximately 7 wk after transplanting) have been found to cause economic damage to tobacco (Olthoff et al., 1973). Damage to flue-cured tobacco caused by *P. penetrans* is more severe when springs are cool and wet, than when they are warm and dry (Townshend & Marks, 1976, 1977).

Cyst Nematodes

Species

A complex of round cyst nematodes in the genus *Globodera* attack various types of tobacco around the world. They have been classified as minor or moderately important pests (Shepherd & Barker, 1990). The taxonomy of these species has been somewhat controversial since their discovery, but they are usually classed as subspecies within a single species based upon biochemical, hybridization, and morphometric data (Mugniery et al., 1992; Bossis & Mugniery, 1993; Mota & Eisenback, 1993a,b,c). *Globodera tabacum tabacum* has been an important pest of shade-grown tobacco in Connecticut since 1951 (Shepherd & Barker, 1990). *Globodera tabacum solanacearum* is an important problem on flue-cured tobacco in Virginia (Shepherd & Barker, 1990), and has been recently detected on Maryland-type tobacco in Maryland (S. Sardanelli, 1995, personal communication) and on flue-cured tobacco in a number of counties in North Carolina (Melton et al., 1991; T. A. Melton, 1995, personal communication). *Globodera tabacum virginiae* is the third subspecies within the complex (Shepherd & Barker, 1990). This nematode occurs in Virginia and North Carolina, but does not reproduce on flue-cured tobacco. *Globodera tabacum virginiae* does, however, reproduce slowly on burley tobacco.

Symptoms

Aboveground symptoms of tobacco caused by cyst nematodes are very similar to those caused by root-knot nematodes (Lucas, 1975) (Fig. 22–9). Symptomatic plants may occur singly or in groups distributed throughout the field in no obvious pattern. The foliage of infected plants may show evidence of nutrient deficiencies. Infected plants are stunted to varying degrees, depending

Fig. 22–9. Stunting of flue-cured tobacco caused by tobacco cyst nematodes (*Globodera tabacum solanacearum*) in Virginia.

upon environmental conditions, and often tend to wilt when soil moisture conditions should be adequate, due to a much smaller root system. Cysts approximately 0.5 mm in diameter and ranging from white to dark reddish-brown in color are usually visible on roots of infected plants (Fig. 22–10).

Fig. 22–10. Cysts of *Globodera tabacum solanacearum* on flue-cured tobacco roots.

Biology

When female tobacco cyst nematodes die, their bodies develop into tough, brown structures, called cysts, that are filled with hundreds of eggs. Cysts protect eggs from soil organisms, drying, and chemical soil treatments. The protection afforded by cysts and the apparent capacity to enter "diapause" enables eggs and juveniles to survive for 11 yr or more in the absence of a host plant. Eggs hatch in response to increasing soil temperature and, to some degree, to exudates from host roots (LaMondia, 1995b).

Second-stage juveniles infect roots and establish permanent feeding sites in developing host vascular tissue by stimulating development of syncytial cells (Fig. 22–11). Feeding nematodes molt twice, fourth-stage male nematodes leave their feeding sites, and are attracted to female nematodes (Seinhorst, 1986), whose bodies have begun to protrude from infected roots, facilitating cross-fertilization (Miller & Gray, 1972). Adult female nematodes are pearly white, but they turn yellow as they mature, and then turn brown after they die. Cysts become free in the soil when they break off at the neck. *Globodera tabacum tabacum* can complete its life cycle within 20 d, while *G. t. solanacearum* and *G. t. virginiae* require from 32 to 82 d to complete their life cycles (Barker & Lucas, 1984). Rate of development, cyst size, and number of eggs and juveniles within cysts all vary with soil temperature (Adams et al., 1982). Numbers of cysts and juveniles remain low early in the growing season, perhaps due to land tillage in May and/or movement of soil populations from the soil to host roots (Reilly & Grant, 1985). Population densities in soil reach their peak at the end of the growing season, in September and October. Populations drop precipitously in early fall and remain low until June or July.

Disease Complexes

Globodera tabacum solanacearum increases development of bacterial wilt (*Pseudomonas solanacearum)* and tobacco black shank *(P. parasitica* Dast. var. *nicotiana)* (Grant, 1984). Interactions between cyst nematodes and tobacco black shank are generally synergistic (Grant, 1984). Although both pathogens must be present on the same root system in order to increase root necrosis, nematode development may accelerate on healthy roots of plants infected by the black shank pathogen (Grant, 1984). Population densities of *G. t. tabacum* as low as 25 juveniles/cm^3 of soil increased Fusarium wilt on wilt-susceptible and wilt-resistant cultivars of broad leaf and shade tobacco in the northeastern USA (LaMondia & Taylor, 1987). *Globodera tabacum tabacum* increased wilt in a localized, rather than in a systemic manner. *Globodera tabacum solanacearum* also may suppress reproduction of mycorrhizal fungi, important in normal development of the tobacco crop (Fox & Spasoff, 1972).

Globodera tabacum tabacum inhibits the reproduction of other nematodes, such as *P. penetrans*, on shade tobacco (Barker & Lucas, 1984). Large populations of *P. penetrans* also can slow the increase of tobacco cyst nematode populations, but moderate to high populations of *G. tabacum tabacum* suppress populations of *P. penetrans* to the point that the latter may become undetectable.

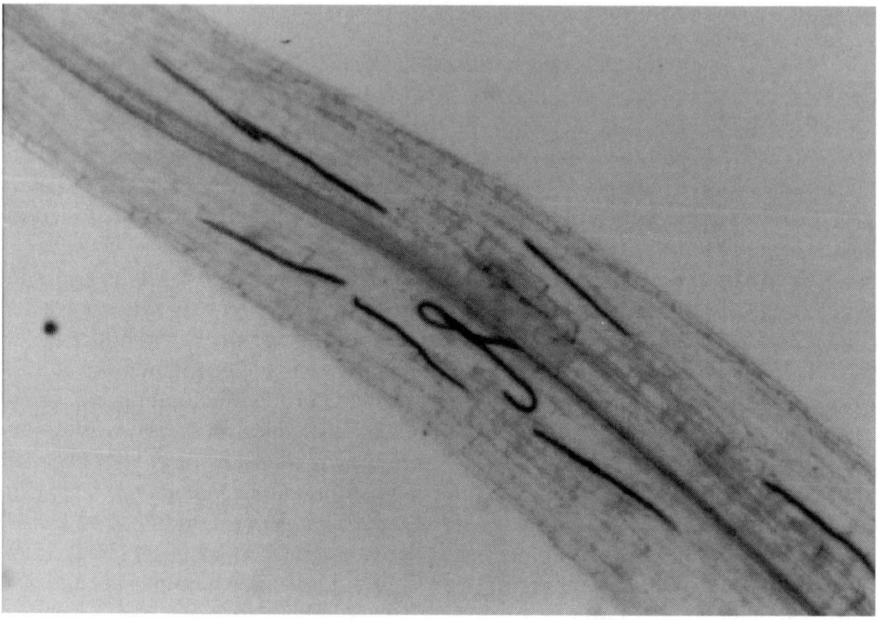

Fig. 22–11. Vermiform juveniles of *Globodera tabacum solanacearum* in roots of flue-cured tobacco.

Economic Importance

The economic importance of tobacco cyst nematodes is increasing as these pests spread within the flue-cured tobacco production areas of Virginia and North Carolina. On average, *G. t. solanacearum* causes yield losses of about 15% in flue-cured tobacco in Virginia (Komm et al., 1983). Losses of 25 to 50% have been estimated from cases involving extremely large population densities, but tobacco may escape significant crop losses when population densities are moderate and environmental conditions favor the crop (Johnson, 1992). Although population densities below 100 juveniles/cm^3 of soil limit yield of shade tobacco less than 5%, 500 to 1000 juveniles/cm^3 can cause maximum yield losses of about 45% (LaMondia, 1995a).

Reniform Nematodes

Species

Rotylenchulus reniformis causes significant damage to tobacco in some countries of the world and under some conditions (Barker & Lucas, 1984). This parasite occurs in Trinidad and in eastern North Carolina, and has been reported to stunt bidi tobacco in greenhouse experiments in India (Shepherd & Barker, 1990). *Rotylenchulus parvus* doesn't parasitize tobacco.

Biology

Female reniform nematodes penetrate parenchyma of tobacco roots intracellularly to reach the pericycle (Patel, 1986). In contrast to root-knot nematodes, only the anterior third of invading females becomes embedded within the host root. When feeding begins, host cells surrounding the head of the nematode are transformed into syncytia by cell-wall dissolution and coalescing of cytoplasm from adjacent cells (Patel et al., 1988). Hypertrophy, hyperplasia, cell wall thickening, granular cytoplasm and enlargement of nuclei and nucleoli are observed in host cells surrounding the feeding site. Damage to phloem and cortical parenchyma kill feeder roots, inhibiting shoot growth of infected plants.

Disease Complexes

Meloidogyne incognita or *M. javanica* alone or in combination with *Rotylenchus reniformis* depress nutrient uptake and translocation more than *R. reniformis* alone (Patel et al., 1988). Similar effects were seen on growth of tobacco (Patel et al., 1989). Both species of root-knot nematode reproduce at a higher rate on tobacco than *R. reniformis*, and either species of root-knot nematode can suppress reproduction of reniform nematodes (Patel & Patel, 1991).

Economic Importance

Rotylenchulus reniformis can limit seed germination, nutrient uptake and growth of tobacco seedlings (Patel et al., 1986). Significant increases in tobacco yield and value also have been reported from control of reniform nematodes (Patel et al., 1987). Control of moderate to high population densities of reniform nematodes also increased yield and value of flue-cured tobacco in the USA, but the increases were not as great as those reported from other countries (Melton & Powell, 1991). Population densities also appeared to decline as the tobacco growing season progressed, suggesting that tobacco may not be a good host for reniform nematodes. Intraspecific differences among reniform nematodes and differences in soil texture between the USA and other tobacco-growing areas were suggested as possible reasons for conflicting results.

Stem and Bulb Nematodes

Species

Ditylenchus dipsaci has been reported on tobacco in many countries. Most of the approximately 20 known host races of the nematode do not reproduce on tobacco (Barker & Lucas, 1984). Isolates extracted from tobacco may not infect other crops, such as wheat (*Triticum aestivum* L.), maize (*Zea mays* L.), or sugarbeet (*Beta vulgaris* L. subsp. *vulgaris*) (Grujicic et al, 1988a). *Ditylenchus destructor* can reproduce to a limited extent on some tobacco genotypes (Basson et al., 1990).

Symptoms

Stem nematodes enter the leaves and stems of tobacco seedlings during wet weather and induce small, yellow swellings ("galls") that may extend 40 cm or more above the soil. Younger plants may be stunted to a greater extent than older plants. Leaves wilt only under severe attack because the vascular system may continue to function for some time after these nematodes have begun their attack. As the number of swellings increase, plant tissue begins to die prematurely. Lower leaves may fall off, and upper leaves may turn yellow. Galls eventually rot, stopping growth of infected plants. Eventually, infected stems break and plants fall over.

Biology

Stem break is usually associated with cool, damp weather and heavy soils. Fourth-stage juveniles of *D. dipsaci* can remain dormant in a cryptobiotic stage for 20 or more years (Barker & Lucas, 1984). Eggs and juveniles can withstand freezing. Low soil moisture and soil temperatures at or below 0°C are optimum for this nematode. A period of 19 to 23 d is usually required for the nematode to complete its life cycle (Lucas, 1975). Reproduction is by amphimixis.

Economic Importance

Yield losses due to "stem break" have only been reported from Holland, France, Germany, Switzerland, and Serbia (Shepherd & Barker, 1990; Grujicic et al., 1988b). Stem break has been reported to cause losses up to 54% on tobacco in northeastern France (Lucas, 1975).

Other Nematodes

Taxa

Species of *Aphelenchoides* have been reported in France, Germany, China, Pakistan, Brazil, and Chile (Shepherd & Barker, 1990). *Aphelenchoides ritzemabosi* has been implicated as the cause of a "checkered leaf disease" in a localized area of France near the Atlantic end of the Pyrenees (Shepherd & Barker, 1990). Various species of *Tylenchorhynchus* have been reported to damage tobacco in Canada, India, New Zealand, and the USA (Shepherd & Barker, 1990). Spiral nematodes such as *Scutellonema brachyurus* are often found in tobacco soils, and have been reported to limit growth of tobacco, but also are only of minor importance on tobacco (Shepherd & Barker, 1990). Trichidorid species such as *Paratrichodorus lobatus* and *Longidorus elongatus* have been reported to cause stunting and severe yield losses in tobacco in several production areas of the world (Barker & Lucas, 1984). Other nematodes associated with tobacco include *Xiphinema americanum*, *Helicotylenchus* spp., *Criconemella* spp., and *Tetylenchus nicotianae* (Barker & Lucas, 1984).

Symptoms

Aphelenchoides ritzemabosi causes polygonal blotches on leaves of infected plants that are bounded by leaf veins, similar to the symptoms caused by this nematode on Max chrysanthemum [*Leucanthemum maximum* (Ramond) DC] (Shepherd & Barker, 1990). The primary symptom of parasitism by ectoparasitic nematodes is a smaller root system that leads to stunting of the overall growth of infected plants (Barker & Lucas, 1984). Ectoparasitic nematodes feed on root tips and kill root hairs, giving the root system a "stubby root" appearance. These nematodes can often be found in the cortical tissue of infected roots. High populations may result in necrotic lesions.

Biology

Xiphinema americanum, *C. ornata*, *H. dihystera*, and *P. minor* reproduce by parthenogenesis (Barker & Lucas, 1984). *Xiphinema americanum* can complete its life cycle one or two times within a year. *Tylenchorhynchus claytoni* is amphimictic, requiring a month to complete its life cycle when temperatures are 30 to 34°C (Barker & Lucas, 1984).

Economic Importance

Ectoparasitic nematodes rarely cause significant problems in tobacco production. Population densities of *L. elongatus* greater than 2500 juveniles/kg of soil were reported to stunt flue-cured tobacco in Canada (Shepherd & Barker, 1990). *Tylenchorhynchus claytoni* has been reported to reproduce on tobacco without affecting yields of the crop (Barker & Lucas, 1984). *Tylenchorhynchus annulatus* has significantly suppressed growth of tobacco seedlings in greenhouse experiments (Shepherd & Barker, 1990). Trichodorid and *Xiphinema* spp. can indirectly suppress tobacco yield and quality by transmitting viruses such as tobacco ringspot and tobacco rattle (Barker & Lucas, 1984). Complexes of stunt nematodes and tobacco rattle virus have been reported to cause yield losses in tobacco in parts of Holland and Germany (Lucas, 1975). Stunt nematodes (*Tylenchorhynchus* spp.) also may enhance the incidence of Fusarium wilt (Barker & Lucas, 1984). *Globodera tabacum tabacum* suppresses infection and survival of *T. claytoni* (Barker & Lucas, 1984).

CONTROL

Host Resistance

Research to develop flue-cured tobacco cultivars with resistance to Races 1 and 3 of *M. incognita* began as early as 1935, and has resulted in the release of many cultivars with resistance to these nematodes (Shepherd & Barker, 1990). Resistance generally doesn't inhibit penetration by *Meloidogyne* species, but prevents successful establishment of a feeding site (Schneider, 1991). Some evidence suggests root exudates of resistant tobacco cultivars also may have nematicidal properties (Shukla et al., 1988). Resistance to Races 1 and 3 of *M. incognita* was transferred into cultivated tobacco from *N. tomentosa* (Shepherd & Barker, 1990). This resistance is controlled by a single pair of dominant genes located on Chromosome G (Rufty et al., 1983). A majority of flue-cured tobacco cultivars planted in the USA currently possess this type of resistance to root-knot nematodes (Johnson, 1989a).

Cultivars resistant to Races 1 and 3 of *M. incognita* also possess limited resistance to *M. arenaria* (primarily host Race 1) and/or *M. javanica*, and are more tolerant to attack from *Meloidogyne* species in general (Barker & Melton, 1990; Schneider, 1991). Sources of resistance to Races 3 and 4 of *M. incognita* have been identified from a cross of *N. repanda* × *N. tabacum* and in the La Quinta accession of *N. otophora*, respectively (Shepherd & Barker, 1990; Reed & Schneider, 1992). Resistance to *M. javanica* has been found in *N. longiflora*, *N. megalosiphon*, and *N. repanda*, as well as several other sources (Davis et al., 1988a,b; Shepherd & Barker, 1990). Resistance to *M. javanica* has been combined with that to *M. incognita* in some breeding lines from Zimbabwe, but some of these breeding lines remain susceptible to Race 3 of *M. incognita* and to *M. arenaria* (Shepherd & Barker, 1990). Egg-mass production by *M. arenaria* was significantly restricted on selections of *N. knightiana*, *N. sanderae*, and *N. velutina* (Davis et al., 1988b). Low egg-mass production by *M. arenaria* and *M. javanica* has been noted on *N. glauca*, *N. longiflora*, *N. nudicaulis*, *N. plumbaginifolia*, and *N. repanda*. Unfortunately, resistance to *M. incognita* and *M. javanica* can break down at extremely high temperatures (30–35°C), and some strains of *M. javanica* have been reported to parasitize resistant cultivars (Shepherd & Barker, 1990).

Recent advances in molecular biology may enable development of tobacco cultivars that prevent *Meloidogyne* spp. from successfully initiating feeding sites without increasing development of other races or species of plant-parasitic nematodes (Opperman & Conkling, 1994). A tobacco root-specific gene (*TobRB7*) has recently been identified whose expression becomes restricted to feeding sites of *M. arenaria*, *M. incognita*, and *M. javanica* as the root system develops (Opperman et al., 1994). This gene construct might enable development of flue-cured tobacco cultivars resistant to all species and races of root-knot nematodes.

High levels of resistance to *Pratylenchus* spp. are not available in released tobacco cultivars. Low reproduction of *Pratylenchus* spp. has been frequently observed in root-knot resistant tobacco germplasm (Barker & Lucas, 1984). Variability in tolerance to *P. brachyurus* also has been observed among selected burley and flue-cured tobacco cultivars (Barker & Lucas, 1984). Certain burley tobacco cultivars possess limited resistance to *P. penetrans* (Shepherd & Barker, 1990).

Few females of *G. t. solanacearum* develop on *N. paniculata*, *N glutinosa*, and *N. plumbaginifolia* (Barker & Lucas, 1984). Crosses of *N. tabacum* with *N. repanda* also have shown resistance to this nematode, as have several plant introductions (Shepherd & Barker, 1990; Hayes, 1995). Resistance to *G. t. solanacearum* and *G. t. tabacum* also has been identified in released flue-cured tobacco cultivars, although this resistance had not been selected for in cultivar development (LaMondia, 1988). This resistance has been linked with resistance to wildfire of tobacco, caused by *Pseudomonas tabaci* (Hayes, 1995). However, limited reproduction has also recently been noted on *N. miersii*, which is susceptible to wildfire (Hayes, 1995). Resistance to tobacco cyst nematodes may have been introduced into flue-cured tobacco germplasm from burley tobacco lines with resistance to wildfire derived from *N. longiflora*. Juveniles penetrate roots of resistant cultivars in similar numbers to those for susceptible cultivars, but fewer develop successfully in the resistant cultivar (Wang, 1996). Unfortunately, *G. t. solanacearum* causes similar reductions in tobacco yield and quality in resistant and susceptible cultivars (Wang, 1996). Rotating resistant and susceptible cultivars in infested fields reduces population densities of *G. t. solanacearum*, but does not increase economic returns to the grower compared to use of a high-yielding susceptible cultivar with a nematicide (Johnson et al., 1989: Johnson, 1990b). Resistance to *G. t. solanacearum* was thought to be multigenic, but a single, dominant gene has been suggested as responsible for resistance to *G. t. tabacum* (LaMondia, 1991). Resistance to tobacco cyst nematodes appears to be relatively stable, since continuous cropping with resistant cultivars over three growing seasons did not increase the reproductive potential of the nematode (Elliott et al., 1986). Resistance and tolerance to *G. t. solanacearum* appear to be inherited independently (Fox & Spasoff, 1976). Unfortunately, expression of *TobRB7* is not induced by *G. t. solanacearum* (Opperman et al., 1994).

Although no tobacco cultivars are available with resistance to reniform nematodes, some variation in host status has been noted among tobacco genotypes (Melton & Powell, 1991).

Cultural

While unable to provide satisfactory control when used alone, cultural practices such as crop rotation and early destruction of tobacco debris provide the foundation for consistently effective nematode management in tobacco (Johnson, 1989a). Most tobacco fields in the USA are rotated on 1- or 2-yr intervals. Longer intervals between tobacco crops would reduce nematode populations more dramatically, but small farm size and the low value of rotation crops relative to tobacco prevents growers from using these systems (Johnson, 1989a). Bare fallow can reduce nematode populations, but exposes the soil to excessive erosion (Shepherd & Barker, 1990). Weed fallow allows build-up of plant-parasitic nematodes on alternative hosts (Tedford & Fortnum, 1988; Madulu & Trudgill, 1993). *Globodera* species and *D. dipsaci* have narrow host ranges that allow greater flexibility in selecting rotation crops, but may require more than 1 or 2 yr between tobacco crops to achieve satisfactory control.

Root-knot and lesion nematodes have wide host ranges that may differ among species within each genus of nematodes, making selection of effective rotation crops even more difficult (Shepherd & Barker, 1990; Fortnum & Currin, 1993, Hirunsalee et al., 1995). Maize, small grains, and pasture grasses (particularly fescue, *Festuca arundinacea* Schreber) are commonly rotated with tobacco and effectively reduce populations of root-knot and cyst nematodes (Johnson, 1989a; Shepherd & Barker, 1990). Grain sorghum [*Sorghum bicolor* (L.) Moench] also has been shown to be a poor host for root-knot nematodes and may be useful in tobacco crop rotation systems (Fortnum & Currin, 1988). Vegetables are often good hosts for nematode pests on tobacco and are, therefore, poor rotation crops for tobacco (Shepherd & Barker, 1990). Lesion nematodes can build up on bluegrass (*Poa* spp.), corn (*Zea mays* L.), rye (*Secale cereale* L.), and many legumes, although barley (*Hordeum vulgare* L.), oat (*Avena sativa* L.) and sweet-potato [*Ipomoea batatas* (L.) Lam.] limit reproduction of some *Pratylenchus* species (Arsenault et al., 1989; Shew & Lucas, 1991). *Crotalaria* species, lovegrass (*Eragrostis* spp.), rhodesgrass (*Chloris gayana* Kunth), and panicgrass (*Panicum* spp.) also can be used to restrict nematode population densities in tobacco fields (Shepherd & Barker, 1990; Madulu et al., 1994). Unfortunately, *Crotalaria* species are often highly toxic to livestock and may persist as weeds in subsequent crops (Polhill, 1982). Ranges of susceptibility to *M. javanica* also have been found within and between strains of rhodesgrass (York, 1990).

Tobacco roots and stalks are destroyed within several weeks of harvest in 95 to 99% of the flue-cured tobacco fields in the USA (Johnson, 1989a). However, this practice is most effective when performed immediately after harvest. Tobacco roots should be completely removed from the soil and killed. In many cases, crop debris is not destroyed completely, or for several weeks or months after harvesting has ended, reducing the effectiveness of this practice.

Planting tobacco on elevated beds or ridges also has been shown to increase control of root-knot nematodes, probably by increasing soil temperatures and reducing soil moistures to levels lethal to juvenile nematodes (Shepherd & Barker, 1990). Delaying transplanting may help control both components of the black shank/root-knot complex (Csinos et al., 1994).

Chemical

In the USA, tobacco seed beds are almost always fumigated with methyl bromide before seeding. This practice has virtually eliminated infected transplants as an initial source of nematode problems in tobacco fields. Tobacco transplants also are increasingly produced in greenhouse hydroponic systems, where nematode parasitism is not a significant concern.

Nematicides are used in 60 to 95% of the flue-cured tobacco fields in Brazil, Canada, USA, and Zimbabwe (Rich et al., 1989). The high cost and toxicity of these products make safety and effectiveness key issues. The proportion of use for fumigant vs. nonfumigant nematicides varies depending on soil type, the range of nematode and other pest species present, and environmental concerns. Due to their greater effectiveness against more aggressive species of root-knot nematodes, fumigants are most widely used in areas where *M. javanica* and P*seudomonas solanacearum* (cause of bacterial wilt) are the predominate target species (Rich et al., 1989; Shepherd & Barker, 1990). Chlorinated fumigants also are reported to provide better control of *Pratylenchus* spp. (Shepherd & Barker, 1990). Fumigants can inhibit the activity of nitrifying bacteria in soil (Shepherd & Barker, 1990; Tu et al., 1995). Under some environmental conditions, this interference of normal soil nitrification processes increases total alkaloids and decreases reducing sugars in cured tobacco leaf, lowering tobacco quality (Shepherd & Barker, 1990).

Fumigants have been most commonly applied in tobacco production as preplant row treatments. A 2- to 3-wk preplant interval is generally required to avoid potential fumigant injury to the crop (Melton et al., 1994). Broadcast fumigation is increasing, particularly in the fall or winter before a field is transplanted (Fortnum et al., 1990). Methyl bromide, 1,3-dichloropropene (1,3-D), and chloropricrin (trichloronitromethane) are the major fumigants used in the USA and Canada, although metham-sodium (sodium N-methyldithiocarbamate) and dazomet (tetrahydro-3,5-dimethyl-1,3,5-thiadiazine-2-thione) also are registered for some uses (Rich et al., 1989; Shepherd & Barker, 1990). Although most fumigants are sold as a liquefied gas or volatile liquid, dazomet is applied in a granular form that reacts with soil water to form the gaseous active ingredient methyl isothiocyanate (Hague & Gowen, 1987). Mixtures of chloropicrin with one of the other fumigants are often used in fields where a complex of soilborne plant pathogens exist. Ethylene dibromide (EDB) (1,2-dibromoethane) is used in some tobacco production areas outside the USA (Rich et al., 1989; Shepherd & Barker, 1990). Satisfactory control results from proper placement of appropriate rates when soil conditions (porosity, moisture content, and temperature) allow fumigants to diffuse throughout the soil, partition between the water and air phases of the soil, and remain long enough to adequately kill the target nematode population (Hague & Gowen, 1987; Lembright, 1990).

The nonfumigant nematicides registered for use on tobacco include aldicarb [2-methyl-2-(methylthio) propionaldehyde-O-(methylcarbamoyl) oxime], carbofuran (2,3-dihydro-2,2-dimethyl-7-benzofuranyl methylcarbamate), ethoprop O-ethyl S,S-dipropyl phosphorordithioate), fenamiphos {ethyl 3-[methyl-4-(methylthio) phenyl] (1-methyl) phosphoramidate}, fensulfothion {O,O-diethyl O-[4-(methulsulfinyl)phenyl] phosphorothioate}, and oxamyl {Methyl N′, N-dimethyl-N-[(methylcarbamoyl)oxy]-1-thiooxamimidate} (Rich et al., 1989; Shepherd & Barker, 1990). Fosthiazate [(RS)-S-sec-butyl-O-ethyl-2-oxo-1,3-thiazolidine-3-yl-phosphonothioate] also has demonstrated activity against nematode pests of tobacco (Johnson, 1995). These products usually possess insecticidal properties, and also may have growth regulating effects on host plants (Johnson, 1989b; Barker & Powell, 1988). Nonfumigant nematicides are usually formulated as granules or liquids and are applied 2 wk to immediately before transplanting (Melton et al., 1994). Liquid nematicide formulations are usually sprayed broadcast, while granular formulations are banded in the tobacco rows when they are bedded (Melton et al., 1994). Oxamyl also can be applied in the transplant water (Reddy, 1976). Banded application of liquid nonfumigant nematicides may enable tobacco producers to reduce nematicide use without sacrificing nematode control (Johnson, 1990a). Nematicides also have performed satisfactorily when applied via impregnated fertilizer granules or through irrigation water (Johnson, 1974; Johnson et al., 1979; Csinos et al., 1986). Outside the USA, nonfumigant nematicides also have been applied as root dips and as foliar sprays or spot treatments (Patel & Patel, 1986; Gowda et al., 1988; Deall, 1991).

Satisfactory performance of nonfumigant nematicides is dependent upon proper application and timing. The activity of nonfumigant nematicides is especially influenced by their degradation, movement, and persistence in soil, as well as by their mode of action and movement in plants (Hague & Gowen, 1987). Differences in these characteristics among nonfumigant nematicides can result in differing levels of activity against different nematode species (Nordmeyer & Dickson, 1985). Recent evidence also suggests that continual use of these compounds can lead to enhanced biodegradation that can reduce nematode control (Davis et al., 1993).

Biological

Root exudates from marigold (*Tagetes* spp.) can reduce populations of *Meloidogyne* and *Pratylenchus* species in tobacco fields, but practical systems to use this activity on a farm scale have yet to be developed (Lucas, 1975). Decomposing cellulose amendments can suppress emergence and root invasion by *G. t. tabacum*, especially when supplemented with N (Miller et al., 1968). Mixtures of chitinous materials with urea and soybean [*Glycine max* (L.) Merr.] meal reduced galling by *M. arenaria* and increased tobacco yields in microplot tests, but had less beneficial effects in field experiments (Melton et al., 1989; Rodriguez-Kabana et al., 1990). Similar amendments had no effect on *G. t. tabacum* in shade tobacco (LaMondia, 1994). Amendments such as oil seed cakes have been tested, but did not significantly suppress galling in tobacco seed beds (Krishnamurthy, 1993). Organic manures can limit parasitism by root-knot nematodes and increase tobacco yields, but balancing nematicidal effects with potential phytotoxicity is a significant concern (d'Errico & Sannino, 1982; Sinha et al., 1985). Essential oils extracted from leaves of *N. tabacum* have shown nematicidal activity against *M. incognita* (Singh et al., 1991).

Foliar application of an antihelminthic antibiotic reduced development and reproduction of *M. javanica* on tobacco seedlings in the greenhouse (Gowda & Setty, 1984). Some forms of avermectins limit parasitism by root-knot nematodes on tobacco, but these materials are generally unsuitable for field use (Nordmeyer & Dickson, 1985).

Although *Paecilomyces lilacinas* does not provide acceptable levels of root-knot nematode control in tobacco, *Pasteuria penetrans* is suppressive to these nematodes (Hewlett et al., 1988; Chen et al., 1994). Use of a combination of *P. lilacinus* and *Pasteuria penetrans* resulted in better control of *M. incognita* than when either was used alone (Dube & Smart, 1987). Incorporating the endomycorrhizal fungus *Glomus fasciculatum* into the soil of tobacco seed beds also suppressed galling by *M. incognita* and improved plant growth (Prassad, 1991).

Other Possible Methods

Burning residues of various plants [tobacco, rice (*Oryza sativa* L.), wheat, etc.,] is a practice called "rabbing" that is used in some countries for control of soilborne pathogens. Rabbing can reduce nematode populations in tobacco seed beds, but does not provide the weed control also obtained from fumigation (Shepherd & Barker, 1990). Flooding, steaming, or solarizing soil before planting also can reduce initial nematode population densities. Steaming can kill many soilborne pathogens, but is only practical for use with tobacco seed-beds (Shepherd & Barker, 1990). Some nematode control has been found due to flooding in areas of the world where tobacco is rotated with rice or where flooding of fields is common (Shepherd & Barker, 1990). Soil solarization also can be used to reduce nematode populations in tobacco plant beds and fields (Madulu & Trudgill, 1994; Mazza et al., 1994).

Integrated Control

Nematode management decisions in tobacco are usually based upon field histories and, to some extent, nematode advisory systems. Current nematode advisory systems are based upon soil assays (Barker & Imbriani, 1984). These systems are generally unable to identify nematode populations beyond the genus level and are used by only a limited number of farmers (Johnson, 1989a). Nematode management decisions in tobacco are usually targeted on a single plant-parasitic nematode species, but nematode management practices are strongly intertwined with those for soil insects and root diseases. Rotation crops and intervals, cultivars, and nematicides are selected based upon the nematode-soil insect-root disease complex thought to be important in a field or on the farm as a whole. Specific practices are chosen for their ability to maximize tobacco yield and quality, given the relative importance of the various components of the soil-borne pest complex.

Nematode control in tobacco is based upon a combination of cultural control, host resistance, and chemical control practices. In the USA, consistent control of plant-parasitic nematodes involves crop rotation, destruction of tobacco roots and stalks as soon as possible after harvest, root-knot resistant cultivars, and nematicides. Rotation crops and intervals provide the foundation for consistent nematode control, but are often determined by government programs and economic factors. Inappropriate crop rotation systems and reliance upon host resistance or a nematicide in continuous tobacco culture can lead to control failures, often because of the diversity of factors and organisms influencing nematode control in tobacco (Fortnum et al., 1984; Davis et al., 1993).

OUTLOOK

The complexity of plant-parasitic nematode communities in tobacco fields is increasing, at least in the USA. Future improvements in nematode control will, therefore, likely involve systems to manage several plant-parasitic nematode species at the same time (Johnson, 1995). Although research continues to identify more effective nematicide products, economic and political pressure to reduce the use of nematicides can be expected to continue. No biological control agents or methods have yet been identified with immediate promise for widescale effective use in tobacco. Unfortunately, economic and political constraints also can be expected to limit intensive use of crop rotation to lower nematode population densities in tobacco fields. Consequently, improved nematode management in tobacco will probably involve developing tobacco cultivars resistant to multiple nematode races and species and improving the efficiency with which tobacco producers use nematicides.

Our improved understanding of plant-nematode interactions at the molecular level is giving rise to new, promising, strategies of host resistance (Opperman et al., 1994). These strategies may improve the durability of nematode resistance, as well as provide multiple nematode species resistance.

Optimizing nematicide use in tobacco will require new application techniques and improved detection and monitoring of plant-parasitic nematode communities. Techniques such as use of enzyme phenotypes, monoclonal antibodies, and DNA analysis (RFLP and RAPD) need to be applied to the "real world" as nematode assays that can differentiate species within a genus and, when necessary, separate biotypes within a species (Evans, 1995). New sampling strategies and recommendation systems also are needed to improve the practicality and usefulness of nematode assay services (Barker & Imbriani, 1984; Barker et al., 1985; Johnson, 1989a). The complex, evolving nature of tobacco cropping systems will increase the need for accurate information on plant-parasitic nematode populations. Improved monitoring of nematode populations will likely be a foundation stone for tobacco culture in the "Information Age."

ACKNOWLEDGMENTS

The author wishes to express his appreciation to Dr. J. D. Eisenback and Dr. K. R. Barker for their help in preparing this publication.

REFERENCES

Adams, H.H., W.W. Osborne, and A.J. Webber. 1982. Effect of temperature on development and reproduction of *Globodera solanacearum*. Nematropica 12:305–311.

Agrios, G.N. (ed.). 1978. Plant pathology. Acad. Press, New York.

Akehurst, B.C. 1981. Tobacco. Human. Press, New York.

Anonymous. 1993. Tobacco disease losses for 1991 and 1992. p. 85–122. *In* B.W. Maw (ed.) Proc. 35th Tobacco Workers Conf., Savannah, GA. 11–14 January. Univ. Georgia, Tifton.

Arens, M.L., and J.R. Rich. 1981. Yield response and injury levels of *Meloidogyne incognita* and *M. javanica* on the susceptible tobacco 'McNair 944'. J. Nematol. 13:196–201.

Arens, M.L., J.R. Rich, and D. W. Dickson. 1981. Comparative studies on root invasion, root galling, and fecundity of three *Meloidogyne* spp. on a susceptible tobacco cultivar. J. Nematol. 13:201–205.

Arsenault, W.J., J. Kimpinski, and J.A. McLeod. 1989. Root lesion nematodes and flue-cured tobacco in a rye-tobacco or soybean-tobacco crop sequence. Tob. Sci. 33:72–73.

Barker, K.R. 1989. Yield relationships and population dynamics of *Meloidogyne* spp. on flue- cured tobacco. Suppl. J. Nematol. 21(4S):597–603.

Barker, J.R., and J.L. Imbriani. 1984. Nematode advisory programs–status and prospects. Plant Dis. 68:735–741.

Barker, J.R., and G.B. Lucas. 1984. Nematode parasites of tobacco. p. 213–242. *In* W.R. Nickle (ed.) Plant and insect nematodes. Marcel Dekker, Inc., New York.

Barker, J.R., and T.A. Melton. 1990. Comparative host sensitivity and efficiency of selected tobacco cultivars to *Meloidogyne* species and populations. Tob. Sci. 34:44–49.

Barker, J.R., and N. T. Powell. 1988. Influence of aldicarb on the growth and yield of tobacco. J. Nematol. 20:432–438.

Barker, J.R., and W. W. Weeks. 1991. Relationships between soil and levels of *Meloidogyne incognita* and tobacco yield and quality. J. Nematol. 23:82–90.

Barker, J.R., D.P. Schmitt, and J.P. Noe. 1985. Role of sampling for crop loss assessment and nematode management. Agric. Ecosyst. Environ. 12:355–396.

Barker, J.R., F.A. Todd, W.W. Shane, and L.A. Nelson. 1981. Interrelationships of *Meloidogyne* species with flue-cured tobacco. J. Nematol. 13:67–79.

Basson, S., D. DeWaele, and A.J. Meyer. 1990. An evaluation of crop plants as hosts for *Ditylenchus destructor* isolated from peanut. Nematropica 20:23–29.

Bird, A.F. 1969. The influence of tobacco ring spot virus and tobacco mosaic virus on the growth of *Meloidogyne javanica*. Nematologica 15:201–209.

Bisessar, S., and K.T. Palmer. 1984. Ozone, antioxidant spray and *Meloidogyne hapla* effects on tobacco. Atmos. Environ. 18:1025–1027.

Bossis, M., and D. Mugniery. 1993. Specific status of six *Globodera* parasites of solanaceous plants studied by means of two-dimensional gel electrophoresis with comparison of gel patterns by a computed system. Fund. Appl. Nematol. 16:47–56.

Bridge, J., S.M. Jordan., and S.L.J. Page. 1984. Combined effects of root-knot nematode and *Orobranche* species on tomato and tobacco. Haustorium 12:1–2.

Chen, S., D.W. Dickson, and E.B. Whitty. 1994. Response of *Meloidogyne* spp. to *Pasteuria penetrans*, fungi, and cultural practices in tobacco. Suppl. J. Nematol. 26(4S):620–625.

Csinos, A.S., A.W. Johnson, and A.M. Golden. 1986. Metalaxyl and fenamiphos applied through the irrigation water to control black shank/root-knot complex on tobacco. Plant Dis. 70:210–213.

Csinos, A.S., N.A. Minton, W.G. Stephenson, and T.P. Gaines. 1994. Effects of planting dates, metalaxyl, and fenamiphos on the tobacco black shank-root knot complex, tobacco quality, and yield. Tob. Sci. 38:82–85.

Davis, E.L., and J.R. Rich. 1987. Nicotine content of tobacco roots and toxicity to *Meloidogyne incognita*. J. Nematol. 19:23–29.

Davis, E.L., J.R. Rich, and R.G. Gwynn. 1988a. Reaction of selected *Nicotiana* spp. x *N. tabacum* crosses grown in microplots to three *Meloidogyne* spp. Nematropica 18:109–115.

Davis, E.L., J.R. Rich, R.G. Gwynn, and V. Sisson. 1988b. Greenhouse evaluation of *Nicotiana* spp. for resistance to root-knot nematodes. Nematropica 18:99–107.

Davis, R.F., A.W. Johnson, and R.D. Wauchope. 1993. Accelerated degradation of fenamiphos and its metabolites in soil previously treated with fenamiphos. J. Nematol. 25:679–685.

Daykin, M.E., and R.S. Hussey. 1985. Staining and histopathological techniques in nematology. p. 39–48. *In* K.R. Barker et al. (ed.). An advanced treatise on *Meloidogyne*. Vol. 2. Methodology. North Carolina State Univ. Graphics, Raleigh.

Deall, M.W. 1991. Studies on extending control of *Meloidogyne javanica* with postplanting application of Nemacur on tobacco. Pflanzenschutz. Nachricht. Bayer 44(62):67–78.

DiVito, M., N. Greco, and A. Carella. 1983. The effect of population densities of *Meloidogyne incognita* on the yield of cantaloupe and tobacco. Nematol. Medit. 11:169–174.

Dube, B., and G.C. Smart. 1987. Biological control of *Meloidogyne* by *Paecilomyces lilacinus* and *Pasteuria penetrans*. J. Nematol. 19:222–227.

Elliott, A.P., P.M. Phipps, and T.R. Terrill. 1986. Effects of continuous cropping of resistant and susceptible cultivars on reproduction potentials of *Heterodera glycines* and *Globodera tabacum solanacearum*. J. Nematol. 18:375–379.

Evans, K. 1995. Invited paper: Closing the gap between molecular biologists and traditional nematologists. Nematologica 41:385–394.

d'Errico, F.P., and L. Sannino. 1982. Effect of two organic manures on the development of *Meloidogyne incognita* on tobacco. Redia 65(Appendix):41–46.

Ferris, H. 1974. Correlation of tobacco yield, value, and root-knot index with early-to-midseason, and postharvest *Meloidogyne* population densities. J. Nematol. 6:75–81.

Fortnum, B.A., and R.E. Currin, III. 1988. Host suitability of grain sorghum cultivars to *Meloidogyne* spp. Ann. Appl. Nematol. 2:61–64.

Fortnum, B.A., and R.E. Currin, III. 1993. Crop rotation and nematicide effects on the frequency of *Meloidogyne* spp. in a mixed population. Phytopathology 83:350–355.

Fortnum, B.A., J. P. Krausz, and N. G. Conrad. 1984. Increasing incidence of *Meloidogyne arenaria* on flue-cured tobacco in South Carolina. Plant Dis. 68:244–245.

Fortnum, B.A., D. T. Gooden, R. E. Currin, III, and S. B. Martin. 1990. Spring or fall fumigation for control of *Meloidogyne* on tobacco. Suppl. J. Nematol. 22(4S):645–660.

Fox, J.A., and L. Spasoff. 1972. Interaction of *Heterodera solanacearum* and *Endogone gigantea* on tobacco. J. Nematol. 4:224–225.

Fox, J.A., and L. Spasoff. 1976. Resistance and tolerance of tobacco to *Heterodera solanacearum*. J. Nematol. 8:284–285.

Goswami, B.K., and S.P. Raychaudhuri. 1973. Host-parasite relationship between tobacco and root-knot nematode, *Meloidogyne javanica* (Treub) Chitwood, influenced by tobacco mosaic virus infection. Phytopathol. Soc. Jpn. Ann. 39:99–102.

Gowda, D.N., and K.G.H. Setty. 1984. Efficacy of the anithelminthic antibiotic (KT-199) for the control of root-knot nematode *Meloidogyne javanica* on tobacco. Nematol. Medit. 12:231–234.

Gowda, D.N., N. Shivanandappa, and C. Karegowda. 1988. Effect of pesticides as root dip treatment in control of root-knot of tobacco. Pesticides 22:28–29.

Graham, T.W., Z.T. Ford, and R.E. Currin. 1964. Response of root-knot resistant tobaccos to the nematode root disease complex caused by *Pratylenchus* spp. and *Meloidogyne incognita acrita*. Phytopathology 54:205–210.

Grant, C.E. 1984. Interaction of a tobacco cyst nematode (*Globodera solanacearum*) and the black shank fungus (*Phytophthora parasitica* var. *nicotianae*) with flue-cured tobacco (*Nicotiana tabacum*). Ph.D. diss. Virginia Polytech. Inst. State Univ., Blacksburg (Diss. Abstr. 84-15951).

Grujicic, G., D. Jovicic, G. Komar, R. Pvalovic, P. Markovic, and D. Jovanovic. 1988a. Study of *Ditylenchus dipsaci* with wheat, maize, sugarbeet and tobacco as host plants. Zast. Bilja 39:5–16.

Grujicic, G., D. Jovicic, G. Komar, R. Pvalovic, P. Markovic, and D. Jovanovic. 1988b. Harmful nematofauna in tobacco fields in Serbia. Zast. Bilja 39:149–157.

Hague, N.G.M., and S.R. Gowen. 1987. Chemical control of nematodes. p. 131–178. *In* R.H. Brown and K.R. Kerry (ed.) Principles and practice of nematode control in crops. Acad. Press, Sydney, Australia.

Hayes, A.J. 1995. Identification and genetic characterization of tobacco accessions possessing resistance to tobacco cyst nematode. M.Sc. thesis. Virginia Polytech. Inst. State Univ., Blacksburg.

Hewlett, T.E., D.W. Dickson, D.J. Mitchell, and M.E. Kannwischer-Mitchell. 1988. Evaluation of *Paecilomyces lilacinus* as a biocontrol agent of *Melodogyne javanica* on tobacco. J. Nematol. 20:578–584.

Hirunsalee, A., K.R. Barker, and M.K. Beute. 1995. Effects of peanut-tobacco rotations on population dynamics of *Meloidogyne arenaria* in mixed race populations. J. Nematol. 27:178–188.

Hussey, R.S. 1989. Disease-inducing secretions of plant-parasitic nematodes. Annu. Rev. Phytopathol. 27:123–141.

Ibrahim, I.K.A. 1987. Interaction between *Meloidogyne arenaria* and *M. incognita* on tobacco. Nematol. Medit. 15:287–291.

Ibrahim, S.K. 1991. Peroxidase isoenzymes from *Meloidogyne* spp. cultured on different hosts. Rev. Nematol. 14:335–344.

Johnson, A.W. 1974. Relative efficacy of selected nonvolatile nematicides for control of root-knot nematodes on flue-cured tobacco. Tob. Sci. 18:132–133.

Johnson, A.W., and C.J. Nusbaum. 1970. Interactions between *Meloidogyne incognita*, *M. hapla*, and *Pratylenchus brachyurus* in tobacco. J. Nematol. 2:334–340.

Johnson, A.W., J.D. Miles, E.W. Hauser, T.P.Gaines, and G.A. Slappey. 1979. Effect of fluid and granular fertilizer-pesticide combinations on nematodes, weeds, and flue-cured tobacco. Tob. Sci. 23:67–69.

Johnson, C.S. 1989a. Managing root-knot on tobacco in the southeastern United States. Suppl. J. Nematol. 21(4S):604–608.

Johnson, C.S. 1989b. Insecticidal activity of products applied for control of tobacco cyst nematodes, 1988. Insect. Acaric. Tests 14:291–292.

Johnson, C.S. 1990a. Band vs. broadcast application of contact nematicides for control of tobacco cyst nematodes in Virginia. Phytopathology 80:671–672.

Johnson, C.S. 1990b. Control of *Globodera tabacum solanacearum* by rotating susceptible and resistant flue-cured tobacco cultivars. Suppl. J. Nematol. 22(4S):700–706.

Johnson, C.S. 1992. Tobacco cyst nematodes can stimulate yield of flue-cured tobacco. Phytopathology 82:1095.

Johnson, C.S. 1995. Use of fosthiazate to control tobacco cyst nematodes in flue-cured tobacco. Tob. Sci. 39:69–76.

Johnson, C.S., and T. D. Reed. 1994. Tobacco. p. 323–336. *In* C.J. Arntzen and E.M. Ritter (ed.) Encyclopedia agriculture. science. Vol. 4. Acad. Press, San Diego, CA.

Johnson, C.S., D.A. Komm, and J.L. Jones. 1989. Control of *Globodera tabacum solanacearum* by alternating use of host resistance and a nematicide. J. Nematol. 21:16–23.

Kincaid, R.R., F.G. Martin, N. Gammon, H.L. Breland, and W.L. Pritchett. 1970. Multiple regression of tobacco black shank, root knot, and coarse root indexes on soil pH, potassium, calcium, and magnesium. Phytopathology 60:1513–1516.

Komm, D.A., J.J. Reilly, and A.P. Elliott. 1983. Epidemiology of a tobacco cyst nematode (*Globodera solanacearum*) in Virginia. Plant Dis. 11:1249–1251.

Krishnamurthy, G.V.G. 1993. Effect of oil seed cakes on the control of root-knot nematode, *Meloidogyne javanica* in tobacco nursery. p. 434–437. *In* M.S. Chari and G. Ramaprasad (ed.) Proc. Natl. Symp. Ind. Soc. Tob. Sci., Rajahmundry, India. 21–22 Jan. 1990. Ind. Soc. Tob. Sci., Rajahmundry, India.

LaMondia, J.A. 1988. Tobacco resistance to *Globodera tabacum*. Ann. Appl. Nematol. 2:77–80.

LaMondia, J.A. 1991. Genetics of tobacco resistance to *Globodera tabacum tabacum*. Plant Dis. 75:453–454.

LaMondia, J.A. 1992. Predisposition of broadleaf tobacco to *Fusarium* wilt by early infection with *Globodera tabacum tabacum* or *Meloidogyne hapla*. J. Nematol. 24:425–431.

LaMondia, J.A. 1994. The effects of a chitin-urea soil amendment on *Globodera tabacum* population changes under shade tobacco. Tob. Sci. 38:18–20.

LaMondia, J.A. 1995a. Shade tobacco yield loss and *Globodera tabacum tabacum* population changes in relation to initial nematode density. J. Nematol. 27:114–119.

LaMondia, J.A. 1995b. Hatch and reproduction of *Globodera tabacum tabacum* in response to tobacco, tomato, or black nightshade. J. Nematol. 27:382–386.

LaMondia, J.A., and G.S. Taylor. 1987. Influence of the tobacco cyst nematode (*Globodera tabacum*) on Fusarium wilt of Connecticut broadleaf tobacco. Plant Dis. 71:1129–1132.

Lembright, H.W. 1990. Soil fumigation: principles and application technology. Suppl. J. Nematol. 22(4S):632–644.

Lucas, G.B. 1975. Diseases of tobacco. Biol. Consult. Assoc. Raleigh, NC.

Madulu, J.D., and D.L. Trudgill. 1993. Weed hosts of *Meloidogyne javanica* in Tanzanian tobacco fields. Pak. J. Nematol. 11:61–64.

Madulu, J.D., and D.L. Trudgill. 1994. Influence of temperature on the development and survival of *Meloidogyne javanica*. Nematologica 40:230–243.

Madulu, J.D., D.L. Trudgill, and M. S. Phillips. 1994. Rotational management of *Meloidogyne javanica* and effects on *Pasteuria penetrans* and tomato and tobacco yields. Nematologica 40:438–455.

Marks, C.F., and R.M. Sayre. 1964. Effect of potassium on the rate of development of the root-knot nematodes *Meloidogyne incognita*, *M. javanica*, and *M. hapla*. Nematologica 10:323–327.

Mazza, S., F. Di Serio, F.P. D'Errico, G. Sannino, L. Barone, and A. Scialla. 1994. Prospects of soil solarization for controlling nematodes on tobacco in Caserta province. Il Tabacco 2:23–30.

McIntyre, J.L., and P.M. Miller. 1976. Competitive interaction of *Tylenchorhynchus claytoni* and *Pratylenchus penetrans* in tobacco roots. Phytopathology. 66:1427–1430.

McIntyre, J.L., and P.M. Miller. 1978. Protection of tobacco against *Phytophthora parasitica* var. *nicotianae* by cultivar-nonpathogenic races, cell-free sonicates, and *Pratylenchus penetrans*. Phytopathology 68:235–239.

Melton, T.A., and N.T. Powell. 1991. Effects of nematicides and cultivars on *Rotylenchulus reniformis* and flue-cured tobacco yield. J. Nematol. 23(4S):712–716.

Melton, T.A., D. Porter, and K. Wood. 1994. Disease management. p. 85–111. *In* G.F. Peedin (ed.) Flue-cured tobacco information 1995. North Carolina Coop. Ext. Ser. Publ. AG-187.

Melton, T.A., D. Porter, K. Wood, and P. Wickham. 1989. Extension-research flue-cured tobacco disease report-1989. North Carolina Coop. Ext. Ser. Publ. AG-191.

Melton, T.A., J. A. Philips, J. L. Imbriani, and K. R. Barker. 1991. First report of *Globodera tabacum solanacearum* on flue-cured tobacco outside Virginia. Plant Dis. 75:1074.

Miller, L. I., and B. J. Gray. 1972. *Heterodera solanacearum* n. sp., a parasite of solanaceous plants. Nematologica 18:404–413.

Miller, P.M., G.S. Taylor, and S.E. Wihrheim. 1968. Effect of cellulosic soil amendments and fertilizers on *Heterodera tabacum*. Plant Dis. Rep. 52:441–445.

Moorman, G.W., J.S. Huang, and N.T. Powell. 1980. Localized influence of *Meloidogyne incognita* on *Fusarium oxysporum* wilt resistance of flue-cured tobacco. Phytopathology 70:969–970.

Mota, M.M., and J.D. Eisenback. 1993a. Morphology of second-stage juveniles and males of *Globodera tabacum*, *G. t. virginiae*, and *G. t. solanacearum* (Nemata: Heteroderinae). J. Nematol. 25:27–33.

Mota, M.M., and J.D. Eisenback. 1993b. Morphology of females and cysts of *Globodera tabacum*, *G. t. virginiae*, and *G. t. solanacearum* (Nemata: Heteroderinae). J. Nematol. 25:136–147.

Mota, M.M., and J.D. Eisenback. 1993c. Morphometrics of *Globodera tabacum*, *G. t. virginiae*, and *G. t. solanacearum* (Nemata: Heteroderinae). J. Nematol. 25:148–160.

Mugniery, D., M. Bossis, and P. Jean-Sebastien. 1992. Hybridization between *Globodera rostochiensis* (Wollenweber), *G. pallida* (Stone), *G. virginiae* (Miller and Gray), *G. solanacearum* (Miller and Gray), et *G. "mexicana"* (Campos-Vela). Description and future of the hybrids. Fund. Appl. Nematol. 15:375–382.

Ng'ambi, T.B.S., R.C. Rufty, and K.R. Barker. 1995. Response of root-knot resistant tobacco to concomitant populations of *Meloidogyne* species. Plant Dis. 79:1008–1013.

Noe, J.P., and K.R. Barker. 1985a. Relation of within-field spatial variation of plant-parasitic nematode population densities and edaphic factors. Phytopathology 75:247–252.

Noe, J.P., and K.R. Barker. 1985b. Overestimation of yield loss in tobacco caused by aggregated spatial pattern of *Meloidogyne incognita*. J. Nematol. 17:245–251.

Noe, J.P., and C. L. Campbell. 1985. Spatial pattern analysis of plant-parasitic nematodes. J. Nematol. 17:86–93.

Nordmeyer, D., and D.W. Dickson. 1985. Management of *Meloidogyne javanica*, *M. arenaria*, and *M. incognita* on flue-cured tobacco with organophosphate, carbamate, and avermectin nematicides. Plant Dis. 69:67–69.

Olthoff, T.H.A., C.F. Marks, and J.M. Elliot. 1973. Relationship between population densities of *Pratylenchus penetrans* and crop losses in flue-cured tobacco in Ontario. J. Nematol. 5:158–162.

Opperman, C.H., and M.A. Conkling. 1994. Nematode-induced plant gene expression and related control strategies. Fund. Appl. Nematol. 17:211–217.

Opperman, C.H., C.G. Taylor, and M.A. Conkling. 1994. Root-knot nematode-directed expression of a plant root-specific gene. Science (Washington, DC) 263:221–223.

Patel, B.A., K.N. Vyas, and D.J. Patel. 1988. Histological changes due to combined inoculation of two root-knot and reniform nematodes in tobacco. Pak. J. Nematol. 6:87–92.

Patel, D.B., D.J. Patel, and H.M. Shah. 1987. Susceptibility of different crop plants to *Rotylenchulus reniformis*. Int. Nematol. Network Newsl. 4:25–27.

Patel, D.J. 1986. Histopathology of tobacco roots infected by *Rotylenchulus reniformis*. Pak. J. Nematol. 4:75–78.

Patel, D.J., and S. K. Patel. 1986. Effect of post-spot application of phenamiphos in management of root-knot disease in bidi tobacco field. p. 53. *In* Proc. 6th Natl. Symp. Tobacco, 13–17 February. Indian Soc. Tob. Sci., Rajahmundry, Andhra Pradesh, India.

Patel, D.J., M. V. Desai, and D. B. Patel. 1986. Effect of *Rotylenchulus reniformis* on tobacco seed germination and seedling development. Int. Nematol. Network Newsl. 3:6–7.

Patel, S.K., and D.J. Patel. 1991. The interaction between *Meloidogyne incognita*, *M. javanica*, and *Rotylenchulus reniformis* in tobacco. Nematol. Medit. 19:41–42.

Patel, S.K., A.J. Patel, and D.J. Patel. 1988. Effect of interaction between root knot (*Meloidogyne incognita* and *M. javanica*) and reniform (*Rotylenchulus reniformis*) nematodes on nutrient uptake by tobacco plant. Tob. Res. 14:106–108.

Patel, S.K., D.J. Patel, and A.J. Patel. 1989. Interaction between root-knot (*Meloidogyne incognita* and *M. javanica*) and reniform (*Rotylenchulus reniformis*) nematodes on bidi tobacco. Tob. Res. 15:96–99.

Polhill, R.M. 1982. Crotalaria in Africa and Madagascar. A. A. Balkema, Rotterdam.

Powell, N.T. 1979. Internal synergisms among organisms inducing disease. p. 113–133. *In* J.G. Horsfall and E.B. Cowling (ed.) Plant disease—an advanced treatise. Vol. 4. Acad. Press, New York.

Powell, N.T., P.L. Melendez, and C.K. Batten. 1971. Disease complexes in tobacco involving *Meloidogyne incognita* and certain soil-borne fungi. Phytopathology 61:1332–1337.

Prassad, K.S. 1991. Influence of a vesicular arbuscular mycorrhiza on the development and reproduction of root-knot nematode affecting flue-cured tobacco. Afro-Asian J. Nematol. 1:130–134.

Rahi, G.S., J.R. Rich, and C. Hodge. 1988. Effect of *Meloidogyne incognita* and *M. javanica* on leaf water potential and water use of tobacco. J. Nematol. 20:516–522.

Rammah, A., and H. Hirschmann. 1990. *Meloidogyne morocciensis* n. sp. (Meloidogynidae), a root-knot nematode from Morocco. J. Nematol. 22:279–291.

Ramakrishnayya, B.V., and S.S. Hussaini. 1987. Preliminary observations on interaction of root-knot nematode and potash in flue-cured tobacco. Indian J. Nematol. 17:27–29.

Reddy, D.D.R. 1976. Control of root-knot nematodes on tobacco with oxamyl through transplant water. Plant Dis. Rep. 60:430–431.

Reed, S.M., and S.M. Schneider. 1992. Evaluation of *Nicotiana otophora* as a source of resistance to *Meloidogyne incognita* race 4 for tobacco. J. Nematol. 24:253–256.

Reilly, J.J., and C.E. Grant. 1985. Seasonal fluctuations of *Globodera tabacum solanacearum* as estimated by two extraction techniques. J. Nematol. 17:354–360.

Rich, J.R., and K.R. Barker. 1984. Flowering delay in flue-cured tobacco infected with *Meloidogyne* species. J. Nematol. 16:402–404.

Rich, J.R., C. Hodge, and J.T. Johnson. 1984. Population development and pathogenicity of *Meloidogyne javanica* on flue-cured tobacco as influenced by ethoprop and DD. J. Nematol. 16:240–245.

Rich, J.R., J.D. Arnett, J.A. Shepherd, and M.C. Watson. 1989. Chemical control of nematodes on flue-cured tobacco in Brazil, Canada, United States, and Zimbabwe. Suppl. J. Nematol. 21(4S):609–611.

Rodriguez-Kabana, R., D. Boube, and R.W. Young. 1990. Chitinous materials from blue crab for control of root-knot nematode. II. Effect of soybean meal. Nematropica 20:153–168.

Rufty, R.C., E.A. Wernsman, and N.T. Powell. 1983. A genetic analysis of the association between resistance to *Meloidogyne incognita* and a necrotic response to infection by a strain of potato virus Y in tobacco. Phytopathology 73:1413–1418.

Sasser, J.N., and D.W. Freckman. 1987. A world perspective on nematology: the role of society. p. 7–14. *In* J.A. Veech and D.W. Dickson (ed.). Vistas on nematology. Soc. Nematol., Inc. Hyattsville, Md.

Schmitt, D.P., and K.R. Barker. 1988. Incidence of plant-parasitic nematodes in the coastal plain of North Carolina. Plant Dis. 72:107–110.

Schneider, S.M. 1991. Penetration of susceptible and resistant tobacco cultivars by *Meloidogyne* juveniles. J. Nematol. 23:225–228.

Seinhorst, J.W. 1986. The development of individuals and populations of cyst nematodes on plants. p. 101–117. *In* F. Lamberti and C.E. Taylor (ed.) Cyst nematodes. Plenum Press, New York.

Shepherd, J.A., and K.R. Barker. 1990. Nematode parasites of tobacco. p. 493–517. *In* M. Luc et al. (ed.) Plant parasitic nematodes in subtropical and tropical agriculture. CAB Int. Wallingford, England.

Shew, H.D., and G.B. Lucas. 1991. Compendium of tobacco diseases. APS Press, St. Paul, MN.

Shukla, Y.M., and M.K. Chakraborty. 1988. Biochemical studies of tobacco and tomato plants to root knot nematode infection. Tob. Res. 14:43–50.

Shukla, Y.M., M.K. Chakraborty, and D.J. Patel. 1988. Biological activity and semiochemicals in the root exudates of host plants resistant to root knot nematodes. Tob. Res. 14:51–57.

Singh, R.P., S.S. Tomar, C. Devakumar, B.K. Goswami, and D.B. Saxena. 1991. Nematicidal efficacy of some essential oils against *Meloidogyne incognita*. Indian Perfum. 35:35–37.

Sinha, R., S.S.L. Dwivedi, and A.K. Pandey. 1985. Performance of chewing tobacco under varying doses and methods of compost application. Tob. Res. 11:71–73.

Sosa-Moss, C., K.R. Barker, and M.E. Daykin. 1983. Histopathology of selected cultivars of tobacco infected with *Meloidogyne* species. J. Nematol. 15:392–397.

Taylor, A.L., and J.N. Sasser. 1978. Biology, identification and control of root-knot nematodes (*Meloidogyne* species). North Carolina State Univ. Graphics, Raleigh.

Taylor, A.L., J.N. Sasser, and L.A. Nelson. 1982. Relationship of climate and soil characteristics to geographical distribution of *Meloidogyne* species in agricultural soils. North Carolina State Univ. Graphics, Raleigh.

Tedford, E.C., and B.A. Fortnum. 1988. Weed hosts of *Meloidogyne arenaria* and *M. incognita* common in tobacco fields in South Carolina. Ann. Appl. Nematol. 2:102–105.

Townshend, J.L. 1971. Influence of edaphic factors on penetration of corn roots by *Pratylenchus penetrans* and *P. minyus* in three Ontario soils. Nematologica 18:201–212.

Townshend, J.L. 1984. Anhydrobiosis in *Pratylenchus penetrans*. J. Nematol. 16:282–289.

Townshend, J.L. 1987. Anhydrobiosis in *Pratylenchus penetrans* and *Tylenchorhynchus* n. sp. in cultivated soils cropped to winter rye. J. Nematol. 19:164–171.

Townshend, J.L., and C.F. Marks. 1976. Available soil moisture and the expression of damage caused by *Pratylenchus penetrans* and crop losses in flue-cured tobacco in Ontario. Nematologica 22:65–70.

Townshend, J.L., and C.F. Marks. 1977. Temperature and the expression of damage caused by *Pratylenchus penetrans* in flue-cured tobacco. Nematologica 23:29–32.

Townshend, J.L., and J. L. Webber. 1971. Movement of *Pratylenchus penetrans* and the moisture characteristics of three Ontario soils. Nematologica 17:47–57.

Tu, C.M., C.F. Marks, P.W. Johnson, and J.M. Elliot. 1995. Effects of nematicides on populations of soil microflora and *Pratylenchus penetrans*, soil enzymes and flue-cured tobacco. J. Environ. Sci. Health B30(1):95–111.

Van Gundy, S.D. 1981. Ecology of *Meloidogyne* spp.—emphasis on environmental factors affecting survival and pathogenicity. p. 177–182. *In* J.N. Sasser and C.C. Carter (ed.) An advanced treatise on *Meloidogyne*. Vol. I. Biology and control. North Carolina State Univ. Graphics, Raleigh.

Wang, J. 1996. Characterizing resistance in flue-cured tobacco to *Globodera tabacum solanacearum*. Ph.D. diss. Virginia Polytech. Inst. State Univ., Blacksburg.

Wheeler, T.W., K.R. Barker, and S.M. Schneider. 1991. Yield-loss models for tobacco infected with *Meloidogyne incognita* as affected by soil moisture. J. Nematol. 23:365–371.

York, P.A. 1990. Range of susceptibility within and between diploid and tetraploid strains of *Chloris gayana* (Rhodes grass) to *Meloidogyne javanica*. Rev. Nematol. 13:45–50.

Yung, C.H., E.A. Wernsman, and G.V. Gooding, Jr. 1991. Characterization of potato virus Y resistance from gametoclonal variation in flue-cured tobacco. Phytopathology 81:887–891.

Zhang, Z.C., and R.H. Zhou. 1989. Influences of root-knot nematode (*Meloidogyne incognita*) on the plant growth and physiological activities of tobacco. Acta Agric. Univ. Henanensis 233:291–296.

Zunke, U. 1990a. Observations on the invasion and endoparasitic behaviour of the root lesion nematode *Pratylenchus penetrans*. J. Nematol. 22:309–322.

Zunke, U. 1990b. Ectoparasitic feeding behavior of the root lesion nematode, *Pratylenchus penetrans*, on root hairs of different host plants. Rev. Nematol. 13:331–337.

23 Peanut

DON W. DICKSON

University of Florida
Gainesville, Florida

Peanut (*Arachis hypogaea* L.) is native to South America where it is believed to have first been domesticated (Hammons, 1982). The plant, an annual legume, has at least two distinct growth habits: erect (bunch or upright) and prostrate (running or trailing) (Wynne & Coffelt, 1982). Two subspecies of peanut are *A. hypogaea* and *A. fastigiata*. The former has a maturation time of 145 to 165 d (spreading-bunching growth habit), and the latter has a shorter maturation time of 125 to 145 d (erect growth habit) (Shokes & Malouk, 1995).

Growth stages (vegetative and reproductive) of peanut have been described (Boote, 1982). Flowers begin to appear 4 to 6 wk after planting. After fertilization, the gynophore, commonly called a peg, develops and elongates. The peg responds to geotropism and grows downward into the soil, where it turns horizontally and forms a swollen ovary that enlarges to form the pod.

Peanut is grown worldwide in more than 50 countries. In the USA, approximately 677 000 ha are planted annually. Worldwide there are 19 389 000 ha planted on four continents as follows: North America and Caribbean—763 000 ha; South America—246 000 ha; Africa—5 180 000 ha; and Asia—13 174 000 ha. Yields per hectare (2941 kg ha^{-1}) in the USA far exceed those in other peanut-producing countries (USDA Crop Stat. Agric. Stat., 1993).

Peanut is a hardy plant that tolerates varying environmental conditions. It grows well in deep, well-drained sandy soil in temperate to tropical environments. Flowering continues throughout the relatively long growing season; thus, pods also continue to form throughout the season. Since these pods are produced on pegs belowground, they are especially vulnerable to nematode-induced diseases. They are exposed to nematode infection for a relatively long period that may last from approximately 45 d to more than 100 d. This relatively long exposure period contributes to severe yield losses and poor quality of peanut as a consequence of nematode parasitism. This growth pattern also exacerbates the problem of nematode management, especially with nematicides. Even when a nematicide is incorporated into the disease management program in nematode problem fields, it is difficult to avoid late increases in nematode population densities that may damage maturing pods.

Root-knot nematodes (*Meloidogyne* spp.) are the most important nematode pathogens of peanut. At least eight other genera also infect peanut. Many other nematode species have been listed in association with peanut, particularly in Africa and Asia (Sharma & McDonald, 1990; Sharma & Subrahmanyam, 1990; Sharma et al., 1991, 1994). However, pathogenesis has not been confirmed. Further research may establish the pathogenicity of these nematodes to peanut and demonstrate economic losses. Other nematodes may simply be associated with peanut roots because of their proximity to other host roots, or these nematodes feed on peanut roots but cause little or no economic damage.

Some of the most important nematode parasites of peanut are widely disseminated, whereas others are limited to specific geographic regions. Annual losses on peanut to nematode diseases are estimated at 12%; thus, the monetary loss is well over U.S. $1 billion (Sasser & Freckman, 1987).

ROOT-KNOT NEMATODES

Species and Distribution

Three *Meloidogyne* spp., *M. arenaria*, *M. hapla*, and *M. javanica*, are recognized as important nematode pathogens of peanut. The first report of root-knot disease on peanut was made by Neal (1889) in north central Florida. He wrote, "The roots of peanuts became masses of knotty roots, the worse case of the disease I ever saw." Machmer (1951) positively identified root-knot on peanut in Georgia in 1947, where he discovered large infested areas in the Chattahoochee Valley of Georgia. He stated, "A field where infestation is general and severe is usually a total loss." The following year Wilson (1948) reported root-knot disease on peanut in Alabama.

Root-knot nematodes infect peanut roots, pegs, and pods throughout the season, stunting the plants and suppressing yields. The direct injury to pods causes reduced quality. There may be significant pod rotting associated with root-knot nematode induced galls. Infection of peanut by root-knot nematode can increase the severity of white mold (*Sclerotium rolfsii*) (Starr et al. 1996) and other soilborne diseases.

Meloidogyne arenaria (peanut root-knot nematode) was described by Chitwood (1949) from diseased peanut in Georgia discovered by Machmer in 1947, and is recognized as the most important nematode pathogen of peanut in Alabama, Florida, Georgia, North Carolina, South Carolina, and Texas (Sturgeon, 1986). It also occurs in Virginia and Oklahoma. The only peanut-producing state in the USA not infested with *M. arenaria* is New Mexico.

The peanut root-knot nematode is known to occur in the major producing countries of India (Sharma & McDonald, 1990; Sakhuja & Sethi, 1985b), China (Zhang, 1985), and several African peanut-producing countries. The first reports in Africa were in Zimbabwe (Martin, 1958) and Egypt (Ibrahim & El-Saedy, 1976). Other country reports include Taiwan (Cheng & Tu, 1980), and Israel (Orion & Cohn, 1975).

Meloidogyne hapla (northern root-knot nematode) was first found on peanut in North Carolina by Cooper (1950). It has subsequently been detected in all peanut-producing states in the USA except Florida. Instances of *M. hapla* infections on crops other than peanut, e.g., strawberry (*Fragaria* × *ananass*), occur in Florida, but the nematode apparently fails to survive during the summer's exposure under tropical conditions. *Meloidogyne hapla* is known to parasitize peanut as far south as southern Alabama (Henry County). This parasite is considered the dominant nematode species infecting peanut in North Carolina, Virginia, and Oklahoma.

The northern root-knot nematode, generally considered a cool-season nematode, can survive hot summer months in southern Alabama, Georgia, and Oklahoma. On a worldwide basis this parasite occurs on peanut in Australia (Colbran, 1958), China (Zhang, 1985), Israel (Minz, 1956), Japan (Mitsui et al., 1976), Korea (Choi, 1978), and South Africa (Van der Linde, 1956).

Martin (1958) reported that *M. javanica* (Javanese root-knot nematode) caused light galling on peanut in Zimbabwe. This finding was thought to be unusual since peanut was considered to be a nonhost of *M. javanica* (Sasser, 1954). In 1968, peanut in south Georgia (Tift County) was found to be infected by *M. javanica* (Minton et al.,1969). Field-grown plants had poor top growth and yields; however, the occurrence of the Javanese root-knot nematode on peanut is believed to be rare in the southeastern USA. However, recently Tomaszewski et al. (1994) discovered this parasite on peanut in Texas. They also reported that 14 isolates of *Meloidogyne* spp., collected from infected peanut roots in four governorates of Egypt, were all identified as *M. javanica*. When these isolates were bioassayed on peanut in the greenhouse, their reproduction rates on 'Florunner' peanut were similar to that of *M. arenaria* Race 1. These *M. javanica* isolates—identified based on morphology, esterase phenotypes, and presence of species-specific restriction fragment length polymorphisms in mtDNA (Powers & Harris, 1993; Tomaszewski et al., 1994)—gives us the most authoritative report to date of *M. javanica* infection on peanut.

Other observations of *M. javanica* infecting peanut have been made in Africa (Ibrahim & El-Saedy, 1976, Rammah & Hirschmann, 1990), India (Patel et al., 1988, Prasad et al., 1964, Sakhuja & Sethi, 1985a), and Brazil (Lordello & Gerin, 1981). Populations of *M. javanica* able to parasitize peanut are apparently more common in Africa and India than in other parts of the world.

Biological Races

The use of host differential plants to separate species of root-knot nematodes was first proposed by Sasser (1954). An outgrowth of Sasser's proposal in 1954 was the concept of host races among the major agriculturally important *Meloidogyne* spp., which includes *M. arenaria*, *M. hapla*, and *M. javanica*. A host race is described as an isolate or population that produces a different reaction (galling and egg-mass production) on a particular host plant than another isolate of the species that is morphologically identical. Taylor and Sasser (1978) proposed four host races of *M. incognita* and two host races of *M. arenaria*. No host races were reported at that time for *M. hapla* or *M. javanica*; however, recently host races for *M. javanica* have been proposed (Rammah & Hirschmann, 1990, Sharma et al., 1995). *Meloidogyne incognita* Races 1 to 4 are separated based on their reaction on *M. incognita*-resistant tobacco ('NC-95') and cotton (*Gossypium hirsutum* L.) ('Deltapine 61'). Race 1 does not infect either host, Race 2 infects NC- 95 tobacco, Race 3 infects Deltapine 61 cotton, and Race 4 infects both differential hosts (Sasser & Carter, 1982). Peanut (Florunner) is a differential host used to separate *M. arenaria* Races 1 and 2. Race 1 infects peanut, and Race 2 does not. *Meloidogyne javanica* Races 1 to 3 are separated on pepper (*Capsicum annuum* L.) ('Early California Wonder') and peanut (Florunner). Race 1 does not infect either host, Race 2 infects pepper, and Race 3 infects peanut.

Triantaphyllou (1985) described two cytological races of *M. hapla*: Races A and B. Race A populations of *M. hapla*, which includes most of those studied, reproduces by facultative meiotic parthenogenesis, i.e., usually by meiotic parthenogenesis and less frequently by cross fertilization. Most of these populations have a haploid chromosomal complement of 17, but some have $n = 16$, and a few have $n = 15$, 14, or 13. In diploid populations of *M. hapla*, some polyploid females were identified with $n = 28$ to 34. Race B populations, which reproduce by obligatory mitotic parthogenesis, have a somatic chromosome number varying between 43 and 48, but some have only 30 to 32 chromosomes.

Reproduction of *M. arenaria* is exclusively by mitotic parthenogenesis (Triantaphyllou, 1985). No distinction is given between host Races 1 and 2 of *M. arenaria*. This species is cytologically very complex because it includes populations with chromosome numbers varying from $2n = 30$ to $2n = 56$. Most commonly, populations of *M. arenaria* are triploid with somatic chromosome numbers larger than 50. Chromosome numbers can be used to separate *M. arenaria* from other major root-knot nematodes, because the other major species have fewer chromosomes.

Signs and Symptoms

The most obvious sign of peanut root-knot nematode is to locate and observe the female in the galls either on roots, pegs, or on pods. The female nematode is approximately the size of a typewritten period on a page (800-μm length by 500-μm width). The female nematodes are pearly white and have a sharply pointed neck and head that is generally visible. Egg masses are attached to females and may extend from the surface of galled roots, pegs, and pods. Egg masses (which are about 1 mm in width) appear as a brownish mass adhering to the galled tissue and contain ≤300 to >500 eggs each. They are generally plentiful along the galled tissue. The egg mass is extruded from the vulva end of the female that is generally at or very near the root surface. This positions the egg mass on the outside surface of galled tissue, which facilitates both egg hatch and secondary infection of roots by the freshly hatched juveniles.

Peanut plants growing in soil infested with the peanut or Javanese root-knot nematode generally show noticeable above- and below-ground symptoms. In heavily infested fields symptoms can be seen 45 to 75 d following planting; however, most severe symptoms are observed after 90 to 120 d. The general characteristics of root knots on peanut are typical of other plants diseased by these nematodes. Plants heavily infected with either *M. arenaria* and *M. javanica* often are severely stunted, showing signs of chlorosis, incipient wilting, nutrient deficiencies, and are distributed in patches of varying sizes (Fig. 23–1A). Plants exhibit a rusty, yellowish, and mottled appearance. Galls on peanut pods and pegs are distinctive (Fig. 23–1C,D; Fig. 23–2A). If drought occurs near the end of the season the severity of root knot is accentuated and weakened plants die (Fig. 23–1B).

Second-stage juveniles infect peanut roots soon after planting, but noticeable galling and egg masses are not apparent on the roots until 55 to 90 d after

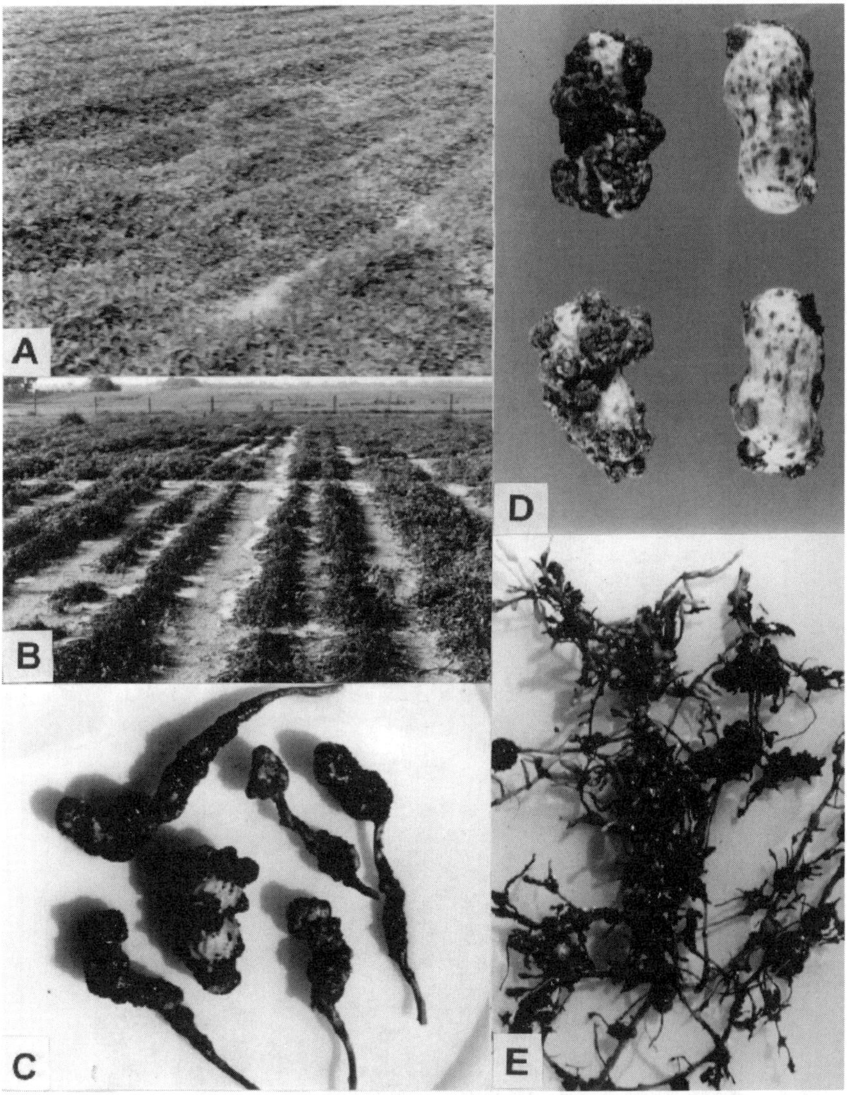

Fig. 23–1. Field symptoms of the peanut root-knot nematode (*Meloidogyne arenaria* Race 1) on peanut. (*A*) Peanut field in north central Florida infested with the peanut root-knot nematode. Symptoms include stunted plants, chlorosis, and patches of varying size plants. (*B*) Peanut plants severely infected by the peanut root-knot nematode. Many plants weakened by drought eventually died. (*C*) Pegs and pods of peanut severely galled by the peanut root-knot nematode. (*D*) Pods of peanut showing varying degrees of galling from infection of the peanut root-knot nematode. (*E*) Roots of peanut galled by the peanut root-knot nematode.

planting. The characteristic symptoms of root knot are the abnormal swelling (galls or knots) on the roots (Fig. 23–1E, 23–2B, 23–3A) Galls are not as easily detected on peanut roots as on many other more succulent root systems, e.g., tomato (*Lycopersicon esculentum* Miller), tobacco (*Nicotiana tabacum* L.), round melon [*Praecitrullus fistulosus* (Stocks) Pangalo], and okra [*Abelmoschus escu-*

Fig. 23–2. Peg, pod, and root symptoms of the peanut root-knot nematode (*Meloidogyne arenaria* Race 1) on peanut. (*A*) Pegs and pods showing typical galling induced by the peanut root-knot nematodes. (*B*) Roots heavily galled by the peanut root-knot nematode.

lentus (L.) Moench]. This is distinct from what happens on tomato, tobacco, cotton, etc., where galling is evident on roots 2 to 3 wk after planting. Peanut root galls are small and generally indiscrete, whereas galls on the other host crops may be large and coalesced. Nematode galls can be distinguished from rhizobium nod-

Fig. 23–3. Root and pod symptoms of the peanut root-knot nematode (*Meloidogyne arenaria* Race 1). (*A*) Portions of a peanut root system showing varying degrees of galling resulting from infection by the peanut root-knot nematode. (*B*) Contrast of pods infected (bottom row) and uninfected (top row) by the peanut root-knot nematode.

ules. Nodules are distinctive, round swellings attached to the root and are easily detached; whereas, nematode galls are swellings that constitute a part of the fibrous root system and can not be removed without destroying the integrity of the root. Also, because second-stage nematodes can infect nodules, in some instances, galls may appear on the nodules and vice versa. Soon after blooming and initiation of pod set (generally about 45 d after planting), pegs and pods also may be infected by root-knot nematode juveniles. Early infection may result in a weakened peg so that pods are lost during harvesting. Pods may become heavily infected and extensively galled by *M. arenaria* (Fig. 23–3B) or *M. javanica* late in the season. This situation results in drastic yield suppression. Galls induced by *M. hapla* on roots are smaller and have a characteristic hairy or matted appearance. Infection by *M. hapla* causes a proliferation of adventitious roots from the infected areas on roots.

Meloidogyne hapla symptoms are generally less severe than those caused by either *M. arenaria* or *M. javanica*. Aboveground symptoms of *M. hapla* may be difficult to detect because this parasite may not cause any stunting or chlorosis. The most severe symptoms generally indicate large population densities of infective juveniles in the soil. Severity over a field varies, depending on the variations in soil population densities and soil type. Sandy areas within a field often show the most severe symptoms. Plants with slight infections usually do not show stunting or chlorosis. Typically, the only indication of root knot on such plants is galls on roots, pegs, and pods.

Biology and Life History

Root-knot nematodes are sedentary endoparasites that have evolved a specialized and complex relationship with host plants (Hussey, 1985). The second-stage juvenile (Fig. 23–4A), which is the infective stage, moves through soil in search of plant roots or, in the case of peanut, roots, pegs, and pods. Dickson and Hewlett, 1988b showed that second-stage juveniles of *M. arenaria* were capable of moving 90 cm vertically, infecting plant roots within 16 d, whereas migrating 120 cm vertically required 29 d.

Second-stage juveniles that locate peanut roots, pegs, and pods often move into them immediately (Hussey, 1985). The juveniles generally penetrate the root tissue directly behind the root cap, then move intercellularly in the cortex to the region of cell differentiation. After moving short distances within the root system the juveniles embed their heads in the periphery of the vascular tissue. The remainder of the body extends into the cortex with the long axis of the body parallel with the root axis. The juveniles begin feeding on vascular cells, inducing the formation of giant cells. These cells become hypertrophied hyperplasia, which result in the formation of the typical root gall (Fig. 23–4C). Galls induced by *M. hapla* are generally smaller and with numerous lateral roots extending from them, whereas *M. arenaria* and *M. javanica* induce large galls and with few lateral roots.

The life cycle of root-knot nematodes includes one egg stage, four juvenile stages, and the adult stage. Once the juveniles feed they begin the process of molting in order to grow. Juveniles pass through four molts before reaching adult-

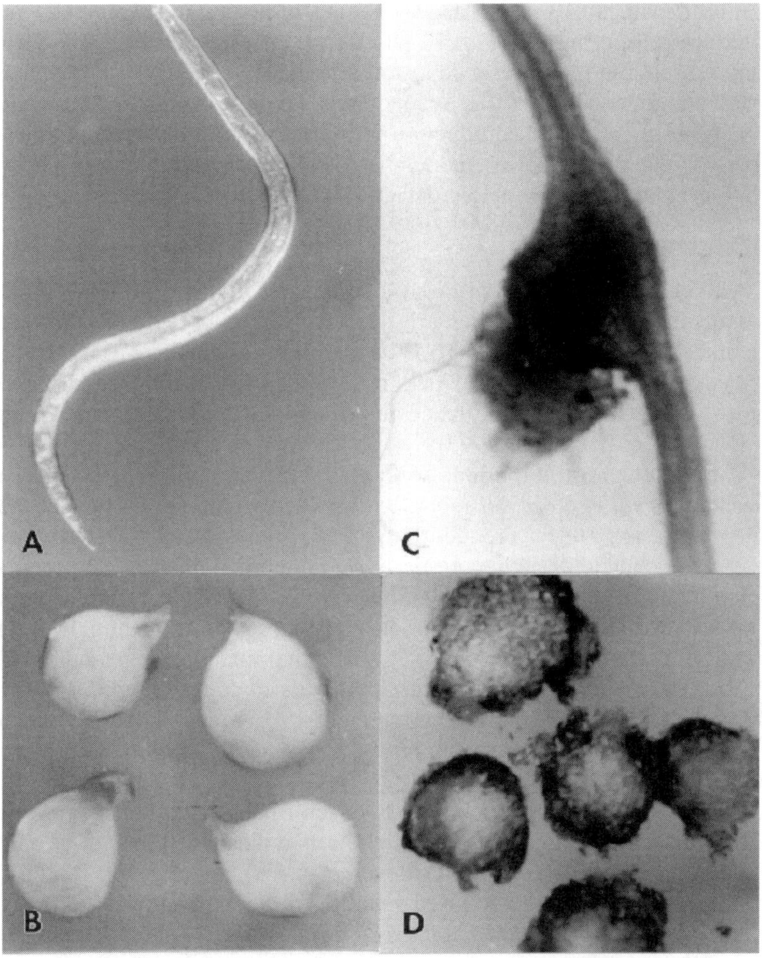

Fig. 23–4. Light microscope photographs of some of the life stages of the peanut root-knot nematode (*Meloidogyne arenaria* Race 1). (*A*) Second-stage juvenile. (*B*) Pyriform shaped females. (*C*) Single gall and egg mass on tomato root. (*D*) Egg masses.

hood. The first-stage juvenile molts the first time within the egg shell before hatch. The females are pyriform in shape (Fig. 23–4B), whereas males are fusiform in shape. The preponderance of juveniles develops into females; however, under adverse condition, e.g., crowding, drought stress, and low light intensity, juveniles may undergo sex reversal producing males with two gonads (Triantaphyllou, 1982). Males seek out females for mating, but do no feed.

Root-knot nematode infested soil contains varying population densities of eggs and infective juveniles. The infective juveniles may be distributed in soil to 75 cm deep, particularly in deep sandy soil in Florida (Martinez, 1976). In Florida, the largest numbers are generally found at the time of planting in the deeper levels (60–75 cm), whereas in North Carolina, most occur in the upper 30 cm of soil (K. R. Barker, 1995, personal communication). In contrast, the lowest population densities of second-stage juveniles at planting are generally found 0 to 15 cm deep. Egg masses (Fig. 23–4C,D) are encountered on roots during the growing season, but their presence in soil after harvest declines rapidly, such that nearly all are gone in 60 or 90 d. Second-stage juveniles constitute the major survival stage in most peanut-production regions and may be encountered in soil anytime of year. Nevertheless, population levels typically decline sharply after crop harvest and may exist below detection limits with current technology.

Disease Complexes

Although a few reports of disease complexes involve root-knot nematodes on peanut, more detailed studies are needed. This is especially true regarding interactions involving *Sclerotium rolfsii* (southern blight) and *Cylindrocladium parasiticum* (Cylindrocladium black rot, CBR), two soilborne diseases that appear to be increasing in importance on peanut in many major peanut-producing states (Melouk & Backman, 1995).

The single most important soilborne fungal pathogen of peanut in the southeastern USA is *S. rolfsii* (southern blight), which causes peanut stem rot (Brenneman et al., 1995). Frequently both *M. arenaria* and *S. rolfsii* are found together in peanut fields. When *M. arenaria* population densities are reduced via nematicide applications or crop rotations, the incidence of southern blight also may be reduced (Rodríguez-Kábana et al., 1982a, 1994). Although these field observations and other studies have suggested that infection of peanut by *M. arenaria* increases the incidence of southern blight, only one attempt has been made to document a disease complex (Starr et al., 1996). These workers were unable to demonstrate an interaction between *M. arenaria* and *S. rolfsii* in microplot experiments.

Significant positive correlations between final populations of microsclerotia of *C. parasiticum* and *M. hapla* in field tests indicate this nematode can affect Cylindrocladium black rot (CBR) development (Diomande & Beute, 1981). Follow-up studies showed increased severity of black rot on susceptible peanut cultivars infected by both *M. arenaria* or *M. hapla* (Culbreath et al., 1992). Disease ratings of roots from four peanut genotypes were higher in microplots infested with either *M. arenaria* or *M. hapla* and *C. parasiticum* than in plots with *C. parasiticum* alone.

Recently, Hirunsalle et al. (1995a) attempted to clarify the interaction of concomitant populations of *M. arenaria* Races 1 and 2, and *M. hapla* on peanut. Race 2 of *M. arenaria* tended to depress *M. arenaria* Race 1 development on peanut, whereas *M. hapla* had little effect on *M. arenaria* Race 1 on this crop. In mixed (1:1) populations of *M. arenaria* Races 1 and 2, Race 1 was dominant on peanut (Hirunsalle et al., 1995b). Crop rotation with a poor host for Race 1 for the previous 2 yr lowered the number of nematodes; however, they increased rapidly when peanut was reintroduced. *M. arenaria* Race 1 had a greater rate of reproduction than *M. arenaria* Race 2 in 2nd-yr peanut following resistant tobacco, and in 3rd-yr peanut following 2nd-yr resistant tobacco. Interaction studies between *M. arenaria* and *M. hapla* on five peanut genotypes showed that *M. arenaria* had greater infection capacity and caused more crop damage than *M. hapla*; however, both species had similar reproductive potentials (Hirunsalle et al.,

1995c). *Meloidogyne arenaria* was more competitive than *M. hapla* on peanut. Infection and reproductive potentials, and crop damage induced by mixed populations were similar to that induced by *M. arenaria* alone.

Peanut-pod decay associated with lesion nematodes (Fig. 23–5A) generally is much less severe than that induced by certain fungi (Fig. 23–5B). Garcia and Mitchell's (1975a,b) studies showed synergistic interactions in peanut pod rot (Fig. 23–5B) and damping off when *Pythium myriotylum* was combined with *Fusarium solani* and *M. arenaria*.

Fig. 23–5. Photograph shows differences in: (*A*) pod lesions caused by infection of the lesion nematode *Pratylenchus brachyurus*, and (*B*) pod rot lesions caused by soilborne fungi.

Economic Importance

In fields heavily infested with root-knot nematodes, large patches of plants often die. In Alabama, Georgia, Florida, and Texas, as many as 40% of the peanut fields are estimated to be infested with this pathogen (Sturgeon, 1986). Where damaging levels of *M. arenaria* occur, as much as 50% of the yield potential can be lost. In three nematicide efficacy trials conducted in peanut fields heavily infested with *M. arenaria*, the most effective nematicide treatment increased yields 65, 83, and 96%, respectively (Dickson & Hewlett, 1988a).

In Florida, the damage threshold under some environmental conditions was estimated to be as low as a single juvenile per 100 cm^3 of soil (McSorley et al., 1992). Similar data on the relationships between *M. arenaria* densities and yield losses have been reported from Alabama (Rodríguez-Kábana et al., 1982b), North Carolina (Koenning & Barker, 1992), and Texas (Wheeler & Starr, 1987). Based on studies in Texas, more than 10% of the peanut fields from five major producing counties have *M. arenaria* population densities that exceed the level that causes a 10% yield suppression (Wheeler & Starr, 1987).

LESION NEMATODE

Steiner (1945, 1949) first reported the lesion nematode (*Pratylenchus brachyurus*) on peanut in Alabama and Virginia. This nematode is now known to occur on peanut throughout the southeastern USA and causes damage to many tropical and subtropical crops (Thames, 1982). Taxonomic separation of species of *Pratylenchus* is difficult because they exhibit little morphological diversity (Roman & Hirschmann, 1969).

Symptoms

Distinct field symptoms in *P. brachyurus*-infested sites are difficult to discern. Severely infected peanut plants may be stunted and chlorotic, but this is rare. The most obvious symptom of lesion nematodes are small, purplish-brown to black lesions that form on the peanut shell (Fig. 23–5A) (Boswell, 1968; Good et al., 1958). These lesions have distinct boundaries and first appear as small brown tunnels in the shell and later coalesce to form large lesions. Secondary soilborne pathogens may enter these lesions causing the lesions to greatly increase in size or the infected pegs and pods may rot. Infection of pegs by *P. brachyurus* has been correlated with a peg rot condition resembling the peg rot disease caused by *S. rolfsii* (Good et al., 1958). Combinations of fungal and nematode-induced lesions may occur, but this has received little study (Good et al., 1958).

Biology and Life History

The lesion nematode is a migratory endoparasite. All life stages except the egg and first-stage juveniles are infective. The second-, third-, and fourth-stage juveniles, and adults are fusiform shaped. They are capable of moving to infection sites on peanut roots, pegs, and pods. Good et al. (1958) found that *P. brachyurus* were most numerous in the mature shells, where they colonized in dark-colored necrotic lesions. The nematodes remain viable in these infected shells and serve as a source of inoculum even after the shells have been cured and stored overwinter. They may penetrate anywhere on roots, pegs, and pods and move from old infection sites to induce new infection sites. Females lay eggs singly inside lesions or outside the plant tissue in soil. Apparently, the pod shell tissue is more favorable for reproduction, with six to eight times greater numbers occurring in it as in equal portions of root tissue (Good et al., 1958). Infected pods may serve as a source of spreading this parasite. Good and Stansell (1965) found that irrigation markedly increased lesion nematode in peanut pegs with 3 to 271 nematodes per gram of shell tissue for nonirrigated and irrigated plots, respectively.

The presence or absence of host races within *P. brachyurus* has not been documented; however, their existence has been suggested by variation in numbers of *P. brachyurus* extracted from roots of *Citrus limon* (L.) Burman f. seedlings when inoculated with different nematode isolates (O'Bannon & Tomberlin, 1970). Field observations also have suggested behavioral differences within *P. brachyurus* populations, particularily on peanut (Dickson & Dunn, 1988, personal observations). However, attempts to separate races of *P. brachyurus* failed to discern behavioral differences among four populations tested on seven species of crop plants (Payan & Dickson, 1988). On these seven crop plants population densities of two populations from peanut were not different from two other populations that originated from soybean [*Glycine max* (L.) Merr.] or corn (*Zea mays* L.).

Economic Importance

Pratylenchus brachyurus is only occasionally associated with severe peanut yield losses. In greenhouse and microplot studies high initial population densities damage peanut substantially; however, the nematode does not increase greatly on Florunner peanut. Damage in the field is probably caused by high initial population densities that carried over on a previous crop, e.g., corn. The nematode does not appear to increase on cotton and soybean cultivars (Thames, 1982). In lesion nematode infested sites up to 19% of the pods have been left in the ground at digging compared to sites where the nematode was controlled (Good et al., 1958). The lesions these nematodes cause on pods also contribute directly to pod rot.

RING NEMATODE

The ring nematode [*Criconemella ornata* (syn. *Mesocriconema ornatum*)] is a common parasite of peanut in the southeastern USA. Relatively large numbers are found frequently in peanut roots, pegs, and pods.

Symptoms

Obvious field symptoms either on the above-ground plant or on the below-ground roots are seldom seen. Machmer (1953) reports a yellowing of peanut vines in soil heavily infested with ring nematodes in Georgia. Barker (1982) induced the "yellows" symptom on peanut grown in microplots with freshly extracted, greenhouse grown nematodes. However, the "yellows" symptom is not common in infested fields. The author has never seen these symptoms in peanut fields with heavy ring nematode infestations. Other reported symptoms such as stunting and small brown necrotic lesions on roots and pods are difficult to detect and are of little practical use in field diagnoses.

Biology and Life History

The ring nematode is a slow-moving, sedentary ectoparasite. It feeds on the outside of the roots by thrusting its long stylet into the root tissue and induces the formation of food cells (Hussey et al., 1992). Ring nematodes were observed to feed up to 8 d from single food cells in the outer root cortex. The nematode's stylet penetrated between epidermal cells and frequently through a subepidermal cortical cell. Ultrastructural studies failed to show any destructive cellular changes in host roots. Apparently, ring nematodes have evolved a highly specialized feeding relationship with cells of its hosts. Although this ultrastructure work was with *C. xenolax*, a parasite of peach (*Prunus persica* L.) trees, it is suspected that *C. ornata* acts in a similar manner on peanut.

Little information is available on the life cycle of ring nematodes primarily because of the difficulty studying their behavior. Some species of ring nematodes complete their life cycle in 25 to 34 d. Females deposit single eggs in the soil at the rate of 4 to 8 eggs/d (Jenkins & Taylor, 1967). Peanut has proven to be an excellent host for *C. ornata* with population increases as great as 970-fold (Barker, 1982). The rate of population increases was inversely related to initial population numbers. Corn is considered an intermediate host (population increase of 264-fold), and tobacco and soybean are poor hosts (Barker, 1982).

Disease Complexes

Diomande and Beute (1981) found generally positive correlations between CBR and *C. ornata*, but there was a less pronounced correlation than that induced by *M. hapla*. The final population densities of *C. ornata* were greater on the CBR-resistant peanut genotype NC 2033 than on the CBR-susceptible 'Florigiant' peanut, but *C. ornata* combined with CBR in microplot tests caused more disease on Florigiant than did CBR in absence of *C. ornata*.

Economic Importance

Ring nematodes are sometimes associated with reduced peanut yields. Minton and Bell (1969) found roots, pods, and pegs of 'Argentine' and 'Starr' severely discolored and pod yields reduced about one-half by ring nematode in microplot tests. Barker (1982) also reported that as few as 178 freshly introduced *C. ornata*/500 cm^3 of soil stunted peanut in microplot tests, whereas much greater residual population densities following tobacco (poor host) had little effect on the growth of peanut. It was suggested that many nematodes in the field following tobacco died.

STING NEMATODE

The sting nematode (*Belonolaimus longicaudatus*) is reported to cause losses on peanut only in localized areas in North Carolina and southern Virginia. It is distributed throughout 17 counties in North Carolina, nine of which are major peanut-producing counties. Sting nematode is distributed throughout the southeastern USA and has been reported in other peanut-producing states including Florida, Georgia, Oklahoma, and Texas. Wheeler and Starr (1987) detected *B. longicaudatus* in 0.8% of samples surveyed in five peanut-producing counties in Texas. The populations in North Carolina and Virginia apparently are unique in that they reproduce on peanut, whereas populations in other southeastern states are not known to attack peanut. The population from North Carolina and Virginia have morphological characteristics that differ from other populations of this species and probably should be reclassified (R. T. Robbins, 1995, personal communication).

Symptoms

Sting nematodes induce severe root growth restrictions and damage. Roots of affected plants become discolored, sparse, and the lateral roots are frequently destroyed. Pods and pegs are affected, and consequently yields are drastically affected. The aboveground plant parts are chlorotic and markedly stunted. Nodule development also may be limited by this pathogen. Plant stand and quality are reduced, and yields are often severely suppressed.

Biology and Life History

Belonolaimus longicaudatus is a migratory ectoparasitic feeding nematode. The nematode has a long stylet, which it uses to pierce deep into the root tissue to obtain nutrition. Sting nematode distribution is limited to the upper 45-cm depth of the soil profile (Potter, 1966) . Populations are encountered only in soils of 90% sand or greater (Perry & Rhoades, 1982). Apparently, soils with small pore spaces are unfit for nematode movement. Sting nematodes are among the largest (2–3 mm long by 50–80 μm wide) of the plant-parasitic nematodes. Reproduction is by cross fertilization, and males and females are present in equal numbers in the soil surrounding peanut roots.

Economic Importance

Sting nematode on peanut is a devastating pathogen causing severe economic losses if not managed, but fortunately the distribution of populations of sting nematodes that attack peanut are limited. Losses are relatively low being reported for North Carolina (0.38%) and Virginia (0.50%) (Anonymous, 1987). In North Carolina yield increases of 67 to over 400% have been attained in nematicide treated plots compared to untreated plots (Cooper et al., 1959; Sasser et al., 1960; Sasser & Cooper, 1961; Cooper & Sasser, 1962).

PEANUT POD NEMATODE

Peanut pod nematode (*Ditylenchus africanus*) was described on the hulls and seeds of peanut in South Africa in 1988 (Jones & De Waele, 1988). The nematode was first reported as *D. destructor* (the potato rot nematode) but was later described as a new species, *Ditylenchus africanus* (Wendt et al., 1995).

Symptoms

No visible symptoms with infection by this nematode are seen on roots; however, seeds show blemishes and premature germination before harvest (De Waele et al., 1989). Infected hulls exhibit brown necrotic tissue at the point of connection with the peg and dark brown to black discoloration along the longitudinal veins. Infected seed are usually shrunken with dark brown to black micopyles and flaccid testa and dark vascular strands. In heavily infested fields about 40 to 60% of pods and seeds show symptoms. The infection results in a lower quality grade and reduced peanut yield.

Biology and Life History

Ditylenchus africanus develops from egg to adult in 8 d at 25°C. Egg hatching at 28 to 30°C starts at around 3 d, and by the sixth day 90% have hatched (C. Venter, 1995, personal communication). The nematode is able to enter a state of anhydrobiosis (Basson et al., 1993). About one-third of the anhydrobiotic nematodes becomes active after rehydration to invade hulls and seeds of a newly planted crop (Jones & De Waele, 1990). Although *D. africanus* is present in the roots of peanut and in soil, 90% of the total population at harvest is found in pods (Basson et al., 1991; Venter et al., 1993). Infestations of up to 97 000 nematodes per pod are not uncommon for peanut grown in the field.

According to Venter et al. (1995), *D. africanus* enters the immature pegs and pods of peanut at the peg-connection; however, the infective stage is unknown. It subsequently invades the parenchymatous region of the hull exocarp and the endocarp, and eventually the seed testa. The nematode causes malformation of the cells of infected tissues, cell wall breakage and cell collapse. Damage appears to be caused by enzymatic activity. The entire parenchyma region in some testa is destroyed. In immature pods, *D. africanus* may move across the fibrous region of the mesocarp into the hull of the endocarp. The lignification of the fibrous mesocarp in mature pods at around 105 d is a barrier, however, to penetration of the inner pod tissues. Nematodes artificially inoculated after 105 d are no longer able to cause damage to the seed. Increased numbers of eggs and anhydrobiotic forms of the nematode are found in the hull tissues and eggs in the seed testa of late harvested pods (about 189 d after planting). Eggs, some containing first-stage juveniles, have been observed in the parenchymatous mesocarp of the hull (Venter et al., 1995). Also, egg numbers increase in the seed testa of late harvested pods. Both occurrences may indicate the onset of survival mechanisms of the nematode. Apparently, all life stages can be found in the hull. It appears that eggs and anhydrobiotes are involved in winter survival in decaying hulls and stubble, whereas eggs are the important survival stage in stored seed.

Economic Importance

About 200 000 ha of peanut are grown annually in South Africa with *D. africanus* being present in all major peanut-production areas of the country (Bolton, 1989). Greenhouse damage-potential studies (Venter et al., 1991) showed that at an initial population density (P_i) of 250 nematodes/3-L pot, 10 to 25% of seeds germinated into second-generation seedlings before harvest, and fresh weight of harvested seed was suppressed 20 to 50%. At P_i 50 or a final population density (P_f) greater than 20/seed, pod disease severity was 3 to 7 (scale of 1–10), and 15 to 80% of seeds were blemished or unsound.

TESTA NEMATODE

The testa nematode (*Aphelenchoides arachidis*) is a facultative endoparasite of peanut in Nigeria. In 1970, the nematode was found in seed testas of kernels being examined for fungal pathogens and it was later described by Bos (1977). This was the first record of a seedborne nematode parasite of peanut. It is not known to be a parasite of peanut outside of Nigeria.

Symptoms

The seed coats of kernels become discolored where there are more than 2000 nematodes/testa (Bridge et al., 1977). Up to 35 000 nematodes/testa have been recovered. However, even healthy looking seed may be infected, varying from 4 to 30% of seed lots tested. Generally, these healthy looking seeds had fewer than 2000 nematodes/testa. Severely infected seeds are light-weight and have a translucent testa, appear light brown and have dark vascular strands within the testa.

Nematodes are found mainly in the subepidermal parachymatus layer and around the tracheids of the testas. Infected testas appeared thicker and more uneven than uninfected testas. Nematodes break down the parenchyma cell walls making the cells appear larger. Pod shells, roots and hypocotyl also harbor the nematode, but there are no noticeable signs of necrosis or other symptoms in these tissues. Heavy infection results in perforation and cracking of the testa and a spongy appearance (McDonald et al., 1979).

Seeds with heavy nematode infection have greater fungal infections than uninfected seed. Nematode infection facilitates the entry of soilborne fungi

(McDonald et al., 1979), thus leading to reduced seedling emergence when infected seeds are planted.

Biology and Life History

Peanut pods are invaded 10 d after the pegs penetrate infested soil (Bridge et al., 1977). Nematode numbers in pods increase rapidly 30 d after infection occurs, with greatest numbers recorded at about 60 d. All nematode developmental stages occur within infected tissue, which includes the pod shells, testas, hypocotyl and roots, but not in the stele. At plant maturity, heavily infected testas contain mainly juvenile stages with few adults, whereas testas showing no external symptoms contain mainly adults and eggs. *Aphelenchoides arachidis* is capable of feeding ectoparasitically on peanut roots and on the hypha of soilborne fungi. The latter probably enable the nematode to survive long periods in the absence of peanut.

Aphelenchoides arachidis can survive desiccation in stored peanut pods; however, the peanut cultivar affects the length of survival. Nematodes survive up to 1 yr in some peanut cultivars, whereas in other cultivars the nematode survival time is reduced. Juveniles have better survival rates than adult nematodes.

Economic Importance

Bridge et al. (1977) reports that *A. arachidis* affects the commercial value of confectionery peanut but had little effect on yield; however, McDonald et al. (1979) found that seed heavily infested with the nematode had higher levels of infection by the more parasitic species of fungi than did visually nematode-free seed. Also, they report that seedling emergence was affected by the nematode infestation.

OTHER NEMATODES

Aphasmatylenchus straturatus

Aphasmatylenchus straturatus was found around roots of peanut in Haute-Volta, West Africa, and described as a new species in 1970 (Germani, 1970). The nematode is a migratory, ecto-endoparasite that causes interveinal chlorosis and stunting of aboveground plant parts, reduces root growth, rhizobium nodulation, and yields (Germani & Dhéry, 1973). Germani and Luc (1982a) refer to the disease as "voltaic chlorosis" of peanut. They demonstrated a direct relationship between nematode density and chlorosis intensity by calculating the value of N_2 assimilation. Nitrogen fixation in chlorotic peanut was about one-fifth that in uninfected peanut.

Distinct dry and wet seasons occur in the West Africa region where this nematode is found. Some nematode species, e.g., *Scutellonema cavenessi* and *S. clathricaudatum*, which are also in close proximity to peanut roots, are anhydrobiotes. However, *A. straturatus* does not enter an anhydrobiotic state (Germani &

Luc, 1982b) and does not infect peanut during the dry season (November–April). During this period the nematode is located 40 to 50 cm deep in soil and in close association with the karité tree (*Vitellaria paradoxa* C.F. Gaertn.). Germani and Luc (1982b) proposed the karité tree as the natural host for the nematode, but as the rainy season begins (June–November) the nematode migrates from the tree roots to the peanut roots where it causes chlorosis and yield reductions. *Aphasmatylenchus straturatus* is not known to infect peanut in other peanut-growing regions of the world.

Stunt Nematode

The stunt nematode (*Tylenchorhynchus brevilineatus*) was first observed in 1976 in the Kalahasti region of Andhra Pradesh State, India (Reddy, 1984). No reports have been made on the life cycle or disease cycle of this nematode on peanut. The nematode induces the disease, "Kalahasti Malady," which is characterized by reduction in pod size and brownish-black discoloration of the pod surface. Small, brownish-yellow lesions also appear on the pegs and developing pods. The margins of the lesions are slightly elevated because of cellular proliferation around the sites of nematode infection. Pegs are greatly reduced in length, and the surface of heavily infected pods becomes completely discolored. The nematode does not affect the seed within pods. Plant roots also are parasitized, but discoloration is less severe than on pods. Aboveground symptoms include stunted plants that have greener foliage than uninfected plants. The disease is generally distributed in patches within fields and is most severe in sandy soils. The nematode has not been reported as a parasite of peanut in other regions of the world.

Needle Nematode

The needle nematode (*Paralongidorus bullatus*) is suspected to be a major pathogen of peanut in the semiarid regions of West Africa (Baujard et al., 1993). Sharma and Siddiqi (1990) report that *P. bullatus* is widely distributed in Niger and that the nematode is associated with peanut crop growth variability problems. The nematode occurs in sandy soil around peanut roots showing poor growth symptoms (Baujard et al., 1993). It is found 50 to 100 cm deep in soil, which makes detection difficult. The nematode had been overlooked in previous sampling of the region because of its deep occurrence in soil and the absence of anhydrobiotic juvenile or adult stages. High soil temperatures of 32 to 36°C and moderate soil moisture of 7 to 11% are optimum for reproduction. In greenhouse studies, the nematode induced a significant reduction of fresh and dry weights of peanut roots and aerial parts. Controlling this nematode in sandy soils often increases yields by more than 1000 kg ha^{-1}.

Spiral Nematodes

Both spiral nematodes (*Scutellonema cavenessi*) and (*S. clathricaudatum*), have been described as parasites on peanut in Sénégal (Germani, 1981; Germani et al., 1985), and Niger (Sharma et al., 1992), respectively. However, according to Baujard and Martiny (1995a), the positive correlation between the initial population level of *S. cavenessi* and yield refuted the earlier research. Field observations have shown that high initial population densities of *S. cavenessi* are often associated with high yields of peanut (Baujard & Martiny, 1995a). Based on additional field and laboratory studies, the pathogenicity of *S. cavenessi* on peanut now appears doubtful. Culturing *S. cavenessi* in the greenhouse has not been possible (Baujard & Martiny, 1995b), thus pathogenicity studies on peanut using pure populations of the nematode have not been conducted. Similarly, no pathogenic effects of *S. clathricaudatum* were shown in laboratory studies on the nematode on peanut (Baujard & Martiny, 1995c).

MANAGEMENT

For nematode management producers must rely on cultural practices and nematicides (Dunn & Dickson, 1995). Currently, no peanut cultivars resistant to nematodes are available (Starr et al., 1995).

Nematode Monitoring and Field Choice

Since peanut is currently one of the most profitable agronomic crops, it should be planted only in fields with the best yield potential. Fields with unusually high risks of nematode or other pathogens and disease problems should be avoided. To determine the suitability of each field for planting peanut, growers should know which nematodes are present and the severity of their infestations. Nematode infestations should be mapped based on symptoms seen on susceptible crops and on laboratory assays of nematode samples in countries where such services are available. Both are needed to have the best information about all kinds of nematodes present and their distributions.

Field Map

The best information about distribution of nematode populations comes from records of previous cropping seasons. Root-knot nematodes are readily monitored on peanut by observing galling of peanut roots, pegs, and pods near the end of the growing season. The galls caused on many plants by root-knot nematodes make it easy to map the incidence of "root knot."

The peanut, northern, and Javanese root-knot nematodes can be identified in the laboratory, but identification of them as pathogens of peanut requires running a bioassay with peanut plants. To ensure that a population of root-knot nematodes can damage peanut, it is necessary to see the galls on roots, pegs, or pods. Host Race 1 of the peanut root-knot nematode, *M. arenaria* that infects peanut cannot be separated easily from the closely related Race 2 of this nematode species. These two races of the peanut root-knot nematode are morphologically identical. Race 1 infects peanut, Race 2 does not infect peanut. Similarly, Races 1 and 2 of *M. javanica* do not attack peanut, whereas Race 3 does attach peanut, yet all three races are morphologically identical. As far as is known all populations of *M. hapla* attack peanut.

The most common species of lesion nematode (*P. brachyurus*) causes distinctive brown lesions on roots, pegs, and pods of peanut. Care should be taken to distinguish pod lesions caused by lesion nematode from those caused by pathogenic fungi. Generally fungal lesions are larger and less discrete than those caused by the lesion nematode. Extraction of nematodes from soil and root samples can be used to identify all kinds of nematodes associated with any crop grown in the field; however, processing samples and identifying nematodes that attack peanut require special equipment, techniques, and expertise in nematode identification.

Sample Analysis

A nematode assay laboratory can help identify kinds of nematodes present and their relative numbers, but no single sample can provide distribution information as efficiently as a field map based on personal experiences. Samples of soil and roots are best taken in the autumn, shortly before or after harvest of crops (autumn–early winter seasons). Samples taken later, up to 30 d before planting, can help identify the kinds of nematodes present in a field, but are less useful for detecting low population densities or estimating the likely levels of damage.

Experiments have been conducted to determine the relationship between yield of peanut and preplant peanut root-knot nematode densities (Rodríquez-Kábana et al., 1982b; Wheeler & Starr, 1987; McSorley et al., 1992). Tests have shown that 1 to 5 nematodes/100 cm^3 of soil caused 10% pod losses; thus, the thresholds for plant damage for this nematode was established at 1 or less than 1 nematode/100 cm^3 of soil. These experiments indicate that the tolerance limits of the peanut root-knot nematode on peanut in most seasons is quite close to the limits of detection. Consequently, the final population density (population levels at or near harvest) of the previous crop provides the best opportunity to detect economic damaging levels of root-knot nematodes that infect peanut.

Land Management Practices

Crop Rotation

Peanut should follow crops such as pasture grasses that are poor or nonhosts for root-knot nematodes. Wheat (*Triticum aestivum* L.) should be avoided because some cultivars are good hosts for the peanut root-knot nematode (Dunn & Dickson, 1995; Dickson & Melouk, 1995). Following soybean, tobacco, or vegetables also should be avoided. Field corn is only fair for reducing root-knot nematode population densities, but is better than continuous peanut or other good hosts of the nematode. Different cultivars of corn have different reactions to root-knot nematode infection (Windham & Williams, 1994). Some species of lesion nematodes increase to high densities on field corn. Grain sorghum [*Sorghum bicolor* (L.) Moench] appears to suppress root-knot nematode population densities two to four times better than corn (McSorley & Gallaher, 1991). Cotton fits well in rotation with peanut in fields infested with *M. arenaria* Race 1, *M. hapla*, or *M. javanica*, since these nematodes do not live on cotton, and the root-knot nematode that infects cotton (*M. incognita* Race 3) does not live on peanut (Sasser & Carter, 1982; Rodríguez-Kábana et al., 1994).

Several years of the tropical forage bahiagrass (*Paspalum notatum* Fluegge) have long been recognized as one of the best rotations to precede peanut (Norton et al., 1977; Rodríguez-Kábana et al., 1994). Long rotations of three or more years out of peanut and other favorable hosts are better than 1- or 2-yr rotations (Dickson & Hewlett, 1989). However, control of weeds in a bahiagrass planting is essential. Hairy indigo (*Indigofera hirsuta* L.), alyceclover [*Alysicarpus vaginalis* (L.) DC], and morning-glory (*Ipomoea* spp.) are a few common weeds that occur frequently in the southeastern USA. Each of these is a good host for the peanut root-knot nematode. Coffee senna [*Senna occidentalis* (L.) Link] and sicklepod (*S. obtusifolia*) are reported as hosts for the peanut root-knot nematode (Machmer, 1951); however, this is variable and needs further study for clarification.

Coastal bermudagrass (*Cynodon dactylon* L.) has been grown in the southeastern USA for decades as a forage and rotation crop for the management of root-knot nematodes in flue-cured tobacco, sweetpotato [*Ipomoea batatas* (L.) Lam.], and other vegetables (Burton & Johnson, 1987). It has been recently proposed as a general solution to problems caused by root-knot nematodes in the southeastern USA (Anonymous, 1989). However, recent data by Rodríguez-Kábana et al. (1994) showed that 3-yr rotations with coastal bermudagrass failed to increase peanut yields and resulted in the highest population densities of *M. arenaria* J2 on peanut. Also, bermudagrass did not reduce the incidence of southern blight (*Sclerotium rolfsii*), whereas both bahiagrass and cotton did.

Some unusual crops, such as partridge pea [*Chamaecrista fasciculata* (Michaux) Greene] and American jointvetch (*Aeschynomene americana* L.) suppress development of *M. arenaria* in soil when either crop was grown for 2 yr in rotation with peanut (Rodriguez-Kabana et al., 1991). Partridge pea reduced *M. arenaria* population densities more than jointvetch. Peanut yield was increased after 2 yr of planting jointvetch.

Crop rotation should not be expected to reduce a root-knot nematode population abruptly. In fact, some of the nematode population will survive the winter without a host and most crop plants can support at least some nematode reproduction. Also, most fields have some weeds that support nematode reproduction. Rotation is a far better tool to help keep relatively low populations from becoming too high, or for gradually reducing high populations over several years. Unfortunately, rotations do not significantly reduce lesion nematode numbers because the nematode lives well on many different crops including most grass family crops, soybean, and most vegetables.

Fallow

Fallow (leaving a field unplanted to reduce nematodes through lack of host plants) is often tried as an alternative to planting and maintaining an unprofitable rotation crop. Theoretically, it should reduce nematodes and other pests or pathogens that must have living host plants for survival. However, fallowing has several limitations as a nematode management practice. First, it is rare that a "fallow" field is truly lacking plants on which nematodes can live, volunteer plants from previous crops and weeds usually appear in the field. Thus, the three factors cited above that reduce the effectiveness of rotation also apply to fallowing. If the field is kept practically free of potential host plants by some combination of chemical herbicides and cultivation, the cost is likely to approach the cost of nematode control by chemical means, and there are likely to be serious losses of soil organic matter and perhaps heavy erosion because of wind storms or heavy rains during prolonged fallow periods.

Crop Destruction

After harvest, roots of the preceding crop should be destroyed as promptly as possible to interrupt reproduction of nematodes and other pests that could otherwise continue to increase their numbers on the old plants and any weeds that are present. Shredding crop residues, as by plowing, disking, or rototilling, also initiates decay of plant roots so nematodes are exposed to natural control agents in the soil and to nematicides.

Winter Cover Crops

A small grain cover crop can help prevent growth of weeds that are hosts of peanut nematodes and will help protect the soil from erosion during the winter. However, small grains that are planted while the soil is sufficiently warm (above 18°C) for root-knot nematodes to infect new roots can support some nematode reproduction before temperatures drop in winter. Rye (*Secale cereale* L.), triticale (× *Triticosecale*), most varieties of wheat, and barley (*Hordeum vulgare* L.) can support root-knot nematodes; many varieties of oat (*Avena sativa* L.) support less or none (Opperman et al., 1988). Root-knot nematodes may not increase their numbers in some of these grains, but overwinter decline of the population may be less than anticipated.

Tillage

Turning the land deeply at least 4 wk before fumigation or before final land preparation buries surface litter, thus reducing some fungus diseases. It also encourages decay of live plant roots (such as those of a winter cover crop and weeds) that could protect nematodes from their natural enemies or nematicides.

Nematicides

Chemical nematicides are available for nematode problems on peanut (Dunn & Dickson, 1995). It is suggested that nematicides and other soil-applied insecticides be alternated in fields that are to be planted in succeeding crop years. Repeated applications of some soil-applied nematicides have led to the build-up of bacteria and other microbes that rapidly degrade them (Ou et al., 1993, 1995).

The soil fumigant, 1,3-D (1,3-dichloropropene), must be applied 1 to 3 wk before planting. Most other products can be applied at the time of planting or immediately before. Nonfumigants (Fig. 23–6A) have been more consistently successful in heavier soils, whereas sandy soils favor fumigation. Many nonfumigants are good soil insecticides. This aspect should be considered when selecting a product to use for nematode control.

Supplemental treatments of nonfumigant nematicides may be applied during peg initiation ("pegging time") to protect young peanut plants when they are most susceptible to nematode damage (Dickson & Waites, 1980; Dickson & Hewlett, 1988a).

Soil Fumigation

The soil fumigant 1,3-D has often been the most effective nematicide in terms of consistency and yield increases in peanut nematode control tests (Kinloch & Dickson, 1991). A few practices that improve the chance for effective peanut nematode control with 1,3-D include the following. It must be used 1 to 3 wk preplant because stand reduction, delayed emergence, and stunting of early growth have all been observed when 1,3-D was applied under the row at or near planting time. Fumigated areas should not be planted until the odor of 1,3-D is gone from the treated zone. Injection should be at least 30 cm below the final soil surface when measured from the bottom of the outlet tube. Row injection at 30 cm deep or greater can lead to treating more soil area with the same amount of applied fumigant. Complete closure or sealing of treated soil is essential (deep chisel plows and subsoiling-bedding equipment have worked well for row treatments, and deep moldboard plow application has worked well for broadcast applications). One effective way of sealing a one- or two-chisel row treatment is to use disk-hillers to throw up a bed 30 to 35 cm high behind the chisel(s), then to remove ("board off") the top of the bed before planting.

An autumn or early winter broadcast application of 1,3-D either by moldboard plow or deep chisels has been tested and found equally effective as a spring application (Kinloch & Dickson, 1991). Seven to ten days following an autumn or winter application, rye or another winter cover crop should be planted over the treated field. Application of 1,3-D should not be made when the soil temperature is below 4.4°C. Soil temperatures that are best for 1,3-D applications lie between 4.4 and 26°C. Soil moisture should be at one-third to one-half of field capacity or near the wilting point. This point can best be described as when the soil is ideal for planting.

Deep placement by moldboard plow is an efficient way to apply 1,3-D (Fig. 23–6B). Fifty-six to eighty-four liters per hectare of 1,3-D have generally given good control from broadcast moldboard plow application. This method of application apparently is more efficient due to the complete lack of furrows or "chimneys," normally left by chisels. Chisel channels provide a conduit through which fumes can escape prematurely.

For heavily infested fields, especially deep sandy soils, a broadcast application of 1,3-D at 55 L ha^{-1} followed by a nonfumigant nematicide application at peg initiation may be the most effective treatment.

Nonfumigant Nematicides for Use at Planting

Formulations of ethoprop (O-ethyl S, S-dipropyl phosphorodithioate), fenamiphos [ethyl 3-methyl-4-(methylthio)phenyl (1-methylethyl)phophorami-

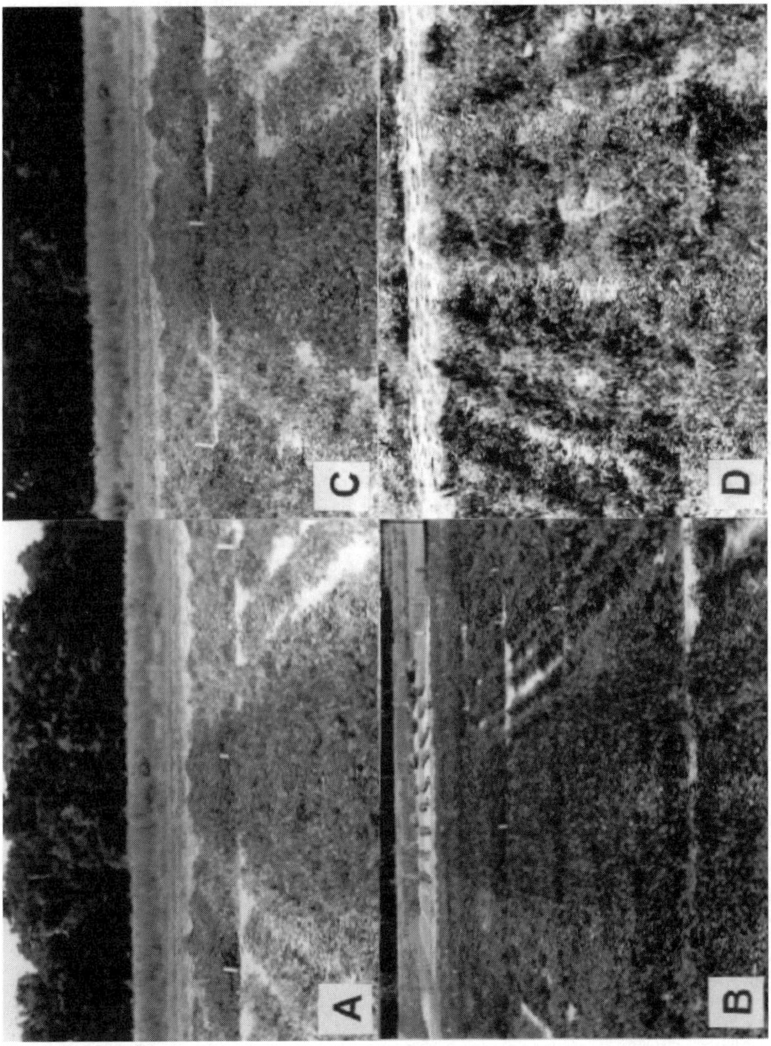

Fig. 23–6. Nematicide test plots in a peanut root-knot nematode (*Meloidogyne arenaria* Race 1) infested field. (*A*) Center two rows treated with a nonfumigant nematicide. Border rows on eithr side untreated. (*B*) Six-row strip through center of field treated with a soil fumigant applied by moldboard plow. (*C*) Center two rows treated with a fumigant nematicide. Border rows on either side untreated. (*D*) Nematicide-treated plots showing damage caused by southern blight (white mold) and Cylindrocladium black rot.

date], aldicarb [2-methyl-2-(methylthio)propionaldehyde O- (methylcarbamoyl) oxime], and oxamyl {methyl N'N'-dimethyl-N-[(methylcarbamoyl)oxy]-1-thio-oxamimidate} are registered for nematode control. When used at planting, aldicarb and fenamiphos have been the most effective of the nonfumigant products according to university research (Fig. 23–6A,C), but these products may fail in the presence of fungal pathogens (Fig. 23–6D). Maximum rates of these products have often produced yields equal to those obtained with 56 to 84 L ha^{-1} of 1,3-D applied in the row.

Nonfumigant Nematicides for Application at Peg Initiation to Supplement the Primary Nematicide

Aldicarb or ethoprop may be applied at peg initiation (about 45 d after planting) to supplement treatment at planting. Such treatments have increased yields 560 to 672 kg ha^{-1}. Nonfumigants should not be applied within 90 d of harvest. Either nematicide may be used at peg initiation following a preplant application of 1,3-D, or at plant application of aldicarb, ethoprop, or fenamiphos. Granules should be applied in a 35- to 47-cm wide band directly over the vines. Foliage should be dry when granules are applied, and a weighted burlap bag or similar drag should be pulled over the vines to dislodge granules from the foliage. The granules should be incorporated by running cultivation sweeps (flat orientation) through the row middles. Granules should be applied only when rain is expected or irrigation can be used to dissolve the active ingredients and carry them into the soil within 7 to 10 d after application. Application should not be made during extended dry periods. Foliar sprays of oxamyl also have been effective when applied at peg initiation in some trials. Label directions should be followed for rates, frequency, and timing of applications.

OUTLOOK

Resistance to Root-Knot Nematodes in Peanut

The most exciting prospects on the horizon for nematode management is the possibility of developing peanut cultivars with root-knot nematode resistance. Just a few years ago this was considered improbable (Miller, 1972; Minton & Hammons, 1975; Holbrook et al., 1983). Thousands of peanut introductions were screened without identifying resistance. In 1986, Baltensperger et al. (1986) reported resistance to *M. arenaria* Race 1 in *A. glabrata* (rhizoma peanut), which may provide a source of resistance should technologies develop that would allow interspecific crossing with domesticated peanut genotypes. However, this may be unnecessary because *M. arenaria* resistance in wild *Arachis* spp. germplasm lines has been reported (Nelson et al., 1989; Holbrook & Noe, 1990; Shew et al., 1993). Nelson et al. (1989) reported resistance to *M. arenaria* in 21 *Arachis* species and two interspecific hybrids. They also detected resistance to *M. hapla* in two *Arachis* species and one of the interspecific hybrids. Castillo et al. (1973) also reported resistance in *M. hapla*. Holbrook and Noe (1992) systematically searched the *A. hypogaea* germplasm collection for useful sources of resistance.

They identified several lines with moderate levels of resistance, none of which had the high levels of resistance observed in the wild species. Recently, two *M. arenaria*-resistant genotypes were developed from wild peanut species (Simpson et al., 1993). Breeding lines with high levels of resistance to *M. arenaria* have been developed (Starr et al., 1995). The source of resistance is readily recoverable from advanced backcross generations.

Three groups in the USA are currently engaged in efforts to further develop resistance to *Meloidogyne* spp. in peanut. Georgia peanut scientists are focusing on a continued systematic search of the available germplasm collection of *A. hypogaea* for useful sources of resistance (Stephenson et al., 1995). Resistance identified to date suppresses population densities of *M. arenaria* by 40 to 60% (Noe et al., 1992). Yield potential of three lines with moderate resistance appears to be equal to that of the susceptible standards in the absence of nematode pressure and superior to the susceptible standard in the presence of damaging levels of *M. arenaria* (Holbrook et al., 1995). North Carolina and Texas scientists are focusing on the introgression of resistance from wild *Arachis* spp. into cultivated peanut. Because *A. hypogaea* is all allotetraploid and most wild species are diploids, introgression of nematode resistance from the wild species into cultivated peanut is a difficult process. The Texas effort has focused on the development of resistance to *M. arenaria* and has followed a diploid route to introgress resistance into *A. hypogaea* (Simpson, 1991). Three wild species were used to develop a nematode-resistant complex hybrid [TxAG-6, (Simpson et al., 1993)] that is cross-compatible with *A. hypogaea.* All three wild species *A. batizocoi, A. cardenasii,* and *A. diogoi* (= *A. chancoensis*) are resistant to *M. arenaria*, and *A. cardenasii* is resistant to *M. hapla* (Nelson et al., 1989). A backcrossing program has been used to introgress the resistance from TxAG-6 into high-yielding cultivars of runner, Spanish, and Virginia market-type peanut (Starr et al., 1995). Seven backcross generations have been completed, and field testing of advanced generation lines for yield potential and other important agronomic traits are projected for the immediate future. These advanced generation lines have retained a level of resistance to *M. arenaria* that provides greater than 90% inhibition of the reproduction of the nematode (Starr et al., 1995).

Three molecular markers, using randomly amplified polymorphic DNA (RAPD), have been used to identify a single resistance derived from *A. cardenasii* (Burow et al., 1996). This gene is the only resistance specifically identified and is the only resistance gene present in several lines examined to date (Choi et al., 1996).

With the recent confirmation of parasitism of peanut by populations of *M. javanica* from Egypt (Thomaszewski et al., 1994), India, and Texas (J.L. Starr, 1996, unpublished data) this nematode also is included in the breeding program. Resistance to *M. javanica* is present in early generation breeding lines TxAG-7 (Thomaszewski et al., 1994) and TP-233 (Abdel-Momen & Starr, 1996, unpublished data). Data from several lines derived from the fourth backcross generation, however, suggest that genes conditioning resistance to *M. javanica* differ from those that condition resistance to *M. arenaria* (J.L. Starr, 1996, unpublished data). Sakhuja and Sethi (1985a) also reported resistance in peanut to *M. javanica*.

In North Carolina, scientists have focused on development of resistance to *M. arenaria* in Virginia market-type peanut, using *A. cardenasii* as the source of resistance (Stalker et al., 1994). High levels of resistance have been introgressed into several breeding lines using a hexaploid route. This effort has reached the point were resistant lines are being screened for yield potential in field tests. Two RAPD markers linked to the resistance genes have been identified (Stalker et al., 1995). In lines with high levels of resistance, there is segregation in a 3:1 ratio indicative of a single dominant gene. Segregation patterns are more complex in lines with moderate levels of resistance (Stalker et al., 1995). The North Carolina program complements the Texas program as it has used a different approach of introgression and may have identified different resistance genes from those in the Texas program. Evidence exists that resistance to *M. arenaria* in *A. cardenasii* is conditioned by multiple, dominant, major genes (Starr & Simpson, 1991).

Several different sources of resistance to *Meloidogne* spp. have been identified and introgressed into numerous peanut genotypes. The availability of multiple genes for resistance to the major species of root-knot nematodes attacking peanut is likely to be an important asset in that it will allow development of gene deployment systems to enhance the durability of the resistance currently being developed. The identification of resistance to *M. javanica* also is likely to be of future importance. Even though *M. javanica* populations parasitic on peanut have been identified from only two locations in the USA, the species is widely distributed, and other populations not yet tested are likely to be parasitic on peanut. As resistance to *M. arenaria* is deployed it is probable that populations of *M. javanica* parasitic on peanut will become more widespread due to their increased competitiveness (Arens et al., 1981). Similar shifts in root-knot nematode species predominance due to introduction of specific resistance genes have been documented for tobacco (Fortnum et al., 1984). Such shifts are similar to those that have been documented for races of the soybean cyst nematode (*Heterodera glycines*) in response to use of specific resistance genes in soybean (Young, 1992).

Biological Control of Root-Knot Nematode on Peanut

Biological control is another area that offers exciting possibilities for the future management of nematode peanut diseases. Recently, *Pasteuria penetrans*, a bacterial endospore forming obligate parasite of nematodes (Fig. 23–7A,B), has been recognized as a suppressive agent for *M. arenaria* Race 1 in peanut fields in Florida and Georgia (Minton & Sayre, 1989; Dickson et al., 1994).

Suppressive Soils

Suppressiveness of soils to plant-parasitic nematodes is a concept that has been studied only during the past two decades. Suppressive soils are defined as those in which disease development is suppressed even though the pathogen is introduced in the presence of a susceptible host (Huber & Schneider, 1982). Only

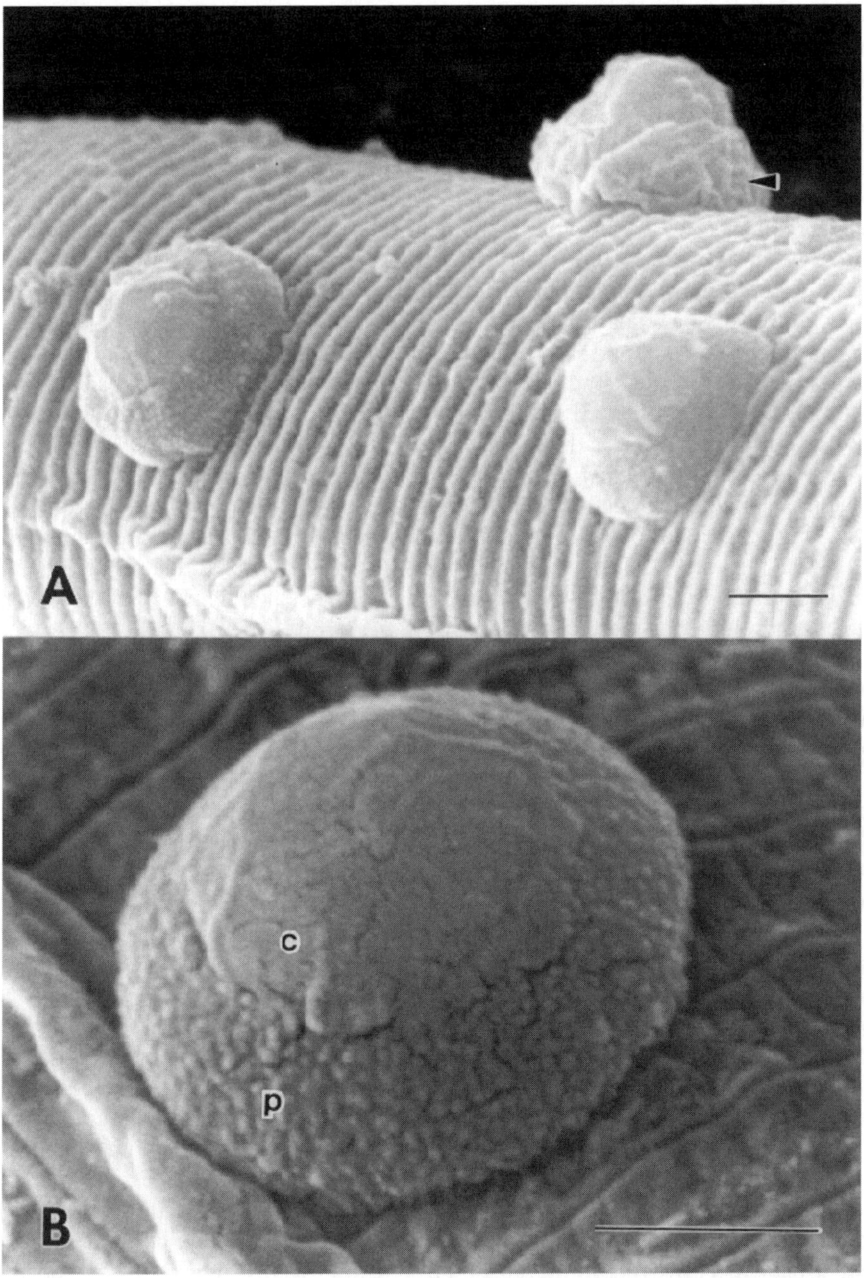

Fig. 23–7. Scanning electron micrographs of *Pasteuria penetrans* endospores attached to the cuticle of a second-stage juvenile of *Meloidogyne arenaria* Race 1. (*A*) Three endospores attached to the anterior region of a second-stage juvenile. One of the endospores retains its sporangium wall (arrow), (scale bar = 1 μm). (*B*) Close-up view of a single endospore. The sporangium wall and exosporium have sloughed off revealing the peripheral fibers (p) and central body (c), (scale bar = 1 μm).

a few documented nematode suppressive soil sites have been reported. Most work to date on suppressive soils has been with a fungal antagonist of nematodes (Jaffee & Zehr, 1982; Kerry, 1982; Stirling & Mankau, 1979; Gair et al., 1969). One of the classical studies was an investigation of cereal monoculture sites in Great Britain that showed a continuous decrease in the population levels of the cereal cyst nematode (*Heterodera avenae*) after a population peak in the 2nd yr. *Nematophthora gynophila* and *Verticillium chlamydosporium*, fungi parasitic on eggs and cysts, were the reported antagonists responsible for the nematode suppressive soil (Kerry, 1982). Stirling and Mankau (1979) reported *Dactylella oviparasitica* suppressing population densities of *Meloidogyne* spp. in California peach orchards on a susceptible 'Lovell' rootstock.

Only a few suppressive soil sites infested with *P. penetrans* and root-knot nematodes have been studied (Dickson et al., 1991; Bird & Brisbane, 1988; Minton & Sayre, 1989; Stirling & White, 1982; Mankau, 1980). A field near Williston, Florida, that had been used for peanut nematode research for several years because of the heavy infestation of *M. arenaria* Race 1 began to show signs that the nematode population density was declining. Gradually over a period of 4 to 5 yr peanut yield increased and root-knot nematode galling on pegs and pods decreased greatly (Dickson et al., 1991). Preliminary examination of the J2 extracted from soil samples taken across the field revealed that 32% of them were infested with one or more endospores of *P. penetrans*.

Peanut was planted in plots randomized across the field and followed by winter crops of rye, vetch (*Vicia villosa* Roth), and bare fallow. Two approaches were taken to determine whether biological agents in the soil caused the suppressiveness. Soil samples were collected from the vetch plots, at 0- to 15- and 15- to 30-cm depth, January 1988 and September 1989. The soil was subjected to four treatments: autoclaving, air drying for 2 wk, drenching with formalin (formaldehyde), or storage at 10°C for 2 wk. The suppressiveness of each soil was bioassayed in pots with tomato inoculated with 0 or 1000 J2 of *M. arenaria* Race 1.

Numbers of galls and egg masses were less on plants in air-dried and stored soil (untreated) collected in January from both depths, compared with autoclaved and formalin-treated soil (Fig. 23–8A). Formalin-treated soil had fewer galls and egg masses compared with autoclaved soil. Soil from the 0- to 15-cm depth was more suppressive than soil from the 15- to 30-cm depth. In September, egg mass counts from the different treatments were not different, but galling was less in soil collected at 0 to 15 cm deep when air dried or stored compared with autoclaved or formalin-treated soil (Fig. 23–8B). The population density of *P. penetrans* endospores in the air-dried soil was determined by adding 1000 J2 per sample. After 3 d, 82 to 100% of the J2 had endospores attached. These data strongly suggest that *P. penetrans* is a major factor in suppressing the population density of *M. arenaria*. However, soilborne fungi parasitic on *M. arenaria* also probably played a role in suppressing the population density of this nematode. In the field at harvest in 1989, 17% of *M. arenaria* females and 50% of the egg masses collected from peanut roots were infected by fungi. The two most common fungi isolated from the females and egg masses were *P. lilacinus* and a *Penicillium* sp., but *V. chlamydosporium*, *Neocosmospora* sp. and *Aspergillus* sp. also were isolated.

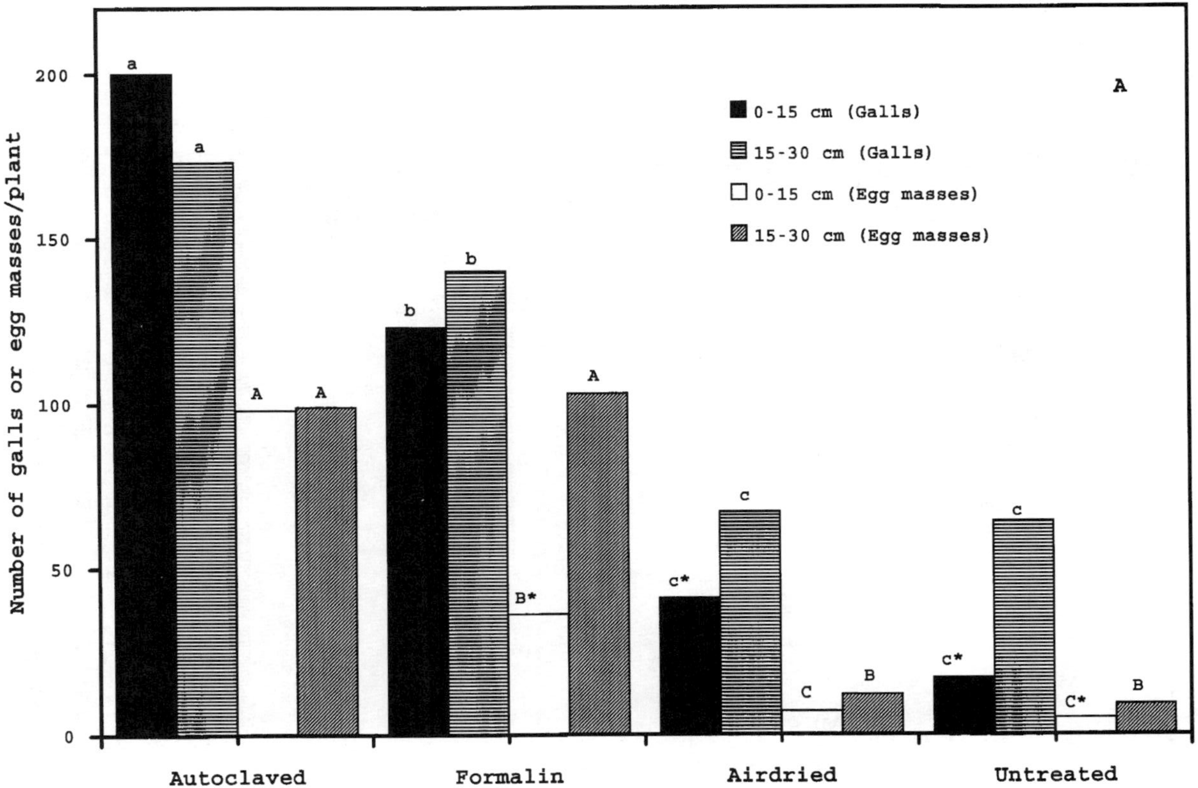

Fig. 23–8. Bar charts showing the number of galls and egg masses on tomato roots grown in autoclaved, formalin, air dried, and untreated soil 597 degree-days after inoculation with healthy second-stage juveniles of *Meloidogyne arenaria* Race 1. The soil was collected from 0 to 15 and 15 to 30 cm deep from a peanut field infested with *M. arenaria* and *Pasteuria penetrans*. (*A*) Winter bioassay, January. Bars with different letters differ significantly across all treatments according to Duncan's multiple-range test ($P \leq 0.05$). Bars with an asterisk differed significantly within a treatment ($P \leq 0.05$).

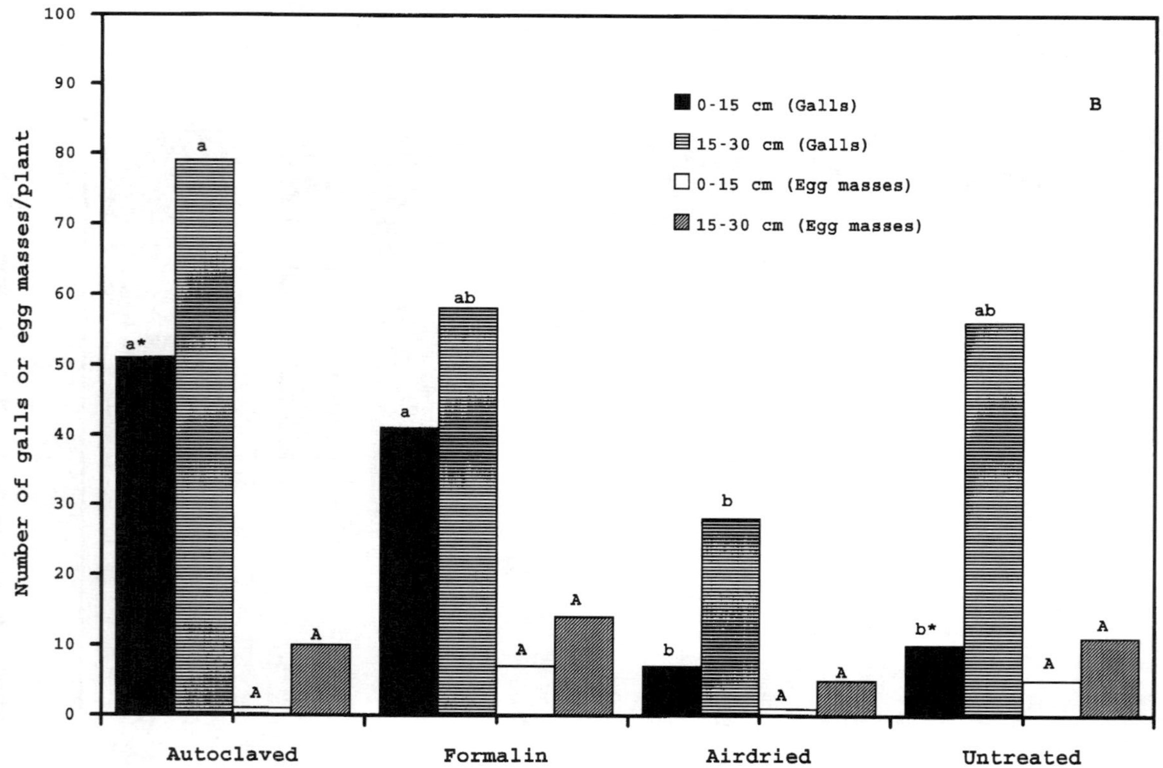

Fig. 23–8. Continued. (*B*) Harvest bioassay, September. Bars with different letters differ significantly across all treatments according to Duncan's multiple-range test ($P \leq 0.05$). Bars with an asterisk differed significantly within a treatment ($P \leq 0.05$).

Development of *Pasteuria penetrans* in Field Soil

Oostendorp et al. (1990, 1991) demonstrated that *P. penetrans* suppressive soil to the peanut root-knot nematode can be developed, which answers one of the most important questions about the potential of *P. penetrans* for biological control, i.e., its efficacy when introduced into nematode infested sites. The organism has been evaluated in only a few tests, which have been mostly of short duration. Dube and Smart (1987) reported control of *M. incognita* on tobacco and pepper with an isolate of *P. penetrans* in greenhouse tests over a 60-d period. This isolate also suppressed *M. incognita* and increased yields of soybean and winter vetch in field microplots over one season. Increased crop yields were reported by Brown et al. (1985) in a 2-yr study in microplots infested with *M. incognita* following application of *P. penetrans*. In at least two other cases where this parasite has been evaluated in short duration tests, control of root-knot nematodes has been reported (Mankau, 1975; Stirling, 1984).

We observed one isolate of *P. penetrans* specific for *M. arenaria* Race 1 in field microplots over a 7-yr period. This afforded us the opportunity to study the population changes of the organism over time and to determine the effect of different winter cropping systems on the parasite. Cropping sequences included Florunner peanut and winter cover crops of rye, hairy vetch or wheat, and bare fallow. *Meloidogyne arenaria* was inoculated into microplots, with and without *P. penetrans*. During the first 2 yr the nematode population densities were high (10 000 J2/100 cm^3 soil) in plots with either *M. arenaria* alone (Fig. 23–9A) or *M. arenaria* plus *P. penetrans* (Fig. 23–9B) (Oostendorp et al., 1991) and peanut yield was greatly suppressed in plots containing nematodes compared with the nematode-free control.

After three growing seasons the peanut yield was significantly higher in *P. penetrans*- infested plots than in plots with *M. arenaria* alone. Root knot on peanut was controlled by *P. penetrans*, although it was not until the third season of peanut before the bacterium had a significant effect on the nematode. Significant yield increases were again achieved in the fourth to sixth peanut crops grown in *P. penetrans*-infested plots (Weibelzahl-Fulton, 1996, unpublished data). The yield in *P. penetrans*-infested plots did not differ from the yield in nematode-free plots. This is definitive evidence that *P. penetrans* can provide control of *M. arenaria*. The rate of endospore attachment to J2 in soil increased continuously during the first 3 yr (Fig. 23–10). At that time the population density of J2 dropped (Fig. 23–9B), with densities so low at peanut harvest that none were recovered with endospores attached (Fig. 23–10), except in the fallowed plots. No root-knot nematode galling was observed on the peanut pods and pegs at harvest in plots with *P. penetrans* over the next 2 yr. Roots also were remarkably free of galling in these plots in the 4th yr and were totally free of visible galling the 5th yr. Obviously, this raises an important question about the possible demise of endospores of *P. penetrans*. Once root-knot nematode J2 population densities are reduced, *P. penetrans* population densities also may diminish due to a decrease in available nematode hosts. However, because of the apparent persistence of endospores, the disappearance or extinction of *P. penetrans* may be delayed.

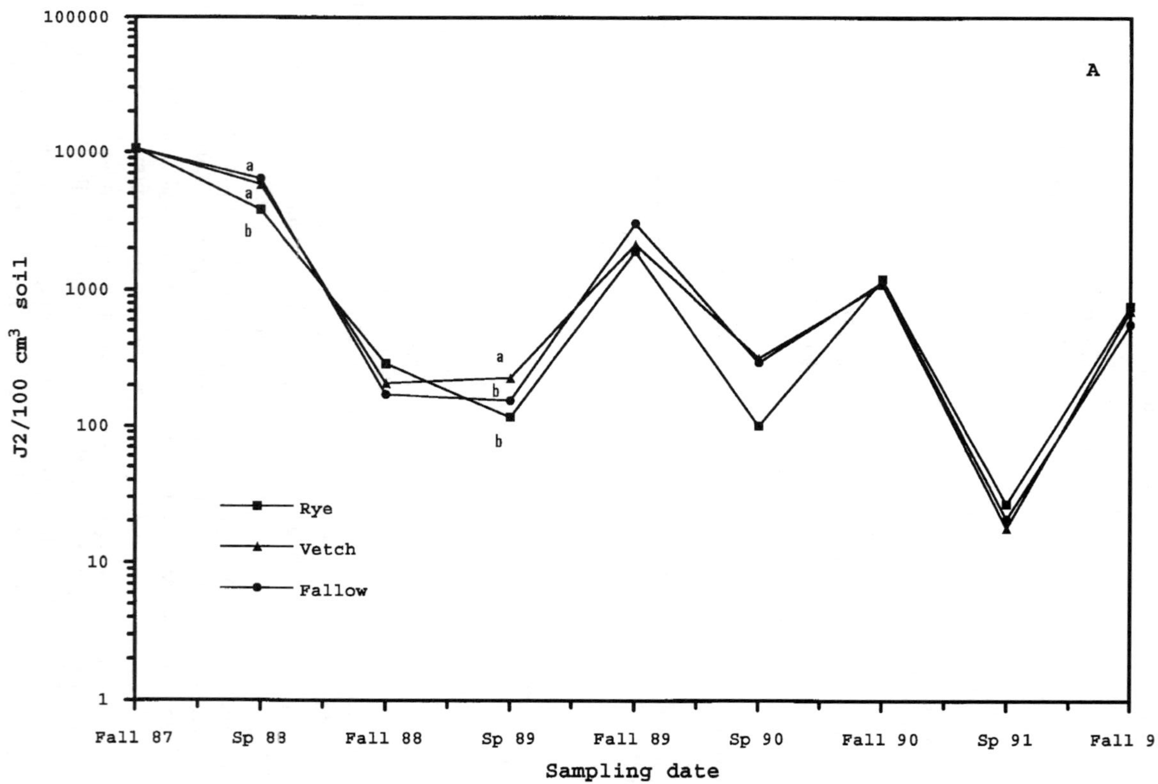

Fig. 23–9. Number of second-stage juveniles (J2) of *Meloidogyne arenaria* Race 1 in soil in microplots planted with peanut in summer and rye, vetch, and bare fallow in winter. (*A*) Without *Pasteuria penetrans*. Means of the cropping sequences at individual sampling dates followed by the same letter are not significantly different according to Duncan's multiple-range test ($P \leq 0.05$), (Sp = spring).

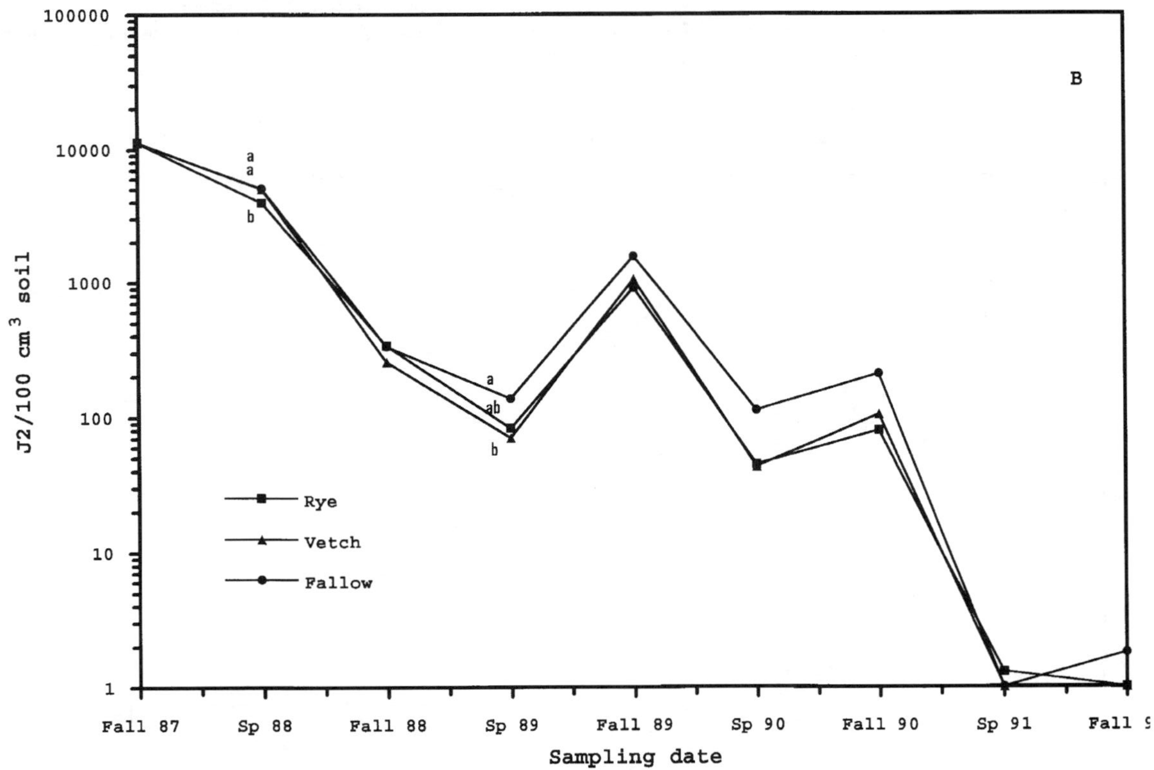

Fig. 23–9. Continued. (*B*) With *P. penetrans*. Means of the cropping sequences at individual sampling dates followed by the same letter are not significantly different according to Duncan's multiple-range test ($P \leq 0.05$), (Sp = spring).

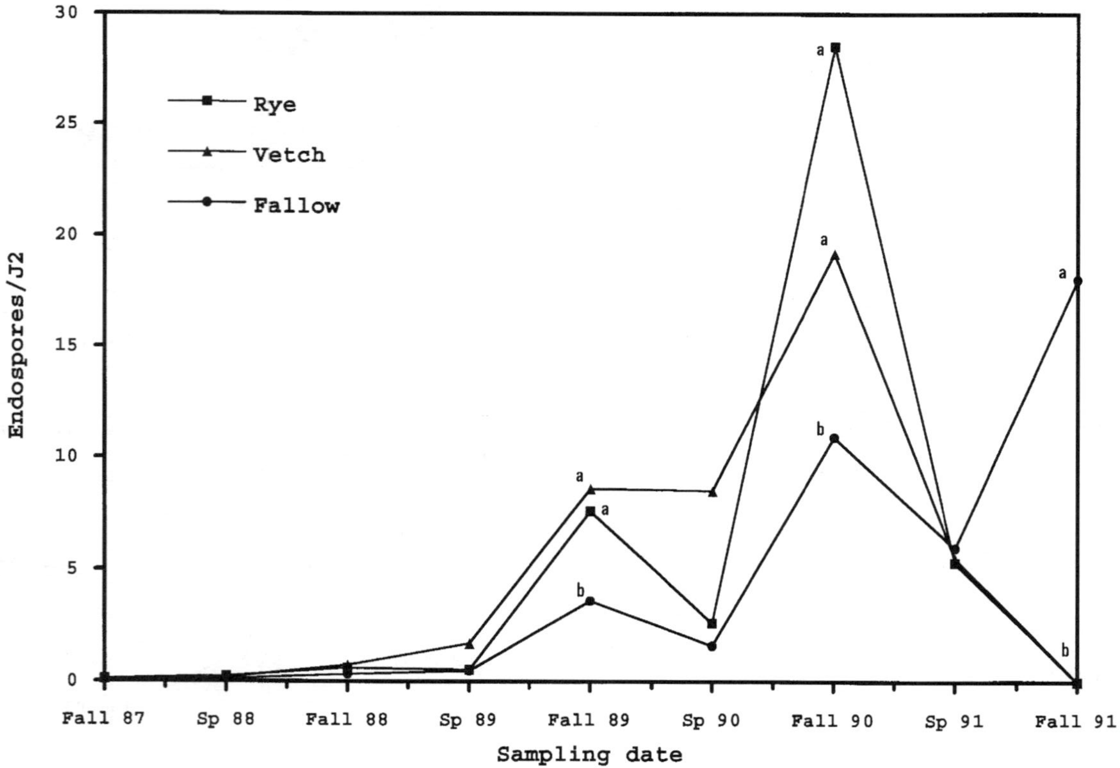

Fig. 23–10. Number of endospores of *Pasteuria penetrans* attached to second-stage juveniles (J2) of *Meloidogyne arenaria* Race 1 in microplots planted with peanut in summer and with rye, vetch, and bare fallow in winter. Means of the cropping sequences at individual sampling dates followed by the same letter are not significantly differed according to Duncan's multiple-range test ($P \leq 0.05$), (Sp = spring).

Fig. 23–11. Biological control of peanut root-knot nematode, *Meloidogyne arenaria* Race 1, using *Pasteuria penetrans*. Endospores were produced in greenhouse pot culture and inoculated in microplots that were preinoculated with *M. arenaria*. Photographs were taken in the second year following inoculation. (*A*) Control plants had few pods and those that formed were severely galled by *M. arenaria*. (*B*) Pods from microplots treated with 100 000 endospore per gram of soil. Pods had few galls.

The number of endospores attached per J2 was affected by the cropping sequence. Endospore attachment was greater on J2 in rye and vetch plots than in fallowed plots (Fig. 23–10). Endospore attachment increased during the 1989 and 1990 winter seasons, particularly in vetch and rye plots. Peanut, which is badly diseased by root-knot nematode when grown in deep (>45 cm) sandy soil (Dickson & Hewlett, 1989), is an extreme test situation for determining the efficacy of *P. penetrans*. Therefore, we were not surprised that it took 2 yr before the impact of *P. penetrans* became apparent on the suppression of root-knot nematodes. Because of technical difficulties in growing *P. penetrans*, only relatively low levels of endospores could be added initially to the microplots.

The winter cover crops affected the population levels of *M. arenaria* (Fig. 23–9A,B). Lower numbers of J2 were recovered from rye than from vetch plots, but in the autumn after peanut this influence was masked. The population density of nematodes dropped sharply after the initial high density of 10 000/100 cm^3 was reached. This reduction was attributed to the massive root damage caused by *M. arenaria*, which destroyed feeding sites. The J2 densities rebounded in the 3rd yr following peanut, and continued to rebound in the nematode-infested plots during each peanut crop over the next 3 yr.

Nematicide Dosages of *Pasteuria penetrans*

When endospores were directly and uniformly applied into nematode-infested soil at relatively high levels, *P. penetrans* suppressed root- and pod-gall development and improved peanut production markedly in a 2-yr field microplot experiment (Chen et al., 1996). Root-gall indices and pod galling were significantly suppressed by 60 to 81% and 90 to 95% for 100 000 endospores per gram of soil, respectively. When 10 000 endospores per gram was applied, root gall indices and pod galling were significantly restricted by 20 to 61% and 65 to 82%, respectively. Peanut growth was improved (Fig. 23–11). Pod yields were not significantly different in the 1st yr but significantly increased by 94% for 100 000 and by 57% for 10 000 endospores per gram soil, respectively, in the 2nd yr. Procedures for identifying naturally suppressive soils need to be developed and used as an alternative sustainable agricultural system for peanut producers.

REFERENCES

Anonymous. 1987. Bibliography of estimated crop losses in the United States due to plant-parasitic nematodes. Ann. Appl. Nematol. (Suppl. J. Nematol.) 1:6–12.

Board of Agriculture, National Research Council. 1989. Alternative agriculture. Natl. Acad. Press, Washington, DC.

Anonymous. 1993. Oil seed situation and outlook. USDA Crop Stat., Agric. Stat., Washington, DC.

Arens, M.L., J.R. Rich, and D.W. Dickson. 1981. Comparative studies on root invasion, root galling, and fecundity of three *Meloidogyne* spp. on a susceptible tobacco cultivar. J. Nematol. 13:201–205.

Baltensperger, D.D., G.M. Prine, and R.A. Dunn. 1886. Root-knot nematode resistance in *Arachis glabrata*. Peanut Sci. 13:78–80.

Barker, K.R. 1982. *Criconemella* in the southeastern United States. p. 150–156 *In* R.D. Riggs (ed.) Nematology in the southern region of the United States. Univ. Arkansas, Fayetteville Agric. Exp. Stn. Bull. 276.

Barker, K.R., D.P. Schmitt, and V.P. Campos. 1982. Response of peanut, corn, tobacco, and soybean to *Criconemella ornata*. J. Nematol. 14:576–581.

Basson, S., D.G.M.A. De Waele, and A.J. Meyer. 1991. Population dynamics of *Ditylenchus destructor* on peanut. J. Nematol. 23:485–490.

Basson, S., D. De Waele, and A.J. Meyer. 1993. Survival of *Ditylenchus destructor* in soil, hulls and seeds of groundnuts. Fund. Appl. Nematol. 16:79–85.

Baujard, P., and B. Martiny. 1995a. Ecology and pathogenicity of the Hoplolaimidae (Nemata) from the sahelian zone of West Africa. 1. Field studies on *Scutellonema cavenessi* Sher, 1964. Fund. Appl. Nematol. 18:261–269.

Baujard, P., and B. Martiny. 1995b. Ecology and pathogenicity of the Hoplolaimidae (Nemata) from the sahelian zone of West Africa. 2. Laboratory studies on *Scutellonema cavenessi* Sher, 1964. Fund. Appl. Nematol. 18:335–345.

Baujard, P., and B. Martiny. 1995c. Ecology and pathogenicity of the Hoplolaimidae (Nemata) from the sahelian zone of West Africa. 3. *Scutellonema clathricaudatum* Whitehead, 1959. Fund. Appl. Nematol. 18:347–353.

Baujard, P., B. Martiny, and A. Traore. 1993. Ecology and pathogenicity of the nematode *Paralongidorus bullatus* (Nemata:Longidoridae) in semi-arid regions of West Africa. Nematropica 23:149–157.

Bird, A.F., and P.G. Brisbane. 1988. The influence of *Pasteuria penetrans* in field soils on the reproduction of root-knot nematodes. Rev. Nématol. 11: 75–81.

Board of Agriculture, National Research Council. 1989. Alternative agriculture. Natl. Acad. Press, Washington, DC.

Bolton, C. 1989. Incidence of *Ditylenchus destructor* (Nematoda) in groundnut seed in South Africa. Phytophylactica 21:111.

Boote, K.J. 1982. Growth stages of peanut (*Arachis hypogaea* L.). Peanut Sci. 9:35–39

Bos, W.S. 1977. *Aphelenchoides arachidis* n. sp. (Nematoda: Aphelenchoidea), an endoparasite of the testa of groundnuts in Nigeria. Z. Pflanzen. Pflanzen. 84:95–99.

Boswell, T.E., 1968. Pathogenicity of *Pratylenchus brachyurus* to Spanish peanut. Ph. D. diss. Texas A&M Univ., College Station. (Diss. Abstr. 68-35708).

Brenneman, T.B., D.R. Sumner, R.E. Baird, G.W. Burton, and N.A. Minton. 1995. Suppression of foliar and soilborne peanut diseases in bahiagrass rotation. Phytopathology 85:948–952.

Bridge, J., W.S. Bos, L.J. Page, and D. McDonald. 1977. The biology and possible importance of *Aphelenchoides arachidis*, a seed-borne endoparasitic nematode of groundnuts from northern Nigeria. Nematologica 23:253–259.

Brown, S.M., J.L. Kepner, and G.C. Smart, Jr. 1985. Increased crop yields following application of *Bacillus penetrans* to field plots infested with *Meloidogyne incognita*. Soil Biol. Biochem. 17: 483–486.

Burow, M.D., C.E. Simpson, A.H. Paterson, and J.L. Starr. 1996. Identification of RAPD markers in peanut (*Arachis hypogaea*) associated with root-knot nematode (*Meloidogyne arenaria*) resistance, derived from *A. cardenasii*. Molec. Breed. 2:369–379.

Burton, G.W., and A.W. Johnson. 1987. Coastal bermudagrass rotations for control of root-knot nematodes. J. Nematol. 19:138–140.

Castillo, M.B., L.S. Morrison, C.C. Russell, and D.J. Banks. 1973. Resistance to *Meloidogyne hapla* in peanut. J. Nematol. 5:281–285.

Chen, Z.X., D.W. Dickson, R. McSorley, D.J. Mitchell, and T.E. Hewlett. 1996. Suppression of *Meloidogyne arenaria* race 1 by soil application of endospores of *Pasteuria penetrans*. J. Nematol. 28:159–168.

Cheng, Y.H., and C.C. Tu. 1980. Identification of *Meloidogyne arenaria* on peanut on Taiwan. J. Agric. Res. China 29:47–53.

Chitwood, B.G. 1949. Root-knot nematodes. I. A revision of the genus *Meloidogyne* Goeldi 1887. Proc. Helminthol. Soc. Wash. 16:90–104.

Choi, Y.E. 1978. A study on the root-knot nematodes affecting economic crops in Korea. Kor. J. Plant Prot. 17:89–98.

Choi, K., C.E. Simpson, M.D. Burow, A.H. Paterson, and J.L. Starr. 1996. Genetics of resistance to root-knot nematodes in peanut. Proc. Am. Peanut Res. Educ. Soc. 28:34.

Colbran, R.C. 1958. Studies of plant and soil nematodes. 2. Queensland host records of root-knot nematodes (*Meloidogyne* spp.). Queensl. J. Agric. Sci. 15:101–136.

Cooper, W.E. 1950. Root knot of peanuts. Phytopathology 40:786.

Cooper, W.E., and J.N. Sasser. 1962. Cynem—an effective control for sting nematodes on peanuts. Plant Dis. Rep. 46:640–642.

Cooper, W.E., J.C. Wells, J.N. Sasser, and T.G. Bowery. 1959. The efficacy of preplant and postplant applications of 1,2-dibromo-3-chloropropane on control of the sting nematode, *Belonolaimus longicaudatus*. Plant Dis. Rep. 43:903–908.

Culbreath, A.K., M.K. Beute, B.B. Shew, and K.R. Barker. 1992. Effects of *Meloidogyne hapla* and *M. arenaria* on black rot severity in new *Cylindrocladium*-resistant peanut genotypes. Plant Dis. 76:352–357.

De Waele, D., B.L. Jones, C. Bolton, and E. Van Den Berg. 1989. *Ditylenchus destructor* in hulls and seeds of peanut. J. Nematol. 21:10–15.

Dickson, D.W., D.J. Mitchell, T.E. Hewlett, M. Oostendorp, and M.E. Kannwischer-Mithcell. 1991. Nematode-suppressive soil from a peanut field. J. Nematol. 23: 526.

Dickson, D.W., and T.E. Hewlett. 1988a. Efficacy of fumigant and nonfumigant nematicides for control of *Meloidogyne arenaria* on peanut. Ann. Appl. Nematol. (Suppl. J. Nematol.) 2:95–101.

Dickson, D.W., and T.E. Hewlett. 1988b. Horizontal and vertical migration of *Meloidogyne arenaria* in soil columns in the field. Nematropica 18:4–5.

Dickson, D.W., and T.E. Hewlett. 1989. Effects of bahiagrass and nematicides on *Meloidogyne arenaria* on peanut. J. Nematol. 21 (Suppl.):671–676.

Dickson, D.W., and H.A. Melouk. 1995. Management of nematode pests. p. 83–86. *In* H.A. Melouk and F.M. Shokes (ed.) Peanut health management. APS Press, St. Paul, MN.

Dickson, D.W., M. Oostendorp, R.M. Giblin-Davis, and D.J. Mitchell. 1994. Control of plant-parasitic nematodes by biological antagonists. p. 575–601. *In* D. Rosen et al. (ed.) Pest management in the subtropics. Biological control—a Florida perspective. Intercept, Andover, England.

Dickson, D.W., and R.E. Waites. 1980. Benefits of additional applications of nematicides on peanut at-pegging for nematode control. Nematropica 10:65.

Diomande, M., and M.K. Beute. 1981. Relation of *Meloidogyne hapla* and *Macroposthonia ornata* populations to Cylindrocladium black rot in peanuts. Plant Dis. 65:339–342.

Dube, B., and G.C. Smart, Jr. 1987. Biological control of *Meloidogyne incognita* by *Paecilomyces lilacinus* and *Pasteuria penetrans*. J. Nematol. 19:222–227.

Dunn, R.A., and D.W. Dickson. 1995. Peanut nematode management. Doc. RF-NG016. Florida Coop. Ext. Ser., Univ. Florida, Gainesville.

Fortnum, B.A., J.P. Krausz, and N.G. Conrad. 1984. Increasing incidence of *Meloidogyne arenaria* on flue-cured tobacco in South Carolina. Plant Dis. 68:244–245.

Gair, R., P.L. Mathias, and P.N. Harvey 1969. Studies of cereal nematode populations and cereal yields under continuous or intensive culture. Ann. Appl. Biol. 63: 503–512.

Garcia, R., and D.J. Mitchell. 1975a. Synergistic interactions of *Pythium myriotylum* with *Fusarium solani* and *Meloidogyne arenaria* in pod rot of peanut. Phytopathology 65:832–833.

Garcia, R., and D.J. Mitchell. 1975b. Interactions of *Pythium myriotylum* with *Fusarium solani, Rhizoctonia solani*, and *Meloidogyne arenaria* in pre-emergence damping-off of peanut. Plant Dis. Rep. 59:665–669.

Germani, G. 1970. *Aphasmatylenchus straturatus* sp. n. (Nematoda: Hoplolaimidae from West Africa. Proc. Helminth. Soc. Wash. 57:48–51.

Germani, G. 1981. Pathogenicity of the nematode *Scutellonema cavenessi* on peanut and soybean. Rev. Nématol. 4:203–208.

Germani, G., P. Baujard, and M. Luc. 1985. Control of phytoparasitic nematodes in the Bassin Arachidier of Senegal. Orstom, Dakar.

Germani, G., and M. Dhéry. 1973. Observations et expérimentations concernant le role des nématodes dans deux affections de l'arachide en Haute Volta: la "chlorose" et le "clump". Oléagineux 28:235–242.

Germani, G., and M. Luc. 1982a. Etudes sur la "chlorose voltaïque" des légumineuses due au nématode *Aphasmatylenchus straturatus* Germani. I. Rev. Nématologie 5:139–146.

Germani, G., and M. Luc. 1982b. Etudes sur la "chlorose voltaïque" des légumineuses due au nématode *Aphasmatylenchus straturatus* Germani. II. Rev. Nématol. 5:195–199.

Good, J.M., L. W. Boyle, and R.O. Hammons. 1958. Studies on Pratylenchus *brachyurus* on peanuts. Phytopathology 48:530–535.

Good, J.M., and J.R. Stansell. 1965. Effect of irrigation, soil fumigation, and date of harvest on *Pratylenchus brachyurus* (Godfrey). Nematologia 11:38–39.

Hammons, R.O. 1982. Origin and early history of the peanut. p. 1–20. *In* H.E. Pattee and C.T. Young (ed.) Peanut science and technology. Am. Peanut Res. Educ. Soc., Yoakum, TX.

Hirunsalle, A., K.R. Barker, and M.K. Beute. 1995a. Infection, reproduction potential, and root galling by root-knot nematode species and concomitant populations on peanut and tobacco. J. Nematol. 27:172–177.

Hirunsalle, A., K.R. Barker, and M.K. Beute. 1995b. Effects of peanut-tobacco rotations on population dynamics of *Meloidogyne arenaria* in mixed race populations. J. Nematol. 27:178–188.

Hirunsalle, A., K.R. Barker, and M.K. Beute. 1995c. Effects of peanut genotypes on *Meloidogyne* species interactions. J. Nematol. 27:189–199.

Holbrook, C.C., D.A. Knauft, and D.W. Dickson. 1983. A technique for screening peanut for resistance to *Meloidogyne arenaria*. Plant Dis. 67:957–958.

Holbrook, C.C., and J.P. Noe. 1990. Resistance to *Meloidogyne arenaria* in *Arachis* spp. and the implications on development of resistant peanut cultivars. Peanut Sci. 17:35–38.

Holbrook, C.C., and J.P. Noe. 1992. Resistance to peanut root-knot nematode (*Meloidogyne arenaria*) in *Arachis hypogaea*. Peanut Sci. 19:35–37.

Holbrook, C.C., J.P. Noe, D.W. Gorbet, and M.G. Stephenson. 1995. Evaluation of peanut breeding lines with resistance to the peanut root-knot nematode. Proc. Am. Peanut Res. Educ. Soc. 27:24.

Huber, D.M., and R.W. Schneider. 1982. The description and occurrence of suppressive soils. p. 1–7. *In* R.W. Schneider (ed.) Suppressive soils and plant disease. APS Press, St. Paul, MN.

Hussey, R.S. 1985. Host-parasite relationship and associated physiological changes. p. 143–153. *In* J.N. Sasser and C.C. Carter (ed.) An advanced treatise on *Meloidogyne*. Vol. 1. Biology and control. North Carolina State Univ. Graphics, Raleigh.

Hussey, R.S., C.W. Mims, and S.W. Westcott, III. 1992. Ultrastructure of root cortical cells parasitized by the ring nematode *Criconemella xenoplax*. Protoplasma 167:55–65.

Ibrahim, I.K.A., and M.A. El-Saedy. 1976. Development and pathogenesis of *Meloidogyne javanica* in peanut roots. Nematol. Medit. 4:231–234.

Jaffee, B.A., and E.I. Zehr. 1982. Parasitism of the nematode *Criconemella xenoplax* by the fungus *Hirsutella rhossiliensis*. Phytopathology 72:1378 –1381.

Jenkins, W.R., and D.P. Taylor. 1967. Plant nematology. Reinhold Publ., New York.

Jones, B.L., and D. De Waele. 1988. First report of *Ditylenchus destructor* in pods and seeds of peanuts. Plant Dis. 72:453.

Jones, B.L., and D. De Waele. 1990. Histopathology of *Ditylenchus destructor* on peanut. J. Nematol. 22:268–272.

Kerry, B.R. 1982. The decline of *Heterodera avenae* populations. Eur. Plant Prot. Org. Bull. 12:491–496.

Kinloch, R.A., and D.W. Dickson. 1991. Comparison of winter and spring soil fumigation with 1,3-D for the management of *Meloidogyne arenaria* on peanut. J. Nematol. 23 (Suppl):682–685.

Koenning, S.R., and K.R. Barker. 1992. Relative damage functions and reproduction potentials of *Meloidogyne arenaria* and *M. hapla* on peanut. J. Nematol. 24:187–192.

Lordello, A.I.L., and M. Gerin. 1981. Nematóide javanes parasitando raízes e nódules de amendoim. Rev. Agric. Pirac. 56:278.

Machmer, J.H. 1951. Root knot of peanut. 1. Distribution. Plant Dis. Rep. 35:364–366.

Machmer, J.H. 1953. *Criconemoides* sp., a ring nematode associated with peanut "yellows." Plant Dis. Rep. 37:156.

Mankau, R. 1975. *Bacillus penetrans* n. comb. causing a virulent disease of plant-parasitic nematodes. J. Invert. Pathol. 26:333–339.

Mankau, R. 1980. Biological control of *Meloidogyne* populations by *Bacillus penetrans* in West Africa. J. Nematol. 12:230.

Martin, G.C. 1958. Root-knot nematodes (*Meloidogyne* sp.) in the Federation of Rhodesia and Nyasaland. Nematologia 3:332–349.

Martinez, R.G. 1976. Vertical distribution and survival stages of *Meloidogyne arenaria* in Florida with peanut as a host crop. M.S. thesis. Univ. Florida, Gainesville.

McDonald, D., W.S. Bos, and M.H. Gumel. 1979. Effects of infestation of peanut (groundnut) seed by the testa nematode, *Aphelenchoides arachidis*, on seed infection by fungi and on seedling emergence. Plant Dis. Rep. 63:464–467.

McSorley, R., D.W. Dickson, E.M. Candanedo-Lay, T.E. Hewlett, and J.J. Frederick. 1992. Damage functions for *Meloidogyne arenaria* on peanut. J. Nematol. 24:193–198.

McSorley, R., and R.N. Gallaher. 1991. Nematode population changes and forage yields of six corn and sorghum cultivars. J. Nematol. 23 (Suppl.):673–677.

Melouk, H.A., and P.A. Backman. 1995. Management of soilborne fungal pathogens. p. 75–82. *In* H.A. Melouk and F.M. Shokes (ed.) Peanut health management. APS Press, St. Paul, MN.

Melouk, H.A., and F.M. Shokes (ed.) Peanut health management. APS Press, St Paul, MN.

Miller, L.I. 1972. Resistance of plant introductions of *Arachis hypogaea* to *Meloidogyne hapla*, *Meloidogyne arenaria*, and *Belonolaimus longicaudatus*. Virginia J. Sci. 23:101.

Minton, N.A., and D.K. Bell. 1969. *Criconemoides ornatus* parasitic on peanuts. J. Nematol. 1:349–351.

Minton, N.A., and R.O. Hammons. 1975. Evaluation of peanut for resistance to the peanut root-knot nematode, *Meloidogyne arenaria*. Plant Dis. Rep. 59:944–945.

Minton, N.A., J.F. McGill, and A.M. Golden. 1969. *Meloidogyne javanica* attacks peanuts in Georgia. Plant Dis. Rep. 53:668.

Minton, N.A., and R.M. Sayre. 1989. Suppressive influence of *Pasteuria penetrans* in Georgia soils on reproduction of *Meloidogyne arenaria*. J. Nematol. 21:574–575.

Minz, G. 1956. The root-knot nematode, *Meloidogyne* spp., in Israel. Plant Dis. Rep. 40:798–801.

Mitsui, Y., T. Yoshida, K. Okamoto, and R. Ishii. 1976. Relationship between nematode-trapping fungi and *Meloidogyne hapla* in peanut field. Jpn. J. Nematol. 6:47–55.

Neal, J.C. 1889. The root-knot disease of the peach, orange, and other plants in Florida, due to the work of the *Anguillula*. USDA Div. Entomol. Bull. 20. U.S. Gov. Print. Office, Washington, DC.

Nelson, S.C., C.E. Simpson, and J.L. Starr. 1989. Resistance to *Meloidogyne arenaria* in *Arachis* sp. germplasm. J. Nematol. 21 (Suppl.):654–660.

Noe, J.P., C.C. Holbrook, and N.A. Minton. 1992. Field evaluation of susceptibility to *Meloidogyne arenaria* in *Arachis hypogaea* plant introductions. J. Nematol. 24 (Suppl.):712–716.

Norton, A.J., V.G. Perry, F.G. Martin, and J. NeSmith. 1977. Effect of age of bahiagrass sod on succeeding peanut crops. Peanut Sci. 4:71–74.

O'Bannon, J.H., and A.T. Tomberlin. 1970. *Pratylenchus* sp. as citrus pathogens. Phytopathology 60:1540.

Oostendorp, M., D.W. Dickson, and D.J. Mitchell. 1990. Host range and ecology of isolates of *Pasteuria* sp. from the southeastern United States. J. Nematol. 22:525–531.

Oostendorp, M., D.W. Dickson, and D.J. Mitchell. 1991. Population development of *Pasteuria penetrans* on *Meloidogyne arenaria*. J. Nematol. 23:58–64.

Opperman, C.H., J.R. Rich, and R.A. Dunn. 1988. Reproduction of three root-knot nematodes on winter small grain crops. Plant Dis. 72:869–871.

Orion, D., and E. Cohn. 1975. Crop rotation for control of certain nematodes. Phytoparasitica 3:69.

Ou, L.-T., J.E. Thomas, and D.W. Dickson. 1993. Enhanced biodegradation of the nematicide fenamiphos in soil. p. 253–260. *In* D.M. Linn et al. (ed.) Sorption and degradation of pesticides and organic chemicals in soil. SSSA Special Publication Number 32. SSSA and ASA, Madison, WI.

Ou, L.-T., K.-Y. Chung, J.E. Thomas, T.A. Obreza, and D.W. Dickson. 1995. Degradation of 1,3-dichloropropene (1,3-D) in soils with different histories of field applications of 1,3-D. J. Nematol. 27:249–257.

Patel, D.J., B.A. Patel, J.C. Chavda, and H.V. Patel. 1988. Record of *Meloidogyne javanica* on groundnut in Gujarat, India. Int. Arach. Newsl. 3:16–17.

Payan, L.A., and D.W. Dickson. 1988. Host specificity of four populations of *Pratylenchus brachyurus*. Ann. App. Nematol. (Suppl. J. Nematol.) 2:140–143.

Perry, V.G., and H.L. Rhoades. 1982. The genus *Belonolaimus*. p. 144–149. *In* R.D. Riggs (ed.) Nematology in the Southern region of the United States. Univ. Arkansas, Fayettville, Agric. Exp. Stn. Bull. 276.

Potter, J.W. 1966. Aspects of the ecology and parasitic relationship to peanuts of several ectoparasitic nematodes in North Carolina. Ph. D. diss. North Carolina State Univ., Raleigh (Diss. Abstr. AA16705696).

Powers, T.O., and T.S. Harris. 1993. A polymerase chain reaction method for identification of five major *Meloidogyne* species. J. Nematol. 25:1–6.

Prasad, S.K., D.R. Dasgupta, and M.C. Mukhopadhyay. 1964. Nematodes associated with commercial crops in north India and host range of *Meloidogyne javanica*. Indian J. Entomol. 26:438–446.

Rammah, A., and H. Hirschmann. 1990. Morphological comparison of three host races of *Meloidogyne javanica*. J. Nematol. 22:56–68.

Reddy, D.D.R. 1984. A nematode disease of peanut caused by *Tylenchorhynchus brevilineatus*. Plant Dis. 68:526–529.

Rodríguez-Kábana, R., N. Kokalis-Burelle, D.G. Robertson, P.S. King, and L.W. Wells. 1994. Rotations with coastal bermudagrass, cotton, and bahiagrass for management of *Meloidogyne arenaria* and southern blight in peanut. J. Nematol. 26 (Suppl.):665–668.

Rodríguez-Kábana, R., D.G. Robertson, P.S. King, and L. Wells. 1991. American jointvetch and partridge pea for the management of *Meloidogyne arenaria* in peanut. Nematropica 21:97–103.

Rodríguez-Kábana, R., R.A. Shelby, P.S. King, and M.H. Pope. 1982a. Application time and effectiveness of four systemic nematicides against *Meloidogyne arenaria* on Florunner peanuts. Nematropica 12:85–96.

Rodríguez-Kábana, R., J.C. Williams, and R.A. Shelby. 1982b. Assessment of peanut yield losses caused by *Meloidogyne arenaria*. Nematropica 12:279–288.

Roman, J., and H. Hirschmann. 1969. Morphology and morphometrics of six species of *Pratylenchus*. J. Nematol. 1:363–386.

Sakhuja, P.K., and C.L. Sethi. 1985a. Screening of groundnut germplasm for resistance to root-knot nematode, *Meloidogyne javanica*. Indian J. Nematol. 15:129–130.

Sakhuja, P.K., and C.L. Sethi. 1985b. Frequency of occurrence of various plant-parasitic nematodes and root-rot fungi on groundnut in Punjab. Indian J. Nematol. 15:191–194.

Sasser, J.N. 1954. Identification and host-parasitic relationships of certain root-knot nematodes (*Meloidogyne* sp.). Agric. Exp. Stn. Tech. Bull. A-77. Maryland.

Sasser, J.N., and C.C. Carter. 1982. Root-knot nematodes (*Meloidogyne* sp.): Identification, morphological and physical variation, host range, ecology, and control. p. 21–32. *In* R.D. Riggs (ed.) Nematology in the southern region of the United States. Arkansas Agric. Exp. Stn., Fayetteville, Bull. 276.

Sasser, J.N., and W.E. Cooper. 1961. Influence of sting nematode control with O,O-diethyl O-2-pyrazinyl phosphorothioate on yield and quality of peanuts. Plant Dis. Rep. 45:173–175.

Sasser, J.N., W.E. Cooper, and T.G. Bowery. 1960. Recent developments in the control of sting nematode, *Belonolaimus longicaudatus*, on peanuts with 1,2-dibromo-3-chloropropane and EN 18133. Plant Dis. Rep. 44:733–737.

Sasser, J.N., and D.W. Freckman. 1987. A world perspective on nematology: The role of the Society. p. 7–14. *In* J.A. Veech and D.W. Dickson (ed.) Vistas on nematology. Soc. Nematol., Hyattsville, MD.

Sharma, S.B., and D. McDonald. 1990. Global status of nematode problems of groundnut, pigeonpea, chickpea, sorghum and pearl millet, and suggestions for future work. Crop Prot. 9:453–458.

Sharma, S.B., and M.R. Siddiqi. 1990. *Paralongidorus bullatus* n. sp. from groundnut soils in Niger and comments on *Xiphinema parasetariae* Luc. J. Nematol. 22:579–584.

Sharma, S.B., M.R. Siddiqi, N.V. Van, and N.X. Hong. 1994. Plant-parasitic nematodes associated with groundnut in North Vietnam. Afro-Asian J. Nematol. 4:185–189.

Sharma, S.B., D.H. Smith, and D. McDonald. 1995. Host races of *Meloidogyne javanica*, with preliminary evidence that the 'groundnut race' is widely distributed in India. Int. Arach. Newsl. 15:43–45.

Sharma, S.B., and P. Subrahmanyam. 1990. Plant parasitic nematodes associated with groundnut in Niger. Trop. Pest Manage. 36:71–71.

Sharma, S.B., F. Waliyar, and B.J. Ndunguru. 1991. Plant-parasitic nematodes associated with groundnut in parts of Benin, Burkina Faso, and Nigeria. Oléagineux 46:119–120.

Sharma, S.B., F. Waliyar, P. Subrahmanyam, and B.J. Ndunguru. 1992. Role of *Scutellonema clathricaudatum* in etiology of groundnut growth variability in Niger. Plant Soil 143:133–139.

Shew, B.B., H.T. Stalker, and M.K. Beute. 1993. Microplot evaluations of resistance to *Meloidogyne arenaria* in *Arachis hypogaea*, *A. cardenasii*, *A. chacoensis*, and genotypes derived from a cross of *A. hypogaea* x *A. cardenasii*. Proc. Am. Peanut Res. Educ. Soc. 25:62.

Shokes, F.M., and H.A. Melouk. 1995. Plant health management in peanut production. p. 1–6. *In* H.A. Melouk and F.M. Shokes (ed.) Peanut health management. APS Press, St. Paul, MN.

Simpson, C.E. 1991. Pathways for introgression of pest resistance into *Arachis hypogaea* L. Peanut Sci. 18:22–26.

Simpson, C.E., S.C. Nelson, J.L. Starr, K.E. Woodard, and O.D. Smith. 1993. Registration of TxAG-6 and TxAG-7 peanut germplasm lines. Crop Sci. 33:14–18.

Stalker, H.T., G.M. Garcia, B.B. Shew, M.K. Beute, T.G. Isleib, and G. Kochert. 1994. Introgression from *A. cardenasii* to *A. hypogaea*. Proc. Am. Peanut Res. Educ. Soc. 26:33.

Stalker, H.T., B.B. Shew, G.M. Garcia, M.K. Beute, K.R. Barker, C.C. Holbrook, J.P. Noe, and G.A. Kochert. 1995. *Meloidogyne arenaria* resistance in advanced-generation *Arachis hypogaea* x *A. cardenasii* hybrids.. Proc. Am. Peanut Res. Educ. Soc. 27:24.

Starr, J.L., M.-Y. Shim, T.A. Lee, Jr., and C.E. Simpson. 1996. Additive effects of *Meloidogyne arenaria* and *Sclerotium rolfsii* on peanut. J. Nematol. 28:99–106.

Starr, J.L., and C.E. Simpson. 1991. Segregation of resistance to *Meloidogyne arenaria* in progeny of interspecific hybrids. Proc. Am. Peanut Res. Educ. Soc. 23:23.

Starr, J.L., C.E. Simpson, and T.A. Lee, Jr. 1995. Resistance to *Meloidogyne arenaria* in advanced generation breeding lines of peanut. Peanut Sci. 22:59–61.

Steiner, G. 1945. Meadow nematodes as the cause of root destruction. Phytopathology 35:935–937.

Steiner, G. 1949. Plant nematodes the grower should know. Proc. Soil Sci. Soc. Florida, 1942. 4-B:72–117.

Stephenson, M.G., C.C. Holbrook, J.P. Noe, and W.F. Anderson. 1995. Evaluation of additional sources of resistance to the peanut root-knot nematode in the cultivated species of peanut. Proc. Am. Peanut Res. Educ. Soc. 27:25.

Stirling, G.R. 1984. Biological control of *Meloidogyne javanica* with *Bacillus penetrans*. Phytopathology 74:55–60.

Stirling, G.R., and R. Mankau. 1979. Mode of parasitism of *Meloidogyne* and other nematode eggs by *Dactylella oviparasitica*. J. Nematol. 11:282–288.

Stirling, G.R., and A.M. White. 1982. Distribution of a parasite of root-knot nematodes in south Australian vineyards. Plant Dis. 66:52–53.

Sturgeon, R.V. 1986. Peanut disease loss estimates for major peanut producing states in the United States in 1984 and 1985. Proc. Am. Peanut Res. Educ. Soc. 18:24–26.

Taylor, A.L., and J.N. Sasser. 1978. Biology, identification and control of root-knot nematodes (*Meloidogyne* species). North Carolina State Graphics, Raleigh.

Thames, W.H. 1982. The genus Pratylenchus. p. 108–126. *In* R.D. Riggs (ed.) Nematology in the Southern region of the United States. Agric. Exp. Stn., Univ. Arkansas, Fayetteville, Bull. 276.

Tomaszewski, E.K., M.A. M. Khalil, A.A. El-Deeb, T.O. Powers, and J.L. Starr. 1994. *Meloidogyne javanica* parasitic on peanut. J. Nematol. 26:436–441.

Triantaphyllou, A.C. 1982. Cytogenetics and sexuality of root-knot and cyst nematodes. p. 71–76. *In* R.D. Riggs (ed.) Nematology in the Southern region of the United States. Agric. Exp. Stn., Univ. Arkansas, Fayetteville, Bull. 276.

Triantaphyllou, A.C. 1985. Cytogenetics, cytotaxonomy and phylogeny of root-knot nematodes. p. 113–126. *In* J.N. Sasser and C.C. Carter (ed.) An advanced treatise on *Meloidogyne*. Vol. 1. North Carolina State Univ. Graphics, Raleigh.

Van der Linde, J.W. 1956. The *Meloidogyne* problem in South Africa. Nematologica 1:177–183.

Venter, C., D. De Waele, and A.J. Meyer. 1991. Reproductive and damage potential of *Ditylenchus destructor* on peanut. J. Nematol. 23:12–19.

Venter, C., D. De Waele, and A.J. Meyer. 1993. Reproductive and damage potential of *Ditylenchus destructor* on six peanut cultivars. J. Nematol. 25:59–62.

Venter, C., G. van Aswegen, A.J. Meyer, and D. De Waele. 1995. Histological studies of *Ditylenchus africanus* within peanut pods. J. Nematol. 27:284–291.

Wendt, K.R., A. Swart, T.C. Vrain, and J.M. Webster. 1995. *Ditylenchus africanus* sp. n. from South Africa; a morphological and molecular characterization. Fund. Appl. Nematol. 18:241–250.

Wheeler, T.A., and J.L. Starr. 1987. Incidence and economic importance of plant-parasitic nematodes on peanut in Texas. Peanut Sci. 14:94–96.

Wilson, C. 1948. Root-knot nematode on peanuts in Alabama. Plant Dis. Rep. 32:443.

Windham, G.L., and W.P. Williams. 1994. Reproduction of *Meloidogyne incognita* and *M. arenaria* on tropical corn hybrids. J. Nematol. 26 (Suppl):753–755.

Wynne, J.C., and T.A. Coffelt. 1982. Genetics of *Arachis hypogaea* L. p. 50–94. *In* H.E. Pattee and C.T. Young (ed.) Peanut science and technology. Am. Peanut Res. Educ. Soc., Yoakum, TX.

Young, L.D. 1992. Problems and strategies associated with long-term use of nematode resistant cultivars. J. Nematol. 24:228–233.

Zhang, Y. 1985. Occurrence and control of peanut root-knot disease in nonirrigated sloping fields of Zhangjiang District. Agric. Sci. Guamgdong Prov. 6:48–49.

24 Potato

BILL B. BRODIE

USDA-ARS
Ithaca, New York

The cultivated potato belongs to genus *Solanum*, which contains over 2000 species that are distributed practically all over the world except for the polar regions. In addition to the potato of commerce (*S. tuberosum* L.), the genus contains at least seven other cultivated species and about 200 wild tuber-bearing species (Hawkes, 1978). The wild tuber-bearing species are important to potato improvement programs because of their resistance to many pests and pathogens and are adapted to environmental extremes.

Evidence indicates that the cultivated potato was once confined to the high Andes of South America and the coastal strip of Chile. It was imported into Spain during the last half of the 16th century and from there spread throughout Europe and most other parts of the world. The exact date potato was introduced to North America is questionable. Some experts believe that the potato arrived in 1621 from England via Bermuda and was grown in the colony of Virginia (Hawkes, 1978). Others believe that it did not arrive until 1719 and was grown in New Hampshire by the Scottish-Irish (Plaisted, 1971). The first introductions to the USA of direct South American origin were in 1848 and 1851. The variety 'Garnet Chili' arose from these introductions and is in the pedigree of many present-day cultivars (Plaisted, 1971).

The subspecies *S. tuberosum* ssp. *tuberosum* was selected from introductions of its South American ancestor (*S. tuberosum* ssp. *andigena*) and subsequently adapted its photoperiodic requirements to long days. From these meager introductions of unadapted plants, potato has evolved from a botanical curiosity to use as livestock feed, then as a delicacy eaten only by royalty, to finally a staple in human diets. Potato is now one of the world's major food crops. Currently, potato production exceeds over 290 Tg annually and ranks in the top five in tonnage produced among the world's major food crops (Horton, 1987).

At least 68 species of plant-parasitic nematodes representing 24 genera are associated with potato (Jensen et al., 1979). The economic importance of many of these species to potato production is yet to be established, whereas others are known major pests. The potato cyst nematodes (*Globodera* spp.) are the most economically important nematode parasites of potato. Species of *Meloidogyne* (root-knot nematodes) can cause severe potato yield losses under suitable soil and environmental conditions. In addition to causing severe damage alone, root-lesion nematodes (*Pratylenchus* spp.) often occur with the fungus *Verticillium dahliae* where they interact causing a disease known as "potato early dying" (Riedel et al., 1985). The false root-knot nematode (*Nacobbus aberrans*) is a serious pest of potato in certain South American countries. The potato-rot nematode (*Ditylenchus destructor*) occurs in isolated areas where it causes severe losses in potato yields. Species of *Trichodorus* and *Paratrichodorus* transmit the tobacco rattle virus to potato causing a disease known as "corky ring spot." Other nematodes such as sting (*Belonolaimus* spp.) and burrowing nematodes (*Radopholus* spp.) occur less frequently in potato soils, but where they do occur they cause losses that are substantial to the potato grower.

IMPORTANT NEMATODE PARASITES

Cyst Nematodes

Potato cyst nematodes (PCN, *Globodera* spp.) are economically important pathogens of potato throughout the temperate regions of the world. Their importance is underscored by the quarantine or regulatory actions imposed against them in most countries where they occur. Actions taken range from banning of host crop production to an integrated management systems that limit their spread (Brodie & Mai, 1989).

Species and Pathotypes

For many years, both species of PCN were classified as a single species, *Heterodera rostochiensis*, with the common names of potato root eelworm or golden nematode. Potato cyst nematodes were later separated into two species, *H. rostochiensis* with golden females and *H. pallida* with white or cream-colored females (Stone, 1973). The subgenus *Globodera* was elevated to generic rank (Behrens, 1975) to which round cyst nematodes were assigned, including both species of PCN (Loof & Bakker, 1992). The current designation of PCN species is *G. rostochiensis* and *G. pallida*.

Each species of PCN is composed of pathotypes or races distinguished by their ability to multiply on *Solanum* clones with different genes for resistance. These clones were used to develop two different schemes to clarify the identification and classification of PCN population variants (Canto-Saenz & de Scurrah, 1977; Kort et al., 1977). Depending upon the classification scheme used, there are five pathotypes or four races of *G. rostochiensis* recognized and three pathotypes or six races of *G. pallida* (Table 24–1). Both of these classification schemes are still cited in the literature but have come under considerable scrutiny (Trudgill, 1985). Recent studies showed that virulence in *G. rostochiensis* to the gene H_1 is controlled by a single major recessive gene and that many populations of *G. rostochiensis* are heterogeneous for this virulence gene (Janssen et al., 1991). Trudgill (1985) suggests that only Ro1 (including Ro4) of *G. rostochiensis* and Pa1 of *G. pallida* can be regarded as distinct pathotypes and that other pathotypes are composed, within each species, of a continuous population of individuals with different virulence genes that should be considered virulence groups.

Table 24–1. Current schemes for identifying and classifying pathotypes or races of potato cyst nematodes.

	Designation of *Globodera rostochiensis* pathotypes					
	Nijboer & Parlevliet (1990)	Ro1		Ro3		Ro5
	Kort et al. (1977)	Ro1	Ro4	Ro2	Ro3	Ro5
Differential host	Canto & de Scurrah (1977)	R_1A	R_1B	R_2A	R_3A	–†
Solanum tuberosum ssp. *tuberosum*		+	+	+	+	+
S. tuberosum ssp. *andigena* (H_1)		–	–	+	+	+
S. kurtzianum KTT/60.21.19		–	+	–	+	+
S. vernei GLKS.58.1642.4		–	+	–	–	+
S. vernei $(VT^n)^2$62.33.3		–	–	–	–	+

Designation of *Globodera pallida* pathotypes†							
Kort et al. (1977)	Pa1	–‡	–‡	–‡	Pa2	Pa3	–‡
Canto & de Scurrah (1977)	P_1A	P_1B	P_2A	P_3A	P_4A	P_5A	P_6A§
Solanum tuberosum ssp. *tuberosum*	+	+	+	+	+	+	+
S. multidissectum (H_2)	–	–	+	+	+	+	+
S. kurtzianum KTT/60.21.19	+	+	–	+	+	+	+
S. vernei GLKS.58.1642.4	+	+	+	–	+	+	+
S. vernei $(VT^n)^2$62.33.3	–	+	–	–	–	+	+
CIP 280090.10¶					–	–	+

† EPPO, 1992 recognizes only Pa1, Trudgill, 1985 and Nijboer and Parlevliet, 1990 recognized virulence groups but not pathotypes.
‡ No comparable pathotype reported.
§ New race identified by Franco and Gonzalez.
¶ New differential clone added by Franco and Gonzalez.

Initially, only *G. rostochiensis* pathotype Ro1 was known to occur in the USA. In 1995, however, populations of *G. rostochiensis*, found in potato fields on Long Island and in western New York, reproduced on Ro1-resistant plants indicating that a second pathotype of *G. rostochiensis* exists in these fields (Brodie, 1995). Nijboer and Parlevliet (1990) suggest that three pathotypes of *G. rostochiensis* are recognizable and should be renamed Ro1 (old Ro1 and Ro4) Ro3 (old Ro2 and Ro3) and Ro5 (old Ro5) (Table 24–1). They were unable to reliably identify pathotypes of *G. pallida*. Recent advances in DNA technology to differentiate PCN populations should prove useful in resolving the pathotype controversy within PCN species and could eventually provide routine classification systems (Burrows, 1988; Folkertsma et al., 1994).

Symptoms

No distinct host symptoms are associated with low PCN population densities, but as population density increases symptoms appear. The first sign of PCN presence is poor growth of plants in small spots that enlarge with continuous growing of potato. Plants in these spots exhibit symptoms of water and/or mineral deficiency stress with a sickly, yellow appearance that wilt at midday and die prematurely. Roots of infected plants contain minute, white bodies, which are the immature females that have erupted through the root epidermis (Fig. 24–1). When the females mature they become dark brown cysts that dislodge from the

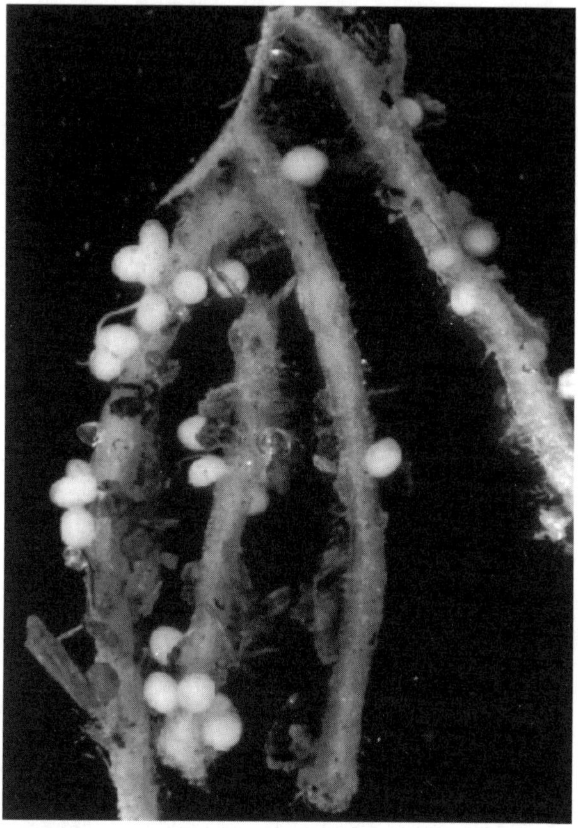

Fig. 24–1. Adult females (white) and cysts (dark) of *Globodera rostochiensis* on potato roots (courtesy W. F. Mai, Cornell Univ.).

roots when the plants are harvested. At extremely high nematode densities, tubers may become infected resulting in the appearance of cysts on their surface.

Biology

Potato cyst nematodes are sedentary endoparasites that exhibit marked sexual dimorphism (Thorne, 1961). The females enlarge to become spherical when mature, but males retain a worm-like shape. The female body dies, and its cuticle forms a protective cyst containing 200 to 500 eggs. At maturity, PCN's go into a state of arrested development and require a hatching stimulant before most of the eggs hatch and juveniles emerge from the cysts (Evans & Stone, 1977). This arrested development may be further specialized as a diapause (Hominick et al., 1985). Diapause can be circumvented by preventing the desiccation of cysts (Janssen et al., 1987). These features serve to synchronize the life cycle of the nematode with that of its host. In the absence of a host, the eggs inside cysts remain dormant with about one-third of them hatching each year, resulting in cysts that contain viable eggs for 20 yr or more (Jones, 1970).

In the presence of host plants, exudates from their roots stimulate 60 to 80% of the eggs to hatch releasing second-stage juveniles (Rawsthorne & Brodie, 1986a). When second-stage juveniles (J2) emerge from the eggs, they locate and enter host roots just behind the growing point. The J2 cut through cell walls, eventually coming to rest with their heads oriented toward the stele (Evans & Stone, 1977), where they feed on pericycle, cortex and endodermis cells. During feeding, they inject saliva into the cell causing its enlargement and cell wall breakdown, which forms a large transfer cell (Jones & Northcote, 1972). The nematodes become sedentary, feed from these cells and molt through the third- and fourth-juvenile stages. After the final molt, males emerge from the root in the wormlike state. Females increase in size and rupture through the root with only their head and neck attached to the root. Females release a substance that attracts many males that mate with them repeatedly (Green et al., 1970). Embryos develop inside the eggs as far as the second-stage juveniles and remain inside the female body until stimulated to hatch by potato root exudates.

Disease Complexes

Limited documentation of PCN involvement in disease complexes suggests that the severe damage caused by these nematodes masks the effects of other potato pathogens. Grainger and Clark (1963) found that *G. rostochiensis*, together with *Rhizoctonia solani*, caused much greater reduction in potato yield than did either organism alone. However, a later study indicated only a slight effect from the combined infection of these organisms on potato (Dunn & Hughes, 1967). A positive interaction was demonstrated between the fungus *Verticillium dahliae* and *G. rostochiensis* that resulted in early death of potato (Harrison, 1971). Both species of PCN appear equally capable of interacting with *V. dahliae* (Storey & Evans, 1987). Corbett and Hide (1971) suggested that female disruption of the endodermis and pericycle provided a pathway for *V. dahliae* invasion, but detailed histopathology studies showed that juveniles of *G. pallida* enhance the capacity of *V. dahliae* to overcome the natural defenses of the roots by opening invasion channels for the fungus (Storey & Evans, 1987). Infection by *V. dahliae* also influences the tolerance of some *G. rostochiensis* resistant cultivars making them intolerant of *G. rostochiensis* attack (Evans, 1983).

Economic Importance

Cyst nematodes are the most economically important nematodes parasitizing potatoes worldwide and are the subject of strict regulations and quarantines in most countries where they occur (Spears, 1968; EPPO, 1992). One or both species have been found in at least 58 countries in Europe, Asia, Africa, North America, Central America, South America, and Oceania (EPPO, 1992). Additionally, infestations of *G. rostochiensis* were found in Israel in 1954 and 1965 but were successfully eradicated (Evans & Brodie, 1980). The most recent discovery of PCN occurred in 1981 in Victoria, Australia (EPPO, 1992). Most countries or areas within countries where PCN do not occur have rigid regulations governing the importation of potato or other products that might carry these pathogens (Evans & Brodie, 1980). Because infested tubers are the chief means of their long-distance spread, PCN's are of great importance to countries that export seed tubers. Countries that import seed tubers have strict requirements on their entrance (Spears, 1968).

Few reliable estimates of yield losses caused by PCN's are available. Where damage due to widespread nematode infestation is limited by various management procedures, up to 9% of the potato yields are lost annually due to PCN's (Evans & Brodie, 1980). In the USA, *G. rostochiensis* has been under some type of state or federal regulations since its discovery in 1941. These regulations have restricted population densities of this pathogen in the USA below plant damaging levels. When PCN population densities are kept extremely low by strict quarantine regulations, no direct loss in potato yields occurs. Nevertheless, associated costs of surveillance and quarantine procedures are considerable (Brodie, 1984; Brodie & Mai, 1989). Where no control of PCN is practiced, potato yields are often less than the amount of tubers planted (Mai, 1977).

Several mathematic models/equations have been advanced to describe the relationship between PCN population density to yield loss of potato (Brodie et al., 1993). Because of variation in damage caused by PCN at different sites (Trudgill, 1991), differences in tolerance of cultivars, and environmental influences on potato yield, these models may not be entirely satisfactory. One model shows a linear relationship between yield of different potato varieties grown on different sites and PCN population densities (Brown &d Sykes, 1983). This model estimates losses in potato yield to average 2.75 t ha^{-1} for every 20 000 eggs kg^{-1} (20 eggs g^{-1}) of soil with a maximum loss of 22 t ha^{-1}.

Root-Lesion Nematodes

Root-lesion nematodes are found associated with poor growth of potatoes in most potato production areas. Both temperate and tropical species are extremely widespread and occur over a broad range of climatic conditions. These nematodes have broad host ranges and are known to parasitize nearly 400 plant species (Mai et al., 1977).

Species and Races

Although at least 15 species of root-lesion nematodes (*Pratylenchus* spp.) are reported to parasitize potato, those most frequently encountered in potato fields include *P. penetrans, P. crenatus, P. neglectus, P. alleni, P. thornei* and *P. scribneri* (Brodie et al., 1993). Considerable movement of *Pratylenchus* with plant material occurs throughout the world causing some species to be almost cosmopolitan (Loff, 1991).

The existence of races of *Pratylenchus* spp. was suggested as early as 1956, but little experimental evidence was presented to support this theory (Slootweg, 1956). Olthof (1968) later identified two races of *P. penetrans* based on host suitability and pathogenicity on tobacco (*Nicotiana tabacum* L.) and celery (*Apium graveolens* v. *dulce*). Morphological characteristics of the tail terminus of females was used to identify two pathotypes of *P. penetrans* (Townshend et al., 1978b). Host suitability of certain potato cultivars to *P. penetrans* is reported to differ with site suggesting the existence of different races (Dunn, 1973; Kotcon et al., 1987). Recent research confirmed the existence of isolates of *P. penetrans* that differ in their ability to reproduce on resistant potato clones (France & Brodie, 1995). Additionally, races of *P. penetrans* may differ in their ability to damage potato tubers (Olthof & Wolynetz, 1991).

Symptoms

Aboveground symptoms are expressed as patches of poor potato growth with occasional yellowing of the foliage (Brodie et al., 1993). Infection by *Pratylenchus* spp. can be more easily determined by examining potato roots. Infected portions of roots turn dark brown to reddish in color (Fig. 24–2) and may be enhanced by invading secondary organisms (Loof, 1991). The discolored areas are known as lesions from which the common name root-lesion nematodes was derived.

Some species of *Pratylenchus* including *P. scribneri, P. penetrans* and *P. neglectus* also infect potato tubers (Olthof & Wolynetz, 1991). Symptoms vary with species from a scabby appearance with sunken lesions (*P. scribneri*) to wart-

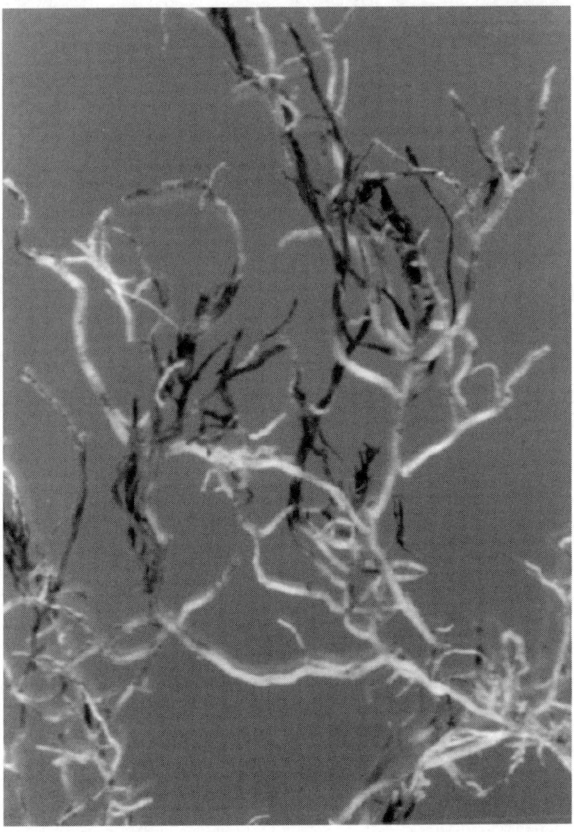

Fig. 24–2. Potato roots infected with *Pratylenchus penetrans* showing typical lesions (courtesy USDA- ARS).

Fig. 24–3. Healthy potato tubers (top) and tubers infected with *Pratylenchus penetrans* (bottom) showing typical protuberances (courtesy W. F. Mai, Cornell Univ.).

like protuberances (*P. penetrans*) varying from brown to black at harvest and turning purple in storage (Fig. 24–3). Lesions on the tubers are shallow, rarely penetrating deeper than 0.5 mm.

Biology

All species of *Pratylenchus* are migratory endoparasites. They usually reside in roots, rhizomes, or tubers but may leave plant tissues and live for some time in the soil. Between 20 to 50% of the *Pratylenchus* population has been reported to reside in the soil of potato fields (MacGuidwin, 1989). When in the soil, they may feed ectoparasitically from root hairs (Zunke, 1990).

All stages of *Pratylenchus* spp. from the second-stage juveniles onward infect roots, usually penetrating the roots at the zone of elongation (Zunke, 1991). During penetration, they puncture epidermal cells that then become attractive to other specimens of *Pratylenchus* spp. that enter through these punctured areas (Zunke, 1991). Once inside the root, they migrate through the root killing the cell through which they pass and feeding on cortical cells. Two types of feeding have been observed; brief feeding of 5 min or less, which causes little or no cell death, and extended feeding of several hours, which results in cell death (Zunke, 1991). Once roots are damaged, *Pratylenchus* spp. leave them in search of a more suitable food source.

Eggs are deposited in clusters inside the roots at a rate of 1 egg d^{-1} at 23°C (Zunke, 1991). Depending upon temperature, generation time is about 6 to 8 wk, allowing several generations each growing season. All stages survive winter temperatures, but fourth-stage juveniles and adults are the most abundant overwintering stages. Temperate species, particularly *P. penetrans*, can survive extremely cold temperatures but survive very poorly at temperatures above 30°C (Mai et al., 1977). Some species undergo anhydrobiosis, which may explain why they can survive extended periods of subzero temperatures (Townshend, 1984).

Disease Complexes

Infection by *Pratylenchus* spp. may predispose potato plants to infection by *Rhizoctonia solani* (Cetas & Harrison, 1963). A more important interaction is that involving *Pratylenchus* spp. with *Verticillium dahliae* causing an economically important disease known as "potato early dying" (Martin et al., 1982; Rowe et al., 1987). The species most frequently found associated with potato early dying is *P. penetrans*. The nematode and fungus interact and cause early dying at population densities at which either pathogen alone have little or no effect on potato (Rowe et al., 1985). The species *P. scribneri* also interacts with *V. dahliae* and causes early dying when plants are stressed by high temperature (Riedel et al., 1985). The species *P. thornei* and *P. penetrans* interact with *V. dahliae* and cause early dying of both *V. dahliae*-susceptible and tolerant potato cultivars (Krikum & Orion, 1977; MacGuidwin & Rouse, 1990). The widely occurring species, *P. crenatus,* does not interact to cause early dying, and the same appears to be true with the less prevalent species *P. neglectus* (Riedel et al., 1985).

Economic Importance

Root-lesion nematodes are serious pathogens of potato throughout the potato production areas of the northeastern USA, the St. Lawrence River basin region, and the Maritime Provinces of Canada (Townshend et al., 1978a). The most economically important species is *P. penetrans*. Population densities of 1000 to 2000 *P. penetrans* per kilogram of soil (damage threshold density) result in measurable reduction in potato yields (Olthof, 1987). The greatest economic importance of *P. penetrans* in potato production is reached when it interacts with Verticillium wilt fungi causing potato early dying, where much lower nematode populations result in economic yield losses (Wheeler et al., 1994). The damage threshold density of *P. scribneri* on potato is about the same as that of *P. penetrans* (Riedel et al., 1985), but that of *P. neglectus* on potato is less (Olthof, 1990). Relatively high population densities of *P. crenatus* can exist in potato fields without causing measurable crop damage (Dickerson et al., 1964; Kimpinski, 1979).

Damage to potato differs with different species of *Pratylenchus* and is influenced by climatic conditions. In experiments where *P. penetrans* was controlled in some plots, a 46% increase in potato yields was recorded (Olthof,

1989a). In microplot studies, an initial level of 186 *P. neglectus* per kilogram of soil restricted marketable potato yield by 19% one season, but 10 times that density was required to restrict marketable yields by 19% the following season (Olthof, 1990), underscoring the importance of climatic conditions to the amount of damage caused by *Pratylenchus* spp.

Root-Knot Nematodes

Root-knot nematodes were first recognized as parasites of potato when Neal (1889) described a root-knot disease of potato in Florida. In general, these nematodes reach their greatest potential as potato pathogens in warm climates. However, some species have adapted to cooler climates and cause severe damage in certain potato production areas. When potato is grown in warm climates, the damage experienced from root-knot is confounded by the interaction of these nematodes with other disease-causing organisms.

Species and Races

The three species of *Meloidogyne* that parasitize potato differ in distribution and amount of damage they cause. Each species has a common or descriptive name that relates to the climate or location in which it occurs. These species are *M. incognita* (southern root-knot nematode), *M. hapla* (northern root-knot nematode), and *M. chitwoodi* (Columbia root-knot nematode).

Four races of *M. incognita* and three of *M. chitwoodi* are distinguishable by their ability to reproduce on different host crops (Sasser & Carter, 1985; Santo & Pinkerton, 1985; Van Meggelen et al., 1994). These races are often referred to as host races and are unrelated to cytological and biological races. Cytological races of *M. incognita* are distinguished by chromosome number (Triantaphyllou, 1985), and biological races are distinguished by reproduction on different genotypes of a single host crop (Canto-Saenz & Brodie, 1986). No host races of *M. hapla* have been reported, but it can be distinguished from the three races of *M. chitwoodi* by its ability to reproduce on different crops (Mojtahedi et al., 1988).

Symptoms

Depending upon nematode population density and environmental conditions, infected plants may show varying degrees of stunting, chlorosis, and wilting. Plants infected with *Meloidogyne* spp. display knots or galls of varying size and shapes on their roots (Fig. 24–4). Gall shape ranges from almost spherical to very rough and irregular. In addition to galling, *M. hapla* infection causes extensive lateral root formation, giving the entire root system a "wiry" appearance. When nematode numbers are high, tubers are infected and display galls that gives the tubers a "warty" appearance (Fig. 24–5). White, pear-shaped females can be found inside the galls. Attached to the females usually are cream-colored to brownish egg masses that protrude outside the root. It is not uncommon for certain *Meloidogyne* spp., such as *M. chitwoodi*, to infect potato roots and not produce galls (Santo et al., 1980).

Fig. 24–4. Potato root infected with *Meloidogyne incognita* showing typical galling.

Biology

Root-knot nematodes are sedentary endoparasites that exhibit sexual dimorphism. The life cycle of root-knot nematodes begins when second-stage juveniles emerge from the egg and enter the roots just behind the root cap. Inside the root, the nematodes become sedentary with their heads near the developing vascular system where they feed, increase in size, and induce alteration in the cells resulting in the formation of galls from which the nematodes derive their food. The nematodes undergo three additional molts, and females become pear-shaped. The males emerge from the root after the fourth molt and are vermiform. Reproduction in these *Meloidogyne* spp. commonly occurs in the absence of males through a process called mitotic parthenogenesis (Bird, 1978). Females produce 300 to 400 eggs that are extruded in a gelatinous matrix outside the root. First-stage juveniles molt inside the eggs to become second-stage juveniles that emerge from the eggs about 10 d later and enter roots to initiate a second generation. The life cycle of all species is essentially the same, but they differ in temperature requirements. *Meloidogyne chitwoodi* and *M. hapla* require about 5°C

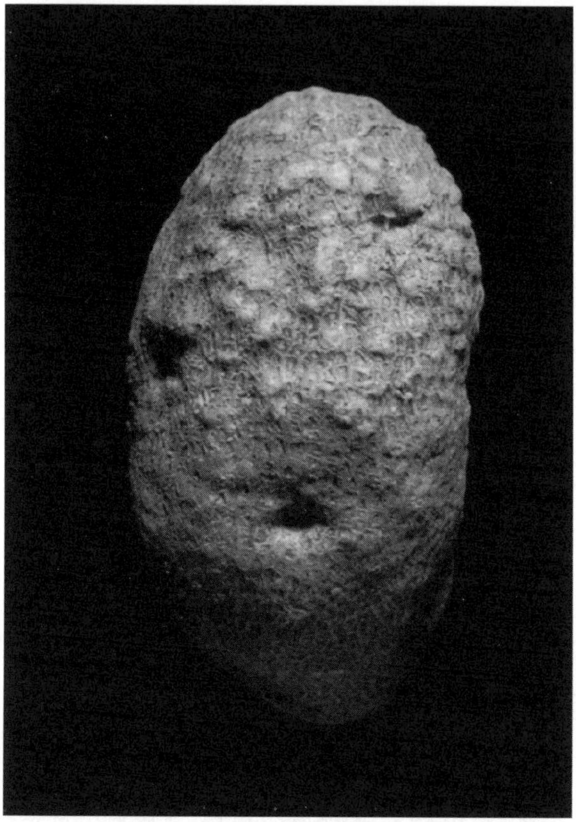

Fig. 24–5. Potato tubers infected with *Meloidogyne hapla* showing typical galling on surface.

less than *M. incognita* for optimum growth (Pinkerton et al., 1991). The number of generations produced depends on temperature. In the temperate zone where potato is grown, generation time is about 4 to 6 wk, and three or four generations are produced each season.

Disease Complexes

Disease complexes involving root-knot nematodes on potato have been recorded for *M. incognita* and *Pseudomonas solanacearum* (Jatala & Martin, 1977), *M. hapla* and *Verticillium albo-atrum* (Jacobsen et al., 1979), and possibly *M. incognita* and *Rhizoctonia solani* (Joubert & Dalmasso, 1972). In the case of *P. solanacearum*, the nematode is thought to provide entry sites for *P. solanacearum* as well as alter the host physiology so that the host is more quickly colonized by the bacterium (Jatala & Martin, 1977). Potato plants infected with both the nematode and bacteria show early and more severe symptoms of wilt and die sooner than those infected with the *P. solanacearum* alone. Similar responses of the plant are noted when both *M. hapla* and *V. albo-atrum* infect potato.

Economic Importance

In tropical climates where root-knot nematodes reach their greatest potential for damage, annual losses to the potato crop are currently estimated to be 24% (Sasser, 1979). However, potato plants escape much of the damage that could occur in the tropics because they are grown at high elevations or during cooler months of the year. The most widely distributed species of root-knot nematodes is *M. incognita*. Potato plants are extremely susceptible to *M. incognita,* but they usually escape severe damage from this species because they are not grown in its presence or are grown during the part of the year when *M. incognita* is inactive. Still potato suffers severe damage when grown at temperatures conducive to *M. incognita* activity. Losses of 25% in potential potato yield occurred when conditions favored *M. incognita* infection of potato (Sitterly & Fassuliotis, 1965).

Because of its lower temperature requirements and ability to survive severe winter temperatures, *M. hapla* causes losses to potato throughout the northern areas of the temperate zone. A relatively large area of the Pacific Northwest (in the states of Washington, Oregon, and Idaho) is infected with *M. chitwoodi.* Taxonomically, this species is closely related to *M. hapla* and was apparently confused with *M. hapla* for many years. It differs from *M. hapla* in that it reproduces well on grain crops and at lower temperatures (15–25°C) than does *M. hapla* (Golden et al., 1980). *Meloidogyne chitwoodi* is the most economically important species of root-knot nematodes attacking potato in the USA. It also is a serious pathogen of potato in the Netherlands and other parts of Europe (Van Meggelen et al., 1994).

Nematodes of Localized Importance

Several ectoparasitic nematodes and other species are important potato parasites in certain production areas. Some of these nematodes, such as the stubby-root and dagger nematodes, are important to potato production not only because of the direct damage they may cause but also because they are known to transmit viruses. Other nematodes such as sting, and potato-rot and stem nematodes (*Ditylenchus* spp.), cause severe damage in localized areas. The false root-knot nematode causes severe damage to potato in South America. The spiral (*Helicotylenchus dihystera*), stunt (*Tylenchoryhnchus claytoni*), ring, burrowing, reniform and foliar nematodes have been associated with damaged potato but lack positive proof of pathogenicity.

Genera and Species

Stubby-root nematodes are in the genera *Trichodorus* and *Paratrichodorus.* Ten species of these nematodes have been reported to parasitize potato (Brodie, 1984). These species include *P. minor, P. pachydermus, P. allius, P. nanus, P. teres, T. cylindrious, T. primitivus, T. proximus, T. similis,* and *T. sparsus.*

The species of sting nematode reported to damage potato is *Belonolaimus longicaudatus.* Sting nematodes cause a serious suppression of potato yield in an isolated area of northeast Florida (Weingartner et al., 1975).

The reniform nematode (*Rotylenchulus reniformis*) was first described from Hawaii where potato was listed as a host crop (Linford & Oliveira, 1940). It was later reported to damage potato in India, Egypt, and in greenhouse trials in the USA (Rebois et al., 1978).

False root-knot nematodes are in the genus *Nacobbus*, which contains the two species *N. aberrans* and *N. batatiformis* (Sher, 1970). Only *N. aberrans* attacks potato.

Two species of *Ditylenchus* are known to damage potato. These are *D. destructor*, which causes potato tuber rot, and *D. dipsaci*, which attacks potato stems and leaves.

Biology

In most cases, our knowledge of the biology of these nematodes is limited to their feeding habits, temperature relations, and time required for egg production. Both stubby-root and sting nematodes feed ectoparasitically on plants, and all stages of these nematodes feed. The absence of stubby-root nematodes from soil during winter suggests that they overwinter in the egg stage, but other stages can survive at lower soil depth. In subtropical climates, all stages of sting nematodes are present throughout the year. Eggs are deposited singly in the soil and hatch within a few days. Some species of stubby-root nematodes complete their life cycle within 21 to 27 d (*P. minor)*, and others take longer (Decraemer, 1991).

First-stage juveniles of reniform nematodes emerge from the egg and undergo three molts before feeding after which they undergo the final molt. Only young females are infective, and they initiate syncytial formation from a single endodermal cell. The adult females embed themselves partly in the root, and the anterior portion enlarges to the familiar kidney shape (reniform). Eggs are deposited in a gelatinous matrix extruded from the nematode. The life cycle is completed in 19 d (Riggs & Niblack, 1993).

The false root-knot nematode (*N. aberrans*) exhibits pronounced sexual dimorphism with the adult female transformed into a swollen, irregularly shaped body, but the male remains a typical worm-shaped nematode with a short tail. Second-stage juveniles emerge from the egg and invade small roots, undergo three additional molts, and increase in size. Some nematodes then leave the roots and others continue feeding. Those that continue feeding initiate gall development and produce eggs, and those that leave the roots complete the final molt and become males or active females. The females re-enter larger roots and initiate gall development. At the surface of the root an opening is formed through which eggs are discharged into a gelatinous matrix exuded by the nematode (Clark, 1967).

All stages of *N. aberrans*, except adult males, may invade tubers penetrating just beneath the surface. They remain in a semiquiescent state in the tubers and may be disseminated over long distances and become active when the tuber is planted (Jatala & de Scurrah, 1975).

The biology of the stem nematode (*D. dipsaci*) and potato-rot nematodes (*D. destructor*) differs only slightly. Both will infect potato tubers through lenticels or eyes and feed on tissue just beneath the surface. During wet weather *D. dipsaci* invades the aerial parts of the plant. Once the nematode penetrates the stem or leaf tissues, multiplication is rapid and the plant is severely stunted and distorted.

Symptoms

Feeding of stubby-root nematodes (*Trichodorus* and *Paratrichodorus* spp.) occurs in the meristematic region of the root tissues causing cessation of root growth that results in numerous stunted roots, giving the root system a "stubby-root" appearance. Although infected plants show stunting and wilting, a more important aspect of these nematodes is their ability to transmit viruses to potato. The viruses that they transmit are in the tobravirus group. The one important to potato is tobacco rattle virus (TRV). This virus causes a disease of potato known as corky ring spot. Symptoms of corky ring spot consists of irregularly shaped tubers with cracks within arc-shaped lesions, and brown concentric rings on the surface of the tubers. Irregularly shaped lesions that have a corky texture appear internally, and deep cracks or shallow corky depressions may develop late in the season that renders the tubers unmarketable.

Symptoms of sting nematode attack consist of dark, shrunken lesions along the root axis and at the tip with lesions sometimes girdling the root causing it to break. Field symptoms consist of the usual decline and death of small plants with stunting and chlorosis of those remaining. Such symptoms are accompanied by a restricted number and size of tubers (Weingartner et al., 1975).

Under conditions of high infestations reniform nematodes destroy the tap root resulting in severe damage to the plant. This nematode can severely reduce tuber yield and quality (Rebois et al., 1978).

False root-knot nematodes induce root galls similar to those produced by *Meloidogyne* spp. but are spherical in shape and occur in a beadlike fashion. Normally the galls lack extension of lateral roots resulting in abnormal fibrous root growth. Massive invasion by juveniles may cause death and deterioration of small roots. The nematode also is capable of attacking potato tubers, but no obvious symptoms or damage result as they penetrate only 1 to 2 mm deep (Jatala & de Scurrah, 1975). The earliest symptoms of potato-rot nematode (*D. destructor*) infection are small, white, chalky or light-colored spots that can be seen just below the skin of the tuber. The infected tissues are dry and granular, and infected spots may coalesce. Affected tissue darkens gradually through grayish to dark brown or black, as secondary organisms, such as bacteria, fungi, or saprophytic nematodes, invade. The skin of the tuber is not attacked but becomes paper thin and cracks as underlying infected tissues dry and shrink.

In contrast to symptoms caused by *D. destructor*, which initially tend to be superficial, symptoms caused by the stem nematode (*D. dipsaci*) may extend throughout the tuber. Tuber tissues infected with *D. dipsaci* become spongy and are yellowish to brown in color. Although distinct differences between the two species may occur in tuber symptoms, diagnosis is often difficult and species identification is necessary. In addition, *D. dipsaci* attacks aboveground parts of the plant, causing typical stunting, thickening, and distortion of stems and leaf petioles.

Economic Importance

The stubby-root nematodes are distributed worldwide with certain species limited by soil type and temperature requirements. In the USA, *P. minor* (formerly *P. christiei*) and *P. allius* are the most common species associated with potato, whereas *P. pachydermus* and *T. primitivus* are most frequently found associated with potato in Europe (Winslow & Willis, 1972). The greatest economic importance of these nematodes is realized in their ability to transmit TRV to potato.

Sting nematodes are limited in distribution and have been reported mainly from tropical and subtropical areas with very sandy soils. They are associated with poor growth of potato in a relatively small area of northeastern Florida (Weingartner et al., 1975). Their lack of widespread importance as potato pests is probably because potato is not grown to a great extent in areas where *B. longicaudatus* exists or at the time of year the nematode is active.

The reniform nematode also is primarily a parasite in tropical and subtropical climates. It has been reported as a parasite of potato in Egypt, Hawaii, and India. The nematode occurs in the southern USA, but its damage potential to potato is based largely on greenhouse studies (Rebois et al., 1978). It is probably not an economically important pest of potato because potato is not grown where high populations of *R. reniformis* exist.

False root-knot nematodes occur in several countries but are known to damage potato only in South America. Because of its rather wide distribution, ability to survive and reproduce under a wide range of temperatures and soil types, and its extensive host range, it is surprising that *N. aberrans* has not been encountered more frequently damaging potato. It appears that host-specific races exist and that the "potato race" has a rather limited geographic distribution, and is confined primarily to South America.

Both the potato-rot and the stem nematodes are found throughout the world, but are of minor importance on potato in most countries. The stem nematode (*D. dipsaci*) is regarded as a serious pest of potato only in Germany, the Netherlands, and parts of the former USSR.

Generally, the potato-rot nematode (*D. destructor*) is a much more important pest of potato with its greatest importance being in temperate regions. The nematode has been reported from many areas of Europe and is especially widespread in the former USSR (Evans & Trudgill, 1992). It also is known to occur in South Africa, some areas of the Mediterranean region, South America, and isolated areas of North America where it has been successfully controlled.

CONTROL

Host Resistance

Host crop resistance is an invaluable resource for control of PCN. The single gene H_1, originally from *S. tuberosum* ssp. *andigena* CPC clone 1673 (Ellenby, 1954), has been bred into commercial potato cultivars throughout the world. In the USA, 27 potato cultivars carrying the H_1 gene have been released (Brodie et al., 1993). This resistance is stable and essentially prevents reproduction of *G. rostochiensis* pathotypes Ro1 and Ro4 but is ineffective against other pathotypes of *G. rostochiensis* and *G. pallida* (Cook & Evans, 1987).

Plants carrying the H_1 gene cause stimulation of PCN hatch similar to that caused by susceptible plants (Turner & Stone, 1984; Rawsthorne & Brodie, 1986b). Unlike susceptible plants, resistant plants limit PCN development by improper establishment of feeding sites (Hoopes et al., 1978) and egression of juveniles from resistant roots (Mullin & Brodie, 1988a). The few females that develop on resistant plants have reduced fecundity, probably resulting from feeding from a poorly developed syncytium (Mullin & Brodie, 1988b).

Where only one pathotype of *G. rostochiensis* (Ro1) is present, growth of cultivars carrying the gene H_1 has successfully managed PCN (Brodie & Mai, 1989). If mixtures of pathotypes exist, growing such cultivars will select out pathotypes that can multiply upon them. This process is particularly rapid in fields where the two species of PCN are present because the two species do not interbreed (Cook & Evans, 1987).

Host resistance has played a much lesser part in the control of *P. penetrans* primarily because of the lack of resistant germplasm to incorporate into commercial cultivars. Early reports indicate that certain commercial cultivars are less suitable for *P. penetrans* multiplication than are others, but the data are inconclusive (Dunn, 1973). Partial resistance to *P. penetrans* was identified in a potato breeding population derived from *S. tuberosum* ssp. *andigena* and *S. vernei* (Brodie & Plaisted, 1991). This resistance has been transferred into more widely adapted potato clones. The beneficial effects of this resistance may be negated by the use of cover crops that are highly susceptible to *P. penetrans* multiplication (Fawole & Mai, 1988).

Although sources of resistance to several species of *Meloidogyne* in wild and cultivated species of *Solanum* have been identified, no commercial cultivars are yet available. Good resistance to *M. incognita*, *M. javanica*, and *M. arenaria* was identified in *S. sparsipilum* (Jatala & Rowe, 1976). Resistance to *M. hapla* was found in *S. tuberosum* subsp. *andigena* (Brodie & Plaisted, 1977). Brown (1989) and Brown et al. (1991) confirmed resistance to Races 1 and 2 of *M. chitwoodi* in the Mexican wild species *S. bulbocastanum* and the hexaploid *S. hougasii*.

Host resistance offers the most practical method of control of false root-knot nematodes. Several resistant cultivars of *S. tuberosum* ssp. *andigena* native to South America have been identified, as well as resistant wild tuber-bearing *Solanum* species (Cahuana et al., 1975; PROINPA, 1991, p. 47–56).

Tolerance

Tolerance is described as the extent to which the plant is able to withstand infection without suffering undue damage (Evans & Haydock, 1990; Trudgill, 1986). Resistance and tolerance are independent attributes of plants, but resistance may confer tolerance (Trudgill, 1991). In general, potato cultivars are tolerant of the PCN species or pathotype to which they are resistant, but there are exceptions. Also, all susceptible cultivars do not react identically as a host of PCN and may exert different influences on nematode populations (Dale & Phillips, 1985). Cultivars that support fewer cysts may provide an important source of breeding material to be combined with tolerance. A disadvantage of true tolerance (no resistance) is that repeated growing of tolerant cultivars may increase the nematode population to a level where even the tolerant cultivar is damaged (Evans & Haydock, 1990).

Based on yield of plants grown in *P. penetrans* infested soil, certain potato cultivars appear to be tolerant of *P. penetrans* infection (Olthof, 1986). 'Russet Burbank' demonstrated an intermediate degree of tolerance, and 'Yukon Gold' appeared highly tolerant to *P. penetrans* damage. Tolerance to infection by *P. penetrans* has been demonstrated for other widely grown potato cultivars in the USA (Kotcon et. al., 1987).

The greatest usefulness of tolerance may be in combination with nematicides or partially resistant cultivars in an integrated approach to maintain profitable potato yields and extend the usefulness of partially resistant cultivars in the presence of mixed populations of nematodes.

Cultural

Because of their limited host range, crop rotation is a practical means for PCN control and is used in most countries where they occur. The length of the gap between potato crops depends upon initial nematode population density. At low population densities a 2- to 4-yr gap between potato crops is sufficient (Chitwood, 1951), but at higher nematode densities, up to a 7-yr gap in potato crops is necessary (Jones, 1970). A wide choice of nonhost crops is available because tomato and eggplant are the only other commercial host crop for PCN, and they are not usually grown in the same areas as potato. Populations of PCN decrease about 30 to 40% when nonhosts are grown, but this figure varies with different soil types and climates (Jones, 1970). This slow rate of PCN decline when nonhosts are grown causes infestations to exist for 20 yr or more and necessitates long rotations (Trudgill et al., 1987).

The wide host range of *Pratylenchus* spp. makes crop rotation less effective in managing these nematodes. Most crops grown in rotation with potato are suitable hosts for multiplication of *P. penetrans* (MacDonald & Mai, 1963). Rye (*Secale cereale* L.) is commonly grown in winter as a cover crop and wheat (*Triticum aestivum* L.), oat (*Avena sativa* L.), and corn (*Zea mays* L.) are frequently grown in rotation with potato. Multiplication of *P. penetrans* varies on these crops but all of them maintain relatively high population densities of this nematode (Dunn & Mai, 1973; Florini & Loria, 1990; Johnston et al., 1994).

Different species of *Sorghum* and their hybrids suppress *P. penetrans* in potato production (Florini & Loria, 1990), but further studies are needed to more clearly determine their ability to suppress population on potato production systems. Marigold (*Tagetes* spp.) is known to suppress populations of *Pratylenchus* spp., and recent reports suggest their possible use to manage *P. penetrans* in potato production (Sieczka et al., 1991).

Root-knot nematodes are hosts for over 2000 plant species (Bird, 1978), which makes the choice of alternate crops in rotations very difficult. Because root-knot species differ somewhat in their host ranges, it is important to correctly identify the species of *Meloidogyne* involved before embarking upon a scheme of control by crop rotation. In the northwestern USA, recommended control of *M. hapla* on potato is rotation with grain crops, but *M. chitwoodi*, the predominant species of root-knot nematodes attacking potato in that area, reproduces well on grains (Santo et al., 1980). Rapeseed (*Brassica napus* cv. Jupiter) and sudangrass (*Sorghum x drummondii* cv. Piper) appear promising as rotational crops for managing *M. chitwoodi* (Santo et al., 1991). Rotation with such crops reduces *M. chitwoodi* populations, but they rebound quickly after a single year of potato production.

Use of healthy, uninfected seed is the essential first step in controlling potato-rot nematodes. Control of weeds that may act as alternative hosts also is important. These two factors have largely eliminated this nematode from intensive potato production areas of the USA. Despite the wide host range of this nematode, crop rotation also has contributed to its control. Kornobis and Stefan (1991) note that 2 to 3 yr of cereals helps eliminate *D. destructor* from potato fields. MacGuidwin and Slack (1991) also record oat as a nonhost.

Rotation studies with nonhost annual crops [e.g., horse bean (*Vicia faba* L.), oat] have increased potato yield without affecting gall formation caused by false root-knot nematodes (Brodie et al., 1993). Weed control (especially *Spergula arvensis*) is very important in limiting damage caused by this nematode.

Chemical

Both fumigant and nonfumigant nematicides have been used in PCN management schemes. Soil fumigants have been traditionally used where a drastic and rapid reduction in PCN population density is required, such as in a quarantine or regulatory setting. Nonfumigant nematicides are used to protect the potato crop from damage and to produce a profitable yield in the presence of the nematode. Because of potential groundwater contamination and other environmental and health concerns, none of these nematicides are currently used in the USA to control PCN (Brodie & Mai, 1989).

The soil fumigants 1,3-dichloropropene or dichloropropene (1,3-dichloropropene)-methylisothiocynate mixtures have been used to control PCN in most countries where they occur and were the key component of the regulatory program to control *G. rostochiensis* in the USA until 1984. Soil fumigants are extremely phytotoxic and must be applied well in advance of planting, often the previous fall. Spring or early summer application often requires that the land be out of potato production for that growing season (Spears, 1968).

The nonfumigants aldicarb [2-methyl-2-(methylthio) propionaldehyde *O*-(methylcarbamoyl) oxime] and oxamyl {methyl *N*′, *N*′-dimethyl-N-[(methylcarbamoyl)oxy]-1-thiooxamimidate} provided effective control of *G. rostochiensis* when applied to the soil surface and incorporated just before planting (Whitehead, 1986) or applied to the seed furrow at the time of planting potato (Brodie, 1983).

Both fumigant and nonfumigant nematicides provide effective control of *P. penetrans* and, in most instances, promote potato yield increases. Application of 1,3-dicholopropene, aldicarb, oxamyl, or metham sodium (sodium methyldithiocarbamate) restricted *P. penetrans* population increases (Kimpinski, 1982; Olthof, 1987, 1989b), but increases in potato yields due to such treatments were more pronounced when plants were under moisture stress than when adequate moisture was available (Kimpinski, 1979). Metham sodium effectively reduced *P. penetrans* populations when applied in irrigation water but did not result in increased yields of potato (Olthof, 1989a). Soaking potato tubers in oxamyl or direct application of oxamyl to tubers provided some measure of *P. penetrans* control (Townshend & Olthof, 1988), but was less effective when applied as a drench at planting or as a foliar spray (Olthof, 1989b). Preplant treatments of 1,3-dicholopropene or a mixture of methylisothiocynate and 1,3-dicholopropene provided good control of *P. penetrans* (Olthof, 1987), and combination treatments of 1,3 dicholopropene and aldicarb also restricted *P. penetrans* numbers and increased potato yields (Olthof, 1987, 1989b).

Soil fumigation with 1,3-dichloropene has been successful in controlling root-knot nematodes on potato. The effectiveness of such treatment is usually independent of root-knot species involved. When nematode densities are high, fumigation usually results in a striking reduction in nematode population densities (99%) and increase in potato yields (55%) (Winslow & Willis, 1972). However, soil fumigants have given inconsistent results with *M. chitwoodi* (Santo & Wilson, 1990; Mojtahedi et al., 1991a).

Nonvolatile nematicides, such as aldicarb, phenamiphos [ethyl 4-(methylthio)-*m*-tolyl isopropylphosphor-amidate], and oxamyl, have shown promise of controlling root-knot nematodes on potato, particularly *M. hapla* (Santo & Qualls, 1984; Ingham et al., 1991). These compounds also have been used effectively as a seed-piece dip to suppress root-knot and root-lesion nematode infection of seed (Rodriguez-Kabana & Ingram, 1977; Townshend & Olthof, 1988).

Soil fumigants and nonfumigant nematicides are the best methods of control for false root-knot nematodes (Inserra et al., 1985). Under some conditions nonfumigant nematicides increase potato yields by up to 40% but fail to decrease nematode galling (Otazú et al., 1985; Bascopé & Lopez, 1987). Some liquid formulations [fenamiphos 40%, ethoprop (*O*-ethyl *S*,*S*-dipropyl phosphorodithioate) 70%] can be used successfully to dip tubers (2000 mg $mL^{-1} \times 10$ min) in order to avoid the dissemination of *N. aberrans*.

Soil fumigation with ethylene dibromide (EDB) (1,2-dibromoethane) was used successfully against heavy infestations of the potato-rot nematode, but because of environmental concerns this nematicide has been withdrawn from usage.

Although both fumigant and nonfumigant nematicides effectively control potato nematodes, their use is heavily restricted due to environmental concerns and possible human health hazards. The use of certain of these nematicides in potato fields have resulted in contamination of the groundwater causing serious concerns over the safety of drinking water (Zaki et al., 1982). Because of their potential for contaminating groundwater under certain conditions, many soil applied nematicides have been banned for use in certain areas. Consideration of possible effects on the environment should be a part of any decision to apply nematicides to control nematodes in potato production.

Biological

Although a number of fungi have been isolated from *Globodera* spp., biological control of PCN with natural enemies has realized limited success. Early surveys indicated a low level of infection of *G. rostochiensis* by fungal pathogens (Wilcox & Tribe, 1974). More recent surveys for natural enemies of PCN in its native habitat, the Andean region of South America, have been only partially successful (LaMondia & Brodie, 1984; Morgan-Jones et al., 1985).

Up to 93% infection of PCN eggs was achieved with *V. suchlasporium*, which was previously isolated from PCN eggs (Dackman, 1990). Up to 63% infection of PCN eggs was experienced with *V. lecani* (Uziel & Sikora, 1992). The egg parasite *Paecilomyces lilacinus* is reported to provide control of *G. rostochiensis* similar to that achieved with aldicarb (Khan, 1988). An unidentified fungus isolated from PCN eggs in Germany infected 70% of *G. rostochiensis* and *G. pallida* eggs (Goswami & Rumpenhorst, 1978). Such fungi may eventually prove to be effective biological control agents of PCN.

A few experiments have focused on potential biological control agents against *Pratylenchus* spp. The fungus *Hirsutella rhossiliensis* is a virulent pathogen of *P. penetrans* and reduced the numbers of *P. penetrans* in soil (Timper & Brodie, 1993). The trapping fungi, *Arthrobotrys dactyloides, A. oligospora, Monacrosporium ellipsosporum* and *M. cionapagum*, trapped and killed most *P. penetrans* added to fungus cultures but have not been tested in soil.

Mycelial and endospore-forming bacteria are known to infect *Pratylenchus* spp. (Sayer & Starr, 1988). The species that is specific for *Pratylenchus* spp. is *Pasteuria thornei* (Starr & Sayer, 1988) and is a proven pathogen of *P. brachyurus* but has not been fully tested against other species of *Pratylenchus*. However, the obligate parasitism exhibited by *Pasteuria* spp. may limit their usefulness as biological control agents as spore load in the soil is reduced because *Pasteuria* spp. are unable to reproduce in the absence of a host. Still, the endospore of *Pasteuria* can withstand environmental extremes, such as temperature and moisture fluctuations, and maintain viability which may enhance its usefulness as a biological control agent of *Pratylenchus* spp. (Sterling, 1991).

The fungus *P. lilacinus* is reported to be parasitic on eggs of *M. incognita*. Reports indicate that introduction of this fungus into soil infested with *M. incognita* reduced production of viable eggs on potato roots by 70% (Jatala et al., 1979).

Biocontrol agents hold some promise for controlling false root-knot since laboratory and greenhouse experiments have shown that *Beauveria brogniartii* will parasitize and reduce multiplication of *N. aberrans* on potato (Balderrama, 1992). The use of biological control agents to control potato nematodes will likely become more important as constraints on the use of pesticides increase.

Integrated Nematode Control

Inherent weaknesses or limitations in most nematode control practices has forced the development of integrated control strategies involving two or more control practices whose effects are complementary. Soil fumigation, application of nonfumigant nematicides, and growing of resistant cultivars have limited the spread of *G. rostochiensis* in the USA (Evans & Brodie, 1980). When nematicides were discovered contaminating groundwater, a system was adopted that consisted of growing resistant cultivars for 2 yr followed by a nonhost crop the 3rd yr and a susceptible cultivar the 4th yr (Brodie & Mai, 1989). In the Netherlands, dichloropropene and metham sodium are used extensively in an integrated approach to control PCN in starch potato (Trudgill et al., 1987). Because PCN population in the USA consists predominately of only *G. rostochiensis* Ro1, such short rotations using resistant cultivars carrying the gene H_1 have been extremely successful (Brodie & Mai, 1989). Integrated control strategies for PCN reduces the risk of a rapid selection for virulent pathotypes. Further, it limits the pesticide load in the environment and shortens considerably the interval between potato crops allowing potato to be grown more frequently on the most productive land (Evans & Brodie, 1980).

Information is insufficient on the individual effects of rotation, weed control and resistant cultivars on nematode population dynamics and potato crop performance to devise integrated control strategies for root-lesion or false root-knot nematodes (Brodie et al., 1993).

OUTLOOK

The potato occupies fourth place behind wheat, rice (*Oryza sativa* L.) and corn among the world's food crops. It is of major significance in human nutrition in the USA where just under 12 billion tons are produced annually. Potato is second only to eggs in nutritional value as a single food source (Swaminathan & Sawyer, 1983). Today's health-conscious society will likely cause an increase in the demand for potato as a nutritional food source. To meet this demand, potato will be grown more frequently on land in current production areas, or potato production will be expanded into nontraditional climates of the world. Either scenario will be accompanied by an increase in production problems caused by nematodes.

Potato cyst nematodes will continue to be a major production and/or regulatory problem in the cooler climates where potato are grown. The Colombia root-knot nematode (*M. chitwoodi*) will continue to be a major constraint to potato production in traditional climates where potato are grown. The expansion of potato production into nontraditional, warmer climates of the world will be accomplished with increased production problems involving root-knot nematodes and their interaction with other plant pathogens. With a decline in usage of chemical nematicides, damage caused by root-lesion nematodes will become more apparent and will more frequently be implicated in early dying of potato.

Chemical nematicides will continue to be a factor in controlling nematodes on potato, but their use will continue to decline. To meet increased production demands, greater emphasis will be placed on host resistance and biological controls of potato nematodes. More attention will be given to integrated nematode management strategies rather than relying on a single control tactic to manage nematode populations. Because of the emergence of new races associated with repeated growing of resistant cultivars, more effort will be devoted to managing nematode-resistant genes to prolong the usefulness of resistant cultivars.

REFERENCES

Balderrama, F. 1992. *Beauvaria brogniartii* como biocontrolador del namatodo del rosario de la papa, *Nacobbus aberrans* (Thorne 1935) Thorne y Allen, 1944. M.S. thesis, Universidad Mayor de San Simon, Cochabamba, Bolivia.

Bascopé, J.B., and B. López. 1987. Ensayo preliminar para el control químico de *Nacobbus* sp. en la zona de Chaqueli. p. 7–10. *In* Control integegrado de plagas. Cochabamba, Bolivia.

Behrens, E. 1975. Globodera Skarbilovic, 1959. Eine eelbatandige Gattung in der Uterfanilie Heteroderinae Skarbilovic, 1947 (Nematoda: Heteroderidae). Vortragst. Aktuell. Probl. Phytonematol. 29:12–26.

Bird, A.F. 1978. Root-knot nematodes in Australia. CSIRO Aust. Div. Hort. Res. Tech. Pap. no. 2. CSIRO, Div. Hort. Res., Adelaide, South Australia.

Brodie, B.B. 1983. Control of *Globodera rostochiensis* in relation to method of applying nematicides. J. Nematol. 15:491–495.

Brodie, B.B. 1984. Nematode parasites of potato. p. 167–212. *In* W. R. Nickle (ed.) Plant and insect nematodes, Marcel Dekker, Inc., New York and Basel.

Brodie, B.B. 1995. The occurrence of a second pathotype of potato cyst nematodes in New York. J. Nematol. 27:493–494.

Brodie, B.B., K. Evans, and J. Franco. 1993. p. 87–132. *In* K. Evans et al. (ed.) Plant parasitic nematodes in temperate agriculture. CAB Int., Wallingford, England.

Brodie, B.B., and W. F. Mai. 1989. Control of the golden nematode in the United States. Annu. Rev. Phytopathol. 27:443–461.

Brodie, B.B., and R. L. Plaisted. 1977. Breeding for resistance to root-knot nematodes in potatoes. Nematropica 7:2.

Brodie, B.B., and R.L. Plaisted. 1991. Resistance in potato to *Pratylenchus penetrans*. J. Nematol. 23:522.

Brown, C.R. 1989. Comparison of reproductive efficiency of *Meloidogyne chitwoodi* on *Solanum bulbocastanum* in soil and in vitro test. Plant Dis. 73:957–959.

Brown, C.R., H. Mojtahedi, and G. S. Santo. 1991. Resistance to Columbia root-knot nematode in *Solanum* spp. and in hybrids of *S. hougasii* with tetraploid cultivated potato. Am. Potato J. 68:445–452.

Brown, E.B., and G.B. Sykes. 1983. Assessment of the losses caused to potatoes by the potato cyst nematodes, *Globodera rostochiensis* and *G. pallida*. Ann. Appl. Biol. 103:271–276.

Burrows, P.R. 1988. The use of nucleic acid probes to identify plant parasitic nematodes. Pest Dis. 2:811–820.

Cahuana, A., O.W. Cornejo, and S.V. Huanco. 1975. Sources of resistance to the false root-knot nematode (*Nacobbus* sp.) among 50 potato clones. Fitopatología 10:71.

Canto-Saenz, M., and B.B. Brodie. 1986. Host efficiency of potato to *Meloidogyne incognita* and damage threshold densities on potato. Nematropica 16:109–116.

Canto-Saenz, M., and M. M. de Scurrah. 1977. Races of the potato cyst nematode in the andean region and a new system of classification. Nematologica 23:340–349.

Cetas, R.C., and M.B. Harrison. 1963. Evaluation of fumigants for control of early maturity wilt of potatoes on Long Island. Phytopathology 53:347–348.

Chitwood, B.G. 1951. The golden nematode of potatoes. USDA Circ. no. 875. U.S. Gov. Print., Washington, DC.

Clark, S.A. 1967. The development and life history of the false root knot nematode, *Nacobbus serendipitious*. Nematologica 13:91–101.

Cook, R., and K. Evans. 1987. Resistance and tolerance. p. 179–231. *In* R. H. Brown and B. R. Kerry (ed.) Principles and practice of nematode control in crops. Acad. Press, London.

Corbett, D.C.M., and G.A. Hide. 1971. Interaction between *Heterodera rostochiensis* Woll. and *Verticillium dahliae* Kleb on potatoes and the effects of CCC on both. Ann. Appl. Biol. 68:71–80.

Dackman, C. 1990. Fungal parasites of the potato cyst nematode *Globodera rostochiensis*: isolation and reinfection. J. Nematol. 22:594–597.

Dale, M.F.B., and M.S. Phillips. 1985. Variation for the degree of susceptibility to the potato cyst nematode (*Globodera pallida*) within *Solanum tuberosum* ssp. *tuberosum*. Potato Res. 28:55–64.

Decraemer, W. 1991. Stubby root and virus vector nematodes: *Trichodorus, Paratrichodorus, Allotrichodorus*, and *Monotrichodorus*. p. 587–625. *In* W.R. Nickle (ed.) Manual of agricultural nematology. Marcel Dekker, New York.

Dickerson, O.J., H.M. Darling, and G.D. Griffin. 1964. Pathogenicity and population trends of *Pratylenchus penetrans* on potato and corn. Phytopathology 54:317–322.

Dunn, E., and W.A. Hughes. 1967. Interaction of *Oospora pustulans, Rhizoctonia solani* and *Heterodera rostochiensis* on potato. Eur. Potato J. 10:327–328.

Dunn, R.A. 1973. Resistance in potato (*Solanum tuberosum*) to *Pratylenchus penetrans*. 2nd Int. Congr. Plant Pathol., St. Paul, MN. 5–12 September. Int. Congr. Plant Pathol.

Dunn, R.A., and W. F. Mai. 1973. Reproduction of *Pratylenchus penetrans* in roots of seven cover crop species in greenhouse experiments. Plant Dis. Rep. 57:728–730.

Ellenby, C. 1954. Tuber forming species and varieties of the genus *Solanum* tested for resistance to the potato root eelworm *Heterodera rostochiensis* Wollenweber. Euphytica 3:195–202.

European Plant Protection Organization. 1992. Distribution of *Globodera rostochiensis* and *G. pallida*. Rep. Serv. 523/16. EPPO, Secretariat, Paris.

Evans, K. 1983. Interactions between *Globodera pallida, G. rostochiensis* and *Verticillium dahliae* on four potato cultivars. Nematologica 28:143–144.

Evans, K., and B. B. Brodie. 1980. The origin and distribution of the golden nematode and its potential in the U.S.A. Am. Potato J. 57:79–89.

Evans, K., and P. P. J. Haydock. 1990. A review of tolerance by potato plants of cyst nematode attack, with consideration of what factors may confer tolerance and methods of assaying and improving it in crops. Ann. Appl. Biol. 117:703–740.

Evans, K., and A. R. Stone. 1977. A review of the distribution and biology of the potato cyst-nematodes *Globodera rostochiensis* and *G. pallida*. PANS 23:178–189.

Evans, K., and D. L. Trudgill. 1992. Pest aspects of potato production. Part 1. The nematode pests of potatoes. p. 438–475. *In* P.M. Harris (ed.) The potato crop. 2nd ed. Chapman and Hall, London.

Fawole, B., and W.F. Mai. 1988. Risk of rye as a cover crop in alternate planting with potato in *Pratylenchus penetrans*-infested soil. Fitopatol. Bras. 13:346–348.

Florini, D.A., and R. Loria. 1990. Reproduction of *Pratylenchus penetrans* on potato and crops grown in rotation with potato. J. Nematol. 22:106–112.

Folkertsma, R.T., J.N.A.M. Rouppe van der Voort, M.P.E. van Gent-Pelzer, K.E. de Groot, W.J. van den Bos, A. Schots, J. Bakker, and F.J. Gommers. 1994. Inter- and intraspecific variation between populations of *Globodera rostochiensis* and *G. pallida* revealed by random amplified polymorphic DNA. Phytopathology 84:807–811.

France, R.A., and B.B. Brodie. 1995. Differentiation of two New York isolates of *Pratylenchus penetrans* based on their reaction on potato. J. Nematol. 27:339–345.

Golden, A.M., J.H. O'Bannon, G.S. Santo, and A.M. Finley. 1980. Description and SEM observations of *Meloidogyne chitwoodi* n. sp. (*Meloidogynidiae*), a root-knot nematode on potato in the Pacific Northwest. J. Nematol. 12:319–327.

Goswami, B.K., and H.J. Rumpenhorst. 1978. Association of an unknown fungus with potato cyst nematodes, *Globodera rostochiensis* and *G. pallida*. Nematologica 24:251–256.

Grainger, J., and M.R.M. Clark. 1963. Interaction of Rhizoctonia and potato root eelworm. Europ. Potato J. 6:131–132.

Green, C.D., D.N. Greet, and F.G.W. Jones. 1970. The influence of multiple mating on the reproduction and genetics of *Heterodera rostochiensis* and *H. schachtii*. Nematologica 16:309–326.

Harrison, J.A.C. 1971. Association between the potato cyst-nematode, *Heterodera rostochiensis* Woll. and *Verticillium dahliae* Kleb in the early-dying disease of potatoes. Ann. Appl. Biol. 67:185–193.

Hawkes, J.G. 1978. History of the potato. p. 1–14. *In* P.M. Harris (ed.) The potato crop. The scientific basis for improvement. Chapman & Hall, London.

Hominick, W.M., J.M.S. Forrest, and A.A.F. Evans. 1985. Dispause in *Globodera rostochiensis* and variability in hatching trials. Nematologica 31:159–170.

Hoopes, R.W., R.E. Anderson, and W.F. Mai. 1978. Internal response of resistant and susceptible potato clones to invasion by potato cyst nematode *Heterodera rostochiensis*. Nematropica 8:13–40.

Horton, D. 1987. Potatoes: production, marketing, and programs for developing countries. Westview Press, Boulder, CO.

Ingham, R.E., M. Morris, and G.B. Newcomb. 1991. Effects of incorporation method of ethoprop and addition of aldicarb on potato tuber infection by *Meloidogyne hapla*. Suppl. J. Nematol. 23:686–692.

Inserra, R.N., G.D. Griffin, and J. Anderson. 1985. The false root-knot nematode *Nacobbus aberrans*. Utah Agric. Exp. St., Bull. 510.

Jacobsen, B.J., D.H. MacDonald, and H.L. Bissonette. 1979. Interaction between *Meloidogyne hapla* and Verticillium wilt disease of potato. Phytopathology 69:288–292.

Janssen, R., J. Bakker, and F.J. Gommers. 1987. Circumventing the diapause of potato cyst nematodes. Netherlands J. Plant Pathol. 93:107–113.

Janssen, R., J. Bakker, and F.J. Gommers. 1991. Mendelian proof for a gene-for-gene relationship between virulence of *Globodera rostochiensis* and the H_1 resistance gene in *Solanum tuberosum* ssp. *andigena* CPC 1673. Rev. Nématol. 14:213–219.

Jatala, P. and M.M. de Scurrah. 1975. Mode of dissemination of *Nacobbus* spp. in certain potato-growing areas of Peru and Bolivia. J. Nematol. 7:324–325.

Jatala, P., J. Franco, A. Vilca, and W. Cornejo. 1979. Nonsolanaceous hosts of *Globodera* in the andes. J. Nematol. 210–211.

Jatala, P., and C. Martin. 1977. Interactions of *Meloidogyne incognita acrita* and *Pseudomonas solanacearum* on field grown potatoes. Proc. Am. Phytopathol. Soc. 4:177–178.

Jatala, P., and P. R. Rowe. 1976. Reaction of 62 tuber-bearing *Solanum* species to the root-knot nematode, *Meloidogyne incognita acrita*. J. Nematol. 8:290.

Jensen, H.J., J. Armstrong, and P. Jatala. 1979. Annotated bibliography of nematode pests of potato. Int. Potato Center, Lima, Peru.

Johnston, H.W., M.J. Celetti, J. Kimpinski, and H.W. Platt. 1994. Fungal pathogens and *Pratylenchus penetrans* associated with preceding crops of clovers, winter wheat, and annual ryegrass and their influence on succeeding potato crops on Prince Edward Island. Am. Potato J. 71:797–808.

Jones, F.G.W. 1970. The control of the potato cyst-nematode. J. R. Soc. Arts 118:179–199.

Jones, M.G.K., and D.H. Northcote. 1972. Nematode induced syncythium: A multinucleate transfer cell. J. Cell Sci. 10:789–809.

Joubert, J., and A. Dalmasso. 1972. Possible interaction between *Rhizoctonia solani* and *Meloidogyne incognita* on potato. Ann. Phytopathol. 4:409.

Khan, M. 1988. Biological control of golden cyst nematode, *Globodera rostochiensis* in potato with the fungus *Paecilomyces lilacinus*. p. 291–294. *In* M.A. Maqbool et al. (ed.) Advances in plant nematology. Natl. Nematol. Res. Centre, Univ. Karachi, Pakistan.

Kimpinski, J. 1979. Root lesion nematodes in potatoes. Am. Potato J. 56:79–86.

Kimpinski, J. 1982. The effect of nematicides on *Pratylenchus penetrans* and potato yields. Am. Potato J. 59:327–335.

Kornobis, S., and K. Stefan. 1991. Plant parasitic nematodes as pests of potatoes in Poland. EPPO Bull. 21:33–34.

Kort, J., H. Ross, H.J. Rumpenhorst, and A.R. Stone. 1977. An international scheme for identifying and classifying pathotypes of potato cyst-nematodes *Globodera rostochiensis* and *G. pallida*. Nematologica 23:333–339.

Kotcon, J.B., R. Loria, and D. J. Wixted. 1987. *Pratylenchus penetrans* population dynamics on three potato cultivars. J. Nematol. 19:361–368.

Krikum, J., and D. Orion. 1977. Studies of the interaction of *Verticillium dahliae* and *Pratylenchus thornei* on potato. Phytoparasitica 5:67.

LaMondia, J.A., and B.B. Brodie. 1984. An observation chamber technique for evaluating potential biocontrol agents of *Globodera rostochiensis*. J. Nematol. 16:112–115.

Linford, M.B., and J.M. Oliveira. 1940. *Rotylenchus reniformis* nov. gen., n. sp., a nematode parasite of roots. Proc. Helminthol. Soc. Washington 7:35–42.

Loof, P.A.A. 1991. The family Pratylenchidae Thorne, 1949. p. 363–421. *In* W. R. Nickle (ed.) Manual of agricultural nematology, Marcel Dekker, New York-Basel-Hong Kong.

Loof, P.A.A., and J. Bakker. 1992. Authorities of specific names in, and transfers to, the nominal genus *Globodera* Skarbilovich, 1959. Nematologica 38:385–386.

MacDonald, D.H., and W.F. Mai. 1963. Suitability of various cover crops as hosts for the lesion nematode, *Pratylenchus penetrans*. Phytopathology 53:730–731.

MacGuidwin, A.E. 1989. Distribution of *Pratylenchus scribneri* between root and soil habitats. J. Nematol. 21:409–415.

MacGuidwin, A.E., and D. I. Rouse. 1990. Role of *Pratylenchus penetrans* in the potato early dying disease of Russet Burbank potato. Phytopathology 80:1077–1082.

MacGuidwin, A.E., and S. A. Slack. 1991. Suitability of alfalfa, corn, oat, red clover and snapbean as hosts for the potato rot nematode, *Ditylenchus destructor*. Plant Dis. 75:37–39.

Mai, W.F. 1977. Worldwide distribution of potato-cyst nematodes and their importance in crop production. J. Nematol. 9:30–34.

Mai, W.F., J. R. Bloom, and T. A. Chen. 1977. Biology and ecology of the parasitic nematode *Pratylenchus penetrans*. Pennsylvania State Univ., Coll. Agric. Bull. 815.

Martin, M.J., R.M. Riedel, and R.C. Rowe. 1982. *Verticillium dahliae* and *Pratylenchus penetrans*: Interactions in the early dying complex of potato in Ohio. Phytopathology 72:640–644.

Mojtahedi, H., G.S. Santo, and J.H. Wilson. 1988. Host tests to differentiate *Meloidogyne chitwoodi* Races 1 and 2 and *M. hapla*. J. Nematol. 20:468–473.

Mojtahedi, H., G.S. Santo, and J. N. Pinkerton. 1991a. Efficacy of ethoprop on *Meloidogyne hapla* and *M. chitwoodi*, and enhanced biodegradation in soil. J. Nematol. 23:372–379.

Morgan-Jones, G., R. Rodríguez-Kábana, and P. Jatala. 1985. Fungi associated with cysts of potato cyst nematodes in Peru. Nematropica 16:21–31.

Mullin, B.A., and B.B. Brodie. 1988a. Effects of host resistance on second-stage juveniles and adult males of *Globodera rostochiensis*. J. Nematol. 20:335–339.

Mullen, B.A., and B.B. Brodie. 1988b. Effects of host resistance on the fecundity of Globodera rostochiensis. J. Nematol. 20:109–112.

Neal, J.C. 1889. The root-knot disease of the peach, orange and other plants in Florida, due to the work of *Anguillula*. USDA Bull. 20. U.S. Gov. Print. Office, Washington, DC.

Nijboer, H., and J.E. Parlevliet. 1990. Pathotype-specificity in potato cyst nematodes, a reconsideration. Euphytica 49:39–47.

Olthof, Th.H.A. 1968. Races of *Pratylenchus penetrans*, and their effect on black root rot resistance of tobacco. Nematologica 14:482–488.

Olthof, Th.H.A. 1986. Reaction of six *Solanum tuberosum* cultivars to *Pratylenchus penetrans*. J. Nematol. 18:54–58.

Olthof, Th.H.A. 1987. Effects of fumigants and systemic pesticides on *Pratylenchus penetrans* and potato yield. J. Nematol. 19:424–430.

Olthof, Th.H.A. 1989a. Control of *Pratylenchus penetrans* on potato with metam-sodium applied in irrigation water. Suppl. J. Nematol. 21:693–696.

Olthof, Th.H.A. 1989b. Effects of fumigant and nonfumigant nematicides on *Pratylenchus penetrans* and yield of potato. Suppl. J. Nematol. 21:645–649.

Olthof, Th.H.A. 1990. Reproduction and parasitism of *Pratylenchus neglectus* on potato. J. Nematol. 22:303–308.

Olthof, Th.H.A., and M.S. Wolynetz. 1991. *Pratylenchus penetrans* and *P. neglectus* in tubers of potato (*Solanum tuberosum*) in Ontario. Can. J. Plant Sci. 71:1251–1256.

Otazú, V., R. Hoopes, G. Cacro, and I. Huayta. 1985. El Rosario de la papa causado por *Nacobbus aberrans* (Thorne 1935) Thorne y Allen 1944. Su efecto en el rendimiento y algunus aspectos que inciden en su propagación y prevalencia en Bolivia. Fitopatologia 20:65–70.

Pinkerton, J.N., G.S. Santo, and H. Mojtahedi. 1991. Population dynamics of *Meloidogyne chitwoodi* on Russet Burbank potatoes in relation to degree-day accumulation. J. Nematol. 23:283–290.

Plaisted, R.L. 1971. 400 years of potato evolution. N.Y. Food Life Sci. Q. 4:24–26.

Programa de Investigacion de la Papa (PROINPA). 1991. Proyecto de Investigación de la Papa. Informe Anual 1990–91. Compendio, Cochabamba, Bolivia.

Rawsthorne, D., and B.B. Brodie. 1986a. Relationship between root growth of potato, root diffusate production, and hatching of *Globodera rostochiensis*. J. Nematol. 18:379–384.

Rawsthorne, D., and B.B. Brodie. 1986b. Root growth of susceptible and resistant potato cultivars and population dynamics of *Globodera rostochiensis* in the field. J. Nematol. 18:501–504.

Rebois, R.V., B.J. Eldridge, and R.E. Webb. 1978. *Rotylenchulus reniformis* parasitism of potatoes and its effect on yields. Plant Dis. Rep. 62:520–523.

Riedel, R.M., R.C. Rowe, and M.J. Martin. 1985. Differential interactions of *Pratylenchus crenatus, P. penetrans,* and *P. scribneri* with *Verticillium dahliae* in potato early dying disease. Phytopathology 75:419–422.

Riggs, R.D., and T.L. Niblack. 1993. Nematode pests of oilseed crops and grain legumes. p. 209–258. *In* K. Evans et al. (ed.) Plant parasitic nematodes in temperate climates. CAB Int., Wallingford, England.

Rodriguez-Kabana, R., and E.G. Ingram. 1977. Treatment of potato seed-pieces with oxamyl for control of plant-parasitic nematodes. Plant Dis. Rep. 61:29–31.

Rowe, R.C., J.R. Davis, M.L. Powelson, and D.I. Rouse. 1987. Potato early dying: Causal agents and management. Plant Dis. 71:482–489.

Rowe, R.C., R.M. Riedel, and M.J. Martin. 1985. Synergistic interactions between *Verticillium dahliae* and *Pratylenchus penetrans* in potato early dying disease. Phytopathology 75:412–418.

Santo, G.N., and J.N. Pinkerton. 1985. A second race of *Meloidogyne chitwoodi* discovered in Washington. Plant Dis. 69:361.

Santo, G.N., and M. Qualls. 1984. Control of *Meloidogyne* spp. on Russet Burbank potato applying metham sodium through center pivot irrigation systems. J. Nematol. 16:159–161.

Santo, G.N., H. Mojtahedi, and J.H. Wilson. 1991. Effect of rapeseed and sudangrass green manure on *Meloidogyne chitwoodi*. J. Nematol. 23:550.

Santo, G.N., J. H. O'Bannon, A.M. Finley, and A.M. Golden. 1980. Occurrence and host range of a new root-knot nematode (*Meloidogyne chitwoodi*) in the Pacific Northwest. Plant Dis. 64:951–952.

Santo, G.S., and J.H. Wilson. 1990. Evaluation of ethoprop and cadusafos for control of *Meloidogyne chitwoodi* on potato. Nematropica 20:137–142.

Sasser, J.N. 1979. Economic importance of *Meloidogyne* species in tropical countries. p. 359–374. *In* F. Lamberti and C.E. Taylor (ed.) Systematics, biology and control, Acad. Press, New York.

Sasser, J.N., and C. C. Carter. 1985. Overview of the International Meloidogyne Project 1975–1984. p. 19–24. *In* J.N. Sasser and C.C. Carter (ed.) An advanced treatise on Meloidogyne. Vol. I. North Carolina State Univ. Graphics, Raleigh.

Sayer, R.M., and M.P. Starr. 1988. Bacterial diseases and antagonisms of nematodes. *In* G.O. Poinar and H.-B. Jansson (ed.) Diseases of nematodes. CRC Press, Boca Raton, FL.

Sher, S.A. 1970. Revision of the genus *Nacobbus* Thorne and Allen, 1944 (Nematoda: Tylenchoidea). J. Nematol. 2:228–235.

Sieczka, J.B., D.D. Moyer, B.B. Brodie, and J.M. Kossowski. 1991. Marigold rotation studies on Long Island. Am. Potato J. 68:633.

Sitterly, W.R., and G. Fassuliotis. 1965. Potato losses in South Carolina due to the cotton root-knot nematode, *Meloidogyne incognita acrita*. Plant Dis. Rep. 49:723.

Slootweg, A.F.G. 1956. Root rot of bulbs caused by *Pratylenchus* and *Hoplolaimus* spp. Nematologica 1:192–201.

Spears, J.F. 1968. The golden nematode handbook. USDA Agric. Handb. 353. U.S. Gov. Print Office, Washington, DC.

Starr, M.P., and R.M. Sayer. 1988. *Pasteuria thornei* sp. nov. and *Pasteuria penetrans* sensu stricto emend., mycelia and endospore-forming bacteria parasitic, respectively, on plant parasitic nematodes of the genera *Pratylenchus* and *Meloidogyne*. Ann. Inst. Pasteur/Microbiol. 139:11–31.

Sterling, G.R. 1991. Biological control of plant parasitic nematodes. CAB Int., Wallingford, England.

Stone, A.R. 1973. *Heterodera pallida* n. sp. (Nematoda: Heteroderidae), a second species of potato cyst nematode. Nematologica 18:591–606.

Storey, G.W., and K. Evans. 1987. Interactions between *Globodera pallida* juveniles, *Verticillium dahliae* and three potato cultivars, with descriptions of associated histopathologies. Plant Pathol. 36:192–200.

Swaminathan, M.S. and R.L. Sawyer. 1983. The potential of potato as a world food. p. 3–4. *In* W.J. Hooker (ed.) Research for the potato in the year 2000. Int. Potato Center, Lima, Peru.

Thorne, G. 1961. Principles of nematology. McGraw-Hill, New York.

Timper, P., and B.B. Brodie. 1993. Infection of *Pratylenchus penetrans* by nematode-pathogenic fungi. J. Nematol. 25:297–302.

Townshend, J.L. 1984. Anhydrobiosis in *Pratylenchus penetrans*. J. Nematol. 16:282–289.

Townshend, J.L., and Th.H.A. Olthof. 1988. Growth of potato and control of *Pratylenchus penetrans* with oxamyl-treated seed pieces in greenhouse studies. J. Nematol. 20:405–409.

Townshend, J.L., J.W. Potter, and C.B. Willis. 1978a. Ranges of distribution of species of *Pratylenchus* in Northeastern North America. Can. Plant Dis. Surv. 58:80–82.

Townshend, J.L., R. Tarte, and W.F. Mai. 1978b. Growth response of three vegetables to smooth- and crenated-tailed females of three species of Pratylenchus. J. Nematol. 10:259–263.

Triantaphyllou, A.C. 1985. Cytogenetics, cytotaxonomy and phylogeny of root-knot nematodes. p. 113–126. *In* J.N. Sasser and C.C. Carter (ed.) An advanced treatise on *Meloidogyne*. Vol. 1. North Carolina State Univ. Graphics, Raleigh.

Trudgill, D.L. 1985. Potato cyst nematodes: A critical review of the current pathotyping scheme. EPPO Bull. 15:273–279.

Trudgill, D.L. 1986. Concepts of resistance, tolerance and susceptibility in relation to cyst nematodes. p. 179–189. *In* F. Lamberti and C.E. Taylor (ed.) Cyst nematodes. Plenum Press, New York and London.

Trudgill, D.L. 1991. Resistance to and tolerance of plant parasitic nematodes in plants. Annu. Rev. Phytopathol. 29:167–192.

Trudgill, D.L., M.S. Phillips, and T.J.W. Alphey. 1987. Integrated control of potato cyst nematode. Outlook Agric. 16:167–172.

Turner, S.J., and A.R. Stone. 1984. Development of potato cyst-nematodes in roots of resistant *Solanum tuberosum* ssp. *andigena* and *S. vernei* hybrids. Nematologica 30:324–332.

Uziel, A., and R.A. Sikora. 1992. Use of non-target isolates of the entomopathogen *Verticillium lecanii* (Zimm.) viegas to control the potato cyst nematode, *Globodera pallida* (Stone). Nematologica 38:123–130.

Van Meggelen, J.C., G. Karssen, G.J.W. Janssen, B. Verkerk-Bakker, and R. Janssen. 1994. A new race of *Meloidogyne chitwoodi* Golden O'Bannon, Santo and Finley, 1980? Fund. Appl. Nematol. 17:93–96.

Weingartner, D.P., J.R. Shumaker, G.C. Smart, and D.W. Dickson. 1975. A new nematode control program for potatoes grown in northeast Florida. Proc. Fla. State Hort. Soc. 88:175–182.

Wheeler, T.A., L.V. Madden, R.M. Riedel, and R.C. Rowe. 1994. Distribution and yield-loss relations of *Verticillium dahliae, Pratylenchus penetrans, P. scribneri, P. crenatus,* and *Meloidogyne hapla* in commercial potato fields. Phytopathology 84:843–852.

Whitehead, A.G. 1986. Chemical and integrated control of cyst nematodes. p. 413–432. *In* F. Lamberti and C.E. Taylor (ed.) Cyst nematodes. Plenum Press, New York and London.

Wilcox, J., and A.T. Tribe. 1974. Fungal parasitism of cysts of Heterodera. I. Preliminary investigations. Trans. Br. Mycol. Soc. 62:585–594.

Winslow, R.D., and R.J. Willis. 1972. Nematode diseases of potato. *In* J.M. Webster (ed.) Economic nematology. Academic, New York.

Zaki, M.H., H.D. Moran, and D. Harris. 1982. Pesticides in ground water: The aldicarb story in Suffolk County, NY. Am. J. Publ. Health 72:1391–1395.

Zunke, U. 1990. Ectoparasitic feeding behaviour of the root lesion nematode, *Pratylenchus penetrans*, on root hairs of different host plants. Rev. Nématol. 13:331–337.

Zunke, U. 1991. Observations on the invasion and endoparasitic behavior of the root lesion nematode *Pratylenchus penetrans*. J. Nematol. 22:309–320.

25 Vegetable Crops

ALVA W. JOHNSON

USDA-ARS, Nematodes, Weeds, and Crops Research Unit
Tifton, Georgia

Most vegetable crops are attacked by one or more species of nematodes and are often involved in complexes including other soilborne plant pathogens that limit crop production. It is difficult, by aboveground symptoms alone, to distinguish between nematode injury and damage caused by other factors, such as soilborne pathogens, inadequate nutrition or irrigation, pests, or herbicides. Under optimum growing conditions, especially ample moisture, fertility, and good weed control, most vegetable crops may produce an acceptable yield while supporting large numbers of nematodes. When present, aboveground symptoms of nematode-infected plants are similar to those caused by conditions limiting efficient functioning of the root system. Erratic stands, chlorosis, stunting, incipient wilting, and premature leaf drop usually are the first visible symptoms of large populations of nematodes.

Although often not noticed, nematodes cause major yield losses in many vegetable crops. Damage caused by plant-parasitic nematodes on 24 vegetable crops in the USA was estimated to be approximately 11%, or 3.8 Tg (Feldmesser et al., 1971). Vegetable crop losses caused by nematodes are unequally distributed across the USA because of regional climatic differences (Noe & Sikora, 1990). Greater losses usually occur in the warmer and more humid regions (Johnson, 1992).

During the past 50 yr, efforts in nematicide development and application, genetic resistance, and cultural practices have yielded a number of means of suppressing nematode damage to vegetable crops. The few currently registered nematicides offer the most immediate short-term solutions for nematode control. However, nematicide use is usually justified only where the total costs [chemical plus application costs, approximately \$24–\$100 ha^{-1} (\$10–\$40 acre^{-1})] can be returned threefold or more in value from increased yield of high cash value crops (Taylor & Sasser, 1978). Modern long-term sustainable nematode management focuses on the interfacing of several practices into an integrated system designed for specific crops and locations. Examples of the many methods used to manage parasitic nematodes on vegetable crops are crop rotation, cultural practices (land preparation, flooding, fallow, time of planting, trap crops, and soil solarization), resistant cultivars, chemical, biological, and the integration of various strategies.

IMPORTANT NEMATODE PARASITES

Root-Knot Nematodes

Species and Races

Root knot has been known as a disease of vegetable crops since 1855, when Berkeley (1855) in England first described the disease on cucumber (*Cucumis sativus* L.) roots (Eisenback & Triantaphyllou, 1991). The causal organism was described as *Heterodera radicicola*. From 1884 to 1949, root-knot nematodes were considered a single species in combination with cyst nematodes and referred to by a number of designations (Johnson & Fassuliotis, 1984). Chitwood (1949) described morphological differences among populations, and he reassigned the root-knot nematodes to the genus *Meloidogyne*. At that time, *M. incognita*, *M. incognita acrita*, *M. arenaria*, *M. javanica*, *M. hapla*, and *M. exiqua* were recognized primarily on the basis of perineal patterns and other morphological characteristics. *Meloidogyne incognita acrita* was later synonymized with *M. incognita* by Triantaphyllou and Sasser (1960). Since 1949, some 90 to 100 species of *Meloidogyne* have been described (Eisenback & Triantaphyllou, 1991).

More than 1000 living populations of *Meloidogyne* species collected from 75 countries of the world have been subjected to the differential host test and identified by morphological study and by chromosome counts (Taylor & Sasser, 1978). More than 95% of the samples were collected from crops, including vegetables, growing in farm fields. When the samples were compiled by species the results were: 52% *M. incognita*, 30% *M. javanica*, 8% *M. arenaria*, and 8% *M. hapla* for a total of 98%. The remaining 2% were *M. exigua* and other species. Based on the differential host test (Taylor & Sasser, 1978), *M. incognita* consist of four races; *M. arenaria* has two races; and *M. javanica* and *M. hapla* show no clearly defined host races. However, both *M. hapla* and *M. arenaria* have two cytological races (Eisenback & Triantaphyllou, 1991). Later, *M. chitwoodi* was reported as an important pest in the Pacific Northwest. This species attacks many hosts including carrot [*Daucus carota* subsp. *sativus* (Hoffm.) Arcang.], turnip (*Brassica rapa* L.), sweet corn (*Zea mays* L.), snapbean (*Phaseolus vulgaris* L.), cowpea [*Vigna unguiculata* (L.) Walp.], garden pea (*Pisum sativum* L.), onion (*Allium cepa* L.), tomato (*Lycopersicon esculentum* Miller), and eggplant (*Solanum melongena* L.) (O'Bannon et al., 1982). Two host races have been described for populations of *M. chitwoodi* (Santo & Pinkerton, 1985). Populations of Race 1 cannot reproduce on alfalfa (*Medicago sativa* L.), and most carrot cultivars are good to moderate hosts. Race 2 populations readily reproduce on alfalfa but not on most cultivars of carrot.

Symptoms

The aboveground symptoms of vegetable plants infected with *Meloidogyne* spp. include erratic stands (Fig. 25–1), stunting (Fig. 25–2), chlorosis, wilting (Fig. 25–3), and premature leaf drop (Fig. 25–4). The belowground symptoms and signs include root-galls (Fig. 25–5) and egg masses. Infected plants also tend

Fig. 25–1. Erratic stand of sweet corn in a field infested with *Meloidogyne* spp.

to have fewer secondary roots than normal, and symptoms may be a combination of galls and coarse roots. The roots of most plants infected with *M. hapla* tend to form branches, and galls often have one or more short lateral rootlets. The number of galls and egg masses that occur on roots of plants may vary with the nematode population densities in the soil, susceptibility of the host, and environmental conditions.

Biology and Life History

Meloidogyne incognita, *M. javanica*, *M. arenaria*, *M. hapla*, and *M. chitwoodi* are the most important root-knot nematode species associated with veg-

Fig. 25–2. Sweet corn growing in soil infested with *Meloidogyne* spp.: left = untreated, right = soil treated with a nematicide.

Fig. 25–3. Cucumber wilting in moist soil infested with root-knot nematodes.

etable crops. The general biology and pathogenicity of these root-knot nematode species are similar on all vegetable crops. The oval-shaped eggs are deposited in a gelatinous egg sac surrounding the posterior end of the pear-shaped female nematode whose head is embedded in the root. The first molt occurs within the egg. Upon hatching, the slender, vermiform second-stage juveniles are attracted to plant roots (Christie, 1936). The infective second-stage juveniles usually enter and move through the growing point of the root and region of cell elongation. They migrate through the undifferentiated root cells until they find the developing vascular system where they begin to feed. Nematodes pierce cell walls with

Fig. 25–4. Premature leaf drop on okra growing in soil infested with root-knot nematodes.

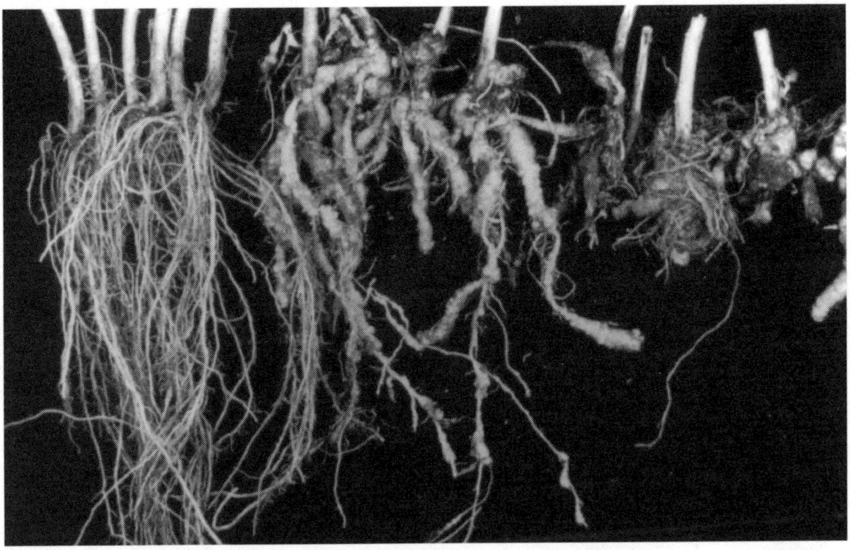

Fig. 25–5. Roots of pole bean from soil infested with root-knot nematodes: left = soil treated with a nematicide + film mulch; center = untreated + film mulch; right = untreated, no film mulch.

their stylet, and secretions from the esophageal glands are extruded through the stylet which alter the plant cells on which the nematodes feed. These "nurse" or "food" cells are called giant cells and grow to various sizes, depending on the host plant and temperature. The chemical nature of nematode secretions and their biological activity are only partially understood (Hussey, 1992).

Nematodes remain sedentary during feeding and enlarge in width but not in length. About 14 d later, the juvenile molts and the stylet and esophageal bulb disappear. Three molts take place between Days 14 and 18 and no further feeding occurs until after the fourth and final molt. Soon after the fourth molt, the stylet and median bulb are regenerated, the female reproductive system is formed, and a perineal pattern is visible (Eisenback & Triantaphyllou, 1991). The female undergoes several changes in shape during development, and after the last molt the female grows rapidly and becomes pear-shaped at maturity.

In the developing male there is rapid metamorphosis; a vermiform nematode develops inside the cuticle, possessing a stylet, esophagus with median bulb, spicules, and sperm in the testis. The male develops faster than the female and emerges from the old cuticle in search of a female even though most species reproduce by parthogenesis (Eisenback & Triantaphyllou, 1991).

Disease Complexes

In vegetable production fields, infection of plants by *Meloidogyne* species alone is improbable; fungi, bacteria, and viruses are most always present and often interact with these nematodes. A number of excellent reviews have been written on the interaction of *Meloidogyne* spp. and other plant-parasitic nematodes with other plant pathogens including those by Johnson and Fassuliotis (1984), Webster (1985), Netscher and Sikora (1990) and Potter and Olthof

(1993). These authors describe many specific interactions. In most interactions involving fungi, the nematode was not essential for the establishment and development of the fungal pathogen. The nematode usually assisted the fungus by altering the number of infection sites and speed of the development of the pathogen, and thus the severity of disease. The major genera of fungi that interact with root-knot nematodes on vegetable crops include *Fusarium*, *Rhizotonia*, *Pythium*, and *Sclerotium*.

The most common complexes include several species of *Fusarium*, *Pythium*, *Rhizoctonia*, or *S. rolfsii* on tomato (Johnson & Fassuliotis, 1984); sweetpotato [*Ipomoea batatas* (L.) Lam.] (Giamalva et al., 1962); celery (*Apium graveolens* L.) (Starr & Mai, 1976); okra [*Abelmoschus esculentus* (L.) Moench] (Johnson et al., 1977); cucumber (*Cucumis sativus* L.) (Johnson et al., 1979b,c); watermelon [*Citrullus lanatus* (Thunb.) Matsum & Nakai] (Sumner & Johnson, 1973); squash (*Cucurbita pepo* var. *melopepo*) (Johnson et al., 1979c); polebean (*Phaseolus vulgaris* L.) (Johnson et al., 1979a; France & Abawi, 1994); snapbean (*Phaseolus vulgaris* L.) and southern pea [*Vigna unguiculata* (L.) Walp.] (Smittle & Johnson, 1982), and other vegetable crops (Sumner et al., 1976, 1978; Johnson & Fassuliotis, 1984). When *M. incognita* and *S. rolfsii* occurred together on eggplant, 25% of the plants wilted compared to 6% wilted when the fungus alone was present (Goswami et al., 1970). Golden and Van Gundy (1975) demonstrated that okra and tomato roots infected with *M. incognita* are highly susceptible to infection by *R. solani*, the fungus is specifically attracted to nematode gall tissue, and that sclerotia are selectively formed on nematode galls. The severe root rot of tomato caused by *M. incognita* and *R. solani* is associated with nutrient mobilization into gall tissue and root exudation (Van Gundy et al., 1977).

The first report associating plant-parasitic nematodes and bacteria in a disease complex was that of Hunger (1901) who showed that tomato became infected by the wilt bacterium (*Pseudomonas solanacearum*) in the presence of the root-knot nematodes, but remained healthy in their absence. Interactions of the root-knot nematode (*M. incognita*) and bacteria were described in a comparison of gall development of tomato seedlings grown under aseptic and septic conditions (Pitcher, 1965). Plants infected by root-knot nematodes and grown under aseptic conditions had massive galls and were firm and did not disintegrate, whereas galls formed under septic conditions became necrotic. Bacterial wilt, caused by *P. solanacearum*, is more severe in root-knot nematode resistant cultivars of tomato and eggplant in the presence of *M. incognita*. Tomato plants inoculated with *M. incognita* plus the bacterium develop wilt 1 to 3 wk earlier than those with the bacterium alone. The combination of the two organisms also reduced the survival rate of wilt-resistant tomato plants by 33 to 36% (Valdez, 1978). A positive interaction of *M. incognita* and tobacco mosaic virus on eggplant has been reported (Goswami & Chenulu, 1974).

Economic Importance

The role *Meloidogyne* plays in total crop loss is difficult to ascertain in cases where crops are suffering from inadequate nutrition or soil moisture or from simultaneous attack by fungi, bacteria, viruses, insects and other nematodes, a situation very common in crop production. *Meloidogyne incognita*, *M. javanica*, *M. arenaria*, and *M. hapla* are the most widespread and common *Meloidogyne* species worldwide and very probably cause more damage to vegetable crops than all other *Meloidogyne* species combined (Sasser & Freckman, 1987), but additional research is needed to support this hypothesis. *Meloidogyne chitwoodi* attacks many vegetable crops in the Pacific Northwest as well as in other countries such as the Netherlands (O'Bannon et al., 1982; Coomans, 1993; Eisenback & Triantaphyllou, 1991). However, the economic impact on vegetable production is unknown.

Cyst Nematodes

Species

Fifty-seven species of *Heterodera* have been described, but only two species are important parasites of vegetable crops (Baldwin & Mundo-Ocampo, 1991). The sugar beet nematode (*H. schachtii*) and the brassica cyst nematode (*H. cruciferae*) both parasitize cruciferous crops. *Heterodera schachtii* was first observed in 1859 on the roots of sugar beet (*Beta vulgaris* L.) in Germany and now occurs in most sugar beet-production regions of the world (Baldwin &d Mundo-Ocampo, 1991). *Heterodera schachtii* occurs in most of the sugar beet-growing areas in the USA from Michigan to California (Thorne, 1952). Isolated infestations have been found on crucifers in New York and Florida.

Heterodera cruciferae was first reported on brussels sprout plants (*Brassica oleracea* var. *gemmifera* DC.) in California (Raski, 1950), the only known distribution of *H. cruciferae* in the USA. *Heterodera carotae* (the carrot cyst nematode) was found in 67% of the carrot production fields surveyed in Michigan (Berney & Bird, 1992). To date, Michigan remains the only area in North or South America to report *H. carotae*, but information on the extent of damage to carrot production is lacking. *Heterodera goettingiana* (the pea cyst nematode) is an important parasite on bean (*Vicia faba* L.) and pea (*Pisum sativum* L.) in many temperate regions including the Pacific Northwest (Stone & Course, 1974; Haglund, 1984). *Heterodera glycines* causes isolated problems, especially among legumes, including snapbean and pea (Baldwin & Mundo-Ocampo, 1991).

Symptoms

Symptoms caused by *H. cruciferae* and *H. schachtii* are similar. *Heterodera cruciferae* has many hosts in common with *H. schachtii* but its host range is more limited. Broccoli (*B. oleracea* var. *botrytis* L.), brussels sprout, cabbage (*B. oleracea* var. *capitata* L.), cauliflower (*B. oleracea* var. *botrytis* L.), kale (*B. oleracea* var. *viridis* L.), radish (*Raphanus sativus* L.), rutabaga [*B. napus* var. *napobrassica* (L.) Reichb.], table beet (*Beta vulgaris* L.), turnip, and similar crops may be severely damaged by *H. cruciferae* or *H. schachtii* (Figs. 25–6, 25–7). Brassica seedlings infected with *H. cruciferae* produce extra lateral roots often bearing small gall-like formations and stunted shoots with either reddened leaves or interveinal chlorosis associated with decline and dying of plants (McCann, 1981). As

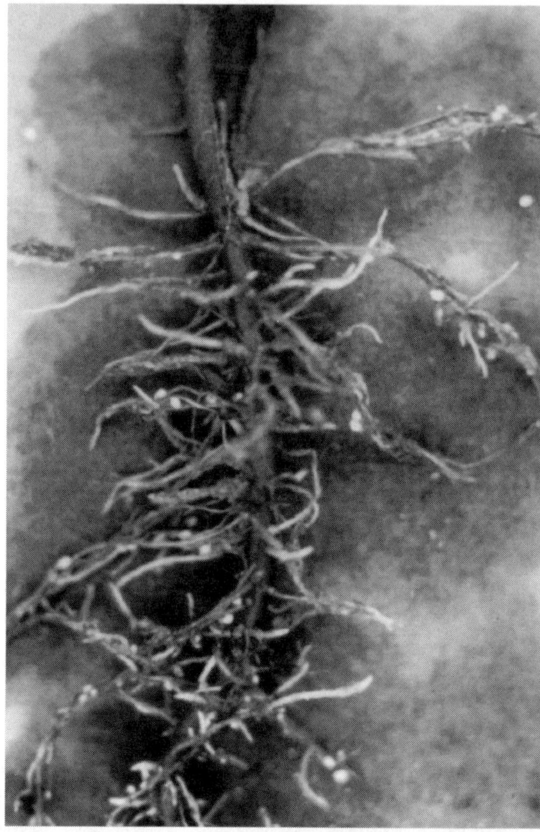

Fig. 25–6. Roots of red table beet cv. Detroit Dark Red infected with *Heterodera schachtii* (courtesy of Dr. John Potter).

Fig. 25–7. Sparse stand of red table beet in soil infested with *Heterodedra schachtii* (courtesy of Dr. John Potter).

the season progresses, areas of sparse and stunted plants may appear pale yellowish green in color, and most older leaves die.

Biology and Life History

The biology and life history of *H. schachtii* and *H. cruciferae* are similar. *Heterodera schachtii* eggs borne within mature females and newly formed cysts hatch in the presence of host plants (Raski, 1950). After hatch and emergence from the cyst, the juveniles enter host plant roots, feed on "nurse cells," develop to maturity, and complete the life cycle in about 30 d (Baldwin & Mundo-Ocampo, 1991). The nematodes are amphimictic. Soon after fertilization, the adult females appear as white lemon-shaped bodies attached to the host roots. A gelatinous matrix, produced in the uterine wall, extruded from the vulval region surrounds the posterior end of the adult female in which 200 or more eggs are deposited. After a short period of growth, the female dies and its cuticular body forms a tough, saclike protective cyst containing from <10 to >600 eggs. The second-stage juveniles are stimulated to hatch by root diffusates from *Brassica* spp., but not from diffusates of other cruciferous hosts (Lewis, 1971).

Disease Complexes

Disease complexes involving cyst nematodes and other root-invading organisms on vegetables are not common (Potter & Olthof, 1993). *Heterodera carotae* occurred in association with *M. incognita* on carrot (Lamberti, 1971). A synergistic suppression of tomato by *H. schachtii* and *M. hapla* was reported by Griffin (1985). However, in California, tomato in rotation with sugar beet for many years are not affected by *H. schachtii* in the presence or absence of root-knot nematodes (Baldwin & Mundo-Ocampo, 1991).

Economic Importance

The worldwide agricultural impact of cyst nematodes on all crops is probably surpassed only by root-knot nematodes (Baldwin & Mundo-Ocampo, 1991). In New York State, *H. schachtii* causes significant yield losses to cabbage and table beet. Marketable yields of both crops are reduced from 23 to 54% by initial population densities of 9 to 64 eggs and juveniles per gram of soil (Abawi & Mai, 1980). Olthof et al. (1974) reported similar losses in Ontario, Canada. A marked reduction in curd quality of cauliflower results when *H. cruciferae* and *H. schachtii* population densities exceed 75 000 eggs/kg (75 eggs/g) of soil (Sykes & Winfield, 1966).

Root-Lesion Nematode

Species and Races

A key and diagnostic compendium published by Handoo and Golden (1989) includes 55 species of *Pratylenchus*. Ten species of *Pratylenchus* have been found in the rhizosphere or roots of vegetable crops (Netscher & Sikora, 1990). *Pratylenchus* spp. have a large geographical distribution with a wide host range on vegetable and rotation crops (Loof, 1991). The most often cited species of *Pratylenchus* that damage vegetable crops include *P. brachyurus*, *P. coffeae*, *P. penetrans*, *P. scribneri*, and *P. vulnus* (Jensen, 1972). These species of *Pratylenchus* should be considered of potential importance when found in root tissue.

Symptoms

The aboveground symptoms of plants infected with *Pratylenchus* spp. include a gradual decline or lack of plant vigor (Fig. 25–8). Shoots of infected plants are stunted and chlorotic and wilt readily.

The belowground symptoms vary for different crop species, but they are characterized by the lack of fibrous roots and root pruning that results from the formation of dark brown elongate lesions containing eggs, juveniles, and adult nematodes. Severely affected plants may be easily pulled from the soil because of almost complete destruction of the root systems. Moderately parasitized plants exhibit varying stages of root survival, but generally the individual roots are discolored, stubby, and surviving roots are found near the soil surface. The affected roots consist of cortical lesions varying from pale yellow to almost black. Injury is caused by mechanical and enzymatic destruction of cells, causing discoloration of nearby intact cells and subsequent accumulation of phenolic compounds. These lesions generally result in complete girdling of the roots followed by loss of the cortical tissues through sloughing off (Loof, 1991). Infected roots are poor absorbers of water and nutrients.

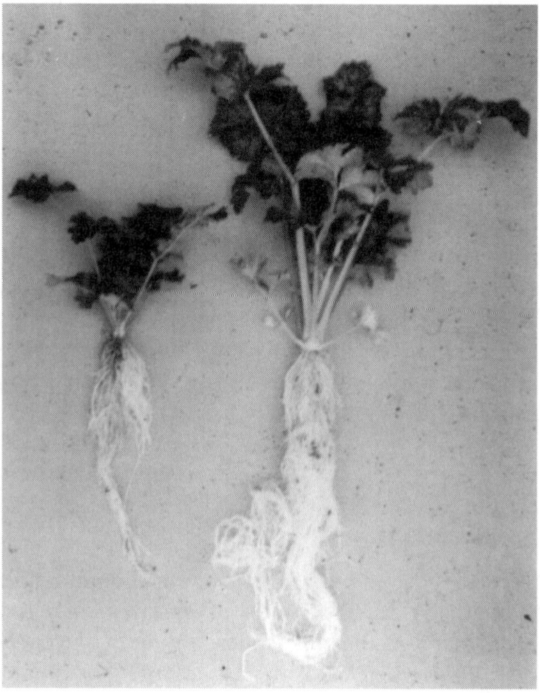

Fig. 25–8. Celery from soil infested with *Pratylenchus penetrans* (left) and noninfested soil (right) (courtesy of Dr. John Potter).

Biology and Life History

The life cycle of *Pratylenchus* spp. is simple, and reproduction is frequently by bisexual cross-fertilization, taking 30 to 90 d for completion, depending on soil temperature (Loof, 1991). Some species of *Pratylenchus* reproduce by amphimixis, whereas others reproduce by parthenogenesis. The female deposits eggs singly in roots or soil. The first molt occurs in the egg, the second-stage juvenile hatches from the egg, and molts three additional times between feeding intervals to become an adult. *Pratylenchus* spp. enter the feeder roots and begin to reproduce soon after the rootlets form. Lesion nematodes are migratory endoparasites, and all juvenile stages and adults can infect plant roots. The nematodes overwinter in soil and roots primarily as adults and fourth-stage juveniles and as eggs in roots. *Pratylenchus penetrans* is most often found in sandy soil, and the nematodes migrate farther in coarse soil rather than in fine-textured soil. All stages of the nematode can be found as deep as 60 cm in sandy soil.

Disease Complexes

Several interactions between *Pratylenchus* species and fungi occur in vegetable crops. *Pratylenchus* spp. are important parasites of pea and are known to form disease complexes with many different soilborne root rotting fungi, thereby increasing root damage (Celetti et al., 1990).

Pratylenchus penetrans is not pathogenic to eggplant at 4000 nematodes per pot; however, in the presence of *Verticillium albo-atrum* a synergistic relationship occurs and severe crop damage results (Taylor, 1979). At low and intermediate levels of the wilt organism, the incidence of wilt was increased in the presence of the nematode. Also, in the presence of the wilt fungus, *P. penetrans* reproduction increased. On pepper (*Capsicum annuum* L.) and tomato, the final population densities of *P. penetrans* were higher in the presence of *Verticillium* than in its absence (Mountain & McKeen, 1962). In pepper, the damage caused by *Verticillium* and *P. penetrans* was additive (Olthof & Reyes, 1969). In addition to interactions with fungi, some bacterial infections are enhanced by *Pratylenchus*, including bacterial wilt in tomato caused by *Pseudomonas* (Dropkin, 1989).

Economic Importance

Pratylenchus spp. are probably the second-most important group of nematodes, following *Meloidogyne* spp., of economic importance to vegetable production in temperate agriculture (Potter & Olthof, 1993). Tomato is a host of at least five species: *P. brachyurus*, *P. coffeae*, *P. penetrans*, *P. scribneri*, and *P. vulnus* (Jensen, 1972). *Pratylenchus* spp. have a cosmopolitan distribution with a wide host range on other vegetable crops including pea, cabbage, cowpea, carrot, pepper, tomato, spinach (*Spinacia oleracea* L.), lettuce (*Lactuca sativa* L.), beet, radish, celery, onion (*Allium cepa* L.), eggplant, and bean (*Phaseolus* spp.). According to Mai et al. (1977, p. 1–64), *P. penetrans* is the most economically important plant-pathogenic nematode in the northeastern USA and also is found in other regions (Johnson & Fassuliotis, 1984).

Bean plants are highly susceptible to several species of *Pratylenchus* (Sikora & Greco, 1990). On 'Kentucky Wonder', pathogenicity tests in pot-grown plants indicated that the nematode increased nearly 75-fold over a 50-d period and affected the shoot growth. 'Fordhook 52' lima bean (*Phaseolus lunatus* L.) is a poor host to *P. scribneri*. Attacked roots were able to overcome the nematode infection by eliciting a hypersensitive response that allowed only a minimal reproduction of the nematode (Thomason et al., 1976). In greenhouse experiments, 'Top Crop' bean was an excellent host of *P. neglectus* (Olthof, 1979). After 120 d, plants that had been initially inoculated with 50 nematodes per pot supported about 30 000 nematodes. Cabbage plants may carry large population densities of *P. penetrans* (Acedo & Rohde, 1971).

Onion yields in northern Indiana are severely restricted by *P. penetrans* (Bergeson, 1962). The nematode initially causes long, narrow, grayish, and opaque root lesions. A single nematode causes a lesion 1 to 3 mm long and one to two cells wide. Onion seedlings show characteristic symptoms of injury when more than 100 *P. penetrans* are present per gram of root (Ferris, 1962).

Bulb and Stem and "Potato-Rot" Nematodes

Species and Races

A recent compilation of nominal species of the genus *Ditylenchus* is reported by Sturhan and Brzeski (1991).

Ditylenchus dipsaci is composed of a complex of races that are differentiated from each other by their respective host ranges (Johnson & Roberts, 1995). Many races have a wide host range, whereas others are host specific. More than 30 biological races of *D. dipsaci* are known, which are named either for their main preferred host or for the plant on which they were first detected (Sturhan, 1971). Mixtures of two or more races may occur in the same field.

Ditylenchus destructor (the potato rot nematode) attacks stolons and tubers. This nematode can cause great losses in potato production, especially in eastern Europe (Sturhan & Brzeski, 1991). A more in-depth coverage of nematodes on potato is presented in Chapter 24 in this monograph (Brodie, 1998).

Symptoms

Symptoms of injury differ for different plant species and different parts of the same plant (Netscher & Sikora, 1990). Infected seedlings become twisted, enlarged, and deformed and frequently die in severely infested areas of the field. The symptoms on infected onion and garlic (*Allium sativum* L.) plants include: erratic stands (Fig. 25–9), stunting, looping, and bending of leaves (Fig. 25–10), pimplelike spots called spikkels, and extensive longitudinal splitting of cotyledons and leaves. Leaves are short and thickened, and frequently contain brown or yellowish spots and stem swelling (Johnson & Roberts, 1995). As the season progresses, onion and garlic foliage collapses and a softening of the bulb begins at the neck and gradually proceeds downward. Scales become soft and light gray (Fig. 25–11). On garlic, symptoms arising from infested seed lots are often not

Fig. 25–9. Erratic stand of garlic in a field infested with *Ditylenchus dipsaci* (courtesy of Dr. P. A. Roberts).

apparent until midseason. Bulbs often become desiccated, shrunken, and light in weight (Fig. 25–12). Bacterial soft rot usually completes the process of destruction and is accompanied by an offensive odor. Many diseased onions decay at the base and contain various types of secondary invaders. Similar decay and secondary infection occurs in infected garlic bulbs. During drying before harvest,

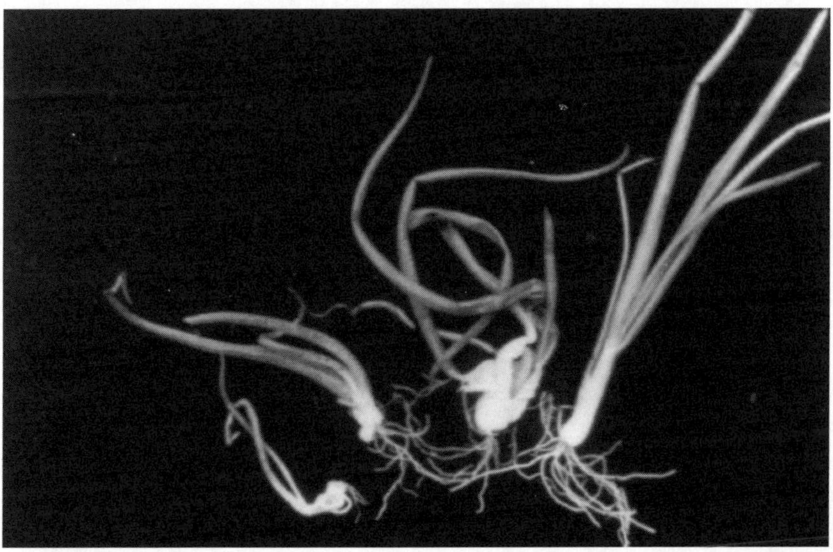

Fig. 25–10. Onion from soil infested with *Ditylenchus dipsaci*: right = healthy; left = infested (courtesy of Dr. R. S. Hussey).

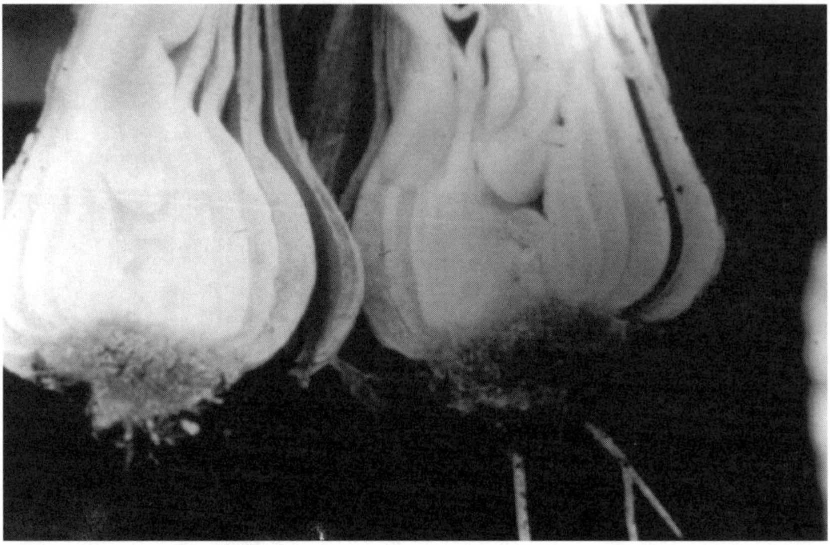

Fig. 25–11. Garlic bulbs from soil infested with *Ditylenchus dipsaci* (courtesy of Dr. P. A. Roberts).

infected garlic bulbs become so light they are seldom harvested. Symptom response is related to nematode population densities.

Biology and Life History

Ditylenchus dipsaci penetrates germinating onion seeds usually before cotyledons emerge from the soil and inhabit the parenchyma tissue beneath the hilum or in the cotyledon (Johnson & Roberts, 1995). While seedlings are young

Fig. 25–12. Garlic bulbs from soil infested with *Ditylenchus dipsaci*: left = healthy, right = infested (courtesy of Dr. P. A. Roberts).

and tissues are soft, *D. dipsaci* may invade any portion of the onion plant (Sturhan & Brzeski, 1991). When plants are more mature, nematodes are likely to enter the base of the bulb where the roots originate. Damage to onion seedlings is attributed to feeding activity of nematodes in the cortical parenchyma as the nematodes migrate through the tissue. Cells are punctured and contents are withdrawn until cells are so weakened that they collapse under the pressure of the nematode's body. There is little evidence that *D. dipsaci* is capable of causing much direct mechanical injury. Most damage appears to result from the effects of salivary secretion of the nematode on the surrounding tissue. In storage tissues, these separated cells become rounded, which usually give the structure a whitish appearance and a mealy texture.

The life cycle of D. *dipsaci* on onion seedlings is completed in 19 to 23 d at 15°C with four molts and four juvenile stages (Yuksel, 1960). The first molt takes place within the egg. About 2 d after oviposition, second-stage juveniles hatch from eggs and within 4 to 5 d develop into adults. The fourth juvenile stage of *D. dipsaci* is the primary invader, penetrating the epidermis or through stomata of stems and leaves (Sturhan & Brzeski, 1991). Mating is necessary for reproduction and the female begins to oviposit 4 d after the final molt. Females deposit 8 to 10 eggs each day for 25 to 50 d for a total of 200 to 500 eggs. Males and females live for 45 to 73 d under normal conditions of temperature and moisture.

Ditylenchus dipsaci is very tolerant to desiccation, and may survive in the dried state for several years by the process of anhydrobiosis. The fourth-stage juvenile is the primary anhydrobiotic stage, although other stages also can enter anhydrobiosis. *Ditylenchus dipsaci* also is very resistant to low temperatures (Evans, 1987). In freezing experiments, some juveniles survived an exposure of −150°C for 18 mo (Sayre & Wang, 1975). Soil moisture appears to affect both nematode movement and longevity. During heavy rains, nematodes become active and move up the plant in a film of water and enter the foliage through stomata. Rain and irrigation water also cause local plant-to-plant spread of the nematode. Vertical migration of juveniles is slower when soil pore spaces are filled with water than when they are empty.

Populations of *D. dipsaci* in infected plants can increase by more than 1000-fold in a season, with as many as 50 000 per plant. Population densities in soil fluctuate, depending on soil type and host plants. In heavy clay soils, low levels of infestation generally increase during the summer and decrease in the winter whether a host or nonhost is grown. Damage occurs in many crops at infestation levels as low as 10 nematodes per 500 g of soil. An aggregated or skewed distribution of *D. dipsaci* occurs in infected garlic seed-clove lots. In field plantings of infected garlic lots, most nematodes are concentrated on a few seeds or cloves. Large populations of the nematode may be recovered from the soil and tissue remnants of bulb scales, leaves, and roots.

Optimum conditions for survival in soil are a combination of low soil moisture and soil temperature near or below freezing. Mortality of *D. dipsaci* in the absence of a host is greater at a soil temperature of 21°C when soil moisture is at field capacity than at the wilting point. Soil temperature of 21°C is optimum for penetration of onion seedlings, movement and reproduction, and symptom severity by these nematodes.

Disease Complexes

Ditylenchus dipsaci interacts with bacteria, fungi, maggots, thrips larvae, bulb mites, and a number of saprophagous nematodes in soft-rot complexes of onion and garlic.

Economic Importance

Ditylenchus dipsaci is among the nematodes of greatest economic impact worldwide (Sturhan & Brzeski, 1991). This pathogen is by far the most important nematode on garlic (Johnson & Roberts, 1995). In some areas of Russia, losses of garlic or onion reached 60% (Kirjanova & Krall, 1980).

The onion-garlic race of *D. dipsaci* attacks onion, garlic, leek (*Allium porrum* L.), chive (*A. schoenoprasum* L.), pea, parsley (*Petroselinum crispum* Miller), celery, lettuce, hairy nightshade (*Solanum* sp.), and salsify (*Tragopogon porrifolius* L.). *Ditylenchus dipsaci* also is an important parasite of broad bean (*Vicia faba* L.), pea, carrot, and parsnip (*Pastinaca sativa* L.). This nematode is widespread from New Hampshire to California and is a potential pathogen in onion- and garlic-production areas. The nematode is spread by planting infected sections of the bulb called cloves (Johnson & Roberts, 1995). Much of the garlic grown in California is treated with hot water before planting to disinfect seed-cloves of *D. dipsaci*.

Reniform Nematode

Species

Only one of nine described species of *Rotylenchulus* feeds on vegetable crops (Jatala, 1991). The reniform nematode (*R. reniformis*) was observed for the first time in the roots of cowpea plants growing in pineapple (*Ananas comosus* L. Merr.) production fields in the Hawaiian Islands (Linford & Oliveira, 1940). Smith and Taylor (1941) first reported *R. reniformis* in the USA, where it was found infecting cotton (*Gossypium hirsutum* L.) and soybean [*Glycine max* (L.) Merr.] near Baton Rouge, Louisiana. Vegetable crops susceptible to *R. reniformis* include sweet potato, beet, kale, chard [*Beta vulgaris* var. *cicla* (L.) Koch], cabbage, Chinese cabbage [*Brassica rapa* ssp. *pekinensis* (Lour.) Hanelt], artichoke (*Cynara cardunculus* L.), bush bean (*Phaseolus vulgaris* L.), lima bean, muskmelon (*Cucumis melo* L.), okra, pea, sweet corn (*Zea mays* L.), cucumber, lettuce, tomato, eggplant, radish, cowpea, squash, pumpkin (*Cucurbita pepo* L.), watermelon [*Citrullus lanatus* (Thunb.)], and cantaloupe (*Cucumis melo* L. var. *cantalupensis*). Resistant or immune vegetable crops include broccoli, carrot, cauliflower, mustard (*Brassica juncea* L.), onion, pepper, spinach, and turnip (Linford & Yap, 1940; Birchfield & Brister, 1962; Heald, 1978; Clark and Wright, 1983; Walters & Barker, 1993).

Symptoms

Symptoms of *R. reniformis* damage on vegetable crops include severe necrosis and pruning of seedling roots, leaf chlorosis and subsequent stunting of plants, a generally unhealthy appearance, and low yields. Cracking and distortions of sweet potato have been associated with damage caused by *R. reniformis* (Clark & Wright, 1983). The kidney-shaped females and their adhering egg masses can be easily observed under the dissecting microscope. Soil adhering to the gelatinous egg masses often give them a dark appearance aiding in detection (Netscher & Sikora, 1990).

Biology and Life History

Varied reports have been made on the feeding site of *R. reniformis*. In cowpea roots, feeding takes place in the cortical region, and galls are not initiated (Linford & Oliveira, 1940). Reniform nematodes do not develop on the fleshy storage roots of sweet potato after they have enlarged beyond a diameter of approximately 5 to 10 mm (Birchfield, 1972). In tomato roots, the nematode feeds near the stele of the root (Sivakumur & Seshadri, 1972) and a cell adjacent to the nematode's head contains a feeding peg enclosing the stylet (Razak & Evans, 1976). Opposite the stylet tip a helical hyaline feeding tube has been observed coiled in the feeding cell. This tube is thought to act as a filter through which the nematode obtains cell solutes. Heald (1975) reported that *R. reniformis* penetrated the cortex perpendicular to the vascular system and rested its head against the endodermis in young roots of cantaloup. Feeding stimulated the pericycle on either side of the endodermal feeding cell and caused cell hypertrophy with enlargement of the nucleoli and granular thickening of the cytoplasm. In older roots where the endodermis had collapsed, the nematode fed directly in the pericycle and caused similar symptoms. Nematode development from eggs is more rapid at 27 than at 21°C, but at both temperatures 24 h after inoculation, juveniles were found attached to roots of plants. At 27°C, maturing females were found in 6 d and began to secrete a gelatinous matrix on the 7th d. Females were completely enveloped with eggs deposited in the matrix on the 10th d. Newly hatched juveniles were present in the matrix on the 13th d. The life cycle can be completed in 24 to 29 d (Sivakumar & Seshadri, 1971).

Disease Complexes

The association of *R. reniformis* with *R. solani* in disease complexes in okra has been studied by a few researchers (Siddiqui et al., 1987) Charcoal rot caused by *Macrophomena phaseolina* on cantaloupe is significantly higher when the roots are infested with reniform nematode (Carter, 1980). The intensity of Brinjal mosaic virus and okra yellow vein mosaic is more severe on eggplant and okra when parasitized by *R. reniformis* (Naqui & Alam, 1975).

Economic Importance

The reniform nematode (*R. reniformis*) attacks over 100 plant species including many vegetable crops and is a limiting factor in vegetable production (Sikora & Greco, 1990). The distribution of the reniform nematode in the continental USA is limited to the southern and southeastern states, where it attacks many vegetable crops (Johnson & Fassuliotis, 1984; Heald & Robinson, 1990).

A recent survey showed that *R. reniformis* is becoming more widespread in Georgia (Richard Baird, 1996, personal communication). Little research has been conducted to determine the effects of the nematode on vegetable crops, although yield losses in the Rio Grande Valley and on cucumber in Georgia have been attributed to the reniform nematode. Heald (1978) reported that yields of okra, tomato, lettuce, and squash were 19, 13, 57, and 69% higher, respectively, from plots treated with 1,3-dichloropropene than those from untreated plots infested with *R. reniformis*.

Stubby-Root Nematode

Species

About 59 species have been described in the genera *Trichodorus* and *Paratrichodorus* (Siddiqi, 1973; Decraemer, 1991). As a result of recent taxonomic studies, *T. christiei* has been synonymized under *P. minor*. *Paratrichodorus minor* appears to be cosmopolitan as a plant pathogen and is responsible for stubby-root injury, root weight loss, and decreased seedling height and weight of more than 100 plant species, including many vegetable crops (Decraemer, 1991). This nematode is considered an important limiting factor on vegetables grown in coarse-textured sandy soils in the subtropical regions of the USA (Perry & Rhoades, 1982).

Symptoms

Symptoms of injury by stubby-root nematodes on vegetable crops consist of general stunting of the entire plant. Yellowing in plants attacked by *P. minor* may be caused by a mineral deficiency. One month after inoculating sweet corn

Fig. 25–13. Sweet corn showing symptoms of magnesium deficiency in soil infested with *Paratrichodorus minor*.

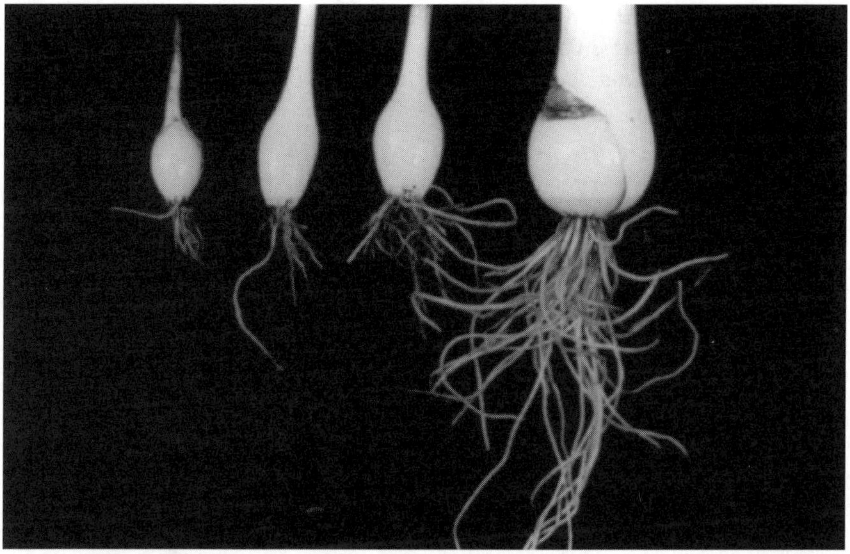

Fig. 25–14. Onion bulbs from soil infested with *Paratrichodorus minor*: left = severe damage, center = moderate damage, right = healthy (courtesy of Dr. W. F. Mai).

plants with 300 *P. minor* per pot, a yellow striping between leaf veins developed similar to that attributed to Mg deficiency (Fig. 25–13) (Johnson, 1967). The primary effect of the feeding is malformation of lateral roots near the base of the stem and devitalization of the root tip, which stops growth (Fig. 25–14). By pruning of the root primordia they cause a cessation of root elongation and an overall reduction in root development, inducing the stubby-root symptoms associated with these nematodes (Fig. 25–15). Usually, this is the "stubby-root" type of

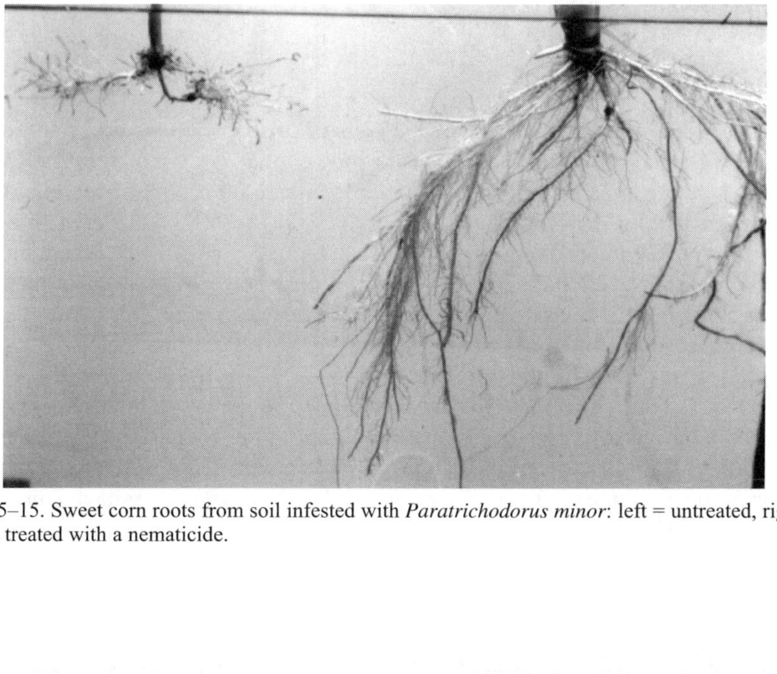

Fig. 25–15. Sweet corn roots from soil infested with *Paratrichodorus minor*: left = untreated, right = soil treated with a nematicide.

abnormality, although the result may sometimes be "coarse root." The root injury caused by stubby-root nematodes may be deceptive because it often lacks discoloration, necrotic lesions, and other conditions usually associated with nematode damage.

Biology and Life History

Adults of *P. minor* are ectoparasites, feeding mostly at root tips, but they may feed along the sides of young succulent roots. Population densities increase rapidly under favorable conditions. The nematode completes its life cycle in 16 to 17 d at 30°C and in 21 to 27 d at 22°C on corn (Rohde & Jenkins, 1957b). Specimens that develop at the lower temperature are larger than those recovered at the higher temperature. The optimal temperature for nematode reproduction is between 20 and 25°C. A conceptual model of the life cycle of *P. minor* consisting of the egg stage, four juvenile stages, and the adult stage was described by Schneider and Ferris (1987). The mode of reproduction is usually by parthenogenesis.

The tolerance limit on tomato, as determined by lateral root growth, is between 500 and 750 nematodes per plant (Högger, 1973). *Paratrichodorus minor* is virulent to eggplant when inoculated with 200 nematodes per plant, drastically inhibiting growth. (Schilt & Cohn, 1975).

Disease Complexes

Trichodorus spp. and *Paratrichodorus* spp. are vectors of tobacco rattle and pea early browning viruses. Pepper ringspot virus has been reported from South America with *P. minor* as vector (Decraemer, 1991). Pea early browning virus occurs naturally in pea and lucerne (*Medicago sativa* L. subsp. *sativa*), and is only recorded from Western Europe, particularly the Netherlands and Great Britain (Decraemer, 1991).

Economic Importance

The economic significance of the genus *Paratrichodorus* was first demonstrated by Christie and Perry (1951), who showed that *Trichodorus christiei*, later named *P. minor*, was responsible for severe damage on roots of beet, celery, and sweet corn in Florida. Rohde and Jenkins (1957a) tested 45 different plants as hosts for *P. minor* and vegetable plants that rated as good to excellent hosts include mustard, cabbage, eggplant, turnip, lettuce, tomato, onion, beet, muskmelon, bean, and sweet corn. *Paratrichodorus minor* may cause a 50% restriction in root growth of tomato (Netcher, 1970). Poor hosts includes cucumber, garden pea, radish, and spinach. Asparagus (*Asparagus officinalis* L.) is a nonhost of *P. minor* (Rohde & Jenkins, 1957a). The nematodes cause severe crop losses to a variety of vegetable crops including onion, tomato, pepper, eggplant, beet, broccoli, brussels sprout, cabbage, cauliflower, Chinese cabbage, radish, rutabaga, turnip, endive (*Cichorium endivia* L.), lettuce, and spinach (Inst. Food Agric. Sci., 1989).

Sting Nematode

Species

The genus *Belonolaimus* was established with *B. gracilis* as the type species (Steiner, 1949). Only nine species of *Belonolaimus* have been described (Smart & Nguyen, 1991). *Belonolaimus longicaudatus* is the species most often reported as associated with damage to vegetable crops.

Symptoms

Symptoms of injury caused by sting nematodes include poor stands, erratic uneven plant growth, chlorosis, severe stunting (Fig. 25–16), and short, coarse roots often malformed with few or no fibrous roots (Fig. 25–17) (Johnson & Fassuliotis, 1984). Nematode feeding induces stubby roots and necrotic lesions with severe damage leading to plant death (Fig. 25–18) (Netscher & Sikora, 1990).

Biology and Life History

Sting nematodes cause damage to vegetable crops by feeding ectoparasitically on or near the root tip and along the sides of succulent roots and other subterranean parts of a plant (Smart & Nguyen, 1991). A study of three *B. longicaudatus* populations from North Carolina and one from Georgia indicated differences among populations (Robbins & Barker, 1973). The North Carolina populations are very similar in host suitability comparisons. The Georgia population differed from the North Carolina populations by the ability to reproduce on cucumber, a nonhost of the North Carolina populations. Also, the reproductive rate of

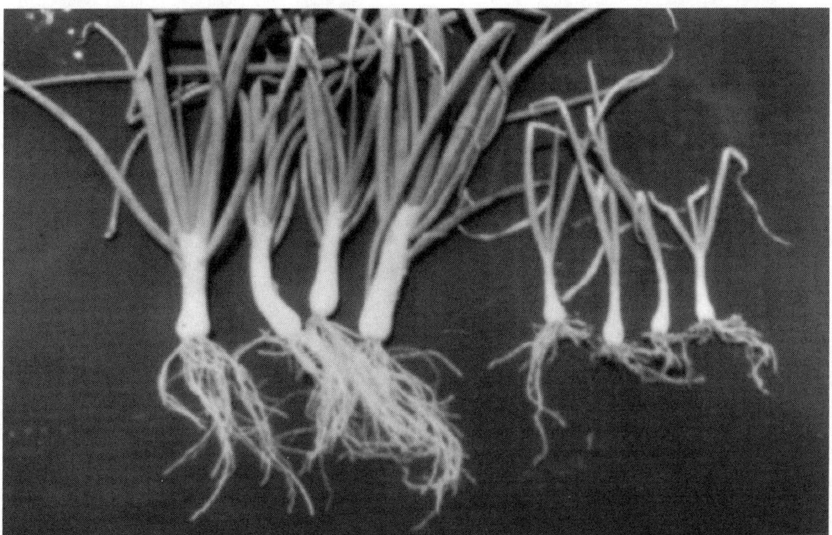

Fig. 25–16. Onions from soil infested with *Belonolaimus longicaudatus*: left = soil treated with a nematicide, right = untreated.

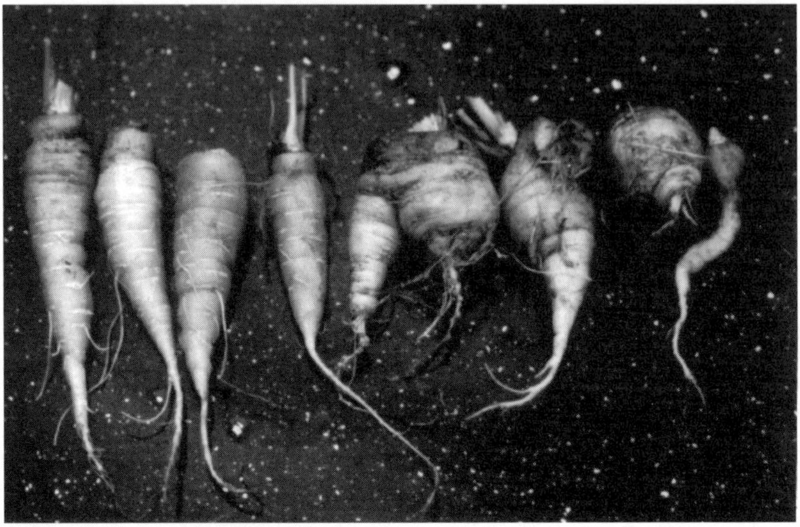

Fig. 25–17. Carrot from soil infested with *Belonolaimus longicaudatus*: left = soil treated with a nematicide, right = untreated (courtesy of Dr. J. O. Becker).

the Georgia population was greater than that of the North Carolina populations. Cucumber, 'Clemson Spineless' okra, asparagus, and watermelon are nonhosts for the North Carolina sting nematode populations. *Belonolaimus longicaudatus* reproduces more readily in sandy soils than in clay and muck soils possibly related to their large size and amphimictic mode of reproduction (Robbins, 1972). The optimum soil conditions for *B. longicaudatus* reproduction are 120- to 170-μm

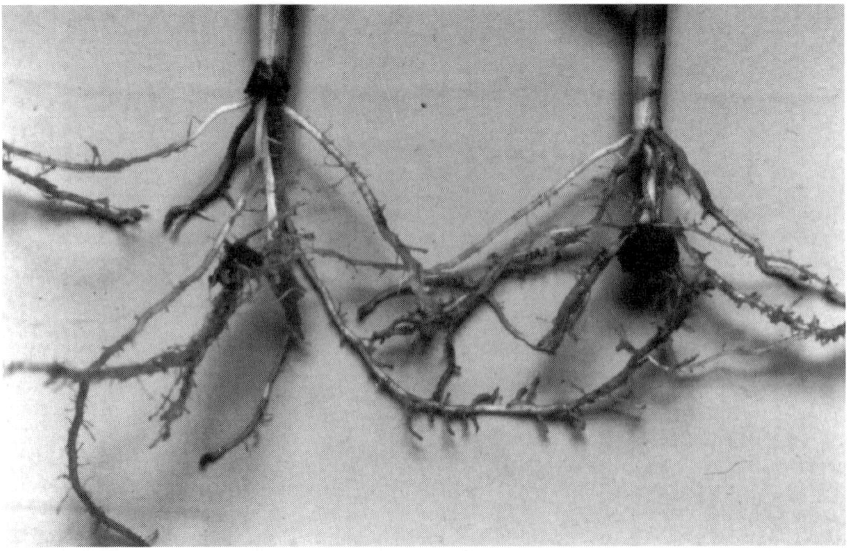

Fig. 25–18. Sweet corn roots from soil infested with *Belonolaimus longicaudatus*.

soil particle size at 25 to 30°C and 7% soil moisture (Robbins, 1972). This nematode completes one generation within 28 d at an optimum temperature of 28 to 30°C.

Disease Complexes

Although disease complex associations with *B. longicaudatus* have been detected on other hosts, only a few have been reported on vegetable crops. This pathogen in association with *Verticillium* suppressed growth of tomato by 69% (Overman & Jones, 1977).

Economic Importance

Sting nematodes appear to be indigenous to the light sandy soils of the southeastern Coastal Plain of the USA (Robbins, 1972). They occur in the Atlantic and Gulf coastal states from Connecticut to Texas as well as Arkansas, California, Kansas, Missouri, New Jersey, Nebraska, and Oklahoma (Smart & Nguyen, 1991). This parasite has a wide host range, and causes serious damage to vegetable crops including carrot, bean, beet, cabbage and other crucifers, celery, sweet corn, cucumber, pumpkin, okra, onion, pea, and pepper (Smart & Nguyen, 1991). These nematodes are highly virulent, and if only one specimen is found in a soil sample, crop damage can be expected. In soil heavily infested with *B. longicaudatus*, yield losses of susceptible vegetable crops may be complete where the nematode is not controlled (Rhoades, 1971).

Other Nematodes

False Root-Knot Nematode

The false root-knot nematode (*Nacobbus aberrans*) is an important nematode parasite on bean, carrot, cucumber, eggplant, lettuce, garden pea, pumpkin, and tomato (Thorne & Schuster, 1956; Ayoub, 1980; Jatala, 1991).

Spiral Nematodes

Rotylenchus robustus has been found in Rhode Island, New Jersey, New York, and California. The injury has been characterized by reduced and decayed roots, accompanied by extreme stunting and yellowing of plants. Losses in yield of lettuce, pea, and carrot have been associated with large populations of *R. robustus* (Seinhorst, 1954; Lear et al., 1969; Fortuner, 1991). In field studies, yields of marketable heads of lettuce were doubled when *R. robustus* was controlled with nematicides. *Rotylenchus robustus* also reproduces on cabbage, red beet, spinach, broccoli, celery, brussels sprout, and cauliflower.

Rotylenchus uniformis is a parasite of carrot (Seinhorst & Kuniyasu, 1969). The tolerance of carrot to the nematode is much lower at 8 than at 18°C. Tolerance to the nematode is considered to be due to increased root density during development of the plants.

Stunt Nematodes

Twenty-two species of stunt nematodes (*Tylenchorhynchus*) have been found associated with vegetable crops (Netscher & Sikora, 1990; Anderson & Potter, 1991). With the exception of *T. brassicae*, none of the other species have been shown to be of significant economic importance on vegetable crops. *Tylenchorhynchus brassicae* is a root parasite that may feed ecto- or endoparasitically. High population densities of *T. brassicae* are often associated with poor areas of growth in cabbage and cauliflower production fields (Khan, 1969). *Tylenchorhynchus brassicae* are confined to the outer cortical layers with their bodies lying parallel to the longitudinal axis of the roots, and the anterior portion of the body curved toward the conducting tissue. When 22 vegetables were inoculated with 1000 *T. brassicae*, only 'Golden Acre' cabbage and 'Snowball' cauliflower were suitable hosts, and significant damage occurred when the reproductive rate was five times above the initial population density. The most favorable temperature for reproduction is 30°C and the most favorable soil moisture is 25 to 30%. *Rhizoctonia solani* and *T. brassicae* are often associated with roots of cabbage and cauliflower. *Rhizoctonia solani* alone can suppress the emergence of cauliflower seedlings by 81%. When both organisms were present, 97% of the seeds failed to germinate (Khan & Saxena, 1969).

Ring Nematodes

Six species of ring nematodes, *Criconemella* (under the names *Criconemoides*, *Macroposthonia*, and *Mesocriconema*), have been detected in the rhizosphere of a wide range of vegetable crops (Netscher & Sikora, 1990). *Criconemella* spp. were found associated with sweet potato; however, pathogenicity was not established (Román, 1978). Data from a 4-yr study on six intensive cropping sequences [turnip (T)-corn (*Zea mays* L.) (C)-snapbean (SNB); T-peanut (*Arachis hypogaea* L.) (P)-SNB; T-C-T; T-P-C; SNB-soybean (SB)-cabbage; and T-cucumber (CU)-southern pea (SP)-T] indicated that numbers of *Criconemella ornata* remained low (< 60/150 cm^3 soil) on vegetable crops, but increased on corn and peanut (Johnson et al., 1983). Numbers of *C. ornata* were consistently suppressed to lowest levels (<15/150 cm^3 soil) in the T-CU-SP-T cropping sequence.

Needle Nematodes

These nematodes are locally important problems on many vegetable crops. They can cause severe damage especially on sandy soils and are probably often overlooked wherever root-knot nematode predominate. *Longidorus africanus* is a pathogen of many vegetable crops in the Imperial Valley of southern California (Radewald et al., 1969). The disease is evident at the time of plant emergence, with nonuniform seedling emergence in some fields and uniform in others. The cotyledon leaves of the stunted seedlings are cupped downward as if they are wilting, though there is ample moisture for normal growth. Leaves of infected seedlings are grayish-green, and the margins are often chlorotic. Taproots of affected seedlings fail to elongate normally and are short with terminal swellings

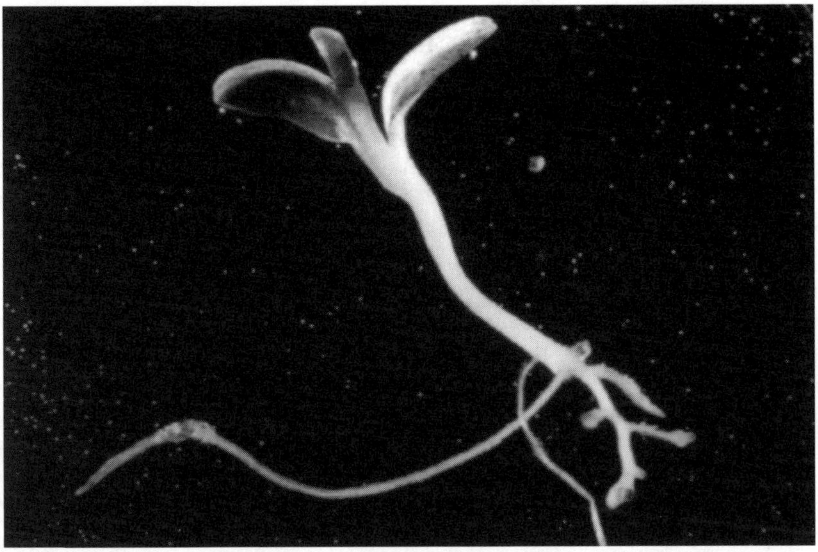

Fig. 25–19. Roots of lettuce damaged by *Longidorus africanus* (courtesy of Dr. J. D. Radewald).

(25–19). Necrotic areas on the swollen root tips are sometimes evident, and it appears that the meristematic tissue is destroyed. Diseased plants frequently compensate for the loss of the main taproot by lateral root proliferation; often these newly formed roots are similarly affected.

Awl Nematodes

Dolichodorus heterocephalus can damage vegetable crops, especially on wet, sandy soils (Smart & Nguyen, 1991). Roots of young tomato and celery plants are almost completely destroyed when grown in soil heavily infested with *D. heterocephalus* (Johnson & Fassuliotis, 1984). The nematode causes stubby-root symptoms and severe root necrosis, indicating a close association with root-rotting fungi (Netscher & Sikora, 1990). This species also attacks the base of the hypocotyl where necrotic tissue can be observed (Johnson & Fassuliotis, 1984).

CONTROL

Host Resistance

The least expensive and often the only practical means of managing some nematode species is with resistant cultivars. Several excellent reviews of resistance to plant nematodes have been published (Roberts, 1982; Lamberti & Taylor, 1983; Cook & Evans, 1987). Resistant cultivars not only prevent nematode damage to the resistant crop, but lower population densities and extend protection to a following susceptible crop. Resistant cultivars usually manage nematode numbers and provide crop yields equal to a nematicide, provided resistance is adequate for the primary nematodes inhabiting the soil (Wyatt et al., 1980). For low value crops, use of a resistant cultivar often is the only practical management method because of the high costs of nematicides. The use of nematode-resistant crop cultivars is often viewed as the foundation of a successful integrated nematode-management system for all high value crops.

"Resistance-breaking" races or pathotypes of root-knot nematodes can limit the continued usefulness of a resistant cultivar. Rotations must be designed to minimize the use of resistant cultivars employing the same resistance genes in order to limit the development of resistance-breaking races. Resistance-breaking races of several nematode species have evolved on a number of resistant cultivars when monocropping was practiced (Sauer & Giles, 1959). The selection of resistance-breaking types is minimized by alternating the growing of resistant with susceptible tomato plants, and after 5 yr, resistance is retained (Giles & Hutton, 1958).

Several lines of cowpea (Fery & Dukes, 1993; Matthews & Roberts, 1994), sweet potato (Jones & Dukes, 1980), and tomato have multiple root-knot nematode resistance genes and are more valuable for planting over a larger geographical area than crops such as bean plants having resistance to only one nematode species. Plant breeders and nematologists have developed tomato cultivars with desirable horticultural traits that are resistant to *M. incognita*, *M. javanica*, and *M. arenaria*, but not to *M. hapla*. Over 50 cultivars and hybrids of tomato are resistant to root-knot nematodes, many of which are readily available from seed catalogs (Sasser & Kirby, 1979). All cultivars of eggplant are susceptible to root-knot nematodes, but a high level of resistance has been identified in a related wild species, *S. sisymbriifolium* (Fassuliotis, 1975).

Peppers express a complete range of galling response from highly resistant to very susceptible. The small-fruited hot peppers (*Capsicum frutescens* L. var. *longum*) are resistant to *M. incognita*, *M. javanica*, and *M. arenaria*, but not to *M. hapla*. Bell and pimento types (*Capsicum frutescens* L. var. *grossum*) lack resistance. The root-knot resistant 'Nemaheart' pimento pepper and 'Mississippi 68' bell pepper were developed over 20 yr ago, but have not been used extensively. A pepper-breeding program at the U.S. Vegetable Laboratory, Charleston, South Carolina, includes root-knot nematode resistance breeding as an objective. A new cultivar, Carolina Cayenne, with resistance to the southern root-knot nematode (*M. incognita*), was released (Fery et al., 1986). More recently, it was reported that Carolina Cayenne was resistant to all races of *M. incognita* (Zamora et al., 1994). A new cayenne pepper cultivar, Charleston Hot, resistant to *M. javanica*, *M. arenaria*, *M. hapla*, and the four races of *M. incognita*, was released by the USDA (1992).

Root-knot nematode juveniles freely enter the roots of some resistant plants such as the common bean, but failed to survive (Barrons, 1940). Resistance to root-knot nematodes (*M. javanica* and *M. arenaria*) was identified in breeding lines of common bean, but its combination with *M. incognita* resistance in commercial bean has not yet been made (Omwega et al., 1990). Both pole-type and bush-type snapbean have been developed with resistance to *M. incognita*. Among the pole types are: 'Alabama No. 1', 'No. 8', 'No. 19', and 'Manoa Wonder'. A nongalling bush-type snapbean cultivar NemaSnap was developed with a higher level of resistance to *M. incognita* than the pole types (Wyatt et al., 1983).

Lima bean cultivars with resistance to *M. incognita* are: Hopi, Nemagreen, Maria, UC92, and White Ventura. A breeding line, L-40215, was developed at the U.S. Vegetable Laboratory primarily for the southern market. In California, effort is underway to combine *M. incognita* resistance with low-galling trait to *M. javanica* in breeding line L-136 (Roberts, 1982). 'Iron' cowpea is resistant to root-knot nematodes (Webber & Orton, 1902). Using this cultivar as a resistant parent, 'Mississippi Silver' with resistance to *M. incognita*, *M. javanica*, and *M. arenaria* was developed (Hare, 1967). A crowder-type cowpea cultivar, Worthmore, developed by crossing Mississippi Silver with 'Pink Eye Purple Hull', is resistant to *M. incognita*, *M. arenaria*, *M. hapla*, and *M. javanica* (Gay, 1976). More recently, other breeding lines and cultivars of cowpea with resistance to root-knot nematodes have been released (Fery & Dukes, 1980, 1982, 1988a,b, 1992; Fery et al., 1994).

Despite the considerable progress being made in developing root-knot nematode resistance in other vegetable crops, none has been developed in the Cucurbitaceae because extensive searching has not produced germplasm with resistance to root-knot nematodes in the commercial-type cucurbits. The only known sources of resistance that may be potentially useful in a breeding program are found in noncultivated species of *Cucumis*, i.e., *C. anguria* (West India gherkin), *C. ficifolius*, *C. longipes*, *C. metuliferus*, and *C. heptadactylus* (Fassuliotis, 1970). 'Sumter' cucumber is more susceptible than the horned cucumber (*C. metuliferus* E. Meyer ex Naudin) to *M. incognita*, *M. javanica*, and *M. arenaria* (Wehner et al., 1991). All cultivars of cucumber are more resistant to *M. hapla* and *M. arenaria* than to either *M. incogita* or *M. javanica*.

No resistance to root-knot nematodes was found in 11 cabbage cultivars (Winstead, 1959). All cultivars of okra in the USA are susceptible to *M. incognita* (Thies & Fery, 1994). Commercial cultivars of onion with suitable resistance to *P. penetrans* to be used as an alternative to chemical control have not been found (Bergeson, 1962). The wild onion, *A. canadense* L., is resistant to *P. penetrans*.

Resistance to root-knot nematodes has been incorporated into several sweetpotato cultivars. The major sources of resistance originally came from the Jersey types and Tinian, a wild type, which was not resistant to all races of *M. incognita*. Recent releases of sweet potato germplasm from the U.S. Vegetable Laboratory are being used by plant breeders throughout the world for improving local cultivars. Many of these lines are highly resistant to *M. javanica*, *M. hapla*, and several races of *M. incognita* (Jones et al., 1980, 1985, 1989). Sweetpotato appears to be a nonhost of *M. arenaria* (Giamalva et al., 1963). None of the sweetpotato selections yet tested is sufficiently resistant or tolerant to *R. reniformis* to offset the damage caused by the reniform nematode under field conditions (Clark & Wright, 1983). Three breeding lines of carrot, MSU 872, 5988, and 1475, are tolerant to *M. hapla* (Yarger & Baker, 1981), and resistance to *M. javanica* and *M. incognita* occurs in some breeding lines (Huang et al., 1986).

Cultural

Fallow

Fallow is currently used to a limited degree during the period between successive crops to reduce nematode population densities in the soil (Noling & Becker, 1994), but is considered too expensive for some growers (McKenry et al., 1994). The most appropriate time to fallow comes immediately after final crop harvest. Summer and winter fallow is effective in managing plant-parasitic nematodes (Johnson & Fassuliotis, 1984). The most effective nematode management is obtained when fallow is maintained during hot, dry weather, either by withholding irrigation (which is easily accomplished in arid areas) or by preventing plant growth, including weeds, by frequent and repeated plowing, disking, or use of nonselective herbicides. In some areas of the USA, *Meloidogyne* population densities in fields can be reduced by plowing at intervals of 2 to 4 wk during the hot, dry season. Plowing the soil exposes nematode eggs, juveniles, and adults to desiccation, and many nematodes in the upper layers of soil are killed. This practice may be sufficient to increase yield of a subsequent susceptible vegetable crop. Many examples of successful nematode management by fallow are available (Johnson & Fassuliotis, 1984; Brown & Kerry, 1987; Netscher & Sikora, 1990; Noling & Becker, 1994) although this method may not be effective in controlling all species of plant-parasitic nematodes (Johnson & Motsinger, 1990). Fallowing, of even short duration, when coupled with crop rotation, crop root destruction, and other management strategies generally gives significant and immediate impact on nematode population densities in the soil (Noling & Becker, 1994).

Flooding

Where water is abundant and fields are level, it is sometimes possible to manage nematodes by flooding the soil to a depth of 10 cm or more for several months (Rhoades, 1982). Flooding does not kill the eggs and juveniles of root-knot nematodes or inhibit infection and reproduction of nematodes on any plants that grow while the field is flooded. Flooding experiments are best evaluated by measuring yields of a subsequent crop, not by survival of nematode juveniles. Juveniles may survive flooding but not be infective (Taylor & Sasser, 1978). Alternating cycles of flooding and drying are more effective than continuous flooding (Overman, 1964). Based on the current attitudes concerning water use and conservation and increased urban demand, it is unlikely that water management agencies will allow or continue to permit flooding for control of nematodes in regions where the tactic is feasible (Noling & Becker, 1994).

Land Preparation

The value of land preparation methods on nematode population densities on vegetable crops has not been fully investigated. Tilling soil in preparation for planting reduces nematode population densities, but provides little long-term crop protection (Good, 1968). The destruction of nematode host plants after harvest reduces nematode population densities and has been recommended for nematode management for many years (Todd, 1971). Tillage practices (disk harrow, subsoil-bed, subsoil-plant, and moldboard) did not significantly affect nematode population densities in intensive cropping systems including several vegetable crops (Johnson et al., 1983). The beneficial effects of non-tillage systems on nematode population densities were not significant (Stinner & Crossley, 1982). Seed bed construction and planting, allowing maximum utilization of dried surface soil, may minimize nematode damage. Plowing soil late in the autumn to expose nematodes to freezing conditions also is effective in reducing root-knot nematode population densities (Potter & Olthof, 1993).

Time of Planting

In long-season, warm-climate regions, certain vegetable crops may be planted during the winter months and harvested before injury occurs in the spring. For example, in Georgia, turnip for greens and snapbean are planted in February or March and harvested in April or May. During this time, root-knot nematodes are less active and cause less damage than in warmer summer and fall months (Sumner et al., 1978; Smittle & Johnson, 1982; Johnson et al., 1983). Carrot may be planted at times when the temperature is below the activity threshold to reduce damage caused by *M. incognita* (Roberts, 1987).

Trap Crops

The use of trap crops has some merit as a method for managing endoparasitic nematodes (Potter & Olthof, 1993). A highly susceptible, rapidly growing crop is planted in infested soil, allowed to grow a short period of time, then plowed under or otherwise destroyed. Management is based on the fact that the nematodes enter the roots and begin to feed and are destroyed along with the plants before reproduction occurs. This method appears valid in theory, but has not been very effective in practice. The practice involves careful timing, for if the crop grows too long the nematode infestation may be increased rather than decreased.

Cover Crops

Cultural practices such as the use of cover crops may reduce nematode population densities (Johnson & Motsinger, 1990; Netscher & Sikora, 1990; Rodriguez-Kabana, 1992; Noling & Becker, 1994). Various cover crops including annual and perennial legumes, small grains, and pasture and forage grasses have been evaluated for nematode management (Rodriguez-Kabana et al., 1989a,b, 1990, 1992; Araya & Caswell-Chen, 1994; Johnson et al., 1995). Many of these crops provide additional benefits to crop production other than nematode suppression. Sorghum is a popular cover crop, which may suppress root-knot nematode population densities and provides large amounts of organic matter to the soil. However, the crop is a suitable host for maintaining or increasing many ectoparasitic nematodes commonly found in the southeastern Coastal Plain (Good, 1972). Control of root-knot nematodes in carrot is obtained by growing

Crotalaria spectabilis, or marigold (*Tagetes* spp.) for several months before the carrot crop (Huang, 1984). However, farmers are reluctant to plant *C. spectabilis* because all parts of the plant are poisonous, whether green or dried in hay. The seeds contain an alkaloid, monocrotaline, that is toxic to chicken, horse, cattle, swine, and other animals (Miller et al., 1974). The effect of marigold as a cover crop on populations of *Meloidogyne* species is highly variable (Belcher & Hussey, 1977). Root galling on lettuce caused by *M. hapla* was less following sudangrass (*Sorghum* × *drummondii*) than following rye (*Secale cereale* L.) and phacelia (*Phacelia leucophylla* Torr.) (Viaene & Abawi, 1994).

Crop Rotation

Use of crop rotations is and will continue to be a primary means of managing nematode population densities on vegetable crops (Nusbaum & Ferris, 1973; Johnson, 1982, 1992; Johnson et al., 1983, 1995; Barker, 1991; Weaver et al., 1994). Management of *Meloidogyne* species by crop rotation usually rotates susceptible crops with nonhost or resistant crops. The susceptible crop is frequently the most profitable and the rotation crops less profitable. For example, tomato is usually a profitable crop, but some cultivars are susceptible to all common species of *Meloidogyne*. After a susceptible tomato crop is harvested, the root-knot nematode population density in the soil has increased to a high level. A second crop of tomato or other susceptible vegetable would be damaged severely in early growth stages. The high population densities are reduced by one or more rotations of nonhost or resistant crops. Uncommon crops such as hairy indigo (*Indigofera hirsuta* L.), horse bean (*Canavalia ensiformis* L. DC.), sesame (*Sesamum indicum* L.), velvetbean [*Mucuna pruriens* var. *utilis* (Wallich ex wight)], castor (*Ricinus communis* L.), and American jointvetch (*Aeschynomene americana* L.) are being introduced into cropping systems in some areas of the southeastern USA to suppress nematode population densities on vegetable crops (McSorley et al., 1994a,b,c).

The resistance, susceptibility, or tolerance of crops to nematodes are the basis for planning crop rotations and developing a systems approach to nematode management. However, before selecting rotations for specific farm situations or for areawide recommendations, the predominant pathogenic nematode species must be determined. Consideration must be given not only to the dominant species, but to other nematode species that can establish dominance and cause economic crop damage under a given rotation system. For example, in a polyspecific community of approximately 10% *M. hapla* and 90% *M. incognita*, population densities of *M. hapla* increased from a minor to a major problem when peanut was planted in multiple-cropping systems with vegetable crops (Johnson, 1985). The host-suitability of many crops to the most important nematode species is sufficiently known to allow planning of vegetable crop rotations to manage complex communities of mixed nematode genera (Johnson et al., 1983; Johnson, 1985, 1992).

The major constraints to the development and use of crop rotation include the wide-host range of many economically important nematodes, occurrence of several nematode species in a given production field, availability of resistant or tolerant cultivars, or the lack of cultivars which are suitably adapted for the specialized cropping practices (Sasser & Kirby, 1979). Other constraints include biotic, climatic, and geographical environments in which crops will be grown or markets for specialized crops and the requirements for specialized equipment to grow and harvest crops (Thomason & Caswell, 1987).

Chemical

Nematicides are one of the most important and reliable means of managing a wide variety of nematodes on vegetable crops. Commercial use of nematicides began about 1945, increased rapidly from 1960 to 1983, and declined from 1985 to 1996. The principal use of nematicides is for control of nematode populations in soil before planting crops. Suitable application methods have been devised for applying about 18 nematicides (Johnson & Feldmesser, 1987; Netscher & Sikora, 1990; Potter & Olthof, 1993). Some of the nematicides that are available on the world market are effective, but are not registered for use on vegetable crops.

Nematicides are used to protect a number of highly susceptible vegetable crops from nematode attack. The action by the U.S. Environmental Protection Agency (EPA) to suspend the use of DBCP (1,2-dibromo-3-chloropropane) on all crops (except pineapple in Hawaii) in 1977 and EDB (ethylene dibromide) (1,2-dibromoethane) in 1983 reduced the options for effective and economical soil fumigants that can be used for control of nematodes on vegetable crops in the USA. Other nematicides are currently under review by the EPA. Based on environmental and health risks, the use of nematicides is discouraged in some countries.

Two broad types of nematicides for control of nematodes on vegetable crops are available. The older nematicides are liquids, which are injected into the soil and volatilize to produce vapors that kill nematodes. These vapors diffuse through the soil, are dissolved in soil water, and enter nematode bodies through the cuticle. Nonfumigant nematicides are incorporated into the soil mechanically or distributed through the soil by percolation of water. They also enter nematode bodies through the cuticle. Other types of nematicides are systemic. These nematicides can be taken up by plants through the roots after application to the soil, or through foliage after spray application. The chemical is translocated to roots where it kills feeding nematodes. New methods for delivery of nematicides have been developed. Chemigation is a method used to apply agricultural chemicals (fertilizers, nematicides, herbicides, fungicides, insecticides, and growth regulators) to the soil via overhead-sprinkler and trickle/drip irrigation systems (Johnson et al., 1986). In the northwestern USA, metham sodium (sodium methyldithiocarbamate) applied through center pivot irrigation systems is effective in controlling nematodes and diseases of vegetable crops and also offers some weed and insect population reduction (Santo & Qualls, 1984). In the southeastern USA, several researchers have applied nematicides and other crop protection chemicals through center pivot irrigation systems to manage pests populations in several vegetable crops (Johnson et al., 1982, 1986; Sumner et al., 1981). Chemical applications in trickle/drip irrigation systems under a plastic mulch are widely used (Overman, 1982; Johnson et al., 1979a,b; Keng & Vander

Gulik, 1984; Garabedian & Van Gundy, 1985). Portable soil drenching devices have been designed and developed to deliver nematicidal agents including plant extracts in the soil, but are not generally used in vegetable production (McKenry & Buzo, 1992).

It is difficult to predict the future for the development of new nematicides. Chemicals derived from plants, microorganisms and industrial byproducts have been, or are being, tested for nematicidal efficacy. Many may have potential for commercial use in specific crops or locations, but at present they do not appear to be generally competitive with presently marketed nematicides (Thomason, 1987).

Biological

The literature on biological control of plant-parasitic nematodes has received extensive review (Sayre, 1971; Webster, 1972, Jatala, 1985; Jones & Rodriguez-Kabana, 1987; Kerry, 1987; Stirling, 1993). These reviews discuss various soil organisms antagonistic to nematodes.

Predators of nematodes include fungi, nematodes, turbellarians, enchytraeids, insects, and mites. Parasites include viruses, protozoa, bacteria, and fungi. A great diversity of microorganisms attack plant-parasitic nematodes under natural conditions in agricultural soils. Soil amendments, to encourage the activity of indigenous soil microorganisms, are one form of biological control by which a multitude of natural enemies suppress nematodes. Currently, no effective nematode biological control agents are available that can be successfully integrated to maintain sustainable vegetable crop production systems, even though some products are commercially available (Meeker, 1988). The agents with the greatest potential for plant-parasitic nematode control are likely to be those that are endemic in agricultural soils and can be induced to proliferate by methods compatible with conventional crop production practices. The increased interest in such fungi as *Glomus fasciculatum* (Kotcon et al., 1985; MacGuidwin et al., 1985), *Meria coniospora* (Janson et al., 1985; Townshend et al., 1989), and *Paecilomyces lilacinus* (Cabanillas et al., 1989; Walters & Barker, 1994), and bacteria such as *Pasteuria penetrans* (Serracin et al., 1994) may lead to use in the biocontrol of some nematode pests on vegetable crops.

Heat

Heat is one of the oldest methods of managing nematode population densities by physical means. Most plant-parasitic nematodes are killed at temperatures between 44 and 48°C (Christie, 1959). Steam, dry heat (Jenkins, 1960), and hot water (Jenkins & Taylor, 1967; Noling, 1994; Johnson & Roberts, 1995) have been used to control plant-parasitic nematodes in soil or plant parts under special usage.

Soil solarization is a recently developed technique for management of nematode population densities (Noling & Becker, 1994). Katan (1981) published an extensive review of the soil solarization technique. Solar heating involves a method of trapping solar energy in the soil by covering the soil surface with polyethylene tarps. Soil solarization is most effective when conducted during the hottest season of the year. A thin transparent polyethylene tarp is more suitable than an opaque tarp. The tarp must be sealed and kept firmly against the soil surface, and the soil must be in good till and kept moist for 4 to 8 wk to attain temperatures lethal to nematodes. More research is needed to fully develop solarization as an effective method used to manage nematode population densities in vegetable crop production systems.

OUTLOOK

The medical and health experts are encouraging our society to eat more fruits and vegetables. Also, the world population continues to increase. Perhaps, in the near future, farmers will depend less on government subsidies for production of major agronomic crops that could result in a greater shift to more vegetable crop production and add diversity to the farming enterprise. These combined factors indicate a need for more and higher quality vegetables in the future. As cropping systems intensify, nematode damage and management strategies also will increase. It is evident that nematodes cause severe damage to vegetable crops. Management of nematode population densities on vegetable crops offers many challenges and opportunities for practical ecological studies, because efficient management of nematodes requires the use of many strategies. The most promising managing techniques must be integrated into management systems. This takes into consideration the total environment, which includes crop cultivars, cultural practices, nematicides, nematodes and other pests, biological enemies, and the effects of environmental and soil conditions on the nematodes and crops. The goal is to develop vegetable crop protection systems that are sustainable, economical, and environmentally sound.

REFERENCES

Abawi, G.S., and W.F. Mai. 1980. Effects of initial population densities of *Heterodera schachtii* on yield of cabbage and table beets in New York State. Phytopathology 70:481–485.

Acedo, J.R., and R.A. Rohde. 1971. Histochemical root pathology of *Brassica oleracea capitata* L. infected by *Pratylenchus penetrans* (Cobb) Filipjev and Schuurmans-Stekhoven (Nematoda:Tylenchidae). J. Nematol. 3:62–68.

Anderson, R.V., and J.W. Potter. 1991. Stunt nematodes: *Tylenchorhynchus*, *Merlinius*, and related genera. p. 529–586. *In* W.R. Nickle (ed.) Manual of agricultural nematology. Marcel Dekker, New York.

Araya, M., and E.P. Caswell-Chen. 1994. Host status of *Crotalaria juncea*, *Sesamum indicum*, *Dolichos lablab*, and *Elymus glaucus* to *Meloidogyne javanica*. J. Nematol. 26:492–497.

Ayoub, S.M. 1980. Plant nematology an agricultural training aid. NemaAid Publ., Sacramento, CA.

Baldwin, J.G., and M. Mundo-Ocampo. 1991. Heteroderinae, cyst- and non-cyst-forming nematodes. p. 275–362. *In* W.R. Nickle (ed.) Manual of agricultural nematology. Marcel Dekker, New York.

Barker, K.R. 1991. Rotation and cropping systems for nematode control: The North Carolina experience—introduction. J. Nematol. 23:342–343.

Barrons, K.C. 1940. Root-knot resistance in beans. J. Hered. 31:35–38.

Belcher, J.V., and R.S. Hussey. 1977. Influence of *Tagetes patula* and *Arachis hypogaea* on *Meloidogyne incognita*. Plant Dis. Rep. 61:525–528.

Bergeson, G.B. 1962. A report on the testing of onion varieties for resistance to lesion nematodes (*Pratylenchus penetrans*). Plant Dis. Rep. 46:535–536.

Berkeley, M.J. 1855. Vibrio forming excrescences on the roots of cucumber plants. Garden Chron. 14:220.

Berney, M.F., and G.W. Bird. 1992. Distribution of *Heterodera carotae* and *Meloidogyne hapla* in Michigan carrot production. Suppl. J. Nematol. 24:776–778.

Birchfield, W. 1972. Differences in host-cell response to the reniform nematode. Phytopathology 62:747.

Birchfield, W., and L.R. Brister. 1962. New hosts and nonhosts of reniform nematode. Plant Dis. Rep. 46:683–685.

Brodie, B.B. 1998. Potato. p. 567–594. *In* K.R. Barker et al. (ed.) Plant and nematode interactions. Agron. Monogr. 36. ASA, CSSA, SSSA, Madison, WI.

Brown, R.H., and B.R. Kerry. 1987. Principles and practice of nematode control in crops. Acad. Press, London.

Cabanillas, E., K.R. Barker, and L.A. Nelson. 1989. Growth of isolates of *Paecilomyces lilacinus* and their efficacy in biocontrol of *Meloidogyne incognita* on tomato. J. Nematol. 21:164–172.

Carter, W.W. 1980. Interaction of *Rotylenchulus reniformis* and *Macrophomina phaseolina* in charcoal rot of cantaloupe. J. Nematol. 12:217.

Celetti, M.J., H.W. Johnston, J. Kimpinski, and H.W. Platt. 1990. Plant-parasitic nematodes and fungi associated with root rot of peas on Prince Edward Island. Suppl. J. Nematol. 22:676–680.

Chitwood, B.G. 1949. Root-knot nematodes. Part I. A revision of the genus *Meloidogyne* Goeldi, 1887. Proc. Helminthol. Soc. Wash. 16:90–104.

Christie, J.R. 1936. The development of root-knot nematode galls. Phytopathology 26:1–22.

Christie, J.R.. 1959. Plant nematodes, their bionomics and control. The H. & W. B. Drew Co., Jacksonville, FL.

Christie, J.R., and V.G. Perry. 1951. A root disease of plants caused by a nematode of the genus *Trichodorus*. Science (Washington, DC) 113:491–493.

Clark, C.A., and V.L. Wright. 1983. Effect of reproduction of *Rotylenchulus reniformis* on sweet potato selections. J. Nematol. 15:197–203.

Cook, R., and K. Evans. 1987. Resistance and tolerance. p. 179–231. *In* R.H. Brown and B.R. Kerry (ed.) Principles and practice of nematode control in crops. Acad. Press, London.

Coomans, A.V. 1993. Development of nematology in the low land countries. p. 11–20. *In* F.J. Gommers and P.W.Th. Maas (ed.) Nematology from molecule to ecosystem. Europ. Soc. Nematol., Inwergrowie, Dundee, Scotland.

Decraemer, W. 1991. Stubby root and virus vector nematodes: *Trichodorus*, *Paratrichodorus*, *Allotrichodorus*, and *Monotrichodorus*. p. 587–625. *In* W.R. Nickle (ed.) Manual of agricultural nematology. Marcel Dekker, New York.

Dropkin, V.H. 1989. Introduction to plant nematology. John Wiley & Sons, Inc., New York.

Eisenback, J.D., and H. H. Triantaphyllou. 1991. Root-knot nematodes: *Meloidogyne* species and races. p. 191–274. *In* W.R. Nickle (ed.) Manual of agricultural nematology. Marcel Dekker, New York.

Evans, A.A.F. 1987. Diapause in nematodes as a survival strategy. p. 180–187. *In* J.A. Veech and D.W. Dickson (ed.) Vistas on nematology. Soc. Nematol., Hyattsville, MD.

Fassuliotis, G. 1970. Resistance of *Cucumis* spp. to root-knot nematode, *Meloidogyne incognita acrita*. J. Nematol. 2:174–178.

Fassuliotis, G.. 1979. Plant breeding for root-knot nematode resistance. p. 425–453. *In* F. Lamberti and C.E. Taylor (ed.) Root-knot nematodes (*Meloidogyne* species), systematics, biology, and control. Acad. Press, London.

Feldmesser, J., D.I. Edwards, J.M. Epps, C.M. Heald, W.R. Jenkins, H.J. Jensen, B. Lear, C.W. McBeth, E.L. Nigh, and V.G. Perry. 1971. Estimated crop losses from plant-parasitic nematodes in the United States. Comm. Crop Losses. Spec. Publ. no. 1. Soc. Nematol., Hyattsville, MD.

Ferris, J.M. 1962. Some observations on the number of root lesion nematodes necessary to cause injury to seedling onions. Plant Dis. Rep. 46:484–485.

Fery, R.L., and P.D. Dukes. 1980. Inheritance of root-knot resistance in the cowpea (*Vigna unguiculata* (L.) Walp.). J. Am. Soc. Hort. Sci. 105:671–674.

Fery, R.L., and P.D. Dukes. 1982. `Colossus 80' southern pea. HortScience 17:263–264.

Fery, R.L., and P.D. Dukes. 1988a. `Kiawah' southern pea. HortScience 23:645–646.

Fery, R.L., and P.D. Dukes. 1988b. US-311 and US-427 southern pea breeding lines. HortScience 23:643–644.

Fery, R.L., and P.D. Dukes. 1992. `Carolina Crowder' southern pea. HortScience 27:1335–1337.

Fery, R.L., and P.D. Dukes. 1993. `Bettergro Blackeye' southern pea. HortScience 28:62–63.

Fery, R.L., P.D. Dukes, and W.L. Ogle. 1986. `Carolina Cayenne' pepper. HortScience 21:330.

Fery, R.L., P.D. Dukes, and J.A. Thies. 1994. Characterization of new sources of resistance in cowpea to the southern root-knot nematode. HortScience 29:678–679.

Fortuner, R. 1991. The Hoplolaiminae. p. 669–719. *In* W.R. Nickle (ed.) Manual of agricultural nematology. Marcel Dekker, New York.

France, R.A., and G.S. Obawi. 1994. Interaction between *Meloidogyne incognita* and *Fusarium oxysporum* f. sp. *phaseoli* on selected bean genotypes. J. Nematol. 26:467–474.

Garabedian, S., and S.D. Van Gundy. 1985. Effects of non-fumigant nematicides applied through low pressure drip irrigation on control of *Meloidogyne incognita* on tomatoes. Plant Dis. 68:138–140.

Gay, J.D. 1976. `Worthmore' southern pea. HortScience 11:621–622.

Giamalva, M.J., W.J. Martin, and T.P. Hernandez. 1962. Relationship of root-knot nematodes to the development of Fusarium wilt in the sweet potato. Phytopathology 52:733.

Giamalva, M.J., W.J. Martin, and T.P. Hernandez. 1963. Sweet potato varietal reaction to species and races of root-knot nematodes (*Meloidogyne* spp.). Phytopathology 53:1187–1189.

Giles, J.E., and E.M. Hutton. 1958. Combining resistance to the root-knot nematode, *Meloidogyne javanica* (Treub) Chitwood, and Fusarium wilt in hybrid tomatoes. Aust. J. Agric. Res. 9:182–192.

Golden, J.K., and S.D. Van Gundy. 1975. A disease complex of okra and tomato involving the nematode, *Meloidogyne incognita*, and the soil-inhabiting fungus, Rhizoctonia. Phytopathology 65:265–273.

Good, J.M. 1968. Soil management practices. p. 113–138. *In* G.C. Smart, Jr., and V.G. Perry (ed.) Tropical nematology. Univ. Florida Press, Gainesville.

Good, J.M. 1972. Management of plant parasitic nematode populations. p. 107–127. *In* E.V. Komarek (ed.) Proc. Tall Timbers Conf. Ecol. Anim. Cont. Habitat Manage., Tallahassee, FL. 24–25 February.

Goswami, B.K., and V.V. Chenulu. 1974. Interaction of root-knot nematode, *Meloidogyne incognita*, and tobacco mosaic virus in tomato. Indian J. Nematol. 4:68–80.

Goswami, B.K., D.V. Singh, and J.N. Gupta. 1970. Studies on association of root-knot nematodes, *Meloidogyne incognita* (Kofoid and White) Chitwood and *Sclerotium rolfsii* Sacc. in brinjal (*Solanum melongena* L.). Indian Phytopathol. 23:587–589.

Griffin, G.D. 1985. Interrelationship of *Heterodera schachtii* and *Meloidogyne hapla* on tomato. J. Nematol. 17:385–388.

Haglund, W.A. 1984. Diseases caused by nematodes. p. 38–40. *In* D.J. Hagedorn (ed.) Compendium of pea diseases. Am. Phytopathol. Soc. Press, St. Paul, MN.

Handoo, Z.A., and A.M. Golden. 1989. A key and diagnostic compendium to the species of the genus *Pratylenchus* Filipjev, 1936 (lesion nematodes). J. Nematol. 21:202–218.

Hare, W.W. 1967. A combination of disease resistance in a new cowpea, Mississippi Silver. Phytopathology 57:460.

Heald, C.M. 1975. Pathogenicity and histopathology of *Rotylenchulus reniformis* infecting cantaloup. J. Nematol. 7:149–152.

Heald, C.M. 1978. Effects of the reniform nematode on vegetable yields. Plant Dis. Rep. 62:902–904.

Heald, C.M., and A.F. Robinson. 1990. Survey of current distribution of *Rotylenchulus reniformis* in the United States. Suppl. J. Nematol. 22:695–699.

Högger, C.H. 1973. Preferred feeding site of *Trichodorus christiei* on tomato roots. J. Nematol. 5:228–229.

Huang, S.P. 1984. Cropping effects of marigolds, corn, and okra on population levels of *Meloidogyne javanica* and on carrot yields. J. Nematol. 16:396–398.

Hunger, F.W.T. 1901. Een Bacterieziekte der tomaat. Meded. Slands Plantent. 48:1–57.

Hussey, R.S. 1992. Secretions of esophageal glands in root-knot nematodes. p. 41–50. *In* F.J. Gommers and P.W.Th. Maas (ed.) Nematology from molecule to ecosystem. Europ. Soc. Nematol., Inwergrowie, Dundee, Scotland.

Institute of Food and Agricultural Sciences. 1989. Florida nematode control guide. Entomol. Nematol. Dep. Coop. Ext. Ser., Univ. Florida, Gainesville.

Janson, H.B., A. Jeyaprakash, and B.M. Zuckerman. 1985. Control of root-knot nematodes on tomato by the endoparasitic fungus *Meria coniospora*. J. Nematol. 17:327–329.

Jatala, P. 1985. Biological control of nematodes. p. 303–308. *In* J.N. Sasser and C.C. Carter (ed.) An advanced treatise on *Meloidogyne*. Vol. 1. North Carolina State Univ. Graphics, Raleigh.

Jatala, P. 1991. Reniform and false root-knot nematodes, *Rotylenchulus* and *Nacobbus* spp. p. 509–528. *In* W.R. Nickle (ed.) Manual of agricultural nematology. Marcel Dekker, New York.

Jenkins, W.R. 1960. Control of nematodes by physical methods. p. 443–446. *In* J.N. Sasser and W.R. Jenkins (ed.) Nematology; fundamentals and recent advances with emphasis on plant parasitic and soil forms. North Carolina Press, Chapel Hill.

Jenkins, W.R., and D.P. Taylor. 1967. Plant nematology. Reinhold Publ. Corp., New York.

Jensen, H. J. 1972. Nematode pests of vegetable and related crops. p. 377–408. *In* J.M. Webster (ed.) Economic nematology. Acad. Press, New York.

Johnson, A.W. 1967. Ecological associations between certain species of plant parasitic nematodes in mixed populations. Ph.D. diss. North Carolina State Univ., Raleigh (Diss. Abstr. 67-2684).

Johnson, A.W. 1982. Managing nematode populations in crop production. p. 113–203. *In* R.D. Riggs (ed.) Nematology in the southern region of the United States. Arkansas Agric. Exp. Stn. South. Coop. Ser. Bull. 276.

Johnson, A.W. 1985. Specific crop rotation effects combined with cultural practices and nematicides. p. 283–301. *In* J.N. Sasser and C.C. Carter (ed.) An advanced treatise on *Meloidogyne*. Vol. 1. North Carolina State Univ. Graphics, Raleigh.

Johnson, A.W. 1992. Nematode management on vegetable crops. p. 234–239. *In* F.J. Gommers and P.W.Th. Maas (ed.) Nematology from molecule to ecosystem. Europ. Soc. Nematol., Inwergrowie, Dundee, Scotland.

Johnson, A.W., G.W. Burton, J.P. Wilson, and A.M. Golden. 1995. Rotations with coastal bermudagrass and fallow for management of *Meloidogyne incognita* and soilborne fungi on vegetable crops. J. Nematol. 27:457–464.

Johnson, A.W., C.C. Dowler, N.C. Glaze, and D.R. Sumner. 1983. Effect of intensive cropping systems and pesticides on nematodes and crop yields. USDA-ARS Agric. Res. Results ARR-S-14. USDA-ARS, New Orleans, LA.

Johnson, A.W., and G. Fassuliotis. 1984. Nematode parasites of vegetable crops. p. 323–372. *In* W.R. Nickle (ed.) Plant and insect nematodes. Marcel Dekker, New York.

Johnson, A.W., and J. Feldmesser. 1987. Nematicides—a historical review. p. 448–454. *In* J. A. Veech and D.W. Dickson (ed.) Vistas on nematology. Soc. Nematol., Hyattsville, MD.

Johnson, A.W., C.A. Jaworski, D.R. Sumner, and R.B. Chalfant. 1979a. Effects of film mulch, trickle irrigation, and soil pesticides on nematodes and yields of polebean. Plant Dis. Rep. 63:360–364.

Johnson, A.W., and R.E. Motsinger. 1990. Effects of planting date, small grain crop destruction, fallow, and soil temperature on the management of *Meloidogyne incognita*. J. Nematol. 22:348–355.

Johnson, A.W., and P.A. Roberts. 1995. Diseases caused by nematodes. p. 35–40. *In* H.F. Schwartz and S.K. Mohan (ed.) Compendium of onion and garlic diseases. Am. Phytopathol. Soc. Press, St. Paul, MN.

Johnson, A.W., D.R. Sumner, and C.A. Jaworski. 1979b. Effect of film mulch, trickle irrigation, and DD-MENCS on nematodes, fungi, and vegetable yields in a multicrop system. Phytopathology 69:1172–1175.

Johnson, A.W., D.R. Sumner, and C.A. Jaworksi. 1979c. Effects of management practices on nematode and fungus populations on cucumber yield. J. Nematol. 11:84–93.

Johnson, A.W., D.R. Sumner, C.A. Jaworski, and R.B. Chalfant. 1977. Effects of management practices on nematode and fungi populations and okra yield. J. Nematol. 9:136–142.

Johnson, A.W., J.R. Young, E.D. Threadgill, C.C. Dowler, and D.R. Sumner. 1986. Chemigation for crop production management. Plant Dis. 70:998–1004.

Jones, A., and P. D.Dukes. 1980. Heritabilites of sweet potato resistances to root knot caused by *Meloidogyne incognita* and *M. javanica*. J. Am. Soc. Hort. Sci. 105:154–156.

Jones, A., P.D. Dukes, and J.M. Schalk. 1985. `Regal' sweet potato. HortScience 20:781–782.

Jones, A., P.D. Dukes, and J.M. Schalk. 1989. `Excel' sweet potato. HortScience 24:171–172.

Jones, A., P.D. Dukes, J.M. Schalk, M.A. Mullen, M.G. Hamilton, D.R. Paterson, and T.E. Boswell. 1980. W-71, W-115, W-119, W-149, and W-154 sweet potato germplasm with multiple insect and disease resistance. HortScience 15:835–836.

Jones, G.M., and R. Rodriguez-Kabana. 1987. Fungal biocontrol for the management of nematodes. p. 94–99. *In* J.A. Veech and D.W. Dickson (ed.) Vistas on nematology. Soc. Nematol., Hyattsville, MD.

Katan, J. 1981. Solar heating (solarization) of soil for control of soilborne pests. Annu. Rev. Phytopathol. 19:211–236.

Keng, J.C.W., and T. Vander Gulik. 1984. Application of chemicals through a trickle system for soil-borne pest control. 1. Derivation of basic physical theory for practical use. Can. Agric. Eng. 27:31–33.

Kerry, B.R. 1987. Biological control. p. 233–263. *In* R.H. Brown and B.R. Kerry (ed.) Principles and practice of nematode control in crops. Acad. Press, London.

Khan, A.M. 1969. Studies on plant parasitic nematodes associated with vegetable crops in Uttar Pradesh. Final Tech. Rep. Grant no. FG-In-225. Project no. A7-CR-65. Aligarh Muslim Univ., Aligarh, India.

Khan, A.M., and S.K. Saxena. 1969. Effect of *Rhizoctonia solani* and *Tylenchorhynchus brassica* on the emergence of cauliflower seedlings. p. 12. *In* Proc. All Indian Nematol. Symp., New Delhi, India. 21–22 August. Indian Agric. Res. Inst., New Delhi, India.

Kirjanova, E.S., and E.L. Krall. 1980. Plant-parasitic nematodes and their control. Amerind Publ. Co., PVT Ltd., New Delhi, India.

Kotcon, J.B., G.W. Bird, L.M. Rose, and K. Dimoff. 1985. Influence of *Glomus fasciculatum* and *Meloidogyne hapla* on *Allium cepa* in organic soils. J. Nematol. 17:55–60.

Lamberti, F.C. 1971. Nematode-induced abnormalities of carrot in southern Italy. Plant Dis. Rep. 55:111–113.

Lamberti, F.C., and C.E. Taylor. 1983. Resistance of crops to nematodes. p. 197–206. *In* F. Lamberti et al. (ed.) Durable resistance in crops. Plenum Press, New York.

Lear, B., D.E. Johnson, and S.T. Miyagawa. 1969. A disease of lettuce associated with an ectoparasitic nematode, *Rotylenchus robustus*. Plant Dis. Rep. 53:952–954.

Lewis, S. 1971. Observations on the development of the Brassica cyst eelworm. Plant Pathol. (London) 20:144–148.

Linford, M.B., and J.M. Oliveira. 1940. *Rotylenchulus reniformis* nov. gen., n. sp., a nematode parasite of roots. Proc. Helminthol. Soc. Wash. 7:35–42.

Linford, M.B., and F. Yap. 1940. Some host plants of the reniform nematode in Hawaii. Proc. Helminthol. Soc. Wash. 7:42–44.

Loof, P.A.A. 1991. The family Pratylenchidae Thorne, 1949. p. 363–421. *In* W.R. Nickle (ed.) Manual of agricultural nematology. Marcel Dekker, New York.

Mai, W.F., J.R. Bloom, and T.A. Chen. 1977. Biology and ecology of the plant-parasitic nematode *Pratylenchus penetrans*. Pennsylvania State Univ. Agric. Exp. Stn. Bull. 815.

MacGuidwin, A.E., G.W. Bird, and G.R. Safir. 1985. Influence of *Glomus fasciculatum* on *Meloidogyne hapla* infecting *Allium cepa*. J. Nematol. 17:389–395.

Matthews, W.C., and P.A. Roberts. 1994. Resistance in cowpea to *Meloidogyne incognita* isolates and *M. javanica* virulent rk gene. J. Nematol. 26:109.

McCann, J. 1981. Threshold populations of *Heterodera cruciferae* and *H. schachtii* causing damage to cabbage seedlings. Plant Dis. Rep. 65:264–266.

McKenry, M.V., and T. Buzo. 1992. Field performance of a soil drenching device. J. Nematol. 24:605–606.

McKenry, M.V., B. Westerdahl, and O. Becker. 1994. Beyond pesticides—some California field examples. J. Nematol. 26:558.

McSorley, R., D.W. Dickson, and J.A. de Brito. 1994a. Host status of selected tropical rotation crops to four populations of root-knot nematodes. Nematropica 24:45–53.

McSorley, R., D.W. Dickson, J.A. de Brito, and R.C. Hochmuth. 1994b. Tropical rotation crops influence nematode densities and vegetable yields. J. Nematol. 26:308–314.

McSorley, R., D.W. Dickson, T.E. Hewlett, and J. J. Frederick. 1994c. Uncommon crops for management of *Meloidogyne arenaria*. J. Nematol. 26:110.

Meeker, D. 1988. Biologicals go commercial. Impact 5:11–13.

Miller, J.F., A.H. Kates, D.E. Davis, and J. McCormack. 1974. Poisonous plants of the southern United States. North Carolina State Univ.Agric. Ext. Ser.

Mountain, W.B., and C.D. McKeen. 1962. Effect of *Verticillium dahliae* on the population of *Pratylenchus penetrans*. Nematologica 7:261–266.

Naqui, S.Q.A., and M.M. Alam. 1975. Influence of brinjal mosaic virus in the populations of *Tylenchorhynchus brassicae* and *Rotylenchulus reniformis* around eggplant roots. Geobios 2:120–121.

Netscher, C. 1970. Les nématodes parasites des cultures maraichéres au Senegal. Cah. O.R.S.T.O.M. Sér. Biol. 11:209–229.

Netscher, C., and R.A. Sikora. 1990. Nematode parasites of vegetables. p. 237–283. *In* M. Luc et al. (ed.) Plant parasitic nematodes in subtropical and tropical agriculture. CAB Int. Inst. Parasitol., Wallingford, England.

Noe, J.P., and R.A. Sikora. 1990. Effects of tropical climates on the distribution and host-parasite relationship of plant parasitic nematodes. p. 583–597. *In* M. Luc et al. (ed.) Plant parasitic nematodes in subtropical and tropical agriculture. CAB Int. Inst. Parasitol., Wallingford, England.

Noling, J.W. 1994. Use of hot water, broad spectrum fumigants, and soil solarization for nematode control. J. Nematol. 26:562.

Noling, J.W., and J.O. Becker. 1994. The challenge of research and extension to define and implement alternatives to methyl bromide. Suppl. J. Nematol. 26:573–586.

Nusbaum, C.J., and H. Ferris. 1973. The role of cropping systems in nematode population management. Annu. Rev. Phytopathol. 11:423–440.

O'Bannon, J.H., G.S. Santo, and A.P. Nyczepir. 1982. Host range of the Columbia root-knot nematode. Plant Dis. 66:1045–1048.

Olthof, T.H.A. 1979. The use of beans and Kentucky blue grass for rearing *Pratylenchus neglectus*, *P. projectus* and *Helicotylenchus digonicus*. Can. J. Plant Sci. 59:897–898.

Olthof, T.H.A., and A.A. Reyes. 1969. Effect of *Pratylenchus penetrans* on *Verticillium* wilt of pepper. J. Nematol. 1:21–22.

Olthof, T.H.A., J.W. Potter, and E.A. Peterson. 1974. Relationship between population densities of *Heterodera schachtii* and losses in vegetable crops in Ontario. Phytopathology 64:549–554.

Omwega, C.O., I.J. Thomason, and P.A. Roberts. 1990. A single dominant gene in common bean conferring resistance to three root-knot nematode species. Phytopathology 80:745–748.

Overman, A.J. 1964. The effect of temperature and flooding on nematode survival in fallow sandy soil. Proc. Soil Crop Sci. Soc. Fla. 34:197–200.

Overman, A. 1982. Soil fumigation via drip irrigation under full-bed mulch culture for row crops. Proc. Soil Crop Sci. Soc. Fla. 41:153–155.

Overman, A., and J.P. Jones. 1977. Effects of *Belonolaimus longicaudatus*, *Criconemoides* sp., and *Meloidogyne incognita* on *Verticillium* wilt of tomato. J. Nematol. 9:279–280.

Perry, V.G., and H.L. Rhoades. 1982. The Trichodorid nematodes. p. 183–186. *In* R.D. Riggs (ed.) Nematology in the southern region of the United States. Arkansas Agric. Exp. Stn. South. Coop. Series Bull. 276.

Pitcher, R.S. 1965. Interrelationships of nematodes and other pathogens of plants. Helminthol. Abstr. 34:1–17.

Potter, J.W., and T.H.A. Olthof. 1993. Nematode pests of vegetable crops. p. 171–207. *In* K. Evans et al. (ed.) Plant parasitic nematodes in temperate agriculture. CAB Int. Inst. Parasitol., Wallingford, England.

Radewald, J.D., J.W. Osgood, K.S. Mayberry, A.O. Paulus, and F. Shibuya. 1969. *Longidorus africanus* a pathogen of head lettuce in the Imperial Valley of southern California. Plant Dis. Rep. 53:381–384.

Raski, D.J. 1950. The life history and morphology of the sugar-beet nematode, *Heterodera schachtii* Schmidt. Phytopathology 40:135–152.

Razak, A.R., and A.A.F. Evans. 1976. An intracellular tube associated with feeding by *Rotylenchulus reniformis* on cowpea. Nematologica 22:182–189.

Rhoades, H.L. 1971. Chemical control of the sting nematode, *Belonolaimus longicaudatus*, on direct-seeded cabbage. Plant Dis. Rep. 55:412–414.

Rhoades, H.L. 1982. Effect of temperature on survival of *Meloidogyne incognita* in flooded and fallow muck soil. Nematropica 12:33–37.

Robbins, R.T. 1972. Morphology and ecology of the sting nematode, *Belonolaimus longicaudatus*. Ph.D. diss. North Carolina State Univ., Raleigh (Diss. Abstr. 72-5605).

Robbins, R.T., and K.R. Barker. 1973. Comparisons of host range and reproduction among populations of *Belonolaimus longicaudatus* from North Carolina and Georgia. Plant Dis. Rep. 57:750–754.

Roberts, P.A. 1982. Plant resistance in nematode pest management. J. Nematol. 14:24–33.

Roberts, P.A. 1987. The influence of planting date of carrot on *Meloidogyne incognita* reproduction and injury to roots. Nematologica 33:335–342.

Rodriguez-Kabana, R. 1992. Cropping systems for the management of phytonematodes. p. 219–233. *In* F.J. Gommers and P.W.Th. Maas (ed.) Nematology from molecule to ecosystem. Europ. Soc. Nematol., Invergowrie, Dundee, Scotland.

Rodriguez-Kabana, R., P.S. King, D.G. Robertson, and C.F. Weaver. 1989a. Crops uncommon to Alabama for the management of *Meloidogyne arenaria* in peanut. Suppl. J. Nematol. 21:712–716.

Rodriguez-Kabana, R., J. Pinochet, D.G. Robertson, and L. Wells. 1992. Crop rotation studies with velvetbean (*Mucuna deeringiana*) for the management of *Meloidogyne* spp. Suppl. J. Nematol. 24:662–668.

Rodriguez-Kabana, R., D.B. Weaver, R. Garcia, D.G. Robertson, and E.L. Carden. 1989b. Bahiagrass for the management of root-knot and cyst nematodes in soybean. Nematropica 19:185–193.

Rodriguez-Kabana, R., C.F. Weaver, D.G. Robertson, P.S. King, and E.L. Carden. 1990. Sorghum in rotation with soybean for the management of *Heterodera glycines* and *Meloidogyne arenaria*. Nematropica 20:111–119.

Rohde, R.A., and W.R. Jenkins. 1957a. Host range of a species of *Trichodorus* and its host- parasite relationships on tomato. Phytopathology 47:295–298.

Rhode, R.A., and W.R. Jenkins. 1957b. Effect of temperature on the life cycle of stubby-root nematodes. Phytopathology 47:29.

Román, J. 1978. Fitonematologia tropical. Estación Exp. Agricola, Univ. Puerto Rico, Rio Piedras.

Santo, G.S., and J.N. Pinkerton. 1985. A second race of *Meloidogyne chitwoodi* discovered in Washington. Plant Dis. 69:361.

Santo, G.S., and M. Qualls. 1984. Control of *Meloidogyne* spp. on Russet Burbank potato by applying metham sodium through center pivot irrigation systems. J. Nematol. 16:159–161.

Sasser, J.N., and D.W. Freckman. 1987. A world perspective on nematology: The role of the society. p. 7–14. *In* J.A. Veech and D.W. Dickson (ed.) Vistas on nematology. Soc. Nematol., Hyattsville, MD.

Sasser, J.N., and M.F. Kirby. 1979. Crop cultivars resistant to root-knot nematodes *Meloidogyne* species. North Carolina State Univ. Graphics, Raleigh.

Sauer, M.R., and J.E. Giles. 1959. A field trial with a root-knot resistant tomato variety. Irrig. Res. Stn. Tech. Pap. no. 3. CSIRO, East Melbourne, Victoria, Australia.

Sayre, R.M. 1971. Biotic influences in soil environment. p. 235–256. *In* B.M. Zuckerman et al. (ed.) Plant parasitic nematodes. Acad. Press, New York.

Sayre, R.M., and S.W. Wang. 1975. Freezing and storing *Ditylenchus dipsaci* in liquid nitrogen. J. Nematol. 7:199–202.

Schilt, H.G., and E. Cohn. 1975. Pathogenicity and population increase of *Paratrichodorus minor* as influenced by some environmental factors. Nematologica 21:71–80.

Schneider, S.M., and H. Ferris. 1987. Stage-specific population development and fecundity of *Paratrichodorus minor*. J. Nematol. 19:395–403.

Seinhorst, J.W. 1954. Een ziefte in erwten, veroorzaakt door het saltje *Hoplolaimus uniformis* Thorne. Tijdschr. Plantenz. 60:262–264.

Seinhorst, J.W., and K. Kuniyasu. 1969. *Rotylenchus uniformis* (Thorne) on carrots. Netherlands J. Plant Pathol. 75:205–223.

Serracin, M., A.C. Schuerger, D.W. Dickson, D.P. Weingartner, and T. Hewlett. 1994. An alternative method for culturing *Pasteuria penetrans*. J. Nematol. 26:565.

Siddiqi, M.R. 1973. Systematics of the genus *Trichodorus* Cobb 1913 (Nematoda: Dorylaimida), with descriptions of three new species. Nematologica 19:259–278.

Siddiqui, M.A., A. Haseeb, and M.M. Alam. 1987. Combined effect of two nematodes and a fungus on the growth and water absorption capability of okra. Indian J. Plant Pathol. 5:83–86.

Sikora, R.A., and N. Greco. 1990. Nematode parasites of food legumes. p. 181–235. *In* M. Luc et al. (ed.) Plant parasitic nematodes in subtropical and tropical agriculture. CAB Int. Inst. Parasitol., Wallingford, England.

Sivakumar, C.V., and A.R. Seshadri. 1971. Life history of the reniform nematode, *Rotylenchulus reniformis*, Linford and Oliveira, 1940. Indian J. Nematol. 1:7–20.

Sivakumar, C.V., and A.R. Seshadri. 1972. Histopathology of infection by the reniform nematode, *Rotylenchulus reniformis* Linford and Oliveira, 1940 on castro, papaya, and tomato. Indian J. Nematol. 2:173–181.

Smart, G.C., Jr., and K.B. Nguyen. 1991. Sting and awl nematodes: *Belonolaimus* spp. and *Dolichodorus* spp. p. 627–667. *In* W.R. Nickle (ed.) Manual of agricultural nematology. Marcel Dekker, New York.

Smith, A.L., and A.L. Taylor. 1941. Nematode distribution in the 1940 regional cotton-wilt plots. Phytopathology 31:771.

Smittle, D.A., and A.W. Johnson. 1982. Effects of management practices on *Meloidogyne incognita* on snapbean yield. J. Nematol. 14:63–68.

Starr, J.L., and W.F. Mai. 1976. Effect of soil microflora on the interaction of three plant-parasitic nematodes with celery. Phytopathology 66:1224–1228.

Steiner, G. 1949. Plant nematodes the grower should know. Proc. Soil Sci. Soc. Fla. (1942) 4-B:72–117.

Stinner, B.R., and D.H. Crossley, Jr. 1982. Nematodes in no-tillage agroecosystems. p. 14–28. *In* D.W. Freckman (ed.) Nematodes in soil ecosystems. Univ. Texas Press, Austin.

Stirling, G.R. 1993. Outlook for biocontrol. p. 257–265. *In* F.J. Gommers and P.W.Th. Maas (ed.) Nematology from molecule to ecosystem. Europ. Soc. Nematol., Invergowrie, Dundee, Scotland.

Stone, A.R., and J.A. Course. 1974. *Heterodera goettingiana*. Set 4. No. 47. Commonw. Inst. Helminthol. St. Albans, England.

Sturhan, D. 1971. Comparative investigations on the host plants of stem eelworms (*Ditylenchus dipsaci*) from beets of different origin. Meded. Land. 30:1468–1474.

Sturhan, D., and M.W. Brzeski. 1991. Stem and bulb nematodes, *Ditylenchus* spp. p. 423–464. *In* W.R. Nickle (ed.) Manual of agricultural nematology. Marcel Dekker, New York.

Sumner, D.R., and A.W. Johnson. 1973. Effect of root-knot nematodes on *Fusarium* wilt of watermelon. Phytopathology 63:857–861.

Sumner, D.R., S.J. Kays, and A.W. Johnson. 1976. Etiology and control of root diseases of spinach. Phytopathology 66:1267–1273.

Sumner, D.R., S.J. Kays, C.A. Jaworski, and R.B. Chalfant. 1978. Influence of film mulches and soil pesticides on root diseases and populations of soil-borne fungi in vegetables. Plant Soil 49:267–283.

Sumner, D.R., S.C. Phatak, D.A. Smittle, A.W. Johnson, and N.C. Glaze. 1981. Control of cucumber foliar diseases, fruit rot, and nematodes by chemicals applied through overhead sprinkler irrigation. Plant Dis. 65:401–404.

Sykes, G.B., and A.L. Winfield. 1966. Studies on Brassica cyst nematode *Heterodera cruciferae*. Nematologica 12:530–538.

Taylor, A.L., and J.N. Sasser. 1978. Biology, identification and control of root-knot nematodes (*Meloidogyne* species). North Carolina State Univ. Graphics, Raleigh.

Taylor, C.E. 1979. *Meloidogyne* in interrelationships with microorganisms. p. 375–398. *In* F. Lamberti and C.E. Taylor (ed.) Root-knot nematodes (*Meloidogyne* species), systemics, biology and control. Acad. Press, London.

Thies, J.A., and R.L. Fery. 1994. Evaluation of okra cultivars for resistance to *Meloidogyne incognita*. J. Nematol. 26:567–568.

Thomason, I. J. 1987. Challenges facing nematology: Environmental risks with nematicides and the need for new approaches. p. 469–476. *In* J.A. Veech and D.W. Dickson (ed.) Vistas on nematology. Soc. Nematol., Hyattsville, MD.

Thomason, I.J., and E.P. Caswell. 1987. Principles of nematode control. p. 87–130. *In* R.H. Brown and B.R. Kerry (ed.) Principles and practice of nematode control in crops. Acad. Press, London.

Thomason, I.J., J.R. Rich, and F.C. O'Melia. 1976. Pathology and histopathology of *Pratylenchus scribneri* infecting snapbean and lima bean. J. Nematol. 8:347–352.

Thorne, G. 1952. Control of the sugar beet nematode. USDA Farm. Bull. 2054. Bur. Plant Indust., Washington, DC.

Thorne, G., and M.L. Schuster. 1956. *Naccobbus batatiformis*, n. sp. (Nematoda:Tylenchidae) producing galls on the roots of sugar beets and other plants. Proc. Helminthol. Soc. Wash. 23:128–134.

Todd, F.A. 1971. System control: A prescription for flue-cured tobacco diseases. North Carolina Agric. Ext. Circ. 530.

Townshend, J.L., M. Meskine, and G.L. Barron. 1989. Biological control of *Meloidogyne hapla* on alfalfa and tomato with the fungus *Meria coniospora*. J. Nematol. 21:179–183.

Triantaphyllou, A.C., and J.N. Sasser. 1960. Variation in perineal patterns and host specificity of *Meloidogyne incognita*. Phytopathology 50:724–735.

U.S. Department of Agriculture. 1992. Notice of release of Charleston Hot cayenne pepper. USDA-ARS, Washington, DC.

Valdez, R.B. 1978. Nematodes attacking tomato and their control. p. 136–152. *In* R. Cowell (ed.) Proc. 1st Int. Symp. Trop. Tomato, AVRDC Publ. 78-59, Shanhua, Taiwan. 23–27 October. Asian Veg. Res. Develop. Center, Shanhua, Taiwan.

Van Gundy, S.D., J.D. Kirkpatrick, and J. Golden. 1977. The nature and role of metabolic leakage from root-knot nematode galls and infection by *Rhizoctonia solani*. J. Nematol. 9:113–121.

Viaene, N.M.M., and G.S. Abawi. 1994. Effects of cover crops on *Meloidogyne hapla* on lettuce grown in organic soil. J. Nematol. 26:123.

Walters, S.A., and K.R. Barker. 1993. Reproductive and damage potentials of two populations of *Rotylenchulus reniformis* on sweet potato and related comparisons with *Meloidogyne javanica* on tomato. Suppl. J. Nematol. 25:830–835.

Walters, S.A., and K.R. Barker. 1994. Efficacy of *Paecilomyces lilacinus* in suppressing *Rotylenchulus reniformis* on tomato. Suppl. J. Nematol. 26:600–605.

Weaver, C.F., R. Rodriguez-Kábana, and D.G. Robertson. 1994. Rotations with winter cover crops for the management of root-knot nematodes in eggplant. J. Nematol. 26:569.

Webber, H.J., and W.A. Orton. 1902. Some diseases of cowpea. II. A cowpea resistant to root-knot (*Heterodera radicicola*). USDA Bull. 17. Bur. Plant Industry, Washington, DC.

Webster, J.M. 1972. Nematodes and biological control. p. 461–496. *In* J.M. Webster (ed.) Economic nematology. Acad. Press, London.

Webster, J.M. 1985. Interaction of *Meloidogyne* with fungi on crop plants. p. 183–192. *In* J.N. Sasser and C.C. Carter (ed.) An advanced treatise on *Meloidogyne*. Vol. I. North Carolina State University Graphics, Raleigh.

Wehner, T.C., S.A. Walters, and K.R. Barker. 1991. Resistance to root-knot nematodes in cucumber and horned cucumber. Suppl. J. Nematol. 23:611–614.

Winstead, N.N. 1959. Reaction of cabbage varieties and clubroot-resistant lines to root-knot nematodes. Plant Dis. Rep. 43:909–912.

Wyatt, J.E., G. Fassuliotis, and A.W. Johnson. 1980. Efficacy of resistance to root-knot nematode in snap beans. J. Am. Soc. Hort. Sci. 105:923–926.

Wyatt, J.E., G. Fassuliotis, J.C. Hoffman, and J.R. Deakin. 1983. 'NemaSnap' snap bean. HortScience 18:776.

Yarger, L.W., and L.R. Baker. 1981. Tolerance of carrots to *Meloidogyne hapla*. Plant Dis. 65:337–339.

Yuksel, H.S. 1960. Observations on the life cycle of *Ditylenchus dipsaci* on onion seedlings. Nematologica 5:289–296.

Zamora, E., P.W. Bosland, and S. Thomas. 1994. 'Carolina Cayenne' as a source of resistance to *Meloidogyne incognita* races 1, 2, 3, and 4. HortScience 29:1184–1185.

26 Fruit and Citrus Trees

ANDREW P. NYCZEPIR

USDA-ARS, Southeastern Fruit and Tree Nut Research Laboratory
Byron, Georgia

J. OLE BECKER

University of California
Riverside, California

This chapter focuses on the recent developments of nematode pests of fruit and citrus trees. Fruit trees herein include stone fruits {apricot [*Prunus armeniaca* L.], cherry [*P. avium* (L.) L., and *P. cerasus* L.], peach/nectarine [*P. persica* (L.) Batsch], and plum [*P. domestica* L., *P. salicina* Lindl. hybrids and *P. cerasifera* Ehrh. hybrids]}, and pome fruits [apple (*Malus domestica* Borkh.) and pear (*Pyrus communis* L.)]. Almond [*Prunus dulcis* (Mill.) Webb] is usually classified with nuts and will not be discussed in this chapter.

Edible citrus include sweet orange [*Citrus sinensis* (L.) Osbeck], sour orange (*C. aurantium* L.), lemon [*C. limon* (L.) Burm. f.], lime [C. *aurantiifolia* (Christm.) Swingle], citron (*C. medica* L.), pummelo [*C. maxima* (Burm.) Merr.], grapefruit (*C.* × *paradisi* Macfad.), mandarin (*C. reticulata* Blanco), and kumquat [*Fortunella margarita* (Lour.) Swingle].

STONE FRUITS

Stone fruits are grown in areas where soil, moisture, and climatic conditions are suitable, regardless of their place of origin. Climatic conditions such as chilling hours, frost-free spring, and water availability determines the limits of species distribution. Stone fruits are generally confined to those areas 23° N and 23° S of the Tropics of Cancer and Capricorn, respectively. The origin and history of stone fruits are well documented (Westwood, 1988) and are, therefore, only briefly discussed in this chapter.

Peach/Nectarine

Peach originated in China and was transported into Persia, Italy, and other temperate areas throughout Europe (Fogel et al., 1974). Peach was introduced into the Western Hemisphere by the Spaniards around the mid-1500s. Over 50 countries produce peach/nectarine throughout the world, with production estimated over 10 Tg. Italy, USA, and Greece are the top three producers, accounting for over 43% of the world's total production (FAO, 1993). Peach is relatively nutritious providing the consumer with a good source of Vitamin A, ascorbic acid, and K. Nematodes cause some of the most important rootstock diseases of peach/nectarine that contribute to reduced yield and vigor, and tree death. Major nematode parasites of peach include *Criconemella* spp., *Meloidogyne* spp., *Pratylenchus* spp., and *Xiphinema* spp.

Cherry

The USA is the world's largest producer of both sweet and sour cherry, followed closely by Germany and Turkey (USDA, 1990). Almost all of the world's fresh cherry come from sweet varieties, with 'Bing' the most popular. Half of the cherries produced by the USA go to the fresh market. *Criconemella* spp., *Meloidogyne* spp., *Pratylenchus* spp., and nematode-vectored viruses are all economically important to cherry production.

Plum

Sixty-two countries produce plum/prune throughout the world, with China, USA, and the Commonwealth of Independent States (CIS, formerly USSR) leading the way, respectively (FAO, 1993). Plum that contains a high-soluble solid content and that can be dried without fermenting is called prune (Ramming & Cociu, 1991). Plum is relatively nutritious, and is utilized in making products such as jams, jellies, and juices as well as going to fresh market. Nematode pests associated with plum decline include *Criconemella* spp., *Meloidogyne* spp., *Pratylenchus* spp., and *Xiphinema* spp.

Apricot

In 1992, some 46 countries were recognized by the FAO (1993) as producers of apricot with Turkey, Spain, and Italy accounting for more then 34% of the world production. Nutritionally, apricot is an excellent source of Vitamins C and A, Ca and Fe. Species of *Criconemella*, *Meloidogyne*, *Pratylenchus*, and *Xiphinema* have been associated with apricot decline.

POME FRUITS

Most of the domestic cultivars of apple available today have been derived from the common apple of Europe (*Malus pumila* Miller). By 1492, there were several hundred known cultivars of apple. Today, only 13 cultivars make up 90% of production in the USA. The climatic adaptability of apple has made it the most widely planted temperate zone fruit (Westwood, 1988).

Domestic pear of Europe has been around since prehistoric times, whereas written records for the domesticated Asian pear exist from 1100 B.C. Like apple, only a few pear cultivars account for most of the world's production.

Apple

Over 43 Tg of apple were produced among 69 countries in 1992, with 33% of the total production coming from the USA, China, and CIS. Apple is not considered a major food crop even though it is nutritious, high in fiber, and ranks second to grape (*Vitis vinifera* L.) in the world as a deciduous fruit crop (Way et al., 1991). Nematode pests of apple include species of *Meloidogyne*, *Pratylenchus*, and *Xiphinema*.

Pear

In 1992, over 10 Tg of pear were produced, with China, Italy, and USA accounting for over 45% of the total production. Pear is higher in fiber content than apple, but is comparable in total carbohydrate composition (Bell, 1991). The major nematode pests of pear are *Pratylenchus* spp.

CITRUS

The citrus family (*Rutaceae*) originated in Southeast Asia where various types are thought to have been cultivated before recorded history. Trade and exploration disseminated many types of citrus throughout the world. Today they are grown in almost every tropical to warm temperate region country. Citrus production will probably exceed 80 Tg by the year 2000 (Davies & Albrigo, 1994). Nematode pests of citrus include *Belonolaimus*, *Hemicycliophora*, *Meloidogyne*, *Paratrichodorus*, *Pratylenchus*, *Radopholus*, *Trichodorus*, *Tylenchulus*, and *Xiphinema* spp.

IMPORTANT NEMATODE PARASITES OF STONE FRUITS

Root-Knot Nematodes

Species and Races

Each of the four major *Meloidogyne* spp. have been shown to cause economic damage to stone fruits in different parts of the world. *Meloidogyne incognita* and *M. javanica* are the predominant species on peach and plum. The USA (Ferris et al., 1976), Israel (Minz. 1963), Spain (Marull et al., 1994), Libya (Siddiqi, 1982), India (Saxena & Chhabra, 1970) and Japan (Kawase, 1972) are just a few countries that have reported both species attacking peach. Additionally, *M. javanica* has been found in Morocco (Esmenjaud et al., 1994) and Turkey (Fernandez et al., 1994) and *M. incognita* and *M. arenaria* in China (Wu et al., 1993) and Italy (Simeone & Di Vito, 1992). *Meloidogyne arenaria* also has been found on peach in Spain (Marull et al., 1994), and *M. hapla* in Japan (Saigusa & Matsumoto, 1961) and Israel (Minz, 1963). Of more limited geographic distribution, *M. hispanica* from Spain (Hirschmann, 1986) and *M. morocciensis* from

Morocco (Rammah & Hirschmann, 1990) also have been reported on peach. On plum, *M. incognita* and *M. javanica* have been found in the USA (Norton et al., 1963) and *M. incognita* from Spain (Fernandez et al., 1994). An Italian population of *M. incognita* was reported parasitizing apricot in Italy (Siniscalco et al., 1976) even though most apricot rootstocks are touted as immune, and *M. hapla* was found on sour cherry from the USA (Mai et al., 1960).

Limited information exists regarding the relationship between races of *Meloidogyne* spp. and their pathogenicity on *Prunus*. In Cyprus, *M. incognita* Race 2 was reported on peach (Philis, 1983), and in the USA a Florida isolate of *M. incognita* Race 3 is known to parasitize Okinawa, Nemared, and Nemaguard peach rootstocks (Sherman et al., 1991).

Biology

The biology of *Meloidogyne* spp. on stone fruits follows the general pattern as described in earlier chapters. Briefly, the nematodes belonging to this genus are sedentary endoparasites and can have several generations in 1 yr on peach if environmental and edaphic conditions are conducive. Typically, root-knot nematode damage is more serious to trees growing in sandy soil compared to finer-textured soils (Stirling, 1975). Soil temperature can alter susceptibility of nematode resistant *Prunus* spp. to *M. incognita* and *M. javanica*. Nemared peach and Myrobalan 29C plum maintained high levels of resistance to *M. incognita* when grown at soil temperatures of 23 and 31°C. The experimental peach-almond hybrid G × N N°22 exhibited a partial loss in resistance at 31°C. The susceptible

Fig. 26–1. Stunting of 1-yr-old Goldprince/Lovell peach trees planted into *Meloidogyne* spp. infested soil (photo supplied by A. P. Nyczepir).

peach-almond hybrid GF-677 exhibited more extensive galling at higher temperatures (Fernandez et al., 1993). A similar loss in resistance also occurred with Nemaguard rootstock, which was moderately galled when grown at 30°C (Wehunt, 1972). High soil temperature (32°C) also increased root-galling and favored nematode reproduction on the peach-almond hybrid, G × N no.1, parasitized by *M. javanica* 90 d after inoculation (Canals et al., 1992).

Certain vesicular-arbuscular mycorrhiza (VAM), like *Gigaspora margarita*, have been shown to suppress reproduction of *M. incognita* on peach (Strobel et al., 1982).

Symptoms

A peach orchard infested with root-knot nematodes usually contains trees of uneven size (Fig. 26–1). Stunting of trees is usually first detected in the first 2 yr following planting. Affected trees also exhibit symptoms that include reduced vigor, growth, and yield, poorly developed foliage, early defoliation and occasionally twig dieback and tree death. These symptoms can be intensified in periods of drought. In the southeastern USA, it is not uncommon to detect bearing trees on Lovell rootstock that are heavily galled as a result of root-knot nematode infection, but show no symptoms of stunted growth. According to Bertrand (1985), these orchards may have been established on infested lands having low initial population densities of *Meloidogyne* spp.

Root galls of various sizes associated with poorly developed secondary and tertiary roots are typical belowground symptoms found on trees heavily infected with *Meloidogyne* spp. (Fig. 26–2).

Fig. 26–2. Galls on peach roots caused by *Meloidogyne* sp. (photo supplied by A. P. Nyczepir).

Fig. 26–3. Severe stunting of Lovell peach tree when grown in soil infested with both *Meloidogyne incognita* (*Mi*) and *Criconemella xenoplax* (*Cx*) as compared to Mi-alone or untreated check (photo supplied by A. P. Nyczepir).

Disease Complexes

Meloidogyne spp. have been implicated in peach disease complexes as a result of interactions with bacteria and other nematodes. Attack of Lovell peach seedlings by the crown gall bacterium (*Agrobacterium tumefaciens*) was increased in the presence of *M. javanica* (Nigh, 1966). In Canada (Dhanvantari et al., 1975), percentage of crown gall on peach roots increased in the presence of *M. hapla* 62 d after inoculation. However, in a follow-up study, the bacterial disease was not affected by the presence of *M. hapla* or *M. incognita* after 157 d. Growth of Lovell peach was suppressed more in the presence of *M. incognita* plus *C. xenoplax* as compared to either nematode alone, 26 mo after inoculation in field microplots (Nyczepir et al., 1993) (Fig. 26–3). Furthermore, *M. incognita* appears to be more pathogenic on Lovell peach than *C. xenoplax* (Fig. 26–4).

Economic Importance

Economic importance of *Meloidogyne* spp. to stone fruit production has been assessed with the use of nematicides and rootstock resistance. In the southeastern USA, preplant fumigation with methyl bromide to control *Meloidogyne* sp increased the 3-yr cumulative yield in the 3rd to 5th growing seasons of Redhaven on Lovell rootstock by 2535 kg ha^{-1} (Sharpe et al., 1993). Additionally, Foster et al. (1972) reported that the use of either a preplant fumigant nematicide [e.g., 1,2-dibromo-3-chloropropane (DBCP) or ethylene dibromide (1,2- dibromoethane), neither of which are available today] or resistant rootstock (S-37) increased peach yields in *Meloidogyne*-infested soil as compared to untreated checks. In the third and final year of the study, greatest returns were obtained from trees on S-37 rootstock. In Libya, stone fruit nurseries established on sites heavily infested with *M. javanica* estimated their loss due to the nematode at 115 500 LD (Libyan Dinars) over a 2-yr period (Siddiqi, 1982).

Ring Nematodes

Species

Like *Meloidogyne* spp., ring nematodes are widely distributed throughout the world with certain species considered to be economically important to the stone fruit industry. Probably two of the most important ring nematode species associated with stone fruit decline diseases include *Criconemella xenoplax* and *C. curvata*. Both species have been found in North America, Europe, India, and Africa (Mulvey, 1960; Williams, 1972; Loof, 1974) and both species have been reported on peach in the USA (Hung & Jenkins, 1969; Nyczepir et al., 1985). *Criconemella xenoplax* also has been found on peach in South America (Carneiro et al., 1993), South Africa (Heyes, 1970; Hugo & Meyer, 1995), Italy (Nicotina, 1990), and Australia (Stirling, 1975). Of lesser importance, *C. rustica* was reported on peach in the USA. (Jaffee et al., 1987b). *Criconemella curvata* and *C. xenoplax* were found on cherry in Canada (Mulvey, 1960) and the USA (Melakeberhan et al., 1993), respectively. *Criconemella xenoplax* also has been found on apricot in India, USA, and Africa (Heyes, 1970; Williams, 1972; Sharma et al., 1991) and on plum in Africa and USA (Heyes, 1970; Lownsbery et al., 1974).

Biology

The life cycle of *C. xenoplax* is completed in 25 to 34 d at 22 to 26°C, under laboratory conditions (Seshadri, 1964). The development of nematode life stages requires 11 to 13 d for eggs, 3 to 5 d for J2, 4 to 7 d for J3, 5 to 6 d for J4 and 2 to 3 d for adult maturation. Adult females lay 8 to 15 eggs over a 2- to 3-d period. Egg deposition occurs in close proximity to host roots or on the root surface. Males are scarce, and reproduction is presumed to be primarily by parthenogenesis. Feeding is required for juveniles to molt and for maturation of oocytes in adult females.

Feeding activity of *C. xenoplax* occurs along peach roots and at the tips. Unlike other ectoparasitic feeders, *C. xenoplax* was observed feeding from a single-root cortical cell for 1 to 8 d with no necrotic tissue development associated with the feeding site (Westcott & Hussey, 1992). Additionally, individual cortical cells of clover (*Trifolium* spp.), carnation (*Dianthus caryophyllus* L.) and tomato (*Lycopersicon esculentum* Miller) parasitized by *C. xenoplax* were modified into discrete "food cells" in order to sustain ingestion of food for long periods of time (Hussey et al., 1992).

Criconemella xenoplax reproduces faster in sandy soils as compared to loam or silty loam soils and prefers soil temperatures ranging from 22 to 26°C (Seshadri, 1964). In another study, 66% of *C. xenoplax* eggs hatched between 13 and 32°C, but died at temperatures >32.5°C (Westcott & Burrows, 1991).

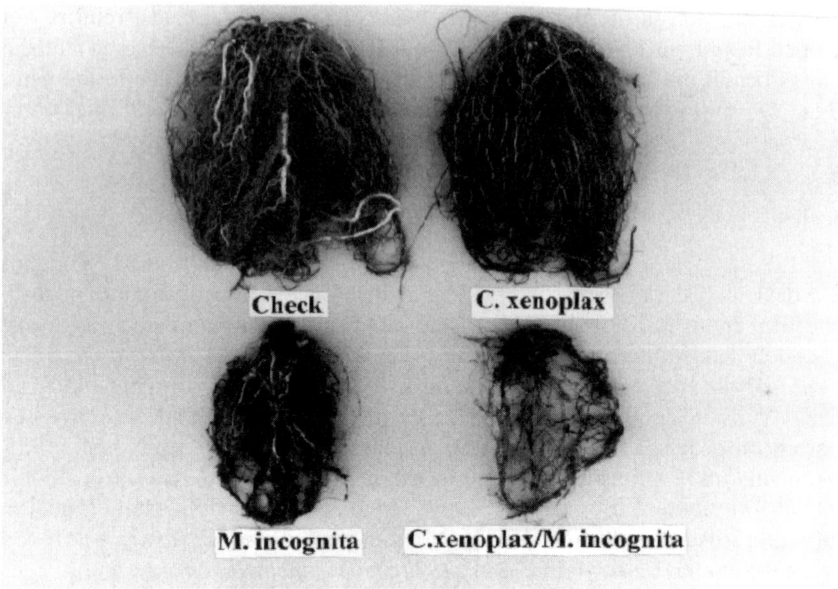

Fig. 26–4. Influence of *Meloidogyne incognita (*alone), *Criconemella xenoplax* (alone), or both nematodes together on Lovell peach root growth in the greenhouse after 150 d (P_i = 2000 eggs and/or nematodes) (photo supplied by A. P. Nyczepir).

Symptoms

Stone fruit trees parasitized by *Criconemella* spp. may exhibit various symptoms depending on age of trees and climatic conditions that influence the presence of certain causal agents of tree death for a particular region of the world. The nematodes belonging to this genus have an ectoparasitic feeding habit. Feeding by *C. curvata* on sterile cultured peach seedlings resulted in extensive pits and lesions on roots (Hung & Jenkins, 1969), whereas parasitism by *C. xenoplax* is usually characterized by lack of feeder roots (Nyczepir et al., 1987).

In the southeastern USA, peach trees parasitized by *C. xenoplax* are predisposed to cold injury, bacterial canker *(Pseudomonas syringae*) or both. This disease complex is termed peach tree short life (PTSL) (Nyczepir, 1990). The rapid fluctuation in late winter/early spring temperatures typically of this region in the USA are associated with the outbreak of PTSL. In California, *C. xenoplax* renders *Prunus* more susceptible to bacterial canker, a disease complex termed Bacterial Canker Complex (McKenry, 1989a).

Generally, aboveground PTSL symptoms express themselves when trees are 3 to 6 yr of age although younger or older trees can be affected. Symptoms are similar to those of any plant deprived of an adequate root system (Fig. 26–5). Typically, initiation of normal tree growth is followed by chlorosis, wilting, and sudden death shortly after bloom (Ritchie & Zehr, 1995). In severe cases, neither flower nor leaf buds open at all. In a study by Nyczepir and Wood (1988), it was reported that autumn leaf senescence was delayed in 2-yr-old Nemaguard trees grown in field microplots in the presence of *C. xenoplax* prior to PTSL symptom development in subsequent years.

Fig. 26–5. Dead peach tree with suckers at the crown during summer in the presence of *Criconemella xenoplax* and the peach tree short life disease complex (photo supplied by A. P. Nyczepir).

Bacterial canker of peach is generally associated with delay of bloom or leaf out of individual limbs, with such limbs usually dying by summer's end. If the infection is severe, entire trees may collapse with branches exhibiting alternate zones of darkened and healthy tissue having a sour sap odor. Dead trunk tissue usually does not extend beneath the soil line, thus leaving the primary root system alive. As a result, suckers are usually produced at the crown of the tree during summer (Fig. 26–5). When cold injury is present, trees will begin to leaf out until additional water is required by the tree. At that time, leaves collapse, and the bark may crack and separate from the scaffold limbs and trunk of the tree. In California, limbs and even entire trees affected by bacterial canker complex die in spring.

Disease Complexes

Replant problems of peach and other stone fruits, defined as the unsatisfactory performance of trees grown in soil from which peach trees were previously removed, have plagued orchardists since the late 1600s. Furthermore, replant problems have been defined as soil sickness, sudden decline or root residue toxicity. In the southeastern USA a specific type of disease complex problem exists, which has been termed PTSL. This disease complex is characterized by the sudden collapse of new growth above the soil line and subsequent death of the aboveground portion of the tree in late-winter/early spring following cold injury and (or) bacterial canker infection. The principal biotic factor that is responsible for predisposing peach trees to bacterial canker or cold injury is *C. xenoplax* (Nyczepir, 1990). Additionally, *C. xenoplax* appears to be the only plant-parasitic nematode associated with PTSL. In an ongoing field microplot study, the dynamics of concomitant populations of *C. xenoplax* and *M. incognita*

Fig. 26–6. Dead 6-yr-old Lovell peach trees in the presence of *Criconemella xenoplax* (alone) (*Cx*) and the peach tree short life disease complex (*Mi* = *Meloidogyne incognita*) (photo supplied by A. P. Nyczepir).

are being studied on Lovell peach (Nyczepir & Wood, 1995) Results after 6 yr indicate that the only trees to die from PTSL are those that are planted in *C. xenoplax*-alone infested soil (Fig. 26–6). Furthermore, severity of PTSL in orchards infested with *C. xenoplax* appears to increase and orchard lifespan decrease with each successive peach planting (Nyczepir & Okie, 1996). In South Africa, peach shoots were defoliated 90 to 120 d (3–4 mo) before natural leaf drop when trees were parasitized by a concomitant infestation of *C. xenoplax* and *M. javanica* (Hugo & Meyer, 1995). In California, *C. xenoplax* and more recently *Criconema mutabile* have been associated with making apricot, peach, and plum trees more susceptible to bacterial canker complex (BCC) (McKenry, 1989a, 1990b). A distinction between BCC and PTSL is that cold injury is not associated with BCC, even though BCC is more likely to occur in "cold-spots" within an orchard.

Economic Importance

Tree loss due to PTSL varies from year to year in the southeastern USA. In a 13-yr tree death survey (1980–1992) in South Carolina alone, a total loss in potential income to growers from PTSL was estimated at $85 million, which represents over $6 million per year (Miller, 1994).

Root-Lesion Nematodes

Species and Races

At least nine species of root-lesion nematode have been found associated with stone fruits throughout the world. *Pratylenchus vulnus* was found on peach and plum in the USA (Fliegel, 1969; Lownsbery et al., 1974). Additionally, *P. vulnus* was found on peach in Italy (Ricciardi et al., 1975), Australia (Stirling, 1975), China (Gao et al., 1989), Brazil (Lordello, 1973), Spain (Fernandez et al., 1992) and France (Scotto La Massese, 1979). Apricot and cherry also were found to be suitable hosts for *P. vulnus* in the USA (Lownsbery & Serr, 1963) and Spain (Pinochet et al., 1992b), respectively. Recently, Pinochet et al. (1993b) confirmed the existence of strains with different damage potentials on GF-677 peach-almond hybrid among geographically separated populations of *P. vulnus*. In another study, Pinochet et al. (1994) identified distinct races of *P. vulnus,* where an apricot and rose (*Rosa* spp.) isolate of *P. vulnus* did not attack citrus, and the walnut isolate did.

Pratylenchus penetrans has been detected on peach in Canada (Mountain & Patrick, 1959) and China (Wu et al., 1993), and on cherry in the USA (Melakeberhan et al., 1993), Canada (Vrain & Yorston, 1987), Poland (Szczygiel & Rebandel, 1985), and Germany (Harr et al., 1976).

Pratylenchus pratensis, *P. brachyurus*, *P. zeae*, and *P. convallariae* were all found on peach in China (Wu et al., 1993). *Pratylenchus zeae* also was detected on peach in the USA (Fliegel, 1969) and Australia (Colbran, 1955).

In Spain (Pinochet et al., 1991), *P. neglectus* and *P. thornei* are frequently detected in peach nurseries and orchards but are not regarded as economically important pests. *Pratylenchus neglectus* also has been found on cherry in Poland in addition to *P. crenatus* (Zepp & Szczygiel, 1985). *Pratylenchus hexincisus* was the dominant species found on apricot trees in Hungary (Dabaj & Jenser, 1991).

Biology

Pratylenchus spp. are migratory endoparasites that enter and move inter- and intracellularly while feeding on cortical cells where they cause extensive damage. All life stages of *P. vulnus* were observed in "large pockets" and cavities of healthy cortical parenchyma cells of the peach-almond hybrid, G × N no. 1 (Marull & Pinochet, 1991) and plum rootstocks (Pinochet et al., 1993c). Active juveniles and adult stages have the capability of entering and leaving the root. Generally, adult females deposit eggs singly within individual root cells. However, some eggs can be found in necrotic tissue. The first molt occurs in the egg with the J2 emerging at hatch. The juvenile molts three more times, between intervals of feeding, to become an adult. The duration of the life cycle for *P. penetrans* is temperature dependent and varies between 30 (30°C) and 92 d (15°C) (Mai et al., 1977). *Pratylenchus penetrans* is usually associated with cooler temperatures and higher elevations, whereas *P. vulnus* is typically more often associated with warmer temperature regions (McKenry, 1989a). Both *P. penetrans* and *P. vulnus* reproduce sexually. *Pratylenchus penetrans* and *P. vulnus* penetrate roots at a faster rate and do more damage to plants growing in sandy loam (coarse) as compared to finer textured soils. On peach in the USA, *P. vulnus* population density was greatest from August to December (Fliegel, 1969). *Pratylenchus penetrans* population density on most crops is highest in late summer and early fall and lowest in late spring and early summer.

Reproduction of *P. vulnus* on Marianna 2624, Myrobalan 605, and St. Julien 655-2 plum (Camprubi et al., 1993) and St. Lucia cherry rootstocks

(Pinochet et al., 1995) was not suppressed in the presence of either arbuscular mycorrhizal *Glomus mosseae* or *Glomus intraradices* fungi, respectively.

Symptoms

The presence of *Pratylenchus* spp. in stone fruit orchards can affect tree establishment, orchard uniformity, and ultimately fruit production. Aboveground symptoms can include tree decline with general characteristics of nutrient deficiency, reduced shoot growth and general tree vigor, and a reduction in fruit size. First-leaf plum trees growing in a field naturally infested with *P. vulnus* were severely stunted; however, damage among older trees was not as discernable (McKenry, 1989a). *Pratylenchus vulnus* was associated with reduced peach tree vigor in Georgia (Fleigel, 1969). In Canada, *P. penetrans* was affiliated with peach tree decline, chlorosis and reduced growth (Mountain & Patrick, 1959), and with impaired cherry tree growth in the USA (Mai et al., 1994). *Pratylenchus hexincisus* was associated with stunting, dieback, and death of apricot trees in Hungary (Dabaj & Jenser, 1991).

Generally, belowground symptoms of *Pratylenchus* spp. damage include a reduction in number of feeder roots, darkening of roots, and necrotic lesions

Fig. 26–7. Influence of *Pratylenchus vulnus* on Myrobalan 29C plum root growth after 2-yr (left = untreated, right = inoculated, P_i = 1000 nematodes per plant) (photo supplied by J. Pinochet).

(McKenry, 1989a) (Fig. 26–7). Rapid root deterioration and discoloration of peach feeder roots was detected with parasitism by *P. vulnus* (Marull & Pinochet, 1991), whereas *P. penetrans* caused necrosis of peach roots in the absence of other microorganisms (Mountain & Patrick, 1959). The extent of tissue darkening is dependent on phenolic content of the cells attacked or the ability of cells to synthesize phenols after injury. Symptoms on sour cherry roots caused by *P. penetrans* include necrotic lesions, dead rootlets, and a reduced and discolored root system (Mai, 1960).

Disease Complexes

Pratylenchus penetrans is associated with the peach replant problem in Canada (Mountain & Patrick, 1959). It causes necrosis of peach roots in the absence of bacteria and fungi, and the necrosis takes place in advance of the invading nematode. The role of the nematode in replant disease is to initiate root degeneration, thus providing sites for secondary infection by other soil microbes. Recently, Melakeberhan et al. (1993) reported that *P. penetrans* had a higher population density and frequency of occurrence than all other plant-parasitic nematodes identified in Michigan sweet cherry orchards in decline. The role of *P. penetrans* in decline of sweet cherry is not well understood, but it was noted that low soil pH was a common abiotic factor associated with this disease complex. In the northeastern USA, young cherry trees growing in soils with a high population of *P. penetrans* were more prone to cold injury than trees growing in soil treated with a preplant fumigant nematicide (Edgerton & Parker, 1958). In Germany (Harr et al., 1976), the cherry replant problem is thought to be the result of damage caused by *P. penetrans*, or the fungus *Thielaviopsis basicola,* or both. Growth reduction caused by combining low levels of both pathogens resulted in additive damage, whereas high concentrations of both organisms resulted in synergistically increased damage. *Pratylenchus vulnus* is associated with peach replant problems in the USA (McKenry, 1989a) and Italy (Ricciardi et al., 1975). In the southeastern USA, *P. vulnus* was the primary root-lesion nematode species most often associated with extensive feeder root deterioration and reduced tree vigor. Apricot decline, on the other hand, has been associated with the presence of *P. hexincisus.*

Pratylenchus vulnus adversely limits plant growth and colonization by the endomycorrhizal fungus (*G. mosseae*) in Marianna 2624 plum (Camprubi et al., 1993). On cherry, internal spore production by the VAM fungus (*G. intraradices*) was significantly reduced in the presence of this nematode (Pinochet et al., 1995).

Economic Importance

Pratylenchus vulnus and *P. penetrans* are the two most economically important root-lesion nematodes on stone fruits in California, with the former species being more important than the latter (McKenry, 1989a). Damage by *P. vulnus* to plum and peach rootstocks is estimated at approximately a 16% reduction in marketable fruit (reduction in fruit size and yield), with peach rootstocks more affected than plum. In the southeastern USA, *P. vulnus* was associated with reduced peach tree vigor, whereas *P. brachyurus* and *P. zeae* were not (Fliegel,

1969). *Pratylenchus vulnus* also was responsible for causing a 21 and 38% suppression in shoot and root growth, respectively, of PSM 101 plum in field microplots (Pinochet et al., 1993c). Additionally, Barker and Clayton (1973) tested the reaction of six *Pratylenchus* species on peach cultivars and found that only *P. penetrans* (California population) and *P. vulnus* (North Carolina population) caused root damage. *Pratylenchus penetrans* also is associated with the peach replant problem in Canada as previously mentioned.

Dagger Nematodes

Species

The geographic distribution and virus-vectoring capabilities of *Xiphinema* spp. were challenged and redefined in 1979 (Lamberti & Bleve-Zacheo, 1979). Prior to this time, *Xiphinema americanum* was considered a cosmopolitan species in Europe, but after careful examination these nematodes are known to be *X. brevicolle* or *X. pachtacium* (Brown & Taylor, 1987).

At least seven *Xiphinema* spp. have been identified as being associated with stone fruit disease. They include *X. americanum* group (*X. americanum*, *X. rivesi*, *X. californicum*, and *X. brevicolle*), *X. diversicaudatum*, *X. basiri,* and *X. vuittenezi*. *Xiphinema* spp. can cause a reduction in tree vigor, but they are more importantly associated with vectoring nepoviruses. *Xiphinema americanum* was found associated with peach, cherry, and plum roots in the USA (Mai et al., 1960; Jaffee et al., 1987a), and also with peach in Canada (Van Driel et al., 1990) and South Africa (Meyer & Hugo, 1994). In India and Western Australia, *X. americanum* was found on apricot (Siddiqi, 1973). *Xiphinema rivesi* was detected on peach in the USA (Jaffee et al., 1987a) and Canada (Van Driel et al., 1990) and found on plum in the USA (Koepsell et al., 1990). *Xiphinema diversicaudatum* was found on peach in Italy (Lamberti et al., 1993), cherry and plum in the Slovak Republic (Liskova et al., 1992), and on plum in Romania (Teodorescu et al., 1986). In Hungary (Jenser, 1985), *X. vuittenezi* was found occurring in the rhizosphere of apricot, peach, plum, and sweet and sour cherry. *Xiphinema basiri* and *X. brevicolle* were detected in the rhizospheres of apricot in India (Sharma et al., 1991) and on plum in Romania (Teodorescu et al., 1986), respectively; whereas *X. californicum* was found on peach in the USA (Jaffee et al., 1987a). This nematode has been found along the western seaboard USA and occasionally from the eastern USA. Additionally, *X. californicum* can coexist in the same peach orchard with *X. americanum* and *X. rivesi* (Jaffee et al., 1987a). *Xiphinema index* and *X. pachtacium* were found in the rhizosphere of peach in Italy (Nicotina, 1990), however, it was not clear as to what hosts these two species were parasitizing.

Biology

Xiphinema spp. are ectoparasites that frequently feed just behind the root tip (Abawi & Mai, 1990). Eggs are laid singly in the soil near the root system of host plants. Nematodes hatch as first-stage juveniles (J1). Recently, it was shown that native North American species, including *X. americanum*, *X. rivesi*, and *X. californicum,* have three juvenile stages while European species such as *X. diversicaudatum* have four (Halbrendt & Brown, 1992). The vermiform juvenile stages can be recognized by the differences in the lengths of their developing stylets. Males are rare in most species, and reproduction is primarily parthenogenic. In most temperate regions of the world only one generation is produced per year. However, *X. diversicaudatum* has been reported to have a life cycle of up to 3 yr from egg to egg (Pitcher et al, 1974). Soil temperature, moisture, and texture affect nematode population density and the efficiency of vectoring nepoviruses. In Hungary (Jenser, 1985), calcareous and noncalcareous loam soil provided favorable conditions for reproduction of *X. vuittenezi* on cherry, peach, plum, and apricot. *Xiphinema diversicaudatum* occurs primarily in sandy and loamy uncarbonated soils in the Slovak Republic where plum was a favored host as compared to other stone fruits grown in the region (Liskova et al., 1992). Population density of *X. americanum* increased at 21°C, but decreased at 16, 27, and 32°C under controlled conditions (Abawi & Mai, 1990). Additionally, *X. americanum* and *X. rivesi* population densities on peach were usually highest in fall and spring and lowest in summer under field conditions in the eastern USA (Jaffee et al., 1987a).

Symptoms

Symptoms and damage caused by *X. vuittenezi* on stone fruits is similar to that caused by other *Xiphinema* spp. (Jenser, 1985). Apricot trees parasitized by *X. vuittenezi* exhibit stunted shoot growth. Belowground the parasitized roots appear swollen, root growth is retarded, and the apical meristem decays leading to lateral root formation. In South Africa (Meyer & Hugo, 1994), peach trees parasitized by *X. americanum* showed various degrees of early leaf drop and dieback, whereas the root system was reduced, discolored, and had a limited number of functional feeder roots.

Prunus spp. infected with nepoviruses, vectored by different *Xiphinema* spp., also are responsible for causing characteristic aboveground symptoms of the tree. Three North American nepoviruses reportedly transmitted by the *Xiphinema americanum* group species (following the 1979 review) include cherry rasp leaf virus, tomato ring spot virus, and peach rosette mosaic virus (Brown et al., 1993). A fourth *Xiphinema* transmitted nepovirus to be discussed is strawberry latent ring spot virus.

Cherry rasp leaf virus (CRLV) is vectored by *X. americanum*, *X. californicum*, and *X. rivesi.* Cherry rasp leaf virus causes the disease known as cherry rasp leaf, which also has been called leaf enation, ruffled leaf, and cockscomb disease. It has been detected in western North America, Europe, and New Zealand. This disease has been reported primarily on sweet cherry and Mazzard seedlings, but also occurs in Montmorency sour cherry. Leaves of infected trees exhibit prominent enations on the underside and are distorted, whereas the upper leaf surface appears rough and coarse textured. Lower branches and spurs of infected trees may die thus reducing fruit production. Trees can die if infected at an early age (Nyland, 1976).

Tomato ring spot virus (TmRSV) is vectored by *X. americanum*, *X. californicum*, *X. rivesi*, and *X. bricolensis*. It causes prunus stem pitting (PSP), prune brown line (PBL), Stanley constriction and decline (SCAD) and yellow bud mosaic (YBMV) diseases. Prunus stem pitting has been found associated with stone fruit diseases in North America and occurs on peach, cherry, apricot, and plum (Mircetich & Fogel, 1976; Mircetich et al., 1977). Peach trees affected by PSP appear girdled, the leaves droop, curl upward lengthwise, and may appear chlorotic. In autumn, leaf color changes to red-purple followed by premature leaf abscission. Fruits on affected trees are smaller, tasteless, more brightly colored, and ripen and drop earlier than on TmRSV-free trees. Below or at ground level, the trunk is enlarged and the bark is thick and spongy. Removal of the bark from the lower portion of the enlarged area reveals pits and grooves in the wood (Fig. 26–8). Trees usually die within 2 to 3 yr after symptoms are initially observed.

Peach rosette mosaic virus (PRMV) is vectored by *X. americanum*. It has been found in the USA (New York and Michigan), with a similar disease reported from Italy (Klos, 1976). Stone fruit hosts of PRMV include peach and plum. Peach trees infected with PRMV are darker green than healthy trees and the internodes are shortened giving rise to rosettes of leaves. Additional symptoms may include delayed foliation, distortion and chlorotic mottling of newly formed leaves.

Strawberry latent ring spot virus (SLRV) is vectored by *X. diversicaudatum* and has been reported from numerous European countries (Taylor & Brown, 1981). Less frequent reports of detection have come from the USA, New

Fig. 26–8. Peach tree with bark removed at the graft union to show symptoms of Prunus stem pitting disease caused by tomato ringspot virus (photo supplied by G. L. Reighard).

Zealand, Asia, and Australia. Strawberry latent ring spot virus has been reported on peach, cherry, and plum (Hussey & McGuire, 1987). Damage to apricot is more serious when associated with other viruses such as necrotic ring spot virus (Crossa-Raynaud & Audergon, 1987).

Disease Complexes

As previously mentioned *Xiphinema* spp. may reduce tree vigor, but they are economically more important as vectors of nepoviruses. See the "Symptoms" section for a discussion of nematode-nepovirus diseases.

Economic Importance

The total economic impact of *Xiphinema* spp. and nepoviruses on the stone fruit industry has not been determined. Only a few reports document the extent of these problems outside the USA. An estimate of the damage caused by *Xiphinema* spp. can be obtained from several surveys, which report the incidence of nepovirus-infected trees in different kinds of stone fruit orchards in the USA. In the western USA, cherry trees affected with PSP in different orchards can range anywhere from a few trees to >90% infection (Mircetich et al., 1977). The proportion of 'Stanley' plum trees affected by PBL in northwestern, west-central, and southwestern regions of Michigan was 18, 32, and 35%, respectively (Ramsdell et al., 1993). In eastern USA, PSP was observed in 20% of the peach orchards surveyed and was the primary cause of tree decline in 8% of those sites. One-hundred percent of the orchards had dagger nematodes present and approximately 90% of these sites contained TmRSV-infected dandelion (*Taraxacum officinale* Wigg.) (Greene, 1986).

Other Nematodes

Lists of nematodes found in the rhizosphere of stone fruit, but not necessarily causing economic damage, are recorded elsewhere (Mai et al., 1960; Wehunt & Nyczepir, 1988). The genus *Paratylenchus* is perhaps the most frequently mentioned in the literature. Species associated with various stone fruits include *P. bukowinensis*, *P. hamatus*, *P. minutus*, *P. neoamblycephalus*, *P. projectus*, and *P. pruni* (Nyczepir & Halbrendt, 1993). Myrobalan plum seedlings parasitized by *P. neoamblycephalus* showed root discoloration and reduction of feeder roots (Braun et al., 1975). Reproduction of *P. neoamblycephalus* was favored at 20°C. *Paratylenchus pruni* suppressed peach growth at an initial population (P_i) of 10 000 per 500 cm^3 soil after 180 d (6 mo) under controlled conditions. Reproduction, determined as final population/initial population (P_f / P_i), of *P. pruni* at P_i 10 000 was 4.7 after 180 d (6 mo) and 97.5 at P_i 10. Spacial distribution studies revealed that the maximum population density of *P. pruni* was from 20-cm depth at 30-cm distance around an 8-yr-old tree (Sharma & Sharma, 1989).

IMPORTANT NEMATODE PARASITES OF POME FRUITS

Root-Lesion Nematodes

Species and Races

At least six *Pratylenchus* spp. have been detected on apple and (or) pear in temperate regions of the world. *Pratylenchus penetrans* has been found on apple and pear in the USA (Mai et al., 1960) and on apple in Canada (Vrain & Yorston, 1987). *Pratylenchus vulnus* has been reported on apple and pear in the USA (Lownsbery & Serr, 1963) and Spain (Fernandez et al., 1992), whereas *P. crenatus* and *P. neglectus* have been found associated with these fruits in Poland (Zepp & Szczygiel, 1985). Recently, Pinochet et al. (1993b) confirmed the existence of strains with different damage potentials on M-26 apple rootstock among geographically separated populations of *P. vulnus*. Furthermore, an apple isolate of *P. vulnus* was distinctly identified because of its ability to attack citrus, whereas an apricot isolate did not (Pinochet et al., 1994). In India (Khan & Sharma, 1990), *P. pratensis* was detected in the rhizosphere of apple trees, whereas *P. coffeae* Zimmerman was found in apple orchards in Australia (Colbran, 1953).

Biology

The life cycle for *Pratylenchus* spp. attacking pome fruits is similar to that described in the stone fruit section and will not be discussed again. Soil type and rootstock influence development of the *P. penetrans* populations in apple feeder roots. Fewer *P. penetrans* occurred in MM 111 apple roots than in MM 106 roots growing in clay soil, but soil pH, organic matter, drainage, and herbicide management did not have an effect (Costante et al., 1985). Soil temperature also influences the pathogenicity of *P. penetrans* on apple where infected roots were more affected at 29°C compared to other temperatures (Mai, 1960).

The presence of the endomycorrhizal fungus (*G. mosseae*) increased the growth of EMLA-26 apple rootstock in the presence of *P. vulnus* more than non-mycorrhizal plants (Pinochet et al., 1993a). Fewer *P. vulnus* also were detected in apple seedling roots in the presence of *G. mosseae*, whereas *P. vulnus* did not affect root colonization by the fungus.

Symptoms

Apple trees parasitized by *P. penetrans* often show obvious aboveground symptoms, while in other fruit trees symptoms and damage may be more difficult to detect (Mai et al., 1994). Stunting and even death are associated with severe infection of young trees, whereas older trees may exhibit leaf chlorosis, poor growth, and reduced vigor and yields. Apple trees are damaged more than pear. Stunting and unthrifty growth of apple seedlings also were correlated with parasitism by *P. coffeae* (Colbran, 1953). In some *P. penetrans* infested apple orchards, root damage can be so extensive that a uniform stand of productive trees cannot be established. Parasitism of apple feeder roots by *P. penetrans* and *P. vulnus* causes root discoloration, lack of feeder roots and necrosis (Fig. 26–9).

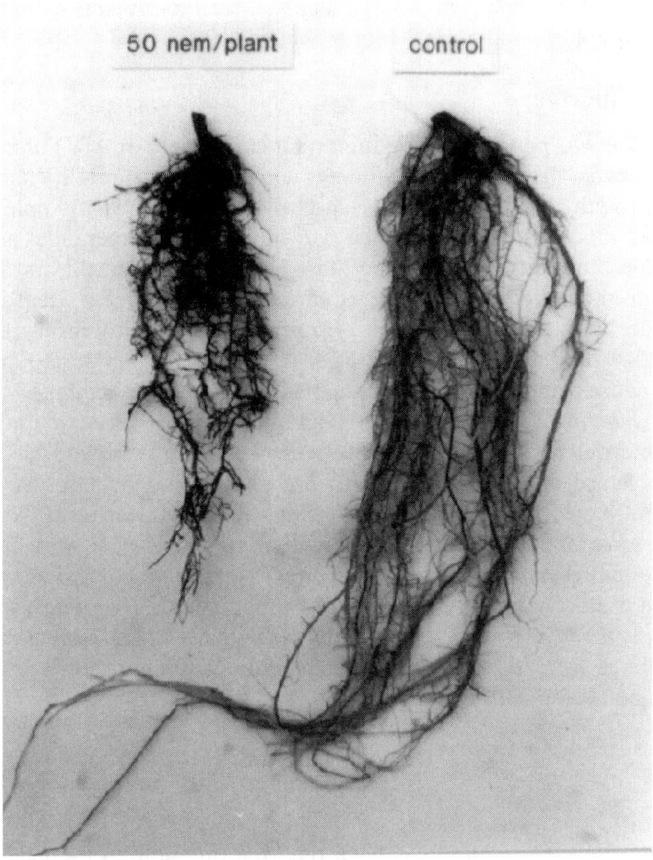

Fig. 26–9. Influence of *Pratylenchus vulnus* on M-26 apple rootstock after 135 d (photo supplied by J. Pinochet).

Apple is a less sensitive host than peach, since discoloration and necrosis takes place within a few hours of nematode invasion in peach vs. 30 or more days in apple (Pitcher et al., 1960). On apple, the belowground symptoms associated with *P. coffeae* include a discolored and abnormal root system. The tap roots of stunted trees are decayed and the discolored fibrous roots concentrated in the top 5 cm of soil (Colbran, 1953).

Disease Complexes

Pratylenchus penetrans is considered one of the primary nematode pathogens of pome fruit trees in the world. This nematode plays an important role in root destruction and as a causal agent of apple replant disease (ARD) in the USA (Mai et al., 1994) and Canada (Pitcher et al., 1960). Roots of young apple trees appear to be damaged more by *P. penetrans* than those of older trees, and severe root damage may result in tree mortality. Root damage is intensified by interactions with other soil microorganisms as well as by soil texture, low soil moisture, and cold temperatures (Mai & Abawi, 1981).

Economic Importance

Pratylenchus penetrans is an important component of ARD in northeastern USA and Canada. Preplant fumigation has decreased tree mortality and increased yield and growth in many apple orchards (Mai et al., 1994). Apple yield was increased as much as 100% with the use of preplant nematicides in an apple orchard. Apple tree growth recorded during the second and third seasons averaged 3.6 cm in trunk circumference in unfumigated soil and 8.5 cm in soil preplant-treated with methyl bromide (bromoethane)-chloropicrin (trichloronitromethane). A damage threshold of 25 to 150 *P. penetrans* per 100 cm^3 soil in sandy soil causes significant growth suppression of apple seedlings.

Pratylenchus vulnus is considered an important pathogen of apple and pear in the Mediterranean region (Fernandez et al., 1992). In vitro studies showed that *P. vulnus* caused extensive damage to both apple and pear rootstocks in the absence of other microorganisms, with visible lesions appearing at 7 d after inoculation. In a greenhouse experiment, growth of apple and pear were significantly reduced in the presence of *P. vulnus* as compared to uninoculated controls. Other species that are economically important to apple include *P. coffeae*, *P. crenatus*, and *P. neglectus* (Colbran, 1953; Zepp & Szczygiel, 1985). Tolerance limits of apple seedlings to *P. crenatus* and *P. neglectus* were approximately 50 and 30 nematodes per 100 g soil, respectively.

Dagger Nematodes

Species

Xiphinema americanum has been reported on apple in the USA (Jaffee et al., 1987a), India (Rai et al., 1988), Canada (Lana et al., 1983), and Romania (Teodorescu et al., 1986) and on pear in the USA (Mai et al., 1960). *Xiphinema americanum* and *X. rivesi* have been known to occur within the same apple orchard in the USA (Rosenberger et al., 1983). *Xiphinema* spp. detected in the rhizosphere of apple include *X. brevicolle* in Romania (Teodorescu et al., 1986), *X. index* in Italy (Coiro et al., 1985), and *X. belmontense* in Portugal (Roca & Pereira, 1992). Virus-vectoring capability and pathogenic habit of the latter three species on pome fruits is unknown. *Xiphinema* spp. do not appear to be a problem on pear.

Biology

The biology for *Xiphinema* spp. attacking pome fruits is similar to that described in the stone fruit section and will not be discussed again.

Symptoms

High populations of *X. americanum* have been associated with unthrifty growth and poor yield of apple trees in the eastern USA (Abawi & Mai, 1990).

Under controlled conditions, apple seedling feeder roots exhibited swollen tips, necrosis, brown lesions, and sloughing off of the cortex.

Xiphinema americanum and *X. rivesi* are known vectors of TmRSV, which is the causal agent of apple union necrosis and decline disease (AUND). This disease is only a problem where the scion variety is resistant to TmRSV and the rootstock is susceptible. This nepovirus disease has been found in eastern and northeastern USA (Rosenberger et al., 1983). Affected trees appear to be girdled. Leaves are pale, dull green, small, and sparse. Twigs exhibit a bunchy appearance because terminal growth is restricted. The best diagnostic symptoms associated with AUND of young bearing apple trees include pitting, invagination, and necrosis in the woody cylinder at the graft union, followed by general decline in growth and productivity, and even death. Diseased trees often break off at the graft union.

Tobacco ring spot virus (TobRSV) is vectored by *X. americanum* and is the causal agent of a union incompatibility condition of apple reported in Canada (Lana et al., 1983). This nepovirus is unique in that it can be vectored by aphids (*Myzus persicae* Sulzer and *Aphis gossypii* Glover), flea-beetles (*Epitrix hirtipennis* Melsheiner), and thrips (*Thrips tabci* Lindeman), in addition to *Xiphinema* spp. (Brown et al., 1993). This disease is characterized by necrosis, stem pitting, and breaking off at the graft union. At the end of the growing season, foliage is sparse, and leaves prematurely turn chlorotic and are diffusely mottled.

Xiphinema americanum also is a vector of CRLV, the causal agent of flat apple disease in the western USA (Parish, 1977). Symptoms occur on fruit and foliage of mainly Golden and Delicious apple cultivars. The fruit on affected limbs looks flat with little indentations of the stem end of the flattened fruit. Prominent lobes also occur at the calyx cavity. Affected branch growth is severely suppressed, and leaves roll upward thus exposing their underside. Some chemical thinners cause flat apples, but these apples, unlike CRLV-infected apples, contain no seeds. Furthermore, symptoms of flat apple disease are different from cherry rasp leaf although it is caused by the same virus (see "Symptoms" under "*Xiphinema*" in the "Stone Fruit" section).

Disease Complexes

As previously mentioned *Xiphinema* spp. cause reduction in pome tree vigor, but they are more importantly associated with vectoring nepoviruses. See the "Symptoms" section for nematode-nepovirus disease discussion.

Economic Importance

As with the stone fruits, the economic importance of *Xiphinema* spp. to the pome fruit industry has been difficult to evaluate. However, in one report TmRSV was detected in the inner bark of 159 of 309 trees indexed in northeastern USA, and in 89% of the trees exhibiting AUND graft-union symptoms (Rosenberger et al., 1983).

Root-Knot Nematodes

Species and Races

Meloidogyne hapla and *M. incognita* are parasitic on pear and apple in Japan, but their economic importance is unknown (Saigusa & Matsumoto, 1961). *Meloidogyne incognita* also has been found on apple in India (Khan & Sharma, 1990). However, *M. mali* is considered economically important to the apple industry in the northern parts of Hokkaido and Honshu, Japan (Inagaki, 1978).

Biology

The life cycle of *M. mali* takes 18 to 20 wk to complete. This nematode has only one generation per year in Japan (Inagaki, 1978). Most J2-stage *M. mali* nematodes were found to be distributed within the soil from 0 to 25 cm in depth.

Symptoms

Apple trees infected with *M. mali* show a restriction in trunk circumference, number of leaves, and number and length of new primary shoots (Inagaki, 1978). Secondary shoots increase in length and number. Belowground symptoms include typical root-galling.

Economic Importance

Growth of infected apple seedlings is reduced by 15 to 43%. On 36- to 50-yr-old apple trees the growth damage is indistinct, but fruit yield is significantly lower on heavily infected trees (Inagaki, 1978).

Ring Nematodes

Criconemella xenoplax and *C. curvata* do not appear to be economically important on pome fruits. *Criconemella xenoplax* was detected on apple in Africa (Heyes, 1970) and in the USA (Mai et al., 1960), whereas pear was classified a nonhost in a greenhouse experiment (Seshadri, 1964).

Other Nematodes

Other nematodes found in association with apple and pear are listed elsewhere (Mai et al., 1960; Wehunt & Golden, 1982). Additional research is needed to determine the economic importance of these nematodes.

IMPORTANT NEMATODE PARASITE OF CITRUS

Although many plant-parasitic nematodes have been detected in the rhizosphere of citrus trees, only a few have been shown to be of economic importance (Duncan & Cohn, 1990). Citrus also is prone to many virus diseases. However, virus transmission by dorylaimid nematode species feeding on citrus has not been demonstrated (Stokes, 1977).

Citrus Nematode

Tylenchulus semipenetrans is regarded as a worldwide problem in citrus production and is found in all citrus growing regions (Van Gundy & Meagher, 1977).

Species and Races

Four physiological races of *T. semipenetrans* (Poncirus, Citrus, Mediterranean, and Grass) were originally described (Inserra et al., 1980). However, populations found on grasses do not parasitize citrus trees and have subsequently been renamed *T. graminis* and *T. palustris* (Inserra et al., 1988). The Citrus race of *T. semipenetrans* (dominant in North America) is parasitic on citrus, grape, olive (*Olea europaea* L.), and persimmon (*Diospyros khaki* Thunb.). Some hybrids of *T. semipenetrans*-resistant trifoliate orange [*Poncirus trifoliata* (L.) Raf.] and *Citrus* spp. are tolerant or resistant, although the degree of resistance to infection and damage tolerance varies (Baines et al., 1969). The Poncirus race has been found in Japan and California (Baines et al., 1969), and recent studies suggested a possible occurrence in Florida (Duncan et al., 1994). It reproduced well on citrus, trifoliate orange and grape but not on olive (O'Bannon & Ford, 1977). The Mediterranean race, found in southern Europe and Africa, appears similar to Citrus but does not reproduce on olive.

Biology

Tylenchulus semipenetrans is able to live and reproduce almost everywhere citrus is produced (Van Gundy & Meagher, 1977). Citrus nematodes complete their life cycle in about 6 to 8 wk under optimum conditions. Freshly hatched female juveniles seek out young feeder roots and start feeding on epidermal cells. They penetrate the root tissues and initiate nurse cell feeding sites deep in the cortex. When the females mature, their posterior enlarges while extending out from the root surface. Approximately 100 eggs are deposited into a gelatinous matrix, which provides protection against desiccation and possibly microbial attack. Males and females undergo four molts, but the males remain outside of the root and do not feed. Citrus nematodes reproduce by amphimixis or mitotic parthenogenesis (Van Gundy, 1959).

Population dynamics of the nematode are closely linked to the phenology of the host. Feeder roots are preferred sites for penetration and nematode development (Cohn, 1964). Distinct fluctuations of seasonal population densities have been reported (Duncan et al., 1993). In general, the highest populations were found during spring and late autumn following flushes of root growth. Carbohydrate availability in roots (Duncan & Eissenstat, 1993), their lignification (Duncan et al., 1993) and attractiveness (Duncan et al., 1993) for citrus nematode juveniles, have been associated with seasonal fluctuations. In addition, various soil factors such as composition and texture (Labanauskas et al., 1965; O'Bannon, 1968), moisture (Van Gundy et al., 1962), temperature (Reynolds et al., 1970), O_2 (Van Gundy et al., 1962) and salt content (Mashela et al., 1992a,b) have been found to affect nematode population development. Infection, growth and reproduction occur between 20 and 30°C (O'Bannon et al., 1966) and are generally favored by high soil moisture in fine-textured soils (Labanauskas et al., 1965). Although citrus nematodes can survive several years in field soil (Baines et al., 1978), the population typically will decrease dramatically during the first few months after trees have been removed (Reynolds et al., 1970). Body food reserves are used up more rapidly in warm and dry aerated soils (Van Gundy et al., 1967).

Symptoms

Damage caused by the citrus nematode is usually referred to as citrus slow decline. This disease is characterized by a gradual starvation of the tree. The true impact of *T. semipenetrans* on tree health and yield remained obscure until demonstrated through the use of chemical control with pre- and postplant nematicides (O'Bannon & Tarjan, 1973). Typically the first aboveground symptoms are smaller leaves and fruits, followed by a deterioration in tree appearance. Leaf chlorosis and curling, and dieback are further consequences of poor root development and feeder root decay as nematodes, pathogenic fungi and secondary microorganisms in concert cause a decline in root health. Parasitized trees also become more sensitive to other biotic or abiotic stresses such as fungal pathogens, waterlogging, drought, salinity and cold. Feeder roots heavily infected with mature females often appear thicker and darker. Soil particles sticking to the gelatinous matrix, which surrounds the nematode egg masses, give the roots a dirty appearance.

Failure of citrus replants to grow as well as the previous plantings or compared to those grown on sites never before planted with citrus is referred to as the citrus replant problem. Most studies indicate that the build-up of citrus nematodes and deleterious root pathogens such as *Phytophthora* spp. during prolonged cropping to citrus is the primary cause for the citrus replant problem (Tsao et al., 1989).

Disease Complexes

Fusarium solani often has been associated with the citrus nematode in diseased roots (Feldmesser et al., 1962). Interactions between *T. semipenetrans* and *Fusarium* spp. have resulted in greater yield reduction than disease caused by either organism alone (O'Bannon et al., 1967). Although direct interactions with *Phytophthora* spp. have not been documented, they appear to be interlinked since both organisms commonly occur with citrus tree decline worldwide (Graham & Timmer, 1992; Duncan et al., 1993). Heavily infected feeder roots are often invaded by secondary organisms, which may exacerbate decay of the cortex and shorten root life (Cohn, 1965).

Economic Importance

Citrus nematodes may adversely affect tree health and reduce fruit quality and yield. However, damage and, consequently, management recommendations are dependent on a number of factors such as the particular nematode population, scion and rootstock, and biotic and abiotic environmental conditions. Management guidelines for citrus nematodes set action thresholds at about 10 000 juveniles per 500 cm^3 soil (Duncan & Cohn, 1990) and 1100 to 1400 females per gram feeder roots (Van Gundy, 1984).

Burrowing Nematode

Burrowing nematodes parasitize a large number of plants and occur throughout most tropical areas. However, in citrus production these nematodes are a problem only in Florida (Poucher et al., 1967; DuCharme, 1968).

Species and Races

Burrowing nematodes in citrus were first identified in 1953 and were thought to be a citrus race of *Radopholus similis* (Suit & DuCharme, 1953). Later it was renamed *R. citrophilus* based on genetic and behavior studies, host preference (Huettel et al., 1984) and morphological distinctions (Huettel & Yaegashi, 1980). Two biotypes of *R. citrophilus* have been differentiated based on their ability to reproduce on certain rootstocks (Kaplan & O'Bannon, 1985).

Biology

Burrowing nematodes only invade young feeder rootlets where they feed, establish burrows, and eventually lay eggs. Only females enter and stay within the roots, where they remain for their entire life as long as the root remains intact and provides sufficient food (DuCharme, 1968). In this shelter, the nematodes can resist periods of adverse environmental conditions whereas they do not survive more than 180 d (6 mo) in sandy soil free of roots (Tarjan, 1961). Reproduction is either by amphimixis (Huettel et al., 1982) or parthenogenesis (Brooks & Perry, 1962). Juveniles hatch out of the eggs within a week and start feeding in the root. Under optimal conditions *R. citrophilus* completes its life cycle within 3 wk (DuCharme & Price, 1966). Decay caused by secondary invaders or high population pressure within a tissue induces migration towards a new site (Poucher et al., 1967).

Symptoms

Damage by the burrowing nematode is called spreading decline. It initially appears in a group of trees within a grove, with the area of decline typically increasing in size annually. *Radopholus citrophilus* is a migratory endoparasite that causes massive damage to root tissues. Affected trees are prone to wilting during dry winter and spring periods when the disease spreads rapidly through infested orchards. Sparse foliage, defoliated branches, twig dieback, and small leaves and fruit are direct effects of malnutrition due to the partial loss of the root system. In contrast to the disease caused by citrus nematodes, most of the root damage occurs at depths below 25 cm (DuCharme, 1968). Large portions of the deep root system may be affected or destroyed. Infected roots develop dark lesions where the nematode penetrated and older roots display the characteristic cavities caused by the tunneling of *R. citrophilus*. Massive invasion of the roots can cause swelling and stunting of the root tips (DuCharme, 1959).

Disease Complexes

Although spreading decline is mainly caused by burrowing nematodes, the disease progress is often accelerated by interactions with various soilborne microorganisms (DuCharme, 1968). The large burrows and cavities provide secondary invaders such as saprophytic fungi and bacteria with the opportunity to utilize normally nonaccessible substrate. The succession of decomposition is further advanced by members of the soil fauna, such as mites, which feed on both microorganisms and decaying root tissues.

Economic Importance

In Florida, economic losses due to *R. citrophilus* have been estimated as much as 70% for orange and grapefruit (DuCharme, 1968). However, it has been suggested that the use of resistant rootstocks, optimum fertilization, and adequate water may have reduced the losses over the past two decades (Duncan & Cohn, 1990).

Root-Lesion Nematodes

Pratylenchus spp. have been found on citrus in various parts of the world (Gotoh & Ohshima, 1963; O'Bannon et al., 1972; Siddiqui et al., 1973; Inserra & Vovlas, 1974; Huang & Chang, 1976; Milne, 1982). Their effect on tree health is sometimes referred to as citrus slump disease (Tarjan & O'Bannon, 1984). The name "root-lesion nematode" refers to the darkened necrotic lesions that appear on parasitized roots. Greenhouse trials have demonstrated the ability of various species such as *P. coffeae*, *P. brachyurus* and *P. vulnus* to reduce citrus plant growth (Feldmesser & Hannon, 1965; Brooks & Perry, 1967; O'Bannon & Tomerlin, 1969; Radewald et al., 1971b; O'Bannon et al., 1976; Inserra & Vovlas, 1977). While both *P. coffeae* (O'Bannon & Tomerlin, 1973) and *P. brachyurus* (O'Bannon et al., 1974) were found to be seriously damaging to young trees, *P. coffeae* also is very injurious to mature trees. They reproduced best at soil temperatures above 25°C, and preferred sandy to sandy loam textured soils. Typically, these migratory endoparasites penetrate the zone of root elongation and invade the cortex but normally do not enter vascular tissues (Radewald et al., 1971a). Rapid population increases have been observed, with *P. coffeae* yielding >10 000 nematodes per gram root (O'Bannon et al., 1976). This destructive parasite can survive in the absence of host roots >120 d (4 mo) (Radewald et al., 1971b). Field observations have suggested that *P. coffeae* and *T. semipenetrans* might be incapable of concomitant parasitism of citrus roots for prolonged periods of time (Kaplan & Timmer, 1982). *Pratylenchus brachyurus* rarely exceeds 200 nematodes per gram fresh root and damage potential usually decreases with plant age (O'Bannon et al., 1974). As with the burrowing nematode, the destruction of tissues by root-lesion nematodes is closely followed by invasion of various secondary organisms that accelerate root decay.

Sting Nematode

The sting nematode is indigenous to the sandy soils of the southeastern USA and has a wide host range including citrus (Smart & Nguyen, 1991). In Florida, 87 of 1147 groves surveyed were infested with this nematode (Esser & Simpson, 1984). However, a survey in progress shows >90% infestation. The discrepancy is probably due to more intensive sampling in the current study (L.W. Duncan, 1995, personal communication). This ectoparasitic nematode feeds preferentially on young feeder roots and can cause reductions of the root system with terminal swellings and lesions (Tarjan & O'Bannon, 1984). Aboveground symptoms include dieback, leaf chlorosis and stunting of trees (Standifer & Perry, 1960; Esser & Simpson, 1984; Kaplan, 1985). Host-range comparisons of three sting nematode populations indicated that physiological races exist that differ widely in their ability to reproduce and cause symptoms on rough lemon.

Root-Knot Nematodes

These sedentary, endoparasitic nematodes are not known to be a major problem in citrus production although root-knot galling on citrus has been reported from various parts of the world (Minz, 1956; Colbran, 1958; Chitwood & Toung, 1960; Gill, 1971; Pan, 1984). In most cases, common species such as *M. incognita*, *M. javanica*, and *M. arenaria* were identified as the causal organisms. Although *M. javanica* was reported to reproduce in roots of several citrus species (Gill, 1971), it appears more often that even when invasion and galling occurred, no giant cell formation followed and, consequently, reproduction failed (Van Gundy et al., 1959; Orion & Cohn, 1975; Inserra et al., 1978). A *Meloidogyne* sp. named the "Asiatic pyroid citrus nema" induced gall formation, bloating of tips, and abnormal branching as well as reproduced on various citrus hosts under controlled conditions (Chitwood & Toung, 1960).

Dagger Nematodes

Dagger nematodes have been reported from many countries to be associated with citrus (Yassin, 1974; Baines et al., 1978; Milne, 1982; Esser et al., 1978). However, the economic importance of *Xiphinema* species to citrus is unknown. *Xiphinema brevicolle* and *X. index* dramatically suppressed growth of citrus seedlings in greenhouse trials (Cohn & Orion, 1970), where *X. brevicolle* was identified with the decline of grapefruit trees (Yassin, 1974). Symptoms caused by this nematode are similar to those of other crops attacked by dagger nematodes. High populations of *X. americanum* can restrict the growth of the feeder root system and lead to short, swollen roots with dark appearance (Baines et al., 1978).

Stubby-Root Nematodes

Trichodorus and *Paratrichodorus* spp. have been found in low numbers in major citrus-growing areas (Baines et al., 1959; Malo, 1961; Colbran, 1965). At high population densities, they may suppress root growth and induce the typical stubby-root symptom. Such symptoms have reportedly been caused by *P. minor* (syn. *T. christiei*) feeding in the meristematic region of grapefruit roots (Standifer & Perry, 1960). Additionally, *P. porosus* (Baines et al., 1978) and *P. lobatus* affected nursery citrus trees in vigor and predisposed them to wilt (Stirling, 1976).

Sheath Nematodes

Although *Hemicycliophora* spp. have frequently been found in samples from citrus orchards, only *H. arenaria* has been studied in detail (Van Gundy, 1959). This ectoparasite is indigenous to inland deserts of southern California (McElroy et al., 1966). It feeds on the epidermal and cortical cells near the root tip and causes enlargements of the terminal and lateral root tips, which resemble small knobs (Van Gundy & Rackham, 1961). Similar symptoms on citrus have been observed as a response to feeding by *Caloosia nudata* (Colbran, 1963). *Hemicycliophora arenaria* preferred sandy soils, and suppressed the growth of inoculated citrus seedlings by more than a third compared to control plants in greenhouse trials (Van Gundy & Rackham, 1961). Despite their wide host range (McElroy et al., 1966), sheath nematodes are considered of minor economic importance to the citrus industry.

MANAGEMENT OF MAJOR NEMATODE SPECIES

In selecting sites for stone and pome fruit production, those areas having a history of nematode problems or replant diseases should be avoided. If such a site is unavailable, the following options should be considered.

STONE FRUITS

Root-Knot Nematodes

Host Resistance

Planting certified nematode-free rootstock, when available, is extremely important as a management strategy to avoid any future problems following orchard establishment. However, compatibility between scion and rootstock and other edaphic and biological pest problems also must be considered. The following rootstocks are generally used in the various stone fruit industries. Sweet and sour cherry are generally grafted onto seedlings of *P. mahaleb* or *P. avium* (Iezzoni et al., 1991). Peach is grown primarily on *P. persica* seedling rootstocks in the USA and on plum, peach or peach × almond rootstocks in Europe (Scorza & Okie, 1991). Plum rootstocks are used for peach and Japanese plum only where their tolerance to wet soils is needed. In Europe, plums are commonly grown on plum rootstocks. However, California plum growers prefer to graft plum scions onto Lovell or Nemaguard peach rootstock when waterlogging is not a problem because the trees grow better (McKenry, 1989b). Rootstocks for apricot include apricot, peach, and plum (Mehlenbacher et al., 1991).

Evaluation of *Prunus* spp. and other deciduous fruits for resistance against *Meloidogyne* spp. was recognized as an important consideration for these industries at the turn of the 20th century. Development and release of resistant germplasm over the years is the primary reason why *Meloidogyne* spp. are no longer a major threat to the stone fruit industry. However, evaluation techniques for nematode resistance is an ongoing process. The objectives are to accelerate the screening process and release of new *Prunus* germplasm that can be used to manage *Meloidogyne* spp. problems in different regions of the world (Esmenjaud et al., 1995). Resistance to *Meloidogyne* spp. occurs in cherry, peach, plum, and apricot. Resistance to *M. incognita* has been reported in *P. avium* and *P. cerasus* and to *M. javanica* in Masto de Montanana cherry (Breton, 1980, p. 21–31; Pinochet et al., 1992a).

In peach, Nemaguard, Nemared, and Okinawa among others are resistant to *M. incognita* and *M. javanica*; with Nemaguard and Nemared being the two most widely used rootstocks today (Ramming & Tanner, 1983). In Florida, an isolate of *M. incognita* Race 3 infects Nemaguard and Nemared rootstock. Breeding efforts have resulted in the rootstock Flordaguard, that is resistant to this nematode (Sherman et al., 1991). Another new commercial peach rootstock, Guardian (=BY520-9), also was found to be resistant to *M. incognita* (Okie et al., 1994a). Other sources of resistance to either *M. incognita* or *M. javanica* have been reported from Italy (Simeone & Di Vito, 1992) and Spain (Marull et al., 1994). Resistance to *M. arenaria* also has been found in peach-almond hybrids in Spain (Fernandez et al., 1994). In plum, Marianna 2624 (*P. cerasifera* × *P. munsoniana*) and Myrobalan (Norton et al., 1963) have been selected for *Meloidogyne* resistance, while Citation plum is highly resistant to *M. javanica* (Pinochet et al., 1992a). Resistance to *M. arenaria*, *M. javanica*, and/or *M. incognita* from other sources have been reported from Europe (Fernandez et al., 1994). Apricot has been reported to be resistant to *M. incognita* and (or) *M. javanica* in Israel and the USA (Minz & Cohn, 1962; Norton et al., 1963).

Cultural Practices

Ground covers that suppress *Meloidogyne* spp. and can be grown in the tree or tractor row middles are of interest. In Georgia, bahiagrass (*Paspalum notatum* Flugge 'Paraguayan-22') shows promise as a nonhost of *Meloidogyne* spp., but it must be planted in row middles and not within the tree row because it is a strong competitor for nutrients and water (Evert & Bertrand, 1993). In California, peach tree growth was increased while the *M. incognita* J2 soil population density increased 10-fold and root galling decreased with use of black polyethylene mulching as a preplant management strategy in orchard establishment (Duncan et al., 1992). Additionally, incorporation of refuse from 'Cahaba' white vetch (*Vicia sativa* L.) followed by irrigation also shows promise as a practice for suppressing a wide variety of nematode species (excluding *M. hapla*) (McKenry, 1995, personal communication).

Nematicides

Preplant chemical treatment provides the most effective control of *Meloidogyne* spp. because it allows trees to have a healthy start by preventing the nematode from causing major root damage (Nyczepir, 1991). Preplant control is more effective than postplant because such treatments involve general biocides that provide protection during the key time of tree establishment.

Other Tactics

Submersion of *Meloidogyne*-infected dormant seedlings of *P. mahaleb* and Lovell peach in a hot water bath at 48°C/30 min, 49°C/20 min, 50°C/10 min, 51°C/5 min, or 51°C/10 min effectively killed nematodes inside the roots (Nyland, 1955).

Ring Nematodes

Host Resistance

Lovell rootstock has been recommended to southeastern USA stone fruit growers for more than 15 yr because it has demonstrated partial tolerance to *C. xenoplax* and survives longer on PTSL sites as compared to Nemaguard (Sharpe et al., 1989). Nemaguard provides control against *M. incognita* and *M. javanica*. Recently, a new peach rootstock Guardian (=BY520-9) that tolerates parasitism by *C. xenoplax* and increases scion longevity on PTSL sites more than Lovell has become commercially available (Okie et al., 1994a,b).

Cultural Practices

A 10-Point Management Program is recommended to peach growers in the southeastern USA to help reduce tree losses from PTSL (Ritchie & Zehr, 1995). This program is not a cure for PTSL, but rather a management system for the disease complex. One of the major points to this program is pre- and postplant soil fumigation as a management strategy for *C. xenoplax*. However, continued removal of chemical nematicides from the market is leaving growers with fewer nematode management options. Alternatives to chemical control for nematode management have been explored over the past 9 yr with some promising results. Several ground covers appear promising as either preplant or postplant management strategies for *C. xenoplax*. In North Carolina, nimblewill (*Muhlenbergia schreberi* J. Gmelin) planted around peach trees suppressed populations of *C. xenoplax* without being highly competitive for nutrients and water (Meyer et al., 1992). However, it does support reproduction by *M. javanica* and *M. arenaria*, and to a lesser extent *M. incognita* (McKenry, 1990a; Nyczepir, 1995 personal communication). Also, establishment of nimblewill in Georgia and South Carolina peach orchards has been extremely difficult. In California, select marigold species (*Tagetes* spp.), California poppy (*Eschscholzia papaver* Cham), and Cahaba white vetch all exhibited nematicidal activity against *C. xenoplax*, but the former two are phytotoxic (Tudor & McKenry, 1989). In Georgia, chemically mowed Paraguayan-22 bahiagrass sod previously grown in orchard row middles infested with *C. xenoplax* suppressed the nematode population density in addition to improving soil structure and increasing organic matter content (Evert & Bertrand, 1993). Additionally, double-cropping wheat (*Triticum aestivum* L. 'Stacy') and sorghum [*Sorghum bicolor* (L.) Moench 'NK2660'] has potential as a preplant management strategy in suppression of *C. xenoplax* soil population density in establishment of peach orchards in the southeastern USA (Nyczepir et al., 1996).

Nematicides

Criconemella xenoplax can be controlled by preplant soil fumigation plus postplant application of nonfumigant nematicides (Nyczepir, 1991). A nematicide treatment threshold of 38 to 83 *C. xenoplax*/100 cm^3 soil is recommended in North Carolina (Ritchie, 1988) and >50 *C. xenoplax*/100 cm^3 soil in South Carolina.

Biological

Hirsutella rhossiliensis is a common fungal endoparasite of *C. xenoplax*. The fungus suppressed *C. xenoplax* in some controlled tests (Eayre et al., 1987). Recently a field site in South Carolina was identified as being suppressive to *C. xenoplax* and incidence of PTSL (Kluepfel et al., 1993). Seven strains of a bacteria that were isolated and found to be suppressive to *C. xenoplax* were all identified as *Pseudomonas aureofaciens*. It also was demonstrated that a selected strain of *P. aureofaciens* and a genetically modified bacteria derivative both effectively inhibited *C. xenoplax* egg hatch under controlled conditions (Westcott & Kluepfel, 1993).

Root-Lesion Nematodes

Host Resistance

Differences in rootstock susceptibility to *Pratylenchus* spp. that are implicated in *Prunus* replant diseases have been reported. Tolerance to *P. vulnus* has been reported in the plum rootstocks Marianna 2624 and Myrobalan 29C, 'Royal' apricot, and sand cherry (*P. besseyi* L. Bailey) (Norton et al., 1963; Culver et al., 1989). In Spain, Marianna 2624 proved to be a suitable host for a *P. vulnus* isolate obtained from rose, whereas Myrobalan 605 supported fewer nematodes (Pinochet et al., 1991). Furthermore, the plum rootstock 'Bruce' exhibited tolerance to *P. vulnus* in greenhouse experiments (Ledbetter, 1992). In California, Nemaguard peach rootstock is considered to be susceptible to *P. vulnus* (McKenry, 1989a), whereas it proved to be an unsuitable host when evaluated against a *P. vulnus* isolate from Spain (Pinochet et al., 1992b). Mahaleb cherry is tolerant to *P. penetrans* in light well-drained soils (Edgerton & Parker, 1958). However, a selection of *Prunus mahaleb*, Santa Lucia 64, was shown to be susceptible to *P. vulnus* in microplot experiments (Pinochet et al., 1995). Tolerance to *P. penetrans* was identified in a peach cross (Okinawa × Nemaguard), also which demonstrated resistance to *Meloidogyne* spp.

Generally, resistance to *P. vulnus* and *P. penetrans* is not available in commercial rootstocks, although a few wild *Prunus* clones have been shown to be resistant (Scotto La Massese, 1975). In contrast to *Meloidogyne* spp., little is known on how the resistant trait(s) is transmitted and what possible defense mechanism(s) is involved. An effort is currently being pursued in Spain and France (Pinochet et al., 1993b, 1994) and the USA (Ledbetter, 1992) to search for and incorporate a source of resistance into *Prunus* rootstocks.

Cultural Practices

Brassica spp. contain glucosinolates that form toxic volatile isothiocyanates upon hydrolysis that are similar in structure and activity to commercial fumigants. Several glucosinolate compounds were evaluated in vitro for control of *P. penetrans* (McFadden-Smith et al., 1993). Allyl glucosinolate [allyl (2-propenyl) glucosinolate] was shown to be the most nematicidal compound in laboratory tests, but had no significant effect on the nematode when applied as a green manure under field conditions.

Nematicides

Preplant soil fumigation and periodic postplant application of a recommended nematicide provided the best management against *Pratylenchus* spp. involved in replant diseases of stone fruits (Nyczepir, 1991). Local sources of information should be consulted for updated nematicide recommendations.

Dagger Nematodes

Host Resistance

Production of virus-free nursery stock (scion and rootstock) is the most important means of reducing virus disease problems in stone fruit orchards (Layne, 1987). Resistance or immunity to TmRSV, which causes PBL, has been identified in Marianna 2624 plum rootstock, whereas Myrobalan plum was susceptible (Hoy & Mircetich, 1984). Additionally, Mazzard cherry remained symptomless after inoculation with TmRSV, whereas Mahaleb rootstock developed typical PSP symptoms (Mircetich et al., 1977). However, no commercially acceptable rootstocks with TmRSV resistance are available for peach or nectarine.

Cultural Practices

Preplant cultural management strategies for PSP include growing grain or grass crops and control of broadleaf weeds (e.g., dandelion, since they serve as reservoirs of TmRSV in the orchard). Postplant strategies include control of the broadleaf weeds in the row middle by establishing a dense sod and within the tree-row middles by utilizing herbicides (Greene, 1986; Halbrendt, 1992). Orchards that have a good broadleaf weed management program are not likely to develop serious TmRSV problems even if dagger nematodes are present. Recently, Halbrendt (1994) reported that utilizing rapeseed (*Brassica napus* L.)

as a green manure or cover crop effectively suppressed the population density of *Xiphinema* sp.

Soil cultivation also can be used to reduce the population density of *Xiphinema* spp. since they are sensitive to desiccation (Abawi & Mai, 1990).

Nematicides

Xiphinema spp. can be effectively managed with most fumigant and non-fumigant nematicides, but only at high treatment rates (Abawi & Mai, 1990).

POME FRUITS

Root-Knot Nematodes

The best defense against *M. mali* is to purchase nematode-free rootstock and use a preplant fumigant nematicide (Inagaki, 1978) .

Root-Lesion Nematodes

Host Resistance

Apple rootstocks that were less sensitive to *P. penetrans* infection on sandy soils in the northeastern USA included M.7 and M.12. Trees on sensitive rootstocks reacted to soil fumigation with increased growth, whereas the resistant rootstocks did not (Abawi & Mai, 1990).

Evaluation of apple and pear rootstocks for host suitability to *P. vulnus*, *P. neglectus*, and *P. thornei* are reported elsewhere (Pinochet et al., 1991, 1992b; Fernandez et al., 1992). The apple rootstock EMLA 26, and pear rootstock OHF-333, were both suitable hosts for *P. vulnus*, but apple was more susceptible than pear. The OHF-333 pear rootstock was rated as a poor to nonhost for *P. neglectus* and *P. thornei.*

Cultural Practices

Removal of old pear and apple roots and preplanting an orchard site with select marigold species (*Tagetes*) or wild oat (*Avena sativa* L. 'Saia') for 1 to 2 yr aids in suppressing *P. penetrans* soil population density (Mai et al., 1994). Furthermore, establishing an orchard ground cover [e.g., tall fescue (*Festuca arundinacea* Schreb.)] that is a poor host to *P. penetrans* and maintaining a weed-free area within tree rows or around trees also is beneficial in reducing nematode damage to trees. Following routine orchard management practices (e.g., fertilization, soil pH) also are encouraged.

Nematicides

Preplant soil fumigation and periodic postplant application of a recommended nematicide provide the best management against *Pratylenchus* spp.

involved in replant diseases of pome fruits (Mai et al., 1994). Local sources of information should be consulted for updated nematicide recommendations.

Dagger Nematodes

Host Resistance

Apple is a host for *Xiphinema* spp., but scion cultivars vary in their resistance or susceptibility to nepoviruses. As stated previously, AUND only seems to be a problem when the rootstock is susceptible to TmRSV and the scion variety resistant. Clones of 'Delicious' grafted on MM.106 rootstock are reported to be more severely affected with AUND than other scion/rootstock combinations. It has been suggested that the incidence of AUND is associated with the release of MM.106 rootstock (Rosenberger et al., 1983). However, not all scion cultivars grafted on MM.106 develop AUND.

Resistant rootstocks to TmRSV include C6, Novole, Ottawa 7 and 11, M.2, M.7, M.4, M.9, M.13, Robusta 5, Bud. 54-146, Bud. 57-490, Bud. 57-491, CG.10, CG.24, Kansas-14, NAC-24, MM.102, OAR-1, P-1, P-13, P-16, and P-22. Rootstocks susceptible to TmRSV include M.26, MM.106, MAC-30, MAC-39 and P-2. The virus was inconsistently recovered from M.27, MM.111, Budagovsky 9, MAC-1, MAC-9, Ottawa 3, and P-18. Scion cultivars most often affected by AUND include Delicious, Jerseymac, Quinte, Rhode Island Greening, and Tydeman's Early. Trees of 'Empire', 'Golden Delicious', 'Rome Beauty', and 'York Imperial' on MM.106 rootstock and established on a known AUND site have not exhibited symptoms (Cummins & Gonsalves, 1982).

Red and Golden Delicious are susceptible to CRLV, and should not be planted on a site known to have a history of this disease (Parish, 1977).

Cultural Practices

Cultural control of *Xiphinema* spp. and TmRSV on pome fruits (AUND), is the same as for stone fruits.

Nematicides

Preplant soil fumigation and most nonfumigant nematicides provide effective control of *Xiphinema* spp. associated with AUND in addition to improving fruit tree growth and yield (Mai & Abawi, 1981).

CITRUS

Most information about nematode control in citrus is directed towards *T. semipenetrans* and *R. citrophilus* due to their economic importance. Ideally, citrus management strategies should be based on nematode-free planting material, nematode resistant or tolerant rootstocks, a nematode-free planting site and avoidance of nematode-infested equipment and run-off water. Preplant nematode management and sanitation measures are the most important means for optimizing young tree growth and securing long-term productivity (Van Gundy, 1984). It is important for sound integrated pest management decisions to understand the dynamic interactions between the host, the pest population, and the biotic and abiotic environment.

Host Resistance

Most commercially produced citrus are propagated on rootstocks because seedling trees usually have a prolonged juvenile period and may not be true-to-type for certain cultivars. Furthermore, it has not been feasible to combine excellent fruit quality with other desirable traits such as disease and pest resistance (Davies & Albrigo, 1994). In general, rootstock breeding focuses mainly on soil-related objectives such as salinity and drought tolerance, resistance to *Phytophthora* diseases as well as to citrus and burrowing nematodes. However, more than 20 horticultural characteristics are influenced by the rootstock and need to be considered depending on regional conditions and demands (Castle, 1987). Breeding for nematode resistance is an ongoing effort as new biotypes of nematodes evolve. Often more than 20 yr are necessary before new rootstocks are released (Castle, 1987). However, recently developed biotechnology techniques have the potential to accelerate this process and to solve some of the citrus rootstock-related nematode problems (Grosser & Gmitter, 1990).

Poncirus trifoliata hybridizes readily with most *Citrus* spp. and has been used as a source of resistance to *T. semipenetrans* for developing new rootstocks (Van Gundy & Meagher, 1977). The resistance is based on several mechanisms including periderm formation at the nematode feeding site, a hypersensitive reaction, toxic compounds and other factors that cause low densities of *T. semipenetrans* in the rhizosphere (Van Gundy & Kirkpatrick, 1964; Kaplan, 1981; Gottlieb et al., 1986). The hybrid cultivar Swingle citrumelo (*C. paradisi* × *P. trifoliata*) combines disease resistance against *Phytophthora parasitica* and Tristeza with good resistance against most populations of the citrus nematode (Wutscher, 1974). The good performance of trifoliate orange and some of its hybrids in the presence of most strains of the citrus nematode and *Phytophthora* spp. has decreased the citrus replant problem.

Milan lemon and Ridge Pineapple sweet orange are currently recommended in Florida for use against *R. citrophilus*, although only Biotype 1 does not reproduce well on these rootstocks (Kaplan & O'Bannon, 1985). Resistant cultivars for other citrus nematode pests are not available or economically important.

Cultural Practices

Cultural management of nematode pests emphasizes a long-term approach to reduce nematode populations to noninjurious levels. It also applies to preventative techniques, which check the increase of nematode populations. The economic threshold for these pests can be quite variable depending on the overall health of the tree. Cultural practices should therefore provide conditions that minimize stress such as disease pressure (especially by *Phytophthora* spp.) or water- and nutrient-related deficiencies. Measures used to aid symptomatic trees include the soil incorporation of various organic amendments (Mankau & Minteer, 1962; Tarjan, 1977; Spiegel et al., 1989), and N_2 fertilizers (Scotto La Massese et al., 1973).

Most nematodes parasitic on citrus do not move very far by themselves (Tarjan & O'Bannon, 1984). Therefore, planting nematode-free rootstocks and avoiding nematode dispersal on implements and via irrigation or run-off water are important for preventing the introduction of nematodes into a new grove or reinfestation of previously fumigated soil. Nursery certification programs have successfully prevented the spread of citrus, root-lesion and burrowing nematodes (Poucher et al., 1967). Hot-water treatments have been used to eradicate burrowing nematodes (Poucher et al., 1967), root-lesion nematodes (Kaplan, 1982) and sting nematodes (Kaplan, 1985) from infected root systems of young citrus nursery stock.

Replanting old citrus orchards requires special precautions because citrus nematodes can survive extended host-free periods (Van Gundy et al., 1967). Several years of rotation with nonhost crops may help to accelerate the decomposition of old citrus roots that harbor nematodes (Van Gundy, 1984).

The major impact of the burrowing nematode is mainly on the deep root system. Frequent short irrigation cycles, optimal fertilization regimes and herbicide use (instead of cultivation for weed management) allow establishment and maintenance of a vigorous shallow root system (Bryan, 1969; Tarjan & O'Bannon, 1977).

Nematicides

Nematicides have historically played an important role in proving that plant-parasitic nematodes cause economic losses in citrus production (O'Bannon & Tarjan, 1973). However, excessive use of nematicides in citrus groves within the USA resulted in groundwater contamination and consequent closure of drinking water wells as well as cancellation of several nematicide registrations (Kaplan, 1988).

In some industries, preplant soil fumigation of orchards with a history of citrus nematode infestation is still an important measure to avoid a quick population build-up on young trees even with nematode tolerant or resistant rootstocks (Baines et al., 1966; Scotto La Massese, 1977; Van Gundy, 1984). Occasionally, soil fumigation has produced undesirable effects such as stunting or malnutrition due to the elimination of mycorrhizae (Kleinschmidt & Gerdemann, 1972; O'Bannon & Nemec, 1978). Mycorrhizal citrus plants are more efficient in uptake of immobile nutrients such as Zn, Cu and particularly P (Kleinschmidt & Gerdemann, 1972; Menge et al., 1978) although rootstocks differ significantly in their dependence on these symbionts (Nemec, 1978). Citrus seedlings inoculated with mycorrhizal fungi produced better growth in fumigated soil (Hattingh & Gerdemann, 1975), making such a technique feasible for replenishing fumigated soils with mycorrhizae (Ferguson & Menge, 1986).

Citrus growers have often been reluctant to use postplant chemical control because of federal- and state-imposed pesticide restrictions and the high cost of most nematicides. Nematode control with nonvolatile nematicides often has been variable (Davis & Wilhite, 1985; Duncan, 1989; Philis, 1989; Lo Giudice et al., 1990) and resulted in inconsistent return on investments (Van Gundy & Meagher, 1977). Some studies indicate that yield or quality improvements can be expected only after two or more years of nematicide treatments (Hamid et al., 1988; Greco et al., 1993). Further improvement of performance will depend on a better understanding of treatment aspects such as timing, placement and retention time of the nematicide (Van Gundy et al., 1981; Duncan & Cohn, 1990) in regards to the biology of the organisms and their host-parasite relationship.

Biological

Nematode-destroying organisms can be easily obtained from the citrus rhizosphere (Gaspard & Mankau, 1986; Walter & Kaplan, 1990). Furthermore, it has been suggested that nematodes and their antagonists have established an effective population equilibrium in California soils that results in relatively little economic damage to citrus by *T. semipenetrans* (Mankau, 1982). However, in-depth studies are lacking, and little is known about pathogens and parasites of maturing females and eggs.

Soil-borne bacteria of the genus *Pasteuria* are obligate, host-specific pathogens of nematodes (Mankau, 1975) and considered as promising candidates for biological control (Stirling, 1991). Several associations of *T. semipenetrans* with this parasite have been reported (Sturhan, 1988; Fattah et al., 1989; Walter & Kaplan, 1990), but no data are available on the efficacy of *Pasteuria* species to control *T. semipenetrans*.

Utilizing antagonists directly or in combination with organic amendments to control plant-parasitic nematodes in citrus has not been successful even though some studies have reported a reduction in nematode populations and/or increase in plant growth (Mankau & Das, 1974; Walter & Kaplan, 1990).

OUTLOOK

The importance of various nematode species and their economic impact on tree fruit and citrus production was demonstrated during the 1950s and 1960s when soil fumigants with selective activity against plant-parasitic nematodes became available. It was not until after the USEPA suspended registration of DBCP as a soil fumigant for all crops in 1981 and eventually other nematicides that nematode management started to evolve from a "one-tool reliance" (nematicides) to "multi-tool reliance" [use of resistant and tolerant rootstocks, cultural technologies (e.g., certification programs for nematode-free nursery trees, crop rotation with antagonistic plants, nematode-suppressive ground covers, and sanitation), and nematicides]. Research in the future is likely to emphasize: (i) interactions among different plant-parasitic nematodes and the effects they have on their host; (ii) interactions between plant pathogens, nematodes, and their host; (iii) continued search for improved rootstock nematode resistance and development of biological management technologies to control nematodes in production field situations; (iv) development of programs on application of cultural and biological technologies for use with existing integrated pest management programs; (v) exploration of the possibility of developing transgenic plants to control virus diseases; (vi) elucidation of factors responsible for nematode control and/or suppression in natural soils, rootstock resistance, and other cultural technologies; and (vii) exploration of different formulations, rates, and placement of new and existing nematicides.

Furthermore, it is important that orchard management practices be optimized to limit the biotic and abiotic stress factors on trees. Proper fertilization (following leaf and soil analysis), use of a nematode resistant/tolerant rootstock (if available), irrigation with tensiometer-driven microsprinkler systems vs. furrow irrigation, and proper pathogen, insect, and weed control are just some of the management practices that need to be followed in order to grow a healthy tree. A healthy tree is important to improve and maintain production, increase tree longevity, and even reduce the impact of a given number of plant-parasitic nematodes that otherwise might affect tree performance. Some of the potential benefits to a multitool approach and understanding are more efficient and safer production methods, reduction in environmental contamination, and reduced costs for the grower and consumer.

REFERENCES

Abawi, G.S., and W.F. Mai. 1990. Dagger nematodes. p. 73–74. *In* A.L. Jones and H.S. Aldwinckle (ed.) Compendium of apple and pear diseases. APS Press, St. Paul, MN.

Baines, R.C., L.J. Klotz, T.A. DeWolfe, R.H. Small, and G.O. Turner. 1966. Nematocidal and fungicidal properties of some soil fumigants. Phytopathology 56:691–698.

Baines, R.C., T. Miyakawa, J. W. Cameron, and R. H. Small. 1969. Biotypes of the citrus nematode. p. 955–956. *In* H.D. Chapman (ed.) Proc. 1st Int. Citrus Symp., Riverside, CA. 16–27 Mar. 1968. Univ. California, Riverside.

Baines, R.C., S.D. Van Gundy, and E.P. DuCharme. 1978. Nematodes attacking citrus. p. 321–345. *In* W. Reuther et al. (ed.) The citrus industry. Univ. California, Berkeley.

Baines, R.C., S.D. Van Gundy, and S.A. Sher. 1959. Citrus and avocado nematodes. Calif. Agric. 13:16–18.

Barker, K.R., and C.N. Clayton. 1973. Nematodes attacking cultivars of peach in North Carolina. J. Nematol. 5:265–271.

Bell, R. 1991. Pears (*Pyrus*). p. 655–697. *In* J.N. Moore and J.R. Ballington (ed.) Genetic resources of temperate fruit and nut crops. Int. Soc. Hort. Sci. Wageningen, The Netherlands.

Braun, A.L., H. Mojtahedi, and B.F. Lownsbery. 1975. Separate and combined effects of *Paratylenchus neoamblycepalus* and *Criconemoides xenoplax* on 'Myrobalan' plum. Phytopathology 65:328–330.

Breton, S. 1980. The rootstocks. *In* S. Breton (ed.) The cherry. (In French.) Centre Tech. Interprof. Fruits Legumes, Paris, France.

Brooks, T.L., and V.G. Perry. 1962. Apparent parthenogenetic reproduction of the burrowing nematode *Radopholus similis*. Proc. Soil Crop Sci. Soc. Fla. 22:160–162.

Brooks, T.L., and V.G. Perry. 1967. Pathogenicity of *Pratylenchus brachyurus* to citrus. Plant Dis. Rep. 51:569–573.

Brown, D. J. F., J.M. Halbrendt, R.T. Robbins, and T.C. Vrain. 1993. Transmission of nepoviruses by *Xiphinema americanum*-group nematodes. J. Nematol. 25:349–354.

Brown, D.J.F., and C.E. Taylor. 1987. Comments on the occurrence and geographical distribution of longidorid nematodes in Europe and the Mediterranean region. Nematol. Medit. 15:33–373.

Bryan, O.C. 1969. Living with the burrowing nematode. Citrus Veg. Mag. 33:29, 38.

Camprubi, A., J. Pinochet, C. Calvet, and V. Estaun. 1993. Effects of the root-lesion nematode *Pratylenchus vulnus* and the vesicular-arbuscular mycorrhizal fungus *Glomus mosseae* on the growth of three plum rootstocks. Plant Soil 153:223–229.

Canals, J., J. Pinochet, and A. Felipe. 1992. Temperature and age of plant affect resistance in peach-almond hybrid infected with *Meloidogyne javanica*. HortScience 27:1211–1213.

Carneiro, R.M.D. G., J.F. Fortes, and M.R.A. Almeida. 1993. Associacao de *Criconemella xenoplax* com a morte do pessegueiro no rio grande do sul. Nematol. Brasileira 17:122–131.

Castle, W.S. 1987. Citrus rootstocks. p. 361–399. *In* R.C. Rom and R.F. Carlson (ed.) Rootstocks for fruit crops. John Wiley & Sons, Inc., New York.

Chitwood, B.G., and M.C. Toung. 1960. Host-parasite interaction of the asiatic pyroid citrus nema. Plant Dis. Rep. 44:848–854.

Cohn, E. 1964. Penetration of the citrus nematode in relation to root development. Nematologica 10:594–600.

Cohn, E. 1965. On the feeding and histopathology of the citrus nematode. Nematologica 11:47–54.

Cohn, E., and D. Orion. 1970. The pathological effect of representative *Xiphinema* and *Longidorus* species on selected host plants. Nematologica 16:423–428.

Coiro, M.I., F. Lamberti, and E. Vindimian. 1985. Susceptibility of three apple rootstocks to *Xiphinema index*. Redia 68:99–100.

Colbran, R.C. 1953. Problems in tree replacement. I. The root-lesion nematode *Pratylenchus coffeae* Zimmerman as a factor in the growth of replant trees in apple orchards. Aust. J. Agric. Res. 4:384–389.

Colbran, R.C. 1955. A preliminary survey of plant nematodes in Queensland. J. Aust. Agric. 21:167–169.

Colbran, R.C. 1958. Studies of plant and soil nematodes. II. Queensland host records of root-knot nematodes (*Meloidogyne* spp.). Queensl. J. Agric. Sci. 15:101–135.

Colbran, R.C. 1963. Studies of plant and soil nematodes. VI. Two new species from citrus orchards. Queensl. J. Agric. Sci. 20:469–474.

Colbran, R.C. 1965. Studies of plant and soil nematodes. IX. *Trichodorus lobatus* n. sp. (Nematoda: Trichodoridae), a stubby-root nematode associated with citrus and peach trees. Queensl. J. Agric. Anim. Sci. 22:273–276.

Costante, J. F., W.F. Mai, and R.M. Klein. 1985. Effects of rootstock and soil type on *Pratylenchus penetrans* populations in apple feeder roots. J. Am. Soc. Hort. Sci. 110:38–41.

Crossa-Raynaud, P., and J.M. Audergon. 1987. Apricot rootstocks. p. 295–320. *In* R.C. Rom and R.F. Carlson (ed.) Rootstocks for fruit crops. John Wiley & Sons, Inc., New York.

Culver, D.J., D.W. Ramming, and M.V. McKenry. 1989. Procedures for field and greenhouse screening of *Prunus* genotypes for resistance and tolerance to root-lesion nematode. J. Am. Soc. Hort. Sci. 114:30–35.

Cummins, J.N., and D. Gonsalves. 1982. Recovery of tomato ringspot virus from inoculated apple trees. J. Am. Soc. Hort. Sci. 107:798–800.

Dabaj, K.H. and G. Jenser. 1991. Preliminary observation on apricot decline. Acta Hort. 293:511–513.

Davies, F.S., and L.G. Albrigo. 1994. Citrus. CAB Int., Wallingford, England.

Davis, R.M., and H.S. Wilhite. 1985. Control of *Tylenchulus semipenetrans* on citrus with fenamiphos and oxamyl. Plant Dis. 69:974–976.

Dhanvantari, B.N., P.W. Johnson, and V.A. Dirks. 1975. The role of nematodes in crown gall infection of peach in southwestern Ontario. Plant Dis. Rep. 59:109–112.

DuCharme, E.P. 1959. Morphogenesis and histopathology of lesions induced on citrus roots by *Radopholus similis*. Phytopathology 49:388–395.

DuCharme, E.P. 1968. Burrowing nematode decline of citrus, a review. p. 20–37. *In* G.C. Smart and V.G. Perry (ed.) Tropical nematology. Univ. Florida Press, Gainesville.

DuCharme, E.P., and W.C. Price. 1966. Dynamics of multiplication of *Radopholus similis*. Nematologica 12:113–121.

Duncan, L.W. 1989. Effect of fenamiphos placement on *Tylenchulus semipenetrans* and yield in a Florida citrus orchard. J. Nematol. 21:703–706.

Duncan, L.W., and E. Cohn. 1990. Nematode parasites of citrus. p. 321–346. *In* M. Luc et al. (ed.) Plant parasitic nematodes in subtropical and tropical agriculture. CAB Int., Wallingford, England.

Duncan, L.W., and D.M. Eissenstat. 1993. Responses of *Tylenchulus semipenetrans* to citrus fruit removal: Implications for carbohydrate competition. J. Nematol. 25:7–14.

Duncan, L.W., J.H. Graham, and L.W. Timmer. 1993. Seasonal patterns associated with *Tylenchulus semipenetrans* and *Phytophthora parasitica* in the citrus rhizosphere. Phytopathology 83:573–581.

Duncan, L.W., R.N. Inserra, J.H. O'Bannon, and M.M. El-Morshedy. 1994. Reproduction of a Florida population of *Tylenchulus semipenetrans* on resistant citrus rootstocks. Plant Dis. 78:1067–1071.

Duncan, R.A., J.J. Stapleton, and M.V. McKenry. 1992. Establishment of orchards with black polyethylene film mulching: Effect on nematode and fungal pathogens, water conservation, and tree growth. J. Nematol. 24 (4S):681–687.

Eayre, C.G., B.A. Jaffee, and E.I. Zehr. 1987. Suppression of *Criconemella xenoplax* by the nematophagous fungus *Hirsutella rhossiliensis*. Plant Dis. 71:832–834.

Edgerton, L.J., and K.G. Parker. 1958. Effect of nematode infestation and rootstock on cold hardiness of Montmorency cherry trees. Proc. Am. Soc. Hort. Sci. 72:134–138.

Esmenjaud, D., J.C. Minot, R. Voisin, J. Pinochet, and G. Salesses. 1994. Inter- and intraspecific resistance variability in Myrobalan plum, peach, and peach-almond rootstocks using 22 root-knot nematode populations. J. Am. Soc. Hort. Sci. 119:94–100.

Esmenjaud, D., J.C. Minot, R. Voisin, G. Salesses, and A. Bonnet. 1995. Effect of cutting age on the resistance of *Prunus cerasifera* (Myrobalan plum) to *Meloidogyne arenaria*. J. Nematol. (4S):634–638.

Esser, R.P., and S.E. Simpson. 1984. Sting nematode on citrus. Nematol. Circ. no. 106. Florida Dep. Agric. Consum. Serv., Div. Plant Ind.

Esser, R.P., G.T. Smith, and J.H. O'Bannon. 1978. An eleven year phytoparasitic nematode survey of Florida citrus groves and their environs. Florida Dep. Agric. Consum. Serv. Bull. 15.

Evert, D.R., and P.F. Bertrand. 1993. Survival and growth of peach trees planted in killed bahiagrass at an old orchard site. HortScience 28:26–28.

Food and Agriculture Organization (FAO). 1993. Production yearbook. Vol. 46. FAO, Rome.

Fattah, F.A., H.M. Saleh, and H.M. Aboud. 1989. Parasitism of the citrus nematode, *Tylenchulus semipenetrans*, by *Pasteuria penetrans* in Iraq. J. Nematol. 21:431–433.

Feldmesser, J., W.A. Feder, and R.V. Rebois. 1962. Movement of *Tylenchulus semi-penetrans* into Rough lemon roots and in soil and its relation to *Fusarium* in the roots. Phytopathology 52:9.

Feldmesser, J., and C.I. Hannon. 1965. Susceptibility of two citrus rootstocks to *Pratylenchus* spp. Plant Dis. Rep. 53:603–607.

Ferguson, J.J., and J.A. Menge. 1986. Response of citrus seedlings to various field inoculation methods with *Glomus deserticola* in fumigated nursery soils. J. Am. Soc. Hort. Sci. 111:288–292.

Fernandez, C., J. Pinochet, and R. Dolcet. 1992. Host-parasite relationship of *Pratylenchus vulnus* on apple and pear rootstocks. Nematropica 22:227–236.

Fernandez, C., J. Pinochet, D. Esmenjaud, G. Salesses, and A. Felipe. 1994. Resistance among new *Prunus* rootstocks and selections to root-knot nematodes in Spain and France. HortScience 29:1064–1067.

Fernandez, C., J. Pinochet, and A. Felipe. 1993. Influence of temperature on the expression of resistance in six *Prunus* rootstocks infected with *Meloidogyne incognita*. Nematropica 23:195–202.

Ferris, H., M.V. McKenry, and H.E. McKinney. 1976. Spacial distribution of nematodes in peach orchards. Plant Dis. Rep. 60:18–22.

Fliegel, P. 1969. Population dynamics and pathogenicity of three species of *Pratylenchus* on peach. Phytopathology 59:120–124.

Fogel, H.W., H.L. Kiel, W.L. Smith, S.M. Mircetich, L.C. Cochran, and H. Baker. 1974. Peach production. USDA-ARS Agric. Handb. 463. U.S. Gov. Print. Office, Washington, DC.

Foster, H.H., C.E. Gambrell, Jr., W.H. Rhodes, and W.P. Byrd. 1972. Effects of preplant nematicides and resistant rootstocks on growth and fruit production of peach trees in *Meloidogyne* spp.-infested soil of South Carolina. Plant Dis. Rep. 56:169–173.

Gao, X., H.R. Cheng, Z.D. Fang. 1989. Identification and determination of pathogenicity of lesion nematodes in peach roots. Acta Phytopathol. Sin. 19:65–68.

Gaspard, J.T., and R. Mankau. 1986. Nematophagous fungi associated with *Tylenchulus semipenetrans*. Nematologica 32:359–363.

Gill, H.S. 1971. Occurrence and reproduction of *Meloidogyne javanica* on three species of citrus in California. Plant Dis. Rep. 55:607–608.

Gotoh, V.A., and Y. Ohshima. 1963. *Pratylenchus*-Arten und ihre geographische Verbreitung in Japan (Nematoda: Tylenchida). Jpn. J. Appl. Entomol. Zool. 7:187–199.

Gottlieb, Y., E. Cohn, and P. Spiegel-Roy. 1986. Biotypes of the citrus nematode (*Tylenchulus semipenetrans* Cobb) in Israel. Phytoparasitica 14:193–198.

Graham, J.H., and L.W. Timmer. 1992. *Phytophthora* diseases in citrus. p. 250–269. *In* J. Kumar et al. (ed.) Plant diseases of international importance. Vol.3. Diseases of fruit crops. Prentice-Hall, Englewood Cliffs, NJ.

Greco, N., M. Basile, T.D'Abbabbo, and A. Brandonisio. 1993. Influence of aldicarb and fenamiphos on *Tylenchulus semipenetrans* population densities and orange yield. J. Nematol. 25:768–772.

Greene, G.M., II. 1986. Save the peaches. Am. Fruit Grower 106:34–36.

Grosser, J.W., and F.G. Gmitter Jr. 1990. Protoplast fusion and citrus improvement. p. 339–374. *In* J. Janick (ed.) Plant breeding reviews. Timber Press, Portland.

Halbrendt, J.M. 1992. Ground cover management: Impact on nematodes and virus. Pennsylvania Fruit News 72:95–96.

Halbrendt, J.M. 1994. Rapeseed as a dagger nematode suppressive rotation crop. J. Nematol. 26:101–102.

Halbrendt, J.M., and D.J.F. Brown. 1992. Morphometric evidence for three juvenile stages in some species of *Xiphinema americanum* sensu lato. J. Nematol. 24:305–309.

Hamid, G.A., S.D. Van Gundy, and C.J. Lovatt. 1988. Phenologies of the citrus nematode and citrus roots treated with oxamyl. p. 993–1004. *In* R. Goren and K. Mendel (ed.) Proc. 6th Int. Citricult. Congr., Tel Aviv, Israel. 6–11 March. Balaban Publ., Rehovot, Israel.

Harr, J., S.A.G. Basel, and J. Klingler. 1976. Single and combined effect of *Pratylenchus penetrans* and *Thielaviopsis basicola* on the growth of cherry tree cuttings. J. Plant Dis. Prot. 83:615–619.

Hattingh, M.J., and J.W. Gerdemann. 1975. Inoculation of Brazilian sour orange seed with an endomycorrhizal fungus. Phytopathology 65:1013–1016.

Heyes, J. 1970. South African Criconematinae. Part 3. More species of *Hemicriconemoides* and *Macroposthonia* (Nematoda). Phytophylactica 2:243–249.

Hirschmann, H. 1986. *Meloidogyne hispanica* n. sp. (Nematoda: Meloidogynidae), the 'Seville' root-knot nematode. J. Nematol. 18:520–532.

Hoy, J.W., and S.M. Mircetich. 1984. Prune brownline disease: Susceptibility of prune rootstocks and tomato ringspot virus detection. Phytopathology 74:272–276.

Huang, C.S., and Y.C. Chang. 1976. Pathogenicity of *Pratylenchus coffeae* on sunki orange. Plant Dis. Rep. 60:957–960.

Huettel, R.N., D.W. Dickson, and D.T. Kaplan. 1982. Sex attractants and behavior in the two races of *Radopholus similis*. Nematologica 28:360–369.

Huettel, R.N., D.W. Dickson, and D.T. Kaplan. 1984. *Radopholus citrophilus* n. sp. (Nematoda), a sibling species of *Radopholus similis*. Proc. Helminthol. Soc. Wash. 51:32–35.

Huettel, R.N., and T. Yaegashi. 1980. Morphological differences between *Radopholus citrophilus* and *similis*. J. Nematol. 20:150–157.

Hugo, H.J., and A.J. Meyer. 1995. Severe nematode damage to peach trees in South Africa. Nematologica 41:310.

Hung, C.L.P., and W.R. Jenkins. 1969. *Criconemoides curvatum* and the peach tree decline problem. J. Nematol. 1:12.

Hussey, R.S., and J.M. McGuire. 1987. Interactions with other organisms. p. 293–328. *In* R.H. Brown and B.R. Kerry (ed.) Principles and practice of nematode control in crops. Acad. Press Australia, Marrickville, Australia.

Hussey, R.S., Mims, C.W., and S.W. Westcott III. 1992. Ultrastructure of root cortical cells parasitized by the ring nematode, *Criconemella xenoplax*. Protoplasma 167:55–65.

Iezzoni, A., H. Schmidt, and A. Albertini. 1991. Cherries (*Prunus*). p. 109–174. *In* J.N. Moore and J.R. Ballington (ed.) Genetic resources of temperate fruit and nut crops. Int. Soc. Hort. Sci. Wageningen, The Netherlands.

Inagaki, H. 1978. Apple root-knot nematode, *Meloidogyne mali*, its taxonomy, ecology, damage and control. The Kasetsart J. 12:25–30.

Inserra, R.N., G. Perotta, N. Vovlas, and A. Catara. 1978. Reaction of citrus rootstocks to *Meloidogyne javanica*. J. Nematol. 10:181–184.

Inserra, R.N., and N. Vovlas. 1974. Damage by *Pratylenchus vulnus* to sour orange in Apulia. Nematol. Medit. 2:183–185.

Inserra, R.N., and N. Vovlas. 1977. Effects of *Pratylenchus vulnus* on the growth of sour orange. J. Nematol. 9:154–157.

Inserra, R.N., N. Vovlas, and J.H. O'Bannon. 1980. A classification of *Tylenchulus semipenetrans* biotypes. J. Nematol. 12:283–287.

Inserra, R.N., N. Vovlas, J.H. O'Bannon, and R.P. Esser. 1988. *Tylenchulus graminis* n. sp. and *T. palustris* n. sp. (Tylenchulidae), from native flora of Florida, with notes on *T. semipenetrans* and *T. furcus*. J. Nematol. 20:266–287.

Jaffee, B.A., M.A. Harrison, R.L. Shaffer, and M.B. Strang. 1987a. Seasonal population fluctuation of *Xiphinema americanum* and *X. rivesi* in New York and Pennsylvania orchards. J. Nematol. 19:369–378.

Jaffee, B.A., A.P. Nyczepir, and A.M. Golden. 1987b. *Criconemella* spp. in Pennsylvania peach orchards with morphological observations of *C. curvata* and *C. ornata*. J. Nematol. 19:420–423.

Jenser, G. 1985. *Xiphinema vuittenezi* Luc, Lima, Weischer & Flegg (Nematoda) associated with apricot roots. Acta Hort. 192:247–249.

Kaplan, D.T. 1981. Characterization of citrus rootstock responses to *Tylenchulus semipenetrans*. J. Nematol. 13:492–498.

Kaplan, D.T. 1982. Eradication of *Pratylenchus coffeae* in 'Carrizo' citrange seedlings. Proc. Florida State Hort. Soc. 95:70–72.

Kaplan, D.T. 1985. Influence of the sting nematode, *Belonolaimus longicaudatus*, on young citrus trees. J. Nematol. 17:408–414.

Kaplan, D.T. 1988. Future considerations for nematode management in citrus. Proc. 6th Int. Citrus Congr. 6:969–975.

Kaplan, D.T., and J.H. O'Bannon. 1985. Occurrence of biotypes in *Radopholus citrophilus*. J. Nematol. 17:158–162.

Kaplan, D.T., and L.W. Timmer. 1982. Effects of *Pratylenchus coffeae—Tylenchulus semipenetrans* interactions on nematode population dynamics in Citrus. J. Nematol. 14:368–373.

Kawase, K. 1972. Resistance of peach rootstocks to the root-knot nematode (*Meloidogyne incognita acrita* Chitwood) in Japan. Kurume Hort. Res. Stn. Bull. Series-D. 7:1–11.

Khan, M.L., and G.C. Sharma. 1990. Effect of temperature and moisture on population fluctuation of nematodes in an apple orchard. Indian J. Nematol. 20:10–13.

Kleinschmidt, G.D., and J.W. Gerdemann. 1972. Stunting of citrus seedlings in fumigated nursery soils related to the absence of endomycorrhizae. Phytopathology 62:1447–1453.

Klos, E.J. 1976. Rosette mosaic. p. 135–138. *In* Virus diseases and noninfectious disorders of stone fruits in North America. USDA-ARS Agric. Handb. 437. U.S. Gov. Print. Office, Washington, DC.

Kluepfel, D.A., T.M. McInnis, and E.I. Zehr. 1993. Involvement of root-colonizing bacteria in peach orchard soils suppressive of the nematode *Criconemella xenoplax*. Phytopathology 83:1240–1245.

Koepsell, P.A., J.W. Pscheidt, R. Forster, S.K. Mohan, O. Maloy, R. Byther, and D. Johnson. 1990. Pacific Northwest plant disease control handbook. Oregon State Univ. Ext. Serv., Corvallis, OR.

Lamberti, F., and T. Bleve-Zacheo. 1979. Studies on *Xiphinema americanum* sensu lato with descriptions of fifteen new species (Nematoda, Longidoridae). Nematol. Medit. 7:51–106.

Lamberti, F., S. Landriscina, A. Ciancio, and L. Catalano. 1993. Control trials against *Xiphinema diversicaudatum*, nematode vector of the strawberry latent ringspot virus on peach in Piedmont. Inform. Fitopatol. 5:57–59.

Lana, A.F., J.F. Peterson, G.L. Rouselle, and T.C. Vrain. 1983. Association of tobacco ringspot virus with union compatibility of apple. Phytopathol. Z. 106:141–148.

Layne, R.E.C. 1987. Peach rootstocks. p. 185–216. *In* R.C. Rom and R.F. Carlson (ed.) Rootstocks for fruit crops. John Wiley & Sons, Inc., New York.

Ledbetter, C.A. 1992. Protecting orchards against nematodes. *In* M. Wood (ed.) Agric. Res. 40 (7):31.

Liskova, M., M. Sabova, and B. Valocka. 1992. Contribution to the finding of the southern border of *Xiphinema diversicaudatum* (Micoletzky, 1927) Thorne, 1939 distribution range in the Slovak Republic. Helminthologia 29:181–184.

Lo Giudice, V., A. Ciancio, S. Landriscina, and F. Lamberti. 1990. Effect of nematicidal treatments on population densities of *Tylenchulus semipenetrans*. Nematol. Medit. 18:123–125.

Loof, P.A.A. 1974. *Macroposthonia curvata*. C. I. H. Descriptions of plant parasitic nematodes. Set 4, no. 58. Commonw. Inst. Helminthol., St. Albans, England.

Lordello, L.G.E. 1973. Nematoide nocivo ao pessegueiro. Rev. Agric. Piracicaba 43:128.

Lownsbery, B.F., E.H. Moody, and A.L. Braun. 1974. Plant-parasitic nematodes in California prune orchards. Plant Dis. Rep. 58:633–636.

Lownsbery, B.F., and E.F. Serr. 1963. Fruit and nut tree rootstocks as hosts for a root-lesion nematode, *Pratylenchus vulnus*. Am. Soc. Hort. Sci. 82:250–254.

Mai, W.F. 1960. Growth of apple seedlings in relation to soil temperature and inoculation with *Pratylenchus penetrans*. Phytopathology 50:237–238.

Mai, W.F., and G.S. Abawi. 1981. Controlling replant diseases of pome and stone fruit in northeastern United States by preplant fumigation. Plant Dis. Rep. 65:859–864.

Mai, W.F., J.R. Bloom, and T.A. Chen. 1977. Biology and ecology of the plant-parasitic nematode *Pratylenchus penetrans*. Pennsylvania Agric. Exp. Stn. Bull. 815.

Mai, W.F., H.W. Crittenden, and W.R. Jenkins. 1960. Distribution of stylet-bearing nematodes in the northeastern United States. New Jersey Agric. Exp. Stn. Bull. 796.

Mai, W.F., I.A. Merwin, and G.S. Abawi. 1994. Diagnosis, etiology and management of replant disorders in New York cherry and apple orchards. Acta Hort. 363:33–41.

Malo, S.E. 1961. Nematode populations associated with citrus roots in central Florida. Plant Dis. Rep. 45:20–23.

Mankau, R. 1975. *Bacillus penetrans* n. comb. causing a virulent disease of plant parasitic nematodes. J. Invert. Pathol. 26:333–339.

Mankau, R. 1982. Attractiveness, predacity and ecological relationships of nematode-trapping fungi associated with the citrus nematode. Nematologica 28:158–159.

Mankau, R., and S. Das. 1974. Effect of organic materials on nematode bionomics in citrus and root-knot nematode infested soil. Indian J. Nematol. 4:138–151.

Mankau, R., and R.J. Minteer. 1962. Reduction of soil populations of citrus nematode by the addition of organic materials. Plant Dis. Rep. 46:375–378.

Marull, J., and J. Pinochet. 1991. Host suitability of *Prunus* rootstocks to four *Meloidogyne* species and *Pratylenchus vulnus* in Spain. Nematropica 21:185–195.

Marull, J., J. Pinochet, A. Felipe, and J.L. Cenis. 1994. Resistance verification in *Prunus* selections to a mixture of thirteen *Meloidogyne* isolates and resistance mechanisms of a peach-almond hybrid to *M. javanica*. Fund. Appl. Nematol. 17:85–92.

Mashela, P., L. Duncan, and R. McSorley. 1992a. Salinity reduces resistance to *Tylenchulus semipenetrans* in citrus rootstocks. Nematropica 22:7–12.

Mashela, P., L.W. Duncan, J.H. Graham, and R. McSorley. 1992b. Leaching soluble salts increases population densities of *Tylenchulus semipenetrans*. J. Nematol. 24:103–109.

McElroy, F.C., S.A. Sher, and S.D. Van Gundy. 1966. The sheath nematode, *Hemicycliophora arenaria*, a native of California soils. Plant Dis. Rep. 50:581–583.

McFadden-Smith, W., J.W. Potter, M. Hawke, and G. Lazarovits. 1993. Use of *Brassica* green manures for control of *Verticillium dahliae* and *Pratylenchus penetrans*. p. 282. *In* Proc. 6th Int. Congr. Plant Pathol., Montreal (Quebec) Canada. 28 July–6 August. Natl. Res. Council Canada, Ottawa, ON.

McKenry, M.V. 1989a. Nematodes. p. 139–147. *In* J.H. LaRue and R.S. Johnson (ed.) Peaches, plums, and nectarines. Growing and handling for fresh market. Publ. 3331. Univ. California Div. Agric. and Nat. Resources, Oakland, CA.

McKenry, M.V. 1989b. Damage and development of several nematode species in a plum orchard. Appl. Agric. Res. 4:10–14.

McKenry, M.V. 1990a. Cover crops and nematode species. p. 67–70. *In* Proc. Int. Conf. Agric. 21st Century, Maui, HI. 12–14 October. Pacific Cult. Center & MOA Found., Honolulu, HI.

McKenry, M.V. 1990b. *Criconema mutabile* associated with bacterial canker and Nemaguard rootstock. Plant Dis. 74:394.

Mehlenbacher, S.A., V. Cociu, and L.F. Hough. 1991. Apricots (*Prunus*). p. 63–107. *In* J.N. Moore and J.R. Ballington (ed.) Genetic resources of temperate fruit and nut crops. Int. Soc. Hort. Sci. Wageningen, The Netherlands.

Melakeberhan, H., A.L. Jones, P. Sobiczewski, and G.W. Bird. 1993. Factors associated with the decline of sweet cherry trees in Michigan: Nematodes, bacterial canker, nutrition, soil pH, and winter injury. Plant Dis. 77:266–271.

Menge, J.A., C.K. Labanauskas, E.L.V. Johnson, and R.G. Platt. 1978. Partial substitution of mycorrhizal fungi for phosphorus fertilization in the greenhouse culture of citrus. Soil Sci. Soc. Am. J. 42:926–930.

Meyer, J.A., and H.J. Hugo. 1994. *Xiphinema americanum* damaging peach trees in South Africa. J. Nematol. 26:111.

Meyer, J.R., E.I. Zehr, R.L. Meager Jr., and S.K. Salvo. 1992. Survival and growth of peach trees and pest populations in orchard plots managed with experimental ground covers. Agric. Ecosyst. Environ. 41:353–363.

Miller, R.W. 1994. Estimated peach tree losses 1980 to 1992 in South Carolina—causes and economic impact. p. 121–127. *In* A.P. Nyczepir et al. (ed.) Proc. 6th Stone Fruit Decline Workshop, Fort Valley, GA. 26–28 Oct. 1992. USDA-ARS, ARS-122, Natl. Tech. Inform. Serv., Springfield, VA.

Milne, D.L. 1982. Nematode pests of citrus. p. 12–18. *In* D.P. Keetch and J. Heyns (ed.) Nematology in Southern Africa. RSA Dep. Agric. and Fish. Sci. Bull.

Minz, G. 1956. The root-knot nematode, *Meloidogyne* spp. in Israel. Plant Dis. Rep. 40:798–801.

Minz, G. 1963. Additional hosts of the root-knot nematode, *Meloidogyne* spp., recorded in Israel during 1960–1962. Israel J. Agric. Res. 13:133–134.

Minz, G., and E. Cohn. 1962. Susceptibility of peach rootstocks to root-knot nematodes. Plant Dis. Rep. 46:531–534.

Mircetich, S.M., and H.W. Fogel. 1976. Peach stem pitting. p. 77–87. *In* Virus diseases and noninfectious disorders of stone fruits in North America. USDA-ARS Agric. Handb. 437. U.S. Gov. Print. Office, Washington, DC.

Mircetich, S.M., W. J. Moller, and G. Nyland. 1977. Prunus stem pitting in sweet cherry and other stone fruit trees in California. Plant Dis. Rep. 61:931–935.

Mountain, W.B., and Z.A. Patrick. 1959. The peach replant problem in Ontario. VII. The pathogenicity of *Pratylenchus penetrans* (Cobb, 1917) Filip. & Stek. 1941. Can. J. Bot. 37:459–470.

Mulvey, R.H. 1960. Some records of plant-parasitic nematodes in Canada in 1960. Can. Plant Dis. Surv. 40:101–103.

Nemec, S. 1978. Response of six citrus rootstocks to three species of *Glomus*, a mycorrhizal fungus. Proc. Florida State Hort. Soc. 91:10–14.

Nicotina, M. 1990. Nematodes living in the rhizosphere of peaches in decline in Lazio. Inform. Agrar. 46:63–64.

Nigh, E.L., Jr., 1966. Incidence of crown gall infection in peach as affected by the Javanese root-knot nematode. Phytopathology 56:150.

Norton, R.A., C.J. Hansen, H.J. O'Reilly, W.H. Hart. 1963. Rootstocks for plums and prunes in California. Univ. California Agric. Exp. Stn. Ext. Serv. Leaflet 158.

Nyczepir, A.P. 1990. Influence of *Criconemella xenoplax* and pruning time on short life of peach trees. J. Nematol. 22:97–100.

Nyczepir, A.P. 1991. Nematode management strategies in stone fruits in the United States. J. Nematol. 23:334–341.

Nyczepir, A.P., and J.M. Halbrendt. 1993. Nematode pests of deciduous fruit and nut trees. p. 381–425. *In* K. Evans et al. (ed.) Plant parasitic nematodes in temperate agriculture, CAB Int., Wallingford, Oxon, England.

Nyczepir, A.P., and W.R. Okie. 1996. Occurrence of peach tree short life on a field site with no history of peach production. HortScience 31:163.

Nyczepir, A.P., and B.W. Wood. 1995. Interactions between *Mesocriconema xenoplax* and *Meloidogyne incognita* and the incidence of peach tree short life. J. Nematol. 27:513.

Nyczepir, A.P., and B.W. Wood. 1988. Peach leaf senescence delayed by *Criconemella xenoplax*. J. Nematol. 20:585–589.

Nyczepir, A.P., P.F. Bertrand, and B.M. Cunfer. 1996. Suitability of a wheat-sorghum, double-crop rotation to manage *Criconemella xenoplax* in peach production. Plant Dis. 80:629–632.

Nyczepir, A.P., P.F. Bertrand, R.W. Miller, and R.W. Motsinger. 1985. Incidence of *Criconemella* spp. and peach orchard history in short-life and non-short-life sites in Georgia and South Carolina. Plant Dis. 69:874–877.

Nyczepir, A.P., C.C. Reilly, and W.R. Okie. 1987. Effect of initial population density of *Criconemella xenoplax* on reducing sugars, free amino acids, and survival of peach seedlings over time. J. Nematol. 19:296–303.

Nyczepir, A.P., M.B. Riley, and R.R. Sharpe. 1993. Dynamics of concomitant populations of *Meloidogyne incognita* and *Criconemella xenoplax* on peach. J. Nematol. 25:659–665.

Nyland, G. 1955. Killing root knot nematodes in some stone fruit tree rootstocks. Plant Dis. Rep. 39:573–575.

Nyland, G. 1976. Cherry rasp leaf. p. 219–221. *In* Virus diseases and noninfectious disorders of stone fruits in North America. USDA-ARS Agric. Handb. 437. U.S. Gov. Print. Office, Washington, DC.

O'Bannon, J.H. 1968. The influence of an organic soil amendment on infectivity and reproduction of *Tylenchulus semipenetrans* on two citrus rootstocks. Phytopathology 58:597–601.

O'Bannon, J.H., and H.W. Ford. 1977. Resistance in citrus rootstocks to *Radopholus similis* and *Tylenchulus semipenetrans* (Nematoda). Proc. Int. Soc. Citricult. 2:544–549.

O'Bannon, J.H., C.R. Leathers, and H.W. Reynolds. 1967. Interactions of *Tylenchulus semipenetrans* and *Fusarium* species on rough lemon (*Citrus limon*). Phytopathology 57:414–417.

O'Bannon, J.H., and S. Nemec. 1978. Influence of soil pesticides on vesicular-arbuscular mycorrhizae in a citrus soil. Nematropica 8:56–61.

O'Bannon, J.H., J.D. Radewald, and A.T. Tomerlin. 1972. Population fluctuation of three parasitic nematodes in Florida citrus. J. Nematol. 4:194–199.

O'Bannon, J.H., J.D. Radewald, A.T. Tomerlin, and R.N. Inserra. 1976. Comparative influence of *Radopholus similis* and *Pratylenchus coffeae* on citrus. J. Nematol. 8:58–63.

O'Bannon, J.H., H.W. Reynolds, and C.R. Leathers. 1966. Effects of temperature on penetration, development and reproduction of *Tylenchulus semipenetrans*. Nematologica 12:483–487.

O'Bannon, J.H., and A.C. Tarjan. 1973. Preplant fumigation for citrus nematode control in Florida. J. Nematol. 5:88–94.

O'Bannon, J.H., A.C. Tarjan, and F.W. Bistline. 1974. Control of *Pratylenchus brachyurus* on citrus and tree response to chemical treatment. Proc. Soil Crop Sci. Soc. Fla. 33:65–67.

O'Bannon, J.H., and A.T. Tomerlin. 1969. Population studies on two species of *Pratylenchus* on citrus. J. Nematol. 1:299–300.

O'Bannon, J.H., and A.T. Tomerlin. 1973. Citrus tree decline caused by *Pratylenchus coffeae*. J. Nematol. 5:311–316.

Okie, W.R., T.G. Beckman, A.P. Nyczepir, G.L. Reighard, W.C. Newall, Jr., and E.I. Zehr. 1994a. BY520-9, a peach rootstock for the southeastern United States that increases scion longevity. HortScience 29:705–706.

Okie, W.R., G.L. Reighard, T.G. Beckman, A.P. Nyczepir, C.C. Reilly, E.I. Zehr, W.C. Newall, Jr., and D.W. Cain. 1994b. Field-screening *Prunus* for longevity in the southeastern United States. HortScience 29:673–677.

Orion, D., and E. Cohn. 1975. A resistant response of citrus roots to the root-knot nematode, *Meloidogyne javanica*. Marcellia 38:327–328.

Pan, C. 1984. Studies on plant-parasitic nematodes on economically important crops in Fujian. I. Species of root-knot nematodes (*Meloidogyne* species) and their host-plants. Acta Zool. Sin. 30:159–167.

Parish, C.L. 1977. A relationship between flat apple disease and cherry rasp leaf disease. Phytopathology 67:982–984.

Philis, J. 1983. Occurrence of *Meloidogyne* spp. and races on the island of Cyprus. Nematol. Medit. 11:13–19.

Philis, J. 1989. Yield loss assessment caused by the citrus nematode *Tylenchulus semipenetrans*. Nematol. Medit. 17:5–6.

Pinochet, J., C. Calvet, A. Camprubi, and C. Fernandez. 1995. Interaction between the root-lesion nematode *Pratylenchus vulnus* and the mycorrhizal association of *Glomus intraradices* and Santa Lucia 64 cherry rootstock. Plant Soil 170:323–329.

Pinochet, J., A. Camprubi, and C. Calvet. 1993a. Effects of the root-lesion nematode *Pratylenchus vulnus* and the mycorrhizal fungus *Glomus mosseae* on the growth of EMLA-26 apple rootstock. Mycorrhiza 4:79–83.

Pinochet, J., L. Cenis, C. Fernandez, M. Douchet, and J. Marull. 1994. Reproductive fitness and random amplified polymorphic DNA variation among isolates of *Pratylenchus vulnus*. J. Nematol. 26:271–277.

Pinochet, J., C. Fernandez, D. Esmenjaud, and M. Douchet. 1993b. Effects of six *Pratylenchus vulnus* isolates on the growth of peach-almond hybrid and apple rootstocks. J. Nematol. 25(4S):843–848.

Pinochet, J., J. Marull, and A. Felipe. 1992a. Response of newly introduced peach, plum, and cherry rootstocks to *Meloidogyne javanica* in Spain. Nematropica 22:99–102.

Pinochet, J., J. Marull, R. Rodriguez-Kabana, A. Felipe, and C. Fernandez. 1993c. Pathogenicity of *Pratylenchus vulnus* on plum rootstocks. Fund. Appl. Nematol. 16:375–380.

Pinochet, J., S. Verdejo-Lucas, and J. Marull. 1991. Host suitability of eight *Prunus* spp. and one *Pyrus communis* rootstocks to *Pratylenchus vulnus, P. neglectus, and P. thornei*. J. Nematol. 23 (4S):570–575.

Pinochet, J., S. Verdejo, A. Soler, and J. Canals. 1992b. Host range of a population of *Pratylenchus vulnus* in commercial fruit, nut, citrus, and grape rootstocks in Spain. J. Nematol. 24 (4S):693–698.

Pitcher, R.S., Z.A. Patrick, and W.B. Mountain. 1960. Studies on the host-parasite relations of *Pratylenchus penetrans* (Cobb) to apple seedlings. I. Pathogenicity under sterile conditions. Nematologica 5:309–414.

Pitcher, R.S., M.R. Siddiqi, and D.J.F. Brown. 1974. *Xiphinema diversicaudatum*. C. I. H. Descriptions of plant-parasitic nematodes. Set 4, no. 60. Commonw. Inst. Helminthol., St. Albans, England.

Poucher, C., H.W. Ford, R.F. Suit, and E.P. DuCharme. 1967. Burrowing nematode in citrus. Florida Dep. Agric. Cons. Serv., Div. Plant Ind. Bull. 7:63.

Radewald, J.D., J.H. O'Bannon, and A.T. Tomerlin. 1971a. Anatomical studies of *Citrus jambhiri* roots infected by *Pratylenchus coffeae*. J. Nematol. 3:409–416.

Radewald, J.D., J.H. O'Bannon, and A.T. Tomerlin. 1971b. Temperature effects on reproduction and pathogenicity of *Pratylenchus coffeae* and *P. brachyurus* and survival of *P. coffeae* in roots of *Citrus jambhiri*. J. Nematol. 3:390–394.

Rai, K.M., R. Joshi, and G.M. Tripathi. 1988. Short note on *Xiphinema americanum* Cobb, a plant parasitic nematode, recorded on apple roots for the first time in U. P. Hills. Prog. Hort. 20:179–180.

Rammah, A., and H. Hirschmann. 1990. *Meloidogyne morocciensis* n. sp. (Meloidogyninae), a root-knot nematode from Morocco. J. Nematol. 22:279–291.

Ramming, D.W., and V. Cociu. 1991. Plums (*Prunus*). p. 233–287. *In* J.N. Moore and J.R. Ballington (ed.) Genetic resources of temperate fruit and nut crops, Int. Soc. Hort. Sci. Wageningen, The Netherlands.

Ramming, D.W., and O. Tanner. 1983. 'Nemared' peach rootstock. HortScience 18:376.

Ramsdell, D.C., V.A. Adler, and C.R. Kesner. 1993. Prune brown line disease occurrence in declining 'Stanley' plum orchards in Michigan and factors related to its incidence. HortTechnology 3:325–329.

Reynolds, H.W., J.H. O'Bannon, A.T. Tomerlin, E.L. Nigh, and D.R. Rodney. 1970. The influence of various ecological factors on survival of *Tylenchulus semipenetrans*. Proc. Soil Crop Sci. Soc. Fla. 30:366–370.

Ricciardi, P., A. Amici, and F. Lalatta. 1975. Ulteriori indagini sul ruolo di *Pratylenchus vulnus* nel problema del reimpianto del pesco. Riv. Ortoflorofrut. It. 59:109–123.

Ritchie, D.F. 1988. Population dynamics of ring nematodes and peach tree short life in North Carolina. p. 34–37. *In* E.I. Zehr (ed.). Proc. 3rd Stone Fruit Decline Workshop, Clemson, SC. 28–29 Oct. 1986. USDA-ARS, Natl. Tech. Inform. Serv., Springfield, VA.

Ritchie, D.F., and E.I. Zehr. 1995. Peach tree short life. p. 45–46. *In* J.M. Ogawa et al. (ed.) Compendium of stone fruit diseases. APS Press, St. Paul, MN.

Roca, F., and M.J. Pereira. 1992. *Xiphinema belmontense* sp. n. (Nematoda: Longidoridae) from Portugal. Fund. Appl. Nematol. 15:251–255.

Rosenberger, D.A., M.B. Harrison, and D. Gonsalves. 1983. Incidence of apple union necrosis and decline, tomato ringspot virus, and *Xiphinema* vector species in Hudson Valley orchards. Plant Dis. 67:356–360.

Saigusa, T., and Y. Matsumoto. 1961. A list of the host plants of the root-knot nematodes, *Meloidogyne incognita* var. *acrita* and *M. hapla*. Plant Prot. Serv. Jpn. Res. Bull. 1:84–88.

Saxena, P.K., and H.K. Chhabra. 1970. Infection of root-knot nematode in peaches (*Prunus persica*)-A new record from India. Labdev J. Sci. Tech. 8:115.

Scorza, R., and W.R. Okie. 1991. Peaches (*Prunus*). p. 175–231. *In* J.N. Moore and J.R. Ballington (ed.) Genetic resources of temperate fruit and nut crops. Int. Soc. Hort. Sci., Wageningen, The Netherlands.

Scotto La Massese, C. 1975. Tests d'hotes de quelques porte-greffes et varietes fruitieres a l egard de *Pratylenchus vulnus* Allen et Jensen. C. R. Acad. d' Agric. France 61:1088–1095.

Scotto La Massesse, C. 1977. Some aspects of host reaction by resistant citrus rootstock (*Poncirus trifoliata*) to *Tylenchulus semipenetrans* infection. p. 833–835. *In* W. Grierson (ed.) Proc. 3rd Int. Soc. Citricult., Orlando, FL. 1–8 May. Int. Soc. Citricult., Orlando, FL.

Scotto La Massese, C. 1979. Possibilites offertes par les "courbes isopathologiques" dans l' appreciation de la pathogenie des parasites illustrees par deux exemples en nematologie fruitiere. Rev. Zool. Agric. 78:97–113.

Scotto La Massesse, C., R. Vassy, and H. Zaouchi. 1973. Elimination de *Tylenchulus semipenetrans* par des apports azotes appliqués a des clémentiniers greffés sur bigaradiers. Nematol. Medit. 1:15–21.

Seshadri, A.R. 1964. Investigations on the biology and life cycle of *Criconemoides xenoplax* Raski, 1952 (Nematoda: Criconematidae). Nematologica 10:540–562.

Sharma, G.C., and N.K. Sharma. 1989. Spacial distribution of *Paratylenchus pruni* in peach rhizosphere. Him. J. Agric. Res. 15:99–100.

Sharma, G.C., N.K. Sharma, and M.L. Khan. 1991. A note on parasitic nematodes associated with apricot (*Prunus armeniaca*). Him. J. Agric. Res. 17:112–114.

Sharpe, R.R., P.L. Pusey, A.P. Nyczepir, and W.J. Florkowski. 1993. Yield and economics of intervention with peach tree short life disease. J. Prod. Agric. 6:241–244.

Sharpe, R.R., C.C. Reilly, A.P. Nyczepir, and W.R. Okie. 1989. Establishment of peach in a replant site as affected by soil fumigation, rootstock, and pruning date. Plant Dis. 73:412–415.

Sherman, W.B., P.M. Lyrene, and R.H. Sharpe. 1991. Flordaguard peach rootstock. HortScience 26:427–428.

Siddiqui, I.A., S.A. Sher, and A.M. French. 1973. Distribution of plant parasitic nematodes in California. Dep. Food Agric., Div. Plant Ind., Sacramento, CA.

Siddiqi, M.R. 1973. *Xiphinema americanum.* C. I. H. Descriptions of plant-parasitic nematodes. Set 2, no. 29. Commonw. Inst. Helminthol., St. Albans, England.

Siddiqi, Z.A. 1982. Nematode problems in fruit tree nurseries in Libya. Libyan J. Agric. 11:147–152.

Simeone, A.M., and M. Di Vito. 1992. Reactions to nematodes of selections of peach rootstocks. Acta Hort. 315:197–202.

Siniscalco, A., F. Lamberti, and R. Inserra. 1976. Reaction of peach rootstocks to Italian populations of root-knot nematodes (*Meloidogyne* spp.). Nematol. Medit. 4:79–84.

Smart, G.C., and K.B. Nguyen. 1991. Sting and awl nematodes: *Belonolaimus* spp. and *Dolichodorus* spp. p. 627–667. *In* W.R. Nickle (ed.) Manual of agricultural nematology. Marcel Dekker, New York.

Spiegel, Y., E. Cohn, and I. Chet. 1989. Use of chitin for controlling *Heterodera avenae* and *Tylenchulus semipenetrans*. J. Nematol. 21:419–422.

Standifer, M.S., and V.G. Perry. 1960. Some effects of sting and stubby root nematodes on grapefruit roots. Phytopathology 50:152–156.

Stirling, G.R. 1975. A survey of the plant-parasitic nematodes in Riverland peach orchards. Agric. Record 2:11–13.

Stirling, G.R. 1976. *Paratrichodorus lobatus* associated with citrus, peach and apricot trees in South Australia. Nematologica 22:138–144.

Stirling, G.R. 1991. Biological control of plant parasitic nematodes: Progress, problems and prospects. CAB Int., Wallingford, England.

Stokes, D.E. 1977. The role of nematodes as vectors of citrus viruses. Proc. Int. Soc. Citriculture 3:848–853.

Strobel, N.E., R.S. Hussey, and R.W. Roncadori. 1982. Interactions of vesicular-arbuscular mycorrhizal fungi, *Meloidogyne incognita*, and soil fertility on peach. Phytopathology 72:690–694.

Sturhan, D. 1988. New host and geographical records of nematode-parasitic bacteria of the *Pasteuria penetrans* group. Nematologica 34:350–356.

Suit, R.F., and E.P. DuCharme. 1953. The burrowing nematode and other parasitic nematodes in relation to spreading decline. Citrus Leaves 33:8–9, 32–33.

Szczygiel, A., and Z. Rebandel. 1985. The effect of root-lesion nematode (*Pratylenchus penetrans*) on the growth of two seedling rootstocks, Antonvka and Mazzard cherry, in field conditions. Roczniki Akadenii Rolniczej W Poznaniu 156:209–214.

Tarjan, A.C. 1961. Longevity of *Radopholus similis* (Cobb) in host free soil. Nematologica 6:170–175.

Tarjan, A.C. 1977. Use of municipal refuse compost on nematode-infected citrus. Citrus Veg. Mag. 40:44–49.

Tarjan, A.C., and J.H. O'Bannon. 1977. Nonpesticidal approaches to nematode control. Proc. Int. Soc. Citricult. 3:848–853.

Tarjan, A.C., and J.H. O'Bannon. 1984. Nematode parasites of citrus. p. 395–433. *In* W.R. Nickle (ed.) Plant and insect nematodes. Marcel Dekker, New York.

Taylor, C.E., and D.J.F. Brown. 1981. Nematode-virus interactions. p. 281–302. *In* B.M. Zuckerman and R.A. Rhode (ed.) Plant parasitic nematodes. Vol. 3. Acad. Press, New York.

Teodorescu, G.M., E. Romascu, and M. Ivan. 1986. Identified helminthological fauna on apple and plum trees and its control in virus free mother plantations. Acta Hort. 193:149–154.

Tsao, P.H., J.P. Martin, and R.M. Davis. 1989. The citrus replant problem. p. 261–279. *In* W. Reuther et al. (ed.) The citrus industry. Univ. California, Statewide Integrat. Pest Manage. Proj., Div. Agric. Natural Resour., Berkeley, CA.

Tudor, M.E., and M.V. McKenry. 1989. Unraveling the various nematicidal activities within aqueous extracts of plant refuse. J. Nematol. 21:592.

U. S. Department of Agriculture, Foreign Agricultural Service. 1990. Hort. Prod. Rev. FHORT 11–90. U.S. Gov. Print. Office, Washington, DC.

Van Driel, J.W. Potter, and B.A. Ebsary. 1990. Distribution of virus vector nematodes associated with peach and other fruit crops in Essex County, Ontario. Can. Plant Dis. 70:23–26.

Van Gundy, S.D. 1959. The life history of *Hemicycliophora arenaria* Raski (Nematoda: Criconematidae). Proc. Helm. Soc. Wash. 26:67–72.

Van Gundy, S.D. 1984. Nematodes. p. 129–131. *In* M.L. Flint (ed.) Integrated pest management for citrus. Univ. California Integrated Pest Manage. Project. Univ. Calif., Berkeley.

Van Gundy, S.D., A.F. Bird, and H.R. Wallace. 1967. Aging and starvation in larvae of *Meloidogyne javanica* and *Tylenchulus semipenetrans*. Phytopathology 57:559–571.

Van Gundy, S.D., S. Garabedian, and E.L. Nigh. 1981. Alternatives to DBCP for citrus nematode control. Proc. Int. Soc. Citricult. 387–390.

Van Gundy, S.D., and J.D. Kirkpatrick. 1964. Nature of resistance in certain citrus rootstocks to citrus nematode. Phytopathology 54:419–427.

Van Gundy, S.D., and J.W. Meagher. 1977. Citrus nematode (*Tylenchulus semipenetrans*) problems worldwide. Proc. Int. Soc. Citricult. 3:823–826.

Van Gundy, S.D., and R.L. Rackham. 1961. Studies on the biology and pathogenicity of *Hemicycliophora arenaria*. Phytopathology 51:393–397.

Van Gundy, S.D., L.H. Stolzy, T.E. Szuszkiewicz, and R.L. Rackham. 1962. Influence of oxygen supply on survival of plant parasitic nematodes in soil. Phytopathology 52:628–632.

Van Gundy, S.D., I.J. Thomason, and R.L. Rackham. 1959. The reaction of three *Citrus* spp. to three *Meloidogyne* spp. Plant Dis. Rep. 43:970–971.

Vrain, T.C., and J.M. Yorston. 1987. Plant-parasitic nematodes in orchards of the Okanagan valley of British Columbia, Canada. Plant Dis. 71:85–87.

Walter, D.E., and D.T. Kaplan. 1990. Antagonists of plant-parasitic nematodes in Florida citrus. J. Nematol. 22:567–573.

Way, R.D., H.S. Aldwinkle, R.C. Lamb, A. Rejman, S. Sansavini, T. Shen, R. Watkins, M.N. Westwood, and Y. Yoshida. 1991. Apples (*Malus*). p. 3–62. *In* J.N. Moore and J.R. Ballington (ed.) Genetic resources of temperate fruit and nut crops, Int. Soc. Hort. Sci., Wageningen, The Netherlands.

Wehunt, E.J. 1972. Influence of temperature on infection of *Meloidogyne incognita acrita* on 'Nemaguard' peach seedlings. Plant Dis. Rep. 56:305–308.

Wehunt, E.J., and A.M. Golden. 1982. Nematodes of pear. p. 408–413. *In* T. van der Zwet and N.F. Childers (ed.) The pear. Hort. Publ., Gainesville, FL.

Wehunt, E.J., and A.P. Nyczepir. 1988. Nematodes on peaches in the U.S.—possible problems. p. 739–750. *In* N. F. Childers and W. B. Sherman (ed.) The peach. Hort. Publ., Gainesville, FL.

Westcott, S.W., III, and P.M. Burrows. 1991. Degree-day models for predicting egg hatch and population increase of *Criconemella xenoplax*. J. Nematol. 23:386–392.

Westcott, S.W., and R.S. Hussey. 1992. Feeding behavior of *Criconemella xenoplax* in monoxenic cultures. Phytopathology 82:936–940.

Westcott, S.W., and D.A. Kluepfel. 1993. Inhibition of *Criconemella xenoplax* egg hatch by *Pseudomonas aureofaciens*. Phytopathology 83:1245–1249.

Westwood, M.V. 1988. Temperate-zone pomology. W. H. Freeman and Co., San Francisco, CA.

Williams, K.J.O. 1972. *Macroposthonia xenoplax*. C. I. H. Descriptions of plant parasitic nematodes. Set 1, no. 12. Commonw. Inst. Helminthol., St. Albans, England.

Wu, S.F., Z. Hu, Z. Shi, and M. Xu. 1993. Investigation and identification of root parasitic nematodes of peach (*Amygdalus persica* L.) in Zhejiang Province. Acta Agric. Zhejiang. 5:245–247.

Wutscher, H.K. 1974. Swingle citrumelo: An ultraresistant rootstock. Citrograph 59:387–388, 391.

Yassin, A.M. 1974. A note on *Longidorus* and *Xiphinema* species from the Sudan. Nematol. Medit. 2:141–147.

Zepp, A.L., and A. Szczygiel. 1985. Pathogenicity of *Pratylenchus crenatus* and *P. neglectus* (Pratylenchidae, Nematoda) to three fruit tree seedling rootstocks. Fruit Sci. Rep. 12:109–118.

27 Small Fruits

JOSEPH ESNARD

Cornell University
Ithaca, New York

BERT M. ZUCKERMAN

University of Massachusetts
Amherst, Massachusetts

Small fruits grown on a commercial level include grape (*Vitis* spp.), raspberry (*Rubus idaeus* L. and *R. occidentalis* L.), blackberry (*Rubus plicatus* Weihe & Nees), strawberry (*Fragaria* spp.), cranberry (*Vaccinium macrocarpon* Aiton), blueberry (*Vaccinium corymbosum* L.), gooseberry (*Ribes uva-crispa* L.), currant (*Ribes nigrum* L. and *R. rubrum* L.) and to a much lesser extent, elderberry (*Sambucus canadensis* L.) and juneberry (*Amelanchier* spp.). They grow as shrubs or vines that are mostly adapted to temperate zones in areas with warm to cool temperatures. Small fruits are used for direct consumption as fresh fruit, as components of many processed foods and beverages, for medicinal purposes, for making dyestuffs and as ornamental plants in gardens. The relative importance of pests and diseases in small fruit cultivation depends on the host species. For example, very few bacterial pathogens attack strawberry, but several nematodes are known to cause dramatic distortions of the foliage.

Nematodes are some of the principle disease-inciting agents in small fruit culture. A wide spectrum is known to cause economically important diseases wherever small fruits are grown. They siphon off plant juices, act as vectors for bacteria and viruses (Taylor & Brown, 1981), provide entry sites for many bacterial and fungal pathogens, and damage host tissue directly or by secretions that alter tissue development. Weakened root systems can not sustain normal top growth of the plants and, consequently, aboveground parts lose color, become stunted and/or wilt. The overall effect is a shortening of plantation lifetime.

Reliable estimates of the percentage or dollar losses caused by nematode-related diseases of small fruits (Raski & Krusberg, 1984) that would clearly dramatize the economic importance of these pathogens are acutely scarce. Unlike other competitors, the presence of nematodes is not correlated with any characteristic external symptoms of the plant during its growth up to harvest. For the few species that induce specific symptoms (such as root knot), correlations with yield loss are hard to establish (Seinhorst, 1981). Symptoms also may be confused with stress caused by other pathogens, mineral deficiency or toxicity, and waterlogging or drought.

Yield losses associated with nematode interactions with small fruit can range from very little to more than 80%. The cost of nematode or viral disease management has not been well documented but damage to small fruit plants is often highly significant. Increases in nematode damage to these and other crops will be dramatic as management options (including chemical pesticides) continually become unacceptable or ineffective (Soc. Nematol., 1994; Zuckerman & Esnard, 1994). This chapter provides an overview of the nature of nematode-related problems and an outlook on future priorities for small-fruit cultivation.

NEMATODE PARASITES OF GRAPE

Root-Knot Nematode

Species

Only 4 of the approximately 60 known root-knot nematode species cause significant economic damage on commercial grape, viz. *Meloidogyne incognita, M. javanica, M. arenaria* and *M. hapla*. In the USA, *M. hapla* damage is significant on the grape grown only in Washington and Michigan. The first three are common in vineyards worldwide. The virulence of different root-knot nematode races on small fruit crops is not well known. *Meloidogyne nataliei* has been reported from Michigan vineyards on *Vitis labrusca* 'Concord'.

Symptoms

Poor vine growth, reduced yields, and off-colored grape are observed in fields infested with root-knot nematodes (Raski, 1988). Although tissue malformations do not appear on above-ground parts, the secondary root systems become severely deformed by the formation of galls at points in the root where the nematodes have invaded and become sedentary. Gall formation reduces nutrient and water uptake tremendously resulting in nonfunctional rootlets, general weakening and yield reduction. Feeder roots are usually killed. Aboveground symptoms of root-knot infection are most pronounced in sandy soils where nutritional deficiencies and water stress are greatest. *Meloidogyne nataliei* parasitizes grape without inducing gall formation but is known only in Michigan vineyards. (Golden et al., 1981; Diamond, 1994; Bird et al., 1994).

Life Cycle and Epidemiology

The life cycles and epidemiology of root-knot nematodes have been described in Chapter 15 (Kinloch, 1998) and Chapter 22 (Johnson, 1998) of this monograph.

Economic Importance

Root-knot, dagger (*Xiphinema* spp.), lesion (*Pratylenchus* spp.) and citrus (*Tylenchulus* spp.) nematodes are the four most damaging nematodes in grape cultivation. According to Raski (1986), 30 yr of trials in California have shown that loss in grape crop value could be attributed to root-knot nematode damage. In 1983, this amounted to $190 million for a $950 million crop. Ten-percent of a $39-million grape crop was lost to root-knot nematode infestations in Washington. The ban of major nematicides has exacerbated the situation even further (Zuckerman & Esnard, 1994) and will highlight the devastating effect root-knot nematodes have on grape production. Unfortunately, root-knot nematodes are difficult to eradicate.

Dagger and Needle Nematodes

Causal Organisms

The most important dagger nematodes reported in association with grape are *Xiphinema index*, *X. americanum*, and *X. italiae* (Israel only). *Xiphinema diversicaudatum* has been reported in European vineyards (Raski, 1986). *Xiphinema mediterraneum* and *X. pachtaicum* are quite widespread. *Xiphinema algeriense*, *X. brevicolle*, *X. turcicum*, and *X. vuittenzi* have been found in association with grape. These species may be found singly or in various combinations.

The following needle nematodes have been found in association with grape: *Longidorus elongatus*, *L. attenuatus*, *L. protae*, *L. diadecturus*, *L. iranicus*, *L. macrosoma*, and *L. sylphus*.

The most destructive longidorid pathogen is *X. index*. It occurs worldwide (Raski, 1988). In addition, *X. index* transmits the grapevine fanleaf virus (GFLV), which makes the nematode even more destructive in a vineyard that is infested with viruliferous nematodes. *Xiphinema americanum* is widespread in vineyards on most continents except Europe. Little is known about the pathogenicity of most of the dagger nematodes reported from vineyards.

Symptoms

Damage occurs only on the root systems. Feeding occurs mostly at the root tips where further development of the rootlet stops. Root bending and swelling occur as a result of hyperplasia and hypertrophy. Prolonged feeding results in necrotic spots that spread over the entire root tip giving a witch's-broom effect as new rootlets appear and are, in turn, damaged (Pinochet et al., 1976; Raski, 1986). The overall effect is a general weakening of vines due to a lack of adequate water and nutrient uptake through the damaged roots. *Xiphinema index, X. diversicaudatum* and *X. americanum* induce such characteristic symptoms but to a lesser extent by the latter (Raski, 1986, 1988). Vines decline until they become totally unproductive.

Life Cycle and Epidemiology

Dagger and needle nematodes feed ectoparasitically on grape root systems by using their long stylets to penetrate root tips. Based on the observations of two species, needle nematodes can cause necrosis and malformations of grape roots. Eggs are laid singly and unprotected in the soil. Each embryo develops into an elongate juvenile that hatches from the egg as a first-stage juvenile (J1). Further growth and molting occur only after feeding on a suitable host. The J1 molts four times in the soil to become the adult. All stages remain vermiform and resemble the adults. Some species reproduce mostly parthenogenetically, rarely produce males, or produce an equal number of each sex (e.g., *X. diversicaudatum*). *Xiphinema index* takes 22 to 27 d to develop from egg to adult in California whereas 210 to 270 d (7–9 mo) elapse for one life cycle of the nematode in Israel (Raski, 1986, 1988). Although the reason for this phenomenal difference is unknown, it might be related to genotype-environmental interactions or to specific grape cultivar responses in the two environments.

Dagger and needle nematodes spread by means of infected root stocks, contaminated irrigation water and other cultural practices (Raski, 1988). *Xiphinema americanum* can be found on grape roots >2 m deep in soil of Michigan vineyards (Bird & Ramsdell, 1985; Ramsdell & Myers, 1974).

Economic Importance

No estimates on the direct losses in grape caused by dagger and needle nematodes are available. *Xiphinema index* or *X. italiae* can transmit GFLV and act synergistically with the virus to compound the damage to grapevines when viruliferous nematodes are involved. The economic life expectancy of a field infested with viruliferous nematodes is about 12 to 20 yr (Raski, 1988).

Lesion Nematodes

Causal Organisms

Lesion nematodes are even more destructive to grape than root-knot nematodes as control measures do not reverse the damage on already declining vines. The most damaging and widespread species is *Pratylenchus vulnus*. It is found on grape in California and Australia. Other species associated with grape production are *P. pratensis* (Tashkent, Russia), *P. brachyurus* (California), *P. scribneri* (Australia, California) and *P. neglectus* (Australia, California). No single species is found worldwide (Raski, 1986), but lesion nematodes exist in nearly all grape vineyards.

Symptoms

Aboveground symptoms of lesion nematode infection are not easy to detect and also are not diagnostic for the disease. However, vines show poor growth and decreased yields but seldom die. Feeder roots appear dead and darkened. New feeder roots that are produced in response to root damage also are killed resulting in a matted, witch's-broom effect in all or parts of the root system (Raski, 1986, 1988).

Life Cycle and Epidemiology

The life cycle and epidemiology of lesion nematodes are discussed in Chapter 16 (Windham, 1998) in this monograph.

Citrus Nematode

Causal Organisms

The citrus nematode (*Tylenchulus semipenetrans*) has been reported on grape in California, Australia, Egypt, India, and the Philippines. It might occur in all places where grape follows citrus cultivation, but may not necessarily be pathogenic on grape.

Symptoms

On aboveground parts, grape does not exhibit distinctive symptoms (Anwar & van Gundy, 1989). Feeder roots die or support large numbers of the nematodes that produce a gelatinous matrix to which soil particles cling (Raski, 1986, 1988). Unlike lesion and root-knot nematodes, no necrosis and malformations are produced. Female juveniles may survive one or more years in the soil.

Life Cycle and Epidemiology

The life cycle of the citrus nematode is described in Chapter 26 (Nyczepir & Becker, 1998) in this monograph.

Economic Importance

Raski (1988) considered the citrus nematode one of the most pathogenic nematodes on grape. Although specific estimates on grape crop losses are not available, it is well known that yields inevitably decline, and vineyards infested with citrus nematodes become uneconomical.

Other Nematodes

Causal Organisms

Other nematode ectoparasites found in vineyards whose pathogenic potential is not fully known are ring nematodes, *Criconemella xenoplax*; pin nematodes, *Paratylenchus hamatus* and *P. neoamblycephalus*; reniform nematodes, *Rotylenchulus* spp.; spiral nematodes, *Helicotylenchus* spp.; lance nematodes, *Hoplolaimus* spp.; stubby-root nematodes, *Paratrichodorus minor*; and stunt nematodes, *Tylenchorhynchus* (syn. *Telotylenchus*) spp. The reniform nematode is widespread in distribution and has been reported on grape in Algeria, Egypt, India and the Philippines (Raski, 1986).

These nematodes do not produce effects with characteristic symptoms on grapevines. More studies are needed to characterize the effects of these ectoparasites on grapevines. Most of the remaining species are ectoparasitic, feeding (without penetrating the roots) by puncturing cells in deep layers of the roots. The general life cycles of nematodes belonging to these genera have been discussed in Chapter 2 (Ferris & Ferris, 1998) in this monograph.

Estimates of losses induced specifically by these nematodes have not been reported. The management methods described for other nematodes are applicable for control of these ecto- and semiendoparasites.

NEMATODE PARASITES OF RASPBERRY AND BLACKBERRY

Root-Lesion Nematodes

Species

Root-lesion nematodes (*Pratylenchus* spp.) are endemic on *Rubus* worldwide. Of the six species reported on raspberry, only *P. penetrans* is important. *Pratylenchus vulnus* is implicated as the cause of red raspberry decline in Utah, and *P. crenatus* occurs to some extent in Europe (McElroy, 1991).

Symptoms

Early symptoms of root-lesion nematode infection of *Rubus* spp. are difficult to detect (Jennings, 1988) because they appear only as small lesions on young roots. However, as the nematode populations increase, feeder roots loose vitality and die resulting in a proliferation of finer roots, which are, in turn, invaded by the pathogen. Eventually what is left of the root system is only the larger older roots, which do not have the absorptive capacity of the finer, feeder roots. Decreased water and nutrient uptake give rise to slight stunting of plants that may go undetected. After several years the canes show reduced diameter, some dieback and weakened growth, which characterizes a slow decline. Infected plants are more susceptible to winter injury. Because buds die in the process, growth at the beginning of the season of subsequent years is generally poor. The decline usually occurs over 3 to 4 yr but plants can die 2 to 3 yr after the decline period depending on stresses related to water, disease, insect, and other factors (McElroy, 1991).

Life Cycle and Epidemiology

The life cycle of lesion nematodes is dealt with in Chapter 16 (Windham, 1998) in this monograph. Although populations of lesion nematodes are active throughout the year in raspberry plantations, their numbers and activity are more reduced during the winter and in areas where soil has dried up. These nematodes were even detected in soil frozen 15 cm deep. Their population densities are highest in the fall when damage potential is greatest (McElroy, 1991).

Dagger and Needle Nematodes

Species

The three dagger nematodes known to be associated with raspberry and blackberry are *Xiphinema bakeri*, *X. americanum*, and *X. diversicaudatum*. *Xiphinema americanum* and *X. diversicaudatum* have their greatest impact on plant health as vectors of nepoviruses. *Xiphinema bakeri* is not known to be a virus vector but among the three dagger nematodes, it has the greatest direct impact on *Rubus*. *Xiphinema americanum* mostly occurs in North America while *X. diversicaudatum* is widely distributed in Europe. There is still controversy over the taxonomy of the *X. americanum* group of nematodes.

Fig. 27–1. Fruits from 'Fairview' red raspberry infected with tomato ringspot virus (ToRSV). Healthy fruit (center) is compared with diseased fruit (left and right) showing reduced droplet set, small size and abnormal shape (after Stace-Smith and Converse, 1987).

Several needle nematodes are associated with *Rubus* spp. *Longidorus elongatus*, *L. macrosoma*, and *L. attenuatus* are often encountered. They do not seriously damage root systems but have their greatest impact when they transmit viruses.

Symptoms

Xiphinema bakeri and *X. diversicaudatum* induce the formation of galls on raspberry and blackberry roots, which halts further primary and lateral root elongation. The feeding activity of *X. bakeri* causes swelling and curling of the root tip (McElroy, 1991). The resulting reduced root system cannot support good cane growth and plants begin to decline significantly. *Xiphinema americanum* does not cause any direct damage of the root systems of *Rubus* spp.

Life Cycle and Epidemiology

Dagger and needle nematodes are ectoparasites that feed by penetrating deep into the root tissue with their long stylets and, in the process, transmit viruses (Fig. 27–1). Their life cycle is similar to that described in Chapter 16 (Windham, 1998) for longidorid nematodes. The adults can live for 2 to 3 yr (McElroy, 1991). They reproduce mostly parthenogenetically. Dagger and needle nematodes are native to raspberry and blackberry plantations and acquire nepoviruses naturally harbored by weeds present during their feeding activities.

Other Nematodes

Other nematodes of the genera *Meloidogyne, Hemicycliophora, Criconema* and *Criconemella* have been found in association with *Rubus* spp., but their pathogenicity has not been confirmed. *Meloidogyne hapla* infects 'Candy' and 'Newburgh' raspberry, but the only significant effect is increased crown gall incidence.

NEMATODE-RELATED DISEASES OF STRAWBERRY

Root-Knot Nematodes

Species

The major species of root-knot nematodes on strawberry is *Meloidogyne hapla*. It is found on strawberry in southern areas such as California and Louisiana, but is mostly tolerant of low temperatures and is widely distributed in the northern two-thirds of the USA. *Meloidogyne hapla* causes the most serious nematode problems in strawberry in the mid-Atlantic regions and is found in most plantings in Queensland and Korea (Choo et al., 1987; Raski & Krusberg, 1984).

Meloidogyne javanica was found on strawberry in Israel imported from Japan. This nematode can cause severe galling. It also was found on strawberry grown in Zambia. Severe damage by *M. incognita* was observed in Florida. *Meloidogyne thamesi* caused root-knot in field strawberry in Florida (Nemec, 1984).

Symptoms

Root-knot nematodes cause generally small, variously shaped galls on the roots of strawberry plants. Symptomatology associated with root-knot nematode infection is described in an earlier section of this chapter and Chapter 15 (Kinloch, 1998).

Life Cycle

The life cycle of root-knot nematodes has been described in Chapter 15 (Kinloch, 1998). Damage caused by these nematodes in strawberry is more severe in sandy soils, than in heavy soils, and during periods of summer drought (Nemec, 1984). Very few studies have characterized interactions between root-knot nematodes and other pathogens. Meagher and Jenkins (1970) reported on the interaction between *M. hapla* and *Verticillium dahliae* and more recently, Szczygiel and Profic-Alwasiak (1989) identified the interaction of *M. hapla* with *V. dahliae, Fusarium oxysporum*, and *Rhizoctonia fragariae*.

Economic Importance

Although reliable estimates have not been published, root-knot nematode damage can prevent sale of plants and reduce berry yields.

Dagger and Needle Nematodes

Species

Dagger and needle nematode transmit nepoviruses that cause important diseases in strawberry (Fig. 27–2). Several species are known to cause direct damage to strawberry, viz., *X. americanum, X. diversicaudatum, X. chambersi, X. bakeri, X. index, X. brevicolle* and *L. elongatus.*

Xiphinema americanum is the principal species occurring on strawberry in North America. *Xiphinema diversicaudatum* is widely distributed on strawberry plants mostly in Europe and to a limited extent in the USA. *Xiphinema chambersi* has been reported on strawberry in the USA while *X. bakeri* is known in the USA and Canada. *Xiphinema* spp. are reported as parasites on strawberry in several other countries such as India, Israel, Sri Lanka, and Hawaii (Nemec, 1984).

Longidorus elongatus has been reported as a parasite of strawberry grown in the Netherlands, Poland, Belgium, Germany, Sweden, England and Canada.

Symptoms

Longidorids feed ectoparasitically and reach the vascular system by penetrating the root epidermis and up to five layers of cortical cells (Trudgill, 1976). *Xiphinema americanum* and *X. chambersi* cause small, reddish-brown, sunken lesions that eventually coalesce and become blackened. The root system is

Fig. 27–2. Chlorotic blotch on leaf of 'Cambridge Vigour' strawberry graft-inoculated with strawberry latent ringspot virus (after Murant & Lister, 1987).

reduced, plants appear stunted, and runner production is reduced. *Xiphinema bakeri* and *X. diversicaudatum* induce the same symptoms in addition to distortion (curling) and galling of the strawberry root tips. The malformations induced by *L. elongatus* are similar to those associated with the feeding of *X. diversicaudatum*.

Life Cycle and Epidemiology

Strawberry production is significantly affected in the 2nd or 3rd yr reflecting a very slow decline. This is related to the relatively long life cycles of longidorid nematodes that may take 22 d to 1 yr from egg to adult (Cotten, 1975; Wyss, 1970). All stages occur in the soil close to the roots. However, soils high in organic matter and extremes of soil wetness or dryness severely affect migration and survival of longidorid nematodes (Nemec, 1984). In heavy clay soil, inoculated populations of *L. elongatus* decline tremendously. These factors also affect the efficiency of virus transmission of these nematodes. *Longidorus elongatus* is known to interact with the fungus *Rhizoctonia fragariae* in affecting strawberry growth (Szczygiel & Profic-Alwasiak, 1989). This interaction might be highly environment-dependent.

Economic Importance

No reliable estimates of the direct damage caused by longidorid nematodes on strawberry production have been reported. However, these nematodes can cause severe damage in strawberry directly and by transmitting virulent nepoviruses.

Root-Lesion Nematodes

Species

Several species of lesion nematodes have been reported in association with strawberry root diseases viz., *Pratylenchus penetrans*, *P. coffeae* (Fig. 27–3), *P. vulnus*, *P. zeae*, *P. pratensis*, and *P. scribneri* (Adams & Hickman, 1970; Chikaoka, 1970; Nemec, 1984). *Pratylenchus penetrans* is the most common, and probably the most important, nematode associated with root diseases of strawberry (Nemec, 1984; Potter & Dale, 1991). Lesion nematodes have been found in association with both wild and cultivated strawberry. Most weeds in strawberry plantations could be infested with lesion nematodes. The interaction of lesion nematodes and fungi on strawberry have been studied (Szczygiel & Profic-Alwasiak. 1989).

Symptoms

Chen and Rich (1962) first demonstrated the pathogenicity of *P. penetrans* to strawberry. In the field, plants infected with root-lesion nematodes show root systems with brown lesions that develop into a black necrosis on most of the roots. Blackening results from the oxidation of phenolic compounds within the

Fig. 27–3. Damage to strawberry plants caused by *Pratylenchus coffeae* (courtesy of Robert D. Riggs, Univ. of Arkansas).

lesions. Plants are stunted, produce shorter, more erect petioles and produce fewer runner plants.

Adult and juvenile nematodes and eggs are found within the cortical tissue of the roots. Penetration of the roots by these nematodes leaves open entry points for fungi that can easily invade and increase the severity of black root-rot disease in strawberry.

Epidemiology

Lesion nematodes have been associated with black root rot of strawberry, a disease complex involving *P. penetrans* and soilborne pathogens such as *Rhizoctonia, Verticillium, Fusarium*, and/or *Pythium* spp. (LaMondia & Martin, 1989; McKinley & Talboys, 1979; Nemec, 1984; Raski & Krusberg, 1984). The nematodes tend to leave the root tissue after it has been invaded by root rot fungi and bacteria. Even Verticillium wilt-resistant cultivars such as Redgauntlet showed some damage in the presence of *P. penetrans*. On both 'Senga Sengana' and *Fragaria vesca* L., a synergistic effect was observed when *Verticillium albo-atrum* and *P. penetrans* were inoculated together (Muller, 1973).

Research in Poland showed that seedlings were more severely affected than runner plants and that older seedlings and runner plants were more susceptible than the younger seedlings or runner plants (Szczygiel, 1986). Szczygiel et al. (1983a) reported a very low damage threshold for *P. penetrans* on strawberry in Poland (viz. <500/kg of soil). The presence of root-rot fungi lowered the limit by about half. In Poland, strawberry plants are more tolerant of *P. penetrans* in the spring than in the fall. Lesion nematode populations are lowest in early spring and reach a peak in late summer.

Pratylenchus penetrans development is favored by acidic soils, which could increase the incidence and severity of root damage to strawberry roots (Szczygiel et al., 1983a). Pathogenesis is enhanced at lower moisture levels and in the lightest (coarse-textured) soils (Szczygiel & Soroka, 1983; Szczygiel et al., 1983b). The activity of *P. penetrans* is adversely affected in soils with high organic matter content (Szczygiel & Zepp, 1983).

Economic Importance

Estimates of losses in dollars are difficult to obtain because of the interaction of lesion nematodes and fungal root pathogens. *Rhizoctonia* alone can cause 30 to 38% root rot, whereas nematode colonization can increase root rot to 70 to 82% at 24°C. Together the lesion nematodes and fungi reduced the length of lateral and main roots by up to 40% and feeder roots by ≤60% (LaMondia & Martin, 1988, 1989).

Although no reliable estimates of economic losses due to root-lesion nematode damage are available, Wolfe et al. (1990) conducted a study to determine if the cost of fumigation would be recovered in the 1st yr of strawberry production when yield potential was limited by the root-rot complex. Although the cause of black root rot was not determined in the study, soil fumigation with MC33 [67% methyl bromide + 33% chloropicrin (trichloronitromethane)] or MC 2 (98% methyl bromide + 2% chloropicrin) increased yields enough to justify its use economically. Methyl bromide, however, has been identified as an environmental pollutant.

Other Nematodes

Other species mostly in the genera *Belonolaimus, Nothotylenchus, Criconemella, Helicotylenchus, Pratylenchus, Rotylenchus* and *Tylenchorynchus* have been reported as associated with strawberry on several continents (Birchfield et al., 1978; Kir'yanova & Krall', 1980; Kondo & Ishibashi, 1977; Scotto la Massese, 1973; Szczygiel & Hasior, 1972).

Although the importance of these nematodes are not fully understood, sting nematodes *Belonolaimus gracilis* and *B. longicaudatus*, both of limited distribution, are of economic importance on strawberry in the southeastern USA but mainly in Florida. These nematodes are ectoparasites and feed mainly on the very young feeder roots, which die in the process. When the fine roots die, only old, coarse roots remain. Plants decline, leaf buds are not produced and death results. During severe infection, the edges of older leaflets turn dark brown and, with the darkening progressing toward the midrib, leaves eventually die.

Both fruiting and nursery plants can be affected. Field infestations cause damage that appear as spots, which slowly enlarge and coalesce. The disease is most severe on plants growing in light sandy soils and in fields that have been continuously planted to strawberry or vegetables in soils of low humus content.

Most control methods used for ectoparasitic nematodes are effective against sting nematodes on strawberry. Soil fumigation with thionazin [*O,O*-diethyl-*O*-diethyl-(2-pyrazinyl) phosphorothioate], phorate [*O,O*-diethyl *S*-(ethylthio)methyl phosphorodithioate], DBCP (1,2- dibromo-3-chloropropane), ethylene dibromide (1,2-dibromoethane) and demeton {phosphorothioic acid *O,O*-diethyl *O*-[2-(ethylthio)ethyl] ester mixture with *O,O*-diethyl *S*-[2-(ethylthio) ethyl]phosphorothioate} suppressed *B. longicaudatus* populations throughout the crop cycle resulting in improved strawberry yields (Nemec, 1984). Most fumigants and non-fumigants have been banned or restricted. In the southern USA, high temperatures can be effective during summer fallow.

In Japan, the most common nematodes attacking aboveground parts of strawberry are *Nothotylenchus acris* and *Aphelenchoides fragariae* (Kondo & Ishibashi, 1977).

Aphelenchids Causing Spring Dwarf

Species

Spring dwarf is caused by *Aphelenchoides fragariae* or *A. ritzemabosi*. In Europe, spring dwarf can be caused by either species whereas *A. fragariae* is the most common cause in the USA. *Aphelenchoides fragariae* is important on strawberry in the eastern regions of Russia, Holland, southern Poland, India, and along the East Coast from New England to Georgia. It is the most important nematode pest on strawberry grown in coastal regions of California. *Aphelenchoides ritzemabosi* is reported on strawberry in Poland, the Netherlands and the USA (Raski & Krusberg, 1984).

Symptoms

Spring dwarf nematodes feed ectoparasitically on the surfaces of buds and leaflets of growing points. They cause the greatest damage early in the growing season when leaves and flowers that expand from infected buds are stunted, deformed and crinkled. As a consequence, little or no fruit production occurs. The hot summer weather lowers the nematode populations, and plants appear to produce healthy looking runners. However, symptoms appear the following spring since these runners are infected. Some plants are killed by the feeding activity of these nematodes.

Aphelenchoides fragariae and/or *A. ritzemabosi* occasionally interact with pathogenic bacteria (*Rhodococcus fascians* = *Lorynebacterium fascians*) and cause cauliflower disease in strawberry. The crowns of diseased plants are stunted with deformed fleshy rosettes of flowers, leaves and stems that resemble tiny "cauliflowers" (Crosse & Pitcher, 1952). Both the nematode and bacterium must be present for induction of cauliflower disease symptoms. Cauliflower disease has not been reported on strawberry in North America (Nemec, 1984).

Epidemiology

Spring dwarf nematodes depend upon a thin film of moisture (from rain, dew or high humidity) for infection and movement. These nematodes do not persist long enough in soil to spread the disease from old to new strawberry plantings. *Aphelenchoides fragariae* persisted in strawberry residue for not more than

90 d (3 mo) under optimum conditions (Szczygiel & Hasior, 1971). The disease is transmitted through infected runner plants. The spring dwarf nematodes, however, can survive winters with subzero temperatures and hot summers, but not for prolonged periods.

In cauliflower disease of strawberry, the nematode involved is essential as a vector that carries *R. fasciens* from infected parts of the same plant or from the soil to the apical meristem. The metabolites produced by the bacteria interfere with development of the apical meristem. The nematode is considered an incitant since mechanical inoculation with the bacterium does not induce cauliflower disease symptoms. *Rhodococcus fasciens* is a ubiquitous epiphyte on many plants. Only a few strains can induce cauliflower-type symptoms and, in their absence, only alaminate leaves appear. *Aphelenchoides* spp. are bisexual nematodes. They reproduce and develop within the buds of the strawberry plants.

Economic Importance

As with virtually all nematode diseases, very few reliable estimates of losses or damage are available. Two years of tests in Poland showed a 32 to 61% yield reduction in seven strawberry varieties parasitized by *A. fragariae* and *A. ritzemabosi* (Szczygiel, 1967). *Aphelenchoides fragariae* could be difficult to eradicate if it gets established in nuclear stock (McNamara et al., 1986).

Summer Dwarf Nematodes

Species

Summer dwarf of strawberry is similar to spring dwarf but is caused by the nematode *Aphelenchoides besseyi*. The disease is distributed mostly in southern regions of the USA up to Arkansas and southern Virginia and southern California, but can survive as far north as Delaware and southern Illinois. It has been reported in North Carolina, Tennessee, Massachusetts, New York, Kentucky, Maryland and Pennsylvania. *Aphelenchoides besseyi* is known in other parts of the world. In Australia, severe losses can occur (McCulloch, 1978). It also has been reported from continental Europe and Great Britain.

Symptoms

Aphelenchoides besseyi induce symptoms similar to spring dwarf on strawberry. The nematode inhabits spaces in the leaf and runner buds and feeds externally on leaflet surfaces. The feeding, excretory and reproductive activity of the nematodes causes the growing points to develop abnormally. Developing leaves appear crinkled, twisted, glossy, stiffer and darker green than normal. Deformed leaves are reduced in size and are sometimes elongate with shortened petioles. The main bud is killed, and multiple crowns are produced. Malformed leaves appear in the center of the plant and are surrounded by unaffected ones. In cooler fall weather expanding leaves may appear normal. Most runner plants are infected, and infected flower buds may produce deformed fruits.

Epidemiology

The life cycle of *A. besseyi* takes about 2 wk for completion. The sexes are similar in shape, and juveniles resemble adults. Development is similar to that of spring dwarf nematodes. In California, *A. besseyi* can survive winter and adverse conditions in soil and plant debris (Raski & Krusberg, 1984). *Aphelenchoides besseyi* requires a thin film of moisture for movement and infection.

Summer dwarf of strawberry differs from spring dwarf in that leaf symptoms in the former develop during summer and early fall. The nematodes are present in buds in spring but only build up to destructive levels in warm weather when temperatures are >23°C. Unlike spring dwarf nematodes, *A. besseyi* can persist long enough to spread the summer dwarf disease from an old to a new planting within a year. In both diseases, the most common means of spread is from infested runner plants. Infestations are most severe during the rainy season and in low areas of infested fields.

Economic Importance

Summer dwarf is important on strawberry grown in several areas. Up to 50% yield loss was recorded in Australia (McCulloch, 1978). In the USA, although few reliable estimates have been published, up to 75% loss has been recorded for southern regions (McElroy, 1972). However, persistent roguing and selection have lessened the impact of this disease in the USA.

Bulb and Stem Nematodes

Species

The bulb and stem nematode *Ditylenchus dipsaci* is a serious pest of strawberry in many countries including parts of Holland, Italy, Russia and the USA where it is restricted to the northern Pacific Coast (Kir'janova & Krall', 1980). It occurs sporadically in Poland but is not economically important there.

The species parasitizes hundreds of plants including many bulb crops all over the world. It is a serious pest of clover (*Trifolium* spp.). The strawberry race of *D. dipsaci* can attack red clover (*T. pratense* L.) and the red clover race can attack strawberry. Four races of *D. dipsaci* from oat (*Avena sativa* L.)-onion (*Allium cepa* L.)-rye (*Secale cereale* L.), red clover, narcissus (*Narcissus* spp.), and teasel (*Dipsacus fullonum* L.) also can attack strawberry. However, at least six physiological races attack strawberry.

Symptoms

Ditylenchus dipsaci feeds endoparasitically and can be found in high numbers inside foliage tissues in the swollen parenchyma. Softening and eventual breakdown of tissues occur. Symptoms appear as swollen petioles, distorted leaves, gall-like swellings on leaf bases, peduncles, fruit clusters and flowers, stunting and poor yields. Red, yellow or brown spots appear in the foliage of infected strawberry plants. Certain symptoms resemble those induced by other foliar nematodes on strawberry making accurate identification of the nematode species necessary.

Life Cycle and Epidemiology

The disease is most severe in cool, wet, spring weather (McGrew, 1978) as exists in the Pacific Northwest. Symptoms usually appear on plant parts, developing early in the growing season. The nematode can be found ectoparasitically on crowns of strawberry, and under wet conditions, it enters and feeds endoparasitically inside young leaf and stem tissues.

Economic Importance

Severe infestations can result in entire loss of strawberry plantings. Yields can decrease 85% in strawberry plantings in eastern Russia (Szczygiel, 1977). However, losses due to foliar nematode infestations of strawberry are more severe in places while persistent roguing and selection are not rigorously practiced, for example, in certain areas of Russia and Europe. Few reports are available that quantify the economic importance of the damage induced by foliar nematodes.

NEMATODE PARASITES OF CRANBERRY

Stubby-Root Nematode

Species

Stubby-root nematode (*Paratrichodorus* spp.), although not the most prevalent nematode found around roots of the American cranberry (*Vaccinium macrocarpon* Aiton), is the most serious pathogen. *Paratrichodorus minor* (syn. *Paratrichodorus christiei, Trichodorus minor, T. christiei)* or *P. nanus* were found in Massachusetts in 42% of the bogs sampled. In New Jersey, *P. minor* occurred in 24% of bogs sampled while small numbers of *T. californicus* were found in two of the bogs sampled in Wisconsin.

Symptoms

Pathogenicity experiments showed that growth of lateral roots was halted, resulting in the typical stubby-root symptoms in inoculated plants (Fig. 27–4). Comparison of nematode-infected plants with uninfected showed drastic suppression in root (Fig. 27–5) and seedling growth. A later study in New Jersey, reported high numbers of *P. minor* in association with a dieback disease of cranberry, but no correlation was found between the nematode and expression of disease symptoms (Myers, 1991).

Population Dynamics

The population dynamics of *P. minor* were studied in Massachusetts bogs (Zuckerman et al., 1964). Although the numbers were low in the bogs selected for sampling, populations were consistently highest in November in 2 yr..

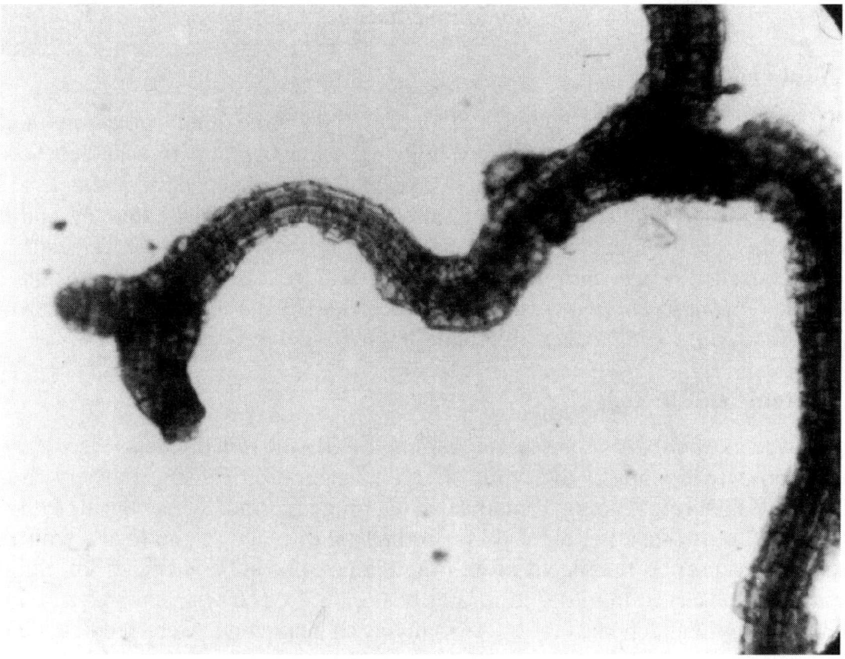

Fig. 27–4. 'Stubby root' caused by the feeding of *Paratrichidorus minor* on cranberry roots (after B. M. Zuckerman; *Nematologica* 6:135–143).

Fig. 27–5. Cranberry seedlings from a greenhouse trial. Those to the left were from soil infested with *Paratrichodorus minor*, as compared to seedlings from uninfested soil (right) (courtesy of B.M. Zuckerman).

Sheath Nematodes

Causal Organisms

Sheath nematodes (*Hemicycliophora* spp.) are found in sandy soils throughout the world, and cranberry bogs are no exception. In Massachusetts, 72.5% of the bogs sampled contained several species of sheath nematodes, with *H. similis*, *H. gracilis* (*H. vaccinii*), and *H. uniformis* the most common, and *H. gigas* the rarest. Other species observed in bogs are *H. zuckermani* in Massachusetts, *H. vaccinii* in New Jersey, and *H. typica* and *H. obtusa* in Wisconsin. *Hemicycliophora similis* and *H. gracilis* exist in bogs in New Jersey. *Hemicycliophora similis* also was found in Oregon (McKewan, 1979).

Symptoms and Biology

Hemicycliophora similis causes small galls on roots (Zuckerman 1961), and root growth stops at the point of stylet penetration causing root curvature (Fig. 27–6). During feeding a small adhesive plug is formed where the stylet penetrates the root (Kisiel et al., 1971). Feeding takes place at one locus, with the stylet often inserted to a depth of half the diameter of the cranberry root. Single nematodes observed in vitro fed for periods up to 48 h. *Hemicycliophora similis* can molt while attached to roots. As many as 16 nematodes were seen clustered

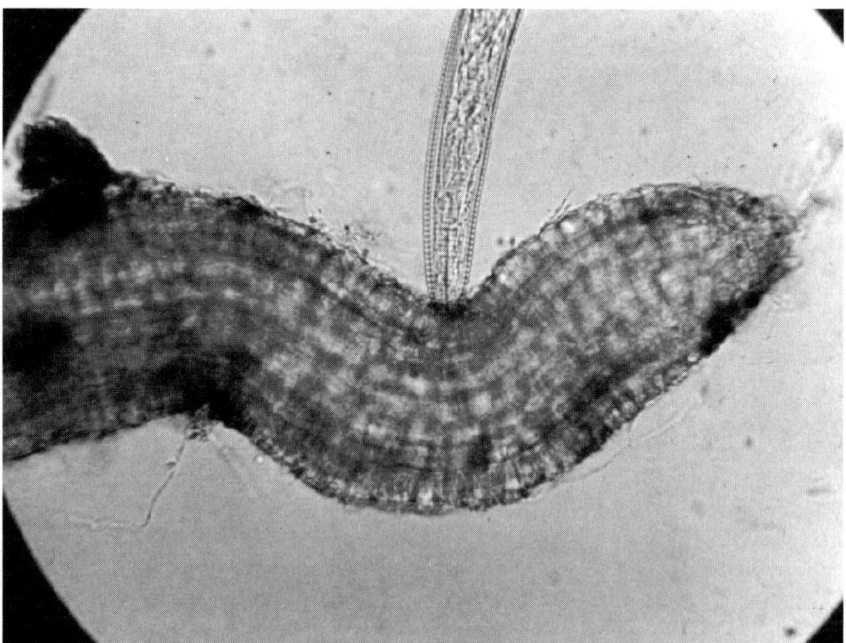

Fig. 27–6. *Hemicycliophora similis* feeding on cranberry root. The curvature of the root shows that cells around the point of penetration of the stylet had stopped growing (×200) (after B. M. Zuckerman, Nematologica 6:135–143).

and feeding on a single root tip. This activity is associated with a significant decrease in overall root growth. Populations of *H. similis* increased on cranberry in Massachusetts, New Jersey, and Oregon (McKewan, 1979).

Epidemiology

Population dynamics of *H. similis* on cranberry were studied throughout the year in Massachusetts. Populations were highest in the spring and fall and lowest when bogs were flooded during the winter months. In New Jersey, *H. similis* persisted at the same levels in flooded model cranberry bogs as they did in the field under similar conditions. In Massachusetts, *H. similis* migrated downwards during a drought period in July (63.9% of the population was found at a 1.2-m depth or above), whereas 48 h after irrigation, 88.1% of the population was found above 1.2 m in depth. Normal root depth in Massachusetts bogs was about 0.9 m, suggesting a decrease in nematode damage during drought periods.

Other Nematodes

Several other nematode species have been studied on cranberry. *Tetylenchus joctus* (syn. *Merlinius joctus*) and *T. productus* were found in 21.9% of the bogs sampled in Massachusetts. *Tetylenchus joctus* also was reported from New Jersey. Parasitism to cranberry seedlings grown in vitro, as well as minor pathogenic symptoms, also were described, but little else is known of the biology of this species on this crop. *Tylenchorhynchus* sp., mostly *T. claytoni* were found in 16.3% of the bogs in Massachusetts and *T. dubius* from New Jersey. *Tetylenchus claytoni* also fed on cranberry roots in vitro and caused collapse of root epidermal cells.

Helicotylenchus spp, mostly *H. erythrinae*, occurred in 17.5% of the bogs examined in Massachusetts. *Helicotylenchus pseudorobustus* was reported from 20% of the area surveyed in Wisconsin and *Helicotylenchus* spp. from New Jersey. Other stylet-bearing nematodes found with some frequency in cranberry bogs were *Criconemella xenoplax* and *Criconemella* spp., *Hoplolaimus* spp., and *Paratylenchus arenarium. Meloidogyne* spp. and *Pratylenchus* spp., reported infrequently in surveys from New Jersey and Massachusetts, were probably from roots of weeds in the bogs (since neither root-knot nor endoparasitic nematodes have been recovered from cranberry roots). The report of a *Xiphinema* sp. from cranberry probably also represents an unusual occurrence.

Atylenchus decalineatus was considered widespread on cranberry in New Jersey and Wisconsin. Nothing is known about the biology of this nematode, except that it failed to reproduce on cranberry.

The sole report on nematode-fungus interactions was that of Myers (1991) on cranberry dieback disease. High numbers of *Hemicycliophora ritteri* and *P. minor* were associated with diseased vines; however, no correlation was found between nematode numbers, yields of berries and disease ratings. Myers (1991) concluded that nematodes apparently were not involved in the etiology of the disease.

NEMATODE PARASITES OF BLUEBERRY

Stubby-Root Nematodes

Species

The most virulent nematodes on blueberry (*Vaccinium corymbosum* L.) are *Paratrichodorus* spp. *Paratrichodorus* spp. associated with blueberry are *P. minor* and *Trichodorus aequalus* in New Jersey, *P. minor* in Michigan and Massachusetts, and *Paratrichodorus* spp. in Massachusetts, North Carolina, Oregon, Michigan and Arkansas (Clark & Robbins, 1994; Converse & Ramsdell, 1982; Milholland et al., 1974; Ramsdell & Myers, 1974).

Symptoms

In in vitro and greenhouse trials, *P. minor* caused typical stubby-root symptoms on blueberry. It specifically stopped root growth and developed enlarged tips. Blueberry cuttings inoculated in the greenhouse with *P. minor* had an 82% reduction in root growth. Observations on blueberry rooting beds in New Jersey showed a correlation between high numbers of *Paratrichodorus* spp. and the failure of cuttings to root.

Tetylenchus

Tetylenchus joctus (syn. *Merlinius joctus*) are frequently found in high numbers in blueberry fields. Surveys reported *T. joctus* from all blueberry-growing areas in the USA. The frequency with which this species is found in association with blueberry led to the speculation that *T. joctus* injury contributed to the stunting of blueberry plants.

In Massachusetts, studies of *T. joctus* showed that the nematode fed on blueberry roots grown in agar plates and reproduced on cuttings grown in the greenhouse. However, no easily detectable root symptoms were observed nor was root growth reduced in greenhouse trails. Detailed studies on the biology of *T. joctus* described the life cycle, and support the conclusion that this nematode is not a serious plant pathogen (Norgren et al., 1968).

Sheath Nematode

Sheath nematodes (*Hemicycliophora* spp.) were found in moderate to high numbers around blueberry roots in several states. Surveys recorded *H. gracilis, H. similis*, and *H. uniformis* in New Jersey; *H. uniformis* and *H. similis* in Massachusetts; *Hemicycliophora* spp. in Michigan; *H. typica* in North Carolina; and *H. zuckermani* in Arkansas (Clark & Robbins, 1994; Milholland et al., 1974).

In Massachusetts, parasitism by *H. similis* was observed on blueberry roots grown in petri plates. Small root galls and reduced root growth occurred in greenhouse pot trials on cuttings, leading to the conclusion that studies on commercial cutting beds could be useful. Studies in New Jersey suggested that nematicides would control the problem of reduced growth caused by nematodes in commercial nursery beds.

Nematode Parasites of Lesser Importance

Other nematode species associated with blueberry are considered as potentially harmful pathogens, but only a limited amount of experimental work has proceeded on this subject. Since awl nematodes were found in low numbers in only 2 of 10 blueberry fields in Massachusetts, the possibility of their being serious pathogens of blueberry is minimal. Transmission electron microscope studies of *D. heterocephalus*, revealed what appeared to be a phenomenon of very rare occurrence in nematodes; viz. the presence of intracellular viruslike particles (Fig. 27–7) within an epithelial cell (probably gonad epithelium). This is the only known case suggesting cellular virus infection in plant-parasitic nematodes (Zuckerman et al., 1973). No attempt was made to isolate and characterize the viruslike particles. The suggestion of possible viral diseases of nematodes remains intriguing.

Root-Knot Nematode

Meloidogyne spp. have been rarely associated with blueberry, and have been observed on only a few occasions in the northeastern USA. These observa-

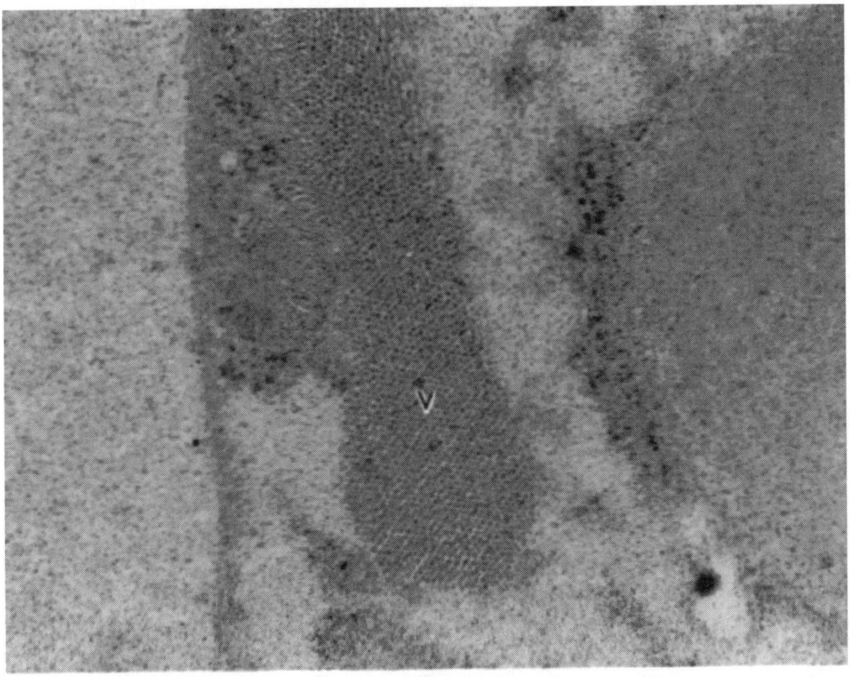

Fig. 27–7. Polyhedral virus (V) within an intestinal epithelial cell of *Dolichodorus heterocephalus* from blueberry (after B. Zuckerman et al.; *Nematologica* 19:117).

tions probably represented infections of weed roots in the samples. Highbush blueberry is resistant to root-knot nematodes. However, a new species of root-knot, described as *M. carolinensis*, was found parasitizing blueberry in North Carolina (Eisenback, 1982). The economic importance of *M. carolinensis* to blueberry production is not described. *Helicotylenchus* species have been reported on blueberry in Arkansas (Clark & Robbins, 1994).

NEMATODE PARASITES OF GOOSEBERRY AND CURRANTS

There have been few reports on nematode diseases of gooseberry and black and red currants over the last two decades. Several viruses, mostly nematode-vectored, are of little economic impact because of their restricted occurrence.

Bud and Leaf Nematode

The only nematode disease reported with some degree of importance is caused by *Aphelenchoides ritzemabosi* (Harmat et al., 1990). It occurs in New Zealand and Europe and affects mainly the buds of currant and gooseberry.

NEMATODE-TRANSMITTED VIRUS DISEASES

A summary of the nematode-related viral diseases associated with small fruits is given in Table 27–1. The reader is referred to more detailed treatment of the different diseases in the references cited and a recent overview by Taylor and Brown (1997).

Grape

Among the major viruses of economic importance in grape production are the following nepoviruses: GFLV, tomato ringspot (ToRSV), tobacco ringspot (TRSV) and peach rosette mosaic (PRMV) viruses (Ahmedullah & Himelrick, 1990). Cultivars sensitive to grapevine fanleaf degeneration could suffer up to 80% losses because of low fruit quality and yield, shortened productive life of vineyards and decreased resistance to adverse environmental conditions. Three distinct reactions to the virus are known (Table 27–1). Fruit set is poor and yield could be almost zero. In both the second and third syndromes, little or no malformations occur. Apart from one recently discovered natural variant, all GFLV isolates are serologically uniform (Martelli & Savino, 1988). Dagger nematodes cannot be eradicated in established vineyards.

Tomato ringspot virus decline symptoms are more severe in colder regions (Gonsalves, 1988b). Despite the injury caused by the virus, infected vines often survive more than 3 yr although unproductive. This nepovirus has a very wide host range, which includes many common vineyard weeds. The natural vectors (*Xiphinema* spp.) can retain the virus for very long periods and can acquire it from feeding on infected weeds and grapevines. The virus can be introduced to a

Table 27–1. Major small fruit viruses as related to hosts, vectors and managment options.

Virus (Acronym)	Modes of transmission	Distribution	Important hosts	Symptoms	Tested management tactics	References
Grapevine fanleaf nepovirus (GFLV)	*Xiphinema index, X. italiae* pollen, sap, infected stock	Worldwide	Grape	Type I: (spring) malformations, fewer branches, poor fruit set; Type II: (spring) chrome yellow mosaic; Type III: (summer) yellow vein banding/ flecking; also, trabeculae in vascular and epidermal cells, inclusions and cell-wall outgrowths	Use of virus-free stock, nematode- or virus-resistant stock, cross protection with ArMV, vector control	Martelli & Savino (1988), Walker et al. (1989)
Tomato ringspot nepovirus (ToRSV)	*X. americanum, X. californicum, X. rivesi;* seedborne, infected stock, grafting, sap	Northeastern USA, Lake Ontario, British Columbia, California, Oregon, Pennsylvania, Washington, eastern Siberia	Grape, red raspberry, blackberry, strawberry, highbush blueberry, elderberry, currant, cherry [*Prunus avium* (L.) L.], peach [*P. persica* (L.) Batsch], apple (*Malus domestica* Borkh.), vineyard weeds	In northeastern USA (Year 1): few leaves mottled in oak leaf pattern; Year 2: weak new growth, small distorted leaves, stunted growth, berry plants sparse and uneven; Year 3: stunted growth in surviving suckers; strawberry: symptomless or stunting, leaf mottling, restricted runner production, crumbly fruit, leaf/plant death; raspberry: new sucker/foliage growth delayed, bronzing in new foliage, yellow rings, vein chlorosis, line patterns, symptoms not clear in summer but weakening and stunting clear; blackberry: decreased fruit yield, yellow veins, oakleaf patterns, yellow spots blotches at leaflet bases; blueberry: asymptomatic or cupped malformed leaves with circular spots, mottled young leaves, necrotic spots on canes, poor fruit quality and yield, severe defoliation, plants die; red currant: <21°C, chlorotic spotting and vein banding, >27°C new growth asymptomatic, 21 to 27°C, only chlorotic spots and rings on young leaves	Working soil 1 yr before planting, weed control, virus-free cuttings, resistant cultivars or rootstocks, preplant fumigation, avoiding vector-infested soils, roguing and fumigation of infection foci areas	Gonsalves (1988b), Converse (1977, 1991, 1995), Stace-Smith & Converse (1987), Forer & Stouffer (1982), Williams et al. (1987)

(continued on next page)

Table 27–1. Continued.

Virus (Acronym)	Modes of transmission	Distribution	Important hosts	Symptoms	Tested management tactics	References
Tobacco ringspot nepovirus (TRSV)	*X. americanum, X. rivesi*(?), sap	Northeastern USA, New Jersey, Michigan, Arkansas, Oregon, Illinois	Grape, weed hosts of ToRSV, *Plantago major*, highbush blueberry, weeds	Similar to ToRSV; blueberry: slow decline, misshapen and rumpled leaves with small necrotic spots that may fall out, small necrotic spots on twigs, or small rosetted leaves, or stem dieback; reduced fruit yields, infection foci widening at about 30 d/yr	Similar to ToRSV; "clean fields," fumigation, virus-/vector-free stock, roguing	Gonsalves (1988b), Meer (1987b)
Peach rosette mosaic nepovirus (PRMV)	*X. americanum* (tentative for blueberry), *Longidorus diadecturus,* grape pomace, sap, seedborne	Michigan, Ontario	Grape, peach, highbush blueberry, perennial weeds	Circular pattern of dead vines expanding at about 30 d/yr, internode crooked and shortened, leaves puckered and misshapened, berry fall off; blueberry: strap- or cresent- or spoon-shaped leaves, no chlorosis or necrosis	Removal of infected plant parts, working soil in summer, soil fumigation in autumn, sampling for vector at new sites, virus-free planting stock	Ramsdell (1988, 1995b), Ramsdell et al. (1995), Ramsdell & Myers (1974)
Arabis mosaic nepovirus (ArMV)	*Xiphinema diversicaudatum,* sap, weed seeds, infected stock	Europe	Raspberry, blackberry, strawberry, black and red currant, gooseberry, weeds, crops	Raspberry yellow dwarf: mottling, yellow speckling, vein yellowing or down curling of leaves, stunting, near-zero fruit production; strawberry: similar to SLRSV; yellow mottle of black currant (*Ribes nigrum* L.): reduced fruit size, bushes and vigor	Virus- and vector-free stock, "clean field", resistant cultivars, preplant soil fumigation, weed control, crop rotation	Forest & Stouffer (1982), Murant (1987)
Strawberry latent ringspot nepovirus (?) (SLRSV)	*Xiphinema diversicaudatum,* sap, weed seeds, infected stock, soilborne	Europe	Raspberry, strawberry, red currant, weeds, many crops	Similar to ArMV; strawberry: stunting, leaf mottling, mosaics, crinkling, twisting or cupping, death within 1–2 yr, symptomless in midsummer but progressively stunted and devitalized	Similar to ArMV	Forer & Stouffer (1982), Ramsdell (1995b)

Raspberry ringspot nepovirus (RpRSV)	*L. elongatus, L. macrosoma*, sap, weed seeds, infected stock, soilborne	Europe, Russia	Raspberry, gooseberry, (Fig. 27–8), red currant, strawberry, weeds, crops [e.g. clover (*Trifolium* spp.)]	Raspberry leaf curl: chlorotic ring spots which disappear during hot summer months, down curling of brittle leaves, dwarfing, reduced vigor or asymptomatic; strawberry: angular chlorotic spots, rings, streaks, stunting death, or new leaves symptomless; gooseberry deterioration: asymptomatic or indistinct mosaic patterns, late berry ripening, small misshapened berries, death in few years; spoon leaf of red currants: bright yellow mosaics, ringspots, leaf deformations with indentations and margin curling	Similar to ArMV, chemical decontamination, soil sterilization for strawberry	Szczygiel & Rebandel (1991), Meer (1987a,c)
Tomato blackring nepovirus (TBRV)	*L. elongatus, L. attenuatus*, sap, weed seeds, infected stock	Europe, Russia	Raspberry, strawberry, weeds, crops (e.g., clover)	Similar to RpRSV but milder, crumbly poor fruit	Similar to ArMV and RpRSV	Forer & Stouffer (1982)
Tobacco rattle tobravirus (TRV)	*Trichodorus primitivus, T. hooperi, T. similis,* sap	Netherlands	Red currants	Leaf pattern: "oakleaf" patterns on certain branches persisting for several years	Control mostly not necessary; use of virus-free stock and "clean fields"	Meer (1987b)
Minor viruses: grapevine chrome mosaic, artichoke Italian latent, grapevine Bulgarian latent (=blueberry leaf mottle) nepoviruses	Mostly *Xiphinema* and *Longidorus* spp.	Europe	Grape, wide host range	Various (nondistinct)	Nematode- and virus-free clones, heat treatments (38°C/≥60 d), soil fumigation, nonhost rootstocks	Stellmach & Goheen (1988)

new vineyard through the use of infected planting material and by seeds dispersed from infected weeds. Because of the wide host range and ease of seed transmission, ToRSV is very likely present in most land being prepared for grape production.

Tomato ringspot virus is generally the most economically important virus infecting fruit trees in the northeastern USA. In susceptible cultivars with desirable agronomic properties such as Cascade, ToRSV can cause fruit production to become uneconomical by the 2nd yr of the disease. Resistant cultivars or rootstocks give good control (Allen et al., 1982). All American (*V. labrusca*) type cultivars and rootstocks are resistant. The following hybrids are known to be susceptible: Cascade, Baco noir, DeChaunac, Chelois, Vidal blanc, Ventura, Dutchess, Siegfriedrebe and Vincent (Gonsalves, 1988b). Only the tolerant cultivars of *V. vinifera* are used in California where the disease is now probably absent. Rootstocks C1616, Clinton, Kober 5BB, 41B Millardet et de Grasset, 44-53 Malegue, 110 Richter, Shakoka and Sanona are reported resistant to graft inoculation (Gonsalves, 1988b) while SO4 and C3309, although susceptible to graft transmission, show good field resistance.

Tobacco ringspot virus is serologically distinct from ToRSV though they are vectored by the same nematode *X. americanum* (Gonsalves, 1988a). Because the disease has been confined largely to *V. vinifera* cultivars that are not grown extensively in northeastern USA, TRSV decline is of minor importance in that part of the country. *Vitis labrusca* is resistant to both TRSV and ToRSV.

Symptoms of PRMV decline could be confused with those caused by boron deficiency or grapevine fanleaf viral infection of *V. vinifera* cultivars. About 9 to 10% of seeds harvested from a diseased vineyard may carry the virus. After a 3- to 4-yr latent period, the disease spreads at a rate of approximately 1 m yr^{-1} to new grape plants. Under field conditions certain cultivars such as Couderc 1616 and 1202, Foch, Seyval and Delaware show good resistance to PRMV (Ramsdell et al., 1995; Ramsdell & Gillett, 1985) while cultivars Concord (Ramsdell & Myers, 1978), Niagara, Vignoles, Teleki 5A and 5C (Ramsdell et al., 1995) and Catawba are susceptible. Couderc 1616 and 1202 and Delaware supported the same population levels of *X. americanum* as Concord. Vine vigor decreases over several years, but the rate of decline may be increased by the effect of winter injury on diseased plants.

In Michigan, the PRMV decline is mainly limited to grape cultivars Concord and Niagara. Peach rosette mosaic virus can cause up to 50-fold losses in infected grapevines compared to healthy vines (Ramsdell et al., 1995). Concord is a popular juice grape and can suffer up to 40% yield loss and up to 60% growth reduction. French-American hybrids also are threatened. Soil fumigation gives effective control but is costly (usually >\$2000 ha^{-1}). Long fallow periods are not effective against PRMV decline since the vector can easily survive for several years on infected surviving vine roots. *Longidorus diadecturus* is not an important vector in Michigan (Ramsdell et al., 1995). Good rootstocks for grafting are important but susceptible cultivars (such as the juice types Concord and Niagara) include Couderc 1616 and the less cold-tolerant Couderc 1202. Excellent scion cultivars include Foch, Seyval and, to a lesser extent Chancellor and Delaware (Ramsdell et al., 1995; Ramsdell & Gillett, 1985).

Grapevines may have dual infections by the leafroll agent and one of several minor nepoviruses (Table 27–1). In Europe, seven nepoviruses in addition to GFLV have been identified serologically: raspberry ringspot virus, arabis mosaic virus, strawberry latent ringspot virus, tomato black ringspot virus, grapevine chrome mosaic virus, artichoke Italian latent virus and grapevine Bulgarian latent virus (= blueberry leaf mottle virus) (Stellmach & Goheen, 1988).

Raspberry

In Europe, four nematode-vectored viruses are known to cause major losses in raspberry. These are arabis mosaic (ArMV), strawberry latent ringspot (SLRSV), raspberry ringspot (RpRSV) and tomato black ring viruses (TBRV) (Table 27–1). None of those four viruses is found naturally in North America. Serologically different strains of these nepoviruses have been detected. ELISA and cDNA probes are the best tools for identifying ArMV, SLRS, RpRSV and TBRV. Vectors of ArMV and SLRSV can become viruliferous within 1 d and transmit the virus acquired within 3 d. In fallow soil the vector can retain either nepovirus for at least 90 d (3 mo). The RpRSV and TBRV are naturally transmitted by three needle nematode species, which can retain the viruses for up to 9 wk.

Strawberry

The most important nematode-vectored viral diseases of strawberry in Europe are caused by the four viruses: RpRSV, TBRV, ArMV, and SLRSV (Fig. 27–2). The RpRSV and TBRV strains are transmitted by the same needle nematode while the same dagger nematode is the vector for ArMV and SLRSV. Raspberry ringspot virus tends to be associated with another soilborne virus related to TBRV. The RpRSV and TBRV do not occur naturally in the Western Hemisphere where they can pose a very serious threat to the strawberry industry because of the extensive cultivation of susceptible cultivars. All common cultivars are susceptible to both RpRSV and TBRV except Huxley, which is immune to TBRV. In the field, outbreaks of the disease appear in patches a few square meters to several hectares in area.

Both RpRSV and TBRV, single or in combination, can cause strawberry production to become uneconomical within 1 to 2 yr of infecting a susceptible cultivar. This presents a serious problem in international exchange of planting stocks. However, in Scotland, because of the localized occurrence of RpRSV and TBRV, and the use of effective management strategies in strawberry production, the diseases are not of general importance. In Britain the spread of these viruses is effectively controlled.

Problems with such viruses also can stem from the fact that they induce few or no symptoms initially making it relatively easy to accidentally introduce the virus to a new area through planting stocks. In the Western Hemisphere, where susceptible cultivars of strawberry are widely grown, strict quarantine measures are taken to keep those viruses from being introduced.

Natural ArMV, SLRSV, RpRSV and TBRV disease outbreaks occur in patches reflecting the horizontal distribution of the nematode vectors in field soil.

'Alpine' strawberry infected with ArMV remains symptomless but shows loss of vigor. In general there seems to be few cultivars that are immune to ArMV.

Since newly infected strawberry plants are usually symptomless, accidental introduction of the virus into soils infested with the nematode can become problematic. Dagger nematodes, once established, become endemic. The ArMV and SLRSV are effectively controlled in Britain but are potential problems in the international movement of planting material. The two viruses can be so severe on certain strawberry cultivars that a crop can be made valueless within 1 to 2 yr of infection. These viruses are of serious economic importance mostly in the southern and southwestern parts of Great Britain where they are prevalent (Murant & Lister, 1987).

Converse (1981) reported an infection rate by ToRSV of 30% in strawberry after 19 mo. All juvenile stages of the nematode including adults can acquire the virus and can transmit it to a susceptible host within 1 h of feeding. Tomato ringspot virus can be eradicated in peach by incubating infected plants at 38°C for at least 21 d. However, eradicative measures have not been necessary in strawberry since no clones have been discovered that were completely infected by the virus (Converse & Stace-Smith, 1987). Several cultivars may be immune to leaf-graft infection (Converse & Stace-Smith, 1987) but the reactions of most strawberry cultivars to ToRSV infection via dagger nematodes are not known [though Converse & Stace-Smith (1987) suggested that immunity might be widespread].

The DNA probes developed for detection of the ToRSV worked well on herbaceous hosts but gave variable results with infected *Rubus* tissue (Converse, 1991). The ELISA, however, is most commonly used for strain detection in *Rubus* spp. (Bitterlin & Gonsalves, 1988). The vector nematode acquires the virus after feeding on an infected host and can transmit the virus for several months. In raspberry fields the virus spreads from plant to plant at a rate of about 2 m yr^{-1}. Experimental transmission of ToRSV has been achieved from diseased blueberry plants to other hosts (Johnson, 1972). However, direct transmission from nematode to blueberry has not been confirmed.

Blueberry

Necrotic ringspot (caused by TRSV) has been reported in highbush, but not in lowbush or rabbit-eye, blueberry (Ramsdell, 1995a). As with most viruses of small fruit, symptoms vary between cultivars and disease increase is slow. All infected cultivars become unproductive. Peach rosette mosaic has been observed mostly in experimental plantings of the highbush cultivars Jersey and Berkeley (Ramsdell, 1995b). Peach rosette mosaic virus, however, is a destructive disease of grape and peach. The epidemiology of PRMV on blueberry is not well studied, but control methods used in grape should be effective in blueberry production. In general, the best approach for control of small fruit viruses is prevention.

Gooseberry and Currants

Although available information is limited, gooseberry may be damaged when infected by the raspberry ringspot virus (Fig. 27–8). No transmission experiments have been conducted in *Ribes* species with ArMV, a pathogen on that crop. However, *X. diversicaudatum* is the confirmed vector of ArMV in other crops (Adams & Thresh, 1987b). Losses are not economically important because the infection is not widespread.

Other viruses that are presumed naturally transmitted by *X. diversicaudatum* are SLRSV and ArMV on red currant (Table 27–1). Little or no information is available on the effect of the two nepoviruses on cropping of red currant. In sites where the nematode vector does not exist, selection and propagation of clean planting material should still be ascertained. No effective treatment is available for infected plants. The effect of nematode control on the disease is not known.

Other nepoviruses isolated from black currant for which little information on nematode-transmission has been gathered are SLRSV in Scotland, RpRSV and ToRSV in the former Soviet Far East, and TBRV in Finland (Adams & Thresh, 1987a). No information is available on natural spread of the virus in currant but *X. americanum* is implicated as the vector which, if present, could lead to the establishment of the virus in weeds and other hosts.

Up-to-date information on the economic importance of ToRSV disease in red currant (also called American currant mosaic) is not available. Yield losses and occasional death of plants may occur. For disease prevention, only red currant stock that has been indexed and certified as disease-free should be used for propagation. Exploration of cultivar resistance may be possible not only against

Fig. 27–8. Gooseberry 'Whynham's Industry' showing deteriorated (lower) shoot infected with raspberry ringspot virus (RpRSV) as compared with a healthy shoot (after Van der Meer, 1987a).

Fig. 27–9. Leaf symptom expression caused by tobacco rattle virus on *Ribes sanguineum* 'King Edward VII' (after F. A. van der Meer, 1987b).

the virus but also for *X. americanum* and related nematode species. To date no information exists on effective treatments for the virus in red currant.

Tobacco rattle virus (TRV) causes a disease on red currants called leaf pattern (Fig. 27–9). The disease only has been observed in three bushes of red currant in the Netherlands (van der Meer, 1987b) and because of this rare occurrence, leaf pattern is considered of little economic importance.

CONTROL

Control methods for nematode-vectored viral diseases are outlined in Table 27–1. An overview of the strategies used for controlling the nematode parasites is given below.

Cultural

Several nonchemical approaches have been used for nematode management in grape and other berry cultivation, such as exclusion, side-dressings, replant strategies, good weed control, and resistant rootstocks. Every possible measure, (for example, the use of virus-free certified planting stock), is taken to prevent introduction of the nematode into "clean" fields. Mulching may not be effective for controlling all nematode populations in the soil. In blueberry, *P. minor* but not *X. americanum* populations were lower in mulched plots (Myers, 1995).

Cultural control measures are similar for certain nematodes (for example, longidorid and lesion nematodes). Exclusion of longidorid nematodes from the growing area is very important because once they become established, the infestation is permanent. The introduction of longidorid nematodes by way of the plant stock material is less likely than for root lesion nematodes since the former are strictly ectoparasitic. The economic threshold level for the non-vector, *X. bakeri*, is approximately 50 nematodes per 500 cm^3 of soil. Even *X. bakeri* initial populations should be reduced to much lower levels in the preplant stages. *Xiphinema bakeri* is, however, easily controlled by leaving the bare soil fallow and repeatedly turning and drying the soil during hot summer periods.

In the preplant period, if the nematode vectors are detected, a shallow-root grass crop grown for 1 to 2 yr will bring the nematodes closer to the surface where they are more effectively controlled by fumigation treatment. The use of a nonhost crop reduces viral initial inoculum and increases the probability that all viruliferous nematodes would be killed by the fumigation treatment. Given the wide host range of the nepoviruses involved, good weed control is of paramount importance especially in view of the fact that the viruses can be seedborne and remain in symptomless association with those weed hosts.

Hot water disinfestation of strawberry plants also has proved useful to a limited extent. However, temperatures that killed the nematodes also delayed fruiting and did not increase yields. Cold storage after hot water treatments gives good control of *M. hapla* and *P. penetrans* and good plant growth. Effective treatments have ranged from 47 to 52.8°C for 2 to 15 min (Bridge, 1975). Other methods of control, such as crop rotations or fallow, are not very practical for management of longidorid nematodes, which have long survival rates in soil and wide host ranges.

Nematodes can be introduced into a field in soil and water by humans and machinery; and once established, they become endemic. For example, tremendous progress has been made in the management of root lesion nematodes on *Rubus* plants. An effective systemic herbicide could be applied to plants still existing in the field to drive nematodes that are inside the roots into the soil where control is easier. Adequate soil preparation is imperative. The plowed crop should be first allowed to decompose sufficiently. Root-lesion nematodes, for example, are difficult to control in the soil. Crop rotation is not practical since *Pratylenchus* spp. are harbored in many common weeds and crop hosts. Where appropriate, sudangrass [*Sorghum* × *drummondii* (Nees ex steudel) Millsp. & Chase], sweet corn (*Zea mays* ssp. *mays* L.) or pumpkin (*Cucurbita pepo* L.) may be used as a nonhost rotation crop for control of *Pratylenchus*. Disinfestation of strawberry roots colonized by lesion nematodes could be achieved by dipping dormant plants in hot water (<54°C). Soil organic amendments, such as composted hay, may reduce lesion nematode populations.

The cloning of strawberry from runners avoided rooting strawberry in nematode-infested soil and, together with fumigation, gave sustained impetus to strawberry production (Potter & Dale, 1991). As with other nematode diseases, the use of nematode-free certified planting stock is of paramount importance in soils that are free of the nematode.

The most effective cultural management strategy for spring dwarf include use of certified nematode free plants, heat treatments of dormant stock plants at 46.1°C for 10 min (Bridge, 1975), and roguing of infested plants as early as possible. Total eradication of the nematodes in nuclear stock can be achieved by excision of affected parts and regeneration of crowns from previously dormant buds (McNamara et al., 1986). At the East Malling Research Station, runners are rooted in artificial medium (Perlite) instead of peat because it is difficult to eradicate the nematodes from the latter. Now, all introduced varieties are tested for nematodes.

Management strategies administered for control of summer dwarf are similar to those used for spring dwarf. Several grass weeds commonly found in strawberry fields in Australia were found to harbor *A. besseyi* (McCulloch, 1978) and could act as sources of inocula when they are plowed under.

The best management strategy for control of bulb and stem nematodes on strawberry has been the use of certified nematode-free planting stock to establish new plantings. Fields with or adjacent to plots with a history of the nematode on susceptible crops should be avoided. If only a few plants are infected, roguing might be effective. Rotations of strawberry with nonhost crops can control bulb and stem nematodes (Nemec, 1984). Successful heat therapy was achieved by dipping dormant strawberry plants in water at 46.1°C for 10 min (Bridge, 1975).

Resistance

Grape

Multiple resistance to most nematodes is not available in any one grape root stock (Raski, 1986). In the USA, 'DogRidge' and 'Salt Creek' (= 'Ramsey') are almost immune to root-knot nematodes and grow vigorously in most soils (except coarse sands). 'Freedom' and 'Harmony' show good resistance to root-knot nematodes and phylloxera but the latter genotype supports large numbers of *X. index* and *M. hapla*. 'Kober 5BB', '1103 Paulsen', '1447 Paulsen', 'SO4', 'Teleki 8B', '1103', '1616C', and '3306' are resistant to *M. arenaria* in the Mediterranean region while the four already mentioned are resistant to *M. javanica* in Australia and South Africa (Brown et al. 1994). Few rootstocks are known with adequate resistance to *X. index* and *X. italiae*. Two rootstocks from a *Vitis rotundifolia* × *V. vinifera* cross showed promising resistance to *X. index* and the GFLV virus (Raski, 1988).

Control methods described for root-knot nematode management also are effective for lesion nematodes. However, because of differences in feeding requirements, *P. vulnus* may infect rootstocks (such as Dogridge) that are resistant to root-knot nematodes (Raski, 1986). Rootstocks developed for resistance to root-knot nematodes also should be tested against lesion nematodes. Little testing has been done on rootstocks or cultivars for resistance to citrus nematodes.

Raspberry and Blackberry

Raspberry and blackberry varieties originally screened for resistance to nepoviruses by graft transmission may not necessarily be resistant to these viruses when grown in soil infested with viruliferous nematodes (Jones et al., 1989).

Strawberry

Although some strawberry cultivars are tolerant or resistant to root-knot nematodes, immunity is rare. Several noncommercial species of strawberry do not support *M. hapla* reproduction and can serve as sources for introgressing resistance genes into desirable cultivars. Resistance to root-knot nematodes is found in cultivars such as Blakemore, Delite, Fairfax, Fletcher, Florida Ninety and Jerseybelle.

In an extensive study Szczygiel (1981) compared 28 cultivars of North American, British and local (European) origin and found that none were immune to *P. penetrans*. No strawberry cultivar is known to have good resistance to lesion nematodes. A local cultivar Senga Sengana plus 'Guardian' and 'Redchief' supported significantly lower numbers of nematodes than 'Midway', 'Grenadier' and 'Vesper' which were excellent hosts for *P. penetrans*. Growth suppression in Midway was not proportional to nematode damage and suggested that both growth inhibition and root damage should be considered in cultivar evaluations. In another study that involved 14 cultivars grown in Canada's Pacific Northwest, no source of resistance was identified (Vrain, 1985).

More recently, in Canada, Potter and Dale (1991) attempted improvement of tolerance to *P. penetrans* in strawberry cultivars through intraspecific crosses of *Fragaria* × *ananassa* cultivars, and by searching among *F. chiloensis* and wild *F. virginiana* clones. Nematode tolerance (measured as the ability to survive large nematode populations with minimal damage) was identified in Midway while nematode suppressiveness (taken as the failure to support nematode population increase) was found in Guardian. The same results had been obtained in Europe (Szczygiel, 1981). Inbreeding of *F.* × *ananassa* might have inadvertently resulted in the loss of tolerance to root-lesion nematodes and other traits (Potter & Dale, 1991; Sjulin & Dale, 1987). Suppressive clones were not found among *F. chiloensis* suggesting that any genetic resistance to *P. penetrans* present in *F. x ananassa* most likely does not originate in *F. chiloensis*. However, several clones among wild *F. virginiana* showed good suppressiveness and vigorous growth habit. Although the study of Potter and Dale (1991) considered a fairly limited selection of strawberry clones, there seems to be some diversity that can be exploited even by way of introgressing resistance genes from unimproved wild germplasm (Scott & Lawrence, 1975; Sjulin & Dale, 1987). Cultivars resistant to *Aphelenchoides* spp. are known and can be used in breeding programs. There is little resistance among strawberry cultivars to stem and bulb nematodes.

Other Small Fruits

Resistance to nematodes has not been identified in any cranberry or blueberry selections. There also is an acute scarcity of information on the varietal response of gooseberry and currant to plant-parasitic nematodes.

Chemical

Grape

The one selectively nematicidal chemical, DBCP was banned from the market in 1977 after being detected in groundwater (Raski, 1988). Nonfumigant, systemic nematicides of the organophosphate and carbamate types have been used, but these nematicides must necessarily be mixed with the soil or water and then carried by water to the target. Their movement is restricted in clay soils and, besides their high cost, they may build up unacceptable levels of residue in the fruit (Raski, 1986).

Declining vineyards that do not recover from side-dressing treatments should be replanted. In the past methyl bromide (MeBr) and 1,3-D (1,3-dichloropropene) have proved to be very effective (Raski, 1988). These fumigants require deep injection for effective control. Methyl bromide is an effective, but expensive, highly toxic general biocide. In the USA, a ban on production and importation of methyl bromide is scheduled to begin in the year 2001. After fumigation, vineyards consistently yielded growth of uniform vigor. Since certain nematodes (such as *Meloidogyne* spp.) are never completely eradicated from a vineyard, early detection of the buildup of nematode populations is crucial for timely side-dressing treatments.

The DBCP gave effective control of *X. index* (free of GFLV) for 2 to 3 yr. Nonfumigant systemic nematicides have not given such effective control. Good control of this nematode has been achieved with MeBr and 1,3-D. Preplant fumigations failed to control *X. index* and could not reverse the decline of newly planted vineyards within 1 to 2 yr (Raski, 1988).

Replant treatments, exclusion tactics and side-dressing of established vineyards with citrus nematodes are similar to those for root-knot nematodes (Raski, 1988). Although the semi-endoparasitic nematode may be more exposed to chemical treatments, the gelatinous matrix still provides some degree of protection to the embedded eggs and females. No chemical is registered for controlling citrus nematodes on grape.

Raspberry and Blackberry

Deep cultivation (0.8–1.5 m) and the working of the soil to a seedbed condition where large clods are broken up are necessary to achieve the most effective fumigation. Soil moisture should be at about 50% field capacity (Pritts & Handley, 1989). Methyl bromide plus chloropicrin (trichloronitromethane) under continuous polyethylene tarp have given effective control of nematodes, fungi, weeds and insects in *Rubus* plantations (McElroy, 1991). An economic threshold density for this crop is approximately 500 nematodes/500 cm^3. A single fumigation, coupled with the use of disease-free planting material and followed by strict soil population monitoring, can give 2 yr of profitable berry production. However, currently used fumigants are expensive and may not be available for use on crops in the field in the near future.

For treatment of an established crop, nematode sampling should be conducted at least every year to ensure timely response to a decline in plant health.

Fenamiphos [ethyl 3-,ethyl-4-(methylthio)phenyl (1-methylethyl) phosphoamidate] is not registered for use on blackberry and is now the only nematicide registered for use on red raspberry. This nonfumigant nematicide disrupts nematode feeding and reproduction and gives some control (Vrain & King, 1986). It does not kill the organism immediately as does fumigant treatments. Fenamiphos must not be applied less than 180 d (6 mo) before harvest. About 7 to 13 kg ha^{-1} of fenamiphos are needed for reversing the effects of root-lesion nematodes already at densities of 1000 per 500 cm^3 of soil (McElroy, 1991).

The detection of only a single vector nematode necessitates preplant fumigation treatment (McElroy, 1991). No effective postplant treatment is currently available for controlling dagger and needle nematodes and the viruses they transmit.

Strawberry

Soil fumigation prior to planting strawberry has long been the most effective means of controlling root-knot nematodes and other soilborne pathogens (Raski & Krusberg, 1984; Wilhelm & Paulus, 1980). However, preplant fumigation is useless if nematode-free planting stocks are not used for establishing the field. Most fumigants have been banned or are under review because of environmental and health concerns. Nonfumigant nematicides, such as fenamiphos and aldicarb {2-methyl-2-(methylthio)propanal *O*-[(methylamino)carbonyl]oxime}, have given effective control in Maryland as side dressings. One effective regimen in Maryland had been a quarterly alternation of applications starting at planting (in fumigated soil).

Most fumigants recommended for controlling other nematodes on strawberry also are effective against longidorid nematodes. In the USA, methyl bromide or DD-Mencs (20% methyl isothiocyanate + 80% chlorinated C_3 hydrocarbons) was most effective against *X. americanum* (Nemec & Malek, 1975). In Switzerland, DD (dichloropropane-dichloropropene), dazomet (tetrahydro-3,5-dimethyl-2*H*-1,3,5-thiadiazine-2-thione) or fenamiphos gave the most effective control for *X. diversicaudatum*. In England, MeBr eliminated 99% of *X. diversicaudatum* to a depth of 70 cm (Raski & Krusberg, 1984). Given the slow rate of development of longidorid nematodes, another fumigation treatment would not be required for several years.

The DBCP applied as a postplant treatment effectively reduced *P. penetrans* populations, but it has been banned (Nemec, 1984). In Poland up to 10-fold increases in yield were obtained with Di-Trapex, which effectively reduced *P. penetrans* and the incidence of *Verticillium dahliae* Kleb. on strawberry (Szczygiel & Rebandel, 1982).

Organophosphate insecticides such as parathion (diethyl-*p*-nitrophenyl monothiophosphate) and demeton {phosphorothioic acid *O*,*O*-diethyl *O*-[2-(ethylthio)ethyl] ester mixture with *O*,*O*-diethyl *S*-[2-(ethylthio)ethyl]phosphorothioate} have been used as sprays or dips. In California, parathion did not give satisfactory control of *A. fragariae* and was phytotoxic to strawberry (Raski & Krusberg, 1984). In Europe, granular nematicides such as aldicarb prevented serious damage but did not eliminate the pest (McNamara et al., 1986). However up to 80% control of *Aphelenchoides* spp. in plant buds is known. The use of nematode-free planting stock has been very successful in California. In Maryland, the routine use of parathion in nurseries might have kept bud and leaf nematodes in check (Raski & Krusberg, 1984). In Russia, several methods are used such as sanitation, heat therapy, chemical dips and sprays with mixtures containing parathion (Kir'janova & Krall', 1980). *Aphelenchoides besseyi* was effectively controlled by dips in zinophos [*O,O*-diethyl-*O*-(2-pyrazinyl)phosphorothioate] but the chemical is no longer used. Control of *A. besseyi* was achieved with one granule of fenamiphos 10G applied to the crown of each strawberry plant that showed symptoms 30 d (1 mo) after planting.

In Italy foliar sprays of fenamiphos, oxamyl {N′,N′-dimethyl-N-[(methylcarbamoyl)oxy]-1-thiooxamimidic acid methyl ester} and thionazin [O,O-diethyl-O-diethyl-(2-pyrazinyl) phosphorothioate] were the most effective for control of *D. dipsaci* and increased fruit yield (Tacconi, 1976). Successful control of bulb and stem nematodes with parathion sprays, dips and drenches were achieved. However, chemical treatments have not been generally effective in Russia (Szczygiel, 1977; Gulyaev & Akinin, 1990).

Cranberry and Blueberry

Nematode control in mature plantings of cranberry and blueberry is not warranted in most cases. Early studies with the nematicide zinophos in cranberry bogs infested with *Paratrichodorus* showed significant yield increases with no phytotoxicity to cranberry vines in 3-yr field trials. Increased yields appeared to be associated with the formation of a higher number of fruit buds. In Massachusetts and New Jersey, zinophos and DBCP increased cranberry yields, reduced nematode populations and increased the percentage of shoot uprights and buds. Ethoprop (O-ethyl *S,S*-dipropylphosphorodithioate), carbofuran (2,3-dihydro-2,2-dimethyl-7-benzofuranylmethylcarbamate), fenamiphos, or oxamyl did not significantly affect yield or numbers of stubby root or sheath nematodes.

Soil fumigation of blueberry cutting beds prior to planting gives satisfactory control. Methyl isothiocyanate, N-methyldithiocarbamate and 1,3-dichloropropene are the current preplant fumigants labeled for nematode control in blueberry. The non-fumigant fenamiphos significantly reduced populations of *P. penetrans* and *X. americanum.* Fenamiphos, however, may not always reduce *P. penetrans* population levels (Myers, 1995).

OUTLOOK

Multiple resistance to most nematode parasites is not available in any one small fruit genotype. There is a scarcity of nematode resistant rootstock among small fruit selections, and many genotypes showing good tolerance to a particular nematode also are highly susceptible to a viral, fungal, or other nematode pathogen. Also, given the fate/shortcomings and impracticality of chemical and other control methods (such as crop rotation) used for nematodes and viral diseases of small fruits, the search for general resistance and alternative nonchemical management tactics will have to be intensified even more.

To aid the development of acceptable alternatives to current nematode control strategies in small fruit culture, more basic information on the ecology, host-vector interaction, and epidemiology is needed. The urgency of that need is exemplified by the observation that cultivars screened for viral resistance by graft transmission could become susceptible to nepoviruses in soil where viruliferous nematodes are present (Jones et al., 1989). There is very little experimentation with biological control agents or organic amendments for control of nematodes on small fruits. As with many crops, there is still an acute scarcity of reliable estimates of crop losses that clearly demonstrate the economic importance of nematode-related diseases on grape and berry crops. The most effective approach, however, to a long-term solution to the complexity of viral diseases is still the culture of tolerant varieties with the adoption of measures that would eliminate viruses from nematode populations. As novel viral, nematode, and host resistance genes are identified in small fruits and other organisms (including pathogens), molecular biologists and plant breeders will have new options to explore for the improvement of desirable small fruit cultivars. Biologically based management practices may provide economically and environmentally acceptable levels of crop protection.

REFERENCES

Adams, R.E., and C.E. Hickman, 1970. Influence of nematicidal treatments and fungicidal sprays on yield of strawberries. Plant Dis. Rep. 54:923–926.

Adams, A.M., and J.M. Thresh. 1987a. Other nepoviruses isolated from black currant. p. 140. *In* R. H. Converse (ed.) Virus diseases of small fruits. USDA Agric. Handb. 631. USDA-ARS, Washington, DC.

Adams, A.M., and J.M. Thresh. 1987b. Yellow mottle of black currant. p. 139–140. *In* R.H. Converse (ed) Virus diseases of small fruits. USDA Agric. Handb. 631. USDA-ARS, Washington, DC.

Ahmedullah, M., and D.G. Himelrick 1990. Grape management. p. 383–471. *In* G.J. Galleta and D.G. Himelrick (ed.) Small fruit crop management. Prentice-Hall, Englewood Cliffs, NJ.

Allen, W.R., H.F. Dias, and J.G. Van Schagen. 1982. Susceptibility of grape cultivars and rootstocks to an Ontario isolate of tomato ringspot virus. Can. J. Plant Pathol. 4:275–277.

Anwar, S.A., and S.D. van Gundy. 1989. Influence of four nematodes on root and shoot growth parameters in grape. J. Nematol. 21:276–283.

Birchfield, W., J.P. Hollis, and W.J. Martin. 1978. A list of nematodes associated with some Louisiana plants. Louisiana State Univ. Agric. Exp. Stn. Bull. 101.

Bird, G., C. Diamond, F. Warner, and J. Davenport. 1994. Distribution and regulation of *Meloidogyne nataliei*. J. Nematol. 26(4S):727–730.

Bird, G.W., and D.C. Ramsdell. 1985. Population trends and vertical distribution of plant parasitic nematodes associated with *Vitis labrusca* L. in Michigan. J. Nematol. 17:100–107.

Bitterlin, M.W., and D. Gonsalves. 1988. Serological grouping of tomato ringspot virus isolates: Implications for diagnosis and cross-protection. Phytopathology 78:278–285.

Bridge, J. 1975. Hot water treatment to control plant parasitic nematodes of tropical crops. Meded. Fac. Landbouwiw. Univ. Gent. 40:249–259.

Brown, D.J.F., J.M. Halbrendt, and A.J. Jones. 1992. Transmission of nepoviruses by populations of four *Xiphinema americanum* group species. Nematologica 38:401–402.

Brown, D.J.F., J.M. Halbrendt, A.J. Jones, T.C. Vrain, and R.T. Robbins. 1994. Transmission of three North American nepoviruses by populations of four distinct species of the *Xiphinema americanum* group. Phytopathology 84:646–649.

Chen, T., and A.E. Rich. 1962. Pathogenicity of *Pratylenchus penetrans* on strawberry and Ladino white clover seedlings. Phytopathology 52:922–923.

Chikaoka, I. 1970. Studies on the control of *Pratylenchus vulnus* infesting strawberry. Bull. Kanagawa Pref. Agric. Res. Inst. 109:61–77.

Choo, H.Y., H.K. Kim, J.C. Park, S.M. Lee, and J.I. Lee. 1987. Studies on the pattern of plastic film house, their growing conditions and disease and pest occurrence on horticultural crops in southern part of Korea. Insects and nematodes associated with horticultural crops and effect of nursery soil conditions on the infection of root-knot nematode. Kor. J. Plant Pathol. 26:195–201.

Clark, J.R., and R.T. Robbins, 1994. Phytoparasitic nematodes associated with three types of blueberries in Arkansas. J. Nematol. 26(4S):761–766.

Converse, R.H. 1981. Infection of cultivated strawberries by tomato ringspot virus. Phytopathology 71:1149–1152.

Converse, R.H. 1991. Nematode-transmitted diseases. p. 47–50. *In* M.A. Ellis et al. (ed.) Compendium of raspberry and blackberry diseases and insects. APS Press, St. Paul, MN.

Converse, R.H. 1995. Tomato ringspot. p. 55–56. *In* F.L. Caruso and D.C. Ramsdell (ed.) Compendium of blueberry and cranberry diseases. APS, St. Paul, MN.

Converse, R.H., and D.C. Ramsdell. 1982. Occurrence of tomato and tobacco ringspot viruses and of dagger and other nematodes associated with cultivated highbush blueberries in Oregon. Plant Dis. 66:710–712.

Converse, R.H., and R. Stace-Smith. 1987. Tomato ringspot virus in strawberry. p. 52–55. *In* R.H. Converse (ed.) Virus diseases of small fruits. USDA Agric. Handb. 631. USDA-ARS, Washington, DC.

Cotten, J. 1975. Virus vector species of *Xiphinema* and *Longidorus* in relation to certification schemes for fruit and hops in England. p. 283–285. *In* F. Lamberti et al. (ed.) Nematode vectors of plant viruses. Plenum, New York.

Diamond, C.J. 1994. Observations on the biology of *Meloidogyne nataliei*: With special references to host parasite relationships. M.S. thesis. Michigan State Univ., East Lansing.

Eisenback, J.D. 1982. Description of the blueberry root-knot nematode, *Meloidogyne carolinensis* n. sp. J. Nematol. 14:303–317.

Ferris, J.M., and V. R. Ferris. 1998. Biology of plant-parasitic nematodes. p. 21–35. *In* K.R. Barker et al. (ed.) Plant and nematode interactions. Agron. Monogr. 36. ASA, CSSA, and SSSA, Madison, WI.

Forer, L.B., and R.F. Stouffer. 1982. *Xiphinema* spp. associated with tomato ringspot virus infection of Pennsylvania fruit crops. Plant Dis. 66:735–736.

Golden, A.M., L.M. Rose, and G.W. Bird. 1981. Description of *Meloidogyne nataliei* n. sp. (Nematoda: Meloidogynidae) from grape (*Vitus labrusca*) in Michigan, with SEM observations. J. Nematol. 13:393–400.

Gonsalves, D. 1988a. Tobacco ringspot virus decline. p. 51. *In* R.C. Pearson and A.C. Goheen (ed.) Compendium of grape diseases. APS Press, St. Paul, MN.

Gonsalves, D. 1988b. Tomato ringspot virus decline. p. 49–50. *In* R.C. Pearson and A.C. Goheen (ed.) Compendium of grape diseases. APS Press, St. Paul, MN.

Gulyaev, A.G., and N.I. Akinin. 1990. Protection of strawberry plantlets against stem nematodes. Sadov. Vinograd. 4:26–27.

Harmat, L., A. Porpaczy, D.G. Himelrick, and G.J. Galletta. 1990. Currant and gooseberry management p. 245–272. *In* G.J. Galletta and D.G. Himelrick (ed.) Small fruit crop management. Prentice-Hall, NJ.

Jennings, D.L. 1988. Diseases caused by nematodes. p. 139–142. *In* D.L. Jennings (ed.) Raspberries and blackberries: Their breeding, diseases and growth. Acad. Press. New York.

Johnson, C.S. 1998. Tobacco. p. 487–522. *In* K.R. Barker et al. (ed.) Plant and nematode interactions. Agron. Monogr. 36. ASA, CSSA, and SSSA, Madison, WI.

Johnson, F. 1972. Tomato ringspot virus of blueberries in Washington. Phytopathology 62:1004.

Jones, A.T., M.J. Michell, and D.J.F. Brown. 1989. Infectibility of some new raspberry with arabis mosaic and raspberry ringspot virus and further evidence for variation in British isolates of these two nepoviruses. Ann. Appl. Biol. 115:57–69.

Kinloch, R.A. 1998. Soybean. p. 317–333. *In* K.R. Barker et al. (ed.) Plant and nematode interactions. Agron. Monogr. 36. ASA, CSSA, and SSSA, Madison, WI.

Kir'janova, E.S., and E.L. Krall'. 1980. Plant-parasitic nematodes and their control. Vol. 2. Academy of Sciences of the USSR, 1971. Am. Publ. Co., PVT. Ltd., New Delhi, India.

Kisiel, M., J. Castillo, and B.M. Zuckerman. 1971. An adhesive plug associated with the feeding of *Hemicycliophora similis* on cranberry. J. Nematol. 3:296–298.

Kondo, E., and N. Ishibashi. 1977. Seasonal occurrence, dissemination, and survival of the strawberry bud nematode, *Nothotylenchus acris*. Jpn. J. Nematol. 7:39–44.

LaMondia, J.A., and S.B Martin. 1988. Plant parasitic nematodes and fungi are an unhealthy alliance against strawberries. Front. Plant Sci. 40:3–5.

LaMondia, J.A., and S.B. Martin. 1989. The influence of *Pratylenchus penetrans* and temperature on black root rot of strawberry by binucleate *Rhizoctonia* spp. Plant Dis. 73:107–110.

Martelli, G.P., and V. Savino. 1988. Fanleaf degeneration. p. 48–49. *In* R.C. Pearson and A.C. Goheen (ed.) Compendium of grape diseases. APS Press, St. Paul, MN.

McCulloch, J. 1978. Strawberry crimp. Queensl. Agric. J. 104:345–347.

McElroy, F.D. 1972. Nematodes of tree fruits and small fruits. p. 335–376. *In* J.M. Webster (ed.) Economic nematology. Acad. Press, London.

McElroy, F.D. 1991. Nematode parasites. p.59–62. *In* M.A. Ellis et al. (ed.) Compendium of raspberry and blackberry diseases and insects. APS Press, St. Paul, MN.

McKewan, J. 1979. Studies on the biology and life history of *Hemicycliophora similis* J. Nematol. 11:307–308.

McKinley, R.T., and P.W. Talboys. 1979. Effects of *Pratylenchus penetrans* on development of strawberry wilt *Verticillium dahliae*. Ann. Appl. Biol. 92:347–357.

McNamara, D.G., J.E. Stickels, and J.J.M. Flegg. 1986. Control of leaf nematodes in nuclear stock strawberries. British Crop Prot. Counc. Monogr. 33: 217–219.

Meagher, J.W., and P.T. Jenkins. 1970. Interaction of *Meloidogyne hapla* and *Verticillium dahliae*, and the chemical control of wilt in strawberry. Aust. J. Exp. Agric. Anim. Husb. 10:493–496.

Milholland, R.D., R. Jones, and C.N. Clayton. 1974. Blueberry diseases in North Carolina. North Carolina Agric. Ext. Serv. Circ. 466.

Muller, J. 1973. On the *Verticillium-Pratylenchus* disease complex in strawberries. Gartenbauwissenschaft 38:43–55.

Murant, A.F., and R.M. Lister. 1987. Nematode-borne diseases. p. 46–52. *In* R.H. Converse (ed.) Virus diseases of small fruits. USDA Agric. Handb. 631. USDA-ARS, Washington, DC.

Myers, R.F. 1991. Nematodes associated with dieback disease of cranberries. J. Nematol. 23(4S):629–633.

Myers, R.F. 1995. Diseases caused by nematodes p. 61–62. *In* F.L. Caruso and D.C. Ramsdell (ed.) Compendium of blueberry and cranberry diseases. APS Press, St. Paul, MN.

Nemec, S. 1984. Nematode diseases p. 121–126. *In* J.L. Maas (ed.) Compendium of strawberry diseases. APS Press, St. Paul, MN.

Nemec, S., and R.B. Malek. 1975. Effects of nematicides and strawberry growth on nematodes, especially *Xiphinema americanum*, in root-rot sites in Illinois. J. Nematol.7:328.

Norgren, R.L., B.M. Zuckerman, and D.H. MacDonald. 1968. Investigations on the biology and monoxenic culturing of *Tetylenchus joctus*. Nematologica 14:565–572.

Nyczepir, A.P., and J. Ole Becker. 1998. Fruit and citrus trees. p. 637–684. In K.R. Barker et al. (ed.) Plant and nematode interactions. Agron. Monogr. 36. ASA, CSSA, and SSSA, Madison, WI.

Potter, J.W., and A. Dale. 1991. Root lesion nematode tolerance in wild and cultivated strawberry. p. 202–208. *In* A. Dale and J.J. Luby (ed.) The strawberry into the 21st century. Timber Press, Portland, OR.

Pritts, M., and D. Handley. 1989. Bramble production guide. Northeast Region. Agric. Eng. Service, Ithaca, NY.

Ramsdell, D.C. 1988. Peach rosette mosaic virus decline. p. 51–52. *In* R.C. Pearson and A.C. Goheen (ed.) Compendium of grape diseases. APS Press, St. Paul, MN.

Ramsdell, D.C. 1995a. Necrotic ringspot. p. 53–54. *In* F.L. Caruso and D.C. Ramsdell (ed.) Compendium of blueberry and cranberry diseases. APS Press, St. Paul, MN.

Ramsdell, D.C. 1995b. Peach rosette mosaic. p. 54–55. *In* F.L. Caruso and D.C. Ramsdell (ed.) Compendium of blueberry and cranberry diseases. APS Press, St. Paul, MN.

Ramsdell, D.C., and J.M. Gillett. 1985. Relative susceptibility of American, French hybrid and European grape cultivars to infection by peach rosette mosaic virus. Phytopathol. Medit. 24:41–43.

Ramsdell, D.C., J.M. Gillett, and G.W Bird. 1995. Susceptibility of American grapevine scion cultivars and french hybrid rootstock and scion cultivars to infection by peach rosette mosaic nepovirus. Plant Dis. 79:154–157.

Ramsdell, D.C., and R.L. Myers. 1974. Peach rosette mosaic virus, symptomatology and nematodes associated with grapevine `degeneration' in Michigan. Phytopathology 64:1174–1178.

Ramsdell, D.C., and R.L. Myers. 1978. Epidemiology of peach rosette mosaic virus in a Concord grape vineyard. Phytopathology 68:447–450.

Raski, D. 1986. Plant-parasitic nematodes that attack grapes. p. 43–57. *In* Plant-parasitic nematodes of bananas, citrus, coffee, grapes and tobacco. Union Carbide Agric. Prod. Co., North Carolina.

Raski, D.J. 1988. Parasites of grapes. p. 55–59. In R.C. Pearson and A.C. Goheen (ed.) Compendium of grape diseases. APS Press, St. Paul, MN.

Raski, D.J., and L.R. Krusberg. 1984. Nematode parasites of grapes and other small fruits. p. 457–506. *In* W.R. Nickle (ed.) Plant and insect nematodes. Marcel Dekker Inc., New York.

Scott, D.H., and F.J. Lawrence. 1975. Strawberries. p. 71–97. *In* J. Janick and J.N. Moore (ed.) Advances in fruit breeding. Purdue Univ. Press, West Lafayette, IN.

Scotto la Massese, C. 1973. Les nématodes phytophages associés a la culture des petits fruits en France. Bull. Tech. SICANAPF 5:6–14.

Seinhorst, J.W. 1981. Methods for generating different pest/yield relationships—nematodes. p. 85–89. *In* L. Chiarappa (ed.) Crop loss assessment methods. Suppl. 3. FAO and CAB, Slough, England.

Sjulin, T.M., and A. Dale. 1987. Genetic diversity of North American strawberry cultivars. J. Am. Soc. Hort. Sci. 112:375–385.

Society of Nematologists. 1994. Soil nematodes: Societal impact and focus for the future. J. Nematol. 26:1–26.

Stace-Smith, R., and R.H. Converse. 1987. Tomato ringspot virus in *Rubus*. p.223–227. *In* R.H. Converse (ed.) Virus diseases of small fruits. USDA Agric. Handb. 631. USDA-ARS, Washington, DC.

Stellmach, G., and A.C. Goheen. 1988. Other virus and virus-like diseases. p. 53–54. *In* R.C. Pearson and A.C. Goheen (ed.) Compendium of grape diseases. APS Press, St. Paul, MN.

Szczygiel, A. 1967. Preliminary estimation of the harmfulness of nematodes of the genus *Aphelenchoides* to strawberries in South Poland (Polish). Prace Inst. Sadownictwa 11:210–224.

Szczygiel, A. 1977. Review of Soviet literature on plant parasitic nematodes associated with strawberries. Res. Inst. Pomol., Brezna, Poland.

Szczygiel, A. 1981. Trials on susceptibility of strawberry cultivars to the root lesion nematode *Pratylenchus penetrans.* Fruit Sci. Rep. 8:115–119.

Szczygiel, A. 1986. Effect of *Pratylenchus penetrans* and *Longidorus elongatus* on strawberry plants as influenced by host age at inoculation. Fruit Sci. Rep. 13:31–37.

Szczygiel, A., and H. Hasior. 1971. Possibility of persistence of leaf and bud nematodes (*Aphelenchoides fragariae*) on strawberry plants and in the soil. Zesz. Probl. Post. Nauk Roln. 121:101–106.

Szczygiel, A., and H. Hasior. 1972. Seasonal variations in populations of plant parasitic nematodes in strawberry plantations. Ekol. Polska 20:507–523.

Szczygiel, A., and H. Profic-Alwasiak. 1989. Studies on the interaction between nematodes and fungi in infecting strawberry plants. Acta Hort. 265:561–565.

Szczygiel, A., and Z. Rebandel. 1982. Control of root nematode and *Verticillium* on strawberries using Di-Trapex and Benlate. Roczn. Nauk Rolniczy., E (Ochrona Roslin) 12:163–174.

Szczygiel, A., and Z. Rebandel. 1991. Effectiveness of chemical decontamination in limiting the adverse effects of soil sickness during strawberry replanting. (In Polish.) Akad. Rolnicza, Poznan. Roczniki. Ogrodnictwo 19:149–159.

Szczygiel, A., and A. Soroka. 1983. Effect of soil moisture level on population and pathogenicity of three plant-parasitic nematodes to strawberry plants. Zeszyty Problemowe Postepow Nauk Rolniczych 278:105–111.

Szczygiel, A., A. Soroka, and A. Zepp. 1983a. Effect of soil temperature on population and pathogenicity of *Meloidogyne hapla, Pratylenchus penetrans* and *Longidorus elongatus* to strawberry plants. Zeszyty Problemowe Postepow Nauk Rolniczych 278:87–94.

Szczygiel, A., A. Soroka, and A. Zepp. 1983b. Effect of soil texture on population and pathogenicity of *Meloidogyne hapla*, *Pratylenchus penetrans* and *Longidorus elongatus* to strawberry plants. Zeszyty Problemowe Postepow Nauk Rolniczych 278:77–86.

Szczygiel, A., and A. Zepp. 1983. Effect of organic matter in soil on population and pathogenicity of *Pratylenchus penetrans* and *Longidorus elongatus* to strawberry plants. Zeszyty Problemowe Postepow Nauk Rolniczych 278:113–122.

Tacconi, R. 1976. Prove di lotta control il *Ditylenchus dipsaci* su fragola. II. Ins. Fitopatol. 26:29–36.

Taylor, C.E., and D.J.F. Brown. 1981. Nematode virus interactions. p. 281–301. *In* B.M. Zuckerman and R.A. Rohde (ed.) Plant parasitic nematodes. Vol. 3. Acad. Press, New York.

Taylor, C.E., and D.J.F. Brown. 1997. Nematode vectors of plant viruses. CAB Int., Wallingford, Oxon, England.

Trudgill, D.L. 1976. Observations of the feeding of *Xiphinema diversicaudatum.* Nematologica 22:417–423.

van der Meer, F.A. 1987a. Gooseberry deterioration. p. 131. *In* R.H. Converse (ed.) Virus diseases of small fruits. USDA Agric. Handb. 631. USDA-ARS, Washington, DC.

van der Meer, F.A. 1987b. Leaf pattern of red currant. p. 150–152. *In* R.H. Converse (ed.) Virus diseases of small fruits. USDA Agric. Handb. 631. USDA-ARS, Washington, DC.

Vrain, T. 1985. Resistance to root-lesion nematode in strawberry. Can. Hort. Council Ann. Rep. 75.

Vrain, T.C., and J.C.W. King. 1986. Application of nonvolatile nematicides through a trickle irrigation system to control *Pratylenchus penetrans* in raspberries. Can. J. Plant Pathol. 8:97–101.

Walker, M.A., J.A. Wolpert, E. Vilas, A.C. Goheen, and L. A. Lider. 1989. Resistant rootstocks may control fanleaf degeneration of grapevines. Calif. Agric. 43(20):13–14.

Wilhelm, S., and A.O. Paulus. 1980. How soil fumigation benefits the California strawberry industry. Plant Dis. 64:264–270.

Williams, H.E., Q.L. Holdemann, and R. Casper. 1987. Tomato ringspot virus in red currant. p. 149–150. *In* R.H. Converse (ed.) Virus diseases of small fruits. USDA Agric. Handb. 631. USDA-ARS, Washington, DC.

Windham, G.L. 1998. Corn. p. 335–357. *In* K.R. Barker et al. (ed.) Plant and nematode interactions. Agron. Monogr. 36. ASA, CSSA, and SSSA, Madison, WI.

Wolfe, D., J.R. Hartman, G.R. Brown, and J. Strang. 1990. The influence of soil fumigation on strawberry yield and economics in black root rot infested fields. Appl. Agric. Res. 5:17–20.

Wyss, U. 1970. Parasitierungsvorgan und Pathogenitat Vorndernder Wurzelnematoden an *Fragaria vesca* var. *semperflorens.* Nematologica 16:55–62.

Zuckerman, B.M. 1961. Parasitism and pathogenesis of the cultivated cranberry by some nematodes. Nematologica 6:135–143.

Zuckerman, B.M., and J. Esnard. 1994. Biological control of plant nematodes—current status and hypotheses. Jpn. J. Nematol. 24:1–13.

Zuckerman, B.M., S. Himmelhoch, and M. Kisiel. 1973. Virus-like particles in *Dolichodorus heterocephalus*. Nematologica 19:117.

Zuckerman, B.M., S. Khera, and A.R. Price. 1964. Population dynamics of nematodes in cranberry soils. Phytopathology 54:654–659.

APPENDIX 1

Authorities and key general references for nematode species discussed in this monograph.

Taxon and authorities	References†
Acrobeloides (Cobb, 1924) Steiner & Buhrer, 1933 (microbivorous nematode)	Thorne, 1961
A. buetschlii (de Man, 1884) Steiner & Buhrer, 1933	Thorne, 1991
Amphlimerlinus Siddiqi, 1976	Nickle, 1991
A. icarus (Wallace & Greet, 1964) Siddiqi, 1976	Nickle, 1991
A. macrurus (Goodey, 1932) Siddizi, 1976	Nickle, 1991
Anguina Scapoli, 1777	Christie, 1959; Nickle 1991
A. agropyronifloris Norton, 1965	(see Chapt. 20; Bernard et al., 1998)
A. agrostis (Steinbuch, 1799) Filipjev, 1936	Nickle, 1991
A. funesta Price, Fisher & Kerr, 1979	Cook, 1997
A. amsinchia (Steiner & Scott, 1935)	Thorne, 1961
A. graminis (Hardy, 1850) Filipjev, 1936	Nickle, 1991
A. tritici (Steinbuch, 1799) Chitwood, 1935	Nickle, 1991
Aphasmatylenchulus Sher, 1965	Luc et al., 1990
A. straturatus Germani, 1970	Luc et al., 1990
Aphelenchoides-Fischer, 1894	Nickle, 1991; Hunt, 1993; F. Shahina, 1996
A. africanus Dassonville & Heyns, 1984	Hunt, 1993
A. arachidis Bos, 1977	Luc et al., 1990
A. besseyi Christie, 1942	Nickle, 1991
A. bicaudatus (Imamura, 1931) Filipjev & Schuurmans Stekhoven, 1941	Hunt, 1993
A. fragariae (Ritzema Bos, 1891) Christie, 1936	Nickle, 1991
A. nechaleos Hooper & Ibrahim, 1994	Shahina, 1996
A. paranechlaeos Hooper & Ibrahim, 1994	Shahina, 1996
A. ritzemabosi (Schwartz, 1911) Steiner & Buhrer, 1932	Christie, 1959; Nickle, 1991
Ascaris	
A. lumbricoides (Human parasite)	
Atylenchus Cobb, 1913	Thorne, 1961
A. decalineatus Cobb, 1914	Thorne, 1961
Belonolaimus-Steiner, 1949	Christie, 1959; Nickle, 1991
B. gracilis Steiner, 1949	Christie, 1959; Nickle, 1991
B. longicaudatus Rau, 1958	Christie, 1959; Nickle, 1991
B. nortoni Rau, 1963	Luc et al., 1990
Bursaphelenchus-Fuchs, 1937	Nickle, 1991
B. xylophilus (Steiner & Buhrer, 1934) Nickle, 1970	Nickle, 1991

Taxon and authorities	References†
Caenorhabditis Maupas, 1899	
C. elegans Maupas, 1899; Daugherty, 1953; Tarjan & Hooper, 1974	
Caloosia Siddiqi & Goodey, 1964	Luc et al., 1990
C. nudata (Colbran, 1963) Brzeski, 1994	Luc et al., 1990
Cacoparus Thorne, 1943	Nickle, 1991
C. pestis Thorne, 1943	Nickle, 1991
Criconema-Hofmanner & Menzel, 1914	Nickle, 1991
C. mutabile (Taylor, 1936) Raskit & Luc, 1984	Raski & Luc, 1984
Criconemella [see *Mesocriconema* (De Grisse & Loof, 1965)]	Luc et al., 1990
C. curvata (Raski, 1952) Luc & Raski, 1981	Luc et al., 1990
C. ornata (Raski, 1958) Luc & Raski, 1981	Luc et al., 1990
C. rustica (Micoletzky, 1915) Luc & Raski, 1981	Luc et al., 1990
C. sphaerocephala (Taylor, 1936) Luc & Raski, 1981	Luc et al., 1990
C. xenoplax (Raski, 1962) Luc & Raski, 1981	Luc et al., 1990
Cryphodera Colbran, 1966	Mai et al., 1996
C. utahensis Baldwin et al., 1983	Mai et al., 1996
Ditylenchus-Filipjev, 1936	Christie, 1959; Nickle, 1991
D. africanus Wendt et al., 1995	(see Chapt. 23; Dickson, 1998)
D. angustus (Butler, 1913) Filipjev, 1936	Luc et al., 1990
D. destructor Thorne, 1945	Christie, 1959; Nickle, 1991
D. dipsaci (Kühn, 1857) Filipjev, 1936	Christie, 1959; Nickle, 1991
Dolichodorus-Cobb, 1914	Christie, 1959; Nickle, 1991
D. heterocephalus Cobb, 1914	Christie, 1959; Nickle, 1991
Dorylaimus-Dujardin, 1845	Christie, 1959; Mai et al., 1996
Dracunculus [Human parasite]	
D. medinensis	
Globodera-Skarbilovich, 1959	Nickle, 1991
G. pallida (Stone, 1973) Behrens, 1975	Nickle, 1991
G. rostochiensis (Wollenweber, 1923) Behrens, 1975	Nickle, 1991
G. tabacum solanacearum (Miller & Gray, 1972) Behrens, 1975	Nickle, 1991
G. tabacum tabacum (Lownsbery & Lownsbery, 1954) Behrens, 1975	Nickle, 1991
G. tabacum virginiae (Miller & Gray, 1968) Behrens, 1975	Luc et al., 1990
Greeffiella	
G. minutum (marine nematode)	
Helicotylenchus-Steiner, 1945	Christie, 1959; Nickle, 1991
H. digonicus Perry, 1959	Thorne, 1961
H. dihystera (Cobb, 1893) Sher, 1961	Christie, 1959; Nickle, 1991

Taxon and authorities	References†
H. erythrinae (Zimmermann, 1904) Golden, 1956	Thorne, 1961
H. multicinctus (Cobb, 1893) Golden, 1956	Christie, 1959; Nickle, 1991
H. pseudorobustus (Steiner, 1914) Golden, 1956	Luc et al., 1990
Hemicycliophora-deMan, 1921	Raski & Luc, 1987; Mai et al., 1996
H. arenaria Raski, 1958	Luc et al., 1990
H. gigas Thorne, 1955	Raski & Luc, 1987
H. gracilis Thorne, 1955	Thorne, 1961
H. obtusa, Thorne, 1955	Raski & Luc, 1987
H. ritteri Brizuela, 1974	Raski & Luc, 1987
H. similis Thorne, 1955	Thorne, 1961
H. typica deMan, 1921	Mai et al., 1996
H. uniformis Thorne, 1955	Raski & Luc, 1987
H. vacinii Reed & Jenkins, 1963	Raski & Luc, 1987
H. zuckermani Brzeski, 1963	Raski & Luc, 1987
Heterodera-Schmidt, 1871	Christie, 1959; Nickle, 1991
H. avenae Wollenweber, 1924	Nickle, 1991
H. bifenestra Cooper, 1955	Nickle, 1991
H. carotae Jones, 1950	Nickle, 1991
H. cruciferae Franklin, 1945	Nickle, 1991
H. daverti Wouts & Sturhan, 1979	Nickle, 1991
H. elachista Ohshima, 1974	Nickle, 1991
H. galeopsidis Goffart, 1936	Nickle, 1991
H. glycines Ichinohe, 1952	Nickle, 1991
H. goettingiana Liebscher, 1892	Nickle, 1991
H. hordecalis Anderson, 1975	Nickle, 1991
H. humuli Filipjev, 1934	Nickle, 1991
H. latipons Franklin, 1969	Nickle, 1991
H. lespedeza Golden & Cobb, 1963	Nickle, 1991
H. mani Mathews, 1971	Nickle, 1991
H. oryzae Luc & Berdon Brizuela, 1961	Nickle, 1991
H. oryzicola Rao & Jayaprakash, 1978	Nickle, 1991
H. pakistanensis Maqbool & Shahina, 1986	Nickle, 1991
H. sacchari Luc & Merny, 1963	Nickle, 1991
H. schachtii A. Schmidt, 1871	Nickle, 1991
H. trifolii Goffart, 1932	Nickle, 1991
H. zeae Koshy, Swarup & Sethi, 1971	Nickle, 1991
Hirschmanniella-Luc & Goodey, 1964	Nickle, 1991
H. gracilis (deMan, 1880) Luc & Goodey, 1964	Luc et al., 1990
H. mucronata (Das, 1960) Luc & Goodey, 1964	Luc et al., 1990
H. oryzae (van Breda de Haan, 1902) Luc & Goodey, 1964	Nickle, 1991
Hoplolaimus-von Daday, 1905	Christie, 1959; Nickle, 1991
H. aegypti Shafiee & Koura, 1970	Luc et al., 1990
H. columbus Sher, 1963	Luc et al., 1990
H. galeatus (Cobb, 1913) Thorne, 1935	Nickle, 1991
H. indicus Sher, 1963	Luc et al., 1990
H. magnistylus Robbins, 1982	Mai et al., 1996

Taxon and authorities	References†
Longidorus-Micoletzky, 1922	Nickle, 1991; Hunt, 1993
L. africanus Merny, 1966	Luc et al., 1990
L. attenuatus Hooper, 1961	Hunt, 1963
L. breviannulatus Norton & Hoffman, 1975	Hunt, 1963
L. diadecturus Eveleigh & Allen, 1982	Mai et al., 1996
L. elongatus (deMan, 1876) Thorne & Swanger, 1936	Christie, 1959; Mai et al., 1996
L. iranicus Sturhan & Barooti, 1983	Mai et al., 1996
L. leptocephalus Hooper, 1961	Luc et al., 1990
L. macrosoma Hooper, 1961	Hunt, 1993
L. protae Lamberti & Bleve-Zacheo, 1977	Hunt, 1993
L. sylphus Thorne, 1939	Hunt, 1993; Thorne, 1961
Meloidodera-Chitwood, Hannon & Esser, 1956	Christie, 1959; Mai et al., 1996
M. floridensis Chitwood, Hannon & Esser, 1959	Christie, 1959; Mai et al., 1996
Meloidogyne-Goeldi, 1887	Christie, 1959; Nickle, 1991
M. acronea Coetzee, 1956	Luc et al., 1990
M. arenaria (Neal, 1889) Chitwood, 1949	Christie, 1959; Nickle, 1991
M. artiellia Franklin, 1961	Luc et al., 1990
M. carolinensis Eisenback, 1982	Nickle, 1991
M. chitwoodi Golden, O'Bannon, Santo, & Finley, 1980	Nickle, 1991
M. coffeicola Lordello & Zamith, 1960	Luc et al., 1990
M. exigua Goeldi, 1887	Christie, 1959; Nickle, 1991
M. graminicola Golden & Birchfield, 1965	Luc et al., 1990
M. graminis (Sledge & Golden, 1964) Whitehead, 1968	Nickle, 1991
M. hapla Chitwood, 1949	Christie, 1959; Nickle, 1991
M. hispanica HIrschmann, 1986	Nickle, 1991
M. incognita (Kofoid & White, 1919) Chitwood, 1949	Christie, 1959; Nickle, 1991
M. javanica (Treub, 1885) Chitwood, 1949	Christie, 1959; Nickle, 1991
M. mali Itoh, Oshima & Ichinohe, 1969	Nickle, 1991
M. marylandi Jepsen & Golden, 1987	Nickle, 1991
M. microtyla Mulvey, Townshend, & Potter, 1975	Nickle, 1991
M. morocciensis Rammah & Hirschmann, 1990	(see Chapt. 26; Nyczepir & Ole Becker, 1998)
M. naasi Franklin, 1965	Nickle, 1991
M. nataliei Golden, Rose, & Bird, 1981	Nickle, 1991
M. oryzae Maas, Sanders, & Dede, 1978	Nickle, 1991
M. salasi Lopez, 1984	Nickle, 1991
M. thamesi Chitwood *in* Chitwood, Specht, & Havis, 1952 (= *M. arenaria* thamesi)	Luc et al., 1990
M. trifoliophila Benard & Eisenback, 1997	Benard & Eisenback, 1997
M. triticoryzae Gaur, Saha & Khan, 1993	(see Chapt. 21; McGawley & Overstreet, 1998)
Merlinius Siddiqi, 1970	Luc et al., 1990; Nickle, 1991
M. brevidens (Allen, 1955) Siddiqi, 1970	Luc et al., 1990
M. nothus Allen, 1955	Tarjan & Hopper, 1974

Taxon and authorities	References†
Mesocriconema Andrássy, 1965 (see *Criconemella*)‡	Loof & deGrisse, 1989; Mai et al., 1996
M. ornatum (Raski) Loof & deGrisse, 1989‡	Loof & deGrisse, 1989
M. xenoplax (Raski) Loof & deGrisse, 1989‡	Loof & deGrisse, 1989
Nacobbus-Thorne & Allen, 1944	Christie, 1959; Nickle, 1991
N. aberrans (Thorne, 1935) Thorne & Allen, 1944	Christie, 1959; Nickle, 1991
N. batatiformis Thorne & Schuster, 1956	Thorne, 1961
N. dorsalis Thorne & Allen, 1944	Christie, 1959; Nickle, 1991
Nothotylenchus Thorne, 1941	Thorne, 1961
Oligaphelenchoides	
O. atrebora (fossilized nematode)	
Panagrellus spp. Fuchs, 1930 (microbivorous nematode)	Thorne, 1961
Paralongidorus§ Siddiqi, Hooper, & Khan, 1963	Hunt, 1993
P. bullatus Sharma & Siddiqi, 1990	Mai et al., 1996
P. nanus (Kirjanova, 1951) Lamberti, 1995	Hunt, 1993
Parascaris	
P. equorum (animal parasite)	
Paratrichodorus-Siddiqi, 1974	Hunt, 1993; Nickle, 1991
P. allius (Jensen, 1963) Siddiqi, 1974	Hunt, 1993
P. anemones (Loof, 1965) Siddiqi, 1974	Hunt, 1993
P. christie (cf. *P. minor*)	
P. lobatus (Colbran, 1965) Siddiqi, 1974	Luc et al., 1990
P. minor (Colbran, 1965) Siddiqi, 1974	Hunt, 1993
P. nanus (Allen, 1957) Siddiqi, 1974	Hunt, 1993
P. pachydermis (Seinhorst, 1954) Siddiqi, 1974	Luc et al., 1990
P. porosus (Allen, 1957) Siddiqi, 1974	Hunt, 1993
P. teres (Hooper, 1962) Siddiqi, 1974	Hunt, 1993
Paratylenchus-Micoletzky, 1922	Nickle, 1991; Raski & Luc, 1987
P. arenarium (Raski, 1956) Geraert, 1966	Esser, 1992
P. bukowinensis Micoletzky, 1922	Cook, 1997
P. hamatus Thorne & Allen, 1950	Luc et al., 1990
P. minutus Linford *in* Linford, Oliveira, & Ishii, 1949	Raski & Luc, 1987
P. neoamblycephalus Geraert, 1965	Raski & Luc, 1987
P. projectus (Jenkins, 1956)	Mai et al., 1996; Raski & Luc, 1987
P. prunii Sharma, Sharma & Khan, 1986	Esser, 1992
Placentonema	
P. gigantissima (parasite of whale)	
Pratylenchus-Filipjev, 1936	Christie, 1959; Nickle, 1991
P. agilis Thorne & Malek, 1968	Nickle, 1991
P. alleni Ferris, 1981	Luc et al., 1990
P. brachyurus (Godfrey, 1929) Goodey, 1959	Christie, 1959; Nickle, 1991
P. coffeae (Zimmerman, 1898) Goodey, 1959	Christie, 1959; Nickle, 1991
P. cornvallariae Seinhorst, 1959	Nickle, 1991
P. crenatus Loof, 1960	Luc et al., 1990

Taxon and authorities	References†
P. fallax Seinhorst, 1968	Luc et al., 1990
P. flakkensis Seinhorst, 1968	Luc et al., 1990
P. hexincisus Taylor & Jenkins, 1957	Luc et al., 1990
P. indicus Das, 1960	Luc et al., 1990
P. mediterranus Corbett, 1984	Nickle, 1991
P. neglectus (Rensch, 1924) Filipjev & Schuurmans Stekhoven, 1941	Luc et al., 1990
P. penetrans (Cobb, 1917) Chitwoodi & Oteifa, 1952	Christie, 1959; Nickle, 1991
P. pratensis (deMan, 1880) Filipjev, 1936	Luc et al., 1990
P. scribneri Steiner, 1943	Christie, 1959; Nickle, 1991
P. sefaensis Fortuner, 1974	Luc et al., 1990
P. sudanensis Loof & Yassin, 1971	Luc et al., 1990
P. thornei Sher & Allen, 1953	Christie, 1959; Nickle, 1991
P. vulnus Allen & Jensen, 1951	Christie, 1959; Nickle, 1991
P. zeae Graham, 1951	Luc et al., 1991
Punctodera Mulvey & Stone, 1976	Luc et al., 1990
P. punctata (Thorne, 1928) Mulvey & Stone, 1976	Luc et al., 1990
P chalcoensis Stone, Sosa Moss & Mulvey, 1976	Nickle, 1991
Pterotylenchus Siddiqi & Lenné, 1984	Mai et al., 1996
P. cecidogenus Siddiqi & Lenné, 1984	Mai et al., 1996
Quinisulcius Siddiqi, 1971	Mai et al., 1996
Q. acutus (Allen, 1955) Siddiqi, 1971	Sultan et al., 1995
Radopholus-Thorne, 1949	Christie, 1959; Nickle, 1991
R. citrophilus Huettel, Dickson & Kaplan, 1984	Nickle, 1991
R. oryzae (von Breda de Haan, 1902) Thorne, 1949	Christie, 1959; Thorne, 1961
R. similis (Cobb, 1893) Thorne, 1949	Nickle, 1991
Rhadinaphelenchus-J. B. Goodey, 1960	Nickle, 1991
R. cocophilus (Cobb, 1919) J. B. Goodey, 1960	Nickle, 1991
Rotylenchulus-Linford & Oliveira, 1940	Christie, 1959; Nickle, 1991
R. reniformis Linford & Oliveira, 1940	Christie, 1959; Nickle, 1991
R. parvus (Williams, 1960) Sher, 1961	Luc et al., 1990
Rotylenchus-Filipjev, 1936	Nickle, 1991
R. buxophilus Golden, 1956	Luc et al., 1990
R. robustus (deMan, 1876) Filipjev, 1936	Nickle, 1991
R. uniformis (Thorne, 1949) Loof & Oostenbrink, 1958	Tarjan & Hooper, 1973
Scutellonema-Andrássy, 1958	Nickle, 1991
S. brachyurus (Steiner, 1938) Andrássy, 1958	Nickle, 1991
S. cavenessi Sher, 1964	Luc et al., 1990
S. clathricaudatum Whitehead, 1959	Luc et al., 1990
Tetradonema	
T. plicans (insect parasite)	
Tetylenchus Filipjev, 1936	Luc et al., 1990
T. nicotianae	
T. joctus Thorne, 1949	Thorne, 1961
T. productus Thorne, 1949	Thorne, 1961

Taxon and authorities	References†
Tobrilus Bastian, 1865 (microbivorous nematode)	Tarjan & Hopper, 1974
T. gracilis Bastian, 1865	Tarjan & Hopper, 1974
T. husmanni	
T. pellucidus (Bastian, 1865) Andrássy, 1959	Cook, 1997
Trichodorus-Cobb, 1913	Hunt, 1993; Nickle, 1991
T. aequalis Allen, 1957	Hunt, 1993
T. californicus Allen, 1957	Hunt, 1993
T. cylindrious Hooper, 1962	Hunt, 1993
T. hooperi Loof, 1973	Hunt, 1993
T. primitivus (deMan, 1880) Micoletzky, 1922	Hunt, 1993
T. proximus Allen, 1957	Hunt, 1993
T. similis Seinhorst, 1963	Hunt, 1993
T. sparsus Szczygiel, 1968	Hunt, 1993
Trophotylenchulus Raski, 1957	Luc et al., 1990
T. obscurus (Colbran, 1961) Cohn & Kaplan, 1983	Luc et al., 1991
Tylenchorhynchus-Cobb, 1913	Nickle, 1991
T. acutoides Thorne & Malek, 1968	Tarjan & Hopper, 1974
T. acutus Allen, 1955	Luc et al., 1990
T. agri Ferris, 1963	Nickle, 1991
T. annulatus (Cassidy, 1930) Golden, 1971	Luc et al., 1990
T. brassicae Siddiqi, 1961	Luc et al., 1990
T. brevilineatus Williams, 1960	Luc et al., 1990
T. clarus Allen, 1955	Luc et al., 1990
T. claytoni Steiner, 1937	Nickle, 1991
T. cylindricus Cobb, 1913	Nickle, 1991
T. dubius (Bütschlii, 1873) Filipjev, 1936	Nickle, 1991
T. ewingi Hopper, 1959	Tarjan & Hopper, 1974
T. fischeri Caveness, 1958	(Caveness, Proc. Helminthol. Soc. Wash. D.C. 25:122–124)
T. lamelliferus (de Man, 1880) Filipjev, 1936	Tarjan & Hopper, 1974
T. latus Allen, 1955	Tarjan & Hopper, 1974
T. maximus Allen, 1955	Nickle, 1991
T. nudus Allen, 1955	Luc et al., 1990
T. robustoides Thorne & Malek, 1968	(see Chapt. 20; Bernard et al., 1998)
T. similis Jairajpuri, 1965	Tarjan & Hopper, 1974
T. striatus Allen, 1955	Tarjan & Hopper, 1974
T. vulgaris Upadhyay, Swarup, & Sethi, 1972	Luc et al., 1990
Tylenchulus-Cobb, 1913	Christie, 1959; Nickle, 1991
T. graminis Inserra, Vovlas, O'Bannon, & Esser, 1988	Luc et al., 1991
T. semipenetrans Cobb, 1913	Christie, 1959; Nickle, 1991
Tylenchus Bastian, 1865	Luc et al., 1990
Wilsonema Cobb, 1913	Tarjan & Hopper, 1974
W. capitatum Cobb, 1913 (syn. *Bitholinema schuurmansstekhoveni*)	Tarjan & Hopper, 1974
W. schuurmansstekhoveni	

Xiphinema-Cobb, 1913	Hunt, 1993; Nickle, 1991
X. algeriense Luc & Kostadinov, 1982	Hunt, 1993
X. americanum Cobb, 1913	Hunt, 1993; Mai et al., 1996
X. bakeri Williams, 1961	Hunt, 1993
X. basiri Siddiqi, 1959	Hunt, 1993
X. belmontense Roca & Pereira, 1992	Hunt, 1993
X. brevicolle Lordello & Da Costa, 1961	Hunt, 1993
X. bricolense Ebsary, Vrain, & Graham, 1989	Hunt, 1993
X. californicum Lamberti & Bleve-Zacheo, 1979	Hunt, 1993
X. chambersi Thorne, 1939	Hunt, 1993
X. diversicaudatum (Micoletzky, 1927) Thorne, 1939	Hunt, 1993; Thorne, 1961
X. index Thorne & Allen, 1950	Hunt, 1993; Thorne, 1961
X. italiae Meyl, 1953	Hunt, 1993
X. meridianum Haynes, 1971	Hunt, 1993
X. pachtaicum (Tulaganov, 1938) Kirjanova, 1951	Hunt, 1993
X. rivesi Dalmasso, 1969	Hunt, 1993
X. turcicum Luc & Dalmasso, 1964	Hunt, 1993
X. vuittenezi Luc, Lima, Weischer, & Flegg, 1964	Hunt, 1993

† Bernard, E.C., and J.D. Eisenback. 1997. *Meloidogyne trifoliophila* n. sp. (Nemata Meloidogynidae), a parasite of clover from Tennessee. J. Nematol. 29:43–54.

Bernard, E.C., K.D. Gwinn and G.D. Griffin. 1998. Forage grasses. p. 427–454. *In* K.R. Barker et al. (ed.) Plant and nematode interactions. Agron. Monogr. 36. ASA, CSSA, and SSSA, Madison, WI.

Christie, J.R. 1959. Plant nematodes: Their bionomics and control. Univ. Florida Agric. Exp. Stn.

Cook, R. (ed.). 1997. Nematologiea—index to volumes 1–42. Nematologica 43(3–4):1–249.

Esser, R.P. 1992. A diagnostic compendium to species included in paratylenchinae Thorne, 1949 and Tylenchocriconematinae Raski & Siddiqi, 1975 (Nematodia: Criconematoidea). Nematologica 38:146–163.

Hunt, D.J. 1993. Aphelenchida, Longidoridae, and Trichodoridae. Their systematics and bionomics. CAB Int., Wallingford, Oxon, England.

Loof, P.A.A., and A. deGrisse. 1989. Taxonomic and nomenclatorial observations on the genus *Criconemella* deGrisse and Loof, 1965 *sensu* Luc & Raski, 1981 (Criconematidae). Med. Fac. LandbWet. (Rijksuniv, Gent) 54:53–74.

Luc, M., R.A. Sikora, and J. Bridge (ed.). 1990. Plant parasitic nematodes in subtropical and tropical agriculture. CAB Int., Wallingford, England.

Mai, W. F., P. G. Mullin, H. H. Lyons, and K. Loeffler. 1996. Plant-parasitic nematodes—a pictorial key to genera. 5th ed. Cornell Univ. Press, Ithaca, NY.

McGawley, E.C., and C. Overstreet. 1998. Rice and other cereals. p. 455–486. *In* K.R. Barker et al. (ed.) Plant and nematode interactions. Agron. Monogr. 36. ASA, CSSA, and SSSA, Madison, WI.

Nickle, W.R. (ed.). 1991. Manual of agricultural nematology. Marcell Dekker, Inc., New York.

Nyczepir, A.P., and J.O. Becker. 1998. Fruit and citrus trees. p. 637–684. *In* K.R. Barker et al. (ed.) Plant and nematode interactions. Agron. Monogr. 36. ASA, CSSA, and SSSA, Madison, WI.

Raski, D.J., and M. Luc. 1987. Rev. Nematol. 10:409–444.

Shahina, F. 1996. A diagnostic compendium of the genus *Aphelenchoides* Fisher, 1894 (Nematoda: Aphelenchida) with some new records of the group from Pakistan. Pak. J. Nematol. 14:1–32.

Sultan, M.S., I. Singh, and P.K. Sakhuja. 1995. Plant-parasitic nematodes of Punjab, india IV. Tylenchorhynchinae Eliava, 1964 with *Quinisulcius domesticus* n. sp. Pak. J. Nematol. 13:69–72.

Tarjan, A.C., and B.E. Hopper. 1974. Nomenclatorial compilation of plant and soil nematodes. Soc. Nematol., Hyattsville, MD.

Thorne, G. 1961. Principles of nematology. McGraw-Hill Book Co., Inc., New York.

‡ Mai et al. (1996) and others use *Criconernella* in lieu of *Mesocriconema* for these species.

§ *Paralongidorus* recently restructured into subgenera (Hunt, 1993).

APPENDIX 2

Major parasitic nematodes and their distribution on selected crop plants.†

Crops	Nematodes—genera-species	Geographic distribution	References‡
Alfalfa	*Meloidogyne arenaria*	Temperate and tropical regions	2, 5
	M. incognita	Temperate and tropical regions	2, 4, 5
	M. hapla	Northern and temperate regions	2, 4, 5
	Ditylenchus dipsaci	Most regions where alfalfa is grown	2, 4, 5
	Pratylenchus penetrans	Temperate and northern regions	2, 5
	Xiphinema americanum	Widespread in North America	2, 4, 5
Apple	*Meloidogyne* spp.	Limited	2, 3, 6
	Pratylenchus penetrans	Europe, Canada, USA	3, 4, 6
	Xiphinema spp.	Widespread	2, 3, 6
Azalea	*Meloidogyne* spp.	Widespread in southeastern USA	
	Tylenchorhynchus claytoni	Widespread in southeastern USA	1, 3
Banana	*Helicotylenchus* spp.	Widespread	3
	Meloidogyne spp.	Worldwide	3
	Pratylenchus spp.	Widespread	3
	Radopholus similis	Widespread	1, 3
	Rotylenchulus reniformis	Widespread	1, 3
Bean	*Meloidogyne incognita*	Widespread	3, 4, 7
	Belonolaimus spp.	Limited to very sandy soils, especially southeastern USA	7
	Pratylenchus spp.	Widespread	3, 7
	Heterodera glycines	Primarily in infested soybean-growing regions	7
Beet (table)	*Heterodera schachtii*	Widespread	4, 7
	Meloidogyne spp.	Widespread	1, 7
	Paratrichodorus spp.	Florida	1, 7
Bent grass	*Belonolaimus longicaudatus*	Southeastern USA	12
	Criconemella spp. (*Mesocriconema*)§	Widespread	3
Cabbage crucifers	*Heterodera cruciferae*	England, Wales, California	1, 3, 7
	H. schachtii	Widespread	3, 4, 7
	Meloidogyne spp.	Widespread	3, 4, 7
Carrot	*Heterodera carotae*	England, Netherlands, Michigan (USA)	1, 7
	Meloidogyne hapla	Widespread	1, 4, 7
Celery	*Belonolaimus* spp.	Florida	1, 3, 4, 7
	Meloidogyne hapla	New York	4
	Paratylenchus spp.	Northeastern USA, the Netherlands	1, 3

Crops	Nematodes—genera-species	Geographic distribution	Refer-ences‡
Citrus trees	*Hemicycliophora* spp.	Australia, USA	3, 6
	Pratylenchus spp.	Widespread	3, 4, 6
	Radopholus citrophilus	Florida	1, 3, 6
	Tylenchulus semipenetrans	All major citrus regions	3, 4, 6
	Xiphinema spp.	Worldwide	3, 6
Clover, white and red	*Meloidogyne* spp.	Widespread	2, 8
	Heterodera trifolii	Widespread	2, 8
	Ditylenchus dipsaci	Temperate regions	2, 8
Coconut palm	*Radopholus similis*	Florida, Jamaica, Sri Lanka, India	3
	Rhadinaphelenchus cocophilus	Caribbean, Central and South America	1, 3
Corn	*Belonolaimus* spp.	Atlantic Coastal Plain of USA	1, 4, 9
	Heterodera zeae	Africa, India, USA (Maryland and Virginia)	1, 3, 4, 9
	Hoplolaimus spp.	Widespread	1, 3, 9
	Longidorus spp.	Temperate and cooler regions	3, 4, 9
	Meloidogyne spp.	Most production regions	3, 4, 9
	Paratrichodorus spp.	Widespread	1, 3, 4, 9
	Pratylenchus spp.	Worldwide, midwestern and eastern USA	1, 3, 4, 9
	Xiphinema spp.	Widespread	3, 9
Cotton	*Belonolaimus* spp.	Atlantic Coastal Plain of USA	1, 3, 4, 10
	Hoplolaimus spp.	Widespread	3, 4, 10
	Meloidogyne incognita	Worldwide	3, 4, 10
	M. acronea	Southern Africa	3, 10
	Paratrichodorus minor	Southeastern USA	3
	Pratylenchus spp.	Widespread	3
	Rotylenchulus reniformis	Tropic and subtropic regions	3, 4, 10
	Scutellonema spp.	Limited	3
Cowpeas	*Belonolaimus* spp.	Southeastern USA (sandy soils)	3
	Heterodera cajani	Egypt, India	3
	Meloidogyne spp.	Most production regions	3, 4
	Pratylenchus spp.	Widespread	3
	Rotylenchulus reniformis	India, USA	3
Cranberry	*Hemicycliophora* spp.	northeastern USA, Wisconsin	4, 11
Fescue	*Pratylenchus* spp.	Widespread	2, 12
	Xiphinema spp.	Widespread	2
Grape	*Meloidogyne* spp.	Widespread	2, 4, 11
	Pratylenchus vulnus	Widespread	4, 11
	Xiphinema spp.	Widespread	4, 11
Mint	*Longidorus* spp.	Oregon	1
	Pratylenchus spp.	Oregon	2
Oats	*Heterodera avenae*	Oregon (USA), Canada, Europe, Japan, Australia, India	1, 2, 3, 4, 13

Crops	Nematodes—genera-species	Geographic distribution	References‡
	Pratylenchus spp.	Widespread	2, 4
Peach	*Criconemella xenoplax* (*Mesocriconema*)§	Widespread	3, 4, 6
	Meloidogyne incognita	Western and southern USA	3, 4, 6
	Pratylenchus vulnus	Canada, Europe, USA	4, 6
Peanut	*Belonolaimus* spp.	Atlantic Coastal Plain of USA (sandy soils)	1, 3, 4, 14
	Criconemella ornata (*Mesocriconema*)§	Africa, USA	3, 4, 14
	Ditylenchus africanus	South Africa	3, 14
	Meloidogyne arenaria	Australia, Japan, North, Central and South America	1, 3, 4, 14
	M. hapla	North Carolina, Virginia, and other cool production areas	1, 3, 4, 14
	Pratylenchus brachyurus	Warmer production areas	3, 4, 14
Pea (garden)	*Heterodera goettingiana*	Oregon (USA), Europe	1, 3, 7
	Meloidogyne spp.	Widespread	3, 7
	Pratylenchus spp.	Widespread	3, 7
Pepper, black	*Meloidogyne* spp.	Widespread	3
	Radopholus similis	India, Indonesia, Malaysia, Thailand, Sri Lanka	3
Pepper (*Capsicum*)	*Belonolaimus* spp.	Florida	1, 4
	Meloidogyne spp.	Widespread	1, 4, 7
	Pratylenchus spp.	Widespread	4, 7
Pineapple	*Meloidogyne* spp.	Most production areas	3
	Rotylenchulus reniformis	Widespread in tropics and subtropics	3
Potato	*Ditylenchus destructor*	Northern Europe, Wisconsin (USA), Prince Edward Island, South America	1, 3, 4
	Globodera rostochiensis	Europe, South America, New York	3, 4
	G. pallida	Europe, South America	3, 4
	Meloidogyne spp.	Widespread	3, 4
	Nacobbus spp.	Russia, South America, USA	3, 4
	Pratylenchus spp.	Widespread	1, 3, 4
Rice	*Aphelenchoides besseyi*	Japan, USA, most rice regions	1, 3, 4, 13
	Criconemella spp. (*Mesocriconema*)§	Occurrence varies with nematode species and habitat	3, 4
	Ditylenchus augustus	Bangladesh, Burma, India, Malaysia, Thailand, Vietnam	3, 13
	Heterodera spp.	West Africa, India, Ivory Coast, Japan, Bangladesh (distribution varies with species)	3
	Hirschmannellia spp.	Widespread in suitable habitats	3, 4, 13
	Meloidogyne spp.	Widespread (species vary with habitat)	3, 4, 13

Crops	Nematodes—genera-species	Geographic distribution	References‡
	Pratylenchus spp.	Worldwide	3, 13
	Tylenchorhynchus spp.	Worldwide	3
	Xiphinema spp.	Ivory Coast, West Africa, Nigeria, India	3
Soybean	*Belonolaimus* spp.	Atlantic Coastal Plain of USA	1, 4, 15
	Heterodera glycines	Brazil, Canada, China, Japan, USA	1, 3, 4, 15
	Hoplolaimus spp.	India, USA	3, 4, 15
	Meloidogyne spp.	Widespread	3, 4, 15
	Pratylenchus spp.	Widespread	3, 4, 15
	Rotylenchulus reniformis	Limited	3, 4, 15
Strawberry	*Aphelenchoides* spp.	Widespread	3, 4, 11
	Belonolaimus spp.	Limited, Florida	1, 11
	Meloidogyne hapla	Widespread	3, 4, 11
	Pratylenchus spp.	Widespread	1, 3, 4, 11
Sugar beet	*Heterodera schachtii*	Europe, Australia, North America	3, 4, 7
Sugarcane	*Helicotylenchus* spp.	Widespread	3, 4
	Meloidogyne spp.	Widespread	3, 4
	Paratrichodorus/*Trichodorus* spp.	Limited	3, 4
	Pratylenchus spp.	Widespread	3, 4
	Tylenchorhynchus spp.	Widespread	3, 4
	Xiphinema spp.	Limited	3
Sweet potato	*Meloidogyne incognita*	Widespread	3
	Nacobbus spp.	Limited (Americas)	3
	Rotylenchulus reniformis	Most production areas	3, 4
	Scutellonema spp.	Limited	3
Tobacco	*Ditylenchus dipsaci*	Europe	1, 3, 4, 16
	Globodera tabacum solanacearum	North Carolina, Virginia, Maryland	3, 4, 16
	G. tabacum	Europe, USA (Connecticut)	3, 4, 16
	Meloidogyne arenaria	Temperate and tropical regions	3, 4, 16
	M. hapla	Temperate and northern regions	3, 4, 16
	M. incognita	Widespread	3, 4, 16
	M. javanica	Temperate and tropical regions	3, 4, 16
	Paratrichodorus spp.	Limited	3, 16
	Pratylenchus spp.	Widespread	3, 4, 16
Tomato	*Meloidogyne* spp.	Widespread	2, 7
	Nacobbus spp.	Limited	2
	Pratylenchus spp.	Widespread	2, 7
	Rotylenchulus reniformis	Limited	2. 7
Wheat	*Anguina tritici*	Worldwide, but usually limited importance	1, 3, 4, 13
	Heterodera avenae	Widespread (Oregon, USA)	3, 4, 13

	Pratylenchus spp.	Widespread	3, 4, 13
Yams	*Meloidogyne* spp.	Africa, Carribean, Japan	3
	Pratylenchus spp.	Widespread	3
	Scutellonema bradys	Africa, Carribean, South America	3

† After Christie (1959), in part [for extensive and specific references for the host-parasite index see: Cook (ed.). 1997. Index of authors and titles, Vols. 1–42. Nematologica 43(3–4)].

‡ 1 = Christie, J.R. 1959. Plant nematodes: Their bionomics and control. Univ. of Florida Agric. Exp. Stn.; 2 = Evans, K., D.L. Trudgill, and J.M. Webster (ed.). 1993. Plant-parasitic nematodes in temperate agriculture. CAB Int., Wallingford, England; 3 = Luc, M., R.A. Sikora, and J. Bridge (ed.). 1990. Plant-parasitic nematodes in subtropical and tropical agriculture. CAB Int., Wallingford, England; 4 = Nickle, W.R. (ed.). 1984. Plant and insect nematodes. Marcel Dekker, Inc., New York; 5 = Griffin, G.D. 1998. Alfalfa. p. 381–397. *In* K.R. Barker et al. (ed.) Plant and nematode interactions. Agron. Monogr. 36. ASA, CSSA, and SSSA, Madison, WI; 6 = Nyczepir, A.P., and J.O. Becker. 1998. Fruit and citrus trees. p. 637–684. *In* K.R. Barker et al. (ed.) Plant and nematode interactions. Agron. Monogr. 36. ASA, CSSA, and SSSA, Madison, WI; 7 = Johnson, A.W. 1998. Vegetable crops. p. 595–635. *In* K.R. Barker et al. (ed.) Plant and nematode interactions. Agron. Monogr. 36. ASA, CSSA, and SSSA, Madison, WI; 8 = Pederson, G.A., and K.H. Quesenberry. 1998. Clovers and other forage legumes. p. 399–425. *In* K.R. Barker et al. (ed.) Plant and nematode interactions. Agron. Monogr. 36. ASA, CSSA, and SSSA, Madison, WI; 9 = Windham, G.L. 1998. Corn. p. 335–357. *In* K.R. Barker et al. (ed.) Plant and nematode interactions. Agron. Monogr. 36. ASA, CSSA, and SSSA, Madison, WI; 10 = Starr, J.L. 1998. Cotton. p. 359–379. *In* K.R. Barker et al. (ed.) Plant and nematode interactions. Agron. Monogr. 36. ASA, CSSA, and SSSA, Madison, WI; 11 = Esnard, J., and B.M. Zuckerman. 1998. Small fruits. p. 685–725. *In* K.R. Barker et al. (ed.) Plant and nematode interactions. Agron. Monogr. 36. ASA, CSSA, and SSSA, Madison, WI; 12 = Bernard, E.C., K.D. Gwinn, and G.D. Griffin. 1998. Forage grasses. p. 427–454. *In* K.R. Barker et al. (ed.) Plant and nematode interactions. Agron. Monogr. 36. ASA, CSSA, and SSSA, Madison, WI; 13 = McGawley, E.C., and C. Overstreet. 1998. Rice and other cereals. p. 455–486. *In* K.R. Barker et al. (ed.) Plant and nematode interactions. Agron. Monogr. 36. ASA, CSSA, and SSSA, Madison, WI; 14 = Dickson, D.W. 1998. Peanut. p. 523–566. *In* K.R. Barker et al. (ed.) Plant and nematode interactions. Agron. Monogr. 36. ASA, CSSA, and SSSA, Madison, WI; 15 = Kinloch, R.A. 1998. Soybean. p. 317–333. *In* K.R. Barker et al. (ed.) Plant and nematode interactions. Agron. Monogr. 36. ASA, CSSA, and SSSA, Madison, WI; 16 = Johnson, C.S. 1998. Tobacco. p. 487–522. *In* K.R. Barker et al. (ed.) Plant and nematode interactions. Agron. Monogr. 36. ASA, CSSA, and SSSA, Madison, WI.

§ Luc et al. (1990), Mai et al. (1996), and others use *Criconemella* in lieu of *Mesocriconema* for these nematode species.

APPENDIX 3

Scientific and common names for plant species discussed in this monograph.

Scientific name	Common name
Abelmoschus esculentus (L.) Moench (syn. *Hibiscus esculentus*)	Okra
Aegilops squarrosa L.	--
A. ventricosa Tausch	--
Aeschynomene afraspera J. Leonard	--
A. aspera L.	Sola-pith plant
A. americana L.	American jointvetch
Agropyron cristatum (L.) Gaertn.	Crested wheatgrass
A. desertorum (Fisch. ex Link) Schultes	Standard crested wheatgrass
Agrostis avenacea J.F. Gmelin	Blown grass, New Zealand wind grass
A. capillaris L. (syn. *A. tennuis*)	Colonial bentgrass
Allium canadense L.	Wild onion
A. cepa L.	Onion
A. porrum L.	Leek
A. sativum L.	Garlic
A. schoenoprasum L.	Chive
Alysicarpus vaginalis (L.) DC.	Alyceclover
Amsinckia intermedia Fish & C.A. May	Fiddleneck
Ananas comosus (L.) Merr.	Pineapple
Andropogon gerardii Vitman	Big bluestem
A. virginicus L.	Boomsedge
Apium graveolens L. var. *dulce* (Miller) Pers.	Celery
Arachis batizocoi	--
A. cardenasii Krapov. & W.C. Gregory	--
A. diogoi Hoehne (syn. *A. chancoensis*)	--
A. glabrata Benth.	Rhizoma perennial peanut
A. hypogaea L.	Peanut
A. hypogaea var. *fastigiata* (Wald.) Krapov. & W.C. Greg. (syn. *A. fastigiata*)	Valencia peanut
Arrhenatherum elatius (L.) J.S. & C. Presl	Tall oatgrass
Asparagus officinalis L.	Asparagus
Avena sativa L.	Oat
Beta vulgaris L.	Sugar beet, table beet
B. vulgaris var. *cicla* L.	Chard
Bouteloua curtipendula (Michx.) Torr.	Sideoats grama
B. gracilis (Kunth) Lag. ex Griffiths	Blue grama
Brassica juncea (L.) Czernj.	Mustard
B. napus L.	Rape
B. napus var. *napobrassica* (L.) Reichb.	Rutabaga
B. oleracea var. *botrytis* L.	Broccoli, cauliflower
B. oleracea var. *capitata* L.	Cabbage
B. oleracea var. *gemmifera* Zenker	Brussels sprout

Scientific name	Common name
B. oleracea var. *viridis* L.	Kale, collards
B. rapa L.	Turnip
B. rapa ssp. *pekinensis* (Lour.) Hanelt (syn. *B. pekinensis*)	Chinese cabbage
Bromus inermis Leyss.	Smooth bromegrass
Buchloe dactyloides (Nutt.) Engelm.	Buffalograss
Canavalia ensiformis (L.) DC.	Horsebean
Capsicum annuum var. *annuum* L.	Bell pepper, green pepper
C. frutescens L. var. *grossum*	Pimento pepper
C. frutescens L. var. *longum*	Hot pepper
Chamoecrista fasciculata (Michx.) Greene	Partridge pea
Chenopodium album L.	Common lambsquarters
Chloris gayana Kunth	Rhodes grass
C. pectina Benth.	--
Cicer arietinum L.	Chickpea
Cichorium endivia L.	Endive
Citrullus lanatus (Thunb.) Matsum. & Nakai (syn. *C. vulgaris*)	Watermelon
Citrus aurantiifolia (Christm.) Swingle	Lime
C. aurantium L.	Sour orange
C. maxima (Burm.) Merr. (syn. *C. grandis*)	Pummelo
C. limon (L.) Burman f.	Lemon
C. medica L.	Citron
Citrus × *paradisi* Macfad.	Grapefruit
C. reticulata Blanco	Mandarin orange, tangerine
C. sinensis (L.) Osbeck	Sweet orange
Cocos nucifera L.	Coconut palm
Crotalaria spectabilis Roth	Showy crotalaria
Cucumis anguria L. (syn. *C. longipes*)	West India gherkin
C. ficifolius A. Rich.	--
C. heptadactylus Naudin	--
C. melo L.	Muskmelon
C. melo var. *cantalupensis* Naudin (syn. var. *reticulatus*)	Cantaloupe
C. metuliferus E. Meyer ex Naudin	Kiwano, horned cucumber
C. sativus L.	Cucumber
Cucurbita pepo L.	Pumpkin
C. pepo var. *melopepo*	Squash
Cynara cardunculus L. (syn. *C. scolymus*)	Artichoke
Cynodon dactylon (L.) Pers.	Bermudagrass
Dactylis glomerata L.	Orchardgrass
Daucus carota L.	Carrot
Dendranthemax × *grandiflorum* (Ramat.) Kitam.	Chrysanthemum
Digitaria eriantha Steud. (syn. *D. decumbens*)	Pangola grass
Diospyros kaki Thunb.	Japanese persimmon
Dipsacus fullonum L.	Wild teasel
Elaeis guineensis Jacq.	African oil palm
Elytrigia intermedia (Host) Nevski (syn. *Thinopyrum intermedium*)	Intermediate wheatgrass

Scientific name	Common name
Eremochloa ophiuroides (Munro) Hack.	Centipede grass
Eschscholzia papaver Cham	California poppy
Fagopyrum esculentum Moench	Buckwheat
Festuca arundinacea Schreb.	Tall fescue
F. pratensis Huds.	Meadow fescue
F. rubra L.	Red fescue
Fortunella margarita (Lour.) Swingle	Kumquat
Fragaria chiloensis (L.) Duchesne	Beach, Chilean strawberry
Fragaria × *ananassa* Duchesne	Garden strawberry
F. vesca L.	European strawberry
F. virginiana Duchesne	Virginia Strawberry
Glycine max (L.) Merr.	Soybean
Gossypium barbadense L.	Long staple, pima cotton
G. hirsutum L.	Cotton, upland cotton
Hordeum vulgare L.	Barley
Humulus lupulus L.	Hop
Indigofera hirsuta L.	Hairy indigo
Ipomoea batatas (L.) Lam.	Sweetpotato
Kummerowia stipulacea (Maxim.) Makino	Korean lespedeza
K. striata (Thunb.) Schindler	Striate lespedeza
Lactuca sativa L.	Lettuce
Lespedeza cuneata (Dum.-Cours.) G. Don	Sericea lespedeza
Leucanthemum maximum (Ramond) DC	Max chrysanthemum
Leymus cinereus (Scribn. & Merr.) A. Love	Great Basin wildrye grass
Linum usitatissimum L.	Flax
Lolium multiflorum Lam.	Annual ryegrass, Italian ryegrass
L. perenne L.	Perennial ryegrass
L. rigidum Gaud.	Wimmera ryegrass
Lotus corniculatus L.	Birdsfoot trefoil
Lycopersicon esculentum Miller	Tomato
L. peruvianum (L.) Mill.	--
Malus domestica Borkh.	Apple
M. pumila Miller	Paradise apple
Medicago sativa L.	Alfalfa
Melilotus albus Medik.	White sweetclover
M. indicus (L.) All.	Indian sweetclover
M. officinalis Lam.	Yellow sweetclover
Mucuna pruriens var. *utilis* (Wallich ex Wight) Baker ex Burck (syn. *M. deeringiana*)	Velvetbean
Muhlenbergia schreberi J.F. Gmelin	Nimblewill
Musa × *paradisiaca* L.	Banana
Nicotiana glauca Graham	Tree tobacco
N glutinosa L.	--
N. knightiana Goodsp.	--
N. longiflora Cav.	--
N. megalosiphon Van Heurck & Mull. Arg.	--
N. miersii J. Remy	--
N. nudicaulis S. Watson	--
N. otophora Griseb.	--

Scientific name	Common name
N. paniculata L.	--
N. plumbaginifolia Viv.	--
N. repanda Willd. ex Lehm.	Wild tobacco
N. × *sanderae* Will. Watson	--
N. tabacum L.	Tobacco
N. tomentosa Ruiz Lopez & Pavon	--
N. velutina Wheeler	--
Olea europaea L.	Olive
Onobrychis viciifolia Scop.	Sainfoin
Oryza sativa L.	Rice
Panicum miliaceum L.	Millet
Pascopyrum smithii (Rydb.) A. Love	Western wheatgrass
Paspalum notatum Flugge	Bahiagrass
Pastinaca sativa L.	Parsnip
Petroselinum crispum (Miller) Nyman ex A.W. Hill	Parsley
Phalaris arundinacea L.	Reed canarygrass
Phaseolus lunatus L.	Lima bean
P. vulgaris L.	Bean
Phleum pratense L.	Timothy
Pinus sylvestris L.	Scotch pine
Piper nigrum L.	Black pepper
Pisum sativum L.	Pea
Plantago major L.	Common plantain, broadleaf plantain
Poa pratensis L.	Kentucky bluegrass
Polypogon monspeliensis (L.) Desf.	Annual beardgrass
Poncirus trifoliata (L.) Raf.	Trifoliate orange
Praecitrullus fistulosus (Stocks) Pangalo	Round melon
Prunus armeniaca L.	Apricot
P. avium (L.) L.	Sweet cherry
P. besseyi L. Bailey	Sand cherry
P. cerasifera Ehrh.	Cherry plum
P. cerasus L.	Sour cherry
P. domestica L.	Plum
P. dulcis (Miller) Webb (syn *P. amygdalus*)	Almond
P. mahaleb L.	Mahaleb cherry
P. munsoniana W. Wight & Hedr.	Wild goose plum
P. persica (L.) Batsch	Peach, nectarine
P. salicina Lindley	Plum, Japanese plum
Pseudoroegneria spicata (Pursh) A. Love	Bluebunch wheatgrass
Pyrus communis L.	Pear
Raphanus sativus L.	Radish
Ribes nigrum L.	Black currant
R. rubrum L.	Red currant
R. uva-crispa L.	Gooseberry, European gooseberry
Ricinus communis L.	Castor bean
Rubus idaeus L.	Red raspberry, European red raspberry
R. occidentalis L.	Black raspberry

Scientific name	Common name
R. plicatus Weihe & Nees	Blackberry
R. strigosus Michaux	American red raspberry
Saccharum officinarum L.	Sugarcane
Sambucus canadensis L.	American elderberry
Schizachyrium scoparium (Michx.) Nash	Little bluestem
S. rhizomatum (Swallen) Gould	Florida bluestem
Secale cereale L.	Rye
Securigera varia (L.) Lassen (syn. *Coronilla varia*)	Crownvetch
Senna obtusifolia (L.) H.S. Irwin & Barneby (syn. *Cassia obtusifolia*)	Sicklepod
S. occidentalis (L.) Link (syn. *Cassia occidentalis*)	Coffee senna
Sesamum indicum L.	Sesame
Sesbania rostrata Bremek. & Oberm.	--
Setaria palmifolia (Koenig) Stapf	Pit-pit plant
Sinapis alba L.	White mustard
Solanum bulbocastanum Dunal	--
S. hougasii Correll	--
S. kurtzianum Bitter & Wittm.	--
S. melongena L.	Eggplant
S. multidissectum Hawkes	--
S. sisymbriifolium Lam.	--
S. sparsipilum (Bitter) Juz. & Buk.	--
S. spegazzinii Bitter	--
S. tuberosum ssp. *andigena* Hawkes	Andigena
S. tuberosum ssp. *tuberosum* L.	Potato
S. vernei Bitter & Wittm.	--
Sorghum bicolor (L.) Moench	Sorghum, milo
S. × *drummondii* (Nees ex Steudel) Millsp. & Chase	Sudangrass
Spergula arvensis L.	Corn spurrey
Spinacia oleracea L.	Spinach
Taraxacum officinale Wigg.	Dandelion
Tragopogon porrifolius L.	Salsify
Trifolium alexandrinum L.	Berseem clover
T. ambiguum Bieb.	Kura clover
T. carolinianum Michx.	Carolina clover
T. fragiferum L.	Strawberry clover
T. hybridum L.	Alsike clover
T. incarnatum L.	Crimson clover
T. nigrescens Viv.	Ball clover
T. pratense L.	Red clover
T. repens L.	White clover, ladino white clover
T. resupinatum L.	Persian clover
T. subterraneum L.	Subterranean clover
T. vesiculosum Savi	Arrowleaf clover
Triticum aestivum L.	Wheat
T. durum Desf.	Durum wheat

(syn. *T. turgidum* var. *durum*)	
Vaccinium corymbosum L.	Highbush blueberry
V. macrocarpon Aiton	Cranberry
Vicia faba L.	Broad bean
V. sativa L.	Common vetch
V. villosa Roth	Hairy vetch
Vigna unguiculata ssp. *unguiculata* (L.) Walp. (syn. = *V. sinensis*)	Cowpea, southern, blackeyed, crowder pea
Vitellaria paradoxa C.F. Gaertn. (syn. *Butyrospermum parkii*)	Karité tree
Vitis labrusca L.	Concord grape, fox grape
V. rotundifolia Michaux	Muscadine grape
V. vinifera L.	European grape, wine grape
Zea diploperennis Iltis et al.	Diploperennial teosinte
Z. mays ssp. *mays* L. (syn. *Z. mays* var. *saccharata*)	Corn, maize, popcorn, sweet corn

SUBJECT INDEX

Common names of nematodes are provided in Table 1–1 and Chapter 3. Scientific names of plant species are given in Appendix 3.